MW00415905

Animal Andrology

Theories and Applications

Peter J. Chenoweth

To my parents (deceased) who instilled the desire for knowledge, and
to Lee who has been my support and counsel.

Steven P. Lorton

To Lynn C. Lorton, my wife, best friend and my best supporter.

To Joseph Curtis (deceased), Neal First, James Mrotek and John Reynolds,
early mentors who sparked my interest in research and sperm in particular,
and who have been my friends for more than 40 years.

Animal Andrology

Theories and Applications

Edited by

Peter J. Chenoweth

ChenoVet Animal Andrology, Wagga Wagga, New South Wales, Australia

Steven P. Lorton

Reproduction Resources, Walworth, Wisconsin, USA

CABI is a trading name of CAB International

CABI
Nosworthy Way
Wallingford
Oxfordshire OX10 8DE
UK

Tel: +44 (0)1491 832111
Fax: +44 (0)1491 833508
E-mail: info@cabi.org
Website: www.cabi.org

CABI
38 Chauncey Street
Suite 1002
Boston, MA 02111
USA

Tel: +1 800 552 3083 (toll free)
Tel: +1 (0)617 395 4051
E-mail: cabi-nao@cabi.org

A catalogue record for this book is available from the British Library, London, UK.

Library of Congress Cataloging-in-Publication Data

Animal andrology : theories and applications / [edited by] Peter J. Chenoweth, Steven P. Lorton.
p. cm.
Includes bibliographical references and index.
ISBN 978-1-78064-316-8 (hbk)
1. Domestic animals--Reproduction--Endocrine aspects. 2. Andrology. 3. Spermatozoa. I. Chenoweth, Peter J., editor of compilation. II. Lorton, Steven P., editor of compilation.
[DNLM: 1. Reproductive Techniques, Assisted--veterinary. 2. Semen Preservation--veterinary. 3. Sperm Retrieval--veterinary. 4. Veterinary Medicine--methods. SF 871]

SF871.A55 2014
636.08'24--dc23

2013042144

ISBN-13: 978 1 78064 316 8

Commissioning editor: Sarah Hulbert / Julia Killick
Editorial assistant: Emma McCann
Production editor: Lauren Povey

Typeset by SPi, Pondicherry, India.
Printed and bound in the UK by CPI Group (UK) Ltd, Croydon, CR0 4YY.

Contents

Contributors

Gary C. Althouse, Department of Clinical Studies, School of Veterinary Medicine, University of Pennsylvania, New Bolton Center, 382 West Street Road, Kennett Square, PA 19348, USA. E-mail: gca@vet.upenn.edu

Sayed Murtaza H. Andrabi, Animal Reproduction Programme, Animal Sciences Institute, National Agricultural Research Centre, Park Road, Islamabad, Pakistan. E-mail: andrabi123@yahoo.com

Abdelhaq Anouassi, Veterinary Research Centre, PO Box 77749, Abu Dhabi, United Arab Emirates. E-mail: anouassi@yahoo.com

Barry A. Ball, Gluck Equine Research Center, Department of Veterinary Sciences, University of Kentucky, Lexington, KY 40546-0099, USA. E-mail: b.a.ball@uky.edu

William E. Berndtson, Department of Biological Sciences, University of New Hampshire, Durham, NH 03824, USA. E-mail: bill.berndtson@unh.edu

Sema Birler, Department of Reproduction and Artificial Insemination, Faculty of Veterinary Medicine, University of Istanbul, 34320, Istanbul, Turkey. E-mail: sbirler@istanbul.edu.tr

Leonardo F.C. Brito, ABS Global, Inc., 1525 River Road, DeForest, WI 53532, USA. E-mail: leo.brito@genusplc.com

Jorge Chacon, Research Program on Applied Animal Andrology, School of Veterinary Medicine, Universidad Nacional (UNA), Heredia, Costa Rica. E-mail: jorge.chacon.calderon@una.cr

Peter J. Chenoweth, ChenoVet Animal Andrology, 22 Peter Street, Wagga Wagga, NSW 2650, Australia. E-mail: peter1@chenovet.com.au

Tonya M. Collop, Missouri Department of Agriculture, 1616 Missouri Boulevard, Jefferson City, MO 65102, USA. E-mail: tcollop@hotmail.com

Lefric Enwall, New Tokyo Medical College, Kolonia, Pohnpei, Federated States of Micronesia. E-mail: lefricenwall@yahoo.com

Bart M. Gadella, Departments of Farm Animal Health and Biochemistry and Cell Biology, Faculty of Veterinary Medicine, Utrecht University, Yalelaan 2, 3584 CM Utrecht, the Netherlands. E-mail: b.m.gadella@uu.nl

William V. Holt, Academic Department of Reproductive and Developmental Medicine, University of Sheffield, Sheffield S10 2SF, UK. E-mail: bill.holt@ioz.ac.uk

Abdullah Kaya, Alta Genetics Inc, N8350 High Road, PO Box 437, Watertown, WI 53094, USA. E-mail: akaya@altagenetics.com

Tamara Keeley, School of Agriculture and Food Sciences, University of Queensland, Gatton, Queensland 4343, Australia. E-mail: zooreproduction@yahoo.com

Julie A. Long, Beltsville Agricultural Research Center, US Department of Agriculture, Animal Research Service, Beltsville, MD 20705, USA. E-mail: julie.long@ars.usda.gov

Steven P. Lorton, Reproduction Resources, Inc., 400 S. Main Street, Walworth, WI 53184, USA. E-mail: splorton@alumni.clarku.edu

Francoise J. McPherson, Charles Sturt University, Boorooma Street, Wagga Wagga, NSW 2678, Australia. E-mail: fmcpherson@csu.edu.au

Erdogan Memili, Department of Animal and Dairy Sciences, Mississippi State University, Mississippi State, MS 39762, USA. E-mail: em149@ads.msstate.edu

Scott T. Norman, School of Animal and Veterinary Sciences, Charles Sturt University, Boorooma Street, Wagga Wagga, NSW 2678, Australia. E-mail: snorman@csu.edu.au

Lisa K. Pearson, Department of Veterinary Clinical Sciences, College of Veterinary Medicine, Washington State University, Pullman, WA 99164-6610, USA. E-mail: pearsonlk@vetmed.wsu.edu

Linda M. Penfold, South-East Zoo Alliance for Reproduction & Conservation, Yulee, FL 32097, USA. E-mail: lindap@wogilman.com

Heriberto Rodriguez-Martinez, Department of Clinical and Experimental Medicine, Faculty of Health Sciences, Linköping University, SE-581 85 Linköping, Sweden. E-mail: heriberto.rodriguez-martinez@liu.se

Margaret V. Root Kustritz, Department of Veterinary Clinical Sciences, College of Veterinary Medicine, University of Minnesota, 1352 Boyd Avenue, St Paul, MN 55108, USA. E-mail: rootk001@umn.edu

Nana Satake, School of Veterinary Sciences, University of Queensland, Gatton Campus, Gatton, Queensland 4343, Australia. E-mail: nanastake@gmail.com

Swanand Sathe, Department of Veterinary Clinical Sciences, Lloyd Veterinary Medical Center, College of Veterinary Medicine, Iowa State University, 1600 S 16th Street Ames, IA 50011, USA. Email: ssathe@iastate.edu

Brian P. Setchell, School of Medical Sciences, University of Adelaide, Adelaide, SA 5005, Australia. E-mail: brian.setchell@adelaide.edu.au

Clifford F. Shipley, College of Veterinary Medicine, University of Illinois, 1008 West Hazelwood Drive, Urbana, IL 61802, USA. E-mail: cshipley@illinois.edu

Rebecca Spindler, Taronga Zoo, Taronga Conservation Society Australia, Bradley's Head Road, Mosman, NSW 2088, Australia. E-mail: rspindler@zoo.nsw.gov.au

Ahmed Tibary, Department of Veterinary Clinical Sciences, College of Veterinary Medicine, Washington State University, Pullman, WA 99164-6610, USA. Email: tibary@vetmed.wsu.edu

Preface

Motivation for this volume on applied animal andrology derives from a number of sources. Firstly, the science of andrology (or male reproduction) is rapidly evolving. Fifty years ago, few would have envisaged today's capabilities, which include identifying specific genomic sites for factors directly affecting male reproduction including those associated with sex-related defects or disease. Reproductive technologies such as cryopreservation, *in vitro* fertilization (IVF) and intracytoplasmic sperm injection (ICSI) have become routine andrological procedures. In the animal world, these complement established technologies such as those used for oestrous detection and synchronization, artificial breeding and embryo transfer. Rapidly evolving technologies such as molecular and cell biology, proteomics and genomics are transforming animal reproduction and livestock production as well as our capabilities to conserve threatened and endangered species. Such progress becomes even more astounding when one considers that frozen semen AI (artificial insemination) has been routinely applied in livestock for only 60 years, the recognition of male factor infertility extends back scarcely 100 years and it is less than 250 years since Antonie van Leeuwenhoek used an 'advanced-magnifier microscope' to describe 'animalcules' within an ejaculate.

An outcome of this accelerating tsunami of knowledge is that it is becoming very challenging for experts in the field to remain abreast of relevant advances in andrology, let alone for those who should benefit from their practical application. Andrology itself is becoming so compartmentalized that the exchange of information across its different sub-disciplines is constrained and it is becoming more and more difficult to maintain an overview. Not too many years ago, there were several pertinent scientific journals only, while today there is a profusion of them. Indeed, Thaddeus Mann and Cecilia Lutwak-Mann could hardly have envisaged the scope of subsequent developments when they wrote the following preface to their landmark book, *Male Reproductive Function and Semen*, in 1980: 'To present a coherent and meaningful survey of scientific research endeavour in an area that has expanded so rapidly as physiology and biochemistry of reproduction in the male is no mean feat these days'. However, despite the accumulation of more and better knowledge, or perhaps because of the volume and complexities involved, it is apparent that practical implementation of many potential benefits is either not occurring or is being unnecessarily delayed. There is an evident need to facilitate the flow of information between animal andrological science and its potential end users.

Such considerations present both challenges and opportunities in attempting to produce a compendium that summarizes current knowledge and wisdom in animal andrology.

Experience from over 40 years of teaching animal reproduction to veterinary and animal science students in both the USA and Australia (PJC) indicates that such a text could represent a useful and relevant resource. Another thread comes from long experience with the livestock artificial breeding industries (SPL), in which warm support for such a book has been expressed.

An additional important consideration is the pressing need to boost animal protein production in developing countries, where a burgeoning human population is causing increased stress on food resources. Here, livestock productivity is often low; a situation compounded by poor reproductive rates. Although it is conceded that a number of factors are involved in this scenario, there is general acceptance of the need for widespread dissemination and adoption of the basic precepts of good reproductive management.

Similar sentiments to those above led to the formation in 1997 of the 'Association for Applied Animal Andrology' (4A; see http://www.animalandrology.org/), which aspires to improve networking and understanding in the discipline of andrology as applied to those animals that are of direct use to mankind. A major founding objective of 4A was to help provide an effective conduit so that current scientific knowledge in andrology can be translated into practices that can directly benefit animal reproduction. This objective is considered to be even more pertinent today, when artificial breeding of many species – e.g. cattle, horses, sheep, deer, dogs, pigs, camelids, chickens, zoo animals – has become so widespread, and superior animal genetics, in the form of liquid or frozen semen, are routinely transported across continents.

In this situation, relevant knowledge and expertise are at a premium, both in developed and developing countries. However, despite an increasing demand for competent animal reproduction/andrology expertise and services, opportunities for appropriate education and training are decreasing. Thus, this text aims to provide useful information for those teaching animal physiology at a tertiary level (and possibly at high secondary level), as well as a reference for those interested in male animal reproductive evaluation (and performance), and in semen evaluation, handling and use for artificial breeding. The book attempts to provide the necessary basic information, and this then leads to informed evaluation of male reproductive function in domestic and exotic species (including semen collection, preservation and evaluation) and newer developments in animal andrology, including advanced reproduction techniques (ART).

As editors, we would be extremely remiss if we did not acknowledge the immense contributions made to animal andrology over many years by dedicated scientists from a number of disciplines, including andrology, gynaecology, biochemistry, physiology, physics, and animal and veterinary science. The list of individuals who deserve appropriate recognition is indeed long, and in the current context we can only attempt to mention some who have made exceptional contributions in terms of applied animal andrology, in the sure knowledge that we have inadvertently omitted worthy candidates. This book is dedicated to several individuals who have had significant influence on either of the editors. Other names that should be duly recognized within this context include: J.O. Almquist; R.P. Amann; L. Ball; A. Bane; J. Bedford; W. Bielanski; A.W. Blackshaw; E. Blom; T. Bonadonna; B. Brackett; M.C. Chang; B.G. Crabo; H.M. Dott; P. Dzuik; D.W. Fawcett; R.H. Foote; D. Galloway; D. Garner; R.M.C. Gunn; J. Hammond; R.A.P. Harrison; A. Iritani; L.A. Johnson; R. Jones; N. Lagerlof; A. Laing; H. Lardy; T. Mann; W.G.R. Marden; D. Mortimer; H.G. Osborne; B.W. Pickett; C. Polge; L.E.A. Rowson; R. Saacke; G.W. Salisbury; B.P. Setchell; S. Solomon; M. Tischner; G.M.H. Waites; W.W. Williams; and R. Yanagimachi.

We believe that the contributing authors to this volume are cut from similar cloth, and represent today's animal andrology leaders in their respective categories. We are immensely proud that such a sterling team, drawn from North and South America, Australasia, Micronesia, Europe, the Middle East and Pakistan, has been assembled for the task. We are most appreciative of their efforts and patience in the preparation of this tome, as we are of the staff of CABI, Sarah Hulbert and Emma McCann, and Connie Clement, our Australian editorial assistant.

Peter J. Chenoweth and Steven P. Lorton

Part I

Animal Andrology Theories

1 Semen and its Constituents

Brian P. Setchell*
University of Adelaide, Adelaide, South Australia

Introduction

Semen, the material that is emitted from the penis at ejaculation, comprises a cellular component, the spermatozoa, and a liquid phase, the seminal plasma. The volume of semen in a single ejaculate varies widely among the domesticated mammals, from about 1 ml in sheep and goats to as much as half a litre in pigs. The density of spermatozoa also varies, being much higher in those species with small ejaculates and lower in those ejaculating large volumes.

Semen is composed of secretions of the ampulla of the ductus deferens, and of the accessory glands, seminal vesicles and prostate, as well as fluid and spermatozoa from the cauda epididymis. Individual components may have different origins.

Composition of Semen

Spermatozoa

The fraction of the semen made up by spermatozoa is known as the spermatocrit, and ranges from more than 30% in sheep to less than 2% in pigs (Table 1.1).

A large amount of information is now available about the structure of the spermatozoa, and for details the reader is referred to the many detailed reviews on this topic (Bishop and Walton, 1960; Phillips, 1975; Bedford and Hoskins, 1990; Gage, 1998; Bedford, 2004; Eddy, 2006). In brief, the spermatozoa of the domestic mammals have spatulate heads containing the nuclear DNA, with an acrosome covering the anterior pole, attached by a specialized neck structure to a midpiece and tail. The midpiece consists of a helix of mitochondria surrounding the central two and surrounding nine fibres, which extend into the tail. The sperm of the domestic mammals are relatively small, at least when compared with those of most rodents, and are similar in size and structure to human sperm. The sperm of most murid rodents are much larger and quite different in shape, being falciform or hook shaped, with the acrosome over one side of the head.

Other cells

As well as spermatozoa, white blood cells (WBC) are often found in semen. In humans, more than 10 WBC/ml semen is often associated

* E-mail: brian.setchell@adelaide.edu.au

Table 1.1. Some details of the composition of the semen of the domestic animals. Based on data from Mann, 1964; Mann and Lutwak-Mann, 1981. Reproduced from Setchell, 1991, with permission from Elsevier.

	Bull	Ram	Goat	Boar	Stallion
Semen					
Dry weight (%)	9.5	14.8	–	4.6	4.3
pH	6.48–6.99	5.9–7.3	–	6.85–7.9	6.2–7.8
Specific gravity	1.035	–	–	–	1.013
Sperm concentration ($\times 10^6$/ml)	300–2000	2000–5000	1000–5000	25–350	30–800
Spermatocrit (%)	10	33	–	2	3
Volume (ml)	2–10	0.5–2	0.5–2.5	150–500	20–300
Seminal plasma					
Acetylglucosaminidase (units/ml)	15,000	16,000	–	–	–
Ascorbic acid (mg/100 ml)	8.7	5	–	–	–
Bicarbonate (mmol/l)	7.1[a]	7.1[a]	–	–	–
Calcium (mmol/l)	9.3	1.9	–	–	6.5
Chloride (mmol/l)	49	18	–	96	–
Citric acid (mg/100 ml)	357–1000	137	–	36–325	8–53
Ergothioneine (mg/100 ml)	Trace	Trace	Absent	6–30	3.5–13.7
Fructose (mg/100 ml)	120–540	150–600	–	20–40	<1
Glutamic acid (mg/100 ml)	35–41	76	–	–	–
Glycerylphosphorylcholine (mg/100 ml)	110–500	1600–2000	1400–1600[a]	110–240	40–110[a]
Inositol (mg/100 ml)	25–46	10–15	–	380–610	19–47
Magnesium (mmol/l)	3.4	2.4	–	–	3.8
Mannosidase (units/ml)	400	50	–	–	–
Potassium (mmol/l)	44	23	–	16	26
Protein (g/100 ml)	3–8	–	–	–	–
Sodium (mmol/l)	117	78	–	122	114
Sorbitol (mg/100 ml)	10–136	26–120	–	6–18	20–60

[a]Whole semen.

with infertility (Wolff, 1995), although this view is now not universally accepted (Aitken and Baker, 1995; Lackner *et al.*, 2010; Tremellen and Tunc, 2010; Henkel 2011). In domestic mammals, WBC are often present in small numbers in semen, although there appears to be no relationship between their numbers and abnormalities of the sperm (Sprecher *et al.*, 1999; Sutovsky *et al.*, 2007; Alghamdi *et al.*, 2010).

Carbohydrates

One of the most remarkable features of semen is that the predominant reducing sugar is not glucose, as in blood, but fructose (Mann 1946a,b), a sugar more usually found in plants. Small amounts of glucose are also present, and boar semen in particular contain large concentrations of inositol, but less fructose than semen from bulls or rams (Mann, 1951). Stallion semen also contains inositol and lower concentrations of fructose (Baronos, 1951; Mann *et al.*, 1963), and other compounds of inositol are also present in some species (Seamark *et al.*, 1968). Fructose in bulls and rams originates in the seminal vesicles, with some from the ampulla, but in the stallion, most comes from the ampulla. Inositol is secreted in the seminal vesicles (Mann and Lutwak-Mann, 1981).

Both glucose and fructose can be utilized by sperm, either by oxidation or glycolysis, although the Michaelis constant (K_m) for glucose is much lower than that for fructose (see Ford and Rees, 1990). The mitochondria, in which oxidative phosphorylation occurs, are arranged as a helix around the midpiece of the sperm, whereas the glycolytic enzymes are concentrated in the principal piece of the tail, while some are bound to the fibrous sheath of the flagellum. However, it is unlikely that glycolysis alone could generate enough ATP for full motility, and while diffusion from the mitochondria may be sufficient in smaller sperm, in larger sperm it is likely that an adenylate kinase shuttle is involved in moving ATP from the mitochondria to the flagellum (Ford, 2006; Miki, 2007; Storey, 2008; Cummins, 2009). There is evidence for the occurrence in sperm of specific glucose transporters that can transport both glucose and fructose (Purcell and Moley, 2009).

Proteins, amino acids and other nitrogen-containing compounds

Seminal plasma contains a variety of proteins and peptides, the total concentration being somewhat less than that in blood plasma (Mann and Lutwak-Mann, 1981). Seminal plasma proteins are derived from the epididymis and the accessory glands, and are involved in several essential steps preceding fertilization, including capacitation, establishment of the oviductal sperm reservoir, modulation of the uterine immune response, sperm transport in the female tract and gamete interaction and fusion (Calvete *et al.*, 1994; Topfer-Petersen *et al.*, 2005; Karekoski *et al.*, 2011).

Some proteins are higher in the semen of fertile bulls, whereas others are more abundant in the semen of bulls of lower fertility (Killian *et al.*, 1993; Bellin *et al.*, 1998; Brandon *et al.*, 1999). In stallions, the abundance of some proteins (kallikrein-1E2, clusterin and seminal plasma proteins 1 and 2 (SP1 and 2) are negatively related to fertility, whereas cysteine-rich secretory protein 3 (CRISP3) is positively related (Novak *et al.*, 2010).

Other proteins are involved in sperm–egg interactions and cell cycle regulation (Gaviraghi *et al.*, 2010). Identified proteins include leptin and insulin-like growth factor I (IGF-I; Lackey *et al.*, 2002) and phospholipid-binding proteins involved in sperm membrane lipid modification during capacitation (Manjunath and Therien, 2002). Seminal plasma from stallions contains SSP-7 (stallion seminal protein 7, also known as horse seminal protein 7 – HSP-7), a member of the spermadhesin protein family that is involved in the sperm binding to the zona pellucida of the oocyte (Reinert *et al.*, 1997), and there are also heparin-binding proteins, which modulate capacitation (Miller *et al.*, 1990; Nass *et al.*, 1990; Bellin *et al.*, 1994). Other proteins inhibit *in vitro* and cooling-induced capacitation (Vadnais and Roberts, 2010) and the ability of sperm to penetrate zona-free oocytes (Henault *et al.*, 1995; Henault and Killian, 1996), as well as sperm transport and elimination (Troedsson *et al.*, 2005), sperm longevity (Karekoski and Katila, 2008) and storage in the oviduct (Gwathmey *et al.*, 2006). Seminal plasma from pigs contains high concentrations of transforming growth factor β (TGF-beta), an important immune deviating agent (Robertson *et al.*, 2002).

Seminal plasma also contains considerable concentrations of free amino acids, particularly glutamic acid in rams and bulls (Setchell *et al.*, 1967; Brown-Woodman and White, 1974) and hypotaurine in boars (Van der Horst and Grooten, 1966; Johnson *et al.*, 1972). Hypotaurine may be important in preventing damage to sperm by reactive oxygen species (Alvarez and Storey, 1983; Bucak *et al.*, 2009).

There are also appreciable concentrations of carnitine in the seminal plasma of rams (Brooks, 1979), bulls (Carter *et al.*, 1980) and stallions (Stradaioli *et al.*, 2004). This substance is involved in fatty acid transport in other tissues, but that present in semen is largely derived from the epididymis (Hinton *et al.*, 1979). Boar semen also contains ergothioneine, the betaine of thiolhistidine, a sulfur-containing reducing base, which comes mainly from the seminal vesicle (Mann and Leone, 1953); it is also present in stallion semen, but in this species, it originates largely from the ampulla (Mann and Lutwak-Mann, 1963).

Semen and seminal plasma from rams, bulls, goats, boars and stallions were found to contain considerable amounts of glycerophosphorylcholine, which originates largely from the epididymis (Dawson *et al.*, 1957; Brooks 1970), as well as glycerylphosphorylinositol.

Lipids

Semen contains considerable amounts of lipid, both neutral lipids and phopholipids, most of which is in the spermatozoa (Hartree and Mann, 1959). In ram semen, the most abundant phospholipid is choline plasmalogen (also known as phosphatidalcholine), whereas in boars, it is lecithin (also known as phosphatidylcholine) and in bull sperm, the two phospholipids are present in approximately equal amounts (see Mann and Lutwak-Mann, 1981). One remarkable feature of these phospholipids is their high concentration of highly unsaturated fatty acids, 22 carbons in length, with six double bonds (22:6) in rams and bulls and five double bonds (22:5) in boars (Johnson *et al.*, 1969; Poulos *et al.*, 1973; Evans and Setchell, 1978). These constituent fatty acids are particularly susceptible to damage from reactive oxygen species. The phospholipids may also be important precursors of platelet activating factor (PAF), which is probably involved in sperm motility, the acrosome reaction and fertilization, and which is found in bull and boar sperm (Parks *et al.*, 1990; Roudebush and Diehl, 2001). Seminal plasma from bulls and stallions contains an acetylhydrolase, which may play a role in regulating autocrine or paracrine functions of PAF (Parks and Hough, 1993; Hough and Parks, 1994).

Semen also contains appreciable concentrations of steroids. In bull semen, the concentrations of several steroids, including progesterone, dihydrotestosterone, androstanediols and oestrogens are much higher than in blood plasma. The oestrogens appear to come from the prostate, whereas the other steroids originate from the epididymis. Testosterone is present in seminal plasma at about the same concentration as in blood plasma, much less than in the rete testis fluid leaving the testis (Ganjam and Amann, 1976).

Prostaglandins were discovered in the 1930s, and were so named because it was thought that they came from the prostate, but in fact they originate largely from the seminal vesicle in rams and bulls. They occur in smaller concentrations in the testis and epididymis (Voglmayr, 1973; Kelly, 1978).

Function of Semen

Transport of spermatozoa

An obvious function of the semen is the transport of the spermatozoa into the female reproductive tract at mating. The site of deposition varies according to species; it is deposited into the vagina in cattle and sheep, but directly into the uterus in pigs and to some extent in horses. (Anderson, 1991). Movements of the female tract – which are probably important in moving the spermatozoa from the site of deposition to that of fertilization – are probably influenced by some seminal constituents, in particular by prostaglandins (Kelly, 1978).

Metabolism of spermatozoa

It has been argued that the spermatozoa are in contact with the seminal plasma for too short a time for metabolism of their constituents to be of major importance, but several hours can elapse between mating and fertilization, so some utilization of metabolites, particularly of sugars, should be possible. However, the sperm probably also utilize cellular constituents, particularly lipids, during this time.

Effects of seminal plasma on the female reproductive tract

The possible involvement of proteins in the seminal plasma in the establishment of oviductal sperm reserves, the capacitation of sperm and the processes of fertilization, including binding to and penetration of the zona pellucida of the oocyte, has already been mentioned. It should be remembered, though, that the conceptus must also be protected from maternal immune attack. This is achieved by the action of molecules in the seminal plasma that bind to receptors on female cells and activate gene expression, leading to modification in cellular composition, structure and function of local and remote tissues, such as the ovaries, spleen and peripheral lymphoid organs (Murray *et al.*, 1983; Mah *et al.*, 1985; Rozeboom *et al.*, 2000; Robertson 2005, 2007).

Seminal plasma induces a state of maternal immune tolerance, probably by mediation of T-regulatory cells (Robertson *et al.*, 2009). Seminal plasma also facilitates early placental development, promotes embryo attachment and implantation, and regulates proliferation, viability and differentiation of embryonic blastomeres. There is also an effect on the interval between the luteinizing hormone (LH) surge and ovulation, and even on subsequent behaviour of the inseminated female (O'Leary *et al.*, 2002, 2004, 2006; Robertson, 2005, 2007; Robertson *et al.*, 2006).

References

Aitken RJ and Baker HWG (1995) Seminal leucocytes: passengers, terrorists or good samaritans *Human Reproduction* **10** 1736–1739.

Alghamdi AS, Funnell BJ, Bird SL, Lamb GC, Rendahl AK, Taube PC and Foster DN (2010) Comparative studies on bull and stallion seminal DNase activity and interaction with semen extender and spermatozoa *Animal Reproduction Science* **121** 249–258.

Alvarez JG and Storey BT (1983) Taurine, hypotaurine epinephrine and albumin inhibit lipid peroxidation in rabbit spermatozoa and protect against loss of motility *Biology of Reproduction* **29** 548–555.

Anderson GB (1991) Fertilization, early development and embryo transfer In *Reproduction in Domestic Animals* 4th edition pp 279–313 Ed PT Cupps. Academic Press, San Diego.

Baronos S (1951) Seminal carbohydrate in boar and stallion *Journal of Reproduction and Fertility* **24** 303–305.

Bedford JM (2004) Enigmas of mammalian gamete form and function *Biological Reviews* **79** 429–460.

Bedford JM and Hoskins DD (1990) The mammalian spermatozoon: morphology, biochemistry and physiology In *Marshall's Physiology of Reproduction* 4th edition Volume 2 pp 379–568 Ed GE Lamming. Churchill Livingstone, London.

Bellin ME, Hawkins HE and Ax RL (1994) Fertility of range beef bulls grouped according to the presence or absence of heparin-binding proteins in sperm membranes and seminal fluid *Journal of Animal Science* **71** 2441–2448.

Bellin ME, Oyarzo JN, Hawkins HE, Zhang H, Smith RG, Forrest DW, Sprott LR and Ax RL (1998) Fertility-associated antigen on bull sperm indicates fertility potential *Journal of Animal Science* **76** 2032–2039.

Bishop MWH and Walton A (1960) Spermatogenesis and the structure of mammalian spermatozoa In *Marshall's Physiology of Reproduction* 3rd edition Volume 1 Part 2 pp 1–129 Ed AS Parkes. Longmans, Green and Co, London.

Brandon CI, Heusner GL, Caudle AB and Fayre-Hosken RA (1999) Two-dimensional polyacrylamide gel electrophoresis of equine seminal plasma proteins and their correlation with fertility *Theriogenology* **52** 863–873.

Brooks DE (1970) Acid-soluble phosphorus compounds in mammalian semen *Biochemical Journal* **118** 851–857.

Brooks DE (1979) Carnitine, acetylcarnitine and the activity of carnitine acyltransferases in seminal plasma of men, rams and rats *Journal of Reproduction and Fertility* **56** 667–673.

Brown-Woodman PDC and White IG (1974) Amino acid composition of semen and the secretions of the male reproductive tract *Australian Journal of Biological Sciences* **27** 415–422.

Bucak MN, Tuncer PB, Sariozkan S, Ulutas PA, Coyan K, Baspinar N and Ozkalp B (2009) Effects of hypotaurine, cysteamine and aminoacids solution on post-thaw microscopic and oxidative stress parameters of Angora goat semen *Research in Veterinary Science* **87** 468–472.

Calvete JJ, Nessau S, Mann K, Sanz L, Sieme H, Klug E and Topfer-Petersen E (1994) Isolation and biochemical characterization of stallion seminal-plasma proteins *Reproduction in Domestic Animals* **29** 411–426.

Carter AL, Hutson SM, Stratman FW and Haning RV (1980) Relationship of carnitine and acetylcarnitine in ejaculated sperm to blood plasma testosterone of dairy bulls *Biology of Reproduction* **23** 820–825.

Cummins J (2009) Sperm motility and energetics In *Sperm Biology: An Evolutionary Perspective* pp 185–206 Ed TR Birkhead, DJ Hosken and S Pittnick. Elsevier, Amsterdam.

Dawson RMC, Mann T and White IG (1957) Glycerophosphorylcholine and phosphorylcholine in semen and their relation to choline *Biochemical Journal* **65** 627–634.

Eddy EM (2006) The Spermatozoon In *Knobil and Neill's Physiology of Reproduction* 3rd edition pp 3–54 Ed JD Neill. Elsevier Academic Press, St Louis.

Evans RW and Setchell BP (1978) The effect of rete testis fluid on the metabolism of testicular spermatozoa *Journal of Reproduction and Fertility* **52** 15–20.

Ford WCL (2006) Glycolysis and sperm motility: does a spoonful of sugar help the flagellum go round? *Human Reproduction Update* **12** 269–274.

Ford WCL and Rees JM (1990) The bioenergetics of mammalian sperm motility In *Controls of Sperm Motility: Biological and Clinical Aspects* pp 175–202 Ed C Gagnon. CRC Press, Boca Raton.

Gage MJG (1998) Mammalian sperm morphometry *Proceedings of the Royal Society London B* **265** 97–103.

Ganjam VK and Amann RP (1976) Steroids in fluids and sperm entering and leaving the bovine epididymis, epididymal tissue and accessory sex gland secretions *Endocrinology* **99** 1618–1630.

Gaviraghi A, Deriu F, Soggiu A, Galli A, Bonacina C, Boniaai L and Roncada P (2010) Proteomics to investigate fertility in bulls *Veterinary Research Communications* **34** S33–S36.

Gwathmey TM, Ignota GG, Mueller JL, Manjunath P and Suarez SS (2006) Bovine seminal plasma proteins PDC-109, BSP-A3 and BSP-30-kDa share functional roles in storing sperm in the oviduct *Biology of Reproduction* **75** 501–507.

Hartree EF and Mann T (1959) Plasmalogen in ram semen, and its role in sperm metabolism *Biochemical Journal* **71** 423–434.

Henault MA and Killian GJ (1996) Effect of homologous and heterologous seminal plasma on the fertilizing ability of ejaculated bull spermatozoa assessed by penetration of zona-free bovine oocytes *Journal of Reproduction and Fertility* **108** 199–204.

Henault MA, Killian GJ, Kavanaugh JF and Griel LC (1995) Effect of accessory gland fluid from bulls of different fertilities on the ability of cauda epididymal sperm to penetrate zona-free bovine oocytes *Biology of Reproduction* **52** 390–397.

Henkel RR (2011) Leucocytes and oxidative stress: dilemma for sperm function and male fertility *Asian Journal of Andrology* **13** 43–52.

Hinton BP, Snoswell AM and Setchell BP (1979) The concentration of carnitine in the luminal fluid of the testis and epididymis of the rat and some other mammals *Journal of Reproduction and Fertility* **56** 105–111.

Hough SR and Parks JE (1994) Platelet-activating factor acetylhydrolase activity in seminal plasma from the bull, stallion, rabbit and rooster *Biology of Reproduction* **50** 912–916.

Johnson LA, Gerrits RJ and Young EP (1969) The fatty acid composition of porcine spermatozoa phospholipids *Biology of Reproduction* **1** 330–334.

Johnson LA, Pursel VG, Gerrits RJ and Thomas CH (1972) Free amino acid composition of porcine seminal, epididymal and seminal vesicle fluids *Journal of Animal Science* **34** 430–434.

Karekoski M and Katila T (2008) Components of stallion seminal plasma and effects of seminal plasma on sperm longevity *Animal Reproduction Science* **107** 249–256.

Karekoski AM, Rivera del Alamo MM, Guvenc K, Reilas T, Calvete JJ, Rodriguez-Martinez H, Andersson M and Katila T (2011) Protein composition of seminal plasma in fractionated stallion ejaculates *Reproduction in Domestic Animals* **46** e79–e84.

Kelly RW (1978) Prostaglandins in semen: their occurrence and possible physiological significance *International Journal of Andrology* **1** 188–200.

Killian GJ, Chapman DA and Rogowski LA (1993) Fertility-associated proteins in Holstein bull seminal plasma *Biology of Reproduction* **49** 1202–1207.

Lackey BR, Gray SL and Henricks DM (2002) Measurement of leptin and insulin-like growth factor-I in seminal plasma from different species *Physiological Research* **51** 309–311.

Lackner JE, Agarwal A, Mahfouz R, de Plessis SS and Schatzl G (2010) The association between leucocytes and sperm quality is concentration dependent *Reproductive Biology and Endocrinology* **8** 12–17.

Mah J, Tilton JE, Williams GL, Johnson JN and Marchello MJ (1985) The effect of repeated matings at short intervals on reproductive performance of gilts *Journal of Animal Science* **60** 1052–1054.

Manjunath P and Therien I (2002) Role of seminal plasma phospholipid-binding proteins in sperm membrane lipid modification that occurs during capacitation *Journal of Reproductive Immunology* **53** 109–119.

Mann T (1946a) Fructose, a constituent of semen *Nature* **157** 79.

Mann T (1946b) Studies on the metabolism of semen 3. Fructose as a normal constituent of seminal plasma. Site of formation and function of fructose in semen *Biochemical Journal* **40** 481–491.

Mann T (1951) Inositol, a major constituent of the seminal vesicle secretion of the boar *Nature* **168** 1043.

Mann T (1964) *The Biochemistry of Semen and of the Male Reproductive Tract.* Methuen, London.

Mann T and Leone E (1953) Studies on the metabolism of semen 8. Ergothioneine as a normal constituent of boar seminal plasma *Biochemical Journal* **53** 140–148.

Mann T and Lutwak-Mann C (1963) Comparative biochemical aspects of animal reproduction *Bulletin de l'Académie Royale de Médecine de Belgique* **3** 563–597.

Mann T and Lutwak-Mann C (1981) *Male Reproductive Function and Semen.* Springer-Verlag, Berlin.

Mann T, Minotakis CS and Polge C (1963) Semen composition and metabolism in the stallion and jackass *Journal of Reproduction and Fertility* **5** 109–122.

Miki K (2007) Energy metabolism and sperm function In *Spermatology* pp 309–325 Ed ERS Roldan and M Gomendio. Nottingham University Press, Nottingham.

Miller DJ, Winer MA and Ax RL (1990) Heparin-binding proteins from seminal plasma bind to bovine spermatozoa and modulate capacitation by heparin *Biology of Reproduction* **42** 899–915.

Murray FA, Grifco P and Parker CF (1983) Increased litter size in gilts by intrauterine infusion of seminal and sperm antigens before mating *Journal of Animal Science* **56** 895–900.

Nass SJ, Miller DJ, Winer MA and Ax RL (1990) Male accessory sex glands produce heparin-binding proteins that bind to cauda epididymal spermatozoa and are testosterone dependent *Molecular Reproduction and Development* **25** 237–246.

Novak S, Smith TA, Paradis F, Burwash L, Dyck MK, Foxcroft GR and Dixon WT (2010) Biomarkers of *in vivo* fertility in sperm and seminal plasma of fertile stallions *Theriogenology* **74** 956–967.

O'Leary S, Robertson SA and Armstrong DT (2002) The influence of seminal plasma on ovarian function in pigs – a novel inflammatory mechanism? *Journal of Reproductive Immunology* **57** 225–238.

O'Leary S, Jasper MJ, Warnes GM, Armstrong DT and Robertson SA (2004) Seminal plasma regulates endometrial cytokine expression, leucocyte recruitment and embryo development in the pig *Reproduction* **128** 237–247.

O'Leary S, Jasper MJ, Robertson SA and Armstrong DT (2006) Seminal plasma regulates ovarian progesterone production, leucocyte recruitment and follicular cell responses in the pig *Reproduction* **132** 147–158.

Parks JE and Hough SR (1993) Platelet-activating factor acetylhydrolase activity in bovine seminal plasma *Journal of Andrology* **14** 335–339.

Parks JE, Hough S and Elrod C (1990) Platelet activating factor activity in the phospholipids of bovine spermatozoa *Biology of Reproduction* **43** 806–811.

Phillips DM (1975) Mammalian sperm structure In *Handbook of Physiology. Male Reproductive System* Section 7 Volume V pp 405–419 Ed DW Hamilton and RO Greep. The American Physiological Society, Bethesda.

Poulos A, Darin-Bennett A and White IG (1973) The phospholipid-bound fatty acids and aldehydes of mammalian spermatozoa *Comparative Biochemistry and Physiology – Part B: Biochemistry & Molecular Biology* **46B** 541–549.

Purcell SH and Moley KH (2009) Glucose transporters in gametes and preimplantation embryos *Trends in Endocrinology and Metabolism* **20** 483–489.

Reinert M, Calvete JJ, Sanz L and Topfer-Petersen E (1997) Immunohistochemical localization in the stallion genital tract, and topography of spermatozoa of seminal plasma protein SSP-7, a member of the spermadhesin family *Andrologia* **29** 179–186.

Robertson SA (2005) Seminal plasma and male factor signalling in the female reproductive tract *Cell and Tissue Research* **322** 43–52.

Robertson SA (2007) Seminal fluid signalling in the female reproductive tract: lessons from rodents and pigs *Journal of Animal Science* **85** E36–E44.

Robertson SA, Ingman WV, O'Leary S, Sharkey DJ and Tremellen KP (2002) Transforming growth factor β – a mediator of immune deviation in seminal plasma *Journal of Reproductive Immunology* **57** 109–128.

Robertson SA, O'Leary S and Armstrong DT (2006) Influence of semen on inflammatory modulators of embryo implantation *Society for Reproduction and Fertility Supplement* **62** 231–245.

Robertson SA, Guerin LR, Moldenhauer LM and Hayball JD (2009) Activating T regulatory cells for tolerance in early pregnancy – the contribution of seminal fluid *Journal of Reproductive Immunology* **83** 109–116.

Roudebush WE and Diehl JR (2001) Platelet-activating factor content in boar spermatozoa correlates with fertility *Theriogenology* **55** 1633–1638.

Rozeboom KJ, Troedsson MH, Hodson HH, Shurson GC and Crabo BG (2000) The importance of seminal plasma on the fertility of subsequent artificial insemination in swine *Journal of Animal Science* **78** 443–448.

Seamark RF, Tate M and Smeaton TC (1968) The occurrence of scyllitol and D-glycerol 1-(L-myoinositol 1-hydrogen phosphate) in the male reproductive tract *Journal of Biological Chemistry* **243** 2424–2428.

Setchell BP (1991) Male reproductive organs and semen In *Reproduction in Domestic Animals* 4th edition pp 221–249 Ed PT Cupps. Academic Press, San Diego.

Setchell BP, Hinks NT, Voglmayr JK and Scott TW (1967) Amino acids in ram testicular fluid and semen and their metabolism by spermatozoa *Biochemical Journal* **105** 1061–1065.

Sprecher DJ, Coe PH and Walker RD (1999) Relationships among seminal culture, seminal white blood cells and the percentage of primary sperm abnormalities in bulls evaluated prior to the breeding season *Theriogenology* **51** 1197–1206.

Storey BT (2008) Mammalian sperm metabolism: oxygen and sugar, friend and foe. *International Journal of Developmental Biology* **52** 427–437.

Stradaioli G, Sylla L, Zelli R, Chiodi P and Monaci M (2004) Effect of L-carnitine administration on the seminal characteristics of oligoasthenospermic stallions *Theriogenology* **62** 761–777.

Sutovsky P, Plummer W, Baska K, Peterman K, Diehl JR and Sutovsky M (2007) Relative levels of semen platelet activating factor-receptor (PAFr) and ubiquitin in yearling bulls with high content of semen white blood cells: implications for breeding soundness evaluation *Journal of Andrology* **28** 92–108.

Topfer-Petersen E, Ekhlas-Hundrieser M, Kirchoff C, Leeb T and Sieme H (2005) The role of stallion seminal proteins in fertilisation *Animal Reproduction Science* **89** 159–170.

Tremellen K and Tunc O (2010) Macrophage activity in semen is significantly correlated with sperm quality in infertile men *International Journal of Andrology* **33** 823–831.

Troedsson MHT, Desvousges A, Alghambi AS, Dahms B, Dow CA, Hayna J, Valesco R, Collahan PT, Macpherson ML, Pozor M and Buhl WC (2005) Components in seminal plasma regulating sperm transport and elimination *Animal Reproduction Science* **89** 171–186.

Vadnais ML and Roberts KP (2010) Seminal plasma proteins inhibit *in vitro*- and cooling-induced capacitation in boar spermatozoa *Reproduction, Fertility and Development* **22** 893–900.

Van der Horst CJG and Grooten HJG (1966) The occurrence of hypotaurine and other sulphur-containing amino acids in seminal plasma and spermatozoa of boar, bull and dog *Biochimica et Biophysica Acta* **117** 495–497.

Voglmayr JK (1973) Prostaglandin F2a concentration in genital tract secretions in dairy bulls *Prostaglandins* **4** 673–678.

Wolff H (1995) The biologic significance of white blood cells in semen *Fertility and Sterility* **63** 1143–1157.

2 Sperm Production and its Harvest

William E. Berndtson*
University of New Hampshire, Durham, New Hampshire, USA

Introduction

Spermatogenesis requires ~60 days in most mammals. It encompasses a series of successive mitotic divisions, two meiotic divisions and the transformation of haploid spermatids into spermatozoa. Spermatogenesis is susceptible to disruption by many physical or chemical agents, which can produce alterations in seminal quality that may be manifested either quickly or weeks thereafter. Recognition of factors that are known to alter spermatogenesis, of the time course for the first appearance of alterations in ejaculated semen, and of subsequent recovery after exposure to disruptive agents is critical for sound reproductive management. Because spermatogenesis proceeds independently of sexual activity, sexual inactivity results in the accumulation of sperm within the extragonadal ducts and subsequent losses via micturition. Consequently, large numbers of spermatozoa must be expected in ejaculates taken after lengthy sexual rest. To harvest the maximal number of spermatozoa, males must be maintained on a regular, frequent ejaculation schedule. Maintaining libido at such ejaculation frequencies can be challenging, and requires the provision of novelty. This can be accomplished for some animals (e.g. the bull) by changing teaser animals or collection locations, or by providing movement by the teaser. This chapter provides an overview of these and other important factors that contribute to the maximum reproductive potential of a given male.

Sperm Production and its Harvest

Successful reproduction requires two major contributions from the male: the production of adequate numbers of viable sperm, and the capacity to mate or to be used for semen collection, so that sperm may be used for artificial insemination. Accordingly, it is important to select and manage the male to maximize sperm production and its harvest.

Spermatogenesis

The germinal cells

The seminiferous tubules contain three major classes of germ cells: spermatogonia, spermatocytes and spermatids. The spermatogonia are the least differentiated of these cells,

* E-mail: bill.berndtson@unh.edu

and they are distinguished from the other types by the fact that they undergo mitotic divisions that increase the number of germ cells, while also providing for stem cell renewal. By classical definition, spermatogonia with nucleoplasm of a smooth, coarse or intermediate texture are designated as type A, type B or type I (intermediate), respectively. Division of these cells yields subtypes, which are identified by subscripts denoting their order of appearance. For example, type A_1 spermatogonia would divide to produce daughter cells of type A_2. Successive divisions might proceed as follows: $A_2 \rightarrow A_3 \rightarrow I \rightarrow B_1 \rightarrow B_2$, etc. The number of divisions and, thus, the number of spermatogonial subtypes, is constant within but differs among individual species. The last in the series of spermatogonial divisions results in the production of primary (1°) spermatocytes.

The testis contains both primary and secondary (2°) spermatocytes, which are distinguished from the other germ cells by the fact that they undergo meiotic divisions. The primary spermatocytes are the most developmentally advanced germ cells to replicate DNA. They undergo the first reduction division to produce secondary spermatocytes, which receive duplicate copies of only one member of each pair of chromosomes. While the autosomes behave similarly, the sex chromosomes (X and Y) are particularly useful for explaining the unique manner in which chromosomes are passed to daughter cells during the meiotic divisions. Each newly formed somatic cell, or spermatogonium, receives both an X and Y chromosome; their DNA is replicated so that upon mitotic division each daughter cell receives one member of each chromosomal pair (i.e. one X plus one Y chromosome). In contrast, the replication of DNA by the primary spermatocytes is followed by the first meiotic division, through which each daughter secondary spermatocyte receives two copies of either the X or the Y chromosome, but not both. The secondary spermatocytes do not replicate DNA; their division involves separation of the identical pairs of chromosomes, resulting in haploid spermatids containing a single X or Y chromosome.

Spermatids are the most developmentally advanced of the three germinal cell types. They do not divide, but undergo transformational events that culminate with their release into the lumen of the seminiferous tubules, at which time they are considered to be spermatozoa.

Kinetics of spermatogenesis

The kinetics of spermatogenesis refers to the number and nature of the various germ cell divisions for a given species, and consists of three major components. Spermatocytogenesis encompasses the cell divisions beginning with the least differentiated spermatogonia and culminating with the production of haploid spermatids. This is followed by spermiogenesis, by which the newly formed spherical spermatids are transformed into more mature forms with a morphology closely resembling that of spermatozoa. Spermiogenesis culminates with spermiation.

Although most spermatogonial divisions yield daughter cells of a more developmentally advanced type, type A_1 spermatogonia must be replenished to prevent the supply of these cells from being exhausted. Mechanisms for accomplishing this stem cell renewal have been investigated in many species. Most proposed models are based on a combination of quantitative and morphological data. A more detailed discussion of stem cell renewal in the boar, bull, goat and ram may be found in references listed in Table 2.1. Finally, the testes contain small numbers of inactive reserve cells designated as type A_0 spermatogonia. These are uncommitted to cell division, but become mitotically active to replenish the type A_1 population when necessary (Clermont, 1962, and others).

Table 2.1. References characterizing spermatogonial stem cell renewal in the boar, bull, goat and ram.

Species	Reference
Boar	Frankenhuis *et al.*, 1980
Bull	Amann, 1962b
Bull	Hochereau-de Reviers, 1970
Bull	Berndtson and Desjardins, 1974
Goat	Bilaspuri and Guraya, 1984
Ram	Ortavant, 1958
Ram	Lok *et al.*, 1982

The cycle of the seminiferous epithelium

The division and maturation of most germ cells proceeds on a schedule that is relatively well timed in normal individuals. Thus, spermatogenesis yields a distinct number of unique combinations of cells, or cellular associations, that are observable together at any given point in time. If one could observe a given cross section of a seminiferous tubule, one would note progressive changes in the cellular associations over time. Ultimately, the cellular association that was present initially would appear once again. The series of changes beginning with the appearance of one cellular association and ending with its reappearance would constitute one cycle of the seminiferous epithelium. In most species, approximately 4.5 cycles are required for the production of sperm cells from the least developmentally advanced spermatogonia. The length of the cycle of the seminiferous epithelium and the duration of spermatogenesis (i.e. the time required to produce spermatozoa from the least developmentally advanced spermatogonia) for several relevant species are given in Table 2.2.

Although spermatogenesis is a continuous process, researchers have found it useful to divide this process into recognizable stages of the cycle of the seminiferous epithelium. The number of stages for any given species is limited only by the ability to discern distinguishing features. Although all staging systems rely on the specific cellular associations that may be observed, the two most common approaches include additional consideration of either general tubular morphology or acrosomal development. Use of the tubular morphology system includes noting whether elongated spermatids are or are not present, and whether the elongated spermatids are embedded deeply within the seminiferous epithelium or line the lumen immediately before spermiation, etc. (see Plate 1); with this system, eight stages are usually recognizable. Staging by the acrosomal system is based primarily on recognizable steps of acrosomal development during spermiogenesis (see Fig. 2.1); for most species, 12–15 stages can be identified by this approach. Publications describing criteria for staging the cycle in several economically important species are listed in Table 2.3.

All stages of the cycle of the seminiferous epithelium can be found within a single seminiferous tubule at any point in time. This feature contributes to the steady, continuous release of sperm from the testis over time. Moreover, these stages are arranged sequentially along the length of the tubule. So a segment containing stage III might be followed by segments in stage IV, V, VI, etc. This arrangement is denoted as the wave of spermatogenesis. Minor disruptions in this pattern, termed modulations, may occur (Perey *et al.*, 1961). An example of this might involve

Table 2.2. The length of the cycle of the seminiferous epithelium and the duration of spermatogenesis in selected species.

Species	Cycle length (days)	Duration of spermatogenesis (days)	Reference
Boar	8.6	–	Swierstra, 1968a
Boar	–	39[a]	Amann and Schanbacher, 1983
Boar	9.0	40.6	Franca and Cardosa, 1998
Boar (wild)	9.05	41[a]	Almeida *et al.*, 2006
Bull	13.3	61	Amann and Almquist, 1962
Bull	13.5	–	Hochereau-de Reviers *et al.*, 1964
Bull	13.5	61[a]	Amann and Schanbacher, 1983
Goat	10.6	47.7	Franca *et al.*, 1999
Ram	10.6	42.3	Cardosa and Queiroz, 1988
Stallion	12.2	–	Swierstra *et al.*, 1974
Stallion	–	55[a]	Amann and Schanbacher, 1983

[a]Estimates based on an assumption that the duration of spermatogenesis requires 4.5 cycles of the seminiferous epithelium.

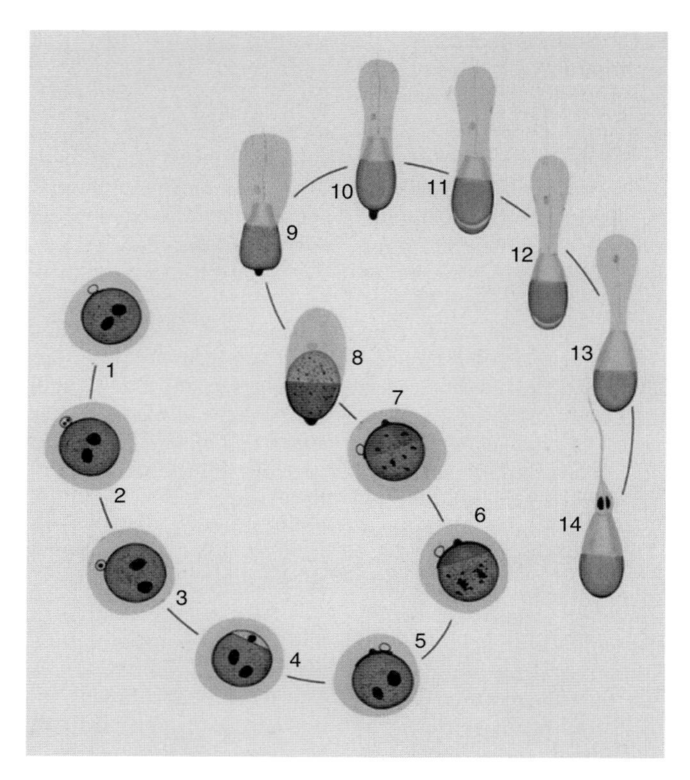

Fig. 2.1. Spermiogenesis in the bull as seen with periodic acid–Schiff (PAS) staining. Fourteen steps in the development of bovine spermatids are depicted. Changes associated primarily with acrosomal development were used by Berndtson and Desjardins (1974) to distinguish these 14 stages of the cycle of the seminiferous epithelium.

Table 2.3. Systems for classifying stages of the cycle of the seminiferous epithelium in selected species.

Species	Staging system	Reference
Boar	Tubular morphology	Swierstra, 1968a
Boar	Acrosomal	Franca *et al.*, 2005
Bull	Tubular morphology	Amann, 1962b
Bull	Acrosomal	Berndtson and Desjardins, 1974
Goat	Tubular morphology	Franca *et al.*, 1999
Goat	Tubular morphology	Onyango *et al.*, 2000
Goat	Acrosomal	Bilaspuri and Guraya, 1984
Ram	Tubular morphology	Ortavant (1959a,b)
Ram	Acrosomal	Clermont and Lebland, 1955
Stallion	Tubular morphology	Swierstra *et al.*, 1974

a spatial distribution of stages along one segment of a seminiferous tubule as follows: stage I → stage II → stage III → stage IV → stage III → stage IV → stage V → stage VI, etc. A wave of spermatogenesis is not present in man. Rather, stages are confined to small, discrete, irregular patches distributed more randomly within the seminiferous tubules (Heller and Clermont, 1964). This arrangement precludes the identification of stages within round seminiferous tubular cross sections as are applied routinely for other mammals.

By examining a large number of round seminiferous tubular cross sections in species other than man, researchers can determine the frequency of appearance for each stage of the cycle of the seminiferous epithelium. This information has several useful applications. Because the timing of spermatogenic events is relatively constant, the frequency of a given stage will be proportional to its relative duration. For example, one cycle of the seminiferous epithelium requires 13.5 days in the bull (Table 2.2). Thus, if a particular stage appeared at a frequency of 10%, it would be apparent that the duration of that stage equalled 10% of one cycle, or 1.35 days. Similarly, estimates of the actual length of one cycle of the seminiferous epithelium or of the duration of spermatogenesis have been based on stage frequency data and the progression of radiolabelled germ cells in virtually every species for which such information has been generated.

Knowledge of the duration of individual stages is useful in predicting the time course over which a treatment that adversely impacted a particular type of cell or cell division would be expected to first become evident in an ejaculate from a treated male, or in estimating the subsequent time course for full recovery upon withdrawal of the causative agent or factor (Foote and Berndtson, 1992). Although the timing of spermatogenic events and the frequency of individual stages appear to be relatively constant in normal individuals, some variability has been reported among normal subjects. In addition, both arrested development and subsequent stage synchronization (i.e. progression of spermatogenesis with most tubular cross sections being at a single, identical stage of development at any given point in time), have been induced experimentally. These findings and their impact on the reliability of stage frequency data have been discussed elsewhere in greater detail (Berndtson, 2011).

Sperm Production Rates

Which factors have an impact on sperm production?

Sperm production is among the most important determinants of the reproductive capacity of an individual male. Indeed, although libido is also quite important with natural mating systems, the number of potential matings to a sire used for artificial insemination (AI) is generally limited only by the number of normal sperm produced per unit of time. It is, therefore, important for andrologists to recognize factors that do or do not have an impact on sperm production.

Testis size

Testis size is highly and positively correlated with sperm production in healthy post-pubertal males (Table 2.4). Accordingly, scrotal circumference or width is an important

Table 2.4. Reported correlations between testis size and the daily sperm production (DSP) of sexually mature males.

Species	Correlation	Reference
Boar	0.90	Swierstra, 1968b
Boar (wild)	0.97	Almeida *et al.*, 2006
Bull	0.81, 0.72, 0.64, 0.40, –0.22[a]	Hahn *et al.*, 1969
Bull	0.71	Berndtson *et al.*, 1987a
Bull	0.72	Berndtson *et al.*, 1987b
Goat	0.94[b]	Jindal and Panda, 1980
Goat	0.50[b]	Ritar *et al.*, 1992
Goat	0.98	Leal *et al.*, 2004
Ram	0.97[c]	Lino, 1972
Ram	0.75–0.80[d]	Alkass *et al.*, 1982
Ram	0.94	Cardosa and Queiroz, 1988
Stallion	0.77	Gebauer *et al.*, 1974a
Stallion	0.89	Johnson *et al.*, 1997

[a]Correlations for Holstein bulls aged 17–22, 34–42, 42–53, 56–69 and 72–150 months of age, respectively.
[b]Correlation between testicular weight and epididymal sperm number.
[c]Correlation between paired testes weight and extragonadal sperm reserves.
[d]Correlation between scrotal circumference and total sperm in the reproductive tract.

component of male breeding soundness examinations (e.g. Shipley, 1999; Eilts, 2005a,b). As discussed subsequently, testis size and sperm production increase, at least up to a point, after puberty. However, this does not diminish the importance of testis size measurements in younger males. Young bulls with relatively small testes tend to develop into adults with testes that are smaller than those of their counterparts, and vice versa (Hahn *et al.*, 1969; Coulter *et al.*, 1975). So breeders should consider testis size before enrolling young bulls in their progeny testing programmes.

Age

Sperm production is influenced by age. For most species, testicular size and sperm production increase to maximal levels over a period of time after puberty, and then remain at that level until ultimately declining as a result of senescence. This pattern is characteristic for the bull, as shown in Table 2.5. In contrast, sperm production appears to increase throughout life in healthy stallions (Table 2.6). Sperm production/g tissue also increases for a period of time after puberty, as illustrated by data in Table 2.6 for the stallion. Thus, young males have testes that are both smaller and less productive/g tissue than those of more sexually mature individuals.

Season

The impact of season on sperm production is species dependent. Seasonal breeding is advantageous for most wild species, in which sperm production may cease entirely during the non-breeding season. In contrast, the management of domesticated animals ensures greater consistency in the availability of feed, protection from severe weather, etc., and this, over time, has presumably diminished seasonal breeding patterns in our domesticated species. Several of our farm animals, such as the bull and boar, produce sperm at a consistent rate throughout the year. In contrast, stallions and the rams of some breeds produce sperm throughout the year, but in greater quantity during the breeding than in the non-breeding season.

Seasonal changes in sperm production are typically associated with decreases in both testicular size and the number of sperm produced per unit of testicular parenchyma. In one study, scrotal width was 11% greater (101 versus 91 mm) at the onset of the breeding season (27 April–26 May) than for the same stallions during the non-breeding season 180 days later (Squires *et al.*, 1981). Based on the imperfect assumption that testes are precise spheres, a 10% difference in testicular diameter would be associated with a corresponding difference of approximately 37% in testicular volume. Testicular tissue must be removed to quantify sperm production by direct methods. This precludes the quantification of sperm production within the same individuals at each season. None the less, sperm production per volume of testicular parenchyma is clearly lower in stallions during the non-breeding season. For example, in one study, daily sperm production (DSP)/g of tissue for stallions ≥4 years old averaged 14.8 versus 18.8 million during non-breeding versus breeding seasons, respectively (Johnson and Thompson, 1983). Thus, although stallions produce sperm throughout the year, sperm production occurs at a markedly reduced rate during the non-breeding season. Hence, season must be considered when estimating the breeding capacity of seasonally breeding males from scrotal measurements.

Table 2.5. Changes in reproductive characteristics with age in the bull. Adapted from Hahn *et al.*, 1969.

	Age (months)				
Variable	17–22	34–42	42–53	56–59	72–150
Scrotal circumference (cm)	35.1	40.0	40.3	42.0	42.6
DSO[a]/male (billions)	4.1	5.9	5.4	6.1	4.0
DSO/g (millions)	8.7	8.4	7.6	7.5	4.9

[a]Daily sperm output.

Table 2.6. Changes in reproductive characteristics with age in the stallion. Adapted from Johnson and Neaves, 1981.

	Age (years)		
Variable	2–3	4–5	13–20
Testicular weight (g)	117	161	213
DSP[a]/testis (billions)	1.27	2.67	3.18
DSP/g (millions)	11.4	18.8	17.0

[a]Daily sperm production.

Environmental factors

Normal spermatogenesis requires a testicular temperature slightly below normal core body temperature. It is for that reason that the testes of all mammals other than the elephant and whale reside within a superficial scrotum. Heat can be dissipated from the scrotal surface, which can be increased or decreased as necessary via the contraction or relaxation of the external cremaster muscles of the spermatic cord and/or the tunica dartos muscle at the base of the scrotum. These muscles also serve to position the testes closer to or further from the rest of the body. In addition, the spermatic cord contains an extensive vasculature of closely intertwined arterial and venous blood vessels known as the pampiniform plexus, which provides a countercurrent exchange mechanism by which warm arterial blood is cooled before reaching the testis by the cooler, returning venous blood, and vice versa.

The importance of testicular thermoregulation is evident in cases of cryptorchidism, in which one or both testes fail to descend into the scrotum. Whereas cryptorchid testes continue to produce near-normal levels of androgens, they do not produce sperm. Although the scrotal testis of a unilaterally cryptorchid individual will continue to produce sperm, cryptorchidism is a heritable condition, and such individuals should not be used for breeding. The disparity in testicular size within one unilaterally cryptorchid stallion is depicted in Plate 2.

Thermoregulatory mechanisms may be incapable of maintaining an appropriate testicular temperature when ambient temperatures are excessive, or during periods of illness accompanied by fever. Shade, fans or other methods for preventing overheating can be very helpful during such times. Beyond that, it is important to recognize the potential for transient depressions in seminal quality or fertility as a result of elevated temperature. As with any insult to the testis, the severity of an effect will reflect both its magnitude and duration. Moreover, the interval required for any resulting negative impacts to be manifested via depressions in the fertility of mated females or the quality of ejaculated semen, and the time required subsequently for full recovery can vary.

A simple assembly line concept is useful for illustrating the reasons for this variability in seminal quality or fertility as a result of elevated temperature. Imagine an assembly line for producing wristwatches consisting of ten stations at which each watch remains for 1 h as different components are added. Now, imagine that a new employee assigned to station number three began to insert components in a manner that would cause the hands of the watch to move counterclockwise rather than clockwise. At the time of the first error, stations 4–10 would contain partially but correctly assembled watches. These would continue to move through the assembly process. Some 7–8 h would elapse before the first defective watch had advanced through the remaining stations and undergone a final quality control inspection. Had an error occurred instead at station 10, defective watches would have been detected soon thereafter, but in this example the problem was not immediately apparent. Indeed, depending on the point in the assembly process at which a problem arose, detection via evaluation of the completed product could take anywhere from 1 to 10 h in this hypothetical example.

By analogy, spermatogenesis requires approximately 60 days in most mammals. Several more days are required for epididymal transit before sperm are available for ejaculation (Table 2.7). Therefore, a problem(s) with spermatogenesis might be reflected in an ejaculate almost immediately, or might require more than 60 days, depending on the particular types of germ cell or spermatogenic events affected. Recovery times are subject to similar variability. Referring once

Table 2.7. Total epididymal transit time.

Species	Total (days)	Reference
Boar	10.2	Swierstra, 1971
Boar	7.94	Franca and Cardosa, 1998
Bull	5.6–8.3	Weisgold and Almquist, 1979
Goat	Unknown	–
Ram	12	Lino, 1972
Stallion	4.9	Gebauer *et al.*, 1974b

again to the wristwatch illustration, at the moment that defective watches were detected, stations 3–10 would already contain watches with components inserted incorrectly. An additional 7–8 h would be required after correction of the problem before properly assembled watches would once again begin coming off the assembly line. Had the problem occurred at station ten, correction would have produced a more immediate remedy.

Once again by analogy, the time required for semen quality to recover after a transient disruption of spermatogenesis could be brief or might require more than 2 months. For AI, it is customary to evaluate every ejaculate for the number and percentage of motile sperm and, occasionally, for spermatozoal morphology. This provides an opportunity for detecting some forms of damage and thereby some safeguard against the use of semen of low quality. Similar evaluations are not performed routinely for males used for natural mating. Although breeding soundness evaluations would certainly be indicated upon detection of herd fertility problems, full recovery during the interval between the unsuccessful matings and the detection of low fertility could render such evaluations inconclusive.

Although severe nutritional deficiencies or imbalances can be disruptive, the nutritional requirements for normal spermatogenesis in sexually mature males do not appear to differ from those for maintaining good general health and body condition (Foote, 1978). The author is unaware of any dietary supplements that are effective in increasing sperm production or seminal quality of normal, healthy males. It should be noted that there has been some recent interest in assessing whether dietary supplementation might prove effective for altering the composition of sperm membranes, thus rendering these cells more resistant to the stresses of cooling or freezing (e.g. Brinsko *et al.*, 2005; Harris *et al.*, 2005). Research in this area is likely to continue.

Exogenous agents

Spermatogenesis is readily disrupted by a variety of exogenous agents. While an extensive description of known anti-spermatogenic agents is beyond the scope of this undertaking, several groups of compounds are of current relevance to the animal industries. As cited previously, one of the main roles of the male is either to mate with receptive females or to ejaculate semen that can be harvested for use via AI. This requires that the male have adequate libido. Because libido is under endocrine control, exogenous hormones have been evaluated as potential treatments for enhancing libido or for treating impotence or other breeding issues (Berndtson *et al.*, 1979; McDonnell, 1999). Such applications are likely to be of greater interest for companion animals of intrinsically high value than for most food-producing species. None the less, from the author's perspective, any use of exogenous hormones should be preceded by a thorough evaluation of the physical and endocrine status of the male. If an endocrine disorder is confirmed, an informed decision would be required to determine whether endocrine therapy might be helpful and advisable.

The administration of testosterone to normal, healthy individuals is not indicated. Exogenous testosterone does not appear to be effective for enhancing the libido of normal males, and higher dosages are clearly detrimental to sperm production (Berndtson *et al.*, 1974, 1979). For example, Berndtson *et al.* (1979) administered 0, 50 or 200 μg of testosterone propionate/kg body weight to normal sexually mature stallions every other day for 88 days. Treatment did not influence the time to erection, interval from first mount to ejaculation or the number of mounts required per ejaculation. However, the highest dosage reduced sperm production/g testis and per stallion by 41 and 57%, respectively, while also reducing sperm motility by ~10 percentage points. In contrast to these findings,

McDonnell *et al.* (2003) suggested that injections of testosterone every other day could be used to boost the sexual arousal of stallions without affecting spermatogenesis. The 80 mg dosage used, which reportedly was without effect on sperm production (McDonnell, 1999), was similar to that eliciting pronounced decreases in sperm production and seminal quality in the aforementioned investigation with normal stallions (Berndtson *et al.*, 1979). The reason for the discrepancy between these two studies is unclear. However, the potential for exogenous androgens to depress sperm production is apparent (Berndtson *et al.*, 1974, 1979), and this should be recognized whenever the administration of exogenous androgens is under consideration.

Anabolic steroids

Anabolic steroids are synthetic hormones with growth-promoting properties. Because they enhance muscle development and physical performance, they have been used to provide a competitive advantage for animals shown at halter or used as athletes. Although such uses are prohibited, anabolic steroid use has probably continued at some level, as it has with professional human athletes. Most, if not all, of the anabolic steroids possess androgenic properties. For the male reproductive system, anabolic steroids produce responses closely mimicking those with exogenous androgens. In one study, in which stallions were injected with Equipoise (the anabolic steroid boldenone undecylenate) at a dosage of 4.4 mg/kg body weight once every 3 weeks over a 15 week period, testis size, sperm production/g testis, sperm production per testis and sperm motility were decreased to approximately 55, 65, 28 and 62% of control values, respectively (Squires *et al.*, 1982). The use of anabolic steroids to provide a competitive advantage should be prohibited for both ethical and physiological justifications.

Exercise

Although the benefits of exercise to male fertility are often cited in the popular literature, results from well-designed scientific studies appear limited. Inconsistencies among the reported benefits may, at least in part, reflect shortcomings in experimental design, such as treatment periods that were shorter than the duration of spermatogenesis and/or the absence of reliable pretreatment baseline data. Other potential contributors include differences in the intensity of exercise, provisions for prior conditioning or the use of field data devoid of suitable controls.

However, Snyder and Ralston (1955) conducted an excellent study to assess seminal characteristics and non-return rates for non-exercised versus exercised dairy bulls. Non-exercised bulls were confined to box stalls. The remaining bulls were exercised on a mechanical walker at a speed of 1.24 miles/h on 6 days/week. Exercise was limited to 15 min/day for the first 35 days, but was increased to 23 min for the next 14 days and to 30 min/day for the remainder of the 6 month study period. Exercise was without effect on seminal characteristics or fertility. The total number of sperm per ejaculate averaged 8.53 versus 7.68 billion in non-exercised versus exercised bulls, respectively. The corresponding values for initial sperm motility, abnormal sperm and 60–90 day non-return rates equalled, 73.2 versus 71.5%, 15.0 versus 13.4% and 65.0 versus 63.8%, respectively. This was a large study involving 32 bulls (16/exercise group), a total of 33,292 inseminations and a treatment period exceeding the duration of spermatogenesis.

One of the first studies to assess the effects of exercise on the stallion was conducted by Dinger and Noiles (1986). This two-phase study involved eight 2-year-old stallions. During phase 1, four stallions were assigned either to confinement in box stalls or to a 6x day/week regimen of either 18 min (12 trotting and 6 walking) or 24 min (16 trotting and 8 walking) of forced exercise. After 16 weeks, treatment assignments were reversed for an additional 16 week period. Semen was collected once every 14 days, at which time libido was scored on a scale of 0–4 (with 4 being highest). Libido declined during the exercise periods. For stallions placed initially in the exercised group, libido scores averaged 2.88, 2.06 and 2.81 before treatment and at the end of the 16 week exercised and non-exercised periods, respectively. For stallions placed first in the non-exercised group, libido scores during the pretreatment, non-exercised and exercised periods averaged 3.00, 3.5 and 2.00, respectively. Treatment was without effect on daily sperm output or seminal quality (Dinger *et al.*, 1986).

Based on the foregoing studies, moderate exercise does not appear to have a substantive impact on spermatogenesis or on the seminal quality of bulls (Snyder and Ralston, 1955) or stallions (Dinger and Noiles, 1986; Dinger *et al.*, 1986). However, whereas intense exercise would be unexpected for most farm animals, its effects are of interest relative to stallions used simultaneously for breeding and athletic competition.

Janett *et al.* (2006) assessed the impact of repeated strenuous exercise on the seminal characteristics of 11 stallions 7–19 year old. They collected one ejaculate/week during each of four consecutive 4 week pretreatment (Period 1), exercise (Period 2) and post-exercise (Periods 3 and 4) periods. Exercise consisted of two sessions/week for a total of eight sessions. For the first two sessions, stallions walked on a treadmill for 5 min at 1.5 m/s at 0% inclination, followed by two intervals of trotting at 3.5 m/s at inclinations of 3 and 6%, respectively. For the subsequent sessions, stallions trotted for 3 min at 3.5 m/s and walked for 1 min at 1.5 m/s as the inclination was increased sequentially from 0 to 9%. At the 9% inclination, trotting speed was increased to 4 and 4.5 m/s. This regimen resulted in increases in average heart rate values ranging from 70.4 to 97.4 beats per minute (bpm) before exercise to between 166.8 and 194.0 bpm immediately thereafter. Strenuous exercise had a negative effect on semen quality; the incidence of acrosome defects and nuclear vacuoles began to rise 3 weeks after the beginning of the exercise period, while the percentage of viable frozen–thawed sperm decreased from 53.8% during the pretreatment period to 49.2% during the exercise period. The number of sperm per ejaculate averaged 7.0, 6.8, 6.1 and 6.6 billion during Periods 1–4, respectively. The decline in sperm number per ejaculate during the 4 week period subsequent to the discontinuation of exercise (i.e. during Period 3) is consistent with a negative effect of strenuous exercise on spermatogenesis, which requires approximately 2 months for completion.

Conflicting reports on the effects of intensive training can be found (e.g. Lange *et al.*, 1997; Davies Morel and Gunarson, 2000). For example, Davies Morel and Gunarson (2000) reported average fertility rates of 74.1, 63.7 and 66.9% for Icelandic stallions receiving intensive, moderate or no training, respectively. However, these findings were based on survey data obtained from a variety of breeding associations and individual stallion owners, rather than from a well-controlled experiment. Intense exercise training has also been associated with negative reproductive consequences in humans (e.g. Arce and De Souza, 1993). Accordingly, intense, stressful exercise is probably detrimental to male fertility. As a practical matter, it is possible that moderate physical conditioning might have a beneficial impact on levels of mating activity for males in a natural mating system, although it does not appear to have a beneficial effect on seminal quality per se.

Frequency of ejaculation

Frequency of ejaculation appears to be without effect on quantitative rates of sperm production. For example, Carson and Amann (1972) conducted a well-replicated study in which daily sperm production of ~8-month-old rabbits was assessed after 40 days of sexual rest or ejaculation either daily or twice daily on an every other day basis. Selected data from that study are presented in Table 2.8. Treatment was without effect on testicular weight or DSP/male or DSP/g of testicular tissue.

Amann (1962a) also studied the effect of frequency of ejaculation on spermatogenesis in dairy bulls. For this study, nine young Holstein bulls were assigned to treatments consisting of either sexual rest or ejaculation 2× or 8×/wk for 20 wk. Five older bulls were either sexually rested or ejaculated 6–8×/week. The results of this study are summarized in Table 2.9.

Frequency of ejaculation was without effect on testicular weight, the percentage of the testicular parenchyma occupied by seminiferous tubules, or sperm production as judged from the number of germ cells/stage I seminiferous tubular cross section. Although the frequency of ejaculation does not appear to influence sperm production rates, it does exert a profound effect on the number of sperm per ejaculate, as described later.

How to measure daily sperm production?

A number of approaches may be used to measure sperm production. Choosing from among these requires consideration of several factors, including whether the testes can be removed or whether sperm production is to be estimated in a prospective breeding animal. In the former case, one should also consider whether one needs to know how many sperm are being produced per unit of time, or whether it is sufficient to simply assess relative changes in sperm production due to an experimental treatment. In addition to this, every method for quantifying sperm production requires one or more technical assumptions. Readers with an interest in quantifying sperm production should familiarize themselves with these and other considerations of importance during the selection of an evaluation method. Such considerations have been presented elsewhere in greater detail (Amann, 1981; Berndtson, 2011). Several useful and popular methods for quantifying sperm production are described briefly below.

Table 2.8. Daily sperm production of sexually rested (SR) versus frequently ejaculated rabbits. Adapted from Carson and Amann, 1972.

	Frequency of ejaculation[a]		
Characteristic	1×/24 h	2×/48 h	SR
Paired testes weight (g)	6.29	6.58	6.40
DSP[b]/male (10^6)	225	247	247
DSP/g testis	37.4	38.9	39.0

[a]Rabbits were sexually rested for 30 days, followed by 40 days of once daily ejaculation, twice every other day ejaculation or sexual rest.
[b]Daily sperm production.

Enumeration of homogenization-resistant spermatids

During spermiogenesis, the chromatin of elongating spermatids condenses and becomes resistant to homogenization (Amann and Almquist, 1961a). These nuclei remain intact during homogenization, while other components of the tissue are destroyed. Thus, after homogenization of a given quantity of tissue in a known volume of fluid, haemocytometry can be used to determine the number of resistant

Table 2.9. Testicular characteristics and sperm production of sexually rested (SR) or ejaculated bulls. Adapted from Amann, 1962a.

				Germ cells/cross section			
Treatment	No. of bulls	Testes weight (g)	Seminiferous tubules (% testis)	A[a]	Young (1°)[b]	Old (1°)[b]	Spermatids
			Young bulls[c]				
SR[d]	3	301.5	76.5	2.75	55.5	54.5	214.0
2×/wk[d]	3	286.5	77.5	3.30	54.5	52.5	186.0
8×/wk[d]	3	290.5	77.0	2.85	52.5	49.5	188.5
			Mature bulls				
SR[e]	2	411.0	73.8	2.90	55.0	52.0	189.0
6–7×/wk[f]	3	396.3	71.7	3.70	64.0	62.0	226.0

[a]Type A spermatogonia (smooth).
[b]Primary spermatocytes.
[c]Animals 36 months old when sacrificed.
[d]Treatment duration 20 weeks.
[e]5 or 20 weeks sexual rest.
[f]Individual bulls ejaculated 6×/wk for 26 wk, ≥6×/wk for 40 wk, followed by 7×/wk for 7 wk and 8×/wk for 17 wk, respectively.

spermatids/g tissue, or per testis or male. The numbers of such spermatids can be compared directly to provide a relative comparison of sperm production rates in control versus treated subjects (Berndtson, 1977). Alternatively, if an estimate of actual daily sperm production is needed, this can be determined by dividing the spermatid number by a time divisor, equivalent to the number of days of sperm production represented by these cells. For example, if spermatids became resistant to homogenization 6.5 days before spermiation, the total number of homogenization-resistant spermatids would be divided by a time divisor of 6.5 days to obtain an estimate of daily sperm production. Table 2.10 reports time divisors used with the homogenization method for several farm species.

Germ cells per Sertoli cell or per round seminiferous tubular cross section

In addition to germinal cells, seminiferous tubules contain somatic cells known as Sertoli cells. These perform functions that include the formation of a blood–testis barrier, secretion of androgen-binding protein and others (Russell and Griswold, 1993). Sertoli cells form at an early age, and were once believed to remain as a numerically stable population thereafter (Gondos and Berndtson, 1993). It is now apparent that the number of Sertoli cells can change in response to age and/or season in some species (Johnson and Thompson, 1983; Johnson and Nguyen, 1986; Johnson *et al.*, 1991). None the less, these cells appear to be quite resistant to harsh treatments, as evident from their persistence after exposure to some treatments that cause almost complete obliteration of the germ cells (Oakberg, 1959). Because of this resistance, the germ cell:Sertoli cell ratio can be used to assess relative changes in sperm production. For example, if a treatment reduced sperm production by 50%, this would be expected to cause a corresponding 50% reduction in the spermatid:Sertoli cell ratio. With this technique, direct counts of the number of germ cells and Sertoli cells are made in a predetermined number of round seminiferous tubular cross sections. For most studies, one stage, usually containing spherical spermatids, is chosen as representing spermatogenesis as a whole.

The histological specimens that are examined contain some nuclei residing entirely within the section, and also fragments produced by sectioning. The proportion of fragments will increase as the section thickness is decreased, and it will be greater for nuclei of larger versus smaller diameter. Therefore, the resulting crude counts, which include both whole and partial nuclei, must be converted to true counts or whole-cell equivalents. Several equations have been developed for this purpose. That used by the author is Abercrombie's equation (Abercrombie, 1946), which is as follows:

True count = Crude count × (section thickness/section thickness + nuclear diameter)

Abercrombie's equation and other similar equations are only applicable for spherical nuclei. Thus, it is customary to select tubules at a stage containing spherical spermatids for counting; irregularly shaped elongated spermatids are not counted. The shape of the Sertoli nuclei is also irregular, so it is also customary to count only those Sertoli nuclei containing a nucleolus. Abercrombie's formula can be applied to the spherical nucleoli to obtain a true count for the Sertoli nuclei. True counts are then used to determine germ cell:Sertoli cell ratios. Alternatively, some investigators calculate the average number of Sertoli cells per tubular cross section in the control subjects or for all animals in

Table 2.10. Reported time divisors used for estimating daily sperm production from the number of homogenization-resistant spermatids.

Species	Time divisor (days)	Reference
Boar	5.86	Okwun *et al.*, 1996
Bull	3.27	Amann and Almquist, 1962
Goat	3.56	Ritar *et al.*, 1992
Ram	5.0	Cardosa and Queiroz, 1988
Stallion	6.0	Johnson and Neaves, 1981

the experiment. The number of germ cells in this average is then expressed as the number of germ cells per tubular cross section. This technique is useful for assessing relative changes in sperm production due to treatment, but it does not provide an estimate of the actual number of sperm being produced/day per unit of tissue or per male.

Volume density approaches for estimating sperm production

For the volume density approach, the volume of the testicular tissue is first recorded. This is usually determined by measuring its fluid displacement or, alternatively, is simply assumed to equal testis weight minus the weight of the testicular capsule, because the specific gravity of the testis in mammals is very close to 1.0 (Swierstra, 1966; Gebauer *et al.*, 1974a; de Jong and Sharpe, 1977; Johnson and Neaves, 1981; Johnson *et al.*, 1981; Mori *et al.*, 1982). The tissue is then processed for histological evaluation. Tissue shrinkage should be recorded, to permit adjustment as necessary (Berndtson, 2011). Next, the eyepiece of a microscope is fitted with a fixed pointer or pointers. The slide of testicular tissue is moved at random, after which the identity of the structure at the tip of the pointer is recorded. This process is repeated many times. With sufficient sampling, the frequency with which a particular structure is 'hit' will be proportional to its volume density. For example, if the nuclei of the spherical spermatids occupied 10% of the tissue, one would expect these nuclei to be 'hit' 10% of the time.

With these data, the total volume of the nuclei of each type of germ cell can be determined as the product of its volume density and the total testicular parenchymal volume. By dividing the total volume of these nuclei by the volume of a single nucleus, an estimate is obtained of the total number of cells of each type. The volume of spherical nuclei is usually determined by entering nuclear diameter into the equation for calculating the volume of a sphere. Reconstructon of serial sections and other more sophisticated procedures have been used to estimate the volume of nuclei with an irregular outline (e.g. elongated spermatids and Sertoli cells; Johnson *et al.*, 1984; Sinha Hikim *et al.*, 1988). Once the total number of cells of a given type has been determined, the data can be used to calculate germ cell:germ cell or germ cell:Sertoli cell ratios. Alternatively, an estimate can be obtained of daily sperm production by dividing the total number of spermatids by a time divisor (the number of days of sperm production represented by these cells). DSP may also be estimated from the numbers of younger germ cells, but this requires correction for subsequent cell divisions. Such estimates are subject to greater potential errors due to normal or treatment-induced cellular attrition.

Estimating daily sperm production (DSP) from daily sperm output (DSO)

Sexual rest results in sperm losses in the urine (Holtz and Foote, 1972), but such losses are minimal for males maintained on a regular, frequent schedule of ejaculations (Amann, 1981). Thus, sperm output that is frequently collected from a male provides an excellent method for estimating DSP. It offers the advantage of not requiring the removal of testicular tissue. In addition, by collecting semen continuously over time, there is the opportunity to assess temporal changes in sperm output and seminal quality that might follow acute, experimental exposure of males to an agent or experimental treatment of interest (e.g. assessment of seminal quality after a single dosage with an anthelmintic).

Typical sperm output of economically important species

Sperm production varies greatly among individuals both within and among species. Several factors contributing to such variability within species have already been described, including inherent differences in testis size and the influences of variables such as age, season (in some species), environmental factors, drugs, etc. The typical sperm production of breeding-age males of several species is summarized in Table 2.11.

Most male mammals of reproductive age other than humans produce and ejaculate

Table 2.11. Daily sperm production (DSP) of sexually mature males (range in parentheses).

Species	DSP/g (million)	DSP/male (billion)	Reference
Boar	–	31.3	Kennelly and Foot, 1964
Boar	24.7	16.5	Swierstra, 1968b
Bull	17.7	11.5	Amann and Almquist, 1962
Bull	16.9	5.3	Swierstra, 1966
Bull	11.2 (7.25–16.67)	–	Berndtson *et al.*, 1987a
Bull	9.31	4.30	Berndtson and Igboeli, 1988
Bull	–	3.79 (1.99–7.86)	Berndtson and Igboeli, 1989
Goat	23.8	4.0–6.4	Ritar *et al.*, 1992
Goat	30.3	5.54	Leal *et al.*, 2004
Ram	~27[a]	12.9	Schanbacher and Ford, 1979
Ram		8.06[b]	Dacheux *et al.*, 1981
Ram	22.8	4.4	Cardosa and Queiroz, 1988
Stallion	21.2	8.0	Gebauer *et al.*, 1974a
Stallion	11.9[c]	3.6[c]	Berndtson *et al.*, 1979
Stallion	17.0–18.8	5.3–6.4	Johnson and Neaves, 1981

[a]Estimate calculated by the author by dividing DSP by testicular weight, without correction for the weight of the tunica albuginea.
[b]Ile-de-France rams.
[c]Estimates derived by dividing the number of homogenization-resistant spermatids by a time divisor of 6.0 days.

sperm in great excess of the numbers required for normal fertility. This does not diminish the importance of considering sperm production during an examination of breeding soundness or during the selection of males for breeding. As discussed later, sperm number declines in successive ejaculates taken on a single day (Amann, 1981). Thus, the potential for sperm numbers per mating to decrease to below an optimal level is always possible for males required to breed large numbers of females within a short period of time via natural mating. However, the latter possibility should be less for males producing large numbers of sperm than for those for which sperm production is more limited. The potential for sperm number to become limited may also differ among species. Foote (1978) reported that rams tend to ejaculate a relatively smaller number of sperm per semen sample than bulls or boars, and that they are capable of ejaculating repeatedly (11×/day in the study of Salamon, 1962). In contrast, boars ejaculate a large number of sperm in each ejaculate, and can deplete their epididymal sperm reserves quite rapidly (Foote, 1978).

Sperm Maturation and Transport through the Excurrent Ducts

Sperm transit time

Upon release from the testis, sperm undergo transit through the efferent ducts and the epididymis. The time required for epididymal transit was summarized in Table 2.7 for several of the economically important species. Whereas the transit time through the caput epididymis and corpus epididymis is relatively constant, transit through the cauda epididymis is more rapid in males ejaculating at a higher frequency (e.g. daily; Amann, 1981).

Maturational changes and the acquisition of fertility

Spermatozoa are not fertile upon release from the testis, but undergo maturation within the epididymis. Epididymal secretions appear to play a critical role in the maturation process, which includes the acquisition of motility, changes in sperm membranes that permit binding to the zona pellucida of the oocyte, nuclear decondensation and other changes (Cooper, 1995). The sperm of most species do not attain fertility until reaching the corpus epididymis.

Extragonadal sperm reserves (EGR)

The epididymides and vas deferentia typically contain the equivalent of several days of sperm production, which collectively constitute the EGR. It is from these reserves, and especially from the tail of the epididymis, that sperm are emitted during ejaculation. During periods of sexual rest, sperm accumulate and reach maximal levels. Degeneration and resorption of aged sperm appears to be minimal in normal males (Amann, 1981). Rather, as additional sperm leave the testis and enter the EGR, others enter the urethra and are flushed from the body via the urine (Holtz and Foote, 1972). The size of the EGR in sexually rested males of several economically important species is summarized in Table 2.12.

Only a portion of the EGR is available for ejaculation. Some sperm reside within the head and initial segment of the body of the epididymis, from which they are unavailable for ejaculation, while the lining of the epididymal duct can also preclude complete removal. Based on studies with rats, bulls and stallions, Amann (1981) estimated that about 55–65% of the sperm within the cauda epididymis and vas deferens of sexually rested males can be removed by harvesting 5–20 ejaculates in succession on a single day.

Table 2.12. Extragonadal sperm reserves (EGR) of sexually rested males.

Species	EGR (billion)	Reference
Boar	166[a]	Swierstra, 1971
Bull	69	Almquist and Amann, 1961
Goat	17.5	Jindal and Panda, 1980
Goat	47.8	Ritar *et al.*, 1992
Ram	93.5	Lino, 1972
Stallion	89	Amann *et al.*, 1979

[a]Values 72 h post depletion of EGR via the collection of one ejaculate in the morning, a second in the afternoon and a third on the following morning.

Several important management concepts are easy to understand by drawing an analogy between the EGR and a reservoir from which water might be obtained. As sperm are released from the testis and enter the extragonadal ducts at a constant rate, one could imagine that water is also entering the reservoir at a constant rate. During times when water is not being used, its level will continue to rise until, upon reaching maximal capacity, it simply overflows the dam. This would be comparable to the filling of the EGR during periods of sexual rest, which would continue until sperm overflowed into the urinary tract. At other times, water might be drawn from a reservoir at a rate exceeding that at which it was being replenished. This might continue for a while, but would be accompanied by a progressive decrease in the level of water behind the dam. This would be analogous to the collection of several successive ejaculates from a male after sexual rest. Each ejaculate would reduce the number of residual sperm within the EGR. Ultimately, if water continued to be drawn from a reservoir at a rate exceeding its rate of replenishment, the reservoir would be depleted. At that point, the rate at which additional water could be drawn would be limited by and equal to its rate of replenishment. Similarly, if a male continues to ejaculate at a high frequency, the EGR will ultimately be depleted. At that point, DSO will be limited by and equal to DSP. These concepts have important implications relative to conventional breeding soundness examinations, and for the management of males to maximize the sperm harvest, which are discussed below.

Sperm Harvest – Maximising DSO and Seminal Quality

Seminal quality and sperm number in ejaculates taken after sexual rest

The first ejaculate(s) taken after sexual rest usually contains very large numbers of sperm. The data of Squires *et al.* (1979) illustrate the decline in sperm output in successive ejaculates taken after sexual rest. In their study, the sperm output of 2–3 year old stallions averaged 4.5, 2.4, 0.9, 0.6 and 0.5 billion for the first to fifth ejaculates, respectively. The corresponding sperm output for 4–6 year old stallions averaged 9.5, 3.5, 2.6, 1.3 and 1.1 billion, while that for 9–16 year old stallions averaged 11.4, 5.5, 2.4, 1.8 and 1.2 billion, respectively. A decline in sperm output in successive ejaculates is to be expected, based on the concept of the EGR described above. For this reason, the number of sperm within the first few ejaculates taken after sexual rest does not provide a reliable estimate of the average DSO to be expected once a male is used on a regular frequent basis. This is an important point relative to routine breeding soundness examinations, for which only one or two ejaculates are often collected after a lengthy period of sexual inactivity. Such collections are useful, because they demonstrate a willingness or ability of the male to ejaculate, while confirming recent spermatogenic activity and providing an opportunity to assess seminal quality. However, they are inadequate for characterizing actual levels of sperm production. Unless there are the time and resources with which to maintain the male on a regular, frequent seminal collection schedule, at which DSO would approach DSP, measurements should be made of scrotal width or circumference, as these allow an estimate of testis size, which is highly correlated with DSP (Table 2.4).

While initial ejaculates taken after sexual rest usually contain large numbers of sperm, the quality of these first ejaculates may be reduced, although this appears species dependent. For example, in one study (Pickett and Voss, 1973), initial motility averaged 56% for ejaculates from stallions collected either once or six times/wk. In contrast, Almquist (1973) collected two ejaculates from bulls on 1 day/wk. Initial motility averaged 48 versus 60% for first and second ejaculates of beef bulls collected without sexual preparation. The corresponding motility for dairy bulls collected after one false mount averaged 63 versus 68%, respectively. Much greater decreases in semen quality are possible during extended periods of inactivity. With respect to a breeding soundness examination, retesting at a later date and/or the collection of additional ejaculates on the same date would be advisable if the quality of an initial ejaculate(s) was questionable. It is equally important to consider that spermatogenesis requires approximately 2 months in most mammals, and that additional time is required for sperm transport through the epididymis (Table 2.7). Thus, the semen that is harvested on any given day represents the culmination of a process that for most species extended over at least the previous 60–70 days. Low semen quality could reflect problems with spermatogenesis or epididymal function arising at any time during that period, from which recovery could require a similar length of time.

The relationship between frequency of ejaculation and DSO

As cited previously, sperm production is not influenced by frequency of ejaculation. During sexual rest, sperm may be lost in the urine (Holtz and Foote, 1972). Higher frequencies of ejaculation can eliminate such losses, while at any frequency above that level, increasing frequencies of ejaculation will reduce the number of sperm per ejaculate. The results of several representative studies conducted to examine the influence of frequency of ejaculation on the DSO of dairy bulls and stallions are given in Tables 2.13 and 2.14, respectively. For most

Table 2.13. Frequency of ejaculation and daily sperm output (DSO) of Holstein bulls.

Frequency of ejaculation	Sperm/ ejaculate (billion)	DSO (billion)	Reference
1×/wk	10.9[a]	1.56[a]	Almquist, 1982
6×/wk	5.1[a]	4.41[a]	
1×/wk	17.8	2.5	Hafs *et al.*, 1959
7×/wk	4.8	4.8	
2×/day on M, W, F	5.3	4.6	Amann and Almquist, 1961b
2×/day on Th, F	7.6	4.3	Seidel and Foote, 1969

[a]Means for 5–9-year old bulls.

species, daily or every other day ejaculation will enable essentially all of a male's sperm production to be harvested.

Sexual preparation and sperm output

To collect semen with an artificial vagina, one need only provide conditions that will arouse the male and permit him to mount. However, the number of sperm per ejaculate from rams (Knight, 1974), bulls (Hale and Almquist, 1960) and boars (Hemsworth, 1979) can be increased by appropriate sexual preparation. Sexual preparation entails administering procedures such as teasing, false mounting or active restraint before ejaculation. Such procedures cause a release of oxytocin and possibly other hormones, which, in turn, enhance sperm transport or emission from the extragonadal ducts (Sharma and Hays, 1973, Berndtson and Igboeli, 1988).

False mounting consists of allowing the male to mount a teaser animal or dummy while the penis is deflected to prevent penetration of the reproductive tract or artificial vagina. Without tactile stimulation, the male will dismount without ejaculating. This procedure is usually repeated 2–3 times. Active restraint involves providing conditions that encourage the male to mount, but restraining him when he attempts to do so. The latter is called teasing when applied to the stallion, for which it is customary to provide a physical barrier between the mare and stallion to prevent injury. The impact of sexual preparation on the sperm output of dairy bulls is illustrated by the data in Table 2.15. Sexual arousal does not increase sperm output from stallions, although teasing increases gel and total seminal volume (Pickett and Voss, 1973).

While sexual preparation is effective in increasing the number of sperm per ejaculate in most species, it is important to recall that ejaculation does not affect sperm production. Rather, by increasing the efficiency of ejaculation, sexual preparation simply allows maximal DSO to be achieved via ejaculation at a lower frequency than would be required in its absence. This is beneficial for the AI industry, because it reduces the total time required to collect and process the maximal quantity of semen for a given male.

Table 2.14. Frequency of ejaculation and daily sperm output (DSO) of stallions.

Frequency of ejaculation	Sperm/ejaculate (billion)	DSO (billion)	Reference
Daily	4.5	4.5	Gebauer *et al.*, 1974a
Every other day	7.0	3.5	Swierstra *et al.*, 1975
1×/wk	13.5	1.93	Pickett and Voss, 1973
3×/wk	12.7	5.43	
6×/wk	7.2	6.16	
1×/wk	11.4	1.63	Pickett *et al.*, 1975
3×/wk	11.7	5.03	
6×/wk	5.9	5.04	

Table 2.15. Increase in sperm output attributable to sexual preparation of bulls.

Ejaculations/wk	No. of false mounts	Active restraint (min)	Sperm/first ejaculation (% increase)	Reference
1	1	2–3	36	Collins *et al.*, 1951
2	1	0	50	Branton *et al.*, 1952
2	2	0	67	
4	0	1	129	Crombach, 1958
4	10	0	147	
4	5	1	251	
6	3	–	~30	Almquist *et al.*, 1958

Stimulus pressure to maintain libido

Although the frequencies of ejaculation required to maximize sperm harvest are not excessive, males maintained on such schedules can experience reduced libido. This is often manifested via increases in reaction time – the time between presentation to a teaser animal and mounting or ejaculation.

This issue was examined extensively for dairy bulls in the classical studies of Hale and Almquist (1960), who developed strategies for maintaining libido that remain the standard of the bovine AI industry. Their research showed that the key to maintaining good libido is to provide novelty, or what they defined as stimulus pressure. In one of their studies, bulls were maintained on a 1×/wk collection schedule. Identical procedures were followed for each collection, and reaction times were recorded. This process was repeated each week until the reaction time for a bull exceeded 10 min. On the subsequent week, the bull was presented to the same teaser animal, which had simply been moved a distance of 3 ft from the location used previously. Reaction time decreased to an average of ~2 min. A similar response was obtained without moving the location of the teaser animal, but by simply rocking the teaser back and forth in place. In another experiment, these authors (Hale and Almquist, 1960) allowed bulls to mount and ejaculate at will for 1 h, during which time individual bulls ejaculated an average of 10.6 times. The interval between successive ejaculates increased during this time, and most bulls were satiated by the end of the first hour. The authors then introduced a new teaser animal, again allowing the bulls to mount at will. During the second hour, bulls ejaculated an additional 8.4 times. These data demonstrated that satiation was associated with a specific teaser animal or set of conditions, rather than with ejaculation per se. The novelty created by a change of teasers, movement and/or a change of locations is exploited routinely by the AI industry.

Almquist and Hale (1956) demonstrated further that the frequency with which teasers, locations, etc. needs to be changed to maintain good libido is related to the frequency of ejaculation. This is evident from the data in Table 2.16. In this study, new teasers were introduced as necessary to maintain minimal reaction times, which averaged ~6 min. The frequency at which new teaser animals needed to be introduced was three times greater for bulls maintained on a 6× versus a 2×/wk collection schedule. Novelty appears to have a similar beneficial impact on libido in many other species (Price, 1985).

Conclusions

Spermatogenesis is a truly remarkable process, resulting in levels of sperm production in most species that enable successful mating with large numbers of females or the potential for insemination of an even greater number of females via AI. Because sperm production is not influenced by frequency of ejaculation, higher frequencies of ejaculation are accompanied by corresponding decreases in the number of sperm per ejaculate. The implications of this relationship for the management of males used for natural mating or for AI have been described. In addition to this, libido must be maintained if producers are to make maximal utilization of valuable sires. Novelty appears to be a key factor affecting libido in bulls, and presumably also in many other species. Several approaches for providing novelty have been described. Judicious selection and management are critical for allowing each male to realize its maximum reproductive potential.

Table 2.16. Stimulus pressure to maintain sexual activity of bulls ejaculated 2× or 6×/wk. From Almquist and Hale, 1956.

Reaction time/Stimulus pressure	Frequency of ejaculation	
	2×/wk	6×/wk
Reaction time (min)	6.4	6.2
Stimulus pressure[a]		
Per 24 wk period	4.5	12.5
Per 48 ejaculates	4.5	4.2

[a]No. stimulus animals or combinations of stimulus animals presented.

References

Abercrombie M (1946) Estimation of nuclear population from microtome sections *Anatomical Record* **94** 238–248.

Alkass JE, Bryant MJ and Walton JS (1982) Some effects of level of feeding and body condition upon sperm production and gonadotropin concentrations in the ram *Animal Production* **34** 265–277.

Almeida FFL, Leal MC and Franca LR (2006) Testis morphometry, duration of spermatogenesis, and spermatogenic efficiency in the wild boar (*Sus scrofa scrofa*) *Biology of Reproduction* **75** 792–799.

Almquist JO (1973) Effects of sexual preparation on sperm output, semen characteristics and sexual activity of beef bulls with a comparison to dairy bulls *Journal of Animal Science* **36** 331–336.

Almquist JO (1982) Effect of long term ejaculation at high frequency on output of sperm, sexual behavior, and fertility of Holstein bulls; Relation of reproductive capacity to high nutrient allowance *Journal of Dairy Science* **65** 814–823.

Almquist JO and Amann RP (1961) Reproductive capacity of dairy bulls. II. Gonadal and extra-gonadal sperm reserves as determined by direct counts and depletion trials; dimensions and weight of genitalia *Journal of Dairy Science* **44** 1668–1678.

Almquist JO and Hale EB (1956) An approach to the measurement of sexual behavior and semen production of dairy bulls In *Third International Congress on Animal Reproduction (Cambridge) Plenary papers* **50** (cited by Hale and Almquist, 1960).

Almquist JO, Hale EB and Amann RP (1958) Sperm production and fertility of dairy bulls at high-collection frequencies with varying degrees of sexual preparation *Journal of Dairy Science* **41** 733 [Abstract].

Amann RP (1962a) Reproductive capacity of dairy bulls. III. The effect of ejaculation frequency, unilateral vasectomy, and age on spermatogenesis *American Journal of Anatomy* **110** 49–67.

Amann RP (1962b) Reproductive capacity of dairy bulls. IV. Spermatogenesis and testicular germ cell degeneration *American Journal of Anatomy* **110** 69–78.

Amann RP (1981) A critical review of methods for evaluation of spermatogenesis from seminal characteristics *Journal of Andrology* **2** 37–58.

Amann RP and Almquist JO (1961a) Reproductive capacity of dairy bulls. I. Technique for direct measurement of gonadal and extra-gonadal sperm reserves *Journal of Dairy Science* **44** 1537–1543.

Amann RP and Almquist JO (1961b) Reproductive capacity of dairy bulls. V. Detection of testicular deficiencies and requirements for experimentally evaluating testis function from semen characteristics *Journal of Dairy Science* **44** 2283–2291.

Amann RP and Almquist JO (1962) Reproductive capacity of dairy bulls. VIII. Direct and indirect measurement of testicular sperm production *Journal of Dairy Science* **45** 774–781.

Amann RP and Schanbacher BD (1983) Physiology of male reproduction *Journal of Animal Science* **57(Supplement 2)** 380–403.

Amann RP, Thompson DL Jr, Squires EL and Pickett BW (1979) Effects of age and frequency of ejaculation on sperm production and extragonadal sperm reserves in stallions *Journal of Reproduction and Fertility Supplement* **27** 1–6.

Arce JC and De Souza MJ (1993) Exercise and male factor infertility *Sport Exercise* **15** 146–169.

Berndtson WE (1977) Methods for quantifying mammalian spermatogenesis: a review *Journal of Animal Science* **44** 818–833.

Berndtson WE (2011) The importance and validity of technical assumptions required for quantifying sperm production rates: a review *Journal of Andrology* **32** 2–14.

Berndtson WE and Desjardins C (1974) The cycle of the seminiferous epithelium and spermatogenesis in the bovine testis *American Journal of Anatomy* **140** 167–180.

Berndtson WE and Igboeli G (1988) Spermatogenesis, sperm output and seminal quality of Holstein bulls electroejaculated after administration of oxytocin *Journal of Reproduction and Fertility* **82** 467–475.

Berndtson WE and Igboeli G (1989) Numbers of Sertoli cells, quantitative rates of sperm production, and the efficiency of spermatogenesis in relation to the daily sperm output and seminal quality of young beef bulls *American Journal of Veterinary Research* **50** 1193–1197.

Berndtson WE, Desjardins C and Ewing LL (1974) Inhibition and maintenance of spermatogenesis in rats implanted with polydimethylsiloxane capsules containing various androgens *Journal of Endocrinology* **62** 125–135.

Berndtson WE, Hoyer JH, Squires EL and Pickett, BW (1979) Influence of exogenous testosterone on sperm production, seminal quality and libido of stallions *Journal of Reproduction and Fertility Supplement* **27** 19–23.

Berndtson WE, Igboeli G and Parker WG (1987a) The numbers of Sertoli cells in mature Holstein bulls and their relationship to quantitative aspects of spermatogenesis *Biology of Reproduction* **37** 60–67.
Berndtson WE, Igboeli G and Pickett BW (1987b) Relationship of absolute numbers of Sertoli cells to testicular size and spermatogenesis in young beef bulls *Journal of Animal Science* **64** 241–246.
Bilaspuri G S and Guraya SS (1984) The seminiferous epithelial cycle and spermatogenesis in goats (*Capra hircus*) *Journal of Agricultural Science (Cambridge)* **103** 359–368.
Branton C, D'Arsenbourg G and Johnston JE (1952) Semen production, fructose content of semen and fertility of dairy bulls as related to sexual excitement *Journal of Dairy Science* **35** 801–807.
Brinsko SP, Varner DD, Love CC, Blanchard TL, Day BS and Wilson ME (2005) Effect of feeding a DHA-enriched nutriceutical on the quality of fresh, cooled and frozen stallion semen *Theriogenology* **63** 1519–1527.
Cardosa FM and Queiroz GF (1988) Duration of the cycle of the seminiferous epithelium and daily sperm production of Brazilian hairy rams *Animal Reproduction Science* **17** 77–84.
Carson WS and Amann RP (1972) The male rabbit. VI. Effects of ejaculation and season on testicular size and function *Journal of Animal Science* **34** 302–309.
Clermont Y (1962) Quantitative analysis of spermatogenesis of the rat: a revised model for the renewal of spermatogonia *American Journal of Anatomy* **111** 111–129.
Clermont Y and Leblond CP (1955) Spermiogenesis of man, monkey, ram and other mammals as shown by the "periodic acid–Schiff" technique *American Journal of Anatomy* **96** 229–253.
Collins WJ, Bratton RW and Henderson CR (1951) The relationship of semen production to sexual excitement of dairy bulls *Journal of Dairy Science* **34** 224–227.
Cooper TG (1995) The epididymal influence on sperm maturation *Reproductive Medicine Review* **4** 141–161.
Coulter GH, Larson LL and Foote RH (1975) Effect of age on testicular growth and consistency of Holstein and Angus bulls *Journal of Animal Science* **41** 1383–1389.
Crombach JJML (1958) The effect of the preparation of A. I. bulls before service on semen production and conception *Tijdschrift Diergeneesk* **83** 137 (cited by Hale and Almquist, 1960).
Dacheux JL, Pisselet C, Hochereau-de-Reviers M-T and Courot M (1981) Seasonal variations in rete testis fluid secretion and sperm production in different breeds of ram *Journal of Reproduction and Fertility* **61** 363–371.
Davies Morel MCG and Gunarson (2000) A survey of the fertility of Icelandic stallions *Animal Reproduction Science* **64** 49–64.
de Jong FH and Sharpe RM (1977) The onset and establishment of spermatogenesis in rats in relation to gonadotropin and testosterone levels *Journal of Endocrinology* **75** 197–207.
Dinger JE and Noiles EE (1986) Effect of controlled exercise on libido in two-year-old stallions *Journal of Animal Science* **62** 1220–1223.
Dinger JE, Noiles EE and Hoaglund TA (1986) Effect of controlled exercise on semen characteristics in two-year-old stallions *Theriogenology* **25** 525–535.
Eilts BE (2005a) Bull Breeding Soundness Examination pp 222–235 Society for Theriogenology, Montgomery. Available at: http://www.vetmed.lsu.edu/eiltslotus/theriogenology-5361/bull.htm (accessed 22 August 2013).
Eilts BE (2005b) Ram Breeding Soundness Examination (pp 575–584) Society for Theriogenology, Montgomery. Available at: http://www.vetmed.lsu.edu/eiltslotus/theriogenology-5361/ram_2.htm (accessed 22 August 2013).
Foote RH (1978) Factors influencing the quantity and quality of semen harvested from bulls, rams, boars and stallions *Journal of Animal Science* **47(Supplement 2)** 1–11.
Foote RH and Berndtson WE (1992) Reversibility in male reproductive risk assessment: the germinal cells In *Male Reproductive Risk Assessment* pp 1–55 Ed AR Scialli and ED Clegg. CRC Press, Inc., Boca Raton.
Franca L R and Cardosa FM (1998) Duration of spermatogenesis and sperm transit time through the epididymis in the Piau boar *Tissue and Cell* **30** 573–582.
Franca LR, Becker-Silva SC and Chiarini-Garcia H (1999) The length of the cycle of the seminiferous epithelium in goats (*Capra hircus*) *Tissue and Cell* **31** 274–280.
Franca LR, Avelar GF and Almeida FFL (2005) Spermatogenesis and sperm transit through the epididymis in mammals with emphasis on pigs *Theriogenology* **63** 300–318.
Frankenhuis MT, de Rooij DG and Kramer MF (1980) Spermatogenesis in the pig In *Proceedings of the 9th International Congress on Animal Reproduction and Artificial Insemination*, 16th–20th June 1980 **Volume III (free communications)** p 265 (cited by Amann *et al.*, 1983) Editorial Garsi Madrid (cited by

Amann *et al.*, 1983 as *Proceedings of the 9th International Congress on Animal Reproduction and Artificial Insemination* **5** 17).
Gebauer MR, Pickett BW and Swierstra EE (1974a) Reproductive physiology of the stallion. II. Daily production and output of sperm *Journal of Animal Science* **39** 732–736.
Gebauer MR, Pickett BW and Swierstra EE (1974b) Reproductive physiology of the stallion. III. Extragonadal transit time and sperm reserves *Journal of Animal Science* **39** 737–742.
Gondos B and Berndtson WE (1993) Postnatal and prepubertal development In *The Sertoli Cell* pp 115–154 Ed LD Russell and MD Griswold. Cache River Press, Clearwater.
Hafs HD, Hoyt RS and Bratton RW (1959) Libido, sperm characteristics, sperm output, and fertility of mature dairy bulls ejaculated daily or weekly for thirty-two weeks *Journal of Dairy Science* **42** 626–636.
Hahn J, Foote RH and Seidel GE Jr (1969) Testicular growth and related sperm output in dairy bulls *Journal of Animal Science* **29** 41–47.
Hale EB and Almquist JO (1960) Relation of sexual behavior to germ cell output in farm animals *Journal of Dairy Science* **(Supplement 43)** 145–169.
Harris MA, Baumgard LH, Arns MJ and Webel SK (2005) Stallion spermatozoa membrane phospholipid dynamic following dietary *n*-3 supplementation *Animal Reproduction Science* **89** 234–237.
Heller CG and Clermont Y (1964) Kinetics of the germinal epithelium in man *Recent Progress in Hormone Research* **20** 545–575.
Hemsworth PH (1979) The effect of sexual stimulation on the sperm output of the domestic boar *Animal Reproduction Science* **2** 387–394.
Hochereau-de Reviers MT (1970) Etudes des divisions spermatogoniales et du renouvellement de la spermatogonie souche chez le taureau. Doctor of Science thesis, University of Paris, Paris.
Hochereau-de Reviers MT, Courot M and Ortavant R (1964) Durée de la spermatogenese chez le taureau étude par autoradiographie testiculaire In *5th International Congress on Animal Reproduction and Artificial Insemination* **Volume III** 541–546 (cited by Amann and Schanbacher, 1983).
Holtz W and Foote RH (1972) Sperm production, output and urinary loss in the rabbit *Proceedings of the Society for Experimental Biology and Medicine* **141** 958–962.
Janett F, Burkhardt C, Burger D, Imboden I, Hassig M and Thun R (2006) Influence of repeated treadmill exercise on quality and freezability of stallion semen *Theriogenology* **65** 1737–1749.
Jindal SK and Panda JN (1980) Epididymal sperm reserves of the goat (*Capra hircus*) *Journal of Reproduction and Fertility* **59** 469–471.
Johnson L and Neaves WB (1981) Age-related changes in the Leydig cell population, seminiferous tubules, and sperm production in stallions *Biology of Reproduction* **24** 703–712.
Johnson L and Nguyen HB (1986) Annual cycle of the Sertoli cell population in adult stallions *Journal of Reproduction and Fertility* **76** 311–316.
Johnson L and Thompson DL Jr (1983) Age-related and seasonal variation in the Sertoli cell population, daily sperm production and serum concentrations of follicle-stimulating hormone, luteinizing hormone and testosterone in stallions *Biology of Reproduction* **29** 777–789.
Johnson L, Petty CS and Neaves WB (1981) A new approach to quantification of spermatogenesis and its application to germinal cell attrition during human spermatogenesis *Biology of Reproduction* **25** 217–226.
Johnson L, Zane RS, Petty CS and Neaves WB (1984) Quantification of the human Sertoli cell population: its distribution, relation to germ cell numbers, and age-related decline *Biology of Reproduction* **31** 785–795.
Johnson L, Varner DD, Tatum ME and Scrutchfield WL (1991) Season but not age affects Sertoli cell number in stallions *Biology of Reproduction* **45** 404–410.
Johnson L, Blanchard TL, Varner DD and Scrutchfield WL (1997) Factors affecting spermatogenesis in the stallion *Theriogenology* **48** 1199–1216.
Kennelly JJ and Foote RH (1964) Sampling boar testes to study spermatogenesis quantitatively and to predict sperm production *Journal of Animal Science* **23** 160–167.
Knight TW (1974) The effect of oxytocin and adrenalin on the semen output of rams *Journal of Reproduction and Fertility* **39** 329–336.
Lange J, Matheja S, Klug E, Aurich C and Aurich JE (1997) Influence of training and competition on the endocrine regulation of testicular function and on semen parameters in stallions *Reproduction in Domestic Animals* **32** 297–302.
Leal MC, Becker-Silva SC, Chiarini-Garcia H and Franca LR (2004) Sertoli cell efficiency and sperm production in goats (*Capra hircus*) *Animal Reproduction* **1** 122–128.

Lino BF (1972) The output of spermatozoa in rams. II. Relationship to scrotal circumference, testis weight, and the number of spermatozoa in different parts of the urogenital tract *Australian Journal of Biological Science* **25** 359–366.
Lok D, Weenk D and de Rooij DG (1982) Morphology, proliferation, and differentiation of undifferentiated spermatogonia in the Chinese hamster and the ram *Anatomical Record* **203** 83–99.
McDonnell SM (1999) Stallion sexual behavior In *Equine Breeding Management and Artificial Insemination* pp 53–56 Ed J Sampler. W B Saunders Co, Philadelphia (as cited by McDonnell *et al.*, 2003).
McDonnell SM, Turner RM, Love CC and LeBlanc MM (2003) How to manage the stallion with a paralyzed penis for return to natural service or artificial insemination In *Proceedings of the 49th Annual Convention of the American Association of Equine Practitioners, New Orleans, Louisiana, USA, 21–5 November 2003 AAEP Proceedings* **49** 291–292. American Association of Equine Practitioners (AAEP), Lexington.
Mori H, Shimizu D, Fukunishi R and Christensen AK (1982) Morphometric analysis of testicular Leydig cells in normal adult mice *Anatomical Record* **204** 333–339.
Oakberg EF (1959) Initial depletion and subsequent recovery of spermatogonia of the mouse after 20 R of gamma rays and 100, 300 and 600 R of X-rays *Radiation Research* **11** 700–719.
Okwun OE, Igboeli G, Ford JJ, Lunstra DD and Johnson L (1996) Number and function of Sertoli cells, number and yield of spermatogonia, and daily sperm production in three breeds of boar *Journal of Reproduction and Fertility* **107** 137–149.
Onyango DW, Wango EO, Otiang'a-Owiti GE, Oduor-Okelo D and Werner G (2000) Morphological characterization of the seminiferous cycle in the goat (*Capra hircus*): a histological and ultrastructural study *Annals of Anatomy* **182** 235–241.
Ortavant R (1958) Le cycle spermatogénétique chez le belier. Doctor of Science thesis, University of Paris, Paris.
Ortavant R (1959a) Déroulement et durée du cycle spermatogénétique chez le belier. Première partie *Annales de Zootechnie* **8** 183–244 (cited by Courot *et al.* (1979) *Journal of Reproduction and Fertility Supplement* **26** 165–173).
Ortavant R (1959b) Déroulement et durée du cycle spermatogénétique chez le belier. Deuxième partie *Annales de Zootechnie* **8** 271–322 (cited by Courot *et al.* (1979) *Journal of Reproduction and Fertility Supplement* **26** 165–173).
Perey B, Clermont Y and Leblond CP (1961) The wave of the seminiferous epithelium in the rat *American Journal of Anatomy* **108** 47–77.
Pickett BW and Voss JL (1973) Reproductive management of the stallion. *Colorado State University Experiment Station General Series* **934**. Fort Collins.
Pickett BW, Sullivan JJ and Seidel, GE Jr (1975) Reproductive physiology of the stallion. V. Effect of frequency of ejaculation on seminal characteristics and spermatozoal output *Journal of Animal Science* **40** 917–923.
Price EO (1985) Sexual behavior of large domestic farm animals: an overview *Journal of Animal Science* **61(Supplement 3)** 62–74.
Ritar AJ, Mendoza G, Salamon S and White IG (1992) Frequent semen collection and sperm reserves of the male Angora goat (*Capra hircus*) *Journal of Reproduction and Fertility* **95** 97–102.
Russell LD and Griswold MD (Eds) (1993) *The Sertoli Cell* Cache River Press, Clearwater.
Salamon S (1962) Studies on the artificial insemination of merino sheep. III. The effect of frequent ejaculation on semen characteristics and fertilizing capacity *Australian Journal of Agricultural Research* **13** 1137–1150 (cited by Foote, 1978).
Schanbacher BD and Ford JJ (1979) Photoperiodic regulation of ovine spermatogenesis: relationship to serum hormones *Biology of Reproduction* **20** 719–726.
Seidel GE Jr and Foote RH (1969) Influence of semen collection interval and tactile stimuli on semen quality and sperm output in bulls *Journal of Dairy Science* **52** 1074–1079.
Sharma OP and Hays RL (1973) Release of an oxytocic substance following genital stimulation in bulls *Journal of Reproduction and Fertility* **35** 359–362.
Shipley CF (1999) Breeding soundness examination of the boar *Swine Health and Production* **7** 117–120.
Sinha Hikim AP, Bartke A and Russell LD (1988) Morphometric studies on hamster testes in gonadally active and inactive states: light microscopic findings *Biology of Reproduction* **39** 1225–1237.
Snyder JW and Ralston NP (1955) Effect of forced exercise on bull fertility *Journal of Dairy Science* **38** 125–130.
Squires EL, Pickett BW and Amann RP (1979) Effect of successive ejaculation on stallion seminal characteristics *Journal of Reproduction and Fertility Supplement* **27** 7–12.

Squires EL, Berndtson WE, Hoyer JH, Pickett BW and Wallach SJR (1981) Restoration of reproductive capacity of stallions after suppression with exogenous testosterone *Journal of Animal Science* **53** 1351–1359.

Squires EL, Todter GE, Berndtson WE and Pickett BW (1982) Effect of anabolic steroids on reproductive function of young stallions *Journal of Animal Science* **54** 576–582.

Swierstra EE (1966) Structural composition of Shorthorn bull testes and daily spermatozoa production as determined by quantitative testicular histology *Canadian Journal of Animal Science* **46** 107–121.

Swierstra EE (1968a) Cytology and duration of the cycle of the seminiferous epithelium of the boar; duration of spermatozoan transit through the epididymis *Anatomical Record* **161** 171–185.

Swierstra EE (1968b) A comparison of spermatozoa production and spermatozoa output of Yorkshire and Lacombe boars *Journal of Reproduction and Fertility* **17** 459–469.

Swierstra EE (1971) Sperm production of boars as measured from epididymal sperm reserves and quantitative testicular histology *Journal of Reproduction and Fertility* **27** 91–99.

Swierstra EE, Gebauer MR and Pickett BW (1974) Reproductive physiology of the stallion. I. Spermatogenesis and testis composition *Journal of Reproduction and Fertility* **40** 113–123.

Swierstra EE, Gebauer MR and Pickett BW (1975) The relationship between daily sperm production as determined by quantitative testicular histology and daily sperm output in the stallion *Journal of Reproduction and Fertility Supplement* **23** 35–39.

Weisgold AD and Almquist JO (1979) Reproductive capacity of beef bulls. VI. Daily spermatozoal production, spermatozoal reserves and dimensions and weight of reproductive organs *Journal of Animal Science* **48** 351–358.

3 Determinants of Sperm Morphology

Abdullah Kaya,[1]* Sema Birler,[2] Lefric Enwall[3] and Erdogan Memili[4]
[1]*Alta Genetics, Inc. Watertown, Wisconsin, USA;* [2]*University of Istanbul, Istanbul, Turkey;* [3]*New Tokyo Medical College, Kolonia, Pohnpei, Federated States of Micronesia;* [4]*Mississippi State University, Mississippi State, Mississippi, USA*

Introduction

Spermatozoa, or sperm cells, develop within the walls of the seminiferous tubules by a complex and time-consuming process known as spermatogenesis. During this process, primordial diploid 'typical' cells are transformed into unique, morphologically and functionally specific haploid cells. Following this process, and after release into the lumen of the seminiferous tubules and then passage into the epididymides, the spermatozoa undergo further maturational changes, e.g. acquiring the ability to be motile, but still undergo further physiological and biochemical changes in the female reproductive tract to acquire their final ability to fertilize ova.

The unique morphological features of spermatozoa are critical for cell functionality. External influences, e.g. genetics, reactive oxygen species, temperature, hormonal modifications, external chemicals and DNA modifications, may negatively affect the production of normal spermatozoa. These factors are discussed in this chapter, but before doing so, a review of spermatozoa morphology is presented in order to provide an understanding of the significance of these external factors.

Normal Sperm Morphology

Classical studies using light and electron microscopy have provided a basic understanding of the structural and functional features of mammalian spermatozoa. Also, many sperm abnormalities have been documented as associated with male infertility and sterility in most of the species studied.

Although the size and shape of spermatozoa are different among different species, the main morphometric structures are similar (Sullivan, 1978). Mammalian spermatozoa consist of two major functionally dependent parts: the head and the tail. While the sperm head contains the materials necessary for fertilization and paternal DNA, the tail comprises the apparatus necessary for sperm energy production and motility (Sullivan, 1978; Roberts, 1986; Barth and Oko, 1989; Garner and Hafez, 2000).

The head

The sperm head is oval and flattened in shape and it is divided into two segments: the anterior acrosome and the posterior post-acrosomal region. The junction of the anterior and posterior

* E-mail: akaya@altagenetics.com

 Animal Andrology: Theories and Applications (eds P.J. Chenoweth and S.P. Lorton) regions is known as the nuclear ring. Underlying the anterior acrosome and continuing to the base of the head is the nucleus (Fig. 3.1).

The nucleus

The largest component of the sperm head is the nucleus, which contains highly condensed DNA, the result of the involvement of sperm-specific proteins, the protamines (Ward and Coffey, 1991; Hazzouri *et al.*, 2000). This condensed DNA state characterizes the cell as being non-dividing and transcriptionally inactive (Balhorn, 1982; Hazzouri *et al.*, 2000; Johnson *et al.*, 2011). Further, the compacted sperm DNA occupies a very small volume compared with the DNA in the mitotic chromosome (Ward and Coffey, 1991). The nucleus is surrounded by a special cytoskeletal complex of perinuclear theca called the post-acrosomal sheet (Fawcett 1975; Sullivan, 1978; Olson *et al.*, 2002).

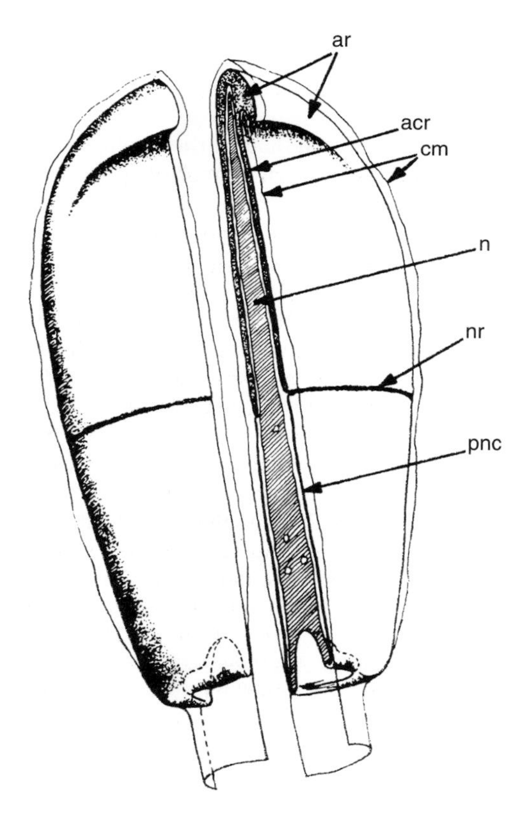

Fig. 3.1. Graphic illustration of the bovine sperm head, indicating the main parts: apical ridge (ar), which is prominent in some ruminants; acrosomal cap (acr); cell membrane (cm); nucleus (n); nuclear ring (nr), which is also known as the equatorial ridge; and the post-nuclear cap (pnc). (Adapted from Saacke RG and Almquist JO (1964a) Ultrastructure of bovine spermatozoa. I. The head of normal ejaculated sperm *American Journal of Anatomy* **115** 144. This material is reproduced with permission of John Wiley & Sons, Inc.)

The acrosome

The acrosome is a unique sperm organelle that develops from the Golgi apparatus during early stages of spermatogenesis. The acrosome is composed of an inner and outer membrane and a protease-filled matrix. In mammals, the acrosome overlays in a cap-like manner about 60% of the apical region of the sperm nucleus and head. It exhibits species-specific differences in shape, size and precise location on the sperm head (Fawcett, 1975; Mann and Lutwak-Mann, 1981).

The acrosomal membrane that just underlies the plasma membrane is known as the outer acrosomal membrane, while that overlying the nucleus is called the inner acrosomal membrane. (Fawcett, 1975; Yanagimachi, 1994; Wassarman, 1999; Eddy, 2006). The posterior region of the acrosome is known as the nuclear ring or the equatorial segment. Because the acrosome develops during spermatogenesis from the Golgi apparatus, it has been referred to as a specialized lysosome. The posterior region of the sperm head (the post-acrosomal region) undergoes changes during post-testicular maturation in the epididymides. These maturational changes play a significant role in sperm-specific binding and fusion with the oolemma (plasma membrane of the oocyte) after sperm have penetrated through the corona radiata and zona pellucida following the acrosome reaction (Sullivan, 1978; Barth and Oko, 1989; Yanagimachi, 1994; Ellis *et al.*, 2002 Olson *et al.*, 2002). Enzymes that have been reported to be present in the acrosome include: hyaluronidase, proacrosin, acrosin, esterases, neuraminidase, acid phosphatase, phospholipases, aryl phosphatase, β-*N*-acetylglucosaminidase, arylamidase, collagenase and corona penetrating enzyme (CPE) (Allison and Hartree, 1970; Zaneveld and Williams,1970; Mann and Lutwak-Mann, 1981; Barth and Oko, 1989; Eddy, 2006).

The specific functions of some of these enzymes are well documented, while those of others are not. Hyaluronidase is released from the acrosome and plays a role in either the dispersal or the digestion of the cumulus oophorus extracellular matrix of the oocyte. CPE, which is also associated with the outer acrosomal membrane, plays a role in penetrating through the corona radiata of the oocyte. Acrosin is present in the acrosome as proacrosin; it is associated with the inner acrosomal membrane and is converted into acrosin. Acrosin is a trypsin-like enzyme that functions to allow penetration of spermatozoa through the zona pellucida during fertilization. Neuraminidase also evidently enhances passage of spermatozoa through the zona pellucida and the plasma membrane of the oocyte (Allison and Hartree, 1970; Zaneveld and Williams, 1970; Mann and Lutwak-Mann, 1981; see Gadella, 2013, and Chapter 4 for in-depth discussions on the physiology and biochemistry of sperm–oocyte interactions).

The tail

The sperm tail consists of four segments: the neck (connecting piece), the midpiece, the principal piece and the endpiece. Each of these four segments is surrounded by a common cell membrane (Fawcett, 1975; Sullivan, 1978; Barth and Oko, 1989; Garner and Hafez, 2000; Olson *et al.*, 2002; Eddy 2006). The primary structural parts of the mammalian sperm tail are the axoneme, the mitochondrial sheath, the outer dense fibres and the fibrous sheath. The centrally located axoneme is composed of nine evenly spaced microtubule doublets and a central pair of singlet microtubules. These extend throughout the length of the tail. The outer dense fibres have a very remarkable cytoskeletal structure, consisting of nine fibres that surround the axoneme and extend through the neck, the midpiece and principal piece of the mammalian sperm tail (Sullivan, 1978; Roberts 1986; Haidl *et al.*, 1991; Mortimer, 1997; Garner and Hafez, 2000; Olson *et al.*, 2002). Haidl *et al.* (1991) also reported that defects of the outer dense fibres result in abnormal sperm morphology in men. The helically arranged mitochondrial sheath surrounds the outer dense fibres in the midpiece, whereas only the fibrous sheath surrounds the outer dense fibres in the principal piece. Only the microtubules terminate in the endpiece (Fawcett, 1975; Mortimer,1997; Eddy, 2006).

The neck

The neck is a short connecting segment between the head and the tail of the spermatozoon (Fig. 3.2). The sperm head and tail are connected via the basal plate at the caudal end of the nucleus, and the capitulum, which adheres to the basal plate of the implantation fossa of the nucleus (Fawcett, 1975; Mortimer, 1997). The neck and the axoneme are formed by a pair of centrioles that are composed of nine circularly arranged microtubular triplets. These two centrioles are present in the spermatid at the time of sperm tail formation. The distal centriole forms the axoneme, whereas the proximal centriole is associated with the formation of capitulum (Barth and Oko, 1989; Mortimer 1997; Eddy, 2006). Thus, the connecting piece has a dual origin from both the proximal and distal centrioles (Gordon, 1972; Fawcett, 1975).

The midpiece

The midpiece (Fig. 3.3) is the region of the tail between the neck and the annulus (Jensen's ring). The annulus is a structural element of the mammalian sperm tail, which connects the midpiece and the principal piece. In the midpiece segment, the outer dense fibre–axoneme complex is surrounded by a helically-wrapped mitochondrial sheath, which extends from the neck to the annulus (Fawcett, 1975; Sullivan, 1978; Mortimer, 1997; Olson *et al.*, 2002; Eddy, 2006).

The mitochondria of the midpiece generate energy in the form of ATP, which is used for sperm locomotion. The inner mitochondrial membrane is the site of energy production (Mortimer, 1997). The elongated mitochondrial helix surrounds approximately 80% of the

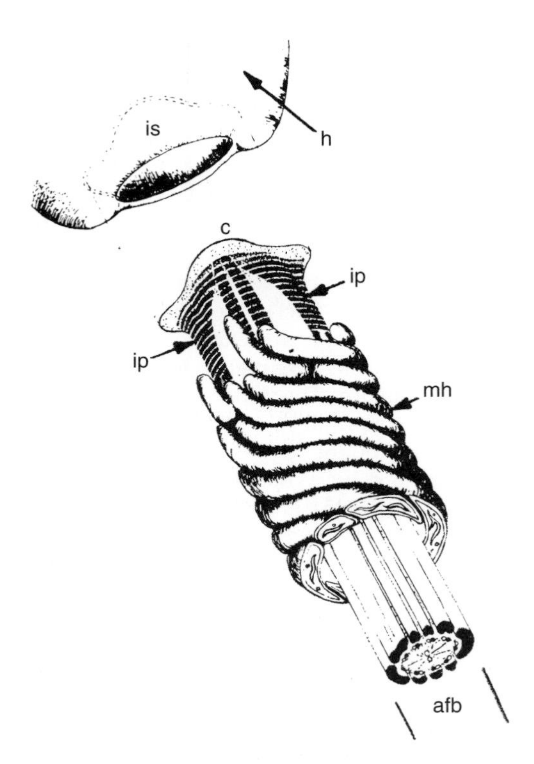

Fig. 3.2. Graphic illustration of the neck region of the bovine sperm, shown with the cell membrane removed and the tail separated from the head (h). Part of the mitochondrial helix (mh) is cut away in order to reveal the axial filament bundle (afb), which extends down the tail, and is composed of 20 fibres arranged in a 9+9+2 array. The outer ring of the nine course fibres merge to form the laminated columns in the neck. The larger laminated columns are known as the implantation plates (ip), and in the neck form the capitulum (c), which fits into the recess in the head, known as the implantation socket (is). (Adapted from Saacke RG and Almquist JO (1964b) Ultrastructure of bovine spermatozoa. II. The neck and tail of normal ejaculated sperm *American Journal of Anatomy* **115** 170. This material is reproduced with permission of John Wiley & Sons, Inc.)

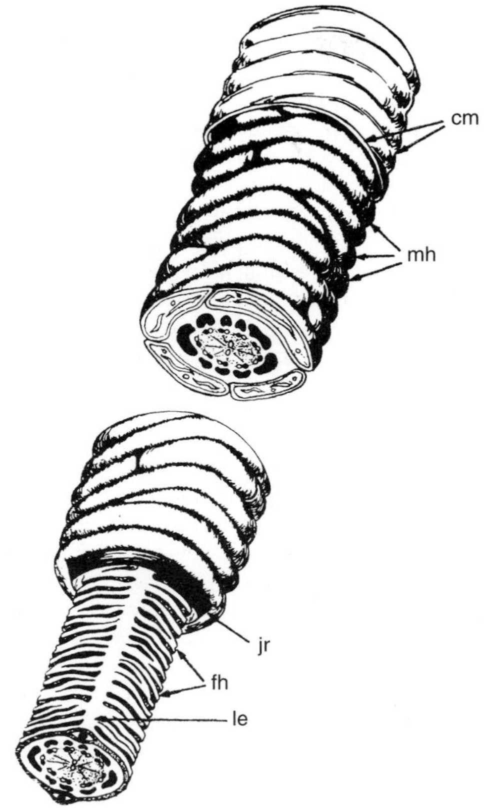

Fig. 3.3. Graphic illustration of the midpiece and anterior portion of the principal piece of the bovine sperm. The cell membrane (cm) is partially removed and the midpiece cut to reveal the internal structure: mitochondrial helix (mh); Jensen's ring (jr), also known as the annulus; fibrous helix (fh); and the longitudinal element (le). (Adapted from Saacke RG and Almquist JO (1964b) Ultrastructure of bovine spermatozoa. II. The neck and tail of normal ejaculated sperm *American Journal of Anatomy* **115** 165. This material is reproduced with permission of John Wiley & Sons, Inc.)

midpiece. There is a wide range of variation among mammals in the length of the midpiece and in the number of mitochondria present (Fawcett, 1975; Sullivan, 1978; Eddy, 2006).

The principal piece

The principal piece is the longest part of the sperm flagellum and extends from the annulus to the terminal piece (Fig. 3.4). Due to the termination of mitochondria in the midpiece, the diameter of the tail in the principal piece is reduced. The principal piece is characterized by the presence of a fibrous sheath, which provides stability for the contractile elements of the tail (Fawcett, 1975; Sullivan, 1978; Mortimer, 1997; Garner and Hafez, 2000; Eddy, 2006).

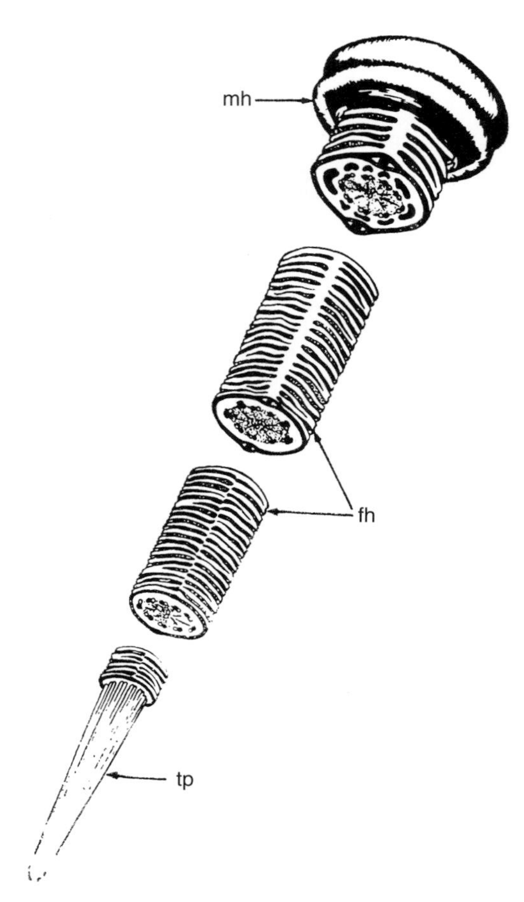

Fig. 3.4. Graphic illustration of the principal piece and terminal piece (tp) of the bovine sperm. The cell membrane is removed and the principal piece cut at several locations to show the change in the axial filament bundle as it passes through the fibrous helix (fh) and the mitochondrial helix (mh). (Adapted from Saacke RG and Almquist JO (1964b) Ultrastructure of bovine spermatozoa. II. The neck and tail of normal ejaculated sperm *American Journal of Anatomy* **115** 169. This material is reproduced with permission of John Wiley & Sons, Inc.)

The endpiece

The endpiece is the region beyond the distal end of the fibrous sheath. This region contains only the terminal segment of the axoneme, surrounded only by the cell membrane of the sperm tail (Fawcett, 1975; Mortimer, 1997; Garner and Hafez, 2000).

Abnormal Sperm Morphology

It has been well documented that a variety of sperm structural modifications are present in ejaculates of all species. The morphometric shapes and/or size of the sperm cells that are different from the normal structure that is characteristic of the individual species are referred to as abnormal or teratozoic spermatozoa. The distinct morphological deviations from normal structure can be usually identified during routine clinical evaluation of semen quality. The importance of morphologically abnormal spermatozoa to fertility was first brought to the attention of other investigators by Williams (1920). Williams and Savage (1925), and then Lagerlof (1934), reported the production of abnormal spermatozoa by infertile and sterile bulls. For more than 70 years, there have been extensive studies involving the identification of abnormal sperm morphology and its relationship with fertility (Lagerlof, 1934; Salisbury *et al.*, 1942; Cupps and Briggs, 1965; Blom, 1972; Lorton *et al.*, 1983; Barth and Oko, 1989; Söderquist *et al.*, 1991; Foote *et al.*, 1992; Howard *et al.*, 1993; Saacke, 2008).

Abnormal sperm cells have long been associated with sub-fertility or sterility, depending on the type or frequency of the morphological abnormalities. However, the prediction of fertilizing ability is still largely a mystery owing to the fact that abnormal sperm cells coexist along with normal sperm cells, and the presence (i.e. number) of normal cells can be adequate for normal fertility. Still, when a very high frequency of several morphological abnormalities is found in the ejaculate, fertility of the male is expected to decline (Salisbury *et al.*, 1942; Barth and Oko, 1989). Thus, relatively arbitrary thresholds of the percentage of morphological abnormalities have been established by andrologists in the animal breeding industry in order to consider the ejaculate as worthy of use in animal breeding. The economic impact of sub-fertility is of the utmost importance, even in light of superior genetics. Therefore, periodic sperm morphological evaluations during either natural or artificial breeding programmes serve as a valuable tool for maintaining optimal fertility levels.

Classification of Abnormal Sperm Morphology

There are various morphological schemes that are used by andrologists to characterize sperm abnormalities and the importance of each abnormality with respect to fertility (Salisbury *et al.*, 1942; Blom, 1972; Barth and Oko, 1989; Söderquist *et al.*, 1991; Saacke *et al.*, 1994). These schemes include those described below.

The origin of the abnormalities

Earlier researchers classified morphological abnormalities as primary, secondary or tertiary according to their origin (Williams and Savage, 1925; Lagerlof, 1934; Salisbury *et al.*, 1942; Blom, 1950; Garner and Hafez, 2000). The primary abnormalities were considered to be developmental abnormalities occurring within the seminiferous epithelium as a result of abnormal spermatogenesis before spermiation. These abnormalities include sperm head and acrosome abnormalities, midpiece abnormalities, e.g. the Dag defect, proximal cytoplasmic droplets and primordial cells. Secondary abnormalities were considered to be those originating after the sperm cells exit the testis. These abnormalities were attributed to altered epididymal maturation, prolonged retention of the sperm cells in the genital tract and abnormal composition of the seminal plasma introduced during ejaculation (Elmore, 1985; Roberts, 1986; Garner and Hafez, 2000). Secondary abnormalities include distal cytoplasmic droplets, simple bent tails, free normal heads and detached acrosomes. Tertiary abnormalities were characterized as being the result of improper semen handling during or after collection. Examples include fast cooling, high ambient temperature, contamination with urine or water and improper slide preparation for examination (Mitchell *et al.*, 1978; Elmore, 1985; Roberts, 1986; Garner and Hafez, 2000).

Impact on fertility: major and minor sperm abnormalities

Although the classification of sperm abnormalities based on their origin has been widely accepted, it is often difficult to distinguish the anatomical origin of those abnormalities, even with the use of the electron microscope. For example, although the Dag defect was considered to be a primary defect, it also was reported to occur during the caput epididymal transit, although it was normal in the testis (Blom, 1972, 1973). Blom (1972) reported that some sperm head abnormalities were not related to improper testicular function, but occurred after the cells exited the testes. Due to the fact that the origin-based classification was observed to be not consistent, Blom (1972) suggested a new scheme in which a sperm abnormality is either associated with a major adverse impact on male fertility, i.e. 'major defects', or is associated with a minor effect on male fertility, i.e. 'minor defects'. A high incidence of major defects is associated with impaired fertility or sterility and probably results from abnormal conditions in the testis or epididymis, or from hereditary genetic defects. The minor defects are considered to be less important for male fertility, unless they are present in a large percentage within the ejaculate.

Localization of abnormalities

Sperm morphological abnormalities have more simply and traditionally been classified based on where the abnormalities are localized on the spermatozoa. Thus, the abnormalities are categorized as being of the acrosome, the head, the neck or midpiece, or the tail (Söderquist *et al.*, 1991; Al-Makhzoomi *et al.*, 2008).

Compensable and uncompensable sperm defects

Spermatozoal abnormalities have also been classified as compensable or uncompensable based upon the ability to overcome the presence of the abnormalities and achieve optimal fertility by increasing number of spermatozoa per breeding dose (Saacke *et al.*, 1994; den Daas *et al.*, 1998; Saacke *et al.*, 2000). It should also be noted that non-motile sperm and sperm with gross abnormal morphologies are prevented by the female reproductive tract from reaching the site of fertilization; so an insufficient number of spermatozoa may then be present to fertilize the oocyte, and decreased fertility is the result (Saacke *et al.*, 1998).

The incidence of compensable sperm defects can be overcome by increasing the number of normal spermatozoa in an artificial breeding dose. Therefore, the fertility of the bull will be at the maximum level for that individual. Increasing the number of normal spermatozoa further will not affect (i.e. increase) fertility. It should also be noted that different individuals exhibit different maximum fertility levels (Pace *et al.*, 1981; Saacke *et al.*, 1994; Saacke, 2008).

Increasing the number of spermatozoa does not overcome the negative effects of uncompensable sperm defects, which may be of molecular or genetic origin. Uncompensable sperm defects are likely to be identified as normal or slightly irregular during routine laboratory evaluations. These sperm are often capable of reaching the site of fertilization, and penetrating and activating the oocyte, but fail to support zygotic and embryonic development (Saacke *et al.*, 1998; Evenson, 1999; Saacke, 2008).

Genetic sperm defects

Genetic sperm defects occur without environmental effects and at a constant rate; they can be transmittable to future generations (Chenoweth, 2005). The diagnosis of such defects is especially important as, otherwise, impaired fertility occurs in both current and future generations (Sullivan, 1978; see Chapter 7 (Chenoweth and McPherson) for a complete discussion of sperm genetic abnormalities).

Reactive Oxygen Species (ROS) and Abnormal Sperm Production

ROS are important modulators of sperm capacitation and function (Agarwal *et al.*, 2005). At high concentrations, ROS can be detrimental to sperm viability; however, at low concentrations, they are important to cell signalling (de Lamirande *et al.*, 1997; Griveau and Le Lannou, 1997; Rivlin *et al.*, 2004; Sanocka and Kurpisz, 2004; Roy and Atreja, 2008). ROS include superoxide anions, hydrogen peroxide, nitric oxide and lipid peroxides. Intracellular protein peroxides are created through photo-oxidation of organic molecules by ultraviolet (UV) and visible light, creating singlet oxygen (Wright *et al.*, 2002). O'Flaherty *et al.* (2006) demonstrated that ROS modulate phosphorylation events at multiple points in the phosphorylation pathways during the capacitation of human sperm. Tyrosine phosphorylation is one such major pathway (Naz and Rajesh, 2004), and this has also been demonstrated in boar (Flesch *et al.*, 1999; Bravo *et al.*, 2005) and equine sperm (Baumber *et al.*, 2003).

Physiologically, an increase in ROS has been shown to have a detrimental effect on sperm function. Gong *et al.* (2012) found that decreased levels of peroxiredoxin enzymes that scavenge free radicals are correlated with increased levels of DNA damage and with decreases in human sperm quality. Indeed, increased levels of ROS are commonly associated with human male infertility (Aitken *et al.*, 2012). Wang and Liu (2012) found increased levels of 8-hydroxy-2′-deoxyguanosine (8-OHdG), a product of DNA oxidation, in sperm from human males clinically diagnosed with idiopathic asthenozoospermia.

Increases in ROS production can also be related to observed sperm morphological alterations, such as seen in sperm from infertile men (Aziz *et al.*, 2004), stallions (Ball *et al.*, 2001) and dogs (Cassani *et al.*, 2005). Such differences in ROS production are also commonly seen in sperm from males of the same species, either as an endogenous trait responding to environmental conditions, as demonstrated in *Bos taurus* and *B. indicus* when responding to seasonal changes (Nichi *et al.*, 2006), or in response to other stressors such as advancing disease and pathology in the human male reproductive tract (Vicari *et al.*, 2006). An increase in ROS in ejaculated sperm has been associated with increases in DNA damage in humans and rams (Fedder and Ellerman-Eriksen, 1995; Hughes *et al.*, 1998;

Twigg *et al.*, 1998a,b; Wang *et al.*, 2003; Said *et al.*, 2005a,b; Kasimanickam *et al.*, 2006).

An important component of mitochondria is cytochrome C (cytC), which, in conjunction with Complex IV, participates in establishing the electrochemical gradient of mitochondria and the subsequent production of ATP. When the structural integrity of the mitochondrial inner membrane is compromised, cardiolipin (CL) is released and binds to cytC (Kagan *et al.*, 2009). The resultant cytC/CL complex produces hydroperoxides (Kagan *et al.*, 2006).

Other critical proteins, such as apoptosis inducing factor (AIF), also emerge from the deteriorating mitochondrial membrane (Susin *et al.*, 1999; Candé *et al.*, 2002; Fleury *et al.*, 2002; van Loo *et al.*, 2002a,b; Philchenkov, 2004) and are present in sperm (Martin *et al.*, 2004; Paasch *et al.*, 2004; Grunewald *et al.*, 2005a,b). Mature sperm are terminally differentiated cells that are also transcriptionally silent (Monesi, 1965), unlike somatic cells, which are transcriptionally active during apoptosis. The caspase proteins, which are part of the signalling pathways to programmed cell death, i.e. apotosis, are absent from mature sperm cells. Their presence in sperm then is indicative of immature or damaged cells, for example after the cryopreservation and thawing processes. The presence of activated apotosis signalling, which has been termed 'apoptotic like', is indicative of decreased fertilizing ability (Grunewald *et al.*, 2008, 2009).

Not all research groups have noted a clear correlation between increased ROS production and negative fertilization effects or morphological abnormalities (Moilanen *et al.*, 1998). Zan-Bar *et al.* (2005) found that ram sperm were more susceptible to light-induced ROS generation than were tilapia sperm. It is also possible that the ability to discern differing levels of ROS production could be compromised, as ROS production has been shown to decline over time during the assay period (Kobayashi *et al.*, 2001); this could potentially skew the results.

Since ejaculates are comprised of millions to billions of individual sperm cells, variability in the overall quality of the ejaculate is reflected in the total number of individual defects. Sperm with defects in the nuclear chromatin or the protamine milieu have been demonstrated to cause problems to sustained embryo development (Ward *et al.*, 2000; Seli *et al.*, 2004; Lewis and Aitken, 2005; Nasr-Esfahani *et al.*, 2005; Suganuma *et al.*, 2005). Excess lipid peroxidation has also been shown to be correlated with midpiece abnormalities in human semen (Rao *et al.*, 1989). However, mitochondrial abnormalities cannot be too severe or the resulting pathology will be reflected in the overall midpiece structure, thereby ablating or seriously attenuating the motility and subsequent fertilizing ability. The ability of sperm to progress though cervical mucus is seriously compromised in sperm with midpiece deformities, while sperm of normal morphology are favoured (Jeulin *et al.*, 1985). It would therefore be logical to believe that midpiece deformities prevent sperm from reaching the site of fertilization. Because of the role of mitochondria in creating and regulating ROS during oxidative phosphorylation (Brookes *et al.*, 2004), and because ROS may have a detrimental effect on sperm, significant disturbances in the functional physiology of the midpiece, which are not necessarily reflected in the morphology, are still of great concern.

The Environment and Abnormal Sperm Production

Temperature

Major disruptions to any of the mechanisms regulating testicular temperature can be detrimental to spermatogenesis. In mammals, the spermatogenic apparatus of the testicles functions optimally at a temperature 2–4°C below that of the core body temperature. This cooler temperature range is maintained in most terrestrial mammals by a combination of sweating and evaporative cooling of the scrotal epidermis (Waites, 1991), by changes in testicular position relative to the body as adjusted by the tunica dartos muscle (Maloney and Mitchell, 1996) and by the countercurrent exchange system of the pampiniform plexus (Coulter and Kastelic, 1994). Waites and Voglmayr (1962) also provided evidence that in rams the apocrine sweat glands of the scrotal epidermis are stimulated by adrenergic sympathetic nerves in response to elevated temperature. Johnson *et al.* (1969) reported a decrease in the testicular weights of rams exposed to an increase in ambient air temperature. The weight changes corresponded to increased levels of lipids and free cholesterol, possibly due to decreased androgen production.

Ambient heat is relevant to sperm quality issues not only because it can be reproduced for scientific study, but also because animals are indeed subjected to elevated temperature and humidity indices. Extremes of heat can compromise sperm quality, as evidenced by an increase in the number of abnormal sperm (Austin *et al.*, 1961; Igboeli and Rakha, 1971; Kumi-Diaka *et al.*, 1981; Parkinson, 1987; Mathevon *et al.*, 1998). Spermatogenesis in animals occurs in a progressive, cyclical manner, known as the cycle of the seminiferous epithelium. The time required for these progressive changes is the duration of the cycle, and differs by species. When scrotal insulation is applied under research conditions, specific patterns of sperm cell abnormalities arise in relation to the temporal characteristics of spermatogenesis (Barth and Oko, 1989; Vogler *et al.*, 1993; Karabinus *et al.*, 1997; Zhu and Setchell, 2004; Enwall, 2009; Rahman *et al.*, 2011).

Cryptorchid animals are a 'natural' example of the effects of heat on spermatogenesis. The temperature of the cryptorchid testis reflects that of the body temperature. Bilaterally cryptorchid individuals are infertile (Setchell, 1998), but spermatogenesis in cryptorchid pigs (Frankenhuis and Wensing, 1979) and rats (Karpe *et al.*, 1984) has been restored if the testicles are removed from the warmer body cavity and moved into the cooler scrotum. Warming the scrotum in a heated water bath (Sailer *et al.*, 1997), heating the neck of the scrotum (Kastelic *et al.*, 1996), or insulating the entire scrotum (Vogler *et al.*, 1991, 1993; Karabinus *et al.*, 1997; Brito *et al.*, 2003) can also induce changes to normal spermatogenesis via heat stress. When using semen from rams that had been scrotally insulated for 16 h/day on 21 consecutive days, Mieusset *et al.* (1992) noted that, while pregnancy rates in ewes did not decrease, embryonic mortality was significantly increased.

Spermatogenesis in the bovine bull occurs over a 65 day process whereby the spermatogonial stem cells undergo mitosis and meiosis, and subsequently physiological and morphological alterations, to produce mature sperm (Johnson *et al.*, 1994). When dairy bulls were subjected to 2 or 3 wk long periods of elevated ambient air temperature (in excess of 37°C), sperm concentration, motility and total numbers were decreased and spermatogenesis was almost ablated (Casady *et al.*, 1953). When scrotal insulation is applied for 48 h, specific patterns of sperm cell abnormalities arise in relation to the temporal characteristics of spermatogenesis (Vogler *et al.*, 1993; Brito *et al.*, 2003; Walters *et al.*, 2005a; Enwall, 2009).

Dutt and Hamm (1957) reported that in unshorn rams subjected to an ambient temperature of 32°C for 1 wk, spermatogenesis was negatively affected 5 wk later. In male rabbits exposed to an air temperature of 43°C for 1 h, an effect was seen on fertilization but not on sperm motility (El-Sheikh and Casida, 1955). In roosters, a decrease of sperm numbers in the ejaculates occurred after a 3 h heat stress event, while sperm number recovered within 2 wk following the stress (Boone and Huston, 1963).

Heat stress to the scrotum not only alters the morphology of testicular spermatogenic cells (Kumi-Diaka *et al.*, 1981; Malmgren and Larsson, 1989), but also affects the morphology of ejaculated sperm (Parkinson, 1987; Vogler *et al.*, 1991, 1993; Barth and Bowman, 1994; Kastelic *et al.*, 1996; Sailer *et al.*, 1996; Enwall, 2009). The resulting altered sperm morphology has also been linked to decreased fertility (Skinner and Louw, 1966; Thundathil *et al.*, 1998, 1999; Ostermeier *et al.*, 2000; Jung *et al.*, 2001; Karaca *et al.*, 2002; Brito *et al.*, 2003; Walters *et al.*, 2005a,b, 2006; Enwall, 2009). This is not surprising given that alterations in the nuclear structure of sperm have been shown to correspond to decreased fertility with or without heat stress (Sailer *et al.*, 1996; Evenson, 1999; Ostermeier *et al.*, 2001).

Nutrition

Spermatogenesis, like other metabolic processes, requires energy as well as the micronutrient resources required by enzymatic reactions. It would then be reasonable to think that irregularities in the dietary intake of an organism might have an effect on an energy intensive process such as spermatogenesis.

For instance, Schmid *et al.* (2012) found that older men with a higher intake of micronutrients, such as vitamin E, β-carotene, zinc, folate and, in particular, vitamin C, had less sperm DNA damage than those with lower micronutrient intake. In rams and goat bucks, poor nutrition has been shown to result in smaller testicular size and reduced sperm output, although evidently the effects were not associated with gonadotrophin levels (Martin and Walkden-Brown, 1995; Martin *et al.*, 2010). Brito *et al.* (2007) and Barth *et al.* (2008) reported that improved diet in young calves not only increased testicular size but also produced a sustained secretion of luteinizing hormone (LH) and increased insulin-like growth factor I (IGF-I). In Menz ram lambs, improved dietary intake resulted in better overall growth and testicular size (Mukasa-Mugerwa and Ezaz, 1992). In roosters, the addition of L-carnitine to the diet improved the number of sperm as well as reducing lipid peroxidation (Zhai *et al.*, 2007). When α-tocopherol was added to the diets of roosters, increased levels of polyunsaturated fatty acids and antioxidant capacity of semen were observed (Surai *et al.*, 1997).

Selenium also seems to play a critical role in animal spermatogenesis. There are at least 11 known selenoproteins, including glutathione peroxidase, phospholipid hydroperoxide glutathione peroxidase, deiodinase and thioredoxin reductase (Holben and Smith, 1999). Contri *et al.* (2011) noted a positive correlation between supplements of organic selenium, vitamin E and zinc and progressive sperm motility in stallions. Slowińska *et al.* (2011) provided evidence that selenium-enriched diets improved sperm concentration and the number of sperm in turkey ejaculates. Selenium supplementation has also been shown to result in similar effects in Egyptian cockerels (Ebeid, 2009), mice (Ibrahim *et al.*, 2012; Ren *et al.*, 2012) and boars (Speight *et al.*, 2012). Because selenium is critical to the function of glutathione peroxidase activity, deficiencies in the element can lead to decreased enzyme function and potential decreased sperm quality (Sławeta *et al.*, 1988). Croxford *et al.* (2011) found that a zinc deficiency in mice resulted in decreased spermatogenesis. Zinc deficiency in humans has also been linked to decreased fertility (Hunt *et al.*, 1992; Colagar *et al.*, 2009).

Gossypol, which is a lipid soluble polyphenolic pigment found in the cotton plant – which is often used as animal feed – increases the percentage of abnormal spermatozoa. The abnormalities are apparent as segmental aplasia of the mitochondrial sheath. Daily sperm production and sperm motility are also decreased, although testicular circumference and weight do not decrease (Chenoweth *et al.*, 1994, 2000). In rats fed gossypol, Hoffer (1983) reported dramatic defects on the mitochondrial sheath of stage 18 and 19 spermatids. Kennedy *et al.* (1983) suggested that the negative effects of gossypol on fertility were via inhibition of the conversion of proacrosin to acrosin, as demonstrated by *in vitro* studies using spermatozoa ejaculated from boar and human (Kennedy *et al.*, 1983). Aflatoxins, which are mycotoxins that can be found in animal feed after storage under certain conditions, also affect male reproduction, and in such cases testes have been observed to be atrophied and spermatogenesis totally arrested. The effects observed are dependent upon the levels of aflatoxins in the feed, so in some animals only increased sperm abnormalities were detected (Ortatatli *et al.*, 2002; Akbarsha *et al.*, 2011).

The presence of heavy metals or pesticides in the diet has been linked to detrimental effects on sperm and fertility. Selvaraju *et al.* (2011) tested the effects of cadmium, lead, chlorpyrifos and endosulfan on buffalo sperm *in vitro* and noted decreased cell integrity and reduced fertilization rates.

Endocrinology of Sperm Production and Physiology

Sperm production is a complex process that is regulated by the effects of many hormones and some key regulators. The primary organs that produce the hormones and regulators involved in sperm production are the hypothalamus, the pituitary gland and the testes. As well as some of the pituitary hormones, e.g. follicle-stimulating hormone (FSH) and LH, the gonadal hormones – androgens and oestrogens (Carreau, 2001; Carreau *et al.*, 2006) – are considered to be the primary hormones for spermatogenesis (Baarends, 1995; Nicholls *et al.*, 2011). Oestrogens, which are converted from androgens by cytochrome P450 aromatase and act via oestrogen receptors (Carreau, 2001) also have important roles during spermatogenesis (Carreau, 2001; Carreau *et al.*, 2006).

The testes have two main functions: sperm and hormone production. Spermatogenesis begins with mitotic divisions of the spermatogonia and, after two meiotic divisions activated by retinoic acid at puberty (Matson *et al.*, 2010), spermatids are released into the lumen of the seminiferous tubules by the process of spermiation. Spermiation is regulated by FSH and testosterone (Saito *et al.*, 2000; Beardsley and O'Donnell, 2003; Beardsley *et al.*, 2006; Nicholls *et al.*, 2011). Serum Anti-Müllerian hormone (AMH), which is stimulated by FSH, is high prepuberty, and declines at puberty as a result of LH-driven androgen suppression of FSH (Young *et al.*, 2005). AMH is not only responsible for the regression of the Müllerian ducts during male sexual differentiation, but also for the regulation of androgen production in the postnatal testes (Spiraman *et al.*, 2001).

Sertoli cells are unique somatic cells that extend from the basement membrane to the lumen of the seminiferous tubules and perform important roles during spermatogenesis and spermiation. These cells produce androgen binding protein (ABP), which is necessary to maintain the required testosterone levels in spermatogenic cells. The extra-tubular somatic cells, the Leydig cells, are under the control of the gonadotrophin LH and produce the hormone testosterone, which is strictly required for spermatogenesis. Testosterone levels in the testes have been reported to be up to 100× higher than in blood serum (Verhoeven *et al.*, 2010).

Endocrine modifiers and their influence on abnormal sperm production

In recent years, social and scientific concerns about endocrine modifiers or endocrine disrupting compounds (EDCs) have been raised. Environmental compounds may have a significant impact on organisms. Phytoestrogens, diethylstilboestrol (DES), bisphenol A (BPA), dichlorodiphenyltrichloroethane (DDT) and vinclozolin are important examples of EDCs. Sources of exposure to EDCs are primarily food, water and air (Diamanti-Kandarakis *et al.*, 2009; Zhang and Ho, 2011). By modulating endocrine signalling pathways, and activating or inhibiting oestrogen and androgen receptors (Safe, 2004, 2005; Tabb and Blumberg, 2006; Diamanti-Kandarakis *et al.*, 2009; Zama and Uzumcu, 2010; Hochberg *et al.*, 2011; Miyagawa *et al.*, 2011; Zhang and Ho, 2011), endocrine modifiers are thought to be responsible for some important male and female reproductive problems, as well as other health problems (Mantovani *et al.*, 1999; Diamanti-Kandarakis *et al.*, 2009; Hochberg *et al.*, 2011).

The primary problem for male reproduction thought to be caused by EDCs, is a decrease in sperm count (Safe, 2005). Gametogenesis can be affected by postnatal and/or adult exposures (Hochberg *et al.*, 2011). As male sexual differentiation is androgen and possibly oestrogen dependent (Mantovani *et al.*, 1999; Diamanti-Kandarakis *et al.*, 2009), EDCs may have different actions in males than in females (Diamanti-Kandarakis *et al.*, 2009). In males, they may cause lowered fertility rates (Tripathi *et al.*, 2009). Safe (2005) reviewed a meta-analysis of sperm count studies and pointed out a dramatic decrease in human sperm count from 1940 to 1990. Apoptotic cell death in germ cells occurs normally during spermatogenesis, but increases due to external stimuli (Tripathi *et al.*, 2009), for example exposure to ethane-1,2-dimethanesulfonate, nitrobenzene, high levels of BPA, lindane and endocrine disrupters (reviewed by Tripathi *et al.*, 2009). Developing organisms are more sensitive to EDCs than adults (Mantovani *et al.*, 1999; Markey *et al.*, 2003; Evans *et al.*, 2004; Diamanti-Kandarakis *et al.*, 2009; Hochberg *et al.*, 2011). Exposure of male embryos to synthetic oestrogens during pregnancy causes changes in the external genitalia of mice (Kim *et al.*, 2004).

According to Skakkebaek *et al.* (2001), testicular dysgenesis syndrome in men, which may be characterized by reduced sperm quality, testicular cancer and/or genital abnormalities, can be the result of adverse effects of environmental endocrine disrupters. Although some epigenetic changes (see next section) can be modified by hormone stimulation, they may also be transmitted to a new generation (Diamanti-Kandarakis *et al.*, 2009; Zama and Uzumcu, 2010; Hochberg *et al.*, 2011; Zhang and Ho, 2011).

Epigenetics of Sperm Abnormalities

Epigenetics is defined as the non-genetic factors that influence gene expression in the absence of changes in the genetic code. Two major epigenetic mechanisms influence sperm, namely, DNA methylation and chromatin structure.

Chromatin structure

Sperm DNA undergoes replication and remodelling during the mitotic and meiotic cell divisions of spermatogenesis. Major changes in chromatin structure involve dynamic transitions of the nuclear proteins, histones and protamines. Before the spermatid stage, sperm DNA is wrapped around somatic histones (namely H1, H2A, H2B, H3, H4), but in the spermatids, the histones are first replaced by transition nuclear proteins (TNP) (Grootegoed *et al.*, 2000; Rajender *et al.*, 2011), and then the TNPs are replaced by protamines. Interestingly, depending on the mammalian species, 1–15% of the sperm nuclear proteins remain as histones (van der Heijden *et al.*, 2008). Following sperm penetration into the oocyte, protamines are replaced by the maternal histones of the oocyte (Corzett *et al.*, 2002).

What is the functional significance of the dynamic chromatin structures during spermatogenesis? Spermatozoa are approximately 10× smaller in size than somatic cells. As a result, sperm genomic DNA must be condensed into a much smaller volume. The tight compaction of the sperm nuclear DNA is achieved by the protamines (Miller *at al.*, 2010; Rajender *et al.*, 2011). The dynamics of sperm chromatin structure, which is an epigenetic phenomenon, is important for sperm functions including fertilization, egg activation and embryo development (Ward, 2010).

The precise regulation of nucleohistone and nucleoprotamine transitions is thought to be important for the protection of sperm DNA and for sperm to gain and retain their fertilizing ability (Ward, 2010). The three structural domains of mature spermatozoa, i.e. large amounts of DNA coiled into toroids by protamines, small amounts of DNA wrapped around histones, and the matrix attachment regions, are also important for sperm quality (Ward, 2010).

The structural integrity of these unique chromatin structures will help to ensure the transmission of a high-integrity paternal genome that is suitable for gene expression following fertilization. For example, it is thought that the ultracompact structure of the sperm has a protective function against DNA damage or viral invasion, and that regions of the retaining histones in the spermatozoa may encode genes whose expression is important for early embryonic development (Hammoud *et al.*, 2009). Indeed, it has been shown that sperm histones have been transmitted to the oocytes. Histones have been detected in the male pronucleus after fertilization (van der Heijden *et al.*, 2008). Posttranslational modifications of the sperm histones are also believed to be important for the sperm epigenome (Carrell and Hammoud, 2010) and abnormalities might affect fertilization and embryonic development. For example, acetylation of sperm histone 4 (H4) is thought to be important in histone replacement via decompaction of the sperm DNA (Govin *et al.*, 2004). Similarly, sperm histones are known to be methylated, and perturbations of sperm DNA are associated with infertility (Lee *et al.*, 2005; Jenkins and Carrell, 2011).

The primary nucleoproteins present in spermatozoa are protamine 1 (PRM1) and protamine 2 (PRM2) in most species, including humans and mice. In the case of bovines, the European hamster, Norwegian rat, and *Rattus tunneyi* (pale field rat or Tunney's rat), only PRM1 is detected (Corzett *et al.*, 2002). It has been demonstrated that deviations from a PRM1/PRM2 ratio of 0.8–1.2 are associated with low fertility in humans (Balhorn *et al.*, 1988; Corzett *et al.*, 2002; de Mateo *et al.*, 2009). Cho *et al.* (2001, 2003) provided evidence that both PRM1 and PRM2 are essential for fertility in mice and that PRM1 deficiency causes sperm head abnormalities, while PRM2 protects the sperm DNA from damage.

DNA methylation

DNA methylation plays an important role in the regulation of gene expression, silencing of endogenous retroviruses, suppression of homologous recombinations and defence against the mutagenic effects of transposable elements in mammalian genomes (Yoder *et al.*, 1997; Bestor, 2000).The enzymes DNMT1 (DNA (cytosine-5)-methyltransferase 1) and DNMT3a and b catalyse reactions that add a methyl group on to the cytosine residues at CpG (cytosine–guanine) dinucleotides on newly synthesized hemimethylated DNA strands during replication, and on hemimethylated and unmethylated DNA (Bird, 1986; Okano *et al.*, 1998). The genes whose promoters are methylated are not expressed because transcription factors and other DNA-binding proteins are not able to bind to the methylated DNA (Bestor *et al.*, 1992).

DNA methylation is a dynamic process during mammalian development. The global DNA methylation of the male and female genomes then decreases during early embryogenesis and then increases again as the embryo becomes a blastocyst (Reik *et al.*, 2001). Abnormalities in the DNA methylation of sperm have been implicated in male-factor infertility cases. For example, Houshdaran *et al.*, (2007) demonstrated that DNA methylation was increased in many regions in poor-quality human sperm. In the same study, significant associations between DNA methylation and sperm concentration, motility and morphology were determined for important genes such as *NTF3*, *MT1A*, *PAX8* and *PLAGL1*. Associations between sperm quality and DNA methylation and gene expression were also demonstrated by Pacheco *et al.* (2011), who provided evidence that DNA methylation and gene transcription were associated with low sperm motility.

Although recent research, as discussed above, has shown that alterations of sperm DNA methylation are associated with decreased fertility and suggest important roles for DNA methylation in fertility, many questions still await answers. For example, what are the mechanisms by which specific methylation profiles of sperm DNA influence male fertility? What do the methylation profiles of sperm DNA tell us; do they present molecular events that took place during spermatogenesis, or do they have any roles once the paternal genome enters the cytoplasm of the oocyte? How heterogeneous are the DNA methylation dynamics among sperm from the same individual and, importantly, to what extent can DNA methylation profiles of sperm be used as a fertility predictor for males?

Conclusion

The continuous increase in the global population results in an increased need for higher quantity and quality meat and milk from food animals. To respond to this important need, science-based solutions need to be developed and applied to increase the reproductive efficiency of livestock. Animal reproductive biotechnology will benefit from a comprehensive understanding of the male gamete and of factors influencing gamete quality. The implementation of advanced technologies for the prediction of male fertility will then greatly promote increased food quality and quantity. The net result will be a significant positive impact on animal producers, the public and the economy. The cellular and molecular attributes of spermatozoa play major roles in the ability of spermatozoa to fertilize and activate the egg, and then sustain continued embryonic development. Recent and cutting edge knowledge of sperm cellular morphology and molecular integrity, for example the methylation of sperm DNA and the post-translational changes of sperm proteins, are exciting advances in the field of animal reproduction that will lead to sustainable animal agriculture in this very crowded and challenged world.

Acknowledgement

The authors wish to thank Dr Steven Lorton, who kindly provided guidance and editorial assistance in the preparation of this chapter.

References

Agarwal A, Gupta S and Sharma R (2005) Oxidative stress and its implications in female infertility – a clinician's perspective *Reproductive BioMedicine Online* **11** 641–50.

Aitken RJ, Jones KT and Robertson SA (2012) Reactive oxygen species and sperm function – in sickness and in health *Journal of Andrology* **33** 1096–1106.

Akbarsha MA, Kunnathodi F and Alshatwi AA (2011) A comprehensive review of male reproductive toxic effects of aflatoxin In *Aflatoxins – Biochemistry and Molecular Biology* pp 177–202 Ed RG Guevara-Gonzalez. InTech, Rijeka. Available at: http://www.intechopen.com/books/aflatoxins-biochemistry-and-molecular-biology/a-comprehensive-review-of-male-reproductive-toxic-effects-of-aflatoxin (accessed 23 August 2013).

Allison AC and Hartree EF (1970) Lysosomal enzymes in the acrosome and their possible role in fertilization *Journal of Reproduction and Fertility* **21** 501–515.

Al-Makhzoomi A, Lundeheim N, Håård M and Rodriguez-Martinez H (2008) Sperm morphology and fertility of progeny-tested AI dairy bulls in Sweden *Theriogenology* **70** 682–691.

Austin JW, Hupp EW and Murphree RL (1961) Effect of scrotal insulation on semen of Hereford bulls *Journal of Animal Science* **20** 307–310.

Aziz N, Saleh RA, Sharma RK, Lewis-Jones I, Esfandiari N, Thomas AJ Jr and Agarwal A (2004) Novel association between sperm reactive oxygen species production, sperm morphological defects, and the sperm deformity index *Fertility and Sterility* **81** 349–354.

Baarends WM (1995) Anti-Müllerian hormone and androgens: regulation of receptors during sex differentiation and gonadal development. Thesis, Erasmus University Rotterdam, Rotterdam.

Balhorn R (1982) A model for the structure of chromatin in mammalian sperm *Journal of Cell Biology* **93** 298–305.

Balhorn R, Reed S and Tanphaichitr N (1988) Aberrant protamine 1/protamine 2 ratios in sperm of human infertile males *Experientia* **44** 52–55.

Ball BA, Vo AT and Baumber J (2001) Generation of reactive oxygen species by equine spermatozoa *American Journal of Veterinary Research* **62** 508–515.

Barth AD and Bowman PA (1994) The sequential appearance of sperm abnormalities after scrotal insulation or dexamethasone treatment in bulls *Canadian Veterinary Journal* **34** 93–102.

Barth AD and Oko RJ (1989) *Abnormal Morphology of Bovine Spermatozoa*. Iowa State University Press, Ames.

Barth AD, Brito LFC and Kastelic JP (2008) The effect of nutrition on sexual development of bulls *Theriogenology* **70** 485–494.

Baumber J, Sabeur K, Vo A and Ball BA (2003) Reactive oxygen species promote tyrosine phosphorylation and capacitation in equine spermatozoa *Theriogenology* **60** 1239–1247.

Beardsley A and O'Donnell L (2003) Characterization of normal spermiation and spermiation failure induced by hormone suppression in adult rats *Biology of Reproduction* **68** 1299–1307.

Beardsley A, Robertson DM and O'Donnell L (2006) A complex containing $\alpha_6\beta_1$-integrin and phosphorylated focal adhesion kinase between Sertoli cells and elongated spermatids during spermatid release from the seminiferous epithelium *Journal of Endocrinology* **190** 759–770.

Bestor TH (2000) The DNA methyltransferases of mammals *Human Molecular Genetics* **9** 2395–2402.

Bestor TH, Gundersen G, Kolsto AB and Prydz H (1992) CpG islands in mammalian gene promoters are inherently resistant to *de novo* methylation *Genetic Analysis, Techniques and Applications* **9** 48–53.

Bird AP (1986) CpG-rich islands and the function of DNA methylation *Nature* **321** 209–213.

Blom E (1950) Interpretation of spermatic cytology in bulls *Fertility and Sterility* **1** 223–238.

Blom E (1972) The ultrastructure of some characteristic sperm defects and a proposal for a new classification of the bull spermiogram. In *Atti del VII Simposio Internationale di Zootecnia, Milano 15–17 Aprile 1972* pp 125–139. Società Italiana per il Progresso della Zootecnia, Milan.

Blom, E (1973) The ultrastructure of some characteristic sperm defects and a proposal for a new classification of bull spermiogram *Nordisk Veterinærmedicin* **25** 383–391.

Boone MA and Huston TM (1963) Effects of high temperature on semen production and fertility in the domestic fowl *Poultry Science* **42** 670–676.

Bravo MM, Aparicio IM, Garcia-Herreros M, Gil MC, Peña FJ and Garcia-Marin LJ (2005) Changes in tyrosine phosphorylation associated with true capacitation and capacitation-like state in boar spermatozoa *Molecular Reproduction and Development* **71** 88–96.

Brito LFC, Silva AEDF, Barbosa RT, Unanian MM and Kastelic JP (2003) Effects of scrotal insulation on sperm production, semen quality, and testicular echotexture in *Bos indicus* and *Bos indicus* × *Bos taurus* bulls *Animal Reproduction Science* **79** 1–15.

Brito LFC, Barth AD, Rawlings NC, Wilde RE, Crews DH Jr, Mir PS and Kastelic JP (2007) Effect of improved nutrition during calfhood on serum metabolic hormones, gonadotropins, and testosterone concentrations, and on testicular development in bulls *Domestic Animal Endocrinology* **33** 460–469.

Brookes PS, Yoon Y, Robotham JL, Anders MW and Sheu SS (2004) Calcium, ATP, and ROS: a mitochondrial love–hate triangle *American Journal of Physiology: Cell Physiology* **4** C817–C833.

Candé C, Cecconi F, Dessen P and Kroemer G (2002) Apoptosis-inducing factor (AIF): key to the conserved caspase-independent pathways of cell death? *Journal of Cell Science* **115** 4727–4734.

Carreau S (2001) Germ cells: a new source of estrogens in the male gonad *Molecular and Cellular Endocrinology* **178** 65–72.

Carreau S, Delalande C, Silandre D, Bourguiba S and Lambard S. (2006) Aromatase and estrogen receptors in male reproduction *Molecular and Cellular Endocrinology* **246** 65–68.

Carrell DT and Hammoud SS (2010) The human sperm epigenome and its potential role in embryonic development *Molecular Human Reproduction* **16** 37–47.

Casady RB, Myers RM and Legates JE (1953) The effect of exposure to high ambient temperature on spermatogenesis in the dairy bull *Journal of Dairy Science* **36** 14–23.

Cassani P, Beconi MT and O'Flaherty C (2005) Relationship between total superoxide dismutase activity with lipid peroxidation, dynamics and morphological parameters in canine semen *Animal Reproduction Science* **86** 163–73.

Chenoweth PJ (2005) Genetic sperm defects *Theriogenology* **64** 457–468.

Chenoweth PJ, Risco CA, Larsen RE, Velez J, Tran T, and Chase CC Jr (1994) Effects of gossypol on aspects of semen quality, sperm morphology and sperm production in young Brahman bulls *Theriogenology* **42** 1–13.

Chenoweth PJ, Chase CC, Risco CA and Larsen RE (2000) Characterization of gossypol-induced sperm abnormalities in bulls *Theriogenology* **53** 1193–1203.

Cho C, Willis WD, Goulding EH, Jung-Ha H, Choi YC, Hecht NB and Eddy M (2001) Haploinsufficiency of protamine-1 or -2 causes infertility in mice *Nature Genetics* **28** 82–86.

Cho C, Jung-Ha H, Willis DW, Goulding EH, Stein P, Xu Z, Schultz RM, Hecht NB and Eddy EM (2003) Protamine 2 deficiency leads to sperm DNA damage and embryo death in mice *Biology of Reproduction* **69** 211–217.

Colagar AH, Marzony ET and Chaichi MJ (2009) Zinc levels in seminal plasma are associated with sperm quality in fertile and infertile men *Nutrition Research* **29** 82–88.

Contri A, De Amicis I, Molinari A, Faustini M, Gramenzi A, Robbe D and Carluccio A (2011) Effect of dietary antioxidant supplementation on fresh semen quality in stallion *Theriogenology* **75** 1319–26.

Corzett M, Mazrimas J and Balhorn R (2002) Protamine 1: protamine 2 stoichiometry in the sperm of eutherian mammal *Molecular Reproduction and Development* **61** 519–527.

Coulter GH and Kastelic JP (1994) Testicular thermoregulation in the bull In *Proceedings 14th Technical Conference on Artificial Insemination and Reproduction* pp 29–34. National Association of Animal Breeders, Columbia, Missouri.

Croxford TP, McCormick NH and Kelleher SL (2011) Moderate zinc deficiency reduces testicular Zip6 and Zip10 abundance and impairs spermatogenesis in mice *Journal of Nutrition* **141** 359–65.

Cupps PT and Briggs JR (1965) Changes in the epididymis associated with morphological changes in the spermatozoa *Journal of Dairy Science* **48** 1241–1244.
de Lamirande E, Jiang H, Zini A, Kodama H and Gagnon C. (1997) Reactive oxygen species and sperm physiology *Reviews of Reproduction* **1** 48–54.
de Mateo S, Gazquez C, Guimera M, Balasch J, Meistrich ML, Ballesca JL and Oliva R (2009) Protamine 2 precursors (Pre-P2), protamine 1 to protamine 2 ratio (P1/P2), and assisted reproduction outcome *Fertility and Sterility* **91** 715–721.
den Daas JHG, DeJong G, Lansbergen LMTE and van Wagtendonk-de Leeuw AM (1998) The relationship between the number of spermatozoa inseminated and the reproductive efficiency of individual bulls *Journal of Dairy Science* **81** 1714–1723.
Diamanti-Kandarakis E, Bourguignon JP, Giudice LC, Hauser R, Prins GS, Soto AM, Zoeller RT and Gore AC (2009) Endocrine-disrupting chemicals: an Endocrine Society Scientific Statement *Endocrinology Reviews* **30** 293–342.
Dutt RH and Hamm PT (1957) Effect of exposure to high environmental temperature and shearing on semen production of rams in winter *Journal of Animal Science* **16** 328–334.
Ebeid TA (2009) Organic selenium enhances the antioxidative status and quality of cockerel semen under high ambient temperature *British Poultry Science* **50** 641–647.
Eddy EM (2006) The spermatozoon In *Physiology of Reproduction* pp 1–54 Ed JD Neill. Elsevier Academic Press, San Diego.
Ellis DJ, Shadan S, James PS, Henderson RM, Edwardson JM, Hutchings A and Jones R (2002) Post-testicular development of a novel membrane substructure within the equatorial segment of ram, bull, boar, and goat spermatozoa as viewed by atomic force microscopy *Journal of Structural Biology* **138** 187–198.
Elmore, RG (1985) Evaluating bulls for breeding soundness: sperm morphology *Veterinary Medicine* **80** 90–95.
El-Sheikh AS and Casida LE (1955) Motility and fertility of spermatozoa as affected by increased ambient temperature *Journal of Animal Science* **16** 1146–1150.
Enwall L (2009) Investigating nuclear shape, morphology, and molecular changes in semen from scrotal-insulated Holstein bulls using Fourier harmonic analysis, flow cytometry, and *in vitro* fertility evaluations. PhD thesis, University of Wisconsin, Madison.
Evans NP, North T, Dye S and Sweeney T (2004) Differential effects of the endocrine-disrupting compounds bisphenol-A and octylphenol on gonadotropin secretion in prepubertal ewe lambs *Domestic Animal Endocrinology* **26** 61–73.
Evenson DP (1999) Loss of livestock breeding efficiency due to uncompensable sperm nuclear defects *Journal of Reproduction and Fertility* **11** 1–15.
Fawcett DW (1975) The mammalian spermatozoon *Developmental Biology* **44** 394–436.
Fedder J and Ellerman-Eriksen S (1995) Effect of cytokines on sperm motility and ionophore-stimulated acrosome reaction *Archives of Andrology* **35** 173–85.
Flesch FM, Colenbrander B, van Golde LM and Gadella BM (1999) Capacitation induces tyrosine phosphorylation of proteins in the boar sperm plasma membrane *Biochemical and Biophysical Research Communications* **262** 787–92.
Fleury C, Mignotte B and Vayssière JL (2002) Mitochondrial reactive oxygen species in cell death signaling *Biochimie* **84** 131–41.
Foote RH, Hough SR, Johnson LA and Kaproth M (1992) Electron microscopy and pedigree study in an Ayrshire bull with tail-stump sperm defects *Veterinary Record* **130** 578–579.
Frankenhuis MT and Wensing CJG (1979) Induction of spermatogenesis in the naturally occurring cryptorchid pig *Fertility and Sterility* **31** 428–433.
Gadella BM (2013) Dynamic regulation of sperm interactions with the zona pellucida prior to and after fertilization *Reproduction, Fertility and Development* **25** 26–37.
Garner DL and Hafez ESE (2000) Spermatozoa and seminal plasma In *Reproduction in Farm Animals* 7th edition pp 96–109 Ed B Hafez and ESE Hafez. Lippincott, Philadelphia.
Gong S, San Gabriel MC, Zini A, Chan P and O'Flaherty C (2012) Low amounts and high thiol oxidation of peroxiredoxins in spermatozoa from infertile men *Journal of Andrology* **33** 1342–1351.
Gordon M (1972) The distal centriole in guinea pig spermiogenesis *Journal of Ultrastructural Research* **39** 364–388.
Govin J, Caron C, Lestrat C, Rousseaux S and Khochbin S (2004) The role of histones in chromatin remodeling during mammalian spermiogenesis *European Journal of Biochemistry* **271** 3459–3469.
Griveau JF and Le Lannou D (1997) Reactive oxygen species and human spermatozoa: physiology and pathology *International Journal of Andrology* **20** 61–69.

Grootegoed JA, Siep M and Baarends WM (2000) Molecular and cellular mechanisms in spermatogenesis *Best Practice and Research Clinical Endocrinology and Metabolism* **14** 331–343.

Grunewald S, Paasch U, Said TM, Sharma RK, Glander HJ and Agarwal A (2005a) Caspase activation in human spermatozoa in response to physiological and pathological stimuli *Fertility and Sterility* **83(Supplement 1)** 1106–12.

Grunewald S, Paasch U, Wuendrich K and Glander HJ (2005b) Sperm caspases become more activated in infertility patients than in healthy donors during cryopreservation *Archives of Andrology* **51** 449–460.

Grunewald S, Said TM, Paasch U, Glander HJ and Agarwal A (2008) Relationship between sperm apoptosis signalling and oocyte penetration capacity *International Journal of Andrology* **31** 325–330.

Grunewald S, Sharma R, Paasch U, Glander HJ and Agarwal A (2009) Impact of caspase activation in human spermatozoa *Microscopy Research and Technique* **72** 878–888.

Haidl G, Becker A, and Henkel R (1991) Poor development of outer dense fibers as a major cause of tail abnormalities in the spermatozoa of asthenoteratozoospermic men *Human Reproduction* **6** 1431–1438.

Hammoud SS, Nix DA, Zhang H, Purwar J, Carrell DT and Cairns BR (2009) Distinctive chromatin in human sperm packages genes for development *Nature* **460** 473–478.

Hazzouri M, Rousseaux S, Mongelard F, Usson Y, Pelletier R, Faure AK, Vourc'h C and Sèle B (2000) Genome organization in the human sperm nucleus studied by FISH and confocal microscopy *Molecular Reproduction and Development* **55** 307–315.

Hochberg Z, Feil R, Constancia M, Fraga M, Junien C, Carel JC, Boileau P, Le Bouc Y, Deal CL, Lillycrop K, Scharfmann R, Sheppard A, Skinner M, Szyf M, Waterland RA, Waxman DJ, Whitelaw E, Ong K and Albertsson-Wikland K (2011) Child health, developmental plasticity, and epigenetic programming *Endocrine Reviews* **32** 159–224.

Hoffer A (1983) Effects of gossypol on the seminiferous epithelium in the rat: a light and electron microscope study *Biology of Reproduction* **28** 1007–1020.

Holben DH and Smith AM (1999) The diverse role of selenium within selenoproteins: a review *Journal of the American Dietetic Association* **99** 836–843.

Houshdaran S, Cortessis VK, Siegmund K, Yang A, Laird PW and Sokol RZ (2007) Widespread epigenetic abnormalities suggest a broad DNA methylation erasure defect in abnormal human sperm *PLoS One* **2** e1289.

Howard JG, Donoghue AM, Johnston LA and Wildt DE (1993) Zona pellucida filtration of structurally abnormal spermatozoa and reduced fertilization in teratospermic cats *Biology of Reproduction* **49** 131–139.

Hughes CM, Lewis SE, McKelvey-Martin VJ and Thompson W (1998) The effects of antioxidant supplementation during Percoll preparation on human sperm DNA integrity *Human Reproduction* **13** 1240–1247.

Hunt CD, Johnson PE, Herbel J and Mullen LK (1992) Effects of dietary zinc depletion on seminal volume and zinc loss, serum testosterone concentrations, and sperm morphology in young men *American Journal of Clinical Nutrition* **56** 148–157.

Ibrahim HA, Zhu Y, Wu C, Lu C, Ezekwe MO, Liao SF and Huang K (2012) Selenium-enriched probiotics improves murine male fertility compromised by high fat diet *Biological Trace Element Research* **147** 251–260.

Igboeli G and Rakha AM (1971) Seasonal changes in the ejaculate characteristics of Angoni (short horn Zebu) bulls *Journal of Animal Science* **33** 651–654.

Jenkins TG and Carrell DT (2011) The paternal epigenome and embryogenesis: poising mechanisms for development *Asian Journal of Andrology* **13** 76–80.

Jeulin C, Soumah A, and Jouannet P (1985) Morphological factors influencing the penetration of human sperm into cervical mucus *in vitro International Journal of Andrology* **8** 215–23.

Johnson AD, Gomes WR and VanDemark NL (1969) Effect of elevated ambient temperature on lipid levels and cholesterol metabolism in the ram testis *Journal of Animal Science* **29** 469–475.

Johnson GD, Lalancette C, Linnemann AK, Leduc F, Boissonneault G and Krawetz SA (2011) The sperm nucleus: chromatin, RNA, and the nuclear matrix *Reproduction* **141** 21–36.

Johnson L, Wilker CE and Cerelli JS (1994) Spermatogenesis in the bull In *Proceedings 15th Technical Conference on Artificial Insemination and Reproduction* pp 9–27. National Association of Animal Breeders, Columbia, Missouri.

Jung A, Eberl M and Schill W-B (2001) Improvement of semen quality by nocturnal scrotal cooling and moderate behavioural change to reduce genital heat stress in men with oligoasthenoteratozoospermia *Reproduction* **121** 595–603.

Kagan VE, Tyurina YY, Bayir H, Chu CT, Kapralov AA, Vlasova II, Belikova NA, Tyurin VA, Amoscato A, Epperly M, Greenberger J, Dekosky S, Shvedova AA and Jiang J (2006) The "pro-apoptotic genies" get out of mitochondria: oxidative lipidomics and redox activity of cytochrome c/cardiolipin complexes *Chemico-Biological Interactions* **163** 15–28.

Kagan VE, Bayir HA, Belikova NA, Kapralov O, Tyurina YY, Tyurin VA, Jiang J, Stoyanovsky DA, Wipf P, Kochanek PM, Greenberger JS, Pitt B, Shvedova AA and Borisenko G (2009) Cytochrome c/cardiolipin relations in mitochondria: a kiss of death *Journal of Free Radicals in Biology and Medicine* **46** 1439–1453.

Karabinus DS, Vogler CJ, Saacke RG and Evenson DP (1997) Chromatin structural changes in sperm after scrotal insulation of Holstein bulls *Journal of Andrology* **18** 549–555.

Karaca AG, Parker HM, Yeatman JB and McDaniel CD (2002) The effects of heat stress and sperm quality classification on broiler breeder male fertility and semen ion concentrations *British Poultry Science* **44** 621–628.

Karpe B, Ploen L and Ritzen EM (1984) Maturation of juvenile rat testis after surgical treatment of cryptorchidism *International Journal of Andrology* **7** 154–161.

Kasimanickam R, Pelzer KD, Kasimanickam V, Swecker WS and Thatcher CD (2006) Association of classical semen parameters, sperm DNA fragmentation index, lipid peroxidation and antioxidant enzymatic activity of semen in ram-lambs *Theriogenology* **65** 1407–1421.

Kastelic JP, Cook RB, Coulter GH and Saacke RG (1996) Insulating the scrotal neck affects semen quality and scrotal/testicular temperatures in the bull *Theriogenology* **45** 935–942.

Kennedy WP, Van der Ven HH, Straus JW, Bhattacharyya AK, Waller DP, Zaneveld LJD and Polakoski KL (1983) Gossypol inhibition of acrosin and proacrosin, and oocyte penetration by human spermatozoa *Biology of Reproduction* **29** 999–1009.

Kim KS, Torres CR Jr, Yucel S, Raimondo K, Cunha GR and Baskin LS (2004) Induction of hypospadias in a murine model by maternal exposure to synthetic estrogens *Environmental Research* **94** 267–275.

Kobayashi H, Gil-Guzman E, Mahran AM, Sharma RK, Nelson DR, Thomas AJ Jr and Agarwa A (2001) Quality control of reactive oxygen species measurement by luminol-dependent chemiluminescence assay *Journal of Andrology* **22** 568–574.

Kumi-Diaka J, Nagaratnam V and Rwuaan JS (1981) Seasonal and age-related changes in semen quality and testicular morphology of bulls in a tropical environment *Veterinary Record* **108** 13–15.

Lagerlof N (1934) Morphological studies on the changes in the sperm structure and in the testes of bulls with decreased or abolished fertility *Acta Pathologica et Microbiologica Scandinavica* **19** 254–267.

Lee MG, Wynder C, Cooch N and Shiekhattar R (2005) An essential role for CoREST in nucleosomal histone 3 lysine 4 demethylation *Nature* **437** 432–435.

Lewis SE and Aitken RJ (2005) DNA damage to spermatozoa has impacts on fertilization and pregnancy *Cell and Tissue Research* **322** 33–41.

Lorton SP, Wu ASH, Pace MM, Parker WG and Sullivan JJ (1983) Fine structural abnormalities of spermatozoa from a subfertile bull: a case report *Theriogenology* **20** 585–599.

Malmgren L and Larsson K (1989) Experimentally induced testicular alterations in boars: histological and ultrastructural findings *Journal of Veterinary Medicine* **36** 3–14.

Maloney SK and Mitchell D (1996) Regulation of ram scrotal temperature during heat exposure, cold exposure, fever, and exercise *Journal of Physiology* **496** 421–430.

Mann T and Lutwak-Mann C (1981) *Male Reproductive Function and Semen* pp 55–159. Springer-Verlag, Berlin.

Mantovani A, Stazi AV, Macri C, Maranghi F and Ricciardi C (1999) Problems in testing and risk assessment of endocrine disrupting chemicals with regard to developmental toxicology *Chemosphere* **39** 1293–1300.

Markey CM, Rubin BS, Soto AM and Sonnenschein C (2003) Endocrine disruptors: from wingspread to environmental developmental biology *Journal of Steroid Biochemistry and Molecular Biology* **83** 235–244.

Martin G, Sabido O, Durand P and Levy R (2004) Cryopreservation induces an apoptosis-like mechanism in bull sperm *Biology of Reproduction* **71** 28–37.

Martin GB and Walkden-Brown SW (1995) Nutritional influences on reproduction in mature male sheep and goats *Journal of Reproduction and Fertility Supplement* **49** 37–49.

Martin GB, Blache D, Miller DW and Vercoe PE (2010) Interactions between nutrition and reproduction in the management of the mature male ruminant *Animal* **4** 1214–1226.

Mathevon M, Buhr MM and Dekkers JCM (1998) Environmental, management, and genetic factors affecting semen production in Holstein bulls *Journal of Dairy Science* **81** 3321–3330.

Matson CK, Murphy MW, Griswold MD, Yoshida S, Bardwell VJ and Zarkower D (2010) The Mammalian doublesex homolog DMRT1 is a transcriptional gatekeeper that controls the mitosis versus meiosis decision in male germ cells *Developmental Cell* **19** 612–624.

Mieusset R, Quintana Casares P, Sanchez Partida LG, Sowerbutts SF, Zupp JL and Setchell BP (1992) Effects of heating the testes and epididymides of rams by scrotal insulation on fertility and embryonic mortality in ewes inseminated with frozen semen *Journal of Reproduction and Fertility* **94** 337–343.

Miller D, Brinkwort M and Iles D (2010) Paternal DNA packaging in spermatozoa: more than the sum of its parts? DNA, histones, protamines and epigenetics *Reproduction* **139** 287–301.

Mitchell JR, Hanson RD and Fleming WN (1978) Utilizing differential interference microscopy for evaluating abnormal spermatozoa In *Proceedings 7th Technical Conference on Artificial Insemination and Reproduction* pp 64–68. National Association of Animal Breeders, Columbia, Missouri.

Miyagawa S, Sato M and Iguchi T (2011) Molecular mechanisms of induction of persistent changes by estrogenic chemicals on female reproductive tracts and external genitalia *Journal of Steroid Biochemistry and Molecular Biology* **127** 51–57.

Moilanen JM, Carpén O and Hovatta O (1998) Flow cytometric light scattering analysis, acrosome reaction, reactive oxygen species production and leukocyte contamination of semen preparation in prediction of fertilization rate *in vitro Human Reproduction* **13** 2568–2574.

Monesi V (1965) Differential rate of ribonucleic acid synthesis in the autosomes and sex chromosomes during male meiosis in the mouse *Chromosoma* **17** 11–21.

Mortimer ST (1997) A critical review of the physiological importance and analysis of sperm movement in mammals *Human Reproduction Update* **3** 403–39.

Mukasa-Mugerwa E and Ezaz Z (1992) Relationship of testicular growth and size to age, body weight and onset of puberty in Menz ram lambs *Theriogenology* **38** 979–988.

Nasr-Esfahani MH, Salehi M, Razavi S, Anjomshoa M, Rozbahani S, Moulavi F, and Mardani M (2005) Effect of sperm DNA damage and sperm protamine deficiency on fertilization and embryo development post-ICSI *Reproductive BioMedicine Online* **11** 198–205.

Naz RK and Rajesh PB (2004) Role of tyrosine phosphorylation in sperm capacitation/acrosome reaction *Reproductive Biology and Endocrinology* **2**:75.

Nichi M, Bols PE, Züge RM, Barnabe VH, Goovaerts IG, Barnabe RC and Cortada CN (2006) Seasonal variation in semen quality in *Bos indicus* and *Bos taurus* bulls raised under tropical conditions *Theriogenology* **66** 822–828.

Nicholls PK, Harrison CA, Walton KL, McLachlan RI, O'Donnell L and Stanton PG (2011) Hormonal regulation of Sertoli cell micro-RNAs at spermiation *Endocrinology* **152** 1670–1683.

O'Flaherty C, de Lamirande E and Gagnon C (2006) Reactive oxygen species modulate independent protein phosphorylation pathways during human sperm capacitation *Journal of Free Radicals in Biology and Medicine* **40** 1045–1055.

Okano M, Xie S and Li E (1998) Dnmt2 is not required for *de novo* and maintenance methylation of viral DNA in embryonic stem cells *Nucleic Acids Research* **26** 2536–2540.

Olson GE, NagDas SK and Winfrey VP (2002) Structural differentiation of spermatozoa during post-testicular maturation In *The Epididymis: From Molecules to Clinical Practice* pp 371–387 Ed B Robaire and BT Hinton. Kluwer Academic/Plenum Publishers, New York.

Ortatatli M, Ciftci MK, Tuzcu M and Kaya A (2002) The effects of aflatoxin on the reproductive system of roosters *Research in Veterinary Science* **72** 29–36.

Ostermeier GC, Sator-Bergfelt R, Susko-Parrish JL and Parrish JJ (2000) Bull fertility and sperm nuclear shape *AgBiotechNet* **2** 1–6.

Ostermeier GC, Sargeant GA, Yandell BS, Evenson DP and Parrish JJ (2001) Relationship of bull fertility to sperm nuclear shape *Journal of Andrology* **22** 595–603.

Paasch U, Grunewald S, Dathe S and Glander HJ (2004) Mitochondria of human spermatozoa are preferentially susceptible to apoptosis *Annals of the New York Academy of Science* **1030** 403–409.

Pace MM, Sullivan JJ, Elliott FI, Graham EF and Coulter GH (1981) Effects of thawing, temperature, number of spermatozoa, and spermatozoal quality on fertility of bovine spermatozoa packaged in 0.5ml French straws *Journal of Animal Science* **53** 693–701.

Pacheco SE, Houseman EA, Christensen BC, Marsit CJ, Kelsey KT, Sigman M and Boekelheide K (2011) Integrative DNA methylation and gene expression analyses identify DNA packaging and epigenetic regulatory genes associated with low motility sperm *PLoS One* **6** e20280.

Parkinson TJ (1987) Seasonal variations in semen quality of bulls: correlations with environmental temperature *Veterinary Record* **120** 479–82.

Philchenkov A (2004) Caspases: potential targets for regulating cell death *Journal of Cell and Molecular Medicine* **8** 432–444.

Rahman MB, Vandaele L, Rijsselaere T, Maes D, Hoogewijs M, Frijters A, Noordman J, Granados A, Dernelle E, Shamsuddin M, Parrish JJ and Van Soom A (2011) Scrotal insulation and its relationship to abnormal morphology, chromatin protamination and nuclear shape of spermatozoa in Holstein–Friesian and Belgian Blue bulls *Theriogenology* **76** 1246–1257.

Rajender S, Avery K and Agarwal A (2011) Epigenetics, spermatogenesis and male fertility *Mutation Research* **72** 62–71.

Rao B, Soufir JC, Martin M and David G (1989) Lipid peroxidation in human spermatozoa as related to midpiece abnormalities and motility *Gamete Research* **24** 127–134.

Reik W, Dean W and Walter J (2001) DNA methylation and mammalian epigenetics *Science* **293** 1089–1093.

Ren XM, Wang GG, Xu DQ, Luo K, Liu YX, Zhong YH and Cai YQ (2012) The protection of selenium on cadmium-induced inhibition of spermatogenesis via activating testosterone synthesis in mice *Food and Chemical Toxicology* **50** 3521–3529.

Rivlin J, Mendel J, Rubinstein S, Etkovitz N and Breitbart H (2004) Role of hydrogen peroxide in sperm capacitation and acrosome reaction *Biology of Reproduction* **70** 518–522.

Roberts SJ (1986) Infertility in male animals In *Veterinary Obstetrics and Genital Disease (Theriogenology)* 3rd edition, pp 751–893. David and Charles, North Pomfret.

Roy SC and Atreja SK (2008) Effect of reactive oxygen species on capacitation and associated protein tyrosine phosphorylation in buffalo (*Bubalus bubalis*) spermatozoa *Animal Reproduction Science* **107** 68–84.

Saacke RG (2008) Sperm morphology: its relevance to compensable and uncompensable traits in semen *Theriogenology* **70** 473–478.

Saacke RG and Almquist JO (1964a) Ultrastructure of bovine spermatozoa. I. The head of normal ejaculated sperm *American Journal of Anatomy* **115** 143–162.

Saacke RG and Almquist JO (1964b) Ultrastructure of bovine spermatozoa. II. The neck and tail of normal ejaculated sperm *American Journal of Anatomy* **115** 163–184.

Saacke RG, Nadir S. and Nebel RL (1994) Relationship of semen quality to sperm transport, fertilization, and embryo quality in ruminants *Theriogenology* **41** 45–50.

Saacke RG, DeJarnette JM, Bame JH, Karabinus DS and Whitman S (1998) Can spermatozoa with abnormal heads gain access to the ovum in artificially inseminated super- and single-ovulating cattle? *Theriogenology* **51** 117–128.

Saacke RG, Dalton JC, Nadir S, Nebel RL and Bame JH (2000) Relationship of seminal traits and insemination time to fertilization rate and embryo quality *Animal Reproduction Science* **60–61** 663–677.

Safe S (2004) Endocrine disruptors and human health: is there a problem? *Toxicology* **205** 3–10.

Safe S. (2005) Clinical correlates of environmental endocrine disruptors *Trends in Endocrinology and Metabolism* **16** 139–144.

Said TM, Aziz N, Sharma RK, Lewis-Jones I, Thomas AJ Jr and Agarwal A (2005a) Novel association between sperm deformity index and oxidative stress-induced DNA damage in infertile male patients *Asian Journal of Andrology* **7** 121–126.

Said TM, Agarwal A, Sharma RK, Thomas AJ Jr and Sikka SC (2005b) Impact of sperm morphology on DNA damage caused by oxidative stress induced by beta-nicotinamide adenine dinucleotide phosphate *Fertility and Sterility* **83** 95–103.

Sailer BL, Jost LK and Evenson DP (1996) Bull sperm head morphometry related to abnormal chromatin structure and fertility *Cytometry* **24** 167–173.

Sailer BL, Sarkar LJ, Bjordahl JA, Jost LK and Evenson DP (1997) Effects of heat stress on mouse testicular cells and sperm chromatin structure *Journal of Andrology* **18** 294–301.

Saito K, O'Donnell L, McLachlan RI and Robertson DM (2000) Spermiation failure is a major contributor to early spermatogenic suppression caused by hormone withdrawal in adult rats *Endocrinology* **141** 2779–2785.

Salisbury GW, Willett EL and Seligman J (1942) The effect of the method of making semen smears upon the number of morphologically abnormal spermatozoa *Journal of Animal Science* **1** 199–205.

Sanocka D and Kurpisz M (2004) Reactive oxygen species and sperm cells *Reproductive Biology and Endocrinology* **2**:12.

Schmid TE, Eskenazi B, Marchetti F, Young S, Weldon RH, Baumgartner A, Anderson D and Wyrobek AJ (2012) Micronutrients intake is associated with improved sperm DNA quality in older men *Fertility and Sterility* **98** 1130–1137.

Seli E, Gardner DK, Schoolcraft WB, Moffatt O and Sakkas D (2004) Extent of nuclear DNA damage in ejaculated spermatozoa impacts on blastocyst development after *in vitro* fertilization *Fertility and Sterility* **82** 378–383.

Selvaraju S, Nandi S, Gupta PS and Ravindra JP (2011) Effects of heavy metals and pesticides on buffalo (*Bubalus bubalis*) spermatozoa functions *in vitro Reproduction in Domestic Animals* **46** 807–813.

Setchell BP (1998) The Parkes Lecture: heat and the testis *Journal of Reproduction and Fertility* **114** 179–194.

Skakkebaek NE, Rajpert-De Meyts E and Main KM (2001) Testicular dysgenesis syndrome: an increasingly common developmental disorder with environmental aspects *Human Reproduction* **16** 972–978.

Skinner JD and Louw GN (1966) Heat stress and spermatogenesis in *Bos indicus* and *Bos taurus* cattle *Journal of Applied Physiology* **21** 1784–1790.

Sławeta R, Laskowska T and Szymańska E (1988) Lipid peroxides, spermatozoa quality and activity of glutathione peroxidase in bull semen *Acta Physiologica Polonica* **39** 207–214.

Slowińska M, Jankowski J, Dietrich GJ, Karol H, Liszewska E, Glogowski J, Kozłowski K, Sartowska K and Ciereszko A (2011) Effect of organic and inorganic forms of selenium in diets on turkey semen quality *Poultry Science* **90** 181–190.

Söderquist L, Janson L, Larsson K and Einarsson S (1991) Sperm morphology and fertility in A.I. bulls *Zentralblatt für Veterinärmedizin, Reihe A* **38** 534–543.

Speight SM, Estienne MJ, Harper AF, Crawford RJ, Knight JW and Whitaker BD (2012) Effects of dietary supplementation with an organic source of selenium on characteristics of semen quality and *in vitro* fertility in boars *Journal of Animal Science* **90** 761–770.

Spiraman V, Niu E, Matias JR, Donahoe PK, Maclaughlin DT, Hardy MP and Lee MM (2001) Müllerian inhibiting substance inhibits testosterone synthesis in adult rats *Journal of Andrology* **22** 750–758.

Suganuma R, Yanagimachi R and Meistrich ML (2005) Decline in fertility of mouse sperm with abnormal chromatin during epididymal passage as revealed by ICSI *Human Reproduction* **20** 3101–3108.

Sullivan JJ (1978) Morphology and motility of spermatozoa In *Physiology of Reproduction and Artificial Insemination of Cattle* 2nd edition pp 286–328 Ed GW Salisbury, NL Van Demark and JR Lodge. W.H. Freeman Company, San Francisco.

Surai PF, Kutz E, Wishart GJ, Noble RC and Speake BK (1997) The relationship between the dietary provision of alpha-tocopherol and the concentration of this vitamin in the semen of chicken: effects on lipid composition and susceptibility to peroxidation *Journal of Reproduction and Fertility* **110** 47–51.

Susin SA, Lorenzo HK, Zamzami N, Marzo I, Snow BE, Brothers GM, Mangion J, Jacotot E, Costantini P, Loeffler M, Larochette N, Goodlett DR, Aebersold R, Siderovski DP, Penninger JM and Kroemer G (1999) Molecular characterization of mitochondrial apoptosis-inducing factor *Nature* **397** 441–446.

Tabb MM and Blumberg B. (2006) New modes of action for endocrine-disrupting chemicals *Molecular Endocrinology* **20** 475–482.

Thundathil J, Palasz AT, Barth AD and Mapletoft RJ (1998) Fertilization characteristics and *in vitro* embryo production with bovine sperm containing multiple nuclear vacuoles *Molecular Reproduction and Development* **50** 328–333.

Thundathil J, Palasz AT, Mapletoft RJ and Barth AD (1999) An investigation of the fertilizing characteristics of pyriform-shaped bovine spermatozoa *Animal Reproduction Science* **57** 35–50.

Tripathi R, Mishra DP and Shaha C (2009) Male germ cell development: turning on the apoptotic pathways *Journal of Reproductive Immunology* **83** 31–35.

Twigg J, Fulton N, Gomez E, Irvine DS and Aitken RJ (1998a) Analysis of the impact of intracellular reactive oxygen species generation on the structural and functional integrity of human spermatozoa: lipid peroxidation, DNA fragmentation and effectiveness of antioxidants *Human Reproduction* **13** 1429–1436.

Twigg JP, Irvine DS and Aitken RJ (1998b) Oxidative damage to DNA in human spermatozoa does not preclude pronucleus formation at intracytoplasmic sperm injection *Human Reproduction* **13** 1864–1871.

van der Heijden GW, Ramos L, Baart EB, van der Berg IM, Derijck AA, van der Vlag J, Martini E and de Boer P (2008) Sperm derived histones contribute to zygotic chromatin in humans *BMC Developmental Biology* **8**:34.

van Loo G, Saelens X, van Gurp M, MacFarlane M, Martin SJ and Vandenabeele P (2002a) The role of mitochondrial factors in apoptosis: a Russian roulette with more than one bullet *Cell Death and Differentiation* **9** 1031–1042.

van Loo G, Saelens X, Matthijssens F, Schotte P, Beyaert R, Declercq W and Vandenabeele P (2002b) Caspases are not localized in mitochondria during life or death *Cell Death and Differentiation* **9** 1207–1211.

Verhoeven G, Willems A, Denolet E, Swinnen JV and Gendt KD (2010) Androgens and spermatogenesis: lessons from transgenic mouse models *Philosophical Transactions of the Royal Society of London B* **365** 1537–1556.

Vicari E, La Vignera S, Castiglione R and Calogero AE (2006) Sperm parameter abnormalities, low seminal fructose and reactive oxygen species overproduction do not discriminate patients with unilateral or bilateral post-infectious inflammatory prostato-vesiculo-epididymitis *Journal of Endocrinological Investigation* **29** 18–25.

Vogler CJ, Saacke RG, Bame JH, DeJarnette JM and McGillard ML (1991) Effects of scrotal insulation on viability characteristics of cryopreserved bovine semen *Journal of Dairy Science* **74** 3827–3835.

Vogler CJ, Bame JH, DeJarnette JM, McGilliard ML and Saacke RG (1993) Effects of elevated testicular temperature on morphology characteristics of ejaculated spermatozoa in the bovine *Theriogenology* **40** 1207–1219.

Waites GM (1991) Thermoregulation of the scrotum and testis: studies in animals and significance for man *Advances in Experimental Medicine and Biology* **286** 9–17.

Waites GM and Voglmayr JK (1962) Apocrine sweat glands of the scrotum of the ram *Nature* **196** 965–967.

Walters AH, Saacke RG, Pearson RE and Gwazdauskas FC (2005a) Assessment of pronuclear formation following in vitro fertilization with bovine spermatozoa obtained after thermal insulation of the testis *Theriogenology* **65** 1016–1028.

Walters AH, Eyestone WE, Saacke RG, Pearson RE and Gwazdauskas FC (2005b) Bovine embryo development after IVF with spermatozoa having abnormal morphology *Theriogenology* **63** 1925–1937.

Walters AH, Saacke RG, Pearson RE and Gwazdauskas FC (2006) The incidence of apoptosis after IVF with morphologically abnormal bovine spermatozoa *Theriogenology* **64** 1404–1421.

Wang, X and Liu N (2012) Sperm DNA oxidative damage in patients with idiopathic asthenozoospermia *Zhong Nan Da Xue Xue Bao Yi Xue Ban (Journal of Central South University)* **37** 100–105.

Wang X, Sharma RK, Sikka SC, Thomas AJ Jr, Falcone T and Agarwal A (2003) Oxidative stress is associated with increased apoptosis leading to spermatozoa DNA damage in patients with male factor infertility *Fertility and Sterility* **80** 531–535.

Ward WS (2010) Function of sperm chromatin structural elements in fertilization and development *Molecular Human Reproduction* **16** 30–36.

Ward WS and Coffey DS (1991) DNA packaging and organization in mammalian spermatozoa: comparison with somatic cells *Biology of Reproduction* **44** 569–574.

Ward WS, Kishikawa H, Akutsu H, Yanagimachi H and Yanagimachi R (2000) Further evidence that sperm nuclear proteins are necessary for embryogenesis *Zygote* **8** 51–56.

Wassarman PM (1999) Mammalian fertilization: molecular aspects of gamete adhesion, exocytosis, and fusion *Cell* **96** 175–183.

Williams WW (1920) Technique of collection semen for laboratory examination with a review of several diseased bulls *The Cornell Veterinarian* **10** 87–94.

Williams WW and Savage A (1925) Observation of the seminal micropathology of bulls *The Cornell Veterinarian* **15** 353–375.

Wright A, Bubb WA, Hawkins CL and Davies MJ (2002) Singlet oxygen-mediated protein oxidation: evidence for the formation of reactive side chain peroxides on tyrosine residues *Photochemistry and Photobiology* **76** 35–46.

Yanagimachi, R (1994) Mammalian fertilization In *The Physiology of Reproduction* pp 189–317 Ed E Knobil and JD Neill. Raven Press, New York.

Yoder JA, Soman NS, Verdine GL and Bestor TH (1997) DNA (cytosine-5)-methyltransferases in mouse cells and tissues. Studies with a mechanism-based probe *Journal of Molecular Biology* **270** 385–395.

Young J, Chanson P, Salenave S, Noel M, Brailly S, O'Flaherty M, Schaison G and Rey R (2005) Testicular anti-Müllerian hormone secretion is stimulated by recombinant human FSH in patients with congenital hypogonadotropic hypogonadism *Journal of Clinical Endocrinology and Metabolism* **90** 724–728.

Zama AM and Uzumcu M (2010) Epigenetic effects of endocrine-disrupting chemicals on female reproduction: an ovarian perspective *Frontiers in Neuroendocrinology* **31** 420–439.

Zan-Bar T, Bartoov B, Segal R, Yehuda R, Lavi R, Lubart R and Avtalion RR (2005) Influence of visible light and ultraviolet irradiation on motility and fertility of mammalian and fish sperm *Photomedicine and Laser Surgery* **23** 549–555.

Zaneveld LJ and Williams WL (1970) A sperm enzyme that disperses the corona radiata and its inhibition by decapacitation factor *Biology of Reproduction* **2** 363–368.

Zhai W, Neuman SL, Latour MA and Hester PY (2007) The effect of dietary L-carnitine on semen traits of White Leghorns *Poultry Science* **86** 2228–2235.

Zhang X and Ho SM. (2011) Epigenetics meets endocrinology *Journal of Molecular Endocrinology* **46** R11–R32.

Zhu BK and Setchell BP (2004) Effects of paternal heat stress on the *in vivo* development of pre-implantation embryos in the mouse *Reproduction Nutrition Development* **44** 617–629.

4 Sperm Preparation for Fertilization

Bart M. Gadella*
Utrecht University, Utrecht, the Netherlands

Introduction

Fertilization is a decisive moment in life and enables the combination of the DNA from two gametes to ultimately form a new organism. The sperm surface, especially the head area, has distinguishable functional areas that are involved in distinct fertilization processes. It is known that the sperm surface undergoes constant remodelling during its migration, especially in the epididymis (maturation) and in the oviduct (capacitation). Altogether, these changes serve to prepare sperm to meet the oocyte and bind to the zona pellucida. To this end, not only must sperm acquire hyperactivated motility but also three independent, successive events are required to establish monospermic fertilization. Firstly, the sperm cell must secrete its acrosome contents (the acrosome reaction), thus allowing the sperm to penetrate the zona pellucida and to reach the oocyte plasma membrane – the site of fertilization. Next, the sperm cell binds to the oocyte plasma membrane, and the two different cells fuse together. Finally, the just fertilized oocyte must protect itself from polyspermic fertilization, which is established by the secretion of the cortical granule contents over the entire oocyte cell surface into the perivitelline space (the cortical reaction). Altogether, these events lead to the activation of the oocyte and the formation of one male and one female haploid pronucleus. After syngamy, the resulting diploid zygote is ready to develop into an embryo and a new organism.

This chapter provides an overview of the current understanding of the preparative processes of sperm that are required to achieve mammalian fertilization. In addition, the consequences of semen processing, e.g. for semen storage or for flow cytometry sorting, for sperm integrity and its fertilization capacity are discussed.

Sperm Ergonomics

In the last moments of spermatogenesis, sperm cells are released from the Sertoli cells and from the syncytia of synchronously formed spermatids into the lumen of the seminiferous tubules. The released sperm cells are starting a complex and lengthy process, which should bring them eventually to the cumulus oocyte complexes in the oviducts of the female. During the transit through the male and female reproductive

* E-mail: b.m.gadella@uu.nl

tracts, alterations of spermatozoal metabolism, motility and surface properties occur that allow the sperm to properly interact with the extracellular vestments of the oocyte and to eventually bind to and fuse with the oocyte. After fertilization, the sperm cell also provides a trigger to the oocyte to prevent polyspermic fertilization. Some of the ergonomics of a typical mammalian sperm cell are already apparent at the moment they are released in the testis. Typically, spermatozoa consist of a head, a midpiece and a tail (for details on sperm structure, see Eddy and O'Brien, 1994)

The sperm head

Within the sperm head, two large intracellular compartments can be distinguished: the acrosome and the nucleus. The latter contains the male haploid genome, which is highly condensed, with protamines having replaced histones almost completely during spermatogenesis (for a review see Dadoune, 2003). This results in the chromatin being approximately 10× more compact than somatic cell chromatin, so the proportions of the sperm head are minimized. Moreover, as a haploid cell, the sperm contains only half the complement of DNA of somatic cells. However, the dimensions of the nucleus as seen under a microscope can be misleading: the somatic nucleus is round and ball shaped, while the sperm nucleus of farm animals and equines is extremely flattened (oval and pancake shaped). Furthermore, the extremely compacted DNA is less vulnerable to enzymes (apoptotic related), toxins or oxidant-mediated damage. The nucleus is then 'protected' and compact, both of which are requirements for the optimal transport of the spermatozoa.

The acrosome is located at the apical edge of the sperm head, while the flagellum develops at the distal part of the sperm head. The acrosome is a cap-shaped organelle with an acidic pH and is filled with hydrolytic, glycosidic and proteolytic enzymes (Moreno and Alvarado, 2006). These enzymes are condensed into the so-called acrosomal matrix in the lumen of this organelle. The contents of the acrosome are enclosed by one continuous membrane, although the part directly overlying the nuclear envelope is referred to as the inner acrosomal membrane, while the part associated with the sperm plasma membrane is referred to as the outer acrosomal membrane. After the acrosome reaction, the contents of the acrosome are involved in sperm binding to the zona pellucida of the oocyte and its penetration (Gadella and Evans, 2011), and possibly sperm–cumulus association as well (Sun *et al.*, 2011); the acrosome contents may also contain factors involved in sperm–oocyte binding and fusion (Sachdev *et al.*, 2012).

The sperm midpiece

The sperm midpiece is connected by a connective element to the sperm head and contains a battery of less than 100 mitochondria that are coiled around the flagellum and elongate further down the tail (Bahr and Engler, 1970). It is in the mitochondria of the midpiece that aerobic metabolism occurs (the head and tail are devoid of mitochondria). The ATP produced probably serves to provide the energy for the motility machinery and for housekeeping processes through the sperm membrane. The constantly changing environments of the spermatozoa discussed later in this chapter probably result in adaptive ionic and metabolic processes that are important for maintenance of the intracellular milieu (Shivaji *et al.*, 2009; Ramió-Lluch *et al.*, 2011).

The sperm tail

The tail is the mitochondria-free distal part of the flagellum. This part of the flagellum allows sperm movement. The microtubules of the sperm tail effectively slide over each other, resulting in bending of the tail. Microtubule sliding is an ATP-consuming process, fed by ATP produced either in the midpiece (aerobically) and/or in the tail itself by glycolysis. The central flagellum is surrounded by a fibrous sheath, in which there is high-capacity glycolytic machinery. This sheath provides the tail elastic properties, which are required for the forward-propelling properties of the sperm tail (for a review see Eddy, 2007).

Other ergonomic aspects of sperm

The sperm head, midpiece and tail have their own specializations and ergonomic adaptations that allow the cell to optimally reach and fertilize the oocyte. One aspect of this is that sperm cells have lost various somatic cell processes and organelles. For example, during spermiation, the sperm cell loses its capacity for transcription and translation (Boerke *et al.*, 2007). Once the remaining and specialized organelles are organized as mentioned above, sperm cells also completely lose membrane transport capabilities via vesicles, e.g. they have no endocytosis, exocytosis or inter-organelle transport. A number of other 'typical' cellular components are also absent, including ribosomes (no translation), Golgi complexes, endoplasmic reticulum, peroxisomes and lipid droplets. Sperm cells only contain a minimal amount of cytosol as well (for a review see Eddy and O'Brien 1994). All these features are in fact signs of the minimalization of sperm volume and size, and provide evidence that the cell is terminally differentiated and prepared for only one last task, namely the fertilization of the oocyte. After spermiation, the sperm cell in the testis is not competent to fertilize the oocyte, but it already shares all of these ergonomic features with the activated sperm in the oviduct, which has at that point become competent to fertilize the oocyte. In the testis, the sperm cell is not motile, nor does it have the appropriate surface organization or proteins to recognize the cumulus–oocyte complex (Boerke *et al.*, 2008). The sperm cell must undergo a series of posttranslational and environmental modifications that will result in its ability to undergo the fertilization tasks. The current understanding of the post-testicular sperm remodelling will be reviewed here, and suggestions made for future research efforts to fill the gaps in our current knowledge.

The Sperm Surface

This section focuses on the domained structure of the sperm surface and discusses its importance for fertilization. Functionally, the sperm head surface can be divided into three regions: apical, equatorial and post-acrosomal.

The apical area of the sperm head is involved in sperm–zona recognition (primary zona binding) and consequently exclusively contains the primary zona-binding proteins that are recruited to this apical surface area only shortly before binding takes place (in the oviduct or in *in vitro* fertilization (IVF) media) (van Gestel *et al.*, 2005, 2007). The series of events that are induced after sperm–zona binding are depicted in Plate 3.

After the sperm has bound to the zona pellucida, the acrosome reaction occurs and results in the exposure of the acrosomal enzymatic matrix to the zona pellucida (the secondary zona-binding proteins). At this stage, zona binding also results in zona digestion. To ensure maximal exposure of the enzymatic matrix to the zona pellucida, simultaneous multi-point membrane fusion occurs between the outer acrosomal membrane and the plasma membrane. This causes the formation and release from the sperm cell proper of hybrid vesicles of these two membranes. The resulting hybrid vesicles are sometimes referred to as the 'acrosomal shroud'. Only two-thirds of the apical side of the outer acrosomal membrane is involved in this very characteristic fusion process (Tsai *et al.*, 2012). It is of note that it has recently been established that boar sperm capacitation involves the docking of the sperm acrosome to the sperm surface (Tsai *et al.*, 2010) and that the initiation of the acrosome reaction may be earlier than at zona binding, at least in the mouse (Jin *et al.*, 2011; Sun *et al.*, 2011).

The remaining unfused part of the sperm surface and the outer acrosomal membrane forms the so-called 'equatorial segment'. In this area, the plasma membrane has become continuous with the one-third section of the outer acrosomal membrane, which was already continuous with the inner acrosomal membrane (Tsai *et al.*, 2010). This continuous membrane structure takes over the surface function of the released apical plasma membrane, thereby ensuring the maintenance of an intact intracellular milieu. The resulting equatorial segment has recruited a specific machinery of proteins that will be involved in sperm–oolemma binding and fertilization fusion. In part, these proteins originate from the acrosomal membrane (Ito *et al.*, 2010; for a review see also Gadella and Evans, 2011).

Thus, the sperm head surface contains at least three functional domains involved in gamete interaction: one involved in zona recognition (apical area), one in the acrosomal fusions (the apical and pre-equatorial area) and one in the fertilization fusion (the equatorial segment). The function of the post-acrosomal region of the sperm plasma membrane during the process of fertilization is not known. In the testicular sperm cell, many of the components required for fertilization are not present and need to be recruited from the post-testicular environments (van Gestel *et al.*, 2007; Leahy and Gadella, 2011; Caballero *et al.*, 2012). Once attached to the sperm surface, these components also need to concentrate at their specific surface areas in order to give each domain its specific tasks (van Gestel *et al.*, 2005; Zitranski *et al.*, 2010; Watanabe and Kondoh, 2011). Also, the midpiece and tail have a dynamic domained structure, and probably modifications thereof, that are related to sperm motility characteristics. Testicular sperm cells are immotile, whereas ejaculated sperm have gained straight-forward movement, following their maturation in the epididymes. The sperm will obtain full hyperactivated motility characteristics after capacitation in the oviduct (Suarez and Pacey, 2006; Suarez, 2008).

These dynamic, changing sperm features are accomplished gradually through the influence of the constantly changing extracellular environment encountered en route to fertilization. In the next section, these processes will be followed at the level of the sperm surface. Of considerable cell biological and membrane biochemical interest is the question as to how the domained structure of the sperm surface and its dynamic reordering (Gadella *et al.*, 1995) are accomplished. A number of possible dynamic alterations in the sperm plasma membrane and interacting components of the cytoskeleton and the extracellular matrix may be involved (Fig. 4.1).

General Mechanisms of Sperm Surface Modification by the Extracellular Environment

The actual membrane phenomena underlying plasma membrane domain dynamics at the sperm surface are of interest and may add to the understanding of how and why certain treatments in sperm handling, storage and IVF are good or bad. A hypothetical scheme of possible interactions between the sperm surface and the components of the male or female genital tract is depicted in Fig. 4.2, and the five relevant processes shown in this figure are described below.

From the various epithelia of the male and female genital tract, blebbing cytosol-filled vesicles are released into the genital fluids and then interact and exchange surface components with sperm (see process 1 in Fig. 4.2). It is unlikely that these vesicles fuse with the sperm, as this would dramatically increase sperm volume (which has already been reduced maximally in order to obtain an ergonomically designed cell optimally suited for fertilization). Blebbing of vesicles has been demonstrated in the epididymal duct and in the epithelia of the vesicular gland and the prostate of the boar (Tsai and Gadella, 2009). The function of the blebbed vesicles is largely unknown, although hypothetically they may exchange components with the sperm surface and/or modulate the repressive activity of white blood cells.

Serum components are released into the genital fluids by transcytosis (Cooper *et al.*, 1988; see process 2a in Fig. 4.2). Lipoprotein particles may invade the surroundings of the sperm and facilitate exchange at the sperm surface. Fluid-phase secretion and adsorption alter the extracellular matrix (ECM) of sperm by a variety of sticking molecules and enzymes that cause posttranslational modifications of ECM molecules (process 2b in Fig. 4.2).

Apocrine secretion of exosomes may be involved in altering the sperm surface and function as well. Exosomes are secreted by the epididymis (epididymosomes), by the prostate (prostasomes) and by the uterus

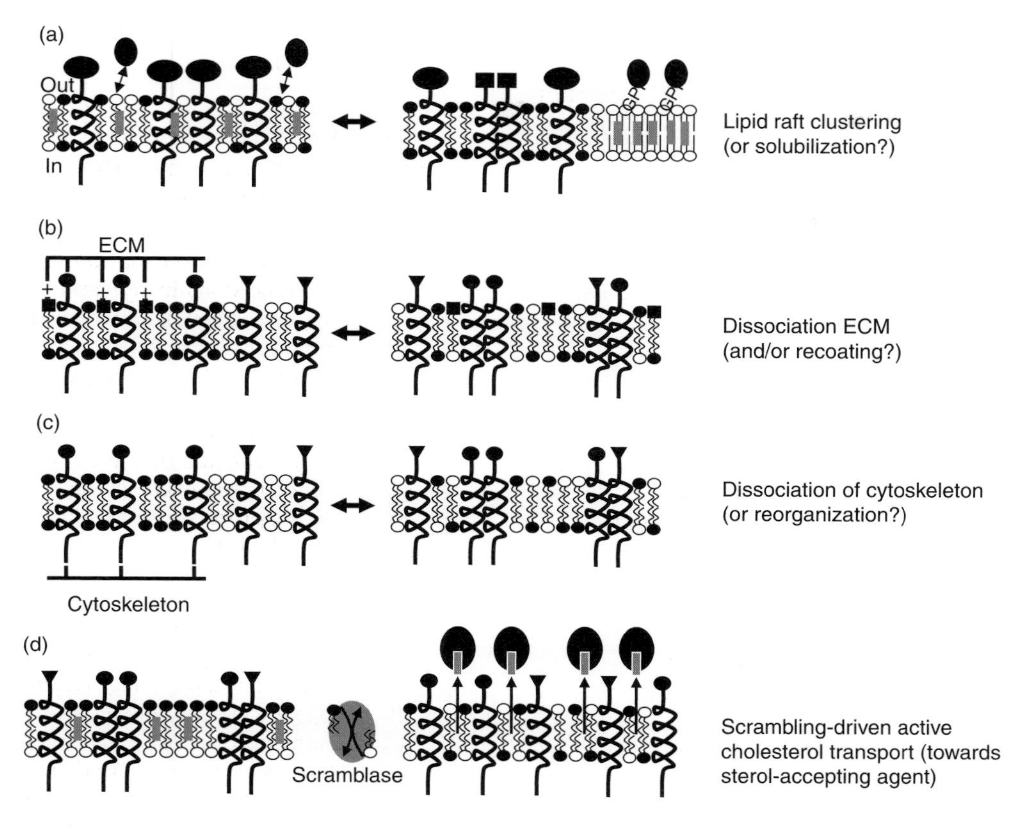

Fig. 4.1. Current models for lateral polarity dynamics upon sperm capacitation. (a) GPI (glycosylphosphatidylinositol)-anchored peripheral membrane proteins can induce the formation of lipid ordered micro-domains in the membrane that exclude freely diffusible transmembrane proteins. GPI-anchored proteins may be recruited from the fluid-phase membrane area or from lipoprotein particles and exosomes that surround the migrating sperm. The raft area is presumably enriched in saturated phospholipids, cholesterol gangliosides and sphingomyelin. (b) The extracellular matrix (ECM) surrounding the sperm is heterogeneous, so the ECM may assist in the creation of membrane domains via electrostatic interactions with glycolipids or with membrane proteins. Disruption or reordering of the sperm's ECM as it travels from the testis towards the isthmus of the oviduct where it fertilizes the egg might allow redistribution of lipids and membrane proteins. (c) A similar scenario to that described for (b) may be valid for the heterogeneous organization of the sperm cytoskeleton. (d) Processes (a–c) may result from an increase in disordered lipid packing in the sperm plasma membrane that allows cholesterol efflux. (Modified from Flesch and Gadella, 2000, and Gadella, 2008.)

(uterosomes) (Gatti *et al.*, 2005; Griffiths *et al.*, 2008; Thimon *et al.*, 2008) (process 3 in Fig. 4.2). Further, exosomes may provide sperm with tetraspanins, a group of membrane proteins involved in the tethering of proteins into protein complexes. The addition of CD9 (an exosome marker) originating from the oocytes on to the sperm surface by membrane particles occurs even when sperm reach the perivitelline space (Barraud-Lange *et al.*, 2007a,b), although the function of exosomes in fertilization fusion is unclear (Miyado *et al.*, 2008; Barraud-Lange *et al.*, 2012).

A change in the sperm's surface properties may alternatively be elicited by sperm interactions with ciliated epithelial cells in the epididymis, uterus and oviduct (Zhang and Martin-Deleon, 2003; Gatti *et al.*, 2004; Sostaric *et al.*, 2008) (process 4 in Fig. 4.2). The interaction at the level of the epididymal epithelia guides

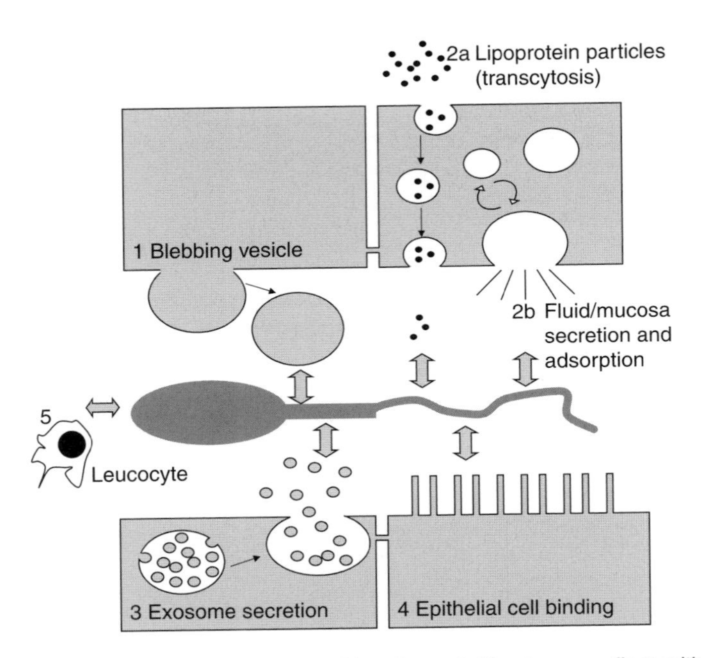

Fig. 4.2. Hypothetical scheme of possible male and female genital tract surroundings with the sperm surface. Details are provided in the text. (1) Membrane blebbing. (2a) Transcytosis of fluid/mucosa components from the circulation into the luminal fluid. (2b) Fluid phase secretion and or adsorption of either fluid or mucosa components into or from the luminal fluid. (3) Apocrine secretion of exosomes into the luminal fluid. (4) Sperm binding to ciliated cells, which form the epithelial lining of the luminal fluid compartment. (5) Interaction with immunological active cells such as leucocytes.

the sperm through a maturation process during which a continuous modification of the sperm surface takes place, and by which the sperm cells obtain their motility characteristics and the removal of their redundant cytoplasmic droplets. The interaction with the oviduct has a physiological role in *in situ* capacitation, but this is not understood at the molecular level (Chang and Suarez, 2012). The importance of sperm interactions with other ciliated epithelial cells of the female and male genital tract is also not fully understood, but could be crucial for sperm surface remodelling and sperm function.

When sperm enter the uterus they, together with seminal plasma components, evoke an immunological response (process 5 in Fig. 4.2) (Schuberth *et al.*, 2008). For instance, leucocytes will infiltrate into the uterine fluid (Taylor *et al.*, 2008) and affect the surface of sperm. This probably serves to cleanse the uterus of degenerated sperm and to prevent an inflammatory response in it (Woelders and Matthijs, 2001). The sperm involved in fertilization colonize the lower parts of the oviduct, where no leucocyte infiltration takes place.

Sperm Surface Modification in the Male Genital Tract

As discussed above, the domained surface of sperm is already apparent in spermatids in the seminiferous tubules of the testis before spermiation (Eddy and O'Brien, 1994). The molecular dynamics involved in the establishment of surface specialization upon spermatogenesis is largely unknown. Generally, it is clear that the polar organization of the extracellular matrix components of the cytoskeleton and the sperm cell organelles are involved in its domained and heterogeneous surface (Plate 3). As already noted, in mature sperm cells, the amount of cytosol is minimal and, indeed, the observed surface domains mirror the organelle organization located just under the sperm surface: at the apical area, the acrosome, the post-acrosomal area, the nuclear envelope, the midpiece surface covering the mitochondria and, in the tail, at the surface covering the fibrous sheath.

Moreover, once liberated in the lumen of the seminiferous tubules, the sperm cells will start their travel through the male and female genital tracts and will meet a sequence of different environments. During this voyage, surface remodelling takes place, most likely at many sites in the two genital tracts: (i) upon somatic maturation in the epididymis (Gatti *et al.*, 2004; Dacheux *et al.*, 2005); and (ii) by recoating and decoating events induced by the accessory fluids added at ejaculation, which probably stabilize the sperm for their further journey in the female genital tract (Gwathmey *et al.*, 2006; Girouard *et al.*, 2008). These changing environments may cause sperm surface remodelling and thus may influence their potential to fertilize the oocyte.

After release from the Sertoli cells into the lumen of seminiferous tubules, sperm cells are equipped with a number of proteins that are reported to be involved in zona binding. At its surface, the sperm has transmembrane proteins belonging to the ADAMs family. These were initially thought to be involved in the fertilization process, but are now reported to be involved in sperm–zona binding. ADAM-2, also known as fertilin β, has such a function in boar sperm (van Gestel *et al.*, 2007). Other testicular sperm proteins, e.g. sperm lysosomal-like protein (SLLP1; Herrero *et al.*, 2005), and sperm acrosomal membrane proteins (SAMP14 and 32; Vjugina and Evans, 2008) and Sp56 (Buffone *et al.*, 2008), are involved in secondary zona binding, as they are localized in the acrosome and only become exposed to the zona pellucida after the induction of the acrosome reaction.

Some secretory proteins, e.g. CRISP (cysteine-rich secretory proteins), are also involved in sperm–zona adhesion, sperm–oolemma binding or fertilization fusion (Busso *et al.*, 2007a; Cohen *et al.*, 2007; Da Ros *et al.*, 2007). CRISP2 is from testicular origin, while CRISP1 and CRISP4 originate from the epididymis (Busso *et al.*, 2007b). The exact mechanism of CRISP's association with the sperm surface is not yet known, although CRISP1 is one of the abundant proteins in epididymosomes (Thimon *et al.*, 2008). Epididymosomes are also reported to influence the lipid composition of the sperm surface (Rejraji *et al.*, 2006). Other proteins that have been shown to be added to the sperm surface in the epididymis are SED1 (or P47). These proteins are known to have a role in oviduct and zona binding (Shur *et al.*, 2006; Nixon *et al.*, 2011). Angiotensin converting enzyme (ACE), which becomes glycosylphosphatidylinositol (GPI) – anchored to the sperm surface upon epidiymal maturation, is involved in the proper orienting of ADAM proteins (Yamaguchi *et al.*, 2006; Nixon *et al.*, 2011).

During sperm capacitation, phospholipases and ACE serve to remove a proportion of GPI-anchored proteins and are believed to play a role in the remodelling of lipid rafts (Kondoh *et al.*, 2005; Inoue *et al.*, 2010; Watanabe and Kondoh, 2011). The addition of GPI-anchored proteins has been reported to occur in the epididymis where the sperm adhesion molecule SPAM-1 is secreted by the epithelial cells into the luminal fluid and later associates with the sperm surface (Zhang and Martin-Deleon, 2003). Likewise, in the pig, the spermadhesin AQN-3 and carbonyl reductase are added to the sperm surface (van Gestel *et al.*, 2007). Inhibition of hamster carbonyl reductase activity results in a decreased affinity for the zona pellucida, while the sperm remain motile and intact (Montfort *et al.*, 2002). Even under very stringent detergent conditions, AQN-3 is still able to bind to the zona pellucida (van Gestel *et al.*, 2007).

Data derived by proteomics has been useful for the identification of proteins that do not have a direct function in sperm–zona binding but are associated with the zona binding protein complex formed at the sperm surface (van Gestel *et al.*, 2007; Gadella, 2008). Some of these proteins (such as protein phosphatases) are involved in sperm signalling processes, while others are involved in the maintenance of the redox balance (peroxiredoxin 5) (see Gadella *et al.*, 2008). The latter include a potassium channel, which might induce membrane hyperpolarization by K^+ efflux. This hyperpolarization may, in turn, open voltage-dependent Ca^{2+} channels that enable Ca^{2+}-dependent processes in the capacitating sperm (Tsai and Gadella, 2009). An interesting observation has been that the major zona binding proteins listed above tend to aggregate in capacitating sperm (under IVF conditions) at the surface area that is involved in zona binding (van Gestel *et al.*, 2005, 2007), and this coincides with the attraction of soluble *N*-ethylmaleimide-sensitive factor attachment protein receptors (SNAREs) involved in the acrosome reaction (Tsai *et al.*, 2007, 2010, 2012). The fact that both outer acrosomal SNAREs and plasma membrane SNAREs are concentrated at the apical ridge area of the sperm head after capacitation has led to the assumption that the lipid-ordered membrane aggregation is a preparative step for the acrosome reaction (as proposed in Fig. 4.3).

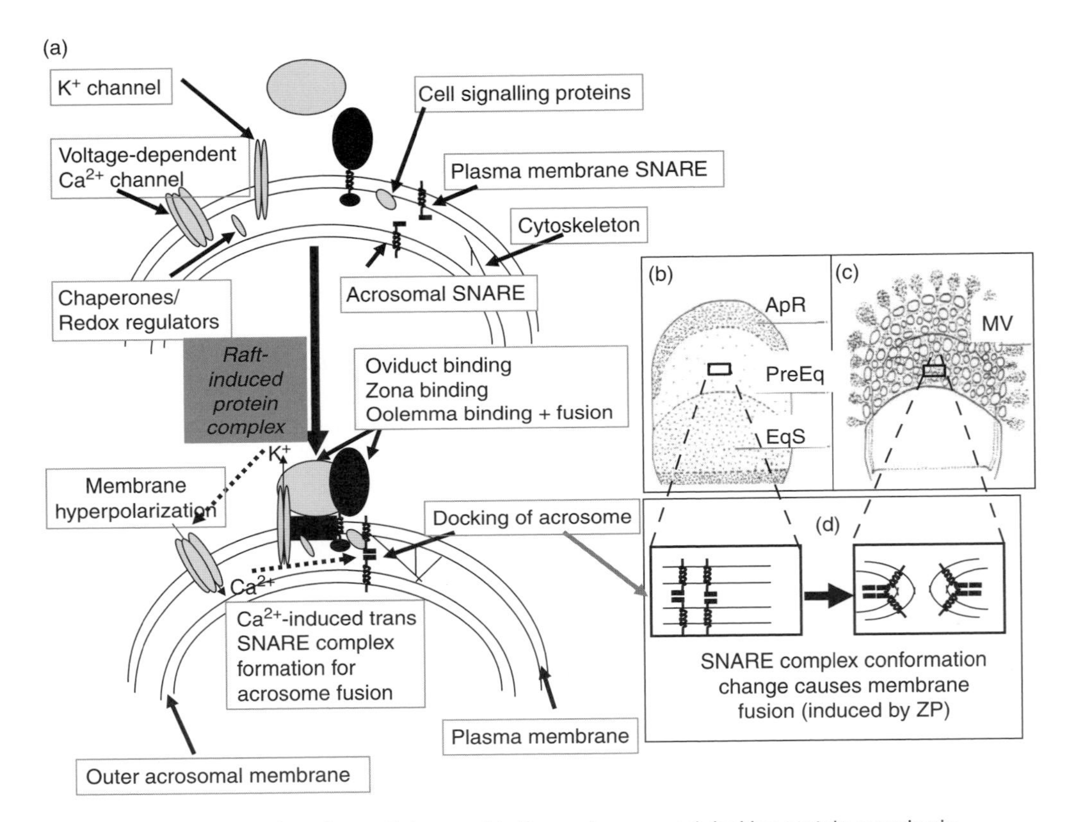

Fig. 4.3. Putative formation of a multiple zona binding and acrosomal docking protein complex in capacitating porcine sperm. (a) A hypothetical model of a sperm zona pellucida-binding complex formed during sperm capacitation by raft-induced protein clustering. This results in a multifunctional protein complex that plays a role in the diverse processes leading to fertilization. SNARE, soluble *N*-ethylmaleimide-sensitive factor attachment protein receptor; for further explanation see text. Scheme based on biochemical and proteomic investigations on epididymal and ejaculated sperm before and after *in vitro* fertilization (IVF) incubations (no sperm surface remodelling by the female genital tract *in situ* is taken into account). (b) Depicts the diverse membrane domains of the sperm head (ApR, apical ridge; EqS, equatorial segment; PreEq, pre-equatorial domain). (c) Depicts a surface view of the sperm head that has initiated the acrosome reaction and in which the ApR and PreEq domains are forming mixed vesicles (MV) with the outer acrosomal membrane. (d) Depicts how in capacitated sperm the plasma membrane and outer acrosomal membrane are positioned with SNARE docking into a transmembrane trimeric protein complex of syntaxin 1B, SNAP 23 and VAMP 3; note that the acrosome remains intact. After binding of the zona pellucida (ZP), a concomitant rise in cytosolic Ca^{2+} allows the multiple membrane fusions involved in the acrosome reaction that are exclusive for the anterior part of the sperm head surface (see Tsai *et al.*, 2012). The *cis*-SNARE membrane complex transition allows these membrane fusions.

For an extensive overview of sperm–zona binding in ruminants, see Gadella (2010). The identified zona interacting protein complex thus is not only involved in sperm-zona binding but may link this event with preparative steps for the acrosome reaction. In fact, we recently provided evidence that capacitation led to docking of the sperm plasma membrane and the outer acrosomal membrane, and that this is exclusive for the area later involved in the acrosome reaction (Tsai *et al.*, 2010). The stably docked acrosome needs an increase in intracellular Ca^{2+} to allow the conformational change of SNARE proteins to subsequently allow the multiple fusions between the apical plasma membrane and the outer acrosomal membrane (Tsai *et al.*, 2012). The author and co-workers have also shown that the docked *trans*-SNARE complex that is formed is stabilized by complexin, and that a Ca^{2+}-dependent removal alters interactions between the plasma membrane and the outer acrosomal membrane SNARE proteins (Tsai *et al.*, 2012). This allows the formation of a new SNARE complex that induces the acrosome reaction after a conformational shift from *trans*- to *cis*-SNARE complexes, which cause the acrosomal fusion. A number of proteins are associated with this multimeric protein complex, which is involved in zona binding and acrosomal fusions. It is likely that recruitment of these components together in the apical ridge area of the sperm head is instrumental for the future fertilization tasks of this specific sperm surface.

In the laboratory, Gadella *et al.* only identified the major zona binding proteins of the sperm plasma membrane, but did not attempt to identify minor proteins or hidden proteins (covered on 2-D IEF-SDS-PAGE gels by the glycosylated zona proteins). Therefore, the possibility cannot be eliminated that some of the sperm surface proteins involved in sperm–zona interactions have not been identified. As no proteins originating from any of the accessory male sex glands were identified, it cannot be concluded that secretions from these glands do not also play a role in sperm surface modification during ejaculation. For instance, the possibility that prostasomes, uterosomes and exosomes originating from the oviduct, or even from the oocyte, could fuse with sperm (Burden *et al.*, 2006; Griffiths *et al.*, 2008; Barraud-Lange *et al.*, 2012) is interesting but needs to be experimentally validated. Another option is that such micro-vesicles exchange components with the sperm surface. It is not clear whether such interactions stabilize the sperm surface and provide decapacitation factors or, alternatively, induce capacitation and destabilize the sperm surface. Both possibilities are described in the literature (for a review, see Tsai and Gadella, 2009). At any rate, to date, no prostate-derived proteins involved in sperm–zona interaction have been described, although prostatic fluids are known to have an influence on sperm surface organization and protein composition (Russell *et al.*, 1984).

Sperm Surface Modification in the Female Genital Tract

Far less is known about the contributions of the female genital tract to the zona binding properties of sperm. Sperm reside for hours to days in the cervix, uterus and, eventually, the isthmus of the oviduct (time and site of deposition is species dependent). Although there are no data concerning the role of the cervical epithelia and secretory products on sperm modifications, it is well established that a reduced number of sperm can be inseminated when deposition is deeper in the cervix or directly into the uterine body (Roberts and Bilkei, 2005; Behan and Watson, 2006). Moreover, a very low dose of sperm can be used for deep intrauterine insemination (for reviews, see Vazquez *et al.*, 2005, 2008). Some studies indicate that, after sperm deposition in the female genital tract, a removal of an extracellular glycoprotein coating (release of decapacitation factors) and further remodelling by cervical, uterine and oviduct secretions is involved in the activation of sperm so that they are capable of penetrating the oocyte (*in vivo* capacitation) (Suarez and Pacey, 2006; Rodriguez-Martinez, 2007).

Sperm also interact with cumulus cells and remaining follicular fluid components. The follicular fluid and cumulus cells impregnate the zona pellucida with secretory components and may also influence sperm behaviour with cumulus-free oocytes (Getpook and Wirotkarun, 2007; Gil *et al.*, 2008). Even the perivitelline space, i.e. the space between the zona pellucida and the oolemma, is filled with components that may influence sperm–oolemma interactions (Barraud-Lange *et al.*, 2007a,b). In an *ex vivo* study, binding of sperm to the uterine epithelium was established. This binding was carbohydrate dependent, as shown by studies with a lectin competition binding assay using uterine segments (Taylor *et al.*, 2008).

However, the proteins involved in this binding are not yet identified, nor is it clear whether this binding is selective and is thus reducing the numbers of sperm capable of migrating deeper into the uterus. Another possibility is that the binding allows modifications of the sperm surface that are important in processes leading to fertilization. The composition of uterine fluid and its effects on the sperm surface or sperm functioning are topics that have been almost completely neglected by researchers. The only study of the effects of porcine uterine fluids on sperm reports lipid modifications in sperm membranes (Evans *et al.*, 1987). Porcine sperm contain non-genomic progesterone receptors at the plasma membrane (Jang and Yi, 2005). Most likely, hormone binding at the sperm surface is part of the *in vivo* sperm capacitation process. The uterine fluids of mice have been shown to contain exosomes, and these particles (uterosomes) contain SPAM-1 and other GPI-linked proteins that can be exchanged with caudal epididymal sperm (Griffiths *et al.*, 2008). It is possible that this exchange functionally improves the capacity of sperm to fertilize the oocyte.

An immunological response is elicited when sperm enter the uterus (Schuberth *et al.*, 2008). This response can be observed by a migration of leucocytes (predominantly polymorphonuclear neutrophils) into the uterine fluid (Taylor *et al.*, 2008). It is not clear whether this infiltration will effect sperm that later enter the oviducts, although leucocytes clearly reduce (by phagocytosis) the amount of sperm that enter the oviducts (Woelders and Matthijs, 2001). It is also uncertain whether phagocytosis is selective (for aberrant sperm) or whether it only non-selectively depletes the amount of sperm that migrate further to the oviduct. As already noted, the oviducts are free of leucocytes and are the site where sperm are finally capacitated *in vivo* in order to fertilize the oocyte (Suarez, 2008).

A number of studies have shown that fluids from the oviduct stimulate sperm capacitation and induce hyperactivated sperm motility. One of the factors involved in this sperm activation is bicarbonate, which is commonly used during IVF (Rodriguez-Martinez, 2007). Oviduct-specific glycoproteins (OSG) as well as osteopontin have been shown to support fertilization in the cow and are secreted by the oviduct (Killian, 2004). A sperm-binding glycoprotein from the oviductal fluid has been shown to induce porcine sperm capacitation (Teijeiro *et al.*, 2008). The lower part of the oviduct is now considered to function as the sperm activation site, making sperm capable of meeting and fertilizing the oocyte (Suarez, 2008). In the isthmus, small numbers of sperm are bound, become capacitated, and are then released after ovulation, thereby enabling them to migrate to the upper part of the oviduct (the ampulla) and to fertilize the passing oocyte(s) (for a review see Suarez, 2008).

The binding and release of sperm to and from the oviduct has recently been shown to be a more complex mechanism than that detailed above (Chang and Suarez, 2012). What has become clear from bovine studies is that the oviductal epithelia and fluids contain sperm-binding factors as well as sperm-releasing factors that cause sperm adhesion and release in the correct time sequence around the time of ovulation (Sostaric *et al.*, 2008). Most likely, spermadhesins, such as AQN-1, are involved in the formation of the oviductal sperm reservoir, as they are involved in sperm binding to this specific epithelium (Ekhlasi-Hundrieser *et al.*, 2005). Note that spermadhesins are added to the sperm during the epididymal sperm maturation phase and/or at ejaculation (Song *et al.*, 2010).

Oviduct-specific glycoproteins have been shown to modulate sperm–zona pellucida interaction and to control the polyspermic fertilization rates in pigs (Hao *et al.*, 2006; Coy *et al.*, 2008; Avilés *et al.*, 2010).

These authors have also pointed out that polyspermy, which is a well-recognized problem in pig IVF, can be prevented by the addition of oviductal fluid components to the IVF media. Oviductal fluid has also been shown to support the early embryonic development of fertilized pig oocytes (Hao *et al.*, 2008; Lloyd *et al.*, 2009). Other oviductal secretory products, e.g. catalases, may protect sperm from peroxidation damage, as has been demonstrated in the cow (Lapointe *et al.*, 1998). Oviduct epithelial annexins have been suggested to immobilize bovine sperm by binding bovine sperm proteins (BSP; Ignotz *et al.*, 2007). Annexin A2 has also been implicated to be involved in sperm–oviduct binding in the sow (Teijero *et al.*, 2009). In the bovine, this interaction is reversed by oviductal fluid factors, such as catalase (Lapointe and Sirard, 1998). The interplay of various glycoproteins at the surface of sperm and oviduct epithelia or in the oviductal fluid, as well as the varying amounts and composition of glycosidases, probably orchestrate proper sperm activation around the time of ovulation in the pig (Carrasco *et al.*, 2008, Töpfer-Petersen *et al.*, 2008).

However, the effect of oviduct and uterine proteins on sperm–zona binding, as well as their putative association with the sperm surface, is not yet established. Of course, it is very possible that secreted products of the female tract enhance sperm–zona binding and that additional female protein candidates may be added to the surface of the zona interacting sperm. Further, the zona pellucida itself, in addition to being a binding target, may also add proteins to the sperm surface. The cumulus cells and the zona pellucida are impregnated with follicular fluid, and remnants of this fluid probably remain attached to the cumulus oocyte complex. For example, different growth factors and extracellular matrix components have been considered to be involved in interactions of sperm with the cumulus oocyte complex (for a review, see Einspanier *et al.*, 1999). Sperm that interact with these structures may respond to these fluid components, in a similar fashion as occurs with extracted follicular fluid (hyperactivated motility; Getpook and Wirotkarun, 2007; Gil *et al.*, 2008).

Finally, an interesting observation has been made that membrane remodelling occurs after the acrosome reaction when sperm reach the perivitelline space but before fertilization fusion. Within the perivitelline space, membrane fragments containing CD9 are added to the sperm surface (Barraud-Lange *et al.*, 2007a,b). If correct, this observation clearly demonstrates the 'feed forward principle' that may exist in mammalian reproduction, wherein the oocyte facilitates the first sperm coming into the perivitelline space to fertilize by adding functional tethering proteins to the surface of the sperm cells. It remains to be established though whether such a process also enables sperm that enter the oviduct to bind to the zona pellucida (Barraud-Lange *et al.*, 2012). It may also be mentioned that sperm proteins are involved in oolemma binding and fertilization fusion in the mouse and human (Ellerman *et al.*, 2006; Vjugina and Evans, 2008; Gadella and Evans, 2011), but data for boar, bull and equine sperm are absent. A list of sperm proteins involved in mouse sperm–egg interaction is published, but the origin and topology of these proteins in and on sperm has not been scrutinized (Stein *et al.*, 2006). Moreover, because mouse sperm are aspirated from epididymes, later post-epididymal sperm surface modifications were not considered. A number of these proteins were recovered in the detergent-resistant membrane (DRM) fraction of mouse sperm (Nixon *et al.*, 2009). This is similar to the high enrichment of zona binding proteins recovered in the DRM fraction of porcine sperm (van Gestel *et al.*, 2005).

Identification of Sperm Proteins Involved in the Cascade Leading to *In Vitro* Fertilization

It is very difficult to study the sperm surface alterations described above *in situ*, although for many mammalian species, including humans, specific sperm handling and incubation media have been optimized for efficient *in vitro* fertilization purposes. In general, mammalian sperm are activated in a medium that is similar to oviductal fluid and contains capacitation factors, e.g. high concentrations of bicarbonate, free calcium ions and lipoproteins (albumin) (Flesch and Gadella, 2000). In some species, specific glycoconjugates (Mahmoud and Parrish, 1996) or phosphodiesterase inhibitors are included *in vitro* for additional sperm activation (Barkay *et al.*, 1984). All strategies are designed to evoke capacitation *in vitro*. This implies that the researcher can evaluate the relevant induced sperm surface reorganization under *in vitro* fertilization conditions.

Sperm membrane composition, as well as the ordering of membrane components, can be compared with that found under control conditions (media without the capacitation factors) or with the membrane ordering of newly collected sperm. Collected sperm is washed through a discontinuous density gradient to remove aberrant sperm and non-sperm particles, seminal plasma and factors delaying sperm capacitation (Harrison *et al.*, 1996). The pelleted cells are resuspended in *in vitro* capacitation media. The surface reordering of membrane proteins and lipids in the sperm head have been studied extensively under these *in vitro* capacitation conditions (for reviews, see Flesch and Gadella, 2000; Gadella and Visconti, 2006). Most relevant for fertilization is that sperm surface proteins that are entrapped into small lipid-ordered domains (lipid rafts) are clustered into the area that is specifically involved in sperm–zona binding (van Gestel *et al.*, 2005) as well as in the docking of the sperm plasma membrane to the outer acrosomal membrane.

It is important to stress the importance of the sperm surface reordering and changes in composition of membrane components by diverse extracellular factors. The induced lateral redistribution of membrane components appears also to be instrumental in the assembly of a functional sperm protein complex that is involved in sperm–zona binding, as well as for the zona induction of the acrosome reaction (Fig. 4.3; van Gestel *et al.*, 2005, 2007; Tsai *et al.*, 2007, 2010, 2012; Ackermann *et al.*, 2008; Zitranski *et al.*, 2010). Therefore, in addition to the composition of sperm surface proteins, one needs to study how these proteins are organized and whether they are functional for their physiological role in fertilization. Moreover, the relatively simple defined *in vitro* capacitation system described above probably does not provide all the information about sperm surface reorganization *in utero* or in the oviducts, where hormones and other bioactive non-protein components probably regulate sperm physiology in different ways. Nevertheless, a number of proteins involved in zona binding and sperm plasma membrane docking with the acrosome have been identified using *in vitro* capacitated pig sperm (see van Gestel *et al.*, 2007; Tsai *et al.*, 2012).

It is surprising that similar approaches have not been performed using bull or stallion sperm, where proteomic research has focused on identifying markers for predicting male fertility (Peddinti *et al.*, 2008; D'Amours *et al.*, 2010; Novak *et al.*, 2010) rather than on the identification of proteins functionally involved in gamete interaction and fertilization. None the less, members of the CRISP protein family functionally known to be involved in oocyte fertilization and/or sperm–zona binding have been shown to correlate with *in vivo* stallion fertility (Novak *et al.*, 2010). The relative abundance of phosphoethanol binding protein or of BSP1 in highly fertile bulls has been compared with that of a lower fertility group (D'Amours *et al.*, 2010). Also, mouse studies utilizing zona pellucida ghosts and purified apical plasma membranes are lacking, although functional proteomic studies for sperm–egg interaction and DRM fractions have resulted in the identification of functional proteins (Stein *et al.*, 2006; Nixon *et al.*, 2009). The results of similar studies using human sperm have recently been published (Nixon *et al.*, 2011). In similar studies, Redgrove *et al.* (2011) used human sperm to identify the composition of multimeric protein complexes involved in capacitation-dependent zona affinity.

Sperm Handling

As is mentioned above, sperm interact with their immediate environment and thus are subject to a continuous surface remodelling. However, in assisted reproductive technologies, the processing of sperm may well frustrate these processes. In principle, sperm handling may alter the sperm surface by dilution and shearing forces that may cause the removal of extracellular matrix components, such as decapacitation factors. Dilution may also alter the antioxidant capacity of seminal plasma, thereby leaving the sperm more vulnerable to oxidative stress. The handling involved in cryostorage and flow cytometric sorting notoriously reduces the integrity of sperm and their fertilizing capacity. These processes lead to reduced pregnancy rates and a loss of production of offspring. In general, the industry has focused on establishing methods to stabilize sperm during storage or sorting. These attempts have been successful in achieving maximum sperm survival, although as reviewed extensively elsewhere (Leahy and Gadella, 2011), it is questionable whether the maximal stabilization of sperm is beneficial for its post-handling fertilization competence. The capacitating sperm faces a series of destabilizing events that may be irreversibly inhibited by these strategies for optimizing sperm survival. Further understanding of sperm surface changes under physiological conditions may lead to improved strategies for sperm storage or sorting.

Conclusions

The continuous sperm surface remodelling that occurs during sperm transit from the rete testis to the oviducts, and possibly even within the perivitelline space, and the physiological role of these surface kinetics is to a large extent terra incognita. The identification of different complex proteins systems within the male and female genital tract is promising, although their role and function in events associated with sperm–oocyte interactions are still difficult to identify (Boerke *et al.*, 2008).

Wild and genotypically knockout mice have been studied to validate the function of certain proteins purported to play a role in sperm–zona binding, the acrosome reaction, oolemma binding and sperm fusion with the oolema (for reviews, see Vjugina and Evans, 2008; Gadella and Evans, 2011). Such studies have provided valuable information on the potential impact of certain proteins involved in mammalian fertilization. However, the molecular intervention of transcription and translation in gametes is hampered by the fact that, in sperm, both processes are silenced, while in the oocyte, almost all mRNA is stored for post-fertilization translation. So it is not possible to intervene with the molecular processes involved in gamete interaction and fertilization; intervention should rather be in either earlier gametogenic processes or in later post-fertilization development processes.

An example of this is a mutation in spermatogenic cells of the syntaxin 2/epimorphin gene. This protein seems to play a role in the acrosome reaction (Tsai *et al.*, 2007), but the mutation caused a defect in the transition from spermatocyte to spermatids. Hence, a phenotypic knockout of the syntaxin 2 gene cannot be used to study the effect of this protein on fertilization, simply because the knockout phenotype fails to produce sperm (Akiyama *et al.*, 2008). Furthermore, in many cases, homologous genetic recombination applications have shown that knocking out the expression of phenotypic factors that were previously believed to be essential for fertilization were found to be dispensable to this process (Okabe and Cummins, 2007). In part, this could be explained by the fact that biological systems contain redundancies and compensatory mechanisms, and both processes are believed to play a prominent role in the evolution of gamete interaction and, therefore, in speciation (Herlyn and Zischler, 2008; Turner and Hoekstra, 2008). Nevertheless, the results of genomic approaches devoted to studying the molecular mechanisms involved in mammalian fertilization may provide support for substantial modifications of classical/current fertilization models (Okabe and Cummins, 2007).

The production of knockout husbandry pigs is very expensive and time-consuming. Gene-specific silencing of protein translation is possible with interference RNA technology, so the specific role of proteins involved in fertilization and in sperm surface kinetics can be studied. The big problem though is that the treatment of explants and cells causes dedifferentiation and results in the alteration of their interaction with sperm (Sostaric *et al.*, 2008).

The chapter has shown that after leaving the testis, sperm are subjected to a series of events causing continuous sperm surface remodelling, and that these events are relevant for their final fertilization task (Yanagimachi, 1994). The kinetics of the sperm surface proteins have also been reviewed. It can be noted that difficulties in molecular intervention approaches, as well as the difficulties in studying sperm surface remodelling *in situ*, are now compensated by high throughput proteomic technologies that allow the identification of proteins in low abundance. In combination with off-gel full LC-MS/MS (liquid chromatography–mass spectrometry/mass spectrometry) platforms, sperm surface isolation and purification technologies and isobaric tagging strategies for peptides (Zieske, 2006; Ernoult *et al.*, 2008), it will be possible in the near future – when the entire porcine and bovine genome and their annotation are available to the public (Archibald *et al.*, 2010; Reese *et al.*, 2010) – to identify the entire sperm surface proteome (Gadella, 2009; Brewis and Gadella, 2010).

Acknowledgements

This chapter is a rewritten and completely updated revised version of the following paper: Tsai P-S and Gadella BM (2009) Molecular kinetics of proteins at the surface of porcine sperm before and during fertilization. In *Control of Pig Reproduction VIII Proceedings of the Eighth International Conference on Pig Reproduction May 31–June 4 2009 The Banff Centre, Banff Alberta Canada* Eds H Rodriguez-Martinez, JL Vallet and AJ Ziecik *Society for Reproduction and Fertility Supplement* **66** 23–36. Nottingham University Press, Nottingham. In the current version, a broader coverage of ruminants and other mammalian species is provided. Copyright permission was provided by Nottingham University Press. The contribution of Dr P.S. Tsai in the original version, which has been useful for preparing this new book chapter, is greatly appreciated.

References

Ackermann F, Zitranski N, Heydecke D, Wilhelm B, Gudermann T and Boekhoff I (2008) The Multi-PDZ domain protein MUPP1 as a lipid raft-associated scaffolding protein controlling the acrosome reaction in mammalian spermatozoa *Journal of Cellular Physiology* **214** 757–768.

Akiyama K, Akimaru S, Asano Y, Khalaj M, Kiyosu C, Masoudi AA, Takahashi S, Katayama K, Tsuji T, Noguchi J and Kunieda T (2008) A new ENU-induced mutant mouse with defective spermatogenesis caused by a nonsense mutation of the syntaxin 2/epimorphin (*Stx2/Epim*) gene *Journal of Reproduction and Fertility* **54** 122–128.

Archibald AL, Bolund L, Churcher C, Fredholm M, Groenen MA, Harlizius B, Lee KT, Milan D, Rogers J, Rothschild MF, Uenishi H, Wang J, Schook LB, Swine Genome Sequencing Consortium (2010) Pig genome sequence – analysis and publication strategy *BMC Genomics* **11**:438.

Avilés M, Gutiérrez-Adán A and Coy P (2010) Oviductal secretions: will they be key factors for the future ARTs? *Molecular Human Reproduction* **16** 896–906.

Bahr GF and Engler WF (1970) Considerations of volume, mass, DNA, and arrangement of mitochondria in the midpiece of bull spermatozoa *Experimental Cell Research* **60** 338–340.

Barkay J, Bartoov B, Ben-Ezra S, Langsam J, Feldman E, Gordon S and Zuckerman H (1984) The influence of *in vitro* caffeine treatment on human sperm morphology and fertilizing capacity *Fertility and Sterility* **41** 913–918.

Barraud-Lange V, Naud-Barriant N, Bomsel M, Wolf JP and Ziyyat A (2007a) Transfer of oocyte membrane fragments to fertilizing spermatozoa *FASEB Journal* **21** 3446–3449.

Barraud-Lange V, Naud-Barriant N, Saffar L, Gattegno L, Ducot B, Drillet AS, Bomsel M, Wolf JP and Ziyyat A (2007b) Alpha6beta1 integrin expressed by sperm is determinant in mouse fertilization *BMC Developmental Biology* **7**:102.

Barraud-Lange V, Chalas Boissonnas C, Serres C, Auer J, Schmitt A, Lefevre B, Wolf JP and Ziyyat A (2012) Membrane transfer from oocyte to sperm occurs in two CD9-independent ways that do not supply the fertilizing ability of Cd9-deleted oocytes *Reproduction* **144** 53–66.

Behan JR and Watson PF (2006) A field investigation of intra-cervical insemination with reduced sperm numbers in gilts *Theriogenology* **66** 338–343.

Brewis IA and Gadella BM (2010) Sperm surface proteomics: from protein lists to biological function *Molecular Human Reproduction* **16** 68–79.

Boerke A, Dieleman SJ and Gadella BM (2007) A possible role for sperm RNA in early embryo development *Theriogenology* **68** 147–155.

Boerke A, Tsai P-S, Garcia-Gil N, Brewis IA and Gadella BM (2008) Capacitation-dependent reorganization of microdomains in the apical sperm head plasma membrane: functional relationship with zona binding and the zona-induced acrosome reaction *Theriogenology* **70** 1188–1196.

Buffone MG, Zhuang T, Ord TS, Hui L, Moss SB and Gerton GL (2008) Recombinant mouse sperm ZP3-binding protein (ZP3R/sp56) forms a high order oligomer that binds eggs and inhibits mouse fertilization *in vitro Journal of Biological Chemistry* **283** 12438–12445.

Burden HP, Holmes CH, Persad R and Whittington K (2006) Prostasomes – their effects on human male reproduction and fertility *Human Reproduction Update* **12** 283–292.

Busso D, Goldweic NM, Hayashi M, Kasahara M and Cuasnicú PS (2007a) Evidence for the involvement of testicular protein CRISP2 in mouse sperm–egg fusion *Biology of Reproduction* **76** 701–708.

Busso D, Cohen DJ, Maldera JA, Dematteis A and Cuasnicu PS (2007b) A novel function for CRISP1 in rodent fertilization: involvement in sperm–zona pellucida interaction *Biology of Reproduction* **77** 848–854.

Caballero J, Frenette G, D'Amours O, Belleannée C, Lacroix-Pepin N, Robert C and Sullivan R (2012) Bovine sperm raft membrane associated glioma pathogenesis-related 1-like protein 1 (GliPr1L1) is modified during the epididymal transit and is potentially involved in sperm binding to the zona pellucida *Journal of Cellular Physiology* **227** 3876–3886.

Carrasco LC, Romar R, Avilés M, Gadea J and Coy P (2008) Determination of glycosidase activity in porcine oviductal fluid at the different phases of the estrous cycle *Reproduction* **136** 833–842.

Chang H and Suarez SS (2012) Unexpected flagellar movement patterns and epithelial binding behavior of mouse sperm in the oviduct *Biology of Reproduction* **86** 1–8.

Cohen DJ, Da Ros VG, Busso D, Ellerman DA, Maldera JA, Goldweic N and Cuasnicú PS (2007) Participation of epididymal cysteine-rich secretory proteins in sperm–egg fusion and their potential use for male fertility regulation *Asian Journal of Andrology* **9** 528–532.

Cooper TG, Yeung CH and Bergmann M (1988) Transcytosis in the epididymis studied by local arterial perfusion *Cell and Tissue Research* **253** 631–637.

Coy P, Cánovas S, Mondéjar I, Saavedra MD, Romar R, Grullón L, Matás C and Avilés M (2008) Oviduct-specific glycoprotein and heparin modulate sperm–zona pellucida interaction during fertilization and contribute to the control of polyspermy *Proceedings of the National Academy of Sciences of the United States of America* **105** 15809–15814.

D'Amours O, Frenette G, Fortier M, Leclerc P and Sullivan R (2010) Proteomic comparison of detergent-extracted sperm proteins from bulls with different fertility indexes *Reproduction* **139** 545–556.

Da Ros V, Busso D, Cohen DJ, Maldera J, Goldweic N and Cuasnicu PS (2007) Molecular mechanisms involved in gamete interaction: evidence for the participation of cysteine-rich secretory proteins (CRISP) in sperm–egg fusion *Society of Reproduction and Fertility Supplement* **65** 353–356.

Dacheux JL, Castella S, Gatti JL and Dacheux F (2005) Epididymal cell secretory activities and the role of proteins in boar sperm maturation *Theriogenology* **63** 319–341.

Dadoune JP (2003) Expression of mammalian spermatozoal nucleoproteins *Microscopy Research and Technique* **61** 56–75.

Eddy EM (2007) The scaffold role of the fibrous sheath *Society of Reproduction and Fertility Supplement* **65** 45–62.

Eddy EM and O'Brien DA (1994) The spermatozoon In *The Physiology of Reproduction* pp 29–78 Eds E Knobil and JD Neild. Raven Press, New York.

Einspanier R, Gabler C, Bieser B, Einspanier A, Berisha B, Kosmann M, Wollenhaupt K and Schams D (1999) Growth factors and extracellular matrix proteins in interactions of cumulus–oocyte complex, spermatozoa and oviduct *Journal of Reproduction and Fertility Supplement* **54** 359–365.

Ekhlasi-Hundrieser M, Gohr K, Wagner A, Tsolova M, Petrunkina A and Töpfer-Petersen E (2005) Spermadhesin AQN1 is a candidate receptor molecule involved in the formation of the oviductal sperm reservoir in the pig *Biology of Reproduction* **73** 536–545.

Ellerman DA, Myles DG and Primakoff P (2006) A role for sperm surface protein disulfide isomerase activity in gamete fusion: evidence for the participation of ERp57 *Developmental Cell* **10** 831–837.

Ernoult E, Gamelin E and Guette C (2008) Improved proteome coverage by using iTRAQ labelling and peptide OFFGEL fractionation *Proteome Science* **6**:27.

Evans RW, Weaver DE and Clegg ED (1987) Effects of *in utero* and *in vitro* incubation on the lipid-bound fatty acids and sterols of porcine spermatozoa *Gamete Research* **18** 153–162.

Flesch FM and Gadella BM (2000) Dynamics of the mammalian sperm plasma membrane in the process of fertilization *Biochimica et Biophysica Acta* **1469** 197–235.

Gadella BM (2008) The assembly of a zona pellucida binding protein complex in sperm *Reproduction in Domestic Animals* **43(Supplement 5)** 12–19.

Gadella BM (2009) Sperm surface proteomics In *Immune Infertility: the Impact of Immunoreactions on Human Infertility* pp 38–48 Eds WK Krause and RK Naz. Springer-Verlag, Berlin.

Gadella BM (2010) Interaction of sperm with the zona pellucida during fertilization *Society for Reproduction and Fertility Supplement* **67** 267–287.

Gadella BM and Evans JP (2011) Membrane fusions during mammalian fertilization *Advances in Experimental Medicine and Biology* **713** 65–80.

Gadella BM and Visconti PE (2006) Regulation of capacitation In *The Sperm Cell: Production, Maturation, Fertilization, Regeneration* pp 134–169 Ed C de Jonge and C Barratt. Cambridge University Press, Cambridge.

Gadella BM, Lopes-Cardozo M, van Golde LM, Colenbrander B and Gadella TW Jr (1995) Glycolipid migration from the apical to the equatorial subdomains of the sperm head plasma membrane precedes the acrosome reaction. Evidence for a primary capacitation event in boar spermatozoa *Journal of Cell Science* **108** 935–946.

Gadella BM, Tsai P-S, Boerke A and Brewis IA (2008) Sperm head membrane reorganisation during capacitation *International Journal of Developmental Biology* **52** 473–480.

Gatti JL, Castella S, Dacheux F, Ecroyd H, Metayer S, Thimon V and Dachuex JL (2004) Post-testicular sperm environment and fertility *Animal Reproduction Science* **82–83** 321–339.

Gatti JL, Métayer S, Belghazi M, Dacheux F and Dacheux JL (2005) Identification, proteomic profiling, and origin of ram epididymal fluid exosome-like vesicles *Biology of Reproduction* **72** 1452–1465.

Getpook C and Wirotkarun S (2007) Sperm motility stimulation and preservation with various concentrations of follicular fluid *Journal of Assisted Reproduction and Genetics* **24** 425–428.

Gil PI, Guidobaldi HA, Teves ME, Uñats DR, Sanchez R and Giojalas LC (2008) Chemotactic response of frozen–thawed bovine spermatozoa towards follicular fluid *Animal Reproduction Science* **108** 236–246.

Girouard J, Frenette G and Sullivan R (2008) Seminal plasma proteins regulate the association of lipids and proteins within detergent-resistant membrane domains of bovine spermatozoa *Biology of Reproduction* **78** 921–931.

Griffiths GS, Galileo DS, Reese K and Martin-Deleon PA (2008) Investigating the role of murine epididymosomes and uterosomes in GPI-linked protein transfer to sperm using SPAM1 as a model *Molecular Reproduction and Development* **75** 1627–1636.

Gwathmey TM, Ignotz GG, Mueller JL, Manjunath P and Suarez SS (2006) Bovine seminal plasma proteins PDC-109, BSP-A3, and BSP-30-kDa share functional roles in storing sperm in the oviduct *Biology of Reproduction* **75** 501–507.

Hao Y, Mathialagan N, Walters E, Mao J, Lai L, Becker D, Li W, Critser J and Prather RS (2006) Osteopontin reduces polyspermy during *in vitro* fertilization of porcine oocytes *Biology of Reproduction* **75** 726–733.

Hao Y, Murphy CN, Spate L, Wax D, Zhong Z, Samuel M, Mathialagan N, Schatten H and Prather RS (2008) Osteopontin improves *in vitro* development of porcine embryos and decreases apoptosis *Molecular Reproduction and Development* **75** 291–298.

Harrison RA, Ashworth PJ and Miller NG (1996) Bicarbonate/CO_2, an effector of capacitation, induces a rapid and reversible change in the lipid architecture of boar sperm plasma membranes *Molecular Reproduction and Development* **45** 378–391.

Herlyn H and Zischler H (2008) The molecular evolution of sperm zonadhesin *The International Journal of Developmental Biology* **52** 781–790.

Herrero MB, Mandal A, Digilio LC, Coonrod SA, Maier B and Herr JC (2005) Mouse SLLP1, a sperm lysozyme-like protein involved in sperm–egg binding and fertilization *Developmental Biology* **284** 126–142.

Ignotz GG, Cho MY and Suarez SS (2007) Annexins are candidate oviductal receptors for bovine sperm surface proteins and thus may serve to hold bovine sperm in the oviductal reservoir *Biology of Reproduction* **77** 906–913.

Inoue N, Kasahara T, Ikawa M and Okabe M (2010) Identification and disruption of sperm-specific angiotensin converting enzyme-3 (ACE3) in mouse *PLoS One* **5** e10301.

Ito C, Yamatoya K, Yoshida K, Maekawa M, Miyado K and Toshimori K (2010) Tetraspanin family protein CD9 in the mouse sperm: unique localization, appearance, behavior and fate during fertilization *Cell and Tissue Research* **340** 583–594.

Jang S and Yi LS (2005) Identification of a 71 kDa protein as a putative non-genomic membrane progesterone receptor in boar spermatozoa *Journal of Endocrinology* **184** 417–425.

Jin M, Fujiwara E, Kakiuchi Y, Okabe M, Satouh Y, Baba SA, Chiba K and Hirohashi N (2011) Most fertilizing mouse spermatozoa begin their acrosome reaction before contact with the zona pellucida during *in vitro* fertilization *Proceedings of the National Academy of Sciences of the United States of America* **108** 4892–4896.

Killian GJ (2004) Evidence for the role of oviduct secretions in sperm function, fertilization and embryo development *Animal Reproduction Science* **82–83** 141–153.

Kondo G, Tojo H, Nakatani Y, Komazawa N, Murata C, Yamagata K, Maeda Y, Kinoshita T, Okabe M, Taguchi R and Takeda J (2005) Angiotensin-converting enzyme is a GPI-anchored protein releasing factor crucial for fertilization *Nature Medicine* **11** 160–166.

Lapointe S and Sirard MA (1998) Catalase and oviductal fluid reverse the decreased motility of bovine sperm in culture medium containing specific amino acids *Journal of Andrology* **19** 31–36.

Lapointe S, Sullivan R and Sirard MA (1998) Binding of a bovine oviductal fluid catalase to mammalian spermatozoa *Biology of Reproduction* **58** 747–753.

Leahy T and Gadella BM (2011) Sperm surface changes and physiological consequences induced by sperm handling and storage *Reproduction* **142** 759–778.

Lloyd RE, Elliott RM, Fazeli A, Watson PF and Holt WV (2009) Effects of oviductal proteins, including heat shock 70 kDa protein 8, on survival of ram spermatozoa over 48 h *in vitro Reproduction, Fertility and Development* **21** 408–418.

Mahmoud AI and Parrish JJ (1996) Oviduct fluid and heparin induce similar surface changes in bovine sperm during capacitation: a flow cytometric study using lectins *Molecular Reproduction and Development* **43** 554–560.

Miyado K, Yoshida K, Yamagata K, Sakakibara K, Okabe M, Wang X, Miyamoto K, Akutsu H, Kondo T, Takahashi Y, Ban T, Ito C, Toshimori K, Nakamura A, Ito M, Miyado M, Mekada E and Umezawa A (2008) The fusing ability of sperm is bestowed by CD9-containing vesicles released from eggs in mice *Proceedings of the National Academy of Sciences of the United States of America* **105** 12921–12926.

Montfort L, Frenette G and Sullivan R (2002) Sperm–zona pellucida interaction involves a carbonyl reductase activity in the hamster *Molecular Reproduction and Development* **61** 113–119.

Moreno RD and Alvarado CP (2006) The mammalian acrosome as a secretory lysosome: new and old evidence *Molecular Reproduction and Development* **73** 1430–1434.

Nixon B, Bielanowicz A, McLaughlin EA, Tanphaichitr N, Ensslin MA and Aitken RJ (2009) Composition and significance of detergent resistant membranes in mouse spermatozoa *Journal of Cellular Physiology* **218** 122–134.

Nixon B, Mitchell LA, Anderson AL, McLaughlin EA, O'Bryan MK and Aitken RJ (2011) Proteomic and functional analysis of human sperm detergent resistant membranes *Journal of Cellular Physiology* **226** 2651–2665.

Novak S, Smith TA, Paradis F, Burwash L, Dyck MK, Foxcroft GR and Dixon WT (2010) Biomarkers of *in vivo* fertility in sperm and seminal plasma of fertile stallions *Theriogenology* **74** 956–967.

Okabe M and Cummins JM (2007) Mechanisms of sperm–egg interactions emerging from gene-manipulated animals *Cellular and Molecular Life Sciences* **64** 1945–1958.

Peddinti D, Nanduri B, Kaya A, Feugang JM, Burgess SC and Memili E (2008) Comprehensive proteomic analysis of bovine spermatozoa of varying fertility rates and identification of biomarkers associated with fertility *BMC Systems Biology* **2**:19.

Ramió-Lluch L, Fernández-Novell JM, Peña A, Colás C, Cebrián-Pérez JA, Muiño-Blanco T, Ramírez A, Concha II, Rigau T and Rodríguez-Gil JE (2011) '*In vitro*' capacitation and acrosome reaction are concomitant with specific changes in mitochondrial activity in boar sperm: evidence for a nucleated mitochondrial activation and for the existence of a capacitation-sensitive subpopulational structure *Reproduction in Domestic Animals* **46** 664–673.

Redgrove KA, Anderson AL, Dun MD, McLaughlin EA, O'Bryan MK, Aitken RJ and Nixon B (2011) Involvement of multimeric protein complexes in mediating the capacitation-dependent binding of human spermatozoa to homologous zonae pellucidae *Developmental Biology* **356** 460–474.

Reese JT, Childers CP, Sundaram JP, Dickens CM, Childs KL, Vile DC and Elsik CG (2010) Bovine Genome Database: supporting community annotation and analysis of the *Bos taurus* genome *BMC Genomics* **11** 645.

Rejraji H, Sion B, Prensier G, Carreras M, Motta C, Frenoux JM, Vericel E, Grizard G, Vernet P and Drevet JR (2006) Lipid remodeling of murine epididymosomes and spermatozoa during epididymal maturation *Biology of Reproduction* **74** 1104–1113.

Roberts PK and Bilkei G (2005) Field experiences on post-cervical artificial insemination in the sow *Reproduction in Domestic Animals* **40** 489–491.

Rodriguez-Martinez H (2007) Role of the oviduct in sperm capacitation *Theriogenology* **68** 138–146.

Russell LD, Peterson RN, Hunt W and Strack LE (1984) Posttesticular surface modifications and contributions of reproductive tract fluids to the surface polypeptide composition of boar spermatozoa *Biology of Reproduction* **30** 959–978.

Sachdev M, Mandal A, Mulders S, Digilio LC, Panneerdoss S, Suryavathi V, Pires E, Klotz KL, Hermens L, Herrero MB, Flickinger CJ, van Duin M and Herr JC (2012) Oocyte specific oolemmal SAS1B involved in sperm binding through intra-acrosomal SLLP1 during fertilization *Developmental Biology* **363** 40–51.

Schuberth HJ, Taylor U, Zerbe H, Waberski D, Hunter R and Rath D (2008) Immunological responses to semen in the female genital tract *Theriogenology* **70** 1174–1181.

Shivaji S, Kota V and Siva AB (2009) The role of mitochondrial proteins in sperm capacitation *Journal of Reproductive Immunology* **83** 14–18.

Shur BD, Rodeheffer C, Ensslin MA, Lyng R and Raymond A (2006) Identification of novel gamete receptors that mediate sperm adhesion to the egg coat *Molecular and Cellular Endocrinology* **250** 137–148.

Song CY, Gao B, Wu H, Wang XY, Chen GH and Mao J (2010) Spatial and temporal expression of spermadhesin genes in reproductive tracts of male and female pigs and ejaculated sperm *Theriogenology* **73** 551–559.

Sostaric E, Dieleman SJ, van de Lest CH, Colenbrander B, Vos PL and Garcia-Gil N and Gadella BM (2008) Sperm binding properties and secretory activity of the bovine oviduct immediately before and after ovulation *Molecular Reproduction and Development* **75** 60–74.

Stein KK, Go JC, Lane WS, Primakoff P and Myles DG (2006) Proteomic analysis of sperm regions that mediate sperm–egg interactions *Proteomics* **6** 3533–3543.

Suarez SS (2008) Regulation of sperm storage and movement in the mammalian oviduct *The International Journal of Developmental Biology* **52** 455–462.

Suarez SS and Pacey AA (2006) Sperm transport in the female reproductive tract *Human Reproduction Update* **12** 23–37.

Sun TT, Chung CM and Chan HC (2011) Acrosome reaction in the cumulus oophorus revisited: involvement of a novel sperm-released factor NYD-SP8 *Protein and Cell* **2** 92–98.

Taylor U, Rath D, Zerbe H and Schuberth HJ (2008) Interaction of intact porcine spermatozoa with epithelial cells and neutrophilic granulocytes during uterine passage *Reproduction in Domestic Animals* **43** 166–175.

Teijeiro JM, Cabada MO and Marini PE (2008) Sperm binding glycoprotein (SBG) produces calcium and bicarbonate dependent alteration of acrosome morphology and protein tyrosine phosphorylation on boar sperm *Journal of Cellular Biochemistry* **103** 1413–1423.

Teijeiro JM, Ignotz GG and Marini PE (2009) Annexin A2 is involved in pig (*Sus scrofa*) sperm–oviduct interaction *Molecular Reproduction and Development* **76** 334–341.

Thimon V, Frenette G, Saez F, Thabet M and Sullivan R (2008) Protein composition of human epididymosomes collected during surgical vasectomy reversal: a proteomic and genomic approach *Human Reproduction* **23** 1698–1707.

Töpfer-Petersen E, Ekhlasi-Hundrieser M and Tsolova M (2008) Glycobiology of fertilization in the pig *The International Journal of Developmental Biology* **52** 717–736.

Tsai P-S and Gadella BM (2009) Molecular kinetics of proteins at the surface of porcine sperm before and during fertilization In *Control of Pig Reproduction VIII Proceedings of the Eighth International Conference on Pig Reproduction May 31–June 4 2009 The Banff Centre, Banff Alberta Canada* Eds H Rodriguez-Martinez, JL Vallet and AJ Ziecik *Society of Reproduction and Fertility* **(Supplement 66)** 23–36.

Tsai P-S, De Vries KJ, De Boer-Brouwer M, Garcia-Gil N, Van Gestel RA, Colenbrander B, Gadella BM and Van Haeften T (2007) Syntaxin and VAMP association with lipid rafts depends on cholesterol depletion in capacitating sperm cells *Molecular Membrane Biology* **24** 313–324.

Tsai P-S, Garcia-Gil N, van Haeften T and Gadella BM (2010) How pig sperm prepares to fertilize: stable acrosome docking to the plasma membrane *PLoS One* **5** e11204.

Tsai P-S, Brewis IA, van Maaren J and Gadella BM (2012) Involvement of complexin 2 in docking, locking and unlocking of different SNARE complexes during sperm capacitation and induced acrosomal exocytosis *PLoS One* **7** e32603.

Turner LM and Hoekstra HE (2008) Causes and consequences of the evolution of reproductive proteins *The International Journal of Developmental Biology* **52** 769–780.

van Gestel RA, Brewis IA, Ashton PR, Helms JB, Brouwers JF and Gadella BM (2005) Capacitation-dependent concentration of lipid rafts in the apical ridge head area of porcine sperm cells *Molecular Human Reproduction* **11** 583–590.

van Gestel RA, Brewis IA, Ashton PR, Brouwers JF and Gadella BM (2007) Multiple proteins present in purified porcine sperm apical plasma membranes interact with the zona pellucida of the oocyte *Molecular Human Reproduction* **13** 445–454.

Vazquez JM, Martinez EA, Roca J, Gil MA, Parrilla I, Cuello C, Carvajal G, Lucas X and Vazquez JL (2005) Improving the efficiency of sperm technologies in pigs: the value of deep intrauterine insemination *Theriogenology* **63** 536–547.

Vazquez JM, Roca J, Gil MA, Cuello C, Parrilla I, Vazquez JL and Martínez EA (2008) New developments in low-dose insemination technology *Theriogenology* **70** 1216–1224.

Vjugina U and Evans JP (2008) New insights into the molecular basis of mammalian sperm–egg membrane interactions *Frontiers in Bioscience* **13** 462–476.

Watanabe H and Kondoh G (2011) Mouse sperm undergo GPI-anchored protein release associated with lipid raft reorganization and acrosome reaction to acquire fertility *Journal of Cell Science* **124** 2573–2581.

Woelders H and Matthijs A (2001) Phagocytosis of boar spermatozoa *in vitro* and *in vivo* In *Control of Pig Reproduction VI: Proceedings of the Sixth International Conference on Pig Reproduction, University of Missouri-Columbia, June 2001* Ed RD Geisert, H Niemann and C Doberska *Reproduction* **(Supplement 58)** 113–127.

Yamaguchi R, Yamagata K, Ikawa M, Moss SB and Okabe M (2006) Aberrant distribution of ADAM3 in sperm from both angiotensin-converting enzyme (Ace)- and calmegin (Clgn)-deficient mice *Biology of Reproduction* **75** 760–766.

Yanagimachi R (1994) Mammalian fertilization In *The Physiology of Reproduction* pp 189–317 Eds E Knobil and JD Neild. Raven Press, New York.

Zhang H and Martin-Deleon PA (2003) Mouse epididymal Spam1 (PH20) is released in the luminal fluid with its lipid anchor *Journal of Andrology* **24** 51–58.

Zieske LR (2006) A perspective on the use of iTRAQ reagent technology for protein complex and profiling studies *Journal of Experimental Botany* **57** 1501–1508.

Zitranski N, Borth H, Ackermann F, Meyer D, Viewig L, Breit A, Gudermann T and Boekhoff I (2010) The "acrosomal synapse": subcellular organization by lipid rafts and scaffolding proteins exhibits high similarities in neurons and mammalian spermatozoa *Communicative and Integrative Biology* **3** 513–521.

5 Fundamental and Practical Aspects of Semen Cryopreservation

William V. Holt[1]* and Linda M. Penfold[2]
[1]*University of Sheffield, Sheffield, UK;* [2]*South-East Zoo Alliance for Reproduction & Conservation, Yulee, Florida, USA*

Introduction

Following the discovery in the middle of the last century that glycerol can be used successfully as a cryoprotectant (Polge *et al.*, 1949), thus allowing spermatozoa to be stored for months, years or decades, many aspects of reproductive technology were revolutionized. The most obvious results of this revolutionary discovery include the massive development of selective breeding in agriculture, whereby semen from desirable bulls can be frozen, stored and shipped within a single country or exported around the world, and then used for artificial insemination (AI). While the dairy industry has been a major global beneficiary of this technology, similar benefits have been derived for human infertility treatment, for which semen from fertile men is banked and used for donor insemination programmes both at home and abroad. Other applications of semen freezing technology have also developed in a variety of different contexts, including: (i) the cryopreservation of genetic resources for food fishes (Solar, 2009) and fishes used in biomedical research (i.e. zebra fish and medaka) (Yang and Tiersch, 2009; Tiersch *et al.*, 2011); (ii) the storage of mouse, rat and primate spermatozoa for biomedical research purposes (Shaw and Nakagata, 2002; Mazur *et al.*, 2008); (iii) the establishment of semen banks for the preservation of regional and national agricultural breeds (Woelders *et al.*, 2006); and (iv) the establishment of semen banks for the support of conservation breeding programmes (Holt *et al.*, 1996; Wildt *et al.*, 1997; Watson and Holt, 2001; Bartels and Kotze, 2006).

A number of important reviews have been written about semen cryopreservation, both with focus on a particular species, such as sheep (Salamon and Maxwell, 1995a,b), humans (Royère *et al.*, 1996), dogs (Pena *et al.*, 2006) or mice (Marschall and Hrabe de Angelis, 1999), and with a more general approach (Watson, 1979, 1990; Holt, 2000a,b; Bailey *et al.*, 2003). This list is not exhaustive, but it is surprising that most of the reviews of semen cryopreservation were published more than a decade ago; more recent reviews include those by Chen and Liu (2007), Curry (2007) and Dinnyes *et al.* (2007). Watson's review in the 1990 edition of *Marshall's Physiology of Reproduction* (Watson, 1990) is particularly interesting for its detailed account of the history of semen collection and preservation. He recounts some of the earliest scientific investigations into semen freezing and artificial insemination, which

* E-mail: bill.holt@ioz.ac.uk

were documented in the 18th and 19th centuries. These include, notably: the observations of Lazzaro Spallanzani (1776), an Italian priest, who described the first successful AI procedure in a dog, and also noted that frog, human and stallion semen could be cooled in snow and their motility restored by rewarming; and Paolo Mantegazza (Mantegazza, 1866), who described the effects of temperature changes on human sperm motility. Mantegazza also forecast correctly that one day in the future it would be possible to collect and store semen from soldiers who died in battle, so that it could be used posthumously to inseminate their wives. This procedure has actually been used on a few occasions over the last 20 years, when it has provoked vigorous ethical debates and outright bans in some countries (for a commentary, see Nakhuda, 2010).

Cryobiological Theory

Physical changes during freezing

A great deal has been written about the general theory of cryobiology, especially as the discipline involves a diversity of species and taxonomic groups that include plants (seeds, meristems, etc), bacteria, fungi, as well as a diversity of animal cells and tissues. The general principles governing our understanding of cell survival during freezing and thawing rely on knowledge of water transport through membranes and interactions between solutes, cryoprotectants, low temperature and membranes.

As shown in Fig. 5.1, when the temperature falls, the water in the sample actually remains liquid unless active steps (seeding) are taken to induce ice crystallization; the solution at this stage is said to undergo 'supercooling' (Fig. 5.1: step 1). Ice formation occurs randomly at some point below the freezing point; the randomness of this stage is dictated by the probability that groups of water molecules organize themselves into clusters and initiate a chain reaction. As this is an exothermic reaction, sufficient heat is released to increase the sample temperature to a significant extent (Fig. 5.1: step 2). Depending on the sample being frozen, and the technique being used to cool it, the sample temperature remains static for a few minutes before cooling is resumed (Fig. 5.1: step 3) or can even

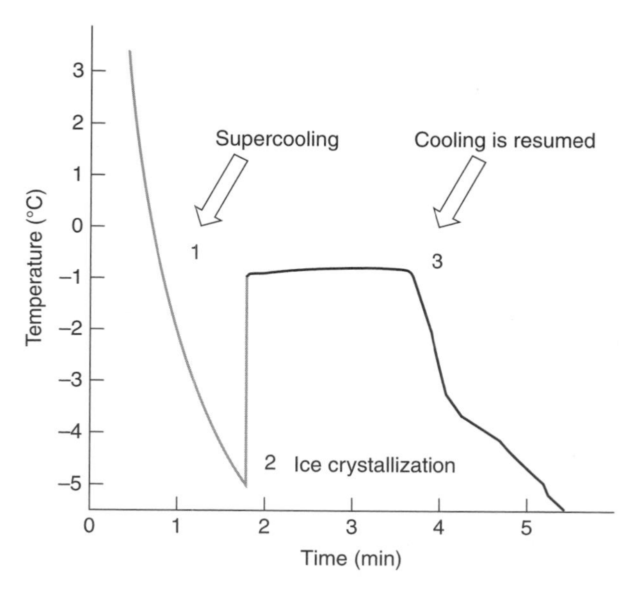

Fig. 5.1. Graph showing the profile of temperature changes that occur when a sample of semen is frozen within a plastic straw. The most significant events are that the temperature first drops to below the usual freezing point (1) (super cooling), then increases again rapidly as the latent heat of fusion is released (2). Dissipation of the latent heat produces a short period when the temperature does not change, after which cooling is finally resumed (3).

result in a transient increase in temperature before cooling resumes. These temperature excursions invariably occur when (plastic) straws of semen are frozen in cold nitrogen vapour (for instance, a typical technique is to place the straw horizontally about 4 cm above the surface of liquid nitrogen), and researchers have been divided as to whether this is a harmful process. To address this issue, directional freezing units that promote seeding of the sample in a linear fashion are thought to minimize these temperature excursions.

When ice begins to form during the cooling process, the dissolved salts and other compounds are excluded from the ice and become concentrated into the ever-decreasing spaces between the developing ice crystals. Biological cells interact osmotically with these unfrozen pockets of high solute concentration and water tends to be drawn out of the cells, resulting in cell shrinkage. This process results in two opposing deleterious effects that counteract each other: if the cooling rate is slow, the cell membranes are exposed for longer periods to the pockets of hypertonic solutions, with possible deleterious effects such as protein and lipid extraction and the generation of reactive oxygen species (ROS); conversely, if the cells are cooled too rapidly their cytoplasmic water content remains high and they run the risk of lethal intracellular ice formation. In practical terms, the optimal cooling rate has to be regarded as a compromise between these opposing effects.

Experimentally it is possible to illustrate some of these effects by setting up idealized conditions using a cryomicroscope. Figure 5.2a shows what happens if some cryopreservation medium lacking cryoprotectant is spiked with a fluorescent dye and frozen at a slow cooling rate (about 5°C/min) to –20°C. The bright lines in the image represent inter-ice crystal regions where the fluorescent dye has become highly concentrated. If cryoprotectant has been added, or if a faster cooling rate had been used, the structure would have been more complex. Figure 5.2b shows a snapshot taken as the ice crystallization front was moving across the field of view (left to right). A number of spermatozoa can be seen near the leading edge of the ice front and can later be seen distributed across different regions of the ice crystal formations (Fig. 5.2c). Figure 5.2d shows a snapshot taken as the sample is being thawed; it is notable that the heterogeneity displayed while the sample is frozen is still evident at the time of thawing.

These observations indicate that individual spermatozoa interact with both ice crystal regions and the interspersed regions of high salt content, and that this level of heterogeneous distribution is established around the time of freezing and then maintained until the sample is thawed. The photographs in Fig. 5.2 were set up to exaggerate the degree of organization visible within the preparations; however, a number of ultrastructural studies have confirmed that spermatozoa within frozen straws of semen become partitioned within regions of high solute concentration, and that they are separated by regions of pure ice (Courtens and Paquignon, 1985; Ekwall *et al.*, 2007).

In addition to the purely osmotic interactions associated with the freezing and thawing processes, the cell membrane lipids undergo partially irreversible phase transitions (Holt and North, 1984), with the appearance of semi-crystalline lipid arrays as the temperature declines during cooling and freezing. This effect is thought to be responsible for the damaging effects of 'cold shock', which occurs when spermatozoa are rapidly cooled in the absence of protective additives (Quinn and White, 1966; Quinn *et al.*, 1969; Moran *et al.*, 1992); cold shock is not only recognizable because of the structural damage caused to the plasma and acrosomal membranes, but is also known to induce loss of cellular homeostasis through inappropriate membrane permeabilization, excessive uptake of calcium and uncontrolled loss of potassium.

In keeping with the long-established fluid mosaic model of membrane structure and organization (Singer and Nicolson, 1972), sperm plasma membrane lipids have long been envisaged as being free to diffuse laterally within the two-dimensional plane of the sperm surface (Wolf and Voglmayr, 1984). This broad view, although a useful conceptual model for thinking about cryopreservation, has been modified somewhat with the realization that membrane lipids and proteins are highly organized into structurally, biochemically and functionally defined regions,

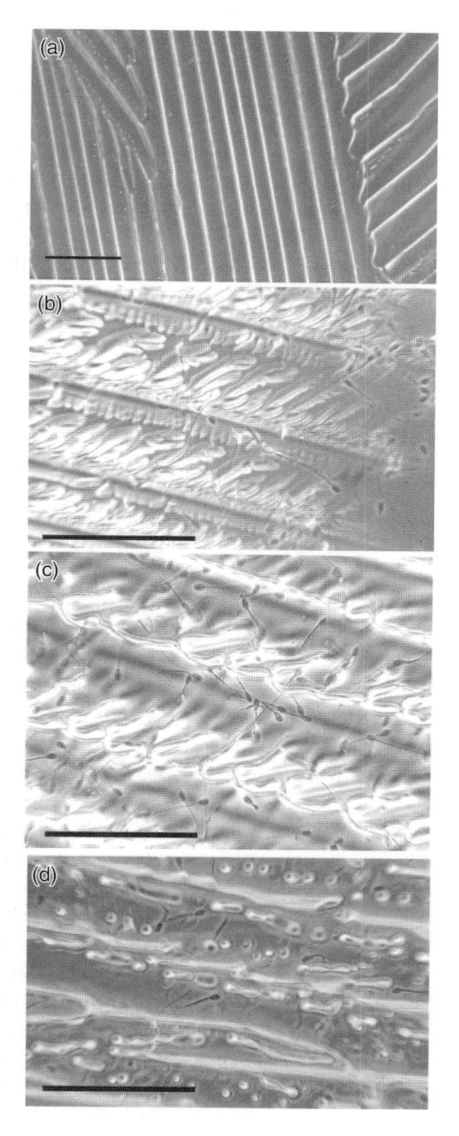

Fig. 5.2. (a) Shows what happens when a thin film of culture media is spiked with a fluorescent dye (fluorescein isothiocyanate) and then frozen to −20°C on a cryomicroscope. The fluorescent dye highlights the regions of high salt content, while the dark areas are formed of pure ice. Bar = 100 μm. (b) Shows a snapshot taken during ice crystal formation in a suspension of spermatozoa. The ice front is travelling from left to right and trapping spermatozoa as it goes. Bar = 100 μm. (c) Shows spermatozoa trapped between the ice crystal formations that developed as shown in (b). Bar = 100 μm. (d) Shows what happens during thawing of the sample illustrated in (c). At this moment, the ice crystal formations have not yet completely disappeared and spermatozoa can be seen lying between them. Bar = 100 μm.

or domains and subdomains, within the plane of the membrane (for reviews, see Zitranski *et al.*, 2010 and Zhu and Inaba, 2011). If anything, gaining further insights into sperm membrane domain organization has strengthened the realization that cryopreservation technology has the potential to cause huge amounts of cellular dysfunction through inappropriate and disruptive phase transitions. Unfortunately, changes in temperature are not the only cause of membrane lipid-phase transitions; they can also be induced by the removal of water, such as occurs during the osmotic fluxes associated with the freezing process or during freeze-drying (Chiantia *et al.*, 2005; Bennun *et al.*, 2008). The occurrence and effects of these drying-induced phase transitions can be offset to some extent by certain sugars, especially trehalose (Rudolph *et al.*, 1986; Chandrasekhar and Gaber, 1988; Crowe *et al.*, 1988). In fact, some authors have reported that the presence of trehalose is beneficial during the cryopreservation of spermatozoa (see, for example, Hu *et al.*, 2010b).

Cryoprotectants

In practice, it is not possible to cryopreserve spermatozoa, and most other cell types, without the presence of cryoprotectants. These chemicals, which belong to several distinct groups of compounds, interact with water molecules and inhibit the formation of the hydrogen bonds that are essential for the formation of ice crystals. The theory of cryoprotectant action is highly complex and multifactorial, but essentially it is understood that these compounds lower the freezing point of water and, thereby, at any given temperature, reduce the extent of ice crystal formation. This has the complementary effect of reducing the formation of unfrozen pockets of high solute concentration and, importantly, permitting these hyperosmotic regions to undergo transition to the glassy phase without necessarily reaching the same high solute concentrations (for a comprehensive review of cryoprotectant action, see Fuller, 2004).

Some cryoprotectants, notably those such as glycerol and dimethyl sulfoxide (DMSO), which are widely used for freezing spermatozoa, permeate through the plasma membrane of the cells in question and exert their cryoprotection by inhibiting the lethal formation of intracellular ice crystals. This allows them to protect cells during freezing by both intracellular actions and extracellular actions (i.e. by reducing the external concentration of solutes). However, there is also another class of cryoprotectants, mostly large macromolecules such as polyvinylpyrrollidone (PVP), hydroxyethyl starch and dextran, which cannot pass through the cell membrane and whose cryoprotective action seems to be brought about by cellular dehydration, freezing point depression and increasing the viscosity of the media at low temperatures, thus inhibiting ice crystal formation. These actions combine to protect the structure of the cell plasma membrane and have much in common with naturally evolved methods of cell preservation *in vivo* that are based on desiccation and the biosynthesis of protective compounds, typically sugars such as trehalose. Small multicellular organisms such as tardigrades and eelworms are able to lose >95% of their body water, remain viable for prolonged periods of up to several years and to withstand extreme heating and cooling (for references about this topic and its centuries-old history, see reviews by Keilin, 1959 and Crowe and Crowe, 1986).

The effectiveness of cryoprotectants varies between cell types, partly because they operate via different mechanisms, but also because they can be differentially toxic to cells. This introduces more variables that need to be considered when examining the way that cryoprotectants work. Even a simple compound such as glycerol has been found to induce sterility in poultry spermatozoa (Hammerstedt and Graham, 1992), where the effects may be due as much to direct alterations of membrane and cytoplasmic structures as to purely anisosmotic damage; it can also cause acrosomal damage in boar spermatozoa if used at concentrations exceeding 3% (w/v) (Hofmo and Almlid, 1991; Gutierrez-Perez *et al.*, 2009). The permeability of the cellular plasma membrane to both water and cryoprotectants is an important variable that governs the optimal design of protocols for cell freezing as well. Some cells behave as perfect osmometers when confronted with solutions of different tonicity and will shrink or expand as appropriate, in ways that can be predicted from biophysical principles. When their surrounding media begin to freeze, with the formation of a hyperosmotic external environment, these cells begin to shrink as their intracellular water is drawn across the membrane. The ideal cooling rates for these cells would be sufficiently fast to minimize exposure to the deleterious effects of unfrozen, extracellular, hypertonic solutions, but not too fast, otherwise the excess water remaining inside the cells would turn into ice crystals, with lethal effects. The optimal cooling rate is indeed predictable for certain, regularly shaped cell types, such as mammalian oocytes, for which measurements of membrane permeability to water and to cryoprotectant can be made.

Spermatozoa are not ideal candidates for modelling their optimal cooling rates because their high degree of structural and functional differentiation has resulted in irregularly shaped cells whose volume is difficult to calculate. A number of researchers have, nevertheless, attempted to measure sperm membrane permeability and predict optimal cooling rates (Gao *et al.*, 1992, 1993; Noiles *et al.*, 1993), but until about 10 years ago these predictions were wildly inaccurate. More accurate predictions have now been obtained for several mammalian and non-mammalian species by applying differential scanning calorimetry and measuring sperm membrane permeability to water and cryoprotectants at sub-zero temperatures, instead of supra-zero temperatures, which would not necessarily be relevant (Devireddy *et al.*, 1999, 2000, 2004, 2006; Thirumala *et al.*, 2006; Alapati *et al.*, 2009; Hagiwara *et al.*, 2009).

Fourier-transform infrared spectroscopy (FTIR) is an alternative biophysical approach to the investigation of sperm membrane phase transitions during cooling, and has been used successfully to investigate cold shock (Drobnis *et al.*, 1993) and freezing (Ricker *et al.*, 2006; Oldenhof *et al.*, 2010). This technique is extremely sensitive to lipid conformational order (Mendelsohn and Moore, 1998) and allows the direct measurement of changes in the CH_2 symmetric stretching frequency (which reflects lipid acyl chain conformation) over a range of temperatures, even in the presence of cryoprotectants and cryoprotective additives. Two recent papers about the cryopreservation of stallion spermatozoa are of considerable interest here: Oldenhof *et al.* (2010) were able to determine the membrane permeability parameters necessary for modelling the optimal cooling and freezing rates for individual animals, and showed that they were not all the same; and Ricker *et al.* (2006) showed that the presence of egg yolk lipoproteins during freezing prevented serious changes in phase transition behaviour that were seen in spermatozoa frozen and thawed in the absence of egg yolk.

Developments in the use of cryoprotective additives

Traditionally, sperm cryopreservation diluents have included various sources of lipid and proteins (e.g. egg yolk, milk proteins, coconut milk, soybean lecithin), as well as sugars such as sucrose and trehalose, to assist with cryoprotection. There have been numerous publications over the last several decades dealing with the merits of various diluent additives as they apply to individual species, and the reader is referred to other publications for further detailed information (Watson, 1979, 1981, 1990; Watson and Holt, 2001). Unfortunately, despite the large amount of research that has been invested in improving the performance of semen cryodiluents, it is clear that the mechanism of action of these additives still remains obscure, which makes it difficult to exploit them further and gain more improvement. For example, the active ingredient of egg yolk was narrowed down to a low-density lipoprotein component (LDL) in the 1970s (Watson, 1976), but even though the use of LDL has recently seen a resurgence as an effective way to enhance semen preservation in several species (Bencharif *et al.*, 2010; Hu *et al.*, 2010a, 2011; Dong *et al.*, 2011; Vera-Munoz *et al.*, 2011), its mode of action is still largely unexplained at the level of the sperm plasma membrane.

The protective effects of exogenous lipids during the cooling and freezing of spermatozoa have been recognized for many years (Evans and Setchell,1978; Graham and Foote, 1987; Holt and North, 1988) and substances such as lecithin (phosphatidylcholine), often of plant origin, form the basis of several commercial cryopreservation media. However, the mechanisms underlying these effects have been difficult to understand, especially as there has been no evidence that the lecithin intercalates into the sperm plasma membrane, with resultant modulations of phase transitions. FTIR studies (Ricker *et al.*, 2006) have, nevertheless, revealed that the presence of soy- or egg yolk-derived lecithin during sperm cryopreservation prevents the deleterious phase transitions that are suffered when spermatozoa are frozen–thawed without the presence of lecithin, and that the lecithin becomes tightly associated with the sperm plasma membrane. These authors also believed that the presence of lecithin as a cryoprotective additive was equally as effective as that of egg yolk lipoprotein for the cryopreservation of stallion spermatozoa. This is consistent with an earlier study (Kolossa and Seibert, 1990), in which the authors developed a chemically defined cryoprotective medium for bovine spermatozoa that had commercial lecithin as the main ingredient, and claimed that it was only slightly less effective than egg yolk-based diluents. These results are especially topical in view of the upsurge of interest in avoiding animal products in semen diluents for the sake of enhanced biosecurity.

Media for semen cryopreservation often contain Equex, an anionic surfactant (sodium triethanolamine lauryl sulfate – SLTS, also known as Orvus ES) (Howard *et al.*, 1986; Pontbriand *et al.*, 1989; Kaplan and Mead 1992; Montfort *et al.*, 1993). This compound is usually used in combination with egg yolk and is thought to emulsify the egg yolk, possibly assisting in its ability to interact with the plasma membrane. It is especially prevalent in diluents for boar semen, but is hardly ever used for bull semen; it has also been tested in a number of wild species, including the African elephant (Howard *et al.*, 1986), and in the domestic dog (Rota *et al.*, 1998). It would be of interest to investigate SLTS in more detail, possibly using FTIR, to find out more about its mechanism of action.

While the cryopreservation of spermatozoa from domestic livestock such as cattle, pigs and sheep typically involves the use of egg yolk or milk as diluent components, there are communities of scientists that focus on other groups of species and have developed different types of cryopreservation media. It is interesting to consider these briefly because there is no doubt that species differences are very important factors in this field. Detailed protocols and discussion about sperm cryopreservation in different species have been collected and published in several reviews (Watson and Fuller, 2001; Watson and Holt, 2001; Lermen *et al.*, 2009).

Interest in cryopreserving mouse spermatozoa from defined genetic lines for biomedical research produced significant advances in appropriate technology in the early 1990s (George *et al.*, 1992; Nakagata and Takeshima 1992; Nakagata *et al.*, 1992; Sztein *et al.*, 1992), and it is interesting that the more successful approaches often employed 18% raffinose, together with skimmed milk. Differences in the post-thaw fertility (from *in vitro* fertilization or tubal insemination) of spermatozoa of various mouse strains have, none the less, remained a problem (e.g. C57BL/6J mice show a 0–20% fertilization rate). Recent advances have shown that supplementing the 18% raffinose medium with L-glutamine and methyl-β-cyclodextrin enormously improves the fertilization rate for this specific mouse strain (bringing it to 69.2%; Takeo and Nakagata, 2010). A similarly dramatic improvement in fertilization rate has also been reported following supplementation of the raffinose medium with the reducing agent monothioglycerol (Ostermeier *et al.*, 2008; Takahashi and Liu, 2010); these authors reported that the fertilization rates of several mouse genetic lines, including C57BL/6, were restored almost to the same level as those of fresh spermatozoa.

The recent findings that cyclodextrin can improve the cryosurvival of mouse spermatozoa are mirrored to some extent by equivalent findings with bull, ram, stallion and goat spermatozoa (Mocé *et al.*, 2010a,b; Moraes *et al.*, 2010; Oliveira *et al.*, 2010; Spizziri *et al.*, 2010). Here, the cyclodextrin is used to load the sperm plasma membrane with cholesterol before cryopreservation. This effectively changes the membrane lipid composition and hence the phase transition behaviour during cooling and freezing. To our knowledge, monothioglycerol has not been tested with domestic livestock, but if it improves cryosurvival by reducing oxidative damage associated with freezing and thawing, it is likely to be successful to some extent. Other thiols have been tested with bull semen (cysteine, *N*-acetyl-L-cysteine and 2-mercaptoethanol; Bilodeau *et al.*, 2001) and at concentrations above 0.5 mM have been shown to have some beneficial effects on post-thaw motility. There are many other studies reporting the inclusion of other antioxidants in semen extenders (liquid diluent added to semen to preserve its fertilizing ability) and cryoprotective media (Pena *et al.*, 2004; Funahashi and Sano, 2005; Gadea *et al.*, 2005; Pagl *et al.*, 2006), but none has been demonstrated to provide beneficial effects matching those seen by adding monothioglycerol to the mouse spermatozoa.

Vitrification

While conventional freezing requires the use of cryoprotectants to prevent damage caused by ice crystal formation, the use of increased cryoprotectant concentrations (up to 6 M) and rapid cooling allows solutions to be cooled to extremely low temperatures without any ice being formed at all. This process is termed vitrification; the name refers to the fact that the solution undergoes a glass transition and forms a stable structure without the presence of ice crystals. Cells that have been vitrified in this way can be returned to ambient temperature without the distortion that normally accompanies conventional freezing. There are, however, a few drawbacks to this technique; the 'glassy' state of vitrified samples tends to be rather fragile and easily damaged but, more importantly, there is a high risk of ice crystal formation during the process of rewarming.

Vitrification techniques have been applied to mammalian oocytes (Shaw *et al.*, 1992; Succu *et al.*, 2008) and embryos (Rall and Fahy, 1985; Rall and Wood, 1994), but success with spermatozoa has been very limited until quite recently. One report (Isachenko *et al.*, 2008) showed that if small (30 μl) droplets of human spermatozoa, suspended in media consisting of human tubal fluid supplemented with 1% human serum and 0.25 M sucrose, were dropped directly into liquid nitrogen; on warming they recovered about 65% progressive motility and retained mitochondrial function. Similar results were also obtained with rainbow trout spermatozoa when 20 μl droplets of spermatozoa, suspended in a mixture containing a standard buffer for fish spermatozoa (Cortland medium), supplemented with combinations of 40% seminal plasma, 1% bovine serum albumin and 0.125 M sucrose, were plunged directly into liquid nitrogen (Merino *et al.*, 2011); these authors reported the recovery of >80% motility upon rewarming and the retention of mitochondrial function.

These results are encouraging because they suggest that alternative approaches to semen cryopreservation are worth investigating. The limitation on droplet size may not turn out to be a serious problem if suitable methods for delivering large numbers of droplets into liquid nitrogen could be found. The more serious downside of this approach might be the high risks of microbial contamination through the liquid nitrogen itself. The research group that has been developing these approaches has evaluated several methods (cryoloops, droplets, open pulled straws – OPS, and open straws) to avoid risks of contamination via liquid nitrogen, and have found that only the OPS method can be considered to be a 'clean' technique (Isachenko *et al.*, 2005a).

Freeze-drying

The advent of intracytoplasmic sperm injection (ICSI) technology, whereby individual spermatozoa are microinjected into the oocyte cytoplasm, has led to considerable interest in the possibility of storing the spermatozoa in the freeze-dried state. Logically, the micro-manipulation technique needed for ICSI can be considered as a replacement for sperm transport and any steps requiring motility or the acrosome reaction, and, at least theoretically, the only sperm component necessary for the production of a zygote is the nucleus. From that argument, it can be concluded that damage to the mitochondria, axoneme and other cytoplasmic components should not be relevant to the developmental potential of freeze-dried spermatozoa.

A number of reports have shown that this approach is feasible (Wakayama and Yanagimachi, 1998; Kaneko *et al.*, 2003a,b; Ward *et al.*, 2003; Liu *et al.*, 2004; Kawase *et al.*, 2007; Kusakabe *et al.*, 2008) and that live offspring have been produced in mice and rabbits following ICSI using freeze-dried nuclei. This mirrors the use of freeze-dried nuclei for nuclear transfer (Loi *et al.*, 2008) after 3 years storage at room temperature.

The principles of anhydrobiosis (i.e. the natural freeze-dried state) mentioned above as occurring in tardigrades and eelworms are very relevant in this context. The presence of trehalose confers protection against DNA fragmentation (Loi *et al.*, 2008), as expected from theoretical principles (Rudolph and Crowe, 1985; Rudolph *et al.*, 1986; Beattie *et al.*, 1997). The disulfide status of the spermatozoa is also important in determining the success of freeze-drying; Kaneko *et al.* (2003b) showed that the ability of cauda epididymal spermatozoa to support normal development could be enhanced if they were first treated with diamide, which oxidises free sulfhydryl (SH) groups to form disulfide bonds. Conversely, if the disulfide bonds were first reduced with dithiothreitol (DTT), their ability to support development was lost. These findings are significant because they directly suggest a crucial role for the nuclear basic proteins – protamine 1 and protamine 2 – as determinants of success in freeze-drying. These protamines package sperm DNA into very compact and highly stable chromatin structures, a function that is largely mediated by the presence of cysteine groups and the formation of disulfide bonds (Balhorn *et al.*, 2000; Balhorn, 2007). Although the number of cysteine residues found in the protamine 1 of eutherian mammals is species specific, and typically varies from five to ten, it is notable that the protamine 1 of most marsupials completely lacks cysteine groups (Retief *et al.*, 1995). It is possible, therefore, that this will have negative implications for any attempts to freeze-dry marsupial spermatozoa.

Genetic Influences on Semen Cryopreservation and Cryoinjury

Semen cryopreservation has been applied successfully in only a relatively small number of species, and despite the commercial species such as cattle, sheep and pigs having attracted large amounts of research investment, considerable and unexplained variation in post-thaw semen quality exists between individuals. Although this phenomenon is a nuisance for those involved in routine semen cryopreservation, it might, nevertheless, be possible to exploit these differences and find a way to refine our knowledge of sperm cryoinjury and, also, a way to improve the between-species situation, which is much less consistent. Recent studies have suggested that there may be a genetic basis for the variation in post-thaw semen quality, and these indicate that modern molecular technologies should enable the identification of markers linked to genes that influence the stability of spermatozoa in the face of extreme stress.

Although there are long-standing hypotheses, especially on sperm membranes and their lipid content (Parks and Lynch, 1992), that attempt to explain interspecies variation in the susceptibility of spermatozoa to cryoinjury, these seem rather unsatisfactory when the issue of within-species variation is considered. If sperm quality and fertility across males of a single species is normally sufficient to achieve successful mating and conception, it is likely that the sperm membrane lipid composition would have to be maintained within certain limits in order to preserve appropriate sperm membrane fluidity and functionality. If that is the case, explanations for inter-male variations in sperm survival during and after cryopreservation may be more complex than previously thought. The existence of inter-male variation in terms of semen freezing is well known (at least anecdotally), and males are often known as 'good' and 'bad' freezers. More formal demonstrations of these differences have been published for a number of species: mice (Songsasen and Leibo 1997); dogs (Yu *et al.*, 2002); rhesus monkeys (Leibo *et al.*, 2007); pigs (Thurston *et al.*, 2002); and stallions (Ortega-Ferrusola *et al.*, 2009).

The mouse studies noted above are of particular interest because they demonstrated that there were major differences in the post-thaw fertility (using *in vitro* fertilization – IVF) of spermatozoa from different genetic lines. Slight but consistent strain-dependent morphological differences in sperm tail and cytoplasmic droplet morphology are known to correlate with *in vitro* fertilization rates (Kawai *et al.*, 2006), although epididymal dysfunction is partly responsible for this effect by inducing tail bending at the sperm neck region in some lines. An earlier series of studies in which different mouse strains were compared (Krzanowska *et al.*, 1991, 1995) had linked the ability of spermatozoa to penetrate the zona pellucida with subtle differences in sperm head shape and, moreover, had shown that these differences were dependent upon genotype. Given that these differences exert influences on fertilizing ability even when no freezing methods are involved, it is not surprising that cryopreservation only amplifies the genetically based variation in fertility.

The observation that the mouse genotype might be responsible for differences in post-thaw fertility variation is consistent with other studies in pigs. In an effort to minimize genetic variation due to other factors, Thurston *et al.* (2002) categorized more than 100 boars from a genetically homogeneous line into 'good', 'bad' and 'intermediate' freezers, based on extensive analysis of their post-thaw semen quality. She then compared the genetic characteristics of the 'good' and 'bad' freezers by using amplified fragment length polymorphism (AFLP) and was able to show that there were at least 16 polymorphic regions of the genomic DNA that correlated with post-thaw semen quality. In a parallel study of sperm head morphology, she also demonstrated systematic variation in pre-freeze sperm head shape that correlated with the post-thaw semen quality (Thurston *et al.*, 2001). A more recent study based on pre-freeze and post-thaw semen quality parameters (Safranski *et al.*, 2011) has used quantitative genetic approaches to implicate genotypic differences in this relationship.

These studies suggest that, in species that have not undergone any selection for good sperm survival during cryopreservation, underlying genotypic differences between individuals could be identified and used as markers. Although routine artificial insemination in pigs uses chilled and non-frozen semen, there are special circumstances in which such markers would be useful. These include the establishment of cryobanks for long-term storage of genetic materials, where it would be advantageous and most efficient to freeze semen from the 'good' freezers only. Biomarkers are probably less useful for the dairy industry, as the bulls are tested for semen freezability early in life and the poor freezers discarded. From a practical point of view, consistent freezability of spermatozoa is less relevant outside the commercial agricultural industry as there is more emphasis on breeding genetically or financially valuable individuals and investing efforts and resources in gaining that specific pregnancy.

Pathogens in Semen, Semen Preservation and Biosecurity

Many bacterial and viral diseases are now known to be transmitted through semen, or physically through breeding, and unless practitioners are constantly vigilant they run the risk of disseminating harmful microorganisms together with their extended or frozen semen samples. Viruses or bacteria may actually be present in the spermatozoa; for example, both bovine and human papilloma viruses enter the spermatozoon (Pao *et al.*, 1996; Lindsey *et al.*, 2009) and the viruses are impossible to remove by the use of washing procedures (Foresta *et al.*, 2011a,b). Alternatively, the contamination may originate from epithelial cells shed into the semen (e.g. foot-and-mouth disease), either directly via seminal plasma, or through external sources. Two recent reviews have summarized the problems of semen contamination and have also presented a series of general recommendations for good AI centre practice (Althouse 2008; Althouse and Rossow, 2011). These are extremely important and relevant to semen collection in many different species and therefore we reproduce them in this chapter (Table 5.1). Disease control measures for semen collection and genetic resource banking have also been extensively reviewed by Philpott (1993) and Kirkwood and Colenbrander (2001). These authors also presented summary lists of diseases that, at the time, were either known or suspected to be transmissible through semen and AI. As these lists are also important, we have updated them to some extent and present them as Table 5.2.

The use of antibiotics and the removal of pathogenic organisms

Antibiotics are routinely added to semen extenders to address bacterial contamination of ejaculates. The seven most common classes of antibiotics used in semen extenders have been summarized by Althouse (2008), who found that while a range of ten antibiotics is commonly used with porcine semen, only three are used with poultry semen. The reason for this is that the antibiotics tend to have deleterious effects on sperm quality parameters, as well as exerting their antimicrobial effects, and it is therefore important to evaluate both the antimicrobial and detrimental effects, and to find antibiotics that offer a compromise between the two. There is an extensive literature about the use of antibiotics in semen (see, for example: Lorton *et al.*, 1988a,b; Shin *et al.*, 1988; Miraglia *et al.*, 2003; Bielanski 2007) and it is not appropriate to review that topic in detail here except to mention that bacterial diseases of particular concern in this context include those caused by mycoplasmas, ureaplasmas, *Histophilus somni*, *Campylobacter fetus* subsp. *venerealis*, tuberculosis and leptospirosis. International standards for the use of antibiotics in frozen semen have been developed by the global body, the World Organisation for Animal Health (OIE), which provides extensive species-specific guidance (in the Terrestrial Animal Health Code, 2013) on its web site (http://www.oie.int/fileadmin/Home/eng/Health_standards/tahc/2010/en_index.htm). For example, the instructions for processing and freezing bovine, porcine and small ruminant semen state the following (Article 4.6.7: 1f):

> A mixture of antibiotics should be included with a bactericidal activity at least equivalent to that of the following mixtures in each ml of frozen semen: gentamicin (250 μg), tylosin (50 μg), lincomycin–spectinomycin (150/300 μg), penicillin (500 IU), streptomycin (500 μg), lincomycin–spectinomycin (150/300 μg), amikacin (75 μg) or divekacin (25 μg).

Table 5.1. General stud hygiene and sanitation recommendations. Reproduced from Althouse, 2008, with permission of John Wiley & Sons, Inc.

Stud personnel

1. Application of good hand hygiene, including appropriate washing and use of protective gloves, should be practised throughout all areas of the stud.
2. Personnel should avoid any contact of bare hands with materials that can later come into contact with semen or extender.
3. Personnel with upper respiratory tract infections should be cognizant of and avoid contamination of materials, semen or extender through aerosolization during sneezing or coughing.
4. Caps and hairnets can be of value if worn by personnel performing the semen collection process and by laboratory personnel as an aid in minimizing hair and dander as a contamination source.
5. Clean protective garments and shoes/boots, provided on-site by the stud, should be available for use by all stud personnel.

Animal housing and handling

1. Animal housing should be put on a regular sanitary maintenance schedule, including removal of organic material and application of a broad spectrum disinfectant.
2. Trimming of hair around the preputial orifice performed on an as-needed basis to eliminate the accumulation of organic matter at this site and its inadvertent introduction into the ejaculate during semen collection.
3. The ventral abdomen should be clean and dry prior to commencing with semen collection.
4. Cleaning of the preputial opening and surrounding area with a single-use disposable wipe should be considered if the area is wet and/or has organic material present.
5. In some species, preputial fluids can have high numbers of bacteria, therefore these fluids should be evacuated immediately prior to the semen collection process.
6. When collecting semen using an artificial vagina or gloved hand, the collector should position the penis in such a way as to minimize gravitation contamination of the semen collection vessel with preputial fluids.
7. If performing gloved hand semen collection, diverting the pre-sperm fraction from the semen collection vessel may aid in reducing ejaculate bacterial load.
8. The semen collection area and any collection equipment should be thoroughly cleaned and disinfected at the end of each collection day.

Laboratory

1. Encourage single-use disposable products when economically feasible to minimize cross-contamination.
2. When using reusable laboratory materials (i.e. glassware, plastic ware, plastic tubing, containers, etc.) that cannot be heat/gas sterilized or boiled, clean these reusables initially using a laboratory-grade detergent (residue free) with water, followed by a distilled water rinse and lastly through a 70% alcohol (non-denatured) rinse. Allow sufficient time and proper ventilation for complete evaporation of residual alcohol on the reusable. Rinse reusables with semen extender prior to their first use of the day.
3. Laboratory-purified water should be checked on a minimum quarterly basis if in-house, and by lot if outsourced. Any bacterial growth should be considered significant and appropriate action taken to identify and eliminate the contaminant source.
4. Disinfect countertops and contaminated laboratory equipment at end of the processing day with a residue-free detergent and rinse.
5. Floor should be mopped at end of each day with a disinfectant.
6. Break down bulk products into smaller daily use quantities immediately after opening.
7. Ultraviolet lighting can be installed to aid in sanitizing reusables and laboratory surfaces; however, safety precautions should be integrated to prevent exposure to personnel.

Table 5.2. Virus diseases that can, or probably could, be transmitted by artificial insemination. Adapted from Kirkwood and Colenbrander, 2001.

Virus disease	Host species
Avian influenza	Birds
Newcastle disease	Birds
Bovine papilloma virus	Bovidae
Bovine viral diarrhoea	Bovidae
Enzootic bovine leucosis	Bovidae
Infectious bovine rhinotracheitis	Bovidae
Epizootic haemorrhagic disease of deer	Deer
Feline infectious enteritis	Felidae
Feline infectious peritonitis	Felidae
Feline leukaemia virus	Felidae
Feline rhinotracheitis	Felidae
African horse sickness	Horse
Equine herpes viruses	Horse
Equine infectious anaemia	Horse
Equine viral arteritis	Horse
Glanders	Horse
Rabies	Mammals
African swine fever	Pigs
Aujesky's disease	Pigs
Blue ear disease	Pigs
Classical swine fever	Pigs
Porcine circovirus	Pigs
Porcine reproductive and respiratory syndrome	Pigs
Swine vesicular disease	Pigs
Teschen disease	Pigs
Vesicular stomatitis	Pigs
Various parvoviruses	Pigs and carnivores
Hepatitis B	Primates
Encephalomyocarditis	Various
Simian, feline and other immunodeficiency viruses	Various
Bluetongue	Various Artiodactyla
Foot-and-mouth disease	Various Artiodactyla
Lumpy skin disease	Various Artiodactyla
Malignant catarrhal fever	Various Artiodactyla
Peste de petite ruminants	Various Artiodactyla
Rift Valley fever	Various Artiodactyla
Rinderpest	Various Artiodactyla
Scrapie	Various Artiodactyla
Sheep pox	Various Artiodactyla
Canine distemper	Various carnivores

A subset of this combination used as a standard procedure for bovine semen storage in the USA, Canada and Europe is known as the GTLS method (gentamicin, tylosin and Linco-Spectin). As described by Shin and Kim (2000), the GTLS method used in the USA involves two steps: a preliminary extension step with antibiotics containing non-glycerolated extender; and a second extension step without antibiotics. Elsewhere, and somewhat also in the USA, a one-step approach is used. Experimental comparison of these approaches revealed that they were equally effective (Shin and Kim, 2000).

Recently, considerable efforts have been made to explore alternative methods of processing semen in order to remove microorganisms. Centrifugation through density gradients is used in some human clinical settings as an effective way of processing HIV-positive semen and rendering it suitable for the insemination of HIV-negative women (Semprini *et al.*, 1992; Loskutoff *et al.*, 2005; Bostan *et al.*, 2008), and similar procedures are being developed for animal semen (Morrell and Geraghty, 2006; Blomqvist *et al.*, 2011; Morrell and Wallgren, 2011). This topic was reviewed in detail by Bielanski (2007), who suggested that the methods currently available are not sufficiently reliable to allow complete confidence in their applicability.

The experimental application of washing procedures to semen contaminated with human papilloma virus (HPV; Czeglédy and Szarka, 2006) failed to eliminate the virus from the spermatozoa of infected patients. Nevertheless, when spermatozoa that tested HPV positive by PCR were used in IVF, healthy children were born. In animal studies, Bielanski *et al.* (1992) tested Percoll gradients in an effort to remove bovine viral diarrhoea virus (BVDV) from infected bull semen, but found that this approach was unsuccessful. In contrast, Morrell and Geraghty (2006) found that by combining density gradient centrifugation with 'swim-up' procedures it was possible to remove equine arteritis virus from 'virus-spiked' stallion semen.

Bielanski (2007) has also cited several alternative approaches to the problem of semen disinfection, which have included: (i) the brief exposure of semen to acidic conditions (e.g. 1 min exposure to pH < 6.0 inactivates bluetongue virus (BTV) and rubella virus), which apparently does not cause sperm damage (Bielanski *et al.*, 1991); (ii) the use of extenders that incorporate immune reactivity against microorganisms (Silva *et al.*, 2000) – in this case hyperimmune egg yolk semen extender was used for the inactivation of bovine herpes virus (BHV-1); (iii) the use of enzymes, such as 0.25% trypsin, to assist the removal of microorganisms from the sperm surface during washing procedures (Loskutoff *et al.*, 2005); and (iv) the use of photosensitive compounds, such as haematoporphyrin (HP) and thiopyronin, which develop antimicrobial properties when irradiated with white light from a halogen lamp. Trials showed that HP was effective against BHV-1 when tested with infected bull semen (Eaglesome *et al.*, 1994), although the effects on semen quality are yet to be determined. In addition to the methods cited by Bielanski (2007), it is also worth noting an earlier report by Schultz *et al.* (1988), in which the authors showed that the addition to egg yolk or milk extenders of gamma globulin isolated from hyperimmune bovine serum could effectively eliminate several significant viruses from infected semen. These included BTV (5 or more logs of infectivity), and both infectious bovine rhinotracheitis virus (IBR) and BVDV (6 or more logs). The authors commented that the direct addition of hyperimmune serum to extender was harmful to the spermatozoa, but that the gamma globulin produced no ill effects. They also suggested that the technique could be adapted to include other viruses if appropriate antibodies were present. Surprisingly, this approach does not seem to have been exploited further, even though it certainly seems to merit further investigation.

Hygienic aspects of sperm storage

The issue of safe and hygienic embryo and sperm cryostorage has recently been the subject of considerable concern (Vajta, 2010). Sample storage in liquid nitrogen containers presents hazards from several sources, including the containers themselves (which may accumulate microorganisms from air and moisture), the non-sterile liquid nitrogen used within the containers, and imperfectly sealed or broken straws or cryovials within the containers. The safety of cryopreserved germplasm became a matter of special concern after the discovery of a case of transmission of human hepatitis B via bone marrow transplants cryopreserved in liquid nitrogen (Tedder *et al.*, 1995). Experimentally, Piasecka-Serafin (1972) was the first to demonstrate the possibility of the translocation of bacteria from infected semen pellets to sterile pellets in liquid nitrogen. Of the sterile samples, 94% became infected with *Escherichia coli* and *Staphylococcus aureus* within 2 h of placing them in a container holding contaminated liquid nitrogen. More recently, Bielanski *et al.* (2000) have demonstrated the possibility of the infection of embryos through liquid nitrogen contaminated with BVDV and BHV-1 if the samples are not sealed properly.

A number of recommendations for good practice have recently been published in a detailed review, which also describes the outcomes of sampling liquid nitrogen tanks for the presence of viruses and bacteria (Bielanski and Vajta, 2009). These and other authors (Isachenko *et al.*, 2005a,b) also considered the emerging issues of hygiene in relation to vitrification, when sperm or embryo samples are directly exposed to potentially contaminated liquid nitrogen. Their suggestion in this instance involved the use of sterile liquid nitrogen for preparative work, and then making sure that the samples are enclosed safely in appropriate containers, straws or vials, for long-term storage. In fact, government protocols from multiple countries mandate the use of virgin liquid nitrogen for the importation of cryopreserved semen from diverse species.

Modern transport methods, security and the international exchange of semen samples

Annually, tens of millions of straws containing cryopreserved semen are shipped internationally and exposed to X-irradiation during security inspections. Current screening methods already employ up to three simultaneous scans on checked bags to build a three-dimensional image of the baggage, and plans are underway to develop X-ray machines that deliver higher doses of radiation to scan large quantities of baggage with a single scan.

Transportation of these biological materials requires that they are exposed to X-irradiation by airport security procedures at ports of entry and exit. A pioneering study published in 1996 (England and Keane, 1996) indicated that the levels of X-irradiation employed at airport security checkpoints do not adversely affect equine sperm quality or fertilizing ability, or embryonic development rates. This was recently confirmed for bovine semen (Hendricks *et al.*, 2010); these authors also failed to find any detrimental effects on sperm chromatin integrity. A similar study in which frozen felid spermatozoa (from a fishing cat and two domestic cats) were exposed to X-irradiation (Gloor *et al.*, 2006) did, however, report adverse effects on post-thaw sperm motility, and an unusually large amount of double-stranded DNA damage. The detrimental effects of X-irradiation on DNA in somatic cells, germ cells and embryos are well recognized (Matsuda *et al.*, 1989; Sailer *et al.*, 1995; Haines *et al.*, 2001), so species differences in sperm chromatin organization and X-ray dose responsiveness may explain the evident discrepancies between these results. Despite the possibility that X-ray induced chromatin damage may be limited, albeit not in all species, or may even be repaired after fertilization, it is important to take into consideration the potential for epigenetic and genetic changes to the original DNA in the embryo formed from X-irradiated sperm. These changes may not be exhibited as a reduction in embryo competence to the blastocyst stage, but as reduced implantation rates, or as increased rates of mid- and late-gestational abnormalities, death or increased susceptibility to disease in the offspring; future studies might include investigation of the effects of X-irradiation on offspring production.

International semen transport for animal conservation

While the general guidance from, for example, the OIE, on semen import/export is designed mainly for routine application in commercial stud situations, there is also considerable interest in the international movement of semen from wild and endangered species in support of zoo-based genetic management and *ex situ* conservation programmes. These activities potentially apply to many different types of species, including mammals, birds, fishes, reptiles and amphibians, all of which are known to carry diseases. In general, international transport of semen from wild species, with its potential to transmit diseases to agricultural livestock, is viewed with considerable caution by national veterinary authorities and is either entirely forbidden (e.g. the importation of some types of semen, especially of ruminants, into the European Union (EU) is currently not permitted at all, because EU countries have yet to agree upon the importation protocols) or is very difficult. Much of the difficulty is centred on establishing species-specific protocols for disease testing, both before and after semen collection. Evidence-based risk analysis models are being developed in order to quantify the risks associated with the movement of whole animals (Sutmoller and Casas Olascoaga 2003; Engel *et al.*, 2006; Travis *et al.*, 2006) or semen from place to place (Loskutoff *et al.*, 2003), and thus assist the decision-making processes that apply to particular situations. Gathering the evidence to build these models is still problematic, however, because they depend heavily on detailed knowledge of the disease status of animals in the relevant country, and also of the specific region of origin within the country. Further, such models depend on the validity of translating the usual disease-testing methods from one species to another, and this is not always possible. The acceptability of these models will also depend on the attitudes of the veterinary authorities in the different receiving countries. Relevant information is typically sparse, so it will be some considerable time before semen from wild ruminants can be transported readily.

Conclusions: The Outstanding Problems

Semen processing and cryopreservation has transformed the dairy industry over the past six or seven decades, and has facilitated major advances in agricultural genetic selection and disease control. Semen cryopreservation has also had a major impact on human clinical medicine, whether for the preservation of donor semen for anonymous AI by a donor (AID), the preservation of spermatozoa from oncology patients about to undergo radiotherapy or chemotherapy, or even from soldiers wounded during battle – exactly as predicted by Mantegazza (1866) in the 19th century.

Nevertheless, the early pioneers of semen cryopreservation technology were rather fortunate in choosing to study bull semen, with which high fertility rates can be achieved by using relatively small numbers of viable spermatozoa (Vishwanath 2003). Had they begun with boar or ram semen, the situation would have been significantly different. Successful inseminations in these species can be achieved with frozen semen, but additional steps are needed to ensure maximum fertility. Conception rates in sheep are excellent if the semen is delivered directly to the uterine horns by the use of surgical methods such as laparoscopy. Similarly, high success rates can be achieved with boar spermatozoa provided that the females are frequently checked for signs of standing heat, and inseminated at a time close to ovulation. These problems and solutions highlight the difficulties associated with the retention of full functionality after cryopreservation. In sheep, the cryopreserved spermatozoa can be deposited intra-cervically during AI, but then they do not possess the ability to progress towards the uterine horns and oviducts. In pigs, although intrauterine insemination is not a problem in itself, the cryopreserved spermatozoa survive for a considerably shorter period than do fresh spermatozoa and, unless ovulation occurs during this critical period, the AI will be unsuccessful. These examples show that although spermatozoa can survive the process of cryopreservation, they tend to be sub-fertile afterwards.

This sub-fertility is a consequence of various pathological changes induced within the spermatozoon by the cryopreservation process. These include inappropriate phosphorylation of flagellar proteins by a process that has been termed 'cryo-capacitation' (Watson, 1995, 1996), the induction of DNA fragmentation (Fraser and Strzezek, 2007; Portas *et al.*, 2009; Thomson *et al.*, 2009) shortly after thawing and, similarly, the induction of 'apoptosis-like' changes in sperm membranes (Pena *et al.*, 2003; Ortega-Ferrusola *et al.*, 2008). As many of these effects have been attributed to the action of ROS, there has been considerable investment in finding measures to counteract these by the use of antioxidants. These endeavours have yet to yield convincing benefits, thus indicating that there is no simple cure for oxidative damage in spermatozoa.

Some recent reports have indicated that exposing spermatozoa to high pressure before commencing cryopreservation improves their survival (Pribenszky and Vajta, 2011; Pribenszky *et al.*, 2011). So far, the apparent success of this approach is difficult to explain, but from a biophysical perspective it may be the case that membranes are forced to undergo a hyperbaric phase transition that alters their membrane permeability characteristics. If this method continues to show promising outcomes, it may be a surprising but effective treatment for the future. Species differences in the biology of spermatozoa also lie at the heart of the cryopreservation problem, especially when the array of wild species is considered. At present, there is no effective cryopreservation method for any marsupial, despite immense efforts from a number of investigators (Rodger *et al.*, 1991; Molinia and Rodger, 1996; Taggart *et al.*, 1996; Holt *et al.*, 2000; Miller *et al.*, 2004; Johnston *et al.*, 2006; Czarny *et al.*, 2009). Similarly, the current and dramatic amphibian extinction crisis has suddenly provoked a demand for sperm freezing techniques in frogs and toads, about which very little is known, except that they may be similar to fish spermatozoa. This multiplicity of demands, coupled with the poor theoretical basis for investigating cryopreservation, underlines the need for detailed theoretical and comparative research into this topic.

References

Alapati R, Stout M, Saenz J, Gentry GT Jr, Godke RA and Devireddy RV (2009) Comparison of the permeability properties and post-thaw motility of ejaculated and epididymal bovine spermatozoa *Cryobiology* **59** 164–170.

Althouse GC (2008) Sanitary procedures for the production of extended semen *Reproduction in Domestic Animals* **43(Supplement 2)** 374–378.

Althouse G and Rossow K (2011) The potential risk of infectious disease dissemination via artificial insemination in swine In *Special Issue: The 7th International Conference on Boar Semen Preservation, Bonn, Germany, 14–17 August 2011 Reproduction in Domestic Animals* **46(Supplement S2)** 64–67.

Bailey J, Morrier A and Cormier N (2003) Semen cryopreservation: successes and persistent problems in farm species *Canadian Journal of Animal Science* **83** 393–401.

Balhorn R (2007) The protamine family of sperm nuclear proteins *Genome Biology* **8**:227.

Balhorn R, Brewer L and Corzett M (2000) DNA condensation by protamine and arginine-rich peptides: analysis of toroid stability using single DNA molecules *Molecular Reproduction and Development* **56** 230–234.

Bartels P and Kotze A (2006) Wildlife biomaterial banking in Africa for now and the future *Journal of Environmental Monitoring* **8** 779–781.

Beattie GM, Crowe JH, Lopez AD, Cirulli V, Ricordi C and Hayek A (1997) Trehalose: a cryoprotectant that enhances recovery and preserves function of human pancreatic islets after long-term storage *Diabetes* **46** 519–523.

Bencharif D, Amirat L, Pascal O, Anton M, Schmitt E, Desherces S, Delhomme G, Langlois ML, Barriere P, Larrat M and Tainturier D (2010) The advantages of combining low-density lipoproteins with glutamine for cryopreservation of canine semen *Reproduction in Domestic Animals* **45** 189–200.

Bennun SV, Faller R and Longo ML (2008) Drying and rehydration of dlpc/dspc symmetric and asymmetric supported lipid bilayers: a combined AFM and fluorescence microscopy study *Langmuir* **24** 10371–10381.

Bielanski A (2007) Disinfection procedures for controlling microorganisms in the semen and embryos of humans and farm animals *Theriogenology* **68** 1–22.

Bielanski A and Vajta G (2009) Risk of contamination of germplasm during cryopreservation and cryobanking in IVF units *Human Reproduction* **24** 2457–2467.

Bielanski A, Eastman P and Hare WCD (1991) Transitory acidification of semen as a potential method for the inactivation of some pathogenic microorganisms. Effect on fertilization and development of ova in superovulated heifers *Theriogenology* **36** 33–40.

Bielanski A, Dubuc C and Hare WCD (1992) Failure to remove bovine diarrhea virus (BVDV) from bull semen by swim up and other separatory sperm techniques associated with *in vitro* fertilization *Reproduction in Domestic Animals* **27** 303–306.

Bielanski A, Nadin-Davis S, Sapp T and Lutze-Wallace C (2000) Viral contamination of embryos cryopreserved in liquid nitrogen *Cryobiology* **40** 110–116.

Bilodeau JF, Blanchette S, Gagnon C and Sirard MA (2001) Thiols prevent H_2O_2-mediated loss of sperm motility in cryopreserved bull semen *Theriogenology* **56** 275–286.

Blomqvist G, Persson M, Wallgren M, Wallgren P and Morrell JM (2011) Removal of virus from boar semen spiked with porcine circovirus type 2 *Animal Reproduction Science* **126** 108–114.

Bostan A, Vannin AS, Emiliani S, Debaisieux L, Liesnard C and Englert Y (2008) Development and evaluation of single sperm washing for risk reduction in artificial reproductive technology (ART) for extreme oligospermic HIV positive patients *Current HIV Research* **6** 461–465.

Chandrasekhar I and Gaber BP (1988) Stabilization of the bio-membrane by small molecules: interaction of trehalose with the phospholipid bilayer *Journal of Biomolecular Structure and Dynamics* **5** 1163–1170.

Chen Y and Liu RZ (2007) [Cryopreservation of spermatozoa] *Zhonghua Nan ke Xue/National Journal of Andrology* **13** 734–738.

Chiantia S, Kahya N and Schwille P (2005) Dehydration damage of domain-exhibiting supported bilayers: an AFM study on the protective effects of disaccharides and other stabilizing substances *Langmuir* **21** 6317–6323.

Courtens JL and Paquignon M (1985) Ultrastructure of fresh, frozen and frozen–thawed spermatozoa of the boar In *Proceedings of the First International Conference on Deep Freezing of Boar Semen* pp 61–87 Ed LA Johnson and K Larsson. Swedish University of Agricultural Sciences, Uppsala.

Crowe JH and Crowe LM (1986) Stabilization of membranes in anhydrobiotic organisms In *Membranes, Metabolism and Dry Organisms* pp 188–209 Ed AC Leopold. Comstock Publishing Associates, Ithaca and London.

Crowe JH, Crowe LM, Carpenter JF, Rudolph AS, Winstrom CA, Spargo BJ and Anchordoguy TJ (1988) Interactions of sugars with membranes *Biochimica et Biophysica Acta* **947** 367–384.

Curry MR (2007) Cryopreservation of mammalian semen *Methods in Molecular Biology* **368** 303–311.

Czarny NA, Harris MS, Iuliis GND and Rodger JC (2009) Acrosomal integrity, viability, and DNA damage of sperm from dasyurid marsupials after freezing or freeze drying *Theriogenology* **72** 817–825.

Czeglédy J and Szarka K (2006) Detection of high-risk HPV DNA in semen and its association with the quality of semen *International Journal of STD and AIDS* **17** 211–212.

Devireddy RV, Swanlund DJ, Roberts KP and Bischof JC (1999) Subzero water permeability parameters of mouse spermatozoa in the presence of extracellular ice and cryoprotective agents *Biology of Reproduction* **61** 764–775.

Devireddy RV, Swanlund DJ, Roberts KP, Pryor JL and Bischof JC (2000) The effect of extracellular ice and cryoprotective agents on the water permeability parameters of human sperm plasma membrane during freezing *Human Reproduction* **15** 1125–1135.

Devireddy RV, Fahrig B, Godke RA and Leibo SP (2004) Subzero water transport characteristics of boar spermatozoa confirm observed optimal cooling rates *Molecular Reproduction and Development* **67** 446–457.

Devireddy RV, Campbell WT, Buchanan JT and Tiersch TR (2006) Freezing response of white bass (*Morone chrysops*) sperm cells *Cryobiology* **52** 440–445.

Dinnyes A, Liu J and Nedambale TL (2007) Novel gamete storage *Reproduction, Fertility and Development* **19** 719–731.

Dong QX, Rodenburg SE, Hill D and Vandevoort CA (2011) The role of low-density lipoprotein (LDL) and high-density lipoprotein (HDL) in comparison with whole egg yolk for sperm cryopreservation in rhesus monkeys *Asian Journal of Andrology* **13** 459–464.

Drobnis EZ, Crowe LM, Berger T, Anchordoguy TJ, Overstreet JW and Crowe JH (1993) Cold shock damage is due to lipid phase-transitions in cell-membranes – a demonstration using sperm as a model *Journal of Experimental Zoology* **265** 432–437.

Eaglesome MD, Bielanski A, Hare WC and Ruhnke HL (1994) Studies on inactivation of pathogenic microorganisms in culture media and in bovine semen by photosensitive agents *Veterinary Microbiology* **38** 277–284.

Ekwall H, Hernandez M, Saravia F and Rodriguez-Martinez H (2007) Cryo-scanning electron microscopy (cryo-SEM) of boar semen frozen in medium-straws and miniflatpacks *Theriogenology* **67** 1463–1472.

Engel G, Hungerford LL, Jones-Engel L, Travis D, Eberle R, Fuentes A, Grant R, Kyes R and Schillaci M (2006) Risk assessment: a model for predicting cross-species transmission of simian foamy virus from macaques (*M. fascicularis*) to humans at a monkey temple in Bali, Indonesia *American Journal of Primatology* **68** 934–948.

England GCW and Keane M (1996) The effect of X-radiation upon the quality and fertility of stallion semen *Theriogenology* **46** 173–180.

Evans RW and Setchell BP (1978) Association of exogenous phospholipids with spermatozoa *Journal of Reproduction and Fertility* **53** 357–362.

Foresta C, Ferlin A, Bertoldo A, Patassini C, Zuccarello D and Garolla A (2011a) Human papilloma virus in the sperm cryobank: an emerging problem? *International Journal of Andrology* **34** 242–246.

Foresta C, Pizzol D, Bertoldo A, Menegazzo M, Barzon L and Garolla A (2011b) Semen washing procedures do not eliminate human papilloma virus sperm infection in infertile patients *Fertility and Sterility* **96** 1077–1082.

Fraser L and Strzezek J (2007) Effect of different procedures of ejaculate collection, extenders and packages on DNA integrity of boar spermatozoa following freezing–thawing *Animal Reproduction Science* **99** 317–329.

Fuller BJ (2004) Cryoprotectants: the essential antifreezes to protect life in the frozen state *CryoLetters* **25** 375–388.

Funahashi H and Sano T (2005) Select antioxidants improve the function of extended boar semen stored at 10°C *Theriogenology* **63** 1605–1616.

Gadea J, Gumbao D, Matas C and Romar R (2005) Supplementation of the thawing media with reduced glutathione improves function and the *in vitro* fertilizing ability of boar spermatozoa after cryopreservation *Journal of Andrology* **26** 749–756.

Gao DY, Mazur P, Kleinhans FW, Watson PF, Noiles EE and Critser JK (1992) Glycerol permeability of human spermatozoa and its activation-energy *Cryobiology* **29** 657–667.

Gao DY, Ashwort E, Watson PF, Kleinhan FW, Mazu P and Critse JK (1993) Hyperosmotic tolerance of human spermatozoa – separate effects of glycerol, sodium-chloride, and sucrose on spermolysis *Biology of Reproduction* **49** 112–123.

George MA, Johnson MH and Vincent C (1992) Use of fetal bovine serum to protect against zona hardening during preparation of mouse oocytes for cryopreservation *Human Reproduction* **7** 408–412.

Gloor KT, Winget D and Swanson WF (2006) Conservation science in a terrorist age: the impact of airport security screening on the viability and DNA integrity of frozen felid spermatozoa *Journal of Zoo and Wildlife Medicine* **37** 327–335.

Graham JK and Foote RH (1987) Effect of several lipids, fatty acid chain length, and degree of saturation on the motility of bull spermatozoa after cold shock and freezing *Cryobiology* **24** 42–52.

Gutierrez-Perez O, Juarez-Mosqueda Mde L, Carvajal SU and Ortega ME (2009) Boar spermatozoa cryopreservation in low glycerol/trehalose enriched freezing media improves cellular integrity *Cryobiology* **58** 287–292.

Hagiwara M, Choi JH, Devireddy RV, Roberts KP, Wolkers WF, Makhlouf A and Bischof JC (2009) Cellular biophysics during freezing of rat and mouse sperm predicts post-thaw motility *Biology of Reproduction* **81** 700–706.

Haines GA, Hendry JH, Daniel CP and Morris ID (2001) Increased levels of comet-detected spermatozoa DNA damage following *in vivo* isotopic- or X-irradiation of spermatogonia *Mutation Research* **495** 21–32.

Hammerstedt RH and Graham JK (1992) Cryopreservation of poultry sperm: the enigma of glycerol *Cryobiology* **29** 26–38.

Hendricks KE, Penfold LM, Evenson DP, Kaproth MT and Hansen PJ (2010) Effects of airport screening X-irradiation on bovine sperm chromatin integrity and embryo development *Theriogenology* **73** 267–272.

Hofmo PO and Almlid T (1991) Recent developments in freezing of boar semen with special emphasis on cryoprotectants In *Boar Semen Preservation II Proceedings, Second International Conference on Boar Semen Preservation* Ed LA Johnson and D Rath *Reproduction in Domestic Animals* **1991(Supplement 1)** pp 111–122.

Holt WV (2000a) Basic aspects of frozen storage of semen *Animal Reproduction Science* **62** 3–22.

Holt WV (2000b) Fundamental aspects of sperm cryobiology: the importance of species and individual differences *Theriogenology* **53** 47–58.

Holt WV and North RD (1984) Partially irreversible cold-induced lipid phase transitions in mammalian sperm plasma membrane domains: freeze-fracture study *Journal of Experimental Zoology* **230** 473–483.

Holt WV and North RD (1988) The role of membrane-active lipids in the protection of ram spermatozoa during cooling and storage *Gamete Research* **19** 77–89.

Holt WV, Bennett PM, Volobouev V and Watson PF (1996) Genetic resource banks in wildlife conservation *Journal of Zoology* **238** 531–544.

Holt WV, Penfold LM, Johnston SD, Temple-Smith P, McCallum C, Shaw J, Lindemans W and Blyde D (2000) Cryopreservation of macropodid spermatozoa: new insights from the cryomicroscope *Reproduction, Fertility and Development* **11** 345–353.

Howard JG, Bush M, Devos V, Schiewe MC, Pursel VG and Wildt DE (1986) Influence of cryoprotective diluent on post-thaw viability and acrosomal integrity of spermatozoa of the African elephant (*Loxodonta africana*) *Journal of Reproduction and Fertility* **78** 295–306.

Hu JH, Li QW, Zan LS, Jiang ZL, An JH, Wang LQ and Jia YH (2010a) The cryoprotective effect of low-density lipoproteins in extenders on bull spermatozoa following freezing–thawing *Animal Reproduction Science* **117** 11–17.

Hu JH, Zan LS, Zhao XL, Li QW, Jiang ZL, Li YK and Li X (2010b) Effects of trehalose supplementation on semen quality and oxidative stress variables in frozen-thawed bovine semen *Journal of Animal Science* **88** 1657–1662.

Hu JH, Jiang ZL, Lv RK, Li QW, Zhang SS, Zan LS, Li YK and Li X (2011) The advantages of low-density lipoproteins in the cryopreservation of bull semen *Cryobiology* **62** 83–87.

Isachenko V, Isachenko E, Montag M, Zaeva V, Krivokharchenko I, Nawroth F, Dessole S, Katkov II and van der Ven H (2005a) Clean technique for cryoprotectant-free vitrification of human spermatozoa *Reproductive Biomedicine Online* **10** 350–354.

Isachenko V, Montag M, Isachenko E, Zaeva V, Krivokharchenko I, Shafei R and van der Ven H (2005b) Aseptic technology of vitrification of human pronuclear oocytes using open-pulled straws *Human Reproduction* **20** 492–496.

Isachenko E, Isachenko V, Weiss JM, Kreienberg R, Katkov II, Schulz M, Lulat AG-MI, Risopatrón MJ and Sánchez R (2008) Acrosomal status and mitochondrial activity of human spermatozoa vitrified with sucrose *Reproduction* **136** 167–173.

Johnston SD, MacCallum C, Blyde D, McClean R, Lisle A and Holt WV (2006) An investigation into the similarities and differences governing the cryopreservation success of koala (*Phascolarctos cinereus*: Goldfuss) and common wombat (*Vombatus ursinus*: Shaw) spermatozoa *Cryobiology* **53** 218–228.

Kaneko T, Whittingham DG and Yanagimachi R (2003a) Effect of pH value of freeze-drying solution on the chromosome integrity and developmental ability of mouse spermatozoa *Biology of Reproduction* **68** 136–139.

Kaneko T, Whittingham DG, Overstreet JW and Yanagimachi R (2003b) Tolerance of the mouse sperm nuclei to freeze-drying depends on their disulfide status *Biology of Reproduction* **69** 1859–1862.

Kaplan JB and Mead RA (1992) Evaluation of extenders and cryopreservatives for cooling and cryopreservation of spermatozoa from the western spotted skunk (*Spilogale gracilis*) *Zoo Biology* **11** 397–404.

Kawai Y, Hata T, Suzuki O and Matsuda J (2006) The relationship between sperm morphology and *in vitro* fertilization ability in mice *Journal of Reproduction and Development* **52** 561–568.

Kawase Y, Hani T, Kamada N, Jishage K and Suzuki H (2007) Effect of pressure at primary drying of freeze-drying mouse sperm reproduction ability and preservation potential *Reproduction* **133** 841–846.

Keilin D (1959) The Leeuwenhoek lecture: The problem of anabiosis or latent life: history and current concept *Proceedings of the Royal Society of London Series B Biological Sciences* **150** 149–191.

Kirkwood JK and Colenbrander B (2001) Disease control measures for genetic resource banking In *Cryobanking the Genetic Resource; Wildlife Conservation for the Future?* pp 69–84 Ed PF Watson and WV Holt. Taylor and Francis, London.

Kolossa M and Seibert H (1990) A chemically 'defined' diluent for cryopreservation of bovine spermatozoa *Andrologia* **22** 445–454.

Krzanowska H, Wabik-Sliz B and Rafinski J (1991) Phenotype and fertilizing capacity of spermatozoa of chimaeric mice produced from two strains that differ in sperm quality *Journal of Reproduction and Fertility* **91** 667–676.

Krzanowska H, Styrna J and Wabik-Sliz B (1995) Analysis of sperm quality in recombinant inbred mouse strains: correlation of sperm head shape with sperm abnormalities and with the incidence of supplementary spermatozoa in the perivitelline space *Journal of Reproduction and Fertility* **104** 347–354.

Kusakabe H, Yanagimachi R and Kamiguchi Y (2008) Mouse and human spermatozoa can be freeze-dried without damaging their chromosomes *Human Reproduction* **23** 233–239.

Leibo SP, Kubisch HM, Schramm RD, Harrison RM and VandeVoort CA (2007) Male-to-male differences in post-thaw motility of rhesus spermatozoa after cryopreservation of replicate ejaculates *Journal of Medical Primatology* **36** 151–163.

Lermen D, Blomeke B, Browne R, Clarke A, Dyce PW, Fixemer T, Fuhr GR, Holt WV, Jewgenow K, Lloyd RE, Lotters S, Paulus M, Reid GM, Rapoport DH, Rawson D, Ringleb J, Ryder OA, Sporl G, Schmitt T, Veith M and Muller P (2009) Cryobanking of viable biomaterials: implementation of new strategies for conservation purposes *Molecular Ecology* **18** 1030–1033.

Lindsey CJ, Almeida ME, Vicari CF, Carvalho C, Yaguiu A, Freitas AC, Becak W and Stocco RC (2009) Bovine papillomavirus DNA in milk, blood, urine, semen, and spermatozoa of bovine papillomavirus-infected animals *Genetics and Molecular Research* **8** 310–318.

Liu JL, Kusakabe H, Chang CC, Suzuki H, Schmidt DW, Julian M, Pfeffer R, Bormann CL, Tian XC, Yanagimachi R and Yang X (2004) Freeze-dried sperm fertilization leads to full-term development in rabbits *Biology of Reproduction* **70** 1776–1781.

Loi P, Matzukawa K, Ptak G, Natan Y, Fulka J Jr and Arav A (2008) Nuclear transfer of freeze-dried somatic cells into enucleated sheep oocytes *Reproduction in Domestic Animals* **43(Supplement 2)** 417–422.

Lorton SP, Sullivan JJ, Bean B, Kaproth M, Kellgren H and Marshall C (1988a) A new antibiotic combination for frozen bovine semen. 2. Evaluation of seminal quality *Theriogenology* **29** 593–607.

Lorton SP, Sullivan JJ, Bean B, Kaproth M, Kellgren H and Marshall C (1988b) A new antibiotic combination for frozen bovine semen. 3. Evaluation of fertility *Theriogenology* **29** 609–614.

Loskutoff NM, Holt WV and Bartels P Eds (2003) *Biomaterial Transport and Disease Risk: Workbook Development.* Conservation Breeding Specialist Group (SSC/IUCN). Omaha, Nebraska.

Loskutoff NM, Huyser C, Singh R, Walker DL, Thornhill AR, Morris L and Webber L (2005) Use of a novel washing method combining multiple density gradients and trypsin for removing human immunodeficiency virus-1 and hepatitis C virus from semen *Fertility and Sterility* **84** 1001–1010.

Mantegazza P (1866) Sullo sperma umano *Rendiconti Reale Istituto Lombardo di Scienze e lettere* **3** 183–196.

Marschall S and Hrabe de Angelis M (1999) Cryopreservation of mouse spermatozoa: double your mouse space *Trends in Genetics* **15** 128–131.

Matsuda Y, Seki N, Utsugi-Takeuchi T and Tobari I (1989) X-ray- and mitomycin c (MMC)-induced chromosome aberrations in spermiogenic germ cells and the repair capacity of mouse eggs for the X-ray and MMC damage *Mutation Research* **211** 65–75.

Mazur P, Leibo SP and Seidel GE (2008) Cryopreservation of the germplasm of animals used in biological and medical research: importance, impact, status, and future directions *Biology of Reproduction* **78** 2–12.

Mendelsohn R and Moore DJ (1998) Vibrational spectroscopic studies of lipid domains in biomembranes and model systems *Chemistry and Physics of Lipids* **96** 141–157.

Merino O, Risopatrón J, Sánchez R, Isachenko E, Figueroa E, Valdebenito I and Isachenko V (2011) Fish (*Oncorhynchus mykiss*) spermatozoa cryoprotectant-free vitrification: stability of mitochondrion as criterion of effectiveness *Animal Reproduction Science* **124** 125–131.

Miller RR Jr, Sheffer CJ, Cornett CL, McClean R, MacCallum C and Johnston SD (2004) Sperm membrane fatty acid composition in the eastern grey kangaroo (*Macropus giganteus*), koala (*Phascolarctos cinereus*), and common wombat (*Vombatus ursinus*) and its relationship to cold shock injury and cryopreservation success *Cryobiology* **49** 137–148.

Miraglia F, Morais ZM, Cortez A, Melville PA, Marvullo MFV, Richtzenhain LJ, Visintin JA and Vasconcellos SA (2003) Comparison of four antibiotics for inactivating leptospires in bull semen diluted in egg yolk extender and experimentally inoculated with *Leptospira santarosai* serovar *guaricura Brazilian Journal of Microbiology* **34** 147–151.

Mocé E, Blanch E, Tomás C and Graham JK (2010a) Use of cholesterol in sperm cryopreservation: present moment and perspectives to future *Reproduction in Domestic Animals* **45(Supplement 2)** 57–66.

Mocé E, Purdy PH and Graham JK (2010b) Treating ram sperm with cholesterol-loaded cyclodextrins improves cryosurvival *Animal Reproduction Science* **118** 236–247.

Molinia FC and Rodger JC (1996) Pellet-freezing spermatozoa of 2 marsupials – the tammar wallaby, *Macropus eugenii*, and the brushtail possum, *Trichosurus vulpecul Reproduction, Fertility and Development* **8** 681–684.

Montfort SL, Asher GW, Wildt DE, Wood TC, Schiewe MC, Williamson LR, Bush M and Rall WF (1993) Successful intrauterine insemination of Eld's deer (*Cervus eldi thiamin*) with frozen–thawed spermatozoa *Journal of Reproduction and Fertility* **99** 459–465.

Moraes EA, Graham JK, Torres CA, Meyers M and Spizziri B (2010) Delivering cholesterol or cholestanol to bull sperm membranes improves cryosurvival *Animal Reproduction Science* **118** 148–154.

Moran DM, Jasko DJ, Squires EL and Amann RP (1992) Determination of temperature and cooling rate which induce cold shock in stallion spermatozoa *Theriogenology* **38** 999–1012.

Morrell JM and Geraghty RM (2006) Effective removal of equine arteritis virus from stallion semen *Equine Veterinary Journal* **38** 224–229.

Morrell JM and Wallgren M (2011) Removal of bacteria from boar ejaculates by single layer centrifugation can reduce the use of antibiotics in semen extenders *Animal Reproduction Science* **123** 64–69.

Nakagata N and Takeshima T (1992) High fertilizing ability of mouse spermatozoa diluted slowly after cryopreservation *Theriogenology* **37** 1283–1291.

Nakagata N, Matsumoto K, Anzai M, Takahashi A, Takahashi Y, Matsuzaki Y and Miyata K (1992) Cryopreservation of spermatozoa of a transgenic mouse *Experimental Animals* **41** 537–540.

Nakhuda GS (2010) Posthumous assisted reproduction *Seminars in Reproductive Medicine* **28** 329–335.

Noiles EE, Mazur P, Watson PF, Kleinhans FW and Critser JK (1993) Determination of water permeability coefficient for human spermatozoa and its activation-energy *Biology of Reproduction* **48** 99–109.

Oldenhof H, Friedel K, Sieme H, Glasmacher B and Wolkers WF (2010) Membrane permeability parameters for freezing of stallion sperm as determined by Fourier transform infrared spectroscopy *Cryobiology* **61** 115–122.

Oliveira CH, Vasconcelos AB, Souza FA, Martins-Filho OA, Silva MX, Varago FC and Lagares MA (2010) Cholesterol addition protects membrane intactness during cryopreservation of stallion sperm *Animal Reproduction Science* **118** 194–200.

Ortega-Ferrusola C, Sotillo-Galan Y, Varela-Fernandez E, Gallardo-Bolanos JM, Muriel A, Gonzalez-Fernandez L, Tapia JA and Pena FJ (2008) Detection of "apoptosis-like" changes during the cryopreservation process in equine sperm *Journal of Andrology* **29** 213–221.

Ortega-Ferrusola C, Garcia BM, Gallardo-Bolanos JM, Gonzalez-Fernandez L, Rodriguez-Martinez H, Tapia JA and Pena FJ (2009) Apoptotic markers can be used to forecast the freezeability of stallion spermatozoa *Animal Reproduction Science* **114** 393–403.

Ostermeier GC, Wiles MV, Farley JS and Taft RA (2008) Conserving, distributing and managing genetically modified mouse lines by sperm cryopreservation *PLoS One* **3** e2792.

Pagl R, Aurich C and Kankofer M (2006) Anti-oxidative status and semen quality during cooled storage in stallions *Journal of Veterinary Medicine A Physiology Pathology and Clinical Medicine* **53** 486–489.

Pao CC, Yang FP and Lai YM (1996) Preferential retention of the E6 and E7 regions of the human papilloma virus type 18 genome by human sperm cells *Fertility and Sterility* **66** 630–633.

Parks JE and Lynch DV (1992) Lipid composition and thermotropic phase behavior of boar, bull, stallion, and rooster sperm membranes *Cryobiology* **29** 255–266.

Pena FJ, Johannisson A, Wallgren M and Rodriguez-Martinez H (2003) Assessment of fresh and frozen-thawed boar semen using an annexin-v assay: a new method of evaluating sperm membrane integrity *Theriogenology* **60** 677–689.

Pena FJ, Johannisson A, Wallgren M and Rodriguez-Martinez H (2004) Antioxidant supplementation of boar spermatozoa from different fractions of the ejaculate improves cryopreservation: changes in sperm membrane lipid architecture *Zygote* **12** 117–124.

Pena FJ, Nunez-Martinez I and Moran JM (2006) Semen technologies in dog breeding: an update *Reproduction in Domestic Animals* **41(Supplement 2)** 21–22.

Philpott M (1993) The dangers of disease transmission by artificial insemination and embryo transfer *British Veterinary Journal* **149** 339–369.

Piasecka-Serafin M (1972) The effect of the sediment accumulated in containers under experimental conditions on the infection of semen stored directly in liquid nitrogen (–196°C) *Bulletin de l'Academie Polonaise des Sciences Serie des Sciences Biologiques* **20** 263–267.

Polge C, Smith AV and Parkes AS (1949) Revival of spermatozoa after vitrification and dehydration at low temperatures *Nature* **164** 166.

Pontbriand D, Howard JG, Schiewe MC, Stuart LD and Wildt DE (1989) Effect of cryoprotective diluent and method of freeze-thawing on survival and acrosomal integrity of ram sperm *Cryobiology* **26** 341–354.

Portas T, Johnston SD, Hermes R, Arroyo F, López-Fernadez C, Bryant B, Hildebrandt TB, Göritz F and Gosalvez J (2009) Frozen-thawed rhinoceros sperm exhibit DNA damage shortly after thawing when assessed by the sperm chromatin dispersion assay *Theriogenology* **72** 711–720.

Pribenszky C and Vajta G (2011) Cells under pressure: how sublethal hydrostatic pressure stress treatment increases gametes' and embryos' performance? *Reproduction, Fertility and Development* **23** 48–55.

Pribenszky C, Horvath A, Vegh L, Huang SY, Kuo YH and Szenci O (2011) Stress preconditioning of boar spermatozoa: a new approach to enhance semen quality In *Special Issue: The 7th International Conference on Boar Semen Preservation, Bonn, Germany, 14–17 August 2011 Reproduction in Domestic Animals* **46(Supplement 2)** 26–30.

Quinn PJ and White IG (1966) The effect of cold-shock and deep-freezing on the concentration of major cations in spermatozoa *Journal of Reproduction and Fertility* **12** 263–270.

Quinn PJ, White IG and Cleland KW (1969) Chemical and ultrastructural changes in ram spermatozoa after washing, cold-shock and freezing *Journal of Reproduction and Fertility* **18** 209–220.

Rall WF and Fahy GM (1985) Ice-free cryopreservation of mouse embryos at –196°C by vitrification *Nature* **313** 573–575.

Rall WF and Wood MJ (1994) High *in vitro* and *in vivo* survival of day 3 mouse embryos vitrified or frozen in a non-toxic solution of glycerol and albumin *Journal of Reproduction and Fertility* **101** 681–688.

Retief JD, Krajewski C, Westerman M and Dixon GH (1995) The evolution of protamine P1 genes in dasyurid marsupials *Journal of Molecular Evolution* **41** 549–555.

Ricker JV, Linfor JJ, Delfino WJ, Kysar P, Scholtz EL, Tablin F, Crowe JH, Ball BA and Meyers SA (2006) Equine sperm membrane phase behavior: the effects of lipid-based cryoprotectants *Biology of Reproduction* **74** 359–365.

Rodger JC, Cousins SJ and Mate KE (1991) A simple glycerol-based freezing protocol for the semen of a marsupial *Trichosurus vulpecula*, the common brushtail possum *Reproduction, Fertility and Development* **3** 119–125.

Rota A, Linde-Forsberg C, Vannozzi J, Romagnoli S and Rodriguez-Martinez H (1998) Cryosurvival of dog spermatozoa at different glycerol concentrations and freezing/thawing rates *Reproduction in Domestic Animals* **33** 355–361.

Royère D, Barthelemy C, Hamamah S and Lansac J (1996) Cryopreservation of spermatozoa: a 1996 review *Human Reproduction Update* **2** 553–559.

Rudolph AS and Crowe JH (1985) Membrane stabilization during freezing. The role of two natural cryoprotectants, trehalose and proline *Cryobiology* **22** 367–377.

Rudolph AS, Crowe JH and Crowe LM (1986) Effects of three stabilizing agents – proline, betaine, and trehalose – on membrane phospholipids *Archives of Biochemistry and Biophysics* **15** 134–143.

Safranski T, Ford J, Rohrer G and Guthrie H (2011) Genetic selection for freezability and its controversy with selection for performance In *Special Issue: The 7th International Conference on Boar Semen Preservation, Bonn, Germany, 14–17 August 2011 Reproduction in Domestic Animals* **46(Supplement 2)** 31–34.

Sailer BL, Jost LK, Erickson KR, Tajiran MA and Evenson DP (1995) Effects of X-irradiation on mouse testicular cells and sperm chromatin structure *Environmental and Molecular Mutagenesis* **25** 23–30.

Salamon S and Maxwell WMC (1995a) Frozen storage of ram semen. I. Processing, freezing, thawing and fertility after cervical insemination *Animal Reproduction Science* **37** 185–249.

Salamon, S and Maxwell, WMC (1995b) Frozen storage of ram semen. II. Causes of low fertility after cervical insemination and methods of improvement. *Animal Reproduction Science* **38** 1–36.

Schultz RD, Kaproth M and Bean B (1988) Immunoextension: a method to eliminate viral infectivity in contaminated semen In *Proceedings of the 11th International Congress on Animal Reproduction and Artificial Insemination Volume 4* Paper 522. University College, Dublin.

Semprini AE, Levi-Setti P, Bozzo M, Ravizza M, Taglioretti A, Sulpizio P, Albani E, Oneta M and Pardi G (1992) Insemination of HIV-negative women with processed semen of HIV-positive partners *Lancet* **340** 1317–1319.

Shaw JM and Nakagata N (2002) Cryopreservation of transgenic mouse lines *Methods in Molecular Biology* **180** 207–228.

Shaw PW, Bernard AG, Fuller BJ, Hunter JH and Shaw RW (1992) Vitrification of mouse oocytes using short cryoprotectant exposure: effects of varying exposure times on survival *Molecular Reproduction and Development* **33** 210–214.

Shin SJ and Kim SG (2000) Comparative efficacy study of bovine semen extension: 1-step vs 2-step procedure In *Proceedings of the Eighteenth Technical Conference on Artificial Insemination and Reproduction* pp 60–62. National Association of Animal Breeders, Milwaukee

Shin SJ, Lein DH, Patten VH and Ruhnke HL (1988) A new antibiotic combination for frozen bovine semen 1. Control of mycoplasmas, ureaplasmas, *Campylobacter fetus* subsp. *venerealis* and *Haemophilus somnus Theriogenology* **29** 577–591.

Silva N, Solana A and Castro JM (2000) Inactivation of bovine herpesvirus 1 in semen using a hyperimmune egg yolk semen extender *Journal of Veterinary Medicine B Infectious Diseases and Veterinary Public Health* **47** 69–75.

Singer SJ and Nicolson GL (1972) The fluid mosaic model of the structure of cell membranes *Science* **175** 720–731.

Solar II (2009) Use and exchange of salmonid genetic resources relevant for food and aquaculture *Reviews in Aquaculture* **1** 174–196.

Songsasen N and Leibo SP (1997) Cryopreservation of mouse spermatozoa. II. Relationship between survival after cryopreservation and osmotic tolerance of spermatozoa from three strains of mice *Cryobiology* **35** 255–269.

Spallanzani L (1776) *Opuscoli de Fisice Animale e Vegetabile Opuscola. II. Observatione e Sperienze Intorno ai Vermicelli Spermatici dell' Homo e degli Animali*. Presso la Societa Tipographica, Modena.

Spizziri BE, Fox MH, Bruemmer JE, Squires EL and Graham JK (2010) Cholesterol-loaded-cyclodextrins and fertility potential of stallions spermatozoa *Animal Reproduction Science* **118** 255–264.

Succu S, Bebbere D, Bogliolo L, Ariu F, Fois S, Leoni GG, Berlinguer F, Naitana S and Ledda S (2008) Vitrification of *in vitro* matured ovine oocytes affects *in vitro* pre-implantation development and mRNA abundance *Molecular Reproduction and Development* **75** 538–546.

Sutmoller P and Casas Olascoaga R (2003) The risks posed by the importation of animals vaccinated against foot and mouth disease and products derived from vaccinated animals: a review *Revue Scientifique et Technique OIE* **22** 823–835.

Sztein JM, Schmidt PM, Raber J and Rall WF (1992) Cryopreservation of mouse spermatozoa in a glycerol/raffinose solution *Cryobiology* **29** 736–737.

Taggart DA, Leigh CM, Steele VR, Breed WG, Temple-Smith PD and Phelan J (1996) Effect of cooling and cryopreservation on sperm motility and morphology of several species of marsupial *Reproduction, Fertility and Development* **8** 673–679.

Takahashi, H and Liu, C (2010) Archiving and distributing mouse lines by sperm cryopreservation, IVF, and embryo transfer. *Methods in Enzymology* **476** 53–69.

Takeo T and Nakagata N (2010) Combination medium of cryoprotective agents containing L-glutamine and methyl-β-cyclodextrin in a preincubation medium yields a high fertilization rate for cryopreserved C57bl/6j mouse sperm *Laboratory Animals* **44** 132–137.

Tedder RS, Zuckerman MA, Brink NS, Goldstone AH, Fielding A, Blair S. Patterson KG, Hawkins AE, Gormon AM, Heptonstall J and Irwin D (1995) Hepatitis B transmission from contaminated cryopreservation tank *The Lancet* **346** 137–140.

Thirumala S, Campbell WT, Vicknair MR, Tiersch TR and Devireddy RV (2006) Freezing response and optimal cooling rates for cryopreserving sperm cells of striped bass, *Morone saxatilis Theriogenology* **66** 964–973.

Thomson LK, Fleming SD, Aitken RJ, De Iuliis GN, Zieschang JA and Clark AM (2009) Cryopreservation-induced human sperm DNA damage is predominantly mediated by oxidative stress rather than apoptosis *Human Reproduction* **24** 2061–2070.

Thurston LM, Watson PF, Mileham AJ and Holt WV (2001) Morphologically distinct sperm subpopulations defined by Fourier shape descriptors in fresh ejaculates correlate with variation in boar semen quality following cryopreservation *Journal of Andrology* **22** 382–394.

Thurston LM, Siggins K, Mileham AJ, Watson PF and Holt WV (2002) Identification of amplified restriction fragment length polymorphism markers linked to genes controlling boar sperm viability following cryopreservation *Biology of Reproduction* **66** 545–554.

Tiersch TR, Yang H and Hu E (2011) Outlook for development of high-throughput cryopreservation for small-bodied biomedical model fishes *Comparative Biochemistry and Physiology C Toxicology and Pharmacology* **154** 76–81.

Travis DA, Hungerford L, Engel GA and Jones-Engel L (2006) Disease risk analysis: a tool for primate conservation planning and decision making *American Journal of Primatology* **68** 855–867.

Vajta G (2010) Biosafety of vitrification. The issue of contamination using an open system In *1st International Congress on Controversies in Cryopreservation of Stem Cells, Reproductive Cells, Tissue and Organs* pp 17–22 Ed A Arav. Medimond, Bologna.

Vera-Munoz O, Amirat-Briand L, Bencharif D, Anton M, Desherces S, Shmitt E, Thorin C and Tainturier D (2011) Effect of low-density lipoproteins, spermatozoa concentration and glycerol on functional and motility parameters of bull spermatozoa during storage at 4°C *Asian Journal of Andrology* **13** 281–286.

Vishwanath R (2003) Artificial insemination: the state of the art *Theriogenology* **59** 571–584.

Wakayama T and Yanagimachi R (1998) Development of normal mice from oocytes injected with freeze-dried spermatozoa *Nature Biotechnology* **16** 639–641.

Ward MA, Kaneko T, Kusakabe H, Biggers JD, Whittingham DG and Yanagimachi R (2003) Long-term preservation of mouse spermatozoa after freeze-drying and freezing without cryoprotection *Biology of Reproduction* **69** 2100–2108.

Watson PF (1976) The protection of ram and bull spermatozoa by the low density lipoprotein fraction of egg yolk during storage at 5°C, and deep freezing *Journal of Thermal Biology* **1** 137–141.

Watson PF (1979) The preservation of semen in mammals In *Oxford Reviews of Reproductive Biology Volume 1* pp 283–350 Ed CA Finn. Clarendon Press, Oxford.

Watson PF (1981) The roles of lipid and protein in the protection of ram spermatozoa *Journal of Reproduction and Fertility* **62** 483–492

Watson PF (1990) Artificial insemination and the preservation of semen In *Marshall's Physiology of Reproduction Volume 2* pp 747–869 Ed G Lamming. Churchill Livingstone, Edinburgh, London, New York.

Watson PF (1995) Recent developments and concepts in the cryopreservation of spermatozoa and the assessment of their post-thawing function *Reproduction, Fertility and Development* **7** 871–891.

Watson PF (1996) Cooling of spermatozoa and fertilizing capacity *Reproduction in Domestic Animals* **31** 135–140.

Watson PF and Fuller BJ (2001) Principles of cryopreservation of gametes and embryos In *Cryobanking the Genetic Resource: Wildlife conservation for the future?* pp 23–46 Ed PF Watson and WV Holt. Taylor and Francis, London.

Watson PF and Holt WV Eds (2001) *Cryobanking the Genetic Resource; Wildlife Conservation for the Future?* Taylor and Francis, London.

Wildt DE, Rall WF, Critser JK, Monfort SL and Seal US (1997) Genome resource banks *Bioscience* **47** 689–698.

Woelders H, Zuidberg CA and Hiemstra SJ (2006) Animal genetic resources conservation in the Netherlands and Europe: poultry perspective *Poultry Science* **85** 216–222.

Wolf DE and Voglmayr JK (1984) Diffusion and regionalization in membranes of maturing ram spermatozoa *Journal of Cell Biology* **98** 1678–1684.

Yang H and Tiersch TR (2009) Current status of sperm cryopreservation in biomedical research fish models: zebrafish, medaka, and *Xiphophorus Comparative Biochemistry and Physiology C Toxicology and Pharmacology* **149** 224–232.

Yu I, Songsasen N, Godke RA and Leibo SP (2002) Differences among dogs in response of their spermatozoa to cryopreservation using various cooling and warming rates *Cryobiology* **44** 62–78.

Zhu L and Inaba K (2011) Lipid rafts function in Ca^{2+} signaling responsible for activation of sperm motility and chemotaxis in the ascidian *Ciona intestinalis Molecular Reproduction and Development* 78 920–929.

Zitranski N, Borth H, Ackermann F, Meyer D, Viewig L, Breit A, Gudermann T and Boekhoff I (2010) The "acrosomal synapse": subcellular organization by lipid rafts and scaffolding proteins exhibit high similarities in neurons and mammalian spermatozoa *Communicative and Integrative Biology* **3** 513–521.

6 Evaluation of Semen in the Andrology Laboratory

Steven P. Lorton*
Reproduction Resources, Walworth, Wisconsin, USA

Introduction

Bartlett (1978) formulated success in artificial breeding using an 'Equation of Reproduction'. According to this formula, success as measured by the percentage of animals born is dependent upon four factors: A, % herd detected in oestrus and inseminated; B, % fertility of the herd; C, % fertility of semen; and D, % efficiency of the inseminator/technician. Therefore:

$$\% \text{ born} = A \times B \times C \times D$$

Because the end result is the cumulative product of the factors, and not their arithmetic average, it is extremely important that each factor remains as high as biologically and practically possible in order to achieve optimal results.

In artificial insemination centres, veterinary clinics and/or in semen collection centres, the quality and quantity of the semen collected must be assessed in order to achieve the optimal level of factor C in Bartlett's Equation of Reproduction. This chapter will discuss the techniques that are currently used to achieve this goal. The significance of these techniques in predicting 'fertility' will also be discussed.

Measurement of Ejaculate Volume

Ejaculates vary significantly in volume among species. Table 6.1 lists typical volume characteristics for several species. As a result of this variation in volume, during collection and subsequent evaluation of ejaculates, suitable devices are used to estimate ejaculate volume. For very small volumes, e.g. for the ram, specialized collection tubes are available (Fig. 6.1). Centrifuge tubes of 15 ml capacity are often used for bull and dog ejaculates, while collection bottles or insulated cups are typically used for the higher volume stallion and boar ejaculates. In many cases, these collection devices have volumetric calibrations, although manufacturers do not often provide accuracy data for these calibrations. Many laboratories use a standard laboratory scale to measure the weight of the ejaculate, after making an allowance for the weight of the collection tube or bottle. Volume is equated with the weight in grams. Kaproth (1988) reported a variation to this gravimetric procedure that employed a regression formula to correct for an increase in specific gravity with concentration.

* E-mail: splorton@alumni.clarku.edu

Table 6.1. Ejaculate characteristics. Adapted from Setchell, 1991.

Species	Ejaculate volume (ml)	Sperm concentration ($\times 10^{-6}$/ml)
Bull	2–10	300–2000
Boar	150–200	25–350
Dog	2–15	60–300
Ram	0.5–2	2000–5000
Stallion	20–300	30–800

Fig. 6.1. A collection device used during semen collection from small animals. Photo courtesy of Pacific Vet Pty Ltd, Melbourne, Australia.

Measurement of Spermatozoal Concentration

Measurement of the concentration of spermatozoa in ejaculates is a fundamental and extremely important step in the evaluation and then in the processing of ejaculates for artificial insemination.

The simplest and most inaccurate method to estimate spermatozoal concentration is by means of visual inspection. This conservative processing technique is used effectively when, for example, a minimum number of insemination doses must be produced and when semen availability is not limited. In the case of liquid boar semen, for example, a fresh ejaculate that has the appearance similar to whole milk, i.e. creamy white, might be extended to yield 6–8 insemination doses; this would compare with 20–24 doses if actual concentration measurements were performed. In contrast, a semi-clear, non-fat or skimmed milk-like appearance of an ejaculate might be extended to yield only 2–3 doses.

The Karras Spermiodensimeter has been used for the determination of sperm concentration in boar and bull ejaculates (see Fig. 6.2). In this case, the measurement is based upon the turbidity of a sperm suspension prepared in normal saline. The concentration is visually estimated by interpretation of the instrument's scale through the turbid solution. Clarity of the scale is an indirect indication of spermatozoal concentration. Vianna *et al.* (2007) compared data obtained using this densimeter with data obtained using the Neubauer haemocytometer. These authors noted that data obtained using the densimeter was consistently an overestimation, which was due especially to the variability in volume of seminal fluids. The authors developed a corrected concentration table for use with the densimeter to provide data more consistent with those obtained with the haemocytometer.

The haemocytometer counting chamber has classically been the 'gold standard' for sperm concentration evaluation. Although variations exist depending upon the manufacturer, this chamber usually consists of a glass slide with two etched grids of precise dimensions. A special coverslip is applied on to support 'rails'; thus creating a chamber of a specific volume over each grid. After filling each chamber with a diluted neat or extended semen sample, the user observes the chamber using a standard microscope and manually counts the sperm within the limits of the grid. (See detailed instructions below.)

Various manufacturers and types of counting chambers are available, which include the Petroff-Hauser, the Improved Neubauer and the Makler® chamber (see Fig. 6.3); however, the Improved Neubauer Haemocytometer is used widely in animal andrology laboratories. Several authors (Belding, 1934; Salisbury *et al.*, 1943; Bane, 1952; Freund and Carol, 1964; Lorton *et al.*, 1984; Seaman *et al.*, 1996; Brazil *et al.*, 2004; Christensen *et al.*, 2005; Payne and

Clark, 2005) have described causes of variation of the data obtained with counting chambers. The variation can be attributed to: type of chamber, technician, number of dilutions and the number of chambers counted. Therefore, it is recommended that a *minimum* of four counts, i.e. both sides of two chambers, are made for each diluted subsample. Additional dilutions, chambers and technicians will reduce variation. Counts of the two sides of each chamber should be within 10% of each other. A standardized method of counting should also be used. The evaluation of samples using haemocytometers is time-consuming – approximately 15–20 min/chamber.

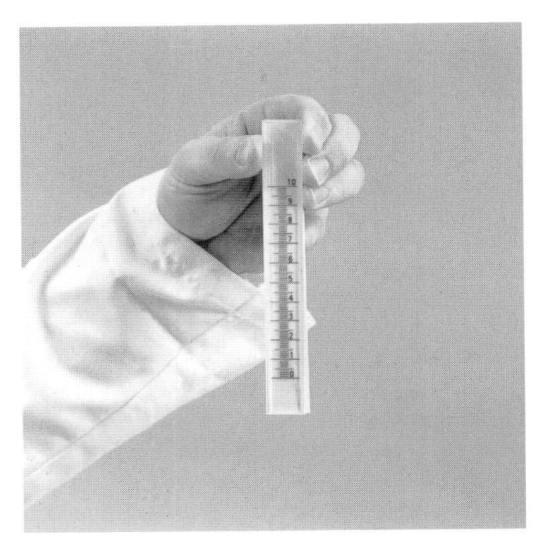

Fig. 6.2. The Karras densimeter. Photo courtesy of Minitüb GmbH Tiefenbach, Germany.

Suggested method for the determination of the concentration of spermatozoa in semen using an Improved Neubauer Haemocytometer

1. Prepare suitable diluted subsamples of the semen. The dilution used is somewhat dependent upon the species being studied. For example, for rams or bulls, ejaculates are typically very concentrated, so a 1:201 dilution could be used. Boar ejaculates are less concentrated, so a 1:101 dilution could be used.

Notes:

(i) It is important that dilutions are properly prepared. The outside surface of sample pipette tips should be carefully wiped to remove adherent cells. Tips should be wiped from the top towards the open end. The open end of the tip should not be touched, as this will result in sample loss by wicking action from inside the tip.

(ii) It is important that dilutions are properly prepared. For example, 0.01 ml (10 μl) sample plus 1.0 ml diluent results in a 1:101 dilution; 0.01ml (10 μl) sample plus 2.0 ml diluent results in a 1:201 dilution.

(iii) For accurate counts, a diluent that results in immotile spermatozoa should be used, e.g. 10% formaldehyde in saline (10 ml of 40% formaldehyde in saline).

(iv) Prepare a dilution that results in a count of approximately 150 spermatozoa in 5 large squares of the counting chamber.

2. Properly position the haemocytometer coverslip on the haemocytometer. After thoroughly mixing the diluted subsample, a drop of semen is placed on the filling 'V' groove (if present)

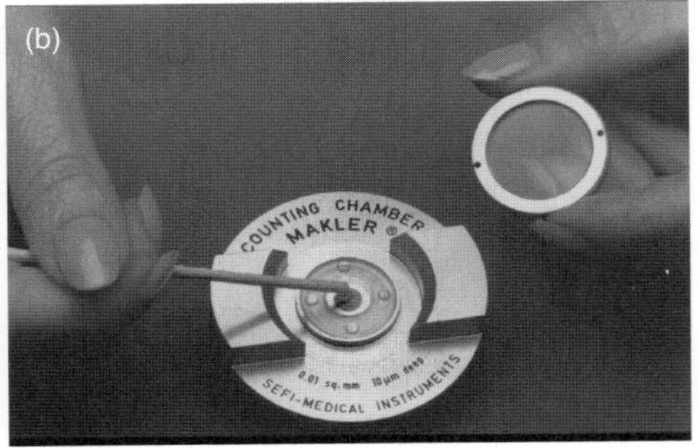

Fig. 6.3. (a) Improved Neubauer Haemocytometer (photo by L. Lorton). (b) Makler® Counting Chamber (photo courtesy of A. Makler, Sefi Medical Instruments, Haifa, Israel).

of the haemocytometer – or it is placed at the intersection of the cover slip and top of the haemocytometer. The liquid should be drawn up to just cover/fill completely the one 'side' of the chamber. This procedure is repeated for the other side of the haemocytometer.

Note: Overfilling or underfilling of the chamber will result in inaccuracy of concentration measurements.

3. Wait approximately 5–10 min for the sperm to settle to the bottom of the haemocytometer. *Do not* allow the haemocytometer to dry out before counting. In order to be sure that the haemocytometer does not dry out, the following procedure can be followed. Place a paper towel large enough to hold the haemocytometer in the bottom of a dish and wet the towel. Then place two microscope slides on the wet paper towel as a base, positioned so that the haemocytometer can be placed on the slides above the wet towel. The slides prevent the bottom of the haemocytometer from getting wet, and avoid any need to manually dry the chamber, as doing so may dislodge the haemocytometer coverslip. Cover the dish and allow the necessary time for the cells to settle.

4. The haemocytometer consists of a grid system. Five large squares should be counted – either each of the corners and the centre square, or from corner to corner diagonally (see Figs 6.4 and 6.5).

Counting suggestions:

(i) Count the sperm heads only, ignore the sperm tails.

(ii) Establish a pattern to use every time.

(iii) Count the sperm heads touching two of the sides of the large squares, e.g. left and bottom. Do not count the sperm heads touching the remaining two sides, e.g. right and top.

(iv) Total counts for the two sides of each individual chamber should be within 10% of each other. If not, these counts should be discarded and another chamber prepared and counted.

5. Add the counts from all five squares of one side of the haemocytometer as one total count, and then count the other side. It is recommended to count at a minimum four sides, using two haemocytometers for each semen sample under evaluation.

6. Calculate the total number and concentration of spermatozoa as follows (steps A–D):

A. Average the haemocytometer counts.

B. Multiply the average of the counts by the dilution factor.

C. Multiply the result of step 'B' by 50,000 (5×10^4). This result is the number of cells/ml ejaculate. (Note: the factor 50,000 is used only if five large squares of the chamber are counted.)

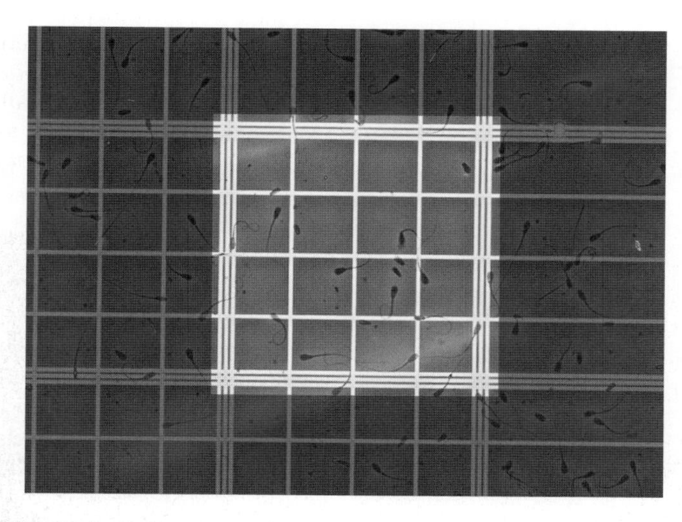

Fig. 6.4. Phase micrograph of a large square (containing 16 small squares in a 4×4 pattern) of the Improved Neubauer Haemocytometer counting grid. Photo courtesy of B. Hasell and the Artificial Breeding Centre, Beef Breeding Services, Primary Industries and Fisheries, Department of Employment, Economic Development and Innovation, Queensland, Australia.

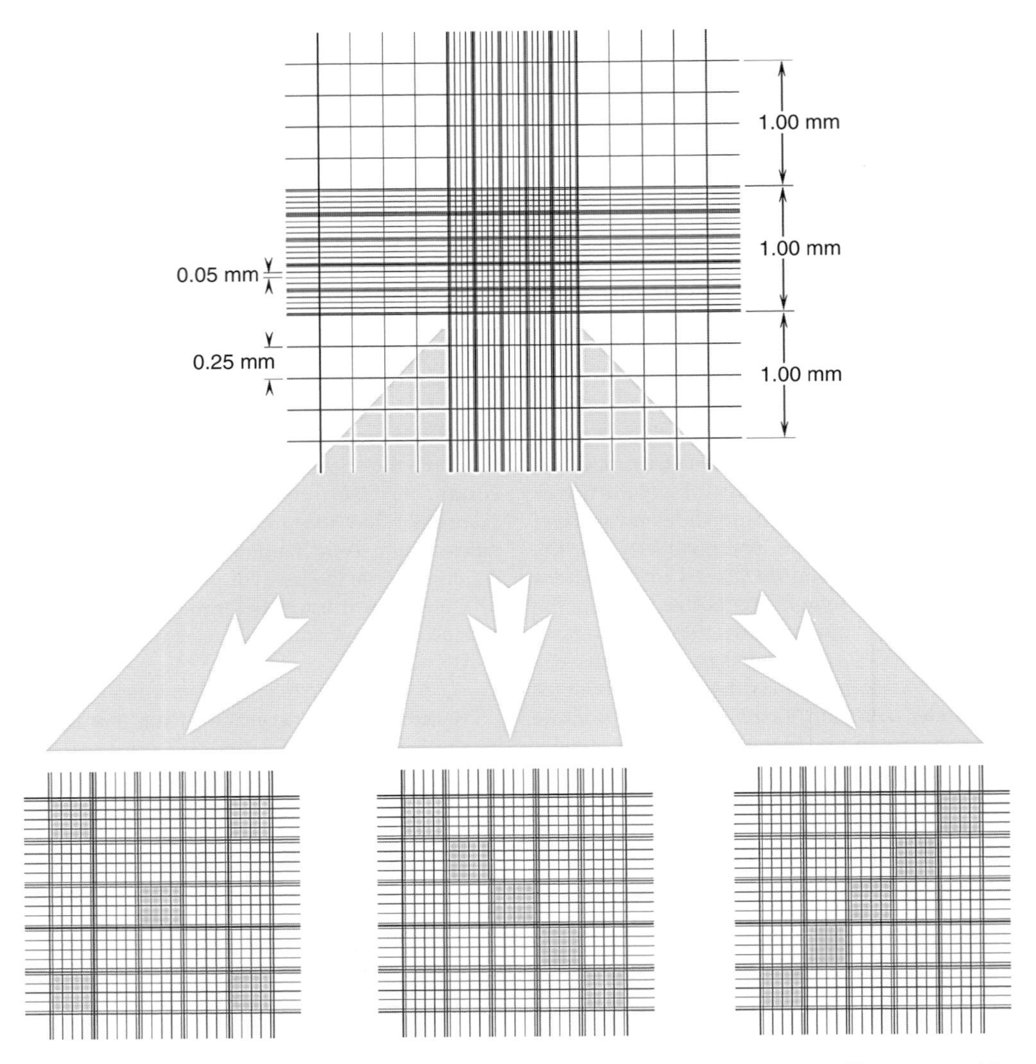

Fig. 6.5. Suggested counting techniques using the Improved Neubauer Haemocytometer. Three acceptable counting patterns are shown. Using one of the three patterns outlined in grey at the bottom, counting the spermatozoa located in the five large squares (16 small squares/large square) will result in a valid representation of the entire chamber. Figure by R. DeMarco, Reproduction Resources.

D. Multiply the result of step 'C' by the volume of the ejaculate. This result is the total number of sperm in the ejaculate.

Example:
Ejaculate volume = 10 ml
Total counts in each of the five large squares are: 161, 168, 164, 167 after a dilution of 101 (counts from two chambers, two sides each).
A. $(161 + 168 + 164 + 167) \div 4 = 660 \div 4 = 165$
B. $165 \times 101 = 16665 = 16.7 \times 10^3$
C. $(16.7 \times 10^3) \times 50{,}000 = (16.7 \times 10^3) \times (5 \times 10^4) = 8.4 \times 10^8$ = no. cells/ml
D. $8.4 \times 10^8 \times 10$ ml $= 8.4 \times 10^9$ = 8.4 billion sperm total in the ejaculate.

7. Clean the haemocytometer and coverslip carefully.

Note: These are coverslips unique to the haemocytometer and cannot be replaced with an 'ordinary' microscope coverslip.

Colorimeters, photometers and spectrophotometers have been widely used in semen production facilities owing to their ease and speed of use. Comstock and Green (1939) described the measurement of light transmission through ram sperm suspensions using the photoelectric colorimeter and suggested that data obtained by this method would be comparable to the mean of two or three haemocytometer counts. Salisbury *et al.* (1943) compared the use of the haemocytometer and the colorimeter using bull semen. Using raw semen samples diluted 1:11 in 0.067 M sodium citrate, these authors reported that a 'straight line relationship existed between the Log (100/reading of sample) and the mean of duplicate haemocytometer counts'. Emik and Sidwell (1947) indicated that pipettes, sample turbidity and, especially debris, contributed to technique error. Type of colorimeter used, and semen species (i.e. concentration of ejaculates), influenced the optimal dilution rate. Willett and Buckner (1951) and Young *et al.* (1960) used regression analysis to estimate the number of sperm/ml using absorption (OD, optical density) measurements, where OD = (2 – log %T), and T is the transmittance. The principles, calibration of and procedures for the photometric measurement of sperm cell concentration in raw (neat) semen have been reviewed – for bovine semen, by Foote (1968, 1972) and Foote *et al.* (1978), and for equine semen by Pickett (1968).

Improved Neubauer Haemocytometer maths

- Counting chamber depth: 0.100 mm.
- Ruling pattern: 1/400 square mm.
- Rulings cover 9 mm^2 (9 squares of 1 × 1mm).
- The centre square is subdivided into 25 large squares separated by triple lines.
- Each large square is subdivided into 16 small squares of 1/400 mm^2.
- The chamber height is 1/10mm, so the volume over each small square is 1/4000 mm^3 or 0.00025 mm^3.

- If 'N' is the number of cells counted in five large (80 small) squares:
- Then: N = no. of cells in 80/4000 mm^3.
- So in 1 mm^3, there are 4000 (N)/80 cells.
- Therefore, in 1 ml, there are 4,000,000 (N)/80 cells.
- If 'D' is the dilution factor, then the number of cells/ml is: 4,000,000 (N) (D)/80 or (5×10^4) (N) (D).

Lorton (1986) described a technique for evaluating initially extended semen using a spectrophotometer. Photometric estimations of sperm concentration typically are simple and relatively quick (30–60 s). Photometers/spectrophotometers must be calibrated, i.e. to give OD versus sperm concentration, when purchased, as well as by periodic checks and after changing the light source. Calibrations change over time, especially in machines equipped with incandescent light sources, although they vary to a lesser degree in machines with light emitting diodes (LED) as the light source. Foote *et al.* (1978) reviewed calibration techniques when using haemocytometer counts. Calibrations must be made for each individual species of sperm under study. The NucleoCounter® SP-100™ (see below) provides more repeatable and more accurate data than haemocytometers and can be used for calibrations instead. The electronic function of photometers/spectrophotometers should be evaluated frequently (daily, weekly) with turbidity standards and/or filters.

Evaluation of concentrations can be made using the flow cytometer (Evenson *et al.*, 1993). Within these machines, individual cells pass in a liquid fluid stream through a laser and are electronically counted. 'Gates' based upon cell size can be established in order to reduce error from debris and cells other than sperm (COTS), and extremely accurate data can be obtained. However, flow cytometers are expensive and relatively difficult to maintain and operate. Few semen production laboratories utilize them for concentration measurements alone, though they are used for more complex morphological and functional analyses (see below).

The NucleoCounter® SP-100™ (ChemoMetec A/S, Allerod, Denmark) is a relatively new device for the measurement of sperm concentration in raw (neat) ejaculates and/or extended semen. After a simple dilution with a reagent that results in disintegration of the sperm cell membranes, the sample (50 μl) is loaded into a special cassette (Fig. 6.6), which is manufactured with the fluorescent staining agent propidium iodide (PI) immobilized within it. Due to the absence/disruption of the cell membranes, the PI can be incorporated into the spermatozoa and selectively binds to the DNA of each cell. The microscope of the SP-100™ utilizes a green light source that causes the PI-DNA complexes to fluoresce red (Fig. 6.7), thus allowing the machine's processing unit to count each complex. Algorithms eliminate COTS, invalid data due to dilution error and/or cassette loading error.

Due to the specificity of the PI-DNA binding and fluorescence, extender, seminal gel, COTS and other debris do not interfere with data analyses. Hansen *et al.* (2006) compared the use of the SP-100™ with the haemocytometer, the spectrophotometer, computer assisted semen analysis (CASA) machines and the flow cytometer (FACS). Repeatability for the SP-100™ and the FACS were similar, at 3.1% and 2.7%, respectively, while it was 7.1%, 10.4%, 8.1% and 5.4% for the haemocytometer, the Corning 254 photometer, and two models of CASA machines. Correlations between the FACS and the SP-100™ were highest (as was the repeatability of these instruments). Figure 6.8 presents examples of regressions of SP-100™ data versus flow cytometer data. Further validations of the SP-100™ instrument for measurement of sperm concentration were done by Comerford *et al.* (2008) and by Comerford (2009).

Anzar *et al.* (2009) concluded that the Nucleocounter® SP-100™ and the flow cytometer could be used with equal confidence for the measurement of sperm concentration. These authors also provided evidence that 13% additional bull artificial insemination doses could be produced when using the SP-100™ rather than haemocytometers for the calibration of spectrophotometers. The SP-100™ has become the instrument of choice

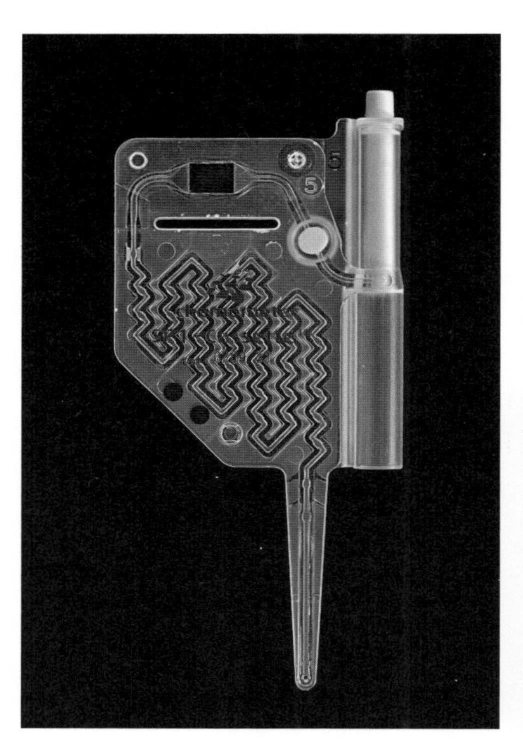

Fig. 6.6. The NucleoCounter® SP-100™ cassette. Note the two black dots on the cassette. During manufacture, the depth of the analysis chamber (upper centre of the cassette) is measured using a laser. The number and location of the black dots on each cassette are evaluated before concentration analysis and represent a final calibration of sample volume in the analysis chamber. Photo by L. Lorton.

Fig. 6.7. The analysis window of the NucleoCounter® SP-100™ as 'seen' by the microscope. Each white dot represents a propidium iodide (PI)–DNA complex, i.e. a sperm cell. Photo courtesy of ChemoMetec A/S, Allerod, Denmark.

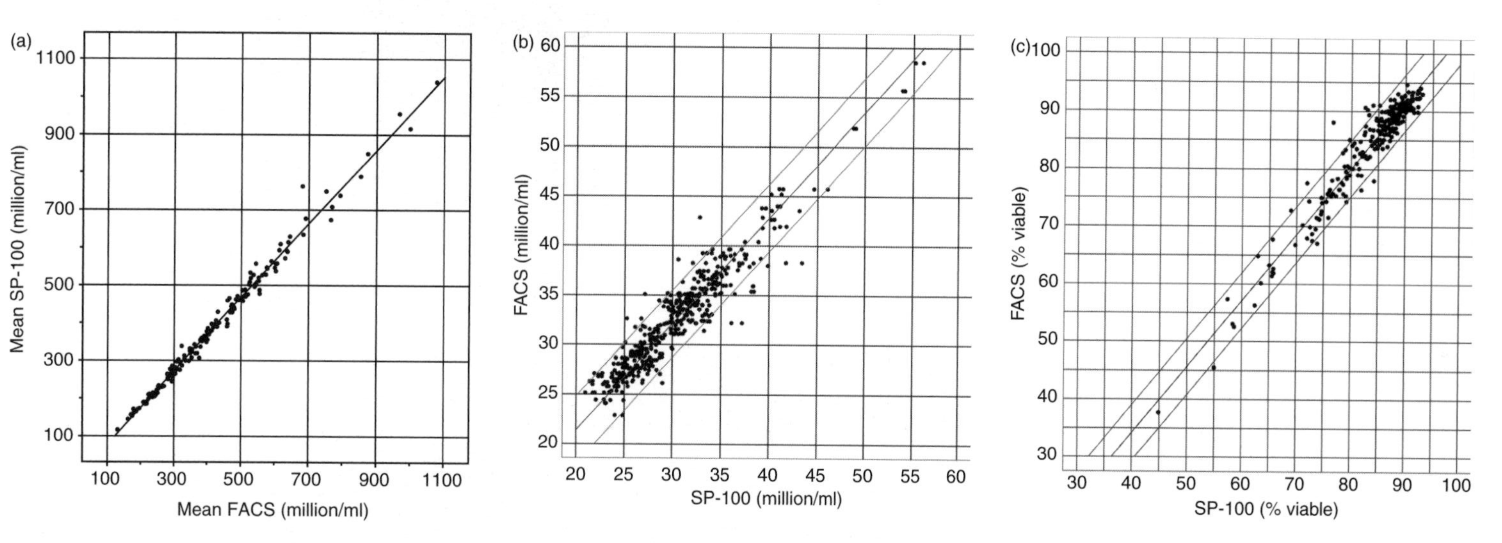

Fig. 6.8. NucleoCounter® SP-100™ versus flow cytometer (FACS) data. (a) Sperm concentration of raw (neat) boar semen. (b) Sperm concentration of extended boar semen. (c) Boar sperm membrane integrity (% viability). The outer lines parallel to the central regression lines in the last two figures indicate the 95% confidence limits for the regressions. Figure courtesy of C. Hansen, Danish Pig Production, Copenhagen, Denmark.

for evaluating sperm concentration in neat and extended semen from several species, including stallions (Comerford *et al.*, 2008; Morrell *et al.*, 2010), bulls (DeJarnette and Lefevre, 2008; S. Lorton, unpublished), pigs, dogs, sheep, deer, wolves (S. Lorton, unpublished) and trout (Nynca and Ciereszko, 2009). Owing to the accuracy and repeatability of the Nucleocounter® SP-100™, it should be considered as the '*new* gold standard' for the measurement of sperm concentration.

Artificial insemination provides the mechanism for the 'extended' use of the male animal, i.e. rather than one natural insemination, multiple inseminations are possible with one collected ejaculate. Sullivan (1970) reviewed early data showing an increase in fertility (non-return rate) with increasing sperm numbers inseminated. The change in fertility was different between bulls of low and high fertility, and, although insignificant due to low numbers in the studies, a decrease in non-return rate was noted after insemination with very high numbers of motile sperm cells. Pace *et al.* (1981) provided evidence of an exponential, asymptotic model relating the number of post-thaw viable spermatozoa that was positively and significantly correlated with the non-return rate. This model confirmed a similar model originally hypothesized by Salisbury and VanDemark in 1961 (see Saacke, 1983).

A later study (S.P. Lorton and M.M. Pace, unpublished) involving approximately 20,000 first inseminations also verified the model and provided additional evidence that the asymptotic value for individual bulls differed significantly. However, it was shown that, across bulls, approximately 10–12 × 10^6 motile spermatozoa per insemination dose resulted in the asymptotic level of fertility (non-return rate). Den Daas *et al.* (1998) provided similar results and also discussed the variability between bulls. Foote and Kaproth (1997) presented evidence that the total sperm number per artificial insemination dose could be reduced to 10 × 10^6 with only a one percentage point reduction in non-return rate. Further, Chenoweth *et al.* (2010) described data showing a strong correlation between the number of progressively motile spermatozoa per insemination dose (1.5, 3.5 and 7.0 × 10^6) and palpated pregnancy rate. A specialized process developed in New Zealand using Caprogen extender and sperm storage at ambient (15–23°C) temperatures results in optimal fertility at levels with approximately 1.5 × 10^6 total (1.0–1.2 × 10^6 motile) spermatozoa per insemination (Shannon *et al.*, 1984). The number of motile spermatozoa inseminated is, therefore, a compensable semen trait. (See Saacke *et al.*, 1991 for a review of compensable and non-compensable semen traits.)

There have been similar, but less extensive, sperm number titration studies in the pig. Baker *et al.* (1968) demonstrated an interaction between the number of sperm inseminated and the volume inseminated. By examining the number of spermatozoa recovered from oviducts of gilts and the number of spermatozoa bound to the zonae pellucidae of ova, these authors concluded that 5–10 × 10^9 total spermatozoa in a volume of approximately 100 ml resulted in optimum fertility. Johnson *et al.* (1981, 1982) demonstrated significant differences among boars in farrowing rate after insemination with 3 × 10^9 total sperm. In contrast, Xu *et al.* (1998) were unable to demonstrate a difference in farrowing rate, while there was a significant difference in total live-born, when the insemination dose was reduced from 3 × 10^9 to 2 × 10^9 total sperm. Flowers (2002) reported a linear relationship of litter size among some boars (32 of 200 boars) versus an asymptotic relationship among most boars (133 of 200 boars) when the number of sperm inseminated was increased from 1 × 10^9 to 9 × 10^9. The authors commented that if additional motile spermatozoa were inseminated, a plateau effect may also have been detected for the animals showing the linear relationship. In contrast, Alm *et al.* (2006) reported a significant decrease in 60-day non-return rate and litter size when the insemination dose decreased from 3 × 10^9 to 2 × 10^9 total sperm. Reicks *et al.* (2008) reported a decrease in litter size when sperm number inseminated was reduced from 4 × 10^9 to 2.5 × 10^9 total sperm, while farrowing rate did not differ. These studies were all performed using 'conventional' cervical artificial insemination (AI) with chilled semen.

There seems to be ample evidence, at least in the species studied, and probably in others, that the following is the case. First, the number of motile, normal spermatozoa inseminated is compensable. Secondly, when performing studies to investigate treatment effects on spermatozoa and the desired end point is a difference in fertility (e.g. non-return rate, pregnancy rate, litter size, etc.), it is critical both to inseminate an accurate number of motile, normal spermatozoa and that the number inseminated should be on the 'upward slope' of the fertility curve (i.e. less than the number when the fertility pattern is asymptotic). Thirdly, the number of inseminations (sample size) per treatment must be adequate to allow the detection of treatment differences; for example, in cattle, one standard deviation around a non-return rate of 70% is 1.5 with 1000 inseminations, while it is 10.2 with 20 inseminations (Oltenacu and Foote, 1976) (see also Pace, 1980; Berndtson, 1991; Amann and Hammerstedt, 2002; Amann, 2005). These details should also be considered when evaluating the literature.

More recently, post-cervical and deep uterine surgical insemination studies in pigs have demonstrated that both insemination volume and number of sperm can be reduced without a reduction in pregnancy rate or litter size, e.g. from 3–3.5 × 10^9 total sperm in 75–80 ml to approximately 1–1.5 × 10^9 total sperm in a volume of 35–40 ml (post-cervical insemination), or even 0.5 ml (deep uterine insemination, using fresh or frozen–thawed semen) (Krueger and Rath, 2000; Rozeboom *et al.*, 2004; Wongtawan *et al.*, 2006; Vazquez *et al.*, 2008 (review); Sonderman *et al.*, 2011).

Evaluation of Spermatozoal Morphology

Our brains, via the eyes, visualize objects and images using colours and/or variations of the grey scale to establish contrast. Using the microscope, we can visualize extremely small objects, many of which have no distinguishable colour or, therefore, contrast. For example, when viewing spermatozoa with the routine bright microscope, the cells appear white with a white/grey background. A wide variety of biological stains and probes are often used to increase the contrast (see Table 6.2), or to identify particular regions or cell structures, often using fluorescent techniques. Many of these stains or probes have recently been used with the flow cytometer to evaluate sperm morphology, cell structures and/or membrane integrity (see Plates 4 and 5). Use of the flow cytometer has the distinct advantage of quickly analysing thousands of cells, compared with only 100–200 cells with the microscope.

The optics of a phase microscope modify the light waves hitting and surrounding the viewed objects, e.g. spermatozoa, to create additional specimen contrast, without the use of biological stains. The result is increased ability by the user to see the objects and differentiate detail. (See Abramowitz (1987) for a detailed discussion of the physics of phase contrast microscopy.) The optics of the phase microscope must be properly set up (aligned) in order to establish the phase image. A description is given below of the suggested method for aligning a phase microscope. Note that this procedure will vary slightly by microscope manufacturer and model.

Assembly, set-up and use of the phase microscope system

The primary components of the phase optical system are the phase turret condenser, the phase objective lenses and the centring telescope (see Fig. 6.9). Note that some microscope manufacturers incorporate the centring device into the body of the microscope and therefore eliminate the use of a separate centring telescope.

Assembly and set up procedure

1. Rotate the coarse focusing knob to move the stage platform to its lowest position.
2. Remove the objective dust caps from the underside of the revolving nosepiece.
3. Screw individual objective lenses into the nosepiece. For convenience, mount the objectives in magnification order i.e. 10×, 20×, 40×, 100×.

Table 6.2. Biological stains and probes used to evaluate spermatozoa (see also Appendix at end of this chapter).

Function	Stain/probe	Reference
General morphology	Alkaline methyl-violet	Hackett and Macpherson, 1965
	Anti-ubiquitin–FITC (fluorescein isothiocyanate)/TRITC (thiol-reactive tetramethylrhodamine-5-(and-6)-isothiocyanate)	Kuster *et al.*, 2004a; Odhiambo *et al.*, 2011
	Belling's iron aceto-carmine	Hackett and Macpherson, 1965
	Casarett stain	Casarett, 1953
	Diff-Quik/Dip Quick®	Root Kustritz *et al.*, 1998; Boersma *et al.*, 2001; Pozor *et al.*, 2012
	Eosin–nigrosin	Hackett and Macpherson, 1965; Brito, 2007; Freneau *et al.*, 2010
	Farelly	Paufler, 1974; Boersma *et al.*, 1999, 2001
	Giemsa	Hancock, 1952; Hackett and Macpherson, 1965
	Harris' haematoxylin	Boersma *et al.*, 2001
	Heidenhain's iron haematoxylin	Hancock, 1952; Hackett and Macpherson, 1965
	India ink	Hackett and Macpherson, 1965
	Papanicolaou	Boersma *et al.*, 1999, 2001
	SpermBlue®	Van der Horst and Maree, 2009
	Williams stain (carbolfuchsin–alkaline methylene blue)	Williams, 1920
General morphology and acrosome	Spermac®	Oettle, 1986; Boersma *et al.*, 2001; Baran *et al.*, 2004
Acrosome	Anti-acrosin–FITC	Garner *et al.*, 1975; Thomas *et al.*, 1997
	Chicago sky blue–Giemsa	Kútvölgyi *et al.*, 2006
	Chlortetracycline (CTC)	Gillan *et al.*, 1997; Maxwell and Johnson, 1997; Green and Watson, 2001; Rathi *et al.*, 2001
	Fast Green FCF–eosin B	Wells and Awa, 1970
	Giemsa	Watson, 1975
	Lectin conjugates – PSA (*Pisum sativum* agglutinin), PNA (peanut agglutinin)	Garner *et al.*, 1986; Graham *et al.*, 1990; Cheng *et al.*, 1996; Gillan *et al.*, 1997; Rathi *et al.*, 2001; Herrera *et al.*, 2002; Nagy *et al.*, 2003; Waterhouse *et al.*, 2006; Celeghini *et al.*, 2007; de Graaf *et al.*, 2007; Garcia-Macias *et al.*, 2007; Petrunkina and Harrison, 2010; Odhiambo *et al.*, 2011; Waberski *et al.*, 2011; Cheuquemán *et al.*, 2012
	Lysotracker Green DND-26 (LYSO-G)	Thomas *et al.*, 1997, 1998
	Merocyanine 540	Rathi *et al.*, 2001; Gadella and Harrison, 2002; Kavak *et al.*, 2003; Peña *et al.*, 2004b; Garcia-Macias *et al.*, 2007
	Naphthol yellow S–aniline blue	Christensen *et al.*, 1994
	Naphthol yellow S–erythrosin b	Christensen *et al.*, 1994
	Nigrosin-eosin-Giemsa	Tamuli and Watson, 1994

Continued

Table 6.2. Continued.

Function	Stain/probe	Reference
Calcium influx/ acrosome reaction	Fluo-3	Parrish *et al.*, 1999; Green and Watson, 2001; Petrunkina *et al.*, 2001; Peña *et al.*, 2004a; Okazaki *et al.*, 2011
	Fura-2	Zhou *et al.*, 1990
DNA	Fuelgen	Hackett and Macpherson, 1965; Ball and Mohammed, 1995
DNA integrity (Comet)	Electrophoresis assay/ethidium bromide/SYBR® Green	Linfor and Meyers, 2002; Fraser and Strzezek, 2004, 2005; Boe-Hansen *et al.*, 2005a
DNA integrity (FISH)	Cy3 probe, DAPI (4′,6-diamidino-2-phenylindole)	Fernández *et al.*, 2003
DNA integrity (SCD – Sperm Chromatin Dispersion test)	DAPI/propidium iodide/ Wright's stain	Fernández *et al.*, 2003; Garcia-Macias *et al.*, 2007
DNA integrity (SCSA® – Sperm Chromatin Structure Assay)	Acridine orange	Ballachey *et al.*, 1987, 1988; Evenson *et al.*, 1994; Gadella and Harrison, 2002; Boe-Hansen *et al.*, 2005b, 2008; Love *et al.*, 2005; Waterhouse *et al.*, 2006; Didion *et al.*, 2009; Teague *et al.*, 2010; Tsakmakidis *et al.*, 2010; Waberski *et al.*, 2011
DNA integrity (TUNEL)	dUTP nick end labelling/ fluorescein/FITC-streptavidin/ propidium iodide	Gadella and Harrison, 2002; Waterhouse *et al.*, 2006; Teague *et al.*, 2010
Mitochondrial function	JC-1	Shaffer and Almquist, 1948; Gravance *et al.*, 2000; Celeghini *et al.*, 2007; Garcia-Macias *et al.*, 2007; Brinsko *et al.*, 2011b; Cheuquemán *et al.*, 2012
	MitoTracker Red®	Gadella and Harrison, 2002; Waterhouse *et al.*, 2006; Celeghini *et al.*, 2007
	MitoTracker Green FA	Shaffer and Almquist, 1948
	Rhodamine 123	Shaffer and Almquist, 1948; Celeghini *et al.*, 2007; de Graaf *et al.*, 2007
Viability/membrane integrity	Annexin-V–FITC	Anzar *et al.*, 2002; Gadella and Harrison, 2002; Peña *et al.*, 2003; Cheuquemán *et al.*, 2012
	Bromophenol blue	Boguth, 1951
	Calcein acetylmethyl ester (CAM)–ethidium homodimer-1 (EH)	Althouse and Hopkins, 1995; Juonala *et al.*, 1999
	Carboxy(methyl) fluorescein diacetate (CFDA)	Garner *et al.*, 1986; Harrison and Vickers, 1990; Donoghue *et al.*, 1995
	Chicago sky blue–Giemsa	Kútvölgyi *et al.*, 2006
	Congo red–nigrosin	Hackett and Macpherson, 1965
	Eosin B–aniline blue	Hackett and Macpherson, 1965; Garner *et al.*, 1997
	Eosin–nigrosin	Hackett and Macpherson, 1965; Graham *et al.*, 1990
	Erythrosin–nigrosin	Hackett and Macpherson, 1965
	Ethidium homodimer-1 (EH)	de Graaf *et al.*, 2007
	Fast Green FCF–eosin B	Hackett and Macpherson, 1965; Wells and Awa, 1970; Aalseth and Saacke, 1986

Continued

Table 6.2. Continued.

Function	Stain/probe	Reference
	Fluorescein diacetate	Garner *et al.*, 1986
	Hoechst 33258	Garner *et al.*, 1986; De Leeuw *et al.*, 1991; Casey *et al.*, 1993; Juonala *et al.*, 1999; Herrera *et al.*, 2002
	Hoechst 33342	Keeler *et al.*, 1983; Celeghini *et al.*, 2007
	Nigrosin–eosin	Hancock, 1952
	Nigrosin–eosin–Giemsa	Tamuli and Watson, 1994
	Propidium iodide (PI)	Graham *el al.*, 1990; Harrison and Vickers, 1990; Papaioannou *et al.*, 1997; Celeghini *el al.*, 2007
	Propidium iodide (PI)–CFDA	Garner *et al.*, 1986; Almid and Johnson, 1988; Tamuli and Watson, 1994; Magistrini *et al.*, 1997
	SNARF-1–PI–FITC-PSA	Peña *et al.*, 1999; Kavak *et al.*, 2003
	SYBR-14/Propidium iodide (PI)	Garner *et al.*, 1994; Garner and Johnson, 1995; Magistrini *et al.*, 1997; Nagy *et al.*, 2003; Garcia-Macias *et al.*, 2007; Boe-Hansen *et al.*, 2008; Petrunkina and Harrison, 2010; Brinsko *et al.*, 2011b; Cheuquemán *et al.*, 2012
	SYTO® 17	Thomas *et al.*, 1997
	Trypan blue–Giemsa	Hackett and Macpherson, 1965; Didion and Graves, 1988; Kovács and Foote, 1992
	YoPro-1	Rathi *et al.*, 2001; Gadella and Harrison, 2002; Kavak *et al.*, 2003; Peña *et al.*, 2004b; Waterhouse *et al.*, 2006

Fig. 6.9. The components of the phase microscope system: phase turret condenser (centre), phase objective lenses (four, bottom) and centring telescope (top right).

4. With the 10× lens in the forward 'use' position', rotate the coarse focusing knob to move the stage platform to its highest position.
5. Loosen the knurled condenser locking screw on the side of the condenser mounting ring (see Figs 6.10 and 6.11).
6. Insert the phase condenser into its mounting ring. The top sleeve of the condenser should be inserted completely 'up' into the mounting ring.
7. Tighten the knurled locking screw to secure the condenser in position.
8. Place a specimen drop on a microscope slide and cover with a coverslip.
9. Place the slide on the microscope stage and adjust the stage so that the edge of cover glass is under the 10× lens.
10. Rotate the phase adjustment control of the condenser so that the bright-field (BF) setting is seen at the front.
11. Close the condenser diaphragm using the control lever on the bottom side of the condenser (see Fig. 6.12).
12. Focus on the edge of the coverslip.
13. Move the slide so that the coverslip is approximately centred under the objective.
14. Remove one eyepiece (ocular) and insert the phase centring telescope.
15. Rotate the condenser adjustment control so that the '10' position is in the forward position.
16. Loosen the knurled locking screw on the side of the centring telescope.

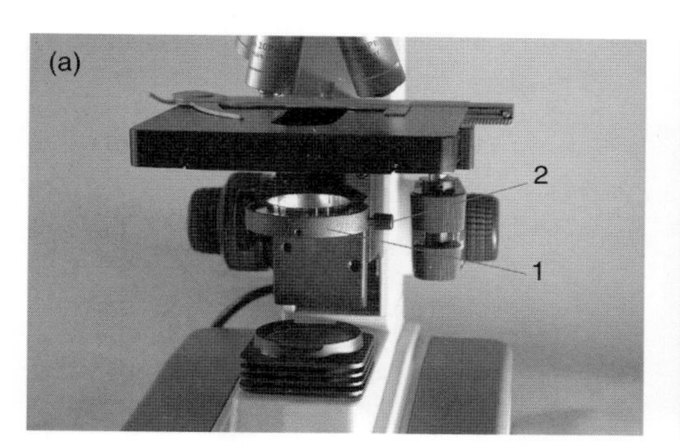

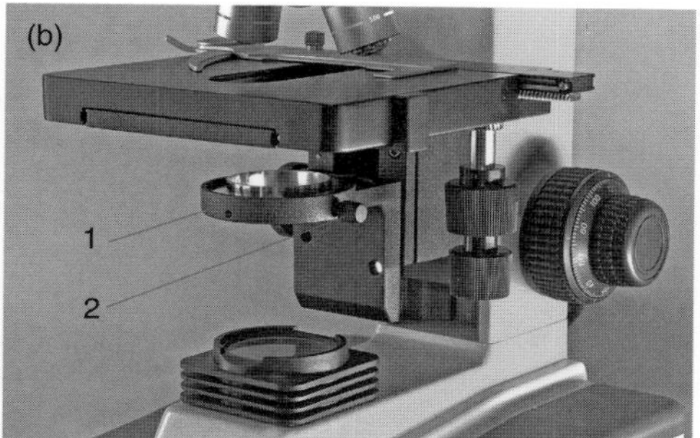

Fig. 6.10. Two views of the phase microscope, both showing the condenser mounting ring (1) and the locking screw (2).

Fig. 6.11. The phase microscope showing the condenser (3) in the mounting ring (1). The locking screw (2) and the condenser diaphragm (4) are also shown.

17. Hold the knurled screw of the centring telescope with one hand, grasp the top of the centring telescope with the other hand and look into the telescope. Slide or rotate the inner sleeve of the centring telescope up or down until the phase rings (round circles) are clearly in focus. Two rings, one black and one white, should be seen. Tighten the telescope knurled locking screw.

18. While looking into the centring telescope, push inwards with each hand the two condenser centring screws that extend outwards from each side of the phase condenser. When completely pushed in, these screws engage centring screws within the condenser assembly.

19. Continue to look into the centring telescope and turn one of the condenser centring screws clockwise while turning the other centring screw counterclockwise. Small adjustments should be made. Note the movement of the white phase ring of the condenser (the black phase ring is actually located in the

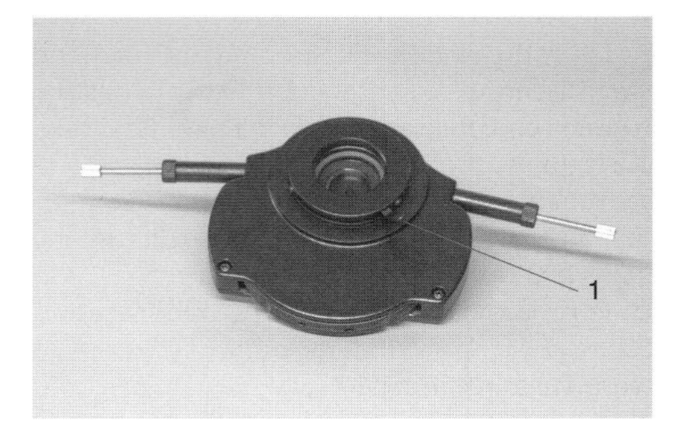

Fig. 6.12. The phase condenser of the phase microscope with the diaphragm control lever (1) shown upside down.

Fig. 6.13. Phase rings of the phase microscope: left – not aligned; right – aligned. Figure by L. Lorton.

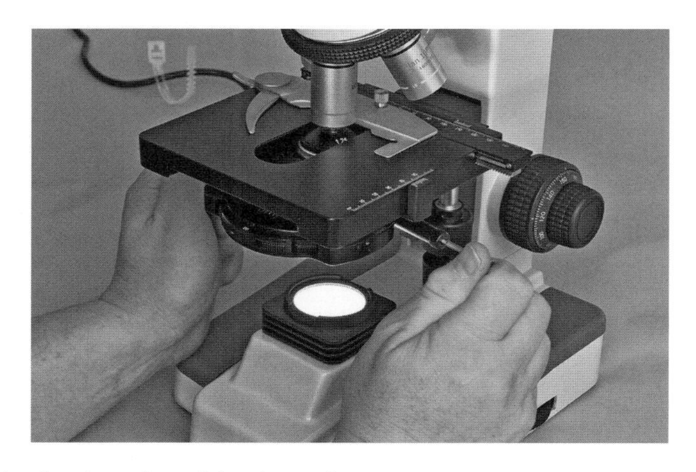

Fig. 6.14. Aligning the phase rings of the phase microscope.

objective and is not moveable). Continue to turn both centring screws in small increments until the white ring is centred directly over the black ring (see Figs 6.13 and 6.14).

20. Disengage the two condenser centring screws by allowing/pulling both to retract to their 'out' position. *Do not* attempt to rotate the condenser adjustment control while the condenser centring screws are in their 'in' position.

21. Rotate the condenser control to position '20' and then rotate the 20× objective to the 'use' position. Repeat the phase rings centring procedure (steps 18–20), if necessary.

22. Repeat the centring steps for the 40× (condenser position '40') and 100× (condenser position '100') objectives, if necessary.

23. Remove the centring telescope and replace with the eyepiece. The microscope is now ready for use.

Note: When the condenser is in the phase positions (10, 20, 40 or 100), the condenser diaphragm should be in the completely open position. When using bright-field microscopy, e.g. with stained specimens, rotate the condenser control to position 'BF' and adjust the diaphragm as desired.

Specimen observation

1. Prepare a slide of the semen sample with a coverslip and place on the microscope stage.

2. Position the slide such that the edge of the coverslip is centred under the 10× objective, with the condenser control positioned at '10'.

3. Focus on the edge of the coverslip.

4. Using the stage movement controls, move the slide so that the centre of the coverslip is approximately under the objective lens.

5. Focus the image, primarily using the 'fine' focus knob.

6. Change the magnification by turning the nosepiece to the desired objective. Set the condenser control to the setting matching the objective.

7. When using phase contrast lenses, the green filter may be fitted on to the recess of the light housing or in a filter holder under the condenser. A blue filter is often employed when using bright-field microscopy, but this may also be used for alternative effects during phase microscopy.

Differential interference contrast (DIC) microscopy is another technique used to enhance contrast of objects using a light microscope. DIC provides an appearance of the 3D structure of viewed objects. The physics of this technique has also been reviewed by Abramowitz (1987). The techniques to set up DIC differ by microscope design and model, are quite complicated, and therefore will not be summarized here. The reader is referred to documentation accompanying individual microscopes.

Wet smears using phase or DIC microscopy may be performed on spermatozoa that are unfixed (Saacke and Marshall, 1968) or fixed with gluteraldehyde or formaldehyde (Hancock, 1957; Pursel and Johnson, 1974; Johnson *et al.*, 1976; Barth and Oko,1989; Richardson *et al.*, 1992), or on spermatozoa that have had their motility inhibited with sodium fluoride (Pursel *et al.*, 1972, 1974). Potential artefacts associated with the staining and preparation of smears are avoided by the use of wet mounts.

Morphological evaluations are performed at 400–1000× magnification, e.g. 40× or 100× (oil) objectives and 10× oculars. Several morphological classification systems have been used. In some laboratories, morphological abnormalities are classified as being 'primary' (originating in the testes), 'secondary' (originating during epidydimal passage) and 'tertiary' (occurring post ejaculation). Blom (1972), as referenced by Barth and Oko (1989), suggested a 'major' and 'minor' classification scheme, according to the importance of the defect with respect to fertility. With these schemes, however, it is often difficult or impossible to classify a particular defect. Often, the origin of the defect is unknown, its importance is unclear, and/or spermatozoa may exhibit more than one defect. A simpler classification may then be used, as follows: abnormal head, abnormal acrosome, abnormal tail, proximal cytoplasmic droplets, distal cytoplasmic droplets and, of course, 'normal'. COTS, debris, etc., should also be noted.

The integrity of the acrosome can be observed using phase contrast or DIC microscopy, and has been correlated with field fertility (Saacke, 1970; Saacke and White, 1972; Pursel *et al.*, 1974; Marshall and Frey, 1976; Mitchell *et al.*, 1978). High magnification is used (e.g. 400–1000×) and 100–200 spermatozoa are evaluated. These evaluations are labour intensive and repetition by individual technicians often causes human eye strain. Some laboratories have discontinued the use of this assay in lieu of SYBR-14 and PI staining using fluorescence microscopes or flow cytometers.

SEM and TEM can also be used to identify and study specific sperm abnormalities. These techniques require specific and meticulous sample preparation. Individual microscopes are expensive. Samples are often transferred to specific laboratories specializing in these techniques.

Irrespective of the classification scheme used, in general a maximum level of 15–20% total abnormal sperm is often considered acceptable. However, whenever performing analyses, it is critical to understand that some abnormalities, e.g. nuclear vacuoles, are extremely significant with respect to fertility. Also, observed abnormalities may be indicative of abnormalities that cannot be observed, e.g. DNA defects. In these cases, the general consideration of 15–20% as acceptable must be reconsidered. In essence, this issue is the basis of the 'major' and 'minor' classification scheme proposed by Blom (1972).

It must also be realized that a morphological evaluation is merely a snapshot of the ability of the male to produce 'normal' spermatozoa. The analysis is performed on a sample collected on a particular day and is indicative only for that particular collection. If the male is young, has been ill, or has been exposed to environmental conditions detrimental to sperm production, a satisfactory semen analysis may indicate adequate/acceptable semen production. Likewise, an unsatisfactory analysis only indicates that on that particular day, the collection consisted of abnormal cells. In both cases, repeated analyses must be done to clearly elucidate the seminal status.

After several satisfactory analyses, repeated in-depth analyses must be done on a periodic basis, depending upon the frequency of semen collection. In between in-depth analyses, laboratory personnel should be aware of abnormal sperm during routine microscopic semen evaluations, e.g. for motility. Salisbury and Mercier (1945) provided evidence that data obtained by examination of 100 spermatozoa on one slide was as reliable as the examination of multiple slides and the examination of up to 500 cells, although differences between technicians may account for greater variation (Root Kustritz *et al.*, 1998; Brito *et al.*, 2011). The proportion of cells examined is, then, extremely low (100–500) compared with the total cells (billions) in the ejaculate. Kuster *et al.* (2004b) discussed the statistical implications of sample size when performing morphological analyses.

Biological stains, DIC, SEM and TEM have all been used to demonstrate various sperm abnormaties: in bulls – Blom (1948), Saacke and Marshall (1968), Saacke (1970), Lorton *et al.* (1983), Barth and Oko (1989), and Foote (1999); dogs – Root Kustritz *et al.* (1998); stallions – Blom (1948) and Brito (2007); and boars – Kojima (1977), Tamuli and Watson (1994) and Briz *et al.* (1996). See also Chapter 3 (Kaya *et al.*) and Chapter 7 (Chenoweth and McPherson) for reviews of environmental and genetically induced abnormalities, respectively. Phase and DIC microscopy of wet smears has been used extensively to characterize acrosomal changes due to sperm ageing *in vitro* (Pursel *et al.*, 1972, 1974; Saacke and White, 1972; Johnson *et al.*, 1976; Aalseth and Saacke, 1986) and after freezing-thawing of ejaculates (Mitchell *et al.*, 1978).

Sperm membrane integrity

Sperm membrane integrity has been used extensively to evaluate sperm quality using three methods: (i) stains – in bulls (Garner *et al.*, 1994; Brito *et al.*, 2003), in boars (Fraser *et al.*, 2001) and in stallions (Brinsko *et al.*, 2011b); (ii) incubation in hyperosmotic media – in rams (Curry and Watson, 1994) and in stallions (de la Cueva *et al.*, 1997); and (iii) incubation in hypo-osmotic media – in bulls (Bredderman and Foote, 1969; Rota *et al.*, 2000; Quintero-Moreno *et al.*, 2008; Rubio *et al.*, 2008; Sahin *et al.*, 2008), in dogs (Hishinuma and Sekine, 2003; Pinto and Kozink, 2008), in donkeys (Rota *et al.*, 2010), in rams (Curry and Watson, 1994), in stallions (Neild *et al.*, 1999; Almin *et al.*, 2010), and in turkeys (Donoghue *et al.*, 1996). The SYBR-14 and PI stain combination is commercially available from Life Technologies (formerly Invitrogen) Molecular Probes® LIVE/DEAD® Sperm Viability Kit for use with the flow cytometer and fluorescence microscope. Johansson *et al.* (2008), Foster (2009), Morrell *et al.* (2010) and Foster *et al.* (2011a,b) have validated the use of the NucleoCounter® SP-100™ for measurement of membrane integrity. Because of its simplicity and assay speed (approx. 2 min for both sample preparation and the total cell and membrane integrity assays), the SP-100™ is easily adapted for use in the andrology laboratory. The flow cytometer has also been used to determine membrane integrity (Thomas *et al.*, 1997; Nagy *et al.*, 2003; Morrell *et al.*, 2010; Foster *et al.*, 2011a; see Plate 5). Flow cytometers are expensive and require specialized training, although Odhiambo *et al.* (2011) demonstrated the availability and accuracy of a relatively simple and easy-to-use instrument.

Sperm membrane integrity assays are often referred to as 'viability' tests. At the time of assay, some spermatozoa may have damaged membranes and, therefore, the assay may indicate loss of membrane integrity, even though some of these cells may be motile and viable. The fact that these particular cells have damaged membranes may, however, be indicative that they will not survive *in vitro* or *in vivo* as long as other cells.

Changes or defects of the spermatozoan's chromatin are non-compensable abnormalities and significantly affect fertility. Evenson and colleagues (Evenson *et al.*, 1980; Ballachey *et al.*, 1987, 1988) developed the SCSA® (Sperm Chromatin Structure Assay, SCSA Diagnostics, Volga, South Dakota, USA), which measures the percentage of sperm with fragmented DNA after denaturation of the DNA and staining with acridine orange. Double-stranded DNA will fluoresce green, while single-stranded DNA fluoresces red (see Plate 5). These authors and others have demonstrated that the amount of DNA fragmentation as measured by SCSA® (or acridine orange staining) is related to the fertility of various species: boars (Evenson *et al.*, 1994; Didion *et al.*, 2000, 2009; Boe-Hansen *et al.*, 2008; Tsakmakidis *et al.*, 2010); bulls (Ballachey *et al.*, 1987, 1988; Sailer *et al.*, 1995; Waterhouse *et al.*, 2006); rams (Sailer *et al.*, 1995); and stallions (Kenney *et al.*, 1995; Sailer *et al.*, 1995; Love and Kenney, 1998; Love, 2005). Evenson (1999) and Marchesi and Feng (2007) have described the potential origin of DNA fragmentation.

The SCSA®/acridine orange technique is considered the 'gold standard', though other techniques for measuring DNA damage have been developed: the Comet assay, which evaluates DNA strand breaks in single cells (Linfor and Meyers, 2002; Boe-Hansen *et al.*, 2005a); the Tunel assay, which measures the amount of 3′-OH ends of DNA using terminal deoxynucleotidyl transferase dUTP nick-end labelling (Anzar *et al.*, 2002; Waterhouse *et al.*, 2006); and the Sperm Chromatin Dispersion (SCD) test (Fernández *et al.*, 2003; Garcia-Macias *et al.*, 2007; de la Torre *et al.*, 2007). These assays have been compared by several authors (Evenson, 1999; Evenson and Wixon, 2005; Chohan *et al.*, 2006; Teague *et al.*, 2010) and have been used to measure various aspects of semen quality, for instance, the effects of: cryopreservation – in chickens (Gliozzi *et al.*, 2011) and in boars (Fraser and Strzezek, 2005); storage of fresh-cooled boar semen (Boe-Hansen *et al.*, 2005b; Waberski *et al.*, 2011); extender differences (Karabinus *et al.*, 1991); age of the male (Karabinus *et al.*, 1990); and experimental scrotal insulation (Karabinus *et al.*, 1997).

Some laboratories now use in-house either the acridine orange assay or the commercial SCSA® assay, but many depend upon subjective microscopic motility measurements and morphology analyses. Numerous morphological abnormalities have been identified (see, for example, Barth and Oko, 1989), but the effects of some of these abnormalities are unknown, especially in cases where the incidence of the abnormality is low (and possibly compensable). Fertility measurements are often deficient and inaccurate as a result of the lack of availability and sufficient quantity of field information. Barth (1989a,b) has reviewed the influence of spermatozoal abnormalities on pregnancy rates in cattle. The influence of several more common abnormalities has also been evaluated with respect to their influence on *in vivo* and *in vitro* fertility, e.g. sperm head shape in cattle (Barth *et al.*, 1992; Saacke *et al.*, 1998; Al-Makhzoomi *et al.*, 2008); presence of cytoplasmic droplets in boar (Waberski *et al.*, 1994; Fischer *et al.*, 2005; Lovercamp *et al.*, 2007a,b), dog (Peña *et al.*, 2007) and bull (Amann *et al.*, 2000) spermatozoa; presence of nuclear vacuoles (Pilip *et al.*,1996); and presence of knobbed acrosomes (Barth, 1986; Thundathill *et al.*, 2000).

Fourier harmonic amplitude (FHA) analysis of sperm nuclear shape

FHA analysis has been used to study the nuclear shape of spermatozoa and its relationship to the sperm morphology and fertility of boars and bulls (Ostermeier, 2000; Ostermeier *et al*, 2001; Enwall, 2009; Rahman *et al.*, 2011; Parrish *et al.*, 2012). These authors have identified boars with reduced fertility, i.e. mean farrowing rate, and significant differences in harmonic amplitudes of the sperm nuclear shape. The sperm of bulls with reduced fertility have that are smaller in overall size and have a longer length component. The authors concluded that some of the differences in nuclear shape between high and low fertility animals may be explained by varying levels of chromatin stability as examined using SCSA. Saravia *et al.* (2007) studied boar sperm head shape using an automated sperm morphology analyser. In this study, sperm head shape was not correlated to chromatin integrity, possibly due to the high quality and low variation of the quality of the semen studied.

Estimation of Spermatozoal Motility

Since the first observations of sperm by Leeuwenhoek and Hartsoeker in the late 17th century (as referenced in the historical review by Afzelius and Bacetti, 1991), andrologists have observed and evaluated semen from various species using the compound microscope. Visual microscopic evaluations are very subjective, as the human eye concentrates on the moving cells. The evaluation of several microscopic fields at 200× magnification, dilution of samples in an iso-osmotic medium, and the use of phase contrast optics, improves observations but does not eliminate subjectivity.

As advancement/modification of earlier work by Rothschild (1953), Elliott *et al.* (1973) and Van Dellen and Elliott (1978) described a 2 s timed-exposure photographic procedure utilizing dark field microscopy for objectively measuring the percentage of progressively motile spermatozoa. Samples were prepared in pre-warmed Petroff-Hausser counting chambers (Fig. 6.15a). Six predetermined positions within each chamber were photographed. The film negative strips were developed and the six fields of each sample were projected on to a white wall or screen for analysis. The projected images represented the 'positive' representations (Fig. 6.15c) of the dark-field microscopic images (Fig. 6.15b). Technicians could easily, quantitatively and repeatably, count the sperm tracks resulting from the 2 s exposures and the non-motile cells. The improved objectivity, compared with visual estimates, of this technique overshadows the increased expense and labour involved. Makler (1978) and Revell and Wood (1978) published modifications of Elliott's photographic procedure. Today, digital photography could be used to advance this technique.

In the early 1980s, in the laboratory of R.H. Foote, analogue video recordings of sperm samples were made (approximately 15 s each of several fields per sample) using standard commercial equipment integrated with a bright-field microscope (see, for example, Parrish and Foote, 1987). Superimposed audio was used to identify samples numerically. Sample numbers were recorded elsewhere along with experimental treatment identifications, etc. Recordings were replayed on to a standard television monitor and observed simultaneously by three to four investigators, who independently and 'blindly' evaluated sample motility. Although also labour intensive, data obtained using this proceedure were

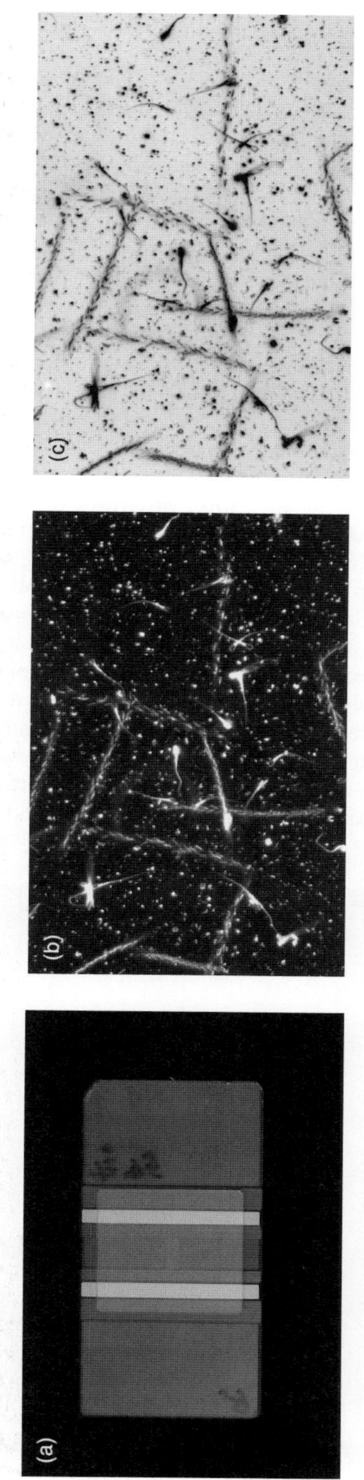

Fig. 6.15. (a) The modified Petroff–Hauser chamber used by Elliott *et al.* (1973) for their timed-exposure photographic tracking procedure for the estimation of sperm motility. (b) Dark field image of 2 s timed-exposure tracking. (c) 'Inverse' or positive image of 2 s timed-exposure tracking. The images were projected on to a blank wall or screen for the analyses. The sperm tracks are easily counted, as are the immotile 'dead' sperm that show no tracks. Figure (a) by L. Lorton; figures (b) and (c) by S. Lorton.

more objective than from individual microscopic evaluations. Liu and Warme (1977) and Amann and Hammerstedt (1980), using video films, and then Budworth *et al.* (1987), using 35mm films, later reported the use of a microcomputer system to analyse digital images of sperm motility.

Other investigators (Rikmenspoel, 1964; Atherton, 1975; Mayevsky *et al.*,1980) reported techniques for measuring spermaozoal motility. These techniques, however, did not progress with time, mostly as a result of the technical advancements and miniaturization of computer systems.

Katz *et al.* (1985) described the real-time assessment and kinemetric analysis of spermatozoal motility using computer-assisted semen analysis (CASA), though the term CASA was not used until later. Mortimer (1990) reviewed in detail the early historical development of CASA sytems, which digitize in multiple frames video images similar to the tracking photographs of Elliott *et al.* (1973) – albeit that the early computer processing units (CPUs) resulted in extremely slow analyses. The identification and characterization of spermatozoa in non-optically clear extenders, e.g. containing egg yolk or milk, resulted in inaccurate analyses. As CPUs advanced, the incorporation of improved microscopic techniques, e.g. negative phase contrast, fluorescence staining of DNA and tail detection, improved the speed of analysis and differentiation of spermatozoa from COTS and media particles (egg yolk and milk).

Davis *et al.* (1992) and Mortimer (2000) described in detail the characteristics of sperm movement as analysed by CASA. In addition to total spermatozoal motility and progressive motility, CASA systems use algorithms to calculate spermatozoal concentration and various velocity parameters. Users establish system set-up parameters to define minimum characteristics of velocity and progressive motility. These parameters may vary between systems and users; therefore, attempts to compare data are often difficult. Different sample chambers are available for use in CASA systems. Chambers of 10–20 μm in depth are preferred to minimize any inability to detect and monitor sperm cell concentration and movement as a result of cells moving in and out of the focal plane. Several manufacturers provide disposable chambers of fixed depth. The Leja® chamber (Leja Products B.V., Netherlands) is widely used with many CASA systems (see Fig. 6.16), although motility variation

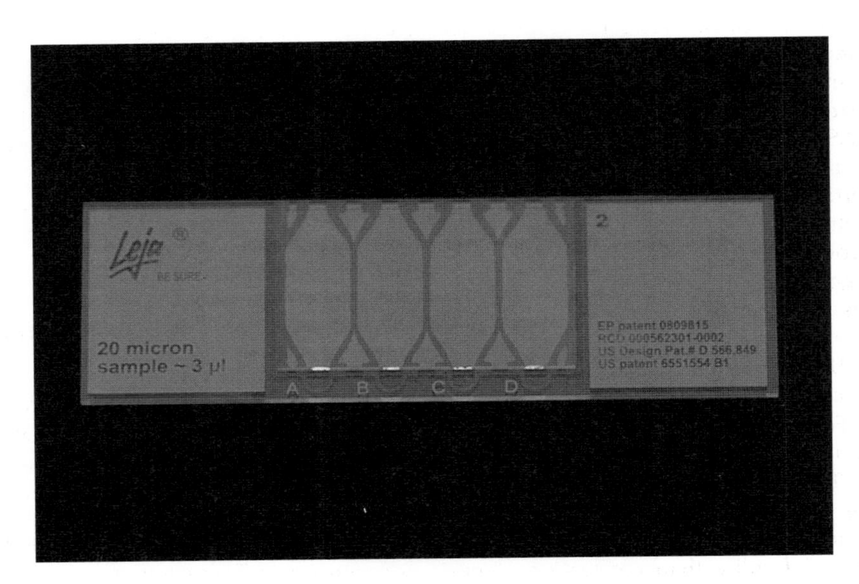

Fig. 6.16. The Leja® counting chamber, which is used with many computer assisted semen analysis (CASA) systems. Photo by L. Lorton.

has been reported when evaluating samples for progressive motility and velocity in different locations within these chambers (Hansen, 2009). Douglas-Hamilton *et al.* (2005a,b) described the Segré–Silberberg (SS) effect, which explains inconsistent filling of sperm in low-volume, capillary-filled counting chambers, and the resulting effects on the estimation of sperm concentration. These authors computed a compensation factor for use in a CASA to correct for the SS effect. Several commercially available CASA systems allow the use of such a factor.

CASA systems have the distinct advantage of increased data objectivity, especially with respect to total and progressive motility, compared with visual microscopic analyses. Many authors have used CASA in attempts to estimate or predict fertility. See, for example: in boars (Evenson *et al.*, 1994; Holt *et al.*, 1997; Hirai *et al.*, 2001; Quintero-Moreno *et al.*, 2004; Ruiz-Sánchez *et al.*, 2006); in bulls (Budworth *et al.*, 1988; Farrell *et al.*, 1998; Palmer and Barth, 2003; Kathiravan *et al.*, 2011 – a review); in stallions (Kirk *et al.*, 2005; Hodder and Liu, 2011; Love, 2011). However, none of the reported analyses of CASA data have identified a technique that clearly differentiates animal fertility levels in the andrology laboratory. This is especially true in the case of dairy bulls in stud, because their fertility levels (as measured by non-return rate) vary only by approximately ten percentage points. Quintero-Moreno *et al.* (2004) and others have suggested that CASA data, as well as other laboratory assays, may be significantly correlated with field fertility.

Holt *et al.* (2007) discussed interpretation and problems associated with CASA data, and suggested that the evaluation of subpopulations of spermatozoa within ejaculates might be more beneficial than evaluating data means. Several authors (Abaigar *et al.*, 1999; Peña *et al.*, 2005; Núñez-Martínez *et al.*, 2006; Ortega-Ferrusola *et al.*, 2009) have identified subpopulations of sperm in ejaculates using CASA. They suggested that these analyses provide additional and better information related to semen quality, semen freezability and animal fertility than conventional analyses. Su *et al.* (2012) have developed a 3D sperm tracking system; these authors can record the movement characteristics of a large number of cells, though the usefulness of their technique awaits further investigations.

Some andrology laboratories perform a post-thaw incubation or 'stress' test as an evaluation of sample quality or 'fertility'. After thawing, samples are evaluated for progressive motility, and then incubated typically at 37°C for up to 2 h, at which time another progressive motility evaluation is performed. The difference between the motility values before and after incubation, or the motility after incubation, are used as a quality assessment. Data in the literature documenting the correlation of these data with fertility are ambiguous (see Saacke, 1983; Vianna *et al.*, 2009). It should be remembered that techniques and media (extender) used for these *in vitro* assays do not simulate the *in vivo* conditions in the female reproductive tract. Individual laboratories have established their own criteria for the characterization of sample quality based upon this assay and the methodology used in the particular laboratory. Saacke and White (1972) and Saacke (1983) described their data showing that spermatozoal quality, i.e. the acrosome integrity, of fresh semen is superior to that of frozen–thawed semen after incubation for up to 4 h at 37°C. The difference observed in semen quality after incubation may be indicative of a loss of membrane integrity, as evidenced by the Saacke and White (1972) data. Freeze–thaw events cause the loss of membrane integrity by some spermatozoa in the sample, and those with the membrane damage may be motile immediately post thaw, but may not survive the incubation test for as long as those without membrane damage.

Spermatozoal Migration *In Vitro*

Kremer (1965) reported a simple *in vitro* technique to measure the ability of spermatozoa to migrate in cervical mucus. Kummerfeld *et al.* (1980, 1981a,b), using cervical mucus collected from cows in oestrus, attempted to correlate *in vitro* spermatozoal migration with bull non-return rate. These authors concluded that *in vitro* spermatozoal migration assays could not be used to differentiate the small range of fertility of bulls in stud. They also demonstrated significant effects of cervical mucus samples and storage time on *in vitro* migration assays, and suggested that a synthetic medium (Lorton *et al.*, 1981) could be used to overcome these effects. Parrish and Foote (1987) used this polyacrylamide medium for migration assays and demonstrated significant correlations between spermatozoal motility and migration distance. The percentage of spermatozoa with intact acrosomes, migration distance in synthetic medium, spermatozoal number after a swim-up procedure and percentage spermatozoal motility were used in a regression model to predict bull non-return rate, an R^2 value of 0.686 was calculated.

Murase *et al.* (1990) stored cervical mucus collected from cows in oestrus at –80°C and subsequently evaluated the *in vitro* migration of freshly collected bull semen. Results from this study indicated a positive correlation of spermatozoal migration and fertility, and semen samples that exhibited more than a 60 mm mucus penetration distance were representative of bulls that had a non-return rate of greater than 60%. Using frozen–thawed sex-sorted and non-sorted ram semen, de Graaf *et al.* (2007) evaluated *in vitro* spermatozoal migration in a commecially prepared sodium hyaluronate medium. Non-sorted spermatozoa migrated in greater numbers to an end position of 1 cm than did sex-sorted spermatozoa. Robayo *et al.* (2008) provided evidence that various velocity parameters measured by CASA were significantly correlated with the ability of ram spermatozoa to migrate *in vitro* through ewe oestrus cervical mucus.

Microbiological Evaluation of Semen

Since the late 1940s, antibiotics have been routinely added to bovine semen (see review by Foote, 1976). Before and after entering studs, bulls are monitored for various microorganisms and viral agents that may be present and could be transferred within semen. Sulfanilamide, penicillin or streptomycin, then penicillin, dihydrostreptomycin and polymixin B and, more recently, gentamicin, tylosin, lincomycin and spectinomycin became the standard cocktail for addition to the raw semen and in the extender (Lorton *et al.*, 1988a,b; Shin *et al.*, 1988).

In the USA, the minimum requirements for the health of bulls and the effective use of antibiotics in frozen semen are under the guidelines of the Certified Semen Services of the National Association of Animal Breeders (see: http://www.naab-css.org/about_css/Final CSS Min Req - Jan 2014.pdf). Similar health and antibiotic requirements exist in other countries, and are typically required for the international trade of semen. The antibiotics serve as a secondary line of defence against semen-borne organisms, especially to combat those that might occur between routine bull health testing. Most, if not all, antibiotics are to semen quality detrimental at some concentration. Further, the level of antibiotics used in semen is not typically sufficient to combat disease, infection or poor management practices. The microbiological evaluation of bovine semen is also routine after observation of COTS and/or when reproductive tract infection is suspected.

International trade in bovine, equine and canine semen requires additional microbiological culture according to individual country requirements. Canine and equine semen are also cultured after observation of COTS and/or when reproductive tract infection is suspected (see reviews by Johnston *et al.*, 2001 and Tibary *et al.*, 2009, respectively). Potassium penicillin and streptomycin are common additives in canine semen extenders, while ticarcillin, amikacin, potassium penicillin or gentamicin are commonly used in equine extenders (Varner *et al.*, 1998; Brinsko *et al.*, 2011a).

Extended boar semen is primarily distributed after cooling to 16–18°C. Gentamicin is the most common antibiotic additive to extenders, although some manufacturers include others in their products (see reviews by Althouse and Lu, 2005 and Althouse *et al.*, 2008). In the USA, it is common practice at boar studs to routinely, e.g. weekly, send subsamples of commercially prepared semen doses to commercial laboratories or veterinary diagnostic facilities for quality assurance analyses. These analyses typically include progressive motility, number of spermatozoa per dose and microbiological culture. In addition, samples are also sent routinely to specific diagnostic laboratories for PRRS (porcine reproductive and respiratory syndrome) virus screening (Rovira *et al.*, 2007; Reicks, 2009).

Conclusion

The overall objective in the andrology laboratory is to produce high-quality artificial insemination doses and to evaluate semen samples as a major part of breeding soundness examinations. To achieve this goal, and therefore to achieve the maximum level of factor 'C' of Bartlett's Equation of Reproduction, a simple, fast, specific and reproducible spermatozoal evaluation procedure, and one that is also highly predictive of fertility potential, is coveted. With frozen semen, e.g. bovine, equine and canine semen, speed of assay is not as critical as in the case of fresh cooled semen, e.g. equine, canine and porcine semen, when processing and evaluation time is minimal owing to the requirement for transport of the semen to the field. Unfortunately, to date, such an assay is not at hand. Instead, laboratories utilize one or more of the available assays and depend to some degree on 'art' rather than on science to achieve their goals.

Acknowledgements

The author wishes to extend his sincere appreciation to Barrett Hasell (Rockhampton, Australia), Mel DeJarnette (Plain City, Ohio, USA), Russ DeMarco (Walworth, Wisconsin, USA), Duane Garner (Reno, Nevada, USA), Claus Hansen (Copenhagen, Denmark), Werner Leidl (Munich, Germany), Amnon Makler (Haifa, Israel), Soren Pederson (Allerod, Denmark), Julie Pryse and Adrienne Pierson (Melbourne, Australia) and Christa Simmet (Tiefenbach, Germany) for permission to use their photographs and figures. Special appreciation is extended to Lewis Lorton (Columbia, Maryland) for his help and expertise in preparing several of the figures in this chapter.

The photo credits for Plate 4 are as follows: a, c, courtesy B. Hasell and the Artificial Breeding Centre, Beef Breeding Services, Department of Employment, Primary Industries and Fisheries, Economic Development and Innovation, Queensland, Australia; b, f, S. Lorton; d, courtesy W. Leidl, Munich, Germany; e, i, reprinted from *Theriogenology* 20(5) 585–599, 1983, with permission from Elsevier; g, h, courtesy D. Garner, GametoBiology Consulting, Graeagle, California, USA.

Appendix

The article 'Some Staining Procedures for Spermatozoa. A Review' by AJ Hackett and JW Macpherson (*Canadian Veterinary Journal*, 1965, pp. 55–62) is reprinted on pp. 136–143 with the kind permission of the Canadian Veterinary Medical Association. This classic review contains many useful procedures and valuable references to older literature.

References

Aalseth EP and Saacke RG (1986) Vital staining and acrosomal evaluation of bovine sperm *Gamete Research* **15** 73–81.

Abaigar T, Holt WV, Harrison RAP and del Barrio G (1999) Sperm subpopulations in boar (*Sus scrofa*) and gazelle (*Gazella dama mhorr*) semen as revealed by pattern analysis of computer-assisted motility assessments *Biology of Reproduction* **60** 32–41.

Abramowitz M (1987) *Contrast Methods in Microscopy: Transmitted Light Volume* 2 pp 12–27. Olympus Publishing Corporation, New York.

Afzelius BA and Bacetti B (1991) History of spermatology In *Comparative Spermatology, 20 Years After* pp 1–10 Ed B Bacetti. Raven Press, New York.

Alm K, Peltoniemi OAT, Koskinen E and Andersson M (2006) Porcine field fertility with two different insemination doses and the effect of sperm morphology *Reproduction in Domestic Animals* **41** 210–213.

Al-Makhzoomi A, Lundeheim N, Håård M and Rodriguez-Martinez H (2008) Sperm morphology and fertility of progeny-tested AI dairy bulls in Sweden *Theriogenology* **70** 682–691.

Almid T and Johnson LA (1988) Effects of glycerol concentration, equilibrium time and temperature of glycerol addition on post-thaw viability of boar spermatozoa frozen in straws *Journal of Animal Science* **66** 2899–2905.

Almin MR, Druart X, and Barrier-Battut I (2010) Osmotic resistance of frozen stallion spermatozoa in relation to fertility *Animal Reproduction Science* **121** S130–S132.

Althouse GC and Hopkins SM (1995) Assessment of boar sperm viability using a combination of two fluorophores *Theriogenology* **43** 595–603.

Althouse GC and Lu KG (2005) Baceriospermia in extended porcine semen *Theriogenology* **63** 573–584.

Althouse GC, Pierdon MS and Lu KG (2008) Thermotemporal dynamics of contaminated bacteria and antimicrobials in extended porcine semen *Theriogenology* **70** 1317–1323.

Amann RP (2005) Weaknesses in reports of "fertility" for horses and other species *Theriogenology* **63** 698–715.

Amann RP and Hammerstedt RH (1980) Validation of a system for computerized measurements of spermatozoal velocity and percentage of motile sperm *Biology of Reproduction* **23** 647–656.

Amann RP and Hammerstedt RH (2002) Detection of differences in fertility *Journal of Andrology* **23** 317–325.

Amann RP, Seidel GE, and Mortimer RG (2000) Fertilizing potential *in vitro* of semen from young beef bulls containing a high or low percentage of sperm with a proximal droplet *Theriogenology* **54** 1499–1515.

Anzar M, He L, Buhr MM, Kroetsch TG and Pauls KP (2002) Sperm apoptosis in fresh and cryopreserved bull semen detected by flow cytometry and its relationship with fertility *Biology of Reproduction* **66** 354–360.

Anzar M, Kroetsch T, and Buhr M (2009) Comparison of different methods for assessment of sperm concentration and membrane integrity with bull semen *Journal of Andrology* **30** 661–668.

Atherton RW (1975) An objective method for evaluating Angus and Hereford sperm motility *International Journal of Fertility* **20** 109–112.

Baker RD, Dzuik PJ and Norton HW (1968) Effect of volume of semen, number of sperm and drugs on transport of sperm in artificially inseminated gilts *Journal of Animal Science* **27** 88–93.

Ball BA and Mohammed HO (1995) Morphometry of stallion spermatozoa by computer assisted image analysis *Theriogenology* **44** 367–377.

Ballachey BE, Hohenboken WD and Evenson DP (1987) Heterogeneity of sperm nuclear chromatin structure and its relationship to fertility of bulls *Biology of Reproduction* **36** 915–925.

Ballachey BE, Evenson DP and Saacke RG (1988) The sperm chromatin structure assay relationship with alternative tests of semen quality and heterospermic performance of bulls *Journal of Andrology* **9** 109–115.

Bane A (1952) A study of the technique of hemocytometric determination of sperm motility and sperm concentration in bull semen *Cornell Veterinarian* **42** 525–530.

Baran A, Sahin BE, Evecen M, Demir K and Ileri IK (2004) Use of Spermac® staining technique in the determination of acrosomal defects in cat semen *Turkish Journal of Veterinary and Animal Sciences* **28** 519–525.

Barth AD (1986) The knobbed acrosome defect in beef bulls *Canadian Veterinary Journal* **27** 379–384.

Barth AD (1989a) Influence of abnormal sperm morphology on conception rates in cattle In *Proceedings of the Annual Meeting, September 29–30, 1989, Coeur D'Alene, Idaho* pp 13–40. Society for Theriogenology, Montgomery.

Barth AD (1989b) Evaluation of frozen bovine semen In *Proceedings of the Annual Meeting, September 29–30, 1989, Coeur D'Alene, Idaho* pp 92–99. Society for Theriogenology, Montgomery.

Barth AD and Oko RJ (1989) *Abnormal Morphology of Bovine Spermatozoa*. Iowa State Press, Ames.

Barth AD, Bowman PA, Bo GA and Mapletoft RJ (1992) Effect of narrow sperm head shape on fertility in cattle *Canadian Veterinary Journal* **33** 31–39.

Bartlett DE (1978) An equation of reproduction In *Proceedings, World Congress of Buiatrics, Mexico* pp 280–284.

Belding DL (1934) Fertility in the male. II Technique of the spermatozoa count *American Journal of Obstetrics and Gynecology* **27** 25–31.

Berndtson WE (1991) A simple, rapid and reliable method for selecting or assessing the number of replicates for animal experiments *Journal of Animal Science* **69** 67–76.

Blom E (1948) Über Medusabilungen (abgelöste Wimpersäume) in Stier- und Hengstsperma und ihre diagnostische Bedeutung. [Medusa formations (detached ciliated borders) in the semen of bulls and stallions and their diagnostic significance] *Gynaecologia* **125** 312–320.

Blom E (1972) The ultrastructure of some characteristic sperm defects and a proposal for a new classification of the bull spermiogram In *Atti del VII Simposio Internationale di Zootecnia, Milano, 15–17 Aprile 1972* pp 125–139. Società Italiana per il Progresso della Zootecnia, Milan (cited in Barth and Oko, 1989).

Boe-Hansen GB, Morris ID, Ersboll AK, Greve T and Christensen P (2005a) DNA integrity in sexed bull sperm assessed by neutral Comet assay and sperm chromatin structure assay *Theriogenology* **63** 1789–1802.

Boe-Hansen GB, Ersboll AK, Greve T and Christensen P (2005b) Increased storage time of extended boar semen reduces sperm DNA integrity *Theriogenology* **63** 2006–2019.

Boe-Hansen GB, Christensen P, Vibjerg D, Nielsen MBF and Hedeboe AM (2008) Sperm chromatin structure integrity in liquid stored boar semen and its relationship with field fertility *Theriogenology* **69** 728–736.

Boersma A, Braun J and Stolla R (1999) Influence of random factors and two different staining procedures on computer assisted sperm head morphometry in bulls *Reproduction in Domestic Animals* **34** 77–82.

Boersma A, Rasshofer R and Stolla R (2001) Influence of sample preparation, staining procedure and analysis conditions on bull sperm head morphometry using the Morphology Analyzer Integrated Visual Optical System *Reproduction in Domestic Animals* **36** 222–229.

Boguth W (1951) Über die Eignung von Färbemethoden zur Bestimmung toter Spermien [Suitability of staining methods to identify dead sperm] *Deutsche Tierärtliche Wochenschritt Suppl. 1: Fortpflanzung und Besamung der Haustiere* **58** 20–21.

Brazil C, Swan SH, Tollner CR, Treece C, Drobnis EZ, Wang C, Redmon JB and Overstreet JW (2004) Quality control of laboratory methods for semen evaluation in a multicenter research study *Journal of Andrology* **25** 646–656.

Bredderman PJ and Foote RH (1969) Volume of stressed bull spermatozoa and protoplasmic droplets, and the relationship of cell size to motility and fertility *Journal of Animal Science* **28** 496–501.

Brinsko SP, Blanchard TL, Varner DD, Schumacher J, Love CC, Hinrichs K and Hartman D (2011a) Semen collection and artificial insemination with fresh semen In *Manual of Equine Reproduction* 3rd edition pp 160–175. Mosby Elsevier, St Louis.

Brinsko SP, Spooner JA, Blanchard TL, Love CC and Varner DD (2011b) Relationships among morphologic characteristics and membrane integrity of stallion sperm *Clinical Theriogenology* **3** 55–59.

Brito LFC (2007) Evaluation of stallion sperm morphology *Clinical Techniques in Equine Practice* **6** 249–264.

Brito LFC, Barth AD, Bilodeau-Goeseels S, Panich PL and Kastelic JP (2003) Comparison of methods to evaluate the plasmalemma of bovine sperm and their relationship with *in vitro* fertilization rate *Theriogenology* **60** 1539–1551.

Brito LFC, Greene LM, Kelleman A, Knobbe M and Turner R (2011) Effect of method and clinician on stallion sperm morphology evaluation *Theriogenology* **76** 745–750.

Briz MD, Bonet S, Pinart B and Camps R (1996) Sperm malformations throughout the boar epididymal tract. *Animal Reproduction Science* **43** 221–239.

Budworth PR, Amann RP and Hammerstedt RH (1987) Microcomputer-photographic method for evaluation of motility and velocity of bull sperm *Journal of Dairy Science* **70** 1927–1936.

Budworth PR, Amann RP and Chapman PL (1988) Relationships between computerized measurements of motion of frozen–thawed bull spermatozoa and fertility *Journal of Andrology* **9** 41–54.

Casarett GW (1953) A one-solution stain for spermatozoa *Stain Technology* **28** 125–127.

Casey PJ, Hillman RB, Robertson KR, Yudin AI, Liu LKM and Drobnis EZ (1993) Validation of an acrosomal stain for equine sperm that differentiates between living and dead sperm *Journal of Andrology* **14** 289–297.

Celeghini ECC, de Arruda RP, de Andrade AFC, Nascimento J and Raphael CF (2007) Practical techniques for bovine sperm simultaneous fluorimetric assessment of plasma, acrosomal and mitochondrial membranes *Reproduction in Domestic Animals* **42** 479–488.

Cheng FP, Fazeli A, Voorhout WF, Marks A, Bevers MM and Colenbrander B (1996) Use of peanut agglutinin to assess the acrosomal status and the zona pellucida-induced acrosome reaction in stallion spermatozoa *Journal of Andrology* **17** 674–682.

Chenoweth P, Zeron Y, Shalit U, Rabinovitch L and Deutsch M (2010) Pregnancy rates in dairy cattle inseminated with different numbers of progressively motile sperm *Clinical Theriogenology* **2** 389.

Cheuquemán C, Bravo P, Treulén F, Giojalas LC, Villegas J, Sánchez R and Risopatrón J (2012) Sperm membrane functionality in the dog assessed by flow cytometry *Reproduction in Domestic Animals* **47** 39–43.

Chohan KR, Griffin JT, Lafromboise M, DeJonge CJ and Carrell DT (2006) Comparison of chromatin assays for DNA fragmentation evaluation in human sperm *Journal of Andrology* **27** 53–59.

Christensen P, Whitfield CH and Parkinson TJ (1994) The use of bright-field microscopy in evaluating bovine acrosome reaction *Theriogenology* **42** 655–662.

Christensen P, Stryhn H and Hansen C (2005) Discrepancies in the determination of sperm concentration using Bürker-Türk, Thoma and Makler counting chambers *Theriogenology* **63** 992–1003.

Comerford KL (2009) Validation of a commercially available fluorescence-based instrument to evaluate stallion spermatozoal concentration and comparison to photometric systems. Master of Science thesis, Texas A&M University, College Station, Texas. Available at: http://repository.tamu.edu/bitstream/handle/1969.1/ETD-TAMU-2009-05-261/1_Comerford.pdf (accessed 28 August 2013).

Comerford KL, Love CC, Brinsko SP, Edmund AJ, Waite JA, Teague SR and Varner DD (2008) Validation of a commercially available fluorescence-based instrument to evaluate stallion spermatozoal concentration *Animal Reproduction Science* **107** 316–317.

Comstock RE and Green WW (1939) Methods for semen evaluation. I. Density, respiration, glycolysis of semen *Journal of Animal Science* **1939** 213–216. Available at: http://www.journalofanimalscience.org/content/1939/1/213.full.pdf+html (accessed 28 August 2013).

Curry MR and Watson PF (1994) Osmotic effects on ram and human sperm membranes in relation to thawing injury *Cryobiology* **31** 39–46.

Davis RO, Niswander PW and Katz DF (1992) New measures of sperm motion. I. Adaptive smoothing and harmonic analysis *Journal of Andrology* **13** 139–152.

de Graaf SP, Evans G, Gillan L, Guerra MMP, Maxwell WMC and O'Brien JK (2007) The influence of antioxidant, cholesterol and seminal plasma on the *in vitro* quality of sorted and non-sorted ram spermatozoa *Theriogenology* **67** 217–227.

DeJarnette JM and Lefevre K (2008) Use of the NucleoCounter in a semen processing laboratory In *Proceedings of the 22nd Technical Conference on Artificial Insemination and Reproduction* pp 89–93. National Association of Animal Breeders, Columbia, Missouri.

de la Cueva FIC, Pujol MR, Rigau T, Bonet S, Miró J, Briz M and Rodriguez-Gill JE (1997) Resistance to osmotic stress of horse spermatozoa: the role of ionic pumps and their relationship to cryopreservation success *Theriogenology* **48** 947–968.

de la Torre J, López-Fernández C, Pita M, Fernández JL, Johnston SD and Gosálvez J (2007) Simultaneous observation of DNA fragmentation and protein loss in the boar spermatozoon following application of the sperm chromatin dispersion (SCD) test *Journal of Andrology* **28** 533–540.

De Leeuw AM, Den Daas JHG and Woelders H (1991) The fix vital stain method. Simultaneous determination of viability and acrosomal status of bovine spermatozoa *Journal of Andrology* **12** 112–118.

Den Daas N, De Jong G, Lansbergen LMTE and Van Wagtendonk-de Leeuw AM (1998) The relationship between the number of spermatozoa inseminated and the reproductive efficiency of individual bulls *Journal of Dairy Science* **81** 1714–1723.

Didion BA and Graves CN (1988) A dual stain technique for detecting viability and acrosomal status of bull spermatozoa In *Proceedings of the 12th Technical Conference on Artificial Insemination and Reproduction* pp 138–139. National Association of Animal Breeders, Columbia, Missouri.

Didion BA, Oleson B, Jost L and Evenson DP (2000) Sperm chromatin structure assay of boar sperm as related to fertility In *Boar Semen Preservation IV. Proceedings, IVth International Conference on Boar Semen Preservation, Beltsville, Maryland* p 243 [abstract] Ed LA Johnson and HD Guthrie. Allen Press, Laurence, Kansas.

Didion BA, Kasperson KM, Wixon RL and Evenson DP (2009) Boar fertility and sperm chromatin structure status: a retrospective report *Journal of Andrology* **30** 655–660.

Donoghue AM, Garner DL, Donoghue DJ and Johnson LA (1995) Viability assessment of turkey sperm using fluorescent staining and flow cytometry *Poultry Science* **74** 1191–1200.

Donoghue AM, Garner DL, Donoghue DJ and Johnson LA (1996) Assessment of the membrane integrity of fresh and stored turkey spermatozoa using a combination of hypo-osmotic stress, fluorescent staining and flow cytometry *Theriogenology* **46** 1543–163.

Douglas-Hamilton DH, Smith NG, Kuster CE, Vermeiden JPW and Althouse GC (2005a) Particle distribution in low-volume capillary-loaded chambers *Journal of Andrology* **26** 107–114.

Douglas-Hamilton DH, Smith NG, Kuster CE, Vermeiden JPW and Althouse GC (2005b) Capillary-loaded particle fluid dynamics: effect on estimation of sperm concentration *Journal of Andrology* **26** 115–122.

Elliott FI, Sherman JK, Elliott EJ and Sullivan JJ (1973) A photographic method of measuring percentage of progressively motile cells using dark-field microscopy In *Atti del VIII Simposio Internationale di Zootecnia* pp 160–168. Società Italiana per il Progresso della Zootecnia, Milan.
Emik LO, and Sidwell GM (1947) Factors affecting the estimation of concentration of sperm in ram's semen by the photoelectrometric method *Journal of Animal Science* **6** 467 – 475.
Enwall L (2009) Investigating nuclear shape, morphology, and molecular changes in semen from scrotal-insulated Holstein bulls using Fourier harmonic analysis, flow cytometry, and *in vitro* fertility evaluations. PhD Dissertation, University of Wisconsin, Madison.
Evenson DP (1999) Loss of livestock breeding efficiency due to uncompensable sperm nuclear defects *Reproduction, Fertility and Development* **11** 1–15.
Evenson DP and Wixon R (2005) Comparison of the Halosperm® test kit with the sperm chromatin structure assay (SCSA®) infertility test in relation to patient diagnosis and prognosis *Fertility and Sterility* **84** 846–849.
Evenson DP, Darzynkiewicz Z and Melamed MR (1980) Relation of mammalian sperm chromatin heterogeneity to fertility *Science* **210** 1131–1133.
Evenson DP, Parks JE, Kaproth MT and Jost LK (1993) Rapid determination on sperm cell concentration in bovine semen by flow cytometry *Journal of Dairy Science* **76** 86–94.
Evenson DP, Thompson L and Jost LK (1994) Flow cytometric evaluation of boar semen by the sperm chromatin structure assay as related to cryopreservation and fertility *Theriogenology* **41** 637–651.
Farrell PB, Presicce GA, Brockett CC and Foote RH (1998) Quantification of bull sperm characteristics measured by computer assisted sperm analysis (CASA) and their relationship to fertility *Theriogenology* **49** 871–879.
Fernández JL, Muriel L, Rivero MT, Goyanes V, Vázquez R and Alvarez JG (2003) The sperm chromatin dispersion test: a simple method for the determination of sperm DNA fragmentation *Journal of Andrology* **24** 59–66.
Fischer KA, Van Leyen K, Lovercamp KW, Manandhar G, Sutovsky M, Feng D, Safranski T and Sutovsky P (2005) 15-lipoxygenase is a component of the mammalian sperm cytoplasmic droplet *Reproduction* **130** 213–222.
Flowers WL (2002) Increasing fertilization rate of boars: influence of number and quality of spermatozoa inseminated *Journal of Animal Science* **80** E47–E53.
Foote RH (1968) Standards for sperm concentration: polystyrene latex particles as an aid in quality control In *Proceedings, 2nd Technical Conference on Artificial Insemination and Reproduction* pp 95–97. National Association of Animal Breeders, Columbia, Missouri.
Foote RH (1972) How to measure sperm cell concentration by turbidity (optical density) In *Proceedings, 4th Technical Conference on Artificial Insemination and Reproduction* pp 57–61. National Association of Animal Breeders, Columbia, Missouri.
Foote RH. (1976) Antibacterial agents for bull semen: do they help? In *Proceedings, 6th Technical Conference on Artificial Insemination and Reproduction* pp 23–30. National Association of Animal Breeders, Columbia, Missouri.
Foote RH (1999) Bull sperm surface "craters" and other aspects of semen quality *Theriogenology* **51** 767–775.
Foote RH and Kaproth MT (1997) Sperm numbers inseminated in dairy cattle and nonreturn rate revisited *Journal of Dairy Science* **80** 3072–3076.
Foote RH, Arriola J and Wall RJ (1978) Principles and procedures for photometric measurement of sperm cell concentration In *Proceedings, 7th Technical Conference on Artificial Insemination and Reproduction* pp 55–61. National Association of Animal Breeders, Columbia, Missouri.
Foster ML (2009) Comparison of methods for assessing viability of equine spermatozoa and effects of seminal plasma on viability and motion characteristics of equine spermatozoa. Master of Science Thesis, Texas A&M University, College Station. Available at: http://repository.tamu.edu/bitstream/handle/1969.1/ETD-TAMU-2009-12-7469/FOSTER-THESIS.pdf?sequence=2 (accessed 28 August 2013).
Foster ML, Varner DD, Hinrichs K, Teague S, LaCaze K, Blanchard TL and Love CC (2011a) Agreement between measures of total motility and membrane integrity in stallion sperm *Theriogenology* **75** 1499–1505.
Foster ML, Love CC, Varner DD, Brinsko SP, Hinrichs K, Teague S, LaCaze K and Blanchard TL (2011b) Comparison of methods for assessing integrity of equine sperm membranes *Theriogenology* **76** 334–341.
Fraser L and Strzezek J (2004) The use of comet assay to assess DNA integrity of boar spermatozoa following liquid preservation at 5°C and 16°C *Folia Histochemica et Cytobiologica* **42** 49–55.

Fraser L and Strzezek J (2005) Effects of freeze–thawing on DNA integrity of boar spermatozoa assessed by the neutral comet assay *Reproduction in Domestic Animals* **40** 530–536.

Fraser L, Gorszczaruk K and Strzezek J (2001) Relationship between motility and membrane integrity of boar spermatozoa in media varying in osmolality *Reproduction in Domestic Animals* **36** 325–329.

Freneau GE, Chenoweth PJ, Ellis R and Rupp G (2010) Sperm morphology of beef bulls evaluated by two different methods *Animal Reproduction Science* **118** 176–181.

Freund M and Carol B (1964) Factors affecting haemocytometer counts of sperm concentration in human semen *Journal of Reproduction and Fertilization* **8** 149–155.

Gadella BM and Harrison RAP (2002) Capacitation induces cyclic adenosine 3′,5′-monophosphate-dependent, but apotosis-unrelated, exposure of aminophospholipids at the apical head plasma membrane of boar sperm cells *Biology of Reproduction* **67** 340–350.

Garcia-Macias V, de Paz P, Martinez-Pastor F, Alvarez M, Gomes-Alves S, Bernardo J, Anel E and Anel L (2007) DNA fragmentation assessment by flow cytometry and Sperm-Bos-Halomax (bright-field microscopy and fluorescence microscopy) in bull sperm *International Journal of Andrology* **30** 88–98.

Garner DL and Johnson LA (1995) Viability assessment of mammalian sperm using SYBR-14 and propidium iodide *Biology of Reproduction* **53** 276–284.

Garner DL, Easton MP, Munson ME and Doane MA (1975) Immunofluorescent localization of bovine acrosin *Journal of Experimental Zoology* **191** 127–131.

Garner DL, Pinkel DP, Johnson LA, and Pace MM (1986) Assessment of spermatozoal function using dual fluorescent staining and flow cytometric analysis *Biology of Reproduction* **34** 127–138.

Garner DL, Johnson LA, Yue ST, Roth BL and Haugland RP (1994) Dual DNA staining assessment of bovine sperm viability using SYBR-14 and propidium iodide *Journal of Andrology* **15** 620–629.

Garner DL, Thomas AC, Joery AC, DeJarnette JM and Marshall CE (1997) Fluor assessment of mitochondrial function and viability in cryopreserved bovine spermatozoa *Biology of Reproduction* **57** 1401–1406.

Gillan L, Evans G and Maxwell WMC (1997) Capacitation status and fertility of fresh and frozen–thawed ram spermatozoa *Reproduction, Fertility and Development* **9** 481–487.

Gliozzi TM, Zaniboni L and Cerolini S (2011) DNA fragmentation in chicken spermatozoa during cryopreservation *Theriogenology* **75** 1613–1622.

Graham JK, Kunze E and Hammerstedt RH (1990) Analysis of sperm cell viability, acrosome integrity and mitochondrial function using flow cytometry *Biology of Reproduction* **43** 55–64.

Gravance CG, Garner DL, Baumber J and Ball BA (2000) Assessment of equine sperm mitochondrial function using JC-1 *Theriogenology* **53** 1691–1703.

Green CE and Watson PF (2001) Comparison of the capacitation-like state of cooled boar spermatozoa with true capacitation *Reproduction* **122** 889–898.

Hackett AJ and Macpherson JW (1965) Some staining procedures for spermatozoa. A review *Canadian Veterinary Journal* **6** 55–62.

Hancock JL (1952) The morphology of bull spermatozoa *Journal of Experimental Biology* **29** 445–453.

Hancock JL (1957) The morphology of boar spermatozoa *Journal of the Royal Microscopy Society* **76** 84–97.

Hansen C (2009) Leja-4 counting chamber exerts negative effect on sperm motility In *VIIIth International Conference on Pig Reproduction, The Banff Centre, Banff, Alberta May 31st–June 4th 2009, Program and Abstract Book* p 129 Poster presentation 252-13. Available at: http://www.porkcrc.com.au/ICPR_2009_Abstract_Book.pdf (accessed 28 August 2013).

Hansen C, Vermeiden T, Vermeiden JPW, Simmet C, Day BC and Feitsma H (2006) Comparison of FACSCount AF System, Improved Neubauer haemocytometer, Corning 254 photometer, SpermVision, Ultimate and NucleoCounter SP-100™ for determination of sperm concentration of boar semen *Theriogenology* **66** 2188–2194.

Harrison RAP and Vickers SE (1990) Use of fluorescent probes to assess membrane integrity in mammalian spermatozoa *Journal of Reproduction and Fertility* **88** 343–352.

Herrera J, Fiero R, Zayas H, Conejo J, Garcia A and Betancourt M (2002) Acrosome reaction in fertile and subfertile boarsperm *Archives of Andrology* **48** 133–139.

Hirai M, Boersma A, Hoeflich A, Wolf E, Föll J, Aumüller R and Braun J (2001) Objectively measured sperm motility and sperm head morphometry in boars (*Sus scrofa*): relation to fertility and seminal plasma growth factors *Journal of Andrology* **22** 104–110.

Hishinuma M and Sekine J (2003) Evaluation of membrane integrity of canine epididymal spermatozoa by short hypoosmotic swelling test with ultrapure water *Journal of Veterinary Medical* Science **65** 817–820.

Hodder ADJ and Liu IKM (2011) Spermatozoal motility In *Equine Reproduction* 2nd edition pp 1292–1296 Ed AO McKinnon, EL Squires, EE Vaala and DD Varner. Wiley-Blackwell, Ames.

Holt C, Holt WV, Moore HD, Reed HCB and Curnock RM (1997) Objectively measured boar sperm motility parameters correlate with the outcomes of on-farm inseminations: results of two fertility trials *Journal of Andrology* **18** 212–223.

Holt WV, O'Brien J and Abaigar T (2007) Applications and interpretation of computer-assisted sperm analyses and sperm sorting methods in assisted breeding and comparative research *Reproduction, Fertility and Development* **19** 709–718.

Johansson CS, Matsson FC, Lehn-Jensen H, Nielsen JM and Petersen MM (2008) Equine spermatozoa viability comparing the NucleoCounter SP-100 and the eosin–nigrosin stain *Animal Reproduction Science* **107** 325–326.

Johnson L, Berndtson WE and Pickett BW (1976) An improved method for evaluating acrosomes of bovine spermatozoa *Journal of Animal Science* **42** 951–954.

Johnson LA, Aalbers JG, Willems CMT and Sybesma W (1981) Use of boar spermatozoa for artificial insemination. I. Fertilizing capacity of fresh and frozen spermatozoa in sows on 36 farms *Journal of Animal Science* **52** 1130–1136.

Johnson LA, Aalbers JG, Willems CMT, Rademaker JHM and Rexroad CE (1982) Use of boar spermatozoa for artificial insemination. III. Fecundity of boar spermatozoa stored in Beltsville Liquid and Kiev extenders for three days at 18 C *Journal of Animal Science* **54** 132–136.

Johnston SD, Root Kustritz MV and Olson PNS (2001) Semen collection, evaluation, and preservation In *Canine and Feline Theriogenology* pp 287–306. Saunders, Philadelphia.

Juonala T, Salonen E, Nurttila T and Andersson M (1999) Three fluorescence methods for assessing boar sperm viability *Reproduction in Domestic Animals* **34** 83–87.

Kaproth MT (1988) Use of the programmable balance in semen processing and quality control In *Proceedings, 12th Technical Conference on Artificial Insemination and Reproduction* pp 142–143 National Association of Animal Breeders Columbia, Missouri.

Karabinus DS, Evenson DP, Jost LK, Baer RK and Kaproth MT (1990) Comparison of semen quality in young and mature Holstein bulls measured by light microscopy and flow cytometry *Journal of Dairy Science* **73** 2364–2371.

Karabinus DS, Evenson DP and Kaproth MT (1991) Effects of egg yolk-citrate and milk extenders on chromatin structure and viability of cryopreserved bull sperm *Journal of Dairy Science* **74** 3836–3848.

Karabinus DS, Vogler CJ, Saacke RG and Evenson DP (1997) Chromatin structural changes in sperm after scrotal insulation of Holstein bulls *Journal of Andrology* **18** 549–555.

Kathiravan P, Kalatharan J, Karthikeya G, Rengarajan K and Kadirvel G (2011) Objective sperm motion analysis to assess dairy bull fertility using computer-aided system – a review *Reproduction in Domestic Animals* **46** 165–172.

Katz DF, Davis RO, Delandmeter BA and Overstreet JW (1985) Real-time analysis of sperm motion using automatic video image digitization *Computer Methods and Programs in Biomedicine* **21** 173–182.

Kavak A, Johannisson A, Lundeheim N, Rodriguez-Martinez A, Aidnik M and Einarsson S (2003) Evaluation of cryopreserved stallion semen from Tori and Estonian breeds using CASA and flow cytometry *Animal Reproduction Science* **76** 205–216.

Keeler KD, Mackenzie NM and Dresser DW (1983) Flow microfluoremetric analysis of living spermatozoa stained with Hoechst 33342 *Journal of Reproduction and Fertility* **68** 205–212.

Kenney RM, Evenson DP, Garcia MC and Love CC (1995) Relationships between sperm chromatin structure, motility, and morphology of ejaculated sperm, and seasonal pregnancy rate In *Equine Reproduction VI, Papers from the 6th International Symposium on Equine Reproduction, Aug. 7–13, 1994, Caxambu, Brazil Biology of Reproduction Monograph Series 1* Ed DC Sharp and FW Bazer pp 647–653. Society for the Study of Reproduction, Madison.

Kirk ES, Squires EL and Graham JK (2005) Comparison of *in vitro* laboratory analyses with the fertility of cryopreserved stallion spermatozoa *Theriogenology* **64** 1422–1439.

Kojima Y (1977) Light and electron microscopic studies of "hairpin curved deformity" on boar spermatozoa *Japanese Journal of Veterinary Science* **39** 265–272.

Kovács A and Foote RH (1992) Viability and acrosome staining of bull, boar and rabbit spermatozoa *Biotechnic and Histochemistry* **67** 119–124.

Kremer J (1965) A simple sperm penetration test *International Journal of Fertility* **10** 209–215.

Krueger C and Rath D (2000) Intrauterine insemination in sows with reduced sperm number *Reproduction, Fertility and Development* **12** 113–117.

Kummerfeld HL, Lorton SP and Foote RH (1980) Relationship of bull fertility to sperm migration and motility In *Proceedings, Fifth Annual Meeting of the American Society of Andrology, Chicago* p 81. American Society of Andrology Schaumburg, Illinois.

Kummerfeld HL, Lorton SP and Foote RH (1981a) Relationship of spermatozoal migration in cervical mucus to bovine fertility *Journal of Andrology* **2**103–107.

Kummerfeld HL, Vosburg JK, Lorton SP and Foote RH (1981b) Influence of bovine cervical mucus samples and storage conditions on sperm migration *in vitro Fertility and Sterility* **35** 218–221.

Kuster CE, Hess RA and Althouse GC (2004a) Immunofluorescence reveals ubiquitination of retained distal cytoplasmic droplets on ejaculated porcine spermatozoa *Journal of Andrology* **25** 340–347.

Kuster CE, Singer RS and Althouse GC (2004b) Determining sample size for the morphological assessment of sperm *Theriogenology* **61** 691–703.

Kútvölgyi G, Stefler J and Kovács A (2006) Viability and acrosome staining of stallion spermatozoa by Chicago sky blue and Giemsa *Biotechnic and Histochemistry* **81** 109–117.

Linfor JJ and Meyers SA (2002) Detection of DNA damage in response to cooling injury in equine spermatozoa using single-cell gel electrophoresis *Journal of Andrology* **23** 107–113.

Liu YT and Warme PK (1977) Computerized evaluation of sperm cell motility *Computers and Biomedical Research* **10** 127–138.

Lorton SP (1986) Determination of spermatozoal concentration in initially extended bovine semen In *Proceedings, 11th Technical Conference on Artificial Insemination and Reproduction* pp 77–79. National Association of Animal Breeders, Columbia, Missouri.

Lorton SP, Kummerfeld HL and Foote RH (1981) Polyacrylamide as a substitute for cervical mucus in sperm migration tests *Fertility and Sterility* **35** 222–225.

Lorton SP, Wu ASH, Pace MM, Parker WG and Sullivan JJ (1983) Fine structural abnormalities of spermatozoa from a subfertile bull: a case report *Theriogenology* **20** 585–599.

Lorton SP, Pace MM and Sullivan JJ (1984) Determination of spermatozoal concentration in initially extended bovine semen In *Proceedings, 10th International Congress Animal Reproduction and Artificial Insemination, University of Illinois at Urbana-Champaign Illinois, USDA, June 10–14 1984, Volume II Brief Communications* pp 60–63.

Lorton SP, Sullivan JJ, Bean B, Kaproth M, Kellgren H and Marshall C (1988a) A new antibiotic combination for frozen bovine semen. 2. Evaluation of seminal quality *Theriogenology* **29** 593–607.

Lorton SP, Sullivan JJ, Bean B, Kaproth M, Kellgren H and Marshall C (1988b) A new antibiotic combination for frozen bovine semen. 3. Evaluation of fertility *Theriogenology* **29** 609–61.

Love CC (2005) The sperm chromatin structure assay: a review of clinical applications *Animal Reproduction Science* **89** 39–45.

Love CC (2011) Relationship between sperm motility, morphology and the fertility of stallions *Theriogenology* **76** 547–557.

Love CC and Kenney RM (1998) The relationship of increased susceptibility of sperm DNA to denaturation and fertility in the stallion *Theriogenology* **50** 955–972.

Love CC, Brinsko SP, Rigby SL, Thompson JA, Blanchard TL and Varner DD (2005) Relationship of seminal plasma level and extender type to sperm motility and DNA integrity *Theriogenology* **63** 1584–1591.

Lovercamp KW, Safranski TJ, Fischer KA, Manandhar G, Sutovsky M, Herring W and Sutovsky P (2007a) High resolution light microscopic evaluation of boar semen quality sperm cytoplasmic droplet retention in relationship with boar fertility parameters *Archives of Andrology* **53** 219–228.

Lovercamp KW, Safranski TJ, Fischer KA, Manandhar G, Sutovsky M, Herring W and Sutovsky P (2007b) Arachidonate 15-lipoxygenase and ubiquitin as fertility markers in boars *Theriogenology* **67** 704–718.

Magistrini M, Guitton E, Levern Y, Nicolle JC, Vidament M, Kerboeuf D and Palmer E (1997) New staining methods for sperm evaluation estimated by microscopy and flow cytometry *Theriogenology* **48** 1229–1235.

Makler A (1978) A new multiple exposure photography method for objective human spermatozoal motility determination *Fertility and Sterility* **30** 192–199.

Marchesi DE and Feng HL (2007) Sperm DNA integrity from sperm to egg *Journal of Andrology* **28** 481–489.

Marshall C and Frey L (1976) Semen evaluation at select sires In *Proceedings, 6th Technical Conference on Artificial Insemination and Reproduction* pp 91–93. National Association of Animal Breeders, Columbia, Missouri.

Maxwell WMC and Johnson LA (1997) Chlortetracycline analysis of boar spermatozoa after incubation, flow cytometric sorting, cooling, or cryopreservation *Molecular Reproduction and Development* **46** 408–418.

Mayevsky A, Bar-Sagie D and Bartoov B (1980) A multi-channel system for the measurement of spermatozoan collective motility *International Journal of Andrology* **3** 436–446.

Mitchell JR, Hanson RD and Fleming WN (1978) Utilizing differential interference contrast microscopy for evaluating abnormal spermatozoa In *Proceedings, 7th Technical Conference on Artificial Insemination and Reproduction* pp 64–68. National Association of Animal Breeders, Columbia, Missouri.

Morrell JM, Johannisson A, Juntilla L, Rytty K, Backgren L, Dalin A-M and Rodriguez-Martinez H (2010) Stallion sperm viability, as measured by the NucleoCounter SP-100™, is affected by extender and enhanced by single layer centrifugation *Veterinary Medicine International* **2010:** 659862.

Mortimer D (1990) Objective analysis of sperm motility and kinematics In *CRC Handbook of the Laboratory Diagnosis and Treatment of Infertility* pp 97–133 Ed Keel BA and Webster BW. CRC Press, Boston.

Mortimer ST (2000) CASA – practical aspects *Journal of Andrology* **21** 515–524.

Murase T, Okuda K and Sato K (1990) Assessment of bull fertility using a mucus penetration test and a human chorionic gonadotrophin stimulaton test *Theriogenology* **34** 801–812.

Nagy S, Jansen J, Topper EK and Gadella BM (2003) A triple-stain flow cytometric method to assess plasma and acrosome membrane integrity of cryopreserved bovine sperm immediately after thawing in presence of egg-yolk particles *Biology of Reproduction* **68** 1828–1835.

Neild D, Chaves G, Flores M, Mora N, Beconi M and Agüero A (1999) Hypoosmotic test in equine spermatozoa *Theriogenology* **51** 721–727.

Núñez-Martínez I, Moran JM and Peña FJ (2006) A three-step statistical procedure to identify sperm kinematic subpopulations in canine ejaculates: changes after cryopreservation *Reproduction in Domestic Animals* **41** 408–415.

Nynca J and Ciereszko A (2009) Measurement of concentration and viability of brook trout (*Salvelinus fontinalis*) spermatozoa using computer-aided fluorescent microscopy *Aquaculture* **292** 256–258.

Odhiambo JF, Sutovsky M, DeJarnette JM, Marshall C and Sutovsky P (2011) Adaption of ubiquitin-PNA based sperm quality assay for semen evaluation by a conventional flow cytometer and a dedicated platform for flow cytometric semen analysis *Theriogenology* **76** 1168–1176.

Oettle EE (1986) Using a new acrosome stain to evaluate sperm morphology *Veterinary Medicine* **81** 263–266.

Okazaki T, Yoshida S, Teshima H and Shimada M (2011) The addition of calcium ion chelator EGTA to thawing solution improves fertilizing ability in frozen–thawed boar sperm *Animal Science Journal* **82** 412–419.

Oltenacu EAB and Foote RH (1976) Monitoring fertility of A.I. programs: can non-return rate do the job? In *Proceedings, 6th Technical Conference on Artificial Insemination and Reproduction* pp 61–68. National Association of Animal Breeders, Columbia, Missouri.

Ortega-Ferrusola C, Garcia BM, Rama VS, Gallardo-Bolaños JM, González-Fernández L, Tapia JA, Rodriguez-Martinez H and Peña FJ (2009) Identification of sperm subpopulations in stallion ejaculates: changes after cryopreservation and comparison with traditional statistics *Reproduction in Domestic Animals* **44** 419–423.

Ostermeier GC (2000) Sperm nuclear morphology and its relationship with bull fertility. PhD dissertation, University of Wisconsin, Madison.

Ostermeier GC, Sargeant GA, Yandell BS, Evensen DP and Parrish JJ (2001) Relationship of bull fertility to sperm nuclear shape *Journal of Andrology* **22** 595–603.

Pace MM (1980) Problems associated in evaluation of spermatozoa in the field In *Proceedings, 8th Technical Conference on Artificial Insemination and Reproduction* pp 67–70. National Association of Animal Breeders, Columbia, Missouri.

Pace MM, Sullivan JJ, Elliott FI, Graham EF and Coulter GH (1981) Effects of thawing temperature, number of spermatozoa and spermatozoal quality on fertility of bovine spermatozoa packaged in 0.5ml French straws *Journal of Animal Science* **53** 693–701.

Palmer CW and Barth AD (2003) Comparison of the BullMate™ sperm quality analyzer with conventional means of assessing the semen quality and breeding soundness of beef bulls *Animal Reproduction Science* **77** 173–185.

Papaioannou KZ, Murphy RP, Monks RS, Hynes N, Ryan MP, Boland MP and Roche JF (1997) Assessment of viability and mitochondrial function of equine spermatozoa using double staining and flow cytometry *Theriogenology* **48** 299–312.

Parrish JJ and Foote RH (1987) Quantification of bovine sperm separation by a swim-up method: relationship to sperm motility, integrity of acrosomes, sperm migration in polyacrylamide gel and fertility *Journal of Andrology* **8** 259–266.

Parrish JJ, Susko-Parrish JL and Graham JK (1999) *In vitro* capacitation of bovine spermatozoa: role of intracellular calcium. *Theriogenology* **51** 461–472.

Parrish J, Schindler J, Willenburg K, Enwall L and Kaya A (2012) Quantitative sperm shape analysis: what can this tell us about male fertility In *Proceedings, 24th Technical Conference on Artificial Insemination and Reproduction* pp 74–80. National Association of Animal Breeders, Milwaukee.

Paufler SK (1974) Die künstliche Besamung beim Rind [Artificial insemination of cattle] In *Künstliche Besamung und Eitransplantation bei Mensch und Tier* pp 45–80 Ed SK Paufler. M und M Schaper, Hannover.

Payne B and Clark S (2005) Accuracy of methods used to determine sperm concentration in extended porcine semen doses In *Proceedings of the Annual Meeting of the American Association of Swine Veterinarians, Phoenix, Arizona* p 37. American Association of Swine Veterinarians, Perry, Iowa.

Peña A, Johannisson A and Linde-Forsberg C (1999) Post-thaw evaluation of dog spermatozoa using new triple fluorescent staining and flow cytometry *Theriogenology* **52** 965–980.

Peña FJ, Johannisson A, Wallgren M and Rodriguez-Martinez H (2003) Assessment of fresh and frozen–thawed boar semen using an Annexin-V assay: a new method of evaluating sperm membrane integrity *Theriogenology* **60** 677–689.

Peña AI, Barrio M, Becerra JJ, Quintela LA and Herradón PG (2004a) Zona pellucida binding ability and responsiveness to ionophore challenge of cryopreserved dog spermatozoa after different periods of capacitation *in vitro Animal Reproduction Science* **84** 193–210.

Peña FJ, Johannisson A, Wallgren M and Rodriguez-Martinez H (2004b) Effect of hyaluronan supplementation on boar sperm motility and membrane lipid architecture status after cryopreservation *Theriogenology* **61** 63–70.

Peña FJ, Saravia F, Garcia-Herreros M, Nuñez-Martínez I, Tapia JA, Johannisson A, Wallgren M and Rodriguez-Martinez H (2005) Identification of sperm morphometric subpopulations in two different portions of the boar ejaculate and its relation to postthaw quality *Journal of Andrology* **26** 716–723.

Peña AI, Barrio M, Becerra JJ, Quintela LA and Herradón PG (2007) Infertility in a dog due to proximal cytoplasmic droplets in the ejaculate: investigation of the significance for sperm functionality *in vitro Reproduction in Domestic Animals* **42** 471–478.

Petrunkina AM and Harrison RAP (2010) Systematic misestimation of cell populations in flow cytometry: a mathematical analysis *Theriogenology* **73** 839–847.

Petrunkina AM, Friedrich J, Drommer W, Bicker G, Waberski D and Töpfer-Petersen E (2001) Kinetic characterization of the changes in protein tyrosine phosphorylation of membranes, cytosolic Ca^{2+} concentration and viability in boar sperm populations selected by binding to oviductal cells *Reproduction* **122** 469–480.

Pickett BW (1968) Collection and evaluation of stallion semen In *Proceedings, 2nd Technical Conference on Artificial Insemination and Reproduction* pp 80–8087. National Association of Animal Breeders, Columbia, Missouri.

Pilip R, Del Campo MR, Barth AD and Mapletoft RJ (1996) *In vitro* fertilizing characteristics of bovine spermatozoa with multiple nuclear vacuoles: a case study *Theriogenology* **46** 1–12.

Pinto CRF and Kozink DM (2008) Simplified hypoosmotic swelling test (HOST) of fresh and frozen–thawed canine spermatozoa *Animal Reproduction Science* **104** 450–455.

Pozor MA, Zambrano GL, Runcan E and Macpherson ML (2012) Usefulness of Dip Quick stain in evaluating sperm morphology in stallions In *Proceedings, 58th AAEP Annual Convention, Anaheim, California* pp 506–510. American Association of Equine Practitioners, Lexington.

Pursel VG and Johnson LA (1974) Gluteraldehyde fixation of boar spermatozoa for acrosome evaluation *Theriogenology* **1** 63–68.

Pursel VG, Johnson LA and Rampacek GB (1972) Acrosome morphology of boar spermatozoa incubated before cold shock *Journal of Animal Science* **34** 278–283.

Pursel VG, Johnson LA and Schulman LL (1974) Acrosome morphology of boar spermatozoa during *in vitro* aging *Journal of Animal Science* **38** 113–116.

Quintero-Moreno A, Rigau T and Rodríguez-Gil JE (2004) Regression analyses and motile sperm subpopulation structure study as improving tools in boar semen quality analysis *Theriogenology* **61** 673–690.

Quintero-Moreno A, Rubio J and González D (2008) Estimating cryodamage on plasma membrane integrity in bull spermatozoa using HOS-ENY test *Reproduction in Domestic Animals* **43(Supplement 5)** 62.

Rahman MB, Vandaele L, Rijsselaere T, Maes D, Hoogewijs M, Frijters A, Noordman J, Granados A, Dernelle E, Shamsuddin M, Parrish JJ and Van Soom A (2011) Scrotal insulation and its relationship to abnormal morphology, chromatin protamination and nuclear shape of spermatozoa in Holstein–Friesian and Belgian Blue bulls *Theriogenology* **76** 1246–1257.

Rathi R, Colenbrander B, Bevers MM and Gadella BM (2001) Evaluation of *in vitro* capacitation of stallion spermatozoa *Biology of Reproduction* **65** 462–470.

Reicks DL (2009) Monitoring boar studs for early detection of PRRS virus to avoid complete depop–repop of studs and infection of downstream breeding herds In *Proceedings, Pre-Conference Seminar, Managing PRRS Introduction into High Risk Populations, 40th Annual Meetings of the American Association of Swine Veterinarians* pp 19–22. American Association of Swine Veterinarians, Perry, Iowa

Reicks DL, Levis DG and Kuster C (2008) The effect of sperm count and bacteria on farrowing rate and total born In *Abstracts: 20th International Pig Veterinary Society Congress, June 2008, Durban, South Africa.* International Pig Veterinary Society, Waterkloof

Revell SG and Wood PDP (1978) A photographic method for the measurement of motility of bull spermatozoa *Journal of Reproduction and Fertility* **54** 123–126.

Richardson GF, Donald AW and MacKinnon CE (1992) Comparison of different techniques to determine the percentage of intact acrosomes in frozen–thawed bull semen *Theriogenology* **38** 557–564.

Rikmenspoel R (1964) Electronic analyzer for measuring velocities and the concentration of spermatozoa *Review of Scientific Instruments* **35** 52–57.

Robayo V, Montenegro V, Valdés C and Cox JF (2008) CASA assessment of kinematic parameters of ram spermatozoa and their relationship to migration efficiency in ruminant cervical mucus *Reproduction in Domestic Animals* **43** 393–399.

Root Kustritz MV, Olson PN, Johnston SD and Root TK (1998) The effects of stains and investigators on assessment of morphology of canine spermatozoa *Journal American Animal Hospital Association* **38** 348–352.

Rota A, Penzo N, Vincenti L and Mantovani R (2000) Hypoosmotic swelling (HOS) as a screening assay for testing *in vitro* fertility of bovine spermatozoa *Theriogenology* **53** 1415–1420.

Rota A, Bastianacci V, Magelli C, Panzani D and Camillo F (2010) Evaluation of plasma membrane integrity of donkey spermatozoa *Reproduction in Domestic Animals* **45** 228–233.

Rothschild L (1953) The movements of spermatozoa In *Mammalian Germ Cells* pp 122–133 Ed GEW Wolstenholme. Little, Brown and Company, Boston.

Rovira A, Reicks D and Munoz-Zanzi C (2007) Evaluation of surveillance protocols for detecting porcine reproductive and respiratory syndrome virus infection in boar studs by simulation modeling *Journal of Veterinary Diagnostic Investigation* **19** 492–501.

Rozeboom KJ, Reicks DL and Wilson ME (2004) The reproductive performance and factors affecting on-farm application of low-dose intrauterine deposit of semen in sows *Journal of Animal Science* **82** 2164–2168.

Rubio J, González D, Valeris R, Osorio C and Quintero A (2008) Can ORT complement routine semen analysis and predict potential fertility in bulls *Reproduction in Domestic Animals* **43(Supplement 5)** 64.

Ruiz-Sánchez AL, O'Donoghue R, Novak S, Dyck MK, Cosgrove JR, Dixon WT and Foxcroft GR (2006) The predictive value of routine semen evaluation and IVF technology for determining relative boar fertility *Theriogenology* **66** 736–748.

Saacke RG (1970) Morphology of the sperm and its relationship to fertility In *Proceedings, 3rd Technical Conference on Artificial Insemination and Reproduction* pp 17–30. National Association of Animal Breeders, Columbia, Missouri.

Saacke RG (1983) Semen quality in relation to semen preservation *Journal of Dairy Science* **66** 2635–2644.

Saacke RG and Marshall CE (1968) Observations on the acrosomal cap of fixed and unfixed bovine spermatozoa *Journal of Reproduction and Fertility* **16** 511–514.

Saacke RG and White JM (1972) Semen quality tests and their relationship to fertility In *Proceedings, 4th Technical Conference on Artificial Insemination and Reproduction* pp 22–27 National Association of Animal Breeders, Columbia, Missouri.

Saacke RG, DeJarnette JM, Nebel RL and Nadir S (1991) Assessing bull fertility In *Proceedings, Annual Meeting of the Society for Theriogenology* pp 56–69.

Saacke RG, DeJarnette JM, Bame JH, Karabinus DS and Whitman SS (1998) Can spermatozoa with abnormal heads gain access to the ovum in artificially inseminated super- and single-ovulating cattle? *Theriogenology* **50** 117–128.

Sahin E, Petrunkina AM, Töpfer-Peterson E and Waberski D (2008) Volume regulation ability in bull epididymal sperm *Reproduction in Domestic Animals* **43(Supplement 5)** 48.

Sailer BL, Jost LK and Evenson DP (1995) Mammalian sperm DNA susceptibility to *in situ* denaturation associated with the presence of DNA strand breaks as measured by the terminal deoxynucleotidyl transferase assay *Journal of Andrology* **16** 80–86.

Salisbury GW and Mercier E (1945) The reliability of estimates of the proportion of morphologically abnormal spermatozoa in bull semen *Journal of Animal Science* **4** 174–178.

Salisbury GW and VanDemark NL (1961) *Physiology of Reproduction and Artificial Insemination of Cattle* p 361. WH Freeman and Co, San Francisco (cited by Saacke, 1983).

Salisbury GW, Beck GH, Elliott I and Willett EL (1943) Rapid methods for estimating the number of spermatozoa in bull semen *Journal of Dairy Science* **26** 69–78.

Saravia F, Nuñez-Martínez I, Morán JM, Soler C, Muriel A, Rodriguez-Martinez H and Peña FJ (2007) Differences in boar sperm head shape and dimensions recorded by computer-assisted sperm morphometry are not related to chromatin integrity *Theriogenology* **68** 196–203.

Seaman EK, Goluboff E, BarCharma N and Fisch H (1996) Accuracy of semen counting chambers as determined by the use of latex beads *Fertility and Sterility* **66** 662–665.

Setchell BP (1991) Male reproductive organs and semen. In *Reproduction in Domestic Animals* 4th edition p 245 Ed PT Cupps. Academic Press, San Diego.

Shaffer HE and Almquist JO (1948) Vital staining of bovine spermatozoa with eosin–aniline blue staining solution *Journal of Dairy Science* **31** 677 [abstract].

Shannon P, Curson B and Rhodes AP (1984) Relationship between total spermatozoa per insemination and fertility of bovine semen stored in Caprogen at ambient temperature *New Zealand Journal of Agricultural Research* **27** 35–41.

Shin SJ, Lein DH, Patten VH and Ruhnke HL (1988) A new antibiotic combination for frozen semen. 1. Control of mycoplasmas, ureaplasmas, *Campylobacter fetus* subsp. *venerealis* and *Haemophilus somnus* *Theriogenology* **29** 577–591.

Sonderman J, Rathje T and Stumpf T (2011) The role of reproductive technologies in genetic improvement at the commercial level In *Proceedings of the Annual Meeting of the American Association of Swine Veterinarians, Phoenix, Arizona* pp 237–239. American Association of Swine Veterinarians, Perry, Iowa..

Su T-W, Xue L, and Ozcan A (2012) High-throughput lens-free 3D tracking of human sperms reveals rare statistics of helical trajectories In *Proceedings of the National Academy of Sciences of the United States of America* **109** 16018–16022.

Sullivan JJ (1970) Sperm numbers required for optimum breeding efficiency in cattle In *Proceedings, 3rd Technical Conference on Artificial Insemination and Reproduction* pp 36–43 National Association of Animal Breeders, Columbia, Missouri.

Tamuli MK and Watson PF (1994) Use of a simple staining technique to distinguish acrosomal changes in the live sperm sub-population *Animal Reproduction Science* **35** 247–254.

Teague SR, Bakker AZ, Brehm LM, Love CC and Varner DD (2010) Comparison of SCSA and TUNEL assays for measurement of stallion sperm DNA quality before and after centrifugation through a silica-particle solution *Special Issue: 10th International Symposium on Equine Reproduction, July 26–30, 2010 in Lexington, Kentucky Animal Reproduction Science* **121(Supplement 1–2)** S205.

Thomas CA, Garner DL, DeJarnette JM and Marshall CE (1997) Fluorescent assessments of acrosomal integrity and viability in cryopreserved bovine spermatozoa *Biology of Reproduction* **56** 991–998.

Thomas CA, Garner DL, DeJarnette JM and Marshall CE (1998) Effect of cryopreservation of bovine sperm organelle function and viability as determined by flow cytometry *Biology of Reproduction* **58** 786–793.

Thundathil J, Meyer R, Palasz AT, Barth AD and Mapletoft RJ (2000) Effect of the knobbed acrosome defect in bovine sperm on IVF and embryo production *Theriogenology* **54** 921–934.

Tibary A, Rodriguez J and Samper JC (2009) Microbiology and diseases of semen In *Equine Breeding Management and Artificial Insemination* 2nd edition pp 99–112 Ed JC Samper. Saunders Elsevier, St Louis.

Tsakmakidis IA, Lymberopoilos AG and Khalifa TAA (2010) Relationship between sperm quality traits and field-fertility of porcine semen *Journal of Veterinary Science* **11** 151–154.

Van Dellen G and Elliott FI (1978) Procedure for time exposure darkfield photomicrography to measure percentage progressively motile spermatozoa In *Proceedings, 7th Technical Conference on Artificial Insemination and Reproduction* pp 53–55. National Association of Animal Breeders, Columbia, Missouri.

Van der Horst G and Maree L (2009) SpermBlue®: a new universal stain for human and animal sperm which is also amenable to automated sperm morphology analysis *Biotechnic and Histochemistry* **84** 299–308.

Varner DD, Scanlan CM, Thompson JA, Brumbaugh GW, Blanchard TL, Carlton CM and Johnson L (1998) Bacteriology of preserved stallion semen and antibiotics in semen extenders *Theriogenology* **50** 559–573.

Vazquez JM, Roca J, Gil MA, Cuello C, Parrilla I, Vazquez JL and Martínez EA (2008) New developments in low-dose insemination technology *Theriogenology* **70** 1216–1224.

Vianna WL, Bruno DG, Namindome A, de Camplos Rosseto AC, Barnabe RC, de Sant'Anna Morelli A (2007) Evaluation of the Karras sperm densimeter in relation to the Neubauer counting chamber for sperm concentration in boar semen *Reproduction in Domestic Animals* **42** 466–470.

Vianna FP, Papa FO, Zahn FS, Melo CM and Dell'Aqua JA Jr (2009) Thermoresistance sperm tests are not predictive of potential fertility for cryopreserved bull semen *Animal Reproduction Science* **113** 279–282.

Waberski D, Meding S, Dirksen G, Weitze KF, Leiding C and Hahn R (1994) Fertility of long-term-stored boar semen: influence of extender (Androhep and Kiev), storage time and plasma droplets in the semen *Animal Reproduction Science* **36** 145–151.

Waberski D, Schapmann E, Henning, H, Riesenbeck A and Brandt H (2011) Sperm chromatin structural integrity in normospermic boars is not related to semen storage and fertility after routine AI *Theriogenology* **75** 337–345.

Waterhouse KE, Haugan T, Kommisrud E, Tverdal A, Flatberg G, Farstad W, Evenson DP and De Angelis PM (2006) Sperm DNA damage is related to field fertility of semen from young Norwegian Red bulls *Reproduction, Fertility and Development* **18** 781–788.

Watson PF (1975) Use of a Giemsa stain to detect changes in acrosomes of frozen ram spermatozoa *Veterinary Record* **97** 12–15.

Wells ME and Awa OA (1970) New technique for assessing acrosomal characteristics of spermatozoa *Journal of Dairy Science* **53** 227–232.

Willett EL and Buckner PJ (1951) The determination of numbers of spermatozoa in bull semen by measurement of light transmission *Journal of Animal Science* **10** 219–225.

Williams WW (1920) Technique of collecting semen for laboratory examination with a review of several diseased bulls *Cornell Veterinarian* **101** 87–94.

Wongtawan T, Saravia F, Wallgren M, Caballero I and Rodriguez-Martinez H (2006) Fertility after deep intra-uterine artificial insemination of concentrated low-volume boar semen doses *Theriogenology* **65** 773–787.

Xu X, Pommier S, Arbov T, Hutchings B, Sotto W and Foxcroft GR (1998) *In vitro* maturation and fertilization techniques for assessment of semen quality and boar fertility *Journal of Animal Science* **76** 3079–3089.

Young DC, Foote RH, Turkheimer AR and Hafs HD (1960) A photoelectric method for estimating the concentration of sperm in boar semen *Journal of Animal Science* **19** 20–25.

Zhou R, Shi B, Chou KC, Oswalt MD and Haug A (1990) Changes in intracellular calcium of porcine sperm during *in vitro* incubation with seminal plasma and a capacitating medium *Biochemical and Biophysical Research Communications* **172** 47–53.

Appendix

THE CANADIAN VETERINARY JOURNAL
LA REVUE VETERINAIRE CANADIENNE

Volume 6 March 1965 Number 3

SOME STAINING PROCEDURES FOR SPERMATOZOA. A REVIEW

A. J. Hackett* and J. W. Macpherson*

STAINING is a useful adjunct to the microscopic examination of spermatozoa. Various dyes react differently when mixed with sperm cells. All cells are stained by rose bengal or bengal red; these dyes are therefore useful for morphological studies. Water soluble halogenated derivatives of fluorescein, eosin B, eosin Y, and erythrosin B will stain dead spermatozoa (28). Recently fluorescent dyes have been used to stain spermatozoa in opaque media (1, 2, 38, 39). Feulgen DNA stain is used to estimate the DNA content of spermatozoa (32).

The purpose of this paper is to review and outline the preparation, procedures and dyes used to stain spermatozoa. For these purposes it is convenient to divide them into morphological, differential, fluorescent and Feulgen DNA stains.

MORPHOLOGICAL STAINS

The most frequently used dye for morphological staining is rose bengal. Bengal red, Belling's iron aceto-carmine, India Ink, Alkaline Methyl Violet Solution, Giemsa or Heidenhain's iron hematoxylin are generally used for more specific purposes. An example is the use of India Ink as a contrast medium to outline the acrosome.

Rose Bengal

Preparation
Rose bengal....................................3 gm.
Formalin 40%..................................1 ml.
Distilled water...............................99 ml.

Dissolve the stain in the water and formalin and sterilize.

Staining Method

Place a drop of semen on a clean slide.

Make a thin film by drawing another slide up to the drop until the semen spreads to the edges of the upper slide, gently push the upper slide across the lower slide at a 30° angle away from the drop.

Dry overnight or fix with low heat.

Stain for five min.

Wash gently in distilled water.

Dry and examine.

This method was used by Herman and Swanson (21) for whole semen. It is possible to see protoplasmic droplets if the slide is examined before drying. This dye is suitable for staining all mammalian spermatozoa.

Bengal red may be used in a similar manner.

Belling's Iron Aceto-carmine

Preparation

Sodium citrate dihydrate in distilled water 2.9%

Bouin's fixative.
Saturated aqueous picric acid 75 ml.
Formalin 40%...........................25 ml.
Acetic acid...............................5 ml.

Belling's iron aceto-carmine

Ferric hydrate solution–small amount of ferric hydrate dissolved in 45% acetic acid.

Prepare ordinary aceto-carmine by adding an excess amount of powdered carmine to a 45% solution of boiling glacial acetic acid. Cool and

* Reproductive Physiology Section, Dept. of Animal Husbandry, University of Guelph, Guelph, Ontario, Canada.

filter. To this add a trace of a solution of ferric hydrate dissolved in 45% acetic acid until the liquid becomes bluish red with no visible precipitate. If the stain is too dark, more aceto-carmine may be added or it may be diluted with 45% acetic acid.

Staining Method

Flood a slide with a mixture of 0.16 ml. whole semen and 15 ml. of 2.9% sodium citrate dihydrate.

Allow the excess to drain off.

Fix the remaining sperm with Bouin's fixative.

Dry and rinse.

Stain in Belling's iron aceto-carmine for 20 minutes or longer.

Cummings (12) used this procedure for staining bull spermatozoa and found it satisfactory for counting abnormal sperm.

Alkaline Methyl–Violet Solution

Preparation

Make up stock solutions as follows in distilled water. Sterilize with heat.

A. Sodium chloride solution 1%.

B. Sodium carbonate solution (Na_2CO_3 sicc.) 1%.

C. Methyl-violet solution (stain quality: 6B) 1%.

While solution C will keep for a year or more, the mixture mentioned below, using solutions B and C, must be used within ten minutes.

Staining Method

The sample is diluted with 1% sodium chloride solution. Best smears can be obtained when the mixture contains about 200,000 sperms per cmm. For the average concentration of bull semen this mixture is attained by mixing about one part semen with four to five parts sodium chloride solution.

After thorough mixing (avoid the formation of scum) a suitable drop is picked up with a glass rod and smeared on a slide free of grease using the rod or the edge of a slide. Dry in air. With some experience and with samples of lower concentration, dilution may be omitted and the semen smeared directly in a thin film by means of a glass rod. Such preparations must be fixed and stained immediately.

The preparation is fixed by being passed several times through a flame, or by passage in alcohol.

Wash the slide with distilled water. Pour off the water.

One part of 1% sodium carbonate solution is mixed thoroughly with nine parts of 1% methyl-violet solution immediately before use. Pour some of this on to the still moist preparation and leave for four to five minutes.

Pour off the stain and wash the preparation with flowing distilled water; three or four vigorous jets from a wash-bottle will suffice; do not wash more than ten to fifteen seconds in all. Blot with filter paper. Dry by passing the slide two or three times through a flame.

Mount with neutral Canada balsam (differentiation).

Various mammalian spermatozoa, which will appear a violet colour, may be stained using this procedure (6).

India Ink

Preparation

India Ink[1]

Staining Method

Place a small drop of semen on a clean slide.

Add about ten times as much India Ink.

Gently mix with a swab stick or by tilting the slide from side to side. Smearing is carried out by placing another slide over the drop and allowing the mixture to spread. Carefully separate the slides. If slides are too dark, use another clean slide and repeat until the right intensity is obtained. Dry and examine.

Some inks have not proven satisfactory. If difficulty is experienced various available brands should be tried.

This technique gives a good morphological picture of various mammalian spermatozoa and outlines the acrosome. No fixing or initial smearing is required.

Giemsa Stain

Preparation

Giemsa (Gurr)..3 ml.
Sorensens phosphate buffer
(pH 7.0) ....2 ml.
Glass distilled water.............................35 ml.

Mix and sterilize.

Staining Method

Make smear.

Dry in air or on warm stage.

Fix in neutral formal-saline (5% formalin) for 30 minutes.

Stain for one and one-half hours.

Some differences in morphology of live and dead cells at time of mixing may be distinguished (18).

Heidenhain's Iron Hematoxylin

Preparation

M/15 phosphate buffer (pH 7.6).

Osmium tetroxide vapour.

Two and one-half per cent ferric alum solution (ammonio-ferric sulphate).

Hematoxylin

DPX[2]

Staining Method

Dilute semen in M/15 phosphate buffer pH 7.6.

Make smears and air dry.

Fix in osmium tetroxide vapour for ten minutes at 37° C.

Wash overnight in running tap water.

Mordant[3] 24 hours in two and one-half per cent ferric alum solution.

Stain for similar period in hematoxylin.

Differentiate in fresh alum solution.

Mount in DPX.

The acrosomes of cells which were living at the time of staining show up less than those which were dead at the time of staining (18).

Differential Stains

As early as 1927, Devereux and Tanner (13) reported that yeast cells varied in colour when stained with congo red, eosin, erythrosin, methyl green, neutral red, safranin (0.5%), gentian violet (0.4%), indigo carmine (3.33%), and methylene blue (0.01 to 0.05%). Congo red and eosin gave a faint pink colour and erythrosin gave a dark pink colour. Except for methylene blue (0.01%), the other dyes were unsatisfactory for differentiating live from dead cells. More recently, Hanks and Wallace (20) stated that live tissue culture cells have the ability to exclude eosin. Phillips and Terryberry (32) have found that tissue culture cells also exclude trypan blue and erythrosin. Katsube and Blobel (22) found that viable leucocytes also exclude trypan blue. Stulberg *et al.* (35) and Stulberg (36) used trypan blue for an indicator of viable strains of animal cells. Wales (41) used trypan blue, eosin and erythrosin as an aid in the estimation of the percentage of "live" human and dog spermatozoa. Blackshaw (5), using congo red, reported better results and stated that up to 15% glycerol in diluted semen has no detrimental effects on the staining results. This is contrary to Mixner and Saroff (31) who reported that levels above 4% glycerol would adversely affect the staining results.

Wales and White (40, 42) postulated that dead cells stain because of a change in permeability which results in a loss of ions and other substances and a disruption of the cell surface. Emmens and Blackshaw (16) and Mayer *et al.* (28) reported that staining of dead cells might not occur if the pH, osmotic pressure or stain concentrations vary within a narrow range. Staining did not occur if stains were not in solutions containing highly ionizable salts.

Eosin B Opal Blue[4]

Preparation

M/8 phosphate buffer: (pH 7.4) 80.4 ml. M/8 Na_2 HPO_4 and 19.6 M/8 KH_2 PO_4

Stain A. Eosin B (water and alcohol soluble) 2% M/8 phosphate buffer (pH 7.3)

Stain B. One part opal blue (undiluted)
One part M/8 phosphate buffer (pH 7.4) } pH 5.7

A staining mixture of one part A and one part B is approximately isotonic with semen and has a pH of near 6.7.

Staining method

Place a drop of stain on a clean glass slide.

Mix the amount of semen which adheres to a glass stirring rod, to the dye.

Make a thin film between the surfaces of two slides (similar to India Ink technique).

Dry on a warm plate at 40° C.

The posterior portion of the sperm head will stain with varied intensity from red to dark purple; the anterior part of the head usually stains a light pink and occasionally not at all. The sperm that do not take the dye appear as a clear outline against a light blue background (23). This staining technique is used for ram spermatozoa but would probably be effective for other mammals also. Macleod (24) was unable to duplicate the results with human sperm. Glynden *et al.* (17) used a similar mixture containing 0.4 gm. water soluble eosin B in 50 ml. isosmotic phosphate buffer and 14 ml. opal blue.

Eosin B–Fast Green F C F

Preparation

M/8 phosphate buffer (pH 7.35) (one part M/8 monosodium phosphate and four parts disodium phosphate)....................100 ml.

Eosin bluish (Eosin B water and alcohol soluble)..0.8 gm.

Fast Green F C F..................................2 gm.

Mix and sterilize by heating. The final pH of the mixture is 7.25.
Staining Method
Use the technique similar to Eosin B Opal Blue.
Dead cells are deep purple to red (16, 28, 29). This procedure is useful in staining human, bull, rabbit, ram and stallion semen. Erythrosin bluish 0.8% (Erythrosin B) may be substituted for eosin B. Sometimes 0.4% erythrosin and 0.4% eosin solutions are used, although best results are obtained using 0.8% solutions.

Eosin B–Aniline Blue

Preparation
M/8 phosphate buffer
(pH 7.3–7.4)..100 ml.
Eosin B (water and alcohol
soluble).. 1 gm.
Aniline blue (water soluble)......................4 gm.
Dissolve the eosin and aniline in the phosphate buffer and sterilize by heating.
Filter when cool.
Staining Method
Smears may be made as previously outlined or by mixing a drop of stain with a drop of semen on a slide and spreading by means of another slide similar to the rose bengal technique.
Shaffer and Almquist (34) used this mixture and found aniline blue to be a good background stain. They noted that the percentage differences of unstained cells due to buffer concentrations were not statistically significant while those due to pH (pH of mixture is 7.2) were significant at the 5% level.
At this laboratory this stain was found unsatisfactory for frozen semen diluted in sterile milk (25). It was toxic to spermatozoa stored overnight at 5° C.
Cascarrett (9) observed good results using eosin-aniline blue dye for staining human and dog semen.
Preparation
Eosin (water and alcohol soluble in distilled water) 5%................................one volume
Phenol in distilled water 1%..........one volume
Aniline-water soluble in distilled water 5%..................................two volumes
Mix and filter.
Fixative–equal parts of alcohol and ether.
Staining Method
Liquefy semen for 30 to 60 minutes.
Make thin even smears.
Fix for three minutes and air dry.
Stain five to seven minutes at 40–60° C.
Wash in distilled water.

Revector Soluble Blue

Preparation
Revector soluble blue 706........................1 gm.
Glucose..1.5 gm.
Sodium chloride....................................0.1 gm.
$Na_2HPO_4 \cdot 12H_2O$................................0.3 gm.
KH_2PO_4...0.0005 gm.
Distilled water...50 ml.
Mix the reagents in distilled water.
This solution can be made up fresh or placed in 1 ml. ampoules and autoclaved for later use.
Counterstain–neutral red 1%.
Fixative–one-half volume saturated solution of mercuric chloride in distilled water and one-half volume absolute alcohol; followed by 90% alcohol containing a few drops of iodine in potassium iodide to give a straw coloured solution.
Staining Method
Place four to five drops of seminal fluid on a clean slide or in a small test tube and add one drop of supravital dye and mix well.
Leave mixture for two to three minutes.
Withdraw one drop.
Make a thin smear on a clean dry slide and air dry.
Fix for 15–20 seconds.
Dip in alcoholic iodine and in 90% alcohol.
Dry.
Counterstain in neutral red.
Rinse in water.
Differentiate *with care* in 90% alcohol.
Dry.
These slides may be examined as prepared or passed through absolute alcohol and xylol and mounted in Canada balsam. The dead nuclei stain blue or purple while the live nuclei are *clear* red. A few non-nucleated forms remain unstained. This stain is used mainly for human semen and does not kill the spermatozoa (11). It is said to show viability and abnormalities of spermatozoa while only taking four minutes to complete.

Eosin Nigrosin

Preparation
A. Eosin (water and alcohol soluble) in distilled water, 5%.
B. Nigrosin (water soluble) in distilled water, 10%.
If A and B are sterilized by heat and remain separated, they may be stable for several years.

Staining Method

Place a small drop of semen on a clean slide.

Place two small drops of eosin on the slide.

Place four small drops of nigrosin on the slide.

With a glass rod, mix the semen and eosin thoroughly till homogeneous, then mix with the nigrosin.

Stir for a few seconds.

Smear on a slide.

Dry over a flame.

The mixing and smearing should not take over one minute. A thin smear is important.

This technique was used by Blom (7) and also by Williams and Pollack (44) when opal blue was unobtainable. In this laboratory, this technique was found unsatisfactory for applications to semen frozen in sterile milk diluent. Five per cent eosin was toxic to spermatozoa although 10% nigrosin was not toxic. Emmens and Blackshaw's observation (16) that eosin tends to diffuse into previously unstained cells was confirmed.

Hancock (18, 19) and also Campbell, *et al.* (8) modified Blom's technique by using eosin Y with different dilutions of semen and stain. Swanson and Beardon (37) found that 1% eosin and 5% nigrosin were better combinations of the dyes. No difference was noted if the percentage of concentration of eosin and nigrosin were changed to between 50 and 200% of the original. A range in pH 6.4–8.5 and dilutions of semen to stain of 1:1 to 20:1 were not harmful. A diluter of three to four parts of 3% sodium citrate dihydrate and one part of egg yolk was used to extend semen and very little differences in the live–dead ratios were observed.

Dott (14) observed the partial staining of spermatozoa using eosin nigrosin. He found no differences when using dyes dissolved in distilled water or citrate buffer as long as the pH range was between six and eight. At this pH, tonicity of the stain had no detrimental effect on the cells. Higher ranges tended to increase the intensity of red.

Preparation

Nigrosin (water soluble)....................10 gm.
Eosin (water and alcohol
soluble).......................................1.67 gm.
Distilled water or citrate
buffers...............................q.s. ad 100 ml.

The citrate buffers consist of either 1 gm., 2.9 gm. or 3.9 gm. of sodium citrate ($Na_3C_6H_5O_7 \cdot 2H_2O$) with the pH adjusted to six or eight by 3% citric acid.

Congo Red–Nigrosin

Preparation

Congo red (water soluble)...................3 gm.
Nigrosin (water soluble).......................5 gm.
Buffered citrate.....................q.s. ad 100 ml.
0.083 M sodium citrate dihydrate
0.010 M $Na_2HPO \cdot 12H_2O$
0.010 M $NaH_2PO_4 \cdot H_2O$

Prepare the citrate buffer by dissolving the chemicals in glass distilled water. Sterilize by heat.

The staining mixture is prepared by dissolving the congo red and nigrosin in the citrate buffer. Either two or three grams of congo red may be used.

Staining Method

For whole semen, add one drop of semen to six drops of dye in a specimen tube.

Incubate for three minutes at 30° C.

Smear on to slides in a thin film.

When extended semen is used, change the ratio of stain to semen to one part stain and ten parts extended semen. Dead cells stain red in a black background. This procedure may be used with ram or bull spermatozoa.

Blackshaw (3, 4, 5) found that congo red gave results comparable to those of eosin. He noted that congo red did not diffuse into the "living" cells as did eosin. Martin and Emmens (26) and Martin (27) had good results using congo red- nigrosin.

Erythrosin Nigrosin

Preparation

Erythrosin (water soluble).................0.5 gm.
Nigrosin (water soluble).......................5 gm.
Sodium citrate dihydrate
(2.6%)................................q.s. ad 100 ml.

Mix the nigrosin in the citrate solution then add the erythrosin. Sterilize by heat if desired. The technique is similar to making smears with congo red-nigrosin.

Wales (41) used 0.5% solution of erythrosin with 5% nigrosin to stain human and dog spermatozoa. In work with tissue cultures, Merchant (30) used a 0.4% erythrosin solution.

Fluorescent Dyes

Opaque media such as milk extender cause difficulties when examining for motility and viability of spermatozoa. When fluorescent dyes are used in concentrations of 1:10,000 to 1:40,000 this handicap may be overcome. At these levels, the dyes are non-toxic. Acridine orange, euchrysin 2 GNX, acriflavine HC1 NF and coriphosphin HK are the best dyes to use for sperm in milk diluent (38, 39). Acridine orange is used routinely in this laboratory at present to examine frozen semen with good results. Attempts to use eosin B and erythrosin have proven unsatisfactory.

Bishop and Smiles (1) used 1:30,000 dilutions of primulin. They noted that dead sperm fluoresce light blue while live sperm were invisible. All the sperm, living and dead, were readily seen with a neutral filter substituted for an ultra-violet filter. When broad-band ultra-violet filters were used, the dead spermatozoa appeared as blue fluorescent images while the live spermatozoa appeared red. Primulin and rhodamine 6 G[5] in equal amounts stained the dead cells light blue and the live ones bright yellow. When rhodamine alone was used both the living and dead cells stained yellow. Two granules at the base of the head were clearly differentiated in the living cells but not in the dead cells.

Preparation

Acridine orange....................................1 gm.
Distilled water........................q.s. ad 100 ml.

Mix dye with water. Filter.

Sterilize by heat.

Staining Method

Using a Pasteur pipette, add one drop of dye to a vial of semen. Make routine preparations for motility examination utilizing fluorescent microscopy with the appropriate barrier filters. These particular filters will vary with the exciter filter used and with each observer.

Feulgen DNA Stain

Recently great interest has been aroused in the study of, and fulfilment of, the genetic potentialities of somatic cells. Feulgen DNA stain was used by Salisbury *et al.* (33) who demonstrated that solutions of different ages have different DNA content.

Preparation

Carnoy's fixative

Acetic acid (glacial)..............................10 ml.
Chloroform..30 ml.
Ethyl alcohol (absolute).......................60 ml.

Combine and mix well.

1N HCl (82.5 ml. HCl in 1000 ml. distilled water)

Schiff's Reagent (D. B. Lilly's modification)

Sodium metabisulphite.....................1.9 gm.
Basic fuschin......................................1.0 gm.
0.15 N HCl (Take 15 ml. of 1 N HCl and water q.s. ad 100 ml.)..................................100 ml.

Mix, shake periodically till straw yellow (two to three hours).

Add 0.3 gm. decolourizing charcoal.

Shake for a few minutes.

Allow to settle and if clear filter; if not clear add more charcoal and refilter.

Store in brown, tightly stoppered bottle in refrigerator.

SO_2 solution

Sodium (meta) bisulphite..................1.9 gm.
Concentrated HCl...............................10 ml.
HCl 0.15N.......................................100 ml.

Staining Method (10)

Centrifuge semen sample at 300 to 500 rpm. for three minutes.

Decant. Add Carnoy fixative and fix for ten minutes, after disturbing the bead, i.e., make into a semen suspension.

Centrifuge for two to three minutes at 300 to 500 rpm.

Decant.

Hydrolize with 1 N HCl for eight minutes at 60° C. using a suspension of semen.

Centrifuge at 300 to 500 rpm for one to two minutes and decant.

Feulgen stain in dark for 45 minutes (Use Schiff's Reagent).

Wash three times in SO_2.

Wash in cold tap water.

Place one to two drops of stained semen suspension on a slide–gently squash with a cover-slip.

Freeze on dry ice to fix for about ten minutes.

Summary

A review of the staining techniques for mammalian spermatozoa is presented. The preparation, procedures and dyes used are outlined. For discussion purposes the stains have been divided into morphological, differential, fluorescent, and Feulgen DNA stains.

Résumé

On présente une revue des procédés de teinture des spermatozoaires chez les mammifères. On explique les grandes lignes des modes de préparation, des méthodes suivies et des teintures utilisées. Pour les fins de la discussion, on a divisé les teintures en quatre groupes: morphologiques, différentielles, fluorescentes et Feulgen DNA.

References

1. Bishop, M. W. H., and Smiles, J. Differentiation between living and dead spermatozoa. Nature. 179: 308. 1957.
2. Bishop, M. W. H., and Smiles, J. Induced fluorescence in mammalian gametes with acridine orange. Nature. 179: 307. 1957.
3. Blackshaw, A. W. The affect of equilibration and the addition of various sugars on the revival of spermatozoa from −79° C. Aust. Vet. J. 31: 124. 1955.
4. Blackshaw, A. W. Factors affecting the revival of bull and ram spermatozoa after freezing to −79° C. Aust. Vet. J. 31:238. 1955.
5. Blackshaw, A. W. The effect of glycerol on supra-vital staining of spermatozoa. Aust. Vet. J. 34: 71. 1958.
6. Blom, Erik. A new and uncomplicated method for the staining of bull sperms. Abstract from Skandinavish Vet. Tidskr. 33: 428. 1943.
7. Blom, Erik. A one minute live dead sperm stain by means of eosin nigrosin. Journal of Fertility and Sterility. 1: 176. 1950.
8. Campbell, R. C., Hancock, J. L., and Rothschild, Lord. Counting live and dead spermatozoa. J. Exp. Biol. 30: 44. 1953.
9. Cascarrett, G. W. A one solution stain for spermatozoa. Stain Technology. 28: 125. 1953.
10. Coubrough, R. I. Personal Communication. 1964.
11. Crooke, A. C., and Mandl, A. M. A rapid supravital staining method for assessing the viability of human spermatozoa. Nature. 159: 749. 1947.
12. Cummings, J. N. Testing Fertility in Bulls. Technical Bulletin 212, University of Minnesota Agricultural Experimental Station. 1954.
13. Devereux, E. D., and Turner, F. W. Observations on the growth of yeasts in pure nutrient solutions. J. Bact. 14: 317. 1927.
14. Dott, H. M. Partial staining of spermatozoa in the nigrosin–eosin stain. Proceedings of the 3rd Inter. Congr. of An. Reprd. (Cambridge). Section III: 42. 1956.
15. Emik, L. O., and Sidwell, G. M. Refining methods for using opal blue stain in evaluating ram semen. J. An. Sci. 6: 67. 1947.
16. Emmens, C. W., and Blackshaw, A. W. Artificial insemination. Physiological Reviews. 36: 277. 1956.
17. Glynden, R., Easley, T., Mayer, D. T., and Bogart, R. Influence of diluters, rate of cooling and storage temperature on survival of bull sperm. A.J.V.R. 3: 338. 1943.
18. Hancock, J. L. Morphology of bull spermatozoa. Nature. 157: 447. 1946.
19. Hancock, J. L. Staining technique for study of temperature shock in semen. Nature. 167: 322. 1951.
20. Hanks, H. J., and Wallace, J. H. Determination of cell viability. Proc. Soc. Exper. Biol. and Med. 98: 88. 1958.
21. Herman, H. A., and Swanson, E. W. Variations in dairy bull semen with respect to its use in artificial insemination. Research Bulletin 326, University of Missouri, College of Agriculture, Agricultural Experimental Station. 1941.
22. Katsube, Y., and Blobel, H. *In vitro* phagocytic activities of leucocytes isolated from the mammary secretions of a cow. A.J.V.R. 25:1090. 1964.
23. Lasley, J. F., Easley, G. T., and McKenzie, F. F. Staining method for the differentiation of live and dead spermatozoa. Anat. Rec. 82: 167. 1942.
24. Macleod, J. An analysis in human semen of a staining method for differentiating live and dead spermatozoa. Anat. Rec. 83: 573. 1942.
25. Macpherson, J. W. Sterile milk as a semen diluent. Can. Vet. J. 1: 551. 1960.
26. Martin, I., and Emmens, C. W. Factors affecting the fertility and other characteristics of deep-frozen bull semen. J. Endocrin. 17: 449. 1958.
27. Martin, I. Effects of lecithin, egg-yolk, fructose, and period of storage at 5° C. on bull spermatozoa deep-frozen to −79° C. J. Reprod. Fertility. 6: 441. 1963.
28. Mayer, D. T., Squiers, D., and Bogart, R. An investigation of the staining principle and the background stain in the differentiation of live from dead spermatozoa. J. An. Sci. 6: 499. 1947.
29. Mayer, D. T., Squiers, D., Bogart, R., and Aloufa, M. M. The technique for characterizing mammalian spermatozoa as dead or living by differential staining. J. An. Sci. 10:226. 1951.
30. Merchant, D. J., Kahn, H., and Murphy, W. H. Handbook of cell and organ culture. 2nd Edition. Minneapolis: Burgess Publishing Company. 1964.
31. Mixner, J. P., and Saroff, J. Interference by glycerol with differential staining of bull spermatozoa as used with semen thawed from the frozen state. J. Dairy Sci. 37: 1094. 1954.
32. Phillips, H. J., and Terryberry, J. E. Counting actively metabolizing tissue cultured cells. Exp. Cell. Res. 13: 341. 1957.
33. Salisbury, G. W., Lodge, J. R., and Baker, F. N. Effects of age of stain, hydrolysis time, and freezing of the cells on the feulgen DNA content of bovine spermatozoa. J. Dairy Sci. 47: 165. 1964.
34. Shaffer, H. E., and Almquist, J. O. Vital staining of bovine spermatozoa with an eosin aniline blue staining mixture. J. Dairy Sci. 31: 677. 1948.

35. STULBERG, C. S., PETERSON, W. D., and BERMAN, L. Quantitative and qualitative preservation of cell stain characteristics. Symposium: Analytic cell culture, Detroit, Michigan, June 6 and 7, 1961.
36. STULBERG, C. S. Preservation and characterization of animal cell strains. Reprint from Culture Collections: Perspectives and Problems. Published by the University of Toronto Press, 1963.
37. SWANSON, E. W., and BEARDON, H. J. An eosin-nigrosin stain for differentiating live and dead bovine spermatozoa. J. An. Sci. 10: 981. 1951.
38. VAN DE MARK, N. L., SCHORR, R., and ESTERGREEN, V. C. The use of fluorescent dyes for studying sperm motility. J. An. Sci. 17: 1216. 1958.
39. VAN DE MARK, N. L., ESTERGREEN, V. C., SCHORR, R., and KUHLMAN, D. E. Use of fluorescent dyes for observing bovine spermatozoa in opaque media. J. Dairy Sci. 42: 1314. 1959.
40. WALES, R. G., and WHITE, I. G. The susceptibility of spermatozoa to temperature shock. J. Endocrin. 19: 211. 1959.
41. WALES, R. G. The differential staining of human and dog spermatozoa. Aust. J. Exper. Biol. and Med. Sci. 37: 433. 1959.
42. WHITE, I. G., and WALES, R. G. The susceptibility of spermatozoa to cold shock. Inter. J. of Fertility. 5: 195. 1960.
43. WARREN, E. D., MAYER, D. T., and BOGART, R. The validity of the live dead differential staining technique. J. An. Sci. 18: 803. 1953.
44. WILLIAMS, W. W., and POLLACK, J. OTAKAR. Study of sperm vitality with the aid of eosin-nigrosin stain. J. Fert. and Ster. 1: 178. 1950.

Notes

[1] Chin Chin Brand, Gunther Wagner, Germany.

[2] A mounting medium, made by dissolving a polystyrene in xylene, obtainable from British Drug Houses, Toronto, Ontario.

[3] Mordant–a substance like alum employed to make a dye bite into the tissue and hold on. The dye combines with the mordant which is itself in high concentration in the structures to be stained. In the Iron Hematoxylin technique the slides are mordanted with iron alum. They are briefly washed in distilled water to remove some of the excess mordant.

[4] Opal blue has been unobtainable since World War II.

[5] British Drug Houses, Toronto, Canada.

7 Genetic Aspects of Male Reproduction

Peter J. Chenoweth[1]* and Francoise J. McPherson[2]
[1]*ChenoVet Animal Andrology, Wagga Wagga, New South Wales, Australia;*
[2]*Charles Sturt University, Wagga Wagga, New South Wales, Australia*

Introduction

Genetic disorders are those caused by abnormalities or defects in genes or chromosomes. Although many genetic disorders are transmitted via parental genes, others are a result of DNA changes or mutations that are not necessarily heritable. Congenital disorders, in contrast, are those that exist at birth (though they may become first evident either earlier or later). These may be caused by a number of factors, including genetic disorders. However, they can also be caused by developmental anomalies, infections, the uterine environment, and metabolic, nutritional and toxic factors, in addition to joint genetic–environmental and epigenetic influences. Those that are proven to be heritable may be transmitted via single gene (or Mendelian) inheritance, or via complex mechanisms that may involve multiple genes as well as environmental effects. For examples of conditions subject to the former mode of inheritance in cattle, readers are referred to the web site (http://dga.jouy.inra.fr/lgbc/mic2000/).

Literally thousands of genes are involved, directly or indirectly, in the production of mammalian spermatozoa, and those genes involved in reproduction are generally less conserved than other genes (Leeb, 2007). It is estimated that 10% of genes in the human genome are associated with spermatogenesis and that genetic factors cause 15–30% of human male infertility (Yoshida *et al.*, 1997). In turn, male genetic factors contribute to an estimated 50% of human infertility (Tanaka *et al.*, 2007). Similar estimates may be assumed for domesticated species, in which there is potential for the widespread dissemination of genetic problems, which may be caused by chromosomal or single gene disorders, Mendelian disorders of sex differentiation, genetic disorders either directly or indirectly affecting male fertility, mitochondrial DNA (mtDNA) mutations, Y chromosome deletions, epigenetic effects and multifactorial disorders (see reviews by Wieacker and Jakubiczka, 1997; Shah *et al.*, 2003; Martin, 2008; Visser and Repping, 2010; and Hofherr *et al.*, 2011).

Developmental Anomalies

The undifferentiated embryo undergoes sex determination, which is described as a developmental decision that sends it along one of two major avenues towards sexual dimorphism.

* E-mail: peter1@chenovet.com.au

Sexual differentiation then occurs, with phenotypic sex being determined under the influence of factors produced by the gonadal precursors (Biason-Lauber, 2010). In mammals, early gonadal differentiation is the major factor determining sexual dimorphism, in which distinguishing differences in structures or features between individuals of different sexes are termed secondary sexual characteristics (McPherson and Chenoweth, 2012). Early gonadal differentiation also marks an important phase in which errors in sexual development may occur, with the status of the *SRY* gene, located on the Y chromosome, playing a major role in establishing 'maleness' and also being implicated in XY female gonadal dysgenesis (Biason-Lauber, 2010). Translocation of the *SRY* gene to the X chromosome is strongly implicated in human cases of XX maleness (Müller, 1993).

Developmental anomalies occurring in domesticated animals that adversely affect male fertility to varying degrees include the following: undescended testes, segmental aplasia/hypoplasia of the Wolffian duct system, testicular hypoplasia, hypospadia, freemartinism and different manifestations of intersex.

Chromosomal and DNA Anomalies

Chromosomal anomalies represent a major cause of infertility in human males (Martin, 2008; Hofherr *et al.*, 2011) and this probably applies to domestic animals as well. Chromosome disorders were found in approximately 2–9% of sub-fertile human males, with Klinefelter's syndrome (47,XXY) being one of the most common (Chandley and Cooke, 1994). Those disorders involving autosomal chromosome rearrangements, such as translocations or inversions, are essentially paternal in origin. This includes approximately 35% of Robertsonian translocations (where two acrocentric chromosomes fuse near the centromere, with the loss of the short arms), which occur in cattle, with resultant reduced fertility of both males and females (Gustaffson, 1979). Reciprocal (non-Robertsonian) translocations have been associated with reduced fertility, abortions and birth defects in humans (Shah *et al.*, 2003) as well as with early embryonic deaths and reduced litter sizes in swine (Feitsma and De Vries, 2006). Relevant reviews on chromosomal defects in domestic animals and their effects include those of Popescu and Tixier (1984) and Ducos *et al.* (2008).

Numerical chromosome abnormalities – such as deletion, trisomy and triploidy – also contribute to early embryonic mortality in domestic animals (Courot and Colas, 1986). A notable example of this is encountered in cats, where the genes for either orange or black colours, but not both together, are located on the X chromosome. Thus, males that exhibit both colours (such as calico or tortoiseshell individuals) are polyploid and hence infertile (see Chapter 8, Root Kustritz). In infertile men with poor semen quality, a direct relationship has been suggested between the impairment of spermatogenesis (as reflected in morphologically and cytogenetically abnormal germ cells) and rates of baseline aneuploidy in normal spermatozoa (Bernadini *et al.*, 1998), with most sperm aneuploidies being associated with lowered fertilization rates as well as reduced embryo survival.

A genetic basis for spermatogenic failure has been associated with deletions on the human Y chromosome (Krausz *et al.*, 1999). Such deletions occur at a relatively high level, indicating a susceptibility of the Y chromosome to loss of genetic material (Aitken and Sawyer, 2003; Shah *et al.*, 2003; Hofherr *et al.*, 2011). In man, Y micro-deletions are notable in the AZF (azoospermia factor) regions, supporting the theory that *DAZ* (deleted in azoospermia) genes play an important role in male fertility (Hofherr *et al.*, 2011).

In recent years, advances in intracytoplasmic sperm injection, or ICSI (Chemes and Rawe, 2003), as well as single sperm typing and synaptonemal complex analysis (Martin, 2008), have enhanced understanding of the relationships between specific characteristics of individual sperm and fertilization, as well as post-fertilization development. More specifically, strong links are now evident between sperm morphological abnormalities and structural chromosomal aberrations, with the latter being significantly elevated in human sperm with head abnormalities (Rosenbusch *et al.*, 1992; Lee *et al.*, 1996; Lewis-Jones *et al.*, 2003; Sun *et al.*, 2006; Enciso *et al.*, 2011).

For instance, chromosomal anomalies were four times more likely in sperm that were amorphous, round or elongated (Lee *et al.*, 1996), and abnormal karyotypes were significantly higher in oocytes injected with severely deformed sperm heads (Kishikawa *et al.*, 1999). Relationships occur between poor chromatin structure, DNA damage, abnormal morphology and fertility in bull sperm (Ballachey *et al.*, 1987; Sailer *et al.*, 1996; Karabinus *et al.*, 1997; Smorag *et al.*, 2000; Ostermeier *et al.*, 2001; Vieytes *et al.*, 2008; Enciso *et al.*, 2011).

As an example, the well characterized diadem/crater defect of sperm, found in many species following spermatogenic stress, is linked with failure of normal chromatin condensation (Larsen and Chenoweth, 1990) and represents part of a choreographed continuum of sperm head damage following stress (Chenoweth, 2005). In humans, large vacuoles in the sperm head have been associated with a failure of chromatin condensation (Boitrelle *et al.*, 2011). If a significant proportion of the sperm population is observed to be morphologically abnormal, it is likely that the causative spermatoxicity has adversely affected other, apparently normal, sperm in the population (Vogler *et al.*, 1993). Due to the limitations of sperm morphology assessment using light microscopy, sperm that appear to be morphologically normal may have defective chromatin or other cellular components (Ballachey *et al.*, 1988; Chemes and Rawe, 2003). These latter authors proposed that certain sperm anomalies (e.g. flagellar defects) have a better fertility prognosis than others (e.g. acrosome, sperm chromatin and neck defects). Similar considerations have contributed to the development of several approaches to the categorization of sperm defects, viz.: (i) compensable and uncompensable, based on observed sperm morphology (Saacke *et al.*, 2000); and (ii) systematic versus non-systematic (or non-specific) sperm abnormalities, based on whether they are considered to be due to pathological and/or environmental causes, or to have a proven or suspected genetic origin, respectively (Chemes and Rawe, 2003).

Sperm with severe head malformations often fail to successfully traverse the female tract and reach the site of fertilization and, therefore, can be regarded as compensable defects. However, those sperm with more subtle (i.e. non-head distorting) forms of the diadem/crater defect can gain access to the ovum, leading to both lowered fertility and embryo quality (Miller *et al.*, 1982; Saacke *et al.*, 1994), thus fulfilling the definition of uncompensable sperm defects. Enciso *et al.* (2011) used the Sperm Chromatin Dispersion test in conjunction with Blom's (1972) major/minor categorization of sperm abnormalities to illustrate the relationship between abnormal bull sperm morphology and DNA fragmentation – a relationship that accounted for 61.34% of all deformed sperm, and 46.88% of major defects. In certain major sperm defects, i.e. double forms, narrow head at base, small heads and proximal droplets, DNA fragmentation occurred in all recognized affected sperm.

It can be concluded, then, that sperm with DNA/chromatin abnormalities often show structural abnormalities, which may be identified during routine assessment of sperm morphology (Martin and Rademaker, 1988; Chemes and Rawe, 2003), although caution should be exercised when attempting to extend this generalization to assume that all damaged sperm can be so recognized (Martin and Rademaker, 1988). This is because many examples of damaged sperm DNA or chromatin integrity do not display obvious morphological abnormalities (Sutovsky *et al.*, 2001; Shah *et al.*, 2003), even though they may cause failure of fertilization or post-fertilization loss (Smorag *et al.*, 2000).

In summary, these findings support the following concepts (Chenoweth, 2011): first, that abnormal sperm head morphology is often associated with DNA damage (Erenpreiss *et al.*, 2006); secondly, that the major cause of DNA damage in the male gamete is oxidative stress (Aitken, 2002; Lewis and Aitken, 2005); and thirdly, that sperm DNA abnormalities are a major cause of male-factor sub-fertility. Finally, routine sperm assessment parameters are only partially successful in identifying such damage.

Potential Adverse Genetic Sequelae of Assisted Reproductive Technologies (ARTs)

Sperm morphology, particularly of the head region, is often linked with *in vitro* fertilization (IVF) success rates, but it is apparent that this relationship is less strong than it is with natural fertilization (Windt and Kruger, 2004). Further, the development of ICSI for the treatment of male-factor infertility has resulted in successful fertilizations using sperm that are seriously impaired, as well as those in which normal buffering mechanisms are deficient (Horsthemke and Ludwig 2005). In both cases, a number of barriers that would normally restrict fertilization by abnormal sperm are bypassed, which increases the possibility of subsequent genetic problems, including chromosomal abnormalities, genetic disease and reproductive problems in offspring from ARTs being implicated in disturbances of epigenetic mechanisms and related disorders of genomic imprinting (Fortier and Trasler, 2011).

Male Fertility

Male effects on reproduction may be positive or negative and obvious or subtle. They include not only direct effects on the ability to mate and to achieve fertilization, but also indirect effects on the quality and viability of the conceptus. Embryo development relies not only upon maternal but also on paternal factors (Duranthon and Renard, 2001). For example, differences have been shown to occur among bulls for embryo survival and development, in both *in vivo* and *in vitro* systems (Hillery *et al.*, 1990; Saacke *et al.*, 2000). Earlier work showed that bulls of low fertility used for artificial insemination (AI) gave higher rates of embryonic loss than did bulls of high fertility (cited by Courot and Colas, 1986). Similarly, with sheep, Maxwell *et al.* (1992) reported that rams of differing genetic lines differed in subsequent rates of embryonic loss despite comparable fertilizing capacity. Differences in bull reproductive success are often not predictable on the basis of conventional semen examination. However, recent advances in technologies such as IVF and ICSI have allowed a vastly improved understanding of the causes of variation in male gamete 'success'. Here, bull differences have been reported for IVF rates, initiation and length of the zygotic S-phase, and embryo cleavage and development (cited by Schneider *et al.*, 1999), as well as in sperm *in vitro* ability to access the ovum and accessory sperm numbers (Nadir *et al.*, 1993).

The effects of the male on early pregnancy loss and abortion have been associated with implicating factors such as elevated ambient and scrotal temperatures, 'out-of-season' breeding, and immature and aged sperm (Saacke *et al.*, 2000). It is interesting and useful to note that many of these stressors induce similar patterns of spermatogenic stress. For example, in bulls, a stereotyped response results in a predictable cascade of identifiable sperm abnormalities such as retained cytoplasmic droplets, head loss (decapitation) and diadem-crater defects, with consequent adverse effects on fertilization or subsequent development (Saacke *et al.*, 2000). In pigs, those sperm characteristics associated with successful *in vitro* penetration of oocytes included all conventional semen parameters (Gadea and Matas, 2000), with most of these being significantly correlated with each other. Flowers (2002, 2008) reported that individuals differed in the insemination dose required to consistently produce the greatest number of pigs. Traditional semen assessments tended to be predictive of litter size up to a point, but beyond this, differences in litter size could not be attributed to observed semen differences. Such results illustrate difficulties in predicting sperm fertility from one or two tests, while reinforcing the need to identify biological markers that reflect unifying mechanisms for many aspects of sperm damage.

Even though the heritability of male fertility in domesticated animals is generally considered to be low, some male traits associated with fertility are relatively heritable. For example, adjusted testicular size (measured as scrotal circumference) is moderately to highly heritable in beef breed bulls, where it is favourably associated with age at puberty in related females (Brinks, 1994). Comparable estimates have been obtained in boars (Toelle *et al.*, 1984; Toelle and Robinson, 1985; Young *et al.*, 1986). In turkeys, selection for increased semen yield doubled sperm numbers within five generations (Nestor, 1976). Semen production also increased when turkeys were specifically selected for increased egg yield, and decreased when selection was for increased body weight at 16 weeks (Nestor, 1977).

In rams, scrotal circumference was favourably linked with other male fertility traits; genetic correlations were 0.54 with individual sperm motility and –0.75 with percentage abnormal sperm (Rege *et al.*, 2000). In French dairy cattle, sperm quality traits such as motility were moderately highly heritable; genetic correlations between volume and semen quality traits tended negative, whereas those between concentration and semen quality traits were the opposite (Druet *et al.*, 2009). Some findings from studies on the heritabilities of semen traits in boars, bucks and bulls are shown in Table 7.1.

Heterosis in pigs boosts semen traits such as semen volume, total sperm and sperm viability (Smital *et al.*, 2004). Genetic lines of boars differed in sperm output, which varied by as much as 30 × 10^9 spermatozoa or more per ejaculate, indicating the possibility for selection of boars on the basis of improved sperm production as well as fertility (Flowers, 2008). In a study using 463,130 litter records from Yorkshire, Duroc, Hampshire and Landrace pigs, estimates of sire effects for traits such as pigs born alive, litter weight at 21 days and number weaned ranged from 0.02 to 0.05. Although sire effects were less than environmental effects (0.03–0.08), they were greater than maternal effects (0.00–0.02) across all breeds, providing encouragement for improving litter traits via sire selection (Chen *et al.*, 2003).

In North American Holstein cattle, sire-predicted transmitted ability for twinning was estimated to be 1.6–8.0%, which is considered sufficient to use sire selection as a means of reducing undesirable twinning in dairy cattle (Johanson *et al.*, 2001). Perinatal calf deaths up to 48 h after birth can also be linked to the sire, though the heritability for this parameter in bulls is low (Freeman, 1984).

Stallions are normally not selected on the basis of fertility, but genetic markers have been identified for stallion fertility (Leeb *et al.*, 2005). Three candidate genes have been identified to allow selection of stallions based on reproductive potential. These candidate genes code for equine cysteine-rich secretory proteins (CRISPs) and together make up six polymorphisms as genotyped in 107 Hanoverian stallions. Stallions heterozygous for the *CRISP3* AJ45996c.+622 G>A single nucleotide polymorphism (SNP) have significantly reduced fertility compared with homozygous stallions, which translates into a 7% reduction in conception rate per cycle (Hamann *et al.*, 2007). Other marker genes affecting equine

Table 7.1. Heritabilities of semen traits in boars, bucks and bulls.

Breed	Boars (9 pure breeds and 10 crossbreeds)	Saanen and Alpine bucks	Holstein bulls (<30 months old)	Holstein bulls (4–6 years old)	Simmental bulls	Angus bulls (yearling)
Semen volume	0.58	0.25–0.29	0.24	0.44	0.18	
Sperm concentration	0.49	0.32–0.34	0.52	0.36	0.14	0.13
Sperm motility	0.38	0.12–0.17	0.31	0.01	0.04	0.13
Total no. of sperm/ ejaculate	0.42	0.15–0.25	0.38	0.54	0.22	0.24
Proportion of abnormal sperm	0.34	–	–	–	0.10	–
Authors	Smital *et al.*, 2005	Furstoss *et al.*, 2009	Mathevon *et al.*, 1998	Mathevon *et al.*, 1998	Gredler *et al.*, 2007	Knights *et al.*, 1984

pregnancy rate per oestrus cycle are *SPATA1* and *INHBA* (Gieseke *et al.*, 2010). *SPATA1* codes for a testis-specific protein involved in shaping the sperm head, and conception rates per oestrus cycle are significantly higher for *SPATA1* heterozygous stallions than for homozygous stallions, with a dominance effect of 4.1% (Gieseke *et al.*, 2009).

In addition to these relationships, a large number of conditions, which are either directly or indirectly genetically influenced, can adversely affect male fertility. These may exert their influence not only via sperm and semen traits, but also in relation to mating ability and sperm delivery capabilities. Reviews of genetic disorders linked with male infertility are provided by Wieacker and Jakubiczka (1997), Shah *et al.* (2003) and Martin (2008). In this discussion, the focus is upon conditions affecting sperm production and semen quality. Readers interested in the other aspects, particularly in relation to bulls, may wish to consult Chapter 12 (Brito), which has descriptions of conditions such as testicular hypoplasia, congenitally short penis, congenitally short retractor penis muscle, persistent penile frenulum, penile spiral deviation, aplasia/hypoplasia of efferent duct systems and congenital erectile dysfunction.

Breed effects

Both libido and mating ability in bulls are influenced by genetic factors (Bane, 1954; Chenoweth, 1983) and differences in sex drive and mating behaviour have been reported among breeds of bulls (Chenoweth and Osborne, 1975; Chenoweth *et al.*, 1979) and rams (Chenoweth, 1981), and also between boar lines (Signoret, 1970). Differences in aspects of sperm production and semen collection have been observed between beef and dairy breeds of bulls in general (Chenoweth, 1983). Bulls of *Bos indicus* derivation have been considered to be more sexually 'sluggish' than their European counterparts (Chenoweth, 1981), although this does not necessarily disadvantage breeding success (Chenoweth, 1994).

In a longitudinal study in which 47 stallions (aged 2–26 years), representing seven breeds, were bred to 1664 mares, breed of stallion as well as paternal age and season significantly influenced mare breeding outcomes (Dowsett and Pattie, 1982). Semen factors that were associated with percentage of pregnancies per mare service were volume, concentration, live sperm and total number of sperm (Dowsett and Pattie, 1982). In another stallion study, sire effects on semi-sibling semen characteristics were significant for sperm motility, semen volume and sperm concentration (Parlevliet *et al.*, 1994).

In chickens, Wyandotte and Delaware roosters were less fertile than their Leghorn counterparts (Kirby *et al.*, 1989), and Wyandotte sperm exhibited less metabolic capacity than either Leghorn or Delaware sperm (Kirby and Froman, 1991). Roosters homozygous for the rose comb allele (*R/R*) were sub-fertile compared with heterozygous roosters of genotypes *R/r* or *r/r*, with this depression of fertility being associated with reduced sperm motility and metabolic activity (McLean and Froman, 1996). The reduced sperm motility of *R/R* roosters has been shown to be associated with a significantly reduced intracellular ATP concentration and reduced glucose uptake (McLean *et al.*, 1997).

Sperm morphology can also vary between breeds, as shown in a study comparing eight breeds of Indian buffalo (Aggarwal *et al.*, 2007). Similarly, differences in sperm midpiece length were encountered when comparing Holstein, Friesian, Belgian Blue, Charolais, Limousin and Aberdeen Angus bulls (Shahani *et al.*, 2010). Here, the Belgian Blue and Limousin breeds displayed the longest average midpiece length, while Angus bulls had the shortest. Furthermore, the two dairy breeds showed a negative correlation between sperm midpiece length and 49 day non-return rate. Finally, sperm head morphometry as determined by computerized Fourier analysis differed significantly among 241 mature rams from 36 different sheep dairy flocks (Maroto-Morales *et al.*, 2010).

Inbreeding

Breeding programmes for domestic livestock generally seek to maximize genetic gain while minimizing inbreeding. Inbreeding adversely affects reproductive traits (such as sperm production and semen quality) in two ways: first by reducing heterozygosity, leading to inbreeding depression in those traits affected by directional dominance; and, secondly, via random genetic drift in which advantageous alleles may be lost (Sonersson, 2002). Earlier work (Salisbury and Baker, 1966b) indicated that inbred bulls had more sperm abnormalities than line-crossed bulls, and that the degree of inbreeding is linked with the proportion of abnormal seminiferous tubules (Carroll and Ball, 1970). More recently, inbreeding in endangered ungulates has been linked with high levels of DNA fragmentation and poor sperm quality (Ruiz-Lopez *et al.*, 2010).

In foxhounds, outbred dogs produced a greater ejaculate volume with significantly (P<0.025) higher sperm concentration and more sperm/ejaculate than did inbred males. There was also a trend towards improved sperm motility, ejaculate volume and testes volume for outbred dogs than for inbred dogs (Wildt *et al.*, 1982). In dachshunds, Gresky *et al.* (2005) showed that inbreeding decreased litter sizes and at the same time increased the number stillborn.

Inbreeding can affect not only sperm quality or quantity, but also the occurrence of reproductive organ disorders. For example, undescended testicles, or cryptorchidism, is a common genetic disorder in dogs (Veronesi *et al.*, 2009). A relatively high incidence of cryptorchidism was detected in a colony of Miniature Schnauzers. Of twelve cases examined, nine of the Miniature Schnauzers were descendants of the same sire, either directly or indirectly. Bilateral cryptorchidism was associated with a higher degree of inbreeding than was unilateral cryptorchidism (Cox *et al.*, 1978).

Inbreeding in horses has been associated with increased embryonic loss in Norwegian trotter mares (Klemetsdal and Johnson, 1989) and reduced conception and foaling rates in Standardbred mares (Cothran *et al.*, 1984), though it is conceded that mare management factors could also be involved. When the inbreeding coefficient in Shetland pony stallions exceeded 2%, there was a significant reduction in sperm motility and the percentage of morphologically normal sperm (van Eldik *et al.*, 2006).

Inbreeding in dairy cattle has resulted from extensive use of artificial insemination, with a small number of elite bulls being used for a large number of inseminations (Soares *et al.*, 2011). An added issue is that AI success is declining in highly selected dairy cattle populations (Karoui *et al.*, 2011). In a study by Hudson and van Vleck (1984), the percentage of inbred bulls for Ayrshire, Guernsey, Holstein, Jersey and Brown Swiss breeds in the north-eastern USA was 4.7, 2.8, 12.1, 3.9 and 7.6%, respectively. The average inbreeding coefficients were 0.39, 0.17, 0.36, 0.14 and 0.26, respectively (Hudson and van Vleck, 1984). A more recent study revealed that mean inbreeding coefficients for US dairy cattle have increased to 0.3–0.4, with some animals recording a value greater than 0.5. Milking Shorthorns recorded the highest annual increase in inbreeding level overall, while Holstein cattle recorded the greatest rate of change in the annual increase in inbreeding level (Wiggans *et al.*, 1995). In beef bulls, semen traits deteriorated with increasing inbreeding coefficients (Smith *et al.*, 1989). Although inbreeding coefficients in Austrian dual-purpose Simmental (Fleckvieh) bulls were low (mean 0.013), semen quality traits such as volume, concentration, motility, number of spermatozoa per ejaculate and percentage of viable sperm all showed evidence of inbreeding depression (Maximini *et al.*, 2011).

Methods for minimizing the adverse effects of inbreeding in selection and genetic conservation schemes for livestock breeding, especially in small (≤50) populations, have been described by Sonersson (2002).

Categorization of Defects of Sperm and Spermatogenesis

Classification systems for abnormal morphology of sperm have included those based on: (i) site of origin – primary or secondary (Blom, 1950); (ii) functional significance – major or minor (Blom, 1972); and (iii) relative quantitative significance – compensable or uncompensable (Saacke *et al.*, 1994), and systematic or non-systematic (Chemes and Rawe, 2003).

The major/minor sperm defect classification proposed by Blom (1972) considered major defects to be those that were proven to impair fertility and minor sperm defects as being of little consequence for male fertility. Suggested criteria (Chenoweth, 2005) for the categorization as major sperm defects (also sometimes termed specific sperm defects) were that:

1. They are characterized as 'primary' sperm defects (i.e. they occur during spermatogenesis).
2. They constitute a substantial proportion (at least 10–15%) of the sperm population.
3. They persist over time.
4. They are linked with male infertility or sterility.
5. They are potentially heritable.

This last criterion is particularly relevant to the present discussion. Under natural selection, there is discrimination against genetic factors that result in infertility, though this is not necessarily the case in modern production systems in which artificial selection permits sub-fertile males to propagate infertility factors that are genetically influenced. These factors include genetic sperm defects, whereby, at least in bulls, 'some types of defects appear in the semen at a fairly constant rate and in a very high proportion of the sperm population without any indication of environmental influence. Such defects may be presumed to be rooted in the bull's genome and the prognosis for future improvements in semen quality would be very poor' (Barth and Oko, 1989).

Terminology for generalized sperm and semen defects

Terms employed to describe generalized characteristics of sperm and semen include:

- oligospermia or oligozoospermia – decreased number of spermatozoa in semen;
- aspermia – a complete lack of semen;
- hypospermia – reduced seminal volume;
- azoospermia – absence of sperm in the ejaculate;
- teratospermia – a majority of sperm with abnormal morphology; and
- asthenozoospermia – low sperm motility.

The conditions of asthenozoospermia, azoospermia and teratospermia are included in the discussions below under the heading 'Systematic sperm defects'.

Systematic sperm defects

Although environmental factors are considered to be the most common factors affecting sperm function and form, there is a growing list of sperm defects in domestic animals that are considered to be of genetic origin.

Genetic sperm defects are identified as consistent defects that are genetically transmitted. A number of sperm defects in different species have been established to be genetic, and this number is increasing with improvements in technologies and knowledge. However, while some sperm defects may be caused either by genetic or environmental influences, and some may be due to an interaction of environment and genetic predisposition, yet others are likely to be of as yet unproven genetic origin. Thus the term 'systematic sperm defects' (Chemes and Rawe, 2003) is used in this chapter to allow the inclusion of all such categories.

In the following discussion, and in the accompanying table (Table 7.2), an attempt is made to identify a number of sperm defects that fall into this category.

Asthenozoospermia

Defective sperm motility, or asthenozoospermia, is a classic example of a trait that can be attributable to either genetic or environmental effects. For example, simple coiled tails (term that includes the distal midpiece reflex) are among the most common of sperm defects and are commonly associated with one or more of a variety of non-genetic aetiologies. In general, the condition of 'non-specific flagellar abnormalities' (NSFAs), which in turn results in asthenozoospermia, is most commonly reversible and is secondary to conditions such as infections, varicocele, immunological

Table 7.2. Systematic sperm defects.[a]

Defect	Sperm location	Species	Genetic	Reference
Azoospermia	Absence of sperm in ejaculate	Bull, camel, dog, stallion	?	Olson *et al.*, 1992; Ansari *et al.*, 1993; Estrada *et al.*, 2003; Al-Qarawi and El-Belely, 2004
Defects of the sperm acrosome				
Knobbed acrosome	Acrosome	Boar, bull, ram, stallion	Yes/No	Teunissen, 1946; Hancock, 1949; Donald and Hancock,1953; Bane, 1961; Wohlfarth, 1961; Blom and Birch-Andersen, 1962; Buttle and Hancock, 1965; Bane and Nicander,1966; Cran and Dott, 1976; Barth, 1986; Soderquist, 1998; Thundathil *et al.*, 2000b, 2002; Chenoweth, 2005; Kopp *et al.*, 2008; Sironen *et al.*, 2010; Encisco *et al.*, 2011
Ruffled and incomplete acrosome	Acrosome	Bull, mouse	No	Rajasekarasetty, 1954; Saacke *et al.*, 1968
'Miniacrosome'	Acrosome	Human	?	Baccetti *et al.*, 1991
'SME' defect	Acrosome/head	Boar	Yes	Bane, 1961; Blom and Birch-Andersen, 1962, 1975; Buttle and Hancock, 1965; Bane and Nicander, 1966; Blom, 1973a; Andersen and Filseth, 1976; Blom and Jensen, 1977
Defects of the sperm head				
Decapitated sperm head, disintegrated, decaudated, headless, acephalic, microcephalic, alteration of head–neck attachment	Head	Bull (Friesian, Guernsey, Hereford, Jersey, Red Danish, Swedish Red and White), human, dog	Yes/No	Hancock and Rollinson, 1949; Hancock, 1955; Jones, 1962; Williams, 1965; Van Rensburg *et al.*, 1966; Settergren and Nicander, 1968; Blom and Birch-Andersen, 1970; Blom, 1977; Perotti and Gioria, 1981; Perotti *et al.*, 1981; Baccetti *et al.*, 1984, 1989b, 2001; Oettlé and Soley, 1985; Chemes *et al.*, 1987a, 1999; Barth and Oko, 1989; Chemes and Rawe, 2003; Chenoweth, 2005; Collodel and Moretti, 2006
Round-headed sperm (globozoospermia)	Head	Bull, dog, human, ram	?	Schirren *et al.*, 1971; Pedersen and Rebbe, 1974; Kullander and Rausing, 1975; Anton-Lamprecht *et al.*, 1976; Baccetti *et al.*, 1977, 2001; Castellani *et al.*, 1978; Vicari *et al.*, 2002; Collodel and Moretti, 2006; Sun *et al.*, 2006

Diadem/crater, pyriform defect	Head	Boar, bull, dog, human, ram, stallion	No (predisposition may occur)	Gledhill, 1965; Bane and Nicander, 1965; Blom, 1972, 1980; Coulter *et al.*, 1978; Truitt-Gilbert and Johnson, 1980; Heath and Ott, 1982; Johnson and Hurtgen, 1985; Baccetti *et al.*, 1989a; Larsen and Chenoweth, 1990; Held *et al.*, 1991; Liu and Baker, 1992; Vogler *et al.*, 1993; Thundathil *et al.*, 1999; Rousso *et al.*, 2002; Chenoweth 2005
Rolled head/nuclear crest/ giant sperm/large heads	Head	Bull, dog, human	?	Blom, 1980; Cran *et al.*, 1982; Escalier 1983; Barth and Oko, 1989; Dahlbom *et al.*, 1997; Chenoweth, 2005
Macrocephalic, multi-nucleated, multiflagellate	Head, principal piece	Human	?	Escalier 1983; Sun *et al.*, 2006
Teratoid Teratospermia?	Head	Bull, cat, dog, stallion	Yes/No	Barth and Oko, 1989; Pukazhenthi *et al.*, 2001; Pukazhenthi *et al.*, 2006; Sun *et al.*, 2006; Rota *et al.*, 2008; Brito *et al.*, 2010
Pin head/megalo pin head/micro-cephalic	Head	Bull, human	?	Blom and Birch-Andersen, 1970; Chemes *et al.*, 1999; Chemes and Rawe, 2003
Narrow head, narrow at base		Bull	?	Savage *et al.*, 1927; Barth *et al.*, 1992; Encisco *et al.*, 2011
Diploid, polyploid	Head	Bull, human, mouse, rabbit	No	Fechheimer, 1961; Gledhill, 1965; Salisbury and Baker, 1966a; Esnault *et al.*, 1967; Beatty and Fechheimer, 1972; Carruthers and Beatty, 1975; Mortimer, 1979; Blom, 1980
Abnormal DNA condensation	Head	Boar, bull, dog, human, ram, red deer	No	Leuchtenberger *et al.*, 1953; Gledhill, 1966; Gledhill *et al.*, 1966; McCosker, 1969; Salisbury *et al.*, 1977; Ballachey *et al.*, 1987; Barth and Oko, 1989; García-Macías *et al.*, 2006; Ardón *et al.*, 2008; Lange-Consiglio *et al.*, 2010; Encisco *et al.*, 2011
Pyriform head	Head	Alpaca, bull	No	Thundathil *et al.*, 1999; Al-Makhzoomi *et al.*, 2008
Defects of the sperm midpiece				
'Corkscrew'	Midpiece	Boar, bull, stallion	No	Blom 1959, 1972, 1973b; Chenoweth *et al.*, 1970; Chenoweth and Burgess, 1972; Settergren and Nicander, 1972; Barth and Oko, 1989
Spindle-shaped body, microtubule mass	Midpiece	Human, rabbit	No	Pedersen, 1969; Wartenberg and Holstein, 1975
'Hairpin'	Midpiece	Boar, human	No	Kojima, 1975
Distal midpiece reflex (DMR)	Midpiece	All	No	Blake, 1945; Kojima, 1975

Continued

Table 7.2. Continued.

Defect	Sperm location	Species	Genetic	Reference
Proximal cytoplasmic droplet	Midpiece	Boar, bull, dog, guinea pig, stallion	No	Retzius, 1909; Redenz, 1924; Hancock, 1956; Bloom and Nicander, 1961; Van Rensburg *et al.*, 1966; Huszar and Vigue, 1994; Gomez *et al.*, 1996; Dahlbom *et al.*, 1997; Keating *et al.*, 1997; Amann *et al.*, 2000; Thundathil *et al.*, 2000a, 2001; Lovercamp *et al.*, 2007; Suzuki-Toyota *et al.*, 2010; Encisco *et al.*, 2011; Carreira *et al.*, 2012
Distal cytoplasmic droplet	Midpiece	Boar, bull, human	No	Retzius, 1909; Redenz, 1924; Hancock, 1956; Van Rensburg *et al.*, 1966; Huszar and Vigue, 1994; Keating *et al.*, 1997
Immotile cilia syndrome, primary ciliary dyskinesis ('Kartagener's syndrome')	Midpiece	Human	Yes/No	Afzelius *et al.*, 1975; Pedersen and Rebbe, 1975; Afzelius and Eliasson, 1979; Baccetti *et al.*, 1979; Leestma and Sepsenwol, 1980; Chemes *et al.*, 1987b; Collodel and Moretti 2006; Siqueira *et al.*, 2010
Abaxial, accessory/ offshoot	Midpiece	Boar, bull, human, ram, stallion	Normal in some species	Blom, 1950; Nicander and Bane, 1962; Aughey and Renton, 1971; Peet and Micke, 1976; Roberts 1986; Peet *et al.*, 1988; Barth, 1989; Barth and Oko, 1989
Disrupted	Midpiece	Buck, bull, stallion	No	Heath *et al.*, 1985; Chenoweth *et al.*, 2000; Molnár *et al.*, 2001
Pseudo-droplet	Midpiece	Bull, dog, stallion	Yes/No	Blom 1968, 1972; Blom and Birch-Andersen 1968; Chenoweth *et al.*, 1970; Aughey and Renton, 1971; Heath *et al.*, 1985
Tail stump	Midpiece	Boar, bull, dog, human, mouse, rabbit, stallion	Yes	Williams and Savage, 1925; Bishop *et al.*, 1964; Coubrough and Barker, 1964; Ross *et al.*, 1971; Baccetti *et al.*, 1975, 1993, 2001; Blom, 1976; Vierula *et al.*, 1987; Williams, 1987; Barth and Oko, 1989; Barthelemy *et al.*, 1990; Peet and Mullins, 1991; Foote *et al.*, 1992; Chacon and Rodriguez-Martinez, 1998; Andersson and Mäkipää, 2000; Revell *et al.*, 2000; Cisale *et al.*, 2001; Chemes and Rawe 2003

Short tail	Midpiece	Boar, bull	?	Baccetti *et al.*, 1993; Andersson and Mäkipää, 2000; Sironen *et al.*, 2002, 2006; Sukura *et al.*, 2002; Chemes and Rawe, 2003; Kopp *et al.*, 2008; Siqueira *et al.*, 2010
Centriolar–mitochondrial defect	Midpiece	Human	?	Williams,1950; Holstein *et al.*, 1986
Dag defect, 'ard' defect, 'Dag-like' defect	Midpiece, tail	Boar, buck, bull, stallion	Yes/No	Blom and Birch-Andersen, 1962, 1966; Blom 1966, 1972; Koefoed-Johnsen and Pedersen, 1971; Van Duijn, 1972; Baccetti *et al.*, 1975; Blom and Wolstrup, 1976; Peet and Micke, 1976; Barth and Oko, 1989; Hellander *et al.*, 1991; Andersen Berg *et al.*, 1996; Molnár *et al.*, 2001
Dysplasia of the fibrous sheath (DFS)	Midpiece	Human	Yes	Chemes *et al.*, 1987b; Chemes, 2000; Rawe *et al.*, 2001; Chemes and Rawe, 2003; Baccetti *et al.*, 2004; Baccetti, 2005; Collodel and Moretti, 2006; Moretti and Collodel, 2006; Moretti *et al.*, 2008
Defects of the sperm tail				
Coiled tails	Tail	Buck, bull, dog, human, mouse, stallion	?	Wood *et al.*, 1986; Love *et al.*, 2000; Molnár *et al.*, 2001; Kawakami *et al.*, 2005; Yeung *et al.*, 2009; Collodel *et al.*, 2011
Multiple tails	Tail	Emu, human	?	Perrin *et al.*, 2008; du Plessis and Soley, 2011
Hair pin sperm defect	Tail	Boar	No	Kojima, 1975

[a]The defects of asthenozoospermia (immotile sperm), necrozoospermia (dead sperm) and teratospermia (the production of abnormal sperm) are not included in this table but are discussed in the text. Otherwise, the order of defects followed in the table is that used in the text, but with extra categories of defects added, in no particular order.

disorders and others (Chemes, 2000). However, pathological defects of flagellar structure are also implicated, as described by Chemes *et al.* (1998), in men with severe asthenozoospermia. A comprehensive review of structural pathologies associated with asthenozoospermia in humans is provided by Chemes (2000) and Chemes and Rawe (2003).

Azoospermia

A complete lack of sperm in the ejaculate can be due to either a failure of spermatogenesis or to causes in sperm transport. With the latter category, sperm may be prevented from contributing to the ejaculate by blockages or occlusions in the extragonadal duct system, or because of an ejaculatory dysfunction. In dogs, for example, the possibility of occlusion being the cause may be determined by measurement of alkaline phosphatase (ALP) in the seminal plasma, as ALP originates from the epididymis and seminiferous tubule epithelium (Kutzler *et al.*, 2003). If the concentration of ALP in seminal fluid is greater than 5000 IU/l, the ejaculated sample contains fluid from those tissues and occlusion may be ruled out.

Testicular causes of azoospermia include failure of spermatogenesis, which can be caused by a number of factors; these include hormonal, immunological, congenital, toxic, pathogenic and traumatic factors. Here, deletions within the AZF region (containing subregions AZFa, AZFb and AZFc) on the Y chromosome play an important role, representing 10–15% of cases of azoo- and oligospermia (Shah *et al.*, 2003). Within the AFZc region, the *DAZ* gene, which is deleted in azoospermia, is considered to be important in spermatogenetic control, as its deletion is commonly detected in infertile men (Shah *et al.*, 2003).

Teratospermia

Teratospermia, or the production of >60% morphologically abnormal sperm, can be due to either genetic or environmental causes, as well as to interactions between these two factors. This defect appears to be especially prevalent in domestic cats and wildcats, and occurs in 70% of felid species (Pukazhenthi *et al.*, 2001). One assumption is that this high level of teratospermia is associated with high levels of inbreeding, although conclusive evidence is still pending. Inbreeding levels tend to be high in cheetahs and other endangered wild cats (O'Brien *et al.*, 1983, 1985) that are encountering population decline due to habitat loss, poaching, disease and other causes (Wildt *et al.*, 1995). In felids, spermatozoa from teratospermic individuals are incapable of fertilizing oocytes owing to an inability to undergo capacitation and the acrosome reaction (Pukazhenthi *et al.*, 2006). Suggestions of a genetic basis for teratospermia have also come from studies of a stallion (Brito *et al.*, 2010) and a dog (Rota *et al.*, 2008).

Defects of the sperm acrosome

KNOBBED ACROSOME (KA). This defect was first observed in a sterile bull of the Friesian breed (Teunissen, 1946), where it was subsequently shown to have an autosomal sex-linked recessive mode of genetic transmission (Donald and Hancock, 1953). It has been associated with infertility in bulls, boars, rams and dogs (Teunissen, 1946; Saacke *et al.*, 1968; Barth, 1986; Soderquist, 1998; Santos *et al.*, 2006). In boars, the defect has been suspected of both dominant (Wohlfarth, 1961) and sex-linked recessive (Bishop, 1972) modes of transmission. In cattle, there are indications that it exists as a genetic defect within the Angus breed in North America (P.J. Chenoweth, unpublished) and there is a suggestion of a genetic linkage in the Charolais breed (Barth and Oko, 1989). Four related male dogs exhibiting high KA levels were incapable of achieving pregnancies with otherwise fertile bitches (Santos *et al.*, 2006).

In the bull, this abnormality is commonly observed as either a refractile, thickened, protruding acrosomal apex or as an indented sperm apex (Thundathil *et al.*, 2000b). Electron microscopy often reveals a cystic region (the 'cystic apical body') containing vesicles with inclusions, as well as abnormal fusion of acrosomal membranes (Bane and Nicander, 1966; Cran and Dott, 1976). There is also often a bending back, or abrupt termination, of the sperm nuclear material (Thundathil *et al.*, 2000b), giving the appearance that the sperm head apex has been cleanly incised.

Elevated levels of knobbed acrosomes in bull semen are associated with either genetic or environmental factors. With the latter, the elevated incidence is usually transitory and occurs in concert with other signs of spermatogenic dysfunction (i.e. increased sperm abnormalities in general, including nuclear vacuoles). A genetic cause is suspected when relatively high proportions (>20%) of sperm exhibit the KA defect in the absence of frequent numbers of other sperm abnormalities, and this persists indefinitely (Thundathil *et al.*, 2000b; Chenoweth, 2005). In Canada, 78 of 1331 (5.9%) bulls had sperm with knobbed acrosomes (Barth and Oko, 1989). In contrast, in the author's laboratory (P.J. Chenoweth, unpublished), a prevalence of 6.74% was observed in bulls in a known affected Angus herd.

In pigs, an incidence for KA of 7.6% was reported in the Finnish Yorkshire breed (Kopp *et al.*, 2008), in which the syndrome was associated with immotile short-tail sperm defects. Two candidate genes, both located on pig chromosome 15, have been linked with knobbed acrosomes: *HECW2* and *STK17b* (Sironen *et al.*, 2010). In comparison with 'normal' boars, those with the KA defect were reported to have seminiferous tubules of smaller diameter, as well as a reduced number of Sertoli cells.

Sperm containing KA are either unable to attach to ova (Buttle and Hancock 1965), or this capability is much reduced (Thundathil *et al.*, 2000b). This defect would, therefore, appear to be eligible for inclusion within the 'compensable' category of sperm defects, in which increasing sperm numbers could compensate for damaged sperm in terms of inseminate fertility. In fact, several studies would appear to support this classification. For example, differences were observed in the fertility of a KA-affected ram when natural service was compared with AI (Soderquist, 1998).

Another thread of evidence for including KA in the compensable category comes from work that showed that the proportion of bull KA sperm decreased during transit through the female genital tract (Mitchell *et al.*, 1985). Despite such findings, *in vitro* studies have shown that apparently normal sperm in ejaculates from animals affected by KA may also have compromised fertility (Thundathil *et al.*, 2000b). Further studies indicate that these apparently morphologically normal sperm have plasma membrane damage, and show premature capacitation, a spontaneous acrosome reaction and impaired chromatin condensation (Thundathil *et al.*, 2002). Thus, the KA defect may encompass both 'compensable' and 'uncompensable' characteristics. As bulls with apparently similar proportions of the defect may vary in infertility, this variation may be either due to such unrecognized causes of sperm dysfunction, or to the numbers of undamaged sperm reaching the fertilization site.

RUFFLED AND INCOMPLETE ACROSOMES. Ruffled and incomplete acrosomes have been reported in sub-fertile bulls (Saacke *et al.*, 1968). Here, ruffled acrosomes had an irregular staining pattern that provided a wrinkled, or ruffled, appearance. Incomplete acrosomes showed an irregular margin, giving the appearance that part of the acrosome was missing or incomplete. A genetic basis was suggested by the occurrence of all three acrosome defects (knobbed, ruffled and incomplete acrosomes) in four sons of a sub-fertile Holstein sire. Some similarity has also been drawn with acrosome abnormalities described in 'genetically-determined quasi-sterile' male mice (Rajasekarasetty, 1954).

Defects of the sperm head

DECAPITATED SPERM DEFECT (HEADLESS FLAGELLAE, DISINTEGRATED, ACEPHALIC, MICROCEPHALIC, PIN HEAD). Separation of the sperm head from its corresponding midpiece and tail can be caused by a number of adverse environmental factors affecting either spermiogenesis or sperm maturation (Barth and Oko, 1989), and also sperm integrity post ejaculation. The attachment of the sperm head and midpiece occurs during spermiogenesis and is mediated by the interaction between the centrioles and the spermatid nucleus (Baccetti *et al.*, 1984; Chemes *et al.*, 1999; Chemes and Rawe, 2003). When the centriole-tail anlage (primordium) fails to attach normally to the nucleus, heads and tails develop independently and separate at spermiation, with the heads often being phagocytosed by Sertoli cells or during epididymal passage

(Chemes *et al.*, 1999; Toyama *et al.*, 2000). A variety of manifestations occur, leading to corresponding and confusing terms, some of which are at odds with their common pathogenesis. These terms include decapitated, headless and acephalic sperm, as well as 'pin heads' (Chemes and Rawe, 2003). The latter term can be misleading as it often refers to a phenotype in which the sperm head is absent, but where a structure is retained that resembles a small sperm head. These structures have been shown to contain no traces of chromatin and are due to structural rearrangements, including retained cytoplasmic material, at the proximal end of the principal piece. In contrast, small sperm heads have been associated with fragmented DNA (Encisco *et al.*, 2011).

A specific, sterilizing form of decapitated sperm has been reported in several cattle breeds (Guernsey, Hereford, Swedish Red and White) in which the majority (80–100%) of sperm are affected, and where the separated tail usually retains motility. Early work indicated that separation was associated with defective development of the sperm head implantation groove and basal plate, and first became evident as sperm traversed the epididymis (Settergren and Nicander, 1968; Blom and Birch-Andersen, 1970). More recent findings (Chemes *et al.*, 1999) provide a more definitive account, in which the centriole-tail anlage fails to connect, as described above. Evidence for the hereditary nature of this defect occurs in bulls (most probably via a sex-limited recessive gene with male and female carriers; Jones, 1962; Van Rensburg *et al.*, 1966) as well as in humans (Baccetti *et al.*, 1984, 1989b, 2001; Chemes *et al.*, 1999). In a group of Hereford bulls, this defect apparently occurred in conjunction with testicular hypoplasia (Williams, 1965).

Blom (1977) suggested the following criteria as being characteristic features of the decapitated sperm defect in bulls:

1. Separation of heads and tails in 80–100% of ejaculated sperm.
2. Active movement of the majority of separated tails.
3. A proximal bending or curling of the sperm midpiece around a retained cytoplasmic droplet in a number of the separated tails.

ROUND-HEADED SPERM (GLOBOZOOSPERMIA). The 'round-headed' sperm defect can occur as a specific defect in the human in which the entire sperm population is affected and where it occurs in association with other features, such as an absence of the acrosome and the post-acrosomal cap (Chemes, 2000), infertility (Schirren *et al.*, 1971; Pedersen and Rebbe, 1974; Baccetti *et al.*, 1977) and various degrees of abnormal chromatin condensation. This defect was reported in four infertile men (two of whom were brothers), where it affected 100% of ejaculated sperm (Kullander and Rausing, 1975). Many of the sperm heads contained vacuole-like structures; none had an acrosome attached. The defect is considered to arise during spermiogenesis, probably due to a malfunction of the Golgi apparatus (Baccetti *et al.*, 1977). In *in vitro* studies, no round-headed sperm showed binding with the zona pellucida.

DIADEM/CRATER DEFECT. Although the diadem/crater sperm defect has been considered as being of genetic origin, there is little evidence to support this contention. On the contrary, it appears that it represents part of a stereotyped response to a wide variety of spermatogenic insults and, as such, occurs in all domestic species. Theories concerning the aetiology of the defect include the deleterious actions of reactive oxygen species (ROS) during a critical phase of spermiogenesis. Here, the severity and length of insult, as well as the time between insult and assessment, determine which sperm malformations are most evident, though there is a consistent sequence in which certain abnormalities predominate in the ejaculate (Vogler *et al.*, 1993), as shown in Fig. 7.1. Thus, many sperm abnormalities formerly considered to be discrete entities can now be regarded as part of a consistent response spectrum of the spermatogenic epithelium to stressors (Larsen and Chenoweth, 1990; Chenoweth, 2005), and represent a failure of normal chromatin condensation (Boitrelle *et al.*, 2011). Such abnormalities include pyriform and 'unripe' (undeveloped) heads, various manifestations of craters and vacuoles. While the diadem/crater defect is not regarded as genetic per se, it is possible that genetic

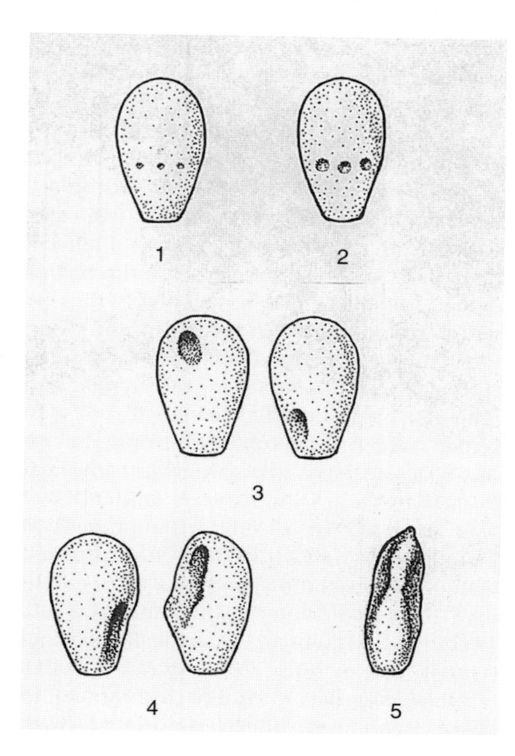

Fig. 7.1. The sequential development (steps 1 to 5) of different forms of the diadem/crater defect following spermatogenic stress. From Chenoweth, 2005 – permission granted by Elsevier.

predispositions to this condition may occur. For example, individuals vary in their capability to regulate testicular temperatures, and this variation could be associated with physical traits that may have a heritable basis.

ROLLED HEAD/NUCLEAR CREST/GIANT HEAD SYNDROME. A number of sperm head abnormalities, i.e. rolled heads, nuclear crests and giant heads, are frequently observed in combination with others, thereby leading to the assumption that they have a common origin (Blom, 1980; Cran *et al.*, 1982). Where such abnormalities occur in significant numbers, there have been suggestions of hereditary linkages (Barth and Oko, 1989), though similar defects have also occurred with ethylene dibromide-induced teratospermia in rams (Eljack and Hrudka, 1979). The ultrastructure of rolled heads and nuclear crests was described by Cran *et al.* (1982) in sperm from a Friesian bull, whose sperm also showed knobbed acrosomes. It is suggested that rolled head and crested sperm both represent a manifestation of the giant sperm head abnormality (Blom, 1980; Barth and Oko, 1989), and that they represent increased ploidy (Gledhill, 1965; Barth and Oko, 1989). Bulls with this defect generally exhibit a consistent spermiogram over time (Barth and Oko, 1989). Although effects on fertility are unclear, it is logical to assume that non-haploid sperm would result in compromised fertility, as encountered, for example, in males with Klinefelter's syndrome (Shah *et al.*, 2003). In work with inbred lines of beef cattle at the San Juan Basin Research Center of Colorado State University, Salisbury and Baker (1966b), concluded that the occurrence of diploid spermatozoa can be genetically influenced.

ABNORMAL DNA CONDENSATION AND INTEGRITY. Differences in sperm DNA content between fertile and infertile males were first described by Leuchtenberger *et al.* (1953) in humans, and by Gledhill (1965) and Gledhill *et al.* (1966) in bulls. Earlier work was conducted using light microscopy in conjunction with DNA-targeting stains; for example, Salisbury *et al.* (1977) used Fuelgen-DNA staining patterns of sperm to conclude that 'differences in spermatozoan DNA exist, not only among individuals of the same species but also among sperm in the same ejaculate'. Signs of abnormal sperm DNA are difficult to identify using routine microscopic techniques, and the task is laborious when using fluorescent microscopy. Here, the use of flow cytometry in concert with DNA-specific fluorochromes allows the rapid assessment of nuclear chromatin characteristics in large numbers of sperm. Abnormal sperm nuclear chromatin can signify disturbances of spermatogenesis, sperm abnormalities and infertility in a number of species (Ballachey *et al.*, 1987).

Defects of the sperm midpiece

'PSEUDO-DROPLET' DEFECT. This defect, which is characterized by a local thickening on the midpiece, was first observed in five related

Friesian bulls in Denmark (Blom, 1968), in which 7–26% of ejaculated sperm were affected. Under normal microscopy, the defect resembles retained cytoplasmic droplets, but it tends to occur in regions where droplets are encountered less frequently (e.g. the middle of the midpiece). They are also more likely to be irregular in shape and more visually dense than cytoplasmic droplets.

A 'mictotubular mass defect' of spermatozoa was reported in the semen of seven Standardbred stallions in which a genetic link was suggested (Heath *et al.*, 1985). Here, irregular masses in the proximal region of the midpiece contained torturous arrays of small abnormal microtubules. Similar structural defects (termed 'knobs') were observed in a sterile stallion (Chenoweth *et al.*, 1970), in bulls that had been diagnosed with a viral disease (bovine ephemeral fever), and also in Brahman bulls fed high levels of gossypol (Chenoweth *et al.*, 1994), where the abnormality was first evident during sperm transit in the epididymis (Chenoweth *et al.*, 2000). Here, ultrastructual examination showed that the abnormal structural features were caused by accumulations of displaced mitochondria.

'CORKSCREW MIDPIECE' DEFECT. The 'corkscrew sperm defect' was first described in the ejaculated sperm of two sterile bulls (Blom, 1959), where it was observed as an irregular distribution ('lumps') of mitochondria that resembled a corkscrew under light microscopy. A genetic cause was suggested, as four of the first five bulls in which the defect was observed were related. The ultrastructural appearance of this defect suggests that it shares a common aetiology with the 'pseudodroplet' defect. This theory is supported by reports in which both defects occurred together: in infertile stallions (Chenoweth *et al.*, 1970; Heath *et al.*, 1985); in bulls following a viral disease (bovine ephemeral fever (Burgess and Chenoweth, 1975); and in bulls being fed gossypol (Chenoweth *et al.*, 2000). Further doubt concerning the genetic basis of this defect came with suggestions that it was associated with preputial ejaculation (Blom, 1973b) and with increased levels of radioactivity (Blom, 1978).

'DAG' DEFECT. Named after the Jersey bull in which it was first identified (Blom, 1966), this defect is characterized by strong folding, coiling and fracture of the distal part of the sperm midpiece (with or without a retained distal cytoplasmic droplet). Because a similar spermiogram was also observed in a full brother to 'Dag', a hereditary basis was suggested, although this defect has also been associated with elevated dietary zinc levels (Blom and Wolstrup, 1976). The defect may reflect disturbance in the testicle or epididymis, and may also be observed at low levels (<4%) in semen, which is regarded as normal. Levels of >50% can have serious fertility implications (Barth and Oko, 1989). A similar defect was reported in a sub-fertile boar (Van Duijn, 1972), as well as in a Standardbred stallion where the majority of sperm exhibited 'Dag-like' defects (Hellander *et al.*, 1991), and also in an infertile Saanenthal (Dutch White) buck (Molnár *et al.*, 2001). Similar irregular midpiece formations in bull semen have been associated with bovine ephemeral fever (Burgess and Chenoweth, 1975) and with gossypol spermatoxicity (Chenoweth *et al.*, 2000).

Defects of the sperm tail

TAIL STUMP DEFECT, SHORT TAIL DEFECT, DYSPLASIA OF THE FIBROUS SHEATH (DFS). The tail stump defect was first reported in bulls (Williams and Savage, 1925) and has since been identified in the mouse, rabbit, dog, stallion and man (Barth and Oko, 1989). Three sterile Canadian bulls (Ayrshire, Shorthorn, Holstein) showed virtual absence of the sperm midpiece and tail in the majority of ejaculated sperm, where they were represented by a small 'stump' or 'stub' (Coubrough and Barker, 1964). In addition, sperm concentration was very low and sperm motility was virtually absent. Monitoring of sperm morphology in these bulls indicated that the percentage of sperm with the defect increased with the age of the bulls. It has been suggested that a prevalence of >25% of this defect in the bull ejaculate is suggestive of a genetic problem (Barth and Oko, 1989). Other reports have linked this defect with sterility in bulls

(Blom, 1976; Williams, 1987; Peet and Mullins, 1991), and suggestion has been made of an inherited mode of transmission (Vierula *et al.*, 1987; Barth and Oko, 1989; Foote *et al.*, 1992). A short tail defect was reported in three Nelore breed bulls in which 34–70% of the sperm population were affected and sperm motility was completely absent (Siqueuira *et al.*, 2010). Encisco *et al.* (2011) noted that short tail sperm in bulls also tended to have fragmented DNA.

A heritable 'short tail' sperm defect associated with infertility has been reported in Yorkshire boars (Andersson *et al.*, 2000). A similar defect has been reported in wild boars (Cisale *et al.*, 2001). A condition termed immotile short tail syndrome (ISTS) has been reported in boars in which sperm have significantly shorter tails (15.4 μm versus 47 μm) and contain significantly greater incidence of proximal cytoplasmic droplets (72.4% versus 6.9%) than do non-affected boars (Sukura *et al.*, 2002).

It is now recognized that a number of systematic sperm defects share a short, thick tail as a common feature, leading to overlaps in recognition and categorization (Baccetti *et al.*, 1975, 1993; Chemes and Rawe, 2003). These include the stump tail and short tail sperm defects and DFS (Chemes *et al.*, 1987a, 1998). Sperm with DFS have characteristic short, thick and irregular flagellae associated with a 'redundant and haphazardly arranged' fibrous sheath (Chemes, 2000; Chemes and Rawe, 2003). Affected sperm display a complete or partial lack of dynein (motor protein) arms and abnormal mitochondrial disposition (Rawe *et al.*, 2001), and increased ubiquination of affected areas (Sutovsky *et al.*, 2001). A complete lack of midpiece mitochondria has been described, which is possibly of genetic origin (Zamboni, 1992).

Care should be taken to differentiate this condition from the 'accessory tail defect' in bulls, which probably shares a common aetiology with the condition of abaxial sperm midpiece, and is reported to have little adverse impact upon bull fertility (Barth, 1989).

PRIMARY CILIARY DYSKINESIA/IMMOTILE CILIA SYNDROME (PCD/ICS). First described in human sperm (Afzelius *et al.*, 1975; Afzelius, 1976; Afzelius and Eliasson, 1979), where it was associated with chronic respiratory disease, this syndrome is now recognized as representing a diverse group of disorders characterized by a structural and generalized abnormality of ciliated cells. In man, a systemic axonemal alteration is associated with Kartagener's syndrome in which the males are infertile and possess immotile spermatozoa. Here, in common with other ciliated cells in the body (such as respiratory epithelial cells), affected sperm (and respiratory tract cilia) have perturbed axonemal structures, e.g. part or complete absence of dynein arms, microtubule disorganization or absent radial spokes, and short tails (Baccetti *et al.*,1975, 1993). This suggests that these structures, which share common organization, have similar genetic coding. In humans, a high incidence of PCD/ICS has been noted in Polynesian populations (Waite *et al.*, 1978), probably associated with autosomal recessive mutation(s) (Chemes and Rawe, 2003).

Similar sperm aberrations occur in animals, although here the link with respiratory diseases has not been adequately pursued. In mice, alterations to sperm tail axonemal complexes were observed in genetically similar individuals that were sterile (Leestma and Sepsenwol, 1980). Later work with mice using gene knockout models has identified specific defects of the sperm flagellum associated with impaired motility (Escalier, 2006), demonstrating that a number of cellular pathways involved with different aspects of the sperm midpiece and tail assembly can be implicated, as well as a number of different genes. It should be noted that the axonemal complex contains over 200 proteins, defects in any of which could be related to problems with genetic coding. For example, *KPL2* gene insertion in porcine chromosome 16 is linked with the 'immotile short-tail sperm defect' of pig sperm, in which sperm production is much reduced (Kopp *et al.*, 2008). Other genes implicated in primary ciliary dyskinesis include those encoding for sperm flagellar protein 2 (Sironen *et al.*, 2011) and for tubulin tyrosine ligase-like 1 protein (Vogel *et al.*, 2010).

Conclusions

There are many genetic influences on male fertility and infertility, and our knowledge and awareness of these is increasing rapidly in step with the development and use of modern technologies. The challenge, as always, is to differentiate between true genetic problems and those that are due to environmental causes – a task made more difficult by interactions between the two. Categorization as a true heritable factor requires a burden of proof that has rigorous rules of application (Chenoweth, 2005), and thus the term 'genetic' should not be used lightly. This is particularly relevant in the context of the livestock purebred breeding industries, where genetics and marketability are often synonymous.

References

Afzelius BA (1976) A human syndrome caused by immotile cilia *Science* **193** 317–319.

Afzelius BA and Eliasson R (1979) Flagellar mutants in man: on the heterogeneity of the immotile-cilia syndrome *Journal of Ultrastructure Research* **69** 43–52.

Afzelius BA, Eliasson R, Johnsen O and Lindholmer C (1975) Lack of dynein arms in immotile human spermatozoa *Journal of Cell Biology* **66** 225–232.

Aggarwal RA, Ahlawat SP, Kumar Y, Panwar PS, Singh K and Bhargava M (2007) Biometry of frozen–thawed sperm from eight breeds of Indian buffaloes (*Bubalus bubalis*) *Theriogenology* **68** 682–686.

Aitken RJ (2002) Active oxygen in spermatozoa during epididymal transit In *The Epididymis from Molecules to Clinical Practice* pp 435–449 Ed B Robaire and B Hinton. Plenum Press, New York.

Aitken RJ and Sawyer DW (2003) The human spermatozoon, not waving but drowning. *Advances in Experimental Medicine and Biology* **518** 85–98.

Al-Makhzoomi A, Lundeheim N, Håård M and Rodriguez-Martinez H (2008) Sperm morphology and fertility of progeny-tested AI dairy bulls in Sweden *Theriogenology* **70** 682–691.

Al-Qarawi AA and El-Belely MS (2004) Intratesticular morphometric, cellular and endocrine changes in dromedary bulls exhibiting azoospermia *The Veterinary Journal* **167** 194–201.

Amann RP, Seidel GE Jr and Mortimer RG (2000) Fertilizing potential *in vitro* of semen from young beef bulls containing a high or low percentage of sperm with a proximal droplet *Theriogenology* **54** 1499–1515.

Andersen K and Filseth O (1976) The occurrence of a "SME"-defect-like abnormality in the head of spermatozoa from a Norwegian Landrace boar *Nordisk Veterinærmedicin* **28** 511–514.

Andersen Berg K, Filseth O and Engeland E (1996) A sperm midpiece defect in a Hereford bull with variable semen quality and freezability *Acta Veterinaria Scandinavica* **37** 367–373.

Andersson M and Mäkipää R (2000) Length of the sperm tail in fertile boars of different breeds and in sterile Yorkshire boars affected with the hereditary "short tail" sperm defect In *Proceedings of the 14th International Congress on Animal Reproduction, Stockholm, Sweden, 2–6 July 2000. Abstracts Volume 1, Section* 2 p 96.

Andersson M, Peloniemi OAT and Mäkinen A (2000) The hereditary "short tail" sperm defect – a new reproductive problem in Yorkshire boars *Reproduction in Domestic Animals* **39** 59–63.

Ansari HA, Jung HR, Hediger R, Fries R, König H and Stranzinger G (1993) A balanced autosomal reciprocal translocation in an azoospermic bull *Cytogenetics and Cell Genetics* **62** 117–123.

Anton-Lamprecht LB, Kotzur B and Schöpf E (1976) Round-headed human spermatozoa *Fertility and Sterility* **27** 685–693.

Ardón F, Helms D, Sahin E, Bollwein H, Töpfer-Petersen, E and Waberski D (2008) Chromatin-unstable boar spermatozoa have little chance of reaching oocytes *in vivo Reproduction* **135** 461–470.

Aughey E and Renton JP (1971) Ultrastructure of abnormal spermatozoa in a stud dog *Journal of Reproduction and Fertility* **25** 303–304.

Baccetti B (2005) Gene deletions in an infertile man with sperm fibrous sheath dysplasia *Human Reproduction* **20** 2790–2794.

Baccetti B, Burrini A and Pallini V (1975) The short tailed human spermatozoa. Ultrastructural alterations and dynein absence *Journal of Submicroscopic Cytology and Pathology* **7** 349–359.

Baccetti B, Renieri T, Rosati F, Selmi RG and Casanova S (1977) Further observations on the morphogenesis of the round headed human spermatozoa *Andrologia* **9** 255–264.

Baccetti B, Burrini AG, Palluni V, Maver A and Reneiri T (1979) 9+0 immotile spermatozoa in an infertile male *Andrologia* **11** 437–443.

Baccetti B, Selmi MG and Soldani P (1984) Morphogenesis of "decapitated" spermatozoa in a man *Journal of Reproduction and Fertility* **70** 395–397.

Baccetti B, Burrini AG, Collodel G, Magnano AR, Oimboni P, Renieri T and Sensini C (1989a) Crater defect in human spermatozoa *Gamete Research* **22** 249–255.

Baccetti B, Burrini AG, Collodel G, Magnano AR, Piomboni P, Renieri T and Sensini C (1989b) Morphogenesis of the decapitated and decaudated sperm defect in two brothers *Gamete Research* **23** 181–188.

Baccetti B, Burrini AG, Collodel G, Piomboni P and Renieri T (1991) A "miniacrosome" sperm defect causing infertility in two brothers *Journal of Andrology* **12** 104–111.

Baccetti B, Burrini AG, Capitani S, Collodel G, Moretti E, Piomboni P and Renieri T (1993) Notulae seminologicae. 2. The 'short tail' and 'stump' defect in human spermatozoa *Andrologia* **25** 331–333.

Baccetti B, Capitani S, Collodel G, Di Cairano G, Gambera L, Moretti E and Plomboni P (2001) Genetic sperm defects and consanguinity *Human Reproduction* **16** 1365–1371.

Baccetti B, Bruni E, Gambera L, Moretti E and Piomboni P (2004) An ultrastructural and immunocytochemical study of a rare genetic sperm tail defect that causes infertility in humans *Fertility and Sterility* **82** 463–468.

Ballachey BE, Hohenboken WD and Evenson DP (1987) Heterogeneity of sperm nuclear chromatin structure and its relationship to bull fertility *Biology of Reproduction* **36** 915–925.

Ballachey BE, Evenson DP and Saacke RG (1988) The sperm chromatin structure assay: relationship with alternate tests of semen quality and heterospermic performance of bulls *Journal of Andrology* **9** 109–115.

Bane A (1954) Studies on monozygous cattle twins XV. Sexual functions in bulls in relation to heredity, rearing intensity and somatic conditions *Acta Agriculturæ Scandinavica* **4** 95–208.

Bane A (1961) Acrosomal defect associated with sterility in boars *In Proceedings of the 4th International Congress on Animal Reproduction and Artificial Insemination, The Hague, 5–9 July 1961* p 810.

Bane A and Nicander L (1965) Pouch formations by invaginations of the nuclear envelope of bovine and porcine sperm as a sign of disturbed spermiogenesis *Nordisk Veterinærmedicin* **17** 628–632.

Bane A and Nicander L (1966) Electron and light microscopical studies on spermateliosis in a boar with acrosome abnormalities *Journal of Reproduction and Fertility* **11** 133–138.

Barth AD (1986) The knobbed acrosome defect in beef bulls *Canadian Veterinary Journal* **27** 379–384.

Barth AD (1989) Abaxial tail attachment of bovine spermatozoa and its effect on fertility *Canadian Veterinary Journal* **30** 656–662.

Barth AD and Oko RJ (1989) *Abnormal Morphology of Bovine Spermatozoa.* Iowa State Press, Ames.

Barth AD, Bowman PA, Bo GA and Mapletoft RJ (1992) Effect of narrow sperm head shape on fertility in cattle *Canadian Veterinary Journal* **33** 31–39.

Barthelemy C, Tharanne MJ, Lebos C, Lecomte P and Lansac J (1990) Tail stump spermatozoa: morphogenesis of the defect. An ultrastructural study of sperm and testicular biopsy *Andrologia* **22** 417–425.

Beatty RA and Fechheimer NS (1972) Diploid spermatozoa in rabbit semen and their experimental separation from haploid spermatozoa *Biology of Reproduction* **7** 267–277.

Bernadini L, Borini A, Preti S, Conte N, Flamigni C, Capitanio GL and Venturini PL (1998) Study of aneuploidy in normal and abnormal germ cells of fertile and infertile men *Human Reproduction* **13** 3406–13.

Biason-Lauber A (2010) Control of sex development *Best Practice and Research Clinical Endocrinology and Metabolism* **24** 163–186.

Bishop MWH (1972) Genetically determined abnormalities of the reproductive system *Journal of Reproduction and Fertility Supplement* **5** 51.

Bishop MWH, Campbell RC, Hancock JL and Walton A (1964) Semen characteristics and fertility in the bull *The Journal of Agricultural Science* **44** 227–248.

Blake (1945) Inheritance of morphological characters in the sperms of cattle *Nature* **155** 631.

Blom E (1950) Interpretation of spermatic cytology of bulls *Fertility and Sterility* **1** 223–238.

Blom E (1959) A rare sperm abnormality: "corkscrew-sperms" associated with sterility in bulls *Nature* **183** 1280–1281.

Blom E (1966) A new sterilizing and hereditary defect (the "Dag-defect") located in the sperm tail *Nature* **209** 739–740.

Blom E (1968) A new sperm defect – "Pseudo-droplets" – in the middle piece of the bull sperm *Nordisk Veterinærmedicin* **20** 279–283.

Blom E (1972) The ultrastructure of some characteristic sperm defects and a proposal for a new classification of the bull spermiogram In *Atti del VII Simposio Internationale di Zootecnia, Milano, 15–17 Aprile 1972* pp 125–139. Società Italiana per il Progresso della Zootecnia, Milan.

Blom E (1973a) Studies on boar semen. 1. A new major defect in the sperm head, the "SME" defect *Acta Veterinaria Scandinavica* **14** 633–635.

Blom E (1973b) Studies on boar semen. II. Abnormal mitochondrial helix "pseudo-corkscrew sperm" due to ejaculation in the preputial cavity *Acta Veterinaria Scandinavica* **14** 636–638.

Blom E (1976) A sterilizing tail stump defect in a Holstein-Friesian bull *Nordisk Veterinærmedicin* **26** 295–298.

Blom E (1977) A decapitated sperm defect in two sterile Hereford bulls *Nordisk Veterinærmedicin* **29** 119–123.

Blom E (1978) The corkscrew sperm defect in Danish bulls – a possible indicator of nuclear fallout? *Nordisk Veterinaermedicin* **30** 1–8

Blom E (1980) Rolled-head and nuclear-crest sperm defects in a rare case of presumed diploidy in the bull *Nordisk Veterinærmedicin* **22** 409–416.

Blom E and Birch-Andersen A (1962) Ultrastructure of sterilising "knobbed sperm" defect in bull *Nature* **194** 989–990.

Blom E and Birch-Andersen A (1966) The ultrastructure of a sterilizing tail defect ("the Dag-defect") in the bull sperm In *Proceedings 10th Nordic Veterinary Congress (Stockholm) Volume 2* 699–705.

Blom E and Birch-Andersen A (1968) The ultrastructure of the "pseudo-droplet" defect in the bull sperm *Proceedings of VIth International Congress on Animal Reproduction and Artificial Insemination, Paris Volume I* pp 117–119.

Blom E and Birch-Andersen A (1970) Ultrastructure of the "decapitated sperm defect" in Guernsey bulls *Journal of Reproduction and Fertility* **23** 67–72.

Blom E and Birch-Andersen A (1975) The ultrastructure of a characteristic spermhead-defect in the boar: the SME-defect *Andrologia* **7** 199–209.

Blom E and Jensen P (1977) Study of the inheritance of the SME seminal defect in the boar *Nordisk Veterinærmedicin* **39** 194–198.

Blom E and Wolstrup C (1976) Zinc as a possible causal factor in the sterilizing sperm tail defect, the "Dag defect" in Jersey bulls *Proceedings of VIIIth International Congress on Animal Reproduction and Artificial Insemination Volume IV* pp 690–693.

Bloom G and Nicander L (1961) On the ultrastructure and development of the protoplasmic droplet of spermatozoa *Zeitschrift für Zellforschung und Mikroskopische Anatomie* **52** 833–844.

Boitrelle F, Ferfouri F, Petit JM, Segretain D, Tourain C, Bergere M, Bailly M, Vialard F, Albert M and Selva J (2011) Large human sperm vacuoles observed in motile spermatozoa under high magnification: nuclear thumbprints linked to failure of chromatin condensation *Human Reproduction* **26** 1650–1658.

Brinks JS (1994) Relationships of scrotal circumference to puberty and subsequent reproductive performance in male and female offspring In *Factors Affecting Calf Crop* pp 363–370 Ed MJ Fields and RS Sand. CRC Press, Boca Raton.

Brito LF, Kelleman A, Greene LM, Raz T and Barth AD (2010) Semen characteristics in a sub-fertile Arabian stallion with idiopathic teratospermia *Reproduction in Domestic Animals* **45** 354–358.

Burgess GW and Chenoweth PJ (1975) Mid-piece abnormalities in bovine semen following experimental and natural cases of bovine ephemeral fever *British Veterinary Journal* **131** 536–544.

Buttle HR and Hancock JL (1965) Sterile boars with "knobbed" acrosomes *The Journal of Agricultural Science* **65** 255–260.

Carreira JT, Mingoti GZ, Rodrigues LH, Silva C, Perri SH and Koivisto MB (2012) Impact of proximal cytoplasmic droplets on quality traits and *in-vitro* embryo production efficiency of cryopreserved bull spermatozoa *Acta Veterinaria Scandinavica* **54** 1–78.

Carroll EJ and Ball L (1970) Testicular changes as affected by mating systems in beef cattle *American Journal of Veterinary Research* **31** 243–254.

Carruthers AD and Beatty RA (1975) The recognition and incidence of haploid and polyploidy spermatozoa in man, rabbit and mouse *Journal of Reproduction and Fertility* **44** 487–500.

Castellani L, Chiara F and Cotelli F (1978) Fine structure and cytochemistry of the morphogenesis of round-headed human spermatozoa *Archives of Andrology* **1** 291–297.

Chacon J and Rodriguez-Martinez H (1998) Tail stump defect as a cause of sterility in an Indobrasil (*Bos indicus*) bull *Reproduction in Domestic Animals* **33** 405–407.

Chandley AC and Cooke HJ (1994) Human male infertility – Y-linked genes and spermatogenesis *Human Molecular Genetics* **3** 1449–1554.

Chemes HE (2000) Phenotypes of sperm pathology: genetic and acquired forms in infertile men *Journal of Andrology* **21** 799–808.

Chemes HE and Rawe YV (2003) Sperm pathology: a step beyond descriptive morphology. Origin, characterization and fertility potential of abnormal sperm phenotypes in infertile men *Human Reproduction Update* **9** 405–428.

Chemes HE, Carizza C, Scarinci F, Brugo Olmedo S, Neuspiler N and Schwarsztein L (1987a) Lack of a head in human spermatozoa from sterile patients: a syndrome associated with impaired fertility *Fertility and Sterility* **47** 310–316.

Chemes H, Brugo Olmedo S, Zanchetti F, Carrere C and Lavieri JC (1987b) Dysplasia of the fibrous sheath: an ultrastructural defect of human spermatozoa associated with sperm immotility and primary sterility *Fertility and Sterility* **48** 664–669.

Chemes HE, Brugo OS, Carrerer C, Oses R, Carizza C, Lesiner M and Blaquier J (1998) Ultrastructural pathology of the sperm flagellum. Association between flagellar pathology and fertility prognosis in severely asthenozoospermic men *Human Reproduction* **13** 2521–2526.

Chemes ET, Puigdomenech C, Carizza S, Brugo Olmedo S, Zanchetti R and Hermes R (1999) Acephalic spermatozoa and abnormal development of the head–neck attachment: a human syndrome of genetic origin *Human Reproduction* **14** 1811–1818.

Chen P, Baas TJ, Mabry JW, Koehler KJ and Dekkers JC (2003) Genetic parameters and trends for litter traits in U.S. Yorkshire, Duroc, Hampshire, and Landrace pigs *Journal of Animal Science* **81** 46–53.

Chenoweth PJ (1981) Libido and mating behavior in bulls, boars and rams. A review *Theriogenology* **16** 155–177.

Chenoweth PJ (1983) Sexual behavior in the bull: a review *Journal of Dairy Science* **66** 173–179.

Chenoweth PJ (1994) Bull behavior, sex-drive and management In *Factors Affecting Calf Crop* pp 319–330 Ed MJ Fields and RS Sands. CRC Press, Boca Raton.

Chenoweth PJ (2005) Genetic sperm defects *Theriogenology* **64** 457–468.

Chenoweth PJ (2011) Male effects on the embryo *Brazilian Journal of Animal Reproduction* **35** 154–159.

Chenoweth PJ and Burgess GW (1972) Mid-piece abnormalities in bovine semen following ephemeral fever *Australian Veterinary Journal* **48** 37–38.

Chenoweth PJ and Osborne HG (1975) Breed differences in the reproductive function of young beef bulls in central Queensland *Australian Veterinary Journal* **51** 405–406.

Chenoweth PJ, Pascoe RR, McDougall HL and McCosker PJ (1970) An abnormality of the spermatozoa of a stallion (*Equus caballus*) *British Veterinary Journal* **126** 476–488.

Chenoweth PJ, Brinks JS and Nett TM (1979) A comparison of three methods of assessing sex-drive in yearling beef bulls and relationships with testosterone and LH levels *Theriogenology* **12** 223–233.

Chenoweth PJ, Risco CA, Larsen RE, Velez J, Chase CC Jr and Tran T (1994) Effect of dietary gossypol on aspects of semen quality, sperm morphology and sperm production in young Brahman bulls *Theriogenology* **42** 1–13.

Chenoweth PJ, Chase CC Jr, Risco CA and Larsen RE (2000) Characterization of gossypol-induced sperm abnormalities in bulls *Theriogenology* **53** 1193–1203.

Cisale HO, Rivola MA and Fernández HA (2001) Tail-stump defect in the semen of a wild boar *Veterinary Record* **149** 682.

Collodel G and Moretti E (2006) Sperm morphology and aneuploidies: defects of supposed genetic origin *Andrologia* **38** 208–215.

Collodel G, Federico MG, Pascarelli NA, Geminiani M, Renieri T and Moretti E (2011) A case of severe asthenozoospermia: a novel sperm tail defect of possible genetic origin identified by electron microscopy and immunocytochemistry *Fertility and Sterility* **95** 289. e11–289.e16.

Cothran EG, MacCluer JW, Weitkamp IR, Pfennig DW and Boyce AJ (1984) Inbreeding and reproductive performance in Standardbred horses *Journal of Heredity* **75** 220–224.

Coubrough RI and Barker CAV (1964) An unusual middle-piece abnormality associated with sterility in bulls *Proceedings of the Vth International Congress on Animal Reproduction and Artificial Insemination, 6–13 September, 1964, Trento* pp 219–229.

Coulter GH, Oko RJ and Costerton JW (1978) Incidence and ultrastructure of the "crater" defect of bovine spermatozoa. *Theriogenology* **9** 165–173.

Courot M and Colas G (1986) The role of the male in embryonic mortality In *Embryonic Mortality in Farm Animals* pp 196–206 Ed JM Greenan and MG Diskin. Martinus Nijhoff, Dordrecht.

Cox VS, Wallace LJ and Jessen CR (1978) An anatomic and genetic study of canine cryptorchidism *Teratology* **18** 233–240.

Cran DG and Dott HM (1976) The ultrastructure of knobbed bull spermatozoa *Journal of Reproduction and Fertility* **47** 407–408.

Cran DG, Dott HM and Wilmington JW (1982) The structure and formation of rolled and crested bull spermatozoa *Nordisk Veterinærmedicin* **32** 409–416.

Dahlbom M, Andersson M, Vierula M and Alanko M (1997) Morphometry of normal and teratozoospermic canine sperm heads using an image analyzer: work in progress *Theriogenology* **48** 687–698.

Donald HP and Hancock JL (1953) Evidence of a gene-controlled sterility in bulls *The Journal of Agricultural Science* **43** 178–181.

Dowsett KF and Pattie WA (1982) Characteristics and fertility of stallion semen *Journal of Reproduction and Fertility Supplement* **32** 1–8.

Druet T, Fritz S, Sellem E, Basso B, Gérard O, Salas-Cortes L, Humblot P, Druart X and Eggen A (2009) Estimation of genetic parameters and genome scan for 15 semen characteristics traits of Holstein bulls *Journal of Animal Breeding and Genetics* **126** 269–277.

Ducos A, Revay T, Kovacs A, Hidas A, Pinton A, Bonnet-Garnier A, Molteni L, Slota E, Switonski M, Arruga MV, Van Haeringen WA, Nicolae I, Chaves R, Guedes-Pinto H, Andersson M and Iannuzzi L (2008) Cytogenetic screening of livestock populations in Europe: an overview *Cytogenetic Genome Research* **120** 26–41.

du Plessis L and Soley JT (2011) Incidence, structure and morphological classification of abnormal sperm in the emu (*Dromaius novaehollandiae*) *Theriogenology* **75** 589–601.

Duranthon V and Renard JP (2001) The developmental competence of mammalian oocytes: a convenient but biologically fuzzy concept *Theriogenology* **55** 1277–1289.

Eljack AH and Hrudka F (1979) Pattern and dynamics of teratospermia induced in rams by parenteral treatment with ethylene dibromide *Journal of Ultrastructural Research* **67** 124–134.

Encisco M, Cisale H, Johnston SD, Sarasa J, Fernandez JL and Gosalvez J (2011) Major morphological sperm abnormalities in the bull are related to sperm DNA damage *Theriogenology* **76** 23–32.

Erenpreiss J, Spano M, Erenpreisa J, Bungum M and Giwercman A (2006) Sperm chromatin structure and male fertility: biological and clinical aspects *Asian Journal of Andrology* **8** 11–29.

Escalier D (1983) Human spermatozoa with large heads and multiple flagella: a quantitative ultrastructural study of 6 cases *Biology of the Cell* **48** 65–74.

Escalier D (2006). Knockout mouse models of sperm flagellum anomalies *Human Reproduction Update* **12** 449–461.

Esnault C, Ortavant R and Nicolle J-C (1967) Origine des spermatozoids diploides présents dans l'éjaculat d'un taureau Charolais *Annales de Biologie Animale Biochimie Biophysique* **7** 25–28.

Estrada A, Samper JC, Lillich JD, Rathi RR, Brault LS, Albrecht BB, Imel MM and Senne EM (2003) Azoospermia associated with bilateral segmental aplasia of the ductus deferens in a stallion *Journal of the American Veterinary Medical Association* **222** 1740–1742.

Fechheimer NS (1961) Poikiloploidy among spermatogenetic cells of *Mus musculus Journal of Reproduction and Fertility* **2** 68–79.

Feitsma H and De Vries AG (2006) Prevalence of reciprocal translocation in A.I. boars and economic justification for culling reciprocal translocation positive boars In *Proceedings of the 19th International Pig Veterinary Society Congress, Copenhagen Volume 1* p 546 Ed JP Nielsen and SE Jorsal. American Association of Swine Veterinarians, Perry, Iowa.

Flowers WL (2002) Increasing fertilization rate of boars: influence of number and quality of spermatozoa inseminated *Journal of Animal Science* **80(E.Supplement 1)** E47–E53.

Flowers WL (2008) Genetic and phenotypic variation in reproductive traits of AI boars *Theriogenology* **70** 1297–1303.

Foote R, Hough S, Johnson L and Kaproth M (1992) Electron microscopy and pedigree study in an Ayrshire bull with tail-stump sperm defects *Veterinary Record* **130** 578–579.

Fortier A and Trasler J (2011) Potential epigenetic consequences associated with assisted reproduction In *Epigenetics and Human Reproduction* pp 3–18 Ed S Rousseaux and S Khochbin. Springer Verlag, Berlin/Heidelberg.

Freeman AE (1984) Secondary traits: sire evaluation and the reproductive complex *Journal of Dairy Science* **67** 449–458.

Furstoss V, David I, Leboeuf B, Guillouet P, Boué P and Bodin L (2009) Genetic and non-genetic parameters of several characteristics of production and semen quality in young bucks *Animal Reproduction Science* **110** 25–36.

Gadea J and Matas C (2000) Sperm factors related to *in vitro* penetration of porcine oocytes *Theriogenology* **54** 1343–1357.

García-Macías V, Martinez-Pastor F, Alvarez M, Garde JJ, Anel E and de Paz P (2006) Assessment of chromatin status (SCSA) in epididymal and ejaculated sperm in Iberian red deer, ram and domestic dog *Theriogenology* **66** 1921–1930.

Gieseke K, Hamann H, Stock KF, Woehlke A, Sieme H and Distl O (2009) Evaluation of *SPATA1*-associated markers for stallion fertility *Animal Genetics* **40** 359–365.

Gieseke K, Sieme H and Distl O (2010) Infertility and candidate gene markers for fertility in stallions: a review *Veterinary Journal* **185** 265–271.

Gledhill BL (1965) Cytophotometry of presumed diploid bull spermatozoa *Nordisk Veterinærmedicin* **17** 328–335.

Gledhill BL (1966) Studies on the DNA content, dry mass and optical area of ejaculated spermatozoa heads from bulls with normal and lowered fertility *Acta Veterinaria Scandinavica* **7** 166–174.

Gledhill BL, Gledhill MP, Rigler R Jr and Ringertz NR (1966) Atypical changes of deoxyribonucleoprotein during spermiogenesis associated with a case of infertility in the bull *Journal of Reproduction and Fertility* **12** 575–578.

Gomez E, Buckingham D, Brindle J, Lanzafame F, Irvine DS and Aitken RJ (1996) Development of an image analysis system to monitor the retention of residual cytoplasm by human spermatozoa: correlation with biochemical markers of the cytoplasmic space, oxidative stress and sperm function *Journal of Andrology* **17** 276–287.

Gredler B, Fuerst C, Fuerst-Waltl B, Schwarzenbacher H and Sölkner J (2007) Genetic parameters for semen production traits in Austrian dual-purpose Simmental bulls *Reproduction in Domestic Animals* **42** 326–328.

Gresky C, Hamann H and Distl, O (2005) Influence of inbreeding on litter size and the proportion of stillborn puppies in dachshunds *Berliner und Münchener Tierärztliche Wochenschrift* **118** 134–139.

Gustaffson I (1979) Distribution and effects of the 1/29 Robertsonian translocation in cattle *Journal of Dairy Science* 62 825–835.

Hamann H, Jude R, Sieme H, Mertens U, Töpfer-Petersen E, Distl O and Leeb T (2007) A polymorphism within the equine *CRISP3* gene is associated with stallion fertility in Hanoverian warmblood horses *Animal Genetics* **38** 259–264.

Hancock JL (1949) Evidence of an inherited seminal character associated with infertility of Friesian bulls *Veterinary Record* **61** 308–310.

Hancock JL (1955) The disintegration of bull spermatozoa *Veterinary Record* **67** 825–826.

Hancock JL (1956) The cytoplasmic beads of boar spermatozoa *Journal of Endocrinology* **14** 38–39.

Hancock JL and Rollinson DHL (1949) A seminal defect associated with sterility of Guernsey bulls *Veterinary Record* **61** 7–12.

Heath E and Ott RS (1982) Diadem/crater defect in spermatozoa of a bull *Veterinary Record* **110** 5–6

Heath E, Aire T and Fujiwara K (1985) Microtubular mass defect of spermatozoa in the stallion *American Journal of Veterinary Research* **46** 1121–1125.

Held JP, Prater P and Stettler M (1991) Spermatozoal head defect as a cause of infertility in a stallion *Journal of the American Veterinary Medical Association* **199** 1760–1761.

Hellander JC, Samper JC and Crabo BG (1991) Fertility of a stallion with low sperm motility and a high incidence of an unusual sperm tail defect *Veterinary Record* **128** 449–451.

Hillery FL, Parrish JJ and First NL (1990) Bull specific effect on fertilization and embryo development *in vitro Theriogenology* **33**:249.

Hofherr EE, Wiktor AE, Kipp BR, Dawson DB and Van Dyke DL (2011) Clinical diagnostic testing for the cytogenetic and molecular causes of male infertility: the Mayo Clinic experience *Journal of Assisted Reproduction and Genetics* **28** 1091–1098.

Holstein AF, Schill WB and Breucker H (1986) Dissociated centriole development as a cause of spermatid malformation in a man *Journal of Reproduction and Fertility* **78** 719–725.

Horsthemke B and Ludwig M (2005) Assisted reproduction: the epigenetic perspective *Human Reproduction Update* **11** 473–482.

Hudson, GFS and Van Vleck LD (1984) Inbreeding of artificially bred cattle in the northeastern United States *Journal of Dairy Science* **67** 161–170.

Huszar G and Vigue I (1994) Correlation between the rate of lipid peroxidation and cellular maturity as measured by creatine kinase activity in human spermatozoa *Journal of Andrology* **15** 71–77.

Johanson JM, Bergert PJ, Kirkpatrick BW and Dentine MR (2001) Twinning rates for North American Holstein sires *Journal of Dairy Science* **84** 2081–2088.

Johnson LA and Hurtgen JP (1985) The morphological and ultrastructural appearance of the crater defect in stallion spermatozoa *Gamete Research* **12** 41–46.

Jones WA (1962) Abnormal morphology of the spermatozoa in Guernsey bulls *British Veterinary Journal* **118** 257–261.

Karabinus D, Vogler CJ, Saacke RG and Evanson DP (1997) Chromatin structural changes in bovine sperm after scrotal insulation of Holstein bulls *Journal of Andrology* **18** 549–555.

Karoui S, Díaz C, Serrano M, Cue R, Celorrio I and Carabaño MJ (2011) Time trends, environmental factors and genetic basis of semen traits collected in Holstein bulls under commercial conditions *Animal Reproduction Science* **124** 28–38.

Kawakami E, Ozawa T, Hirano T, Hori T and Tsutsui T (2005) Formation of detached tail and coiled tail of sperm in a Beagle dog *Journal of Veterinary Medical Science* **67** 83–85.

Keating J, Grundy CE, Fivey PS, Elliot M and Robinson J (1997) Investigation of the association between the presence of cytoplasmic residues in the human sperm midpiece and defective sperm function *Journal of Reproduction and Fertility* **110** 71–77.

Kirby JD and Froman DP (1991) Comparative metabolism of spermatozoa from subfertile Delaware and Wyandotte roosters *Journal of Reproduction and Fertility* **91** 125–130.

Kirby JD, Froman DP, Engel HN Jr and Bernier PE (1989) Decreased sperm survivability in subfertile Delaware roosters as indicated by comparative and competitive fertilization *Journal of Reproduction and Fertility* **86** 671–677.

Kishikawa H, Tateno H and Yanagimachi R (1999) Chromosome analysis of BALB/c mouse spermatozoa with normal and abnormal head morphology *Biology of Reproduction* **61** 809–812.

Klemetsdal G and Johnson M (1989) Effect of inbreeding on fertility in Norwegian trotter *Livestock Production Science* **21** 263–272.

Knights SA, Baker RL, Gianola D and Gibb JB (1984) Estimates of heritabilities and of genetic and phenotypic correlations among growth and reproductive traits in yearling Angus bulls *Journal of Animal Science* **58** 887–893.

Koefoed-Johnsen HH and Pedersen H (1971) Further observations on the Dag-defect of the tail of the bull spermatozoon *Journal of Reproduction and Fertility* **26** 77–83.

Kojima Y (1975) Fine structure of boar sperm abnormality: hairpin curved sperm *Journal of Electron Microscopy* **24** 167–169.

Kopp C, Sironen A, Ijäs R, Taponen J, Vilkki J, Sukura A and Andersson M (2008) Infertile boars with knobbed and immotile short-tail sperm defects in the Finnish Yorkshire breed *Reproduction in Domestic Animals* **43** 690–695.

Krausz C, Quintana-Murci L, Barbaux S, Siffroi J-P, Rouba H, Delafontaine D, Souleyreau-Therville N, Arvis G, Antoine JM, Erdei E, Taar JP, Tar A, Jeandidier E, Plessis G, Bougeron T, Dadoune J-P, Fellous M and McElreavy K (1999) A high frequency of Y chromosome deletions in males with nonidiopathic infertility *The Journal of Clinical Endocrinology and Metabolism* **84** 3606–3612.

Kullander S and Rausing A (1975) On round-headed human spermatozoa *International Journal of Fertility* **20** 33–40.

Kutzler MA, Solter AF, Hoffman WE and Volkmann DH (2003) Characterization and localization of alkaline phosphatase in canine seminal plasma and gonadal tissues *Theriogenology* **60** 299–306.

Lange-Consiglio A, Antonucci N, Manes S, Corradetti B, Cremonesi F and Bizzaro D (2010) Morphometric characteristics and chromatin integrity of spermatozoa in three Italian dog breeds *Journal of Small Animal Practice* **51** 624–627.

Larsen R and Chenoweth PJ (1990) Diadem/crater defects in spermatozoa from two related Angus bulls *Molecular Reproduction and Development* **25** 87–96.

Lee JD, Kamiguchi Y and Yanagimachi R (1996) Analysis of chromosome construction of human spermatozoa with normal and aberrant head morphologies after injection into mouse oocytes *Human Reproduction* **11** 1942–1946.

Leeb T (2007) The horse genome project – sequence based insights into male reproductive mechanisms *Reproduction in Domestic Animals* **42(Supplement 2)** 45–50.

Leeb T, Sieme H and Töpfer-Petersen E (2005) Genetic markers for stallion fertility – lessons from humans and mice *Animal Reproduction Science* **89** 21–29.

Leestma JE and Sepsenwol S (1980) Sperm tail axoneme alterations in the Wobbler mouse *Journal of Reproduction and Fertility* **58** 267–70.

Leuchtenberger C, Schrader F, Weir DR and Gentile DP (1953) The desoxyribosenucleic acid (DNA) content in spermatozoa of fertile and infertile human males *Chromosoma* **6** 61–78.

Lewis SEM and Aitken RJ (2005) DNA damage to spermatozoa has impacts on fertilization and pregnancy *Cell and Tissue Research* **322** 33–41.

Lewis-Jones I, Aziz N, Seshadri S, Douglas A and Howard P (2003) Sperm chromosomal abnormalities are linked to sperm morphologic deformities *Fertility and Sterility* **79** 212–215.

Liu DY and Baker HW (1992) Morphology of spermatozoa bound to the zona pellucida of human oocytes that failed to fertilize *in vitro Journal of Reproduction and Fertility* **94** 71–84.

Love CC, Varner DD and Thompson JA (2000) Intra- and inter-stallion variation in sperm morphology and their relationship with fertility *Journal of Reproduction and Fertility* **56** 93–100.

Lovercamp KW, Safranski TJ, Fischer KA, Manandhar G, Sutovsky M, Herring W and Sutovsky P (2007) High resolution light microscopic evaluation of boar semen quality sperm cytoplasmic droplet retention in relationship with boar fertility parameters *Archives of Andrology* **53** 219–228.

Maroto-Morales A, Ramón M, García-Alvarez O, Soler AJ, Esteso MC, Martínez-Pastor F, Pérez-Guzmán MD and Garde JJ (2010) Characterization of ram (*Ovis aries*) sperm head morphometry using the Sperm-Class Analyzer *Theriogenology* **73** 437–448.

Martin RH (2008) Cytogenetic determinants of male fertility *Human Reproduction Update* **14** 379–390.

Martin RH and Rademaker A (1988) The relationship between sperm chromosomal abnormalities and sperm morphology in humans *Mutation Research* **207** 159–164.

Mathevon M, Buhr MM and Dekkers JC (1998) Environmental, management, and genetic factors affecting semen production in Holstein bulls *Journal of Dairy Science* **81** 3321–3330.

Maximini L, Fuerst-Waltl B, Gredler B and Baumung R (2011) Inbreeding depression on semen quality in Austrian dual-purpose Simmental bulls *Reproduction in Domestic Animals* **46** e102–e104.

Maxwell WMC, Quintana-Casares PI and Setchell BP (1992) Ovulation rate, fertility, and embryo mortality in ewes mated to rams from two different strains *Proceedings of the Australian Society of Animal Production* **19** 192–194.

McCosker PJ (1969) Abnormal spermatozoan chromatin in infertile bulls *Journal of Reproduction and Fertility* **18** 363–365.

McLean DJ and Froman DP (1996) Identification of a sperm cell attribute responsible for subfertility of roosters homozygous for the rose comb allele *Biology of Reproduction* **54** 168–172.

McLean DJ, Jones LG Jr and Froman DP (1997) Reduced glucose transport in sperm from roosters (*Gallus domesticus*) with heritable subfertility *Biology of Reproduction* **57** 791–795.

McPherson FJ and Chenoweth PJ (2012) Mammalian sexual dimorphism *Animal Reproduction Science* **131** 109–122.

Miller D, Hrudka M, Cates WF and Mapletoft R (1982) Infertility in bulls with a nuclear sperm defect *Theriogenology* **17** 611–21.

Mitchell JR, Senger PL and Rosenberger JL (1985) Distribution and retention of spermatozoa with acrosomal and nuclear abnormalities in the cow genital tract *Journal of Animal Science* **61** 956–967.

Molnár A, Sarlós P, Fánci G, Rátky J, Nagy S and Kovács A (2001) A sperm tail defect associated with infertility in a goat – case report *Acta Veterinaria Hungarica* **49** 341–348.

Moretti E and Collodel G (2006) Three cases of genetic defects affecting sperm tail: a FISH study *Journal of Submicroscopic Cytology and Pathology* **38** 137–141.

Moretti E, Pascarelli NA, Federico MG, Renieri T and Collodel G (2008) Abnormal elongation of midpiece, absence of axoneme and outer dense fibers at principal piece level, supernumerary microtubules: a sperm defect of possible genetic origin? *Fertility and Sterility* **90** 1201.e3–1201.e8.

Mortimer D (1979) Functional anatomy of haploid and diploid rabbit spermatozoa *Archives of Andrology* **2** 13–20.

Müller U (1993) The human Y chromosome In *Molecular Genetics of Sex Determination* pp 205–224 Ed SS Wachtel. Academic Press, Elsevier.

Nadir S, Saacke RG, Bame J, Mullins J and Degelos S (1993) Effect of freezing semen and dosage of sperm on number of accessory sperm, fertility and embryo quality in artificially inseminated cattle *Journal of Animal Science* **71** 199–204.

Nestor KE (1976) Selection for increased semen yield in the turkey *Poultry Science* **55** 2363–2369.

Nestor KE (1977) The influence of a genetic change in egg production, body weight, fertility or response to cold stress on semen yield in the turkey *Poultry Science* **56** 421–425.

Nicander L and Bane A (1962) Fine structure of boar spermatozoa *Zeitschrift für Zellforschung und Mikroskopische Anatomie* **57** 390–405.

O'Brien SJ, Wildt DE, Goldman D, Merril CR and Bush M (1983) The cheetah is depauperate in biochemical genetic variation *Science* **221** 459–462.

O'Brien SJ, Roelke ME, Marker L, Newman A, Winkler CW, Meltzer D, Colly L, Everman J, Bush M and Wildt DE (1985) Genetic basis for species vulnerability in the cheetah *Science* **227** 1428–1434.

Oettlé EE and Soley JT (1985) Infertility in a Maltese Poodle as a result of a sperm midpiece defect *Journal of the South African Veterinary Association* **56** 103–106.

Olson PN, Schultheiss, P and Seim HB 3rd (1992) Clinical and laboratory findings associated with actual or suspected azoospermia in dogs: 18 cases (1979–1990) *Journal of Veterinary Medical Science* **201** 478–482.

Ostermeier GC, Sargeant GA, Yandell BS, Evenson D and Parrish JJ (2001) Relationship of bull fertility to sperm nuclear shape *Journal of Andrology* **22** 595–603.

Parlevliet JM, Kemp B and Colenbrander B (1994) Reproductive characteristics and semen quality in maiden Dutch Warmblood stallions *Journal of Reproduction and Fertility* **101** 183–187.

Pedersen H (1969) Ultrastructure of the ejaculated human sperm *Zeitschrift für Zellforschung und Mikroskopische Anatomie* **94** 542–554.

Pedersen H and Rebbe H (1974) Fine structure of round-headed spermatozoa *Journal of Reproduction and Fertility* **37** 51–54.

Pedersen H and Rebbe H (1975) Absence of arms in the axoneme of immotile human spermatozoa *Biology of Reproduction* **12** 541–544.

Peet RL and Micke B (1976) The "Dag" and swollen abaxial midpiece spermatozoan defects in bulls *Australian Veterinary Journal* **52** 476–477.

Peet RL and Mullins KR (1991) Sterility in a poll Hereford bull associated with the 'tail stump' sperm defect *Australian Veterinary Journal* **68** 245 [abstract].

Peet RL, Kluck P and McCarthey M (1988) Infertility in 2 Murray Grey bulls associated with abaxial and swollen midpiece defects *Australian Veterinary Journal* **65** 359–360.

Perotti ME and Gioria A (1981) Fine structure and morphogenesis of "headless" human spermatozoa associated with infertility *Cell Biology International Reports* **5** 113.

Perotti ME, Giarola A and Gioria M (1981) Ultrastructural study of the decapitated sperm defect in an infertile man *Journal of Reproduction and Fertility* **63** 543–549.

Perrin A, Morel F, Moy L, Colleu D, Amice V and De Braekeleer M (2008) Study of aneuploidy in large-headed, multiple-tailed spermatozoa: case report and review of the literature *Fertility and Sterility* **90** 1201.e13–7.

Popescu CP and Tixier M (1984) The frequency of chromosomal abnormalities in domestic animals and their economic consequences *Animal Genetics* **27** 69–72.

Pukazhenthi BS, Wildt DE and Howard JG (2001) The phenomenon and significance of teratospermia in felids *Journal of Reproduction and Fertility Supplement* **57** 423–433.

Pukazhenthi BS, Neubauer K, Jewgenow K, Howard J and Wildt DE (2006) The impact and potential etiology of teratospermia in the domestic cat and its wild relatives *Theriogenology* **66** 112–121.

Rajasekarasetty MR (1954) Studies on a new type of genetically determined quasi-sterility in the house mouse *Fertility and Sterility* **5** 68–97.

Rawe VY, Galaverna GD, Acosta AA, Brugo Olmedo S and Chemes HE (2001) Incidence of tail structure distortions associated with dysplasia of the fibrous sheath in human spermatozoa *Human Reproduction* **16** 879–886.

Redenz E (1924) Versuch einer biologischen Morphologie des Nebenhodens *Archiv für Mikroskopische Anatomie und Entwicklungsmechanik* **103** 593–628.

Rege JE, Toe F, Mukasa-Mugerwa E, Tembely S, Anindo D, Baker RL and Lahlou-Kassi A (2000) Reproductive characteristics of Ethiopian highland sheep. II. Genetic parameters of semen characteristics and their relationships with testicular measurements in ram lambs *Small Ruminant Research* **37** 173–187

Retzius G (1909) Die spermien der Huftiere *Bos taurus Biologische Untersuchungen* **14** 163–178.

Revell S, Cooley W and Cranfield N (2000) A case of sperm tail stump defect in a Holstein bull In *Proceedings of the 14th International Congress on Animal Reproduction, Stockholm, Sweden, 2–6 July 2000. Abstracts Volume 1, Section 2* p 73.

Roberts SJ (1986) *Veterinary Obstetrics and Genital Disease* 3rd edition. David and Charles, Ithaca New York.

Rosenbusch B, Strehler E and Sterzik K (1992) Cytogenetics of human spermatozoa: correlations with sperm morphology and age of fertile men *Fertility and Sterility* **58** 1071–1073.

Ross A, Christie S and Kerr MG (1971) An electron microscope study of a tail abnormality in spermatozoa from an infertile man *Journal of Reproduction and Fertility* **24** 99–103.
Rota A, Manuali E, Caire S and Appino S (2008) Severe tail defects in the spermatozoa ejaculated by an English bulldog *Journal of Veterinary Medical Science* **70** 123–125.
Rousso D, Kourtis A, Mavromatidis G, Gkoutzioulis F, Makedos G and Panidis D (2002) Pyriform head: a frequent but little-studied morphological abnormality of sperm *Systems Biology in Reproductive Medicine* **48** 267–272.
Ruiz-Lopez MJ, Evanson DP, Espeso G, Gomendio N and Roldan ERS (2010) High levels of DNA fragmentation in spermatozoa are associated with inbreeding and poor sperm quality in endangered ungulates *Biology of Reproduction* **83** 332–338.
Saacke RG, Amann RP and Marshall CE (1968) Acrosomal cap abnormalities of sperm from subfertile bulls *Journal of Animal Science* **27** 1391–1400.
Saacke RG, Nadir S and Nebel RL (1994) Relationship of semen quality to sperm transport, fertilization, and embryo quality in ruminants *Theriogenology* **41** 45–50.
Saacke RG, Dalton JC, Nadir S, Nebel RL and Bame JH (2000) Relationship of seminal traits and insemination time to fertilization rate and embryo quality In *Animal Reproduction: Research and Practice II. Proceedings of the 14th International Congress on Animal Reproduction, Stockholm, Sweden, 2–6 July 2000 Animal Reproduction Science* **60–61** 663–677.
Sailer BL, Jost LK and Evanson DP (1996) Bull sperm head morphology related to abnormal chromatin structure and fertility *Cytometry* **24** 167–173.
Salisbury GW and Baker FN (1966a) Frequency of occurrence of diploid bovine spermatozoa *Journal of Reproduction and Fertility* **11** 477–480.
Salisbury GW and Baker FN (1966b) Nuclear morphology of spermatozoa from inbred and linecross Hereford bulls *Journal of Applied Sciences* **25** 476–479.
Salisbury GW, Hart RG and Lodge JR (1977) The spermatozoan genome and fertility *American Journal of Obstetrics and Gynecology* **128** 342–350.
Santos NR, Krekeler N, Schramme-Jossen A and Volkmann DH (2006) The knobbed acrosome defect in four closely-related dogs *Theriogenology* **66** 1626–1628.
Savage A, Williams WW and Fowler NM (1927) A statistical study of the head length variability of bovine spermatozoa and its application to the determination of fertility *Transactions of the Royal Society of Canada Series 3* **21** 425–450.
Schirren CG, Holstein AF and Schirren C (1971) Über die Morphologenese rundköpfiger Spermatozoon des Menschen *Andrologia* **3** 117–125.
Schneider CS, Ellington JE and Wright RW Jr (1999) Effects of bulls with different field fertility on *in vitro* embryo cleavage and development using sperm co-culture systems *Theriogenology* **51** 1085–1098.
Settergren I and Nicander L (1968) Ultrastructure of disintegrated bull sperm *Proceedings of the 6th International Congress Animal Reproduction and AI Paris Volume 1* pp 191–193.
Settergren I and Nicander L (1972) Formation and fine structure of abnormal mitochondrial sheaths in spermatozoa of an infertile boar *Proceedings of the VIIth International Congress Animal Reproduction and Artificial Insemination, Munich Volume I* pp 459–462.
Shah K, Sivapalan G, Gibbons N, Tempest H and Griffin, DK (2003) The genetic basis of infertility *Reproduction* **126** 13–25.
Shahani SK, Revell SG, Argo CG and Murray RD (2010) Mid-piece length of spermatozoa in different cattle breeds and its relationship to fertility *Pakistan Journal of Biological Sciences* **13** 802–808.
Signoret J-P (1970) Swine behavior in reproduction In *Effect of Disease and Stress on Reproductive Efficiency in Swine* pp 28–45. Symposium Proceedings 70-0, Extension Service, University of Nebraska College of Agriculture, Lincoln.
Siqueira JB, Pinho RO, Guimaraes SE, Miranda Neto T and Guimarães JD (2010) Immotile short-tail sperm defect in Nelore (*Bos taurus indicus*) breed bulls *Reproduction in Domestic Animals* **45** 1122–1125.
Sironen AI, Andersson M, Uimari P and Vilkki J (2002) Mapping of an immotile short tail sperm defect in the Finnish Yorkshire on porcine chromosome 16 *Mammalian Genome* **13** 45–49.
Sironen A, Thomsen B, Andersson M, Ahola V and Vilkki J (2006) An intronic insertion in *KPL2* results in aberrant splicing and causes the immotile short-tail sperm defect in the pig *Proceedings of the National Academy of Sciences of the United States of America* **103** 5006–5011.
Sironen A, Uimari P, Nagy S, Paku S, Andersson M and Vilkki J (2010) Knobbed acrosome defect is associated with a region containing the genes *STK17b* and *HECW2* on porcine chromosome 15 *BMC Genomics* **11**:699.

Sironen A, Kotaja N, Mulhern H, Wyatt TA, Sisson JH, Pavlik JA, Miiluniemi M, Fleming MD and Lee L (2011) Loss of *SPEF2* function in mice results in spermatogenesis defects and primary ciliary dyskinesia *Biology of Reproduction* **85** 690–701.

Smital J, De Sousa LL and Mohsen A (2004) Differences among breeds and manifestation of heterosis in AI boar sperm output *Animal Reproduction Science* **80** 121–130.

Smital J, Wolf J and De Sousa LL (2005) Estimation of genetic parameters of semen characteristics and reproductive traits in AI boars *Animal Reproduction Science* **86** 119–130.

Smith BA, Brinks JS and Richardson GV (1989) Estimation of genetic parameters among breeding soundness examination components and growth traits in yearling bulls *Journal of Animal Science* **67** 2892–2896.

Smorag Z, Bocheneck M, Wojdan Z, Sloniewski K and Reklewski Z (2000) The effect of sperm chromatin structure on quality of embryos derived from superovulated heifers *Theriogenology* **53(1)** 206 [abstract].

Soares MP, Gaya LG, Lorentz LH, Batistel F, Rovadoscki GA, Ticinai E, Zabot V, Di Domenico Q, Madureia AP and Pértile SF (2011) Relationship between the magnitude of the inbreeding coefficient and milk traits in Holstein and Jersey dairy bull semen used in Brazil *Genetics and Molecular Research* **10** 1942–1947.

Soderquist L (1998) Reduced fertility after artificial insemination in a ram with a high incidence of knobbed acrosomes *Veterinary Record* **143** 227–228.

Sonersson A K (2002) Managing inbreeding in selection and genetic conservation schemes of livestock. PhD thesis, Animal Breeding and Genetics Group, Wageningen University, Wageningen and Division of Animal Science, Institute for Animal Science and Health, Lelystad.

Sukura A, Mäkipää R, Vierula M, Rodriguez-Martinez H, Sundbäck P and Andersson M (2002) Hereditary sterilizing short-tail sperm defect in Finnish Yorkshire boars *Journal of Veterinary Diagnostic Investigation* **14** 382–388.

Sun F, Ko E and Martin RH (2006) Is there a relationship between sperm chromosome abnormalities and sperm morphology? *Reproductive Biology and Endocrinology* **4**:1.

Sutovsky P, Terada Y and Schatten G (2001) Ubiquitin-based sperm assay for the diagnosis of male factor infertility *Human Reproduction* **16** 250–258.

Suzuki-Toyota F, Ito C, Maekawa M, Toyama Y and Toshimori K (2010) Adhesion between plasma membrane and mitochondria with linking filaments in relation to migration of cytoplasmic droplet during epididymal maturation in guinea pig spermatozoa *Cell and Tissue Research* **343** 429–440.

Tanaka H, Hirose M, Tokuhiro K, Matsoka Y, Miyagawa Y, Tsujimura A and Nishimune Y (2007) Single nucleotide polymorphisms: discovery of the genetic causes of male infertility In *Spermatology: Proceedings of the 10th International Symposium on Spermatology, Held at El Escorial, Madrid, Spain, 17–22 September 2006 Society of Reproduction and Fertility Supplement* **65** 531–534 Ed ERS Roldan and M Gomendio. Nottingham University Press, Nottingham.

Teunissen GHP (1946) Een afwijkning van het acrosom (kopcap) bij de spermatozoiden van een stier *Tijdschrift Voor Diergeneeskunde* **71** 292–303.

Thundathil J, Palasz AT, Mapletoft RJ and Barth AD (1999) An investigation of the fertilizing characteristics of pyriform-shaped bovine spermatozoa *Animal Reproduction Science* **57** 35–50.

Thundathil J, Palasz AT, Barth AD and Mapletoft RJ (2000a) Fertilizing characteristics of bovine sperm with proximal droplets *Theriogenology* **53(1)** 434 [abstract].

Thundathil J, Meyer R, Palasz AT, Barth AD and Mapletoft RJ (2000b) Effect of the knobbed acrosome defect in bovine sperm on IVF and embryo production *Theriogenology* **54** 921–934.

Thundathil J, Palasz AT, Barth AD and Mapletoft RJ (2001) The use of *in vitro* fertilization techniques to investigate the fertilizing ability of bovine sperm with proximal cytoplasmic droplets *Animal Reproduction Science* **65** 181–192.

Thundathil J, Palasz AT, Barth AD and Mapletoft RJ (2002) Plasma membranes and acrosomal integrity in bovine spermatozoa with the knobbed acrosome effect *Theriogenology* **58** 97–102.

Toelle VD and Robison OW (1985) Estimates of genetic correlations between testicular measurements and female reproductive traits in cattle *Journal of Animal Science* **60** 89–100.

Toelle VD, Johnson BH and Robison OW (1984) Genetic parameters for testes traits in swine *Journal of Animal Science* **59** 967–973.

Toyama Y, Iwamoto T, Yajima M, Baba K and Yuasa S (2000) Decapitated and decaudated spermatozoa in man and pathogenesis based on the ultrastructure *International Journal of Andrology* **23** 109–115.

Truitt-Gilbert AJ and Johnson LA (1980) The crater defect in boar spermatozoa: A correlative study with transmission electron microscopy, scanning electron microscopy and light microscopy *Gamete Research* **3** 259–266.

Van Duijn C Jr (1972) Ultrastructural mid-piece defects in spermatozoa from the subfertile Great Yorkshire boar, "Ard" In *Proceedings VIIth International Congress on Animal Reproduction and Artificial Insemination, Milan Volume 1* pp 469–473.

van Eldik P, van der Waai EH, Ducro B, Kooper A.W, Stout TA and Colenbrander B (2006) Possible negative effects of inbreeding on semen quality in Shetland pony stallions *Theriogenology* **65** 1159–1170.

Van Rensburg SWJ, Van Rensburg SJ and de Vos WH (1966) The significance of the cytoplasmic droplet in the disintegration of the semen in Guernsey bulls *Onderstepoort Journal of Veterinary Research* **33** 169–184.

Veronesi MC, Riccardi E, Rota A and Grieco V (2009) Characteristics of cryptic/ectopic and contralateral scrotal testes in dogs between 1 and 2 years of age *Theriogenology* **72** 969–977.

Vicari E, Perdichizzi A, De Palma A, Burrello A, D'Agata R and Calogero AE (2002) Globospermia is associated with chromatin structure abnormalities: case report *Human Reproduction* **17** 2128–2133.

Vierula M, Alanko M, Andersson M and Vanhaperttula T (1987) Tail stump defect in Ayrshire bulls: morphogenesis of the defect *Andrologia* **19** 207–216.

Vieytes AL, Cisale HO and Ferrari MR (2008) Relationship between the nuclear morphology of the sperm of 10 bulls and their fertility *Veterinary Record* **163** 625–629.

Visser L and Repping S (2010) Unravelling the genetics of spermatogenic failure *Reproduction* **139** 303–307.

Vogel P, Hansen G, Fontenot G and Read R (2010) Tubulin tyrosine ligase like 1 deficiency results in chronic rhinosinusitis and abnormal development of spermatid flagella in mice *Veterinary Pathology* **47** 703–712.

Vogler CJ, Bame JH, De Jarnette JM, McGilliard ML and Saacke RG (1993) Effects of elevated testicular temperature on morphology characteristics of ejaculated spermatozoa in the bovine *Theriogenology* **40** 1207–1219.

Waite DA, Wakefield JS, Steele R, Mackay JB, Rose I and Wallace J (1978) Cilia and sperm tail abnormalities in Polynesian bronchiectasis *The Lancet* **312(8081)** 132–133.

Wartenberg H and Holstein A-F (1975) Morphology of the "spindle-shaped body" in the developing tail of human spermatids *Cell and Tissue* Research **159** 435–443.

Wieacker P and Jakubiczka S (1997) Genetic causes of male infertility *Andrologia* **29** 63–69.

Wiggans GR, VanRaden PM and Zuurbier J (1995) Calculation and use of inbreeding coefficients for genetic evaluation of United States dairy cattle *Journal of Dairy Science* **78** 1584–1590.

Wildt D, Pukazhenthi B, Brown J, Monfort S, Howard J and Roth T (1995) Spermatology for understanding, managing and conserving rare species *Reproduction, Fertility and Development* **7** 811–824.

Wildt DE, Baas EJ, Chakraborty PK, Wolfle TL and Stewart AP (1982) Influence of inbreeding on reproductive performance, ejaculate quality and testicular volume in the dog *Theriogenology* **17** 445–452.

Williams G (1965) An abnormality of the spermatozoa of some Hereford bulls *Veterinary Record* **72** 1204–1206.

Williams G (1987) Tail stump defect affecting the spermatozoa of two Charolais bulls *Veterinary Record* **121** 248–250.

Williams WW (1950) Male sterility due to centriolar–mitochondrial disease of the spermatozoa *Journal of Urology* **64** 614–621.

Williams WW and Savage A (1925) Observations on the seminal micropathology of bulls *The Cornell Veterinarian* **15** 353–375.

Windt ML and Kruger TF (2004) The role of sperm morphology in intracytoplasmic sperm injection (ICSI) In *Atlas of Human Sperm Evaluation* pp 19–25 Ed TF Kruger and DR Franken. Taylor and Francis, New York.

Wohlfarth E (1961) Beitrag zum Akrosom-Defekt im Ebersperma [Acrosome defect of boar spermatozoa] *Zuchthygiene, Fortpflanzungsstorungen und Besamung der Haustiere* **5** 268–274.

Wood PD, Foulkes JA, Shaw RC and Melrose DR (1986) Semen assessment, fertility and the selection of Hereford bulls for use in AI *Journal of Reproduction and* Fertility **76** 783–795.

Yeung CH, Tüttelmann F, Bergmann M, Nordhoff V, Vorona E and Cooper TG (2009) Coiled sperm from infertile patients: characteristics, associated factors and biological implication *Human Reproduction* **24** 1288–1295.

Yoshida A, Kazukiyo M and Masufami S (1997) Cytogenetic survey of 1,007 males *Urologia Internationalis* **58** 166–176.

Young LD, Leymaster KA and Lunstra DD (1986) Genetic variation in testicular development and its relationship to female reproductive traits in swine *Journal of Animal Science* **63** 17–26.

Zamboni L (1992) Sperm structure and its relevance to infertility *Archives of Pathology and Laboratory Medicine* **16** 325–344.

Part II

Animal Andrology Applications

8 Applied Small Animal Andrology

Margaret V. Root Kustritz*
University of Minnesota, St Paul, Minnesota, USA

Introduction

There is little in the veterinary literature describing andrology in stud dogs and tomcats compared with the wealth of information that is available for large animal species. The documentation that is available includes information on best practices for the cryopreservation of spermatozoa, and on the semen 'dose' for optimal conception rate and litter size in bitches and queens. Finally, there is very limited information on the biosecurity of shipped semen. This is a review of the current body of literature that is available in canine and feline theriogenology.

The Dog

Puberty

Definition and clinical determination

Puberty, or sexual maturity, is defined as demonstration of normal semen quality and normal breeding behaviours. Male animals require exposure to testosterone before and immediately at the time of birth if they are to show normal development and normal reproductive behaviour as adults (Schulz *et al.*, 2004). The exposure to testosterone is a priming effect; animals not exposed to testosterone at birth cannot respond to the presence of testosterone at puberty with normal breeding behaviour. Exposure to oestrogen at the time of birth also may be important for later reproductive success. In male laboratory animals, lack of oestrogen exposure at birth has been associated with decreased frequency of mounts when breeding (Pereira *et al.*, 2003).

Normal sexual behaviour and normal semen quality do not always develop simultaneously (Corrada *et al.*, 2006; Root Kustritz, 2010a). In one study of male Beagles, the dogs were incapable of ejaculation until they were nearly 8 months old (Takeishi *et al.*, 1975). Determination of puberty onset in male dogs historically required proof through breeding attempts and successful siring of pups. Modern techniques include semen collection and evaluation, but in determining the normality of breeding behaviour, there is no more definitive test than observation after exposure to an oestrous bitch.

Normal male breeding behaviour includes investigation of a bitch's hindquarters, vulvar secretions, anal gland secretions and

* E-mail: rootk001@umn.edu

urine. Inhalation of pheromones and flehmen behaviour in dogs is observed as chattering of the teeth, because the vomeronasal organ is located behind the upper incisors and dogs present pheromonal compounds to that area with rapid flicking of the tongue. Intact male dogs often urinate over the urine of an oestrous female, presumably to 'hide' it from competing males. Male dogs may show mounting and thrusting behaviour as young as 4–6 weeks of age, which is associated with play (Fuller and Fox, 1969; Campbell, 1975; Beaver, 1977). If this play behaviour is restricted, adult mating behaviour may be adversely affected (Beaver, 1977).

Breeding behaviour at puberty is not instinctive; it has been demonstrated that 61% of young male dogs will demonstrate abnormal behaviours such as attempting to mount the bitch's head instead of hindquarters (Beach, 1968).

The normal male will investigate the female and, if she stands and is not hostile, will attempt to mount. The amount of time before the male mounts and the number of mounts he attempts before intromission occurs are variable and are not associated with fertility. The male dog clasps his forelimbs just in front of the bitch's hind limbs and thrusts vigorously for several minutes. Spermatozoa are deposited into the vagina at that time. As the male's penis continues to engorge, it becomes too large for withdrawal from the vagina and vulva, forming the copulatory lock, or tie. The male, still with his penis caught within the vulvar lips, will step one hind limb over the bitch's back, twisting the penis in a horizontal plane such that the bitch and stud are standing facing opposite directions. During this time, vaginal contractions and pulses of prostatic fluid help move spermatozoa cranially in the bitch's reproductive tract. The average duration of the copulatory lock is about 15 min.

Underlying factors

As in many species, puberty occurs when males reach about 80% of adult body weight. Because of the great variation in size of domestic dog breeds, there is a great variation in expected age of puberty onset; there is a positive correlation between age at puberty onset and expected weight at adulthood. Toy and small breed dogs may be sexually mature as early as 4–6 months of age, while large and giant breed dogs may not reach sexual maturity until 18–24 months of age (Beaver, 1977; Root Kustritz, 2010a).

Fertility and determination of breeding soundness

History and pre-breeding testing

When dogs are presented for breeding soundness examination, questions that should be asked include: history of breeding attempts, if any; time since last used for breeding or semen collection; any traumas or surgery, especially to the scrotum, prepuce or penis; history of febrile or systemic disease; drugs or other chemicals administered; brucellosis testing; general historical information, including diet and dietary supplements, and working history. Questions also may be asked about the sire's semen quality; some semen characteristics, such as high sperm motility and low sperm output, have been reported to be highly heritable (England *et al.*, 2010).

Any internal or external insult causing increased intrascrotal temperature may lead to morphological abnormalities of spermatozoa and decreased number of spermatozoa, even if no change of the testis is obvious on palpation. This includes febrile disease, hyperthermia and any direct insult to the scrotum, including frostbite. Drugs implicated in poor semen quality in male dogs include glucocorticoids, anabolic steroids, antifungal agents (including ketoconazole) and chemotherapeutic agents (Johnston *et al.*, 2001a).

Genetic testing takes two forms in stud dogs. One is evaluation for hereditary disease that might preclude use of the animal for breeding. Specific genetic disorders of concern and tests available vary by breed and are best determined by accessing national breed club health information. For example, the national breed clubs of all registered breeds in the American Kennel Club in the USA are required to provide current information about diseases of concern and availability of testing.

Genetic tests include the following: physical tests, which assess for presence or absence of disease directly, such as radiographs to assess for hip or elbow dysplasia; biochemical tests, which identify specific alterations in metabolism associated with hereditary disease; or DNA-based tests, which identify specific chromosomal alterations associated with disease (Patterson, 2009). Some genetic tests identify specific mutations associated with a clinical condition; the mutation present often varies by breed, and more than one mutation may cause a similar condition. Other DNA-based tests are linkage tests, which identify the presence of DNA associated with, but not necessarily causative, of the condition seen. These are less exact tests. New tests become available regularly, and the author encourages veterinarians to help clients find information specific to their breed rather than to try to keep track of all tests available for all breeds.

Genetic testing also may be used to track the use of a given stud dog historically. Cheek swab samples can be collected easily from stud dogs and genetic information stored to permit parentage testing and create a database to better define the incidence of hereditary diseases if information from offspring is submitted to an open registry.

The other test that should be performed in all male dogs before use at stud is for *Brucella canis*, the causative agent of canine brucellosis (Hollett, 2006). Prevalence of this organism is higher among roaming dog populations than in pet populations and is increasing in some parts of the world, possibly because of increased movement of dogs between and within countries. *B. canis* preferentially grows in tissues of the reproductive tract, including the prostate, testes and epididymides (Serikawa *et al.*, 1984). The organism is shed in body fluids, including the urine and semen. Transmission is through ingestion or inhalation, but venereal transmission may also occur. Because the disease is caused by a bacterium, culture is a definitive diagnostic test (Boeri *et al.*, 2008). There are, however, concerns about cultural tests, which include: difficulty in appropriate sampling; overgrowth by normal flora in areas from which samples are drawn, necessitating multiple simultaneous cultures to enhance accuracy; cost; and potential exposure of laboratory personnel to this zoonotic organism. Serological tests include agglutination tests, agar gel immunodiffusion (AGID) tests, enzyme-linked immunosorbent assays (ELISA) and polymerase chain reaction (PCR) tests. At present, agglutination testing is best done for screening, with all positives double checked by AGID testing. When PCR tests become commercially available, they will become the screening tests of choice (Root Kustritz, 2009). PCR testing on semen samples can be performed with exquisite sensitivity and specificity (Keid *et al.*, 2009).

Physical examination findings

PROSTATE. The prostate is the only accessory sex gland in male dogs. It is palpable on transrectal examination as a bilobed, symmetrical structure. It encircles the neck of the urinary bladder and may fall forward into the abdomen with increasing size and weight. If it is not palpable per rectum after this occurs, ability to palpate may be increased by use of the free hand to push the caudal abdominal contents towards the gloved finger in the rectum, or by having an assistant 'wheelbarrow' the dog up on to its hind legs (Plate 6). If the prostate is normal, the dog should not exhibit signs of pain on transrectal palpation.

If the prostate is symmetrically enlarged with age, this provides evidence of benign prostatic hypertrophy/hyperplasia (BPH), which may or may not be associated with changes in semen quality (O'Shea, 1962; Zirkin and Strandberg, 1984; Read and Bryden, 1995). If the dog exhibits signs of pain on palpation of the prostate, acute prostatitis may be present; with chronic prostatitis, dogs rarely show signs of pain.

TESTES. Both testes are completely descended into the scrotum by 10 days of age in most dogs and are easily palpable in the scrotum by 12–16 weeks of age (Gier and Marion, 1969). Testes are unlikely to descend if not in the scrotum by 6 months old – the average age at which the inguinal ring closes. Monorchidism, or the development of one testis only, is extremely uncommon in dogs, so any dog without two descended testes should be assumed to be cryptorchid (Burns and Petersen, 2008). The testes should be symmetrical in size and shape, should have the consistency of a peeled, hard-boiled egg, and should be freely movable in the scrotum.

Size varies with breed. Total scrotal width and total sperm output are positively correlated with body weight (Table 8.1) (Olar *et al.*, 1983; Woodall and Johnstone, 1988a,b). While size of the testes is not necessarily associated with semen quality, the recording of total scrotal width is a useful objective measure to help assess testicular change in dogs with age or disease (Plate 7). Decrease in testicular size may occur as a result of increased intrascrotal temperature, trauma, brucellosis, functional testicular neoplasia, or occlusion of the epididymis or spermatic cord (Carmichael and Kenney, 1968; Vare and Bansal, 1973, 1974).

PENIS/PREPUCE. The penis should be extruded from the prepuce for inspection. A small amount of mucoid or mucopurulent discharge may be present at the preputial orifice or smeared over the surface of the penis in normal dogs. Similarly, a small number of lymphoid follicles may be visible on the bulbus glandis in normal dogs. Abnormalities of the penis that may be associated with poor semen quality and/or abnormal breeding behaviour include balanoposthitis, hypospadias, persistent penile frenulum precluding normal erection, phimosis and penile neoplasia (Ader and Hobson, 1978; Ndiritu, 1979; Rogers, 1997).

Table 8.1. Comparison of body weight with total scrotal width in normal dogs. Adapted from Woodall and Johnstone, 1988b.

Body weight (kg (lbs))	Average total scrotal width (mm)	Acceptable range for total scrotal width (mm)
5 (11)	30	24–35
10 (22)	37	31–45
15 (33)	42	35–50
20 (44)	47	38–56
25 (55)	49	41–60
30 (66)	55	43–64
35 (77)	57	46–68
40 (88)	59	48–74
45 (99)	63	49–76
50 (110)	64	50–77

GENERAL PHYSICAL EXAMINATION. A complete physical examination is recommended. This may include specific testing, such as testing for ophthalmic disorders through the US Canine Eye Registry Foundation (CERF), and cardiac or hip evaluation for submission to the US Orthopedic Foundation for Animals (OFA), or other registries, as proof of worthiness of the dog as a candidate for breeding. For determination of breeding soundness, complete physical examination may allow the detection of signs of systemic diseases or causes of poor libido due to the pain from arthritis or other conditions, and signs of specific conditions such as brucellosis. Obesity has been shown to be associated with poor semen quality in some species due to increased intrascrotal fat and the subsequent increase in intrascrotal temperature, but this has not been reported in dogs. Some people attribute the association of poor semen quality with obesity as a sign of hypothyroidism, although experimental induction of hypothyroidism has not been associated with a decline in semen quality in dogs (Johnson *et al.*, 1999).

Semen collection and libido

Electroejaculation is not routinely used in domestic dogs (Kutzler, 2005), in which manual ejaculation is the method of choice. Semen collection should take place on a non-slip surface. Some practitioners use a consistent rug or room, both to prevent injury and as a training aid. Presence of a teaser bitch, especially if she is in oestrus, increases the number of spermatozoa in the ejaculate (Traas and Root Kustritz, 2004). Other techniques for increasing the number of spermatozoa in the ejaculate include administration of prostaglandin F2alpha (PGF2alpha; 0.1 mg/kg subcut 10–15 min before collection) or gonadotrophin-releasing hormone (GnRH) (1–2 μg/kg IM 60 min before collection) (Purswell and Wilcke, 1993; Root Kustritz and Hess, 2007). These techniques are synergistic; best results are with administration of PGF2alpha in the presence of an oestrous teaser bitch. The teaser bitch should be muzzled or otherwise restrained so that she will not turn on the male dog.

For a collection vessel, cups, bags or commercially available artificial vaginas (AV) may be used. The author prefers to use an enclosed system, such as a rubber AV to which can be attached a centrifuge tube (Fig. 8.1). The advantage of this is twofold; the sample cannot be lost or adulterated as a result of vigorous thrusting by the dog or operator error, and use of the rubber vagina, which tightly encases the erect penis, best mimics natural breeding.

The dog is manually stimulated through the prepuce, briskly and enthusiastically. As erection begins, the prepuce is pushed proximal to the bulbus glandis and the AV introduced. The fingers encircle the penis caudal to the bulbus glandis tightly, stimulating contraction of the constrictor vestibulae muscles during the copulatory lock. Three fractions of semen are ejaculated. First is the clear presperm fraction, which originates in the prostate (England *et al.*, 1990). Next is the cloudy, sperm-rich fraction, which originates in the epididymes and testes, and may be associated with thrusting behaviour. The last fraction is clear prostatic fluid, which may be ejaculated after the dog attempts to step over the collector's arm, mimicking the turn made by male dogs during the copulatory lock, which often is associated with rhythmic anal contractions and urethral pulsations. Once the dog ejaculates the third fraction, collection can be discontinued as there will be no more ejaculation of a sperm-rich fraction during that collection attempt. The grip caudal to the bulbus is released and the AV gently peeled down the penis. The penis should undergo complete detumescence and the prepuce be checked to make sure it has not rolled in along the penis before the dog is kennelled. If penile detumescence is not occurring readily, as may be seen with inexperienced males, the operator should try walking the dog away from the environment where semen collection occurred, gently covering the erect penis with cool, moistened towels or rinsing it with cool water, or distracting the dog with food or other treats.

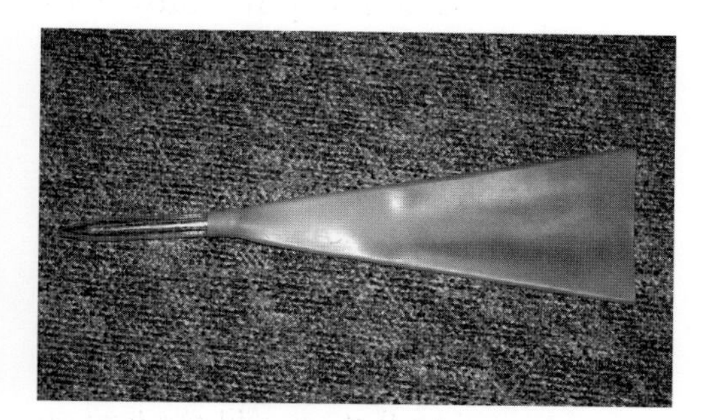
Fig. 8.1. Canine artificial vagina.

Some male dogs are hesitant to breed bitches and rarely show investigative behaviour such as licking or sniffing at urine spots, or flehmen. Established causes of poor libido in male dogs include:

- Lack of appropriate stimulation. This may include bitches being presented before they are in oestrus or at optimal breeding time. The pheromone produced by bitches that identifies them as being in oestrous is methyl-*p*-hydroxybenzoate (Goodwin *et al.*, 1979). A commercial product containing this compound has been marketed for enhancement of male libido.
- Negative breeding experience. This may include threatened or real injury from an aggressive bitch, injury to the penis during mating, or constant discipline by humans for mounting behaviour outside breeding. Some dogs that fail to show normal breeding behaviour as a result of anxiety may respond to use of dog appeasing pheromone (DAP® – Dog Appeasing Pheromone; CEVA Santé Animale, Intervet/Schering Plough Animal Health Canada Inc, Kirkland, Quebec), a compound that originates from the intramammary sebaceous glands of lactating bitches and has been demonstrated to reduce anxiety in adult dogs (Mills *et al.*, 2006). Melatonin has been shown to increase sexual behaviour in some laboratory animal species, but has not been evaluated in dogs (Brotto and Gorzalka, 2000).
- Pain when attempting to mount or ejaculate. This may be due to prostate disease, or to spinal or rear limb disorders.
- Improper breeding environment. Some dogs prefer to breed on their home turf where they presumably feel safe and may be more clearly the dominant male,

while others are comfortable breeding in any environment.

- Mate preference. Some dominant male dogs will breed with the dominant bitch only or with females of their own breed.
- Improper age. The dog is too young or too old.
- Abnormalities of sexual differentiation. This includes hermaphroditism and pseudohermaphroditism, persistent Müllerian duct syndrome and segmental epididymal aplasia (Majeed, 1974; Brown *et al.*, 1976; Hare, 1976; Marshall *et al.*, 1982; Meyers-Wallen and Patterson, 1989; Batista *et al.*, 1998).

Hypothesized causes for poor libido in animals include hyperprolactinaemia, which is associated with poor libido in human males and females, and hypothyroidism. Hyperprolactinaemia is not a demonstrated disorder in dogs, and researchers have also been unable to demonstrate change in libido with experimental induction of hypothyroidism in dogs (Johnson *et al.*, 1999). Low serum testosterone concentration has never been reported as a cause of poor libido in animals, although some males will show more interest in females when treated with testosterone. Routine treatment with testosterone is not recommended; remember that testosterone must be secreted in pulses for the normal production of spermatozoa to occur and if a dog is periodically provided with a large amount of testosterone, that pulsatile release is disrupted and negative feedback exerted on the pituitary, stopping the secretion of luteinizing hormone (LH) and subsequent secretion of testosterone for an unpredictable length of time.

Semen evaluation

GENERAL INFORMATION. There is little information in the veterinary literature linking specifics of semen evaluation to fertility in dogs. One author suggests that one can only make accurate predictions based on canine semen evaluation findings if semen quality is either very good or very poor (Peña Martínez, 2004). Ejaculated spermatozoa have generally not yet undergone capacitation and so do not exhibit the same functions as they would in a bitch's reproductive tract. Spermatozoa attain motility while moving through the epididymis, so testicular or epididymal aspirates will yield immature spermatozoa. Results of the tests that may be performed as described below are affected by sample collection technique, time from sample collection to evaluation, equipment used, skill of the investigator and other factors (Rui *et al.*, 1986; WHO, 1999; Verstegen *et al.*, 2006). Human semen collection is well described. The recommendations below may reflect some of those of the World Health Organization (WHO), which regularly updates a handbook describing quality control measures for human andrology laboratories (WHO, 1999). However, the procedures outlined here are for the evaluation of ejaculated canine spermatozoa (Root Kustritz, 2007).

Canine semen is evaluated at room temperature. Excessive attempts to warm the semen may decrease its quality (Bartlett, 1962; Chatterjee *et al.*, 1976; Threlfall, 2003). The different components that are involved in the manual evaluation of semen are outlined below. Computer assisted sperm analysis (CASA) systems are also available and generate repeatable and accurate results with standardized settings (Schäfer-Somi and Aurich, 2007); they are very useful for some of the evaluations that need to be made (as noted below). Procedures for the use of CASA systems vary with the equipment used.

Semen evaluation findings should be recorded in a legally defensible and retrievable fashion. Appropriate semen evaluation forms are commercially available, such as that used by members of the Society for Theriogenology in the USA (Purswell *et al.*, 2010). General causes of abnormal semen quality in dogs are listed below (Table 8.2).

COLOUR. Normal canine semen is milky or opalescent. All samples that appear cloudy should be evaluated microscopically as some samples containing lipid droplets and no spermatozoa can have this appearance. Clear semen is associated with lack of spermatozoa. A yellow coloration is usually due to urine contamination – in humans, such a yellow discoloration is described in men with icterus

Table 8.2. Causes of abnormal semen quality in dogs.

Abnormality of the semen	Possible causes
Aspermia (no fluid ejaculated)	Poor libido Apprehension Pain in rear limbs or spine Dog very old or very young
Asthenozoospermia (fewer than 70% progressively motile sperm in the ejaculate)	Orchitis Prostatitis Brucellosis Testicular neoplasia Hypothyroidism History of high fever Immotile cilia syndrome Leishmaniasis
Azoospermia (no spermatozoa in the ejaculate)	Abnormal sexual development Hypothyroidism Apprehension Cryptorchidism History of high fever Testicular neoplasia Epididymal occlusion
Oligozoospermia (fewer than 200 to 300 million sperm in the ejaculate)	Orchitis Prostatitis Pain Hypothyroidism Retrograde ejaculation History of high fever
Teratozoospermia (fewer than 80% morphologically normal sperm in the ejaculate)	Orchitis Prostatitis Brucellosis Testicular neoplasia Hypothyroidism History of high fever

or after the ingestion of certain vitamins (WHO, 1999). Green is indicative of a purulent discharge, which could arise either from the penis/prepuce or prostate. Brown is indicative of pooled blood, usually seen with prostate disease (Plate 8). Frank blood may be due to prostate disease or to penile trauma during collection. Occasionally, dogs achieving complete erection for the first time will bleed from the surface vessels on their penis; this is obvious by visual inspection.

VOLUME. Volume is not in and of itself an indicator of semen quality because it is dependent on the operator who collected the semen. If more of the third, or prostatic, fraction is collected, the sample will be larger in volume and more dilute. Volume should be recorded before any samples are withdrawn as that value will be used to calculate total number of spermatozoa in the ejaculate.

PH. Measurement of pH was done traditionally to help clinicians choose an appropriate antibiotic that could penetrate and be trapped within the prostate. Normal pH in non-fractionated canine semen is reported to vary from pH 6.4 to 6.8 (Bartlett, 1962; Chatterjee *et al.*, 1976; Daiwadnya *et al.*, 1995). The inherent inaccuracy of pH measurement with pH paper and the emergence of antibiotics that ionize at multiple pH values have decreased the value of pH testing of canine semen.

PERCENTAGE PROGRESSIVE MOTILITY. Percentage progressive motility is assessed on neat semen. A small volume is placed on a glass slide. A coverslip may or may not be applied; individual clinicians should be consistent in the technique used to make their analyses consistent. A subjective assessment is made of percentage of spermatozoa moving forward. Evaluation also can be made of speed and quality of progression; a canine spermatozoon with normal motility should traverse the microscope field of view at 100× magnification in 2–3 s (Threlfall, 2003). The use of CASA systems permits the identification of details of spermatozoal movement that can be incorporated into equations to produce defined scores, such as a Sperm Motility Index (Rijsselaere *et al.*, 2002). The author is unaware of any published reports associating such values with fertility in dogs. Similarly, details of movement can be used to identify subpopulations of spermatozoa within a sample that may respond differently to cryopreservation or alter fertility (Rigau *et al.*, 2001; Núñez-Martínez *et al.*, 2006; Peña *et al.*, 2006). The author is also unaware of any published reports associating identification of these subpopulations with fertility in dogs.

Normal percentage progressive motility in dogs is 70% or greater. The percentage of progressively motile spermatozoa is not affected by frequency of semen collection (Boucher *et al.*, 1958; England, 1999) and is positively correlated with the percentage of morphologically normal spermatozoa (Ellington *et al.*, 1993; Agarwal *et al.*, 2003). Causes of decreased motility include contaminated collection equipment, testicular disease, prostatic disease and canine brucellosis. If all spermatozoa are non-motile, another ejaculate should be collected using different equipment to ensure that the lack of motility is not an artefact of collection technique. Pathological causes of asthenozoospermia, or poor motility, are the same as those for abnormal morphology, but one cause that is unique to asthenozoospermia is immotile cilia syndrome, or ciliary dyskinesis; this usually is associated with respiratory disease. Another unique cause is infection with *Leishmania chagasi*, which has been associated with asthenozoospermia but with no other changes in semen quality (Assis *et al.*, 2010). Appropriate diagnosis is valuable, as although leishmaniasis is usually spread through an insect vector, affected dogs may shed organisms in their semen and transmit disease to naive bitches (Silva *et al.*, 2009).

CONCENTRATION/TOTAL NUMBER OF SPERM. Sperm concentration is measured with a haemocytometer. Due to the fluid dynamics associated with use of chambers in CASA systems, haemocytometer measurement is more accurate and is the gold standard method (Kuster, 2005). The semen sample is diluted 1:100. This can be accomplished using the Unopette white blood cell counting system (Becton-Dickinson, Rutherford, New Jersey). Semen is drawn up into the 20 µl pipette and dispensed into the diluent chamber. Alternatively, a hand dilution can be made by diluting 1 part semen with 9 parts formal-buffered saline to make a 1:10 dilution, then taking 1 part of the 1:10 dilution and adding 9 parts of formal-buffered saline to make a 1:100 dilution. The diluted semen is dispensed under the glass coverslip of the haemocytometer and permitted to fill that volume by capillary action. The haemocytometer grid is made up of nine large squares. All of the spermatozoa within any one of these large squares are counted. The operator should focus up and down while counting to see all the sperm, some of which will float head up or head down and so not be visible continuously in the grid. The number of spermatozoa counted is the concentration in millions/ml. This value varies with collection technique; as stated above, if more of the third, or prostatic, fraction is collected, the sample will be larger in volume and more dilute.

CASA systems and optical density (OD) measurements can be used to assess the concentration of spermatozoa in dogs (Foote and Boucher, 1964; Gunzel-Apel *et al.*, 1993; Eljarah *et al.*, 2005; Kuster, 2005). Other reported techniques include the use of standardized photographs and of a spermatocrit. In the first method, sperm number is estimated by comparing a neat sample in a 10 µm chamber with photographs of a standardized concentration of spermatozoa; this technique has not been assessed in dogs. In the second method, the concentration of spermatozoa is determined by drawing up a sample in a haematocrit tube and evaluating percentage solids; this technique is not accurate in dogs (Root Kustritz *et al.*, 2007).

The total number of spermatozoa in the ejaculate, unlike the sperm concentration, does not vary with collection technique and is the more valuable parameter for evaluation of semen quality. Total number is calculated by multiplying the concentration (millions/ml) by the volume (ml/ejaculate) to generate the value of millions of spermatozoa in the ejaculate. The normal total number is more than 300 million and is dependent on testicular size. Some toy breeds with very small testes may be fertile with fewer than 300 million spermatozoa. For this reason, some authors suggest that the normal total number of spermatozoa may be as low as 200 million (Verstegen and Onclin, 2003).

Total number of spermatozoa in the ejaculate declines with increasing frequency of semen collection owing to emptying of the epididymal reserves (Foote and Boucher, 1964; Taha *et al.*, 1983; Schäfer *et al.*, 1996; England, 1999). Normal dogs may exhibit oligozoospermia or azoospermia as a result of anxiety or pain (Peña Martínez, 2004). Pathological causes of a decreased total number of spermatozoa include retrograde ejaculation, testicular neoplasia, orchitis, brucellosis, prostatitis, epididymal occlusion, and other causes of occlusion, such as spermatocele. Retrograde flow of a small amount of semen into the urinary bladder during ejaculation is normal. In some dogs, an excessive amount of semen flows into the urinary bladder such that a minimal number of spermatozoa are in the antegrade ejaculate. This most likely occurs as a mismatch of neural signals closing off the bladder neck during emission and ejaculation of semen. Other reasons include: anatomical causes, for example following surgery of the bladder or prostate; neuropathic causes, for example secondary to diabetes mellitus; and psychological causes (Romagnoli and Majolino, 2009). Retrograde ejaculation can be artefactually created by inadequate penile manipulation during manual semen collection (Schnee, 1985). Diagnosis requires collection of a urine sample by cystocentesis after ejaculation. Treatment options include phenylpropanolamine (3 mg/kg p.o. b.i.d. –by mouth twice daily) or pseudoephedrine hydrocholoride (3–5 mg/kg po t.i.d. – by mouth thrice daily; or 3 and 1 h before collection; Romagnoli and Majolino, 2009).

Differentiation of azoospermia due to apprehension from that due to occlusion may be made by measurement of alkaline phosphatase (ALP) in the seminal plasma. ALP in semen arises from the epididymis and seminiferous tubule epithelium (Kutzler, *et al.*, 2003). If the concentration of ALP in the seminal fluid is >5000 IU/l, the ejaculated sample contains fluid from those tissues and occlusion is ruled out.

PERCENTAGE MORPHOLOGICALLY NORMAL SPERMATOZOA. Morphology, or shape, is often evaluated on stained spermatozoa. Phase contrast microscopy permits the evaluation of unstained samples but is not available to most practitioners. Several staining techniques are used to evaluate canine semen. Society for Theriogenology stain (Lane Manufacturing, Denver, Colorado) is an eosin–nigrosin stain. One drop of semen and one drop of stain are placed at one end of a glass slide and another slide used to mix the drops and pull the resultant mixture out into a thick smear. This is allowed to air dry and is then evaluated under oil immersion; the spermatozoa will appear white or pink against a black or dark violet background. A triple stain, such as Diff-Quik™ (Baxter Healthcare, Miami, Florida) can also be used. One drop of semen is placed at one end of a glass slide and another slide used to pull it out as for a blood smear. This is allowed to air dry. The dried slide can be immersed for 5 min in each of the three solutions of the triple stain in sequence, rinsed and allowed to air dry before evaluation under oil immersion.

Another technique is to dip the dried slide into the fixative solution of the triple-stain for 8–10 repetitions, then into the safranin for 8–10 repetitions and finally into the crystal violet for 10–15 repetitions. The slide is not rinsed and the crystal violet is allowed to air dry on the slide before evaluation under oil immersion. The latter technique is quicker but slides must be evaluated immediately as the stain will start to crack over time, obscuring details of spermatozoa morphology. A consistent staining technique should be used, especially for serial evaluations of a given dog; different staining techniques may increase acrosome defects or defects of the head, midpiece and tail, perhaps owing to changes in pH or osmolarity of the stain used (Harasymowycz *et al.*, 1976; Johnson *et al.*, 1991; Root Kustritz *et al.*, 1998).

At least 100 sperm are counted. Precision is increased when higher numbers of spermatozoa are assessed (Christensen *et al.*, 2005). The spermatozoa in the middle third of the slide are those least likely to be affected by preparation technique (Harasymowycz *et al.*, 1976). The parameter of greatest interest is percentage morphologically normal spermatozoa (MNS), which should be 80% or greater in normal dogs. It has been reported that fertility is affected when the percentage of morphologically normal spermatozoa is below 60% in dogs (Oettle, 1993). Although there may be differences in percentage morphology as assessed by different investigators, this is of greatest significance in dogs with a preponderance of abnormal spermatozoa. Abnormalities can be classified as primary (occurring during spermatogenesis), or secondary (occurring after spermatogenesis or as an artefact of sample handling or preparation) (Plate 9). Some authors prefer to designate abnormalities as major or minor, but because there is limited information in dogs on the correlation between given spermatozoal abnormalities and fertility, this author prefers the former terminology (primary/secondary).

Some morphological abnormalities associated with infertility in dogs are proximal cytoplasmic droplets and knobbed acrosomes. One case report documented ultrastructural abnormalities at the sperm neck associated with the formation of proximal droplets, and the inability of affected sperm to undergo the hypermotility associated with capacitation and binding to oocytes (Peña *et al.*, 2007). Four related and somewhat inbred dogs with high percentages of knobbed acrosomes were presented for failing to effect pregnancy when bred to fertile bitches (Santos *et al.*, 2006).

Percentage MNS in a given dog is not altered by frequent semen collection (England, 1999), but samples collected after a sexual rest may contain a higher number of morphologically abnormal spermatozoa, presumably as a result of the presence of a high number of aged spermatozoa from epididymal storage in the sample. Causes of abnormal sperm morphology include testicular disease, prostatic disease, canine brucellosis and immaturity (Corrada *et al.*, 2006).

CYTOLOGY AND CULTURE. Canine semen is naturally dilute, so cytological evaluation of neat semen is often unrewarding. Centrifugation and evaluation of the resultant pellet is more accurate. The presence of fewer than 200 white blood cells/μl of semen is considered normal (Meyers-Wallen, 1991).

Because the distal urethra is not sterile, all semen samples contain some bacteria, but the culture of >100,000 bacteria/ml of any single organism is considered significant. It has been demonstrated that there may be significant growth of bacteria from samples containing unremarkable numbers of white blood cells; if the clinician is suspicious of infection, submission of samples for cultures should not be based solely on presence or absence of inflammatory cells in the ejaculate (Root Kustritz *et al.*, 2005).

SPECIAL TESTS. Specific testing to evaluate cell membrane integrity and fertilizing capability are not commonly performed in dogs. Live–dead staining involves evaluation of the colour of spermatozoa stained with eosin–nigrosin stain. Those with damaged plasma membranes take up stain and appear pink, while those with intact plasma membranes appear white. The author is unaware of any reports in the literature documenting correlation between apparent numbers of (live) spermatozoa and fertility in dogs. Hypo-osmotic swelling tests (HOSTs) also are used for the determination of membrane integrity. Spermatozoa are submerged in a hypo-osmotic medium, such as sodium citrate and fructose (7.35 g and 13.51 g, respectively, in 1000 ml of distilled water) or 100 mM sucrose solution (Inamasu *et al.*, 1999; Pinto *et al.*, 2005; Pinto and Kozink, 2008). Spermatozoa with intact plasma membranes will take in water, as evidenced by coiling of the tail (Peña Martínez, 2004). There are conflicting reports in the literature on correlations between the results of HOSTs and sperm motility and morphology in dogs (England and Plummer, 1993; Kurni-Diaka and Badtram, 1994; Inamasu *et al.*, 1999; Pinto and Kozink, 2008). The author is unaware of reports that correlate the results of HOSTs with fertility in dogs.

Centrifugation gradients may be used either to assess semen quality or to remove abnormal spermatozoa and other cells from the sample. These techniques increase the percentage of morphologically normal and progressively motile spermatozoa in treated samples, but may greatly decrease the total number of spermatozoa in that sample (Mogas *et al.*, 1998). Some authors have attempted to evaluate DNA integrity as a predictor of fertility. Detailed measurement of head shape was used to demonstrate that specific variations in head shape could be associated with more fragmented DNA and decreased fertility (Núñez-Martínez *et al.*, 2007; Lange-Consiglio *et al.*, 2010). Human andrology laboratories may evaluate canine semen samples and report values for DNA fragmentation index and DNA stainability. Reference values have not been reported for the dog. Finally, tests used to assess fertilizing ability, which include zona binding and oocyte penetration assays (Peña Martínez, 2004), are not routinely used in dogs.

INTERPRETATION OF SEMEN EVALUATION RESULTS. No dog should be condemned on the basis of one ejaculate. Examples of factors that may have an impact on semen quality include age, breed and inbred status of the dog. Studies in Dalmatians and Rottweilers identified declines in semen quality after 6 years of age (Schubert and Seager, 1991; Seager and Schubert, 1996). Serial evaluations of populations of Irish wolfhounds have demonstrated lower libido, smaller testes and poorer semen quality compared with control dogs of other breeds (Dahlbom *et al.*, 1997). One study in Foxhounds demonstrated decreased semen quality and a lower number of pups born live to dogs with inbreeding coefficients of 0.125 to 0.558 (Wildt *et al.*, 1982). The reported insemination dose for fresh semen in dogs is 220–250 million morphologically normal spermatozoa (Mickelsen, 1991; Eilts *et al.*, 2003), calculated as the total number of spermatozoa times percentage MNS. Dogs with very high numbers of spermatozoa may achieve that value, even with a relatively low percentage MNS. Owners of bitches to which these males are bred should be counselled that no mathematical calculation can ensure pregnancy in their bitch.

Fertility assessment

Infertility is usually recognized as a lack of pregnancy in bred bitches (Root Kustritz, 2010b). The scheme shown below may be used to evaluate dogs presented for infertility (Fig. 8.2).

The Tomcat

Puberty

Definition and clinical determination

Puberty, or sexual maturity, is defined as the demonstration of normal semen quality and normal breeding behaviours. It is assumed that male cats require exposure to testosterone before and immediately at the time of birth if they are to show normal development and normal reproductive behaviour as adults, as do other species. The exposure to testosterone is a priming effect; animals not exposed to testosterone at birth cannot respond to the presence of testosterone at puberty with normal

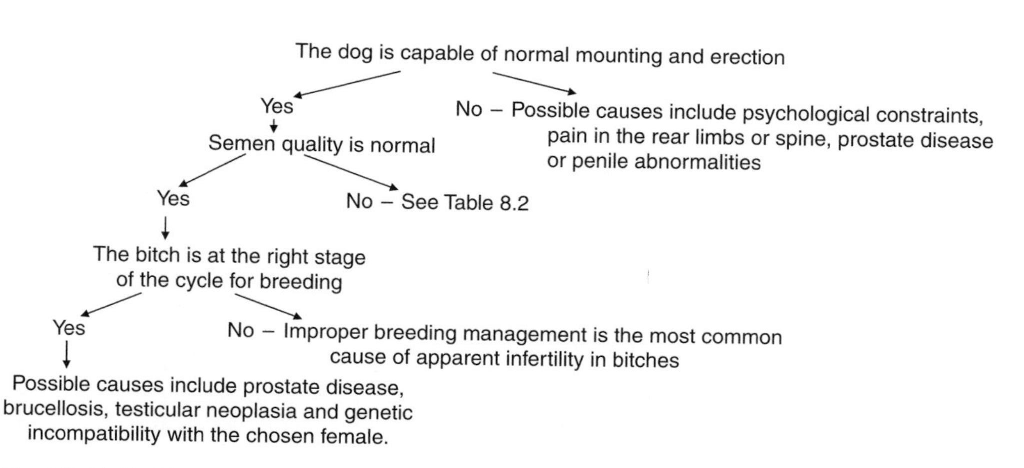

Fig. 8.2. Method for the evaluation of dog infertility.

breeding behaviour. Exposure to oestrogen at the time of birth may also be important for later reproductive success. In male laboratory animals, lack of oestrogen exposure at birth has been associated with decreased frequency of mounts when breeding (Pereira *et al.*, 2003).

Normal sexual behaviour and normal semen quality do not always develop simultaneously. The determination of puberty onset in male cats historically required proof through breeding attempts and successful siring of kittens. Modern techniques include semen collection and evaluation (see below), but in determining normality of breeding behaviour, there is no more definitive test than observation after exposure to an oestrous queen. Normal breeding behaviour includes investigation of the queen and attempts to mount. Inhalation of pheromones and flehmen behaviour in cats is observed as stiffening and extension of the neck after investigation of the queen's hindquarters. The tomcat grasps the scruff of the neck of the queen with his teeth. This grip does not break the skin and is not an aggressive act, but rather serves to stabilize the male and allow him to position the queen for copulation. Intromission and ejaculation last only seconds. The queen undergoes an after reaction consisting of rolling and licking, during which time the male will not approach her. In one study, domestic cats were demonstrated to mate an average of two times an hour, or 15 (or fewer) times in a day (Concannon *et al.*, 1980; Beaver, 1992).

Underlying factors

As in many species, puberty occurs when males reach about 80% of adult body weight, which usually occurs at about 8–10 months of age (Root Kustritz, 2010c). Season has some effect; in temperate latitudes, cats are more likely to achieve puberty during the breeding season, which for most cats includes all months except those with the shortest day length. This effect increases nearer the poles and decreases nearer the equator. The effect is more pronounced in queens than in tomcats (Root Kustritz, 2010c).

Fertility and determination of breeding soundness

History and pre-breeding testing

When cats are presented for breeding soundness examination, questions that should be asked include: history of breeding attempts, if any; any traumas or surgery, especially to the scrotum or penis; history of febrile or systemic disease; and drugs or other chemicals administered.

Genetic testing is not commonly performed in cats. Physical testing may be performed for some specific hereditary disorders, for example renal ultrasound or biopsy for polycystic kidney disease in Persians. Blood typing should be performed, especially in cats of the Abyssinian, Birman, British Shorthair, Cornish Rex, Devon Rex, Himalayan, Persian, Scottish Fold and Somali breeds. Male cats should be bred to queens of the same blood type to prevent the development of neonatal isoerythrolysis (Bucheler, 1999).

Physical examination findings

ACCESSORY SEX GLANDS. The prostate and bulbourethral glands are the accessory sex glands of the cat. They are not palpable per rectum or transabdominally and are rarely diseased, and so are not routinely evaluated in domestic cats.

TESTES. Both testes are completely descended into the scrotum before birth in most cats. Monorchidism, or the development of one testis only, is uncommon in cats, so any cat without two descended testes should be assumed to be cryptorchid. The testes should be symmetrical in size and shape, should have the consistency of a peeled, hard-boiled egg and should be freely movable in the scrotum (Johnston *et al.*, 2001b).

PENIS. The penis of sexually intact cats is encircled with cornified barbs, or spines (Plate 10). These spines are androgen dependent, do not appear until puberty and regress noticeably within 6 weeks of castration (Aronson and Cooper, 1967). The presence of penile spines can be used as a bioassay for testosterone.

Libido

Some male cats are hesitant to breed and rarely show investigative behaviour such as flehmen or approaching the queen. Established causes of poor libido in male cats include:

- Negative breeding experience. This may include injury to the penis during mating.
- Pain when attempting to mount or ejaculate. In cats, this is more commonly due to penile hair rings than to orthopaedic or prostatic pain, as in dogs.
- Chromosomal abnormality. The gene controlling the colours orange and black is on the X chromosome. As male cats should have only one X chromosome (38,XY), they can be orange or black, but not both colours. Male cats demonstrating both colours (calico – orange, black and white; or tortoiseshell – orange and black) must have more than one X chromosome (39,XXY or a 38,XY/38,XY chimera). These cats usually are infertile and may or may not show normal mating behaviour.
- Improper breeding environment. The tomcat is outside his territory.
- Improper age. The tomcat is too young or too old.

Semen collection

Semen collection by three techniques has been described. These are manual ejaculation, electroejaculation and the collection of a urethral sample after anaesthesia.

Manual ejaculation should take place on a safe elevated surface with a non-slip covering. Presence of a teaser queen, either a queen in oestrus or a spayed female cat treated with oestrogen, may increase the chance of successfully collecting an ejaculate (Sojka, 1986). For a collection vessel, most people create an AV with a small centrifuge tube connected to a 2 ml rubber bulb pipette. The tomcat is presented to the queen and the AV introduced over the erect penis to catch the sample as he mounts. It has been reported that about three out of five toms can be trained to ejaculate using this method (Sojka, 1986). This may be valuable in research, but as most practitioners are not likely to see the same tomcat repeatedly, it has little value in clinical practice.

Electroejaculation (EE) is considered by some to be the preferred semen collection technique in cats because it does not require the presence of an oestrous teaser queen or training of the male (Pineda *et al.*, 1984; Zambelli and Cunto, 2006). Electroejaculation requires placement of the cat under general anaesthesia. Reported anaesthetic protocols include ketamine (25–33 mg/kg IM), medetomidine (100–150 μg/kg subcut), and ketamine (5 mg/kg IM) with medetomidine (80 μg/kg subcut) (Platz and Seager 1978; Johnstone, 1984; Howard *et al.*, 1990; Axnér and Linde-Forsberg, 2002; Zambelli *et al.*, 2004). A probe of appropriate size is introduced into the rectum with the electrodes directed ventrally. The nerves are stimulated in a varying pattern using a rheostat (Plate 11). One stimulation protocol is described below (Table 8.3). A small collection tube is placed over the penis to catch the sample.

Retrograde ejaculation is a normal part of the ejaculatory process in cats (Dooley *et al.*, 1991). Semen may be retrieved from the urethra after administration of some anaesthetic agents. The regimen reported to be most successful is medetomidine (130–140 μg/kg subcut); a higher number of sperm is retrieved with medetomidine than with ketamine (Zambelli *et al.*, 2007, 2010; Filliers *et al.*, 2010). A tomcat catheter is passed into the urethra after drug administration to help retrieve the semen sample.

Table 8.3. One reported regimen for electroejaculation in cats (Howard *et al.*, 1990). For each stimulation, the rheostat is controlled to provide a slow rise to the desired voltage over 1 s, held at that voltage for 2–3 s, then abruptly returned to zero for 3 s. The cat is rested for about 30 s between sets.

	Set 1	Set 2	Set 3
Stimulations (no.)	10–10–10[a]	10–10–10	10–10
Voltage (V)	2–3–4	3–4–5	4–5

[a]For example, 10 stimulations at 2 V, 10 at 3 V then 10 at 4V.

Spermatozoa may be retrieved by post-coital vaginal lavage of the bred queen or post-coital cystocentesis of the tom. These methods identify the presence or absence of spermatozoa although they do not permit the evaluation of semen quality.

Semen evaluation

GENERAL EVALUATION. Evaluation to the extent performed in other species is limited by the small volume of the sample. Semen evaluation findings should be recorded in a legally defensible and retrievable fashion.

COLOUR. Normal feline semen is milky or opalescent (Root Kustritz, 2010d). All samples that appear to be cloudy should be evaluated microscopically as some samples containing lipid droplets and no spermatozoa will have this appearance. Clear semen is associated with lack of spermatozoa. Yellow semen is due to urine contamination. Frank blood is most likely to be due to penile trauma during collection.

VOLUME. Volume is not in and of itself an indicator of quality because it is dependent on the semen collection technique. Electroejaculation may be associated with overstimulation of the accessory sex glands and a more dilute sample. Volume should be recorded before any samples are withdrawn as that value will be used to calculate total number of spermatozoa in the ejaculate. Normal ejaculate volume in cats varies from 30–120 μl collected by AV to 114–738 μl collected by electroejaculation, with reported mean values of 45 μl and 194 μl, respectively (Sojka *et al.*, 1970; Platz and Seager, 1978; Platz *et al.*, 1978; Dooley and Pineda, 1986; Howard *et al.*, 1990; Zambelli and Cunto, 2006).

PH. Measurement of pH has no clinical significance in cats and is not routinely performed. If it is measured, be aware that the pH of semen collected by electroejaculation is higher than that of semen collected using an AV, perhaps because of increased contributions from the accessory sex glands (Dooley and Pineda, 1986).

PERCENTAGE PROGRESSIVE MOTILITY. Percentage progressive motility is assessed on neat semen. A small volume is placed on a glass slide. A coverslip may or may not be applied; individual clinicians should be consistent in the technique used to make their analyses consistent. A subjective assessment is made of the percentage of spermatozoa moving forward. Normal percentage progressive motility in cats is 70% or greater (Sojka *et al.*, 1970; Platz and Seager 1978; Platz *et al.*, 1978; Dooley and Pineda, 1986; Howard *et al.*, 1990; Choi *et al.*, 2010; Root Kustritz, 2010d). Evaluation can also be made of the speed and quality of progression. Some suggest that use of the Sperm Motility Index (SMI) promotes the correct emphasis on progressive motility over total motility; SMI is calculated as 0.5 × ((progressive motility × 20) + (% total motility)) (Howard 1992; Howard *et al.*, 1993).

CONCENTRATION/TOTAL NUMBER. Concentration is measured with a haemocytometer. The semen sample is diluted 1:100 as described above for the dog. The haemocytometer grid is made up of nine large squares. All of the spermatozoa within any one of these large squares are counted. The operator should focus up and down while counting to see all of the sperm, some of which will float head up or head down and so not be visible continuously in the grid. The number of spermatozoa counted is the concentration in millions/ml.

The total number of spermatozoa is calculated by multiplying concentration by volume collected. Total number averages 60 million in samples collected using an AV and 117 million in samples collected by electroejaculation (Sojka *et al.*, 1970; Platz and Seager, 1978; Platz *et al.*, 1978; Dooley and Pineda, 1986; Howard *et al.*, 1990; Zambelli and Cunto, 2006). The total number of spermatozoa decreases with daily collection, reaching a constant level of about half that seen on the first day after the fourth day (Sojka *et al.*, 1970).

PERCENTAGE MORPHOLOGICALLY NORMAL SPERMATOZOA. Morphology, or shape, is commonly evaluated on stained spermatozoa. The stain reported to be associated with fewest

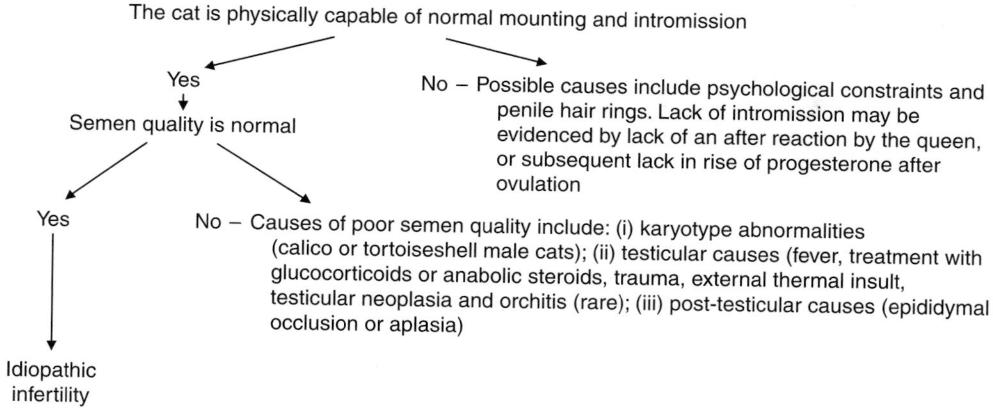

The cat is physically capable of normal mounting and intromission

Yes

Semen quality is normal

No – Possible causes include psychological constraints and penile hair rings. Lack of intromission may be evidenced by lack of an after reaction by the queen, or subsequent lack in rise of progesterone after ovulation

Yes

Idiopathic infertility

No – Causes of poor semen quality include: (i) karyotype abnormalities (calico or tortoiseshell male cats); (ii) testicular causes (fever, treatment with glucocorticoids or anabolic steroids, trauma, external thermal insult, testicular neoplasia and orchitis (rare); (iii) post-testicular causes (epididymal occlusion or aplasia)

Fig. 8.3. Method for the evaluation of cat infertility.

stain-induced artefacts and to permit the greatest differentiation of abnormal morphology in feline semen is Fast Green FCG–rose Bengal staining (Zambelli *et al.*, 1993).

One hundred to 200 sperm are counted under oil immersion. Teratozoospermia is common in domestic and wild felids (Pukazhenthi *et al.*, 2001), but the parameter of greatest interest is the percentage of normal spermatozoa, which has been reported as 44–71% in normal cats (Choi *et al.*, 2010; Root Kustritz, 2010d).

CYTOLOGY AND CULTURE. Because infection of the accessory sex glands is very rare in cats, semen cytology and culture are rarely performed.

SPECIAL TESTS. Specific testing to evaluate cell membrane integrity and fertilizing capability are not commonly performed in cats.

FERTILITY ASSESSMENT. The scheme depicted above may be used to evaluate cats presented for infertility (Fig. 8.3).

References

Ader PL and Hobson HP (1978) Hypospadias: a review of the veterinary literature and a report of three cases in the dog *Journal of the American Animal Hospital Association* **14** 721–727.

Agarwal A, Sharma RK and Nelson DR (2003) New semen quality scores developed by principal component analysis of semen characteristics *Journal of Andrology* **24** 343–352.

Aronson LR and Cooper ML (1967) Penile spines of the domestic cat: their endocrine–behaviour correlations *Anatomical Record* **157** 71–78.

Assis VP, Ribeiro VM, Rachid MA, Santana Castro AC and Ribeiro Valle G (2010) Dogs with *Leishmania chagasi* infection have semen abnormalities that partially revert during 150 days of allopurinol and amphotericin B therapy *Animal Reproduction Science* **117** 183–186.

Axnér E and Linde-Forsberg C (2002) Semen collection and assessment, and artificial insemination in the cat In *Recent Advances in Small Animal Reproduction* Ed PW Concannon, G England, J Verstegen III and C Linde-Forsberg. International Veterinary Information Service, Ithaca, New York. Available at: http://www.ivis.org/advances/concannon/axner/ivis.pdf (accessed 3 September 2013).

Bartlett DJ (1962) Studies on dog semen. I. Morphological characteristics *Journal of Reproduction and Fertility* **3** 173–189.

Batista M, González F, Rodriguez F, Palomino E, Cabrera F, Forga J and Gracia A (1998) Segmental aplasia of the epididymis in a Siberian Husky *Veterinary Record* **142** 250–251.

Beach FA (1968) Coital behavior in dogs. III. Effects of early isolation on mating in males *Behavior* **30** 218–238.

Beaver BV (1977) Mating behavior in the dog *Veterinary Clinics of North America* **7** 723–728.

Beaver BV (1992) Male feline sexual behaviour. In *Feline Behavior: A Guide for Veterinarians* pp 121–139 BV Beaver. WB Saunders Co, Philadelphia.

Boeri E, Escobar GI, Ayala SM, Sosa-Estani S and Lucero NE (2008) Canine brucellosis in dogs in the city of Buenos Aires *Medicina (Buenos Aires)* **68** 291–297.

Boucher JH, Foote RH and Kirk RW (1958) The evaluation of semen quality in the dog and the effects of frequency of ejaculation upon semen quality, libido, and depletion of sperm reserves *The Cornell Veterinarian* **48** 67–86.

Brotto LA and Gorzalka BB (2000) Melatonin enhances sexual behavior in the male rat *Physiology and Behavior* **68** 483–486.

Brown TT, Burek JD and McEntee K (1976) Male pseudohermaphroditism, cryptorchidism and Sertoli cell neoplasia in three Miniature Schnauzers *Journal of the American Veterinary Medical Association* **169** 821–825.

Bucheler J (1999) Fading kitten syndrome and neonatal isoerythrolysis *Veterinary Clinics of North America* **29** 833–870.

Burns JG and Petersen NK (2008) Theriogenology question of the month: monorchidism in a dog *Journal of the American Veterinary Medical Association* **233** 1553–1554.

Campbell WE (1975) Mounting and other sex-related problems *Modern Veterinary Practice* **56** 420–422.

Carmichael LE and Kenney RM (1968) Canine abortion caused by *Brucella canis* *Journal of the American Veterinary Medical Association* **152** 605–616.

Chatterjee SN, Sharma RN and Kar AB (1976) Semen characteristics of normal and vasectomized dogs *Indian Journal of Experimental Biology* **14** 411–414.

Choi EG, Lee YS, Cho SJ, Jeon JT, Cho KW and Kong IK (2010) Semen characteristics of genetically identical male cats cloned via somatic cell nucleus transfer *Theriogenology* **73** 638–644.

Christensen P, Stryhn H and Hansen C (2005) Discrepancies in the determination of sperm concentration using Burker–Turk, Thoma and Makler counting chambers *Theriogenology* **63** 992–1003.

Concannon P, Hodgson B and Lein D (1980) Reflex LH release in estrous cats following single and multiple copulations *Biology of Reproduction* **23** 111–117.

Corrada Y, Hermo G and Gobello C (2006) Theriogenology question of the month: sexual immaturity in a prepubertal dog *Journal of the American Veterinary Medical Association* **228** 855–856.

Dahlbom M, Andersson M, Juga J and Alanko M (1997) Fertility parameters in male Irish Wolfhounds: a two-year follow-up study *Journal of Small Animal Practice* **38** 547–550.

Daiwadnya CB, Huker VB and Sonawane SA (1995) Studies on evaluation of dog semen *Livestock Advisor* **20** 34–37.

Dooley MP and Pineda MH (1986) Effect of method of collection on seminal characteristics of the domestic cat *American Journal of Veterinary Research* **47** 286–292.

Dooley MP, Pineda MH, Hopper JG and Hsu WH (1991) Retrograde flow of spermatozoa into urinary bladder of cat during electroejaculation, collection of semen with artificial vagina, and mating *American Journal of Veterinary Research* **52** 687–691.

Eilts BE, Paccamonti DL and Pinto C (2003) Artificial insemination in the dog In *The Practical Veterinarian: Small Animal Theriogenology* pp 61–95 Ed MV Root Kustritz. Butterworth-Heinemann, St Louis.

Eljarah AH, Chandler JE and Chenevert J (2005) Determining bull semen concentration: a novel design for CASA *Theriogenology* **64** 799 [abstract].

Ellington J, Scarlett J, Meyers-Wallen V, Mohammed HO and Surman V (1993) Computer-assisted sperm analysis of canine spermatozoa motility measurements *Theriogenology* **40** 725–733.

England GCW (1999) Semen quality in dogs and the influence of a short-interval second ejaculation *Theriogenology* **52** 981–986.

England GCW and Plummer JM (1993) Hypo-osmotic swelling of dog spermatozoa *Journal of Reproduction and Fertility Supplement* **47** 261–270.

England GC, Allen WE and Middleton DJ (1990) An investigation into the origin of the first fraction of the canine ejaculate *Research in Veterinary Science* **49** 66–70.

England GCW, Phillips L and Freeman SL (2010) Heritability of semen characteristics in dogs *Theriogenology* **74** 1136–1140.

Filliers M, Rijsselaere T, Bossaert P, Zambelli D, Anastasi P, Hoogewijs M and van Soom A (2010) *In vitro* evaluation of fresh sperm quality in tomcats: a comparison of two collection techniques. *Theriogenology* **74** 31–39.

Foote RH and Boucher JH (1964) A comparison of several photoelectric procedures for estimating sperm concentration in dog semen *American Journal of Veterinary Research* **25** 558–560.

Fuller JL and Fox MW (1969) The behavior of dogs In *The Behavior of Domestic Animals* p 438 Ed ESE Hafez. Williams and Wilkins Co, Baltimore.

Gier HT and Marion GB (1969) Development of mammalian testes and genital ducts *Biology of Reproduction* **1** 1–23.

Goodwin M, Gooding KM and Regnier F (1979) Sex pheromone in the dog *Science* **203** 559–561.

Gunzel-Apel AR, Gunther C, Terhaer P and Bader H (1993) Computer-assisted analysis of motility, velocity and linearity of dog spermatozoa *Journal of Reproduction and Fertility Supplement* **47** 271–278.

Harasymowycz J, Ball L and Seidel GE (1976) Evaluation of bovine spermatozoal morphologic features after staining or fixation *American Journal of Veterinary Research* **37** 1053–1057.

Hare WCD (1976) Intersexuality in the dog *Canadian Veterinary Journal* **17** 7–15.

Hollett RB (2006) Canine brucellosis: outbreaks and compliance *Theriogenology* **66** 575–587.

Howard JG (1992) Feline semen analysis and artificial insemination In *Current Veterinary Therapy XI* pp 929–938 Ed JD Bonagura. WB Saunders Co, Philadelphia.

Howard JG, Brown JL, Bush M and Wildt DE (1990) Teratospermic and normospermic domestic cats: ejaculate traits, pituitary-gonadal hormones, and improvement of spermatozoa motility and morphology after swim up processing *Journal of Andrology* **11** 204–215.

Howard JG, Donoghue AM, Johnston LA and Wildt DE (1993) Zona pellucida filtration of structurally abnormal spermatozoa and reduced fertilization in teratospermic cats *Biology of Reproduction* **49** 131–139.

Inamasu A, Vechi E and Lopes MD (1999) Investigation of the viability of the hypoosmotic test and the relationship with some spermatic characteristics *Revista Brasileira de Reprodução Animal* **23** 302–304.

Johnson C, Jacobs J and Walker R (1991) Morphology-stain induced spermatozoal abnormalities In *Proceedings, Society for Theriogenology,* San Diego, CA, August 16–17, 1991 p 239.

Johnson C, Bari Olivier N, Nachreiner R and Mullaney T (1999) Effect of 131I-induced hypothyroidism on indices of reproductive function in adult male dogs *Journal of Veterinary Internal Medicine* **13** 104–110.

Johnston SD, Root Kustritz MV and Olson PN (2001a) Clinical approach to infertility in the male dog In *Canine and Feline Theriogenology* pp 370–387. WB Saunders Co, Philadelphia.

Johnston SD, Root Kustritz MV and Olson PN (2001b) Disorders of the feline testes and epididymes In *Canine and Feline Theriogenology* pp 525–536. WB Saunders Co, Philadelphia.

Johnstone IP (1984) Electroejaculation in domestic cat *Australian Veterinary Journal* **61** 155–158.

Keid LB, Soares RM, Vasconcellos SA, Megid J, Salgado VR and Richtzenhain LJ (2009) Comparison of agar gel immunodiffusion test, rapid slide agglutination test, microbiological culture and PCR for the diagnosis of canine brucellosis *Research in Veterinary Science* **86** 22–26.

Kurni-Diaka J and Badtram G (1994) Effect of storage on sperm membrane integrity and other functional characteristics of canine spermatozoa: *in vitro* bioassay for canine semen *Theriogenology* **41** 1355–1366.

Kuster C (2005) Sperm concentration determination between hemacytometer and CASA systems: why they can be different *Theriogenology* **64** 614–617.

Kutzler MA (2005) Semen collection in the dog *Theriogenology* **64** 747–754.

Kutzler MA, Solter AF, Hoffman WE and Volkmann DH (2003) Characterization and localization of alkaline phosphatase in canine seminal plasma and gonadal tissues *Theriogenology* **60** 299–306.

Lange-Consiglio A, Antonucci N, Manes S, Corradetti B, Cremonesi F and Bizzaro D (2010) Morphometric characteristics and chromatin integrity of spermatozoa in three Italian dog breeds *Journal of Small Animal Practice* **51** 624–627.

Majeed ZZ (1974) Segmental aplasia of the Wolffian duct; report of a case in a Poodle *Journal of Small Animal Practice* **15** 263–268.

Marshall LS, Oehlert ML, Haskins ME, Selden JR and Patterson DF (1982) Persistent Müllerian duct syndrome in Miniature Schnauzers *Journal of the American Veterinary Medical Association* **181** 798–801.

Meyers-Wallen VN (1991) Clinical approach to infertile male dogs with sperm in the ejaculate *Veterinary Clinics of North America* **21** 609–632.

Meyers-Wallen VN and Patterson DF (1989) Sexual differentiation and inherited disorders of the sexual development of the dog *Journal of Reproduction and Fertility Supplement* **39** 57–64.

Mickelsen WD (1991) Maximizing fertility in stud dogs with poor semen quality In *Proceedings, Society for Theriogenology San Diego, CA, August 16–17, 1991* pp 244–246.

Mills DS, Ramos D, Estelles MG and Hargrave C (2006) A triple blind placebo-controlled investigation into the assessment of the effect of Dog Appeasing Pheromone (DAP) on anxiety related behaviour of problem dogs in the veterinary clinic *Applied Animal Behaviour Science* **98** 114–126.

Mogas T, Rigau T, Piedrafita J, Bonet S and Rodriguez-Gil JE (1998) Effect of column filtration upon the quality parameters of fresh dog semen *Theriogenology* **50** 1171–1189.

Ndiritu CG (1979) Lesions of the canine penis and prepuce *Modern Veterinary Practice* **60** 712–715.

Núñez-Martínez I, Morán JM and Peña FJ (2006) A three-step statistical procedure to identify sperm kinematic subpopulations in canine ejaculates: changes after cryopreservation *Reproduction in Domestic Animals* **41** 408–415.

Núñez-Martínez I, Morán JM and Peña FJ (2007) Identification of sperm morphometric subpopulations in the canine ejaculate: do they reflect different subpopulations in sperm chromatin integrity? *Zygote* **15** 257–266.

Oettle EE (1993) Sperm morphology and fertility in the dog *Journal of Reproduction and Fertility Supplement* **47** 257–260.

Olar TT, Amann RP and Pickett BW (1983) Relationships among testicular size, daily production and output of spermatozoa, and extragonadal spermatozoa reserves of the dog *Biology of Reproduction* **29** 1114–1120.

O'Shea JD (1962) Studies on the canine prostate gland. I. Factors influencing its size and weight *Journal of Comparative Pathology* **72** 321–331.

Patterson EE (2009) Methods and availability of tests for hereditary disorders of dogs In *Kirk's Current Veterinary Therapy XIV* 14th edition pp 1054–1059 Ed JD Bonagura and DC Twedt. Saunders Elsevier, St Louis.

Peña AI, Barrio M, Becerra JJ, Quintela LA and Herradón PG (2007) Infertility in a dog due to proximal cytoplasmic droplets in the ejaculate: investigation of the significance for sperm functionality *in vitro* *Reproduction in Domestic Animals* **42** 471–478.

Peña JF, Núñez-Martínez I and Morán JM (2006) Semen technologies in dog breeding: an update *Reproduction in Domestic Animals* **41(Supplement 2)** 21–29.

Peña Martínez AI (2004) Canine fresh and cryopreserved semen evaluation *Animal Reproduction Science* **82** 209–224.

Pereira OCM, Coneglian-Marise MSP and Garardin DCC (2003) Effects of neonatal clomiphene citrate on fertility and sexual behavior in male rats *Comparative Biochemistry and Physiology, Part A: Molecular and Integrative Physiology* **134** 545–550.

Pineda MH, Dooley MP and Martin PA (1984) Long term study on the effects of electroejaculation on seminal characteristics of the domestic cat *American Journal of Veterinary Research* **45** 1038–1040.

Pinto CRF and Kozink DM (2008) Simplified hypoosmotic swelling test (HOST) of fresh and frozen–thawed canine spermatozoa *Animal Reproduction Science* **104** 450–455.

Pinto CRF, Wrench N and Schramme A (2005) Simplified hypo-osmotic testing of canine spermatozoa *Theriogenology* **64** 811 [abstract].

Platz CC and Seager SW (1978) Semen collection by electroejaculation in the domestic cat *Journal of the American Veterinary Medical Association* **173** 1353–1355.

Platz CC, Wildt DE and Seager SWJ (1978) Pregnancy in the domestic cat after artificial insemination with previously frozen spermatozoa *Journal of Reproduction and Fertility* **52** 279–282.

Pukazhenthi B, Wildt DE and Howard JG (2001) The phenomenon and significance of teratospermia in felids *Journal of Reproduction and Fertility Supplement* **57** 423–433.

Purswell BJ and Wilcke JR (1993) Response to gonadotrophin-releasing hormone by the intact male dog: serum testosterone, luteinizing hormone and follicle-stimulating hormone *Journal of Reproduction and Fertility Supplement* **47** 335–341.

Purswell BJ, Althouse GC, Root Kustritz MV, Pretzer S and Lopate C (2010) Guidelines for using the canine breeding soundness evaluation form *Clinical Theriogenology* **2** 51–59.

Read RA and Bryden S (1995) Urethral bleeding as a presenting sign of benign prostatic hyperplasia in the dog: a retrospective study (1979–1993) *Journal of the American Animal Hospital Association* **31** 261–267.

Rigau T, Farré M, Ballester J, Mogas T, Peña A and Rodríguez-Gil JE (2001) Effects of glucose and fructose on motility patterns of dog spermatozoa from fresh ejaculates *Theriogenology* **56** 801–815.

Rijsselaere T, van Soom A, Maes O and de Kruif A (2002) Use of the sperm quality analyzer (SQA) for the assessment of dog sperm quality *Reproduction in Domestic Animals* **37** 158–163.

Rogers KS (1997) Transmissible venereal tumor *Compendium on Continuing Education for the Practising Veterinarian* **19** 1036–1045.

Romagnoli S and Majolino G (2009) Aspermia/oligozoospermia caused by retrograde ejaculation in the dog In *Kirk's Current Veterinary Therapy XIV* 14th edition pp 1049–1053 Ed JD Bonagura and DC Twedt. Saunders Elsevier, St Louis.

Root Kustritz MV (2007) The value of canine semen evaluation for practitioners *Theriogenology* **68** 329–337.

Root Kustritz MV (2009) Evaluation of clinical usefulness of tests for canine brucellosis by calculation of predictive value *Clinical Theriogenology* **1** 343–346.

Root Kustritz MV (2010a) What is the normal age for puberty onset in bitches and dogs? In *Clinical Canine and Feline Reproduction* p 69. Wiley-Blackwell, Ames.

Root Kustritz MV (2010b) What is the diagnostic approach for infertility of a male dog? In *Clinical Canine and Feline Reproduction* pp 181–183. Wiley-Blackwell, Ames.

Root Kustritz MV (2010c) What is the normal age for puberty onset in queens and toms? In *Clinical canine and feline reproduction* p 211. Wiley-Blackwell, Ames.

Root Kustritz MV (2010d) What are the normal parameters for semen quality in cats? In *Clinical Canine and Feline Reproduction* pp 215–216. Wiley-Blackwell, Ames.

Root Kustritz MV and Hess M (2007) Effect of administration of prostaglandin F2alpha or presence of an estrous teaser bitch on characteristics of the canine ejaculate *Theriogenology* **67** 255–258.

Root Kustritz MV, Olson PN, Johnston SD and Root TK (1998) The effects of stains and investigators on assessment of morphology of canine spermatozoa *Journal of the American Animal Hospital Association* **34** 348–352.

Root Kustritz MV, Johnston SD, Olson PN and Lindeman CJ (2005) Relationship between inflammatory cytology of canine seminal fluid and significant aerobic bacterial, anaerobic bacterial or mycoplasma cultures of canine seminal fluid: 95 cases (1987–2000) *Theriogenology* **64** 1333–1339.

Root Kustritz MV, Kilty C and Vollmer M (2007) Spermatocrit as a measure of the concentration of spermatozoa in canine semen *Veterinary Record* **161** 566–567.

Rui H, Morkas L and Purvis K (1986) Time- and temperature-related alterations in seminal plasma constituents after ejaculation *International Journal of Andrology* **9** 195–200.

Santos NR, Krekeler N, Schramme-Jossen A and Volkmann DH (2006) The knobbed acrosome defect in four closely-related dogs *Theriogenology* **66** 1626–1628.

Schäfer S, Holzmann A and Arbeiter K (1996) The influence of frequent semen collection on the semen quality of Beagle dogs *Tierärztliche Praxis* **24** 385–390.

Schäfer-Somi S and Aurich C (2007) Use of a new computer-assisted sperm analyzer for the assessment of motility and viability of dog spermatozoa and evaluation of four different semen extenders for predilution *Animal Reproduction Science* **102** 1–13.

Schnee CM (1985) Studies on the induction of retrograde ejaculation by alpha-receptor blockage and incorrect semen collection procedure in the dog. Thesis, Tierarztliche Hochschule, Hannover.

Schubert CL and Seager SWJ (1991) Semen collection and evaluation for the assessment of fertility parameters in the male Dalmatian *Canine Practice* **16** 17–21.

Schulz KM, Richardson HN, Zehr JL, Osetek AJ, Menard TA and Sisk CL (2004) Gonadal hormones masculinize and defeminize reproductive behaviors during puberty in the male Syrian hamster *Hormones and Behavior* **45** 242–249.

Seager SWJ and Schubert CL (1996) Semen collection and evaluation for the clinical assessment of fertility parameters in the male Rottweiler *Canine Practice* **21** 30–34.

Serikawa T, Takada H, Kondo Y, Muraguchi T and Yamada J (1984) Multiplication of *Brucella canis* in male reproductive organs and detection of auto-antibody to spermatozoa in canine brucellosis *Developments in Biological Standardization* **56** 295–305.

Silva FL, Oliveira RG, Silva TMA, Xavier MN, Nascimento EF and Santos RL (2009) Venereal transmission of canine visceral leishmaniasis *Veterinary Parasitology* **160** 55–59.

Sojka NJ (1986) Management of artificial breeding in cats In *Current Therapy in Theriogenology* pp 805–808 Ed DA Morrow. WB Saunders Co, Philadelphia.

Sojka NJ, Jemings LL and Hamner CE (1970) Artificial insemination in the cat (*Felis catus* L.) *Laboratory Animal Care* **20** 198–204.

Taha MB, Noakes DE and Allen WE (1983) The effect of the frequency of ejaculation on seminal characteristics and libido in the Beagle dog *Journal of Small Animal Practice* **24** 309–315.

Takeishi M, Toshoyima T and Ryo T (1975) Studies on reproduction in the dog. VI. Sexual maturity of male Beagles *Bulletin of the College of Agriculture and Veterinary Medicine, Nihon University* **32** 213–223.

Threlfall WR (2003) Semen collection and evaluation In *The Practical Veterinarian: Small Animal Theriogenology* pp 97–123 Ed MV Root Kustritz. Butterworth-Heinemann, St Louis.

Traas AM and Root Kustritz MV (2004) Effect of administration of oxytocin or prostaglandin F2α on characteristics of the canine ejaculate *Canadian Veterinary Journal* **45** 999–1002.

Vare AM and Bansal PC (1973) Changes in the canine testis after bilateral vasectomy – an experimental study *Fertility and Sterility* **24** 793–797.

Vare AM and Bansal PC (1974) The effects of ligation of caudal epididymis on the dog testis *Fertility and Sterility* **25** 256–260.

Verstegen J and Onclin K (2003) Male infertility and treatments in the canine species In *Proceedings, Society for Theriogenology, Columbus, OH* pp 221–225.

Verstegen J, Jolin E, Phillips T and Onclin K (2006) Evaluation of the "Minitube Sperm Vision Computer-Based Automated System" for dog semen analysis *Theriogenology* **66** 671 [abstract].

WHO (1999) *WHO Laboratory Manual for the Examination of Human Semen and Sperm–Cervical Mucus Interaction* Published for World Health Organization by Cambridge University Press, Cambridge/New York.

Wildt DE, Baas EJ, Chakraborty PK, Wolfle TL and Stewart AP (1982) Influence of inbreeding on reproductive performance, ejaculate quality and testicular volume in the dog *Theriogenology* **17** 445–452.

Woodall PF and Johnstone IP (1988a) Dimensions and allometry of testes, epididymes and spermatozoa in the domestic dog (*Canis familiaris*) *Journal of Reproduction and Fertility* **82** 603–609.

Woodall PF and Johnstone IP (1988b) Scrotal width as an index of testicular size in dogs and its relation to body size *Journal of Small Animal Practice* **29** 543–547.

Zambelli D and Cunto M (2006) Semen collection in cats: techniques and analysis *Theriogenology* **66** 159–165.

Zambelli D, Bergonzoni ML, de Fanti C and Carluccio A (1993) Tecniche per la valutazione morfologica degli spermatozoi di gatto, coniglio e cane In *Proceedings, SISVET (Società Italiana delle Scienze Veterinarie), Riccione* pp 279–283.

Zambelli D, Baietti B, Prati F and Belluzzi S (2004) The effect of ketamine and medetomidine on quality of domestic cat sperm after electroejaculation In *Proceedings, EVSSAR IV European Congress on Reproduction in Companion, Exotic and Laboratory Animals, June 4–6, Barcelona, 2004* pp 311–312.

Zambelli D, Cunto M, Prati F and Merlo B (2007) Effects of ketamine or medetomidine administration on quality of electroejaculated sperm and on sperm flow in the domestic cat *Theriogenology* **68** 796–803.

Zambelli D, Raccagni R, Cunto M, Andreani G and Isani G (2010) Sperm evaluation and biochemical characterization of cat seminal plasma collected by electroejaculation and urethral catheterization *Theriogenology* **74** 1396–1402.

Zirkin BR and Strandberg JD (1984) Quantitative changes in the morphology of the aging canine prostate *Anatomical Record* **208** 207–214.

9 Applied Andrology in Chickens and Turkeys

Julie A. Long*
Beltsville Agricultural Research Center, US Department of Agriculture Agricultural Research Service, Beltsville, Maryland, USA

Introduction

Modern commercial poultry operations rely heavily on assisted reproduction to maintain the high level of genetic selection that is the backbone of the industry. The growth rate of meat birds and egg production in laying birds are, respectively, the two most economically important traits. Because these traits have negative genotypic and phenotypic relationships, poultry companies either specialize in breeding meat-type birds or have two separate programmes for meat- and egg-producing strains. Primary breeders maintain 50 or more pure lines with relatively small numbers of pedigreed birds. After specific line crosses, the successive generations are multiplied, or produced in greater numbers, until sufficient bird numbers for the final commercial lines are achieved. For the layer and broiler chicken industries, artificial insemination (AI) is used mainly at the primary breeder level. For the turkey industry, all phases of production utilize AI. In this chapter, pertinent aspects of male reproductive biology, semen evaluation and semen storage are reviewed with an emphasis on unique attributes of avian reproductive biology.

Male Reproductive Biology

The reproductive organs of the male bird are completely internalized, unlike the situation in mammals where the testes are externally located. In contrast with the female bird, which has a single functional ovary, avian testes are paired organs anterior to the cranial pole of the kidneys. The male avian reproductive tract consists of the paired testes, epididymal regions and ductus deferens. Avian testes are organized into branching seminiferous tubules containing Sertoli cells, and a relatively small amount of interstitial tissue containing the androgen-secreting Leydig cells. The networks of seminiferous tubules coalesce into a series of ducts, known as the epididymal region, which empties into the ductus deferens. The avian reproductive tract does not contain the characteristically subdivided epididymis found in mammals, in which spermatozoa mature and are stored before ejaculation. Instead, avian spermatozoa mature and are stored in the ductus deferens. Accessory glands found in the mammalian reproductive tract (the seminal vesicle, prostate, etc.) are not present in birds. Males of most avian species do not have a truly intromittent copulatory organ, but rather a phallus on the wall of the cloaca.

* E-mail: julie.long@ars.usda.gov

Avian spermatogenesis is characterized by a relatively short spermatogenic cycle. In mammals, the time to produce a single spermatozoon ranges from 28 to 78 days across species (Hess and Renato de Franca, 2008). In contrast, the time from the onset of meiosis to the end of spermiogenesis is only 14 days in the rooster (de Reviers, 1968), drake (Marchand *et al.*, 1977) and guinea fowl (Brillard, 1981), and 11 days in the quail (Amir *et al.*, 1963). The actual length of the spermatogenic cycle in the turkey has not been confirmed. Using the 5-bromodeoxyuridine (BrdU) technique (Rosiepen *et al.*, 1994), the duration of meiotic prophase in the turkey was estimated to be 4.5 days, and the round spermatid phase had an estimated duration of 2 days; however, no BrdU reaction product was observed on elongated spermatids or testicular spermatozoa (Noirault *et al.*, 2006). Sperm meiosis intervals in the turkey are similar to those reported for the rooster (de Reviers, 1968), drake (Marchand *et al.*, 1977) and quail (Lin *et al.*, 1990), suggesting that the duration of spermatogenesis in the turkey may fall in the range of 11–14 days. In addition to rapid sperm production rates, non-passerine birds appear to produce up to four times the number of spermatozoa per gram of testis than do mammals (Jones and Lin, 1993). It has been concluded that the reproductive strategy of avian species involves the rapid production, maturation and transport of spermatozoa through the reproductive tract, in association with a limited capacity to store spermatozoa for long periods within the male genital ducts (Clulow and Jones, 1982).

Sperm Morphology and Physiology

Morphology

Mature turkey and chicken spermatozoa exhibit similar morphological features (see extensive review by Jamieson, 2007). The spermatozoa are long and narrow with a vermiform appearance, being 0–5–0.8 μm at the widest point. The acrosome of the turkey spermatozoon is 1.8 μm long, reaching 2.5 μm for the rooster acrosome (Marquez and Ogasawara, 1975). A fibrous, actin-containing, rod-like structure, known as the perforatorium or the acrosomal spine (Etches, 1996), is located between the acrosome and the anterior pole of the nucleus in the spermatozoa of non-passerine birds; no corresponding structure is evident in the subacrosomal space of passerine birds (Pudney, 1995) or mammalian spermatozoa, except in rodents (Baccetti *et al.*, 1980). At 1.0 μm in length, the perforatorium of the turkey and rooster is appreciably shorter than that of the guinea fowl (1.9 μm). The role of this structure in bird spermatozoa is suggested to be skeletal in nature, acting as an inner conical support for the acrosome (Baccetti *et al.*, 1980).

The sperm nucleus consists of dense chromatin, forming a concave implantation fossa where it joins the midpiece of the tail. The turkey sperm nucleus is shorter (7–9 μm) than that of guinea fowl or chicken (10–14 μm), and the junction of the nucleus with the midpiece at the neck region is not as conspicuous as in guinea fowl spermatozoa (Thurston and Hess, 1987).

For turkey and chicken spermatozoa, the neck region of the midpiece consists of a proximal and an elongated distal centriole, while guinea fowl spermatozoa contain only a single elongated centriole and associated pericentriolar projections (Thurston and Hess, 1987). Cross sections of the centrioles have the typical 'pinwheel' arrangement of nine triplet microtubules embedded in a cylindrical, dense wall (Jamieson, 2007). Enveloping the distal centriole and extending to the annulus are 25–30 helically arranged mitochondria (Thurston and Hess, 1987). In contrast to the guinea fowl, the cristae of turkey and rooster sperm mitochondria are parallel to the outer membrane (Jamieson, 2007). The absolute size of the midpiece region and the proportion of the midpiece to the head are greater in turkey than in chicken spermatozoa (Lake and Wishart, 1984).

The flagellum ultrastructure consists of the typical 9 + 2 microtubular axonemal complex, but outer dense fibres are absent (Thurston and Hess, 1987). Surrounding the outer doublets is an amorphous sheath (Lake *et al.*, 1968), which defines the principal piece. The portion of the flagellum where the cell membrane is in juxtaposition to the doublet microtubules, without the amorphous sheath, is known as the endpiece (Jamieson, 2007). The flagellum comprises most of the length of the poultry spermatozoon. Turkey and guinea fowl sperm flagellae are usually 60–65 μm long (Marquez and Ogasawara, 1975; Thurston and Hess, 1987), and the flagellum approaches 70 μm in length in rooster spermatozoa (Jamieson, 2007). Thus, the overall length of turkey and guinea fowl spermatozoa (75–80 μm) is also less than that of rooster spermatozoa (90 μm). In all three species, the spermatozoa increase in width from the acrosome to a maximum of 0.5–0.7 μm at the junction of the nucleus with the midpiece. The width then decreases to 0.1–0.2 μm at the end of the flagellum (Thurston and Hess, 1987).

Abnormalities in the morphology of poultry spermatozoa have been widely reported; however, there is no uniform classification system for the different types of sperm defects. For example, the term 'bent sperm' can refer to a bend in the midpiece or the tail. For this reason, the predominance of one type of defect over others is difficult to discern from the literature. Poultry sperm abnormalities can be broadly categorized according to the anatomical region of the defect (e.g. acrosome, head, midpiece, flagellum). Alkan *et al.* (2002) published comprehensive drawings of sperm defects in the Bronze turkey (Fig. 9.1), in which 20 sperm defects are illustrated.

In general, the presence of 15% or fewer abnormal spermatozoa in a given semen sample does not adversely affect fertility (Clark *et al.*, 1984), although the percentage of abnormal spermatozoa in semen can be as high as 24.1% in turkeys (Nestor and Brown, 1971), 46–48.3% in layer chickens (Siudzińska and Łukaszewicz, 2008; Jarinkovićová *et al.*, 2012) and 42.1% in broiler chickens (Edens and Sefton, 2009), illustrating the need for a common classification system for abnormal poultry sperm morphology.

Sperm membrane carbohydrates and lipids

The surface of mammalian spermatozoa comprises a dense coating of carbohydrates forming a 20–60 nm thick glycocalyx (Schröter *et al.*, 1999) that arises from the lipids and proteins forming the plasma membrane. Recently, Long and colleagues have demonstrated that poultry spermatozoa are also covered by carbohydrate residues (Peláez and Long, 2007). The glycocalyx of poultry spermatozoa is extensively sialylated, with sialic acid residues distributed along the

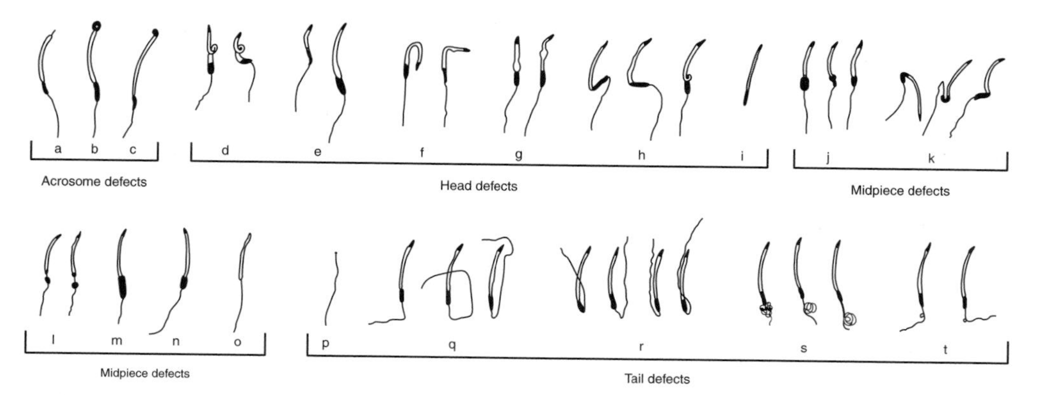

Fig. 9.1. Illustrations of abnormal sperm morphologies associated with the acrosome (a–c), head (d–i), midpiece (j–o) and tail regions (p–t) of turkey spermatozoa: (a) detached acrosome; (b) swollen acrosome; (c) comma-shaped acrosome; (d) knotted heads; (e) micro- and macro-cephalic heads; (f) bent heads; (g) swollen heads; (h) bent or knotted head at neck region; (i) detached head; (j) swollen midpieces; (k) bent midpieces; (l) partially detached midpieces; (m) thickened midpiece; (n) vacuolized midpiece; (o) detached midpiece; (p) detached tail; (q) 90° bent tails; (r) 180° bent tails; (s) curled tails; and (t) knotted tails. Adapted from Alkan *et al.*, 2002.

entire sperm surface (Plate 12a); it also contains residues of α-mannose/α-glucose, α- and β-galactose, α-fucose, α- and β-*N*-acetyl-galactosamine and *N*-acetyl-lactosamine, as well as monomers and dimers of *N*-acetyl-glucosamine. Non-sialic acid carbohydrate residues are segregated among specific morphological zones. For example, α-mannose and α-glucose were detected only in the plasma membrane overlying the head region, whereas, *N*-acetyl-glucosamine residues were distributed mainly along the acrosome region. One notable difference between turkey and chicken spermatozoa was the absence of α-galactose residues on the head region of chicken sperm (Plates 12b and c). *N*-acetyl-glucosamine dimers and *N*-acetyl-lactosamine residues were more prevalent in chicken spermatozoa, but *N*-acetyl-galactosamine residues were more prevalent in turkey spermatozoa.

A major role of glycosylation is the formation of specific molecular surfaces that permit specific intermolecular recognition (Ivell, 1999). The sperm glycocalyx is modified extensively during sperm transport and maturation, and represents the primary interface between the male gamete and its environment. While the function of these glycoconjugates is still largely unknown, in the hen, sialic acid has been implicated for both sperm passage through the vagina (Steele and Wishart, 1996) and sperm sequestration in the sperm storage tubules (Froman and Thursam, 1994), and *N*-acetyl-D-glucosamine is necessary for sperm–egg interaction (Robertson *et al.*, 2000). Long and colleagues have demonstrated that hypothermic storage of turkey semen (4°C) and chicken semen (–196°C) affects the composition of the sperm glycocalyx (Peláez and Long, 2008; Peláez *et al.*, 2011). In particular, terminal sialic acid residues are less abundant after hypothermic storage for both species. Sialic acids act as masking agents on antigens, receptors and other recognition sites of the cell surface, and it appears that the amount of sialic acid in sperm cells is considerably higher than that in somatic cells (Diekman, 2003).

Lipids are a major component and integral part of sperm membranes, and are involved with the series of biochemical and functional changes required for fertilization, such as sperm maturation and the acrosome reaction (Bréque *et al.*, 2003). Phospholipids represent the main lipid type found in avian sperm membranes, with phosphatidylcholine comprising ~40% of the total phospholipid content. Poultry spermatozoa also contain a high proportion of polyunsaturated fatty acids in the plasma membrane (Surai *et al.*, 1998). Polyunsaturated fatty acids are classified as n–3 or n–6 based on the location of the last double bond relative to the terminal methyl end of the molecule. In contrast to the plasma membranes of mammalian spermatozoa, which contain mainly n-3 polyunsaturated fatty acids such as docosahexaenoic acid (22:6n-3; DHA), the phospholipids in avian spermatozoa are enriched mainly with n-6 polyunsaturated fatty acids, including arachidonic (20:4n-6; AA) and docosatetraenoic (22:4n-6; DTA) acids. It has been proposed that the functional significance of this n-6/n-3 dichotomy in the fatty acid profiles of avian and mammalian spermatozoa may represent an adaptation to temperature (Kelso *et al.*, 1997). The higher body temperature of birds (41°C) compared with that of mammals (37°C), in tandem with the fact that avian testes are internalized, requires avian spermatozoa to develop and function in a considerably warmer environment. The major poly-unsaturated fatty acid (PUFA) of avian spermatozoa, DTA, displays the same chain length but fewer double bonds than the mammalian PUFA counterpart DHA; this difference in the degree of polyunsaturation provides a means for maintaining the appropriate biophysical properties of sperm membranes at the different body temperatures.

An interesting feature of the poultry sperm membrane is that neither chicken nor turkey spermatozoa appear to require capacitation (Howarth, 1970; Howarth and Palmer, 1972; Bednarczyk, 1990). Intramagnal insemination of spermatozoa originating from the testis, epididymis and vas deferens resulted in fertility levels of 85–90%; in contrast, vaginal insemination of testicular spermatozoa resulted in a total absence of fertile eggs (Howarth, 1983). In intramagnal insemination of the female bird, spermatozoa are deposited near the site of fertilization, thus bypassing the vagina and the uterovaginal junction, where sperm motility is essential for transport. In mammalian spermatozoa, physiological capacitation occurs in the female reproductive tract and can be induced *in vitro* using species-specific conditions. In either circumstance, capacitation results in membrane destabilization, and this facilitates the acrosome reaction. In contrast, the acrosome reaction of chicken spermatozoa may be induced very rapidly *in vitro* in the presence of the inner perivitelline layer of the egg, or of *N*-linked glycans (derived from the inner perivitelline layer) and extracellular Ca^{2+} (Horrocks *et al.*, 2000). More recently, it was shown that the acrosome reaction occurred within the first minute of incubation with the inner perivitelline layer and Ca^{2+}, with the highest percentages of acrosome reacted spermatozoa occurring after 4–10 min of incubation (Lemoine *et al.*, 2008). The absence of a capacitation-like process before the initiation of the acrosome reaction makes sense in the context of the prolonged stay of poultry spermatozoa in the sperm storage tubules of the hen's reproductive tract, as discussed later in the chapter.

Sperm metabolism and motility

Although the gross morphology of turkey and chicken spermatozoa is similar, ultrastructural differences in the midpiece region appear to be indicative of the different biochemical features of spermatozoa from these two species. The main distinction in energy production is that turkey spermatozoa require oxygen to produce ATP, but chicken spermatozoa can metabolize using either aerobic or anaerobic respiration (Sexton, 1974). Both turkey and chicken spermatozoa use the Embden–Meyerhof–Parnas pathway of glycolysis and the Krebs cycle for oxidative metabolism (Kraft *et al.*, 1978). Turkey spermatozoa have a low glycolytic capacity and require the presence of oxygen for optimal ATP production (Lake and Wishart, 1984). Under anaerobic conditions, chicken spermatozoa form lactate from glucose 13 times faster than turkey spermatozoa (Sexton, 1974; Wishart, 1982) owing, in part, to the 20-fold higher activity of lactate dehydrogenase in chicken spermatozoa (McIndoe and Lake, 1973). Profiles of glycolytic enzymes are similar for spermatozoa from both species; however, chicken sperm enzyme activities are twofold to fourfold higher than those of turkey sperm, with glycerophosphate mutase and lactate dehydrogenase activities 9.5 and 41 times greater, respectively (Wishart and Carver, 1984). Glutamate and glycine are not used by either species as oxidizable substrates, while acetate and succinate increase the rate of oxygen consumption in chicken spermatozoa only (Sexton, 1974).

At the physiological temperature of 40°C, chicken spermatozoa maintain high ATP levels and fertilizing potential for up to 3 h by the aerobic or anaerobic metabolism of glucose (Wishart, 1982). When chicken and turkey spermatozoa were incubated aerobically without glucose, the rate of ATP hydrolysis (a measure of energy utilization) and the rate of oxygen consumption (a measure of the rate of energy production) decreased by 75–80% as the temperature was lowered from 40 to 5°C (Table 9.1). At each temperature, decreases in the rates of oxygen consumption and ATP hydrolysis were equivalent, suggesting a coupling of ATP production to oxygen metabolism at all temperatures, as the sperm ATP levels remained constant (Wishart, 1984). This was not the case with turkey spermatozoa, which, under conditions of anaerobic glycolysis at temperatures between 5 and 40°C, contained <5% of the ATP concentrations exhibited under aerobic conditions (Table 9.1). After storing chicken semen at 5°C under anaerobic conditions for 48 h, no differences in the activity of the glycolytic enzymes aldolase or lactate dehydrogenase were evident, whereas the activities of the two Krebs cycle enzymes fumerase and aconitase were lower in stored than in fresh spermatozoa (Buckland, 1971).

It has been reported that semen ejaculated from the broiler rooster is in an anaerobic state, as there was no free oxygen available in undiluted semen (Parker and McDaniel, 2006). Spermatozoa recovered from different anatomical regions of the male layer rooster reproductive tract had markedly different degrees of sperm motility that corresponded to the degree of sperm maturation: testicular

Table 9.1. ATP concentration in chicken and turkey spermatozoa incubated under aerobic or anaerobic conditions at various temperatures. Adapted from Wishart, 1984.

	ATP concentration (nmol/10^9 spermatozoa)			
	Chicken spermatozoa		Turkey spermatozoa	
Incubation temperature (°C)	Anaerobic conditions	Aerobic conditions	Anaerobic conditions	Aerobic conditions
5	49.0 ± 4.0	44.0 ± 4.0	1.9 ± 0.1	48.0 ± 2.0
10	49.0 ± 3.0	47.0 ± 3.0	1.8 ± 0.7	48.0 ± 4.0
20	48.0 ± 3.0	45.0 ± 3.0	1.6 ± 0.6	50.0 ± 4.0
30	51.0 ± 4.0	48.0 ± 2.0	2.2 ± 0.3	57.0 ± 3.0
40	49.0 ± 5.0	47.0 ± 3.0	1.8 ± 0.3	49.0 ± 2.0

spermatozoa demonstrated the lowest motility, followed by a gradual increase in sperm motility from the epididymal region to the distal ductus deferens (Ahammad *et al.*, 2011). There was no difference in sperm motility between spermatozoa obtained from the distal ductus deferens and ejaculated spermatozoa.

Spermatozoa from some avian species exhibit a unique sperm motility phenomenon of reversible temperature-dependent immobilization (Ashizawa *et al.*, 2010) in which, in simple salt solutions, spermatozoa become immotile at body temperature (40–41°C) but regain motility when the temperature is lowered to 30°C or by the addition of Ca^{2+} at 40°C (Munro, 1938; Wishart and Ashizawa, 1987; Ashizawa *et al.*, 1989; Thomson and Wishart, 1991; Ashizawa *et al.*, 1994). This pattern of temperature-dependent motility occurs in chicken, duck and copper pheasant spermatozoa, but occurs only partially in turkey and not at all in quail or Houbara bustard spermatozoa (Wishart and Wilson, 1999; Ashizawa *et al.*, 2010).

Turkey sperm motility was only partially inhibited at 40°C, while quail and bustard spermatozoa showed no reduction in the percentage of motile spermatozoa at 40°C (Wishart and Wilson, 1999). Ca^{2+} restored the motility of chicken and duck spermatozoa at 40°C and also released the partial inhibition of turkey sperm motility; however, Ca^{2+} had no apparent effect on the proportion of motile bustard spermatozoa and completely inhibited quail sperm motility at 40°C (Wishart and Wilson, 1999). The axoneme and/or accessory cytoskeletal components appear to be directly involved in the temperature-dependent regulatory system, because the motility of de-membranated spermatozoa is similar to that of intact spermatozoa, being negligible at 40°C and restored at 30°C (Ashizawa *et al.*, 2000). This temperature-dependent motility has been shown to be a function of intracellular Ca^{2+} content; intracellular levels of Ca^{2+} were reduced at the higher temperatures, where spermatozoa were immotile, and increased as the temperature was lowered from 40 to 30°C (Thomson and Wishart, 1991). Further, the temperature-dependent retention of Ca^{2+} was a function of the rate of Ca^{2+} efflux rather than influx, and involved a Ca^{2+} ATPase that was relatively inactive at 30°C, but active at 40°C (Thomson and Wishart, 1991). Chicken sperm motility was inhibited by loading with an intracellular Ca^{2+} chelator, and subsequently reactivated by the addition of excess Ca^{2+} to the medium (Ashizawa *et al.*, 1994). Not all avian species demonstrate this temperature-dependent sperm motility phenomenon.

In the chicken and turkey, a sperm motion attribute known as sperm mobility has been positively correlated with fertility in poultry (Donoghue *et al.*, 1999; Froman *et al.*, 1999; King *et al.*, 2000b). Sperm mobility is defined as the net movement of a sperm cell population against resistance at body temperature (Froman and Kirby, 2005), and is assayed by measuring the absorbance of a solution of Accudenz (a cell separation medium) following sperm 'swim down' from an overlaid sperm suspension (Froman and Feltmann, 1998).

Absorbance is proportional to the number of spermatozoa within the sample that have a straight-line velocity of, or greater than, 30 μm/s (Froman and Feltmann, 2000). Simply put, all mobile sperm are necessarily motile, whereas motile sperm are not necessarily mobile (Froman and Kirby, 2005).

When the sperm mobility assay was applied to populations of randomly bred broiler roosters, the phenotype varied among males (Froman *et al.*, 1997), was independent of age (Froman *et al.*, 1999) and determined male fecundity (Froman and Feltmann, 1998). Subsequently, the heritability of this quantitative trait was estimated (Froman *et al.*, 2002), and it was shown that mitochondrial dysfunction accounted for the phenotypic variation (Froman and Kirby, 2005; Froman *et al.*, 2006). In the turkey, the high sperm mobility phenotype was characterized by a straight-line velocity of 45.5 μm/s and an average path velocity of 60.3 μm/s, compared with 28.8 and 45.1 μm/s, respectively, in males exhibiting the low sperm mobility phenotype (Donoghue *et al.*, 1998). Like the sperm mobility phenotype established for the chicken, that in turkeys also varies characteristically among males (King *et al.*, 2000a), is independent of age (Holsberger *et al.*, 1998) and influences paternity (Donoghue *et al.*, 1999). The turkey sperm mobility phenotype does not, however, predict the outcome of sperm function after 24 h of storage at 4°C, as sperm mobility values decline to similar levels irrespective of the male phenotype (Long and Kramer, 2003). Finally, male turkeys characterized as either high or low sperm mobility phenotype exhibit striking differences in the content of carbohydrates in the glycocalyx, including mannose/glucose, *N*-acetylglucosamine and *N*-acetylgalactosamine (Peláez and Long, 2008).

Seminal Plasma

The components of poultry seminal plasma are thought to be derived from the extragonadal duct system within the epididymis, primarily the proximal efferent duct, which appears to be where spermatozoa are mixed with secretions as they are concentrated (Hess *et al.*, 1976; Aire, 1979; al-Aghbari *et al.*, 1992). During natural mating in chickens, a lymph-like, transparent fluid is mixed with the ejaculate. This transparent fluid, derived from the paracloacal vascular bodies (Fujihara, 1992), is chemically distinct from deferent duct fluid (Lake, 1966, 1984). In turkeys, a frothy fluid is secreted from the triangular fold of the cloaca and mixed with the ejaculate during copulation (Fujihara *et al.*, 1985). The chemistry of the frothy fluid resembles that of blood plasma, rather than that of seminal plasma (Fujihara and Nishiyama, 1984). As outlined below, the biochemistry of seminal plasma differs markedly from that of blood plasma. Thus it is not surprising that prolonged exposure of turkey spermatozoa to this frothy fluid adversely affects sperm morphology (Fujihara *et al.*, 1987).

Electrolytes

Compared with mammalian livestock, data on the content of the principal inorganic ions (Na^+, K^+, Cl^-, Ca^{2+}, Mg^{2+}) in the seminal plasma of chicken and turkey semen are sparse, and most reports date from the 1960s (Table 9.2). As in mammals, a reciprocal relationship exists between the amounts of Na^+ and K^+ in avian spermatozoa and seminal plasma, with seminal plasma containing higher Na^+ and lower K^+ concentrations than spermatozoa and, conversely, spermatozoa containing lower Na^+ and higher K^+ than seminal plasma. Chicken seminal plasma, though, contains more Na^+ and less K^+ than has been reported for ram, bull or human seminal plasma (Quinn *et al.*, 1965; Mann and Lutwak-Mann, 1981). Newer results reported by Karaca *et al.* (2002) for inorganic ion concentrations in chicken seminal plasma fell within similar ranges to those in older reports (Hammond *et al.*, 1965; Quinn *et al.*, 1965; El Jack and Lake, 1969; Lake, 1971; Lake and Wishart, 1984), although a recent paper where semen from aged broilers was evaluated after an induced moult reported markedly lower concentrations of Na^+, K^+ and Ca^{2+} in seminal plasma (Khan *et al.*, 2012) than did previous reports (Table 9.2).

Table 9.2. Concentrations of the principal inorganic ions in chicken and turkey seminal plasma (mg/100 ml).

Species	Na^+	K^+	Ca^{2+}	Mg^{2+}	Cl^-	Reference
Chicken	337.0	44.8	7.0	–	497.3	Hammond *et al.*, 1965
	420.0	32.0	8.7	6.2	–	Quinn *et al.*, 1965
	365.2	50.3	5.2	6.2	148.3	El Jack and Lake, 1969
	420.0	32.0	8.7	6.2	190–220	Lake, 1971
	333.5	50.7	5.6	5.6	162.8	Lake and Wishart, 1984
	308.2	39.0	4.0	1.2	237.2	al-Aghbari *et al.*, 1992
	267.9	49.5	4.8	–	215.9	Karaca *et al.*, 2002
	1.2	21.1	1.5	5.5	–	Khan *et al.*, 2012
Turkey	361.1	113.1	–	–	–	Brown *et al.*, 1959
	243.0	79.0	–	7.5	–	Cherms, 1967
	383.1	65.5	4.4	–	108.7	Graham *et al.*, 1971
	312.8	69.8	1.2	10.5	81.4	Lake and Wishart, 1984

Fewer data are available for turkey seminal plasma ion concentrations than for those of chickens, but all reports indicate appreciably higher concentrations of K^+ and lower concentrations of Cl^- in turkey seminal plasma (Brown *et al.*, 1959; Cherms, 1967; Graham, *et al.*, 1971; Lake and Wishart, 1984) than in chicken seminal plasma (Table 9.2).

Amino acids and peptides

Glutamic acid is the principal amino acid in poultry seminal plasma, and is thought to serve as the main anion in place of Cl^- (El Jack and Lake, 1969). Lake and McIndoe (1959) reported that glutamic acid comprises 90% of the free amino acids found in chicken seminal plasma; similarly, a high proportion of glutamic acid is found in turkey seminal plasma (Ahluwalia and Graham, 1966b; Graham *et al.*, 1971). The concentration of glutamic acid in poultry seminal plasma is ten times the total amino acid concentration in bull semen (Lake and McIndoe, 1959), and several 100 times greater than that found in blood plasma (Lake and Wishart, 1984). The functional significance of glutamate, other than in establishing osmolality, is unclear, as poultry spermatozoa lack adequate metabolic capabilities to use glutamate as an energy source (Etches, 1996).

Of the 20 recognized amino acids (excluding glutamic acid), arginine, asparagine, threonine and glycine are the most prevalent in chicken seminal plasma, whereas arginine, asparagine, aspartic acid and serine are the most prevalent in turkey seminal plasma (Ahluwalia and Graham, 1966b). Graham *et al.* (1971) determined that the semen collection frequency in turkeys influenced the amount of free amino acids in seminal plasma; the levels of most amino acids (except for glutamic acid) were higher when semen was collected at 2 day intervals than at 4 or 7 day intervals. It appears that the seminal plasma amino acid profile also varies among genotypes. Froman and Bernier (1987) identified a heritable reproductive disorder in the male rooster in which spermatozoa degenerate prematurely. The inheritance of this trait was attributed to a single dominant gene (Froman *et al.*, 1990), and mutant roosters were characterized by a reduced surface-to-volume ratio within their proximal efferent ducts (Kirby *et al.*, 1990) and a reduced level of glutamic acid in seminal plasma compared with normal roosters (al-Aghbari *et al.*, 1992).

Carnitine plays a vital role in transporting long chain fatty acids across the inner mitochondrial membrane to produce energy through β-oxidation (Bremer, 1983), and may also function as an antioxidant to scavenge free radicals (Agarwal and Said, 2004). Both carnitine and acetyl carnitine are present in poultry seminal plasma (Golan *et al.*, 1982; Lake and Wishart, 1984). While the levels of carnitine (3.2 mmol) and acetyl carnitine (2.1 mmol) are comparable to those reported for mammalian semen (Mann and Lutwak-Mann, 1981), turkey seminal plasma contains about half (1.7 and 0.64 mmol, respectively) of the levels found in chicken seminal plasma (Lake and Wishart, 1984).

Carbohydrates and polyols

Fructose is the principal unbound sugar of seminal plasma for a number of mammalian species (Mann and Lutwak-Mann, 1981), but only trace amounts of fructose are found in poultry seminal plasma (Kamar and Rizik, 1972; Lake and Wishart, 1984; Garner and Hafez, 1993). A few older reports document glucose levels of 80–88 mg/dl in chicken seminal plasma (Mann, 1964; Hammond *et al.*, 1965), though only 3.1 mg/dl was detected by Ahluwalia and Graham (1966a); rather, the primary carbohydrate isolated from chicken seminal plasma in that report was inositol (20.4 mg/dl), along with low amounts of glycerol (2.8 mg/dl). Wishart (1982) concluded that chicken and turkey semen did not contain glycolytic substrates, as no lactate was produced when semen was diluted in buffer without glucose and incubated under anaerobic conditions. A more recent report documented the presence of free oligosaccharides in human seminal plasma – the first report for any species (Chalabi *et al.*, 2002); however, the function of these oligosaccharides remains unknown.

Lipids and phosphodiesters

The lipid composition of poultry seminal plasma has been investigated much more recently than the free carbohydrate or amino acid contents, and differs considerably from the lipid content of poultry spermatozoa. The total lipid content of poultry seminal plasma generally has a lower proportion of phospholipids than are found in spermatozoa and, within the phospholipid class, a lower proportion of phosphatidylcholine (Douard *et al.*, 2000; Zaniboni and Cerolini, 2009). In contrast to spermatozoa, there are significant amounts of cholesterol esters and triglycerides in seminal plasma (Douard *et al.*, 2000). There appear to be species-specific differences in seminal plasma lipids among turkeys, layer chickens and broiler chickens (Table 9.3), a finding reinforced by the results from one laboratory that compared the seminal plasma lipid profiles from all three poultry groups (Cerolini *et al.*, 1997); in this study, the mean total lipid content of turkey, layer chicken and broiler chicken seminal plasma at peak semen production was 1.3, 1.5 and 0.3 mg/ml, respectively. Age of the male may influence the lipid content of seminal plasma,

Table 9.3. Lipid composition of seminal plasma from turkey, broiler chicken and layer chicken semen collected at different ages.

	Turkey		Layer chicken	Broiler		
Lipid category/Age	40 wk[a]	42 wk[b]	23 wk[b]	25 wk[c]	26 wk[b]	60 wk[c]
Total lipid (%)						
Free cholesterol	31.1	48.8	16.6	19.9	32.5	16.0
Cholesterol esters	12.3	6.1	8.8	10.9	9.7	20.8
Triglycerides	6.8	2.1	3.0	22.0	2.3	20.0
Free fatty acids	ND[d]	26.0	34.5	12.9	8.6	14.3
Phospholipids	50.7	40.5	46.2	34.2	46.9	28.1
Phospholipid (%)						
Phosphatidylcholine	4.6	NM[e]	NM	9.8	NM	NM
Phosphatidylethanolamine	31.5	NM	NM	52.6	NM	NM
Sphingomyelin	20.3	NM	NM	19.4	NM	NM
Phosphatidylinositol (PI)/ phosphatidylserine (PS)	14.9	NM	NM	PI ND/PS 18.2	NM	NM
Phosphatidic acid/ diphosphatidylglycerol	8.2	NM	NM	NM	NM	NM
Lysophosphatidylcholine	20.2	NM	NM	NM	NM	NM

[a]Douard *et al.*, 2000; [b]Cerolini *et al.*, 1997; [c]Kelso *et al.*, 1996; [d]not detected; [e]not measured.

as the total lipid content increased from 0.14 to 1.1 mg/ml in broiler males aged 25 and 60 weeks, respectively (Kelso *et al.*, 1996).

Two of the six non-cyclic phosphodiesters are found in semen: glycerol phosphorylcholine in mammalian seminal plasma and serine ethanolamine phosphodiester in avian seminal plasma (Burt and Ribolow, 1994). In chicken seminal plasma, serine ethanolamine phosphodiester was shown to be 23–26% of the total observable phosphate in three males, but in two other males, it was only 7–9% (Burt and Chalovich, 1978). Further, turkey semen collected in the spring showed 'substantial levels' of serine ethanolamine phosphodiester (Burt and Ribolow, 1994), which decreased as males began to stop semen production in late summer (Ribolow and Burt, 1987). Phosphodiesters are a class of compounds that may function as lysophospholipase inhibitors in semen, a role that would decrease the rate of membrane phospholipid turnover and lead to a net sparing of phospholipids (Burt and Ribolow, 1994).

Enzymes and other proteins

Chicken seminal plasma contains 52–77% protein (Blesbois and Hermier, 1990) or 2.0–2.4 g/dl (Harris and Sweeney, 1971; Amen and Al-Daraji, 2011), while turkey seminal plasma averages 1.8 g protein/dl (Thurston *et al.*, 1982). Many proteins in poultry seminal plasma have been identified as enzymes. Autoproteolytic activity has been detected in the seminal plasma of the domestic chicken (Droba, 1986), and serine proteinase inhibitors are present in chicken seminal plasma, with their proposed function being the inactivation of acrosin that is released from dead or damaged spermatozoa (Lessley and Brown, 1978). Turkey seminal plasma contains a unique serine proteinase not found in the chicken or guinea fowl, and the amidase activity of turkey seminal plasma is 23–28 times greater than for chicken or guinea fowl (Thurston *et al.*, 1993). The turkey seminal plasma serine proteases were found to be distinct from the sperm enzyme acrosin (Thurston *et al.*, 1993). The biological roles of these enzymes in birds are unknown as, unlike in mammals, turkey seminal plasma serine proteases cannot participate in semen liquefaction or coagulation because these processes do not occur in birds (Holsberger *et al.*, 2002). More recent research has shown that turkey seminal plasma contains at least three trypsin inhibitors, providing further evidence for increased activities of serine proteinases as a unique characteristic of turkey seminal plasma (Kotłowska *et al.*, 2005).

A number of glycosidases are present in poultry seminal plasma, including β-*N*-acetylglucosaminidase, the α and β forms of mannosidase, galactosidase and glucosidase, and β-glucuronidase (Kannan, 1974; McIndoe and Lake, 1974; Droba and Droba, 1987; Droba and Dzugan, 1993), which hydrolyse a variety of compounds containing terminal, non-reducing carbohydrate residues. Among the glycolytic enzymes present in whole semen, β-*N*-acetylglucosaminidase was the most active, with α-mannosidase ranking next in activity, whereas the activities of β-mannosidase, both forms of galactosidase and glucosidase, and β-glucuronidase, were almost negligible (Kannan, 1974). The activity of acid phosphatases, which catalyse the hydrolysis of phosphate esters, is distinctively high in human and poultry seminal plasma (Wilcox, 1961; Bell and Lake, 1962; McIndoe and Lake, 1974; Hess and Thurston, 1984; Dumitru and Dinischiotu, 1994). The presence of several phospholipases have been confirmed in turkey seminal plasma, including phospholipase A2, phospholipase A1 and lysophospholipase; these have been linked with sperm membrane phospholipid loss during hypothermic semen storage (Douard *et al.*, 2004).

Several antioxidant enzymes are found in poultry seminal plasma, including glutathione peroxidase, superoxide dismutase, paraoxonase, arylesterase and ceruloplasmin (Surai *et al.*, 1998; Khan *et al.*, 2012). The relative activity of superoxide dismutase in the seminal plasma of five different avian species has been ranked as guinea fowl > chicken > goose > duck > turkey; conversely, glutathione peroxidase activity is the highest in the turkey and lowest in the duck and goose (Surai *et al.*, 1998). A potent phospholipid mediator, platelet activating factor

(PAF), has been detected in mammalian spermatozoa from several species (Kumar *et al.*, 1988; Kuzan *et al.*, 1990; Minhas *et al.*, 1991). Platelet activating factor acetylhydrolase, which inactivates PAF, may represent a mechanism for regulating sperm-derived PAF and has been detected in chicken seminal plasma, although the acetylhydrolase activity was lower in rooster seminal plasma than in that of the bull, stallion and rabbit (Hough and Parks, 1994).

Transport proteins, such as albumin, comprise the majority of the protein content of poultry seminal plasma (Blesbois and Caffin, 1992). For example, over 60% of the protein spots from 2D SDS-PAGE gels for chicken and turkey seminal plasma have been identified as serum albumin precursor, ovalbumin or ovotransferrin (J.A. Long, unpublished data). Ovotransferrin is mainly known as an iron-binding glycoprotein found in the egg white (Kurokawa *et al.*, 1999), but it has recently been shown to be a multifunctional protein with a major role in avian natural immunity (Giansanti *et al.*, 2012). Ovalbumin also has been traditionally associated with the egg white and development of fertilized eggs (Smith and Back, 1962; Sugimoto *et al.*, 1999). Serum albumin is the main protein of blood plasma, in which a high binding capacity for water, Ca^{2+}, Na^{+}, K^{+} and fatty acids contributes to the main function of regulating the colloidal osmotic pressure of blood. It seems highly likely that serum albumin precursor has a similar role in seminal plasma.

High density lipoproteins (HDLs) have been detected in chicken seminal plasma; however, the low concentration and the lack of larger lipoproteins suggest that seminal HDLs originate from blood plasma and pass through the blood–testis barrier (Blesbois and Hermier, 1990). As cholesterol carriers, HDLs allow cellular cholesterol efflux (Eisenberg, 1984), and in large amounts could impair the stability of the poultry sperm membrane. Triglyceride-rich lipoproteins (very low, intermediate and low density lipoproteins) were not detected in chicken seminal plasma (Blesbois and Hermier, 1990), but Long and colleagues (unpublished) have identified apovitellenin-1, a low density lipoprotein (LDL), in chicken seminal plasma (Table 9.4). Additionally, these workers identified nine other binding/transport proteins in chicken and/or turkey seminal plasma; vitamin D-binding protein precursor, beta-2-glycoprotein 1, alpha-1-acid glycoprotein 2-like, phosphatidylethanolamine-binding protein 1, nuclear transport factor 2-like protein, calmodulin, retinol binding protein, transthyretin and gamma-glutamyltranspeptidase 1 (Table 9.4). To date, they have identified 29 proteins in the seminal plasma of these two species, with five proteins in common (Table 9.4). One protein common to both chicken and turkey seminal plasma is PIT54 protein precursor; PIT54 is a soluble member of the family of scavenger receptor cysteine-rich proteins whose gene exists only in birds (Wicher and Fries, 2006), and may have an antioxidant role in seminal plasma.

Semen Collection and Evaluation

Semen collection methods for commercial poultry were developed at the Beltsville Agricultural Research Center (Burrows and Quinn, 1937). The method used to manually stimulate ejaculation in gallinaceous birds does not resemble natural mating. The male is placed on his breast and his legs are gently restrained at approximately right angles to the body. The semen collector firmly 'massages' the male's abdomen with one hand while concurrently stroking firmly the back and tail feathers with the other hand. In a well-trained male, this stimulation will cause phallic tumescence within a few seconds. The semen collector then applies gentle pressure to the cloaca, squeezing downwards and inwards with one hand and, at the same time, squeezing upwards and inwards with the hand situated just below the cloaca. This action, referred to as a 'cloacal stroke', results in a partial ejaculation or expulsion of semen from the ductus deferens without obtaining transparent fluid. A second cloacal stroke is required to complete the ejaculation process. No more than two cloacal strokes are used to collect semen, as further strokes may injure the cloaca (Cecil and Bakst, 1985).

Table 9.4. Proteins identified by mass spectrophotometry in the seminal plasma of turkeys and/or chickens. From Long and colleagues (unpublished data).

Protein	Score	Molecular wt	pI	General function	Species
Apovitellenin-1	135	12,016	9.17	Lipoprotein	Chicken
Chain A, chicken calmodulin	354	16,679	4.04	Transport	Chicken/Turkey
Chain A, chicken plasma retinol-binding protein	175	20,409	5.62	Transport	Turkey
Chain A, transthyretin	545	14,209	5.10	Transport	Turkey
Gallinacin-9	144	7,616	8.65	Beta-defensin family	Chicken
Lysozyme C	42	16,741	9.37	Enzyme	Chicken
Ovalbumin	455	43,196	5.19	Storage	Chicken
Ovoglycoprotein	248	22,535	5.11	Transport	Turkey
Ovoinhibitor precursor	205	54,394	6.16	Enzyme	Chicken
Ovotransferrin	1,003	79,552	6.62	Transport	Chicken/Turkey
Phosphatidylethanolamine-binding protein 1	44	21,115	6.96	Transport	Chicken
PIT54 protein precursor	376	52,670	4.61	Antioxidant	Chicken/Turkey
PREDICTED: alpha-1-acid glycoprotein 2-like	460	22,369	5.38	Lipoprotein	Turkey
PREDICTED: apolipoprotein A-I-like	139	13,473	8.87	Lipoprotein	Turkey
PREDICTED: astacin-like metalloendopeptidase-like	157	47,177	6.79	Metalloprotease	Turkey
PREDICTED: beta-2-glycoprotein 1	75	40,069	8.6	Transport	Chicken
PREDICTED: cysteine-rich secretory protein 3-like	322	29,625	6.29	Unknown	Turkey
PREDICTED: gamma-glutamyltranspeptidase 1	104	61,757	6.3	Amino acid transport	Chicken
PREDICTED: haemopexin-like, partial	499	21,673	5.57	Iron binding	Turkey
PREDICTED: hepatocyte growth factor activator	227	62,008	6.96	Unknown	Turkey
PREDICTED: hypothetical protein (*Gallus gallus*)	87	34,637	9.34	Unknown	Chicken
PREDICTED: nuclear transport factor 2-like	175	14,565	5.10	Transport	Turkey
PREDICTED: similar to Rsb-66 protein isoform 1	140	19,531	6.58	Unknown	Chicken
PREDICTED: trypsin inhibitor CITI-1	147	9,405	7.53	Enzyme	Chicken/Turkey
Scavenger receptor cysteine-rich domain-containing protein	93	50,311	4.72	Unknown	Chicken
Serum albumin precursor	218	71,868	5.51	Transport	Chicken/Turkey
Superoxide dismutase (Cu-Zn)	70	15,579	6.1	Enzyme	Chicken
Tyrosine/tryptophan monooxygenase activation protein	262	27,810	4.73	Enzyme	Chicken
Vitamin D-binding protein precursor	358	55,362	6.47	Transport	Chicken

Semen is collected either directly into a conical tube or aspirated into a vial via a mouth pipette. Aspirated semen is often passed through a filter to prevent faecal and/or urate contamination of the semen. Because chicken and turkey semen are characterized by low collection volumes (~0.2–1.0 ml) and high concentrations of spermatozoa (chicken, 6×10^9 spermatozoa/ml; turkey 9×10^9 spermatozoa/ml), semen must be pooled from multiple males to achieve the volumes needed for AI of multiple hens. This requires the semen collector to perform an initial evaluation of semen quality by only pooling semen that is thick and pearly white or cream coloured. Semen that is watery in consistency typically has low sperm concentration. In turkeys, yellow semen will lower the fertilizing potential of the entire semen pool (Kotłowska *et al.*, 2005; Christensen *et al.*, 2010).

Approximately 10% of the commercial turkey population has a condition known as yellow semen syndrome (Thurston and Korn, 1997). Semen from these males has similar volume and sperm concentration to that of normal turkey semen, but yellow semen syndrome has been linked with reduced fertility and hatchability (Saeki and Brown, 1962; Thurston *et al.*, 1992). Seminal plasma from turkeys with yellow semen syndrome is characterized by an elevated protein concentration ~7 g/dl, compared with 1.8 g/dl in normal semen (Thurston *et al.*, 1982). Seminal plasma from these males also shows increased activities of androgen-binding protein (Hess *et al.*, 1984), aspartate aminotransaminase and acid phosphatase, as well as elevated cholesterol (Hess and Thurston, 1984). Yellow semen also contains morphologically abnormal spermatids and high levels of macrophages or spermiophages (Thurston *et al.*, 1975). Histological studies revealed that, in yellow semen syndrome males, the ductuli efferentes epithelia were hypertrophied and filled with lipid-like droplets (Hess *et al.*, 1982); ultrastructural studies showed evidence of cholesterol clefts and excessive apocrine secretion in the epididymal region (Hess *et al.*, 1982). To date, the exact cause of yellow semen syndrome in turkeys is unknown, and breeders typically cull males exhibiting this syndrome from the breeding population.

Sperm concentration

The colour of poultry semen is generally an indication of the density of the ejaculate, ranging from opaque white with a relative high sperm density to clear with declining sperm numbers (Parker and McDaniel, 2006). There are three basic methods used by producers to estimate sperm concentration more precisely: (i) direct counting using a haemocytometer; (ii) the spermatocrit method; and (iii) estimation from optical density.

Direct counting of spermatozoa in a dilute solution using a haemocytometer is the most accurate method, but this method is time-consuming on a commercial scale. Haemocytometer counts are most often used in the one-time preparation of standard curves to correlate the indirect density or volume readings, obtained from instruments such as spectrophotometers or capillary centrifuges, with actual numbers of spermatozoa.

Most producers use the spermatocrit method to estimate sperm concentration. Based on the same principle as determining blood haematocrits, high-speed centrifugation of a sample of poultry semen in a capillary tube will cause the sperm and other cells to become tightly packed. This packed cell volume is then converted to sperm concentration using either a conversion factor or a standard curve, both of which are previously derived by comparing and graphically plotting haemocytometer counts against corresponding packed cell volumes (Bakst, 2010a). The disadvantages of the spermatocrit are that: (i) a high number of other cells (e.g. bacteria, blood cells) can falsely elevate the packed cell volume; and (ii) for semen samples with very low sperm concentrations, the volume of packed cells is poorly correlated with the number of spermatozoa in the sample (Etches, 1996).

Because there is a high correlation between the transmission of light through semen and the concentration of spermatozoa obtained by direct counting (Brillard and McDaniel, 1985), the optical density of a highly diluted semen sample provides an indirect estimate of the sperm concentration when compared with an in-house-generated standard curve, or using the line equation supplied by the manufacturer of the photometer (Long, 2010).

Sperm plasma and acrosome membrane integrity

Sperm viability is determined by measuring the integrity of the sperm membrane with colorimetric or fluorescent stains to provide a ratio of live (e.g. membrane intact) to dead (e.g. membrane compromised) spermatozoa. The most common viability stain for poultry spermatozoa is nigrosin–eosin, which is based on the principle that live spermatozoa have intact membranes that are capable of excluding the eosin stain (Bakst, 2010b); the nigrosin stain provides background contrast to distinguish clear or non-stained spermatozoa (e.g. viable sperm) from purple-stained spermatozoa (non-viable sperm). The size of avian spermatozoa requires the use of an oil immersion lens to accurately distinguish between viable and non-viable cells. An alternative dual stain for poultry spermatozoa that can be used with either a fluorescent microscope (Chalah and Brillard, 1998) or a flow cytometer (Long and Kramer, 2003) uses a combination of SYBR-14® and propidium iodide (PI). SYBR-14® is a membrane permeant (permeating) dye that binds nucleic acids and exhibits bright green fluorescence. PI is impermeant to intact membranes. Upon crossing damaged membranes, PI intercalates with DNA and spermatozoa to exhibit a bright red fluorescence.

Acrosome membrane integrity has been used less frequently than plasma membrane integrity for evaluating semen quality, despite the potential for acrosome membrane disruption during hypothermic semen storage. As in mammals, acrosome-reacted poultry spermatozoa can be detected using fluorescein isothiocyanate-conjugated peanut agglutinin (Donoghue and Walker-Simmons, 1999; Horrocks *et al.*, 2000; Ashizawa *et al.*, 2004). The acrosome of poultry spermatozoa can also be induced to react with inner perivitelline layer fragments isolated from freshly laid eggs (Lemoine *et al.*, 2011), with inner perivitelline derived *N*-linked glycans and extracellular Ca^{2+} (Horrocks *et al.*, 2000), or with inner perivitelline layer fragments and deltamethrin or fenvalerate, specific inhibitors of protein phosphatase type 2B (Ashizawa *et al.*, 2004). The acrosome reaction in poultry spermatozoa can be induced rapidly (within 1 min); however, maximal results were obtained after 4–10 min of incubation (Lemoine *et al.*, 2008). Neither bovine serum albumin nor ovalbumin (its homologue in the hen oviduct), or other components known to stimulate capacitation or the acrosome reaction in mammals (such as $NaHCO_3$ or prostaglandin) stimulate the acrosome reaction in poultry spermatozoa (Lemoine *et al.*, 2008). From all reports to date, the classic hypermotility associated with capacitation/acrosome reaction in mammalian spermatozoa does not occur in poultry.

Sperm motility and mobility

The percentage of motile poultry spermatozoa in a given semen sample is difficult to quantify microscopically unless the sample is diluted to contain 1×10^9 spermatozoa/ml or less. Similarly, the use of computer assisted sperm analysis systems (CASA) requires semen samples to be diluted to contain 25×10^6 spermatozoa/ml (King *et al.*, 2000a). An instrument known as the Sperm Quality Analyzer® uses light beam interference to measure the number and amplitude of poultry sperm movements per second in a capillary tube (McDaniel *et al.*, 1998), although dilutions of chicken and turkey semen greater than fivefold and 20-fold, respectively, resulted in a linear decline in sperm quality index values (McDaniel *et al.*, 1998; Neuman *et al.*, 2002). While the percentage of motile spermatozoa in a semen sample may be high, this is not synonymous with a high percentage of spermatozoa demonstrating progressive, forward motion. In the case of poultry, sperm mobility is defined as the net movement of spermatozoa against resistance at body temperature (Froman *et al.*, 2006). Over a decade of research has shown that sperm mobility is a quantitative trait and is a primary determinant of fertility in commercial poultry (Donoghue *et al.*, 1998; Froman and Feltmann, 1998; Froman *et al.*, 1999; King *et al.*, 2000b; Bowling *et al.*, 2003). An assay has been developed for measuring sperm mobility in chicken and turkey spermatozoa (Froman and McLean, 1996; King *et al.*, 2000a) that also has been validated in boars and stallions (Vizcarra and Ford, 2006). A general schematic for the sperm mobility assay is shown (Fig. 9.2).

Sperm hydrolysing ability

Upon ovulation, the avian ovum is enveloped by a thin, fibrous investment called the inner perivitelline layer, which spermatozoa must digest to reach the germinal disc and fertilize the ovum (Bramwell and Donoghue, 2010). The sperm penetration assay provides a means to quantitatively determine the number of sperm hydrolysis points in the perivitelline layer overlying the germinal disc by dissecting and subjecting this region of the perivitelline layer to background staining with the simple carbohydrate stain of Schiff's reagent. When viewed with 40× magnification, sperm hydrolysis points are easily visualized (Fig. 9.3).

The sperm penetration assay is a predictive measure of fertility when performed with freshly laid eggs (Staines *et al.*, 1998; Kasai *et al.*, 2000; Al-Daraji, 2001; Long *et al.*, 2010). The original procedure, which was developed for chicken eggs (Bramwell *et al.*, 1995), has been adapted for turkey eggs (Donoghue, 1996). An important concept of avian fertilization is that, although typically 50 sperm hydrolysis points are observed in the germinal disc region of the inner perivitelline layer – and are indicative of what might be considered physiological polyspermy – only one spermatozoon is involved in avian syngamy. Auxiliary sperm nuclei undergo degeneration during subsequent embryonic cleavage; however, by quantifying the number of spermatozoa trapped between the inner and outer layers of the perivitelline layer, the phenomenon of physiological polyspermy can be used to predict the duration of fertility and the numbers of spermatozoa residing in the sperm storage tubules

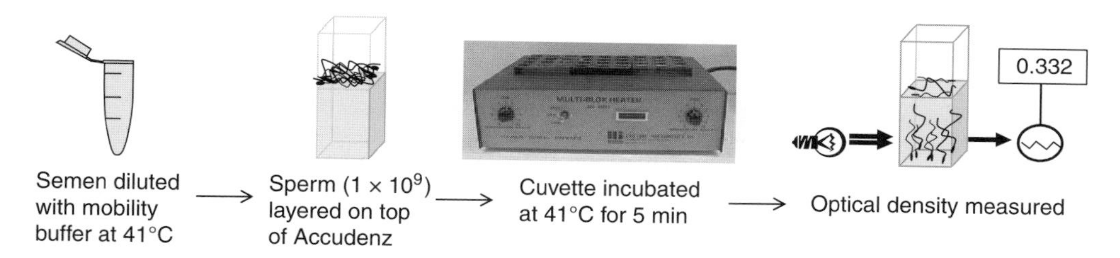

Fig. 9.2. Overview of the sperm mobility assay for poultry spermatozoa.

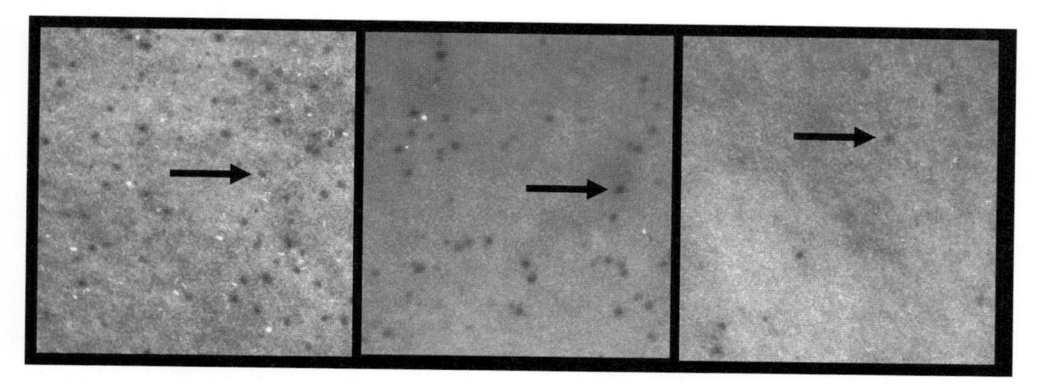

Fig. 9.3. Representative images of hydrolysis points (indicated by arrows) after 5 min of incubation of fresh (left), 6 h stored (centre) or 24 h stored (right) turkey spermatozoa with inner perivitelline layer segments from freshly laid eggs.

(Brillard and Bakst, 1990). Structurally and functionally different from the inner perivitelline layer, the outer perivitelline layer is a fibrous investment that surrounds the ovum shortly after fertilization while it is still within the infundibulum. Those spermatozoa that have not reached the inner perivitelline layer become trapped within the fibrous mesh of the outer perivitelline layer and can be visualized after staining with a fluorescent DNA-binding stain (Wishart and Bakst, 2010).

Artificial Insemination

AI of chickens and turkeys has its roots in the laboratories of the Beltsville Agricultural Research Center (Quinn and Burrows, 1936; Burrows and Quinn, 1937). In those days, poultry AI was not required for reproduction, and natural mating was commonly used for producing offspring. Later, in the 1960s, AI became a critical component of turkey reproduction (Donoghue and Wishart, 2000); then, the advent of broad-breasted turkeys ushered in the current industry standard for reproduction solely through AI. For commercial layer and broiler chickens, AI is used mainly by the primary breeders to produce elite, pedigree lines; thereafter, subsequent generations are produced by natural mating (Etches, 1996). The use of AI, in combination with semen dilution, reduces the number of males needed at each breeding level and enables a high degree of genetic selection.

Semen extenders

As soon as semen is collected from the male, a proportion of the spermatozoa begin to lose integrity and/or functionality. Synthetic diluents have been developed in an attempt to reproduce the *in vivo* environment (Christensen, 1995), and the first chicken semen extender was based on the ionic composition of the seminal plasma (Lake, 1960). There are many variations in diluent composition (Table 9.5); however, the basic goals are to maintain pH and osmolarity, as well as to provide an energy source for metabolism.

Table 9.5. Components of extenders for freshly collected chicken or turkey semen.[a]

	Sexton, 1977[b]; Lake and Ravie, 1982	Sexton, 1982[c]	Lake and Ravie, 1979	
Component (role)	Chicken	Turkey	Chicken	Turkey
Mono-potassium phosphate (buffer)	0.65	0.65	–	–
Di-potassium phosphate (buffer)	12.70	12.70	–	–
BES[d] (buffer)	–	–	30.5	18.9
TES[e] (buffer)	1.95	1.95	–	–
Sodium acetate (osmotic balance)	4.30	4.30	–	2.0
Potassium citrate (osmotic balance)	0.64	0.64	1.28	2.2
Magnesium acetate (osmotic balance)	–	–	0.8	1.05
Magnesium chloride (osmotic balance)	0.34	0.34	–	–
Fructose (metabolic substrate)	5.00	5.00	–	–
Glucose (metabolic substrate)	–	–	6.0	3.6
Sodium glutamate (chelator)	8.67	8.67	19.2	19.2
HCl (pH correction)	–	3.0 ml	–	–
NAOH (pH correction)	–	–	58.0 ml	36.0 ml
pH	7.5	6.5	7.05	7.1
Osmolarity (mOsm/kg)	333.0	350.0	411.0	406.0

[a]Listed values are g/l unless otherwise specified.
[b]Commonly known as Beltsville poultry semen extender.
[c]Commonly known as Beltsville poultry semen extender-II.
[d]*N,N*-bis 2-hydroxyethyl-2-aminoethane sulfonic acid.
[e]*N*-Tris (hydroxymethyl) methyl-2-aminoethane sulfonic acid.

The pH of the semen extender affects both metabolic rate and motility of poultry spermatozoa. For example, research has shown that an alkaline pH increases sperm motility in chickens (Holm and Wishart, 1998) and, similarly, oxygen uptake by turkey spermatozoa has been shown to increase as the pH increased from 7.3 to 7.8 (Pinto *et al.*, 1984). Buffering agents usually include a mixture of phosphates, citrates and/or organic zwitterionic molecules such as *N*,*N*-bis 2-hydroxyethyl-2-aminoethane sulfonic acid (BES) and *N*-Tris (hydroxymethyl) methyl-2- aminoethane sulfonic acid (TES) (Donoghue and Wishart, 2000).

Poultry spermatozoa maintain fertilizing ability across a wide range of osmolarities (Christensen, 1995), although the currently recommended osmolarity of semen extenders is approximately 325–350 mOsm/kg (Graham and Brown, 1971). Under hypo-osmotic conditions, an increased incidence of spermatozoa with bent necks is observed (Clark *et al.*, 1984) – a defect that is frequently found in diluted chicken and turkey semen, and which is negatively correlated with fertility (Donoghue and Wishart, 2000).

Chicken and turkey spermatozoa utilize fructose more efficiently than glucose or inositol (McIndoe and Lake, 1973; Sexton, 1974), so fructose has been the primary sugar in most semen extenders. The high concentration of glutamate and the low incidence of Cl^- in seminal plasma from poultry have prompted inclusion of these components at similar rates in most semen extenders (Etches, 1996).

Sperm storage tubules

Given the absence of a mammalian type oestrous synchronization between multiple ovulations and copulation, female birds have evolved the capacity to store spermatozoa in their oviducts. Anatomically, the sperm storage tubules are tubular invaginations of the luminal surface epithelium restricted to a 2–3 cm wide band at the uterovaginal junction. Spermatozoa within the chicken hen sperm storage tubules retain fertilizing capacity for up to 3 weeks, while turkey hen sperm storage tubules maintain spermatozoa that are capable of fertilization for up to 10 weeks after a single insemination (Brillard, 1993). The extraordinary time frame of this *in vivo* storage phenomenon, however, does not equate to 3 or 4 week intervals between inseminations. To maintain economically acceptable levels of fertility, hens are inseminated weekly during the egg production period to ensure at least 92–96% fertility. Hens ovulate every 24–26 h, and typically produce 5–7 eggs each week for the 5–6 months of egg production. Considering that only 1–2% of inseminated spermatozoa reach the sperm storage tubules (Bakst *et al.*, 1994), a relatively high number must remain viable within the sperm storage tubules to ensure the fertilization of ova that are ovulated between inseminations (Long, 2006).

In Vitro Semen Storage

Poultry semen extenders in use today are modifications of formulas developed over 30 years ago, and intended extending freshly collected semen for only a few hours before insemination. By lowering the pH (Lake and Ravie, 1979; Giesen and Sexton, 1982) of the extender and providing an aerobic environment (Sexton, 1974; Wishart, 1981), viable turkey spermatozoa can be maintained during hypothermic storage at 5–10°C. Under these conditions, commercially acceptable fertility rates are obtained from turkey spermatozoa stored for up to 6 h, though the fertility rates of turkey semen stored for 3–6 h steadily declines after the first 6–7 weeks of egg production. This is in stark contrast with the ability of the sperm storage tubules to maintain viable and functional spermatozoa in the hen for weeks at a time. The consequence for poultry breeders is that semen must be collected and inseminated within 6 h of collection to guarantee high fertility rates throughout production.

Short-term hypothermic (4°C) storage

It has been long recognized that the ability to store turkey semen for 24 h *in vitro* without a significant loss in fertility upon insemination would benefit the commercial turkey industry. A number of recent studies have begun to discern the mechanisms that adversely affect poultry spermatozoa during short-term storage at 4°C. Lipid peroxidation of the plasma membrane is one likely candidate for the poor fertility of stored poultry semen. The high levels of polyunsaturated fatty acids in poultry sperm membranes render them vulnerable to lipid peroxidation (Fujihara and Howarth, 1978; Cecil and Bakst, 1993), a chemical reaction that occurs in the presence of oxygen radicals. Normal by-products of oxidative metabolism form free radicals of oxygen and hydrogen peroxide, which induce the formation of lipid peroxides that are extremely toxic to sperm (Wishart, 1984). For mammalian spermatozoa, lipid peroxidation has been linked with a decline in sperm motility and metabolism *in vitro* (Jones and Mann, 1976). For turkey spermatozoa, the occurrence of lipid peroxidation was demonstrated in samples held at 4°C by measuring malonaldehyde, a by-product of peroxidation (Cecil and Bakst, 1993). More recently, Long and Kramer (2003) demonstrated that the degree of lipid peroxidation varies characteristically for each male, and is a major contributor to the lower fertility rates associated with stored turkey semen; this has profound impacts on the common practice of pooling semen from a large number of tom turkeys for AI.

It has been shown that the phospholipid content of turkey spermatozoa decreases by 30% during 24 h of storage at 4°C, with 20% of the loss occurring between 1 and 4 h (Douard *et al.*, 2003). It has also been demonstrated that membrane-bound phospholipids, especially phosphatidylcholine, were lost during *in vitro* storage of chicken (Blesbois *et al.*, 1999) and turkey spermatozoa (Douard *et al.*, 2000). The inclusion of typical antioxidants, such as vitamin E, in the extender has not been proven to be effective in preventing lipid peroxidation of turkey spermatozoa during semen storage (Donoghue and Donoghue, 1997; Long and Kramer 2003; Douard *et al.*, 2004). However, it has recently been demonstrated that turkey spermatozoa incorporate exogenous phosphatidylcholine into the plasma membrane in a dose-dependent manner during *in vitro* semen storage (Long and Conn, 2012). Higher mean fertility rates were obtained during a 16 week insemination trial when semen was stored with 2.5 mg/ml phosphatidylcholine (66.1%) than without (33.4%), although the fertility levels obtained still are not economically viable for producers (Long and Conn, 2012).

It is likely that other physiological events during short-term semen storage also have an impact on the viability and functionality of poultry spermatozoa. Peláez and Long (2008) have provided evidence that the carbohydrate residues of membrane surface glycoconjugates in turkey spermatozoa undergo quantitative changes during 24 h, hypothermic (4°C) storage of semen. The magnitude and patterns of glycocalyx changes varied with sugar residue type, but the majority of changes in carbohydrate abundance occurred after 8 h of semen storage, which coincides with the dramatic loss in fertility for turkey semen stored longer than 6 h. Because sialic acid is essential for chicken sperm to traverse the vagina and become sequestered within the sperm storage tubules (Froman and Engel, 1989), and removal of terminal sialic acid residues may increase the antigenicity of chicken sperm in the vagina of the hen (Steele and Wishart, 1996), the discovery that hypothermic storage adversely affects the terminal sialic acid content of poultry sperm suggests another potential intervention strategy for improving the fertility of stored semen. In preliminary studies, poultry spermatozoa exhibit maximal binding of exogenous sialic acid within 60 min of incubation in a dose-dependent manner (J.A. Long, unpublished data). Fertility trials were conducted over 12 weeks using semen stored with and without sialic acid, but with phosphatidylcholine. Of the sialic acid doses evaluated, 80 μg/ml, in combination with phosphatidylcholine, supported the highest fertility rates (mean 81.2 ± 9.1%) for the first 6 weeks of insemination (J.A. Long, unpublished data). Using a systematic physiological approach, it is anticipated that new extenders will be developed to support short-term hypothermic storage of poultry semen.

Semen cryopreservation

Over 60 years ago, the discovery of the cryoprotective properties of glycerol pioneered the success of modern cryobiology and led to the development of semen cryopreservation for a wide range of species. Despite the fact that this scientific breakthrough was accomplished with rooster semen (Polge, 1951), the overall fertility rates with frozen–thawed poultry semen are still highly variable and not reliable enough for use in commercial production or for preservation of genetic stocks (Long, 2006). Moreover, significant differences exist among the commercial poultry species in terms of the viability and functionality of sperm after cryopreservation. In particular, the fertility rates from frozen–thawed turkey semen have consistently been lower than cryopreserved chicken semen (Nelson *et al.*, 1980; Sexton, 1981; Kurbatov *et al.*, 1986; Schramm and Hubner, 1988; Wishart, 1989).

Several comprehensive reviews have been published summarizing the empirical studies involving cryoprotectant type and packaging method, as well as freezing and thawing rates, for avian sperm cryopreservation (Lake, 1986; Hammerstedt and Graham, 1992; Donoghue and Wishart, 2000; Blesbois *et al.*, 2008). The two basic methods in use today are: (i) moderately slow freezing rates with glycerol as the cryoprotectant; or (ii) rapid freezing rates with dimethylacetamide (DMA).

Of the cryoprotectants studied to date, glycerol appears to be the most effective for protecting poultry sperm during the cryogenic cycle (Maeda *et al.*, 1984; Bacon *et al.*, 1986; Hammerstedt and Graham, 1992; Donoghue and Wishart, 2000; Peláez *et al.*, 2011). Concentrations of glycerol needed to provide adequate protection (~1 M) are contraceptive in the hen and must be lowered to <0.1 M before the insemination of thawed semen (Neville *et al.*, 1971; Sexton, 1973; Sexton, 1975; Phillips *et al.*, 1996). The precise reason for the lowered fertility of spermatozoa inseminated in the presence of glycerol is unknown, but is most likely related to the osmotic shock following rapid loss of glycerol from spermatozoa in the hen's reproductive tract, and subsequent disruption of the cell membrane (Westfall and Howarth, 1977; Lake *et al.*, 1980). It is well known that glycerolized spermatozoa do not populate the sperm storage tubules. For example, histological sections revealed the absence of spermatozoa in the sperm storage tubules after intravaginal insemination with glycerolized turkey spermatozoa (Marquez and Ogasawara, 1977). Glycerol is believed to adversely affect the sperm cells, rather than the sperm storage tubule ability to store spermatozoa, as timed inseminations of glycerol before, during and after insemination resulted in the greatest contraceptive effect when glycerol and spermatozoa were mixed (Westfall and Howarth, 1977). There are several methods for reducing the glycerol content of frozen–thawed poultry semen, including serial dilution and dialysis; however, these methods can still cause damage to spermatozoa. Long and colleagues reported the feasibility of using a discontinuous gradient of Accudenz (Fig. 9.4) to slowly remove the glycerol

1. Layer 0.5–2.0 ml of thawed semen on gradient.
2. Centrifuge (1250 × *g*; 4°C; 25 min).
3. Recover sperm layer.
4. Dilute with extender prior to insemination.

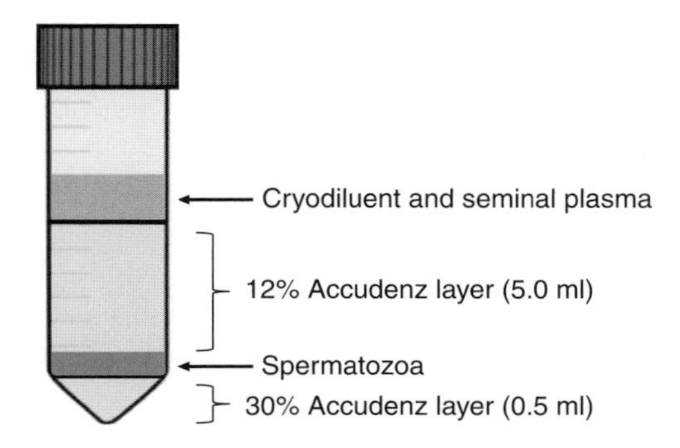

Fig. 9.4. Overview of the Accudenz discontinuous gradient method to remove glycerol from frozen–thawed poultry semen.

without excessively diluting or damaging poultry spermatozoa (Long and Kulkarni, 2004; Long *et al.*, 2010; Peláez *et al.*, 2011).

The post-thaw processing required with glycerol-based cryopreservation procedures has prompted an alternative method using DMA as the cryoprotectant and very rapid freeze–thaw rates (Chalah *et al.*, 1999; Tselutin *et al.*, 1999). In these studies, poultry semen was frozen by the pellet method, whereby semen diluted with a cryodiluent is dropped directly into liquid nitrogen. Although the highest fertility rates were achieved with DMA and the pellet method, this protocol is not ideal for semen banks as it does not confer the high level of safety and clear identification obtained when using glycerol and straw packaging. More recently, a method combining a rapid freeze rate, DMA and straw packaging has been developed for chicken semen (Woelders *et al.*, 2006) that also has shown promise for turkey semen. In preliminary studies, fertile eggs were observed up to 12 weeks after a single insemination of turkey semen frozen using this new strategy (J.A. Long, unpublished data).

These recent advances in poultry semen cryopreservation technology may not provide reliable methodology soon enough for the unique poultry lines that are currently at risk of termination (Delany, 2006; Fulton, 2006), but there may be a temporary solution that takes advantage of the relatively short spermatogenic cycle in poultry. A large body of research clearly demonstrates that the lipid content of the plasma membrane of poultry spermatozoa can be modified by altering the diet (Blesbois *et al.*, 1997; Kelso *et al.*, 1997; Surai *et al.*, 2000; Bréque *et al.*, 2003; Cerolini *et al.*, 2003, 2005, 2006; Blesbois *et al.*, 2004; Dimitrov *et al.*, 2007; Bongalhardo *et al.*, 2009; Zaniboni and Cerolini, 2009; Golzar Adabi *et al.*, 2011). Long-chain polyunsaturated fatty acids of the n-3 and n-6 series cannot be synthesized by vertebrates and therefore must be obtained from the diet, typically from plants in the form of the 18-carbon precursors α-linolenic or linoleic acid, or from the long-chain derivatives (20–22 carbons) found in animal tissues (Cook, 1996). Variation in the plasma membrane lipid composition has been suggested as a key factor for sperm cryotolerance among genetic lines and/or individual males from a wide range of species, including domestic poultry (Blesbois *et al.*, 2005, 2008), silver fox (Miller *et al.*, 2005), horse (Ricker *et al.*, 2006), goat (Chakrabarty *et al.*, 2007), cow (Argov *et al.*, 2007), pig (Waterhouse *et al.*, 2006; Kaeoket *et al.*, 2010) and human (Giraud *et al.*, 2000). New studies are underway in the Beltsville Agricultural Research Center laboratory to determine whether modulating poultry sperm membrane lipids improves the success of cryopreservation.

References

Agarwal A and Said TM (2004) Carnitines and male infertility *Reproductive Biomedicine Online* **8** 376–384.

Ahammad MU, Nishino C, Tatemoto H, Okura N, Kawamoto Y, Okamoto S and Nakada T (2011) Maturational changes in motility, acrosomal proteolytic activity, and penetrability of the inner perivitelline layer of fowl sperm, during their passage through the male genital tract *Theriogenology* **76** 1100–1109.

Ahluwalia BS and Graham EF (1966a) Free carbohydrates in seminal plasma and sperm cells of semen of fowl *Journal of Reproduction and Fertility* **12** 359–361.

Ahluwalia BS and Graham EF (1966b) Free amino acids in the semen of the fowl and the turkey *Journal of Reproduction and Fertility* **12** 365–368.

Aire TA (1979) Micro-stereological study of the avian epididymal region *Journal of Anatomy* **129** 703–706.

al-Aghbari A, Engel HN Jr and Froman DP (1992) Analysis of seminal plasma from roosters carrying the Sd (sperm degeneration) allele *Biology of Reproduction* **47** 1059–1063.

Al-Daraji HJ (2001) Sperm–egg penetration in laying breeder flocks: a technique for the prediction of fertility *British Journal of Poultry Science* **42** 266–270.

Alkan S, Baran A, Özdaş ÕB and Evecen M (2002) Morphological defects in turkey semen *Turkish Journal of Veterinary and Animal Sciences* **26** 1087–1092.

Amen MHM and Al-Daraji HJ (2011) Effect of dietary zinc supplementation on some seminal plasma characteristics of broiler breeder males *International Journal of Poultry Science* **10** 814–818.

Amir D, Braun-Eilon B and Schindler H (1963) Passage and disappearance of labeled spermatozoa in the genital tract of the Japanese quail in segregation or cohabitation *Annales de Biologie Animale, Biochimie, Biophysique* **13** 321–328.

Argov N, Sklan D, Zeron Y and Roth Z (2007) Association between seasonal changes in fatty-acid composition, expression of VLDL receptor and bovine sperm quality *Theriogenology* **67** 878–885.

Ashizawa K, Maeda S and Okauchi K (1989) The mechanisms of reversible immobilization of fowl spermatozoa at body temperature *Journal of Reproduction and Fertility* **86** 271–276.

Ashizawa K, Tomonaga H and Tsuzuki Y (1994) Regulation of flagellar motility of fowl spermatozoa: evidence for the involvement of intracellular free Ca^{2+} and calmodulin *Journal of Reproduction and Fertility* **101** 265–272.

Ashizawa K, Wishart GJ and Tsuzuki Y (2000) Avian sperm motility: environmental and intracellular regulation *Avian Poultry Biology Review* **11** 161–172.

Ashizawa K, Wishart GJ, Ranasinghe AR, Katayama S and Tsuzuki Y (2004) Protein phosphatase-type 2B is involved in the regulation of the acrosome reaction but not in the temperature-dependent flagellar movement of fowl spermatozoa *Reproduction* **128** 783–787.

Ashizawa K, Kawaji N, Nakamura S, Nagase D, Tatemoto H, Katayama S, Narumi K and Tsuzuki Y (2010) Temperature-dependent regulation of sperm motility of Ijima's copper pheasants (*Syrmaticus soemmerringii ijimae*), one of 'near threatened' species *Animal Reproduction Science* **121** 181–187.

Baccetti B, Bigliardi E and Burrini AG (1980) The morphogenesis of vertebrate perforatorium *Journal of Ultrastructure Research* **71** 272–287.

Bacon LD, Salter DW, Motta JV, Crittenden LB and Ogasawara FX (1986) Cryopreservation of chicken semen of inbred or specialized strains *Poultry Science* **65** 1965–1971.

Bakst MR (2010a) Determination of sperm concentration In *Techniques for Semen Evaluation, Semen Storage, and Fertility Determination* 2nd edition pp 26–27 Ed MR Bakst and JA Long. The Midwest Poultry Federation, Buffalo, Minnesota.

Bakst MR (2010b) Sperm viability In *Techniques for Semen Evaluation, Semen Storage, and Fertility Determination* 2nd edition pp 28–34 Ed MR Bakst and JA Long. The Midwest Poultry Federation, Buffalo, Minnesota.

Bakst MR, Wishart GJ and Brillard JP (1994) Oviducal sperm selection, transport, and storage in poultry *Poultry Science Reviews* **5** 117–143.

Bednarczyk M (1990) Penetration of cock spermatozoa into hamster eggs *in vitro* In *Control of Fertility in Domestic Birds* pp 135–140 Ed JP Brillard. INRA Press, Paris.

Bell DJ and Lake PE (1962) A comparison of phosphomonesterase activities in the seminal plasma of the domestic cock, turkey tom, boar, bull, buck rabbit and of man *Journal of Reproduction and Fertility* **3** 363–368.

Blesbois E and Caffin JP (1992) 'Serum like' albumin of fowl seminal plasma and effects of albumin on fowl spermatozoa stored at 4 degrees C *British Journal of Poultry Science* **33** 663–670.

Blesbois E and Hermier D (1990) Effects of high-density lipoproteins on storage at 48°C of fowl spermatozoa *Journal of Reproduction and Fertility* **90** 473–482.

Blesbois E, Lessire M, Grasseau I, Hallouis JM and Hermier D (1997) Effect of dietary fat on the fatty acid composition and fertilizing ability of fowl semen *Biology of Reproduction* **56** 1216–1220.

Blesbois E, Grasseau I and Hermier D (1999) Changes in lipid content of fowl spermatozoa after liquid storage at 2 to 5 degrees C *Theriogenology* **52** 325–334.

Blesbois E, Douard V, Germain M, Boniface P and Pellet F (2004) Effects of n-3 polyunsaturated dietary supplementation on the reproductive capacity of male turkeys *Theriogenology* **61** 537–549.

Blesbois E, Grasseau I and Seigneurin F (2005) Membrane fluidity and the ability of domestic bird spermatozoa to survive cryopreservation *Reproduction* **129** 371–378.

Blesbois E, Grasseau I, Seigneurin F, Mignon-Grasteau S, Saint Jalme M and Mialon-Richard MM (2008) Predictors of success of semen cryopreservation in chickens *Theriogenology* **69** 252–261.

Bongalhardo DC, Leeson S and Buhr MM (2009) Dietary lipids differentially affect membranes from different areas of rooster sperm *Poultry Science* **88** 1060–1069.

Bowling ER, Froman DP, Davis AJ and Wilson JL (2003) Attributes of broiler breeder males characterized by low and high sperm mobility *Poultry Science* **82** 1796–1801.

Bramwell RK and Donoghue AM (2010) Predicting fertility In *Techniques for Semen Evaluation, Semen Storage, and Fertility Determination* 2nd edition pp 91–96 Ed MR Bakst and JA Long. The Midwest Poultry Federation, Buffalo, Minnesota.

Bramwell RK, Marks HL and Howarth B Jr (1995) Quantitative determination of spermatozoa penetration of the perivitelline layer of the hen's ovum as assessed on oviposited eggs *Poultry Science* **74** 1875–1883.

Bremer J (1983) Carnitine – metabolism and functions *Physiological Reviews* **63** 1420–1480.

Bréque C, Surai P and Brillard JP (2003) Roles of antioxidants on prolonged storage of avian spermatozoa *in vivo* and *in vitro Molecular Reproduction and Development* **66** 314–323.

Brillard JP (1981) Influence de la photopériode quotidiènne sur le développement des testicules et sur l'établissement de la spermatogénèse chez la pintade. Thesis, University of Tours, Tours, France.

Brillard JP (1993) Sperm storage and transport following natural mating and artificial insemination *Poultry Science* **72** 923–928.

Brillard JP and Bakst MR (1990) Quantification of spermatozoa in the sperm storage tubules of turkey hens and the relation to sperm numbers in the perivitelline layer *Biology of Reproduction* **43** 271–275.

Brillard JP and McDaniel GR (1985) The reliability and efficiency of various methods for estimating spermatozoa concentration *Poultry Science* **64** 155–158.

Brown KI, Crabo BG, Graham EF and Pace MM (1959) Electrolyte composition and freezing point depression of turkey seminal fluid *Poultry Science* **38** 804–806.

Buckland RB (1971) The activity of six enzymes of chicken seminal plasma and sperm. 1. Effect of *in vitro* storage and full sib families on enzyme activity and fertility *Poultry Science* **50** 1724–1734.

Burrows WH and Quinn JP (1937) The collection of spermatozoa from the domestic fowl and turkey *Poultry Science* **26** 19–24.

Burt CT and Chalovich JM (1978) A new phosphate component of chicken semen *Biochimica et Biophysica Acta – Lipids and Lipid Metabolism* **529** 186–188.

Burt CT and Ribolow H (1994) Glycerol phosphorylcholine (GPC) and serine ethanolamine phosphodiester (SEP): evolutionary mirrored metabolites and their potential metabolic roles *Comparative Biochemistry and Physiology – Part B: Biochemistry and Molecular Biology* **108** 11–20.

Cecil HC and Bakst MR (1985) Volume, sperm concentration, and fertilizing capacity of turkey ejaculates obtained from successive cloacal strokes during semen collection *Poultry Science* **64** 1219–1222.

Cecil HC and Bakst MR (1993) *In vitro* lipid peroxidation of turkey spermatozoa *Poultry Science* **72** 1370–1378.

Cerolini S, Surai P, Maldjian A, Gliozzi I and Noble R (1997) Lipid composition of semen in different fowl breeders *Poultry and Avian Biology Reviews* **8** 141–148.

Cerolini S, Pizzi F, Gliozzi TM, Maldjian A, Zaniboni L and Parodi L (2003) Lipid manipulation of chicken semen by dietary means and its relation to fertility: a review *World Poultry Science Journal* **59** 65–75.

Cerolini S, Surai PF, Speake BK and Sparks NH (2005) Dietary fish and evening primrose oil with vitamin E effects on semen variables in cockerels *British Journal of Poultry Science* **46** 214–222.

Cerolini S, Zaniboni L, Maldjian A and Gliozzi T (2006) Effect of docosahexaenoic acid and alpha-tocopherol enrichment in chicken sperm on semen quality, sperm lipid composition and susceptibility to peroxidation *Theriogenology* **66** 877–886.

Chakrabarty J, Banerjee D, Pal D, De J, Ghosh A and Majumder GC (2007) Shedding off specific lipid constituents from sperm cell membrane during cryopreservation *Cryobiology* **54** 27–35.

Chalabi S, Easton RL, Patankar MS, Lattanzio FA, Morrison JC, Panico M, Morris HR, Dell A and Clark GF (2002) The expression of free oligosaccharides in human seminal plasma *The Journal of Biological Chemistry* **277** 32562–32570.

Chalah T and Brillard JP (1998) Comparison of assessment of fowl sperm viability by eosin–nigrosin and dual fluorescence (SYBR-14/PI) *Theriogenology* **50** 487–493.

Chalah T, Seigneurin F, Blesbois E and Brillard JP (1999) *In vitro* comparison of fowl sperm viability in ejaculates frozen by three different techniques and relationship with subsequent fertility *in vivo Cryobiology* **39** 185–191.

Cherms FL (1967) Elemental analysis of turkey semen *Poultry Science* **46** 1605–1606.

Christensen VL (1995) Diluents, dilution, and storage of poultry semen for six hours In *Proceedings of the First International Symposium on the Artificial Insemination of Poultry* pp 90–106 Ed MR Bakst and GJ Wishart. Poultry Science Association, Savoy, Illinois.

Christensen VL, Bagley L and Long JA (2010) Semen collection and dilution In *Techniques for Semen Evaluation, Semen Storage, and Fertility Determination* 2nd edition pp 7–10 Ed MR Bakst and JA Long. The Midwest Poultry Federation, Buffalo, Minnesota.

Clark RN, Bakst MR and Ottinger MA (1984) Morphological changes in chicken and turkey spermatozoa incubated under various conditions *Poultry Science* **63** 801–805.

Clulow J and Jones RC (1982) Production, transport, maturation, storage and survival of spermatozoa in the male Japanese quail, *Coturnix coturnix Journal of Reproduction and Fertility* **64** 259–266.

Cook HK (1996) Fatty acid desaturation and chain elongation in eukaryotes In *Biochemistry of Lipids, Lipoproteins and Biomembranes* pp 129–152 Ed DE Vance and J Vance. Elsevier, Amsterdam.

Delany ME (2006) Avian genetic stocks: the high and low points from an academia researcher *Poultry Science* **85** 223–226.

de Reviers M (1968) Détermination de la durée des processus spermatogénetiques chez le coq a l'aide de thymidine tritiée *Proceedings, VI International Congress on Animal Reproduction and Artificial Insemination, Paris, France* pp 183–185.

Diekman AB (2003) Glycoconjugates in sperm function and gamete interactions: how much sugar does it take to sweet-talk the egg? *Cellular and Molecular Life Sciences* **60** 298–308.

Dimitrov SG, Atanasov VK, Surai PF and Denev SA (2007) Effect of organic selenium on turkey semen quality during liquid storage *Animal Reproduction Science* **100** 311–317.

Donoghue AM (1996) The effect of 24 hour *in vitro* storage on sperm hydrolysis through the perivitelline membrane of ovipositioned turkey eggs *Poultry Science* **75** 1035–1038.

Donoghue AM and Donoghue DJ (1997) Effects of water- and lipid-soluble antioxidants on turkey sperm viability, membrane integrity, and motility during liquid storage *Poultry Science* **76** 1440–1445.

Donoghue AM and Walker-Simmons MK (1999) The influence of wheat dehydration-induced proteins on the function of turkey spermatozoa after twenty-four-hour *in vitro* storage *Poultry Science* **78** 235–241.

Donoghue AM and Wishart GJ (2000) Storage of poultry semen *Animal Reproduction Science* **62** 213–232.

Donoghue AM, Holsberger DR, Evenson DP and Froman DP (1998) Semen donor selection by *in vitro* sperm mobility increases fertility and semen storage in the turkey hen *Journal of Andrology* **19** 295–301.

Donoghue AM, Sonstegard TS, King LM, Smith EJ and Burt DW (1999) Turkey sperm mobility influences paternity in the context of competitive fertilization *Biology of Reproduction* **61** 422–427.

Douard V, Hermier D and Blesbois E (2000) Changes in turkey semen lipids during liquid *in vitro* storage *Biology of Reproduction* **63** 1450–1456.

Douard V, Hermier D, Magistrini M and Blesbois E (2003) Reproductive period affects lipid composition and quality of fresh and stored spermatozoa in turkeys *Theriogenology* **59** 753–764.

Douard V, Hermier D, Magistrini M, Labbe C and Blesbois E (2004) Impact of changes in composition of storage medium on lipid content and quality of turkey spermatozoa *Theriogenology* **61** 1–13.

Droba M (1986) Autoproteolytic activity in the seminal plasma of the domestic fowl *British Journal of Poultry Science* **27** 103–108.

Droba M and Droba B (1987) Purification and properties of β-galactosidase from chicken seminal plasma *Comparative Biochemistry and Physiology – Part B: Biochemistry and Molecular Biology* **87** 379–384.

Droba B and Dzugan M (1993) Acid alpha-D-mannosidase from turkey seminal plasma *Folia Biologica – Krakow* **41** 49–54.

Dumitru IF and Dinischiotu A (1994) Cock seminal plasma acid phosphatase: active site directed inactivation, crystallization and *in vitro* denaturation–renaturation studies *International Journal of Biochemistry* **26** 497–503.

Edens FW and Sefton AE (2009) Sel-Plex® improves spermatozoa morphology in broiler breeder males *International Journal of Poultry Science* **8** 853–861.

Eisenberg S (1984) High density lipoprotein metabolism *Journal of Lipid Research* **25** 1017–1058.

El Jack MH and Lake PE (1969) The distribution of the principal inorganic ions in semen from the vas deferens of the domestic fowl and the content of carbon dioxide in the seminal plasma *Experimental Physiology* **54** 332–340.

Etches RJ (1996) *Reproduction in Poultry.* Cambridge University Press, Cambridge.

Froman DP and Bernier PE (1987) Identification of heritable spermatozoal degeneration within the ductus deferens of the chicken (*Gallus domesticus*) *Biology of Reproduction* **37** 969–977.

Froman DP and Engel HN Jr (1989) Alteration of the spermatozoal glycocalyx and its effect on duration of fertility in the fowl (*Gallus domesticus*) *Biology of Reproduction* **40** 615–621.

Froman DP and Feltmann AJ (1998) Sperm mobility: a quantitative trait of the domestic fowl (*Gallus domesticus*) *Biology of Reproduction* **58** 379–384.

Froman DP and Feltmann AJ (2000) Sperm mobility: phenotype in roosters (*Gallus domesticus*) determined by concentration of motile sperm and straight line velocity *Biology of Reproduction* **62** 303–309.

Froman DP and Kirby JD (2005) Sperm mobility: phenotype in roosters (*Gallus domesticus*) determined by mitochondrial function *Biology of Reproduction* **72** 562–567.

Froman DP and McLean DJ (1996) Objective measurement of sperm motility based upon sperm penetration of Accudenz *Poultry Science* **75** 776–784.
Froman DP and Thursam KA (1994) Desialylation of the rooster sperm's glycocalyx decreases sperm sequestration following intravaginal insemination of the hen *Biology of Reproduction* **50** 1094–1099.
Froman DP, Kirby JD, Lawler RM and Bernier PE (1990) Onset of spermatozoal degeneration in low-fertility Delaware roosters and test for autoimmune basis *Journal of Andrology* **11** 113–119.
Froman DP, Feltmann AJ and McLean DJ (1997) Increased fecundity resulting from semen donor selection based upon *in vitro* sperm motility *Poultry Science* **76** 73–77.
Froman DP, Feltmann AJ, Rhoads ML and Kirby JD (1999) Sperm mobility: a primary determinant of fertility in the domestic fowl (*Gallus domesticus*) *Biology of Reproduction* **61** 400–405.
Froman DP, Pizzari T, Feltmann AJ, Castillo-Juarez H and Birkhead TR (2002) Sperm mobility: mechanisms of fertilizing efficiency, genetic variation and phenotypic relationship with male status in the domestic fowl, *Gallus gallus domesticus Proceedings of the Royal Society B: Biological Sciences* **269** 607–612.
Froman DP, Wardell JC and Feltmann AJ (2006) Sperm mobility: deduction of a model explaining phenotypic variation in roosters (*Gallus domesticus*) *Biology of Reproduction* **74** 487–491.
Fujihara N (1992) Accessory reproductive fluids and organs in male domestic birds *World's Poultry Science Journal* **48** 39–56.
Fujihara N and Howarth B Jr (1978) Lipid peroxidation in fowl spermatozoa *Poultry Science* **57** 1766–1768.
Fujihara N and Nishiyama H (1984) Addition to semen of a fluid derived from the cloacal region by male turkeys *Poultry Science* **63** 544–557.
Fujihara N, Nishiyama H and Koga O (1985) The mechanism of the ejection of a frothy fluid from the cloaca in the male turkey *Poultry Science* **64** 1377–1381.
Fujihara N, Nishiyama H and Koga O (1987) Effect on turkey spermatozoa of a frothy fluid derived from the cloaca of a male turkey *Theriogenology* **28** 225–235.
Fulton JE (2006) Avian genetic stock preservation: an industry perspective *Poultry Science* **85** 227–231.
Garner DL and Hafez ESE (1993) Spermatozoa and seminal plasma In *Reproduction in Farm Animals* pp 165–187 Ed ESE Hafez. Lea and Febiger, Philadelphia.
Giansanti F, Leboffe L, Pitari G, Ippoliti R and Antonini G (2012) Physiological roles of ovotransferrin *Biochimica et Biophysica Acta – General Subjects* **1820** 218–225.
Giesen AF and Sexton TJ (1982) Beltsville poultry semen extender: 7. Comparison of commercial diluents for holding turkey semen six hours at 15°C *Poultry Science* **62** 379–381.
Giraud MN, Motta C, Boucher D and Grizard G (2000) Membrane fluidity predicts the outcome of cryopreservation of human spermatozoa *Human Reproduction* **15** 2160–2164.
Golan R, Setchell BP, Burrow PV and Lewin LM (1982) A comparative study of carnitine and acylcarnitine concentration in semen and male reproductive tract fluids *Comparative Biochemistry and Physiology – Part B: Biochemistry and Molecular Biology* **72** 457–460.
Golzar Adabi SH, Cooper RG, Kamali MA and Hajbabaei A (2011) The influence of inclusions of vitamin E and corn oil on semen traits of Japanese quail (*Coturnix coturnix japonica*) *Animal Reproduction Science* **123** 119–125.
Graham EF and Brown KI (1971) Effect of osmotic pressure of semen extenders on the fertility and hatchability of turkey eggs *Poultry Science* **50** 836–838.
Graham EF, Schmehl ML, Brown KI, Crabo BG and Ertel G (1971) Some chemical studies of the female reproductive tract and seminal plasma of the male turkey and their relationship to fertility *Poultry Science* **50** 1170–1181.
Hammerstedt RH and Graham JK (1992) Cryopreservation of poultry sperm: the enigma of glycerol *Cryobiology* **29** 26–38.
Hammond M, Boone MA and Barnett BD (1965) Study of the glucose, electrolytes, enzymes and nitrogen components of fowl seminal plasma *Journal of Reproduction and Fertility* **10** 21–28.
Harris GC Jr and Sweeney MJ (1971) Changes in the protein concentration of chicken seminal plasma after rapid freeze–thaw *Cryobiology* **7** 209–215.
Hess RA and Renato de Franca L (2008) Spermatogenesis and cycle of the seminiferous epithelium *Advances in Experimental Medicine and Biology* **636** 1–15.
Hess RA and Thurston RJ (1984) Protein, cholesterol, acid phosphatase and aspartate aminotransaminase in the seminal plasma of turkeys (*Meleagris gallopavo*) producing normal white or abnormal yellow semen *Biology of Reproduction* **31** 239–243.
Hess RA, Thurston RJ and Biellier HV (1976) Morphology of the epididymal region and ductus deferens of the turkey (*Meleagris gallopavo*) *Journal of Anatomy* **122** 241–252.

Hess RA, Thurston RJ and Biellier HV (1982) Morphology of the epididymal region of turkeys producing abnormal yellow semen *Poultry Science* **61** 531–539.

Hess RA, Birrenkott GP Jr and Thurston RJ (1984) Seminal plasma androgen-binding protein activity in turkeys with normal white or abnormal yellow semen *Journal of Reproduction and Fertility* **71** 403–409.

Holm L and Wishart GJ (1998) The effect of pH on the motility of spermatozoa from chicken, turkey and quail *Animal Reproduction Science* **54** 45–54.

Holsberger DR, Donoghue AM, Froman DP and Ottinger MA (1998) Assessment of ejaculate quality and sperm characteristics in turkeys: sperm mobility phenotype is independent of time *Poultry Science* **77** 1711–1717.

Holsberger DR, Rice CD and Thurston RJ (2002) Localization of a proteolytic enzyme within the efferent and deferent duct epithelial cells of the turkey (*Meleagris gallopavo*) using immunohistochemistry *Biology of Reproduction* **67** 276–281.

Horrocks AJ, Stewart S, Jackson L and Wishart GJ (2000) Induction of acrosomal exocytosis in chicken spermatozoa by inner perivitelline-derived *N*-linked glycans *Biochemical and Biophysical Research Communications* **278** 84–89.

Hough SR and Parks JE (1994) Platelet-activating factor acetylhydrolase activity in seminal plasma from the bull, stallion, rabbit, and rooster *Biology of Reproduction* **50** 912–916.

Howarth B Jr (1970) An examination for sperm capacitation in the fowl *Biology of Reproduction* **3** 338–341.

Howarth B Jr (1983) Fertilizing ability of cock spermatozoa from the testis epididymis and vas deferens following intramagnal insemination *Biology of Reproduction* **28** 586–590.

Howarth B Jr and Palmer MB (1972) An examination of the need for sperm capacitation in the turkey, *Meleagris gallopavo Journal of Reproduction and Fertility* **28** 443–445.

Ivell R (1999) Mini symposium: the role of carbohydrates in reproduction *Human Reproduction Update* **5** 277–279.

Jamieson BGM (2007) Avian spermatozoa: structure and phylogeny In *Reproductive Biology and Phylogeny of Birds Volume 6A* pp 349–511 Ed BGM Jamieson. Science Publishers, Enfield, New Hampshire.

Jarinkovičová L, Máchal L, Máchal J, Filipčík R, Tůmová E and Horský R (2012) Relationship of ejaculate quality and selected biochemical parameters of blood in cockerels of three laying lines *Czech Journal of Animal Science* **57** 370–376.

Jones RC and Lin M (1993) Spermatogenesis in birds *Oxford Reviews of Reproductive Biology* **15** 233–264.

Jones R and Mann T (1976) Lipid peroxides in spermatozoa; formation, role of plasmalogen, and physiological significance *Proceedings of the Royal Society B: Biological Sciences* **193** 317–333.

Kaeoket K, Sang-Urai P, Thamniyom A, Chanapiwat P and Techakumphu M (2010) Effect of docosahexaenoic acid on quality of cryopreserved boar semen in different breeds *Reproduction in Domestic Animals* **45** 458–463.

Kamar GA and Rizik MA (1972) Semen characteristics of two breeds of turkeys *Journal of Reproduction and Fertility* **29** 317–325.

Kannan Y (1974) Glycosidases in fowl semen *Journal of Reproduction and Fertility* **40** 227–228.

Karaca AG, Parker HM, Yeatman JB and McDaniel CD (2002) The effects of heat stress and sperm quality classification on broiler breeder male fertility and semen ion concentrations *British Journal of Poultry Science* **43** 621–628.

Kasai K, Izumo A, Inaba T and Sawada T (2000) Assessment of fresh and stored duck spermatozoa quality via *in vitro* sperm–egg interaction assay *Theriogenology* **54** 283–290.

Kelso KA, Cerolini S, Noble RC, Sparks NH and Speake BK (1996) Lipid and antioxidant changes in semen of broiler fowl from 25 to 60 weeks of age *Journal of Reproduction and Fertility* **106** 201–206.

Kelso KA, Cerolini S, Noble RC, Sparks NH and Speake BK (1997) The effects of dietary supplementation with docosahexaenoic acid on the phospholipid fatty acid composition of avian spermatozoa *Comparative Biochemistry and Physiology – Part B: Biochemistry and Molecular Biology* **118** 65–69.

Khan RU, Rahman ZU, Javed I and Muhammad F (2012) Effect of vitamins, probiotics and protein on semen traits in post-molt male broiler breeders *Animal Reproduction Science* **135** 85–90.

King LM, Holsberger DR and Donoghue AM (2000a) Correlation of CASA velocity and linearity parameters with sperm mobility phenotype in turkeys *Journal of Andrology* **21** 65–71.

King LM, Kirby JD, Froman DP, Sonstegard TS, Harry DE, Darden JR, Marini PJ, Walker RM, Rhoads ML and Donoghue AM (2000b) Efficacy of sperm mobility assessment in commercial flocks and the relationships of sperm mobility and insemination dose with fertility in turkeys *Poultry Science* **79** 1797–1802.

Kirby JD, Froman DP, Engel HN Jr, Bernier PE and Hess RA (1990) Decreased sperm survivability associated with aberrant morphology of the ductuli efferentes proximales of the chicken (*Gallus domesticus*) *Biology of Reproduction* **42** 383–389.

Kotłowska M, Kowalski R, Glogowski J, Jankowski J and Ciereszko A (2005) Gelatinases and serine proteinase inhibitors of seminal plasma and the reproductive tract of turkey (*Meleagris gallopavo*) *Theriogenology* **63** 1667–1681.

Kraft LA, Foley CW, Howarth B Jr, Blum MS and Johnson AD (1978) Radiorespirometric studies of carbohydrate metabolism by washed spermatozoa of various species *Comparative Biochemistry and Physiology – Part B: Biochemistry and Molecular Biology* **60** 233–238.

Kumar R, Harper MJK and Hanahan DJ (1988) Occurrence of platelet-activating factor in rabbit spermatozoa *Archives of Biochemistry and Biophysics* **260** 497–502.

Kurbatov AD, Narubina L, Bubliaeva G, Tselutin KV, Mavrodina T, Tur BK and Kosianenko SV (1986) Cryopreservation of fowl semen In *Proceedings of the 7th European Poultry Conference* pp 998–1001 Ed M Larbier. World Poultry Science Association, Paris.

Kurokawa H, Dewan JC, Mikami B, Sacchettini JC and Hirose M (1999) Crystal structure of hen apo-ovotransferrin. Both lobes adopt an open conformation upon loss of iron *The Journal of Biological Chemistry* **274** 28445–28452.

Kuzan FB, Geissler FT and Henderson WR Jr (1990) Role of spermatozoal platelet-activating factor in fertilization *Prostaglandins* **39** 61–74.

Lake PE (1960) Studies on the dilution and storage of fowl semen *Journal of Reproduction and Fertility* **1** 30–35.

Lake PE (1966) Physiology and biochemistry of poultry semen In *Advances in Reproductive Physiology Volume 1* pp 93–123 Ed A McLaren. Academic Press, London.

Lake PE (1971) The male in reproduction In *Physiology and Biochemistry of the Domestic Fowl Volume 3* pp 1411–1447 Ed DJ Bell and BM Freeman. Academic Press, New York.

Lake PE (1984) The male in reproduction In *Physiology and Biochemistry of the Domestic Fowl Volume 5* pp 381–405 Ed BM Freeman. Academic Press, London.

Lake PE (1986) The history and future of the cryopreservation of avian germplasm *Poultry Science* **65** 1–15.

Lake PE and McIndoe WM (1959) The glutamic acid and creatine content of cock seminal plasma *Biochemical Journal* **71** 303–306.

Lake PE and Ravie O (1979) Effect on fertility of storing fowl semen for 24 h at 5 degrees C in fluids of different pH *Journal of Reproduction and Fertility* **57** 149–155.

Lake PE and Ravie O (1982) Effect on fertility of storing turkey semen for 24 hours at 10 degrees C in fluids of different pH *British Journal of Poultry Science* **23** 41–47.

Lake PE and Wishart GJ (1984) Comparative physiology of turkey and fowl semen In *Reproductive Biology of Poultry* pp 151–160 Ed FJ Cunningham, PE Lake and D Hewitt. British Poultry Science, Harlow.

Lake PE, Smith W and Young D (1968) The ultrastructure of the ejaculated fowl spermatozoon *Quarterly Journal of Experimental Physiology and Cognate Medical Sciences* **53** 356–366.

Lake PE, Buckland RB and Ravie O (1980) Effect of glycerol on the viability of fowl spermatozoa – implications for its use when freezing semen *CryoLetters* **1** 299–304.

Lemoine M, Grasseau I, Brillard JP and Blesbois E (2008) A reappraisal of the factors involved in *in vitro* initiation of the acrosome reaction in chicken spermatozoa *Reproduction* **136** 391–399.

Lemoine M, Mignon-Grasteaue S, Grasseaua I, Magistrinia M and Blesbois E (2011) Ability of chicken spermatozoa to undergo acrosome reaction after liquid storage or cryopreservation *Theriogenology* **75** 122–130.

Lessley BA and Brown KI (1978) Purification and properties of a proteinase inhibitor from chicken seminal plasma *Biology of Reproduction* **19** 223–234.

Lin M, Jones RR and Blackshaw AW (1990) The cycle of the seminiferous epithelium in the Japanese quail (*Coturnix coturnix japonica*) *Journal of Reproduction and Fertility* **88** 481–490.

Long JA (2006) Avian semen cryopreservation: what are the biological challenges? *Poultry Science* **85** 232–236.

Long JA (2010) Determination of sperm concentration In *Techniques for Semen Evaluation, Semen Storage, and Fertility Determination* 2nd edition pp 23–25 Ed MR Bakst and JA Long. The Midwest Poultry Federation, Buffalo, Minnesota.

Long JA and Conn TL (2012) Use of phosphatidylcholine to improve the function of turkey semen stored at 4°C for 24 hours *Poultry Science* **91** 1990–1996.

Long JA and Kramer M (2003) Effect of vitamin E on lipid peroxidation and fertility after artificial insemination with liquid-stored turkey semen *Poultry Science* **82** 1802–1807.

Long JA and Kulkarni G (2004) An effective method for improving the fertility of glycerol-exposed poultry semen *Poultry Science* **83** 1594–1601.

Long JA, Bongalhardo DC, Pelaéz J, Saxena S, Settar P, O'Sullivan NP and Fulton JE (2010) Rooster semen cryopreservation: effect of pedigree line and male age on postthaw sperm function. *Poultry Science* **89** 966–973.

Maeda T, Terada T and Tsutsumi Y (1984) Comparative study of the effects of various cryoprotectants in preserving the morphology of frozen and thawed fowl spermatozoa *British Journal of Poultry Science* **25** 547–553.

Mann T (1964) *The Biochemistry of Semen and of the Male Reproductive Tract* Methuen and Company Ltd, London.

Mann T and Lutwak-Mann C (1981) Biochemistry of seminal plasma and male accessory fluids: application to andrological problems In *Male Reproductive Function and Semen* pp 269–336 Ed T Mann and C Lutwak-Mann. Springer-Verlag, New York.

Marchand CR, Gomot L and de Reviers M (1977) Autoradiography and tritiated thymidine labeling during spermatogenesis of the Barbary duck (*Cairina moschata* L.). *Comptes Rendus des Séances de la Societé de Biologie et des ses Filiales* **171** 927–931.

Marquez BJ and Ogasawara FX (1975) Scanning electron microscope studies of turkey semen *Poultry Science* **54** 1139–1142.

Marquez BJ and Ogasawara FX (1977) Effects of glycerol on turkey sperm cell viability and fertilizing capacity *Poultry Science* **56** 725–731.

McDaniel CD, Hannah JL, Parker HM, Smith TW, Schultz CD and Zumwalt CD (1998) Use of a sperm analyzer for evaluating broiler breeder males. 1. Effects of altering sperm quality and quantity on the sperm motility index *Poultry Science* **77** 888–893.

McIndoe WM and Lake PE (1973) Proceedings: Aspects of energy metabolism of avian spermatozoa *Journal of Reproduction and Fertility* **35** 592–593.

McIndoe WM and Lake PE (1974) The distribution of some hydrolytic enzymes in the semen of the domestic fowl, *Gallus domesticus Journal of Reproduction and Fertility* **40** 359–365.

Miller RR Jr, Cornett CL, Waterhouse KE and Farstad W (2005) Comparative aspects of sperm membrane fatty acid composition in silver (*Vulpes vulpes*) and blue (*Alopex lagopus*) foxes, and their relationship to cell cryopreservation *Cryobiology* **51** 66–75.

Minhas BS, Kumar R, Ricker DD, Robertson JL and Dodson MG (1991) The presence of platelet-activating factor-like activity in human spermatozoa *Fertility and Sterility* **1** 1–2.

Munro SS (1938) Fowl sperm immobilisation by a temperature–media interaction and its biological significance *Quarterly Journal of Experimental Physiology* **27** 281–287.

Nelson DS, Graham EF and Schmehl MKL (1980) Fertility after dialysing frozen turkey semen In *Proceedings of the 9th International Congress on Animal Reproduction and AI* Volume 3 pp 417–419 Madrid, Spain.

Nestor KE and Brown KI (1971) Semen production in turkeys *Poultry Science* **50** 1705–1712.

Neuman SL, McDaniel CD, Frank L, Radu J, Einstein ME and Hester PY (2002) Utilisation of a sperm quality analyser to evaluate sperm quantity and quality of turkey breeders *British Journal of Poultry Science* **43** 457–464.

Neville W J, MacPherson JW and Reinhart B (1971) The contraceptive action of glycerol in chickens *Poultry Science* **50** 1411–1415.

Noirault J, Brillard JP and Bakst MR (2006) Spermatogenesis in the turkey (*Meleagris gallopavo*): quantitative approach in immature and adult males subjected to various photoperiods *Theriogenology* **65** 845–859.

Parker HM and McDaniel CD (2006) The immediate impact of semen diluent and rate of dilution on the sperm quality index, ATP utilization, gas exchange, and ionic balance of broiler breeder sperm *Poultry Science* **85** 106–116.

Peláez J and Long JA (2007) Characterizing the glycocalyx of poultry spermatozoa: I. Identification and distribution of carbohydrate residues using flow cytometry and epifluorescence microscopy *Journal of Andrology* **28** 342–352.

Peláez J and Long JA (2008) Characterizing the glycocalyx of poultry spermatozoa: II. *In vitro* storage of turkey semen and mobility phenotype affects the carbohydrate component of sperm membrane glycoconjugates *Journal of Andrology* **29** 431–439.

Peláez J, Bongalhardo DC and Long JA (2011) Characterizing the glycocalyx of poultry spermatozoa: III. Semen cryopreservation methods alter the carbohydrate component of rooster sperm membrane glycoconjugates *Poultry Science* **90** 435–443.

Phillips JJ, Bramwell RK and Graham JK (1996) Cryopreservation of rooster sperm using methyl cellulose *Poultry Science* **75** 915–923.

Pinto O, Amir D, Schindler H and Hurwitz S (1984) Effect of pH on the metabolism and fertility of turkey spermatozoa *Journal of Reproduction and Fertility* **70** 437–442.

Polge C (1951) Functional survival of fowl spermatozoa after freezing at –79 degrees C *Nature* **167** 949–950.

Pudney J (1995) Spermatogenesis in nonmammalian vertebrates *Microscopy Research and Technique* **32** 459–497.

Quinn JP and Burrows WH (1936) Artificial insemination of fowls *Journal of Heredity* **27** 31–37.

Quinn JP, White IG and Wirrick BR (1965) Studies of the distribution of the major cations in semen and male accessory secretions *Journal of Reproduction and Fertility* **10** 379–388.

Ribolow H and Burt CT (1987) Analysis of lipid metabolism in semen and its implications In *Magnetic Resonance of the Reproductive System* pp 128–136 Ed S McCarthy and F Hazeline. Slack Inc, Thorofare, New Jersey.

Ricker JV, Linfor JJ, Delfino WJ, Kysar P, Scholtz EL, Tablin F, Crowe JH, Ball BA and Meyers SA (2006) Equine sperm membrane phase behavior: the effects of lipid-based cryoprotectants *Biology of Reproduction* **74** 359–365.

Robertson L, Wishart GJ and Horrocks AJ (2000) Identification of perivitelline *N*-linked glycans as mediators of sperm–egg interaction in chickens *Journal of Reproduction and Fertility* **120** 397–403.

Rosiepen G, Weinbauer GF, Schlatt S, Behre HM and Nieschlag E (1994) Duration of the cycle of the seminiferous epithelium, estimated by the 5-bromodeoxyuridine technique, in laboratory and feral rats *Journal of Reproduction and Fertility* **100** 299–306.

Saeki Y and Brown KI (1962) Effect of abnormal spermatozoa on fertility and hatchability in the turkey *Poultry Science* **41** 1096–1100.

Schramm GP and Hubner R (1988) Effects of different cryoprotectives and deep freeze techniques on the reproductive potential of turkey sperm after prolonged storage *Monatshefte für Veterinärmedizin* **43** 426–427.

Schröter S, Osterhoff C, McArdle W and Ivell R (1999) The glycocalyx of the sperm surface *Human Reproduction Update* **5** 302–313.

Sexton TJ (1973) Effect of various cryoprotective agents on the viability and reproductive efficiency of chicken spermatozoa *Poultry Science* **52** 1353–1357.

Sexton TJ (1974) Oxidative and glycolytic activity of chicken and turkey spermatozoa *Comparative Biochemistry and Physiology – Part B: Biochemistry and Molecular Biology* **48** 59–65.

Sexton TJ (1975) Relationship of the method of addition and temperature of cryoprotective agents to the fertilizing capacity of cooled chicken spermatozoa *Poultry Science* **54** 845–848.

Sexton TJ (1977) A new poultry semen extender. 1. Effects of extension on the fertility of chicken semen *Poultry Science* **56** 1443–1446.

Sexton TJ (1981) Development of a commercial method for freezing turkey semen: 1. Effect of prefreeze techniques on the fertility of frozen and frozen–thawed semen *Poultry Science* **60** 1567–1572.

Sexton T J (1982) Beltsville poultry semen extender. 6. Holding turkey semen for six hours at 15°C *Poultry Science* **61** 1202–1208.

Siudzińska A and Łukaszewicz E (2008) Effect of semen extenders and storage time on sperm morphology of four chicken breeds *The Journal of Applied Poultry Research* **17** 101–108.

Smith MB and Back JF (1962) Modification of ovalbumin in storage eggs detected by heat denaturation *Nature* **193** 878–879.

Staines HJ, Middleton RC, Laughlin KF and Wishart GJ (1998) Quantification of a sperm–egg interaction for estimating the mating efficiency of broiler breeder flocks *British Journal of Poultry Science* **39** 273–277.

Steele MG and Wishart GJ (1996) Demonstration that the removal of sialic acid from the surface of chicken spermatozoa impedes their transvaginal migration *Theriogenology* **46** 1037–1044.

Sugimoto Y, Sanuki S, Ohsako S, Higashimoto Y, Kondo M, Kurawaki J, Ibrahim HR, Aoki T, Kusakabe T and Koga K (1999) Ovalbumin in developing chicken eggs migrates from egg white to embryonic organs while changing its conformation and thermal stability *The Journal of Biological Chemistry* **274** 11030–11037.

Surai PF, Blesbois E, Grasseau I, Chalah T, Brillard JP, Wishart GJ, Cerolini S and Sparks NH (1998) Fatty acid composition, glutathione peroxidase and superoxide dismutase activity and total antioxidant activity of avian semen *Comparative Biochemistry and Physiology – Part B: Biochemistry and Molecular Biology* **120** 527–533.

Surai PF, Noble RC, Sparks NH and Speake BK (2000) Effect of long-term supplementation with arachidonic or docosahexaenoic acids on sperm production in the broiler chicken *Journal of Reproduction and Fertility* **120** 257–264.

Thomson MF and Wishart GJ (1991) Temperature-mediated regulation of calcium flux and motility in fowl spermatozoa *Journal of Reproduction and Fertility* **93** 385–391.

Thurston RJ and Hess RA (1987) Ultrastructure of spermatozoa from domesticated birds: comparative study of turkey, chicken and guinea fowl *Scanning Microscopy* **1** 1829–1838.

Thurston RJ and Korn N (1997) Semen quality in the domestic turkey: the yellow semen syndrome *Poultry and Avian Biology Reviews* **8** 109–121.

Thurston RJ, Hess RA, Biellier HV, Adldinger HK and Solorzano RF (1975) Ultrastructural studies of semen abnormalities and Herpesvirus associated with cultured testicular cells from domestic turkeys *Journal of Reproduction and Fertility* **45** 235–241.

Thurston RJ, Hess RA, Froman DP and Biellier HV (1982) Elevated seminal plasma protein: a characteristic of yellow turkey semen *Poultry Science* **61** 1905–1911.

Thurston RJ, Hess RA and Korn N (1992) Seminal plasma protein concentration as a predictor of fertility and hatchability in large white domestic turkeys *Journal of Applied Poultry Research* **1** 335–338.

Thurston RJ, Korn N, Froman DP and Bodine AB (1993) Proteolytic enzymes in seminal plasma of domestic turkey (*Meleagris gallopavo*) *Biology of Reproduction* **48** 393–402.

Tselutin K, Seigneurin F and Blesbois E (1999) Comparison of cryoprotectants and methods of cryopreservation of fowl spermatozoa *Poultry Science* **78** 586–590.

Vizcarra JA and Ford JJ (2006) Validation of the sperm mobility assay in boars and stallions *Theriogenology* **66** 1091–1097.

Waterhouse KE, Hofmo PO, Tverdal A and Miller RR Jr (2006) Within and between breed differences in freezing tolerance and plasma membrane fatty acid composition of boar sperm *Reproduction* **131** 887–894.

Westfall F D and B Howarth Jr (1977) Duration of the antifertility effect of glycerol in the chicken vagina *Poultry Science* **56** 924–925.

Wicher KB and Fries E (2006) Haptoglobin, a hemoglobin-binding plasma protein, is present in bony fish and mammals but not in frog and chicken *Proceedings of the National Academy of Sciences of the United States of America* **103** 4168–4173.

Wilcox FH (1961) Phosphatases in chicken semen *Journal of Reproduction and Fertility* **2** 148–151.

Wishart GJ (1981) The effect of continuous aeration on the fertility of fowl and turkey semen stored above 0°C *British Journal of Poultry Science* **22** 445–450.

Wishart GJ (1982) Maintenance of ATP concentrations in and of fertilizing ability of fowl and turkey spermatozoa *in vitro Journal of Reproduction and Fertility* **66** 457–462.

Wishart GJ (1984) Metabolism of fowl and turkey spermatozoa at low temperatures *Journal of Reproduction and Fertility* **70** 145–149.

Wishart GJ (1989) Physiological changes in fowl and turkey spermatozoa during *in vitro* storage *British Journal of Poultry Science* **30** 443–454.

Wishart GJ and Ashizawa K (1987) Regulation of the motility of fowl spermatozoa by calcium and cAMP. *Journal of Reproduction and Fertility* **80** 607–611.

Wishart GJ and Bakst MR (2010) Predicting fertility In *Techniques for Semen Evaluation, Semen Storage, and Fertility Determination* 2nd edition pp 88–91 Ed MR Bakst and JA Long. The Midwest Poultry Federation, Buffalo, Minnesota.

Wishart GJ and Carver L (1984) Glycolytic enzymes of fowl and turkey spermatozoa *Comparative Biochemistry and Physiology – Part B: Biochemistry and Molecular Biology* **79** 453–455.

Wishart GJ and Wilson YI (1999) Temperature-dependent inhibition of motility in spermatozoa from different avian species *Animal Reproduction Science* **57** 229–235.

Woelders H, Zuidberg CA and Hiemstra SJ (2006) Animal genetic resources conservation in the Netherlands and Europe: poultry perspective *Poultry Science* **85** 216–222.

Zaniboni L and Cerolini S (2009) Liquid storage of turkey semen: changes in quality parameters, lipid composition and susceptibility to induced *in vitro* peroxidation in control, n-3 fatty acids and alpha-tocopherol rich spermatozoa *Animal Reproduction Science* **112** 51–65.

10 Applied Andrology in Sheep, Goats and Selected Cervids

Swanand Sathe[1*] and Clifford F. Shipley[2]
[1]*Iowa State University, Ames, Iowa, USA;* [2]*University of Illinois, Urbana, Illinois, USA*

Introduction

An understanding of both normal and abnormal male sexual development in small ruminants and cervids represents an essential basis for implementing breeding soundness evaluations in these species, and interpreting the results obtained. This includes factors, such as seasonality, that affect the attainment of puberty, as well as other reproductive traits. Variations occur among the different species discussed with reference to semen collection, handling and evaluation.

Male Sexual Development

Normal development

The male reproductive system in small ruminants and cervids is similar in most respects and comprises a pair of extra-abdominal testes suspended within the scrotum, which is covered by fine hair, the excurrent duct system, a fibroelastic penis and accessory sex glands. The fetal testes migrate from the posterior pole of the kidney and pass through the abdominal wall to eventually reach the scrotum (Hutson *et al.*, 1997). At the time of birth, the testis is already lodged in the scrotum in sheep and goats (Gier and Marion, 1970). Each testis is oriented vertically in the scrotum with the head of the epididymis located at the proximal pole of the testis and its tail distal to and extending beyond the corresponding two caudal poles of the testis. In the ram, the testicles weigh between 200 and 250 g each, whereas in the buck they weigh between 130 and 160 g each. The epididymides in small ruminants are oriented caudo-medially and the epididymal canal can be as long as 40 to 60 mm.

The penis in small ruminants and cervids is fibroelastic in nature and remains rigid even in the quiescent stage, without the erectile tissue undergoing much engorgement. The penis of the ram and buck has a sigmoid flexure located caudal to the spermatic cord, which straightens during erection. In contrast, the sigmoid flexure is absent in most cervid species. The penis in sheep and goats also has a urethral process that projects about 2 to 3 cm beyond the glans penis. In juvenile or immature males, there

* E-mail: ssathe@iastate.edu

are natural adhesions between the urethral process, glans penis and prepuce, thus rendering these animals incapable of copulation (Plate 13).

The urethral process is absent in cervids and the urethra, which runs along the ventral surface of the penis throughout its length, curves dorsally at its extremity. This permits the ejection of urine and a few sperm in an upward direction, which is important during the 'thrash-urination' activity indulged in by rutting wapiti and red deer males as they spray upwards almost at right angles on to their ventral abdomens, neck and throat (Haigh, 2007). In these species, the prepuce also has well-developed muscles that are responsible for the remarkable rapid forward and backward movement, often called palpitation, of this region seen during the rut. This behaviour, and the degree of development of these muscles, has not been reported in any other genus (Haigh, 2007).

Rams and bucks have a full complement of accessory sex glands, of which the seminal vesicles are the largest and lie on each side of the caudal part of the dorsal surface of the bladder. The disseminated prostate completely surrounds the wall of the pelvic urethra in small ruminants, whereas in red deer and wapiti, it is a discrete, bilobed body lying anterior to and continuous with the disseminate prostate. The epithelium of the body of the prostate undergoes marked changes according to the season, but few or no changes are seen in the epithelial architecture of the disseminate part. The paired bulbourethral glands are found lying on the dorsal surface of the urethra opposite the ischial arch and are covered by a thick layer of dense connective tissue and by the proximal part of the thick bulbospongiosus muscle (Haigh, 2007).

Abnormal development

Several disorders of sexual development have been studied in depth and documented for domestic small ruminants, but very little information is available for the cervid species. The nature of these disorders is usually complex, involving several factors, and is generally discussed in association with: (i) abnormalities of chromosomal sex, which include syndromes related to sex chromosome monosomy, trisomy, mosaicism, chimerism and translocation; (ii) abnormalities of gonadal sex, which include XX and XY sex reversal syndromes and true hermaphrodites; and (iii) abnormalities of phenotypic sex, which include true hermaphrodites, female and male psuedohermaphrodites, and cryptorchidism (Mastromonaco *et al.*, 2011).

Intersexuality

Intersex animals, which include true and pseudohermaphrodites, may exhibit decreased reproductive efficiency owing to altered sexual behaviour, ambiguous external genitalia and decreased production of viable gametes (Ladds, 1993).

Intersexuality has been well documented in goats, with an association shown between the natural absence of horns and the intersex condition. An increased prevalence is seen in polled dairy breeds (i.e. those lacking in horns) such as the Saanen, Toggenburg, Alpine and Damascus (Basrur and Kochhar, 2007), while the condition is rare or not reported for breeds such as the Nubian and Angora, as it is likely that these breeds have a different mode of horn inheritance (Crepin, 1958). Hermaphroditism and horned traits in goats are controlled by recessive genes, and these two loci are close to each other on the same chromosome (i.e. they are linked) (Mickelsen and Memon, 2007). The polled condition is a result of mutation at the horn locus. The polled trait (*P*) is dominant to the horned trait (*p*) and appears together with hermaphroditism (*h*) in *PPhh* or *Pphh* goats because the two loci are linked (Asdell, 1944). Intersexes are thus seen mainly among polled animals and rarely among horned (*pp*) animals, where they can occur because of occasional crossing over between the two loci. Most polled caprine intersexes are genetically female (XX), as demonstrated by karyotyping, and the breeding history usually indicates that they are homozygous for the polled trait.

In the female caprine fetus that is homozygous for the polled gene, two manifestations of the *P* gene divert the process of sexual differentiation towards the male despite the presence of two X chromosomes (Wachtel *et al.*, 1978). Polled intersex goats have been shown to be negative for the gene *SRY* on the Y chromosome; this sex-determining gene is one of the major genes responsible for testicular induction (Just *et al.*, 1994). Recently, an 11.7 kb deletion on chromosome 1 of the goat known as the PIS (polled trait related intersex syndrome) deletion has been identified as causing the polled condition (Basrur and Kochhar, 2007). Most caprine intersexes appear female-like at birth, with the presence of small teats and a shortened penis or a bulbous clitoris, but as sexual maturity approaches they start displaying secondary male sexual characteristics such as masculine head appearance and erect hair on the neck. The clitoris may become visibly enlarged. They may behave like bucks, as shown by their aggressive behaviour towards other goats and people. In the presence of a doe in heat, such animals may display a pronounced male libido. Most intersexes usually have intra-abdominal testes, although in some the testes may partially or completely descend. The principal hormone produced by caprine intersexes is usually testosterone (Smith and Sherman, 1994). They may be used as teaser animals as they are sterile. Goats that are true hermaphrodites are often whole-body chimeras (Basrur and Kochhar, 2007). Bhatia and Shanker (1992) have reported the occurrence of a single fertile Saanen × Beetal buck goat with both XY and XXY cell lines in its blood. However, the mechanism by which this mosaic was created could not be determined and at least four of the buck's 16 progeny were culled for breeding problems.

Intersexuality is rare in sheep and is caused by whole-body chimeras due to freemartinism. Unlike the situation in cattle, where the incidence of freemartins in female heterosexual twins is close to 90% (Ruvinsky and Spicer, 1999), in sheep only 1% of heterosexual twins are diagnosed as intersex (Smith *et al.*, 1998). Sheep diagnosed with intersex may show an XX/XY karyotype, which may be the result of mosaicism due to failure of cell division of the zygote, or chimerism. The latter may occur when a polar body is accidentally fertilized by a Y chromosome spermatozoon and the corresponding oocyte is fertilized by an X spermatozoon, with the occurrence of a mixture of genetic material of two diploid cells resulting in an individual with two cell populations (Cribiu and Chaffaux, 1990). Intersexuality in sheep results in masculinization of the reproductive tract (Wilkies *et al.*, 1978) and associated abnormalities. These could range from a complete absence of a cervix and bilateral presence of female and male gonads (ovary and testis), to variable degrees of development of the epididymis accompanied by marked agenesis of the deferent ducts and other structures arising from the mesonephric ducts. Extreme cases may sometimes be accompanied by the presence of a pseudoprepuce and hypertrophy of the clitoris.

Reports of sexual abnormalities in wild ruminants are still scarce (Bunch *et al.*, 1991), although a recent report of a case of true hermaphroditism has been described for an Iberian roe deer (Pajares *et al.*, 2009). The animal showed ovarian-like structures with follicles on the surface and the presence of testosterone-producing testes on necropsy. Based on lack of evidence for *SRY* sequences and the PIS deletion, this roe deer was diagnosed as possessing the *SRY*-negative XX hermaphroditism syndrome. Similar cases have been documented for captive bred red-fronted gazelles (*Gazella rufifrons laevipes*) at the San Diego Zoo in California, where three animals were evaluated for presence of abnormal genitalia (Mastromonaco *et al.*, 2011). One of the gazelles assessed to be a female showed the presence of a vagina, immature uterus and ovaries, and a lack of testicular tissue, but had a male karyotype (58,XY). It was hypothesized that this animal may have lacked the essential genes for promoting male sex determination, which could be due to a mutation or deletion in the *SRY* or a related gene, such as SRY-box 9 (*SOX-9*). These genes have been demonstrated to play a role in case of sex reversal in domestic species. The other two gazelles were diagnosed as a true hermaphrodite and a male psuedohermaphrodite based on their phenotypic ambiguities.

Sperm granulomas

Unlike polled female intersexes, polled male bucks do not generally show any obvious malformation of the external genitalia and undergo normal growth and sexual maturation. However, because of defects in their gonadal ductal system, these animals often suffer from a stenosis of the epididymis (Hamerton *et al.*, 1969). This results in the retention of spermatozoa in large masses in the lumen of the duct. Sometimes, there is a rupture of the seminiferous tubules close to the rete testes, leading to extravasation of sperm in the interstitium, which, in turn, leads to an inflammatory granulomatous reaction often described as a sperm granuloma. This type of epididymal stenosis leading to sterility is detected in less than 30% of polled bucks resulting from polled-to-polled matings, but it is believed that the afflicted bucks are homozygotes for the polled gene (Basrur and Kochhar, 2007).

Thus, homozygosity for the polled mutation (PIS deletion) appears to be disadvantageous to both sexes as in the male it causes poor differentiation of the androgen-dependent duct system and leads to sterility, and in the female it allows the expression of genes in the testicular differentiation pathway leading to the masculinization of the gonad, gonaduct and external genitalia. The mechanism leading to the interruption of Wolffian duct differentiation in X*YPP* animals is not yet understood.

Cryptorchidism

Cryptorchidism may be defined as the failure of one or both testes to descend into the scrotum at the appropriate time for the species of interest, and is considered to be one of the most frequent abnormalities of male sex differentiation (Thonneau *et al.*, 2003). The retained testicle is located anywhere along the normal path of descent, or is diverted to an ectopic location. Unilateral cryptorchidism may or may not adversely affect semen quality, whereas bilateral cryptorchidism results in sterility. Among unilateral cryptorchids, the right testis is retained in the abdomen in about 90% of the animals (Ott and Memon, 1980). Cryptorchidism in goats is usually associated with the intersex condition, except in the case of Angora bucks, where it may also be related to a recessive trait (Skinner *et al.*, 1961). Persistent Müllerian Duct syndrome (PMDS), which is a condition distinct from cryptorchidism, is also known to cause bilateral cryptorchidism. Affected animals are karyotypically male with a well-developed penis and are sterile. Failure of the Müllerian duct to regress in PMDS males is attributed either to arrested or delayed function of the fetal Sertoli cells, which normally produce the high molecular weight glycoprotein dimer, Anti-Müllerian hormone (AMH), during the differentiation of the embryonic ducts. Müllerian duct regression could also be impeded by the absence of receptors for binding this hormone on the embryonic paramesonephric derivatives (Basrur and Kochhar, 2007). This anomaly in goats is inherited probably as an autosomal recessive trait and is most often not recognized as such because it is mistaken for the intersex condition (Haibel and Roijko, 1990).

Cryptorchidism in rams is chiefly attributable to recessive genetic factors and usually occurs bilaterally (Wendt *et al.*, 1960). A recent study (Williams *et al.*, 2007) has provided evidence that this defect in sheep could be associated with the insulin-like hormone 3 (*INSL3*) gene, which is involved in testicular descent and development of external genitalia. Owing to the heritable nature of this condition, it is highly desirable to cull affected animals from the flock.

Beside genetic and chromosomal abnormalities, endocrine disrupting chemicals (EDCs) have also been shown to influence the occurrence of cryptorchidism, as noted in a field study of Sitka black-tailed deer (*Odocoileus hemionus sitkensis*) on Aliulik Peninsula, Alaska. These animals were initially being examined for antler deformities. The incidence of unilateral or bilateral cryptorchidism in this deer population was estimated at 75%, and was associated with testicular lesions and antler deformities (Veeramachaneni *et al.*, 2006). Based on the lesions observed, it was hypothesized that it was more likely that this testis–antler dysgenesis resulted from continuing exposure of pregnant females to an oestrogenic environmental agent(s), thereby transforming testicular cells, affecting the development of primordial antler pedicles and blocking transabdominal descent of fetal testes.

Hypospadias

Hypospadias is an abnormality of the external genitalia in the male that is characterized by a congenital defect of the urethra, which fails to fuse along its entire length, so that it opens at some point along the ventral aspect of the penis or in the perineum. This condition is most commonly seen in humans and has also been reported in various species of domestic animals. The aetiology of hypospadias is not well understood; it seems to be multifactorial and may be related to genetic, endocrinological and environmental factors (Silver, 2000). Familial clustering of hypospadias among first-degree relatives, as well as twin studies and segregation analysis, have supported a strong heritable component in this disorder in humans (Fredell *et al.*, 2002).

In goats, hypospadias is usually associated with the intersex condition and affected animals should be culled. In an Iranian study, 24 cases of caprine hypospadias from eight different flocks were evaluated (Sakhaee and Azari, 2009) within a span of 7 months. All affected animals were confirmed to be males based on external phenotypic characteristics, with no true or pseudohermaphroditism detected. However, no karyotyping was performed on these animals. Penile and testicular hypoplasia was detected in all cases, with one animal even showing the presence of an ectopic penis located between the anus and scrotum. The external urethral opening was usually located along the ventral aspect of the penile shaft, except for one case in which it was seen in the scrotal region. Hypospadias in goats may be asymptomatic, and careful examination of the exteriorized penile shaft is often necessary to diagnose the condition. Depending on the location of the defect, some animals may show signs of urine scalding and leakage in the subcutaneous tissue.

In rams, hypospadias is a relatively rare condition, and occurs in animals with a genetic predisposition, as seen in certain breeds such as the Merino (Dennis, 1979). A large abattoir survey in the UK gave a reported incidence rate of 0.23% (Smith *et al.*, 2006). Hypospadias in sheep may or may not be accompanied by other abnormalities of the reproductive tract, such as a divided scrotum or a shortened penis with absence of a sigmoid flexure. In some cases, the urethral process is found to be absent or tightly adhered to the galea. As in goats, the presence and severity of clinical signs depends on the location of the external urethral opening, with severe ulceration of the scrotum seen especially in rams that have a periscrotal opening.

Puberty

Puberty may be defined as that stage in the process of sexual maturation when the animal becomes capable of reproduction. In males, it is usually indicated by the presence of viable spermatozoa in the ejaculate as well as the display of mating behaviour. Several factors, such as the season of birth, body weight at weaning, plane of nutrition and growth rate, have been shown to influence the onset of puberty. Most small ruminants and cervids are seasonal breeders, and hence age at puberty may be influenced by the season of birth, with spring-born animals reaching puberty faster than autumn-born bucks/kids. As puberty approaches, increases in serum testosterone are preceded by increases in the plasma concentration of luteinizing hormone (LH). Most pubertal animals will start displaying mounting behaviour, flehmen, interest in females in the flock or herd and attempts to gain intromission. The semen quality at this stage is, however, poor and it is highly recommended that full sexual maturity be attained before introducing these animals as potential sires into a breeding flock. At this stage, the semen has a high proportion of sperm abnormalities and low sperm motility. Semen quality improves significantly within a few months of puberty, with the rate of transition from poor to good semen being dependent on breed (Court, 1976), and more fertile or prolific breeds achieving these parameters earlier. Crossbreeding of breeds of low prolificacy with those of high prolificacy and out-of season-breeding ability may in fact hasten puberty in F_1 crosses, as well as help in improving reproductive performance (Kridli *et al.*, 2006).

In goats, photoperiod does not seem to influence testicular growth and spermatogenesis significantly as bucks can attain puberty during long or short day lengths. Instead, age, body weight and nutrition seem to play important roles. The average age of puberty in bucks often depends on the breed of the animal, with the pygmy breeds achieving puberty earlier (at 2 to 3 months) than Nubian or Boer bucks (4 to 5 months) (Edmondson *et al.*, 2012), whereas Damascus bucks achieve puberty by the age of 24 to 48 months (Epstein and Herz, 1964). Boer goat kids have been found to be capable of spermatogenesis when they are just 84 days old, with sperm found in the epididymis at 140 days of age (Skinner, 1970). In the northern hemisphere, most goat breeds usually attain puberty by the age of 4 to 5 months; natural adhesions between the prepuce and the penis, which preclude the buck from successful mating before puberty, start separating under the influence of testosterone a month earlier (Plate 13). By the age of 4 to 5 months, it is advisable to start keeping the male and female kids separately, as buck kids will start displaying libido and fertile mating may be possible, although bucks belonging to the Nubian or Boer goat breeds do not produce good quality semen until they are approximately 8 months old.

The average age for achieving puberty in rams is around 6 months, depending on the breed of the animal and season of birth. Puberty in rams, unlike that in goats, is greatly influenced by the season of birth, with ram lambs born in the spring and receiving adequate nutrition achieving puberty at 5 to 6 months in the subsequent autumn. In contrast, autumn-born lambs do not achieve puberty until they are about 10 to 12 months old as the long days of the subsequent spring and summer delay its onset. Similarly, rams exposed periodically to cycling ewes also tend to reach puberty faster and display an earlier onset of mating activity than non-exposed rams. Spring-born rams from temperate climates show a gradual increase in testicular size that parallels changes in growth rate, with a more rapid phase of testicular growth occurring in the autumn. Changes in gonadotrophin-releasing hormone (GnRH)-induced LH secretion drive the final maturation of the testes, including the stimulation of testosterone secretion (Fitzgerald and Morgan, 2007). Reproductive efficiency in sheep is improved by various methods, such as selection, the introduction of new sires and crossbreeding. Thus, breeds such as the Romanov, which are characterized by early sexual maturity (3 months old), high prolificacy and out-of-season breeding ability, can be crossed with late-maturing breeds such as the Awassi, which does not achieve puberty until 8 to 9 months old. In rams, body weight rather than chronological age is the more important factor regulating onset of puberty. Hence, optimal post-weaning nutrition management has a strong influence on lamb weight gain, which, in turn, is related to testicular growth and puberty (Mukasa-Mugerwa and Ezaz, 1992). Nutritionally induced fetal growth restriction also has a significant impact on the onset of sexual maturation (Da Silva *et al.*, 2001). High maternal nutritional intakes throughout gestation in adolescent pregnant sheep, resulting in rapid maternal growth rates and significant reductions in both placental and fetal mass at term, can cause a significant delay in the time of onset of puberty in male lambs.

Almost all cervids, like sheep, have a defined breeding season (rut) related to photoperiod, which greatly influences the attainment of puberty and secretion of reproductive hormones. The breeding season among cervids is strongly linked to the development of secondary sexual characteristics such as increases in testicular diameter and the musculature of the neck, as well as antler growth and size. Attainment of puberty is evidenced by steady increases in the serum level of testosterone, which reaches a peak associated with the beginning of reproductive ability. In red deer, testicular size starts to increase in the autumn, when the male is usually around 3 months old; this increase is arrested in the subsequent spring, though the actual size does not decrease. The increase in testicular size resumes in the second autumn of life, by which time the male has attained a threshold live weight. Thus, onset of puberty in red deer extends from 9 to 15 months of age (Lincoln, 1971). In contrast, white tailed deer (WTD) can reach puberty as early as 6 months, with males known to breed successfully by the age of 1 year (Schultz and Johnson, 1992). With the onset of puberty, WTD buck fawns can display small calcified antler buttons in the first winter of life. Reindeer calves can also attain puberty and sexual maturity by the age of 6 to 8 months, and in the absence of an adult male, larger male calves can take over a dominant role and successfully breed with receptive females. Among tropical species of deer, such as the chital or spotted deer, male fawns attain puberty by the age of 14 months, with the first rut occurring after the first hardening of the second antlers. Male sexual development of the endangered male Pampas deer of South America is usually determined based on the antler characteristics (Ungerfeld *et al.*, 2008). These males are known to grow small, single-spiked, 2–8 cm long antlers and begin displaying courtship behaviour by 5 to 6 months of age. Puberty is thought to occur by the age of 1 year although fawns born in the spring are not sexually mature by the following autumn.

Breeding Soundness Evaluation

Breeding soundness evaluation (BSE) of the ram or buck is an overall clinical assessment of the capacity to impregnate a certain number of healthy females (a common standard is 50 females for a mature ram/buck) during a defined breeding season, usually the fall (autumn) in North America. Most importantly, BSE provides criteria for the identification and culling of potential sub-fertile/infertile males, as they could account for major genetic changes in a flock. A positive correlation has been found between use of a male proven satisfactory on a BSE and both the increased percentage of pregnant females and the size of the offspring crop at the end of the breeding season. The use of proven rams has also resulted in more lambs being born early in the lambing season, which, indirectly, results in heavier lamb weights at weaning. Identifying satisfactory breeders also helps to rule out this parameter as a potential cause of infertility in the herd or flock. Besides confirming a suspected case of male infertility, the result of the examination may allow the owner to receive compensation or a replacement animal from the seller if the ram or buck is insured. A complete BSE includes the evaluation of anatomical and structural correctness, testing for freedom from disease, optimal body condition score, examination of external and internal genitalia, and semen collection and evaluation. Ideally, a BSE should be performed at least 30 to 60 days before a breeding season, thereby allowing time to recheck or replace sub-fertile rams and bucks.

The traditional BSE performed in rams and bucks is usually not performed in farmed male cervids because of the obvious difficulty of handling and examining the conscious male, especially in the rutting season. However, male cervids can be restrained in a crush or squeeze chute or tranquillized/anaesthetized for the purpose of semen collection and evaluation.

Physical examination

The first portion of the BSE is a thorough physical examination for general health (Ott and Memon, 1980). The physical examination should include observation of all characteristics that may interfere with a ram's or buck's ability to locate ewes and does in heat, and to successfully breed with them. A body condition score (BCS) of 3–3.5 out of 5 is usually recommended for rams and bucks when they are entering the breeding season. To be classified as a satisfactory potential breeder, a ram or buck should be in good flesh. Thin or excessively fat animals should be avoided, but a little reserve flesh is desirable, as males can be expected to lose weight during the breeding period. Red deer stags are recommended to have a BCS as close as possible to 5 (on a scale of 1 to 9), as they will lose 25–30% of their pre-rut weight even if fed well (Haigh and Hudson, 1993). It has been reported that shearing rams before they are used for breeding in late summer and early autumn results in greater fertility (Hulet, 1977). Shearing apparently helps to alleviate the detrimental effects of high ambient temperature on fertility in rams. In addition to body condition, mucous membranes and sleekness of the hair coat should be evaluated for evidence of parasitism, malnutrition or chronic infections. Deworming before the breeding season has been suggested as a prudent practice.

The ram/buck should be free of known genetic or possibly hereditary defects such as hernias, jaw malformations or supernumerary teats. Bucks especially should not be phenotypically polled because of the association of this trait with the intersex condition. The presence of extra teats apparently has no relationship to the fertility of the buck, but if the trait is passed on to the next generation, the buck's female offspring may be less suitable for milking (Schönherr, 1956). Teeth, eyes, feet and joints should be in good condition so that the ram/buck can continue to eat well and can follow and mount the ewe/doe in oestrus. Males with severe structural defects, such as post-leggedness, tend to exert more pressure on their pasterns, resulting in abnormal growth of hooves. Such animals require more frequent trimming of their feet. Similarly, arthritic conditions, foot rot, foot abscesses and overgrown hoofs may severely impair the ram or buck from satisfactorily performing as per expectations. Structural and physical conditions that cannot be corrected should warrant culling of the animal to prevent future economic losses.

In addition to the physical examination, it is important to consider an animal's breeding history and review records of the past breeding performance, such as length of lambing or kidding period. A thorough medical history, with emphasis on recently treated disease conditions; especially fever, may raise testicular temperature and render the male temporarily infertile for up to 60 days. Lastly the optimal age of the animal should be considered: the optimal breeding age is over 6 months and up to 4 years old.

Disease testing

Ovine epididymitis caused by *Brucella ovis* is an important cause of orchitis and infertility in rams. Similar symptoms have been reported in the male red deer in New Zealand. *B. ovis* is occasionally associated with abortion in ewes, and can cause increased perinatal mortality in lambs. Experimental infections have been reported in goats and cattle, but there is no evidence that these species are infected in nature. *B. ovis* is often transmitted via homosexual activities, or via the ewe during the breeding season. Rams often become persistently infected, and many of these animals shed *B. ovis* intermittently in the semen for 2 to 4 years or longer. Direct non-venereal ram-to-ram transmission is poorly understood and may occur by a variety of routes, including oral transmission. Shedding has been demonstrated in the urine as well as in semen and genital secretions. The demonstration of the existence of genital lesions (unilateral, or, occasionally, bilateral epididymitis) by palpating the testicles of rams may be indicative of the presence of this infection in a given flock. However, this clinical diagnosis is not sensitive enough because only about 50% of rams infected with *B. ovis* present with epididymitis. Moreover, the clinical diagnosis is extremely non-specific due to the existence of many other bacteria causing clinical epididymitis. The most frequently reported isolates causing epididymitis in rams include *Actinobacillus seminis*, *A. actinomycetemcomitans*, *Histophilus ovis*, *Haemophilus* spp., *Corynebacterium pseudotuberculosis ovis*, *B. melitensis* and *Chlamydophila abortus* (formerly *Chlamydia psittaci*). It must be emphasized that many palpable epididymal lesions in rams are sterile, trauma-induced spermatic granulomas (OIE, 2013).

Serological testing of all breeding males is, therefore, strongly recommended before the start of the breeding season, and may be incorporated as an annual exercise, preferably during the periods of lowest sexual activity. In single-sire flocks, the ram should be tested at purchase and retested 30 days later. The complement fixation test (CFT), agar gel immunodiffusion (AGID) test and indirect ELISA (I-ELISA) using soluble surface antigens obtained from *B. ovis*, are currently available, and are preferred for routine diagnosis. The sensitivities of the AGID test and I-ELISA are similar, and sometimes the I-ELISA has higher sensitivity than CFT. A combination of the AGID test and I-ELISA seems to give the best results in terms of sensitivity, but in terms of simplicity and cost, the AGID test alone is the most practicable test for the diagnosis of *B. ovis*. Despite this, because of the lack of standardized methods recognized at the international level for I-ELISA and AGID, the prescribed test for international trade remains the CFT.

Direct diagnosis of *B. ovis* infection is made by means of bacteriological staining (Stamp's method) or isolation of *B. ovis* from semen samples or tissues of rams, or vaginal discharge and milk of ewes, on adequate selective media (modified Thayer–Martin's medium). Molecular biological methods have been developed that could be used for complementary identification based on specific genomic sequences. PCR methods provide additional means of detection. Rams that test positive should be immediately culled. As *B. ovis* has the potential of being in a latent stage and not showing a positive test in a multi-sire flock, the entire group of rams should also be retested at the end of the breeding season/at shearing, once again culling all positive rams.

Examining the scrotum and measuring its circumference

Scrotal examination includes palpation of the scrotum and its contents as well as assessment of testicular size by measuring the scrotal circumference. The testes and the epididymides should be examined by visual assessment and palpation for tone and symmetry. Testicular tone may be scored quantitatively in bulls and rams as softness or flabbiness may indicate testicular dysfunction or degeneration. Likewise, irregular contours, lumps or an excessively hard texture may be due to testicular fibrosis or calcification secondary to degeneration and inflammation (Plate 14). As additional procedures for evaluation of the scrotal contents, thermography and testicular ultrasonography are fast gaining popularity, even under field conditions. The caput (head), corpus (body) and cauda (tail) can easily be palpated in the ram and buck. The examination of the cauda epididymis in rams is particularly important in view of the infectious causes of epididymitis in this particular species. The scrotal neck and the spermatic cord should be palpated for presence of vasectomy scars as well as for signs of scrotal hernia. Lastly the scrotal skin should be examined for signs of dermatitis, mange and trauma.

Scrotal circumference, testicular volume and daily sperm output are highly correlated with each other. Measuring the scrotal circumference is an integral part of BSE in bulls and The Society for Theriogenology has developed scoring methods for evaluating the bull and ram. Although scrotal circumference or testicular diameter charts are lacking for most breeds of goats, it is reasonable to assume that, as for bulls and rams, the male with the larger testicular size at a given age is likely to produce more sperm. There is a positive correlation between scrotal circumference and body weight in bucks.

Similarly, there is considerable evidence to suggest that testicular size serves as an indicator of ram fertility, is positively related to ewe fertility under heavy breeding pressure (Schoeman *et al.*, 1987) and, when measured at puberty, is a more accurate indication of ovulation rate in female relatives than either prepubertal or post-pubertal size. There is a significant correlation between scrotal circumference (SC) and body weight (BW) in rams of all breeds (Braun *et al.*, 1980). The (US) Western Regional Coordination Committee on Ram Epididymitis and Fertility recommends that ram lambs over 70 kg (150 lb) have an SC of greater than 30 cm and that the SC of yearling rams (12–18 months) should measure >33 cm. Producers of seed or breeding stock should set higher standards. Rams weighing 115 kg (250 lb.) or more should have an SC > 36 cm (Kimberling and Parsons, 2007). As an approximate guideline for dairy goat breeds in the USA, bucks weighing more than 40 kg should have an SC of at least 25 cm. Bucks of British breeds had a mean scrotal circumference of 24 cm when they reached sexual maturity at

5½ months, but the value had decreased by several centimetres in the following January and February (Ahmad and Noakes, 1996a,b). Rams and bucks failing to meet these criteria should be culled, as they are likely to be incapable of producing enough semen to service multiple females during the breeding season.

Measurements of SC should be recorded by pulling the testes firmly down into the lower part of the scrotum and placing a measuring tape around the widest point. Scrotal circumference may vary with season and body condition. Photoperiod effects appear to account for a 2 cm difference in SC in Alpine and Nubian bucks kept in environmental chambers (Nuti and McWhinney, 1987). In contrast, a study of Creole goats found that nutrition, rather than photoperiod, determined SC in this breed as managed locally (de la Vega *et al.*, 2006). In Jordan, Damascus goats had the largest SC in the spring, when day length was increasing; this is their normal breeding season (Al-Ghalban *et al.*, 2004). SC in cervids also increases markedly and peaks at about the time of onset of the rutting season (Table 10.1).

A threefold increase in scrotal size between early summer and autumn has been noted in red deer stags, and the SC of male elk increases by about 50% during this period (Haigh *et al.*, 1984). After this, it declines fairly steadily until it reaches its nadir in spring. Several types of measuring tapes intended for measuring SC are available on the market (Plate 15). Some of these also have a small tension marker, which indicates the adequate tension required to obtain an accurate measurement. In others, the tension applied to the measuring tape should be just sufficient to cause a slight indentation in the skin of the scrotum. Placement of the thumb of the hand holding the neck of the scrotum between the cords should be avoided as it will cause separation of the testes and an inaccurate measurement.

Examination of the external and internal genitalia

The prepuce and the penis should be visually inspected closely as well as palpated for signs of injury, pizzle rot (ulcerative posthitis) and adhesions. The penis should be extended either by restraining the animal in its dock or during electroejaculation. The tip of the penis, especially the urethral process, should be examined for signs of adhesions or ulceration. Occasionally, mineral deposits/crystals adhered to the prepuce might be observed; such animals should be observed closely for signs of urinary calculi and obstruction.

Palpation of the accessory sex glands is either not possible or not as easy to perform as it is in bulls. However, the prostate and the seminal vesicles can be felt in most rams and bucks by inserting a gloved, lubricated finger into the rectum. This can be done in conjunction with cleaning out any faecal pellets from the rectum just before electroejaculation. Signs of pain, asymmetry or abnormal texture of these glands might give an indication of an underlying disease process.

Semen collection

Semen collection and evaluation comprises the final part of a BSE. A semen sample may

Table 10.1. Average scrotal circumferences of white tailed deer (WTD).[a]

Age (years)	No. of animals	Live[b]	Testicles[c]	Average scrotal circumference (SC, cm)
1.5	7	6	1	18.91
2.5	11	11	–	18.04
3.5	2	2	–	18.50
4.5	2	2	–	20.25

[a]Preliminary unpublished data from the authors.
[b]Deer examined live (and semen collected by electroejaculation).
[c]Deer measured when submitted for post-mortem epididymal flushing.

be satisfactorily collected from rams, bucks and wild cervids using either an artificial vagina or by electroejaculation.

The artificial vagina

The artificial vagina (AV) is the method of choice for collecting semen in small ruminants because it gives the most reliable and representative semen sample for laboratory evaluation. At the same time, it also requires adequate training of the males and the presence of an oestrous female for optimal stimulation. During the breeding season, most bucks will be willing to mount any restrained female. It may be beneficial to induce oestrus in cycling does or ewes with 5 mg of prostaglandin F2 alpha, or to treat them with oestrogen (1 mg oestradiol administered 1–2 days previously) (Memon *et al.*, 1986) so as to have a more willing mount. The training period for rams and bucks for semen collection using an AV can be very brief, but may extend up to 3 weeks depending on the individual animal and season (Terrill, 1940). The AV consists of the following parts: (i) a rubber tube that is 8 inches long and 2 inches in diameter, and is equipped with an air and water valve; (ii) a rubber pressure bulb; (iii) a latex inner liner; (iv) a latex semen collection funnel; (v) two wide rubber bands; (vi) a 15 ml graduated Pyrex® collection vial; and (vii) a non-drying, non-spermicidal lubricant (Memon *et al.*, 2007). The casing of the AV is filled with warm water to provide an internal temperature of approximately 39°C at the time of collection. Air is then added to create a pressure to simulate the cervix; the AV is lubricated immediately before collection. After two false mounts, the buck or ram is allowed to serve the AV. The penis is never handled directly, but is directed by grasping the sheath or the prepuce. The buck or ram will then seek the AV and thrust into it.

Electroejaculation

Electroejaculation (EE) is frequently carried out in untrained bucks and rams as the presence of an oestrous female is not required to stimulate ejaculation during this procedure. It is particularly suited for captive cervid species, as it can be successfully carried out in restrained, sedated or anaesthetized animals. EE, although convenient, also has its own drawbacks. Some animals do not respond well to the electrical stimulus, especially if a second or third collection is desired. They may fail to ejaculate, or they may contaminate the semen sample with urine. Also, greater concentrations of sodium and potassium are usually found in both the sperm and seminal plasma obtained by EE. Further, EE yields ejaculates of larger volumes and lower sperm concentration than those obtained by AV owing to stimulation of the accessory sex glands, though it does not affect the motility of spermatozoa (Akusu *et al.*, 1984). As seminal plasma has been found to be generally detrimental to the storage of spermatozoa (Nunes, 1982), EE is generally not a preferred method of semen collection, especially for the purpose of cryopreservation. However, recent research in Guirra rams has shown that high-quality semen can be successfully collected via EE without noticeable differences from semen collected using an AV. The semen collected by EE was also shown to have a higher quality even after cryopreservation. Quinn *et al.* (1968) suggested that spermatozoa collected by AV were more resistant to cold shock than those from EE. The post-thaw results for both methods yielded similar seminal quality as measured by both microscopic analysis and computer assisted sperm analysis (CASA). In this study, a significant effect on capacitation status and acrosomal integrity was detected: the percentages of non-capacitated viable and acrosome-intact viable cells were higher in samples collected by EE; the numbers of acrosome-reacted viable spermatozoa were also lower from EE than via AV.

EE can be performed in certain larger cervids, such as the wapiti and the red deer, in a standing position using adequate restraint, such as a squeeze chute or crush, and without drugs, but if these species are in full antler they have to be sedated or anaesthetized. Small species of cervids, such as WTD and the sika deer, are preferably anaesthetized to prevent undue excitement and symptoms relating to capture myopathy (Plate 16). The authors' preferred method for field anaesthesia of WTD and mule deer for semen collection uses a combination of Telazol® plus xylazine. Dosage for this combination is 1mg/lb of each drug IM. This regime usually allows more than adequate time for semen collection. Action of the xylazine may be reversed using agents such as tolazine, atipamazole or yohimbine at the end of the procedure. Tolazine (1–2 mg/kg) may be given IM to decrease the cardiovascular problems encountered with IV administration. For more detailed information on sedation and anaesthesia in cervids, readers are encouraged to refer to specific texts, such as the *Handbook of Wildlife Chemical Immobilization* (4th edition) (Kreeger and Arnemo, 2012), and also to access web sites such as Safe Capture International (www.safecapture.com). Some practitioners prefer to use inhalation agents, but these can be difficult to administer in the field. Supplemental oxygen and intubation may be indicated in some situations. Rams and bucks can be restrained against a wall, chute or in lateral recumbency, and semen then collected satisfactorily.

The process of EE has two phases and involves the stimulation by electric current of emission, erection and ejaculation. The emission of semen from the ampullae and vasa deferentia into the pelvic urethra is a sympathetic response effected by contraction of smooth muscle in response to stimulation of the lumbar sympathetic nerves that form the hypogastric nerve. In contrast, erection and ejaculation are controlled by sacral parasympathetic nerves that form the pelvic and internal pudendal nerves located at the level of the pelvic urethra and the body of the prostate gland (Ball 1986).

Several models and types of electroejaculators are available in the market (Plate 17). Most electroejaculators available for small ruminants can be used for smaller species of cervids, whereas hand-crafted probes may be used for red deer and Wapiti (size 4 cm diameter for red deer and 6 cm diameter for wapiti). These usually have two or three ventrally placed electrodes. The instrument from Lane Manufacturing Inc. (Denver, Colorado) has a commercial wapiti probe that may be attached to its Pulsator IV ejaculator unit and used on larger cervids. The older commercial electroejaculation probes had ring electrodes, which surrounded the barrel of the probe. These electrodes stimulated nerves other than those required for EE. In particular, nerves located dorsal to the rectum and supplying muscles in the hind limb were stimulated, resulting in strong contractions of the muscles of the legs, thighs and back. These contractions are severe enough with some types of probes to cause haemorrhage and bruising of affected muscles and stiffness for a few days. Newer probes have longitudinal electrodes on the ventral side that concentrate the electrical stimulation in the area where the relevant nerves are located, with less stimulation of motor nerves that supply skeletal muscles (Plates 17 and 18). This simple modification of probe design has decreased the intensity of the physical response to electrical stimulation significantly. In general, sine wave currents appear to be more effective than other forms of electrical stimuli in achieving ejaculation (Carter *et al.*, 1990).

Procedure

Semen collection using an electroejaculator can be performed in cervids, rams and bucks with the animal either standing or laying on its side. Anaesthetized cervids are usually collected in the lateral recumbency with an assistant holding the back legs and another assistant holding the head (Haigh and Hudson, 1993). A cleaner collection can be obtained by initially snaring the penis with a cotton bandage and directing it into a collection vessel, so as to avoid preputial contamination. For all three animals (cervids, rams and bucks), care is taken to collect the ejaculate in fractions, so that every collection can yield some viable sperm.

Semen in rams and bucks can be collected satisfactorily without manually extending the penis if it is for a BSE only; in such cases the animal can be restrained against a wall in a standing position. However, if semen is to be collected for extension and freezing, it is generally recommended to manually extend the penis in order to avoid contamination during collection from debris falling from the body or from the prepuce. In this case, the ram or buck is then set on its rump or restrained in a lateral recumbency. The shaft of the penis is grasped through the wall of the prepuce and gradually pushed out, so that the glans penis can be grasped and held with a gauze sponge. A lubricated finger is inserted into the rectum to remove faecal pellets. The lubricated probe is then inserted into the rectum, with downward pressure on the front of the probe; the area of the seminal vesicles is massaged several times. With the probe held on to the area of the seminal vesicles, an electrical charge is applied for 3–4 s. The electrostimulation is stopped briefly (3–4 s) while further massage is applied with the probe. This cycle is repeated until a suitable (usually 0.5–2 ml) sample of semen is collected (usually 2–3 electrostimulations for rams and bucks). The semen sample is usually best collected into a warm, 17 × 100 mm plastic test tube, but a whirl-pack bag can also be used. The sample is placed into a heat block to maintain a temperature of near 37°C until the BSE is completed. The urethral process usually twirls rapidly during ejaculation and sprays the semen into the test tube. An occasional ram or buck may have a urethral defect that allows semen to flow out of the side of the head of the penis. This is not normal, and hence not desirable, though the actual effect of this type of defect on conception is not known.

In anaesthetized cervids, more stimulation is needed than in standing collection for the ram or buck. Anaesthetized males are also more likely to spoil the semen collection with urine contamination or accessory sex gland contamination. Also in cervids, particular care needs to be taken to collect each fraction of the ejaculate separately because some of the fractions may be toxic to sperm or make semen processing difficult, as described below. The different fractions of the semen collected are as follows. The pre-ejaculate, which contains seminal fluid, is clear and watery, whereas the sperm-rich fraction is usually cloudy. Yellow vesicular fluid is sometimes seen early in the breeding season. This fraction is toxic to sperm and, accordingly, care needs to be taken to exclude it from the sperm-rich fraction. If contamination from urine or accessory sex glands occurs, then the affected fraction should be discarded. A very thick 'honey like' fraction is sometimes noted in cervids. The origin of this fraction is unclear, but it does not seem to have any detrimental effect on semen quality; however, it does make the evaluation or processing of semen more difficult (authors' personal observations).

Although EE enables the collection of semen from untrained animals very effectively, it also raises some concerns about animal welfare. Improper techniques, equipment and use of high voltages often lead to strong muscular spasms, urination and vocalization. Confident, sexually active bucks usually do not resist entering the collection compound and show no signs of aversion to a repetitive EE programme (Ball, 1986). In contrast, individuals with temperament difficulties may not adjust to EE and may display signs of aversion to repeated treatments. Samanta *et al.* (1990) recorded respiration, pulse rate, body temperature and haematological parameters in two groups of bucks on which EE was used every 3 days and every 7 days for a month. No significant difference was noted between the groups. Recently, Orihuela *et al.* (2009b) showed that the electric stimulus during EE increased the heart rate and cortisol concentrations in rams, and recommended that the use of IM anaesthesia with ketamine and xylazine might improve their welfare during the procedure (Orihuela *et al.*, 2009a). Accordingly, techniques to minimize the stress response to EE should be developed for practical application.

Many modern probes used for EE have a pre-programmable current pattern starting at the lowest voltage level and increasing gradually until ejaculation occurs, at which stage the current can be stabilized. This allows for more control over the unwanted muscular reactions of the back and limbs typically seen at the voltage required for ejaculation. Probes with longitudinally placed electrodes produce better results than do probes with ring electrodes. Sedation or analgesia may be considered as an option to reduce anxiety and discomfort associated with the procedure. Similarly, prolonged stimulation of animals during EE should be avoided. In short, every effort should be aimed at adopting the least stressful technique that is efficient enough to get the desired results. Semen collection in

WTD in a restraining chute and sedated pre-collection with haloperidol has been reported as an alternative to anaesthesia (M.R. Woodbury, personal communication).

Epididymal flushing

Epididymal flushing is an alternative method of semen collection usually reserved for post-mortem sperm recovery from valuable animals. Although the procedure is quite simple and can be performed with minimal equipment, care has to be taken to ensure that the testes harvested from the animal are processed as soon as possible, as there is a sharp decline in the number of viable sperm by 24 h post-mortem. Owners sending testes to the authors' laboratory are recommended to do so by packaging them in a plastic bag kept over ice in a Styrofoam box for insulation. The source of post-mortem spermatozoa is the cauda epididymis, as the maturation stage and fertility of spermatozoa stored in this part of the epididymis are similar to those of ejaculated spermatozoa. In the laboratory, the testes are cleaned, and the cauda epididymis and vas deferens are dissected out. Sperm recovery from the cauda epididymides may be accomplished by two methods: (i) by cannulating and retrograde flushing of the vas deferens with the preferred commercial extender; or (ii) by a flotation technique in which 10–15 cuts/slashes are made horizontally in the distal epididymis and vas deferens with a No. 10 scalpel blade. The epididymides are placed in a 50 ml conical tube or a Petri dish and covered with approximately 5 ml of the same commercial semen extender. The samples are then agitated and incubated at room temperature for 10 min. The retrograde flushing method is preferred for sample collection as this results in recovery of a less contaminated (with blood) sample and has also proven to be superior when the semen is cryopreserved (Martínez-Pastor *et al.*, 2006). Sperm recovery using the flotation and cutting technique is quicker and easier to perform and may be used in certain instances where it is difficult to perform the retrograde flush.

Semen Handling and Processing

Sperm, though capable of being vigorous and motile, are also fragile cells. They can be easily damaged by changes in several environmental conditions. Of particular importance are exposure to toxic chemicals and thermal stress.

It is important to keep equipment used for semen collection clean; however, the use of soap and strong disinfectants can prove disastrous as these chemicals are potent spermicidals. Hot running water is usually sufficient to clean most AV liners but, if the use of soap becomes necessary, it should be of a non-residual type such as Alconox (Alconox Inc., New York). Disinfectant solutions or soaps containing disinfectants (e.g. povidone–iodine or chlorhexidine) may damage spermatozoa. Povidone–iodine concentrations as low as 0.05% have been shown to render spermatozoa completely immotile within 1 min of contact (Brinsko *et al.*, 1990). Rubber products such as AV liners are particularly vulnerable to harbouring spermicidal residues, which can cause varying degrees of damage to sperm. Some kill sperm immediately, some slowly. Hence, the damage is not always detected under a microscope during semen evaluation, but may reflect as a decreased pregnancy rate. Rinsing the rubber liner with tap water and then keeping it immersed in an alcohol bath (70% isopropyl alcohol) is usually sufficient to get rid of most debris. Sometimes, it may be worthwhile to rinse off the liners and other equipment with deionized water. The liner is then hung in a cabinet for drying. Gas sterilization with ethylene oxide can be used, provided that the liners are allowed to air for 48–72 h before use. Irrespective of the method of collection, all storage vessels, dilution solutions, glassware and pipettes that come into contact with semen should be warmed to 37°C to prevent cold shock, which will kill sperm and reduce motility estimates. Equipment can be warmed on a slide warmer or kept in an incubator until just immediately before usage. A heated microscope stage is an invaluable tool while observing wet mounts. When the weather is cool, the manipulation and evaluation of semen should be done in a controlled environment to avoid the effects of cold shock. Some possible sites for the microscope and the other laboratory equipment needed include a heated office or shed, a pickup cab, the back seat of a car or a mobile trailer. To prevent contamination or compromised sperm quality, it is best to use disposable plastic tubes, syringes and pipettes.

Semen Evaluation

Semen evaluation should be done immediately after collection. Semen colour, odour and appearance should not be ignored as these may provide valuable information. In small ruminants and cervids, semen samples are usually very concentrated owing to the small volume that is ejaculated (0.5–2 ml). The semen is usually creamier and thicker in highly concentrated samples. Dilute low-concentration samples may look watery and grey. Occasionally, the semen sample may get contaminated with urine, which can give it a characteristic odour. Contamination with urine can kill spermatozoa. Haemospermia may cause the semen to appear pink or red in colour, whereas seminal vesiculitis can result in flocculent clumps of purulent material.

A microscopic examination should be performed to assess sperm motility, morphology and to look for the presence of white blood cells. White blood cells indicate infection, often by *B. ovis*, which causes epididymitis in older rams. In young virgin rams, white blood cells (WBCs) often indicate infection with *Actinobacillus* spp. or *Histophilus* spp. A ram with >10 WBCs per 20× field and no palpable lesions should be considered for testing and re-evaluated. If *B. ovis* is indicated in a mature ram, the animal should be tested and culled if found positive. Other conditions that can reduce fertility and can be detected by the presence of WBCs are caseous lymphadenitis and spermatic granulomata.

Motility

As done routinely, sperm motility is a very subjective measurement and is affected by many things, such as the diluent, equipment (glassware, pipettes, etc.), prostatic fluid, seminal pH and ion composition and, most importantly, by temperature. Care must be taken to keep all equipment coming into contact with semen at 37°C to prevent cold shock, which will kill sperm or reduce motility estimates. Although it is possible to effectively evaluate sperm motility by using bright-field microscopy, it is preferable to use phase contrast microscopy and to employ a thermo-controlled heating stage to reduce the chances of sperm cold shock. The semen sample collected is usually evaluated for gross and individual sperm motility. A small (~10 μl) drop of fresh, undiluted semen is placed on a pre-warmed (37° C) glass slide without a coverslip, and wave motion is observed using a 10× phase contrast objective (magnification × 100). An estimate of the vigour of the wave motion is graded as follows: very good (++++, vigorous swirls), good (+++, slow swirls), fair (++, no swirls, but generalized oscillation) or poor (+, sporadic swirls) (Memon *et al.*, 2007).

Progressive motility of individual spermatozoa may be evaluated by diluting a fresh drop of raw semen with a drop of pre-warmed phosphate-buffered saline or 2.9% sodium citrate solution (pH 7.4). A small drop of this diluted semen is then examined under a 40× phase contrast objective (magnification ×400) using a coverslip. It is common practice to observe at least five different fields randomly in order to be able to subjectively evaluate the number of progressively motile spermatozoa (to the nearest 5%) in each field. Progressive motility in bucks ranges from 70 to 90% (average 80%). Rams should have more than 30% progressively motile cells for a 'satisfactory' rating and more than 70% for an 'exceptional' rating (Bulgin, 1992; Kimberling and Parsons, 2007). Sperm subjected to cold or chemical shock may often exhibit circular or reverse motion. However, reverse motion may also be observed in sperm that have a high percentage of midpiece abnormalities. Motility can be depressed outside the breeding season. Similar values are used by the authors for evaluation of cervid semen.

Subjective assessment of sperm motility by standard microscopy can be highly variable, even among highly trained personnel. In an attempt to eliminate these variations, many objective techniques for evaluating sperm motility have been developed; these include computerized analysis, time-lapse photomicrography, frame-by frame playback videomicrography and spectrophotometry. CASA analyses the movement of a large number of sperm and, from this, determines a number of parameters, including the percentage of motile sperm, percentage of progressively motile sperm, amplitude of lateral head displacement, average path velocity and curvilinear velocity. CASA thus provides a rapid, precise and validated objective sperm motion characteristic (Holt and Palomo, 1996), and measurements can be more closely related to fertility than are subjective motility measurements (Farrell *et al.*, 1998). This technique has been applied to the study of short-term (Joshi *et al.*, 2001) and long-term preservation of ram spermatozoa (Edward *et al.*, 1995; Bag *et al.*, 2002a,b; Joshi *et al.*, 2005, 2008). For example, in a recent study, CASA-derived sperm motion characteristics revealed that the semen quality of native Malpura rams was superior to that of crossbred Bharat Merino rams during a major breeding season in a semi-arid tropical climate (Kumar *et al.*, 2010). In cervids, CASA systems have been used to analyse sperm motility parameters, and it has been possible to identify subpopulations defined by motility patterns, and their changes under different circumstances (Martínez-Pastor *et al.*, 2005a,b). In a recent study, four well-defined sperm subpopulations could be identified by CASA in Florida goat buck ejaculates, showing that the spermatozoa of each buck had different motility patterns (Dorado *et al.*, 2010).

The main disadvantage of using these sophisticated systems lies in the equipment and cost, thereby limiting the use of these techniques to universities, research laboratories or commercial studs. Subjective assessment still remains common practice for evaluating semen motility in the field.

Morphology

The assessment of sperm morphology is probably the most useful and important aspect of the semen examination. The goal of evaluating sperm morphology is not only to identify normal or abnormal sperm, but to also place these abnormalities into different categories. Sperm morphology can be assessed by diluting the semen sample with a fixative, such as buffered formalin, in order to render the spermatozoa non-motile, while maintaining their essential structures. The idea is to get a sample of non-motile spermatozoa that are sufficiently separated and spread across the microscopic field so as to enable their proper visualization and evaluation. These samples may then be evaluated by directly placing a drop of unstained semen on a glass slide with a coverslip on and observing with high-power (either 40× or, preferably, 100×) phase contrast or differential interference contrast microscopy microscope.

Evaluation of sperm morphology on a wet mount of fixed cells viewed by phase contrast or differential interference contrast microscopy eliminates artefacts seen in stained smears (Ball *et al.*, 1971), such as distal midpiece reflexes, which can be caused by incorrect osmolality of the stain. These types of microscopes are seldom available to veterinary practitioners, and in that case, samples can also be observed with a bright-field microscope, provided that they are stained with an appropriate stain that assists in viewing individual sperm. Eosin–nigrosin is probably the most commonly used stain as it allows assessment of the morphology as well as the membrane integrity of spermatozoa. The nigrosin stain produces a dark background on which the sperm stand out as lightly coloured objects. Normal live sperm exclude the eosin stain and appear white in colour, whereas 'dead' sperm (i.e. those with loss of membrane integrity) take up eosin and appear pinkish in colour. Toluidine blue is another stain that can be used for staining spermatozoa; however, the staining process is more difficult and time-consuming than with eosin–nigrosin. For the latter, a drop of diluted semen is placed on a warmed microscope slide (at 37°C) and mixed with a drop of the eosin–nigrosin stain. Then a thin smear of the mixture is drawn across the slide, and the slide is air dried. A coverslip is applied to the preparation and sample is examined at either 40× or 100×, with the latter requiring an oil immersion objective. Typically 100–200 spermatozoa are counted and classified as normal, having primary abnormalities or secondary abnormalities.

Primary abnormalities are considered to be those that occur during spermatogenesis, hence originating in the testis, and secondary abnormalities as being due to incomplete sperm maturation or to environmental and handling effects during collection and processing of the semen sample. Defects of the head and midpiece are usually classified as primary (proximal droplets, midpiece defects, tightly coiled or strongly folded tails, severe head abnormalities), while those associated with the tail are regarded as secondary (distal droplets, simple bent tails, tailless sperm, presence of the acrosomal reaction, bodies other than spermatozoa – such as red blood cells, white blood cells, bacteria, squamous epithelial cells or spermatozoal clumps). Rams should have at least 70% normal spermatozoa to be classified as satisfactory potential breeders, or more than 80–90% to be classified as exceptional. Bucks, in contrast, should have at least 80% normal cells to be classified as satisfactory breeders, and animals with lower values may require re-evaluation after 8 weeks rest, or need to be culled.

Sperm morphology can be influenced by age, season and diseases of the reproductive tract and should, therefore, be interpreted after taking these factors into consideration. If collections are made during the summer, the percentage of abnormalities can be expected to be higher. Rams exposed to moderate heat can show a marginal degree of abnormal sperm morphology, which is closely associated with a poor conception rate and high incidence of early embryonic death (Rathore, 1968). Skalet *et al.* (1988) demonstrated in one study that young Nubian bucks had 65% of abnormal sperm at the onset of puberty, which decreased to 12% by 8 months of age, with only midpiece abnormalities being common. Dead sperm have been observed to be more common in the semen of young bucks in spring and summer than in autumn and winter (Ahmad and Noakes, 1996b). In older bucks, it has been observed that season does not influence sperm abnormalities, as shown in a study on bucks of native breeds in India (Sahni and Roy, 1972). A significantly lower fertilizing ability and increase in sperm abnormalities in ejaculates collected in the spring has been reported for ram semen used in artificial insemination (AI), although motility of the spermatozoa was not impaired (Colas and Courot, 1977). Elk and red deer have a high percentage of abnormal cells in July and August in the northern hemisphere, but by the beginning of September they usually start having a higher proportion of normal sperm. At the beginning of the breeding season, a higher percentage of secondary abnormalities can be seen in the semen sample and, similarly, individual males will display a higher percentage of secondary abnormalities towards the end of the breeding season as well. In some instances, an excess of 50% or more sperm in one ejaculate can show distal droplets before the cessation of viable sperm production (Bringans *et al.*, 2007). WTD will typically produce good-quality semen after achieving hard antler status and up to 2 weeks or so after antler shedding, but these findings may vary between individual bucks (authors' personal observations).

Semen quality and morphology can also be affected by various infections and disease conditions. The most common morphological changes of sperm associated with *B. ovis* infection in rams are detached heads and bent tails (Cameron and Lauerman, 1976). Infection in red deer stags is associated with detached heads as well, and also with a significant reduction in sperm motility (Ridler and West, 2002). Detached heads, tightly coiled tails and thickened midpieces were prominent in semen samples from an 18-month-old infertile LaMancha buck with testicular degeneration (Refsal *et al.*, 1983); detached heads (70% of the spermatozoa) were also observed in the semen of a Nubian buck with histologically diagnosed seminal vesciculitis (Ahmad *et al.*, 1993).

Acrosomal staining

The acrosome reaction of mammalian spermatozoa is an essential contributor to fertilization because only acrosome-reacted spermatozoa can penetrate the zona pellucida and fuse with an oocyte. This event is an exocytotic reaction that involves multiple sites of fusion between the sperm plasma membrane and the outer acrosomal membrane, with subsequent vesiculation and release of the acrosomal contents (reviewed by Yanagimachi, 1994). Sperm that have undergone an acrosomal reaction are no longer able to participate in fertilization, and hence visualization of sperm acrosome integrity may be beneficial in the evaluation of semen quality, as male infertility could be caused by a lack of sperm with intact acrosomes at ejaculation. The most accurate method of assessing acrosomal status is transmission electron microscopy (TEM), but because of the time and expense required to conduct this assay, other methods such as the use of specific acrosomal stains and light microscopy have been used in different species (Cross and Meizel, 1989). The integrity of the acrosomal membrane can be assayed by fluorescence microscopy using fluorochrome-labelled lectins, e.g. pea or peanut agglutinins that bind to acrosomal contents.

In vitro, the acrosome reaction can be induced by incubating spermatozoa with the calcium ionophore A23187 (Zhang *et al.*, 1991), or with physiological inducers such as the zona pellucida (Ellington et al., 1993), follicular fluid (Tarlatzis *et al.*, 1993) or progesterone (Meyers *et al.*, 1995). Although chlortetracycline (CTC) fluorescence can be used to differentiate acrosome-intact from acrosome-reacted spermatozoa, the most commonly used fluorescent probes are fluorescein isothiocyanate (FITC, green fluorescent fluorophore) conjugated peanut agglutinin (FITC–PNA) or FITC-conjugated *Pisum sativum* agglutinin (FITC–PSA), which offer the advantage over CTC that they can be analysed by flow cytometry. These indicators of acrosomal status are usually combined with a viability probe, such as propidium iodide (PI), so that four groups of spermatozoa can be identified: live acrosome intact, live acrosome reacted, dead acrosome intact and dead acrosome reacted. In rams, FITC–RCA (*Ricinus communis* agglutinin)–lectin binding has been demonstrated to be a simple, reliable and useful method for assessing the acrosomal status of fresh ram spermatozoa in suspension; the acrosomal status of unfixed ram spermatozoa can be assessed by this method (Martí *et al.*, 2000). Using Trypan blue in a modified triple-stain technique (TST), it has been possible to evaluate sperm viability and acrosomal status in fixed deer spermatozoa, and this process helps in differentiating spermatozoa that have undergone a true acrosome reaction from those that have undergone a false acrosome reaction (Garde *et al.*, 1997).

Sperm concentration

The sperm concentration of ram or buck semen can be determined by haemocytometric techniques or by the measurement of optical density after calibrating the instruments for a particular species.

The haemocytometer technique, while time-consuming, is probably the most commonly used technique for counting sperm numbers. This technique offers a direct visualization of sperm and requires minimal equipment. For routine haemocytometer counting of sperm, the platelet/white blood cell Unopette system (Becton-Dickinson, Franklin Lakes, New Jersey) with 20 μl capillary pipettes can be used if available. The capillary pipette is filled with semen, which is transferred to the Unopette (providing a 1:100 dilution). After thorough mixing, both sides of the coverslipped haemocytometer chamber are loaded, and a few minutes are allowed for sperm to settle on the haemocytometer grid. The central grid is identified under the microscope (20× objective). The central grid is a 5 × 5 arrangement of squares, with each square subsequently divided into 4 × 4 smaller squares. Sperm within all 25 squares that form the central grid are counted. The count is then repeated on the second chamber, and the mean represents the number of sperm in the original raw semen sample in millions/ml.

The CASA technique can also be utilized for determining sperm concentration using chambers of a known volume and depth. The software can then calculate the total sperm, dilution volume and dose volume for preparing semen for AI provided that it is calibrated for that particular species.

Several photometers, such as the SDM 6 and SDM 1 (Minitüb, Germany), are available with preloaded programs for analysing small ruminant semen. Though relatively expensive, these offer the advantage of accurate measurement of live sperm concentration, as well as providing calculations of the extender volume required and the number of doses.

The NucleoCounter® SP-100® (ChemoMetec A/S, Denmark) is a relatively newer technique available for estimating ovine and cervid semen. It involves diluting semen samples into a detergent solution and loading into specially designed SP-1 cassettes containing a microfluid network. The channels in the microfluid network also contain PI, which intercalates with DNA and stains the nuclei of the spermatozoa. The integrated fluorescence microscope within the NucleoCounter® detects the fluorescent signal from each sperm nucleus, and the automated semen analysis software, SemenView®, converts this information into a value for sperm concentration.

Electron microscopy

TEM is useful for detecting subtle lesions of the plasma membrane, acrosome, nuclear chromatin, mitochondria or axoneme in cases of unexplained infertility.

Fertility

Service capacity testing

Reproductive performance in rams is highly variable, so the selection of high-performance rams is greatly desired as it can improve the flock fertility, especially when breeding intensity is great (Stellflug *et al.*, 2006). To identify these high-performance rams it is greatly beneficial to perform service capacity tests. These tests are the most reliable predictor of the adequacy of libido in an individual animal. In a service capacity test, rams are exposed to oestrous ewes for a specified time period and their subsequent breeding activity is recorded over a period of 2 weeks or more. Individual rams can be placed with 2–4 cycling, unrestrained ewes for a period of 20–40 min and their sexual activity can be recorded with an emphasis on the number of mating attempts. Selection of high-performing rams will result in higher lambing percentages and greater numbers of live-born lambs per exposed ewe. Service capacity testing may also help in determining proper ram to ewe ratios, as well as in producing a shorter, more uniform lambing season. It appears that rams born co-twin to male siblings have higher service capacities than those born co-twin to females (Yarney and Sanford, 1993). Usually, adult rams achieving 4–6 or more matings within 30 min are desirable, although those achieving 2–3 matings may still be acceptable.

Any ram that appears to be sexually inactive may be tested by leaving it in the pen overnight with ewes that have different colours painted on their rumps. The ram should be examined the next day for signs of paint on its chest. A multitude of factors can influence an individual service capacity test, such as unfamiliarity with test pens and oestrous ewes (Zenchak and Anderson, 1980). A study by Stellflug and Berardinelli (2002) shows that six 30 min individual-ram performance tests were required to obtain 95% reliability for serving capacity test scores. Hence, individual serving capacity tests (ISCT) may sometimes be time-consuming and laborious. A cohort serving capacity test (COSCT), which includes multiple males competing for oestrous ewes may be an alternative, as it will result in a more efficient use of time. As rams remain with other cohorts, separation anxiety is minimal and competition among rams helps to separate high-performance rams from the lower performing animals. A single three-ram, 30 min COSCT is a reliable and efficient alternative to a series of ISCTs for characterizing sexual activity in rams, and the use of multiple cohort tests can provide some protection against the unnecessary culling of highly sexually active rams (Stellflug *et al.*, 2008).

A ram with good-quality semen, adequate testicular size, and good libido can breed 100 ewes in a 17 day breeding season (Fitzgerald and Perkins, 1991; Kimberling and Parsons, 2007). Most producers in North America use 3–3.5 rams per 100 ewes. Yearlings and mature rams can be expected to service 35–50 ewes; whereas ram lambs should be expected to service only 15–25 ewes (Edmondson *et al.*, 2012). Adjustments should be made for multiple sire breeding units. It is desirable to always have more than three rams in a multiple sire unit, because this tends to alleviate some of the territorial fighting among rams (Edmondson *et al.*, 2012). Mature fertile wapiti and red deer male can service 25–50 females, and young males (1.5–2.5 years old) may service 10–20 females (Haigh and Hudson, 1993). Young (1.5 years old) WTD bucks may service 5–10 females, while mature (2.5 years or older) may service 10–20 females. Because cervids are very seasonal in their breeding patterns and WTD bucks 'tend' does that are in oestrus, a tightly synchronized breeding group may lead to decreased first service conception rates and a strung-out calving/fawning season (authors' personal observations).

Seasonality

Seasonality is a circannual (approximately yearly) phenomenon encountered in small ruminants and cervids, which is characterized by large seasonal variations in sexual behaviour, testicular and antler size, and growth, as well as quantitative and qualitative changes in the sperm output. Seasonal reproductive activity is regulated by an endogenous hormonal rhythm, synchronized by environmental stimuli, with photoperiod playing the most important role (Zarazaga *et al.*, 2010). In small ruminants and cervids, decreasing day length stimulates the onset of reproductive activity by increasing the secretion of melatonin from the pineal gland. In contrast, the end of the breeding season is thought to occur as a result of refractoriness that develops to the previously stimulatory short days (Lincoln, 1980). Melatonin pulses during shorter day length are longer with higher amplitude, but during the summer, release of melatonin is negligible. This increased frequency and amplitude of melatonin secretion in turn leads to the activation of the hypothalamo–pituitary–gonadal axis, resulting in increased testosterone production, spermatogenesis and display of mating behaviour (rut). Melatonin has also been proven to be involved in the regulation of semen quality and the antioxidant enzyme activity that affects the reproductive performance of rams, as seasonal variations of fertility in the ram involve interplay between melatonin and the antioxidant defence system (Casao *et al.*, 2010). Prolactin also plays a major role in testicular recrudescence (Sanford and Dickson, 2008), as well as in androgen production, and in seasonal breeders such as rams, has been shown to facilitate the central activation of gonadotrophin secretion. Studies inducing hypoprolactinaemia in rams have been shown to diminish testosterone secretion by the developing (Lincoln *et al.*, 2001; Sanford and Dickson, 2008) and redeveloped (Regisford and Katz, 1993) testes.

Sheep and goat breeds originating from temperate climates display a distinct seasonal pattern, with males showing dramatic changes in sexual behaviour, testicular weight and qualitative and quantitative sperm production, coincident with decreasing day length. The testis weight in the adult Ile-de-France ram varied from 180–190 g in the late winter–early spring to 300–320 g in late summer and autumn (Ortavant *et al.*, 1988). The increase in testis weight started before the beginning of summer and its regression began before winter in rams not trained for mating. Furthermore, the quality of semen (indicated by the percentage of normal spermatozoa) and its fertility were lower in spring than in autumn. Similar tendencies have been observed in many other breeds, even if there were variations between them or between individuals within a breed. Dorset, Rambouillet and Finn sheep are generally considered to be less seasonal than other breeds, while Suffolk, Hampshire and Columbia breed sheep generally are poor out-of-season breeders.

Seasonality patterns in goats are similar to those in sheep. Breeds such as the Saanen and Alpine show the highest levels of serum testosterone and maximal testicular weight during the months of late summer and early autumn, with the lowest values recorded in winter and spring. Seasonality in breeds originating from the tropics may not be as pronounced as in those originating from temperate climates. Damascus bucks produce the best quality semen during increasing daylight in the summer and spring, which coincides with their breeding season in native Jordan (Al-Ghalban *et al.*, 2004). In contrast, West African Dwarf (Bitto and Egbunike, 2006) and Creole bucks (Delgadillo *et al.*, 1999) in their natural environment do not display seasonality and are capable of producing viable spermatozoa throughout the year. Besides photoperiod, the nutritional plane can affect testicular development and the production of spermatozoa. In genotypes that are highly responsive to photoperiod, such as the Suffolk, nutritional stimuli can reinforce the response to changes in day length, but photoperiod is the completely dominant factor, and will override nutritional signals to the point where animals can be losing weight at the same time as gaining testicular mass (and vice versa). In contrast, Merino rams respond to photoperiod, but their reproductive system is dominated by nutrition to the point where animals can be losing testicular mass at the onset of the breeding season when photoperiod would otherwise be driving the testes to grow (Martin *et al.*, 1999, 2002; Hötzel *et al.*, 2003). In temperate breeds maintained under environmental conditions similar to those from which they originate, intra-breed variability also exists. Some reproductive traits, such as the onset, offset and duration of the breeding season have a hereditary basis and can therefore be used for genetic selection (Quirke *et al.*, 1986; Hanrahan, 1987; Smith *et al.*, 1992; Al-Shorepy and Notter, 1997).

Efforts to control the initiation of the reproductive cycle in small ruminants have received much interest. Of these, photoperiodic variation and melatonin treatments have been found to be most effective. These methods allow semen collection throughout the year instead of just being confined to the 6 month breeding season. Artificial reversal of the annual rhythm of photoperiodic variations induces reversal of the periods of recrudescence and regression of testicular size and sperm production in rams, with the alternation of 3- or 4-month periods of constant long (16h light–8h dark) and constant short (8h light–16h dark) days seen to induce testicular growth (Lincoln and Davidson, 1977). In Alpine and Saanen bucks kept in a lightproof building and subjected to 1 month of long days (16 h light/day) and 1 month of short days (8 h light/day), testicular weight increased progressively 5 months after the beginning of the study. Similarly, fresh semen quality, as well as the fertility of frozen semen, were comparable to values obtained normally in the natural sexual season in control animals (Delgadillo *et al.*, 1991, 1992; Chemineau *et al.*, 1999). Treatment with exogenous melatonin mimicking the effect of short days generally stimulates reproductive activity in short-day breeders. In rams and bucks undergoing melatonin treatment, it is advisable before the insertion of implants to impose a light treatment composed of real long days or long days mimicked by 1 h of extra light during the photosensitive phase (Chemineau *et al.*, 1996). Combined treatment of photoperiod variation and melatonin has been shown to be more efficient in the induction and maintenance of out-of-season sexual activity than a succession of long days alone (Zarazaga *et al.*, 2010).

Seasonality in cervids from temperate climates follows the same pattern as that in rams and bucks, and shows marked annual cycles of testicular involution and recrudescence, including the transition between totally arrested and highly active spermatogenesis (Asher *et al.*, 1999). The family Cervidae represents close to 40 species and 200 subspecies, and due to their wide geographical distribution, no single species can be considered to represent a 'typical' deer in terms of reproductive function. Besides the annual cycle of testicular regression and recrudescence, the gonadal hormones also influence the development of secondary sexual characteristics, such as a long neck mane, an enlarged neck and especially the regrowth of antlers. In addition to the gonadotrophins, prolactin has been shown to modulate the spermatogenic and the steroidogenic function in testes of seasonally breeding cervids such as red deer (Jabbour *et al.*, 1998).

Temperate species such as wapiti (*Cervus canadensis*) and red deer (*C. elaphus*) stags start showing a marked increase in GnRH output around the summer solstice, and 2 months later have markedly high serum LH levels (Haigh, 2007). This stimulates testosterone production, and levels reach a peak around the autumnal equinox, corresponding to velvet shedding and antler cleaning. Testosterone levels, scrotal circumference and percentage of normal sperm increase sharply, reaching a peak just before the rut itself. Within a month of the autumnal equinox, testosterone drops to a baseline level, where it stays for the rest of winter. In red deer and wapiti, a second, subtle and transient elevation in serum testosterone may be seen following the winter solstice (Fennessy *et al.*, 1988). This is seen especially in well-nourished males, and can be accompanied by rutting behaviour. Similar changes have been reported for other species of cervids as well (Giménez *et al.*, 1975). Fallow deer (*Dama dama*) also exhibit a highly seasonal pattern of reproduction, with the rut (peak of mating activity) occurring in October/November in the northern hemisphere and in April/May in the southern hemisphere. Efforts to advance the breeding season in this species by administering melatonin implants have shown that rutting activity, testis development and neck muscle hypertrophy can be advanced by as much as 6–8 weeks (Asher and Peterson, 1991).

In contrast tropical and subtropical cervids may or may not display the pronounced seasonality observed in the temperate species, as their reproductive cycles may be governed by the pattern of annual rainfall. Eld's deer (*C. eldi thamin*) a subtropical species native to eastern India and southern China, begins with pituitary activation during autumn and winter. Peak testis size, testosterone secretion, rutting behaviour and maximal sperm production in this species occur during winter and spring, as day lengths increase (Monfort *et al.*, 1993). However, despite seasonal phase differences, the interactive dynamics of the pituitary–gonadal axis in Eld's deer is remarkably similar to that in red deer stags and fallow deer bucks.

Certain species, such as the male muntjacs (*Muntiacus* spp.), despite demonstrating an annual antler cycle, do not exhibit 'annual puberty', and remain fertile throughout the year. Despite low concentrations of testosterone from early May to early August, muntjacs are capable of producing fertile spermatozoa, although in lower numbers, unlike deer from the temperate regions (Pei *et al.*, 2009). Studies performed in Texas on the spotted deer or chital (*Axis axis*), also indicate that, while seasonal changes in testis sperm content, morphology and antler pattern do occur in the stag, these changes are less pronounced than in more temperate species of cervids, and the stags are able to maintain testicular function throughout the year, regardless of antler status (Willard and Randel, 2002). This is in agreement with studies performed later in India, which confirmed that spotted deer have a physiological breeding season in the months of March–May and when the stags are in hard antler (Umapathy *et al.*, 2007). Staggered antler patterns have been observed in many species of deer, both temperate and subtropical, such as the sika deer (*C. nippon*), red deer and spotted deer (Raman, 1998). Juvenile males have been observed to come into hard antler associated with a peak of testosterone level and sexual activity a few months later than adults of the same species. This phenomenon may be a natural strategy to avoid intra-sexual conflict among males, and may apply in areas that have a high density of males.

References

Ahmad N and Noakes DE (1996a) Sexual maturity in British breeds of goat kids *British Veterinary Journal* **152** 93–103.

Ahmad N and Noakes DE (1996b) Seasonal variations in the semen quality of young British goats *British Veterinary Journal* **152** 225–236.

Ahmad N, Noakes DE and Middleton DJ (1993) Seminal vesiculitis and epididymitis in an Anglo-Nubian buck *Veterinary Record* **133** 322–323.

Akusu MO, Agiang EA and Egbunike GN (1984) Ejaculate and plasma characteristics of West African Dwarf (WAD) buck In *Proceedings, 10th International Congress on Animal Reproduction and Artificial Insemination, June 10–14, University of Illinois Volume 2* pp 50–52.

Al-Ghalban AM, Tabbaa MJ and Kridli RT (2004) Factors affecting semen characteristics and scrotal circumference in Damascus bucks *Small Ruminant Research* **53** 141–149.

Al-Shorepy SR and Notter DR (1997) Response to selection for fertility in a fall-lambing sheep flock *Journal of Animal Science* **75** 2033–2040.

Asdell SA (1944) The genetic sex of intersexual goats and a probable linkage with the gene for hornlessness *Science* **99** 124.

Asher GW and Peterson AJ (1991) Pattern of LH and testosterone secretion of adult male fallow deer (*Dama dama*) during the transition into the breeding season *Journal of Reproduction and Fertility* **91** 649–54.

Asher GW, Monfort SL and Wemmer C (1999) Comparative reproductive function in cervids: implications for management of farm and zoo populations *Journal of Reproduction and Fertility Supplement* **54** 143–156.

Bag, S, Joshi A, Naqvi SMK, Rawat PS and Mittal JP, (2002a) Effect of freezing temperature, at which straws were plunged into liquid nitrogen, on the post-thaw motility and acrosomal status of ram spermatozoa *Animal Reproduction Science* **72** 175–183.

Bag S, Joshi A, Rawat PS and Mittal JP (2002b) Effect of initial freezing temperature on the semen characteristics of frozen–thawed ram spermatozoa in a semi-arid tropical environment *Small Ruminant Research* **43** 23–29.

Ball L (1986) Electroejaculation In *Applied Electronics for Veterinary Medicine and Animal Physiology* pp 395–441 Ed WR Klemm. CC Thomas, Springfield Illinois.

Ball L, Pickett BW and Gerbauer MR (1971) Staining technique and stallion sperm morphology *Journal of Animal Science* **33** 248.

Basrur PK and Kochhar HS (2007) Inherited sex abnormalities in goats In *Current Therapy in Large Animal Theriogenology* 2nd edition pp 590–594 Ed RS Youngquist and WR Threllfall. WB Saunders, Philadelphia.

Bhatia S and Shanker V (1992) First report of a XX/XXY fertile goat buck *Veterinary Record* **130** 271–272.

Bitto II and Egbunike GN (2006) Seasonal variations in sperm production, gonadal and extragonadal sperm reserves in pubertal West African dwarf bucks in their native tropical environment *Livestock Research for Rural Development* **18**:134.

Braun WF, Thompson JM and Ross CV (1980) Ram scrotal circumference measurements *Theriogenology* **13** 221–229.

Bringans MJ, Plante C and Pollard J (2007) Cervid semen collection and freezing In *Current Therapy in Large Animal Theriogenology* 2nd edition pp 982–986 Ed RS Youngquist and WR Threllfall. WB Saunders, Philadelphia.

Brinsko SP, Varner DD, Blanchard TL and Meyers SA (1990) The effect of post breeding uterine lavage on pregnancy rate in mares *Theriogenology* **33** 465–475.

Bulgin MS (1992) Ram breeding soundness examination and SFT form In *Proceedings of the Annual meeting of the Society for Theriogenology, San Antonio, Texas* pp 210–215. Society for Theriogenology, Nashville, Tennessee.

Bunch TD, Callan RJ, Maciulis A, Dalton JC, Figueroa MR, Kunzler R and Olson RE (1991) True hermaphroditism in a wild sheep: a clinical report *Theriogenology* **36** 185–190.

Cameron RDA and Lauerman LH (1976) Characteristics of semen changes during *Brucella ovis* infection in rams *Veterinary Record* **99** 231–233.

Carter PD, Hamilton PA and Duffy JH (1990) Electroejaculation in goats *Australian Veterinary Journal* **67** 91–93.

Casao A, Cebrián I, Asumpção ME, Pérez-Pé R, Abecia JA, Forcada F, Cebrián-Pérez JA and Muiño-Blanco T (2010) Seasonal variations of melatonin in ram seminal plasma are correlated to those of testosterone and antioxidant enzymes *Reproductive Biology and Endocrinology* **8**:59.

Chemineau P, Malpaux B, Pelletier J, Leboeuf B, Delgadillo JA, Deletang F, Pobel T and Brice G (1996) Emploi des implants de mélatonine et des traitements photopériodiques pour maîtriser la reproduction saisonnière chez les ovins et les caprins (Use of melatonin implants and photoperiodic treatments to control seasonal reproduction in sheep and goats) *INRA Productions Animales* **9** 45–60.

Chemineau P, Baril G, Leboeuf B, Maurel MC, Roy F, Pellicer-Rubio M, Malpaux B and Cognié Y (1999) Implications of recent advances in reproductive physiology for reproductive management of goats *Journal of Reproduction and Fertility Supplement* **54** 129–142.

Colas G and Courot M (1977) Production of spermatozoa, storage of semen and artificial insemination in the sheep In *Management of Reproduction in Sheep and Goats: Proceedings, Symposium, University of Wisconsin, Madison, Wisconsin, July 24–25, 1977* pp 31–40 Ed CE Terill. University of Wisconsin, Madison.

Court M (1976) Semen quality and quantity in the ram In *Sheep Breeding* 2nd edition pp 495–503, Ed GJ Tomes, DE Robertson and RJ Lightfoot. Butterworths, London.

Crepin, P (1958) Les cornes dans l'espèce caprine (Horns in the goat) *Mouton* **13** 79.

Cribiu EP and Chaffaux S (1990) L'intersexualité chez les mammifères domestiques [Intersexuality in domestic mammals] *Reproduction Nutrition Development* **30(Supplement 1)** 51s–61s.

Cross NL and Meizel S (1989) Methods for evaluating the acrosomal status of mammalian sperm *Biology of Reproduction* **41** 635–641.

Da Silva P, Aitken RP, Rhind SM, Racey PA and Wallace JM (2001) Influence of placentally mediated fetal growth restriction on the onset of puberty in male and female lambs *Reproduction* **122** 375–383.

de la Vega AC, Morales P, Zimerman M and Wilde O (2006) Annual variation of scrotal circumference in male Creole goats *Archivos de Zootecnia* **55** 113–116.

Delgadillo JA, Leboeuf B, and Chemineau P (1991) Decrease in the seasonality of sexual behaviour and sperm production in bucks by exposure to short photoperiodic cycles *Theriogenology* **36** 755–770.

Delgadillo JA, Leboeuf B and Chemineau P (1992) Abolition of seasonal variations in semen quality and maintenance of sperm fertilizing ability by short photoperiodic cycles in goat bucks *Small Ruminant Research* **9** 47–59.

Delgadillo JA, Canedo GA, Chemineau P, Guillaume D and Malpaux B (1999) Evidence for an annual reproductive rhythm independent of food availability in male Creole goats in subtropical northern Mexico *Theriogenology* **52** 727–737.

Dennis SM (1979) Hypospadias in merino lambs *Veterinary Record* **105** 94–96.

Dorado J, Molina I, Muñoz-Serrano A and Hidalgo M (2010) Identification of sperm subpopulations with defined motility characteristics in ejaculates from Florida goats *Theriogenology* **74** 795–804.

Edmondson MA, Roberts JF, Baird AN, Bychawski S and Pugh DG (2012) Theriogenology of sheep and goats In *Sheep and Goat Medicine* 2nd edition pp 150–230 Ed DG Pugh and AN Baird. Elsevier Saunders, Missouri.

Edward AY, Windsor DP, Purvis W, Sanchez-Partida LG and Maxwell WMC (1995) Distribution of variance associated with measurement of post-thaw function in sperm *Reproduction, Fertility and Development* **7** 129–134.

Ellington JE, Ball BA and Yang X (1993) Binding of stallion spermatozoa to the equine zona pellucida after coculture with oviductal epithelial cells *Journal of Reproduction and Fertility* **98** 203–208.

Epstein H and Herz A (1964) Fertility and birth weights of goats in a subtropical environment *Journal of Agricultural Science* **62** 237–244.

Farrell PB, Presicce GA, Brockett CC and Foote RH (1998) Quantification of bull sperm characteristics measured by computer-assisted sperm analysis (CASA) and the relationship to fertility *Theriogenology* **49** 871–879.

Fennessy PF, Suttie JM, Crosbie SF, Corson ID, Elgar HJ and Lapwood KR (1988) Plasma LH and testosterone responses to gonadotrophin-releasing hormone in adult red deer (*Cervus elaphus*) stags during the antler cycle *Journal of Endocrinology* **117** 35–41.

Fitzgerald J and Morgan G (2007) Reproductive physiology of the ram In *Current Therapy in Large Animal Theriogenology* 2nd edition pp 617–620 Ed RS Youngquist and WR Threllfall. WB Saunders, Philadelphia.

Fitzgerald J and Perkins A (1991) Serving capacity tests for rams In *Handbook of Methods for Study of Reproductive Physiology in Domestic Animals* Ed PJ Dziuk and M Wheeler. University of Illinois Press, Urbana, Illinois.

Fredell L, Kochum I, Hansson E, Holmner S, Lundquist L, Lackgren G, Pedersen J, Stenberg A, Westbacke G and Nordenskjold A (2002). Heredity of hypospadias and the significance of low birth weight *Journal of Urology* **167** 1423–1427.

Garde JJ, Ortiz N, García A and Gallego L (1997) Use of a triple-stain technique to detect viability and acrosome reaction in deer spermatozoa *Archives of Andrology* **39** 1–9.

Gier HT and Marion GB (1970) Development of the mammalian testis In *The Testis* pp 1–45 Ed AD Johnson, WR Gomes and NL Vandemark. Academic Press New York.

Giménez T, Barth D, Hoffmann B and Karg H (1975) Blood levels of testosterone in the roe deer (*Capreolus capreolus*) in relationship to the season *Acta Endocrinologica (Copenhagen)* **Supplementum 193** 59.

Haibel GK and Roijko JL (1990) Persistent Müllerian duct syndrome in a goat *Veterinary Pathology* **27** 135–137.
Haigh JC (2007) Reproductive anatomy and physiology of male wapiti and red deer In *Current Therapy in Large Animal Theriogenology* 2nd edition pp 932–936 Ed RS Youngquist and WR Threllfall. WB Saunders, Philadelphia.
Haigh JC and Hudson RJ (1993) *Farming Wapiti and Red Deer.* Mosby, St Louis.
Haigh JC, Cates WF, Glover GJ and Rawlings NC (1984) Relationships between seasonal changes in serum testosterone concentrations, scrotal circumference and sperm morphology of male wapiti (*Cervus elaphus*) *Journal of Reproduction and Fertility* **70** 413–418.
Hamerton JL, Dickson JM, Pollard CE, Grieves SA and Short RV (1969) Genetic inter-sexuality in goats *Journal of Reproduction and Fertility Supplement* **7** 25–51.
Hanrahan JP (1987) Genetic variation in seasonal reproduction in sheep In *Proceedings of the 38th Annual Meeting of the European Association for Animal Production, September 27–October 1, 1987, Lisbon, Portugal* pp 904 [abstract].
Holt WV and Palomo MJ (1996) Optimization of a continuous real-time computerized semen analysis system for ram sperm motility assessment and evaluation of four methods of semen preparation *Reproduction Fertility and Development* **8** 219–230.
Hötzel MJ, Walkden-Brown SW, Fisher JS and Martin GB (2003) Determinants of the annual pattern of reproduction in mature male Merino and Suffolk sheep: responses to a nutritional stimulus in the breeding and non-breeding seasons *Reproduction Fertility and Development* **15** 1–9.
Hulet CV (1977) Prediction of fertility in rams: factors affecting fertility and collection, testing and evaluation of semen *Veterinary Medicine Small Animal Clinician* **72** 1363–1367.
Hutson JM, Hasthorpe S and Heyns CF (1997) Anatomical and functional aspects of testicular descent and cryptorchidism *Endocrine Reviews* **18** 259–280.
Jabbour HN, Clarke LA, McNeilly AS, Edery M and Kelly PA (1998) Is prolactin a gonadotrophic hormone in red deer (*Cervus elaphus*)? Pattern of expression of the prolactin receptor gene in the testis and epididymis *Journal of Molecular Endocrinology* **20** 175–182.
Joshi A, Bag S, Naqvi SMK, Sharma RC, Rawat PS and Mittal JP (2001) Effect of short-term and long-term preservation on motion characteristics of Garole ram spermatozoa: a prolific microsheep breed of India *Asian–Australasian Journal of Animal Sciences* **14** 1527–1533.
Joshi A, Bag S, Naqvi SMK, Sharma RC and Mittal JP (2005) Effect of post-thawing incubation on sperm motility and acrosomal integrity of cryopreserved Garole ram semen *Small Ruminant Research* **56** 231–238.
Joshi A, Kumar D, Naqvi SMK and Maurya VP (2008) Effect of controlled and uncontrolled rate of cooling, prior to controlled rate of freezing, on motion characteristics and acrosomal integrity of cryopreserved ram spermatozoa *Biopreservation and Biobanking* **6** 277–284.
Just W, De Almeida CC, Goldshmidt B and Vogel W (1994) The male pseudohermaphrodite XX polled goat is Zfy and Sry negative *Hereditas* **120** 71–75.
Kimberling CV and Parsons GA (2007) Breeding soundness evaluation and surgical sterilization of the ram In *Current Therapy in Large Animal Theriogenology* 2nd edition pp 620–628 Ed RS Youngquist and WR Threllfall. WB Saunders, Philadelphia.
Kreeger TJ and Arnemo JM (2012) *Handbook of Wildlife Chemical Immobilization* 4th edition. Privately published. Details available at: http://www.vetconsult.no/pdfs/Handbook.pdf (accessed 5 September 2013). Download available at: http://www.gobookee.net/handbook-of-wildlife-chemical-immobilization/ (accessed 5 September 2013).
Kridli RT, Abdullah AY, Shaker MM and Al-Momani AQ (2006) Age at puberty and some biological parameters of Awassi and its first crosses with Charollais and Romanov rams *Italian Journal of Animal Science* **5** 193–202.
Kumar D, Joshi A and Naqvi SMK (2010) Comparative semen evaluation of Malpura and Bharat Merino rams by computer-aided sperm analysis technique under semi-arid tropical environment *International Journal of Animal and Veterinary Advances* **2** 26–30.
Ladds PW (1993) Congenital abnormalities of the genitalia of cattle, sheep, goats and pigs *Veterinary Clinics of North America Food Animal Practice* **9** 127–144.
Lincoln, GA (1971) Puberty in a seasonally breeding male, the red deer stag (*Cervus elaphus L.*) *Journal of Reproduction and Fertility* **25** 41–54.
Lincoln, GA (1980) Photoperiodic control of seasonal breeding in rams. The significance of short-day refractoriness In *Proceedings of the V International Congress of Endocrinology, February 10–16, 1980, Melbourne, Australia* pp 283–287.

Lincoln GA and Davidson W (1977) The relationship between sexual and aggressive behaviour, and pituitary and testicular activity during the seasonal sexual cycle of rams, and the influence of photoperiod *Journal of Reproduction and Fertility* **49** 267–276.

Lincoln GA, Townsend J and Jabbour HN (2001) Prolactin actions in the sheep testis: a test of the priming hypothesis *Biology of Reproduction* **65** 936–943.

Martí JI, Cebrián-Pérez JA and Muiño-Blanco T (2000) Assessment of the acrosomal status of ram spermatozoa by RCA lectin-binding and partition in an aqueous two-phase system *Journal of Andrology* **21** 541–548.

Martin GB, Tjondronegoro S, Boukhliq R, Blackberry MA, Briegel JR, Blache D, Fisher JS and Adams NR (1999) Determinants of the annual pattern of reproduction in mature male Merino and Suffolk sheep: modification of endogenous rhythms by photoperiod *Reproduction, Fertility and Development* **11** 355–366.

Martin GB, Hötzel MJ, Blache D, Walkden-Brown SW, Blackberry MA and Boukhliq R (2002) Determinants of the annual pattern of reproduction in mature male Merino and Suffolk sheep: modification of responses to photoperiod by an annual cycle in food supply *Reproduction, Fertility and Development* **14** 165–175.

Martínez-Pastor F, Diaz-Corujo AR, Anel E, Herraez P, Anel L and de Paz P (2005a) Post mortem time and season alter subpopulation characteristics of Iberian red deer epididymal sperm *Theriogenology* **64** 958–974.

Martínez-Pastor F, Diaz-Corujo AR, Anel E, Herraez P, Anel L and de Paz P (2005b) Sperm subpopulations in Iberian red deer epididymal sperm and their changes through the cryopreservation process *Biology of Reproduction* **72** 316–327.

Martínez-Pastor F, Garc-Macías V, Álvarez M, Chamorro C, Herráez P, de Paz P and Anel L (2006) Comparison of two methods for obtaining spermatozoa from the cauda epididymis of Iberian red deer *Theriogenology* **65** 471–485.

Mastromonaco GF, Houck ML and Bergfelt DR (2011) Disorders of sexual development in wild and captive exotic animals In *Disorders of Sex Development in Domestic Animals* pp 84–95 Ed DAF Villagómez, L Iannuzzi and WA King. S Karger AG, Basel.

Memon MA, Bretzlaff KN and Ott RS (1986) Comparison of semen collection techniques in goats *Theriogenology* **26** 823–827.

Memon MA, Micklesen DW and Goyal HO (2007) Examination of the reproductive tract and evaluation of potential breeding soundness in the buck In *Current Therapy in Large Animal Theriogenology* 2nd edition pp 515–518 Ed RS Youngquist and WR Threllfall. WB Saunders, Philadelphia.

Meyers SA, Overstreet JW, Liu IK and Drobnis EZ (1995) Capacitation *in vitro* of stallion spermatozoa: comparison of progesterone-induced acrosome reactions in fertile and subfertile males *Journal of Andrology* **16** 47–54.

Mickelsen DW and Memon MA (2007) Infertility and diseases of the reproductive organs of bucks In *Current Therapy in Large Animal Theriogenology* 2nd edition pp 519–523 Ed RS Youngquist and WR Threllfall. WB Saunders, Philadelphia.

Monfort SL, Brown JL, Wood TC, Wemmer C, Vargas A, Williamson LR and Wildt DE (1993) Seasonal patterns of basal and GnRH-induced LH, FSH and testosterone secretion in Eld's deer stags (*Cervus eldi thamin*) *Journal of Reproduction and Fertility* **98** 481–8.

Mukasa-Mugerwa E and Ezaz Z (1992) Relationship of testicular growth and size to age, body weight and onset of puberty in Menz ram lambs *Theriogenology* **38** 979–988.

Nunes, J, (1982) Etude des effets du plasma seminal sur la survie *in vitro* des spermatozoides de bouc (Study of the effects of seminal plasma on the *in vitro* survival of goat spermatozoa). Thèse de Doctorat (Doctoral Thesis), Université Pierre et Marie Curie, Paris.

Nuti LC and McWhinney DR (1987) Photoperiod effects on reproductive parameters in two breeds of dairy goat bucks (*Capra hircus*) In *Proceedings of 4th International Conference on Goats, Brasilia, Brazil, 8–13 March, 1987 Volume 2* pp 1508–1509.

OIE (2013) Ovine epididymitis (*Brucella ovis*) (2011) In *OIE Manual of Diagnostic Tests and Vaccines for Terrestrial Animals 2013 Volume 2 Twentieth Edition* Chapter 2.7.9. World Organisation for Animal Health, Paris Available at: http://www.oie.int/fileadmin/Home/eng/Health_standards/tahm/2.07.09_OVINE_EPID.pdf (accessed 5 September 2013).

Orihuela A, Aguirre V, Hernandez C, Flores-Perez I and Vazquez R (2009a) Effect of anaesthesia on welfare aspects of hair sheep (*Ovis aries*) during electro-ejaculation *Journal of Animal and Veterinary Advances* **8** 305–308.

Orihuela A, Aguirre V, Hernandez C, Flores-Perez I and Vazquez R, (2009b) Breaking down the effect of electro-ejaculation on the serum cortisol response, heart and respiratory rates in hair sheep (*Ovis aries*) *Journal of Animal and Veterinary Advances* **8** 1968–1972.

Ortavant R, Bocquier F, Pelletier J, Ravault JP, Thimonier J and Volland-Nail P (1988) Seasonality of reproduction in sheep and its control by photoperiod *Australian Journal of Biological Science* **41** 69–85.

Ott RS and Memon MA (1980) Breeding soundness examination of rams and bucks, a review *Theriogenology* **13** 155–164.

Pajares G, Balseiro A, Pérez-Pardal L, Gamarra JA, Monteagudo LV, Goyache F and Royo LJ (2009) Sry-negative XX true hermaphroditism in a roe deer *Animal Reproduction Science* **112** 190–197.

Pei KJ, Foresman K, Bing-Tsan Liu B, Long-Hwa Hong L and Yu JY (2009) Testosterone levels in male Formosan Reeve's muntjac: uncoupling of the reproductive and antler cycles *Zoological Studies* **48** 120–124.

Quinn PJ, Salamon S and White IG (1968) The effect of cold shock and deep-freezing on ram spermatozoa collected by electrical ejaculation and by an artificial vagina *Australian Journal of Agricultural Research* **19** 119–128.

Quirke JF, Hanrahan JP, Loughnane W and Triggs R (1986) Components of the breeding and non-breeding seasons in sheep: breed effects and repeatability *Irish Journal of Agricultural Research* **25** 167–172.

Raman RS (1998) Antler cycles and breeding seasonality of the chital (*Axis axis*) in southern India *Journal of the Bombay Natural History Society* **95** 377–380.

Rathore AK (1968) Effects of high temperature on sperm morphology and subsequent fertility in Merino sheep In *Proceedings of the Australian Society of Animal Production* **7** 270–274.

Refsal KR, Simpson DA and Gunther JD (1983) Testicular degeneration in a male goat: a case report. *Theriogenology* **19** 685–691.

Regisford EGC and Katz LS (1993) Effects of bromocriptine-induced hypoprolactinaemia on gonadotrophin secretion and testicular function of rams (*Ovis aries*) during two seasons *Journal of Reproduction and Fertility* **99** 529–537.

Ridler AL and West DM (2002) Effects of *Brucella ovis* infection on semen characteristics of 16-month-old red deer stags *New Zealand Veterinary Journal* **50** 19–22.

Ruvinsky A and Spicer LJ (1999) Developmental genetics: sex determination and differentiation In *The Genetics of Cattle* pp 456–461 Ed R Fries and A Ruvinsky. CAB International, Wallingford.

Sahni KL and Roy A (1972) A note on seasonal variation in the occurrence of abnormal spermatozoa in different breeds of sheep and goat under tropical conditions *Indian Journal of Animal Sciences* **42** 501–504.

Sakhaee E and Azari O (2009) Hypospadias in goats *Iranian Journal of Veterinary Research* **10** 298–301.

Samanta AK, Bhattacharyya B and Moitra DN (1990) Electroejaculatory stress of breeding bucks II. Changes in some biochemical constituents of blood *Indian Journal of Animal Health* **29** 123–125.

Sanford LM and Dickson KA (2008) Prolactin regulation of testicular development and sexual behavior in yearling Suffolk rams *Small Ruminant Research* **77** 1–10.

Schoeman SJ, Els HC and Combrink GC (1987) A preliminary investigation into the use of testis size in cross-bred rams as a selection index for ovulation rate in female relatives *South African Journal of Animal Science* **17** 144–147.

Schönherr S (1956) Die Unfruchtbarkeit der Ziegenböcke, ihre Verbreitung, frühzeitige Erkennung und Bekämpfung (The sterility of the goats, their dissemination, early detection and control) *Zeitschrift für Tierzüchtung und Züchtungsbiologie (Journal of Animal Breeding and Breeding Biology)* **66** 209–234 and 381–416.

Schultz SR and Johnson MK (1992) Breeding by male white tailed deer fawns *Journal of Mammalogy* **73** 148–150.

Silver RI (2000) What is the etiology of hypospadias? A review of recent research *Delaware Medical Journal* **72** 343–347.

Skalet LH, Rodrigues HD, Goyal HO, Maloney MA, Vig MM and Noble RC (1988) Effects of age and season on the type and occurrence of sperm abnormalities in Nubian bucks *American Journal of Veterinary Research* **49** 1284–1289.

Skinner JD (1970) Post-natal development of the reproductive tract of the male Boer goat *Agroanimalia* **2** 177–180.

Skinner JD, Van Heuden JAH and Goris EJ (1961) A note on cryptorchidism in Angora goats *South African Journal of Animal Science* **2** 93–95.

Smith JF, Johnson DL and Reid TC (1992) Genetic parameters and performance of flocks selected for advanced lambing date In *Proceedings of the New Zealand Society for Animal Production* **52** 129–132.

Smith KC, Long SE and Parkinson TJ (1998) Abattoir survey of congenital reproductive abnormalities in ewes *Veterinary Record* **143** 679–685.

Smith KC, Brown P and Parkinson TJ (2006) Hypospadias in rams *Veterinary Record* **158** 789–795.

Smith MC and Sherman DM (1994) Reproductive system In *Goat Medicine* 3rd edition p 439 Ed CC Cann and DA DiRienzi. Lea and Febiger, Philadelphia.

Stellflug JN and Berardinelli JG (2002) Ram mating behaviour after long-term selection for reproductive rate in Rambouillet ewes *Journal of Animal Science* **80** 2588–2593.

Stellflug JN, Cockett NE and Lewis GS (2006) The relationship between sexual behaviour classifications of rams and lambs sired in a competitive breeding environment *Journal of Animal Science* **84** 463–468.

Stellflug JN, Lewis GS, Moffet CA and Leeds TD (2008) Evaluation of three-ram cohort serving capacity tests as a substitute for individual serving capacity tests *Journal of Animal Science* **80** 2024–2031.

Tarlatzis BC, Danglis J, Kolibianakis EM, Papadimas J, Bontis J, Lagos S and Mantalenakis S (1993) Effect of follicular fluid on the kinetics of human sperm acrosome reaction *in vitro Archives of Andrology* **31** 167–175.

Terrill C (1940) Comparison of ram semen collection obtained by three different methods for artificial insemination *Journal of Animal Science* **1940** 201–207.

Thonneau PF, Candia P and Mieusset R (2003) Cryptorchidism: incidence, risk factors, and potential role of environment; an update *Journal of Andrology* **24** 155–162.

Umapathy G, Sontakke SD, Reddy A and Shivaji S (2007) Seasonal variations in semen characteristics, semen cryopreservation, estrus synchronization, and successful artificial insemination in the spotted deer (*Axis axis*) *Theriogenology* **67** 1371–1378.

Ungerfeld R, González-Pensado S, Bielli A, Villagrán M, Olazabal D and Pérez W (2008) Reproductive biology of the pampas deer (*Ozotoceros bezoarticus*): a review *Acta Veterinaria Scandinavica* **50**:16.

Veeramachaneni DNR, Amann RP and Jacobson JP (2006) Testis and antler dysgenesis in Sitka black-tailed deer on Kodiak Island, Alaska: sequela of environmental endocrine disruption? *Environmental Health Perspectives* **114** 51–59.

Wachtel SS, Basrur PK and Koo GC (1978) Recessive male determining genes *Cell* **15** 279–281.

Wendt K, Pohl I and Mrosk H (1960) Der Kryptorchismus des Schafes und seine wirtschaftliche Bedeutung (Cryptorchidism in sheep and its economic importance). *Archiv Tierzucht (Archives of Animal Breeding)* **3** 440–458.

Wilkies PR, Munro IB and Wijeratne WVS (1978) Studies on a sheep freemartin *Veterinary Record* **102** 140–142.

Willard ST and Randel RD (2002) Testicular morphology and sperm content relative to age, antler status and season in axis deer stags (*Axis axis*) *Small Ruminant Research* **45** 51–60.

Williams GA, Ott TL, Michal JJ, Gaskins CT, Wright RW Jr, Daniels TF and Jiang Z (2007) Development of a model for mapping cryptorchidism in sheep and initial evidence for association of *INSL3* with the defect *Animal Genetics* **38** 189–191.

Yanagimachi R (1994) Mammalian fertilization In *Physiology of Reproduction* pp 189–317 Ed E Knobil and JD Neill. Raven Press, New York.

Yarney TA and Sanford LM (1993) Pubertal development of ram lambs: physical and endocrinological traits in combination as indices of postpubertal reproductive function *Theriogenology* **40** 735–744.

Zarazaga LA, Gatica MC, Celi I, Guzmán JL and Malpaux B (2010) Effect of artificial long days and/or melatonin treatment on the sexual activity of Mediterranean bucks *Small Ruminant Research* **93** 110–118.

Zenchak JJ and Anderson GC (1980) Sexual performance levels of rams (*Ovis aries*) as affected by social experiences during rearing *Journal of Animal Science* **50** 167–174.

Zhang JJ, Muzs LZ and Boyle MS (1991) Variations in structural and functional changes of stallion spermatozoa in response to calcium ionophore A23187 *Journal of Reproduction and Fertility Supplement* **44** 199–205.

11 Applied Andrology in Horses

Barry A. Ball*
University of Kentucky, Lexington, Kentucky, USA

Introduction

Reproductive evaluation of the stallion is a reasonably common procedure in equine veterinary practice, and the techniques to evaluate potential fertility of the stallion have improved considerably over the past several years. The evaluation typically includes physical examination, semen collection and evaluation, examination of the internal and external genitalia, and some evaluation of mating behaviour and libido. Ancillary diagnostic tests, such as endoscopy or ultrasonography of the reproductive tract may be used to provide additional information. Genetic testing, endocrine evaluation and diagnostic testing for specific infectious diseases are increasingly common as part of reproductive evaluation of the stallion. Ultimately, however, the evaluation of the stallion can only provide an estimate of stallion fertility. In many cases, reproductive evaluation is more effective in determining potential sub-fertile stallions than in accurately identifying future stallion fertility, which may be influenced by a number of factors that are not considered as part of the routine reproductive evaluation.

Stallion Reproductive Evaluation

Reproductive evaluation of the stallion attempts to provide an estimate of a stallion's future fertility based on evaluation of history, physical examination, semen evaluation and other diagnostic procedures. Many other factors, such as management, have a large impact on stallion fertility, and the fertility of a stallion may change over time. Therefore, the reproductive evaluation is in many cases an attempt to provide an estimate of a stallion's potential fertility and should not be interpreted as an absolute measure of a particular stallion's fertility. The typical reproductive evaluation – or breeding soundness evaluation (BSE) – is composed of the following components:

- history;
- general physical examination, including examination of the external and internal genitalia;
- evaluation of libido and mating behaviour;
- semen collection and evaluation; and
- ancillary diagnostic procedures.

All of the components in the list above are covered in the first part of this chapter. The second part of the chapter is devoted to other

* E-mail: b.a.ball@uky.edu

specific topics in andrology that are related to a stallion's breeding soundness, and expand on some of the subjects covered in the first half. These include the following: oxidative stress in normal and abnormal functioning of equine spermatozoa; endocrinological evaluation of prospective and active breeding stallions; testicular biopsy; diseases of the scrotum and testis; diseases of the scrotum, tunica vaginalis and spermatic cord; diseases of the excurrent duct system; diseases of the accessory sex glands; diseases of the penis and prepuce; behavioural dysfunction; and ejaculatory dysfunction.

History

A detailed history is one of the most important aspects of the BSE but may be difficult to obtain. On many occasions, the owner or agent may not have direct knowledge of the stallion's past use or fertility and, in some cases, may be unwilling to provide a detailed history. It is also important to be aware of the reason for the evaluation being made. Emphasis on different portions of the examination may differ between a stallion that is being evaluated for infertility and one on which a pre-purchase examination is being made.

It is important that the clinician positively identifies the stallion that is examined. In addition to the signalment (recording of peculiar, appropriate, or characteristic marks), this should include lip tattoos, photographs, and markings or scars or microchip numbers. These should become part of the permanent record that is kept with each BSE. It may become important later to be able to refer to the record and positively identify the stallion that was examined on a particular date.

The history should include prior and intended future use of the stallion, as well as establishing previous ownership. The owner or agent may be able to provide little useful information if the stallion has recently been acquired. General health should be queried, including past illnesses or injuries, previous or current medication, type of housing, feeding programme and routine health maintenance. Therapy with steroids or gonadotrophins is of particular concern.

The history should attempt to detail past breeding performance of the stallion. This should include the number of years at stud, size of the stallion's book, frequency of use, type of breeding management (natural cover versus artificial insemination), and the date the stallion was last bred or semen was collected. If the stallion has been used for breeding, the history should elicit breeding shed behaviour and management techniques that may influence libido. If possible, fertility should be expressed as services per conception (Kenney *et al.*, 1971), or as per cycle conception rate (calculated as the number of mares that were detected pregnant/number of cycles mated or inseminated) (Love, 2003). The detected incidence of embryonic or fetal loss should be recorded as well, along with any other unusual occurrences, such as the number of mares noted with endometritis or shortened oestrus cycles after mating. The distribution of the stallion's book (maiden versus foaling versus barren) should also be noted. Reproductive management of the mare band, such as teasing method, ovulation detection, hormonal therapy, vaccination programme and method of pregnancy detection should be discussed. The status of the stallion relative to equine infectious anaemia, equine viral arteritis and contagious equine metritis (CEM) should be noted. The incidence of possible genetic defects, such as cryptorchidism or parrot mouth, in the stallion's progeny should also be recorded (Kenney *et al.*, 1983).

Physical Examination of the Stallion

A general physical examination should be included in the routine BSE of the stallion. The purpose of this examination is to identify defects that might possibly be of genetic origin and that will adversely affect the ability of the stallion to serve. In particular, the visual, cardiopulmonary and locomotor systems should receive special attention in the stallion.

Evaluation of the penis, prepuce and urethral process should be done when the stallion's penis is first washed for semen collection. The relative size of the erect penis should be assessed. The skin of the penis and prepuce should be carefully examined for lesions such as habronemiasis or neoplasia (e.g. squamous cell carcinoma), and for scars indicating prior trauma or infection with equine herpesvirus 3 (which causes coital exanthema). The fossa glandis surrounding the urethral process should be examined for accumulations of smegma 'bean'.

Evaluation of the scrotum and testes is best conducted after semen collection when the stallion is usually more tractable. The scrotal skin should be thin and pliable. The number, size, orientation and consistency of both testes should be noted. The stallion's testes normally are positioned horizontally within the scrotum with the tail of the epididymis oriented caudally. The length, width and height of each testis should be measured. The testes should be roughly symmetrical in size and consistency. They should not be overly firm or soft (the normal consistency approximates that of the fleshy portion of the hand at the base of the thumb with the thumb extended). The testes should be free within the scrotum (Plate 19). The epididymis originates at the head located on the cranio-dorso-lateral pole of the testis and continues along the dorso-lateral aspect of the testis as the thin body. The epididymis terminates caudally as the prominent tail. A remnant of the gubernaculum, known as the scrotal ligament, can often be palpated as a firm structure attached to the tail of the epididymis of the stallion. The spermatic cord should be examined from its origin at the cranio-dorsal pole of the testis until it enters the external inguinal ring. Rotation of the testis may occasionally be noted with up to an 180° rotation noted without apparent effects. If the testis is rotated more than 180°, vascular compromise with clinical signs may occur secondarily to torsion.

Total scrotal width is used as one estimate of testis mass in the stallion (Gebauer *et al.*, 1974). Testis mass, in turn, is correlated with daily sperm production and output (DSO) in the stallion. Total scrotal width and DSO increase with age, though the relationship between total scrotal width and DSO appears less valid in aged stallions; in one study, the correlation between total scrotal width and DSO was 0.55 (Thompson *et al.*, 1979). Total scrotal width is measured by placing one hand above the testes to position both testes in the ventral aspect of the scrotum and then measuring the widest portion across both testes. The accuracy of the measurement is increased by the use of calipers and by taking the average of three measurements. Stallions with a total scrotal width of less than 8 cm should be strongly suspected of testicular hypoplasia or degeneration (Kenney *et al.*, 1983; Pickett *et al.*, 1987). The shape of the testes can influence the relationship between total scrotal width and testis mass; therefore, measurements of length, width and thickness of individual testes are often included in the BSE. A more accurate determination of testis volume can be made based upon ultrasonographic determination of testis width, height and length:

$$\text{volume} = 0.52 \times \text{height} \times \text{width} \times \text{length}$$

and DSO can be estimated as:

$$\text{DSO (billions)} = (0.024 \times \text{volume}) - 1.26$$

(Love *et al.*, 1991)

Comparison of projected DSO based upon testis volume with measured DSO based upon semen collection provides a useful means to assess the efficiency of spermatogenesis in the stallion. When large disparities occur between estimated DSO based upon measured testis volume and actual DSO based upon semen collection, this suggests that efficiency of spermatogenesis may be reduced (Blanchard *et al.*, 2001).

As with the external genitalia, examination of the internal genitalia of the stallion is best conducted after semen collection. Restraint of the stallion and protection of the examiner are important considerations, because most stallions are not accustomed to examination per rectum. Usually, with a slow careful approach, most will tolerate the examination. The internal genitalia of the stallion include the bulbourethral gland, pelvic urethra, prostate gland, vesicular glands and ampullae. The paired bulbourethral glands lie at the root of the penis at the ischial arch and are not palpable. The pelvic urethra is the best landmark for palpation of the internal genitalia, and is identified as a cylindrical object lying on the floor of the pelvis with a diameter of 3 to 4 cm. At the cranial extent of the pelvic urethra, the prostate gland is palpated as a firm, 2 × 4 cm glandular structure, with lobes located on either side of the urethra. The ampullae are located along the midline just cranial to the prostate as muscular ducts approximately 1 to 2 cm in diameter and 10 to 20 cm in length. The vesicular glands are sac-like structures located lateral to the ampullae that are often difficult to palpate unless they are filled as a result of sexual stimulation. The internal inguinal rings are palpated just off midline approximately 10 cm cranio-ventral to the pelvic brim.

Bacteriological Evaluation

Microbiological culture of the stallion is frequently included in the BSE (Kenney *et al.*, 1983; Pickett *et al.*, 1989). It is important to realize, however, that the penis and prepuce of the stallion have a number of bacteria isolated that represent 'normal' microflora, although some may also represent potential pathogens. Cultures are typically taken from the urethra after the penis is washed with water and cotton, and again immediately after semen collection, by passing a culturette up the distal urethra. Pre-ejaculatory cultures should be taken after the stallion has been teased so that some pre-ejaculatory fluid is present. Cultures are also taken from the prepuce and semen. Semen cultures require that a sterile semen receptacle be used. Swabs are held in transport media and are submitted for routine aerobic culture.

Interpretation of the microbiological portion of the examination should be done carefully. The organisms most commonly associated with venereal transmission in the stallion include *Taylorella equigenitalis* (the causative agent of CEM), *Pseudomonas aeruginosa*, *Klebsiella pneumoniae* (particularly capsule types 1 and 5) and, rarely, haemolytic streptococci and *Escherichia coli* (Pickett *et al.*, 1989). Because all of these organisms (except for *T. equigenitalis*) may also be isolated from stallions without fertility problems, interpretation of the microbiological findings requires consideration of other findings from the BSE. The organisms are considered as potential venereal pathogens if recovered in moderate or heavy growth in pure culture from several ejaculates, or if associated with increased leucocytes in the semen, or with an increased incidence of post-coital endometritis in mares bred to that stallion. The source of these organisms is typically urethritis, although vesiculitis, ampullitis or epididymitis are rarely associated with these isolates.

Evaluation of Sexual Behaviour and Mating Ability

The ability of the stallion to respond to a mare in oestrus, achieve erection, mount, seek the vulva or artificial vagina, thrust and ejaculate are critical to his use in a natural service or artificial insemination (AI) programme (McDonnell, 1986). Therefore, careful observation of the stallion's sexual behaviour during the BSE is important. Reaction time (interval from presentation of the mare until erection) and number of mounts per ejaculate should be recorded. The agility of the stallion in mounting the mare, seeking the vulva or artificial vagina (AV), thrusting and dismounting the mare are subjectively assessed. Evidence of pain or reluctance to mount may indicate musculoskeletal abnormalities that may impair the stallion's breeding performance. Overly aggressive behaviour towards the handlers or mare should be noted. Aggression and libido are not synonymous in the stallion.

Interpretation of sexual behaviour and mating ability is somewhat subjective. Age, experience of the stallion and season can affect reaction time and libido. Young, inexperienced stallions require careful, patient handling to successfully collect semen and not unduly disturb the development of normal sexual behaviour. Experience (learning) plays an important role in the sexual behaviour of the stallion. Stallions may become conditioned to respond to events that are not normally related to sexual behaviour, such as phantom mares, handlers, the presence of the AV, etc. Likewise, the stallion may have impaired sexual behaviour owing to negative conditioning associated with breeding-related injuries or excessive discipline in a sexual context.

Semen Collection and Handling

The number of ejaculates examined from a stallion will determine the reliability of the estimates made of semen parameters. Authors differ as to the number of ejaculates that should be examined as part of the routine stallion BSE. Currently, the Society for Theriogenology guidelines recommend a minimum of two ejaculates collected an hour apart if they meet the criteria of representativeness (Kenney *et al.*, 1983). The two ejaculates are considered representative if volumes of the ejaculates are similar, and the second ejaculate contains approximately half the number of spermatozoa as the first ejaculate, and has comparable or increased sperm motility. If the two ejaculates do not meet these criteria, the examiner should consider that the collections are not representative. For example, stallions with prolonged sexual rest before evaluation may have much higher numbers of spermatozoa in the first ejaculate than in the second. During the breeding season, it is recommended that the two ejaculates be collected during the regular breeding or collection schedule of the stallion.

Other investigations have recommended that five daily ejaculates be assessed in order to provide reliable estimates of semen parameters (Rousset *et al.*, 1987), and an average of 4.7 days was required to stabilize extragonadal sperm reserves after periods of prolonged sexual rest (Thompson *et al.*, 2004). In some cases, semen may be collected from stallions for at least 5 to 7 days to deplete the extragonadal sperm reserves and to provide an estimate of DSO (Gebauer *et al.*, 1974; Thompson *et al.*, 2004). Practical considerations tend to limit these types of examinations to very valuable stallions, or to those in which numbers of motile, morphologically normal spermatozoa are subnormal or questionable with the standard two collections made an hour apart.

Evaluation of semen for the BSE should be conducted after collection with the AV (see next section on semen evaluation). Other methods of collection, such as the condom, provide an inferior sample, particularly from the microbiological standpoint. Care should be taken to protect the sample from temperature shock, light and excessive agitation/oxygenation, which can adversely affect spermatozoa. All materials that contact the semen should be clean, dry and warmed to body temperature (35 to 37°C) (Kenney *et al.*, 1983).

Immediately after collection, the sample is evaluated for colour, clarity and foreign debris. Normal stallion semen has a skimmed milk appearance that depends on the concentration of spermatozoa present. The volume of the ejaculate is recorded and gel, if present, is removed by filtration through a milk filter or by aspiration. The volume of gel and gel-free semen is also recorded. Filtration also acts to remove any gross debris present in the ejaculate. It should be noted that appreciable numbers of spermatozoa may be lost in the collection equipment and filters (up to 25%), but this number is typically not accounted for in the routine BSE.

Immediately after removal of the gel, aliquots of semen are removed for the assessment of motility, sperm concentration, morphology and pH (Kenney *et al.*, 1971). Subsamples of semen should be removed after mixing the semen by gently swirling the sample, because spermatozoa sediment. Care should be used in handling the sample, such that all pipettes are warmed and clean and no cross-contamination of the sample with chemical fixatives, such as formol-buffered saline (BFS), occurs. Initial estimates of motility and pH should be made within 5 min of collection. Subsamples for concentration and morphological investigations should be taken soon after collection so as to reduce the occurrence of agglutination of spermatozoa in raw semen and also to reduce the morphological artefacts that may occur with time.

Semen Evaluation

Biochemical analysis of seminal fluids

Alkaline phosphatase (ALP) activity in seminal fluid from the stallion is relatively high, with a reported range of 1640–48,700 IU/l

(Turner and McDonnell, 2003). This activity appears to be derived primarily from epididymal secretions, with contributions from the testis. ALP activity can be a useful marker in cases of azoospermia, to help confirm ejaculation and the contribution of epididymal fluids. The activity is low (<90 IU/l) in pre-seminal fluid and in cases of bilateral obstructive azoospermia secondary to ampullar blockage (Turner and McDonnell, 2003). It may be useful to establish reference values for seminal plasma from normal stallions if ALP activity is to be used for the evaluation of azoospermia.

The pH of raw semen should be measured with a pH meter soon after collection. The normal pH of raw semen ranges from 7.2 to 7.7, with a slight increase in pH between the first and second ejaculates (Kenney *et al.*, 1983). An elevated pH may indicate incomplete ejaculation (pre-ejaculatory fluids have a pH of 7.8 to 8.2), urine contamination, inflammation within the genital tract or equipment contamination with soap. Samples that are incubated for a period of time after collection tend to have a lower pH as a result of the accumulation of metabolic by-products (lactic acid).

The osmolality of seminal plasma varies considerably in the stallion. In one study of five ejaculates each from ten stallions, osmolality was 336 ± 10.5 mOsm (± SEM; B.A. Ball, unpublished data). Elevations of osmolality can be used to diagnose urine contamination of semen samples (Griggers *et al.*, 2001). Alternatively, excessive use of water-soluble lubricants based on sodium methylcellulose for the lubrication of AVs during semen collection may result in contamination of the semen sample and elevation of measured osmolality (Devireddy *et al.*, 2002). This increased osmolality may, in turn, adversely affect both sperm motility and freezability.

Urine contamination of semen (urospermia) can be detected in some ejaculates by a gross change of colour or odour, or by the presence of urine crystals on microscopic examination (Fig. 11.1). In other cases, urospermia may be more difficult to detect and requires the use of assays for urea or creatinine. The use of rapid tests for urea nitrogen (Azostix) allowed the detection of urospermia (Althouse *et al.*, 1989); a colour change (yellow to green) was noted within 10 s in ejaculates that had urea nitrogen concentrations greater than 39 mg/dl. Alternatively, measurement of creatinine concentrations in semen can be used for more specific determination of urine contamination; concentrations >2.0 mg/dl are indicative of urine contamination (Dascanio and Witonsky, 2005).

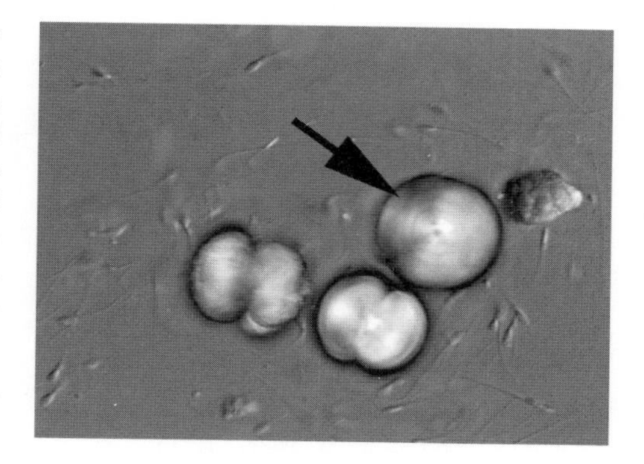

Fig. 11.1. Photomicrograph of urine crystals (arrow) in a semen sample from a stallion with urospermia.

Evaluation of spermatozoa

The normal equine spermatozoon differs from that of other large domestic animals in several respects. Distinguishing characteristics include abaxial attachment of the midpiece, asymmetry of the head, small acrosomal volume and small head size. The percentage of normal sperm in the stallion appears to be lower than that of other domestic animals, with most studies citing between 50 and 60% normal spermatozoa (Dowsett and Pattie, 1982; Jasko *et al.*, 1990; Kenney *et al.*, 1995). Based on the current Society for Theriogenology guidelines, only the total number of normal spermatozoa is considered; the differential distribution of abnormal spermatozoa is not considered. In a recent publication, per cycle pregnancy rates for 88 stallions were correlated with sperm morphological parameters (Love, 2011a). There was a moderate correlation between the percentage of normal sperm and per cycle pregnancy rate (r = 0.42), whereas morphological defects, including abnormal heads ($r = -0.22$), proximal droplets ($r = -0.34$), abnormal midpieces ($r = -0.30$) and coiled tails ($r = -0.35$) had a moderate-to-weak negative correlation with per cycle pregnancy rates (Love, 2011a). Although the percentage of normal spermatozoa is ultimately used for the BSE, the data above suggest that consideration of the type of morphological defect present may be useful in the interpretation of sperm morphology.

The method used to fix and stain spermatozoa can influence the morphological artefacts encountered in assessing stallion sperm (Hurtgen and Johnson, 1982). Cold shock of sperm before fixation and staining can lead to artefactual changes in the acrosome and midpiece. Ideally, raw semen should be fixed (1:10) in BFS immediately after collection. The evaluation of wet-mount specimens with phase-contrast microscopy or differential interference contrast microscopy (DIC) (≥1000× magnification) provides the best assessment of morphology, particularly of the acrosome, midpiece and cytoplasmic droplets.

If phase contrast or DIC microscopy is not available, then eosin–nigrosin stained smears are probably the next best alternative for the evaluation of sperm morphology. Eosin–nigrosin stain is mixed with an equal volume of raw semen and smeared on to a glass slide. Note that stain, semen and slides should be at 35–37°C to avoid inducing artefacts in sperm morphology. While eosin–nigrosin staining has been used to assess live/dead ratios in semen, this test is not highly repeatable under field conditions, and the stain is best used as a counterstain to assess morphology.

The assessment of sperm morphology with either wet mounts of BFS specimens or eosin–nigrosin smears should be conducted on a good-quality microscope under oil immersion (≥1000×). A minimum of 100 (preferably 200) cells should be counted and classified. Eosin–nigrosin stained smears can be held indefinitely, and BFS-fixed samples can be held for extended periods in well-sealed containers.

Neither BFS-fixed samples nor eosin–nigrosin stained smears are suitable for assessing types of cells other than spermatozoa, such as inflammatory cells, red blood cells or spermatocytes in a semen sample. Air-dried smears of semen can be stained with a routine blood stain such as Wright's (Diff-Quik), Giemsa or new methylene blue to observe somatic cells. Inflammatory cells should be noted only rarely (less than 1 per 5–10 high power fields) in normal stallions. Round cells (spermatids, spermatocytes and spermatogonia) should also be present only infrequently (<1–2% of all cells) in ejaculates from normal stallions (Plate 20).

Sperm concentration

The product of the volume of gel-free semen and the sperm concentration/ml is used to assess the total number of spermatozoa in the ejaculate. There are several methods used to assess the concentration of spermatozoa. These include the use of counting chambers (the haemocytometer), spectrophotometry, electronic particle (Coulter) counters, flow cytometery and image-based particle counters (the NucleoCounter SP100®).

The haemocytometer method is the least expensive and most time-consuming way to determine the concentration of spermatozoa. With this method, an aliquot of well-mixed raw semen is diluted (1:100), mixed again and then used to fill a standard Neubauer-ruled haemocytometer. The chamber is allowed to sit for 5 min to allow all sperm to settle to the same plane of focus. The chamber is then examined with either a phase contrast microscope or with a bright-field microscope adjusted for maximum contrast. Spectrophotometric methods are quicker than this and provide acceptable accuracy if properly calibrated. A standard spectrophotometer can be used after constructing a standard curve determined from duplicate haemocytometer counts of semen. There are also a number of commercially available units to determine sperm concentration in stallion semen based upon optical density. Electronic particle counters (Coulter) and flow cytometers can be used to determine sperm concentration, but the cost of obtaining, operating and maintaining this equipment precludes its use for routine semen evaluation. More recently, an image-based method for the determination of sperm concentration has been introduced commercially (Johansson *et al.*, 2008). This unit allows the determination of both sperm number and sperm viability, based upon the use of propidium iodide (PI), a fluorescent, DNA-binding dye.

Sperm motility

Reliable estimation of the motility of stallion sperm requires that all materials that contact the semen are clean, dry and warmed to 35 to 37°C. The initial estimation of motility should be made within 5 min of collection in both raw semen and extended semen. Raw stallion semen tends to agglutinate with both head-to-head agglutination and agglutination to the microscope slide. However, estimates of gross motility in raw semen may be useful for identifying potential technical problems with the extender or possible adverse effects of the extender on a particular semen sample. The repeatability of estimates of sperm motility is generally better with extended than with raw semen (62 versus 41%) (Rousset *et al.*, 1987). For these reasons, both the author and colleagues estimate motility in both raw and extended semen, but rely more heavily on estimates based on extended semen.

The estimation of motility includes both total motility (TM) and progressive motility (PM). TM is an estimation of the percentage of sperm that show any movement, while PM includes only those sperm that are actively moving forward or are moving in large circular paths. Ideally, motility should be evaluated using a standard dilution (25×10^6/ml) and volume of semen for microscopy. This can typically be accomplished by diluting raw semen with a skimmed milk–glucose extender at a ratio of >1:5. Aliquots of semen for evaluation should be taken after gently mixing the semen by swirling the container, because sperm sediment as the sample is stationary. Extender, microscope slides, coverslips, and pipettes should be pre-warmed on a warming tray or in an incubator. A small drop (6–10 μl) of extended semen is placed on a slide, coverslipped and examined with a microscope.

Phase contrast microscopy with a heated microscope stage is ideal for motility assessment. If phase contrast is not available, the microscope condenser should be lowered and the iris diaphragm of the bright-field microscope closed to enhance contrast of the specimen. The examiner should be careful not to move too close to the edge of coverslip, where drying of the specimen occurs rapidly, and estimates of TM and PM should be made after examining several fields. Extended semen from some stallions may demonstrate a high proportion of sperm with circular motility immediately after dilution. In these stallions, incubation for a period of 5 to 10 min may be required before accurate estimates of progressive motility can be obtained. The evaluation of motility should be performed quickly, particularly if the microscope stage is not heated. If necessary, new slides should be made rather than continuing to look at repeated fields on the same slide. Samples in which there are no motile sperm should alert the examiner to possible contamination of the semen with spermicidal compounds, such as alcohol or soaps used to clean the AV or the glassware.

Computer assisted sperm analysis (CASA) systems have been available for many years as a method to provide more objective analysis of sperm motility parameters. These systems provide a host of kinematic data regarding sperm motion and velocity parameters and offer the ability to obtain more repeatable estimates of sperm motility than do subjective estimates obtained using only a microscope. In most cases though, data obtained from CASA does not substantially improve estimates of fertility of a particular sample compared with subjective estimates obtained using a good-quality microscope and heated microscope stage.

Although the concept of PM has long been a part of the routine BSE, its repeatability across different examiners is not good, and the variation in measurement of TM is typically less (Love, 2011b). When sperm motility parameters determined by CASA were compared based upon per cycle pregnancy rates in stallions (91% versus 56% versus 32% per cycle pregnancy rate), the parameters of TM and PM, and of path velocity and progressive velocity were significantly lower for stallions with low seasonal pregnancy rates (Love, 2011a). Not surprisingly, many of the measured motility parameters were highly correlated, including PM and TM and mean progressive velocity, as determined by CASA (Love, 2011a). Correlation coefficients between TM and PM and per cycle pregnancy rates were 0.59 and 0.52, respectively (Love, 2011a).

Ultrastructure of sperm

Transmission electron microscopy (TEM) of equine spermatozoa can provide detailed resolution of the ultrastructure of ejaculated sperm, including defects in the plasma or acrosomal membrane, mitochondria and axoneme, as well as showing chromatin condensation (Veeramachaneni *et al.*, 2006).

Sperm function testing

There are a number of *in vitro* assays that have been used to infer different functional compartments of equine sperm, including the plasma membrane, acrosome, mitochondria and a variety of sperm-specific proteins, but there are relatively little data that associate these *in vitro* assays with stallion fertility. The relative lack of information relating various sperm function tests with fertility in the stallion limits the ultimate use of many of these assays in clinical applications. None the less, a considerable body of information on the normal function of equine spermatozoa has accumulated through such studies, and may ultimately be useful in accurately predicting the fertility of equine semen based solely upon *in vitro* assays. Varner (2008) provides an excellent overview of many of these assays as applied to equine sperm.

A number of fluorescence probes have been used to assess different functional compartments of sperm. Integrity of the plasma membrane can be assessed by the ability of membrane-impermeant, DNA-binding dyes (including PI, ethidium and others) to stain nuclear DNA and thereby assess sperm viability. Likewise, a number of cationic fluorescent probes (rhodamine 123 and derivatives) are preferentially taken up by mitochondria and may be useful for detecting mitochondrial membrane potential. One probe, JC-1, has the advantage of differentially labelling mitochondria with low versus high membrane potential, thereby providing quantitative information on mitochondrial function (Plate 21) (Gravance *et al.*, 2000). Other probes can be used to detect oxidative damage to the sperm membrane (e.g. C-11-Bodipy 581/591) (Ball and Vo, 2002) or the generation of reactive oxygen species during sperm metabolism (Ball *et al.*, 2001b; Sabeur and Ball, 2006; Burnaugh *et al.*, 2007; Ball, 2008). Other changes, such as alterations in sperm membrane structure (Thomas *et al.*, 2006) and apoptosis (Brum *et al.*, 2008) have also been characterized for equine sperm. Many of these fluorescent probes can be detected using fluorescence microscopy; however, more accurate quantitation can be achieved using flow cytometry, which offers the opportunity to assess a relatively large number of sperm (10^4) quickly.

The sperm acrosome, the membrane-bound vesicle covering the rostral sperm head, contains enzymes that are important in the process of sperm penetration through the zona pellucida of the oocyte during fertilization. The acrosome can be damaged during freezing and thawing, and failure of normal acrosomal exocytosis has been identified as a cause of infertility in stallions (Bosard *et al.*, 2005). The acrosome of equine sperm is relatively small and difficult to image using standard microscopy; detection of acrosomal exocytosis is typically based upon either TEM or other specialized staining techniques. Fluoresceinated lectins, such as peanut agglutinin (PNA), have been used to preferentially label the acrosomal membrane and thereby improve detection of the equine acrosome (see Plate 22, which shows equine sperm stained with fluoresceinated pea lectin). FITC (fluorescein isothiocyanate)-PNA is typically combined with a viability probe (PI) to differentiate sperm that have undergone a true acrosome reaction from those that have undergone cell death with subsequent loss of the acrosome. Alternatively, the equine acrosome can be identified with bright-field microscopy after staining with Commassie blue, although this methodology does not allow the simultaneous detection of sperm viability (Brum *et al.*, 2006).

During transit through the mare's reproductive tract, equine sperm associate with the epithelial cells lining the isthmic portion of the oviduct to form a functional sperm reservoir (Thomas *et al.*, 1994). The ability of sperm to interact with oviductal epithelial cells (OEC) *in vitro* has been used as a method to assess sperm function (Dobrinski *et al.*, 1995). Likewise, the ability of equine sperm to bind the zona pellucida *in vitro* has been characterized and a relationship with fertility proposed for the horse (Meyers *et al.*, 1996).

Evaluation of sperm chromatin

Changes in the susceptibility of sperm chromatin (which is composed of highly compacted DNA and nuclear proteins) to damage, and thereby the relationship of sperm chromatin structure with fertility, has been known for several years in a variety of animals, including the stallion (Evenson *et al.*, 2002). To measure such changes in sperm chromatin structure, sperm are exposed to acid to denature chromatin, and then stained with a dye (acridine orange) that differentially stains single- versus double-stranded DNA as a means to detect DNA strand breaks. Stained sperm are examined by flow cytometry and the amount of DNA fragmentation is determined (Evenson *et al.*, 2002). Changes in the stability of sperm chromatin have been associated with changes in stallion fertility (Love and Kenney, 1998). In most cases, the aetiology of differences in susceptibility to the denaturation of sperm chromatin are unknown; however, thermal trauma to the testis may disrupt normal spermatogenesis, with alterations in sperm chromatin stability detected following acute thermal injury to the testis (Love and Kenney, 1999). Other assays have been used to assess changes in sperm chromatin or DNA, including single-cell gel electrophoresis (the comet assay; Fig. 11.2) (Baumber *et al.*, 2002b) and the TdT (terminal deoxynucleotidyl transferase)-mediated-dUTP nick end labelling (TUNEL) assay (Brum *et al.*, 2008); no data are presently available to relate the outcomes of these assays to fertility in the stallion.

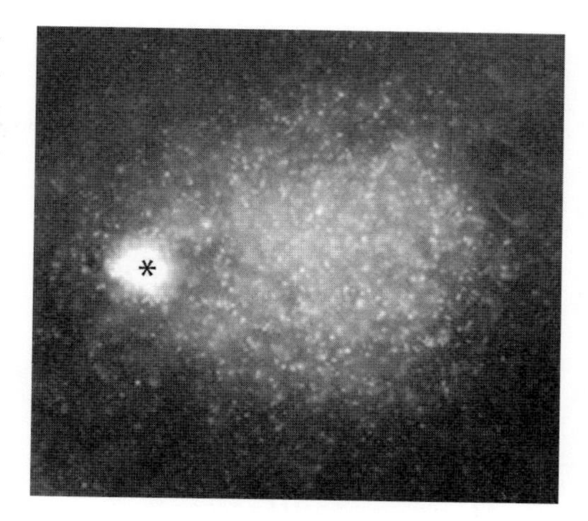

Fig. 11.2. Equine sperm comet assay showing the sperm nucleus (*) and the tail of fragmented DNA that has migrated out from the nucleus.

Ancillary Diagnostic Procedures

Ultrasonographic and endoscopic examination of the stallion's reproductive organs[1]

Ultrasonography

Evaluation of the scrotum, testis, epididymis and spermatic cord by B-mode ultrasonography has become a routine part of reproductive evaluation of the stallion (Pozor, 2005), and the normal ultrasonographic anatomy of the stallions' reproductive tract has been described (Little and Woods, 1987; Love, 1992; Ginther, 1995). In addition, Doppler ultrasonography has been used for the evaluation of both the internal and external genitalia of the stallion (Pozor and McDonnell, 2004; Ginther, 2007; Pozor, 2007).

Lesions of scrotum, epididymis and testis

The ultrasonographic appearance of the normal equine testicular parenchyma is homogeneous with the exception of the central vein of the testis (Fig. 11.3) (Love, 1992).

Changes that may be detected within the testis parenchyma by ultrasound include the presence of tumours, parenchymal oedema, increased echogenicity of the testis parenchyma associated with late-stage testicular

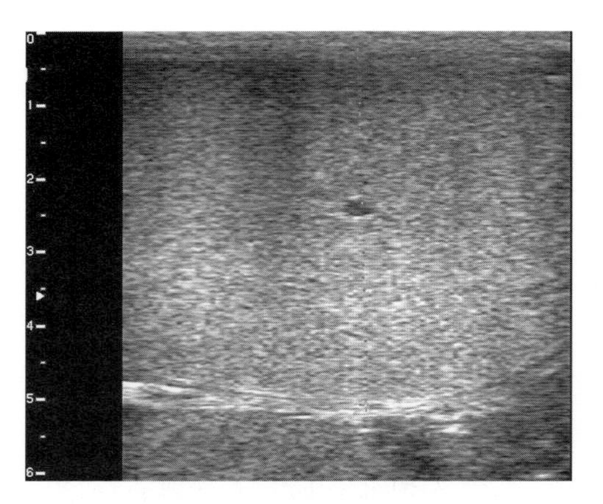

Fig. 11.3. Ultrasonogram of a normal testis showing the central vein.

degeneration and haematomas or abscesses within the testis parenchyma. Testicular tumours are relatively rare in stallions, although seminomas, Sertoli cell tumours, Leydig cell tumours, teratomas and mast cell tumours, among others, have been reported (McEntee, 1990; Brown *et al.*, 2008; Edwards, 2008). Seminomas and Leydig cell tumours occur most often in aged stallions and may be present in either normally descended or cryptorchid testes. Teratomas are more often reported in retained testes (McEntee, 1990). The ultrasonographic appearance of these tumours varies, and relatively few reports have characterized their appearance in the stallion testis. In one report, the ultrasonographic appearance of a seminoma was characterized as diffusely hypoechoic with heterogeneous areas of increased echogenicity (Fig. 11.4) (Beck *et al.*, 2001). Seminomas can metastasize and examination of the spermatic cord and abdominal cavity may be useful for detecting metastatic disease (Fig. 11.5). In another report, a testicular mastocytoma (mast cell tumour) appeared as a heterogenic flocculent mass within the testis (Brown *et al.*, 2008). Routine examinations of the testes of the stallion are likely to yield better diagnostic information on the frequency of testis neoplasia in stallions, along with an earlier detection during the course of the disease.

Fluid accumulations in the vaginal tunic (or tunica vaginalis, the serous covering of the testis), such as hydrocele (an accumulation of serous fluids) and haematocele (an accumulation of blood), are readily detected by ultrasonographic examination of the scrotum and testis in the stallion. Hydrocele can accompany scrotal oedema or may result because of transfer of peritoneal fluid from the abdomen to the vaginal cavity. Hydrocele may also occur during periods of high ambient temperature and resolve as environmental temperatures moderate (Varner *et al.*, 1991). Such fluid accumulations are typically characterized by an anechoic fluid without evidence of increased cellularity or fibrin. In contrast, haematocele (Fig. 11.6) is characterized by the accumulation of blood in the vaginal cavity often associated with trauma to the scrotum or possibly with haemoperitoneum. Ultrasonographically, haematocele is characterized by the presence of increased echoic debris within the tunics, including the presence of fibrin strands.

Scrotal ultrasonography can be very useful in examining the stallion with acute scrotal enlargement (Morresey, 2007). Acute torsion of the spermatic cord, with accompanying testicular oedema, congestion and ischaemia may result in ultrasonographic signs, including increased oedema (reduced echogenicity of the testis parenchyma) (Fig. 11.7), along with evidence of reduced blood flow in the spermatic cord (Pozor, 2007). Inguinal or scrotal hernias typically involve passage of a loop of small intestine through the vaginal ring into the inguinal or scrotal portion of the vaginal tunics. Ultrasound provides a ready method to identify the presence of intestine within the scrotum and greatly facilitates accurate diagnosis.

In addition to torsion, other lesions of the spermatic cord include varicoceles, which represent a venous engorgement or enlargement of the veins of the pampiniform plexus. Although common in human males, varicoceles appear to be rare in the stallion (Varner *et al.*, 1991). The presence of a grossly enlarged venous plexus can be detected by both B-mode and Doppler ultrasonography in the region of the spermatic cord (Pozor, 2005). In human males, the presence of varicocele is associated with increased abnormalities in the spermiogram; anecdotal evidence suggests that stallions with varicocele may also have a suppression of semen quality.

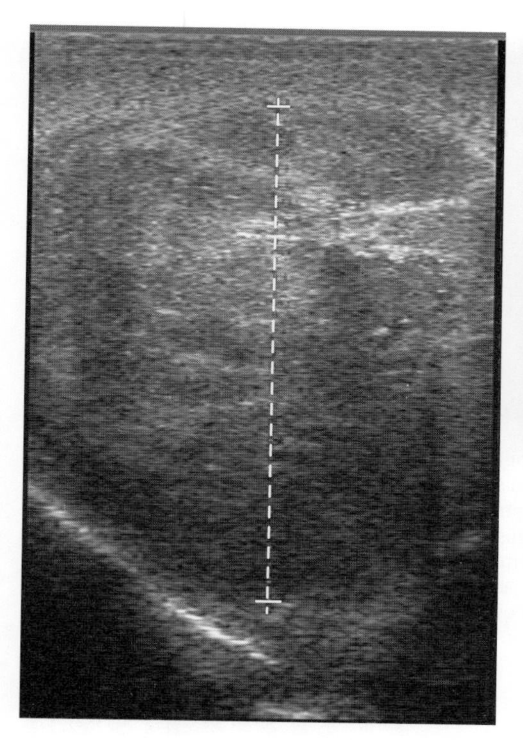

Fig. 11.4. Ultrasonogram of a seminoma in a stallion showing a mixed echoic and hypoechoic pattern.

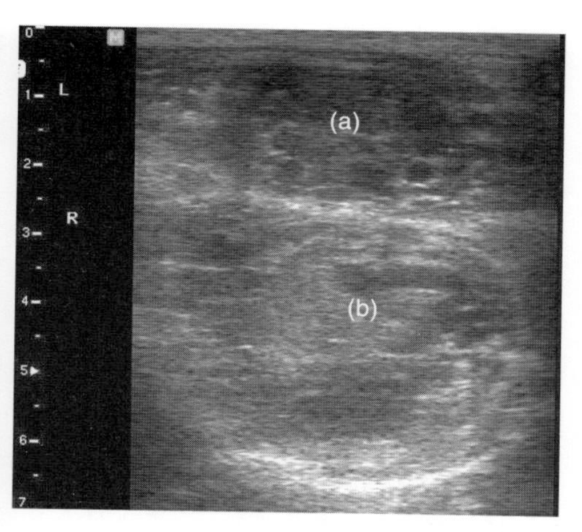

Fig. 11.5. Ultrasonogram of spermatic cord with seminoma. The normal cross section of the spermatic cord (a) presents multiple cross sections of blood vessels, while the affected spermatic cord (b) shows an enlarged cross section with loss of vascular detail and increased echogenicity.

The epididymis can be imaged along its entire length by transcutaneous ultrasonography. Epididymal cysts have been detected ultrasonographically in the caput epididymis of the stallion, and anecdotal evidence suggests that these cysts may be associated with ejaculatory problems (Pozor, 2005). The origin of these cysts has not been defined, but they may be derived from blind-ended efferent ductules that fail to fuse in formation of the epididymal duct, or possibly represent cysts of the appendix epididymis, a remnant of the mesonephric duct (McEntee, 1990; Love, 1992). These cysts may be noted also in stallions in which no decline in semen parameters has been noted; however, rupture of blind-ending efferent ductules can also lead to sperm granuloma formation in the caput epididymis, possibly resulting in obstructive lesions of the epididymal duct. Epididymitis is relatively uncommon in the stallion, but most often involves either unilateral or bilateral disease of the cauda epididymis. These lesions are typically characterized by the presence of epididymal enlargement, along with enlargement of the epididymal lumen (Fig. 11.8). In the acute phase, epididymitis may be accompanied by painful enlargement of the epididymis, also with the presence of pyospermia (or leucocytospermia, a condition in which there is an unusually high number of white blood cells in the semen). As the condition becomes chronic, there may be a reduction in pain associated with the lesion.

Evaluation of penis and prepuce

Acute trauma to the penis, often associated with breeding or collection injuries, may result in haematomas of the corpus cavernosum, which may result in abnormal erection (Hyland and Church, 1995). Such injuries may be characterized ultrasonographically by the appearance of an increased echogenicity (Fig. 11.9) within the corpus cavernosum, accompanied by penile deviation (Fig. 11.10). Kicks to the inguinal region of the stallion may result in penile, or more often preputial, injuries, with haematomas forming in the large venous plexus located dorsal to the prepuce; again, ultrasound imaging may aid the detection and characterization of these lesions.

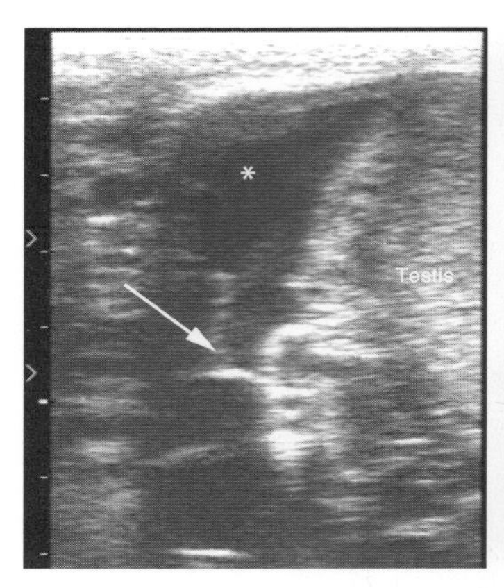

Fig. 11.6. Ultrasonogram of a haematocele in a stallion. Fibrin tags (arrow) are noted between the testis and scrotum, and free blood (∗) is noted within the vaginal tunic.

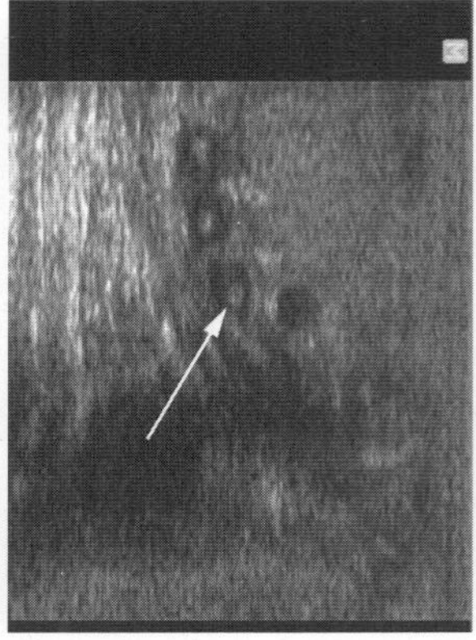

Fig. 11.7. Chronic epididymitis in a stallion. An ultrasonogram of the cauda epididymis reveals numerous sperm granulomas (arrow) that were confirmed on histopathology.

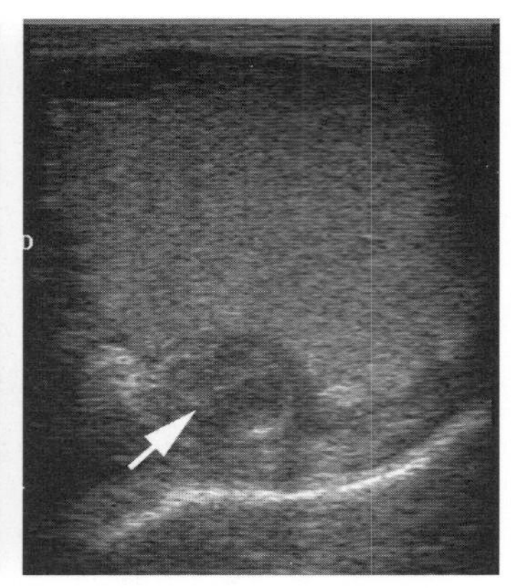

Fig. 11.8. Epididymitis in the stallion. Ultrasound examination of the corpus epididymis revealed an enlargement (arrow) that corresponded to a thickened, fibrotic lesion of the epididymis.

Evaluation of internal genitalia

The internal genitalia of the stallion are readily imaged by transrectal ultrasonography, and changes in the appearance of the accessory sex glands before, during and after ejaculation have been characterized (Weber *et al.*, 1990; Ginther, 1995).

Lesions of the terminal portion of the ductus deferens (ampulla) appear to be the most commonly identified abnormality of the internal genitalia of the stallion. Some stallions appear to have an abnormal retention of spermatozoa within the excurrent duct system, with an accumulation of spermatozoa (Varner *et al.*, 2000). Retained spermatozoa appear to undergo degenerative changes within the excurrent duct system, and ejaculates from these stallions may be characterized by a high sperm concentration (>500 million/ml), low motility and a high percentage of detached heads (Fig. 11.11). In some cases, sperm and epithelial cell debris may form casts that appear within the ejaculate (Fig. 11.12). Occasionally, spermatozoa may form obstructive plugs that are retained within the distal ductus deferens, resulting in unilateral or bilateral obstruction of the ductus deferens. In these stallions, transrectal ultrasonography of the accessory glands may reveal a dilation of one or both ampullar lumen due to obstruction (Fig. 11.13). In most cases, these lesions are readily identified by transrectal ultrasonography.

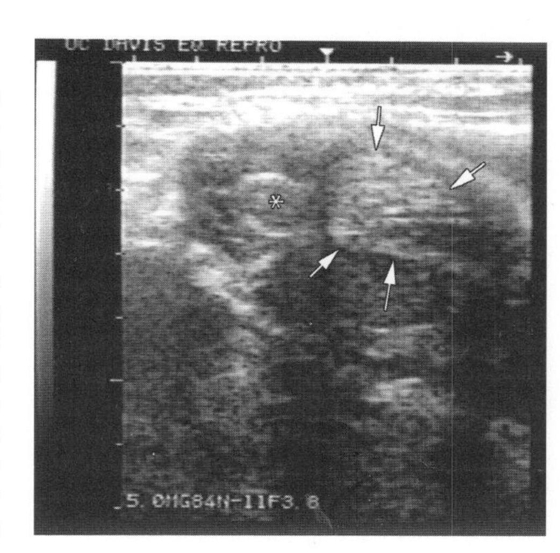

Fig. 11.9. Ultrasonogram of penile hematoma in the corpus cavernosum (arrows) of a stallion, which appears as a hyperechoic region. The urethra and corpus spongiosum is marked (∗).

Fig. 11.10. Marked penile deviation in stallion at the time of ejaculation.

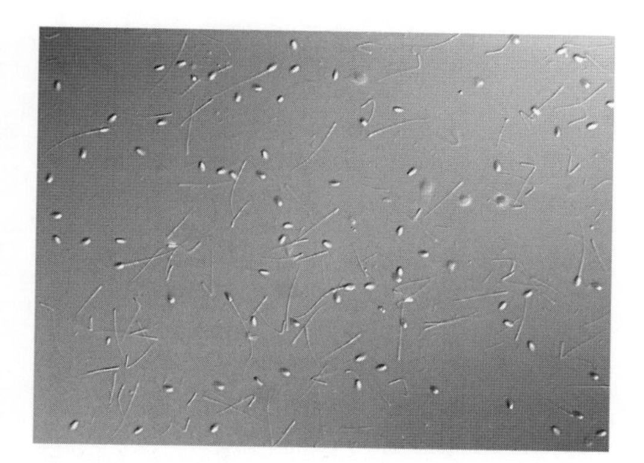

Fig. 11.11. Photomicrograph of a semen sample with a high percentage of detached heads.

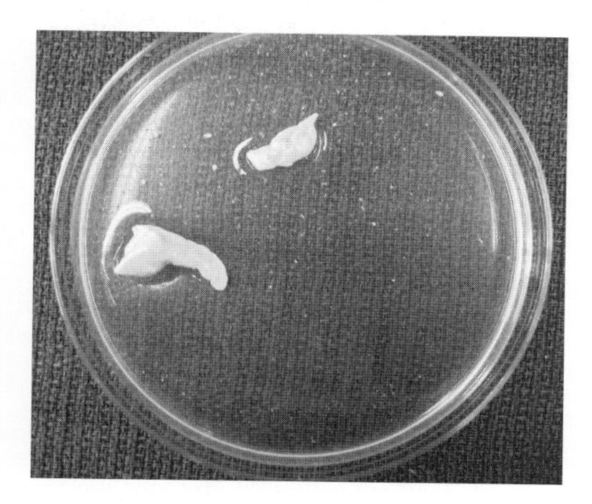

Fig. 11.12. Sperm 'plugs' collected from the semen filter after collection from a stallion with ampullar obstruction (spermiostasis). Grossly, these plugs appear as amorphous masses.

In addition to obstructive lesions of the ampulla, segmental aplasia of the terminal portion of the ductus deferens has been described in the stallion, associated with an apparent failure of fusion of the mesonephric ducts and the urogenital sinus during development (Estrada *et al.*, 2003). Segmental aplasia of the ductus deferens characterized by the presence of dilated terminal cysts within the ampulla near the pelvic urethra has been observed by Ball and colleagues. Although rare, aplasia of different regions of the excurrent duct of the stallion has been reported and should be considered in cases of infertility associated with azoospermia.

Cystic structures associated with the internal genitalia of the stallion have been described based upon their ultrasonographic appearance. One of these, the uterus masculinis, represents a remnant of the paramesonephric duct in the male, which is located in the urogenital fold between the two ampullae of the ductus deferens as they enter the pelvic urethra at the colliculus seminalis (Plate 23). In some stallions, the uterus masculinis may be detected ultrasonographically as a hypoechoic cystic structure lying between the ampullae. While uterus masculinis is frequently found in normal stallions, cystic distension of this structure has been associated with ejaculatory problems by some authors (Pozor, 2005). Urethral cysts have also been identified during ultrasonographic examination of the pelvic urethra of the stallion, though the significance of these lesions remains undetermined (Pozor, 2005).

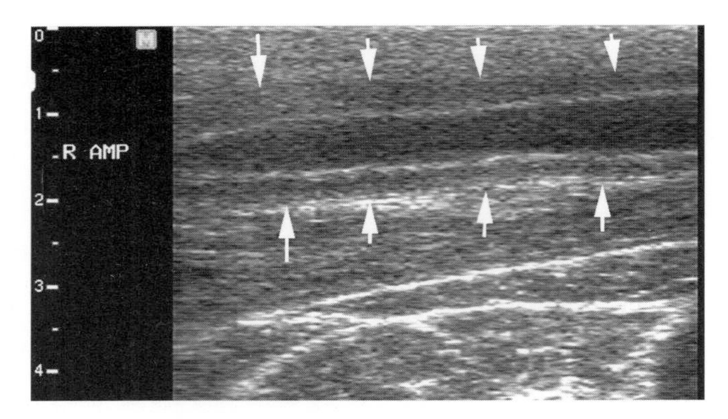

Fig. 11.13. Ultrasonogram of the ampulla of an obstructed stallion showing the luminal dilatation with obstruction of the ampulla, which is shown in longitudinal section and outlined with arrows.

Inflammation of the vesicular glands in the stallion is uncommon, but seminal vesiculitis can present an important source of inflammatory cells in the ejaculate when it occurs. The gross appearance of a semen sample may show a large amount of flocculent debris with evidence of mild haemorrhage (Plate 24). Changes in the ultrasonographic appearance of the vesicular glands may not be detected in vesiculitis but, in some cases, the fluid content of the vesicular glands has been noted to change from a normal anechoic fluid to a more hyperechoic fluid, along with a detectable thickening of the wall of the vesicular gland (Fig. 11.14). Phase contrast microscopy of a semen sample collected from a stallion with vesiculitis showed sperm and many somatic cells (Plate 25), which were identified after Wright's staining as neutrophils.

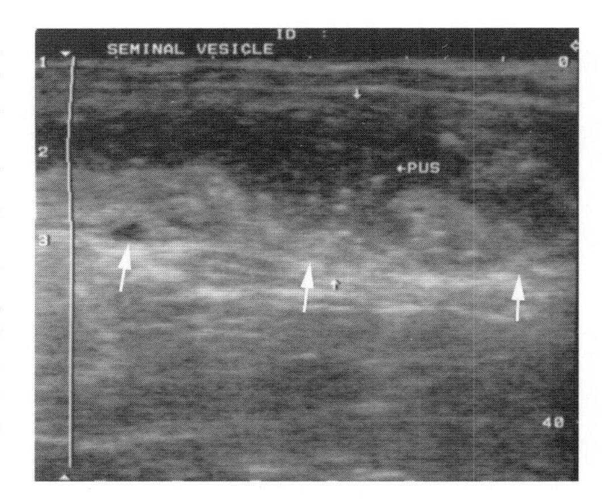

Fig. 11.14. Seminal vesiculitis in a stallion. On ultrasound, the vesicular gland is distended with thickened walls (arrows) and there are echoic particles within the fluid.

Evaluation of retained testes

Ultrasonography can be a useful adjunct to identify testes retained within the inguinal canal or abdomen of the cryptorchid stallion (Jann and Rains, 1990; Schambourg *et al.*, 2006). Identification of the retained testis within the inguinal canal is relatively straightforward using ultrasound, although localization of the abdominal testis by transrectal ultrasonography can require more patience and experience (Jann and Rains, 1990; Schambourg *et al.*, 2006). A technique for the identification of abdominal testes by a systematic transabdominal ultrasound examination with a relatively high sensitivity and specificity has been described (Jann and Rains, 1990; Schambourg *et al.*, 2006).

Endoscopy

Urethroscopy of the stallion is most useful for delineating lesions of the urethra associated with bacterial urethritis or haemospermia (the occurrence of blood in the ejaculate) (Sullins *et al.*, 1988). A 100 cm flexible video endoscope is adequate for imaging the length of the urethra into the pelvis. Lesions identified during endoscopy of the urethra include urethritis, strictures of the urethra, varicosities and urethral rents.

Urethral rents that communicate with the corpus spongiosum may result in haemospermia and are typically located near the ischial arch. Within the pelvic urethra, the termination of the ductus deferens and the openings of the vesicular glands can be imaged dorsally at the colliculus seminalis. The openings of the vesicular glands can be cannulated with a small (3–5 French) polyethylene catheter to facilitate sampling of the content of the vesicular gland. The openings of the ampullae, however, are not readily cannulated via an endoscopic approach.

Genetic testing

Karyotypic analysis is frequently useful in examining individuals whose sexual phenotype is uncertain (i.e. intersex conditions). Such an analysis may also be useful in some undiagnosed cases of infertility or subfertility in stallions. Autosomal abnormalities and sex-chromosome mosaics (i.e. 63,XO/64,XY) have been described as causes of reduced fertility or infertility in stallions (Kenney *et al.*, 1991). More frequent application of cytogenetic studies is likely to reveal other abnormalities that are associated with reduced fertility in the stallion.

Completion of genomic sequencing for the horse, along with an improved understanding of the molecular mechanisms responsible for successful reproduction, offers the opportunity for an improved understanding of the genetics of stallion fertility (Leeb, 2007). As an example of such application, equine cysteine-rich secretory protein (CRISP3) is a major secretory protein in seminal plasma, which has three non-synonymous single nucleotide polymorphisms (SNPs) that are associated with fertility in the stallion (Hamann *et al.*, 2007). Future studies will likely provide an improved understanding of the genetics of stallion fertility, through the identification either of candidate genes or of markers related to fertility (Giesecke *et al.*, 2010).

A number of genetic diseases have been characterized for the horse, including several defects in which the specific genetic mutation underlying the disease have been characterized. A growing number of these diseases have genetic tests that are available to screen for such mutations. Depending upon the breed of stallion being examined, these tests will become increasingly important in the elimination of potential genetic diseases from the breeding population. Some of the diseases characterized to date for which specific mutations have been identified include those listed in Table 11.1.

Interpretation of Findings of the BSE

A final recommendation on breeding capabilities based on the result of the BSE of the stallion requires consideration of all aspects of the examination. Because many factors in addition to the stallion itself potentially affect its breeding performance, the BSE does not provide a precise measure of the future fertility of a given stallion and may have more utility in the identification of potentially sub-fertile stallions. Based upon current knowledge, stallions that have poor sperm quality are more likely to have reduced fertility, though the converse may not be true (Love, 2011b). Stallions with normal sperm quality may have associated good fertility but other factors, such as management, may adversely affect this

Table 11.1. Genetic diseases in horses caused by specific mutations.

Genetic disease	Abbreviation	Breed association (if any)
Hereditary equine regional dermal asthenia	HERDA	American Quarter Horse
Hyperkalemic periodic paralysis	HYPP	American Quarter Horse
Glycogen branching enzyme deficiency	GBED	American Quarter Horse
Junctional epidermolysis bullosa	JEB	Belgian Draft
Polysaccharide storage myopathy	PSSM	–
Severe combined immunodeficiency	SCID	Arabian

fertility, thereby reducing the predictive basis of sperm quality for fertility (Love, 2011b).

The Society for Theriogenology's guidelines (Kenney *et al.*, 1983) on fertility evaluation, which were formulated almost 30 years ago, provided recommendations for a stallion with a full book of 40 mares for natural service or 120 mares for AI (40/120 mares). Over the intervening years, the number of mares bred to stallions with natural service and AI (including a larger proportion of mares bred with transported or frozen–thawed semen) has increased dramatically in some situations, and the utility of these guidelines has diminished.

In the guidelines promulgated by the Society for Theriogenology, the determinant factor for semen quality of the stallion is the number of morphologically normal, progressively motile spermatozoa (Kenney *et al.*, 1983). As noted by Love (2011b), the use of this parameter as the determining factor to assess semen quality may be subject to several problems of interpretation. In particular, the criteria for the determination of progressive motility are not standardized and the repeatability of progressive motility measurements is often lower than that for total motility. In addition, the use of a combined parameter for progressive motility and normal morphology may be overly restrictive, because the parameters of motility and normal morphology are often reasonably correlated. Furthermore, the classification scheme that utilizes only normal sperm morphology, without any attention to the differential analysis of morphologically abnormal spermatozoa, may unduly weight some types of morphological abnormalities (e.g. distal cytoplasmic droplets) that do not have known associations with fertility.

The Society for Theriogenology's guidelines for assessing the potential fertility of a stallion are based on the following criteria:

- The stallion demonstrates good libido and mating ability.
- The penis is normal sized without inflammatory lesions.
- Bacteria recovered from the semen and urethral swabs are mixed in type and decrease after ejaculation.
- The test for equine infectious anaemia is negative.
- The stallion has two scrotal testes and epididymes of normal size, shape and consistency. Total scrotal width is greater than 80 mm.
- The stallion produces a minimum of 1 billion morphologically normal, progressively motile spermatozoa in the second of two ejaculates after correction for season.

In addition, if serological testing for equine arteritis virus is positive in the absence of defined vaccination history, then virus isolation from semen should be conducted to rule out the possibility that the stallion may be shedding the virus in semen.

If a stallion fails to meet these criteria, he is classified as a questionable or unsatisfactory prospective breeder for a full book of mares. The ultimate classification of a stallion will depend on the severity of the problems that have led to an unsatisfactory classification. However, many stallions are not used for 40/120 mares, and may be able to settle a proportionately smaller book of mares with adequate reproductive management.

Oxidative Stress in Normal and Abnormal Function of Equine Spermatozoa

Introduction

Oxidative stress has long been known as an important process in mammalian spermatozoa, having been part of the original description of oxidative damage to mammalian cells (MacLeod, 1943). Although reactive oxygen species (ROS) may form as a normal consequence of oxidative metabolism, specific generating mechanisms within particular cell types, such as leucocytes, may also result in the formation of ROS. From an abundance of literature, it is evident that low-level generation of ROS plays an important role in normal sperm function and that elevated ROS concentrations resulting from an imbalance in either the production or degradation of ROS may adversely affect sperm. During sperm storage, the effects of oxidative stress may be even more important in situations where much of seminal plasma is removed from a semen sample because a great deal of the antioxidant capacity in semen resides in the seminal plasma, and not in sperm themselves, which have relatively little cytoplasm.

Generation of ROS

Two potential mechanisms appear to account for generation of ROS by equine sperm. A sperm-specific NADPH oxidase (NOX5) present in the plasma membrane of the sperm head may be responsible for low-level ROS production, whereas electron leakage from sperm mitochondria appears to produce ROS attributed to sperm metabolism (Sabeur and Ball, 2006, 2007). The superoxide anion ($O_2^{-\cdot}$) is the primary ROS generated by equine sperm, but this molecule rapidly dismutates to form hydrogen peroxide (H_2O_2) (Ball *et al.*, 2001b; Burnaugh *et al.*, 2007). Hydrogen peroxide, in turn, is the ROS that accounts for the major cytotoxic effect in sperm (Baumber *et al.*, 2000). The production of ROS is increased in the presence of cryodamaged, dead or morphologically abnormal sperm, and sperm with residual cytoplasm or abnormal midpieces appear to produce greater quantities of ROS (Ball *et al.*, 2001b). Under these conditions, the generation of ROS is principally driven by electron leakage from the mitochondrial electron transport chain, with subsequent reduction of molecular oxygen to form the superoxide anion (Sabeur and Ball, 2006).

Subsequent to cryopreservation, damage to sperm mitochondria probably results in the generation of superoxide, which contributes to the oxidative damage of frozen–thawed equine sperm. The evaluation of sperm after cryopreservation frequently reveals morphological changes in the sperm midpiece that are characterized by moderate-to-marked-swelling of the mitochondria, suggesting that sperm mitochondria are a significant site of cryodamage, with consequent uncoupling of normal oxidative metabolism, generation of ROS and the induction of degenerative processes such as apoptosis (Brum *et al.*, 2008).

In human semen, contaminating leucocytes are an important source of ROS (Aitken *et al.*, 1994). Ball and colleagues examined the influence of neutrophil (polymorphonuclear leucocyte – PMN) addition to equine sperm to assess the relative effect of PMNs on equine sperm *in vitro* (Baumber *et al.*, 2002a). Equine sperm were separated from seminal plasma by density-gradient centrifugation, and isolated equine PMNs were purified from whole blood. The addition of PMNs activated with phorbol ester to the separated sperm increased hydrogen peroxide generation, and significantly decreased sperm motility (Baumber *et al.*, 2002a). However, this effect was only noted when the ratio of sperm:PMNs was 5:1 or 2.5:1, which suggests that a relatively large contamination of equine semen with PMNs was required to induce notable adverse effects on equine sperm motility.

ROS scavengers in seminal plasma

As already noted, because of the relatively small cytoplasmic volume of the mature spermatozoon, most of the antioxidant scavengers present in semen reside in the seminal plasma, where the primary scavengers include catalase, superoxide dismutase (SOD) and glutathione peroxidase (GPx). There is a wide species variation in the activity of these scavengers in seminal plasma. In equine seminal plasma, the activity of catalase was 98.7 ± 29.2 IU/mg protein, the activity of SOD was 29.15 ± 6.64 IU/mg protein and the activity of GPx was 0.87 ± 0.06 μM NADPH oxidized/mg protein each minute (Ball *et al.*, 2000; Baumber and Ball, 2005). Most of the catalase activity appeared to originate from the prostate gland; whereas SOD was derived primarily from the prostate gland and ampullae; GPx was present in the highest activity in the testis and in the cauda epididymis (Baumber and Ball, 2005). The activity of both catalase and SOD in equine seminal plasma was relatively high compared with that in other species, and there was significant variation between stallions in the activities of these scavengers. Based upon these observations, semen processing that entails the removal of seminal plasma may increase the susceptibility of equine sperm to oxidative damage due to the removal of these enzyme scavengers.

A number of other components of seminal plasma provide antioxidant activity and help to prevent oxidative damage to sperm. Low-molecular weight factors such as lactate, urate, taurine, hypotaurine, pyruvate, ascorbic acid, tocopherol, ergothioneine and albumin are present in seminal plasma and are capable of removing certain ROS (Mann *et al.*, 1963; Alvarez and Storey, 1983; Kovalski *et al.*, 1992; Halliwell and Gutteridge, 1999). There is little research in equine semen to examine the importance of these low molecular weight antioxidants; however, one study suggested that these components may constitute most of the antioxidant capacity of semen (Thiele *et al.*, 1995).

Effects of ROS on normal equine sperm function

The low-level generation of ROS by equine sperm is stimulated by calcium ions, and it appears that the membrane-associated NOX5 (see above) is responsible for the production of the superoxide anion as part of cell signalling events in equine sperm. The generation of superoxide anion, in turn, is important in the induction of capacitation, which is associated with an increase in tyrosine phosphorylation (Baumber *et al.*, 2003b; Burnaugh *et al.*, 2007). During the cryopreservation of equine sperm, Ball and colleagues have hypothesized that the low-level generation of ROS may lead to capacitation-like processes that contribute to shortened sperm longevity after insemination. This is supported by the observation that cryopreserved sperm have an increased intracellular calcium concentration, an increased generation of ROS and a reduced antioxidant capacity because of removal of seminal plasma. Other studies in the same laboratory have supported the observation that the 'cryocapacitation' of sperm results in similar but not identical changes in the sperm membrane to those that are detected during capacitation *in vitro* (Thomas *et al.*, 2006).

Cytopathic effects of ROS on equine sperm

Increased generation of ROS by equine sperm may occur in the presence of large numbers of morphologically abnormal or damaged sperm in a sample, and this may adversely affect the remaining viable sperm via oxidative stress. Although the superoxide anion appears to be the primary product generated by sperm, research by Ball and colleagues demonstrates that the less polar hydrogen peroxide is the most important ROS that results in damage to equine sperm (Baumber *et al.*, 2000). The generation of ROS by the xanthine/xanthine oxidase (X-XO) system results in the production of a superoxide anion that rapidly dismutates to H_2O_2. The increased H_2O_2 results in a decrease in sperm motility with no detectable decrease in viability, acrosomal integrity or mitochondrial membrane potential (Baumber *et al.*, 2000). The addition of catalase (which catabolizes H_2O_2), but not of SOD (which catabolizes the superoxide anion), maintained normal motility secondary to this induced oxidative stress. Based upon these studies, sperm motility appears to be a sensitive indicator of oxidative stress and may be one of the first parameters affected during oxidative stress.

DNA damage is another well-known cytopathic effect of ROS. In equine sperm, exposure to increasing concentrations of ROS generated by X-XO resulted in a dose-dependent increase in DNA damage as detected by the comet assay (Baumber *et al.*, 2003a). This DNA damage was blocked in the presence of catalase or reduced glutathione (GSH), but not in the presence of SOD, which indicates that H_2O_2 was the major ROS responsible for DNA damage in these cells (Baumber *et al.*, 2003a). Damage to sperm DNA appears to be initiated at levels of oxidative stress comparable to those previously shown to affect sperm motility. Aitken *et al.* (1998) proposed that DNA damage may be initiated at relatively low levels of oxidative stress, which might be consistent with induction of capacitation without any change in the motility of human sperm. Therefore, low-levels of oxidative stress may allow sperm with DNA damage to fertilize the oocyte. Sperm have limited to no ability to repair DNA damage, and studies from other species indicate that while fertilization may occur, the rate of subsequent embryonic development is reduced and the rate of early embryonic death is increased in situations in which fertilization is initiated by DNA-damaged sperm (Ahmadi and Ng, 1999; Morris *et al.*, 2002). If this is also true in equine sperm, some of the damage to equine sperm DNA may be present in motile sperm that do not present any evidence of alteration in the standard parameters evaluated as part of routine semen analysis.

During the storage of equine sperm, there is also a measurable increase in DNA damage as detected by the comet assay in sperm from both cooled (Linfor and Meyers, 2002) and frozen storage (Baumber *et al.*, 2003a). Unfortunately, the addition of antioxidants (α-tocopherol, GSH, ascorbic acid) or enzyme scavengers (catalase, SOD) to cryopreservation extenders did not reduce the level of DNA fragmentation subsequent to the freezing and thawing of equine sperm cells (Baumber *et al.*, 2005). In fact, the addition of SOD to cryopreservation extender significantly increased DNA fragmentation, suggesting again that H_2O_2 is the primary ROS resulting in DNA damage to equine sperm – in this case due to the conversion of O_2^- to H_2O_2.

Effect of ROS on sperm membrane damage

Sperm membranes, including those of the horse, are characterized by a high concentration of polyunsaturated fatty acids (Parks and Lynch, 1992), which are susceptible to peroxidative damage (Aitken, 1995). Neither H_2O_2 nor O_2^- are energetic enough to initiate lipid peroxidation on their own, and a transition metal catalyst is required to cause the chain reaction leading to this (Ball and Vo, 2002). The presence of a transition metal catalyst such as Fe^{++} can initiate lipid peroxidation; this results in the formation of lipid peroxides and of cytotoxic malondialdehyde, as well as of the more potent 4-hydroxynonenal (Aitken, 1995). The change in the sperm membrane resulting from lipid peroxidation alters membrane fluidity, and this can affect its ability to fuse during acrosomal exocytosis.

Although lipid peroxidation is well characterized for mammalian sperm, equine spermatozoa appear relatively more resistant to membrane peroxidation than the sperm of other domestic animals (Baumber *et al.*, 2000; Neild *et al.*, 2005). The cryopreservation of equine sperm, however, increases lipid peroxidation, particularly over the region of the sperm midpiece (Neild *et al.*, 2005). Storage of liquid semen at 5°C for 24 to 48 h also resulted in a detectable increase in lipid peroxidation in equine sperm (Ball and Vo, 2002; Raphael *et al.*, 2008). The addition of α-tocopherol significantly reduced lipid peroxidation in equine sperm exposed to ferrous (Fe^{++}) promoters, and the presence of Fe^{++} during cooled storage of equine sperm significantly increased lipid peroxidation and decreased sperm motility (Ball and Vo, 2002). The vitamin E analogue tocopherol succinate is more water soluble than native vitamin E and appears to load more readily into mitochondria. Experimentally, tocopherol succinate was superior to α-tocopherol in preventing the lipid peroxidation of equine sperm, but it did suppress the motility of equine sperm to a greater extent than did α-tocopherol, which limits its practical application (Almeida and Ball, 2005).

Ball and colleagues have also evaluated the effect of the addition of the ROS scavenger catalase, as well as that of lipid and water-soluble antioxidants, on the maintenance of equine sperm motility during cooled storage (Ball *et al.*, 2001a). The addition of catalase to non-fat skimmed milk extenders did not improve the maintenance of motility during 72 h of storage at 5°C. Several lipid-soluble antioxidants were evaluated, including butylated hydroxytoluene (BHT), α-tocopherol and the synthetic antioxidant, Tempo. BHT significantly reduced progressive motility during storage, and there were no positive treatment effects of either α-tocopherol or Tempo on maintenance of motility. In a final experiment, some water-soluble antioxidants were evaluated, including the vitamin E analogue, Trolox, vitamin C and bovine serum albumin, none of which had a positive effect on the maintenance of sperm motility during cooled storage. Aurich *et al.* (1997) observed a positive effect of the addition of ascorbic acid on the preservation of the membrane integrity of cooled equine sperm; however, the addition of catalase under similar conditions had no effect. In conclusion, the addition of catalase or of a variety of lipid- or water-soluble antioxidants did not improve the maintenance of motility during short-term cooled storage of equine sperm in the presence of skimmed milk-based extenders.

A number of investigators have examined the effects of the addition of antioxidants to cryopreserved equine sperm without demonstrating clear-cut positive results on post-thaw parameters or fertility. In studies by Ball and colleagues, the addition of the enzyme scavengers catalase or SOD, or the addition of low molecular weight antioxidants such as GSH, ascorbic acid or α-tocopherol, did not decrease DNA fragmentation, nor increase mitochondrial membrane potential, viability or the motility of frozen equine sperm after thawing (Baumber *et al.*, 2005). In contrast, Aguero *et al.* (1995) reported a positive effect of adding α-tocopherol to cryopreserved equine sperm. In cattle, Foote *et al.* (2002) evaluated multiple antioxidants and combinations of antioxidants for addition to both liquid and frozen bovine sperm, and concluded that they were generally not beneficial. Interestingly, there were interactions between extender type (skimmed milk-based versus egg yolk-based extenders) and the addition of antioxidants. Foote *et al.* (2002) suggest that the addition of casein and other milk proteins in extenders may provide abundant ROS scavenging capability to many extender formulations and may, therefore, obviate the need for the addition of lipid or water-soluble antioxidants to semen extenders containing milk products. As noted above, this notion may also apply to equine semen, as evidenced by the lack of positive effect of the addition of antioxidants in many studies that utilized skimmed milk-based extenders. There may also be important differences between different species, as reports from species other than the horse suggest that the addition of some enzyme scavengers – BHT, GSH and tocopherol or its analogues – had a positive effect on the post-thaw parameters of boar sperm (Großfeld *et al.*, 2008).

Although the literature concerning the addition of vitamin E (α-tocopherol) to semen extenders appears to have, at best, a variable effect on the maintenance of sperm function and fertility, a number of studies suggest that dietary supplementation with vitamin E may have a positive impact on semen quality and the maintenance of sperm during storage. In turkeys and chickens, dietary supplementation with vitamin E reduced the lipid peroxidation of sperm membranes in turkeys and chickens; furthermore, the addition of organic selenium to the diet also increased the activity of selenium-dependent glutathione GPx in avian seminal plasma (Breque *et al.*, 2003). Surprisingly, supplementation of the female with dietary vitamin E and organic selenium also appeared to improve fertility, perhaps through effects on sperm storage in the oviduct (Breque *et al.*, 2003). Dietary levels of vitamin E and selenium in boars affected the percentage of motile, morphologically normal sperm present in the ejaculate as well as the fertilization rate in mated gilts (Marin-Guzman *et al.*, 1997).

A report from studies of the stallion demonstrated an improved maintenance of sperm motility during cooled semen storage after the dietary addition of 3000 IU vitamin E/day for 14 weeks (Gee *et al.*, 2008). Another study demonstrated the favourable effects of the dietary addition of a rice oil supplement to stallions; increases in sperm concentration, motility and total antioxidant capacity of the semen were noted during treatment with this supplement (Arlas *et al.*, 2008). These studies suggest that the dietary addition of antioxidants should be explored further as a means of altering oxidative stress in equine semen that might be associated with cooled or frozen storage of semen or possibly associated with reduced fertility.

Summary and conclusions

Oxidative stress has been associated with the perturbation of normal sperm function, including damage to chromatin, proteins and membrane lipids, but it is important to consider that the low-level generation of ROS appears to have an important role in intracellular signalling events in sperm. Seminal plasma is a rich source of enzyme scavengers and low molecular weight antioxidants, whereas the sperm cell has a limited antioxidant capacity that is related to its small cytoplasmic volume. Many features of semen processing, including the removal of seminal plasma, centrifugation, cooling and re-warming probably contribute to the oxidative damage to sperm during storage. Unfortunately, existing studies on the addition of antioxidants to equine sperm during storage do not demonstrate clear-cut positive effects, and more research needs to be conducted to assess these treatments. Dietary addition of both vitamin E and selenium in the stallion may have positive effects on semen; again, more research should be conducted to address this possibility.

Endocrinological Evaluation of Prospective and Active Breeding Stallions

Diagnostic endocrine testing

Diagnostic endocrine testing in stallions has had limited application in veterinary practice other than for the diagnosis of suspected cryptorchidism in previously castrated stallions. The lack of widely available standardized assays for equine gonadotrophins has hindered their diagnostic use in sub-fertile stallions (Brinsko, 1996), and the relative paucity of information on the diagnostic interpretation of endocrine values has made the clinical application of reproductive endocrinology difficult. Endocrine concentrations in stallions must be interpreted in terms of the stallion's age and the season of the year, because both of these variables have significant effects on the concentrations of luteinizing hormone (LH) and testosterone in plasma (Douglas and Umphenour, 1992). Sexual stimulation can also cause significant increases in the peripheral concentrations of testosterone and should be avoided when obtaining baseline determinations (McDonnell, 1995). Because the secretion of gonadotrophins in the stallion is pulsatile, samples taken over the course of 3 days may be more representative than a single sample. Moreover, there appears to be a diurnal variation in hormone concentrations, which are highest around midday, and some authors suggest that endocrine sampling should be conducted before 9.00 a.m. (Roser, 2001).

The primary indication for endocrine diagnostics in the stallion is presumed testicular degeneration. In stallions, this occurs most often as an age-related, idiopathic condition, often at or near the time of highest genetic and/or monetary value of the breeding sire. Testicular function in the stallion depends upon the normal function of the hypothalamic-pituitary-testicular axis, with a carefully regulated endocrine control, as well as the local regulation of cell function, by autocrine and paracrine factors within the testis. The initiating cause for most cases of testicular degeneration is unknown, though it has been suggested that the primary deficit in idiopathic testicular degeneration is related to changes within the testis that subsequently result in alterations in spermatogenesis and circulating endocrine parameters (Varner *et al.*, 2000; Roser, 2001, 2008).

As in man, testicular degeneration in the stallion is frequently associated with elevated follicle-stimulating hormone (FSH) concentrations, whereas LH and basal testosterone are more variable but are often normal in sub-fertile stallions (Burns and Douglas, 1985; Douglas and Umphenour, 1992; Brinsko, 1996). In stallions with end-stage testicular degeneration, both FSH and LH may be increased, and basal testosterone declines late in the disease process, if at all (Burns and Douglas, 1985; Roser and Hughes, 1992b). Douglas and Umphenour (1992) reported that a FSH increase and a decline in total oestrogens in the stallion were most useful in predicting a decline in fertility, and that increased FSH and decreased total oestrogens were most often highly correlated.

Roser and Hughes (1992a) reported that sub-fertile stallions had increased FSH and LH compared with fertile stallions; however, conjugated oestrogens were not different between these groups. In another study, Roser and Hughes (1992b) reported that although sub-fertile stallions had similar basal testosterone concentrations to fertile stallions, sub-fertile stallions had a significantly lower increase in testosterone following gonadotrophin-releasing hormone (GnRH) stimulation than did fertile stallions. In a group of fertile, sub-fertile and infertile stallions, Roser (1995) also reported that infertile stallions had significantly lower inhibin and oestradiol, as well as significantly higher LH and FSH (basal levels), compared with fertile stallions. While basal testosterone was not different between fertile and infertile stallions, infertile stallions had a lower testosterone increase after the administration of human chorionic gonadotrophin (hCG) than did fertile stallions. The clinical observations across these studies suggest that FSH increases relatively early with alteration (degradation) in testis function and that total oestrogens decline; in contrast, increases in LH are more variable, and basal testosterone does not decline until relatively late in testicular degeneration (Roser, 2001). A reduction in testosterone response to the administration of either GnRH or hCG suggests that there may be a primary defect at the level of the testis in some sub-fertile stallions, but the nature of this defect remains undetermined.

Response testing

Response testing has been advocated to further evaluate changes in the hypothalamic-pituitary-testicular axis. Both fertile and sub-fertile stallions appear able to respond to GnRH stimulation with an elevation in LH and FSH, which implies that the pituitary in many of these stallions is capable of releasing gonadotrophins in response to GnRH. A single GnRH pulse test has been advocated to further differentiate pituitary versus testis changes in the stallion. In this protocol, a single dose of GnRH (25 μg IV) is administered at 9.00 a.m., with blood samples taken at 30 min before administration, at administration and 30, 60 and 120 min later (Roser, 2001). Samples are assayed for both testosterone and LH. Abnormal responses are characterized by failure to elicit an increase in both LH and testosterone after GnRH administration. A 50% increase over baseline values at 30 min post GnRH for LH and a 100% increase over baseline in testosterone at 2 h post GnRH have been suggested as minimal parameters for assessing the response test (Blanchard *et al.*, 2000b).

The administration of hCG has also been used to assess the ability of the testis to produce steroids, thereby bypassing the hypothalamic-pituitary axis (Roser, 1995). Roser (2001) suggested that testosterone response to hCG should be measured every 30 min, beginning 1 h before to 6 h after administration of 10,000 IU hCG IV.

Other endocrine markers

In addition to gonadotrophins, other endocrine markers may be useful for evaluating reproductive function in stallions. Inhibin is a dimeric glycoprotein that is secreted by the Sertoli cells of the testis, primarily in response to FSH stimulation (Stewart and Roser, 1998). Inhibin, in turn, feeds back on the anterior pituitary gland to decrease FSH secretion in a negative feedback loop (Roser *et al.*, 1994). Concentrations of inhibin vary with season (Stewart and Roser, 1998), and circulating inhibin concentrations may be a reasonable marker of Sertoli cell number and function. Stewart and Roser (1998) reported that inhibin concentrations were lower in infertile stallions, and inhibin does appear to have clinical utility in evaluating the sub-fertile stallion. At present, the Clinical Endocrinology Laboratory at UC Davis provides inhibin immunoassays (the normal range in stallions is 2.2–3.4 ng/ml). Decreases in serum inhibin concentrations have been noted in stallions early in the course of testicular degeneration, and measurements of basal serum concentrations of oestrogen, inhibin and FSH appear to be the most useful in predicting changes in fertility; changes in basal LH and testosterone tend to occur late during the course of the condition.

Recently, Ball and colleagues examined another hormone that is produced by the Sertoli cells in the equine testis. Anti-Müllerian hormone (AMH) is a glycoprotein hormone that is first produced by Sertoli cells in the fetal testis; it is responsible for regression of the paramesonephric (Müllerian) duct (Claes *et al.*, 2011). Secretion of AMH by the Sertoli cells continues in the post-natal testis, and the hormone may be important in regulating Leydig cell differentiation and testosterone production (Ball *et al.*, 2008). AMH is produced by Sertoli cells in the adult stallion and is detectable in both blood and the seminal plasma. Because AMH is a specific endocrine product of Sertoli cells, it may be a useful marker for Sertoli cell and testis function. In human males, AMH has been used as a marker for the detection of testis tissue in intersex and cryptorchid conditions (Lee *et al.*, 2003; Demircan *et al.*, 2006), and studies in men have attempted to relate AMH concentrations in seminal plasma to fertility (Fenichel *et al.*, 1999; Sinisi *et al.*, 2008). Future studies should address the application of the AMH assay for similar uses in the stallion.

Testicular Biopsy in the Stallion

Introduction

Descriptions of a variety of techniques for testis biopsy in the stallion have appeared in the veterinary literature for the past 40+ years (Galina, 1971), yet the technique is still not widely applied in either routine or specialist clinical practice in the horse. Techniques described for obtaining a testis biopsy in the stallion include open techniques (DelVento *et al.*, 1992), percutaneous needle biopsy using either manual split-needles (Tru-cut or Vim-Silverman needles) (Galina, 1971; Carluccio *et al.*, 2003), or spring-loaded, self-firing biopsy instruments (Bard – Biopty™) (Faber and Roser, 2000; Roser and Faber, 2007), or fine-needle aspiration (Leme and Papa, 2000). These techniques have been conducted with both general anaesthesia (DelVento *et al.*, 1992; Bartmann *et al.*, 1999) and as standing procedures under sedation (Faber and Roser, 2000; Leme and Papa, 2000; Roser and Faber, 2007).

As with many biopsy techniques, one of the challenges in obtaining testis biopsy samples is obtaining a representative or diagnostic sample for histopathological interpretation. Of the techniques described for testis biopsy, open biopsy procedures provide a larger tissue sample with fewer artefacts than samples obtained by needle biopsy or aspiration (Roser and Faber, 2007). Needle biopsy techniques using manual split needles or spring-loaded biopsy instruments can also yield interpretable samples, though the diameter of the biopsy needle appears to have a large impact on sample size and interpretation. In general, it appears that samples obtained with the spring-loaded biopsy instruments using a 14 gauge biopsy needle provide a superior sample with reduced risk of damage to the testis parenchyma (Roser and Faber, 2007). Fine-needle aspiration of the testis yields an aspirate for cytological evaluation, which may be useful for diagnosis of testicular neoplasia or inflammation. However, for the evaluation of spermatogenesis in the stallion, experience in the interpretation of such cytological preparations is limited (Leme and Papa, 2000), and more information is required on the distribution of cells from the seminiferous epithelium to provide useful diagnostic information from fine-needle aspirates of the testis.

Techniques for testis biopsy

Selection of the biopsy location is an important consideration in obtaining a useful and interpretable sample as well as in minimizing post-biopsy complications, particularly haemorrhage. Ultrasound evaluation of the testis may be a useful adjunct to help define focal lesions in the testis (such as suspect neoplasia) to be sampled by biopsy (Bartmann *et al.*, 1999). Practically speaking, in many cases of testicular neoplasia, an excisional biopsy of the testis obtained at orchiectomy may be the preferred method for diagnosis. In other situations, focal changes in the testis, or changes in the texture or ultrasonographic appearance of the testis, may dictate the site for biopsy.

Knowledge of testis blood supply is critical to avoiding major vessels and reducing the risk of haemorrhage subsequent to biopsy (Smith, 1974; Pozor, 2007). In approximately 70 to 80% of stallions, there is a single testicular artery coursing along the caudal pole of the testis; the remaining 20 to 30% of stallions have two testicular arteries, one along the caudal pole of the testis and another artery (sometimes two) coursing laterally along the surface of the testis in the mid-to-cranial portion of the testis. Venous drainage of the testis is conducted centrally through the testis via the relatively large, central vein, which is readily imaged via ultrasonography (Pozor, 2005). Because most of the arterial supply to the testis is located along the caudal and ventral aspect of the testis, the cranial-lateral pole of the testis is typically selected for the biopsy site (Faber and Roser, 2000), taking care to avoid the head of the epididymis. Because the arterial supply to the testis is variable in some stallions, ultrasound (± Doppler) evaluation of the testis may be useful for identifying those stallions in which a second or third testicular artery arborizes over the lateral aspect of the testis.

Because open biopsy techniques have more critical requirements for general anaesthesia, surgical preparation and possible postoperative complications, they are infrequently used and will not be considered as part of the discussion of techniques for biopsy of the stallion testis. Likewise, the manual split-needle techniques appear to provide more variable sample quality and a higher rate of complications and will not be further considered. The remaining two techniques, spring-loaded biopsy instruments and fine-needle aspiration have been described in detail (Faber and Roser, 2000; Leme and Papa, 2000; Roser and Faber, 2007), and will only be briefly reviewed here.

In general, testis biopsy using a spring-loaded biopsy instrument, such as the Biopty™ instrument, can be performed in the standing stallion. If suitable stocks are available that allow access to the scrotum, the stallion can be restrained in stocks for the procedure. Generally, sedation using a combination of detomidine HCl and butorpanol will provide adequate restraint. As noted above, ultrasound evaluation of the testis may be useful for identifying focal lesions and also in helping to identify lateral branches of the testicular artery to be avoided. Once the site of the biopsy on the cranial-lateral pole of the testis has been identified, the scrotal skin is surgically prepared and infiltrated with 2% lidocaine for local anaesthesia. The surgeon should make a small skin incision over the biopsy site to facilitate introduction of the Biopty™ needle. Once the skin is incised, the armed Biopty™ instrument is placed through the skin incision, the testis is pulled down into the scrotum, and the instrument is fired along the long axis of the testis to obtain the biopsy sample. If desired, the skin incision may be closed with a single interrupted suture; however, this is not necessary.

Careful handling of the biopsy sample is critical to avoid unnecessary artefacts, and careful fixation is required to preserve the biopsy. Although Bouin's solution may provide optimum fixation of testicular tissue (Threlfall and Lopate, 1993), it is sometimes not widely available and may not be preferred by all pathology laboratories. If Bouin's solution is used, the fixed tissue should be transferred to 70% ethanol within 24 h after fixation in Bouin's solution. Alternatively, samples may be fixed in neutral-buffered formalin, though shrinkage and other artefacts may be a problem with this method of fixation.

Fine-needle aspirates of the testis have also been used in the stallion (Threlfall and Lopate, 1993; Leme and Papa, 2000; Roser and Faber, 2007). Preparation of the stallion is similar to that noted above for needle biopsy. Again, the sample should be taken from the cranio-lateral aspect of the testis. For fine-needle aspirates, a 22 or 23 gauge 1½ in needle connected to a 5 ml syringe is introduced directly through the scrotal skin. Two to three needle passes are made through the testis parenchyma while applying negative pressure to the syringe. The pressure on the syringe is released, and the needle is withdrawn from the scrotum. The material obtained will be typically within the needle or needle hub and should be gently expelled onto a clean glass microscope slide, air dried and stained with Giemsa (Leme and Papa, 2000). The cytological evaluation of testis fine-needle aspirates is well described for humans (Meng *et al.*, 2001), but there is relatively little information available for the interpretation of such samples from the stallion, which currently limits the diagnostic use of this technique.

Complications of testis biopsy

Potential complications of testis biopsy include haemorrhage, infection, adhesions

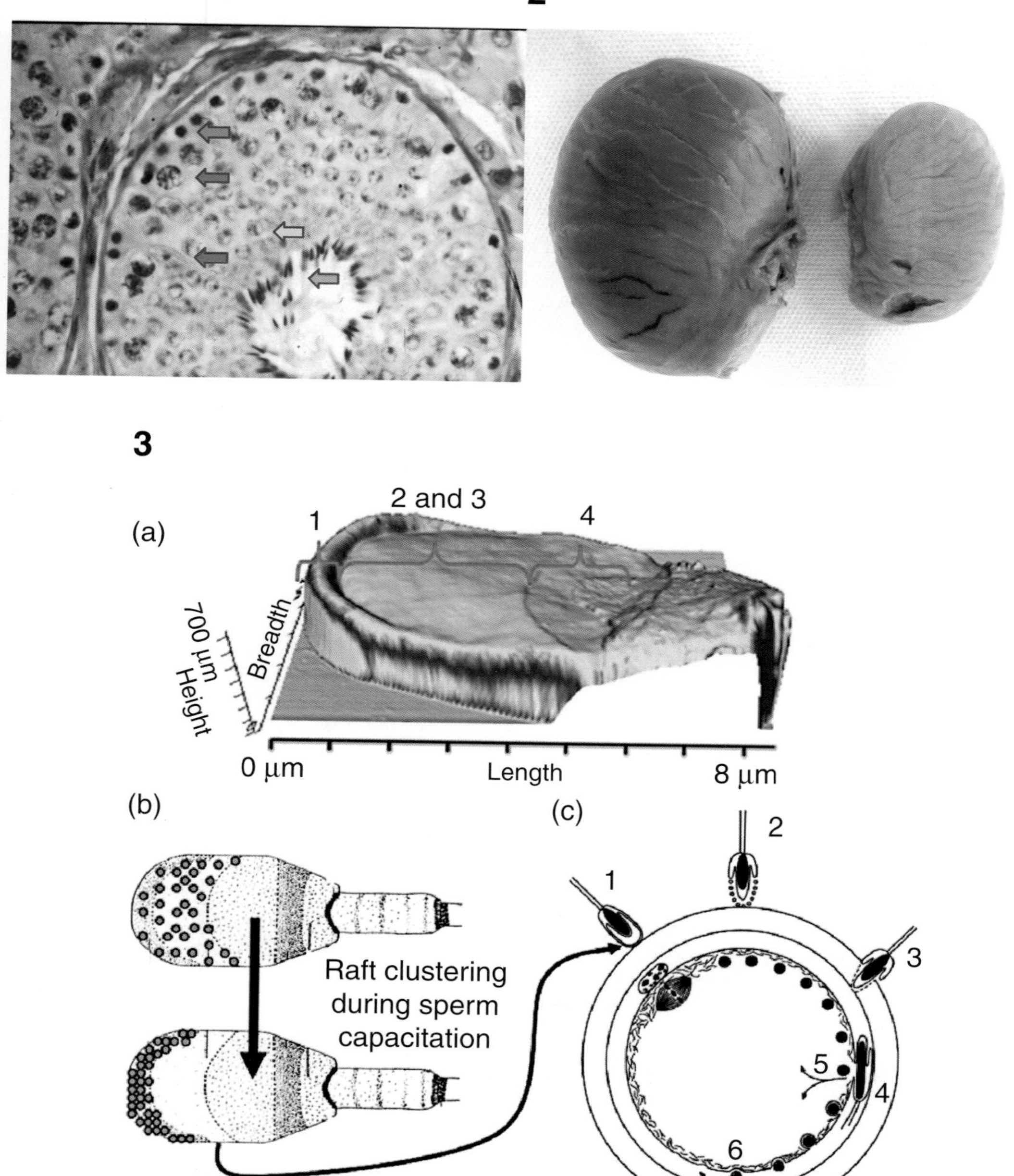

Plate 1. A stage VIII seminiferous tubular cross section from the stallion. Coloured arrows denote nuclei of individual cells as follows: red – Sertoli cell; blue – young primary spermatocyte; green – old (pachytene phase) primary spermatocyte; yellow – young spherical spermatid; orange – elongated spermatid.

Plate 2. Disparity in the size of the scrotal versus cryptorchid testes of an approximately 4-year-old unilaterally cryptorchid stallion.

Plate 3. Protein kinetics at the sperm surface. (a) An atomic force microscopic surface view of a porcine sperm head. (b) Lipid ordered microdomains at the sperm surface cluster into the apical ridge area of the porcine sperm during *in vitro* capacitation. (c) Sperm-zona pellucida and sperm-oocyte interaction leading to fertilization. 1. Zona binding, 2. acrosome reaction, 3. zona drilling, 4. oolemma binding and fertilization, 5. activation of pronucleus formation and oocyte activation, 6. induction of a blockade for polyspermic fertilization. (Numbers in panel A refer to the specific sperm surface areas where these interactions take place.)

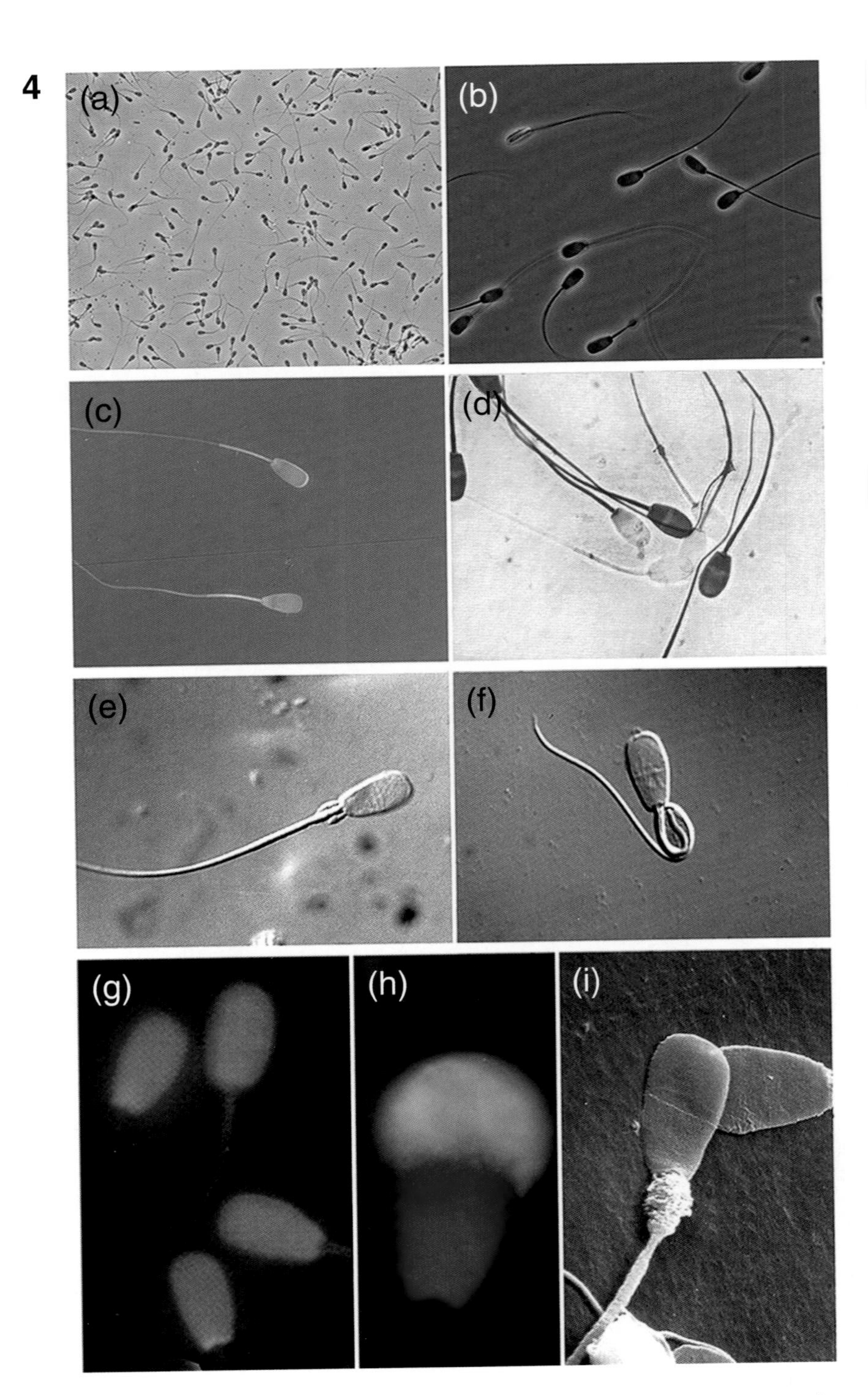

Plate 4. Examples of microscopy techniques: (a) bovine spermatozoa, phase contrast microscopy, low magnification; (b) porcine spermatozoa, phase contrast microscopy, high magnification; (c) bovine spermatozoa, bright field microscopy, eosin–nigrosin stain; (d) ovine spermatozoa, bright field microscopy, bromophenol blue stain; (e) bovine spermatozoa, differential interference contrast (DIC) microscopy; (f) bovine spermatozoa, DIC microscopy; (g) bovine spermatozoa, fluorescence microscopy, SYBR-14 and PI (propidium iodide) staining; (h) bovine spermatozoa, fluorescence microscopy, rabbit anti-acrosin antibody labelled with anti-rabbit immunoglobulin conjugated to FITC (fluorescein isothiocyanate)/PI stain; and (i) bovine spermatozoa, scanning electron microscopy (SEM). (For photo credits, see Chapter 6, p. 123.)

5

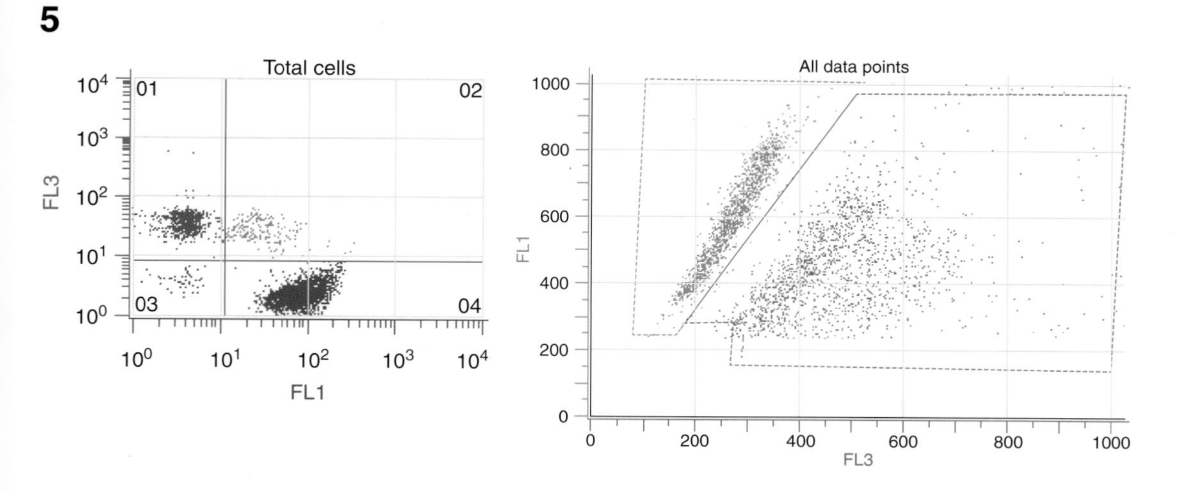

6 7 8

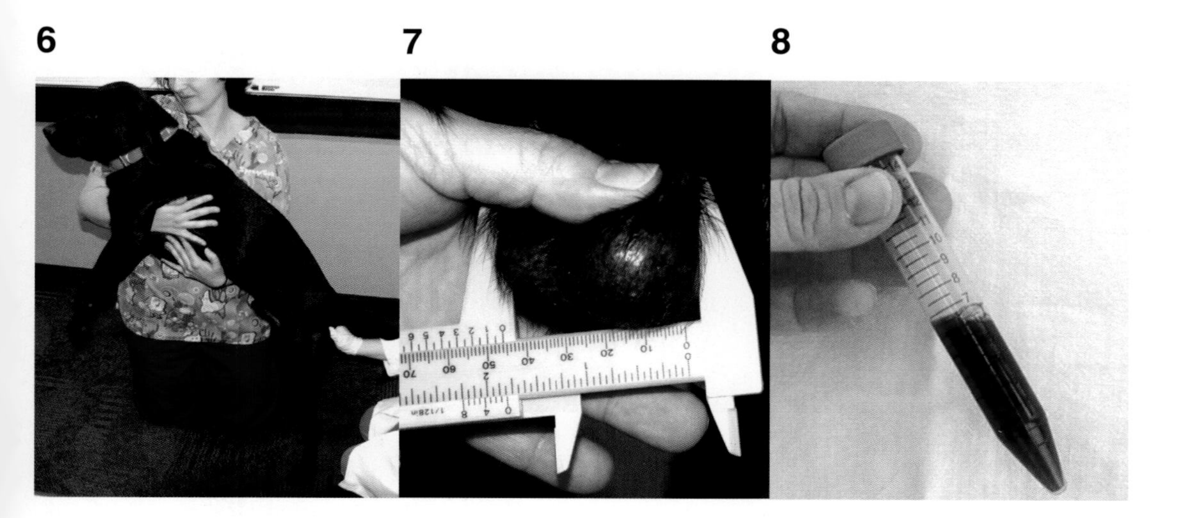

Plate 5. Histograms of flow cytometric analyses of bovine spermatozoa. Left: SYBR-14 and PI (propidium iodide) stains; upper left quadrant – PI stained, dead spermatozoa, i.e. without intact membranes; upper right quadrant – damaged cells, partial staining with SYBR-14 and PI; lower left quadrant – debris; lower right quadrant – SYBR-14 stained, viable spermatozoa, i.e. with intact membranes. Right: acridine orange stained cells – green fluorescent (FL1) cells with double-stranded DNA, and red fluorescent (FL3) cells with single-stranded DNA. Figure courtesy of Select Sires, Plain City, Ohio, USA.

Plate 6. Raising the front of the dog (wheel-barrowing) to facilitate palpation of the prostate.

Plate 7. Measuring total scrotal width of the dog.

Plate 8. Semen from a dog with prostate disease. Reprinted with permission from *Clinical Canine and Feline Reproduction: Evidence-Based Answers*, Wiley-Blackwell, ISBN: 978-0-8138-1584-8.

9 **10** **11**

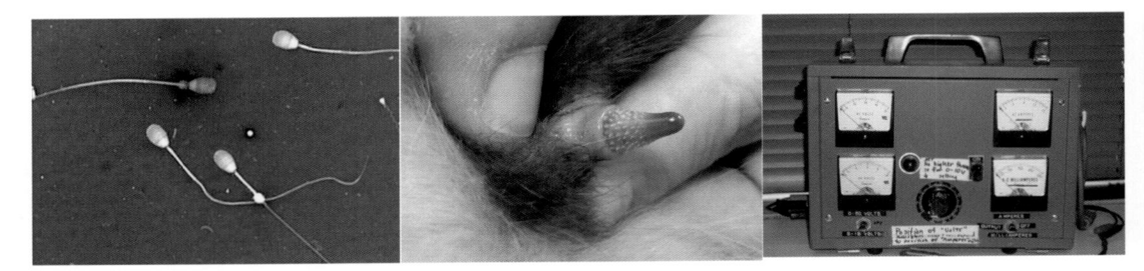

12

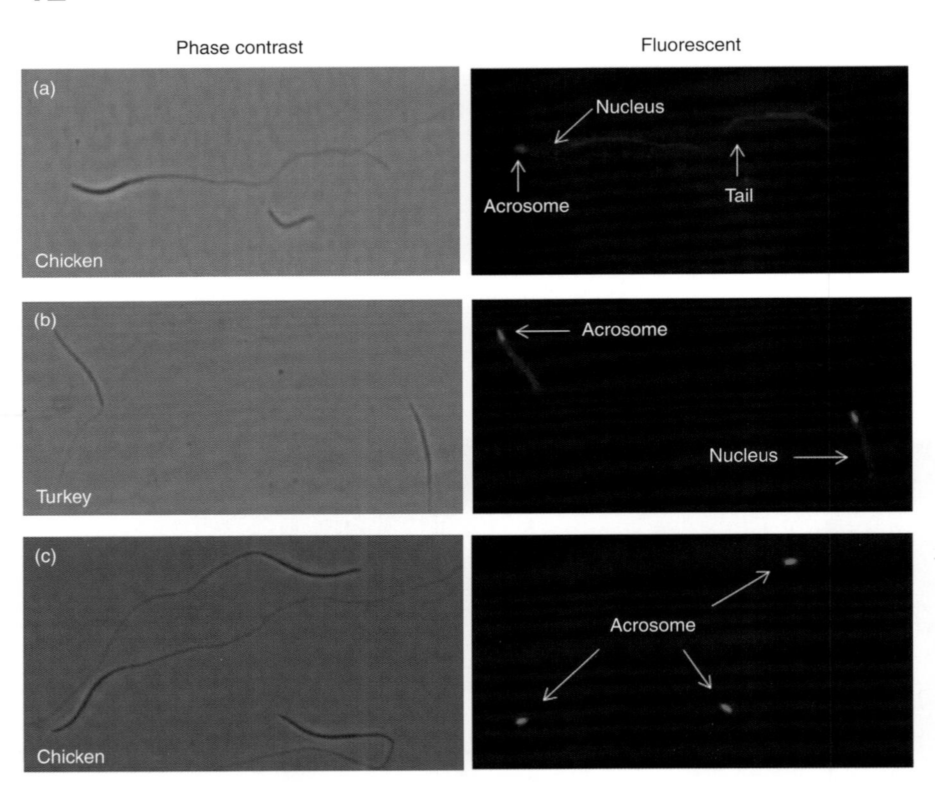

Plate 9. Stained canine spermatozoa for morphology assessment. Note two normal spermatozoa, one with a proximal cytoplasmic droplet (primary defect) and one with a distal cytoplasmic droplet (secondary defect). Reprinted with permission from *Clinical Canine and Feline Reproduction: Evidence-Based Answers*, Wiley-Blackwell, ISBN: 978-0-8138-1584-8.

Plate 10. Penile spines in the cat. Reprinted with permission from *Clinical Canine and Feline Reproduction: Evidence-Based Answers*, Wiley-Blackwell, ISBN: 978-0-8138-1584-8.

Plate 11. Electroejaculation equipment for use in domestic cats. Reprinted with permission from *Clinical Canine and Feline Reproduction: Evidence-Based Answers*, Wiley-Blackwell, ISBN: 978-0-8138-1584-8.

Plate 12. Representative images of poultry spermatozoa stained with carbohydrate-specific lectins and viewed with phase contract and fluorescence microscopy. (a) Sialic acid localization on the glycocalyx of chicken spermatozoa. (b) Species-specific localization of α-galactose residues on the acrosome and nuclear regions of turkey spermatozoa. (c) Species-specific localization of α-galactose residues only on the acrosome regions of chicken spermatozoa.

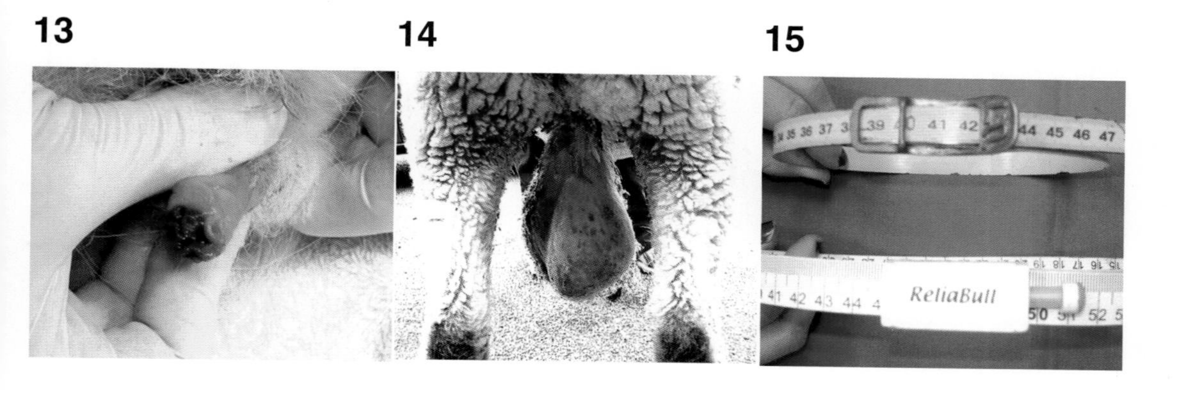

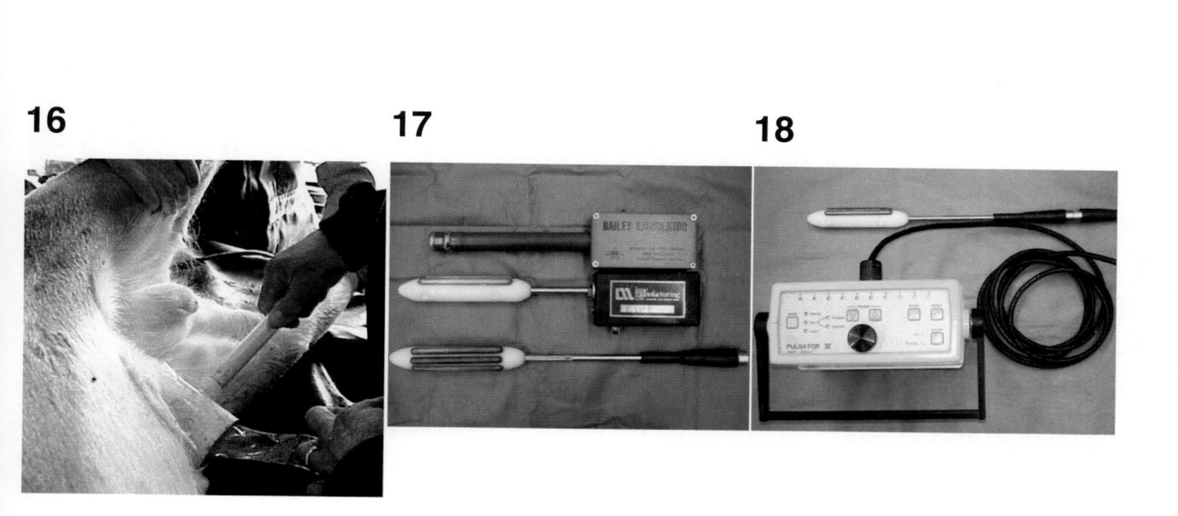

Plate 13. Peri-pubertal goat penis. Notice that the glans and urethral process are still attached to the prepuce.

Plate 14. Ram with right enlarged testicle detected during a breeding soundness examination. Upon ultrasound and surgery, this was found to be a haematoma of the testicle itself, probably due to trauma.

Plate 15. Two different scrotal measuring tapes. The ReliaBull™ has a built-in tensioning system to ensure repeatability of measurement.

Plate 16. White tailed deer buck anaesthetized and positioned in a lateral recumbency in preparation for electro-ejaculation.

Plate 17. Three different probes for electroejaculation of small ruminants. Top: Bailey Ejaculator with circular contacts. Middle: Lane hand-held model with both a high and a low setting for rams and bucks (goats). Bottom: Lane probe for small ruminants that can be attached to a power source.

Plate 18. A three ventral electrode Lane probe for small ruminants attached to an external power source such as the Pulsator IV (Lane Manufacturing Inc. Denver, Colorado).

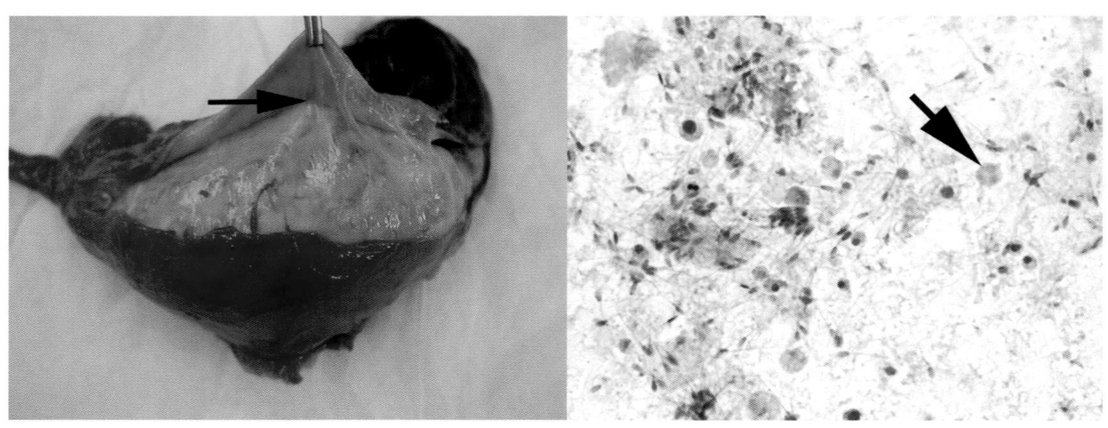

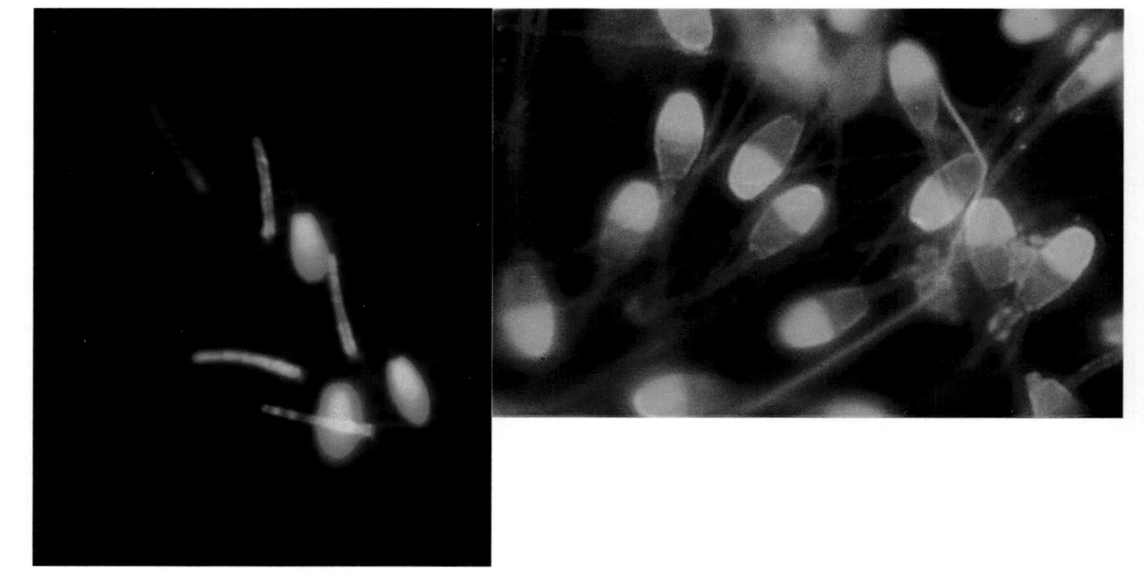

Plate 19. Scrotal adhesions in a stallion secondary to chronic epididymitis. Note the numerous fine, fibrous adhesions between the parietal and visceral vaginal tunics (arrow).
Plate 20. Photomicrograph of a Diff-Quik stained semen smear showing numerous somatic cells, including multinucleate cells from the seminiferous epithelium (arrow).
Plate 21. Fluorescent micrograph showing equine sperm co-stained with JC-1 and propidium iodide (PI). The midpiece of sperm, with high a mitochondrial membrane potential, is stained orange, and the nucleus of non-viable sperm is stained red with PI.
Plate 22. Fluorescent micrograph showing equine sperm stained with fluoresceinated *Pisum sativum* lectin showing the acrosome over the rostral sperm head.

23 **24** **25**

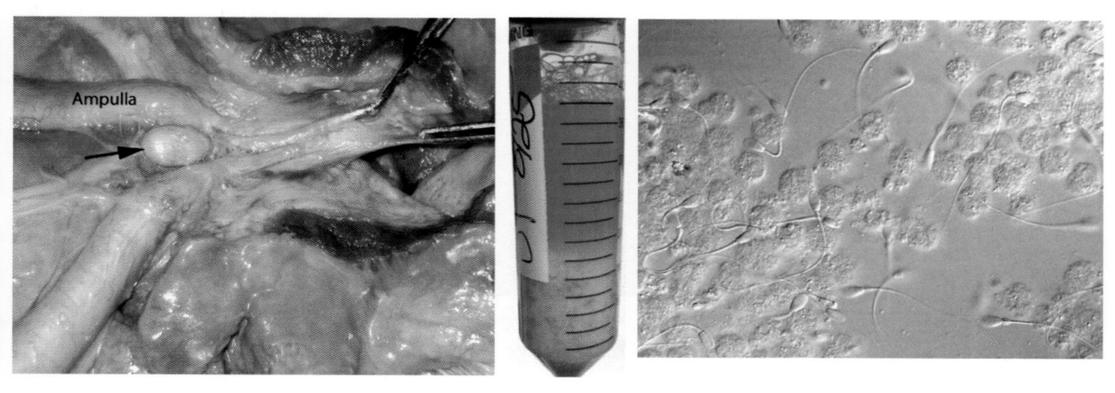

26

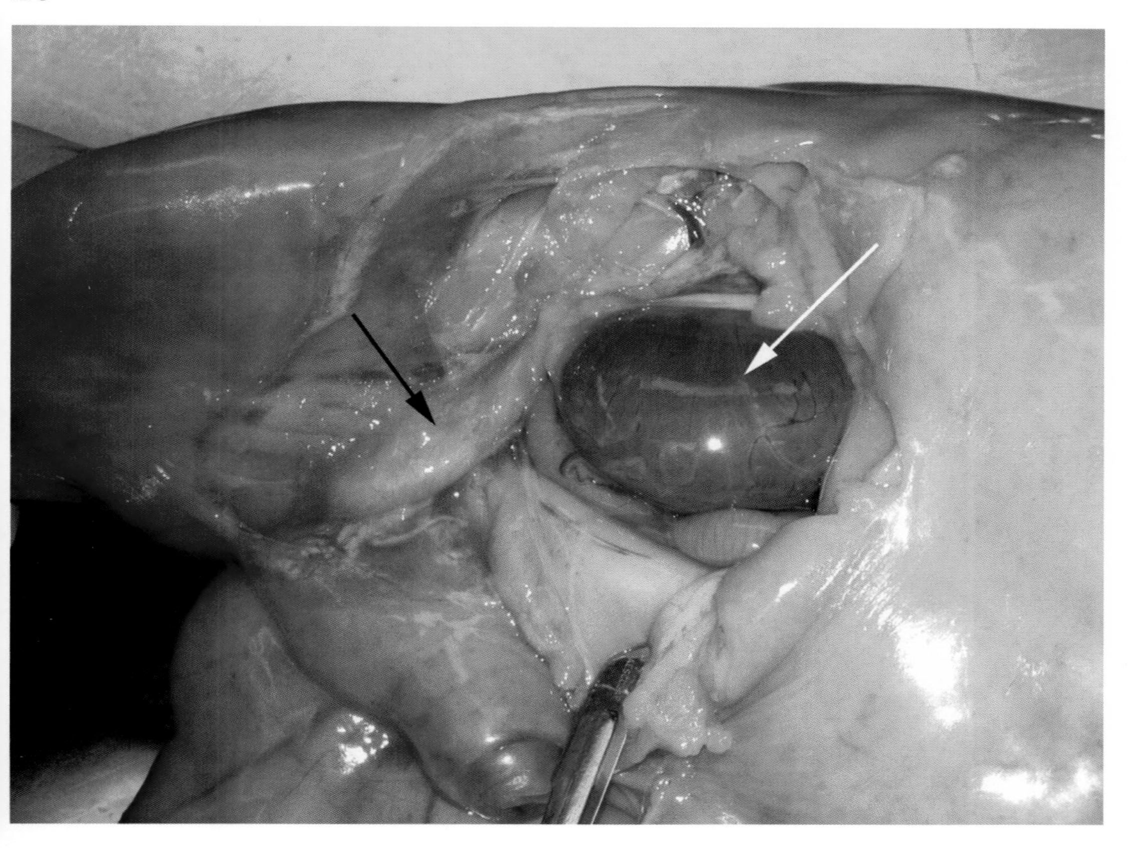

Plate 23. Anatomical specimen from a stallion showing an example of uterus masculinus (arrow). The structure lies between the terminal portion of the ductus deferens (ampullae).

Plate 24. Semen sample from a stallion with seminal vesiculitis. Note the large amount of flocculent debris with evidence of mild haemorrhage.

Plate 25. Phase contrast microscopy of a semen sample from a stallion with seminal vesiculitis, showing sperm and many somatic cells.

Plate 26. Anatomical specimen of a 5-month gestational age equine fetus showing the enlarged gubernaculum (black arrow) in the inguinal canal, and the fetal testis (white arrow) in the caudal abdomen.

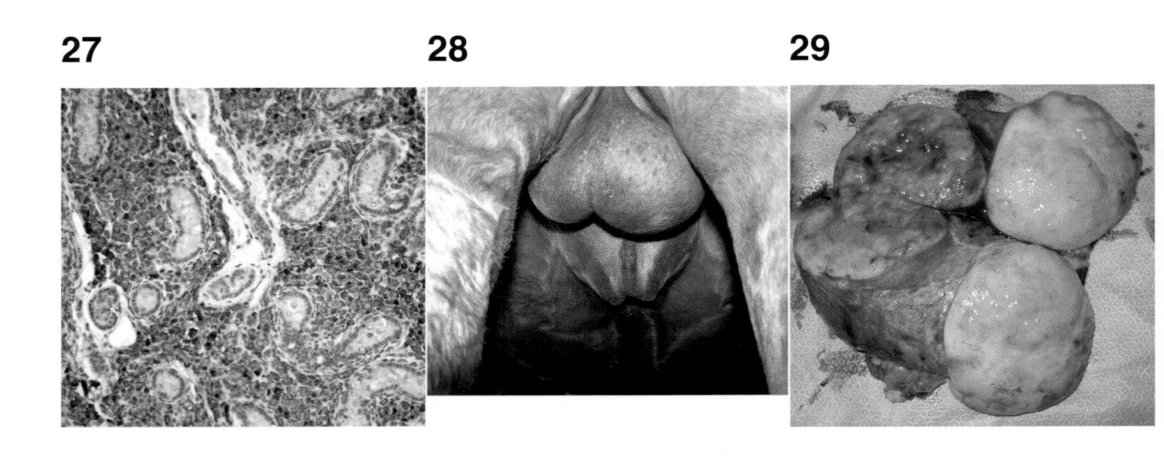

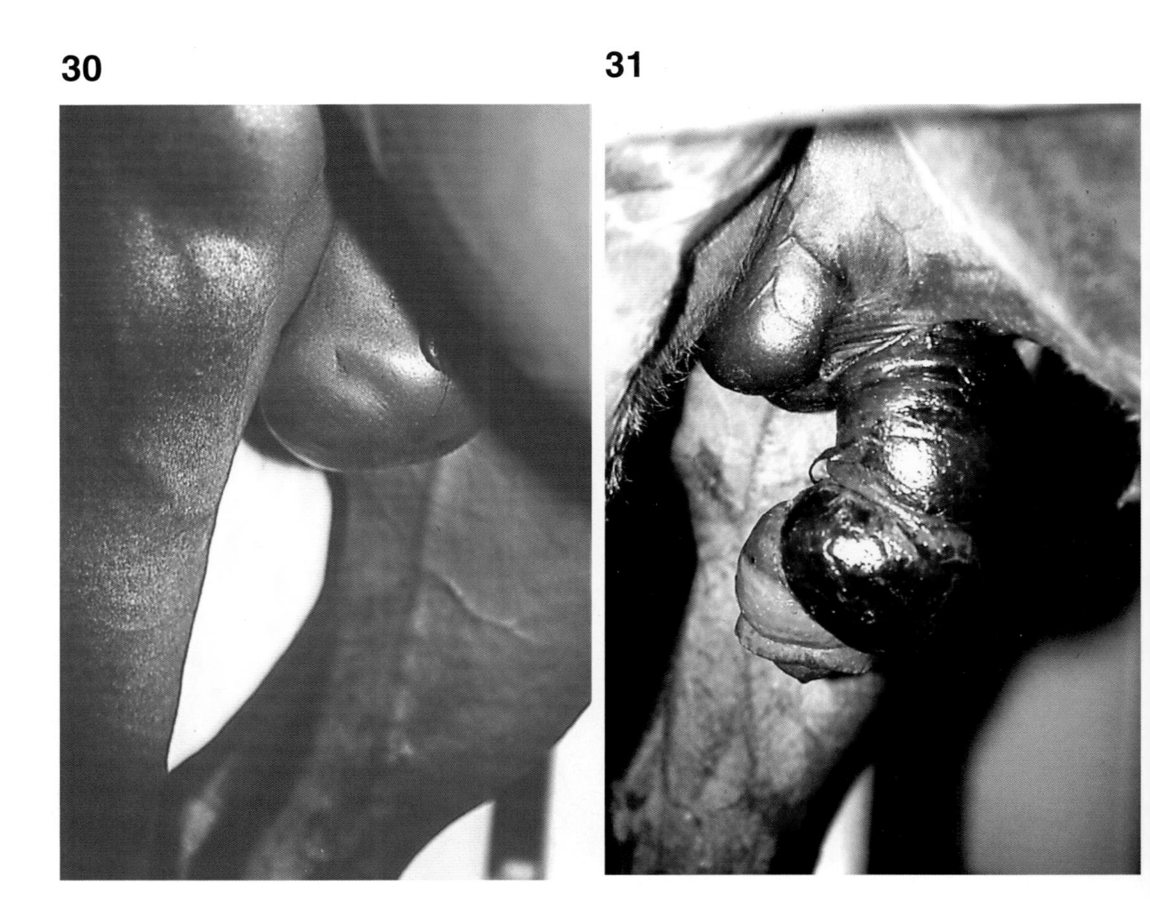

Plate 27. Immunohistochemistry of a cryptorchid equine testis showing labelling for 17-α hydroxylase in Leydig cells and abnormal seminiferous tubules. Testosterone production continues in the absence of normal spermatogenesis.
Plate 28. Seminoma in a stallion. Unilateral scrotal enlargement can be noted as a non-painful enlargement of the right testis.
Plate 29. At necropsy, multiple seminomas were detected in the right testis.
Plates 30 and 31. Breeding-related injuries in a stallion. **Plate 30.** Pitting scrotal oedema. **Plate 31.** Penile trauma with paraphimosis.

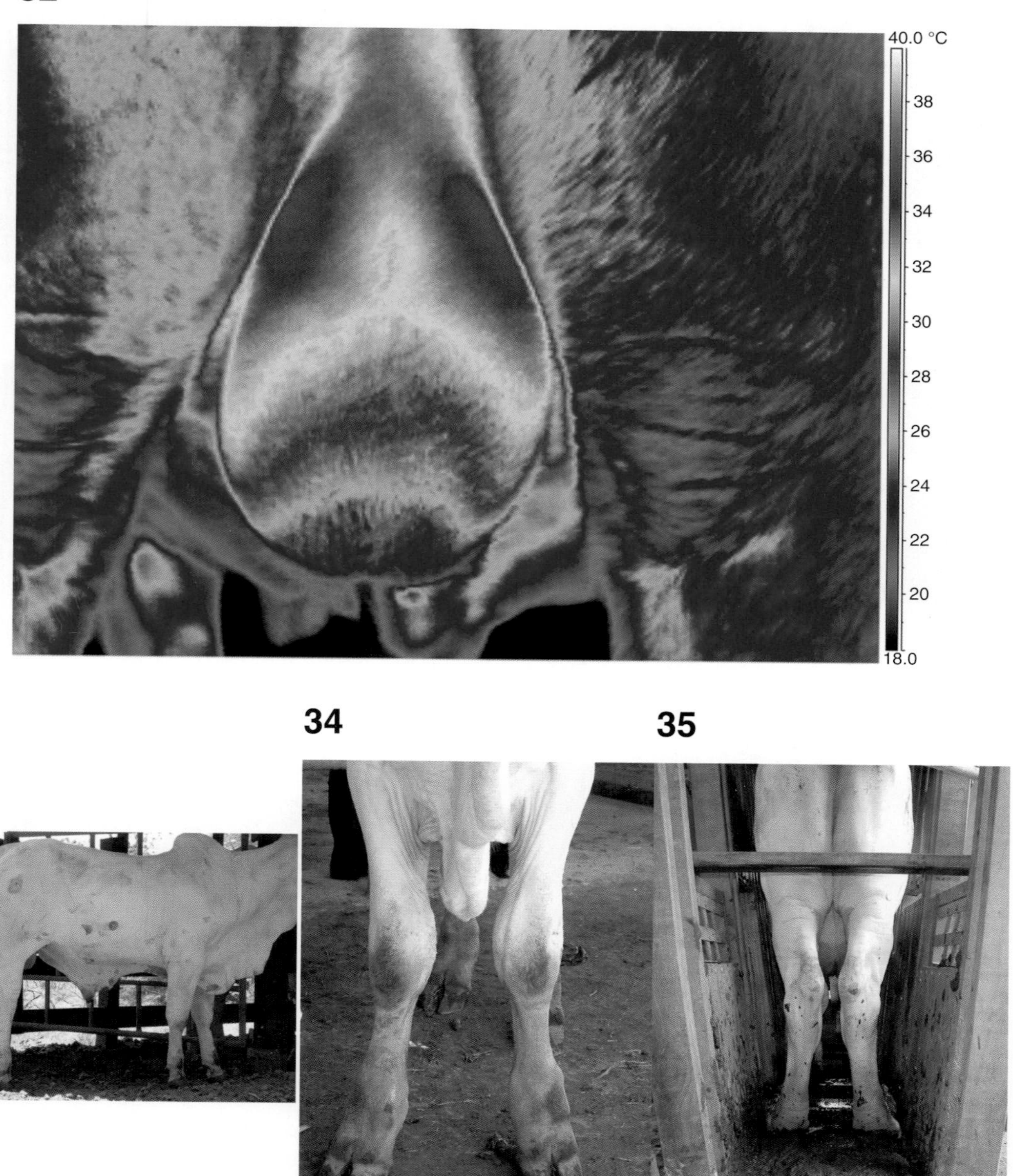

Plate 32. Infrared thermography image of the scrotum of a normal Holstein bull. Note the left-to-right symmetry and the decreasing surface temperature from the top to the bottom.

Plates 33–35. Abnormal conformation of rear feet and limbs are commonly found in extensively reared Brahman bulls, all of which will be classified as unsound for breeding during routine breeding soundness evaluation (BSE).

Plate 33. Post-leg conformation in a full blood 20-month-old sire. Note the plane angle at the hock and stifle joints.

Plate 34. A 10-month-old Brahman bull showing toed-out forelegs. This problem provokes stress on the carpal joint and pastern area. **Plate 35.** Hind limbs of a 30-month-old bull showing the 'cow-hock' (toed-out) defect. This defect is usually seen in conjunction with sickle-hock conformation.

36

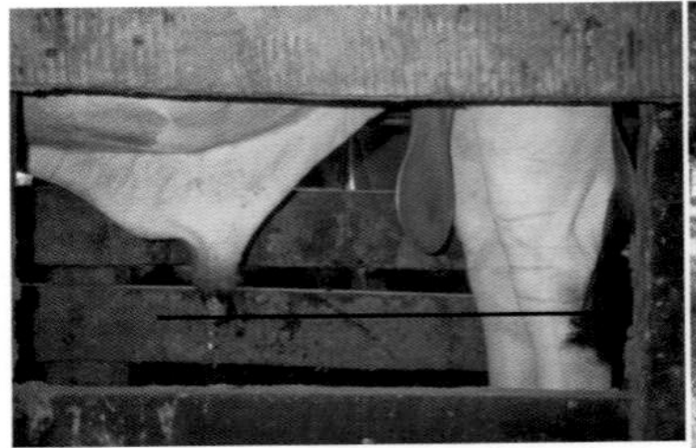

37

38

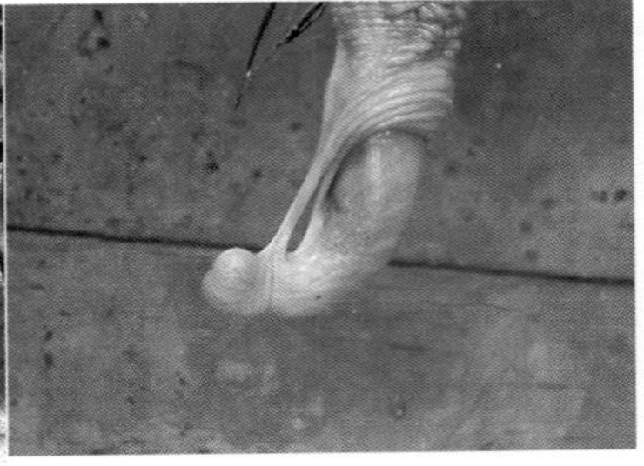

39

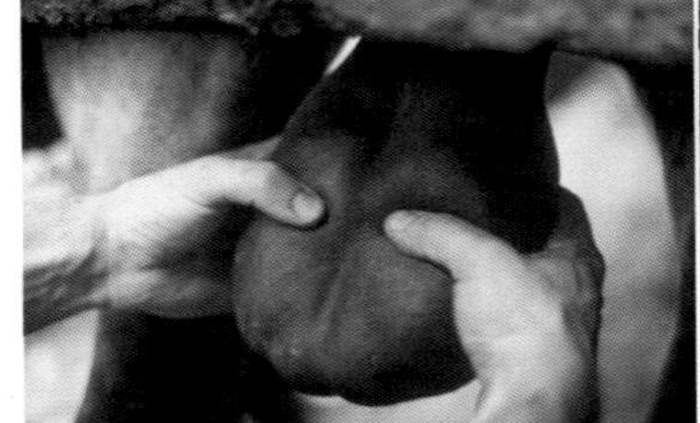

40

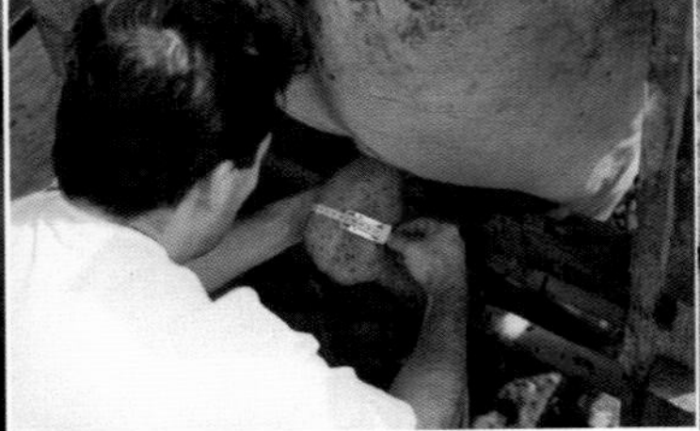

41

42

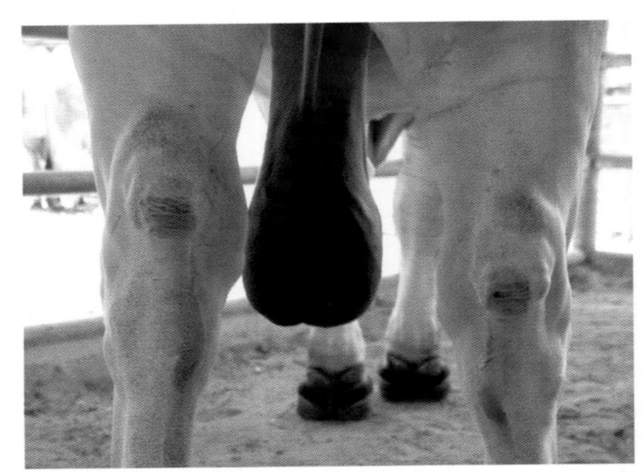

43

Plate 36. Normal preputial conformation. Note the preputial tip length in relation to the hock joint (black line) and the cranial preputial fold that is synonymous with functional anterior preputial muscles.
Plate 37. Elongated preputial sheath in a 3-year-old Brahman sire. Note the lack of the cranial fold giving a 'V-shaped' appearance.
Plate 38. A persistent penile frenulum in a 3-year-old Brahman bull.
Plate 39. Assessment of testicular consistency. Note how both testicles are held firmly, while pushing them to the bottom of the sac; consistency is determined with the thumb from the top to the bottom of the gonad.
Plate 40. Measurement of scrotal circumference (SC). Note the arrangement of the restraint, which allows the practitioner to gently pull the scrotum out of reach of the bull's rear limbs.
Plate 41. Testicular asymmetry in a young (16-month-old) Brahman bull, which persisted into adulthood.
Plate 42. Pendulous scrotum in a 5-year-old Brahman sire. Note the bottom of the sac below the level of the hock joint.
Plate 43. Transrectal examination of internal genitalia in a *Bos indicus* bull. Note the simple characteristics of the chute and the easy immobilization strategy that is routinely used, and also the position of the upper rear restraint at the height of the bull's thighs in order to avoid injuries in case the animal lies down during the examination.

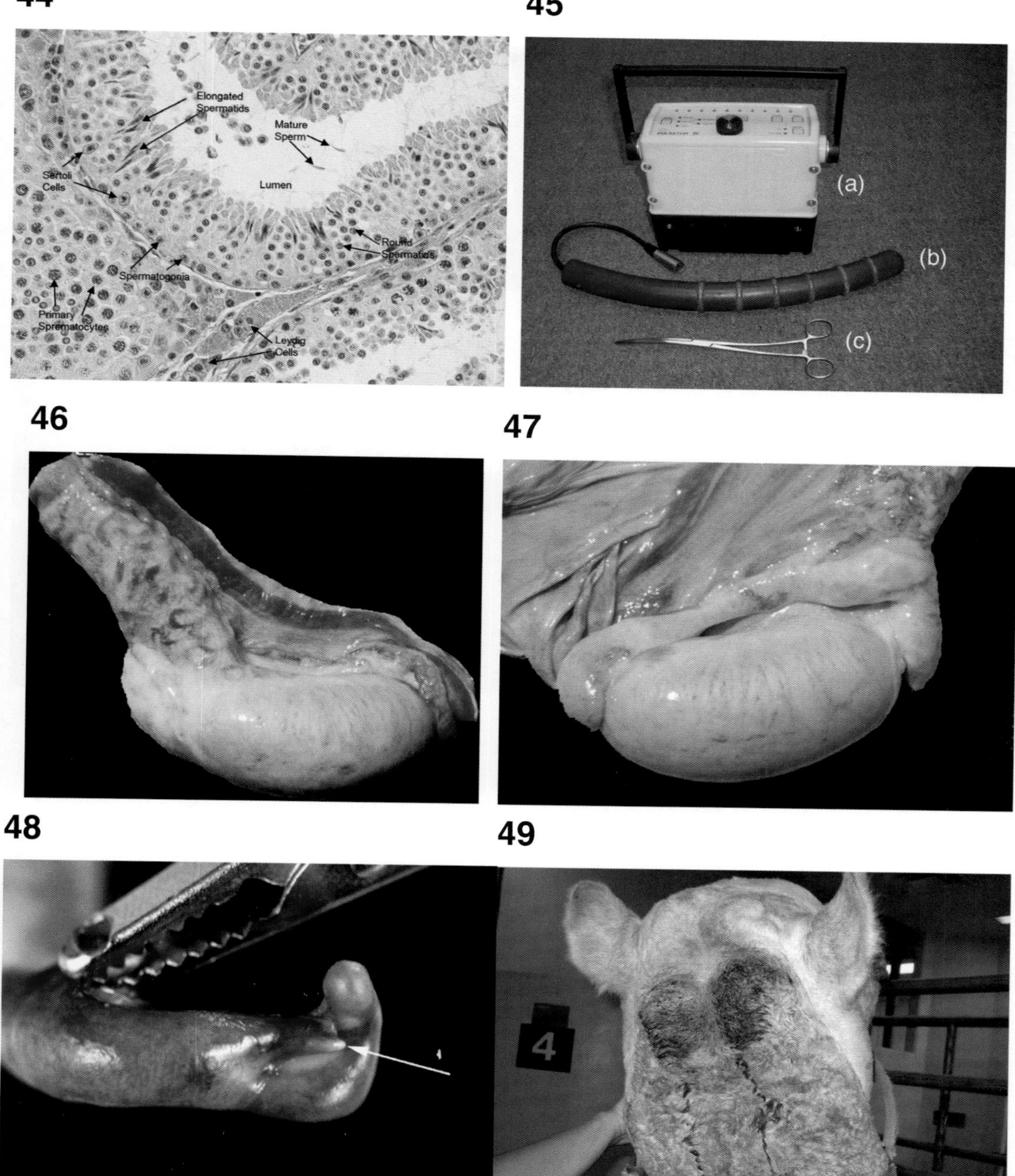

Plate 44. Cross section (×1050) of testis parenchyma from a mature boar stained with haematoxylin and eosin, and identifying prominent cellular structures in the seminiferous tubule and peri-tubular areas.
Plate 45. Equipment used for obtaining a semen collection via electroejaculation (EE) in the boar: (a) electroejaculator; (b) rectal probe; and (c) atraumatic dressing forceps for exteriorizing the penis.
Plate 46. Testicle from a 4-year-old dromedary, lateral view. Note the relatively small size of the cremaster muscle.
Plate 47. Medial view of testicle from a 4-year-old dromedary. Note the relatively small size of the cauda epdidymis.
Plate 48. Glans penis of an alpaca. Note the cartilaginous process, which extends beyond the urethral orifice (arrow).
Plate 49. Active poll gland in a rutting male dromedary.

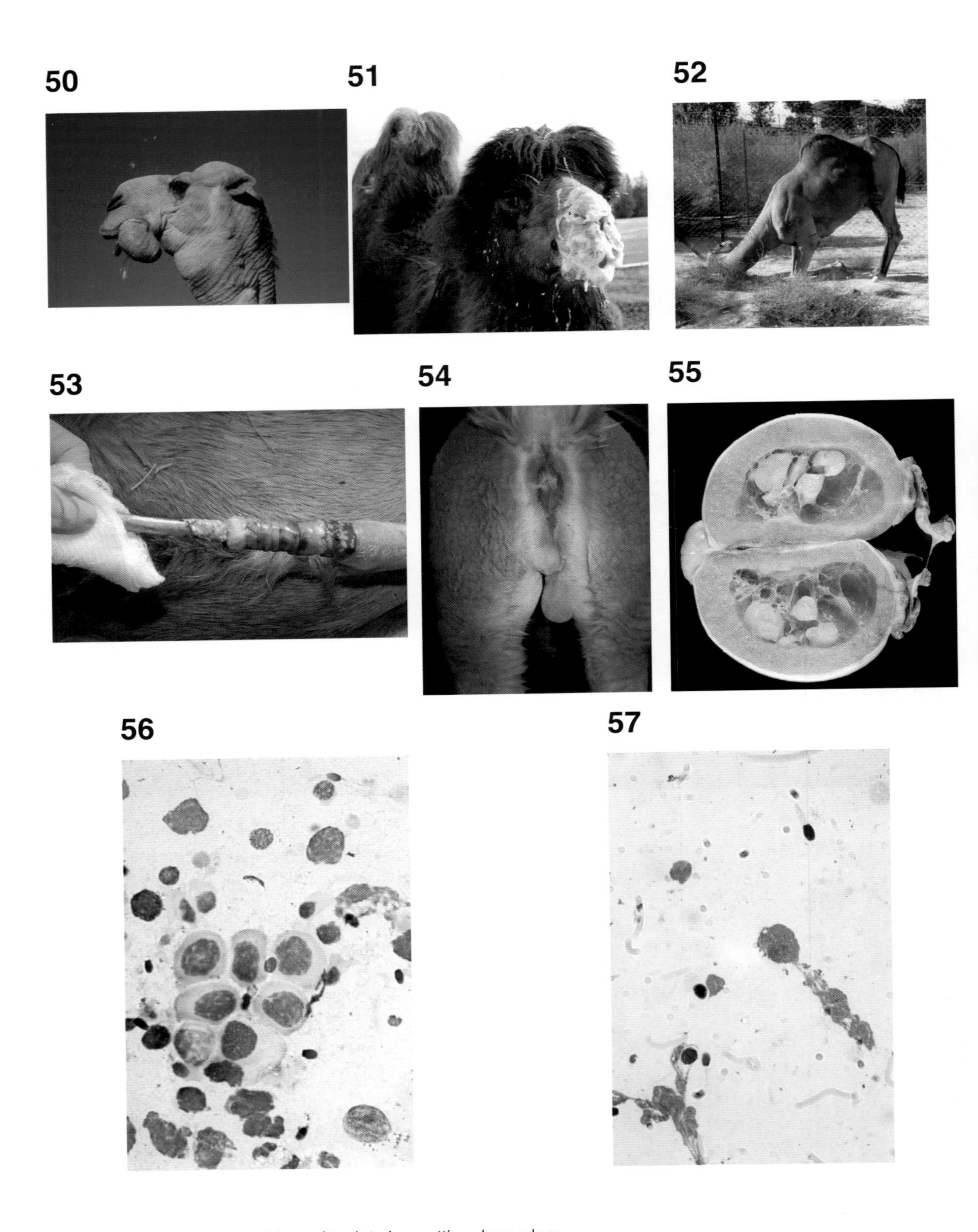

Plate 50. Exteriorization of the soft palate in a rutting dromedary.
Plate 51. Excessive salivation and frothing shown by a rutting Bacterian camel.
Plate 52. Marking behaviour (rubbing of the poll gland secretions) of a male dromedary.
Plate 53. Preputial adhesions, abscessation and severe balanoposthitis in a male alpaca presenting with phimosis.
Plate 54. Ectopic testis in an alpaca.
Plate 55. Testicular (rete testis) cysts in an alpaca (post-castration specimen).
Plate 56. Fine-needle aspirate showing normal spermatogenesis (high cellularity) in an alpaca.
Plate 57. Fine-needle aspirate showing poor spermatogenesis (low cellularity) in an alpaca.

58

59

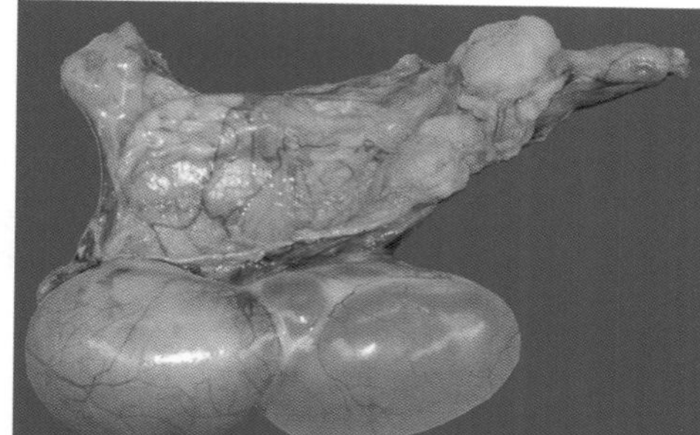

60

61

62

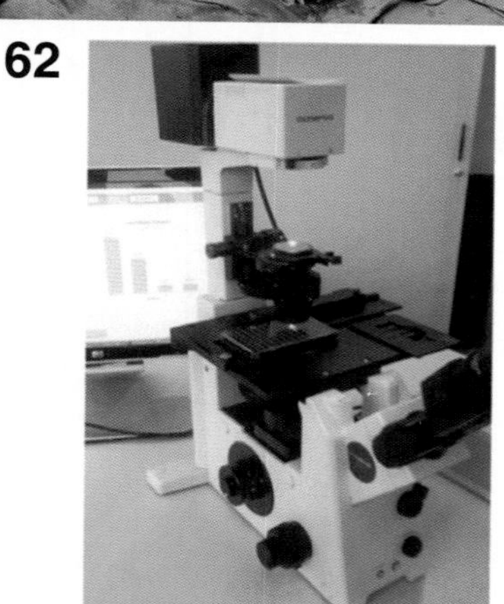

63

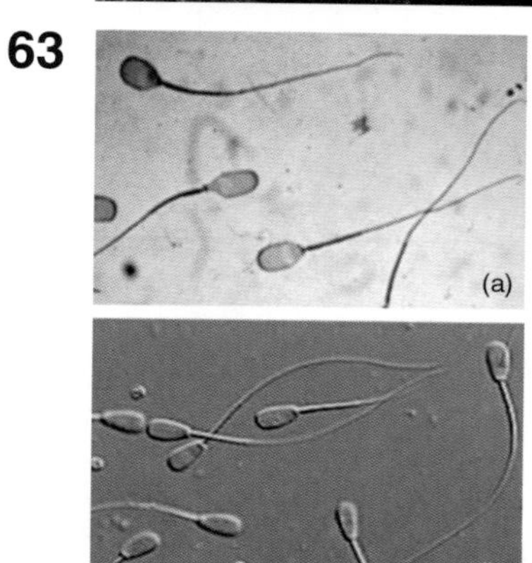

Plate 58. Histological view of testicular degeneration in an alpaca. Note the vacuolated aspect of the seminiferous tubules and the absence of spermatogenic activity.
Plate 59. A large epididymal cyst in an alpaca.
Plate 60. Use of a battery-operated electroejaculation unit in the field. These units are available on a custom-made basis and should be assessed for voltage and amperage delivery before use.
Plate 61. The bifid glans penis of the echidna (*Tachyglossus aculeatus*) is divided into four urethral branches that terminate in 'flower-like rosettes' (Carrick and Hughes, 1978). Photo courtesy of Nana Satake.
Plate 62. Using the Qualisperm™ system (Biophos AG, Geneva, Switzerland), sperm motility can be automatically assessed using novel algorithms to detect single particles (spermatozoa) in confocal volume elements that project individual spermatozoa on to a pixel grid of a CMOS camera and analyse the number of fluctuations using a correlation function. There is a high throughput – four fields/min, and >2000 spermatozoa are usually analysed per sample (courtesy of Prof. R. Rigler).
Plate 63a–b. Sperm morphology can be studied in smears that are either stained (a) or examined using differential interference contrast microscopy (DIC) (b).

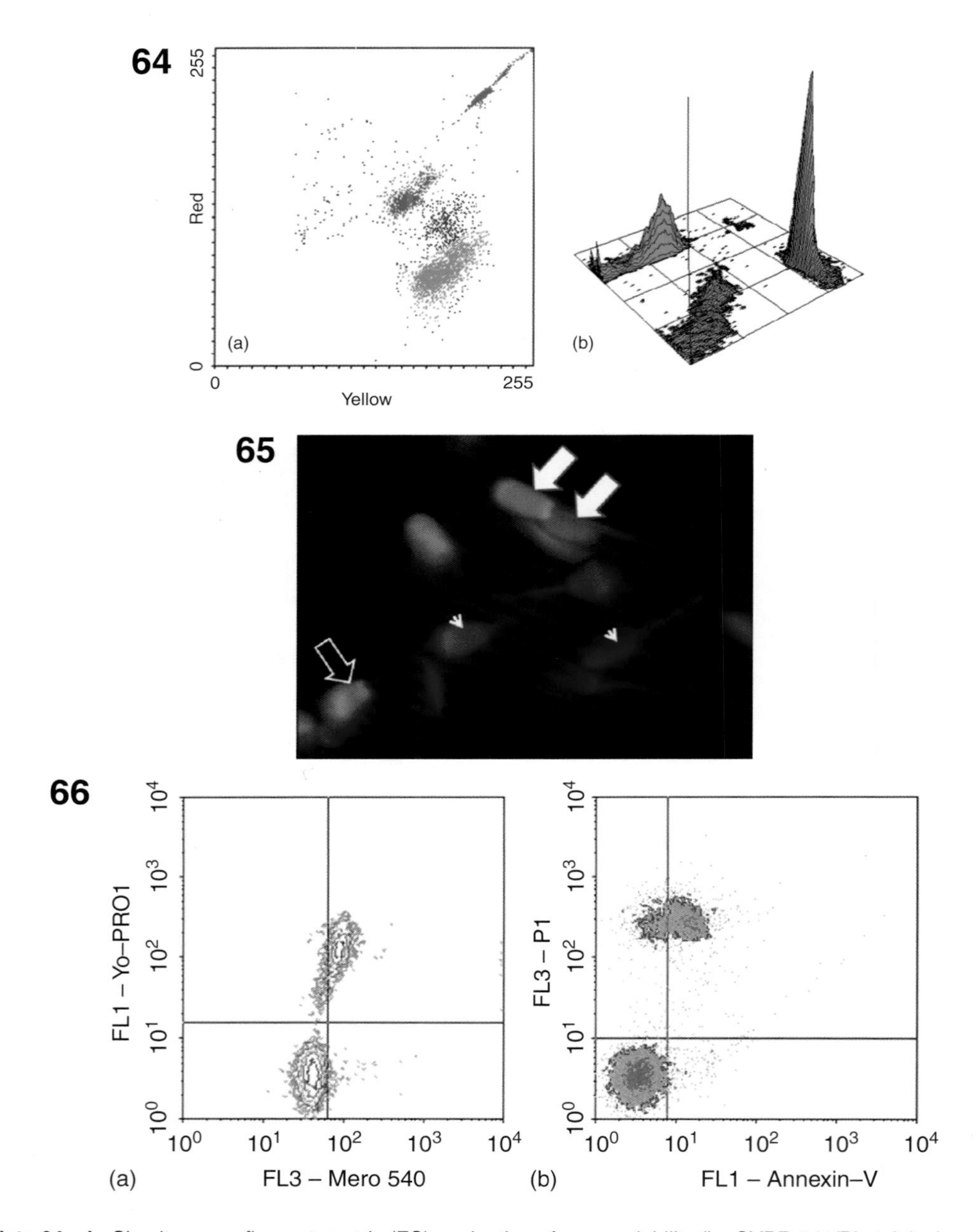

Plate 64a–b. Simultaneous flow cytometric (FC) evaluation of sperm viability (by SYBR 141/PI staining) and cell concentration. (a) A classical dot plot depicting viable (green), dead (red) and dying (blue) sperm dots (relative amounts). The magenta dots are the grouping of fluorescent micro beads (a standard control). Another representation is provided by (b), in which sperm viability and the presence of cell debris are shown in tri-dimensional form. From Hossain *et al.* (2011) *Asian Journal of Andrology* 13, 406–419, with permission.

Plate 65. Changes in sperm membrane permeability can be microscopically determined using a combination of fluorescent probes (YO-PRO-1, SNARF-1, ethidium homodimer). The YO-PRO-1 positive cells (green, thick arrow) indicate increased membrane permeability among viable, stable cells (SNARF-1 positive: brown, arrow heads) and those undergoing membrane damage (SNARF-1/ethidium homodimer positive: green/red, hollow arrow) (courtesy of Prof. Dr F.J. Peña Vega, Cáceres, Spain).

Plate 66a–b. Plasma membrane asymmetry of spermatozoa, as a consequence of lipid destabilization, can be either assessed by flow cytometry using Merocyanine (Mero)-540/YO-PRO-1 labelling (a; contour plots) or by Annexin V/PI (propidium iodide) labelling (b; density plots). In both (a) and (b), viable cells with stable/unaltered or unstable/phosphatidylserine (PS)-exposed plasma membrane are located in the lower quadrants, while dead spermatozoa appear in the upper quadrants. From Hossain *et al.* (2011) *Asian Journal of Andrology* 13, 406–419, with permission.

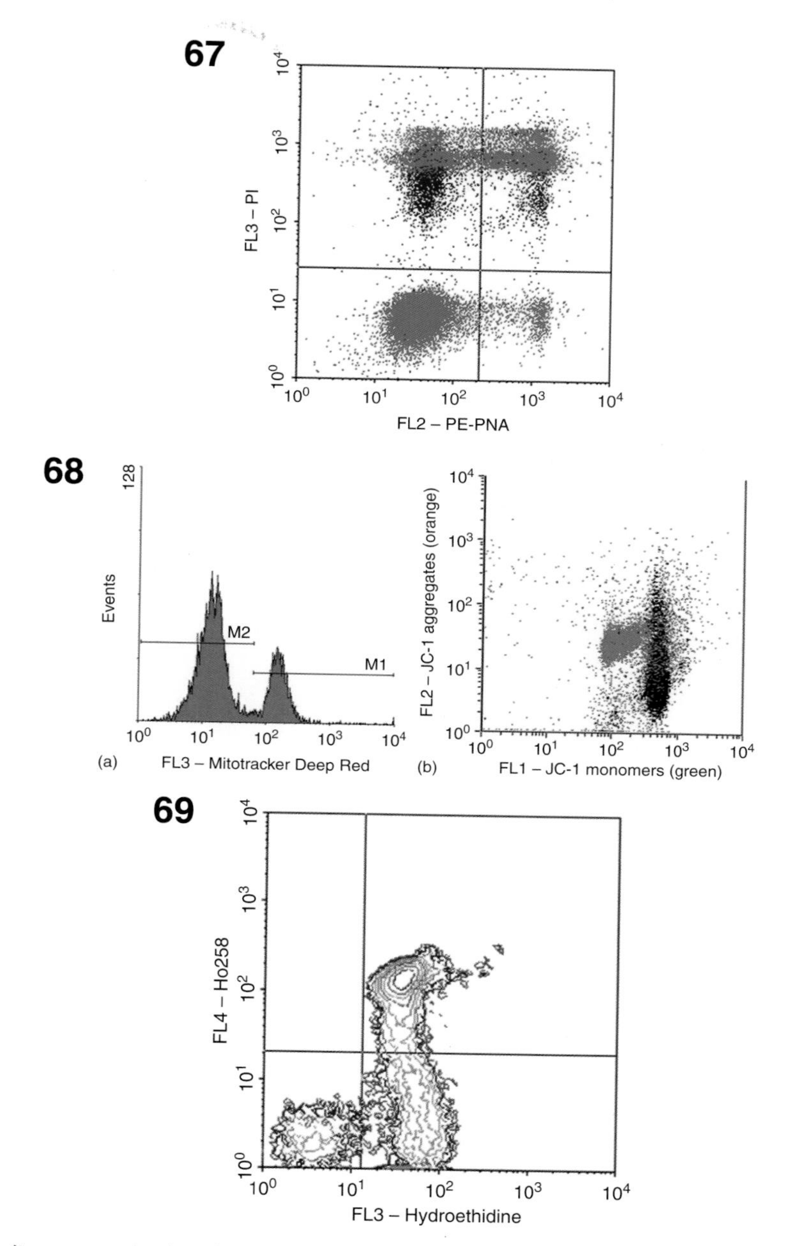

Plate 67. Simultaneous evaluation of sperm viability (membrane integrity) and acrosome integrity, determined, respectively, by SYBR-14/PI (propidium iodide) and PE-PNA (phycoerythrin conjugated peanut agglutinin) labelling. Colour dot plot shows spermatozoa that are: viable, acrosome-intact (lower left quadrant); viable, acrosome-ruptured (lower right quadrant); dead, acrosome intact (upper left quadrant); and dead, acrosome-ruptured (upper right quadrant). Green indicates viable cells, red indicates dead cells, blue indicates dying spermatozoa (evaluated as dead). From Hossain *et al.* (2011) *Asian Journal of Andrology* 13, 406–419, with permission.

Plate 68a–b. Evaluation of mitochondrial membrane potential depicted as: (a), histogram of events when using flow cytometry of spermatozoa loaded with Mitotracker Deep Red and showing either high (M1) or low (M2) membrane potential; or (b), a dot plot of spermatozoa loaded with the JC-1 mitochondrial probe (5,59,6,69-tetrachloro-1,19,3,39-tetraethylbenzimidazolyl-carbocyanine iodide). Sperm aliquots were examined before (red) or after (black) unprotected plunge into liquid N_2. From Hossain *et al.* (2011) *Asian Journal of Andrology* 2011, 13, 406–419, with permission.

Plate 69. Contour plot depicting the production of reactive oxygen species (ROS) in spermatozoa after being loaded with hydroethidine and Hoechst 33258. The quadrants on the right show either viable (lower) or damaged (upper) spermatozoa. The spermatozoa in the lower left quadrant are membrane intact and negative for ROS production. From Hossain *et al.* (2011) *Asian Journal of Andrology* 13, 406–419, with permission.

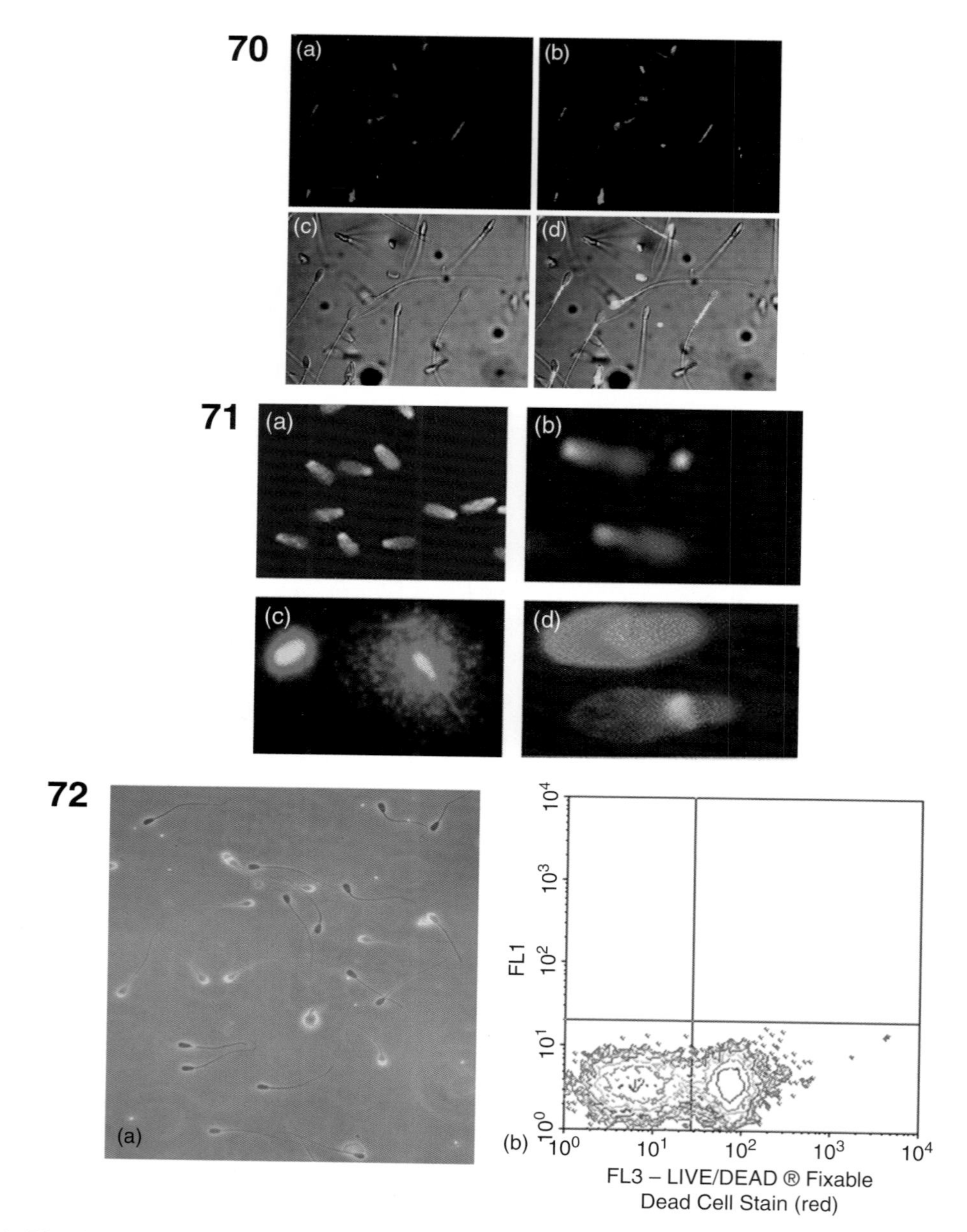

Plate 70a–d. Confocal laser microscopy images of stallion spermatozoa loaded with BODIPY-C_{11} (a probe from the 5-iodoacetamidofluorescein family) showing spermatozoa with areas of lipid peroxidation (LPO). (a) LPO-free spermatozoa, non-oxidized probe, red fluorescence; (b) green oxidized probe; (c) transmission image; (d) overlaid image – areas of lipid oxidation appear yellow (Bio-Rad MRC confocal microscope, 60× magnification). From Ortega Ferrusola *et al.* (2009) *Reproduction* 138, 55–63, with permission.

Plate 71a–d. Spermatozoa from various species undergoing exploration of sperm DNA damage: (a) SCSA (sperm chromatin structure assay); (b) COMET (single-cell gel electrophoresis assay); (c) SCD Halo (sperm chromatin dispersion test); and (d) TUNEL (terminal deoxynucleotidyl transferase-mediated fluorescein-dUTP nick-end labelling). Courtesy of Prof. Dr J. Gosalvez, Madrid, Spain.

Plate 72a–b. Identification of the entire plasmalemma and of acrosome damage in bull spermatozoa using the Live/Dead Fixable Red Dead Cell Stain Kit (Invitrogen L-23102) and Alexa Fluor488 PNA (L21409, green). In (a), fluorescence and differential interference contrast (DIC) microscopy are merged to discriminate between spermatozoa with an intact plasma membrane and intact acrosome (not fluorescent) and those with a disrupted plasma membrane (red tails and/or heads) or a ruptured acrosome (green fluorescence). (b) Flow cytometric contour plots of these populations, with intact spermatozoa in the lower left quadrant and dead spermatozoa in the lower right quadrant. From Hossain *et al.* (2011) *Asian Journal of Andrology* 13, 406–419, with permission.

within the tunics, the formation of anti-sperm antibodies and, possibly, testicular degeneration associated with increased temperature due to inflammation secondary to the surgical procedure. Relatively little information exists on these complications in the stallion. In one study of open biopsy techniques (DelVento *et al.*, 1992), nine stallions were examined for approximately a month after an open biopsy procedure. One month after biopsy, the stallions were castrated and both the control and biopsied testis were examined by histopathology. Granulation tissue and inflammation were noted at the biopsy site, and biopsied testes had increased degeneration in spermatocytes and B spermatogonia in the region of the biopsy compared with control testes. This change was attributed to inflammation at the biopsy site. Figure 11.15 shows a testicular haematoma subsequent to testicular biopsy in a stallion.

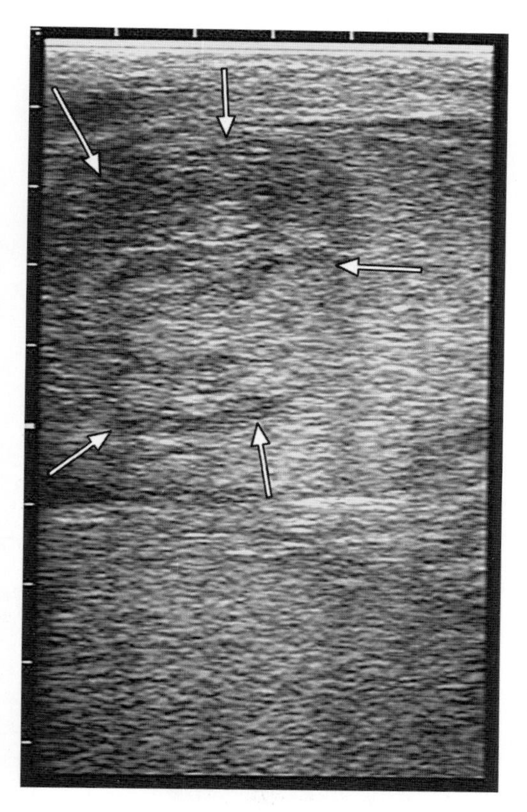

Fig. 11.15. Ultrasonogram of testicular hematoma subsequent to testicular biopsy in a stallion (arrows).

In another study following needle biopsy of the testis in seven stallions (Faber and Roser, 2000), there were no significant changes in semen parameters or fertility in stallions subsequent to testis biopsy. These investigators also examined both serum and seminal plasma for the presence of potential anti-sperm antibody and did not detect changes in either IgG in serum or IgA anti-sperm antibody in the seminal plasma (Faber and Roser, 2000). Immediately after needle biopsy of the testis, local oedema was noted in six of seven stallions, and in one there was a diffuse oedema and increased local temperature, which resolved over the subsequent 5 days. These authors concluded that needle biopsy of the stallion's testis using the Biopty™ instrument was a safe procedure without deleterious effects on fertility in the stallion. In another study, local oedema and haematomas were noted transiently (3–7 days duration) in stallions undergoing testicular biopsy (Pearson *et al.*, 2011).

In the author's experience, needle biopsy of the testis in stallions using the Biopty™ instrument can be associated with scrotal oedema, haematocele and testicular haematomas at a frequency that warrants caution in the selection of cases to be biopsied. Because of these complications, veterinarians should select cases for testicular biopsy under carefully considered indications, as discussed below.

Diagnostic indications for testis biopsy

Testis biopsy is well characterized and described in human andrology (Cerilli *et al.*, 2010). In humans, it is frequently used for the diagnosis of testicular neoplasia and, in azoospermic men, to help evaluate infertility. In addition, testis biopsy may be used in azoospermic men who are candidates for assisted reproductive techniques, such as intracytoplasmic sperm injection (ICSI) using testicular haploid germ cells (Cerilli *et al.*, 2010).

In stallions, diagnostic criteria for testis biopsy are not as clearly defined as in men. In cases of azoospermia, it is important to exclude bilateral obstructive lesions of the excurrent ducts as a cause of azoospermia. The determination of ALP in seminal fluids from azoospermic stallions may be useful in helping this diagnosis. ALP activity in seminal fluid from the stallion is relatively high, with a reported range of 1640–48,700 IU/l (Turner and McDonnell, 2003); the enzyme appears to be derived primarily from epididymal secretions, with some contributions from the testis. The activity of ALP is low (<90 IU/l) in pre-seminal fluid and in cases of bilateral obstructive azoospermia secondary to ampullar blockage (Turner and McDonnell, 2003). If ALP activity in seminal fluids from azoospermic stallions is normal, then primary lesions of the seminiferous epithelium or bilateral obstruction of the epididymal duct are more likely causes to consider.

In stallions that have significant changes in the spermiogram, such as oligospermia (decreased number of sperm), asthenozoospermia (low sperm motility) or teratozoospermia (majority of sperm with abnormal morphology), a careful examination of ejaculated spermatozoa may yield useful diagnostic information on the seminiferous epithelium (Veeramachaneni and Sawyer, 1996; Veeramachaneni *et al.*, 2006). The determination of serum concentrations of gonadotrophins, particularly FSH, may also provide useful information on the hypothalamic-pituitary-gonadal axis in the stallion.

In addition to routine histological evaluation, other assessments of testis function have been performed using tissue derived from biopsy. For example, determinations of the intra-testicular concentrations of oestradiol, testosterone and inhibin have been performed from testis biopsies in the stallion (Faber and Roser, 2000). The evaluation of gene expression from mRNA derived from testicular biopsies in the stallion could provide information about testis function, and in pursuit of this, Das *et al.* (2010) have described methodologies for the isolation of total RNA from stallion sperm and testis biopsies. At present, however, the evaluation of testis biopsies in the stallion is limited primarily to routine histopathological evaluation (inflammation, neoplasia, etc.) and the assessment of spermatogenesis.

Conclusions

Although testis biopsy using needle biopsy techniques has been well characterized for many years in the stallion, the technique is used infrequently in clinical veterinary practice. Possible complications secondary to the technique include scrotal oedema, haematocele and testis haematoma formation, and such complications should be considered before undertaking the procedure. Indications for testis biopsy in the stallion include suspected testicular neoplasia and cases of azoospermia in which obstructive lesions have been excluded; information to determine whether or not spermatogenesis is occurring may be useful in future prognosis or management of the case.

Diseases of the Scrotum and Testis

Cryptorchidism

Cryptorchidism – the failure of one or both testes to descend properly into the scrotum – is a relatively common condition in the stallion and as such, a short discussion of the normal descent of the testis is in order (Bergin *et al.*, 1970). Differentiation of the gonad is complete near the end of the embryonic period (day 40). At this point, the fetal testis lies near the caudal pole of the kidney, and is supported by a fold of mesentery called the mesorchium. An evagination of this peritoneal fold (vaginal process) forms at about this time and will give rise to the inguinal ring. The testis is connected to the abdominal wall by a cord of mesenchyme called the gubernaculum (Plate 26). The gubernaculum attaches to both the caudal pole of the testis and to the mesonephric duct, and will form the ligament of the tail of the epididymis, as well as the proper ligament of the testis. The gubernaculum appears to expand the inguinal canal to facilitate passage of the testis through the canal in late gestation or during the early postpartum period. Around day 150, the developing cauda epididymis is drawn towards the inguinal ring, but the fetal testis is too large to pass through it at this stage. During the second half of gestation, the testis decreases in size and typically passes through the inguinal canal during the last month of gestation, or shortly after birth.

Some 2 to 8% of male horses may be affected by cryptorchidism (Stickle and Fessler, 1978; Cox *et al.*, 1979). The condition appears to be related to a reduction of insulin-like peptide (INSL3), testosterone and/or their respective receptors during testicular descent (Basrur, 2006). It has also been correlated with insufficient intra-abdominal pressure or reduction of testis size (Amann and Veeramachaneni, 2007). Once the cryptorchid condition is established, the undescended testis is subjected to elevated temperature, which negatively influences the development and differentiation of testicular somatic cells and germ cells (Pinart *et al.*, 2000, 2002; Huff *et al.*, 2001).

The testis may be retained within the abdomen, within the inguinal canal (high flanker) or subcutaneously outside the scrotum (ectopic testis). The retention of the testis within the inguinal canal is usually unilateral, with the right and left testes affected with equal frequency. If the testis is retained within the abdomen, the epididymis may or may not extend into the inguinal canal. With abdominal retention, the left testis is retained more frequently than the right (2:1). This pattern does not appear to apply to ponies, in which right-sided inguinal retention appears to be more common.

In addition to the disruption of normal spermatogenesis, cryptorchid testes appear to be at greater risk of neoplastic transformation (Edwards, 2008), and torsion may also occur in abdominally retained testes (Hunt *et al.*, 1990). A genetic predisposition to cryptorchidism has been implied for many years, but there are no good studies on the genetics of the condition. There are breed predilections (the Standardbred, Quarter Horse and Percheron have a higher incidence), and this suggests some type of heritable basis (Hayes, 1986). The actual mechanism of inheritance (autosomal recessive versus polygenic, etc.) has not been determined if, indeed, one exists. New tools in molecular biology are likely to prove very useful in assessing possible heritable factors associated with cryptorchidism in the stallion based upon linkage analysis.

In males in which there is a history of a previous attempt to castrate the animal, or in those in which the history is unknown, other diagnostic tests may be required to identify a retained testis (Cox *et al.*, 1986; Cox, 1999). Although spermatogenesis is arrested in the retained testis as a result of elevated temperature, the Leydig cells continue to secrete testosterone, and these animals will develop secondary sex traits (Plate 27). A number of endocrine assays have been used to distinguish cryptorchids from geldings. Elevated levels of oestrone sulfate are reported to be highly accurate in males over 3 years old for distinguishing a retained testis (Carneiro *et al.*, 1998). Single samples for testosterone determination may also be used for diagnosis in some cases (geldings, 20 to 50 pg/ml, versus cryptorchids, 100–500 pg/ml), but there remains some debate over the use of these single samples. Further, the limit of sensitivity for the detection of testosterone in some laboratories may mean that these values cannot be distinguished. Therefore, the use of an hCG (or GnRH) stimulation test is often recommended. For this test, concentrations of testosterone are compared before and 0.5–2 h after the administration of hCG. A more than twofold rise in testosterone is typically indicative of the presence of testicular tissue.

Orchitis

Bacterial orchitis is rare in stallions but has been reported secondary to infections by *Streptococcus* spp., *Salmonella abortus equi, Brucella abortus* and *Corynebacterium pseudotuberculosis* (McEntee, 1990; Gonzalez *et al.*, 2008). More commonly, orchitis results secondarily to a trauma to the testis (either a blunt or penetrating trauma) with contamination by a number of possible bacteria. Increased testicular temperature, ischaemia, infarction, pressure necrosis and abscessation of the testis may occur as a result of orchitis. Viral orchitis has been reported secondary to equine infectious anaemia, equine viral arteritis and influenza, and vascular lesions associated with these infections may result in infarction of the testis. Aberrant migrating strongyle (redworm) larvae can also produce focal inflammatory lesions of the testis in stallions (Marino *et al.*, 2009).

The developing primary spermatocytes, spermatids and spermatozoa are normally sequestered from the immune system beyond the blood–testis barrier. Damage to the testis or to the epididymis may allow extravasation of sperm and the exposure of sperm antigens, with subsequent formation of anti-sperm antibodies in the male (Kenney *et al.*, 2000). This immune response may lead to the formation of sperm granulomas with testicular degeneration if the process is severe enough.

Trauma to the scrotum and testis

Trauma to the external genitalia of the stallion is a common sequela to breeding accidents in which the stallion is kicked by the mare at mating. Such trauma often involves the prepuce and penis in addition to the scrotum and testis. Even if the trauma does not directly involve the testis, the resulting swelling and oedema in the region may act to insulate the scrotum, with marked effects on spermatogenesis as a result of thermal injury (Plate 30). Initially, frequent cold-water hydrotherapy should be used to reduce swelling. If the tunica albuginea is ruptured, then severe damage to the testis with haemorrhage and necrosis may result. If the tunica albuginea is not ruptured, then formation of a testicular haematoma may occur, with subsequent effects on the remaining normal seminiferous epithelium. Often, the end result of traumatic or bacterial orchitis and periorchitis is marked testicular degeneration. Treatment will include hydrotherapy, anti-inflammatories, antibiotics and unilateral orchiectomy – if only one testis is affected.

Torsion of the testis

Acute torsion (often greater than 180°) may result in ischaemic injury to the testis. Affected stallions may show evidence of colic and scrotal swelling. If the rotation is 180°, the cranial dislocation of the tail of the epididymis may identify it. Manual reduction of the torsion may be possible, but severe cases, with ischaemic damage to the testis, may result in venous congestion, oedema, increased intra-testicular pressure and ultimately, testicular degeneration. Acute testicular torsion typically mandates orchiectomy because the affected testis is usually severely damaged as a result of vascular occlusion. A variety of 'pexy' procedures have been used in attempts to prevent vascular occlusion in stallions with testis rotation, but with variable success (Threlfall *et al.*, 1990). In some stallions, there appears to be an asymptomatic 180° rotation, which may occur transiently. Seminal parameters have been reported to be reduced in these stallions.

Testicular degeneration

Testicular degeneration can result from a variety of factors and often, by the time a diagnosis is made, the aetiology is not known. The degeneration may be unilateral or bilateral, and may be focal or widespread. Typically, there is a generalized loss of germ cells, vacuolization, mineralization and thickening of the basement membrane of the seminiferous tubules. In advanced testicular degeneration, the Sertoli cells and Leydig cells may also be lost. There are a variety of inciting factors that may be associated with the condition; these include thermal damage, trauma, radiation, nutritional imbalances, heavy metals, anabolic steroids, vascular lesions, neoplasia and age-associated degeneration. It may be difficult clinically to distinguish testicular degeneration from hypoplasia. There may be a moderate-to-marked reduction in testis size in either condition. Over time, testicular degeneration may be associated with fibrotic changes that result in a very firm, small testis. In such cases, calcification can be confirmed by ultrasonography. Testicular biopsy may be useful in confirming testicular degeneration.

Thermal injury to the testis has been well characterized for a number of species. In the stallion, the acute application of thermal damage (by scrotal insulation) results in a rapid decrease in the number of normal, motile spermatozoa in the ejaculate. If the thermal insult becomes chronic, the ability of the testis to recover normal function may be impaired and testicular degeneration may occur.

Exogenous anabolic steroids have a significant negative effect on spermatogenesis in most males, including the stallion. Although such agents may be used to improve muscling or performance, they also have a profound negative effect on spermatogenesis (Blanchard, 1985; Garcia *et al.*, 1987). By elevating the levels of circulating androgens, there is a negative feedback on the hypothalamic-pituitary axis, with a reduction in production/release of gonadotrophins. The reduction in LH secretion results in a reduced production of testosterone by Leydig cells and a reduction in the high local levels of testosterone in the seminiferous tubules. Normally, the local concentration of testosterone in the testis is 100× higher than the circulating levels, but the administration of exogenous steroids may reduce circulating LH concentrations by 50%, with a concomitant reduction in the locally high levels of testosterone needed to support normal spermatogenesis. The result is a reduction in testis mass and in daily sperm output in stallions that receive exogenous steroids. These effects may persist for some time after the administration of steroids and it may require 6 to 12 months for recovery of normal spermatogenesis after the steroid therapy is discontinued.

Testicular neoplasia

There are a variety of neoplasms reported in the testis of stallions. These include seminomas, teratomas, Sertoli cell tumours, Leydig cell tumours, dysgerminomas and mast cell tumours (Smyth, 1979; Stick, 1980; Schönbauer and Schönbauer-Längle, 1983; Trigo *et al.*, 1984; Hunt *et al.*, 1990; Brinsko, 1998; Pollock *et al.*, 2002; Edwards, 2008). There is no good data on the relative frequency of these tumours in the stallion, because most male equids are castrated at an early age. However, seminomas and teratomas appear to be the most common. Most testicular tumours are unilateral and present with a painless swelling of one testis. The seminal parameters may be normal or abnormal depending on the influence of the enlarged testis on the remaining normal testis. Seminomas and Leydig cell tumours occur most often in aged stallions and may be present in either normally descended or cryptorchid testes. Teratomas are more often reported in retained testes (McEntee, 1990). The ultrasonographic appearance of these tumours varies, and relatively few reports have characterized the appearance of these tumours in the stallion testis. In one report, the ultrasonographic appearance of a seminoma was characterized as diffusely hypoechoic with heterogeneous areas of increased echogenicity (Beck *et al.*, 2001). Seminomas frequently metastasize, and examination of the spermatic cord and abdominal cavity may be useful for detecting metastatic disease (Plates 28 and 29). Ultrasonography (at a high resolution of 7.5 mHz) may be useful in the detection of some of these tumours earlier than is possible by palpation alone. Biopsy or fine-needle aspiration of the testis may be useful diagnostically.

Diseases of the Scrotum, Tunica Vaginalis and Spermatic Cord

Inflammatory conditions of the testis may often extend to involve the surrounding tunics, resulting in a periorchitis. If such lesions are extensive, significant adhesions may result between the testis and tunica vaginalis, which may interfere with thermoregulation of the testis, even after the inflammation has resolved.

Hydrocele

A hydrocele is an abnormal collection of fluid (typically serous fluid) between the visceral and parietal layers of the tunica vaginalis. It may result from either inflammatory or non-inflammatory conditions. Because the vaginal cavity communicates directly with the peritoneal cavity, ascites may result in secondary hydrocele. In some cases, a hydrocele develops without an obvious aetiology. Transient hydrocele may be noted in stallions exposed to high environmental temperatures. Diagnosis of a hydrocele may be made by palpation if there is a significant accumulation of fluid. Ultrasonography may be very useful for localizing the swelling present in the scrotum to a focal accumulation of hypoechoic material within the vaginal tunics. Careful aseptic aspiration of the fluid accumulation may be useful for diagnosis based on a cytological examination of the fluid recovered, although great care is required to prevent the introduction of bacteria, which may result in sepsis. Persistent accumulation of fluid within the tunica vaginalis may result in reduced fertility because of abnormal thermoregulation of the affected testis.

Inguinal/scrotal hernia

Hernias involving the inguinal canal and scrotum of the stallion typically involve passage of abdominal contents (intestine or mesentery) through the vaginal ring and into the vaginal cavity. Most inguinal hernias in the stallion appear to be congenital and occur as a result of an enlarged vaginal ring. Some authors have suggested a possible heritable basis for such hernias, but this has not been well established. Many of these hernias appear soon after birth and are often easily reduced manually. Some of them appear to be self-correcting by the time the foal is 3 to 6 months of age. The presenting complaint in these cases is typically a soft, non-painful, unilateral inguinal or scrotal swelling. Unless the condition persists, there is typically no need for treatment. In a few cases, these hernias may become incarcerated and require surgical correction.

Acquired inguinal or scrotal hernias may occur in breeding-age stallions and are often associated with acute colic resulting from an incarcerated intestine. The scrotum is enlarged and the viscera may be palpated per rectum as they enter the inguinal canal. Non-surgical reduction of such hernias may be attempted, but surgery may be necessary to reduce the hernia if there is strangulated intestine present (Schneider *et al.*, 1982; Van der Velden, 1988). Acquired hernias may be associated with stallions that have enlarged internal inguinal rings; there appears to be a predisposition to herniation during mating by the stallion.

Varicocele

Varicocele refers to an abnormal distension of the pampiniform plexus of the testis. In man, such lesions have been associated with reduced fertility, apparently due to abnormal thermoregulation. Though common in human males, varicoceles appear to be rare in the stallion (Varner *et al.*, 1991). The presence of a grossly enlarged venous plexus can be detected by both B-mode and Doppler ultrasonography in the region of the spermatic cord (Pozor, 2005). In human males, the presence of varicocele is associated with increased abnormalities in the spermiogram, and anecdotal evidence suggests that stallions with varicocele may also have a suppression of semen quality (Pozor, 2007).

Diseases of the Excurrent Duct System

The efferent ductules (10–23 per testis) conduct spermatozoa from the rete testis to the head of the epididymis. In some cases, these ductules may be blind ending and they may occasionally become distended with secretion in the area of the head of the epididymis. Epididymal cysts have been detected ultrasonographically in the caput epididymis of the stallion, and anecdotal evidence suggests that these cysts may be associated with ejaculatory problems (Pozor, 2005). The origin of the cysts has not been defined, but they may be derived from blind-ended efferent ductules that fail to fuse during formation of the epididymal duct, or possibly represent cysts of the appendix epididymis, a remnant of the mesonephric duct (McEntee, 1990; Love, 1992). The cysts may also be noted in stallions in which no decline in semen parameters has been noted; however, rupture of blind-ending efferent ductules can also lead to sperm granuloma formation in the caput epididymis, possibly resulting in obstructive lesions of the epididymal duct (Pozor and Garncarz, 2006).

While rare, segmental aplasia of the excurrent duct system has been reported in stallions with failure of formation of the epididymis (McEntee, 1990; Blanchard *et al.*, 2000a) and vas deferens (Estrada *et al.*, 2003) occurring either unilaterally or bilaterally.

Epididymitis

Bacterial epididymitis is rare in stallions, although it has been reported (Held *et al.*, 1990; Traub-Dargatz *et al.*, 1991; Brinsko *et al.*, 1992). It is likely to be present in many cases of infectious orchitis in the stallion. Acute epididymitis may be associated with pain. Chronic cases may also be associated with abscessation and adhesion formation. Diagnosis may be based on clinical signs and palpation. Inflammatory cells are typically present in the ejaculate. Therapy may be based on antibiotic treatment after culture and sensitivity testing. Severe cases often result in hemiorchiectomy.

Sperm accumulation syndrome (spermiostasis)

Some stallions appear to have an abnormal retention of spermatozoa within the excurrent duct system, with an accumulation of spermatozoa (Love *et al.*, 1992; Varner *et al.*, 2000). Retained spermatozoa appear to undergo degenerative changes within the excurrent duct system, and ejaculates from these stallions may be characterized by a high sperm concentration (>500 million/ml), low motility and a high percentage of detached heads. In some cases, sperm and epithelial cell debris may form casts that appear within the ejaculate. Occasionally, spermatozoa may form obstructive plugs that are retained within the distal ductus deferens resulting in unilateral (or bilateral) obstruction of the ductus deferens. In these stallions, transrectal ultrasonography of the accessory glands may reveal an asymmetric dilation of one ampullar lumen due to obstruction (if the lesion is unilateral).

In cases of azoospermia, it may be useful to evaluate the levels of ALP in the collected fluid. This enzyme is secreted in high concentrations by the epididymis and testis. As such, it serves as a marker for the contribution of the epididymis to the fluid collected from an azoospermic stallion (Turner and McDonnell, 2003), and may be used to distinguish azoospermia secondary to ejaculatory failure from a primary azoospermia associated with the testis. Management of these cases may include the administration of 10 to 20 IU oxytocin immediately before semen collection in an attempt to dislodge the obstruction (Love *et al.*, 1992). Alternatively, prostaglandin F2α analogues (cloprostenol 50 μg IM), given approximately 15 min before collection or ejaculation, may also be used (D.D. Varner, personal communication). The ejaculate collected immediately after relief of the obstruction is characterized by a high sperm number, a high percentage of detached heads and the appearance of sperm plugs or casts (Card, 2005). More frequent or regular semen collection or mating schedules may be needed to prevent spermiostasis in these stallions.

Diseases of the Accessory Sex Glands

Diseases of the accessory glands of the stallion appear to be rare compared with those in other males, such as the bull. Bacterial infections have been reported and include many of the common bacterial pathogens such as *Pseudomonas*, *Klebsiella*, *Streptococcus* and *Staphylococcus* spp. These infections may involve any of the accessory glands, but those most commonly involved appear to be the vesicular glands. In the acute phase, the stallion may have pain and refuse to ejaculate. There may be evidence of gross contamination of the ejaculate with pus or blood. More often, these infections are chronic and the stallion does not show evidence of pain on ejaculation or palpation of the internal genitalia. In these cases, there is often a general deterioration of semen quality, with PMNs detected in the ejaculate. Further differentiation of the source of the infection requires a careful physical examination to distinguish orchitis or epididymitis from involvement of the accessory glands. Urethroscopic examination of the stallion may allow direct sampling of the vesicular gland for cytology and culture (Varner *et al.*, 2000). Therapy of these chronic cases is often difficult. Antibacterial agents do not reach many of the accessory glands in high levels. Drugs such as trimethoprim/sulfa or fluoroquinolone antibiotics may be reasonable choices if bacterial sensitivity indicates that they may be effective. Surgical vesiculectomy has been conducted in the stallion; however, it remains uncertain as to the effectiveness of this method in correcting chronic infections, partly because other accessory glands may be involved in addition to the vesicular glands.

Diseases of the Penis and Prepuce

Paraphimosis

Paraphimosis refers to the failure to retract the penis into the prepuce as a result of preputial injury or disease. Because this condition in the stallion often is due to trauma (such as a kick injury during mating), in many cases there is injury to both the penis (balanitis and/or phallitis) and to the prepuce (posthitis). After injury, oedema and swelling of the prepuce may prevent retraction of the penis into the prepuce (Plate 31). After a few hours, impaired venous and lymphatic drainage from the penis result in further oedema and swelling of the penis. The corpus cavernosum of the penis may become engorged with blood. If this condition persists, the skin of the prepuce becomes friable, desiccated and, eventually, gangrenous due to vascular stasis.

Therapy of this condition following an injury to the stallion requires prompt attention to prevent the self-perpetuating cycle of oedema and penile injury. If possible, the penis should be replaced into the prepuce and retained with a modified stallion supporter, or possibly with stay sutures. Stay sutures appear to be a less desirable method because of their tendency to cause further oedema of the prepuce and also because they often 'cut out' over a period of days. Swelling of the penis may be reduced by manual massage, by short-term application of an elastic bandage, or by application of a pneumatic bandage. Cold water hydrotherapy may also be used, along with anti-inflammatories to reduce inflammation in the acute phases. If the prolapsed penis can be manually reduced, the prognosis for recovery is often fair to good. If the penis is chronically prolapsed, then severe damage with necrosis and fibrosis often results.

Priapism

Priapism refers to a persistent penile tumescence without sexual arousal. The cause is often unknown, but it has been associated with the administration of some phenothiazine tranquillizers. These result in relaxation of the smooth muscles of the retractor penis muscles and also in enlargement of the vascular spaces within the corpus cavernosum. It is believed that priapism results from a sludging of blood within the vascular sinuses accompanied by a reduction in venous outflow. In the acute phase, the use of the cholinergic blocker benzotropine mesylate (8 mg IV) has been suggested to relieve the problem (Sharrock, 1982; Wilson *et al.*,1991). Surgical treatment has been used also to lavage the corpus cavernosum with heparinized saline to remove the sludged blood or to establish vascular shunts between the corpus cavernosum and corpus spongiosum. The prognosis for return to service of the stallion is typically poor.

Infectious conditions

The skin of the penis and prepuce of the stallion often has a number of microorganisms that may represent a normal microflora of this region. In addition, a number of potentially infectious agents have been isolated from the prepuce and penis of the stallion. Potential infections that may be transmitted venereally include: *P. aeruginosa*, *K. pneumoniae*, *Streptococcus* spp., *E. coli*, *T. equigenitalis* (the causative agent of CEM), equine arteritis virus (EAV; the causative agent of equine viral arteritis – EVA), equine herpesvirus 3

(EHV-3; the causative agent of equine coital exanthema) and *Trypanasoma equiperdum* (which causes dourine).

Bacterial infections

Both *P. aeruginosa* and *K. pneumoniae* have been established as true venereal pathogens that may be transmitted from the stallion to the mare at mating. It has been suggested that other bacteria (such as *Streptococcus* spp. and *E. coli*) may also be carried by the stallion and transmitted during coitus. Colonization of the penis or prepuce of the stallion with *P. aeruginosa* or *K. pneumoniae* is typically not associated with any clinical signs in the stallion. In some cases, these infections may become very persistent, and eliminating the causative organisms can be difficult. The most common problem associated with *P. aeruginosa* and *K. pneumoniae* is the appearance of post-breeding endometritis caused by these organisms in mares bred by stallions that harbour them.

Wet environments appear to be associated with persistence of *P. aeruginosa*. Likewise, it has been shown that frequent use of disinfectants or antibacterial soaps may alter the normal microflora of the penis, allowing colonization by *P. aeruginosa* or *K. pneumoniae*. Therefore, it is routinely recommended to prepare the stallion for semen collection or mating by washing with clean water only and cotton wool without soaps or disinfectants. Heavily encapsulated strains of *K. pneumoniae* (K5) have been implicated more frequently as venereal pathogens in the stallion, and capsule typing may be useful to identify a potentially venereal type of *K. pneumoniae*. Control of these venereal infections can be accomplished via AI with antibiotic-treated semen extenders in breeds that allow AI. In breeds that do not allow AI, the use of minimum contamination techniques may help to reduce transmission. Parenteral antibiotic treatment of the stallion is typically not very effective in eliminating these infections. Topical treatments of the penis have been used with some success. For *P. aeruginosa*, acidification of the penile skin may be useful. This can be accomplished with dilute HCl (10 ml of reagent grade (38%) HCl in 1 gal, i.e. 4.5 l, water) applied daily for 2 weeks. For *K. pneumoniae*, a topical application of dilute sodium hypochlorite (45 ml of 5.25% bleach in 1 gal water) for 2 weeks may be effective. Careful hygiene during the examination or insemination of mares is important because these organisms can also be transmitted mechanically.

Contagious equine metritis (CEM)

CEM is a venereally transmitted bacterial infection caused by *T. equigenitalis*, which was first described as an outbreak in 1977. In mares, the infection results in an acute endometritis with a prominent mucopurulent vulvar discharge about 8 to 10 days after mating with an infected stallion. There are no clinical signs in the stallion. Although most infected mares clear the organism, a few remain persistently infected and may also serve as a reservoir of the infection. The clitoral sinuses appear to be a frequent residual site for *T. equigenitalis* in mares, while the organism appears to persist on the penis and urethral fossa of the stallion. *T. equigenitalis* is difficult to culture without special conditions. A number of serological assays have been used (including complement fixation, agglutination and ELISA); however, these assays are not reliable in the stallion, apparently because the superficial nature of the infection does not elicit a consistent immune response. Control schemes for this infection rely on a number of techniques to limit its spread. Cultures are also taken from the clitoris and uterus of the mare. In stallions, test matings to known negative mares are used, and the mares are sampled by culture and by serology to detect transmission. Cultures of the penis are taken and topical antibacterial therapy (with chlorhexidine solution, followed by nitrofurazone ointment) is also used in the stallion to eliminate the infection.

Coital exanthema

The lesions of equine coital exanthema are typically limited to the external genitalia of the mare and stallion; transmission is usually by the venereal route or, potentially, by mechanical transmission. The disease is most often self-limiting and is characterized by the formation of typical herpesvirus lesions consisting of vesicles and pustules that may become secondarily infected. Treatment consists of enforced sexual rest and possibly topical antibiotics to reduce secondary bacterial infections. Infection by EHV-3 is not associated with abortion disease in the horse.

Behavioural Dysfunction in the Stallion

In the domesticated environment, the stallion is considerably removed from the natural behavioural responses that are associated with reproductive behaviour in feral horses. For example, the stallion is frequently isolated from all mares except at mating, and may not be allowed the opportunity to interact with mares at all except at the time of mating. Often, human intervention determines when a mare is ready to be mated and, in some cases, the stallion is not exposed to mares at all when semen is collected for AI via a phantom mount. Young stallions that are in training may be negatively conditioned or disciplined when displaying sexual arousal at an inappropriate time; mechanical aids such as a stallion ring or brush may be used to discourage erection or masturbation in the stallion. These alterations from the normal 'harem' breeding units that are established in feral horses present a potential problem in behavioural dysfunction in stallions that has been most intensely studied by Dr S.M. McDonnell at the University of Pennsylvania (McDonnell *et al.*, 1987; McDonnell, 1992, 2001).

The sexual behaviour of domestic stallions is strongly influenced by experience, and negative sexual experiences by a young stallion may have profound detrimental effects on their sexual behaviour at later times. Established routines become very important, and the stallion often responds best when this routine is maintained in the breeding shed.

McDonnell and co-workers have characterized sexual behaviour problems in stallions (McDonnell *et al.*, 1987; McDonnell, 1992, 2001). Approximately 50% of their cases involved problems with sexual arousal/interest (poor libido) while another 25% were related to ejaculatory dysfunction. Aggressiveness and self-mutilation were less frequent problems. Slow arousal is a common problem in novice stallions, and often all that is required is careful and patient handling with a quiet mare in good oestrus. Mature, experienced stallions may also suffer from poor libido and, classically, this has been associated with overuse, injury or pain during the breeding season. Other stallions may show preferences (or dislikes) for a particular handler or mare. Stallions that have a very high libido may also present a problem in a hand mating or AI setting. These stallions may present difficulty in teasing, washing or in charging the mare to be bred. Experienced stallion handlers appear to be the best option for dealing with such stallions.

Therapy for sexual behavioural disorders in stallions has been addressed only recently. Behavioural modification through careful training or retraining of stallions may be the most effective. Pasture breeding with a quiet mare in strong oestrus may be a useful method in many cases, particularly with a novice stallion. Another behavioural therapy recommended by McDonnell and co-workers includes the 'novelty' effect. A change of routine is used in an attempt to affect either a positive or negative trait. Sometimes the 'voyeur' effect can be used, in which a stallion is allowed to watch an experienced stallion during mating. In general, androgen therapy should be avoided because of the strong negative feedback effect on spermatogenesis by exogenous androgens. McDonnell cites some effect in slow breeding stallions through the use of GnRH therapy (50 μg GnRH) 2 h before breeding and again at 1 h before. The use of anxiolytic agents (diazepam 0.05 mg/kg IV – up to a maximum of 20 mg) a few minutes before breeding has also been useful in some slow breeding novice stallions.

Ejaculatory Dysfunction

A stallion that shows good libido, mounting and intromission but fails to ejaculate even after repeated mounts can be even more frustrating and perplexing than a stallion with poor libido. The two most common ejaculatory dysfunctions described by Dr McDonnell include failure of ejaculation and urine contamination of semen during ejaculation. Failure of ejaculation may occur with apparently normal libido, mounting and intromission. If the condition persists, the stallion eventually becomes frustrated and may also begin to show other behavioural abnormalities. These 'psychogenic' cases of ejaculatory dysfunction should be distinguished from ejaculatory dysfunction due to some physical cause. Factors such as hind limb lameness, back pain, pleuritis or aorto-iliac thrombosis (McDonnell *et al.*, 1992) may lead to poor intromission, or other signs of pain, during mating. Some of these stallions may vocalize during thrusting or may dismount suddenly with a complete erection.

Persistent urospermia may present with either gross contamination of the semen with urine or with more subtle changes in which the ejaculate manifests poor motility without the noticeable presence of urine. The pattern of urine contamination of the ejaculate appears to vary; however, contamination often appears to occur after ejaculation is complete. In some stallions, the possibility of neurological disease should be considered particularly if other signs of hind limb paresis are present. Imipramine therapy has been used successfully in the therapy of persistent urospermia (Turner *et al.*, 1995).

Therapy for ejaculatory dysfunction requires some attempt to characterize the underlying disorder. If physical pain is present, then the use of anti-inflammatories (phenylbutazone) may be useful. Cases that do not appear to have any physical pain present a greater challenge. Some stallions appear to benefit from careful retraining, as mentioned earlier. Pharmacological therapy may also be useful. Alpha-adrenergic agonists such as noradrenaline (0.01 mg/kg IM) or phenylpropanolamine have been used with some success. More recently, xylazine (which has both $\alpha 1$ and $\alpha 2$ effects) has been used to stimulate ejaculation without copulation in stallions. Tricyclic antidepressants have received considerable attention as a means of treating ejaculatory dysfunction in stallions. Imipramine (500 to 800 mg IV or 100–500 mg orally/day) has been used with some success for stallions with ejaculatory dysfunction (McDonnell, 2001). Both prostaglandins and oxytocin have been used on an anecdotal basis for therapy of ejaculatory dysfunction, but neither appear to be particularly useful except perhaps for stallions with ejaculatory failure due to spermiostasis and sperm accumulation in the ampulla.

Note

[1] From: Ball, BA. Diagnostic methods for evaluation of stallion subfertility: a review. *Journal of Equine Veterinary Science*. 28:650–665, 2008.

References

Aguero A, Miragaya MH, Mora NG, Chaves MG, Neild DM and Beconi MT (1995) Effect of vitamin E addition on equine sperm preservation *Comunicaciones Biologicas* **13** 343–356.

Ahmadi A and Ng SC (1999) Fertilizing ability of DNA-damaged spermatozoa *Journal of Experimental Zoology* **284** 696–704.

Aitken RJ (1995) Free radicals, lipid peroxidation and sperm function *Reproduction, Fertility and Development* **7** 659–668.

Aitken RJ, West K and Buckingham D (1994) Leukocytic infiltration into the human ejaculate and its association with semen quality, oxidative stress, and sperm function *Journal of Andrology* **15** 343–352.

Aitken RJ, Gordon E, Harkiss D, Twigg J, Milne P, Jennings Z and Irvine DS (1998) Relative impact of oxidative stress on the functional competence and genomic integrity of human spermatozoa *Biology of Reproduction* **59** 1037–1046.

Almeida J and Ball BA (2005) Effect of alpha-tocopherol and tocopherol succinate on lipid peroxidation in equine spermatozoa *Animal Reproduction Science* **87** 321–337.

Althouse GC, Seager SWJ, Varner DD and Webb GW (1989) Diagnostic aids for the detection of urine in the equine ejaculate *Theriogenology* **31** 1141–1148.

Alvarez JG and Storey BT (1983) Taurine, hypotaurine, epinephrine and albumin inhibit lipid peroxidation in rabbit spermatozoa and protect against loss of motility *Biology of Reproduction* **29** 548–555.

Amann RP and Veeramachaneni DNR (2007) Cryptorchidism in common eutherian mammals *Reproduction* **133** 541–561.

Arlas TR, Pederzolli CD, Terraciano PB, Trein CR, Bustamante-Filho IC, Castro FS and Mattos RC (2008) Sperm quality is improved feeding stallions with a rice oil supplement In *Special Issue: Proceedings of the 5th International Symposium on Stallion Reproduction* Ed T Katila *Animal Reproduction Science* **107** 306.

Aurich JE, Schonherr U, Hoppe H and Aurich C (1997) Effects of antioxidants on motility and membrane integrity of chilled–stored stallion semen *Theriogenology* **48** 185–192.

Ball BA (2008) Oxidative stress, osmotic stress and apoptosis: impacts on sperm function and preservation in the horse *Animal Reproduction Science* **107** 257–267.

Ball BA and Vo A (2002) Detection of lipid peroxidation in equine spermatozoa based upon the lipophilic fluorescent dye C-11-BODIPY581/591 *Journal of Andrology* **23** 259–269.

Ball BA, Gravance CG, Medina V, Baumber J and Liu IKM (2000) Catalase activity in equine semen *American Journal of Veterinary Research* **61** 1026–1030.

Ball BA, Medina V, Gravance CG and Baumber J (2001a) Effect of antioxidants on preservation of motility, viability and acrosomal integrity of equine spermatozoa during storage at 5°C *Theriogenology* **56** 577–589.

Ball BA, Vo AT and Baumber J (2001b) Generation of reactive oxygen species by equine spermatozoa *American Journal of Veterinary Research* **62** 508–515.

Ball BA, Conley AJ, Grundy SA, Sabeur K and Liu IKM (2008) Expression of anti-Müllerian Hormone (AMH) in the equine testis *Theriogenology* **69** 624–631.

Bartmann CP, Schoon HA, Lorber K, Brickwedel I and Klug E (1999) Testicular sonography and biopsy in the stallion – indication, techniques and diagnostic relevance *Pferdeheilkunde* **15** 506–514.

Basrur PK (2006) Disrupted sex differentiation and feminization of man and domestic animals *Environmental Research* **100** 18–38.

Baumber J and Ball BA (2005) Determination of glutathione peroxidase and superoxide dismutase-like activities in equine spermatozoa, seminal plasma, and reproductive tissues *American Journal of Veterinary Research* **66** 1415–1419.

Baumber J, Ball BA, Gravance CG, Medina V and Davies-Morel MCG (2000) The effect of reactive oxygen species on equine sperm motility, viability, acrosomal integrity, mitochondrial membrane potential and membrane lipid peroxidation *Journal of Andrology* **21** 895–902.

Baumber J, Vo A, Sabeur K and Ball BA (2002a) Generation of reactive oxygen species by equine neutrophils and their effect on motility of equine spermatozoa *Theriogenology* **57** 1025–1033.

Baumber J, Ball BA, Linfor JJ and Meyers SA (2002b) Reactive oxygen species and cryopreservation promote deoxyribonucleic acid (DNA) damage in equine sperm *Theriogenology* **58** 301–302.

Baumber J, Ball BA, Linfor JJ and Meyers SA (2003a) Reactive oxygen species and cryopreservation promote DNA fragmentation in equine spermatozoa *Journal of Andrology* **24** 621–628.

Baumber J, Sabeur K, Vo A and Ball BA (2003b) Reactive oxygen species promote tyrosine phosphorylation and capacitation in equine spermatozoa *Theriogenology* **60** 1239–1247.

Baumber J, Ball BA and Linfor JJ (2005) Assessment of the cryopreservation of equine spermatozoa in the presence of enzyme scavengers and antioxidants *American Journal of Veterinary Research* **66** 772–779.

Beck C, Charles JA and MacLean AA (2001) Ultrasound appearance of an equine testicular seminoma *Veterinary Radiology and Ultrasound* **42** 355–357.

Bergin WC, Gier HT, Marion GB and Coffman JR (1970) A developmental concept of equine cryptorchism *Biology of Reproduction* **3** 82.

Blanchard TL (1985) Some effects of anabolic steroids – especially on stallions *Compendium on Continuing Education for the Practicing Veterinarian* **7** S372–S381.

Blanchard TL, Woods JA and Brinsko SP (2000a) Theriogenology question of the month: azoospermia attributable to bilateral epididymal hypoplasia *Journal of the American Veterinary Medical Association* **217** 825–826.

Blanchard T, Varner D, Miller C and Roser J (2000b) Recommendations for clinical GnRH challenge testing of stallions *Journal of Equine Veterinary Science* **20** 678–737.

Blanchard TL, Johnson L, Varner DD, Rigby SL, Brinsko SP, Love CC and Miller C (2001) Low daily sperm output per ml of testis as a diagnostic criteria for testicular degeneration in stallions *Journal of Equine Veterinary Science* **21** 11–35.

Bosard T, Love C, Brinsko S, Blanchard T, Thompson J and Varner D (2005) Evaluation and diagnosis of acrosome function/dysfunction in the stallion *Animal Reproduction Science* **89** 215–217.

Breque C, Surai P and Brillard JP (2003) Roles of antioxidants on prolonged storage of avian spermatozoa *in vivo* and *in vitro Molecular Reproduction and Development* **66** 314–323.

Brinsko SP (1996) GnRH therapy for subfertile stallions *Veterinary Clinics of North America: Equine Practice* **12** 149–160.

Brinsko SP (1998) Neoplasia of the male reproductive tract *Veterinary Clinics of North America: Equine Practice* **14** 517–533.

Brinsko SP, Varner DD, Blanchard TL, Relford RL and Johnson L (1992) Bilateral infectious epididymitis in a stallion *Equine Veterinary Journal* **24** 325–328.

Brown JA, O'Brien MA, Hodder ADJ, Peterson T, Claes A, Liu IKM and Ball BA (2008) Unilateral testicular mastocytoma in a Peruvian Paso stallion *Equine Veterinary Education* **20** 172–175.

Brum AM, Thomas AD, Sabeur K and Ball BA (2006) Evaluation of Coomassie blue staining of the acrosome of equine and canine spermatozoa *American Journal of Veterinary Research* **67** 358–362.

Brum AM, Sabeur K and Ball BA (2008) Apoptotic-like changes in equine spermatozoa separated by density-gradient centrifugation or after cryopreservation *Theriogenology* **69** 1041–1055.

Burnaugh L, Sabeur K and Ball BA (2007) Generation of superoxide anion by equine spermatozoa as detected by dihydroethidium *Theriogenology* **67** 580–589.

Burns PJ and Douglas RH (1985) Reproductive hormone concentrations in stallions with breeding problems: case studies *Journal of Equine Veterinary Science* **5** 40–42.

Card C (2005) Cellular associations and the differential spermiogram: making sense of stallion spermatozoal morphology *Theriogenology* **64** 558–567.

Carluccio A, Zedda MT, Schiaffino GM, Pirino S and Pau S (2003) Evaluations of testicular biopsy by Tru-cut in the stallion *Veterinary Research Communications* **27** 211–213.

Carneiro GF, Liu IKM, Illera JC and Munro CJ (1998) Enzyme immunoassay for the measurement of estrone sulfate in cryptorchids, stallions and donkeys *Proceedings of the American Association of Equine Practitioners* **44** 3–4.

Cerilli LA, Kuang W and Rogers D (2010) A practical approach to testicular biopsy interpretation for male infertility *Archives of Pathology and Laboratory Medicine* **134** 1197–1204.

Claes A, Ball, BA, Almeida J and Conley AJ (2011) Detection of serum Anti-Müllerian hormone concentrations as a method for diagnosis of cryptorchidism in the horse *Proceedings of the American Associaton of Equine Practitioners* **57** 56 [abstract].

Cox JE (1999) Disturbed testicular descent in horses – principles, diagnosis and therapy *Pferdeheilkunde* **15** 503–505.

Cox JE, Edwards GB and Neal PA (1979) An analysis of 500 cases of equine cryptorchidism *Equine Veterinary Journal* **11** 113–116.

Cox JE, Redhead PH and Dawson FE (1986) Comparison of the measurement of plasma testosterone and plasma oestrogens for the diagnosis of cryptorchidism in the horse *Equine Veterinary Journal* **18** 179–182.

Das PJ, Paria N, Gustafson-Seabury A, Vishnoi M, Chaki SP, Love CC, Varner DD, Chowdhary BP and Raudsepp T (2010) Total RNA isolation from stallion sperm and testis biopsies *Theriogenology* **74** 1099–1106.

Dascanio JJ and Witonsky SG (2005) Theriogenology question of the month: urospermia *Journal of the American Veterinary Medical Association* **227** 225–227.

DelVento VR, Amann RP, Trotter GW, Veeramachaneni DNR and Squires EL (1992) Ultrasonographic and quantitative histologic assessment of sequelae to testicular biopsy in stallions *American Journal of Veterinary Research* **53** 2094–2101.

Demircan M, Akinci A and Mutus M (2006) The effects of orchiopexy on serum anti-Müllerian hormone levels in unilateral cryptorchid infants *Pediatric Surgery International* **22** 271–273.

Devireddy RV, Swanlund DJ, Alghamdi AS, Duoos LA, Troedsson MHT, Bischof JC and Roberts KP (2002) Measured effect of collection and cooling conditions on the motility and the water transport parameters at subzero temperatures of equine spermatozoa *Reproduction* **124** 643–648.

Dobrinski I, Thomas PGA and Ball BA (1995) Cryopreservation reduces the ability of equine spermatozoa to attach to oviductal epithelial cells and zonae pellucdiae *in vitro Journal of Andrology* **16** 536–542.

Douglas RH and Umphenour N (1992) Endocrine abnormalities and hormonal therapy *Veterinary Clinics of North America: Equine Practice* **8** 237–249.

Dowsett KF and Pattie WA (1982) Characteristics and fertility of stallion semen *Journal of Reproduction and Fertility Supplement* **32** 1–8.

Edwards JF (2008) Pathologic conditions of the stallion reproductive tract *Animal Reproduction Science* **107** 197–207.

Estrada A, Samper JC, Lillich JD, Rathi R, Brault LS, Albrecht BA, Imel MM and Senne EM (2003) Azoospermia associated with bilateral segmental aplasia of the ductus deferens in a stallion *Journal of the American Veterinary Medical Association* **222** 1740–1742.

Evenson DP, Larson KL and Jost LK (2002) Sperm chromatin structure assay: its clinical use for detecting sperm DNA fragmentation in male infertility and comparisons with other techniques *Journal of Andrology* **23** 25–43 [review].

Faber NF and Roser JF (2000) Testicular biopsy in stallions: diagnostic potential and effects on prospective fertility *Journal of Reproduction and Fertility Supplement* **56** 31–42.

Fenichel P, Rey R, Poggioli S, Donzeau M, Chevallier D and Pointis G (1999) Anti-Müllerian hormone as a seminal marker for spermatogenesis in non-obstructive azoospermia *Human Reproduction* **14** 2020–2024.

Foote RH, Brockett CC and Kaproth MT (2002) Motility and fertility of bull sperm in whole milk extender containing antioxidants *Animal Reproduction Science* **71** 13–23.

Galina CS (1971) An evaluation of testicular biopsy in farm animals *Veterinary Record* **88** 628.

Garcia MC, Ganjam VK, Blanchard TL, Brown E, Hardin K, Elmore RG, Youngquist RS, Loch WE, Ellersieck MR and Balke JM (1987) The effects of stanozolol and boldenone undecylenate on plasma testosterone and gonadotropins and on testis histology in pony stallions *Theriogenology* **28** 109–119.

Gebauer MR, Pickett BW, Voss JL and Swierstra EE (1974) Reproductive physiology of the stallion: daily sperm output and testicular measurements *Journal of the American Veterinary Medical Association* **165** 711–713.

Gee EK, Bruemmer JE, Siciliano PD, McCue PM and Squires EL (2008) Effects of dietary vitamin E supplementation on spermatozoal quality in stallions with suboptimal post-thaw motility *Animal Reproduction Science* **107** 324–325.

Giesecke K, Sieme H and Distl O (2010) Infertility and candidate gene markers for fertility in stallions: a review *The Veterinary Journal* **185** 265–271.

Ginther OJ (1995) Stallion In *Ultrasonic Imaging and Animal Reproduction: Horses – Book 2* pp 309–336 Ed OJ Ginther. Equiservices, Cross Plains, Wisconsin.

Ginther OJ (2007) Blood flow of male genitalia In *Ultrasonic Imaging and Animal Reproduction: Color-doppler ultrasonography – Book 4* pp 205–224 Ed OJ Ginther. Equiservices, Cross Plains, Wisconsin.

Gonzalez M, Tibary A, Sellon DC and Daniels J (2008) Unilateral orchitis and epididymitis caused by *Corynebacterium pseudotuberculosis* in a stallion *Equine Veterinary Education* **20** 30–36.

Gravance CG, Garner DL, Baumber J and Ball BA (2000) Assessment of equine sperm mitochondrial function using JC-1 *Theriogenology* **53** 1691–1703.

Griggers S, Paccamonti DL, Thompson RA and Eilts BE (2001) The effects of pH, osmolarity and urine contamination on equine spermatozoal motility *Theriogenology* **56** 613–622.

Großfeld R, Sieg B, Struckmann C, Frenzel A, Maxwell WMC and Rath D (2008) New aspects of boar semen freezing strategies *Theriogenology* **70** 1225–1233.

Halliwell B and Gutteridge JMC (1999) *Free Radicals in Biology and Medicine* 3rd edition. Oxford University Press, New York.

Hamann H, Jude R, Sieme H, Mertens U, Topfer-Petersen E, Distl O and Leeb T (2007) A polymorphism within the equine *CRISP3* gene is associated with stallion fertility in Hanoverian warmblood horses *Animal Genetics* **38** 259–264.

Hayes HM (1986) Epidemiological features of 5009 cases of equine cryptorchidism *Equine Veterinary Journal* **18** 468–471.

Held JP, Adair S, McGavin MD, Adams WH, Toal R and Henton J (1990) Bacterial epididymitis in two stallions *Journal of the American Veterinary Medical Association* **197** 602–604.

Huff DS, Fenig DM, Canning DA, Carr MG, Zderic SA and Snyder HM III (2001) Abnormal germ cell development in cryptorchidism *Hormone Research in Paediatrics* **55** 11–17.

Hunt RJ, Hay W, Collatos C and Welles E (1990) Testicular seminoma associated with torsion of the spermatic cord in two cryptorchid stallions *Journal of the American Veterinary Medical Association* **197** 1484–1486.

Hurtgen JP and Johnson LA (1982) Fertility of stallions with abnormalities of the sperm acrosome *Journal of Reproduction and Fertility Supplement* **32** 15–20.

Hyland J and Church S (1995) The use of ultrasonograph in the diagnosis and treatment of a haematoma in the corpus cavernosum penis of a stallion *Australian Veterinary Journal* **72** 468–469.
Jann HW and Rains JR (1990) Diagnostic ultrasonography for evaluation of cryptorchidism in horses *Journal of the American Veterinary Medical Association* **196** 297–300.
Jasko DJ, Lein DH and Foote RH (1990) Determination of the relationship between sperm morphologic classifications and fertility in stallions: 66 cases (1987–1988) *Journal of the American Veterinary Medical Association* **197** 389–394.
Johansson CS, Matsson FC, Lehn-Jensen H, Nielsen JM and Petersen MM (2008) Equine spermatozoa viability comparing the NucleoCounter SP-100 and the eosin–nigrosin stain *Animal Reproduction Science* **107** 325–326.
Kenney RM, Kingston RS, Rajamannon AH and Ramberg CF (1971) Stallion semen characteristics for predicting fertility *Proceedings of the Annual Convention of the American Association of Equine Practitioners* **53** 53–67.
Kenney RM, Hurtgen JP, Pierson R, Witherspoon DM and Simons J (1983) Clinical fertility evaluation of the stallion *Journal of the Society for Theriogenology* **9** 1–100.
Kenney RM, Kent MG, Garcia MC and Hurtgen JP (1991) The use of DNA index and karyotype analyses as adjuncts to the estimation of fertility in stallions *Journal of Reproduction and Fertility Supplement* **44** 69–75.
Kenney RM, Evenson DP, Garcia MC and Love C (1995) Relationship between sperm chromatin structure, motility, and morphology of ejaculated sperm and seasonal pregnancy In *Biology of Reproduction Monograph Series 1: Equine Reproduction VI (Papers from the Sixth International Symposium on Equine Reproduction held in Caxambu, Brazil, August 7–13, 1994)* pp 647–654 Ed DC Sharp. Society for the Study of Reproduction, Madison.
Kenney RM, Cummings MR, Teuscher C and Love CC (2000) Possible role of autoimmunity to spermatozoa in idiopathic infertility of stallions *Journal of Reproduction and Fertility Supplement* **56** 23.
Kovalski NN, De Lamirande E and Gagnon C (1992) Reactive oxygen species generated by human neutrophils inhibit sperm motility: protective effect of seminal plasma and scavengers *Fertility and Sterility* **58** 809–816.
Lee MM, Misra M, Donahoe PK and MacLaughlin DT (2003) MIS/AMH in the assessment of cryptorchidism and intersex conditions *Molecular and Cellular Endocrinology* **211** 91–98.
Leeb T (2007) The Horse Genome Project – sequence based insights into male reproductive mechanisms *Reproduction in Domestic Animals* **42** 45–50.
Leme DP and Papa FO (2000) Cytological identification and quantification of testicular cell types using fine needle aspiration in horses *Equine Veterinary Journal* **32** 444–446.
Linfor J and Meyers SA (2002) Assessing DNA integrity of cooled and frozen–thawed equine sperm using single cell gel electrophoresis *Journal of Andrology* **23** 107–113.
Little TV and Woods GL (1987) Ultrasonography of the accessory sex glands in the stallion *Journal of Reproduction and Fertility Supplement* **35** 87–94.
Love CC (1992) Ultrasonographic evaluation of the testis, epididymis, and spermatic cord of the stallion *Veterinary Clinics of North America: Equine Practice* **8** 167–182.
Love CC (2003) The role of breeding record evaluation in the evaluation of the stallion for breeding soundness In *Proceedings of the Annual Meeting of Society for Theriogenology 2003* pp 68–77. Available at: http://st.omnibooksonline.com/data/papers/2003/20.pdf (accessed 6 September 2013).
Love CC (2011a) Relationship between sperm motility, morphology and the fertility of stallions *Theriogenology* **76** 547–557.
Love CC (2011b) The stallion breeding soundness evaluation: revisited *Clinical Theriogenology* **3** 294–300.
Love CC and Kenney RM (1998) The relationship of increased susceptibility of sperm DNA to denaturation and fertility in the stallion *Theriogenology* **50** 955–972.
Love CC and Kenney RM (1999) Scrotal heat stress induces altered sperm chromatin structure associated with a decrease in protamine disulfide bonding in the stallion *Biology of Reproduction* **60** 615–620.
Love CC, Garcia MC, Riera FR and Kenney RM (1991) Evaluation of measures taken by ultrasonography and caliper to estimate testicular volume and predict daily sperm output in the stallion *Journal of Reproduction and Fertility Supplement* **44** 99–105.
Love CC, Reira FL, Oristaglio-Turner RM and Kenney RM (1992) Sperm occluded (plugged) ampullae in the stallion In *Proceedings of the Annual Meeting of the Society for Theriogenology 1992* pp 117–127.
MacLeod J (1943) The role of oxygen in the metabolism and motility of human spermatozoa *American Journal of Physiology* **138** 512–518.

Mann T, Minotakis CS and Polge C (1963) Semen composition and metabolism in the stallion and jackass *Journal of Reproduction and Fertility* **5** 109–122.

Marin-Guzman J, Mahan DC, Chung YK, Pate JL and Pope WF (1997) Effects of dietary selenium and vitamin E on boar performance and tissue responses, semen quality, and subsequent fertilization rates in mature gilts *Journal of Animal Science* **75** 2994–3003.

Marino G, Zanghi A, Quartuccio M, Cristarella S, Giuseppe M and Catone G (2009) Equine testicular lesions related to invasion by nematodes *Journal of Equine Veterinary Science* **29** 728–733.

McDonnell SM (1986) Reproductive behaviour of the stallion *Veterinary Clinics of North America: Equine Practice* **2** 535–555.

McDonnell SM (1992) Ejaculation. Physiology and dysfunction *Veterinary Clinics of North America: Equine Practice* **8** 57–70.

McDonnell SM (1995) Stallion behaviour and endocrinology: what do we really know? In *Proceedings of the 41st Annual Meeting of the American Association of Equine Practitioners, Lexington, Kentucky, December 1995* pp 18–19.

McDonnell SM (2001) Oral imipramine and intravenous xylazine for pharmacologically-induced ex copula ejaculation in stallions *Animal Reproduction Science* **68** 153–159.

McDonnell S, Garcia M and Kenney R (1987) Imipramine-induced erection, masturbation, and ejaculation in male horses *Pharmacology Biochemistry and Behaviour* **27** 187–191.

McDonnell SM, Love CC, Martin BB, Reef VB and Kenney RM (1992) Ejaculatory failure associated with aortic-iliac thrombosis in two stallions *Journal of the American Veterinary Medical Association* **200** 954–957.

McEntee KE (1990) *Reproductive Pathology of Domestic Animals.* Academic Press, San Diego.

Meng MV, Cha I, Ljung BM and Turek PJ (2001) Testicular fine-needle aspiration in infertile men: correlation of cytologic pattern with biopsy histology *The American Journal of Surgical Pathology* **25** 71–79.

Meyers SA, Liu IKM, Overstreet JW, Vadas S and Drobnis EZ (1996) Zona pellucida binding and zona-induced acrosome reactions in horse spermatozoa: comparisons between fertile and subfertile stallions *Theriogenology* **46** 1277–1288.

Morresey PR (2007) The enlarged scrotum *Clinical Techniques in Equine Practice* **6** 265–270.

Morris ID, Ilott S, Dixon L and Brison DR (2002) The spectrum of DNA damage in human sperm assessed by single cell gel electrophoresis (Comet assay) and its relationship to fertilization and embryo development *Human Reproduction* **17** 990–998.

Neild DM, Brouwers JFHM, Colenbrander B, Aguero A and Gadella BM (2005) Lipid peroxide formation in relation to membrane stability of fresh and frozen thawed stallion spermatozoa *Molecular Reproduction and Development* **72** 230–238.

Parks JE and Lynch DV (1992) Lipid composition and thermotropic phase behaviour of boar, bull, stallion, and rooster sperm membranes *Cryobiology* **29** 255–266.

Pearson LK, Rodriguez JS and Tibary A (2011) How to obtain a stallion testicular biopsy using a spring-loaded split-needle biopsy instrument In *Proceedings of the 57th Annual Convention of the American Association of Equine Practitioners, San Antonio, Texas, USA, 18–22 November 2011* pp 219–225. American Association of Equine Practitioners, Lexington.

Pickett BW, Voss JL, Bowen RA, Squires EL and McKinnon AO (1987) Seminal characteristics and total scrotal width (T.S.W.) of normal and abnormal stallions In *Proceedings of the 33rd Annual Convention of the American Association of Equine Practitioners, New Orleans, Louisiana, November 29th Through December 2nd, 1987* pp 487–518. American Association of Equine Practitioners, Lexington.

Pickett BW, Amann RP, McKinnon AO, Squires EL and Voss JL (1989) *Management of the Stallion for Maximum Reproductive Efficiency, II*, 2nd edition. Colorado State University, Fort Collins.

Pinart E, Sancho S, Briz MD, Bonet S, Garcia N and Badia E (2000) Ultrastructural study of the boar seminiferous epithelium: changes in cryptorchidism *Journal of morphology* **244** 190–202.

Pinart E, Bonet S, Briz M, Pastor LM, Sancho S, García N, Badia E and Bassols J (2002) Histochemical study of the interstitial tissue in scrotal and abdominal boar testes *The Veterinary Journal* **163** 68–76.

Pollock PJ, Prendergas M, Callanan JJ and Skelly C (2002) Testicular teratoma in a three-day-old thoroughbred foal *Veterinary Record* **150** 348–350.

Pozor M (2005) Diagnostic applications of ultrasonography to stallion's reproductive tract *Theriogenology* **64** 505–509.

Pozor MA (2007) Evaluation of testicular vasculature in stallions *Clinical Techniques in Equine Practice* **6** 271–277.

Pozor MA and Garncarz M (2006) Bilateral epididymal nodules in a stallion *Equine Veterinary Education* **18** 63–66.

Pozor MA and McDonnell SM (2004) Color Doppler ultrasound evaluation of testicular blood flow in stallions *Theriogenology* **61** 799–810.

Raphael CF, Andrade AFC, Nascimento J and Arruda RP (2008) Effects of centrifugation on membrane integrity and lipid peroxidation of equine cooled spermatozoa *Animal Reproduction Science* **107** 344–345.

Roser JF (1995) Endocrine profiles in fertile, subfertile and infertile stallions: testicular response to human chorionic gonadotropin in infertile stallions In *Biology of Reproduction Monograph Series 1: Equine Reproduction VI (Papers from the Sixth International Symposium on Equine Reproduction held in Caxambu, Brazil, August 7–13, 1994)* pp 661–669 Ed DC Sharp. Society for the Study of Reproduction, Madison.

Roser JF (2001) Endocrine diagnostics for stallion infertility. In *Recent Advances in Equine Reproduction* Ed BA Ball. International Veterinary Information Service (IVIS), Ithaca, New York. Available at: http://www.ivis.org/advances/reproduction_ball/stallion_fertility_roser/ivis.pdf (accessed 6 September 2013).

Roser JF (2008) Regulation of testicular function in the stallion: an intricate network of endocrine, paracrine and autocrine systems *Animal Reproduction Science* **107** 179–196.

Roser JF and Faber NF (2007) Testicular biopsy In *Current Therapy in Equine Reproduction* pp 205–211 Ed JC Samper, JF Pycock and AO McKinnon. WB Saunders, St Louis.

Roser JF and Hughes JP (1992a) Seasonal effects on seminal quality, plasma hormone concentrations, and GnRH-induced LH response in fertile and subfertile stallions *Journal of Andrology* **13** 214–223.

Roser JF and Hughes JP (1992b) Dose-response effects of gonadotropin-releasing hormone on plasma concentrations of gonadotropins and testosterone in fertile and subfertile stallions *Journal of Andrology* **13** 543–550.

Roser JF, McCue PM and Hoye E (1994) Inhibin activity in the mare and stallion *Domestic Animal Endocrinology* **11** 87–100.

Rousset H, Chanteloube P, Magistrini M and Palmer E (1987) Assessment of fertility and semen evaluations of stallions *Journal of Reproduction and Fertility Supplement* **35** 25–31.

Sabeur K and Ball BA (2006) Detection of superoxide anion generation by equine spermatozoa *American Journal of Veterinary Research* **67** 701–706.

Sabeur K and Ball BA (2007) Characterization of NADPH oxidase 5 in equine testis and spermatozoa *Reproduction* **134** 263–270.

Schambourg MA, Farley JA, Marcoux M and Laverty S (2006) Use of transabdominal ultrasonography to determine the location of cryptorchid testes in the horse *Equine Veterinary Journal* **38** 242–245.

Schneider RK, Milne DW and Kohn CW (1982) Acquired inguinal hernia in the horse: a review of 27 cases *Journal of the American Veterinary Medical Association* **180** 317–320.

Schönbauer M and Schönbauer-Längle A (1983) Seminome beim Pferd Eine Retrospektivuntersuchung [Seminomas in the horse: a retrospective study] *Zentralblatt für Veterinärmedizin A* **30** 189–198.

Sharrock AG (1982) Reversal of drug-induced priapism in a gelding by medication *Australian Veterinary Journal* **58** 39–40.

Sinisi AA, Esposito D, Maione L, Quinto MC, Visconti D, De BA, Bellastella A, Conzo G and Bellastella G (2008) Seminal anti-Müllerian hormone level is a marker of spermatogenic response during long-term gonadotropin therapy in male hypogonadotropic hypogonadism *Human Reproduction* **23** 1029–1034.

Smith JA (1974) Biopsy and the testicular artery of the horse *Equine Veterinary Journal* **6** 81–83.

Smyth GB (1979) Testicular teratoma in an equine cryptorchid *Equine Veterinary Journal* **11** 21–23.

Stewart BL and Roser JF (1998) Effects of age, season, and fertility status on plasma and intratesticular immunoreactive (IR) inhibin concentrations in stallions *Domestic Animal Endocrinology* **15** 129–139.

Stick JA (1980) Teratoma and cyst formation of the equine cryptorchid testicle *Journal of the American Veterinary Medical Assocociation* **176** 211–214.

Stickle RL and Fessler JF (1978) Retrospective study of 350 cases of equine cryptorchidism *Journal of the American Veterinary Medical Assocociation* **172** 343–346.

Sullins KE, Bertone JJ, Voss JL and Pederson SJ (1988) Treatment of hemospermia in stallions: a discussion of 18 cases *Compendium on Continuing Education for the Practicing Veterinarian* **10** 1395–1403.

Thiele JJ, Friesleben HJ, Fuchs J and Ochsendorf FR (1995) Ascorbic acid and urate in human seminal plasma: determination and interrelationships with chemiluminescence in washed semen *Human Reproduction* **10** 110–115.

Thomas AD, Meyers SA and Ball BA (2006) Capacitation-like changes in equine spermatozoa following cryopreservation *Theriogenology* **65** 1531–1550.

Thomas PGA, Ball BA, Miller PG, Brinsko SP and Southwood L (1994) A subpopulation of morphologically normal, motile spermatozoa attach to equine oviduct epithelial cells *in vitro Biology of Reproduction* **51** 303–309.

Thompson DL Jr, Pickett BW, Squires EL and Amann RP (1979) Testicular measurements and reproductive characteristics of stallions *Journal of Reproduction and Fertility Supplement* **27** 13–17.

Thompson JA, Love CC, Stich KL, Brinsko SP, Blanchard TL and Varner DD (2004) A Bayesian approach to prediction of stallion daily sperm output *Theriogenology* **62** 1607–1617.

Threlfall WR and Lopate C (1993) Testicular biopsy In *Equine Reproduction* pp 943–949 Ed AO McKinnon and JL Voss. Saunders, Philadelphia.

Threlfall WR, Carleton CL, Robertson J, Rosol T and Gabel A (1990) Recurrent torsion of the spermatic cord and scrotal testis in a stallion *Journal of the American Veterinary Medical Association* **196** 1641–1643.

Traub-Dargatz JL, Trotter GW, Kaser-Hotz B, Bennett DG, Kiper ML, Veeramachaneni DNR and Squires E (1991) Ultrasonographic detection of chronic epididymitis in a stallion *Journal of the American Veterinary Medical Association* **198** 1417–1420.

Trigo FJ, Miller RA and Torbeck RL (1984) Metastatic equine seminoma: report of two cases *Veterinary Pathology* **21** 259–260.

Turner RMO and McDonnell SM (2003) Alkaline phosphatase in stallion semen: characterization and clinical applications *Theriogenology* **60** 1–10.

Turner RM[O], Love CC, McDonnell SM, Sweeney RW, Twitchell ED, Habecker PL, Reilly LK, Pozor MA and Kenney RM (1995) Use of imipramine hydrochloride for treatment of urospermia in a stallion with a dysfunctional bladder *Journal of the American Veterinary Medical Association* **207** 1602–1606.

Van der Velden MA (1988) Surgical treatment of acquired inguinal hernia in the horse: a review of 51 cases *Equine Veterinary Journal* **20** 173–177.

Varner DD (2008) Developments in stallion semen evaluation *Theriogenology* **70** 448–462.

Varner DD, Schumacher J, Blanchard TL and Johnson L (1991) *Diseases and Management of Breeding Stallions.* American Veterinary Publications, Goleta.

Varner DD, Blanchard TL, Brinsko SP, Love CC, Taylor TS and Johnson L (2000) Techniques for evaluating selected reproductive disorders of stallions *Animal Reproduction Science* **60–61** 493–509.

Veeramachaneni DN and Sawyer HR (1996) Use of semen as biopsy material for assessment of health status of the stallion reproductive tract *Veterinary Clinics of North America: Equine Practice* **12** 101.

Veeramachaneni DNR, Moeller CL and Sawyer HR (2006) Sperm morphology in stallions: ultrastructure as a functional and diagnostic tool *Veterinary Clinics of North America: Equine Practice* **22** 683–692.

Weber JA, Geary RT and Woods GL (1990) Changes in accessory sex glands of stallions after sexual preparation and ejaculation *Journal of the American Veterinary Medical Association*, **196** 1084–1089.

Wilson DV, Nickels FA and Williams MA (1991) Pharmacologic treatment of priapism in two horses *Journal of the American Veterinary Medical Association* **199** 1183–1184.

12 Applied Andrology in Cattle (*Bos taurus*)

Leonardo F.C. Brito*
ABS Global, Inc., DeForest Wisconsin, USA

Introduction

Application of knowledge on bovine andrology can increase reproductive efficiency, which is an imperative for the cattle sector sustainability in a world with increasing protein demand and ever more scarce resources. Within this context, the objective of this chapter is to provide a summary of the available literature to aid dedicated professionals in their quest to feed the world.

Sexual Development

Hormonal control of sexual development

The process of sexual development in bulls involves complex maturation mechanisms of the hypothalamic-pituitary-testicular axis. Sexual development can be divided into three periods according to changes in gonadotrophins and testosterone concentrations, namely the infantile, prepubertal and pubertal periods. The infantile period is characterized by low gonadotrophin and testosterone secretion and extends from birth to approximately 2 months of age. A transient increase in gonadotrophin secretion occurs from approximately 2 to 6 months of age; this so-called 'early gonadotrophin rise' characterizes the prepubertal period, during which testosterone concentrations also begin to rise. The pubertal period corresponds to the period of accelerated reproductive development that occurs after 6 months of age and until puberty. During this period, gonadotrophin secretion decreases, whereas testosterone secretion continues to increase (Rawlings *et al.*, 1978; Lacroix and Pelletier, 1979; McCarthy *et al.*, 1979a,b; Amann and Walker, 1983; Amann *et al.*, 1986).

Although the timing of sexual development is determined primarily by the hypothalamus and gonadotrophin-releasing hormone (GnRH) secretion, the mechanisms regulating GnRH secretion during sexual development in bulls are poorly understood. Gonadotrophin concentrations during the infantile period are low mainly due to reduced GnRH secretion. Maturation changes within the hypothalamus increase the pulse secretion of GnRH, driving the transition from the infantile to the prepubertal period of development through increased secretion of gonadotrophins. Increased GnRH secretion is dependent on either the development of central stimulatory inputs or the removal of inhibitory inputs. The weight of the hypothalamus and its GnRH content do not increase during the infantile period, but

* E-mail: leo.brito@genusplc.com

 Animal Andrology: Theories and Applications (eds P.J. Chenoweth and S.P. Lorton) hypothalamic concentrations of oestradiol receptors decrease after 1 month of age, leading to suggestions that reduced sensitivity to sex steroids could be involved in the augmented GnRH secretion in bulls (Amann *et al.*, 1986).

However, the hypothesis that gonadotrophin secretion is low during infancy due to elevated sensitivity of the hypothalamus to the negative feedback of sex steroids (the gonadostat hypothesis) has been questioned in bulls, because castration did not alter the pulse frequency of luteinizing hormone (LH) or its mean concentrations before 2 months of age (Wise *et al.*, 1987). Nevertheless, as GnRH secretion into hypophyseal portal blood was not necessarily accompanied by LH secretion during the infantile period in bulls, experiments that use LH concentrations to infer GnRH secretion patterns during the infantile period in calves should be interpreted with caution (Rodriguez and Wise, 1989). Another possibility is that removal of opioid inhibition and/or increased dopaminergic activity may be involved in triggering the increase in GnRH secretion during the infantile period. Opioid inhibition of LH pulse frequency during the infantile period was demonstrated by increased LH secretion between 1 and 4 months of age in Hereford calves treated with naxolone, an opioid receptor competitive antagonist (Evans *et al.*, 1993), whereas concentrations of noradrenaline, dopamine and dopamine metabolites increased twofold to threefold in the anterior hypothalamic-preoptic area in Holstein calves between 0.5 and 2.5 months of age (Rodriguez *et al.*, 1993).

The direct evaluation of blood samples from the hypophyseal portal system demonstrated that GnRH pulsatile secretion increased linearly from 2 weeks (3.5 pulses/10 h) to 12 weeks of age (8.9 pulses/10 h) in bull calves. Although GnRH secretion into hypophyseal portal blood was detected at 2 weeks, pulsatile LH secretion was not detected in jugular blood samples before 8 weeks of age. In addition, GnRH pulses were not necessarily accompanied by LH secretion until 8–12 weeks of age, when all GnRH pulses resulted in LH pulses. The increase of pulsatile GnRH release from 2 to 8 weeks of age without a concomitant increase in LH secretion may represent a reduced ability of the pituitary gland to respond to stimulus by GnRH (Rodriguez and Wise, 1989). The period in which GnRH pulses did not stimulate LH secretion corresponded to the period during which there was an increase in pituitary weight, GnRH receptor concentration and LH content (Amann *et al.*, 1986). Moreover, frequent GnRH treatments during the infantile period in calves increased pituitary mRNA for the LH beta subunit (LH-β), LH content and GnRH receptors, with resulting increases in LH pulse frequency and mean concentrations (Rodriguez and Wise, 1991), indicating that increased GnRH pulse frequency results in increased pituitary sensitivity to GnRH. With time, the increased GnRH secretion results in the increased LH pulse frequency observed during the prepubertal period.

During the early gonadotrophin rise that is characteristic of the prepubertal period, there is a transient increase in LH and follicle-stimulating hormone (FSH) concentrations from approximately 2 to 6 months of age (Figs 12.1 and 12.2). Concentrations decrease thereafter and remain at levels only slightly greater than those observed during the infantile period. The main factor responsible for increased gonadotrophin concentrations is the increase in GnRH pulse secretion, as demonstrated by a dramatic increase in LH pulse frequency. The number of LH pulses increased from less than one a day at 1 month of age to approximately 12/day ($\geq$1 pulse/2 h) at approximately 4 months of age. Changes in pulse amplitude during this period were not consistent among reports; amplitude was either reduced, unchanged or augmented. Pulsatile discharges of FSH have been observed in bulls, but are much less evident than those of LH (Amann and Walker, 1983; Amann *et al.*, 1986; Evans *et al.*, 1995, 1996; Aravindakshan *et al.*, 2000).

LH binding sites in testicular interstitial tissue were present in bulls at birth and at 4 months of age, and pulsatile LH secretion is an essential requirement for Leydig cell proliferation and differentiation, and for the maintenance of fully differentiated structure and function (Schanbacher, 1979). Hypophysectomy, the suppression of gonadotrophins by steroid administration, or neutralization of

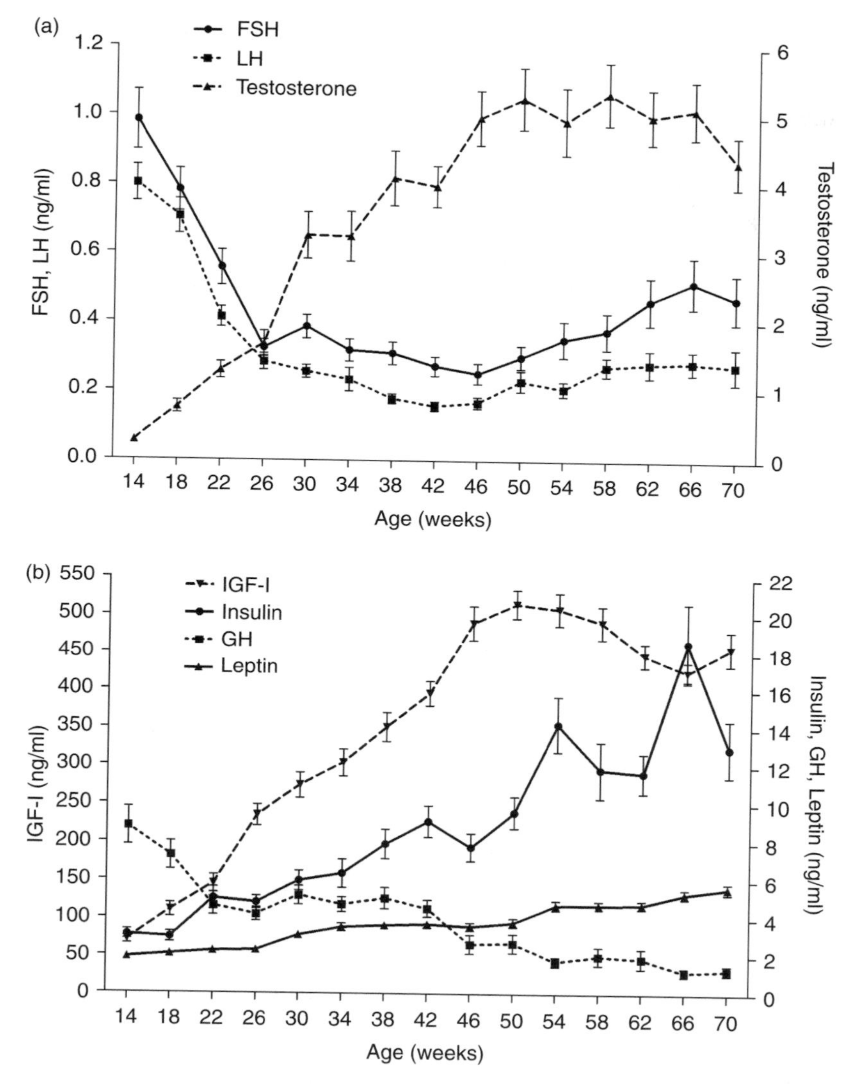

Fig. 12.1. (a) Mean (± SEM) gonadotrophins (FSH, LH) and testosterone, and (b) metabolic hormones (IGF-I, insulin, GH and leptin) concentrations during sexual development in Angus and Angus × Charolais bulls receiving adequate nutrition. FSH (follicle-stimulating hormone), LH (luteinizing hormone), testosterone, IGF-I (insulin growth factor 1) and insulin: *n* (no. of bulls) = 39–62 depending on age, GH (growth hormone) and leptin: *n* = 22–46 depending on age (L. Brito, unpublished data).

GnRH/LH by specific antibodies, caused Leydig cell atrophy and loss of cellular volume, reduction in the number of LH receptors and steroidogenic enzyme activity, and a decrease in the ability to secrete testosterone in response to LH. The characteristic pulsatile nature of LH secretion is important for testosterone production, because continuous exposure of Leydig cells to LH resulted in reduced steroidogenic responsiveness owing to the downregulation of LH receptors (Saez, 1994). The initiation of Leydig cell steroidogenesis is characterized by increased androstenedione secretion, which decreases as the cells complete maturation and begin secreting testosterone. During the first 3 to 4 months of age, testosterone concentrations

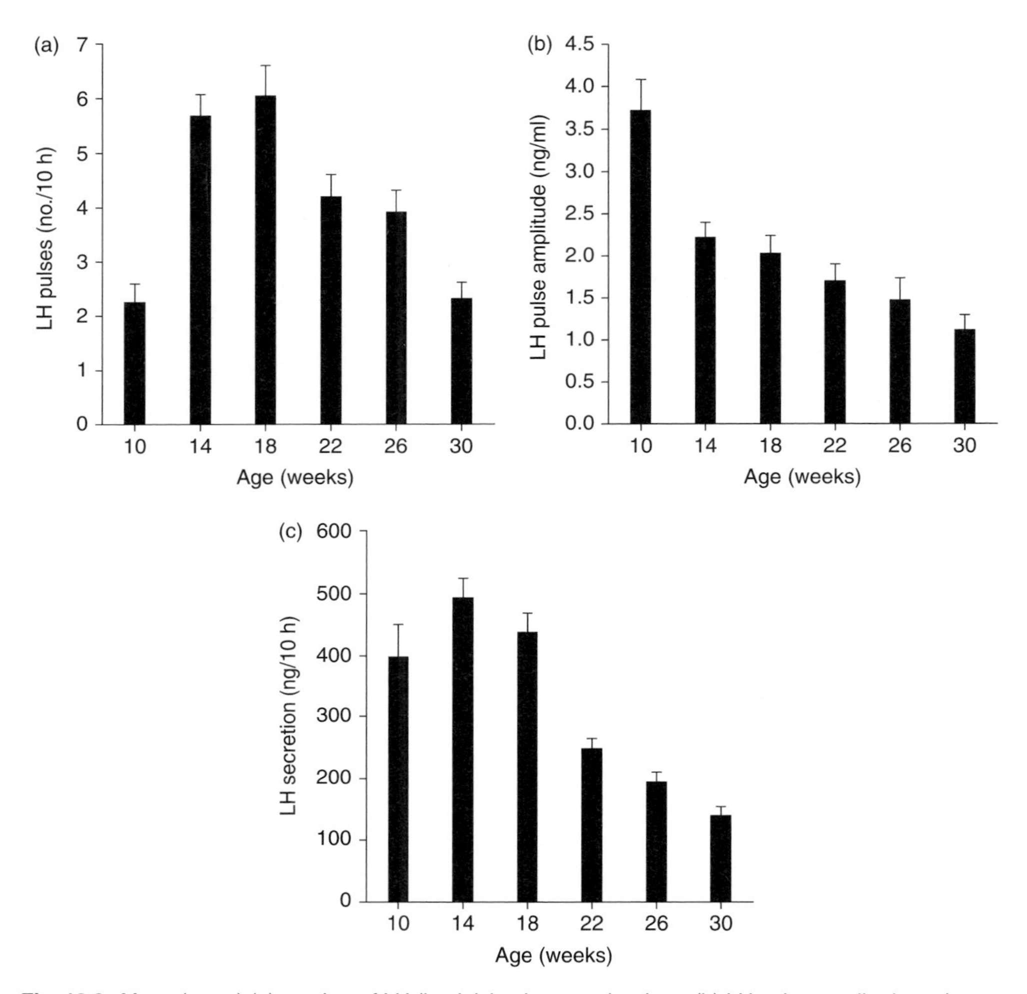

Fig. 12.2. Mean (± SEM) (a) number of LH (luteinizing hormone) pulses, (b) LH pulse amplitude and (c) total LH secretion over 10 h during sexual development in Angus and Angus × Charolais bulls receiving adequate nutrition; *n* (no. of bulls) = 15–39 depending on age (L. Brito, unpublished data).

are low and secretion does not necessarily accompany LH pulses. After this age, LH pulses are followed by testosterone pulses and mean testosterone concentrations began to increase (Fig. 12.1). Various studies have shown that the number of testosterone pulses increased from 0.3–2.3 pulses/24 h at 1–4 months of age to 9–7.5 pulses/24 h at 5 months of age (Rawlings *et al.*, 1972, 1978; Lacroix and Pelletier, 1979; McCarthy *et al.*, 1979b).

FSH binding sites in seminiferous tubules were detected in bull calves at birth and at 4 months of age, and increased FSH concentrations stimulated the proliferation of undifferentiated Sertoli cells and gonocytes (Schanbacher, 1979). While there is considerable evidence that FSH is essential for normal Sertoli cell function, the period of Sertoli cell differentiation coincides with the initiation of testosterone secretion by the Leydig cells, indicating that testosterone may be involved in promoting the maturation of undifferentiated Sertoli cells. Maturation of Sertoli cells and increased testosterone secretion are probably also involved in the differentiation of gonocytes into spermatogonia. The end of the prepubertal period is marked by completion of Sertoli cell differentiation, with the establishment of the blood–testis barrier, formation of the tubular lumen and initiation of germ cell meiosis (Amann, 1983). In Holstein calves, circulating inhibin concentrations either increased around 3 months of age before decreasing, or decreased continuously after birth until 6.5–8 months of age (MacDonald *et al.*, 1991; Matsuzaki *et al.*, 2001).

The crucial role of the early gonadotrophin rise (especially the LH secretion pattern) in regulating sexual development in bulls has been demonstrated in several studies, using a variety of approaches. Prolonged treatment with a GnRH agonist in calves from 1.5 to 3.5 months of age decreased LH and FSH pulse frequency, pulse amplitude and mean concentrations at 3 months of age, delayed the peak mean LH concentration from 5 to 6 months of age, and reduced FSH and testosterone concentrations from 3.5 to 4.5 months of age. These hormonal alterations were associated with delayed puberty and reduced paired testes weight and number of germ cells in tubular cross sections at 11.5 months of age. Conversely, treatment with GnRH every 2 h to mimic pulsatile secretion from 1 to 1.5–2 months of age increased LH pulse frequency and mean concentration during the treatment period, and also resulted in greater scrotal circumference (SC), paired testes weight, seminiferous tubule diameter, and number of germ and Sertoli cells in tubular cross sections at 12 months of age (Chandolia *et al.*, 1997a,b; Madgwick *et al.*, 2008). The LH secretion pattern during the prepubertal period is associated with age at puberty in bulls raised in contemporary groups, suggesting that this is the physiological mechanism by which genetics affects sexual development. Studies have shown that LH pulse frequency was greater at around 2.5 to 5 months of age and that mean LH concentrations increased earlier and reached greater maximum levels in early- than in late-maturing Hereford bulls (age at puberty 9.5 and 11 months, respectively) (Evans *et al.*, 1995; Aravindakshan *et al.*, 2000). Moreover, other studies have indicated that the effects of nutrition on sexual development are mediated through effects on LH secretion patterns (Brito *et al.*, 2007b,c,d).

Reduced gonadotrophin secretion marks the end of the prepubertal period and the beginning of the pubertal period. The rapidly increasing testosterone secretion and, possibly, increased hypothalamic sensitivity to negative feedback from androgens are probably responsible for the decrease in LH secretion, whereas inhibin produced by Sertoli cells may act on the gonadotrophs to limit FSH secretion, because immunization with inhibin antiserum resulted in a marked increase in FSH concentrations in prepubertal bulls (Kaneko *et al.*, 1993; Rawlings and Evans, 1995). Testosterone pulse frequency did not increase after the peri-pubertal period and remained at approximately 4.5 to 6.8 pulses/24 h from 6 to 10 months of age. However, pulse amplitude increased during the pubertal period, with consequent increase in testosterone mean concentrations until approximately 12 months of age. Elevated testosterone secretion is essential for increasing the efficiency of spermatogenesis that eventually leads to the appearance of sperm in the ejaculate (Rawlings *et al.*, 1972, 1978; McCarthy *et al.*, 1979a,b; Rodriguez and Wise, 1991).

The mechanisms controlling reproduction and energy balance are intrinsically related and have evolved to confer reproductive advantages and guarantee the survival of species. The neural apparatus, which is designed to gauge metabolic rate and energy balance, has been denoted the body 'metabolic sensor'. This sensor translates signals provided by circulating (peripheral) concentrations of specific hormones into neuronal signals that ultimately regulate the GnRH pulse generator and control the reproductive process. Metabolic indicator hormones may serve as signs to the hypothalamic-pituitary-gonadal axis and affect sexual development. The patterns of some of these hormones have been studied in growing beef bulls (Fig. 12.1). In contrast to those species in which circulating concentrations of growth hormone (GH) continue to increase until after puberty, GH concentrations decreased during the pubertal period in bulls (Brito *et al.*, 2007a,d). Differences in the stage of body development at which each species attains puberty are likely to be responsible for the different GH profiles among species. Accordingly, the GH profile in bulls seems to indicate that a relatively advanced stage of body development must be attained before the gonads produce sperm efficiently. The differences in GH secretion among species may be due to the regulatory role of steroids on GH secretion. In some other species, steroids stimulate GH secretion, but GH concentrations did not differ between intact bulls and castrated steers (Lee *et al.*, 1991). Furthermore, decreasing GH concentrations during the sexual development of bulls are observed along with increasing testosterone concentrations, indicating that steroids do not have a positive feedback on GH secretion in bulls, as they do other species (Brito *et al.*, 2007a,d).

Circulating concentrations of insulin-like growth factor I (IGF-I) in calves increased continuously and only reached a plateau (or decreased slightly) after sexual development was mostly completed after 12–14 months of age; increasing circulating concentrations of IGF-binding protein 3 and decreasing concentrations of IGF-binding protein 2 were also observed during sexual development (Renaville *et al.*, 1993, 1996, 2000; Brito *et al.*, 2007a,b,c,d). The concomitant decrease in circulating GH concentrations and increase in IGF-I concentrations during sexual development in bulls indicates that there are either drastic changes in liver sensitivity to GH or that other sources are responsible for IGF-I production. A possible IGF-I source might be the testes, as Leydig cells are capable of secreting this hormone in other species. Observations that intact bulls tended to have greater IGF-I concentrations than castrated steers at 12 months of age further support the hypothesis that the testes might contribute substantially to circulating IGF-I concentrations during the peri-pubertal and pubertal periods in bulls (Lee *et al.*, 1991). Close temporal associations observed in a series of nutrition studies strongly suggest that circulating IGF-I might be involved in regulating the GnRH pulse generator and the magnitude and the duration of the early gonadotrophin rise in beef bulls (Brito *et al.*, 2007b,c,d).

A possible effect of IGF-I on testicular steroidogenesis in bulls has also been suggested. Leydig and Sertoli cells both produce IGF-I, indicating the existence of paracrine/autocrine mechanisms of testicular regulation involving IGF-I (Bellve and Zheng, 1989; Spiteri-Grech and Nieschlag, 1992). It is assumed that most of the IGF-I in the testes is produced locally, and that circulating IGF-I may play a secondary role in regulating testicular development and function. However, the temporal patterns and strong associations among circulating IGF-I concentrations, testicular size and testosterone secretion observed in bulls receiving different nutrition argue for a primary role for this hormone (Brito *et al.*, 2007b,c,d). This primary role of increased circulating IGF-I during the pubertal period may be to promote the increase in testosterone concentrations by regulating Leydig cell multiplication, differentiation and maturation. Because testosterone has been shown to upregulate IGF-I production and IGF-I receptor expression by Leydig and Sertoli cells (Cailleau *et al.*, 1990), the establishment of a positive feedback loop between IGF-I secretion and testosterone production may be important for sexual development.

Circulating leptin and insulin concentrations also increased during the pubertal period in bulls. Notwithstanding, developmental and nutritional differences in LH pulse frequency were not related to differences in leptin or insulin concentrations in beef bulls (Brito *et al.*, 2007b,d). Other studies have also demonstrated that leptin did not stimulate *in vitro* GnRH secretion from hypothalamic explants or gonadotrophin secretion from adenohypophyseal cells collected from bulls and steers maintained at an adequate level of nutrition (Amstalden *et al.*, 2005). These results indicate that the role of these hormones in regulating GnRH secretion, if any, might be purely permissive in bulls.

Postnatal testicular development and establishment of spermatogenesis

The testicular growth curve in bulls is sigmoidal with an initial period of little growth followed by a rapid growth phase and then by a plateau. Figure 12.3 illustrates the changes in testicular length and width, as well as in SC in young beef (Angus and Angus cross) bulls on adequate nutrition. Figure 12.4 shows regression

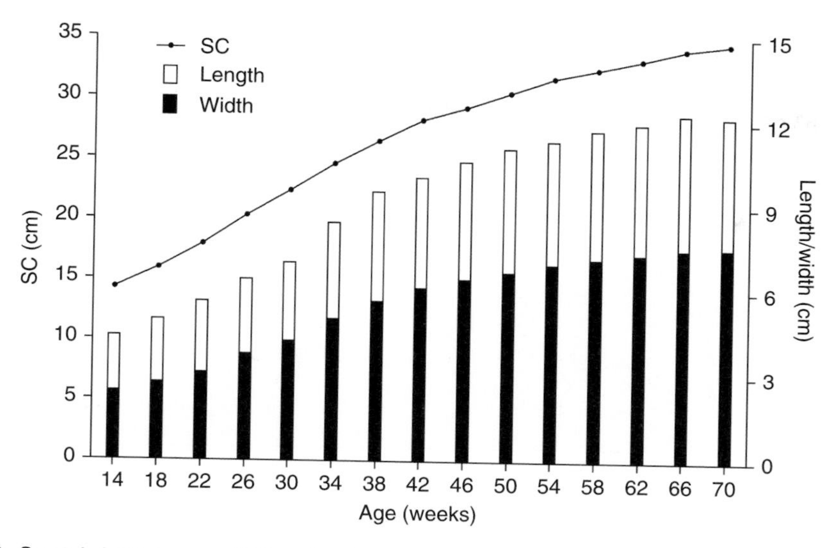

Fig. 12.3. Scrotal circumference (SC) and testicular dimensions according to age in Angus and Angus × Charolais bulls (n = 111) receiving adequate nutrition (L. Brito, unpublished).

curves for SC and paired testes weight (PTW) in dairy (Holstein and Jersey) bulls.

Although the overall pattern of testicular growth is similar in all breeds, the characteristics of the growth curve are greatly affected by genetics. In general, the rapid growth phase is shorter and testicular growth plateaus sooner in bulls from breeds that mature faster (reach puberty earlier) than in bulls from late-maturing breeds, resulting in marked differences in the slope of the curve. In addition, the asymptotic value of the curve, i.e. adult testicular size, also differs considerably among breeds (Coulter and Keller, 1982; Coulter *et al.*, 1987b; Gregory *et al.*, 1991; Pratt *et al.*, 1991; Johnson *et al.*, 1995; Bell *et al.*, 1996; Lunstra and Cundiff, 2003). These same differences can be observed within breed between early- and late-maturing bulls, which emphasizes the effects of genetics on bull testicular growth (Evans *et al.*, 1995).

Testicular growth is accompanied by marked histological changes. The testicular intertubular cell population is composed of mesenchymal-like cells, fibroblasts, Leydig cells, peri-tubular cells and mononuclear cells. At around 1 to 2 months of age, mesenchymal-like cells made up the majority of the cells in the testicular interstitial tissue. These pluripotent cells proliferated by frequent mitoses and were the precursors of Leydig cells, contractile peri-tubular cells and fibroblasts. Approximately 20–30% of all intertubular cells at all ages were mononuclear cells, including lymphocytes, plasma cells, monocytes, macrophages and light intercalated cells (monocyte-derived, Leydig cell-associated typical cells of the bovine testis). The thickness of the tubular basal lamina was approximately 3 μm at 4 months of age, but decreased continuously to 1.2 μm at 5 months of age. Mesenchymal-like cells transformed into peri-tubular cells with elongated nuclei at 4 months of age, while these, in turn, transformed into contractile myofibroblasts at 6 months of age (Wrobel *et al.*, 1988).

Leydig cells are present in the testes from birth to adulthood. At approximately 1 month of age, most (70%) intertubular cells were mesenchymal-like cells with a high mitotic rate. Typical Leydig cells constituted about 6% of all intertubular cells and a number of these cells were found in an advanced degenerative state, probably as remnants of the fetal Leydig cell population. Degenerating fetal and newly formed Leydig cells coexisted until 2 months of age, but only Leydig cells formed postnatally were observed thereafter. At 2 months of age, Leydig cells accounted for approximately 20% of all intertubular cells and comprised 10% of the total testicular volume. At approximately 4 months of age, the mesenchymal-like cells ceased proliferation

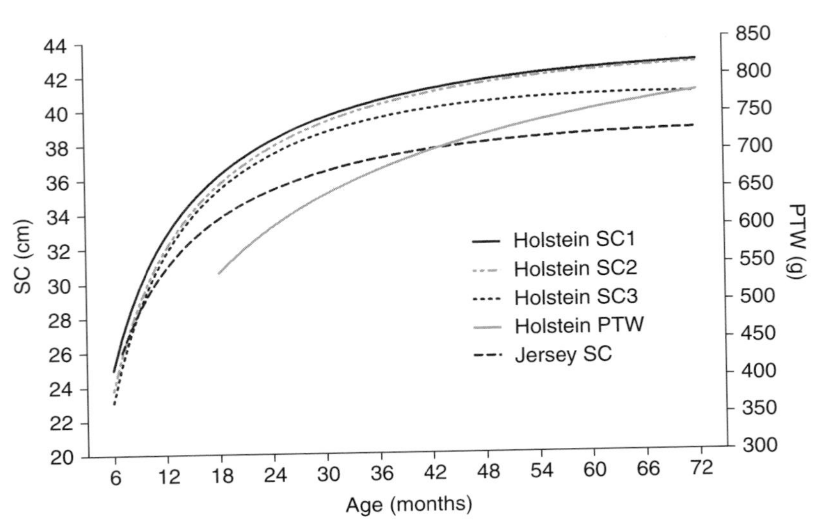

Fig. 12.4. Regression curves for scrotal circumference (SC) and paired testes weight (PTW) according to age in Holstein and Jersey bulls. The relationship of SC to testicular weight with age is best described by quadratic regressions using log age as the independent variable. Holstein SC1: measurements (n = 551) were obtained from bulls 7–148 months old; SC = −7.65 + (50.10 × (log age)) − (12.44 × (log age)^2) (Hahn *et al.*, 1969). Holstein SC2: measurements (n = 5909) were obtained from bulls 6–180 months old; SC = −4.67 + (47.26 × (log age)) − (11.74 × (log age)^2) (Coulter *et al.*, 1975). Holstein SC3: measurements (n = 9614) were obtained from bulls 6–77 months old; SC = −11.75 + (56.7 × (log age)) − (15.3 × (log age)^2) (ASB Global Inc., unpublished). Holstein PTW: measurements were obtained from bulls (n = 250) 19–184 months old; PTW = −368.8 + (952.2 × (log age)) − (180.3 × (log age)^2) (Coulter and Foote, 1976). Jersey SC measurements (n = 1,038) were obtained from bulls 7–75 months old; SC = −0.6814 + (40.26 × (log age)) − (10.27 × (log age)^2) (ABS Global Inc., unpublished data).

and transformed into contractile peri-tubular cells or Leydig cells, thus decreasing the proportion of mesenchymal-like cells to approximately 20%. At 5 months of age, undifferentiated mesenchymal-like cells were rare and the Leydig cell population increased by mitotic proliferation. Newly differentiating, intact and degenerating Leydig cells were observed in close proximity from 4 to 7 months of age. The Leydig cell population found in the adult bull was present at approximately 7 months of age, as mitosis after this age was rare (Wrobel, 1990).

Leydig cell mass increased from 0.15 g/testis at 1 month to 5.8 g/testis at 7–8 months of age. After this, Leydig cell mass increased slowly but continuously to reach about 10 g in the young adult testis at 24 months of age. Nuclear and whole Leydig cell volumes increased from 1 to 4 months of age, but then remained unchanged up to 12 months of age. However, from approximately 12 to 24 months of age there was a considerable increase in nuclear and whole Leydig cell volumes. Leydig cell numbers per testis increased from 1 to 7 months of age (0.42 to 6 billion, respectively) and remained unchanged after that. Therefore, the increase in Leydig cell mass after 7 months of age was a result of hypertrophy and not hyperplasia. Leydig cell mitochondrial mass increased from 1 to 4 months of age and remained relatively constant up to 10 months of age, but then more than doubled from that age until approximately 24 months of age (Wrobel, 1990). Functional maturation of Leydig cells was observed to involve the expression of steroidogenic enzymes and the production of testosterone, although the age-dependent expression of steroidogenic enzymes did not parallel changes in the numbers of Leydig cells in the testis (Waites *et al.*, 1985).

Undifferentiated Sertoli cells (also referred to as undifferentiated supporting cells or pre-Sertoli cells) were present in the seminiferous tubules at birth and were the predominant intra-tubular cells from birth until approximately 4 months of age. Undifferentiated Sertoli cells had little mitotic activity until 1 month of age, but cell multiplication was maximal between 1 and 2 months of age, and subsequently decreased until approximately 4 months of age. At this age, undifferentiated Sertoli cells entered the G0-phase of the cell cycle for the rest of the bull's life. With the end of the proliferative phase, undifferentiated Sertoli cells began to transform into adult-type Sertoli cells. The differentiation process of Sertoli cells during sexual development included distinct morphological changes in the cell shape, nucleus and cellular organelles, as well as an increase in surface specialization and subsequent interaction with other Sertoli cells and germ cells (Sinowatz and Amselgruber, 1986).

At approximately 1 month of age, undifferentiated Sertoli cells were characterized by round or oval nuclei with numerous flakes of heterochromatin dispersed throughout the nucleoplasm. Lateral cell membranes of neighbouring presumptive Sertoli cells contained few interdigitations and no special junctional complexes. Between 4 and 5 months of age, opposing cell membranes of adjacent undifferentiated Sertoli cells started to develop extended junctional complexes above the spermatogonia and in the basal portion of the tubules. The nuclei of these cells became more elongated and irregular, a vacuolar nucleolus characteristic of the mature cell developed, and the clumps of heterochromatin present at younger ages disappeared. Most Sertoli cells had completed their morphological differentiation and attained adult structure by 6 to 7 months of age. Junctional complexes consisting of many serially arranged points or lines of fusion involving neighbouring Sertoli cell membranes could be observed. These junctions formed a functional blood–testis barrier and divided the tubular epithelium into a basal compartment containing spermatogonia and an adluminal compartment (connected to the lumen) containing germ cells at later stages of spermatogenesis (Abdel-Raouf, 1960; Sinowatz and Amselgruber, 1986; Wrobel, 2000). In Holstein bulls, the number of adult-type Sertoli cells increased dramatically from 202 to 8862 million cells/testis between 5 and 8 months of age (Curtis and Amann, 1981).

Intense germ cell proliferation occurred from 50 to 80 days post conception, but the germ cells entered a prolonged G1- or G0-phase and no mitotic activity was observed until after birth in bulls. At birth, the germ cell population was composed solely of gonocytes (also referred to as prespermatogonia or prepubertal spermatogonia). Gonocytes were usually centrally located and had a large nucleus (~12 μm in diameter) with a well-developed nucleolus (Fig. 12.5). Germ cell proliferation slowly resumed between 1 and 2 months of age, when seminiferous tubule diameter was approximately 50–80 μm. Gonocytes were gradually displaced to a position close to the basal lamina and divided by mitosis, originating A-spermatogonia. Differentiation and degeneration resulted in the complete disappearance of gonocytes from the seminiferous tubules by 5 months of age.

Germ cell proliferation reached a maximum between 4 and 8 months of age (tubule diameter 80–120 μm), representing the expansion of the spermatogonial stem cell. In Holstein bulls, the number of spermatogonia increased from 1.8×10^6 cells/testis at 4 months of age to 3.8×10^9 cells/testis at 8 months of age; the number of spermatogonia continued to increase until approximately 12 months of age (Curtis and Amann, 1981). A-spermatogonia divided mitotically to form In- and B-spermatogonia, which, in turn, entered meiosis at around 4–5 months of age, when primary spermatocytes were first observed. The numbers of primary spermatocytes increased slowly until 8 months of age, when they exceeded the number of spermatogonia. Secondary spermatocytes and round spermatids first appeared at approximately 6–7 months of age, whereas elongated spermatids appeared at around 8 months of age. The number of spermatids increased rapidly and, after 10 months of age, spermatid numbers exceeded the numbers of any other germ cell. Mature sperm appeared in the seminiferous tubules at approximately 8–10 months of age (Fig. 12.5). Testes weighing more than 100 g in Swedish Red-and-White bulls, or more

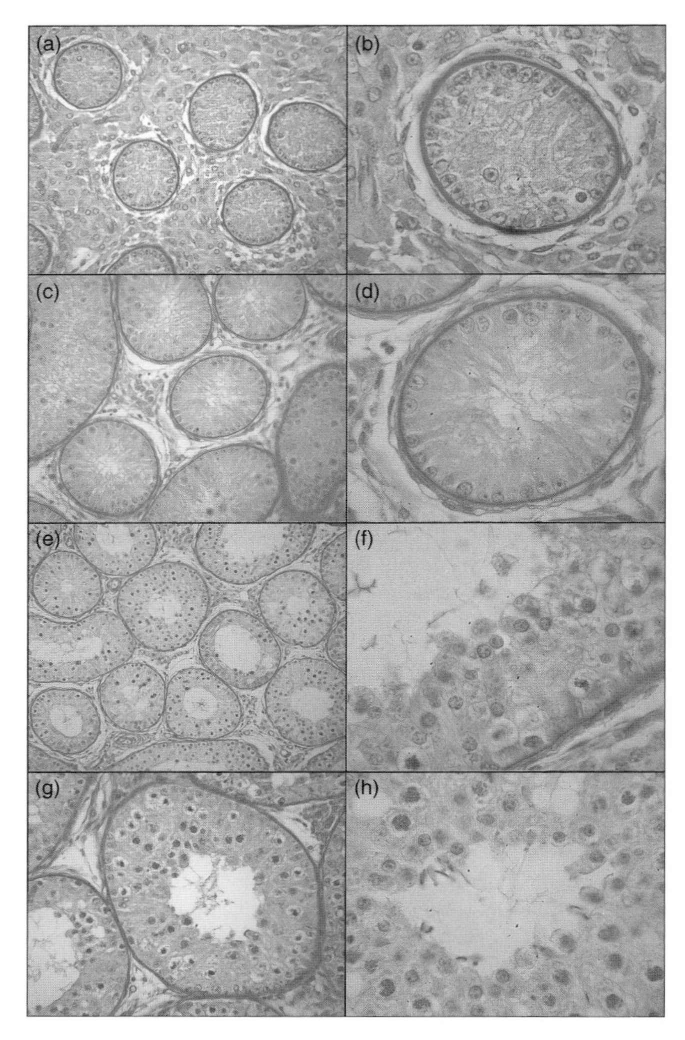

Fig. 12.5. Testicular development and initiation of spermatogenesis in bulls. At 2 months of age, the interstitial tissue occupies most of the testicular parenchyma (a), seminiferous cords are lined by undifferentiated Sertoli cells, and centrally located gonocytes with large nuclei can be observed (b). Between 4 and 6 months of age a cord lumen begins to develop, and gonocytes migrate towards the basement membrane, differentiating into spermatogonia (c, d). Formation of the tubular lumen is evidence of a functional blood–testis barrier (e) and precedes the appearance of primary spermatocytes and spermatids between 6 and 8 months of age (f). With continuous increase in diameter, seminiferous tubules occupy most of the testicular parenchyma, and mature sperm appear in the seminiferous tubules at approximately 8 to 10 months of age (g, h).

than 80 g in Holstein bulls, were likely to be producing sperm (Abdel-Raouf, 1960; Curtis and Amann, 1981; Evans *et al.*, 1996; Wrobel, 2000; Bagu *et al.*, 2006).

Testicular cellular development was accompanied by a progressive increase in the proportion of the testicular parenchyma occupied by seminiferous tubules until approximately 8 months of age. Seminiferous tubule diameter increased gradually from 2 to 5 months of age, and more rapidly from 6 to 10 months of age, increasing fivefold from birth until adulthood. Around 6 months of age, 'cracking' of the tubular cytoplasm was first detected, indicating formation of the tubular lumen. Formation of the tubular lumen was evidence of a functional blood–testis barrier and preceded the appearance of primary spermatocytes and more advanced germ cells. Tubular lumen diameter continued to increase until approximately 8 months of age. Total seminiferous tubule length increased from 830 m/testis at 3 months of age to 2010 m/testis at 8 months of age in Holstein bulls. Therefore, increases in the proportion of parenchyma that was occupied by seminiferous tubules, as well as increases in the tubular diameter and the total length of seminiferous tubules per testis, accounted for initial testicular growth up to 8 months of age; thereafter, testicular growth was primarily the result of increasing total seminiferous tubule length (Abdel-Raouf, 1960; Curtis and Amann, 1981). Testicular echogenicity increased at about 6 and 10 months of age in beef bulls, accompanying the testicular histological changes. When data were analysed according to age at puberty, testicular echogenicity started to increase 16 to 12 weeks before puberty and reached maximum values 4 weeks before or at puberty. These observations indicate that a certain developmental stage of the testicular parenchyma must be reached before puberty and that the composition of the parenchyma remained consistent after puberty (Brito *et al.*, 2012a).

The accelerated testicular growth observed after 6 months of age in bulls occurs when circulating gonadotrophin concentrations are decreasing, which points to the existence of important GnRH-independent mechanisms regulating testicular development. The period of accelerated testicular growth coincides with increasing circulating IGF-I and leptin concentrations, and strong associations between these hormones and testicular size have been observed in growing beef bulls (Brito *et al.*, 2007a,d), indicating that metabolic hormones may be involved in regulating GnRH-independent testicular development. As there was no association between circulating metabolic hormones and gonadotrophin concentrations, the possible effects of metabolic hormones on testicular growth were apparently direct and independent of the hypothalamus and pituitary. Although IGF-I and leptin concentrations were associated with testicular size, there were no associations between these hormones and seminiferous tubule diameter and area, seminiferous epithelium area, or volume occupied by the seminiferous tubules (L. Brito, unpublished). These observations suggest that increased circulating IGF-I and leptin concentrations were associated with increased length of the seminiferous tubules and probably with overall increases in the total number of testicular cells. Considering the cellular events in the testis during the pubertal period, the temporal patterns of metabolic hormone concentrations in bulls indicated that circulating IGF-I and leptin could be involved in regulating Leydig cell multiplication and maturation, Sertoli cell maturation and germ cell multiplication during the period of accelerated GnRH-independent testicular growth.

Testicular concentrations of LH and FSH receptors in beef bulls decreased around 5 to 6 months of age, but increased thereafter until at least approximately 13 months of age; this might act to increase the sensitivity of Leydig and Sertoli cells to the low concentrations of gonadotrophins that occur during the rapid testicular growth phase (Bagu *et al.*, 2006). Other mechanisms that might be associated with GnRH-independent testicular growth include changes in testicular concentrations and bioavailability of growth factors such as transforming growth factor (TGF-alpha and TGF-beta 1, 2 and 3) and interleukins (IL-1 alpha, IL-1 beta and IL-6) (Bagu *et al.*, 2010a,b). In addition, experiments evaluating the effect of nutrition on sexual development have demonstrated that the impact of gonadotrophins on target tissues during the prepubertal period have long-term effects on testicular development in bulls. Bulls with either greater LH pulse frequency or more sustained increased LH pulse frequency during the early gonadotrophin rise had a more prolonged period of increased testicular growth and greater testicular size at 15–16 months of age, even when no differences in metabolic hormones or testosterone concentrations were observed after 6 months of age. These results indicate that the putative effects of circulating metabolic hormones, gonadotrophins, local growth factors and other unknown factors during the period of accelerated testicular growth might be dependent on the previous LH exposure during the prepubertal period. The LH secretion pattern during the early gonadotrophin rise seems to 'prime' testicular development, dictating maximum adult testicular size in bulls (Brito *et al.*, 2007b,c,d).

Puberty

After spermatogenesis is established, there was a gradual increase in the number of testicular germ cells supported by each Sertoli cell and an increase in the efficiency of the spermatogenesis, i.e. an increase in the number of more advanced germ cells resulting from the division of precursor cells. The yields of different germ cell divisions, low during the onset of spermatogenesis, increased progressively to the adult level (Macmillan and Hafs, 1968; Curtis and Amann, 1981). The eventual appearance of sperm in the ejaculate was the result of the increasing sperm production efficiency after initiation of spermatogenesis. In general terms, puberty is defined as the process of changes by which a bull becomes capable of reproducing. This process involves the development of the gonads and secondary sexual organs, and the development of the ability to breed. For research purposes however, puberty in bulls is usually defined as an event instead of a process. Most researchers define attainment of puberty by the production of an ejaculate containing ≥50 million sperm with ≥10% motile sperm (Wolf *et al.*, 1965). The interval between the first observation of sperm in the ejaculate and puberty as defined by these criteria was approximately 30 to 40 days (Lunstra *et al.*, 1978; Jiménez-Severiano, 2002).

Age at puberty determined experimentally can be affected by the age that semen collection attempts are performed, the interval between attempted collections, the method of semen collection (artificial vagina or electroejaculator), the response of the bull to the specific semen collection method and the experience of the collector(s). Moreover, age at puberty is affected by management, nutrition (see below) and genetics. Table 12.1 describes age, weight and SC at puberty in different breeds. Although data from large trials comparing bulls of different breeds raised as contemporary groups are scarce, some liberties could be taken to make some generalizations from these data. Dairy bulls usually mature faster and attain puberty earlier than beef bulls. Bulls from continental beef breeds, with the possible exception of Charolais, usually attain puberty later than bulls from British beef breeds, especially Angus bulls. Bulls from double-muscled breeds are notorious for being late maturing. Puberty is also delayed in bulls from tropically adapted *B. taurus* breeds and in non-adapted bulls raised in the tropics.

There is a large variation in age and body weight at puberty across breeds and within breeds. Although, on average, bulls attain puberty at an SC of between 28 and 30 cm, regardless of the breed, the fact that there is still considerable variation in SC at puberty is sometimes overlooked. Interesting observations have been reported in studies evaluating differences between early- and late-maturing bulls. Bulls that attained puberty earlier were generally heavier and had greater SC than bulls that attained puberty later; however, both weight and SC were smaller at puberty in early-maturing bulls (Evans *et al.*, 1995; Aravindakshan *et al.*, 2000). These observations not only indicate that precocious bulls develop faster, but also suggest that sexual precocity is not simply related to the earlier attainment of threshold body or testicular development. In fact, these thresholds seem to be lower in early-maturing bulls, and late-maturing bulls must reach a more advanced stage of body and testicular development before puberty is attained. Similar observations have also been reported in *B. taurus* × *B. indicus* crossbred bulls (Brito *et al.*, 2004b).

Semen quality in pubertal bulls was generally poor with a gradual improvement characterized by an increase in sperm motility and a marked reduction in morphological sperm abnormalities being observed after puberty. The most prevalent sperm defects observed in pubertal bulls were proximal cytoplasmic droplets and abnormal sperm

Table 12.1. Age, weight and scrotal circumference (SC) at puberty (ejaculate with ≥50 million sperm and ≥10% sperm motility) in different breeds. For crossbred bulls, the first breed noted indicates the sire's breed.

Breed	No. of bulls	Age (months)[a]	Weight (kg)	SC (cm)	Reference
Angus	10	10.1	309	30.0	Wolf *et al.*, 1965
Angus and Angus × Hereford	16	9.7	272	28.3	Lunstra *et al.*, 1978
Angus × Hereford and Angus × composite	24	7.8	323	28.1	Lunstra and Cundiff, 2003
Angus and Hereford × composite	37	8.8	–	28.0	Casas *et al.*, 2007
Belgian Blue × Hereford, Angus, and composite	44	9.0	347	27.9	Lunstra and Cundiff, 2003
Brown Swiss	5	8.7	295	27.2	Lunstra *et al.*, 1978
Brown Swiss	9	10.2	233	25.9	Jiménez-Severiano, 2002
Charolais	22	9.4	396	28.8	Barber and Almquist, 1975
Hereford	10	10.3	322	31.2	Wolf *et al.*, 1965
Hereford	5	10.7	261	27.9	Lunstra *et al.*, 1978
Hereford[b]	27	11.7	342–401	31.3–32.7	Pruitt *et al.*, 1986
Hereford[b]	28	9.6–11.1	357–391	28.5–30.0	Evans *et al.*, 1995
Hereford × Angus and composite	32	8.6	339	27.9	Lunstra and Cundiff, 2003
Hereford × Charolais	20	10.0	–	28.0	Aravindakshan *et al.*, 2000
Holstein	20	9.4	303	–	Killian and Amann, 1972
Holstein	4	10.9	276	28.4	Jiménez-Severiano, 2002
Red Poll	5	9.3	258	27.5	Lunstra *et al.*, 1978
Romosinuano	13	14.2	340	28.8	Chase *et al.*, 1997
Simmental[b]	27	10.6–11.4	328–419	30.6–34.0	Pruitt *et al.*, 1986

[a]Transformed from days or weeks in the original reports.
[b]Ranges indicate differences among groups within an experiment.

heads (approximately 30–60% and 30–40% at puberty, respectively) (Lunstra and Echternkamp, 1982; Evans *et al.*, 1995; Aravindakshan *et al.*, 2000). In a series of studies with beef bulls, age at puberty and age at satisfactory semen quality (≥30% sperm motility, ≥70% morphologically normal sperm) were 308 and 372 days, respectively. Moreover, 10% of the bulls did not have satisfactory semen quality by the end of the experimental period at 15–16 months of age (L. Brito, unpublished data and Fig. 12.6). The results of another study with beef bulls in Western Canada indicated that the proportions of bulls with satisfactory sperm morphology at 11, 12, 13 and 14 months of age were approximately 40, 50, 60 and 70%, respectively (Arteaga *et al.*, 2001). Similarly, only 48% of beef bulls 11 to 13 months old in Sweden had <15% proximal cytoplasmic droplets and <15% abnormal sperm heads (Persson and Söderquist, 2005). These observations have profound implications for the ability of producers to use yearling bulls and the ability of artificial insemination (AI) centres to produce semen for progeny testing at the youngest possible age.

The epididymis continued to grow until at least 6 years of age in Holstein bulls, with epididymal weight increasing from 9 g at 8 months of age to 15, 23, 27 and 38 g at 12, 18, 25–48 and 73–96 months of age, respectively (Almquist and Amann, 1961; Killian and Amann, 1972). The tube-like vesicular glands in newborn calves increased in length and became lobulated during development. The weight of the vesicular glands increased until approximately 4 years of age in Holstein bulls – from 13 g at 8 months of age to 26, 35, 54 and

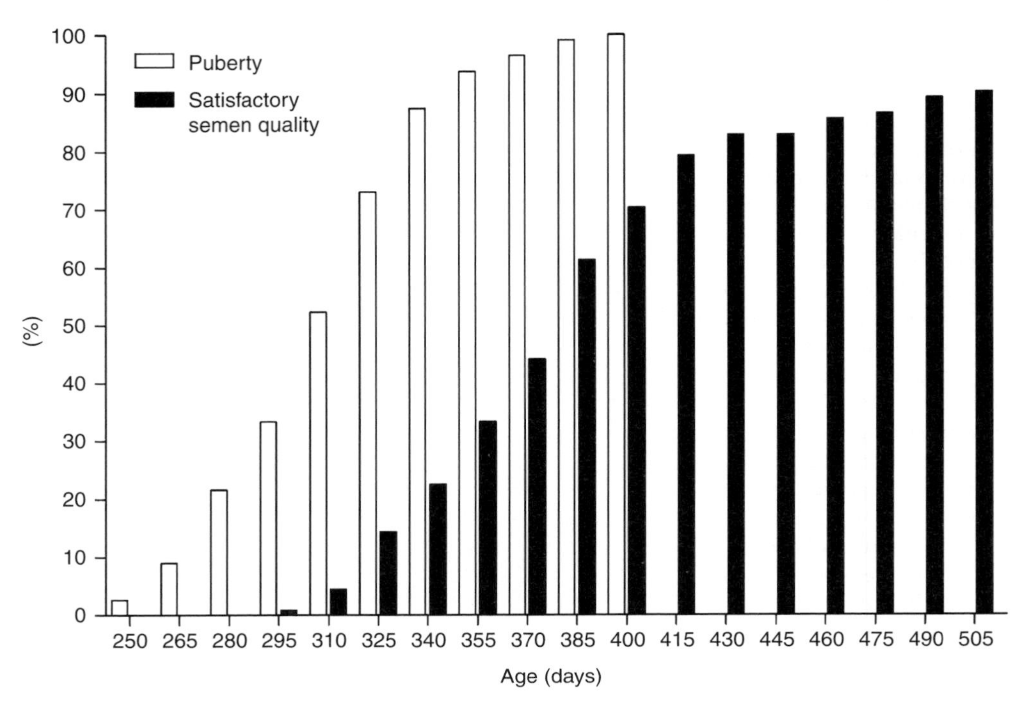

Fig. 12.6. Proportion (%) of Angus and Angus × Charolais bulls receiving adequate nutrition that reach puberty (ejaculate with ≥50 million sperm and ≥10% sperm motility) and satisfactory semen quality (ejaculate with ≥30% sperm motility and ≥70 morphologically normal sperm) according to age; *n* = 111 (L. Brito, unpublished data).

78 g at 12, 18, 25–48 and 73–96 months of age, respectively (Almquist and Amann, 1961; Killian and Amann, 1972). The sigmoid flexure of the penis began to develop at about 3 months of age. Penis length increased by up to five times by the onset of puberty, and length continued to increase until sexual maturity (Coulter, 1986). The penis in Friesian bulls 13 to 19 months old measured 73 to 89 cm (Ashdown *et al.*, 1979a); in contrast, the penis in Holstein bulls ≥25 months old measured 95 to 106 cm (Almquist and Amann, 1961). First protrusion of the penis during mounting was observed at approximately 8 months of age, with complete separation of the penis and sheath observed at approximately 8.5 months of age in Angus, Charolais and Hereford bulls (Wolf *et al.*, 1965; Barber and Almquist, 1975). Complete sheath–penile detachment evaluated during electroejaculation (EE) was observed at around the same time of puberty, whereas first completed service – when evaluated during libido testing – was only observed approximately a month after puberty, indicating that bulls started producing sperm before complete sheath–penile detachment (Lunstra *et al.*, 1978).

Effects of nutrition on sexual development

Very few studies have evaluated the effect of nutrition from birth to maturity on sexual development and reproductive function in bulls. Compared with Holstein bulls receiving control nutrition (100% of requirements), bulls receiving low nutrition (approximately 60–70% of requirements) from birth were older at the time the first ejaculates containing motile sperm were collected and also had smaller testes, but bulls receiving high nutrition (approximately 160% of requirements) had earlier puberty and larger testes

(Bratton *et al.*, 1959; Flipse and Almquist, 1961). These observations have been corroborated and expanded in a series of more recent experiments that have also shown that the most pronounced effects of nutrition occur during the prepubertal period. The recent studies have demonstrated that the adverse effects of low nutrition during the prepubertal period cannot be compensated for by improved nutrition during the pubertal period, and that the beneficial effects of high nutrition during the prepubertal period are sustained even if maintenance diets are fed thereafter. Low nutrition (75% of requirements) during the peri-pubertal period in beef bulls reduced LH pulse secretion, delayed the increase in circulating testosterone concentrations, delayed puberty and resulted in decreased testicular size at 16 months of age; high nutrition (125% of requirements) produced the opposite results (Brito *et al.*, 2007b,c,d).

The effects of nutrition on the sexual development and reproductive function of bulls were mediated through the hypothalamic-pituitary-testicular axis. Nutrition affects the GnRH pulse generator in the hypothalamus, because differences in LH pulse secretion in bulls receiving different nutrition can be observed even in the absence of differences in pituitary LH secretion capability, as determined by GnRH challenge (Brito *et al.*, 2007c,d). Interestingly, when low nutrition was imposed on bulls by limiting the amount of nutrients in a ration fed *ad libitum*, only reduced LH pulse frequency was observed; but when nutrition was restricted by restricting food availability, reduced LH pulse frequency, mean and peak concentrations and secretion were observed after GnRH challenge, (Brito *et al.*, 2007b). These results seem to indicate that the inhibitory effects of limited nutrient availability on LH secretion appear to be exerted on the hypothalamus only, while the combination of limited availability of nutrients combined with hunger sensation experienced by bulls with restricted intake affected both hypothalamic and pituitary function, producing a much more severe inhibition of LH secretion. The effect of nutrition on Leydig cell number and/or function in bulls receiving different diets was demonstrated by differences in testosterone secretion after GnRH challenge, even in the absence of differences in LH secretion after the challenge (Brito *et al.*, 2007c,d).

Differences in yearling SC due to age of the dam in beef bulls could also be interpreted as an indication that nutrition during the pre-weaning period affects sexual development, although possible *in utero* effects cannot be completely ruled out. SC in beef bulls increases as age of the dam increases up to 5 to 9 years of age and then decreases as dams get older. Adjustment factors of 0.7–1.4 cm, 0.2–1.0 cm, 0.1–1.0 cm and 0.3–0.75 cm for yearling SC have been suggested for bulls raised by 2-, 3-, 4- and ≥10-year old dams, respectively (Bourdon and Brinks, 1986; Nelsen *et al.*, 1986; Lunstra *et al.*, 1988; Kriese *et al.*, 1991; Evans *et al.*, 1999; Crews and Porteous, 2003). In these studies, the inclusion of weight as a covariate in the models describing SC resulted in decreased effects of age of the dam, indicating that the effect of dam age on testicular growth seems to be primarily the result of dam age effects on the bull's body weight, probably related to differences in milk production. This theory is also supported by reports that, as observed in bulls receiving low nutrition, LH secretion after GnRH challenge was greater from 3.5 to 6 months of age in bulls raised by multiparous rather than by primiparous females (Bagu *et al.*, 2010c).

Several studies have described the effects of nutrition during the pubertal period only, i.e. after the initial hormonal changes regulating sexual development have occurred. In general, these studies indicate that low nutrition has adverse effects on growth and sexual development. In one study, bulls receiving one third of the amount supplied to their twin controls had lower body and vesicular gland weights, vesicular gland fructose and citric acid contents, and circulating and testicular testosterone concentrations, whereas circulating androstenedione concentrations were increased (Mann *et al.*, 1967). In other experiments, beef bulls 8 to 12 months old receiving diets with low levels of crude protein (8, 5 and 1.5%) for periods of 3 to 6 months had markedly reduced weights of the testes, epididymis and seminal glands compared with control bulls fed diets containing 14%

crude protein. Moreover, seminiferous tubule diameter and seminiferous epithelium thickness were smaller in bulls that were on restricted protein intake (Meacham *et al.*, 1963, 1964).

Though low nutrition during the pubertal period has adverse effects on reproductive function, the potential beneficial effects of high nutrition after weaning are questionable, at best. Effects of energy on sexual development were not consistent in a study with Simmental and Hereford bulls fed diets with low, medium or high energy content (approximately 14, 18 and 23 Mcal/bull daily, respectively) from 7 to 14 months of age. Dietary energy affected sexual development in Simmental bulls, but not in Hereford bulls. Simmental bulls in the high-energy group were heavier and had greater SC and testosterone concentrations than bulls in the low-energy diet group (in general, the medium energy group had intermediate values). However, increased dietary energy did not hasten age at puberty. The only semen trait affected by dietary energy was semen volume, which was depressed in Simmental bulls in the medium-energy group. Serving capacity was greater for Hereford bulls in the high-energy diet; in contrast, medium- and high-energy diets were associated with a decrease in the number of services between two testing periods in Simmental bulls. There was a trend for lower sperm motility and a lower proportion of normal sperm in Simmental bulls fed the low-energy diet (Pruitt and Corah, 1985; Pruitt *et al.*, 1986).

In Holstein bulls producing semen for AI, high energy intake was associated with visual evidence of weakness of the feet and legs and increased reaction time after 3 years of age (Flipse and Almquist, 1961). Under field conditions, post-weaning high-energy diets are frequently associated with impaired reproductive function in bulls, which is most likely related to altered testicular thermoregulation as a result of excessive fat deposition above and around the testes in the scrotum. In one report, sperm motility decreased and the proportion of sperm defects increased with age in Hereford bulls fed to gain >1.75 kg/day, and significantly differed from bulls fed to gain approximately 1 kg/day (control). Even after the high nutrition diet was changed to a control diet, bulls previously receiving high nutrition continued to have lower semen quality. There was greater deposition of fat around the testicular vascular cone in the scrotal neck in bulls in the high nutrition group, and the difference between body and testes temperature was reduced in this group compared with that of bulls in the control group. This difference was still present after the high-energy diet was changed and the bulls had lost a considerable amount of weight, indicating that fat that has accumulated in the scrotum is more difficult to lose than other body fat (Skinner, 1981).

In another series of experiments, Angus, Hereford and Simmental bulls were fed high nutrition (80% grain and 20% forage) or medium nutrition (forage only) from approximately 6.5 until 12–24 months of age. In general, bulls receiving high nutrition had greater body weight and back fat than did the medium nutrition group, although PTW was not affected by diet. Furthermore, bulls receiving high nutrition had lower daily sperm production and epididymal sperm reserves, and a greater proportion of sperm abnormalities. These authors indicated that increased dietary energy may adversely affect sperm production and semen quality owing to fat deposition in the scrotum, which reduces the amount of heat that can be radiated from the scrotal skin, thereby increasing the temperature of the testes and scrotum (Coulter and Kozub, 1984; Coulter *et al.*, 1987a, 1997; Coulter and Bailey, 1988). Observations from another study indicated that bulls fed high-nutrition diets had greater SC and scrotal weight than did bulls fed medium-nutrition diets, though PTW was not different between the two groups (Seidel *et al.*, 1980). Growth rate between 6 and 16 months of age did not affect sexual development and reproductive function in Angus and Angus × Charolais bulls. However, greater body weight at various ages was associated with reduced age at puberty and maturity, and with larger testes at 16 months of age, indicating that improved nutrition might be beneficial, but only when offered before 6 months of age. Average daily gains of 1 to 1.6 kg/day did not result in excessive fat accumulation in the scrotum, or in increased scrotal temperature or reduction in sperm production and semen quality, and could be considered 'safe' targets for growing beef bulls (Brito *et al.*, 2012b).

This summary of the literature supports the intuitive assumption that low nutrition has adverse effects on sexual development and reproductive function regardless of a bull's age. Then again, most research seems to indicate that high nutrition is only beneficial during the first 6 months of life, which presents a challenge to bull producers. Beef bull calves are often nursed until 6 to 8 months of age, with very little attention being paid to their nutrition, whereas nutrition offered to dairy bull calves is often suboptimal. Efforts to obtain maximum weight gain during the first months after birth by offering highly nutritious diets and adopting management practices such as creep feeding should be compensated by reduced age at puberty and greater sperm production capacity in adult bulls. It is also clear that, although high nutrition diets after 6 months of age might be associated with greater SC, this effect is likely to be caused by fat accumulation in the scrotum and is not actually the result of greater testicular size. Moreover, sperm production, semen quality and serving capacity are all compromised in bulls receiving excessive nutrition after this age. Adjusting diets accordingly to maximize growth, but prevent over-conditioning after the peri-pubertal period, is advisable.

Testicular Thermoregulation

Testes in bulls are maintained at 4–5.5°C below body temperature and this lower temperature is essential for normal spermatogenesis (Kastelic *et al.*, 1995, 1997). Adequate testicular temperature is maintained by complex physiological mechanisms that involve the scrotum, the testicular vascular cones and the testes proper. The temperature is determined primarily by the heat conveyed to the testes by the arterial blood, and the temperature of the arterial blood reaching the testes is determined by the amount of heat lost in the testicular vascular cone. This, in turn, is largely dependent on the temperature of the venous blood leaving the testes. Because the testicular veins run on the surface of the testes just below the tunica albuginea, the temperature of the testicular venous blood is greatly affected by the temperature of the scrotum. Therefore, the scrotum, which is equipped with thermoreceptors, is responsible for actively regulating testicular temperature by controlling vascular flow, sweat gland activity and muscle contractility (Waites, 1970; Setchell, 1978; Sealfon and Zorgniotti, 1991).

Increased testicular temperature results in reduced vasoconstrictor tone of the smooth muscles of the skin arteries, which increases blood flow to the scrotum and heat loss by irradiation. Increased testicular temperature also results in increased scrotal sweat production and heat loss by evaporation; scrotal sweat gland density and sweat production are greater than in the skin of other parts of the body (Blazquez *et al.*, 1988). The cremaster muscle and the dartos tunic become completely relaxed when testicular temperature rises, thus maintaining the testes as far away as possible from the abdomen, maximizing the contact of the testes with the scrotum and increasing the scrotal surface area for heat irradiation. Scrotal surface temperature is lower than both body and intra-testicular temperatures, and a positive temperature gradient is also observed between the proximal and distal parts of the scrotum (Plate 32). Scrotal surface temperatures of approximately 30–31°C at the proximal portion and 28–29°C at the distal portion have been reported in beef bulls (Cook *et al.*, 1994; Kastelic *et al.*, 1995, 1996a, 1997).

The testicular vascular cone is a specialized anatomical structure that plays an important role in cooling testicular arterial blood before it reaches the testes and, conversely, in warming venous blood before it re-enters the abdomen. The cone is formed mainly by the very sinuous testicular artery juxtaposed between the fine networks of testicular veins that form the pampiniform plexus. Heat transfer in the testicular vascular cone occurs through a countercurrent mechanism that involves heat transfer between fluids of different temperatures flowing in opposite directions inside adjacent vessels. The degree of heat exchange is determined exclusively by the temperature gradient between the arterial and venous blood, but the efficiency of the mechanism is determined by the length and volume of the artery, the area of the artery's surface contacted by veins, and the distance between the arterial and venous blood (Sealfon and Zorgniotti, 1991). Heat loss by irradiation through the scrotal skin in the vascular cone area is also important, because this is the warmer part of the scrotum and insulation of this area results in increased testicular temperature (Kastelic *et al.*, 1996b).

Anatomical changes of the testicular vascular cone as the bull ages and starts to produce sperm are indications that the efficiency of the countercurrent mechanism needs to improve in order to couple with increasing testicular metabolism. In Angus and Angus × Charolais bulls, the testicular vascular cone diameter measured by ultrasonography increased until approximately 13.5 months of age, or until 1 to 8 weeks before the SC reached a plateau (Brito *et al.*, 2012c). In crossbred beef bulls, the length and diameter of the testicular artery in the vascular cone increased from 6 to 12 months of age (1.8 m and 1.9 mm versus 3.1 m and 3.5 mm, respectively), but did not increase significantly thereafter (Cook *et al.*, 1994). In 15-month-old Angus bulls, the testicular artery length and volume in the vascular cone were 1.6 m and 6 ml, respectively (Brito *et al.*, 2004a). Other studies have reported that the testicular vascular cone length was approximately 10 to 15 cm and that the testicular artery length varied from 1.2 to 4.5 m in adult beef bulls of several breeds (Kirby, 1953; Kirby and Harrison, 1954; Hofmann, 1960).

In addition to the lengthening of the testicular artery, the distance between the arterial and venous blood in the testicular vascular cone also decreased with age. This occurred via thinning of the arterial wall (from 317 to 195 μm thick at 6 and 36 months of age, respectively) and reduction of the distance between the artery and the closest veins (Cook *et al.*, 1994). Another interesting observation was that the artery wall thickness and distance between the artery and the veins also decreased from the proximal to the distal portions of the testicular vascular cone (Hees *et al.*, 1984; Cook *et al.*, 1994; Brito *et al.*, 2004a). The reduction of the distance between the warmer arterial blood and the cooler venous blood may facilitate heat transfer and compensate for the gradual reduction in the arterial–venous temperature gradient that occurs as blood flows through the vascular cone. Evaluation of intra-artery temperature on the dorsal portion of the testis indicated that the arterial blood was cooled down by 3–4.5°C after passing through the testicular vascular cone in 15–18 month-old beef bulls (Kastelic *et al.*, 1997; Brito *et al.*, 2004a).

The testicular artery reaches the dorsal pole of the testis and runs under the tunic albuginea, in close relation to the epididymal body, until it ramifies at the ventral pole. After ramification, several arterial branches run dorsally under the tunic until they penetrate into the testicular parenchyma. The temperature of the arterial blood decreases only slightly between the testicular dorsal pole and the ventral pole, but decreases significantly between the ventral pole and the point of penetration. This arrangement results in a negative sub-tunic temperature gradient between the dorsal and ventral portions of the testis. The combination of the opposing 'top-to-bottom' temperature gradients of the scrotum (positive gradient) and the testicular surface (negative gradient) results in a constant temperature throughout the entire testicular parenchyma. Intra-testicular temperature in beef bulls 15–18 months old was approximately 33.5–35°C (Kastelic *et al.*, 1996a, 1997; Brito *et al.*, 2004a).

Sperm Production and Semen Quality

Daily sperm production is defined as the total number of sperm produced per day by the testes, whereas spermatogenesis efficiency is the number of sperm produced per gram of testicular parenchyma. Spermatogenesis efficiency increased with age and reached adult levels at approximately 12 months of age in Holstein bulls (Macmillan and Hafs, 1968; Killian and Amann, 1972). Individual variation in spermatogenesis efficiency was relatively small and was not affected by ejaculation frequency, with values between 10 and 14 million sperm/g parenchyma being reported for dairy and beef bulls (Macmillan and Hafs, 1968; Killian and Amann, 1972; Amann *et al.*, 1974; Weisgold and Almquist, 1979; Almquist, 1982; Johnson *et al.*, 1995; Lunstra and Cundiff, 2003). As sperm production per gram of testicular parenchyma is somewhat constant among bulls, the daily sperm production of a bull is largely dependent on the weight of the testes. Considering testicular weight at different ages, yearling bulls are expected to produce around 4–5 billion sperm per day, whereas adult bulls are expected to produce around 7–9 billion sperm per day. Sperm output (number of sperm in the ejaculate) in bulls ejaculated frequently (i.e. when extragonadal reserves are stabilized) is essentially the same as sperm production (Amann *et al.*, 1974). One important difference between young and older bulls is the capacity of the epididymis to store sperm. Evaluation of sperm numbers in the tail of the epididymis of 15–17 month-old Holstein bulls demonstrated that sperm available for ejaculation corresponded to approximately 1.5–2 days of sperm production; in older (2–12 year-old) bulls, stored sperm numbers corresponded to the larger amount of approximately 3.5–5 days of sperm production (Amann and Almquist, 1976; Amann, 1990). These observations are especially important for AI centres and indicate that more frequent semen collection is necessary to maximize sperm harvest from young bulls; in older bulls, semen collection intervals of less than 2–3 days have smaller effects on increasing sperm harvest. Sperm output increases with increased ejaculation interval up to the number of days required for epididymal storage capacity to reach its limit. Sperm that are not ejaculated are eliminated with urine or during masturbation.

No differences were found in testicular development, spermatogenesis efficiency or extragonadal sperm reserves at 7 years of age between Holstein bulls that had been submitted to semen collection frequencies of either once or six times a week from 1 year of age using artificial vaginas, indicating that ejaculation at high frequency was not detrimental to either testicular development or sperm physiology. More importantly, the lack of adverse effects on sperm production and semen quality supported the long-term use of a high rate of ejaculation to maximize sperm harvest from bulls in AI centres (Almquist, 1982). With increased weekly ejaculation frequency there was a decrease in ejaculate volume, though sperm concentration was not affected (Almquist and Cunningham, 1967; Almquist *et al.*, 1976). Although ejaculate volume either remained unchanged or decreased with multiple ejaculations on the same day, there was a gradual decrease in sperm concentration and total sperm per ejaculate. In Angus bulls from which seven consecutive ejaculates were obtained, total sperm number decreased continuously from the first to the third ejaculate, but did not change significantly thereafter. The first, first two and first four ejaculates contained 31, 55 and 77%, respectively, of the total sperm output obtained in the seven ejaculates (Foster *et al.*, 1970). A small but significant increase in the proportion of motile sperm in consecutive ejaculates was observed in some studies (Almquist and Cunningham, 1967; Foster *et al.*, 1970; Almquist *et al.*, 1976). In 10–18 month-old Holstein bulls, volume and concentration of the first and second ejaculates were 4 and 3.3 ml and 1.14 and 0.63 billion/ml, respectively (Diarra *et al.*, 1997), whereas in Holstein bulls of various ages, volume, sperm concentration and total sperm in the first and second ejaculates were 8.2 and 7.3 ml, 1.5 and 0.95 billion/ml and 12.0 and 6.8 billion, respectively (Everett and Bean, 1982). Ejaculate volume, sperm concentration and total sperm in dairy and beef *B. taurus* bulls in Brazil were 6.9 to 8.2 ml, 1.2 billion/ml and 8.2 to 9.1 billion, respectively (Brito *et al.*, 2002a,b).

Sperm output is greatly affected by sexual preparation, which has been defined as prolonging the period of sexual stimulation beyond that adequate for mounting and ejaculation (Amann and Almquist, 1976). Optimal sexual preparation ensures that the ejaculate contains the greatest possible number of sperm. The techniques used for sexual preparation include restraint and false mounts, i.e. mounting without ejaculation. While restraint can be effective, it is important to ensure that the bull is actually stimulated and not simply standing close to the stimulus animal; in practice, this can be a subjective assessment. In contrast, false mounting is evident to the observer and therefore is always recommended, except for bulls with severe physical disability. Sperm output increases with both increased restraint time and number of false mounts. In one study with dairy bulls, total sperm numbers in the ejaculate were 7.1, 11.7, 14.2 and 13.5 billion with zero, one, two and three false mounts, respectively. When 5 min of restraint was also used, total sperm numbers were 14.1, 13.8, 15.1 and 17.4 billion, respectively. Greater sperm output was mainly associated with increased semen volume and less so with increased sperm concentration (Hafs *et al.*, 1962). However, dairy and beef bulls respond differently to sexual preparation. In one study, sperm harvest from Angus and Hereford bulls using three false mounts as sexual preparation did not differ from that obtained by using one false mount only and 5 min of restraint, or using no false mounts at all (Foster *et al.*, 1970). Another study demonstrated that total sperm output in two ejaculates from Angus bulls increased significantly if three false mounts were used before collection of the first ejaculate, although the increment in output was minimal when the same sexual preparation was used for the second ejaculate (Almquist, 1973).

A good understanding of the dynamics of spermatogenesis is crucial for understanding abnormal sperm production. Total duration of spermatogenesis is 61 days in bulls. After spermiation, sperm are transported through the epididymis in a process that takes approximately another 10 days. Thus, sperm present in the ejaculate began being produced 71 days earlier and semen quality is a reflection of those events in the previous 2 months that influenced spermatogenesis, sperm maturation and transport. Different testicular germ cell lines have different sensitivities to different insults. For example, spermatocytes and spermatids are particularly sensitive to increased testicular temperature, but spermatogonia are more resistant (Setchell, 1998). Therefore, particular changes in semen quality are manifested after an interval that varies according to the developmental stage of the germ cells at the time of the insult, and the time required for the damaged cells to be released into the seminiferous tubules and transported through the epididymis. If exposure to the insult is limited, a consistent sequence of appearance of sperm defects in the ejaculate is expected, whereas if the exposure is prolonged, a variety of sperm defects may be simultaneously present in the ejaculate. Semen quality improves as the spermatogonia that resisted the insult enter mitosis and meiosis and restart generating normal sperm.

The mildest form of testicular degeneration produces no grossly detectable changes in the testicular tissue and is manifested exclusively by increased production of abnormal sperm. Elevated testicular temperature and endocrine disruption are probably the most common causes of mild testicular degeneration. Normal spermatogenesis in bulls depends upon maintenance of optimal testicular temperature, with increased testicular temperature having detrimental effects on sperm production and semen quality. Metabolic rate and oxygen demand increase as a result of augmented testicular temperature. However, the long and extremely coiled testicular artery limits the blood supply to the testes. As blood flow to the testes either does not increase at all, or does not increase sufficiently to match the increased metabolic rate of the heated tissue, testicular hypoxia develops (Setchell, 1998). Causes of increased testicular temperature include increased whole body temperature (high ambient temperature, pyrexia), increased local temperature (scrotal trauma or dermatitis, orchitis, periorchitis, epididymitis), decreased local heat irradiation (hydrocele, scrotal oedema, fat accumulation around the spermatic cords), and alteration of normal testicular mobility (tunic adhesions, inguinal and scrotal hernias).

Scrotal insulation for more than 24 h results in decreased sperm motility and an increased proportion of abnormal sperm in the ejaculate 7–12 days afterwards. The proportion of abnormal sperm peaks at around 18 to 21 days after insulation, and sperm motility and morphology return to pre-insulation levels only after about 45 days (Fig. 12.7). The sequential appearance of sperm defects after

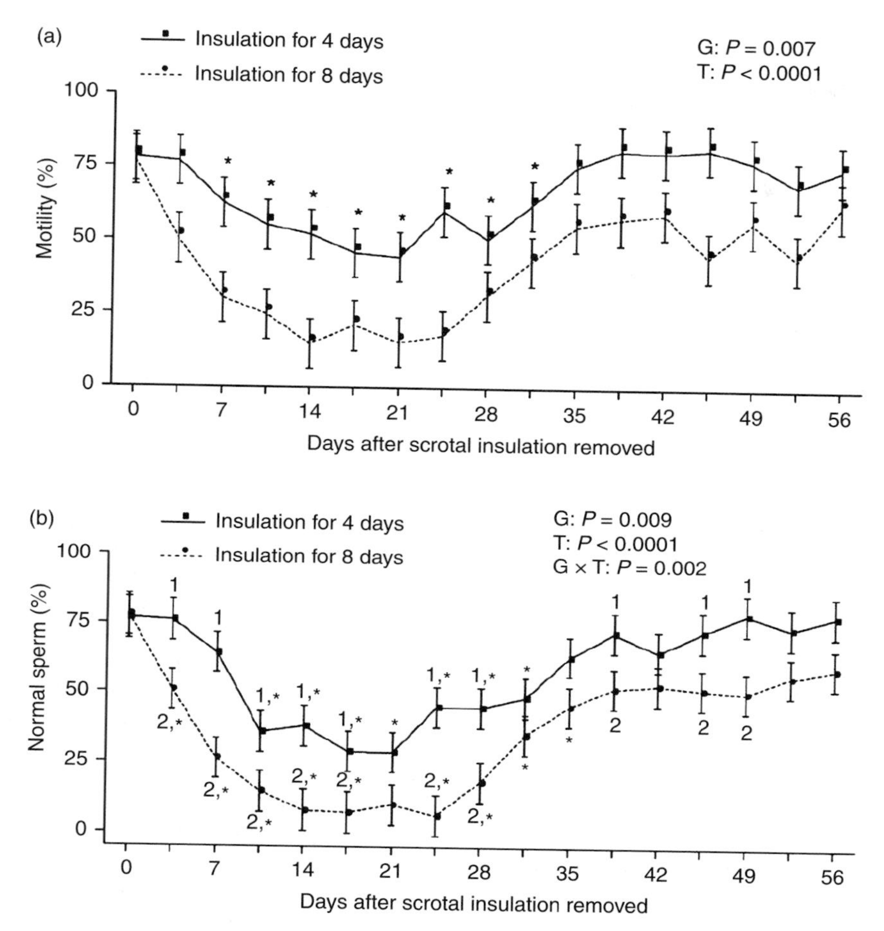

Fig. 12.7. Mean (± SEM) proportion of (a) motile sperm and (b) normal sperm after scrotal insulation for 4 or 8 days (n = 8/group) in Angus, Charolais and Hereford bulls. G = group effect; T = time effect; G × T = group-by-time interaction effect. On each graph, '*' indicates values within the group that differ ($P < 0.05$) from the day before insulation (day 0); '1' and '2' indicate that values differ ($P < 0.05$) between groups within the examination day. Adapted from Arteaga *et al.*, 2005, with permission from Elsevier.

scrotal insulation included an increase of detached sperm heads between 7 to 14 days, midpiece defects between 7 and 21 days, cytoplasmic droplets between 7 and 28 days, nuclear vacuoles between 18 and 21 days, and abnormal sperm head and acrosome defects between 21 and 28 days. There was usually marked individual variation in the proportion of specific sperm defects that were observed after scrotal insulation (Austin *et al.*, 1961; Vogler *et al.*, 1993; Barth and Bowman, 1994). While the increase in abnormal sperm seems to be mainly a reflection of insult to spermatids, decreased sperm production is also often observed after longer periods of scrotal insulation, indicating that spermatocytes are also affected by elevated temperature, but are more likely to degenerate than to develop into abnormal sperm. Exposure of bulls to high temperatures in controlled-environment chambers has profound physiological effects that might also interfere with testicular function. Nevertheless, the observed changes in semen quality resemble those occurring after scrotal insulation (Casady *et al.*, 1953; Skinner and Louw, 1966). The deleterious effect of increased testicular temperature was 'dose dependent', with more prolonged insult resulting in greater decreases of semen quality and sperm production as well as increased time required for recovery (Skinner and Louw, 1966; Arteaga *et al.*, 2005).

The maintenance of adequate intratesticular testosterone concentration is essential for normal spermatogenesis. Endocrine disruption leading to decreased testosterone production results in increased production of abnormal sperm. Stress and the associated release of cortisol is the most common endocrine disruptor causing abnormal sperm production in bulls. Cortisol inhibits GnRH, LH and, ultimately, testosterone production. Illnesses, drastic changes in nutrition or management, changes in social dominance, show ring travelling and long-distance transport are some of the events that may result in stress, with increased production of cortisol and resultant abnormal sperm. Clearly, stress must persist for a long enough period for decreased testosterone concentration to affect semen quality. For example, the proportion of normal sperm decreased from 85 to 20% approximately 20 days after a bull developed lameness, but another bull with a prior history of satisfactory semen quality presented only 45% normal sperm following dehorning a month earlier; recovery (i.e. 85% normal sperm) was observed on a follow-up examination 2 weeks later (L. Brito, personal observations). The deleterious effects of cortisol on spermatogenesis have also been directly demonstrated by treatment of beef bulls with corticosteroids. Interestingly, the dynamic changes in the appearance of sperm defects in the ejaculate, the types of defects observed after treatment, the severity of impairment to spermatogenesis and the time required for recovery after cessation of treatment were virtually identical to those observed after increased testicular temperature induced by scrotal insulation (Barth and Bowman, 1994).

The effects of season on sperm production and semen quality are probably multifactorial and difficult to define. Some of these factors might include temperature, humidity, photoperiod, housing, nutrition and management. Studies conducted at AI centres are the most informative, as at least some of these factors are controlled, but even these have reported inconsistent results. In temperate regions, temperature is less likely to have a significant negative impact on sperm production, and photoperiod might have a more significant effect than in tropical areas, although the mechanisms involved in photoperiodic regulation of sexual function in bulls have not been well studied. In one study in the USA, the most extreme temperatures observed (−24 to −19°C and 27 to 32°C) only caused a 0.3 ml decrease in semen volume and a small decrease in the total number of sperm in the ejaculate (0.3 billion) when compared with what was defined as optimal temperatures (16–21°C) (Taylor *et al.*, 1985). However, while some studies in the UK and the USA observed increased sperm production in Holstein bulls during the summer (Parkinson, 1985; Taylor *et al.*, 1985), other studies in the USA and Canada have reported a decrease in sperm production during this season (Everett and Bean, 1982; Mathevon *et al.*, 1998). In another study in Canada, an inverse seasonal variation in ejaculate volume and sperm concentration (increase of the former and decrease of the latter) from spring to the summer resulted in non-significant variation in sperm production over the year in Holstein bulls (Diarra *et al.*, 1997).

Elevated temperature is more likely to affect sperm motility and morphology, even in temperate regions. In both the UK and Canada, Holstein bulls have been shown to produce sperm with lesser motility and a greater proportion of morphological defects during the summer (Parkinson, 1985; Mathevon *et al.*, 1998), but contradictory observations have been reported in Swedish Red and White and Holstein bulls in Sweden; in one report, the greatest proportion of sperm defects was observed during spring and summer (Söderquist *et al.*, 1996), but in another report, more sperm defects were observed during the winter and spring (Söderquist *et al.*, 1997). In the tropics, sperm production decreased and sperm abnormalities increased in South Devon and Holstein bulls during the warmer seasons when compared with the wet, cool seasons (Kumi-Diaka *et al.*, 1981). In Brazil, season did not affect sperm production in Limousin and Simmental bulls, but sperm abnormalities increased from 18% during the winter to 44%

during the summer (Koivisto *et al.*, 2009). Conversely, month accounted for less than 2% of the variation in sperm production and semen quality over the year in dairy and beef bulls in another study in AI centres in Brazil (Brito *et al.*, 2002b).

Semen quality also differs substantially among mature dairy, young dairy and beef bulls. In addition, season seems to affect these categories of bulls differently. Young dairy and beef bulls usually produce semen with a larger proportion of abnormal sperm than mature dairy bulls, which might be related to the lower selection pressure on the former two bull populations. In addition, a much more pronounced decrease in the production of normal sperm is observed during the summer in beef bulls than in young or mature dairy bulls (Fig. 12.8). An evaluation of records over a period of 5 years in the USA revealed that, whereas only 6.3% of semen batches from mature dairy bulls (n = 26,899) were discharged as a result of poor sperm morphology (<75% morphologically normal sperm), 31.3% of batches from young dairy bulls (n = 26,209) and 32.8% of batches from beef bulls (n = 15,800) were discharged for the same reason (ABS Global, Inc., unpublished).

Sexual Behaviour

A study of the development of sexual behaviour in Hereford bulls indicated that mounting in response to an oestrous female was first observed by 3, 6 and 9 months of age in 18.5, 26 and 48% of the bulls, respectively. Some 59% of the bulls achieved their first ejaculations by 12 months of age, whereas by 15 months of age 78% registered a complete service. The number of services increased with age until 18 months old (Price and Wallach, 1991a). In Angus, Brown Swiss, Hereford, Angus × Hereford and Red Poll bulls the first completed service was observed around 11 months of age (Lunstra *et al.*, 1978). Rearing seems to affect sexual development behaviour, as Hereford bull calves raised in individual pens had a greater number of services when tested for the first time than bulls raised in groups, although these differences quickly disappeared once the bulls were grouped together (Lane *et al.*, 1983). In another study, the influence of the presence of females during bull rearing was evaluated. In the first 2 h of being exposed to females in oestrus, males raised with females had 73% more services than bulls raised in isolation.

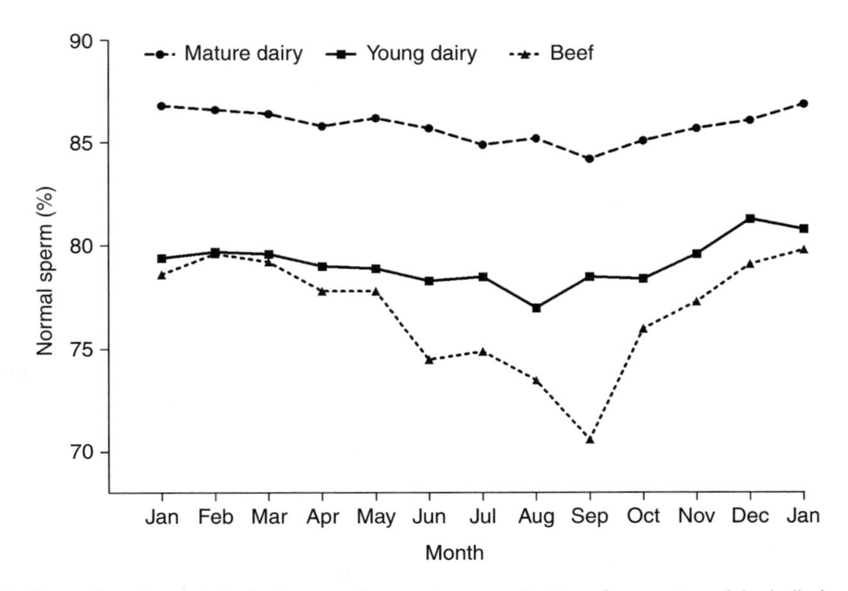

Fig. 12.8. Proportion of morphologically normal sperm in semen batches from mature dairy bulls (n = 7048), young dairy bulls (≤24 months old; n = 6767) and beef bulls (n = 4176) during one year in the USA (ABS Global Inc., unpublished data).

Bulls reared in isolation did compare favourably in subsequent tests though, illustrating that a learning process occurred rather quickly after exposure to females (Price and Wallach, 1990).

The hormonal control of sexual behaviour in bulls has not been very well studied. Normal sexual behaviour requires a threshold of testosterone, but sexual activity was not affected by testosterone (or LH) levels beyond the threshold (Price *et al.*, 1986). In contrast, basal testosterone concentration and total testosterone secretion during 5 h after GnRH challenge have been positively and significantly correlated with libido score (Chase *et al.*, 1997). There are also suggestions that circulating prolactin concentrations and the oestrogen:testosterone concentration ratio are greater in adult Holstein bulls with lower libido, i.e. those with greater reaction time and latency time between successive ejaculates (Henney *et al.*, 1990). There is considerable evidence that sexual behaviour traits in bulls might be heritable (Chenoweth, 1997).

Female cattle play a major role in soliciting bulls. Sexually active females in pro-oestrus and oestrus distance themselves from the rest of the herd, form groups and engage in frequent homosexual mounting activity. The behaviour of the sexually active group offers visual cues to stimulate bulls. Once attracted to the group, bulls investigate individual females for receptivity, initially by sniffing and licking the vulva. These actions are often followed by the flehmen response – a curling upwards of the upper lip that helps to expose and draw molecules to the vomeronasal organ for the detection of pheromones. The bull then tests the female tolerance to mounting by resting its chin on the female's back and making real or sham mounting attempts. Females demonstrate acceptance by remaining quiet and immobile during the investigation. The greatest single stimulus for a bull to mount and attempt service seems to be immobility of the mounting subject, as bulls will mount restrained non-oestrous females, or even bulls or steers, providing that adequate immobility is displayed. During the mount, the bull lifts its forequarters off the ground and clasps the rear of the female between his front limbs. Mounting and clasping allow the tip of the erect penis to access the vagina, accompanied by a series of pelvic oscillations (seeking movements) that culminate with the tip of the penis reaching and penetrating the vaginal orifice. This is immediately followed by two simultaneous events. There is a sudden lengthening of the penis with straightening of the sigmoid flexure and a body thrust, which bring the bull's pelvic region into direct apposition with the cow's perineum, resulting in maximum intromission and ejaculation (Blockey, 1976; Chenoweth, 1983, 1986).

Bulls usually give undivided attention to one oestrous female while ignoring others in the immediate vicinity. However, after serving the female a few times the bull seems to lose interest and then turns its attention to a different female. In some experiments, bulls were shown to rarely breed with the same female more than once and to evenly distribute services; this pattern seems to be dependent on the bull's serving capacity, with high-capacity bulls breeding with the same female repeatedly and more often than lower capacity bulls (Blockey, 1975). Others' reports have shown that bulls repeatedly breed with individual females until female sexual receptivity became attenuated and/or bulls approached sexual satiety (Bailey *et al.*, 2005b). Most of a group of Angus bulls individually exposed to ten females in oestrus until the completion of six services, bred with three to five females only; the bulls bred with at least one female more than once in 92% of the tests, and in 67% of the tests at least one female was bred twice in succession (de Araujo *et al.*, 2003). Bulls are capable of intense sexual activity, and in natural breeding situations it is not uncommon for them to breed 30–35 times over a day. Bulls exposed to oestrus-synchronized females have been observed to breed 14–101 times within a 30 h period (Chenoweth, 1986). After breeding, the bull shows a period of satiation during which there seems to be no sexual interest. The time it takes for a bull to show interest and mount a female again increases as the number of services increase, but decreases with the introduction of novel stimuli (the Coolidge effect). However, other physiological mechanisms and/or physical fatigue also seem to be involved, because breeding activity was greatly reduced for certain periods after exposure to oestrous females regardless of the presentation of novel females (Bailey *et al.*, 2005a).

In AI centres, where semen is routinely collected from individual bulls multiple times every week, maintaining proper sexual stimulation to efficiently obtain the number of desired ejaculates can be a challenge. Reaction time, i.e. the interval between presentation of the bull to the stimulus animal and the initial mount, is considered to be the most objective behavioural measure of bull sexual stimulation (Amann and Almquist, 1976). The length of reaction time is usually greater for beef than for dairy bulls. In one study, the reaction times before the first and second ejaculates for Angus and Hereford bulls were 13 and 6 min, respectively, but for Holstein bulls, the times were 1 and 3.5 min, respectively (Almquist, 1973). Reaction time also gradually increases with repeated exposure to the same stimulus situation; here, presentation of the same stimulus animal at a different location, presentation of a new stimulus animal, or a combination of both, are management options that can be used to shorten the reaction time. The number of combinations of stimulus animals and/or locations necessary to obtain an ejaculate has been defined as stimulus pressure. The more frequently semen is collected from a bull, the greater the stimulus pressure required. In AI centres, it is common to change the stimulus situation every 5–10 min if the bull response is poor (Amann and Almquist, 1976; Amann, 1990).

Knowledge of social behaviour is also important when considering housing, management and use of bulls for natural breeding or for AI. Social interactions between animals can be classified into amicable or agonistic. Amicable behaviours are seen most commonly in groups of young bull calves without a stable social order, and include actions such as play behaviour – including sham fighting and butting, mounting, and licking of the head, neck and preputial regions. Agonistic behaviour becomes more evident in older animals, especially during the formation or re-establishment of the social order of a group. Agonistic behaviours include pawing the ground, rubbing the head and neck in the ground and directing the lowered head toward the opponent. A threat is followed by submission and avoidance, or a threat in return, invariably resulting in butting, head-to-head pushing or blows to the body with the forehead. Physical aggression continues until one of the bulls avoids contact. Interestingly, physical aggression without preceding threats is more common in younger bulls that are establishing their social order, whereas overt aggression is much less common in socially stable groups of older bulls. In these older groups, sparring involves mostly a ritualized sequence of threats, uncommonly followed by contact. In addition, subtle body movements, sometimes almost unnoticeable, indicate threats that result in avoidance behaviour by subordinate bulls (Blockey, 1975).

The outcomes of agonistic interactions between bulls determine the dominance–submission relationships between them. Forcing bulls to enter into each other's personal space or to compete for space at the water trough and determining the 'winner' of the encounter are some methods that have been used to determine the social dominance hierarchy in groups of bulls (Blockey, 1975; Godfrey and Lunstra, 1989). The hierarchy of the herd is established within a few hours of when bulls are grouped. There are indications that, once formed, the hierarchy of the herd is maintained for a considerable period as long as the same animals remain within the herd. In most groups, bulls establish a linear hierarchy order, although linear-tending orders and, more infrequently, complex hierarchy orders are also observed. Age is a major determinant of dominance, and seniority (i.e. relative time with the group) is also important. In one experiment, bulls of similar age and weight newly introduced to the herd 'lost' approximately 95% of their encounters with bulls introduced to the herd a year previously (Blockey, 1975). Other factors that influence dominance are physical size and horns. Bulls establish dominance over females when they are between 1 and 2.5 years of age, and they are less likely to check females for oestrus if they have not yet established dominance.

Sexual behaviour in multi-sire breeding situations is largely dependent on the stability of the social hierarchy. In groups of young bulls in which the hierarchy is unstable and not well defined, bulls are more likely to engage in aggressive behaviour and to overtly compete for females. Two-year-old bulls exposed to oestrous females in groups did not differ in the number of mounts and services that they performed, regardless of their predetermined social hierarchy. In contrast, the social hierarchy is usually stable in older mixed-age groups where bulls tend to be much more observant of their status. In these groups, a subordinate bull will usually abandon a female if approached by a dominant bull. Therefore, the breeding activity of subordinate bulls is much reduced compared with that of dominant bulls. Studies have shown that in three- or four-sire breeding groups, the oldest or second oldest bull sired 60% or more of the calf crop, whereas the youngest bull sired 15% or less. Observations of breeding behaviour in groups that contained two young Angus bulls (2- and 3-years old) and one older bull (5 years old) indicated that the younger, subordinate bulls spent considerably less time with the sexually active group of females than did the older dominant bull (91 versus 53%, respectively). When three or fewer females were in oestrus, the dominant bull(s) successfully prevented subordinate bulls from breeding (Blockey, 1975). Hence, maintaining separate groups of yearling, 2-year-old and older bulls in multiple-sire breeding situations is recommended.

Breeding Soundness Evaluation

Bull breeding soundness evaluation (BSE) procedures have been developed as a means by which veterinarians use their clinical competence and broad knowledge of animal anatomy, physiology and pathology to: (i) identify bulls with a greater probability of being subfertile or infertile: and to (ii) diagnose fertility problems in bulls. The BSE is a time- and cost-effective method that allows veterinarians and bull owners, managers and buyers to make appropriate bull selection and management decisions. The earliest development of a systematic approach that typifies recommended procedures now adopted worldwide can be credited to the Society for Theriogenology (SFT) in North America and the organizations from which the society evolved (Rocky Mount Society for the Study of Breeding Soundness of Bulls and, later, the American Veterinary Society for the Study of Breeding Soundness). Since its inception, the SFT has periodically revised and published guidelines for bull BSE.

In general terms, evaluation of bulls for breeding soundness involves a general physical examination, a specific examination of the reproductive system (including measurement of SC), and a semen examination. Minor changes to the guidelines had been introduced over the years until a major change was adopted with the guidelines published in 1993. Before this date, the guidelines included a scoring system for SC, sperm motility and morphology. The final classification as satisfactory, questionable or unsatisfactory was based on the combined score of all three parameters. Thus bulls with clear deficiencies in one parameter could still obtain a satisfactory classification as long as the scores in other parameters were sufficiently high to compensate for this. In addition, the score system was open to misinterpretation, as it appeared to rank bulls in terms of their potential reproductive performance. Recognition that deficiencies in one category, such as SC, sperm motility or morphology, could not be compensated by high values in another, led to the introduction of a system based on minimum threshold requirements for each of these parameters and the abandonment of the earlier scoring system (Hopkins and Spitzer, 1997). Use of the new system resulted in 5–10% fewer bulls obtaining a satisfactory classification, indicating that the adopted system was indeed stricter than the scoring system that preceded it (Spitzer and Hopkins, 1997; Higdon *et al.*, 2000).

According to the SFT 1993 guidelines, bulls must be free of any relevant physical abnormalities, have a minimum SC of 30–34 cm depending on age (Table 12.2), and have ≥30% progressively motile sperm and ≥70% morphologically normal sperm in order to be classified as satisfactory potential breeders. Unsatisfactory potential breeders are bulls that fail to meet the criteria for satisfactoriness in one or more parameters. In addition, a category termed 'classification deferred' is used for bulls that are considered likely to improve with time or therapy after consideration of signalment, history, test results and nature of the deficiencies. Many bulls classified in this category are yearlings that may not have reached the suggested minima for SC and semen quality by the time of the examination. It is important to note that observation of the bull's ability to successfully breed with females is not a required component of the BSE, so the veterinarian should recommend close observation of bulls during breeding for the identification of problems that cannot be detected during the usual BSE. In addition, tests for venereal diseases (e.g. campylobacterosis and trichomoniasis) and other additional tests may be recommended in certain cases.

Some points in the SFT guidelines warrant consideration. It is important to understand that the minimum requirements for SC, sperm motility and sperm morphology were established to select 'satisfactory' potential breeders and not necessarily 'superior' breeders. Moreover, the thresholds were selected from a physiological perspective and were not intended to be used as tools for genetic selection.

Table 12.2. Minimum scrotal circumference (cm) for breeding soundness classification of bulls at different ages according to the guidelines from the Society for Theriogenology (Hopkins and Spitzer, 1997).

Age (months)	Scrotal circumference (SC)
≤15	30
>15 ≤ 18	31
>18 ≤ 21	32
>21 ≤ 24	33
>24	34

The minimal threshold for SC was selected because the vast majority of bulls with an SC ≥30 cm should be pubertal, and all bulls of all breeds are expected to attain this SC at or before 15 months of age when raised on an adequate nutrition plane. The relatively low threshold for sperm motility was selected after consideration of the inconsistent associations of this parameter with fertility after natural breeding reported in the literature and the often difficult environmental conditions encountered for the evaluation of sperm motility in the field. However, a sperm motility <30% may indicate that a representative ejaculate was not obtained and further collection attempts may be warranted. While it was recognized that further research may require changes to the procedures for sperm morphological evaluation, the minimum recommendation based on the proportion of normal sperm only was chosen to lessen the emphasis and debate on the significance of specific sperm defects (P.J. Chenoweth and J.C. Spitzer, personal communication).

Although the earlier and outdated score classification system from the SFT is still used in some countries, such as Brazil (CBRA, 1998), the general philosophy of the 1993 SFT guidelines has been adopted in several countries other than the USA, such as Canada, South Africa and Australia. The main differences in these countries are the minimum requirements adopted for SC, sperm motility and sperm morphology, and how the relevant results are reported.

The guidelines from the Western Canadian Association of Bovine Practitioners, for example, contain different recommendations for minimum SC that are based on breed and age (Table 12.3), and the minimum sperm

Table 12.3. Minimum scrotal circumference (cm) for breeding soundness classification of bulls of different breeds at different ages according to the guidelines from the Western Canadian Association of Bovine Practitioners (Barth, 2000).

Age (months)	Angus, Brown Swiss, Gelbvieh, Pinzgauer, Simmental	Charolais, Hereford, Holstein Maine Anjou, Red Poll, Salers, Shorthorn	Blonde d'Aquitaine, Galloway, Limousin, Texas Longhorn
12	32	30	29
13	33	31	30
14	34	32	31
15–20	35	33	32

motility threshold is 60%. Under this scheme, bulls can be classified as satisfactory, unsatisfactory or questionable potential breeders, or as 'decision deferred' (Barth, 2000). The guidelines from the South African Veterinary Association contain similar minimum recommendations for SC to those of the SFT for *B. taurus* bulls, but the minimum requirements for motile and morphologically normal sperm are 70 and 75%, respectively. In addition, only bulls considered satisfactory receive a 'certificate of breeding soundness'; evaluation findings and recommendations for bulls deemed not satisfactory are communicated either verbally or in a different written report, and there are no intermediary classifications categories such as 'questionable' or 'decision deferred' (Irons *et al.*, 2007). The guidelines from the Australian Cattle Vets (ACV) allow minimum requirements for SC to be modified to meet breed association requirements, even though there is a large degree of consensus. Sperm motility and normal morphology should be ≥30% and ≥50%, respectively, for bulls used for natural breeding, and ≥60% and ≥70%, respectively, for bulls from which semen is to be frozen for AI. In addition, there are recommendations for maximum tolerated percentage of specific sperm defects, and veterinarians need to be certified to evaluate sperm morphology. The examining veterinarian makes the final determination as to whether a bull should be categorized as a satisfactory potential breeder or not, using established standards for physical soundness, SC, sperm motility and sperm morphology as highly important considerations, in conjunction with other factors such as individual bull, herd and client histories (Fordyce *et al.*, 2006).

Several studies describing the results of BSEs indicated that only 65 to 85% of beef bulls are considered satisfactory prospective breeders (Carroll *et al.*, 1963; Elmore *et al.*, 1975; Spitzer *et al.*, 1988; Bruner *et al.*, 1995; Carson and Wenzel, 1997; Spitzer and Hopkins, 1997; Higdon *et al.*, 2000; Barth and Waldner, 2002; Kennedy *et al.*, 2002; Godfrey and Dodson, 2005; Sylla *et al.*, 2007). The ability to detect and eliminate this relatively large proportion of bulls that are expected to achieve substantially lower fertility rates stresses the importance and the value of performing regular evaluations. However, the results of each specific study are obviously affected by the characteristics of the bull population examined (breeds, age groups, numbers of bulls and herds, management and geographical location), and care must be taken to avoid generalizations.

The prevalence and type of physical abnormalities, for example, vary considerably among studies. In the USA, physical abnormalities were observed in 1.5 to 9.5% of examined bulls (Elmore *et al.*, 1975; Spitzer *et al.*, 1988; Bruner *et al.*, 1995; Carson and Wenzel, 1997; Higdon *et al.*, 2000; Kennedy *et al.*, 2002). Reports vary in describing the most common abnormalities as those associated with eye lesions and feet and leg problems (Carson and Wenzel, 1997), penile problems (fibropapilloma and persistent frenulum) (Bruner *et al.*, 1995) and vesicular adenitis (Spitzer *et al.*, 1988; Kennedy *et al.*, 2002). Similarly, physical abnormalities were observed in 4.3% of Italian beef bulls, although the most common abnormalities were testicular problems (hypoplasia, orchitis and cryptorchidism) (Sylla *et al.*, 2007). In contrast, a much greater prevalence of physical abnormalities (24.3%) was reported in beef bulls in Western Canada; these were mostly associated with a high prevalence of scrotal frostbite and feet and leg problems (Barth and Waldner, 2002). Despite variations among studies, the presence of physical abnormalities is usually the third most common cause of unsatisfactory classifications after small SC and poor sperm morphology.

The effect of age and the interaction of age with breed need to be considered when evaluating the results of bull breeding soundness studies. Evaluation of yearling bulls between 10 and 15 months of age is very common, and bulls at these ages are still going through the pubertal changes associated with rapid testicular growth and improvement in semen quality. So it is not surprising that the proportion of yearling beef bulls classified as satisfactory potential breeders increased from approximately 55% at 10 months of age, to 72% at 11 months and to 78% at 12 months of age as more bulls reach the minimum requirements for SC, sperm motility and sperm morphology (Higdon *et al.*, 2000; Kennedy *et al.*, 2002). As breed differences occur in sexual development, a larger proportion of bulls from early-maturing breeds are expected to obtain satisfactory classification at a younger age than of bulls from late-maturing breeds. An extreme example was found with tropically adapted Senepol bulls; the proportion of satisfactory yearling bulls was only 25–50%, but this increased to 70–85% at 22–26 months of age (Chenoweth *et al.*, 1996; Godfrey and Dodson, 2005). It has also been reported that the proportion of satisfactory bulls decreased after 5–6 years of age as bulls become more prone to develop physical and reproductive problems (Carson and Wenzel, 1997; Barth and Waldner, 2002). Therefore, yearling bulls were more likely to receive an unsatisfactory or deferred classification owing to small SC (Higdon *et al.*, 2000; Kennedy *et al.*, 2002), whereas older bulls were more likely to receive the same classification as a result of poor sperm morphology (Carson and Wenzel, 1997). The proportion of bulls that are classified as unsatisfactory exclusively because of poor sperm motility is usually low.

Scrotal circumference and bull selection

SC is highly correlated with testicular weight (Fig. 12.9) and is a moderately heritable trait in cattle, with yearling heritability estimates ranging from 0.32 to 0.71 (Table 12.4), which indicates that selection can have a very significant impact on this trait. For example, testicular weight at weaning was greater in the progeny sired by Limousin bulls with high expected progeny difference (EPD) for SC when compared with progeny sired by bulls with average or low EPD (Moser *et al.*, 1996). Several studies have also demonstrated moderate-to-high phenotypic correlations between SC and growth traits (Coulter and Foote, 1977; Bourdon and Brinks, 1986; Kriese *et al.*, 1991; Crews and Porteous, 2003), and estimates of the genetic correlations between SC and growth traits are generally positive (Table 12.5), indicating that selection for SC is associated with positive gains in growth traits.

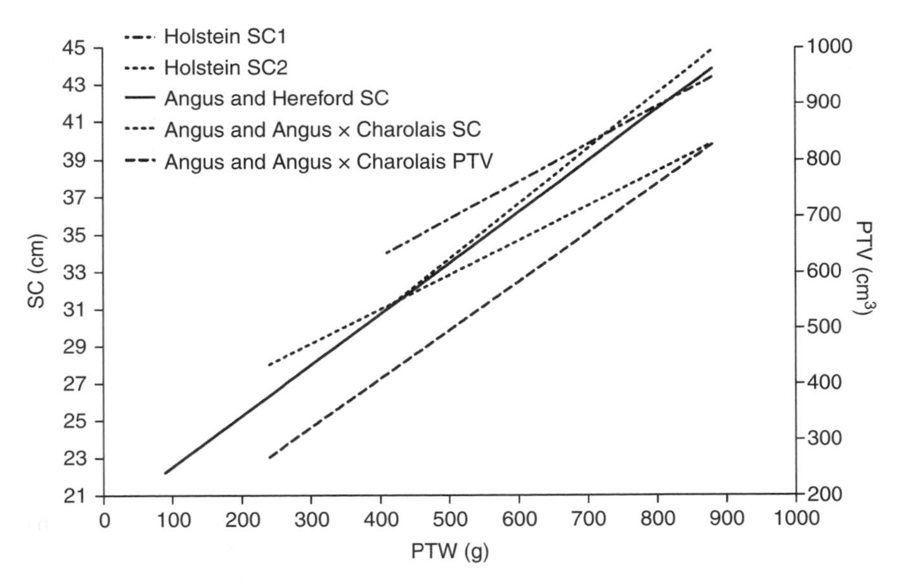

Fig. 12.9. Regression lines for paired testes weight (PTW) according to scrotal circumference (SC) or paired testes volume (PTV = 0.0396 × SC × (mean testicular length)^2). Holstein SC1: measurements obtained from mature Holstein bulls (n = 35); PTW = –1298.5 + (50.2 × SC) (Hahn *et al.*, 1969). Holstein SC2: measurements obtained from Holstein bulls (n = 47) 19–184 months old; PTW = –654.4 + (34 × SC) (Coulter and Foote, 1976). Angus and Hereford SC: measurements obtained from Hereford (n = 199) and Angus (n = 136) bulls 11–30 months old; PTW = –722.28 + (36.53 × SC) (Coulter and Keller, 1982). Angus and Angus × Charolais SC and PTV: measurements obtained from Angus and Angus × Charolais bulls (n = 111) 14–16 months old; PTW = –1274 + (54.04 × SC) or PTW = –67.89 + (1.147 × PTV) – better fit observed for PTV (r^2 = 0.78) than for SC (r^2 = 0.61) (L. Brito, unpublished data).

Table 12.4. Heritability estimates (h^2) for yearling scrotal circumference in bulls.

Breed	h^2	Reference
Angus	0.36–0.55	Latimer *et al.*, 1982; Knights *et al.*, 1984; Gipson *et al.*, 1987; Meyer *et al.*, 1991; Garmyn *et al.*, 2011
Composite	0.66	Mwansa *et al.*, 2000
Hereford	0.40–0.71	Neely *et al.*, 1982; Bourdon and Brinks, 1986; Nelsen *et al.*, 1986; Gipson *et al.*, 1987; Kriese *et al.*, 1991; Meyer *et al.*, 1991; Evans *et al.*, 1999; Crews and Porteous, 2003; Kealey *et al.*, 2006
Holstein	0.67	Coulter *et al.*, 1976
Limousin	0.46	Keeton *et al.*, 1996
Red Angus	0.32	McAllister *et al.*, 2011
Simmental	0.48	Gipson *et al.*, 1987
Various breeds	0.40–0.53	Lunstra *et al.*, 1988; Rose *et al.*, 1988; Smith *et al.*, 1989b; Martínez-Velázquez *et al.*, 2003

Table 12.5. Genetic correlations (r_g) between scrotal circumference and growth traits in bulls.

Breed	Growth trait	r_g	Reference
Angus	Yearling weight	0.24–0.68	Knights *et al.*, 1984; Meyer *et al.*, 1991
Composite	Yearling weight	0.40–0.43	Mwansa *et al.*, 2000
Hereford	Weaning weight	0.08–0.86	Neely *et al.*, 1982; Bourdon and Brinks., 1986; Nelsen *et al.*, 1986; Kriese *et al.*, 1991; Meyer *et al.*, 1991; Crews and Porteous, 2003
	Yearling weight	0.30–0.52	
	Weaning–yearling ADG[a]	0.22–0.35	
Limousin	Weaning weight	0.14	Keeton *et al.*, 1996
Red Angus	Yearling intramuscular fat	0.05	McAllister *et al.*, 2011
	Yearling carcass marbling score	0.01	
Various	Birth–weaning ADG	0.02	Lunstra *et al.*, 1988; Smith *et al.*, 1989b
	Yearling weight	0.10–0.63	
	Weaning weight	0.56	
	Weaning–yearling ADG	0.59	

[a]ADG, average daily gain.

The general trend of increasing SC in certain beef breeds over time seems to be the result of the combination of direct selection for SC and indirect selection for growth traits by breeding stock producers (Fig. 12.10). Although the heritability of semen traits is generally low, SC is positively associated with sperm production and semen quality (Gipson *et al.*, 1987; Moser *et al.*, 1996; Barth and Waldner, 2002), and genetic correlations between SC and semen traits are generally favourable (Table 12.6). This suggests that direct selection for SC would be more effective in bringing about improvement in sperm production and semen quality than would direct selection pressure on semen traits themselves.

Several studies have reported an association between sire SC and daughter age at puberty. In Hereford cattle, genetic correlations between yearling SC and daughter ages at first breeding and at first calving were –0.39 and –0.38, respectively (Toelle and Robison, 1985). In another study with beef cattle, favourable relationships were demonstrated by negative correlation coefficients between greater sire SC and heifer progeny ages at puberty and at first calving (Smith *et al.*, 1989a). In a population of composite beef cattle, correlation coefficients among parental breed group means for SC and percentage of pubertal females at 452 days of age was 0.95, while the correlation with female age at puberty was –0.91 (Gregory *et al.*, 1991).

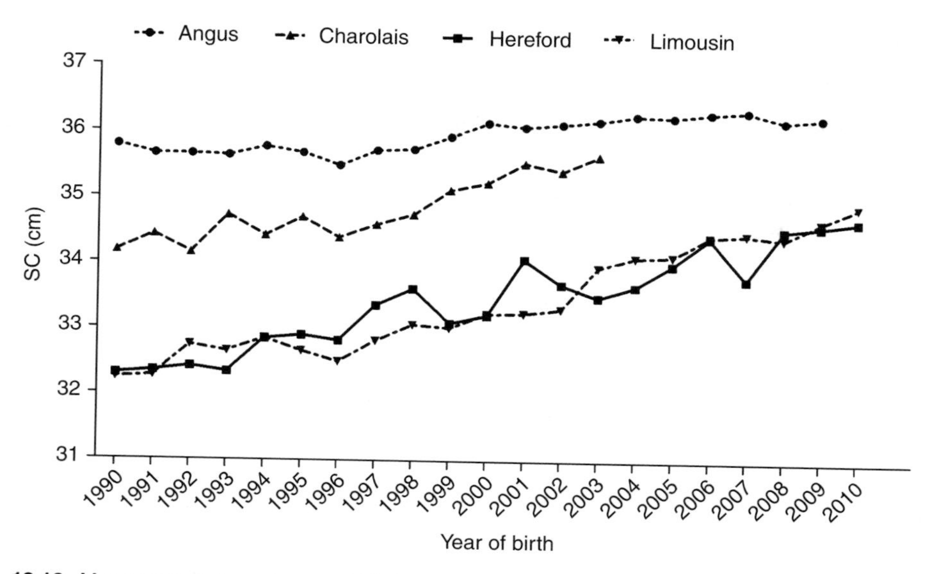

Fig. 12.10. Mean scrotal circumference (SC) in registered yearling bulls according to year of birth. Data for Limousin (n = 73,757; 1184–5200/year) and Angus (unspecified number of bulls) are measurements adjusted to 365 days of age (From North American Limousin Foundation and American Angus Association). Data for Hereford (n = 5553; 360–536/year) and Charolais (n = 6984; 121–997/year) are unadjusted measurements obtained between 321 and 421 days of age (from Canadian Hereford Association and Canadian Charolais Association).

Table 12.6. Genetic correlations (r_g) between scrotal circumference and sperm production and semen quality traits in bulls.

Breed	Trait	r_g	Reference
Angus	Sperm concentration	0.54	Knights *et al.*, 1984; Garmyn *et al.*, 2011
	Sperm motility	0.36	
	Total sperm defects	−0.23	
Hereford	Sperm concentration	0.77	Kealey *et al.*, 2006
	Sperm motility	0.34	
	Normal sperm	0.33	
Hereford and Simmental	Sperm concentration	0.20	Gipson *et al.*, 1987
	Sperm motility	0.11	
	Total sperm number	0.19	

Females sired by Limousin bulls with a larger SC were 7 days younger at puberty than females sired by bulls with smaller SC, though this difference was not significant. In addition, a significantly greater proportion of females had reached puberty at 11 and 13 months of age when sired by Limousin bulls with high EPD for SC than females sired by bulls with low or average EPD for SC. These results suggest that sire selection using EPD for SC might be more effective in reducing age of daughters at puberty than selection based on phenotypic measurements (Moser *et al.*, 1996).

Although sire SC is associated with daughter puberty, genetic correlations between SC and pregnancy rates have often been low and insignificant (Martínez-Velázquez *et al.*, 2003; McAllister *et al.*, 2011). This may be due to a non-linear relationship between these traits. A study with Hereford cattle indicated that the effect of SC breeding values on heifer pregnancy exhibits a threshold relationship. As SC increases in value, there is a diminishing return for improved heifer pregnancy, suggesting that selection for a high SC breeding value may not be an advantage for increased heifer pregnancy over selection for a moderate SC breeding value (Evans *et al.*, 1999). However, even though it would appear that the favourable genetic relationship between SC and heifer offspring age at puberty does not completely translate into heifer pregnancy rates, it is important to note that experimental design might have confounded some of these results. This is because, if an entire group of heifers reach puberty before exposure to breeding, then those heifers reaching puberty at younger ages would show no advantage in conception over those reaching puberty at older ages. Moreover, end-of-season pregnancy rates were used in these studies as opposed to per cycle pregnancy rates, and the value of having heifers conceiving early rather than late in the season might have been lost.

Heritability estimates for SC vary according to age. Studies have demonstrated that heritability estimates increase with age until approximately 1 year old (Coulter *et al.*, 1976; Neely *et al.*, 1982; Mwansa *et al.*, 2000), whereas estimates in yearlings were greater than in 2-year-old bulls (Coulter *et al.*, 1987b; Meyer *et al.*, 1991). Therefore, selection based on yearling SC is recommended over selection based on measurements obtained at other ages. Yearling SC is commonly recorded for performance evaluation programmes in beef bulls, but age effects are very pronounced around this age when testicular growth is rapid. Accordingly, significant differences can be observed in bull SC between 11 and 14 months of age (Table 12.7). In order to adjust SC measurements to 365 days of age, the Beef Improvement Federation (BIF) recommends the use of the age adjustment factors described in Table 12.8. Breed effects on SC are pronounced at either 1 or 2 years of age in beef bulls (Table 12.7 and Fig. 12.11). Adjusting SC for differences in body weight at 368 days of age using regression analysis resulted in reduction in variation among breeds, but significant breed differences in SC were still present, indicating that differences in SC among breeds are only partially attributable to breed differences in yearling weight (Lunstra *et al.*, 1988; Gregory *et al.*, 1991). In Angus and Hereford bulls, correlation coefficients between SC at 1 year of age, and SC and PTW at 2 years of age, were 0.76 and 0.65, respectively, demonstrating that a bull with relatively small or large testes as a yearling will generally have comparable testes size as a 2 year old (Coulter and Keller, 1982).

Attempts to establish guidelines for the selection of bulls at weaning based on the likelihood of attainment of a certain minimum yearling SC have produced mixed results. In one study, it was recommended that the minimum SC in Angus and Simmental bulls 198 to 291 days old should be 23 or 25 cm to ensure an SC of 30 or 32 cm at 365 days of age, respectively. The same recommendations for Hereford bulls were 26 and 28 cm (Pratt *et al.*, 1991).

Table 12.7. Scrotal circumference (cm) according to breed and age in yearling bulls (n = no. bulls sampled).

Breed	10 months	11 months	12 months	13 months	14 months
Charolais[a]	32.3 (n = 246)	33.5 (n = 1,068)	34.7 (n = 2,622)	35.7 (n = 2,504)	36.0 (n = 791)
Chianina[b]	30.6 (n = 455)	32.2 (n = 455)	33.3 (n = 455)	33.8 (n = 455)	34.5 (n = 455)
Hereford[c]	31.1 (n = 77)	31.7 (n = 601)	32.5 (n = 2,510)	33.5 (n = 2,101)	34.4 (n = 341)
Holstein[d]	29.8 (n = 1,004)	31.0 (n = 482)	31.8 (n = 226)	32.8 (n = 162)	33.4 (n = 225)
Marchigiana[b]	30.4 (n = 415)	31.7 (n = 415)	33.0 (n = 415)	33.9 (n = 415)	33.9 (n = 415)
Romagnola[b]	31.5 (n = 425)	33.2 (n = 425)	33.8 (n = 425)	35.1 (n = 425)	35.3 (n = 425)
Simmental[e]	34.6 (n = 129)	35.7 (n = 460)	36.6 (n = 1,400)	37.3 (n = 2,276)	–
Simmental[f]	34.4 (n = 1,609)	35.8 (n = 2,022)	37.3 (n = 1,532)	37.8 (n = 510)	37.1 (n = 120)

Standard deviations were all between 2.1 and 3.0 cm.
Sources: [a]Canadian Charolais Association; [b]Sylla *et al.*, 2007; [c]Canadian Hereford Association; [d]ABS Global Inc.; [e]Canadian Simmental Association; [f]American Simmental Association.

Table 12.8. Age adjustment factors for scrotal circumference (SC) at 365 days of age according to breed; 365-day SC = actual SC + ((365 – age) × age adjustment factor) (BIF, 2010).

Breed	Age adjustment factor
Angus	0.0374
Charolais	0.0505
Gelbvieh	0.0505
Hereford	0.0425
Limousin	0.0590
Red Angus	0.0324
Simmental	0.0543

In another study, differences between bulls that attained a minimum yearling SC of 34 cm and bulls that did not were observed for adjusted SC at 200 days of age (23.3 versus 20.5 cm, respectively). Based on these results, it was suggested that SC at weaning could be used to select bulls for breeding, and 23 cm was proposed as the minimum SC standard at 200 days (Coe and Gibson, 1993). However, this study included bulls from several breeds of known differences in patterns of testicular growth and mature size, while using a singular and very strict yearling SC minimum. Scrotal circumference at 240 days of age could be used as a tool to select bulls with a high probability of meeting the minimum requirements for SC at 365 days of age (i.e. Simmental, 32 cm; Angus, Charolais and Red Poll, 31 cm; Hereford, 30 cm; Limousin, 29 cm); sensitivity and specificity analysis for determining cut-off values indicated that the probability of Charolais bulls with SC ≥ 24 cm, Simmental and Limousin bulls with SC ≥ 22 cm, and Angus, Hereford and Red Poll bulls with SC ≥ 21 cm attaining minimum requirements was greater than 80%. SC at weaning was not useful as a culling tool though, as a large portion of bulls, irrespective of breed, met the minimum requirements at 365 days of age even when SC was <21 cm at 240 days of age (Barth and Waldner, 2002).

Libido and serving capacity tests

Libido has been defined as the willingness and eagerness of a male to mount and service a female, whereas serving capacity is the ability to complete a service. In practical terms, libido and serving capacity tests are very similar. The main difference is that sexual behaviours such as sniffing and licking genitals, flehmen response, chin resting, attempt to mount, incomplete mounting and complete mounting without intromission are all taken into consideration for libido test scoring, but only complete services are quantified in serving capacity tests. The serving capacity test as originally developed included the use of two or three non-oestrous females restrained in service crates in a small pen about 5–7 m apart. Bulls to be tested were allowed to observe the sexual activities of other bulls for 10 min before being exposed in groups of five to the females. Serving capacity test score represented the number of services attained in 40–60 min of observation (Blockey, 1975). The recommended set-up for the libido test is basically the same as that for the serving capacity test, except that bulls are tested in pairs (1:1 bull-to-female ratio). The libido test score is based on the observed sexual behaviour during 10 min of observation (Table 12.9). Retesting bulls a few days apart in different pair combinations is also recommended if results are indeterminate (Chenoweth *et al.*, 1979; Chenoweth, 1986). In order to properly perform behaviour tests, bulls should be handled quietly and sources of distractions should be avoided. Bulls should not be excessively apprehensive or agitated, and conducting tests after stressful handling (EE, vaccination or other processing) or during extreme weather conditions is not recommended.

Bulls have been shown to perform better in libido and serving capacity tests (i.e. proportion of bulls with completed service and number of services) when exposed to non-oestrous, restrained females than when exposed to oestrous, unrestrained females (Blockey, 1975; Chenoweth *et al.*, 1979). There is no evidence that restrained oestrous females are more attractive to bulls than restrained non-oestrous females. During 20 min tests with Hereford bulls, oestrous and non-oestrous females received similar numbers of mount intentions, mount attempts, mounts without ejaculations and ejaculations. Time spent with oestrous and non-oestrous females

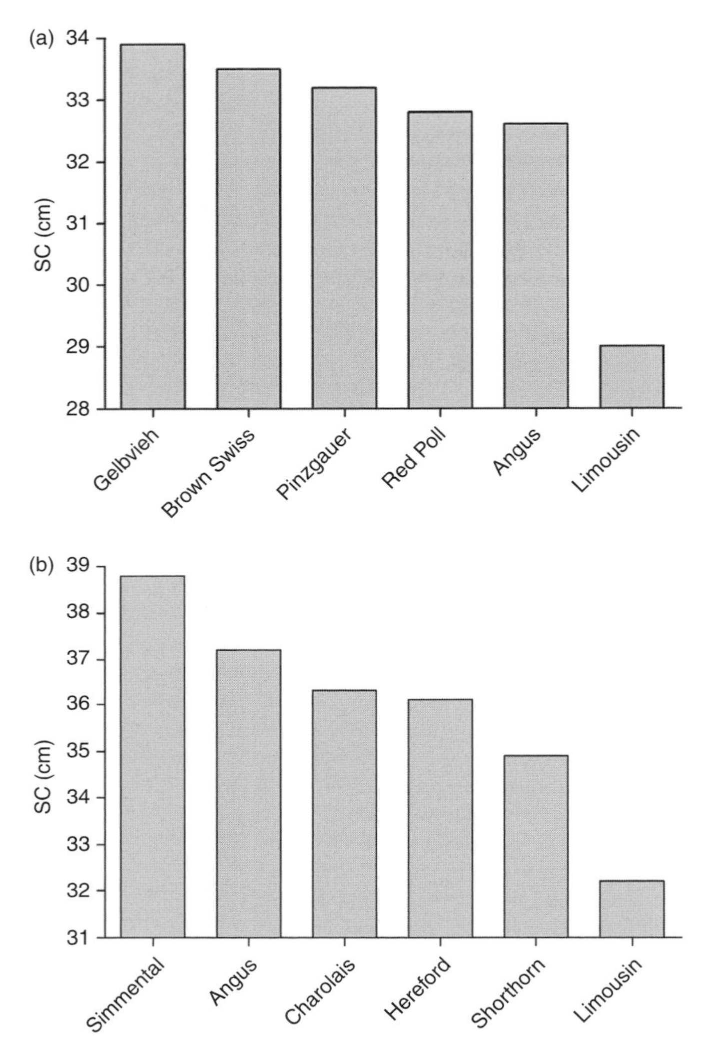

Fig. 12.11. (a) Values for mean scrotal circumference (SC) adjusted for constant age (days) or body weight and reported for 12-month-old bulls of various breeds in post-weaning performance tests in the USA and Canada (Lunstra *et al.*, 1988; Gregory *et al.*, 1991; Pratt *et al.*, 1991; Bell *et al.*, 1996; Barth and Ominski, 2000; Arteaga *et al.*, 2001). Angus, (combined) $n = 1796$; Brown Swiss, $n = 245$; Gelbvieh, $n = 599$; Limousin, $n = 747$; Pinzgauer, $n = 359$; Red Poll, $n = 632$. (b) Scrotal circumference in 2-year-old bulls of various breeds examined at show and sale events in Canada, corrected across breeds for the effects of location, year and sire, and adjusted to a common bull age of 730 days (Coulter *et al.*, 1987b): Angus, $n = 629$; Charolais, $n = 499$; Hereford, $n = 3{,}769$; Limousin, $n = 80$; Shorthorn, $n = 231$.

and latencies to abandon initially chosen females to investigate an alternative female were also similar (Wallach and Price, 1988). Allowing bulls to observe the sexual activity of other bulls provides sexual stimulation and improves their sexual performance during testing. This method of sexual stimulation decreased the time to first mount and first service, and the interval between services, and increased the total number of services in 15–30 min tests when compared with no prior sexual stimulation, stimulation by restraining

Table 12.9. Libido scoring system based on 10 min test (Chenoweth, 1986).

Score	Behaviour
0	Bull showed no sexual interest
1	Sexual interest showed only once
2	Positive sexual interest on more than one occasion
3	Active pursuit of female with persistent sexual interest
4	One mount or mount attempt with no service
5	Two mounts or mount attempts with no service
6	More than two mounts or mount attempts with no service
7	One service followed by no further sexual interest
8	One service followed by sexual interest, including mounts or mounting attempts
9	Two services followed by no further sexual interest
10	Two services followed by sexual interest, including mounts or mounting attempts

in close proximity to a female, or stimulation by being observed by another bull during the test (Mader and Price, 1984). Observation of breeding activity for 10 min is as effective for sexual stimulation as observation for 60 min (Blockey, 1975).

Age and experience have major effects on libido and serving capacity test results. An overall increase in test scores has been observed in several studies with yearling bulls when tests were conducted repeatedly within relatively short intervals, indicating that a maturing and/or learning process occurs rapidly after exposure of bulls to females. When low serving capacity yearling bulls were subsequently re-exposed to oestrous females within a week, 85% of them moved into medium or high serving capacity categories (Boyd and Corah, 1988). No significant correlations were obtained between the numbers of services achieved by Hereford bulls at 12 months of age and their sexual performance at older ages. Only when bulls reached 18 months old were individual differences in serving capacity subsequently consistent, i.e. at 21 and 24 months of age (Price and Wallach, 1991c). In one study, only 53% of 113 yearling bulls completed a service during the 10 min libido test (Chenoweth *et al.*, 1979), whereas with 2-year-old Angus and Hereford bulls, the variance of serving capacity or libido scores obtained from eight tests conducted over 2 months was 69 and 23%, respectively (Landaeta-Hernández *et al.*, 2001). These observations suggest that attempts to predict sexual performance are likely to be unproductive until bulls can fully express their inherent sexual behaviour and serving ability. When adult Angus bulls were evaluated, the time required to complete six services when exposed individually to ten oestrous females increased with age from 31 min in 2-year-old bulls to 43, 55 and 67 min in 3-, 4-, and 5-year-old bulls, respectively (de Araujo *et al.*, 2003). However, no differences in libido score or service capacity test results between 2- and 3-year-old Angus and Hereford bulls were observed in another study (Chenoweth *et al.*, 1984).

The bull-to-female ratio (BFR) and number of bulls tested at the same time might also affect the results of libido and serving capacity tests. When tested in groups, BFRs of 5:2 or 5:4 did not affect the results of a 60 min service capacity test when 2-year-old bulls were evaluated (Blockey, 1975). Hereford bulls 12, 18 and 24 months old tested in large groups (13–14 bulls) had a lower number of mounts and services than bulls tested individually or in small groups (3–4 bulls) when exposed to three restrained females. Likewise, when Angus bulls were evaluated in the same manner, bulls tested individually achieved better test results than bulls evaluated in a group (Price and Wallach, 1991b), which confirmed similar observations from a different study evaluating yearling Hereford bulls (Lane *et al.*, 1983). Sexual behaviour can be depressed by relatively frequent or prolonged bouts of aggression in large group tests in which group size and the BFR are too high. The same might occur if females are restrained too closely together. Considering that age and unfamiliarity tend to maximize the frequency and intensity of agonistic interactions, sexual performance in group tests might be considerably lower when older (≥3 years old), mixed age and(or) unfamiliar bulls are tested. Although individual testing seems to be preferable, the practicality of this approach must also be considered.

Greater libido and serving capacity test scores generally translate into more sexual activity during pasture breeding, but the correlation of test results with fertility is inconsistent. In one study, higher serving capacity bulls, which were subsequently placed with females at a BFR of 1:40 for 6 weeks, served more females, and served them more often, than bulls of lower serving capacity (Blockey, 1975). Though pregnancy rates at the end of the breeding season were not different in heifers exposed to high or medium serving capacity bulls, those heifers bred to the former group conceived significantly earlier than those placed with the latter. This suggests that high serving capacity bulls were more efficient in handling heavy mating loads early in the breeding season, which indicates an economic benefit in selecting and using such bulls (Blockey, 1975).

When composite yearling bulls classified as being of low and high serving capacity were exposed to 25 naturally cycling cows for 3 days, the percentage of cows served zero, one or two, and more than two times were 44, 17 and 39% for low capacity bulls, and 32, 20 and 48% for high capacity bulls, respectively. When these same bulls were exposed to nine oestrous cows for 1 day, there were no differences between groups in the total number of services or the average services per cow. Moreover, pregnancy rates did not differ between service capacity groups in either mating load situation (Boyd *et al.*, 1989). In a different study, the number of services achieved by high serving capacity bulls in double-sire tests was almost double the number achieved by the low capacity bulls, but there was no difference between groups in the single-sire tests. Pregnancy rates after exposure of bulls to 15 oestrous females for 6 h also did not differ between high and low serving capacity bulls (Godfrey and Lunstra, 1989). Angus and Hereford bulls of various ages with a high libido test score achieved a greater number of services (approximately 21/bull) during 30–48 h of exposure to oestrous females than did bulls of medium score, but pregnancy rates after exposure for approximately 1 week did not differ between the high and medium libido groups (Farin *et al.*, 1989). In an extensive evaluation of fertility in 109 Angus, Hereford and Senepol single-sire herds using BFRs of 1:12 to 1:36, correlations of libido test score of bulls ≥2-year-old with (overall) conception rate, conception rate during the first 21 days of the breeding season and calving date were all low and non-significant (Larsen *et al.*, 1990).

Controlling all the variables that affect the temporal expression of bull sexual behaviour is very difficult and represents a challenge to the development of tests that accurately reflect how bulls and females interact during natural breeding and how this interaction affects fertility. Although there is no doubt that libido and serving capacity affect bull reproductive potential, the limitations of the tests, combined with the labour intensity required for conducting them, and the contradictory observations on the association of the test results with fertility, have limited the routine application of these tests to breeding bulls. However, it is always important to remember that observing a bull complete a service is very valuable information.

Selected Congenital Reproductive Disorders

Cryptorchidism

Cryptorchidism is characterized by failure of a testis to descent to its normal puberal position within the scrotum. The pathogenesis is poorly understood, but may involve anomalies in gonadotrophins, testosterone and Anti-Müllerian hormone levels (Ladds, 1993; Amann and Veeramachaneni, 2007). A 26 year survey of North American Veterinary Schools revealed an overall incidence of 0.17% for bull cryptorchidism (St Jean *et al.*, 1992), with the highest levels in Polled Herefords and Shorthorns (approximately 1.0%), and with polled bulls being 40% more represented than horned bulls, regardless of breed. Unilateral cryptorchidism was most common, with the left side being twice as likely to be affected. Cryptorchidism was diagnosed in 0.43% of Chianina, Marchigiana and Romagnola bulls (Sylla *et al.*, 2007), and in 1.4 and 0.9% of 12- and 16-month-old Senepol bulls, respectively (Godfrey and Dodson, 2005).

While the affected testis may be located at any position along the normal trajectory of descent, it is twice as frequently found in the inguinal canal than inside the abdominal cavity. Cryptorchidic testes inside the abdominal cavity are usually located close to the internal inguinal rings (McEntee, 1990). Diagnosis is performed by attempting to identify the retained testis by palpation and/or ultrasonography of the inguinal canal, and per rectum. Cryptorchidic testes are susceptible to disturbed thermoregulation, hypoplasia and aplasia, as well as degeneration. Sperm production and semen quality are reduced or absent in unilaterally affected bulls, whereas those affected bilaterally are sterile (Roberts, 1986; Parkinson, 2001). There is strong evidence to suggest that cryptorchidism is a heritable condition in cattle, and affected bulls should be eliminated from breeding programmes (Roberts, 1986; Steffen, 1997). Absence of one testis (monorchia) and subcutaneous (ectopic) testes have also been reported in bulls (McEntee, 1990).

Testicular hypoplasia

Testicular hypoplasia is a gross diagnosis that covers a number of different conditions with different histological characteristics, including germ cell deficiencies, germ cell weakness and arrested spermatogenesis (McEntee, 1990). The condition occurs owing to a partial or complete lack of germ cells in the seminiferous tubules. Failure of development of the primordial germinal cells in the yolk sac, disrupted migration of these cells to the gonads, and abnormal multiplication or excessive degeneration after the cells reach the gonads might be involved in the pathogenesis of the condition (Roberts, 1986). At least in some cases, testicular hypoplasia does not appear to be caused by deficiencies in gonadotrophin and/or testosterone secretion, suggesting that impaired gonadal responses to these hormones, abnormal secretion of intra-testicular factors or genetic defects are involved. The low number of germ cells and small diameter of the seminiferous tubules are probably responsible for the smaller size of hypoplastic testes (Moura and Erickson, 2001). There appears to be no precise definition of testicular hypoplasia based on testicular measurement, though the condition is suspected when one or both testes are much smaller than expected for the age and the breed. Hypoplastic testes are usually small and firm with a smooth outline. In most cases, the other reproductive organs and sexual behaviour are normal (albeit the ipsilateral epididymis might also be underdeveloped), and the reproductive capability of the affected bull is associated with its ability to produce normal sperm. The degree of either unilateral or bilateral testicular hypoplasia varies from nearly complete bilateral hypoplasia to only slightly and often unsuspected hypoplasia. Semen characteristics and fertility are equally variable from nearly normal to severe oligospermia (oligozoospermia) or azoospermia – which result in infertility or complete sterility (Gledhill, 1973; Roberts, 1986; Parkinson, 2001).

The most comprehensive characterization of testicular hypoplasia has been conducted in Swedish Highland cattle. An increased selection of animals with less pigmentation resulted in increased incidence of gonadal hypoplasia. The left testis was affected in approximately 82% of cases and the condition was bilateral in 14.5% of the cases. Histologically, there was generalized germ cell deficiency, with most seminiferous tubules devoid of germ cells and lined by Sertoli cells only. Pedigree studies indicated that testicular hypoplasia in Swedish Highland bulls is caused by an autosomal recessive gene with incomplete penetrance (Gledhill, 1973; McEntee, 1990). The condition has also been extensively studied in Swedish Red and White cattle. In contrast to that observed in Swedish Highland bulls, affected testes of Swedish Red and White bulls contained at least some (but) degenerating germinal cells. The history of some affected bulls indicated a gradual deterioration of sperm production and semen quality. Thus, testicular hypoplasia in Swedish Red and White bulls seemed to be caused by germ cell weakness and was associated with a progressive, uneven loss of germinal cells. Pedigree analysis provided strong evidence that the condition is heritable in this breed as well (McEntee, 1990; Settergren and McEntee, 1992).

Arrested spermatogenesis has been reported in Swedish Friesian and Hereford bulls. In these cases, both testes were affected, although only one became noticeably smaller with age. Affected bulls had watery ejaculates that might have contained multinucleated giant cells and free spermatids. Histological examination revealed that, whereas germ cells could develop to spermatids, subsequent spermiogenesis was halted, with no sperm development (McEntee, 1990). A similar condition has also been reported in one Angus bull in which histological examination of the testes revealed that germ cell development was arrested during meiosis, and no development occurred past primary spermatocytes (Moura and Erickson, 2001). Trisomy (61,XXY or Klinefelter's syndrome) also results in testicular hypoplasia. Affected bulls reportedly have normal libido and reproductive organs, but they are azoospermic with very small testes. Histologically, the vast majority of seminiferous tubules consisted of irregularly shaped masses of collagenous tissue; a few of these contained Sertoli cells, though none contained germ cells (Logue *et al.*, 1979; Dunn *et al.*, 1980; Schmutz *et al.*, 1994). Testicular hypoplasia and the presence of large foci of seminiferous tubules deficient in germ cells, with germ cells and multinucleated giant cells present in the tubule lumen have also been reported in XY/XX chimeras (Dunn *et al.*, 1979; Bongso *et al.*, 1981).

Regardless of its pathogenesis and pathological characteristics, testicular hypoplasia has been reported in various breeds, and several reports suggest that the condition is more common in highly inbred cattle populations and might also be inherited (McEntee, 1990; Steffen, 1997). Double muscling, which is a characteristic also inherited as an autosomal recessive trait, has also been associated with a high incidence of bilateral testicular hypoplasia (Hoflack *et al.*, 2006). Bulls with small testes or marked testicular asymmetry suggesting hypoplasia should be eliminated from breeding programmes.

Blind-ending efferent ductules and aplasia of the epididymis

Blind-ending efferent ductules usually result as a failure of the mesonephric tubules to connect to the rete tubules. Even so, the results from one study demonstrated that while eight out of 25 bulls had one to five blind-ending ductules, either this population of bulls was more predisposed to the problem or, in most cases, the presence of a few blind ductules has no deleterious consequences. However, blind-ending ductules may become over-filled with sperm and rupture with the increasing pressure. The release of sperm into the interstitial space results in an immune reaction that leads to formation of sperm granulomas in the caput epididymis (Gledhill, 1973; McEntee, 1990). If most or all of the efferent ductules are blind, the rete tubules become distended and the testis may become enlarged and oedematous.

Aplasia of the epididymis has been reported in several breeds of cattle. The condition was observed in 11 of approximately 2000 Red Danish Milk bulls and in 41 of 828 bulls of unspecified breeds in the Netherlands. It has also been reported in Angus, Guernsey, Friesian, Holstein and Simmental bulls. Aplasia might be complete or segmental, and the condition might be bilateral or unilateral (Roberts, 1986; McEntee, 1990; Williams *et al.*, 2010). As observed with blind-ending efferent ductules, the rete tubules in the testis ipsilateral to the abnormal epididymis may become distended and the testis might become enlarged and oedematous. The observation that four of 19 sons of a Red Danish Milk bull with unilateral aplasia also presented with this condition, and that almost a third of 60 fattening calves and newborn calves with aplasia of the epididymis were sired by one affected bull, provides circumstantial evidence that the condition is heritable (McEntee, 1990).

Knobbed acrosome sperm defect

The appearance of the knobbed acrosome defect under light microscopy varies from a bead-like thickening on the sperm head apex to indentation and flattening of the apex. Contrary to the appearance that the nomenclature suggests, protrusion of the acrosome from the head ridge is actually an uncommon presentation of the defect in bulls (Fig. 12.12). An excess of acrosomal matrix and folding of the acrosome over the apex of the sperm head is observed by electron microscopy (EM). Membranous vesicles containing granular or membranous inclusions are commonly observed entrapped in the acrosomal matrix (Blom and Birch-Andersen, 1962; Cran and Dott, 1976; Barth, 1986; Barth and Oko, 1989). Although the knobbed acrosome defect can be caused by environmental factors (e.g. increased testicular temperature, stress), it can also be of genetic origin in bulls. Genetically affected bulls consistently produce a large percentage of knobbed acrosome sperm without apparent cause and without significant changes in other sperm defects (Barth and Oko, 1989; Chenoweth, 2005).

Pedigree analysis of affected bulls indicated that the defect is probably caused by an autosomal recessive gene, and knobbed acrosomes of likely genetic origin have been described in Angus, Charolais, Friesian and Holstein bulls (Donald and Hancock, 1953; Hancock, 1953; Barth, 1986; Chenoweth, 2005). In Western Canada, approximately 0.6% of range beef bulls (10/1758) examined for breeding soundness produced >25% sperm with knobbed acrosomes; eight of these bulls were Charolais (Barth, 1986). A much greater prevalence (>6%) has been reported in a knowingly affected Angus herd (Chenoweth, 2005). *In vitro* studies have suggested that sperm with knobbed acrosomes have altered plasmalemma function that predisposes to premature sperm capacitation and a spontaneous acrosome reaction. Sperm with knobbed acrosomes are unable to bind and penetrate the zona pellucida. Moreover, other genetic defects

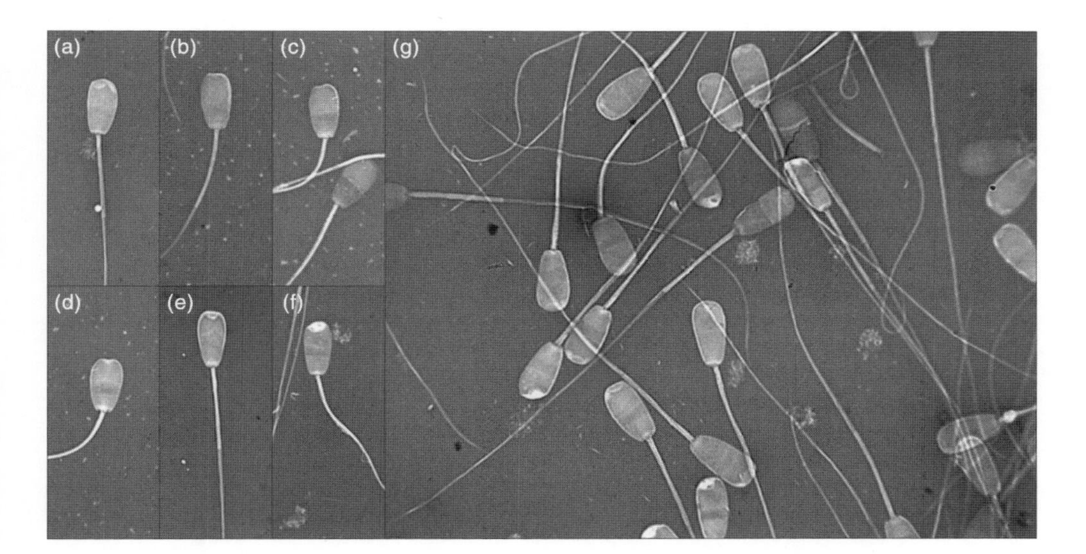

Fig. 12.12. The knobbed acrosome defect might appear as a slight folding of the acrosome (a), as a flattened or indented apex of the head (b–d) or as heavy coiling forming a 'bead' on the head (e, f). Protrusion of the acrosome from the head ridge is actually not a common presentation of this defect in bulls. Note that in (g) virtually all of the sperm are affected and although a few other abnormalities might be observed, knobbed acrosome is by far the most prevalent sperm defect in the sample. In this case, the proportion of sperm with knobbed acrosome defect also did not change with time, indicating that the abnormality was caused by a genetic defect. The images were obtained under 100× magnification from an eosin–nigrosin-stained sample.

in otherwise morphologically normal sperm capable of fertilizing the oocytes probably contributed to the impaired embryonic development observed *in vitro* after the use of semen from bulls producing genetically knobbed acrosomes (Thundathil *et al.*, 2000, 2001, 2002). Bulls that produce a large percentage of knobbed acrosome sperm are infertile or sterile (Donald and Hancock, 1953; Hancock, 1953; Barth, 1986).

Rolled head, nuclear crest, giant head sperm syndrome

The presence of sperm with rolled heads, nuclear crests or giant heads (macrocephalic) can occasionally be observed in very low proportions (<1.5%) in bull ejaculate. However, a syndrome that is characterized by the production of large proportions of all three types of defects has been documented. In rolled sperm, the edges of the head are curved along the long axis to varying degrees. On EM, the plasmalemma bridges the gap between the two edges and does not follow the contour of the 'U'-shaped nucleus. A roughened line extending to variable lengths along the long axis of the sperm head characterizes the sperm nuclear crest. On EM, a typical crested nucleus presents three arms inclined at 120° to each other, though variations, including 'Y'-shaped nuclei, also occur. Giant sperm heads are 1.5–2 times larger than normal sperm heads and are frequently associated with multiple tails and knobbed acrosomes in bulls (Fig. 12.13) (Cran *et al.*, 1982; Barth and Oko, 1989). The syndrome has been observed in Brown Swiss, Friesian, Holstein and South Devon bulls. Reports of its occurrence in closely related bulls suggest a genetic influence on this condition, but the mode of inheritance is not clear. Giant head sperm probably have an abnormal DNA content (e.g. diploid) and are usually immotile. Consequently, the fertility of affected bulls depends on the proportion of normal sperm in the ejaculate (Barth and Oko, 1989).

Stump tail sperm defect

Sperm affected with the stump tail defect have a short tail stump or rudimentary tail when observed under light microscopy. The stump is often covered by cytoplasmic, droplet-like material attached to the sperm's neck (Fig. 12.14). Care must be taken when examining semen samples to not confuse these sperm with simple detached sperm heads. EM shows that many of the droplet-like swellings contain membranous structures, mitochondria, elements of fibrous sheath and individual fibrils. The axoneme is disorganized and does not display the normal tubular, fibre and sheath construction. The defect appears to be the result of an anomaly of development of the distal centriole (Blom, 1976; Blom and Birch-Andersen, 1980; Vierula *et al.*, 1983; Peet and Mullins, 1991; Revell *et al.*, 2000). Although no conclusive data is available, an autosomal recessive mode of inheritance has been suggested based on pedigree analysis of two

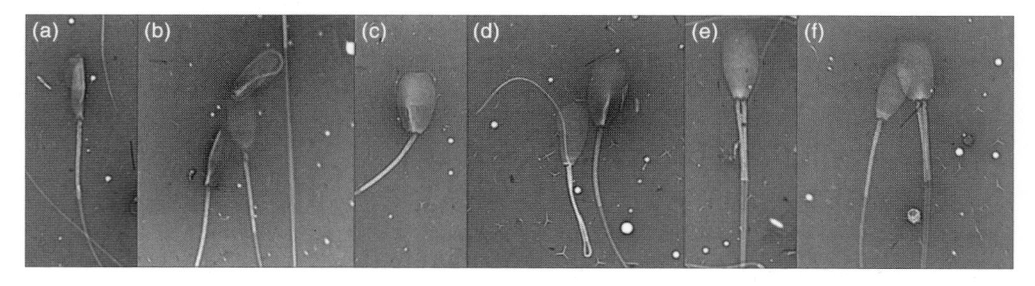

Fig. 12.13. Rolled heads (a, b), nuclear crests (c, d) and giant heads (b–f) are abnormal sperm forms present in the ejaculates of bulls affected by the rolled head/nuclear crest/giant head syndrome. Double tails and acrosome defects are also common and most affected sperm are immotile (dead). The images were obtained under 100× magnification from an eosin–nigrosin-stained sample.

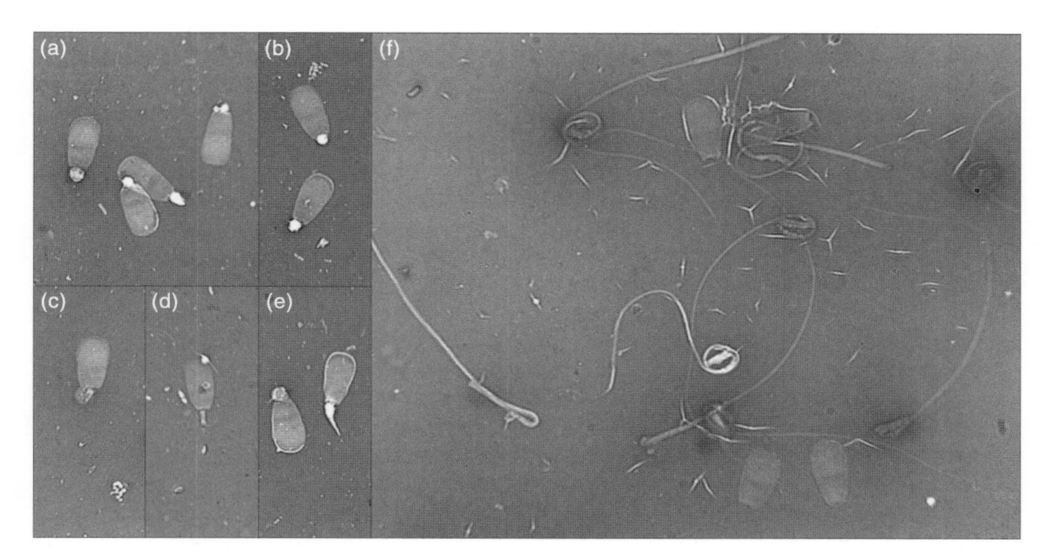

Fig. 12.14. The stump tail defect is sometimes overlooked, as affected sperm appear similar to simple detached heads. However, closer examination reveals the presence of a short tail stump in most of them. The stump is often covered by cytoplasmic, droplet-like material (a–e). Although detached sperm heads are commonly observed, detached tails are usually observed less frequently. In contrast, the genetic form of decapitated sperm is characterized by the presence of a large proportion of both detached heads and detached tails (f). In the stump tail defect, the proximal portion of the midpiece is usually bent or coiled and the tails are motile when unfixed samples are observed; care must be taken not to confuse those with microcephalic sperm. The images were obtained under 100× magnification from an eosin–nigrosin-stained sample.

affected Ayrshire bulls (Foote *et al.*, 1992). The defect has also been described in Charolais, Friesian, Hereford, Holstein, Shorthorn and Swedish Red and White bulls. Total sperm number in affected bulls is usually much lower than normal and >60% of the sperm can be affected. Most sperm are dead and immotile; consequently, affected bulls are usually sterile (Blom, 1976; Williams, 1987; Barth and Oko, 1989; Peet and Mullins, 1991; Revell *et al.*, 2000).

Decapitated sperm defect

Detached sperm heads are commonly observed in the ejaculate of bulls and might result from abnormal spermatogenesis or from ageing of the sperm in the reproductive tract. The decapitated sperm defect is a specific form of detached, tailless sperm. Two characteristics that assist in identifying this defect as a specific syndrome are active movements of the free tails when live sperm are evaluated, and the formation of a loop in many of the detached midpieces (Fig. 12.14). Moreover, >80% of sperm are usually affected and this proportion does not change with semen collection frequency or time between collections. EM shows this defect to be characterized by malformation of structures at the base of the head, especially the implantation fossa and the basal lamella, resulting in 'weak' anchorage of the capitulum (Blom and Birch-Andersen, 1970; Blom, 1977). The defect has been described in Chianina, Guernsey, Hereford and Swedish Red and White bulls, and is most likely associated with a recessive gene, at least in the Guernsey breed. Fertility depends on the proportion of affected sperm in the ejaculate, but is usually low (Barth and Oko, 1989).

'Dag' sperm defect

On light microscopy, the Dag defect is characterized by strongly coiled tails associated with rough, incomplete mitochondrial sheaths and usually accompanied by fractures of the axonemal fibres (Fig. 12.15). On EM

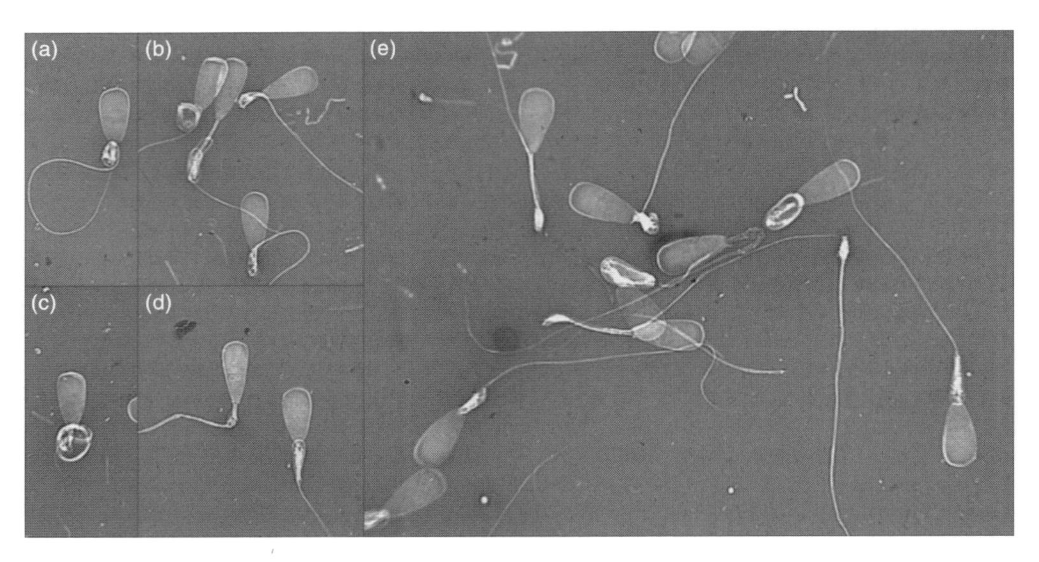

Fig. 12.15. The Dag defect is characterized by midpiece abnormalities that include bending, coiling or breaking. The difference from other forms of tail defects is the roughed, swollen appearance of the midpiece (a–d) and the observation that virtually all sperm in the ejaculate are affected (e). The images were obtained under 100× magnification from an eosin–nigrosin-stained sample.

examination, transverse sections of the tail usually contain more than one axoneme and all elements are enclosed in a common plasmalemma. The mitochondrial sheath appears irregular and disorganized, and the axoneme is abnormal with one or more doublets missing. These changes in the axoneme are accompanied by an absence or displacement of the corresponding dense fibres (Koefoed-Johnsen and Pedersen, 1971; Blom and Wolstrup, 1976; Andersen Berg *et al.*, 1996). The Dag defect is named after a Jersey bull in which the defect was first identified and has since been described in Hereford and Swedish Red and White bulls. The defect is probably caused by an autosomal recessive gene. Genetically affected bulls produce a large percentage of sperm with the defect (>40%), have very low numbers of motile sperm and are severely sub-fertile or sterile (Blom, 1966; Koefoed-Johnsen and Pedersen, 1971; Blom and Wolstrup, 1976; Barth and Oko, 1989). A small percentage of similar ('Dag-like') defects can sometimes be observed in association with other defects in cases of disrupted spermatogenesis, though in these cases the proportion of affected sperm is usually low (<5%) and changes with time.

Short tail sperm defect

The short tail sperm defect has been described in Finnish Yorkshire boars. Affected sperm are immotile and are characterized by malformation of the tail, in which the axonemal structure is abnormal, although the structure and function of other ciliated cells in the animal do not appear to be affected (Sukura *et al.*, 2002). In pigs, the defect is inherited as an autosomal recessive condition caused by an insertion in the *KPL2* gene in porcine chromosome 16 (Sironen *et al.*, 2002, 2006). A similar defect has been observed in two half-brother Red Charolais bulls in Canada (A.D. Barth, personal communication). On light microscopy, only the initial segment of the principal piece appeared to have formed. In some sperm, abnormally thin segments of the principal piece seemed to bend over the midpiece, whereas in others a distal cytoplasmic droplet was retained and was either associated with a distal midpiece reflex or not (Fig. 12.16). The defect has also been recently described in three related Nelore (*B. indicus*) bulls, providing further suggestion that the condition might be genetically influenced in bulls (Siqueira *et al.*, 2010).

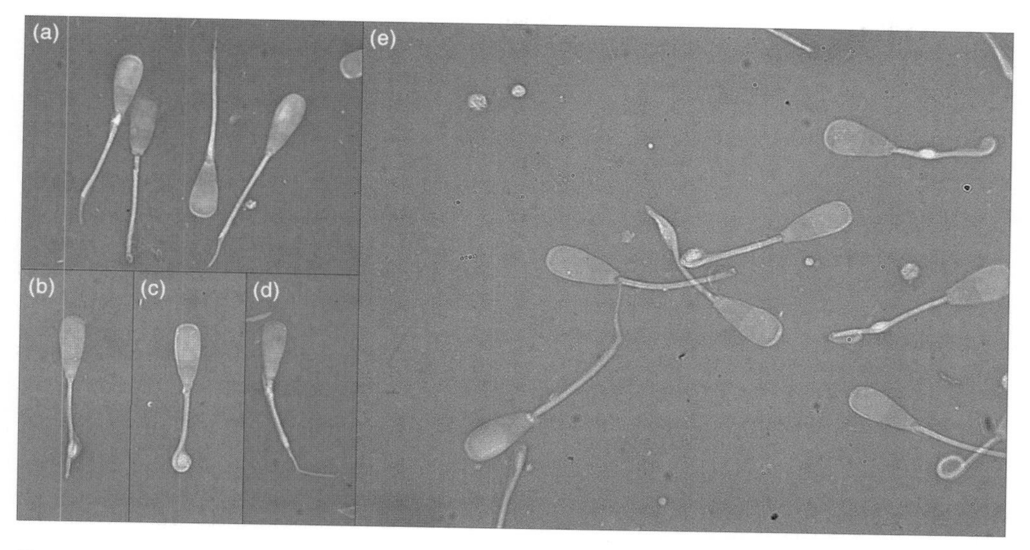

Fig. 12.16. The short tail defect is a genetic sperm defect that has been well documented in boars but has also been observed in bulls. It is characterized by malformation or absence of the principal piece (a). Distal cytoplasmic droplets are observed in some sperm, associated or not with a distal midpiece reflex (b, c). In other cases, the rudimentary principal piece seems to be composed only of the axoneme (d) and virtually all sperm in the ejaculate are affected (e). The images were obtained under 100× magnification from an eosin–nigrosin-stained sample.

Congenital erectile dysfunction

During erection in the bull, the ischiocavernosus muscles contract and relax rhythmically, pumping blood into the corpus cavernosum penis (CCP), while occluding the veins that drain this space, thereby increasing the blood pressure within the CCP and 'stiffening' the penis. Erectile dysfunction results from abnormal blood influx into, or efflux from, the CCP. In either case, the increased blood pressure required to produce erection cannot be achieved and the penis becomes semi-erect only or remains flaccid. Congenital failure of occlusion of the veins that drain the CCP during fetal life results in abnormal venous drainage into the dorsal vein of the penis and erectile dysfunction. Abnormal venous drainage of the CCP has been reported in Devon, Charolais, Friesian and Hereford bulls aged from 18 months to 5 years in which normal service had never occurred, despite adequate libido. The flaccid penis protruded from the preputial orifice in only one bull. Contrast radiography revealed that, contrary to the situation observed in a normal bull, numerous veins directly drained the fine cavernous spaces and the ventral canals of the CPP into the dorsal venous system in these bulls (Ashdown *et al.*, 1979a; Glossop and Ashdown, 1986).

Congenital blockage of the longitudinal canals of the CCP also prevents consistent increased blood pressure from occurring throughout the CCP, also resulting in erectile dysfunction. This condition has been reported in Hereford, Friesian and Sussex bulls 2 to 2.5 years old with the same presentation as those with abnormal venous drainage of the CCP. Radiographic examination in these cases revealed that contrast media injected into the CCP does not spread throughout the entire length of the penis. Post-mortem examinations revealed that the longitudinal canals of the CCP were occluded by dense, translucent fibrous tissue (Ashdown *et al.*, 1979b; Parkinson, 2001). Diagnosis of erectile dysfunction is made by the observation of erection failure during a test of serving ability. Palpation of a mostly flaccid penis during stimulation with an electroejaculator may help to reinforce the diagnosis. A history of previous successful breeding also helps to determine whether the condition is congenital or acquired.

Congenital short penis and short retractor penis muscle

Congenitally short penis has been reported in 20 sons of a Hereford bull and in two closely related Guernsey bulls (Roberts, 1986); the problem has also been observed in bulls of several other breeds (A.D. Barth, personal communication). In affected bulls, the erect penis does not protrude more than 10 to 15 cm from the preputial orifice, even though the retractor penis muscle is normal and there are no adhesions of the penis to the prepuce preventing normal extension. In most cases, the sigmoid flexure does not form a sharp 'S' shape at rest as in normal bulls. Short retractor penis muscle has been reported as a probable recessive hereditary condition in Friesian bulls (Roberts, 1986). The presentation is very similar to that of congenitally short penis, except that the sigmoid flexure is normal but fails to distend during copulation or when manually forced during EE.

Young bulls with either a short penis or short retractor penis muscle may be able to breed heifers, although copulation becomes impossible as the bulls gets older, larger and less agile as the tip of the penis cannot reach the vulva. Personnel in AI centres should be trained to identify these conditions as semen from affected bulls can often still be collected with use of an artificial vagina. Myotomy of the retractor penis muscle to correct the defect is not advised and is illegal in certain countries (Gledhill, 1973). Because there is strong evidence that these conditions are heritable, affected bulls should be eliminated from breeding programmes.

Persistent penile frenulum

At birth, the penis is ventrally united to the prepuce by a band of connective tissue called the frenulum. The frenulum gradually breaks down as the bull matures, and separation of the penis and prepuce is completed between 8 and 11 months of age. The persistent frenulum may be represented as just a thin band of tissue or as a thick and broad band that contains a prominent vessel. Persistence of the penile frenulum results in deviation of the glans penis and prevents normal copulation (Roberts, 1986; McEntee, 1990; Steffen, 1997). One study has demonstrated that the presence of a persistent frenulum in 11-month-old bulls (10.5%) was significantly greater than in older bulls (4.4 to 1.9%), probably due to spontaneous separation while attempting to breed (Bruner *et al.*, 1995). Other studies have reported a prevalence of 3.6 to 5% in yearling bulls (Carroll *et al.*, 1964). Some studies indicate that certain breeds are more commonly affected; others have reported increased prevalence in highly inbred Hereford and Angus herds (Carroll *et al.*, 1963, 1964; Elmore *et al.*, 1978; Bruner *et al.*, 1995). The prevalence of 7.1% found in young Belgian Blue bulls seemed higher than previously reported in other breeds (Hoflack *et al.*, 2006). Transection of the frenulum is often easy and completely resolves this problem, but knowledge of the heritable nature of this condition would suggest caution in using this approach in bulls destined to produce breeding stock (Steffen, 1997).

Spiral deviation of the penis

Spiral twisting of the penis during ejaculation, ranging from a slight twist to a 360° spiral, was observed in over half of bulls from which semen was collected using a transparent artificial vagina, and thus may be considered to be a normal phenomenon (Seidel and Foote, 1969). However, spiral deviation of the penis ('corkscrew penis') occurring before intromission prevents normal copulation and is considered abnormal. The pathogenesis of the abnormality may involve the breakdown or stretching of the attachment of the dorsal ligament of the penis to the tunica albuginea, and/or neuromuscular disorders. In one scenario, the dorsal ligament of the penis slips in position to the left, as the glans bends down and to the right, pulling the tip back into a spiral. Spiralling presumably occurs when the length of the body of the penis exceeds that of its enveloping integument, involves only the glans and is always counterclockwise. Some bulls may experience one or more seasons of normal breeding before developing the problem and others may invariably display premature spiralling when attempting service (Ashdown and Pearson, 1973; Blockey and Taylor, 1984; Roberts, 1986; McEntee, 1990; Parkinson, 2001) Spiralling of the penis can also be elicited during EE, but this is not necessarily associated with the abnormal condition.

Spiral deviation of the penis has been reported in several breeds, with a higher prevalence noted in polled than in horned bulls (Ashdown and Pearson, 1973; Blockey and Taylor, 1984). In the USA, an overall prevalence of 2% has been reported, with spiral deviation being the most common penile defect described (Carroll *et al.*, 1963). A much higher prevalence (10%) was observed in 1083 bulls aged 2–11 years in Australia. Prevalence was greater (16%) in polled bulls (Angus, Polled Hereford, Polled Shorthorn and Red Poll) than in horned Hereford bulls (1%), and increased after 2 years of age. An overall herd prevalence of 60% was observed, with 25 of 42 herds having at least one affected bull. Over a period of 4 years, 25 of 44 Angus bulls from one herd developed the condition. Of these, 22 bulls had a common grandsire. Nine affected Angus bulls from another herd were bred by a common sire and a common grandsire was also identified for three affected Red Poll bulls (Blockey and Taylor, 1984). Even though the mode of transmission has not been elucidated, there is a strong suspicion that the condition is heritable, and affected bulls should be eliminated from breeding programmes (Steffen, 1997).

References

Abdel-Raouf M (1960) The postnatal development of the reproductive organs in bulls with special reference to puberty (including growth of the hypophysis and the adrenals) *Acta Endocrinologica (Copenhagen)* **34(Supplement 49)** 1–109.

Almquist JO (1973) Effects of sexual preparation on sperm output, semen characteristics and sexual activity of beef bulls with a comparison to dairy bulls *Journal of Animal Science* **36** 331–336.

Almquist JO (1982) Effect of long term ejaculation at high frequency on output of sperm, sexual behavior, and fertility of Holstein bulls; relation of reproductive capacity to high nutrient allowance *Journal of Dairy Science* **65** 814–823.

Almquist JO and Amann RP (1961) Reproductive capacity of dairy bulls. II. Gonadal and extra-gonadal sperm reserves as determined by direct counts and depletion trials; dimensions and weight of genitalia *Journal of Dairy Science* **44** 1668–1678.

Almquist JO and Cunningham DC (1967) Reproductive capacity of beef bulls. I. Postpubertal changes in semen production at different ejaculation frequencies *Journal of Animal Science* **26** 174–181.

Almquist JO, Branas RJ and Barber KA (1976) Postpubertal changes in semen production of Charolais bulls ejaculated at high frequency and the relation between testicular measurements and sperm output *Journal of Animal Science* **42** 670–676.

Amann RP (1983) Endocrine changes associated with onset of spermatogenesis in Holstein bulls *Journal of Dairy Science* **66** 2606–2622.

Amann RP (1990) Management of bulls to maximize sperm output In *Proceedings of the 13th Technical Conference on Artificial Insemination and Reproduction, April 20–21, 1990, Milwaukee, Wisconsin* pp 84–91 National Association of Animal Breeders, Columbia, Missouri.

Amann RP and Almquist JO (1976) Bull management to maximize sperm output In *Proceedings of the Sixth Technical Conference on Artificial Insemination and Reproduction, Milwaukee, USA, 20–21 February 1976* pp 1–10. National Association of Animal Breeders, Columbia, Missouri.

Amann RP and Veeramachaneni DN (2007) Cryptorchidism in common eutherian mammals *Reproduction* **133** 541–561.

Amann RP and Walker OA (1983) Changes in the pituitary-gonadal axis associated with puberty in Holstein bulls *Journal of Animal Science* **57** 433–442.

Amann RP, Kavanaugh JF, Griel LC Jr and Voglmayr JK (1974) Sperm production of Holstein bulls determined from testicular spermatid reserves, after cannulation of rete testis or vas deferens, and by daily ejaculation *Journal of Dairy Science* **57** 93–99.
Amann RP, Wise ME, Glass JD and Nett TM (1986) Prepubertal changes in the hypothalamic-pituitary axis of Holstein bulls *Biology of Reproduction* **34** 71–80.
Amstalden M, Harms P, Welsh T, Randel R and Williams G (2005) Effects of leptin on gonadotropin-releasing hormone release from hypothalamic–infundibular explants and gonadotropin release from adenohypophyseal primary cell cultures: further evidence that fully nourished cattle are resistant to leptin *Animal Reproduction Science* **85** 41–52.
Andersen Berg K, Filseth O and Engeland E (1996) A sperm midpiece defect in a Hereford bull with variable semen quality and freezability *Acta Veterinaria Scandinavia* **37** 367–373.
Aravindakshan JP, Honaramooz A, Bartlewski PM, Beard AP, Pierson RA and Rawlings NC (2000) Pattern of gonadotropin secretion and ultrasonographic evaluation of developmental changes in the testis of early and late maturing bull calves *Theriogenology* **54** 339–354.
Arteaga A, Baracaldo M and Barth AD (2001) The proportion of beef bulls in western Canada with mature spermiograms at 11 to 15 months of age *Canadian Veterinary Journal* **42** 783–787.
Arteaga AA, Barth AD and Brito LF (2005) Relationship between semen quality and pixel-intensity of testicular ultrasonograms after scrotal insulation in beef bulls *Theriogenology* **64** 408–415.
Ashdown RR and Pearson H (1973) Studies on "corkscrew penis" in the bull *Veterinary Record* **93** 30–35.
Ashdown RR, David JS and Gibbs C (1979a) Impotence in the bull: (1) Abnormal venous drainage of the corpus cavernosum penis *Veterinary Record* **104** 423.
Ashdown RR, Gilanpour H, David JS and Gibbs C (1979b) Impotence in the bull: (2) occlusion of the longitudinal canals of the corpus cavernosum penis *Veterinary Record* **104** 598–603.
Austin JW, Hupp EW and Murphree RL (1961) Effect of scrotal insulation on semen of Hereford bulls *Journal of Animal Science* **20** 307–310.
Bagu ET, Cook S, Gratton CL and Rawlings NC (2006) Postnatal changes in testicular gonadotropin receptors, serum gonadotropin, and testosterone concentrations and functional development of the testes in bulls *Reproduction* **132** 403–411.
Bagu ET, Gordon JR and Rawlings NC (2010a) Post-natal changes in testicular concentrations of interleukin-1 alpha and beta and interleukin-6 during sexual maturation in bulls *Reproduction in Domestic Animals* **45** 336–341.
Bagu ET, Gordon JR and Rawlings NC (2010b) Postnatal changes in testicular concentrations of transforming growth factors-alpha and-beta 1, 2 and 3 and serum concentrations of insulin like growth factor I in bulls *Reproduction in Domestic Animals* **45** 348–353.
Bagu E, Davies K, Epp T, Arteaga A, Barrett D, Duggavathi R, Barth A and Rawlings N (2010c) The effect of parity of the dam on sexual maturation, serum concentrations of metabolic hormones and the response to luteinizing hormone releasing hormone in bull calves *Reproduction in Domestic Animals* **45** 803–810.
Bailey JD, Anderson LH and Schillo KK (2005a) Effects of novel females and stage of the estrous cycle on sexual behavior in mature beef bulls *Journal of Animal Science* **83** 613–624.
Bailey JD, Anderson LH and Schillo KK (2005b) Effects of sequential or group exposure to unrestrained estrual females on expression of sexual behavior in sexually experienced beef bulls *Journal of Animal Science* **83** 1801–1811.
Barber KA and Almquist JO (1975) Growth and feed efficiency and their relationship to pubertal traits of Charolais bulls *Journal of Animal Science* **40** 288–301.
Barth AD (1986) The knobbed acrosome defect in beef bulls *Canadian Veterinary Journal* **27** 379–384.
Barth AD (2000) *Bull Breeding Soundness Evaluation*. The Western Canadian Association of Bovine Practitioners, Saskatoon.
Barth AD and Bowman PA (1994) The sequential appearance of sperm abnormalities after scrotal insulation or dexamethasone treatment in bulls *Canadian Veterinary Journal* **35** 93–102.
Barth AD and Oko R (1989) *Abnormal Morphology of Bovine Spermatozoa*. Iowa State University Press, Ames.
Barth and Ominski KH (2000) The relationship between scrotal circumference at weaning and at one year of age in beef bulls *Canadian Veterinary Journal* **41** 541–546.
Barth AD and Waldner CL (2002) Factors affecting breeding soundness classification of beef bulls examined at the Western College of Veterinary Medicine *Canadian Veterinary Journal* **43** 274–284.
Bell DJ, Spitzer JC, Bridges WC Jr and Olson LW (1996) Methodology for adjusting scrotal circumference to 365 or 452 days of age and correlations of scrotal circumference with growth traits in beef bulls *Theriogenology* **46** 659–669.

Bellve AR and Zheng W (1989) Growth factors as autocrine and paracrine modulators of male gonadal functions *Journal of Reproduction and Fertility* **85** 771–793.

BIF (2010) *Guidelines for Uniform Beef Improvement Programs*. Beef Improvement Federation, Raleigh.

Blazquez NB, Mallard GJ and Wedd SR (1988) Sweat glands of the scrotum of the bull *Journal of Reproduction and Fertility* **83** 673–677.

Blockey MA (1975) Studies on the social and sexual behaviour of bulls. PhD thesis, University of Melbourne.

Blockey MA (1976) Sexual behaviour of bulls at pasture: a review *Theriogenology* **6** 387–392.

Blockey MA and Taylor EG (1984) Observations on spiral deviation of the penis in beef bulls *Australian Veterinary Journal* **61** 141–145.

Blom E (1966) A new sterilizing and hereditary defect (the "Dag defect") located in the bull sperm tail *Nature* **209** 739–740.

Blom E (1976) A sterilizing tail stump sperm defect in a Holstein–Friesian bull *Nordisk Veterinærmedicin* **28** 295–298.

Blom E (1977) A decapitated sperm defect in two sterile Hereford bulls *Nordisk Veterinærmedicin* **29** 119–123.

Blom E and Birch-Andersen A (1962) Ultrastructure of the sterilizing knobbed sperm defect in the bull *Nature* **194** 989–990.

Blom E and Birch-Andersen A (1970) Ultrastructure of the "decapitated sperm defect" in Guernsey bulls *Journal of Reproduction and Fertility* **23** 67–72.

Blom E and Birch-Andersen A (1980) Ultrastructure of the tail stump sperm defect in the bull *Acta Pathologica et Microbiologica Scandinavia Section A – Pathology* **88** 397–405.

Blom E and Wolstrup C (1976) Zinc as a possible causal factor in the sterilizing sperm tail defect, the 'Dag-defect', in Jersey bulls *Nordisk Veterinærmedicin* **28** 515–518.

Bongso TA, Jainudeen MR and Lee JY (1981) Testicular hypoplasia in a bull with XX/XY chimerism *The Cornell Veterinarian* **71** 376–382.

Bourdon RM and Brinks JS (1986) Scrotal circumference in yearling Hereford bulls: adjustment factors, heritabilities and genetic, environmental and phenotypic relationships with growth traits *Journal of Animal Science* **62** 958–967.

Boyd GW and Corah LR (1988) Effect of sire and sexual experience on serving capacity of yearling beef bulls *Theriogenology* **29** 779–790.

Boyd GW, Lunstra DD and Corah LR (1989) Serving capacity of crossbred yearling beef bulls. I. Single-sire mating behavior and fertility during average and heavy mating loads at pasture *Journal of Animal Science* **67** 60–71.

Bratton RW, Musgrave SD, Dunn HO and Foote RH (1959) *Causes and Prevention of Reproductive Failure in Dairy Cattle: II. Influence of Underfeeding and Overfeeding from Birth to 80 Weeks of Age on Growth, Sexual Development, and Semen Production in Holstein Bulls.* New York State College of Agriculture, Ithaca.

Brito LF, Silva AE, Rodrigues LH, Vieira FV, Deragon LA and Kastelic JP (2002a) Effect of age and genetic group on characteristics of the scrotum, testes and testicular vascular cones, and on sperm production and semen quality in AI bulls in Brazil *Theriogenology* **58** 1175–1186.

Brito LF, Silva AE, Rodrigues LH, Vieira FV, Deragon LA and Kastelic JP (2002b) Effects of environmental factors, age and genotype on sperm production and semen quality in *Bos indicus* and *Bos taurus* AI bulls in Brazil *Animal Reproduction Science* **70** 181–190.

Brito LF, Silva AE, Barbosa RT and Kastelic JP (2004a) Testicular thermoregulation in *Bos indicus*, crossbred and *Bos taurus* bulls: relationship with scrotal, testicular vascular cone and testicular morphology, and effects on semen quality and sperm production *Theriogenology* **61** 511–528.

Brito LF, Silva AE, Unanian MM, Dode MA, Barbosa RT and Kastelic JP (2004b) Sexual development in early- and late-maturing *Bos indicus* and *Bos indicus* × *Bos taurus* crossbred bulls in Brazil *Theriogenology* **62** 1198–1217.

Brito L, Barth A, Rawlings N, Wilde R, Crews D Jr, Mir P and Kastelic J (2007a) Circulating metabolic hormones during the peripubertal period and their association with testicular development in bulls *Reproduction in Domestic Animals* **42** 502–508.

Brito LF, Barth AD, Rawlings NC, Wilde RE, Crews DH Jr, Boisclair YR, Ehrhardt RA and Kastelic JP (2007b) Effect of feed restriction during calfhood on serum concentrations of metabolic hormones, gonadotropins, testosterone, and on sexual development in bulls *Reproduction* **134** 171–181.

Brito LF, Barth AD, Rawlings NC, Wilde RE, Crews DH Jr, Mir PS and Kastelic JP (2007c) Effect of improved nutrition during calfhood on serum metabolic hormones, gonadotropins, and testosterone concentrations, and on testicular development in bulls *Domestic Animal Endocrinology* **33** 460–469.

Brito LF, Barth AD, Rawlings NC, Wilde RE, Crews DH Jr, Mir PS and Kastelic JP (2007d) Effect of nutrition during calfhood and peripubertal period on serum metabolic hormones, gonadotropins and testosterone concentrations, and on sexual development in bulls *Domestic Animal Endocrinology* **33** 1–18.

Brito L, Barth A, Wilde R and Kastelic J (2012a) Effect of growth rate from 6 to 16 months of age on sexual development and reproductive function in beef bulls *Theriogenology* **77** 1398–1405.

Brito L, Barth A, Wilde R and Kastelic J (2012b) Testicular ultrasonogram pixel intensity during sexual development and its relationship with semen quality, sperm production, and quantitative testicular histology in beef bulls *Theriogenology* **78** 69–76.

Brito L, Barth A, Wilde R and Kastelic J (2012c) Testicular vascular cone development and its association with scrotal temperature, semen quality, and sperm production in beef bulls *Animal Reproduction Science* **134** 135–140.

Bruner KA, McCraw RL, Whitacre MD and Van Camp SD (1995) Breeding soundness examination of 1,952 yearling beef bulls in North Carolina *Theriogenology* **44** 129–145.

Cailleau J, Vermeire S and Verhoeven G (1990) Independent control of the production of insulin-like growth factor I and its binding protein by cultured testicular cells *Molecular and Cellular Endocrinology* **69** 79–89.

Carroll EJ, Ball L and Scott JA (1963) Breeding soundness in bulls – a summary of 10,940 examinations *Journal of the American Veterinary Medical Association* **142** 1105–1111.

Carroll EJ, Aanes WA and Ball L (1964) Persistent penile frenulum in bulls *Journal of the American Veterinary Medical Association* **144** 747–749.

Carson RL and Wenzel JG (1997) Observations using the new bull-breeding soundness evaluation forms in adult and young bulls *Veterinary Clinics of North America: Food Animal Practice* **13** 305–311.

Casady RB, Myers RM and Legates JE (1953) The effect of exposure to high ambient temperature on spermatogenesis in the dairy bull *Journal of Dairy Science* **36** 14–23.

Casas E, Lunstra DD, Cundiff LV and Ford JJ (2007) Growth and pubertal development of F1 bulls from Hereford, Angus, Norwegian Red, Swedish Red and White, Friesian, and Wagyu sires *Journal of Animal Science* **85** 2904–2909.

CBRA (1998) *Manual para Exame Andrólogico e Avaliação de Sêmen Animal.* Colegio Brasileiro de Reprodução Animal, Belo Horizonte.

Chandolia RK, Evans AC and Rawlings NC (1997a) Effect of inhibition of increased gonadotrophin secretion before 20 weeks of age in bull calves on testicular development *Journal of Reproduction and Fertility* **109** 65–71.

Chandolia RK, Honaramooz A, Bartlewski PM, Beard AP and Rawlings NC (1997b) Effects of treatment with LH releasing hormone before the early increase in LH secretion on endocrine and reproductive development in bull calves *Journal of Reproduction and Fertility* **111** 41–50.

Chase CC Jr, Chenoweth PJ, Larsen RE, Olson TA, Hammond AC, Menchaca MA and Randel RD (1997) Growth and reproductive development from weaning through 20 months of age among breeds of bulls in subtropical Florida *Theriogenology* **47** 723–745.

Chenoweth PJ (1983) Sexual behavior of the bull: a review *Journal of Dairy Science* **66** 173–179.

Chenoweth PJ (1986) Reproductive behavior of bulls In *Current Therapy in Theriogenology* pp 148–152 Ed DA Morrow. WB Saunders Company, Philadelphia.

Chenoweth PJ (1997) Bull libido/serving capacity *Veterinary Clinics of North America: Food Animal Practice* **13** 331–344.

Chenoweth PJ (2005) Genetic sperm defects *Theriogenology* **64** 457–468.

Chenoweth PJ, Brinks JS and Nett TM (1979) A comparison of three methods of assessing sex-drive in yearling beef bulls and relationships with testosterone and LH levels *Theriogenology* **12** 223–233.

Chenoweth PJ, Farin PW, Mateos ER, Rupp GP and Pexton JE (1984) Breeding soundness and sex drive by breed and age in beef bulls used for natural mating *Theriogenology* **22** 341–349.

Chenoweth PJ, Chase CC Jr, Thatcher MJ, Wilcox CJ and Larsen RE (1996) Breed and other effects on reproductive traits and breeding soundness categorization in young beef bulls in Florida *Theriogenology* **46** 1159–1170.

Coe PH and Gibson CD (1993) Adjusted 200-day scrotal size as a predictor of 365-day scrotal circumference *Theriogenology* **40** 1065–1072.

Cook RB, Coulter GH and Kastelic JP (1994) The testicular vascular cone, scrotal thermoregulation, and their relationship to sperm production and seminal quality in beef bulls *Theriogenology* **41** 653–671.

Coulter GH (1986) Puberty and postpubertal development of beef bulls In *Current Therapy in Theriogenology* pp 142–148 Ed AD Morrow. WB Saunders Company, Philadelphia.

Coulter GH and Bailey DRC (1988) Epididymal sperm reserves in 12-month-old Angus and Hereford bulls: effects of bull strain plus dietary energy *Animal Reproduction Science* **16** 169–175.

Coulter GH and Foote RH (1976) Relationship of testicular weight to age and scrotal circumference of Holstein bulls *Journal of Dairy Science* **59** 730–732.

Coulter GH and Foote RH (1977) Relationship of body weight to testicular size and consistency in growing Holstein bulls *Journal of Animal Science* **44** 1076–1079.

Coulter GH and Keller DG (1982) Scrotal circumference of young beef bulls: relationship to paired testes weight, effect of breed, and predictability *Canadian Journal of Animal Science* **62** 133–139.

Coulter GH and Kozub GC (1984) Testicular development, epididymal sperm reserves and seminal quality in two-year-old Hereford and Angus bulls: effects of two levels of dietary energy *Journal of Animal Science* **59** 432–440.

Coulter GH, Larson LL and Foote RH (1975) Effect of age on testicular growth and consistency of Holstein and Angus bulls *Journal of Animal Science* **41** 1383–1389.

Coulter GH, Rounsaville TR and Foote RH (1976) Heritability of testicular size and consistency in Holstein bulls *Journal of Animal Science* **43** 9–12.

Coulter GH, Carruthers TD, Amann RP and Kozub GC (1987a) Testicular development, daily sperm production and epididymal sperm reserves in 15-mo-old Angus and Hereford bulls: effects of bull strain plus dietary energy *Journal of Animal Science* **64** 254–260.

Coulter GH, Mapletoft RJ, Kozub GC and Cates WF (1987b) Scrotal circumference of two-year-old bulls of several beef breeds *Theriogenology* **27** 485–491.

Coulter GH, Cook RB and Kastelic JP (1997) Effects of dietary energy on scrotal surface temperature, seminal quality, and sperm production in young beef bulls *Journal of Animal Science* **75** 1048–1052.

Cran DG and Dott HM (1976) The ultrastructure of knobbed bull spermatozoa *Journal of Reproduction and Fertility* **47** 407–408.

Cran DG, Dott HM and Wilmington JW (1982) The structure and formation of rolled and crested bull spermatozoa *Gamete Research* **5** 263–269.

Crews D Jr and Porteous DJ (2003) Age of dam and age at measurement adjustments and genetic parameters for scrotal circumference of Canadian Hereford bulls *Canadian Journal of Animal Science* **83** 183–188.

Curtis SK and Amann RP (1981) Testicular development and establishment of spermatogenesis in Holstein bulls *Journal of Animal Science* **53** 1645–1657.

de Araujo JW, Borgwardt RE, Sween ML, Yelich JV and Price EO (2003) Incidence of repeat-breeding among Angus bulls (*Bos taurus*) differing in sexual performance *Applied Animal Behavior Science* **81** 89–98.

Diarra MS, Pare JP and Roy G (1997) Genetic and environmental factors affecting semen quality of young Holstein bulls *Canadian Journal of Animal Science* **77** 77–85.

Donald HP and Hancock JL (1953) Evidence of gene-controlled sterility in bulls *Journal of Agricultural Science* **43** 178–181.

Dunn HO, McEntee K, Hall CE, Johnson RHJ and Stone WH (1979) Cytogenetic and reproductive studies of bulls born co-twin with freemartins *Journal of Reproduction and Fertility* **57** 21–30.

Dunn HO, Lein DH and McEntee K (1980) Testicular hypoplasia in a Hereford bull with 61,XXY karyotype: the bovine counterpart of human Klinefelter's syndrome *The Cornell Veterinarian* **70** 137–146.

Elmore RG, Bierschwal CJ, Martin CE and Youngquist RS (1975) A summary of 1,127 breeding soundness examinations in beef bulls *Theriogenology* **3** 209–218.

Elmore RG, Breuerb J, Youngquist RS, Lasley JF and Bierschwal CJ (1978) Breeding soundness examinations in 18 closely related inbred Angus bulls *Theriogenology* **10** 355–363.

Evans AC, Currie WD and Rawlings NC (1993) Opioidergic regulation of gonadotrophin secretion in the early prepubertal bull calf *Journal of Reproduction and Fertility* **99** 45–51.

Evans ACO, Davies FJ, Nasser LF, Bowman P and Rawlings NC (1995) Differences in early patterns of gonadotrophin secretion between early and late maturing bulls, and changes in semen characteristics at puberty *Theriogenology* **43** 569–578.

Evans ACO, Pierson RA, Garcia A, McDougall LM, Hrudka F and Rawlings NC (1996) Changes in circulating hormone concentrations, testes histology and testes ultrasonography during sexual maturation in beef bulls *Theriogenology* **46** 345–357.

Evans JL, Golden BL, Bourdon RM and Long KL (1999) Additive genetic relationships between heifer pregnancy and scrotal circumference in Hereford cattle *Journal of Animal Science* **77** 2621–2628.

Everett RW and Bean B (1982) Environmental influences on semen output *Journal of Dairy Science* **65** 1303–1310.

Farin PW, Chenoweth PJ, Tomky DF, Ball L and Pexton JE (1989) Breeding soundness, libido and performance of beef bulls mated to estrus synchronized females *Theriogenology* **32** 717–725.

Flipse RJ and Almquist JO (1961) Effect of total digestible nutrient intake from birth to four years of age on growth and reproductive development and performance of dairy bulls *Journal of Dairy Science* **44** 905–914.

Foote RH, Hough SR, Johnson LA and Kaproth M (1992) Electron microscopy and pedigree study in an Ayrshire bull with tail-stump sperm defects *Veterinary Record* **130** 578–579.

Fordyce G, Entwistle K, Norman S, Perry V, Gardiner B and Fordyce P (2006) Standardising bull breeding soundness evaluations and reporting in Australia *Theriogenology* **66** 1140–1148.

Foster J, Almquist JO and Martig RC (1970) Reproductive capacity of beef bulls. IV. Changes in sexual behavior and semen characteristics among successive ejaculations *Journal of Animal Science* **30** 245–252.

Garmyn AJ, Moser DW, Christmas RA and Minick Bormann J (2011) Estimation of genetic parameters and effects of cytoplasmic line on scrotal circumference and semen quality traits in Angus bulls *Journal of Animal Science* **89** 693–698.

Gipson TA, Vogt DW, Ellersieck MR and Massey JW (1987) Genetic and phenotypic parameter estimates for scrotal circumference and semen traits in young beef bulls *Theriogenology* **28** 547–555.

Gledhill BL (1973) Inherited disorders causing infertility in the bull *Journal of the American Veterinary Medical Association* **162** 979–982.

Glossop CE and Ashdown RR (1986) Cavernosography and differential diagnosis of impotence in the bull *Veterinary Record* **118** 357–360.

Godfrey RW and Dodson RE (2005) Breeding soundness evaluations of Senepol bulls in the US Virgin Islands *Theriogenology* **63** 831–840.

Godfrey RW and Lunstra DD (1989) Influence of single or multiple sires and serving capacity on mating behavior of beef bulls *Journal of Animal Science* **67** 2897–2903.

Gregory KE, Lunstra DD, Cundiff LV and Koch RM (1991) Breed effects and heterosis in advanced generations of composite populations for puberty and scrotal traits of beef cattle *Journal of Animal Science* **69** 2795–2807.

Hafs HD, Knisely RC and Desjardins C (1962) Sperm output of dairy bulls with varying degrees of sexual preparation *Journal of Dairy Science* **45** 788–793.

Hahn J, Foote RH and Seidel GE Jr (1969) Testicular growth and related sperm output in dairy bulls *Journal of Animal Science* **29** 41–47.

Hancock JL (1953) The spermatozoa of sterile bulls *Journal of Experimental Biology* **30** 50–56.

Hees H, Leiser R, Kohler T and Wrobel KH (1984) Vascular morphology of the bovine spermatic cord and testis. I. Light- and scanning electron-microscopic studies on the testicular artery and pampiniform plexus *Cell and Tissue Research* **237** 31–38.

Henney SR, Killian GJ and Deaver DR (1990) Libido, hormone concentrations in blood plasma and semen characteristics in Holstein bulls *Journal of Animal Science* **68** 2784–2792.

Higdon HL 3rd, Spitzer JC, Hopkins FM and Bridges WC Jr (2000) Outcomes of breeding soundness evaluation of 2,898 yearling bulls subjected to different classification systems *Theriogenology* **53** 1321–1332.

Hoflack G, Van Soom A, Maes D, de Kruif A, Opsomer G and Duchateau L (2006) Breeding soundness and libido examination of Belgian Blue and Holstein Friesian artificial insemination bulls in Belgium and the Netherlands *Theriogenology* **66** 207–216.

Hofmann R (1960) Die Gefäbarchitektur des bullenhodens, Zugleich eim Versuch ihrer funktionellen Deutung *Zentralblatt für Veterinärmedizin* **7** 59–93.

Hopkins FM and Spitzer JC (1997) The new Society for Theriogenology breeding soundness evaluation system *Veterinary Clinics of North America: Food Animal Practice* **13** 283–293.

Irons PC, Nothling JO and Bertschinger HJ (2007) Bull breeding soundness evaluation in southern Africa *Theriogenology* **68** 842–847.

Jiménez-Severiano H (2002) Sexual development of dairy bulls in the Mexican tropics *Theriogenology* **58** 921–932.

Johnson WH, Thompson JA, Kumi-Diaka J, Wilton JW and Mandell IB (1995) The determination and correlation of reproductive parameters of performance-tested Hereford and Simmental bulls *Theriogenology* **44** 973–982.

Kaneko H, Yoshida M, Hara Y, Taya K, Araki K, Watanabe G, Sasamoto S and Hasegawa Y (1993) Involvement of inhibin in the regulation of FSH secretion in prepubertal bulls *Journal of Endocrinology* **137** 15–19.

Kastelic JP, Coulter GH and Cook RB (1995) Scrotal surface, subcutaneous, intratesticular, and intra-epididymal temperatures in bulls *Theriogenology* **44** 147–152.

Kastelic JP, Cook RB and Coulter GH (1996a) Contribution of the scrotum and testes to scrotal and testicular thermoregulation in bulls and rams *Journal of Reproduction and Fertility* **108** 81–85.

Kastelic JP, Cook RB, Coulter GH and Saacke RG (1996b) Insulating the scrotal neck affects semen quality and scrotal/testicular temperatures in the bull *Theriogenology* **45** 935–942.

Kastelic JP, Cook RB and Coulter GH (1997) Contribution of the scrotum, testes, and testicular artery to scrotal/testicular thermoregulation in bulls at two ambient temperatures *Animal Reproduction Science* **45** 255–261.

Kealey CG, MacNeil MD, Tess MW, Geary TW and Bellows RA (2006) Genetic parameter estimates for scrotal circumference and semen characteristics of Line 1 Hereford bulls *Journal of Animal Science* **84** 283–290.

Keeton LL, Green RD, Golden BL and Anderson KJ (1996) Estimation of variance components and prediction of breeding values for scrotal circumference and weaning weight in Limousin cattle *Journal of Animal Science* **74** 31–36.

Kennedy SP, Spitzer JC, Hopkins FM, Higdon HL and Bridges WC, Jr. (2002) Breeding soundness evaluations of 3,648 yearling beef bulls using the 1993 Society for Theriogenology guidelines *Theriogenology* **58** 947–961.

Killian GJ and Amann RP (1972) Reproductive capacity of dairy bulls. IX. Changes in reproductive organ weights and semen characteristics of Holstein bulls during the first thirty weeks after puberty *Journal of Dairy Science* **55** 1631–1635.

Kirby A (1953) Observations on the blood supply of the bull testis *British Veterinary Journal* **109** 464–472.

Kirby A and Harrison RG (1954) A comparison of the vascularization of the testis in Afrikaner and English breeds of bull In *Studies on Fertility. Proceedings of the Society for Study of Fertility Volume VI* pp. 129–131. Blackwell, Oxford.

Knights SA, Baker RL, Gianola D and Gibb JB (1984) Estimates of heritabilities and of genetic and phenotypic correlations among growth and reproductive traits in yearling Angus bulls *Journal of Animal Science* **58** 887–893.

Koefoed-Johnsen HH and Pedersen H (1971) Further observations on the Dag-defect of the tail of the bull spermatozoon *Journal of Reproduction and Fertility* **26** 77–83.

Koivisto MB, Costa MTA, Perri SHV and Vicente WRR (2009) The effect of season on semen characteristics and freezability in *Bos indicus* and *Bos taurus* bulls in the southeastern region of Brazil *Reproduction in Domestic Animals* **44** 587–592.

Kriese LA, Bertrand JK and Benyshek LL (1991) Age adjustment factors, heritabilities and genetic correlations for scrotal circumference and related growth traits in Hereford and Brangus bulls *Journal of Animal Science* **69** 478–489.

Kumi-Diaka J, Nagaratnam V and Rwuaan JS (1981) Seasonal and age-related changes in semen quality and testicular morphology of bulls in a tropical environment *Veterinary Record* **108** 13–15.

Lacroix A and Pelletier J (1979) Short-term variations in plasma LH and testosterone in bull calves from birth to 1 year of age *Journal of Reproduction and Fertility* **55** 81–85.

Ladds PW (1993) The male genital system In *Pathology of Domestic Animals* pp 471–529 Ed KVF Jubb, PC Kennedy and N Palmer. Academic Press, San Diego.

Landaeta-Hernández AJ, Chenoweth PJ and Berndtson WE (2001) Assessing sex-drive in young *Bos taurus* bulls *Animal Reproduction Science* **66** 151–160.

Lane SM, Kiracofe GH, Craig JV and Schalles RR (1983) The effect of rearing environment on sexual behavior of young beef bulls *Journal of Animal Science* **57** 1084–1089.

Larsen RE, Littell R, Rooks E, Adams EL, Falcon C and Warnick AC (1990) Bull influences on conception percentage and calving date in Angus Hereford, Brahman and Senepol single-sire herds *Theriogenology* **34** 549–568.

Latimer FG, Wilson LL, Cain MF and Stricklin WR (1982) Scrotal measurements in beef bulls: heritability estimates, breed and test station effects *Journal of Animal Science* **54** 473–479.

Lee CY, Hunt DW, Gray SL and Henricks DM (1991) Secretory patterns of growth hormone and insulin-like growth factor-I during peripubertal period in intact and castrate male cattle *Domestic Animal Endocrinology* **8** 481–489.

Logue DN, Harvey MJ, Munro CD and Lennox B (1979) Hormonal and histological studies in a 61XXY bull *Veterinary Record* **104** 500–503.

Lunstra DD and Cundiff LV (2003) Growth and pubertal development in Brahman-, Boran-, Tuli-, Belgian Blue-, Hereford- and Angus-sired F1 bulls *Journal of Animal Science* **81** 1414–1426.

Lunstra DD and Echternkamp SE (1982) Puberty in beef bulls: acrosome morphology and semen quality in bulls of different breeds *Journal of Animal Science* **55** 638–648.

Lunstra DD, Ford JJ and Echternkamp SE (1978) Puberty in beef bulls: hormone concentrations, growth, testicular development, sperm production and sexual aggressiveness in bulls of different breeds *Journal of Animal Science* **46** 1054–1062.

Lunstra DD, Gregory KE and Cundiff LV (1988) Heritability estimates and adjustment factors for the effects of bull age and age of dam on yearling testicular size in breeds of bulls *Theriogenology* **30** 127–136.

MacDonald RD, Deaver DR and Schanbacher BD (1991) Prepubertal changes in plasma FSH and inhibin in Holstein bull calves: responses to castration and(or) estradiol *Journal of Animal Science* **69** 276–282.

Macmillan KL and Hafs HD (1968) Gonadal and extra gonadal sperm numbers during reproductive development of Holstein bulls *Journal of Animal Science* **27** 697–700.

Mader DR and Price EO (1984) The effects of sexual stimulation on the sexual performance of Hereford bulls *Journal of Animal Science* **59** 294–300.

Madgwick S, Bagu ET, Duggavathi R, Bartlewski PM, Barrett DM, Huchkowsky S, Cook SJ, Beard AP and Rawlings NC (2008) Effects of treatment with GnRH from 4 to 8 weeks of age on the attainment of sexual maturity in bull calves *Animal Reproduction Science* **104** 177–188.

Mann T, Rowson LE, Short RV and Skinner JD (1967) The relationship between nutrition and androgenic activity in pubescent twin calves, and the effect of orchitis *Journal of Endocrinology* **38** 455–468.

Martínez-Velázquez G, Gregory KE, Bennett GL and Van Vleck LD (2003) Genetic relationships between scrotal circumference and female reproductive traits *Journal of Animal Science* **81** 395–401.

Mathevon M, Buhr MM and Dekkers JC (1998) Environmental, management, and genetic factors affecting semen production in Holstein bulls *Journal of Dairy Science* **81** 3321–3330.

Matsuzaki S, Uenoyama Y, Okuda K, Watanabe G, Kitamura N, Taya K, Cruzana MB and Yamada J (2001) Prepubertal changes in immunoreactive inhibin concentration in blood serum and testicular tissue in Holstein bull calves *Journal of Veterinary Medical Science* **63** 1303–1307.

McAllister CM, Speidel SE, Crews DH Jr and Enns RM (2011) Genetic parameters for intramuscular fat percentage, marbling score, scrotal circumference, and heifer pregnancy in Red Angus cattle *Journal of Animal Science* **89** 2068–2072.

McCarthy MS, Convey EM and Hafs HD (1979a) Serum hormonal changes and testicular response to LH during puberty in bulls *Biology of Reproduction* **20** 1221–1227.

McCarthy MS, Hafs HD and Convey EM (1979b) Serum hormone patterns associated with growth and sexual development in bulls *Journal of Animal Science* **49** 1012–1020.

McEntee K (1990) *Reproductive Pathology of Domestic Mammals*. Academic Press, San Diego.

Meacham TN, Cunha TJ, Warnick AC, Hentges JF Jr and Hargrove DD (1963) Influence of low protein rations on growth and semen characteristics of young beef bulls *Journal of Animal Science* **22** 115–120.

Meacham TN, Warnick AC, Cunha TJ, Hentges JF Jr and Shirley RL (1964) Hematological and histological changes in young beef bulls fed low protein rations *Journal of Animal Science* **23** 380–384.

Meyer K, Hammond K, Mackinnon MJ and Parnell PF (1991) Estimates of covariances between reproduction and growth in Australian beef cattle *Journal of Animal Science* **69** 3533–3543.

Moser DW, Bertrand JK, Benyshek LL, McCann MA and Kiser TE (1996) Effects of selection for scrotal circumference in Limousin bulls on reproductive and growth traits of progeny *Journal of Animal Science* **74** 2052–2057.

Moura AA and Erickson BH (2001) Testicular development, histology, and hormone profiles in three yearling Angus bulls with spermatogenic arrest *Theriogenology* **55** 1469–1488.

Mwansa PB, Kemp RA, Crews DH Jr, Kastelic JP, Bailey DR and Coulter GH (2000) Comparison of models for genetic evaluation of scrotal circumference in crossbred bulls *Journal of Animal Science* **78** 275–282.

Neely JD, Johnson BH, Dillard EU and Robison OW (1982) Genetic parameters for testes size and sperm number in Hereford bulls *Journal of Animal Science* **55** 1033–1040.

Nelsen TC, Short RE, Urick JJ and Reynolds WL (1986) Heritabilities and genetic correlations of growth and reproductive measurements in Hereford bulls *Journal of Animal Science* **63** 409–417.

Parkinson TJ (1985) Seasonal variation in semen quality of bulls and correlations with metabolic and endocrine parameters *Veterinary Record* **117** 303–307.

Parkinson TJ (2001) Fertility and infertility in male animals In *Arthur's Veterinary Reproduction and Obstetrics* 8th edition pp 695–750 Ed DE Noakes, TJ Parkinson and GCW England. WB Saunders (Harcourt), London.

Peet RL and Mullins KR (1991) Sterility in a poll Hereford bull associated with the 'tail stump' sperm defect *Australian Veterinary Journal* **68** 245.

Persson Y and Söderquist L (2005) The proportion of beef bulls in Sweden with mature spermiograms at 11–13 months of age *Reproduction in Domestic Animals* **40** 131–135.

Pratt SL, Spitzer JC, Webster HW, Hupp HD and Bridges WC Jr (1991) Comparison of methods for predicting yearling scrotal circumference and correlations of scrotal circumference to growth traits in beef bulls *Journal of Animal Science* **69** 2711–2720.

Price EO and Wallach SJ (1990) Short-term individual housing temporarily reduces the libido of bulls *Journal of Animal Science* **68** 3572–3577.

Price EO and Wallach SJ (1991a) Development of sexual and aggressive behaviors in Hereford bulls *Journal of Animal Science* **69** 1019–1027.

Price EO and Wallach SJ (1991b) Effects of group size and the male-to-female ratio on the sexual performance and aggressive behavior of bulls in serving capacity tests *Journal of Animal Science* **69** 1034–1040.

Price EO and Wallach SJ (1991c) Inability to predict the adult sexual performance of bulls by prepubertal sexual behaviors *Journal of Animal Science* **69** 1041–1046.

Price EO, Katz LS, Moberg GP and Wallach SJ (1986) Inability to predict sexual and aggressive behaviors by plasma concentrations of testosterone and luteinizing hormone in Hereford bulls *Journal of Animal Science* **62** 613–617.

Pruitt RJ and Corah LR (1985) Effect of energy intake after weaning on the sexual development of beef bulls. I. Semen characteristics and serving capacity *Journal of Animal Science* **61** 1186–1193.

Pruitt RJ, Corah LR, Stevenson JS and Kiracofe GH (1986) Effect of energy intake after weaning on the sexual development of beef bulls. II. Age at first mating, age at puberty, testosterone and scrotal circumference *Journal of Animal Science* **63** 579–585.

Rawlings NC and Evans AC (1995) Androgen negative feedback during the early rise in LH secretion in bull calves *Journal of Endocrinology* **145** 243–249.

Rawlings NC, Hafs HD and Swanson LV (1972) Testicular and blood plasma androgens in Holstein bulls from birth through puberty *Journal of Animal Science* **34** 435–440.

Rawlings NC, Fletcher PW, Henricks DM and Hill JR (1978) Plasma luteinizing hormone (LH) and testosterone levels during sexual maturation in beef bull calves *Biology of Reproduction* **19** 1108–1112.

Renaville R, Devolder A, Massart S, Sneyers M, Burny A and Portetelle D (1993) Changes in the hypophysial–gonadal axis during the onset of puberty in young bulls *Journal of Reproduction and Fertility* **99** 443–449.

Renaville R, Massart S, Sneyers M, Falaki M, Gengler N, Burny A and Portetelle D (1996) Dissociation of increases in plasma insulin-like growth factor I and testosterone during the onset of puberty in bulls *Journal of Reproduction and Fertility* **106** 79–86.

Renaville R, Van Eenaeme C, Breier B, Vleurick L, Bertozzi C, Gengler N, Hornick J, Parmentier I, Istasse L, Haezebroeck V, Massart S and Portetelle D (2000) Feed restriction in young bulls alters the onset of puberty in relationship with plasma insulin-like growth factor-I (IGF-I) and IGF-binding proteins *Domestic Animal Endocrinology* **18** 165–176.

Revell S, Cooley W and Cranfield N (2000) A case of the sperm tail stump defect in a Holstein bull In *Proceedings of the 14th International Congress on Animal Reproduction, Stockholm, Sweden, 2–6 July 2000 Volume 2–7* p 73.

Roberts SJ (1986) *Veterinary Obstetrics and Genital Diseases* 3rd edition. Published by the author, Woodstock, Vermont. Distributed by David and Charles, Newton Abbot.

Rodriguez RE and Wise ME (1989) Ontogeny of pulsatile secretion of gonadotropin-releasing hormone in the bull calf during infantile and pubertal development *Endocrinology* **124** 248–256.

Rodriguez RE and Wise ME (1991) Advancement of postnatal pulsatile luteinizing hormone secretion in the bull calf by pulsatile administration of gonadotropin-releasing hormone during infantile development *Biology of Reproduction* **44** 432–439.

Rodriguez RE, Benson B, Dunn AM and Wise ME (1993) Age-related changes in biogenic amines, opiate, and steroid receptors in the prepubertal bull calf *Biology of Reproduction* **48** 371–376.

Rose EP, Wilton JW and Schaeffer LR (1988) Estimation of variance components for traits measured on station-tested beef bulls *Journal of Animal Science* **66** 626–634.

Saez JM (1994) Leydig cells: endocrine, paracrine, and autocrine regulation *Endocrine Reviews* **15** 574–626.

Schanbacher BD (1979) Relationship of *in vitro* gonadotropin binding to bovine testes and the onset of spermatogenesis *Journal of Animal Science* **48** 591–597.

Schmutz SM, Barth AD and Moker JS (1994) A Klinefelter bull with a 1;29 translocation born to a fertile 61,XXX cow *Canadian Veterinary Journal* **35** 182–184.
Sealfon AI and Zorgniotti AW (1991) A theoretical model for testis thermoregulation In *Temperature and Environmental Effects on the Testis* pp 123–135 Ed AW Zorgniotti. Academic Press, New York.
Seidel GE Jr and Foote RH (1969) Motion picture analysis of ejaculation in the bull *Journal of Reproduction and Fertility* **20** 313–317.
Seidel GE Jr, Pickett BW, Wilsey CO and Seidel S (1980) Effect of high level of nutrition on reproductive characteristics of Angus bulls In *Proceedings of the 9th International Congress on Animal Reproduction and Artificial Insemination, 16th–20th June, 1980, Madrid, Spain* p 359. Editorial Garsi, Madrid.
Setchell BP (1978) *The Mammalian Testis.* Cornell University Press, Ithaca.
Setchell BP (1998) The Parkes Lecture. Heat and the testis *Journal of Reproduction and Fertility* **114** 179–194.
Settergren I and McEntee K (1992) Germ cell weakness as a cause of testicular hypoplasia in bulls *Acta Veterinaria Scandinavia* **33** 273–282.
Sinowatz F and Amselgruber W (1986) Postnatal development of bovine Sertoli cells *Anatomy and Embryology* **174** 413–423.
Siqueira JB, Pinho RO, Guimaraes SE, Miranda Neto T and Guimaraes JD (2010) Immotile short-tail sperm defect in Nelore (*Bos taurus indicus*) breed bulls *Reproduction in Domestic Animals* **45** 1122–1125.
Sironen AI, Andersson M, Uimari P and Vilkki J (2002) Mapping of an immotile short tail sperm defect in the Finnish Yorkshire on porcine chromosome 16 *Mammalian Genome* **13** 45–49.
Sironen A, Thomsen B, Andersson M, Ahola V and Vilkki J (2006) An intronic insertion in *KPL2* results in aberrant splicing and causes the immotile short-tail sperm defect in the pig *Proceedings of the National Academy of Sciences of the United States of America* **103** 5006–5011.
Skinner JD (1981) Nutrition and fertility in pedigree bulls In *Environmental Factors in Mammals Reproduction* pp 160–168 Ed D Gilmore B and Cook. Macmillan, London
Skinner JD and Louw GN (1966) Heat stress and spermatogenesis in *Bos indicus* and *Bos taurus* cattle *Journal of Applied Physiology* **21** 1784–1790.
Smith BA, Brinks JS and Richardson GV (1989a) Relationships of sire scrotal circumference to offspring reproduction and growth *Journal of Animal Science* **67** 2881–2885.
Smith BA, Brinks JS and Richardson GV (1989b) Estimation of genetic parameters among breeding soundness examination components and growth traits in yearling bulls *Journal of Animal Science* **67** 2892–2896.
Söderquist L, Janson L, Håård M and Einarsson S (1996) Influence of season, age, breed and some other factors on the variation in sperm morphological abnormalities in Swedish dairy AI bulls *Animal Reproduction Science* **44** 91–98.
Söderquist L, Rodriguez-Martinez H, Håård MG and Lundeheim N (1997) Seasonal variation in sperm morphology in proven Swedish dairy AI bulls *Reproduction in Domestic Animals* **32** 263–265.
Spiteri-Grech J and Nieschlag E (1992) The role of growth hormone and insulin-like growth factor I in the regulation of male reproductive function *Hormone Research* **38(Supplement 1)** 22–27.
Spitzer JC and Hopkins FM (1997) Breeding soundness evaluation of yearling bulls *Veterinary Clinics of North America: Food Animal Practice* **13** 295–304.
Spitzer JC, Hopkins FM, Webster HW, Kirkpatrick FD and Hill HS (1988) Breeding soundness examination of yearling beef bulls *Journal of the American Veterinary Medical Association* **193** 1075–1079.
St Jean G, Gaughan EM and Constable PD (1992) Cryptorchidism in North American cattle: breed predisposition and clinical findings *Theriogenology* **38** 951–958.
Steffen D (1997) Genetic causes of bull infertility *Veterinary Clinics of North America: Food Animal Practice* **13** 243–253.
Sukura A, Makipaaa R, Vierula M, Rodriguez-Martinez H, Sundback P and Andersson M (2002) Hereditary sterilizing short-tail sperm defect in Finnish Yorkshire boars *Journal of Veterinary Diagnostic Investigation* **14** 382–388.
Sylla L, Stradaioli G, Borgami S and Monaci M (2007) Breeding soundness examination of Chianina, Marchigiana, and Romagnola yearling bulls in performance tests over a 10-year period *Theriogenology* **67** 1351–1358.
Taylor JF, Bean B, Marshall CE and Sullivan JJ (1985) Genetic and environmental components of semen production traits of artificial insemination Holstein bulls *Journal of Dairy Science* **68** 2703–2722.
Thundathil J, Meyer R, Palasz AT, Barth AD and Mapletoft RJ (2000) Effect of the knobbed acrosome defect in bovine sperm on IVF and embryo production *Theriogenology* **54** 921–934.

Thundathil J, Palomino J, Barth A, Mapletoft R and Barros C (2001) Fertilizing characteristics of bovine sperm with flattened or indented acrosomes *Animal Reproduction Science* **67** 231–243.

Thundathil J, Palasz AT, Barth AD and Mapletoft RJ (2002) Plasma membrane and acrosomal integrity in bovine spermatozoa with the knobbed acrosome defect *Theriogenology* **58** 87–102.

Toelle VD and Robison OW (1985) Estimates of genetic correlations between testicular measurements and female reproductive traits in cattle *Journal of Animal Science* **60** 89–100.

Vierula M, Alanko M, Remes E and Vanha-Perttula T (1983) Ultrastructure of a tail stump sperm defect in an Ayrshire bull *Andrologia* **15** 303–309.

Vogler CJ, Bame JH, DeJarnette JM, McGilliard ML and Saacke RG (1993) Effects of elevated testicular temperature on morphology characteristics of ejaculated spermatozoa in the bovine *Theriogenology* **40** 1207–1219.

Waites GMH (1970) Temperature regulation and the testis In *The Testis* pp 241–279 Ed AD Johnson, WR Gomes and NL Vandermark. Academic Press, London.

Waites GM, Speight AC and Jenkins N (1985) The functional maturation of the Sertoli cell and Leydig cell in the mammalian testis *Journal of Reproduction and Fertility* **75** 317–326.

Wallach SJ and Price EO (1988) Bulls fail to show preference for estrous females in serving capacity tests *Journal of Animal Science* **66** 1174–1178.

Weisgold AD and Almquist JO (1979) Reproductive capacity of beef bulls. VI. Daily spermatozoal production, spermatozoal reserves and dimensions and weight of reproductive organs *Journal of Animal Science* **48** 351–358.

Williams G (1987) 'Tail-stump' defect affecting the spermatozoa of two Charolais bulls *Veterinary Record* **121** 248–250.

Williams HJ, Revell SG, Scholes SF, Courtenay AE and Smith RF (2010) Clinical, ultrasonographic and pathological findings in a bull with segmental aplasia of the mesonephric duct *Reproduction in Domestic Animals* **45** e212–e216.

Wise ME, Rodriguez RE and Kelly CM (1987) Gonadal regulation of LH secretion in prepubertal bull calves *Domestic Animal Endocrinology* **4** 175–181.

Wolf FR, Almquist JO and Hale EB (1965) Prepubertal behavior and pubertal characteristics of beef bulls on high nutrient allowance *Journal of Animal Science* **24** 761–765.

Wrobel KH (1990) The postnatal development of the bovine Leydig cell population *Reproduction in Domestic Animals* **25** 51–60.

Wrobel KH (2000) Prespermatogenesis and spermatogoniogenesis in the bovine testis *Anatomy and Embryology* **202** 209–222.

Wrobel KH, Dostal S and Schimmel M (1988) Postnatal development of the tubular lamina propria and the intertubular tissue in the bovine testis *Cell and Tissue Research* **252** 639–653.

13 Applied Andrology in Cattle (*Bos indicus*)

Jorge Chacón*
Universidad Nacional (UNA), Heredia, Costa Rica

Introduction

The conduct of a breeding soundness evaluation (BSE) on extensively managed *Bos indicus* bulls, particularly when this is done in tropical regions, presents particular challenges for both the handling of the examination and the interpretation of results. The former consideration includes poor facilities in combination with potentially fractious bulls; the latter includes a relative lack of relevant reference data on which to base decisions. Here, it is evident that reliance on the large amount of information collected on *Bos taurus* bulls in temperate environments is insufficient owing to differences that occur in genotype, environment and management. In this chapter, an attempt is made to describe a number of these differences and to place them within the context of conducting and interpreting the BSE of *B. indicus* bulls in tropical regions.

Overview

Bos indicus is one of the two major species of the genus *Bos*, which probably originated from a common non-humped ancestor. For many people, *B. indicus* and zebu cattle denote similar animals, although the term zebu specifically refers to the humped cattle apparently developed by selection from non-humped *B. indicus* in India and Pakistan. *B. indicus* cattle have also been referred to as 'indigenous' cattle. Because most *B. indicus* genotypes throughout the world are humped, the terms zebu and *B. indicus* will be used interchangeably throughout the course of this chapter. Nevertheless, it should be recognized that as indigenous non-humped cattle exist in various parts of the world, the arbitrary use of the word zebu to denote *B. indicus* cattle might cause confusion.

Within the India/Pakistan subcontinent, there are at least 30 separate *B. indicus* breeds, which are divided into six major groups (I, II, III, IV, V, VI) according to a classification by Joshi and Phillip in 1953 (Sanders, 1980). Of those foundation strains, groups I, II and III, which are represented by Guzerat, Nellore and Gir breeds, respectively, have had by far the most influence on zebu cattle breeding in Asia, Australia, North America and Latin America. Other composite breeds were developed more recently through various breed crosses. This is the case for the Indu-Brazil, which was developed in Brazil from a base that involved primarily Gir, Guzerat and Nellore foundations; it is also the case for

* E-mail: jorge.chacon.calderon@una.cr

the Brahman, which was developed in the USA by breeding Guzerat and Nellore (Gray Brahman), or Gir and Indu-Brazil, with some Guzerat influence (Red Brahman), and the probable inclusion of some influence from criollo type cattle. While phenotypic and productive features (including milk attributes) differ widely among breeds, the major use of *B. indicus* cattle in tropical and subtropical areas in the western world has been for beef purposes, either as pure or as crossbred animals.

Features commonly associated with *B. indicus* genotypes include heat and parasite tolerance and the ability to successfully adapt to difficult nutritional conditions, such as those prevailing in many tropical areas. Additional, less favourable, features include delayed puberty and sexual maturity, poor calf survival, prolonged calving–conception intervals and, in general, reduced fertility in comparison with *B. taurus* genotypes. Further, *B. indicus* males have developed a reputation in some quarters for having smaller testicles, lower libido and poorer semen quality than their *B. taurus* equivalents. Consequently, low productive performance has been historically associated with *B. indicus* cattle.

However, this reputation for poor reproductive capability raises questions about the relative roles of genetics, environment and management practices. The central question is whether *B. indicus* cattle are inherently of low genetic fertility or whether their poor reproductive performance is the consequence of a number of external influences, including environment and mismanagement. In this chapter, the latter opinion is supported. The extensive and stressful conditions under which *B. indicus* cattle are managed have confounded environmental and genetic factors and allowed the myth to develop that they can achieve maximum production with minimal input.

The chapter will first address general physiological aspects of *B. indicus* bulls before leading into practical approaches for evaluating their andrological status, taking into consideration the genetic, social and management factors that may influence findings and their interpretation.

Physiological Characteristics of Zebu Bulls

Significant physiological particularities have been determined in *B. indicus* males. The ability of this genus to withstand the environmental conditions prevailing in the tropics has been widely recognized for many decades (Cartwright, 1955; Turner, 1980; Kumi-Diaka *et al.*, 1981), although much of the original focus of study was on their capacity to resist the effects of climate. It is now recognized though that *B. indicus* cattle possess superior capabilities to withstand other adverse effects on cattle production in tropical areas as well, such as low quality forages, and diseases especially caused by haemoprotozoa and other parasites (Cartwright, 1980).

B. indicus bulls have a greater capacity than *B. taurus* bulls to regulate their body temperature in a tropical environment. Studies by Carvalho *et al.* (1995) in Brazil showed that *B. indicus* sires have more sweat glands per unit area of the skin and greater sweat production than *B. taurus* bulls. The difference in sweat gland efficiency is thought to be the result of the distinct glandular shape in *B. indicus* versus native or imported Simmental (*B. taurus*) bulls (baggy and tubular shaped respectively). This characteristic is positively correlated ($P < 0.01$) with the perimeter of the sweat glands (540 ± 19 μm versus 382 ± 27 μm and 497 ± 17 μm, respectively). In addition, the number of layers of the epidermis (epithelial strata) was greater in *B indicus* than in native or imported Simmental (14.93 versus 7.15 and 4.5, respectively), and this was negatively correlated with body temperature during resting or exertion (Carvalho *et al.*, 1995). The higher number of epithelial strata in *B indicus* may be involved in the maintenance of body temperature. Consequently, the sweating ability of these animals is greater and increases more quickly in a hot environment, thus allowing *B indicus* bulls to lose more heat by skin evaporation than European breeds. As a result, signs of heat stress are rarely seen in *B. indicus* bulls exposed to temperatures up to 37°C and high humidity (Da Silva and Casagrande, 1976; Carvalho *et al.*, 1995). The smooth hair coat of zebu bulls has also been found to provide good resistance to solar radiation (Turner, 1980; Finch, 1986; Carvalho *et al.*, 1995).

B. indicus cattle are further recognized for their ability to better utilize low-quality forages and their lower nutritional requirements for maintenance than *B. taurus* cattle (Turner, 1980; Wildeus and Entwistle, 1984). Rekwot *et al.* (1994) suggested that *B. indicus* bulls are able to lower their metabolic rate during the dry season when good quality pastures are scarce. In general, *B. indicus* cattle exhibit a higher resistance to ticks and fly infestations than *B. taurus* cattle (Riek, 1962; Turner, 1980; Fordyce *et al.*, 1996) as well. The skin thickness and its local defence mechanisms of the species could explain this feature.

Haematological differences between *B. indicus* and *B. taurus* cattle have been reported too. Turner (1980) found higher red blood cell counts and haemoglobin levels in Brahman (*B. indicus*) than in Hereford (*B. taurus*) bulls in Florida, which concurred with earlier findings (Evans 1963). This condition, together with their mechanisms of heat loss, may explain the lower body temperatures and respiration rates observed in *B. indicus* during rest or after walking in a tropical climate (Cartwright, 1955; Turner, 1980; Finch, 1986; Carvalho *et al.*, 1995).

On testicular and scrotal thermoregulation, Turner (1980) reported that the scrotal skin in *B. indicus* bulls is hairless and less thick than that in *B. taurus* sires. These features may facilitate heat loss from superficial testicular vessels, and so act in favour of testicular thermoregulation. According to Brito *et al.* (2004), the ratio of testicular artery length and volume to testicular volume ratios were significantly larger in *B. indicus* and related crossbreeds than in *B. taurus* bulls. *B. indicus* sires also showed smaller testicular artery wall thickness than crossbred and *B. taurus* bulls (averages 192.5, 229.0 and 290.0 μm, respectively) and arterial to venous distances (averages 330.5, 373.7 and 609.4 μm, respectively) (Brito *et al.*, 2004). Consequently, the proximity between the arterial and venous blood vessels in *B. indicus* males is more efficient in lowering the temperature of the arterial blood reaching the testicles. These physiological attributes might explain the lower prevalence of testicular degeneration – presumably resulting from environmental heat stress in the tropics – in *B. indicus* than in *B. taurus* bulls (Vale Filho *et al.*, 1980; Kumi-Diaka *et al.*, 1981; Wildeus and Entwistle, 1983, 1984; Crabo, 1988; Ohashi *et al.*, 1988; Chacón, 2000).

Puberty and Sexual Maturity

Factors involved in the modulation of puberty and sexual maturity have received scant attention in *B. indicus* compared with *B taurus* bulls. Moreover, the available information has been generated from limited bull populations. Although the chain of endocrine and other events associated with puberty attainment are similar in the two species, there is general agreement that *B. indicus* bulls reach puberty and sexual maturity at a later age than do *B. taurus* males, even when they are raised under similar conditions (Igboeli and Rakha, 1971a; Fields *et al.*, 1979, 1982; Vale Filho *et al.*, 1980; Wildeus and Entwistle, 1982; Silva-Mena, 1997). Reports on age at puberty and sexual maturity in *B. indicus* bulls differ according to the breed involved, the conditions of the study (particularly the nutritional management of the experimental animals) and how puberty and sexual maturity achievement were defined. Many studies have employed the definition of puberty of Wolf *et al.* (1965) as the age at which a bull is first capable of producing an ejaculate containing at least 50×10^6 spermatozoa with a minimum of 10% progressive motility. Others have defined puberty by recognizing histological indicators of the onset of spermatogenesis in preparations of testicular tissue. While the latter method has the potential of achieving greater accuracy than the former, it suffers from the disadvantage of being unable to be repeated in the same animal.

Scrotal circumference (SC) is the best predictor and indicator of age at puberty and sexual maturity in zebu bulls, regardless of breed, and even more than body weight or chronological age. In tropical Costa Rica, ongoing studies on a large population of Brahman (n = 323) and Nellore (n = 98) bull calves that are grass fed and supplemented with minerals have shown that all of those that were able to reach 28.0 cm SC at 16 months of age were sexually mature at this time. In addition, differences ($P < 0.0001$) in SC with age (27.5 versus 26.4 cm and 18.8 versus 18.0 months, respectively) were found at puberty in both breeds (J. Chacón, unpublished). This feature supports the use of SC as an indicator of puberty and as a selection criterion in young zebu sires, particularly as it also has other attributes, which include a significant correlation with sperm production, semen quality and fertility.

Using Wolf's definition, the age at puberty in supplemented pasture-raised White Fulani (n = 6) and Sokoto zebu bulls (n = 6) in Nigeria was reported as 15.5 and 17 months respectively (Oyedipe *et al.*, 1981), whereas in similarly raised Brahman bulls (n = 12) in Yucatán, México, the mean age at puberty was reported to be 17 months, with a mean body weight and SC of 374 ± 22.5 kg and 28.6 ± 0.6 cm, respectively (Silva-Mena, 1997). Supplemented hay-fed Brahman males (n = 10), which were monitored from 8 to 20 months of age in Florida, attained puberty at around 15.9 months of age, weighing 432 ± 16 kg and with an SC of 33.4 ± 1.2 cm (Fields *et al.*,1982). This relatively large testicular size in contrast with a number of comparable studies could be attributed to this being a closed herd that had been selected and managed optimally for some years. In another study in Florida, and in the same breed, Morris *et al.* (1989) suggested that puberty was attained between 14 and 15 months of age. However, no semen collection was performed in this study, with age at puberty being estimated based on the period of accelerated testicular growth. In tropical Venezuela, Guzerat (n = 159) and Nellore (n = 60) supplemented pasture-raised bulls reached puberty at 18.0 ± 2.0 and 18.5 ± 2.7 months of age, with an SC of 25.6 ± 2.2 versus 23.6 ± 0.2 cm and body weight of 310 ± 42 versus 268.1 ± 12.1 kg, respectively (Trocóniz *et al.*, 1991). Other studies carried out in the same location with Brahman bulls (n = 4) reported puberty at an age of 21.3 months with a mean body weight and SC of 287.3 ± 67.8 kg and 24.9 ± 2.4 cm, respectively (Ocanto *et al.*, 1984). In considering these results and others from different world regions, it is apparent that both genetic and environmental influences can strongly affect the age at puberty in *B. indicus* males.

Puberty, though, is not synonymous with sexual maturity or even acceptable fertility. For example, in yearling Tabapuâ (n = 246) zebu bulls (a *B. indicus* × *B. Taurus* crossbreed developed in Brazil), Corrêa *et al.* (2006) reported values for sperm motility and total spermatozoa per ejaculate (× 10^6) ranging from 14.0 to 42.0% and 91 to 315.8, respectively. Despite such values being consistent with evidence of puberty attainment, they would not permit these bulls to pass a BSE or to achieve satisfactory levels of breeding fertility. Those hurdles were not reached in these same bulls until they achieved a mean SC and body weight of 33.5 cm and 490 kg, respectively, which, in some bulls, was not until they were 2 years old.

Studies based on histological analysis of the testis parenchyma and epididymal contents of young zebu males showed that only Sertoli and spermatogenic stem cells were found in the seminiferous tubules of 7-month-old grass-fed Angoni (short horn *B. indicus*) bulls in Nigeria (Igboeli and Rakha, 1971a). The onset of spermatogenesis was detected at 9 months of age, leading to the presence of first spermatozoa in the cauda epididymis at 11 months of age. Puberty was estimated to have occurred by approximately 14 months old in this breed (Igboeli and Rakha, 1971a). Similar results were found in two different studies by Aire and Akpokodje (1975) and Aponte *et al.* (2005), in Nigeria and Venezuela, respectively. In the first study, conducted with supplemented pasture fed White Fulani (n = 20), primary spermatocytes were found at 9 months of age and spermatozoa were present in large numbers at 15 months of age in both the seminiferous tubules and all regions of the epididymis. In the second study, conducted with similarly raised Brahman bulls (n = 20), the onset of spermatogenesis was estimated to occur at an average at 9 months of age.

Another event that occurs around the age of puberty in zebu sires is the detachment of the preputial mucosa from the glans penis. This is a gradual phenomenon starting at around 7 months of age in most zebu breeds and in Guzerat and Nellore bulls under tropical conditions is completed at around 16 months of age (Trocóniz *et al.*, 1991). Ocanto *et al.* (1984) reported the total separation of the preputial mucosa between 18 and 22 months of age in Brahman bulls; as this phenomenon is reportedly dependent on testosterone levels, which, in turn, are increased by puberty, the nutritional status of the bulls may explain the delayed age reported for this event in this study.

Nutrition plays a significant role in the onset of puberty and attainment of sexual maturity in *B. indicus* bulls (Igboeli and Rakha, 1971a; Crabo, 1988; Vale Filho *et al.*, 1996). For example, supplemented pasture-fed Boran bulls (n = 27) in Ethiopia reached puberty earlier (16.4 versus 17.8 months, respectively, $P < 0.05$), were heavier (208.6 ± 6.0 versus 193.5 ± 7.6 kg, respectively) and had larger SC (24.5 ± 0.6 versus 23.3 ± 0.6 cm, respectively) than control animals that were not supplemented (Tegegne *et al.*, 1992). Underfed young *B. indicus* bulls show impaired body growth, delayed puberty and a retarded development of their endocrine system and reproductive tract (Rekwot *et al.*, 1994). Deficiencies of essential micronutrients, such as zinc, selenium and vitamins A and E can negatively affect gonadotrophin release by the pituitary gland, thus impairing the replication and functionality of Sertoli cells, as well as hormone steroid synthesis by Leydig cells (Dunn and Moss, 1992). Because the critical process of Sertoli cell division occurs during the peri-pubertal period only, nutritional deprivation at this stage can seriously and irreversibly impair the potential reproductive performance of young zebu bulls.

Breeding Management of *B. indicus* bulls in Tropical Areas

Beef cattle production in tropical regions is largely dependent upon pure or crossbred *B. indicus* cattle that are extensively managed on natural pasture. Breeding of the herd is carried out by means of natural mating by using one or several bulls per herd. Data from tropical Costa Rica indicate that natural mating accounts for at least around 95% of the cows sired in beef farms. In general, *B. indicus* bulls are first bred at not earlier than 24 months. Breeding bulls are managed using continuous (i.e. the male mixed with the herd the whole year) or temporal mating seasons with one sire (single sire system) or several sires (multiple sire system). Data from the Andrology Section of the Veterinary School of the Universidad Nacional (UNA) of Costa Rica, based on 1100 BSEs of zebu bulls from the dry Pacific area, showed that 47.6% of the males (n = 524) were in a continuous mating system, and 49.5% of them were in multiple sire systems in which mixed-age or mixed-breed groups were common (Chacón, 2000). Additionally, around 7% of the bulls examined were over 7 years old.

The mean of bulls per group in multiple sire systems in Costa Rica was 4.0 (range 3–11) breeding sires for an average of 101.8 (range 15–450) cows. The mean number of cows in a single sire mating system was 29.3 animals (Chacón, 2000). These data call attention to the fact that, regardless of the system used, there is a bull/cow ratio of approximately 1:25, which is similar to data reported from other tropical countries (McCosker *et al.*, 1989). This management strategy is used by most farmers, as it is believed to represent the maximum breeding capacity of a sire in natural mating, and because it compensates for bull sub-fertility.

Fertility in *B. indicus* bulls

The reproductive efficiency of breeding herds in tropical regions, where beef cattle production is based upon *B. indicus* breeds extensively managed on natural pastures, has often been reported to be low compared with those in more temperate areas, where *B. taurus* cattle predominate.

Müller (1990) estimated that, in Central America, the yearly calving rate in beef cattle does not reach 50%, and stressed that poor reproductive efficiency of the breeding males contributes largely to this scenario – a situation that is still evident (Chacón, 2009). Similar calving rates reported from other tropical areas, such as in Mexico (Lamothe Zavaleta, 1990) and Colombia (Murgueitio, 1990), confirm this observation. How much of the problem is attributable to the sire's role is, as yet, undetermined. However, a lack of periodic andrological control to determine the suitability of bulls for natural breeding certainly contributes to the poor reproductive performance that is often encountered. This observation is reinforced by the fact that, when andrological monitoring is applied, it often results in the identification of bulls with poor reproductive capabilities. It is considered that routine identification and elimination of problem males from breeding herds would help to counter the probably erroneous view held by some authors that *B. indicus* bulls have an inherently low fertility.

Fertility is a multifactorial status to which both males and females contribute. Males must produce a fertile ejaculate and be competent in delivering this to the female. Females must be able to produce fertile oocytes and provide a reproductive system compatible with sperm transport, capacitation, fertilization of the ova, and embryo and fetal development.

The fertility of naturally mating *B. indicus* breeding bulls, when measured by means of their pregnancy or calving rates, must embrace a wide range of decisive factors, which can include climatic stressors, cyclicity and fertility of females, nutritional constraints, sexual behaviour and a number of management considerations that are particularly pertinent for this species. For example, *B. indicus* bulls tend to spend more time in courtship than do European breeds, and this is possibly associated with females tending to spend less time in oestrus. Observations that Brahman bulls may miss mating opportunities owing to focus on a particular female (Aguilar Chavarria, 2008) have yet to receive scientific validation.

Under natural mating conditions, the fertility of *B. indicus* bulls differs between studies. Chenoweth and Osborne (1975) studied the performance of Brahman bulls 16–31 months old in Queensland, Australia. They found lower libido scores in this breed associated with its lower pregnancy rate (54.9%) compared with Shorthorn–Hereford (*B. taurus*) (64.4%) and the Africander crossbred (*B. indicus*) (79.1%) after breeding with 20 to 35 cows for 7 weeks individually. Although no differences in semen quality were noted between breeds, it is noticeable in this study that Brahman bulls were not selected before the mating period, so they also had the highest prevalence of presumed testicular hypoplasia (11.1% versus 1.4% overall), which could have accounted for these fertility differences. In fact, individually, many Brahman bulls performed as well as the best sires from the other breeds. Moreover, the use of some young Brahman bulls (16 months old) probably influenced the differences found as a result of sexual immaturity and lack of mating experience compared with the other bulls tested.

Crockett *et al.* (1978) reported the pregnancy records from mature (4 year old) Angus (*B. taurus*), Hereford and Brahman bulls in Florida, which were naturally mated with groups of 30 heifers for 90 days. Here, breed group affected pregnancy rates, being lower for the Brahman genotype (72%) compared with the other breeds (89.9%). It should be noted, though, that these bulls were selected on the basis of phenotype records as the main criteria.

Silva-Mena *et al.* (2000) studied the sexual behaviour and pregnancy rates of Brahman and Nellore bulls ($n = 15$), which were bred individually with groups of 16 oestrus-synchronized heifers in Mexico. The mean pregnancy rate under these conditions was 59.2 ± 5.8%, which was similar to values reported for European breeds by Pexton *et al.* (1989).

The conception rate in 38 adult zebu bulls individually mated with groups of 24.0 ± 13.6 cows in the southern area of tropical Costa Rica showed significant differences according to the andrological status of the bulls tested (Navarro *et al.*, 2008). Bulls ranked as sound for breeding achieved a higher pregnancy rate compared with unsound sires (82.5 ± 12.1% versus 28.4 ± 30.5%, respectively, $P < 0.05$). Although no statistical differences were found for the conception rate between the sound and deferred (or questionable) sire groups, the latter category tended to show a lower pregnancy rate (77.4 ± 27.2%). These data support the contention that, if *B. indicus* bulls are selected on the basis of attributes related to fertility, they can perform as efficiently as any other sire group regardless of genotype under natural mating conditions. This theory is supported by McCosker *et al.* (1989) who recorded low pregnancy rates (below 40%) in the Brahman herds studied in northern Australia. This was attributed to a high proportion of unsound sires detected after BSEs were performed; a situation compounded by the widespread use of aged bulls (>8 years old), many of whom were positive carriers of venereal diseases (e.g. *Tritrichomonas foetus*). The routine use of mixed age groups of breeding bulls was also considered to be a factor that contributed to the fertility problem observed. After BSE was applied to the Brahman bull stock as selection criteria, together with sanitary controls of the herd stock, subsequent pregnancy rates increased markedly (≥90%).

Bull Breeding Soundness Evaluation (BSE)

General background of the BSE in *Bos indicus*

Many decades of use and refinement have confirmed that the BSE is an essential tool in diagnosing and eliminating bulls of potential low fertility, thereby contributing to the reproductive efficiency of cattle herds. Much of this benefit has been obtained with *B. taurus* cattle in temperate regions. However, it is evident that even greater benefits are possible if widespread use of the BSE is adopted within *B. indicus* beef herds managed under extensive conditions in the tropics. This is because beef cattle in the tropics are generally managed either as herds under extensive grazing conditions or as small groups attached to local communities. Both systems are usually distinguished by the absence of a proscribed breeding period or season, as well as by a lack of adequate records for either productive or reproductive parameters. In addition, veterinary assistance is absent or infrequent, with the consequence that there is inadequate identification and elimination of those animals that have undesirable breeding characteristics. Bulls are often managed in multiple sire systems under extensive conditions, rendering it impossible to identify males of questionable fertility by means of their individual performance records. Therefore, bulls with reproductive problems often go unnoticed and are retained in the breeding herd, thus causing considerable economic losses. If such problem bulls are dominant within multiple sire breeding groups, the adverse effects on fertility can be compounded. Lack of a finite breeding period tends to obscure the extent of unsound breeding bulls in the herd as well.

It is apparent that the importance of the bull's role in ensuring herd reproductive efficiency, particularly in systems based on extensive management, has been neglected in the tropics (McCosker *et al.*, 1989; Müller, 1990; Chacón, 2000). A contributing factor is that most *B. indicus* bulls are selected as breeding sires based solely on phenotypic characteristics, which are not necessarily related to their reproductive performance under natural mating conditions (Chacón, 2000). This was apparent in a survey of beef farmers in tropical Costa Rica, who were asked which were the most important factors that they considered when purchasing a breeding bull (Fig. 13.1). In this study, most bulls were selected based on muscle composition, frame, breed-specific features, pedigree and other characteristics associated with 'type'. In contrast, reproductive aspects of the bulls, such as testicle size (i.e. SC), which are critically important for bull fertility, were underestimated in importance as selection criteria (Chacón, 2009).

When questioned on the utilization of the BSE in *B. indicus* herds, only 20% of beef farmers considered this examination to be a prerequisite to purchasing a bull (Chacón, 2009). Moreover, 88% of farms had never practised a periodic BSE on their bulls. This confirms that the BSE is not a common routine procedure in extensive *B. indicus* operations, as corroborated by observations in Australia (Chenoweth and Osborne, 1975; McCosker *et al.*, 1989). Furthermore, traditional ranchers, who are most often the breeders of pure *B. indicus* sires, are often more reluctant to accept a BSE on their bulls as a result of perceptions that range from potential economic disadvantage to irrational

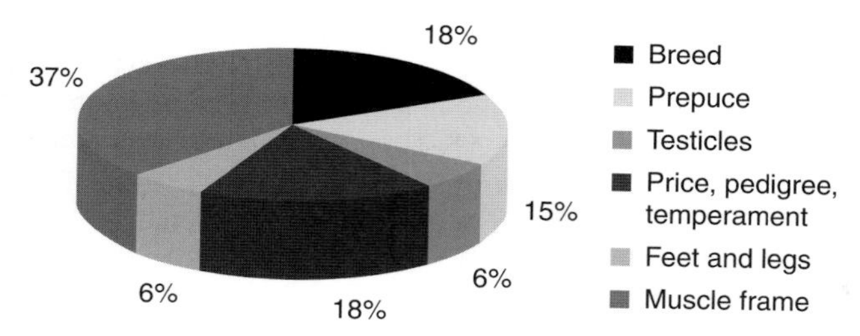

Fig. 13.1. Selection criteria (%) used by beef farmers (n = 1050) at the time of purchasing a *Bos indicus* sire from a survey in tropical Costa Rica. From Chacón, 2009.

myths (Chacón, 2000). Such perceptions must be countered with rational scientific information if the bull BSE is to become accepted as an essential and routine management practice for beef producers in the tropics.

Unfortunately, veterinarians are often not aware of how important a complete bull BSE is under these conditions, and continue to base their evaluations solely on aspects of semen quality and, in particular, on the assessment of sperm mass activity of the collected semen. In general, this practice does not include a complete clinical examination of the sire and, therefore, has led people to wrongly consider a bull BSE as just a 'semen examination' (Chacón, 2000). A similar picture emerges from other tropical regions (McCosker *et al.*, 1989).

The tendency to undervalue the importance of the bull in herd reproductive performance is reflected in the relatively minimal attention given to studies in the tropics of the breeding performance of *B. indicus* bulls and the factors that interfere with their efficiency. An indicator of this imbalance is the ratio of publications concerning reproductive aspects in the male (20%) compared with the female (80%), and the low emphasis given to studying the reproductive aspects of *B. indicus* (19%) compared with *B. taurus* males (38%) (Galina and Russell, 1987; Galina and Arthur, 1991). Further, 42% of published research was associated with aspects of artificial insemination (AI) compared with 10% on natural mating (Galina and Russell, 1987).

Thus there is a strong case for promoting much higher acceptance of bull BSEs in extensively managed beef cattle herds where *B. indicus* type cattle are predominant. Its performance should both provide relatively accurate information and be economically feasible. The ensuing sections will describe the different components of a thorough and practical andrological examination of *B. indicus* sires under field farm conditions.

General information and sire's history

Although semen collection and evaluation are important aspects of the bull BSE, it should be emphasized that other aspects are of equal importance.

At the outset, date of evaluation, farm name, owner and location should be recorded, as well as an unambiguous identification of the bull being examined. One example of the latter is a clearly visible brand number on the upper lateral region of the femoral area. Ear or horn tattoos, and permanent ear tags (using either legible letters/numbers or electronic labelling), are also used. On some farms, additional information can include the bull's birth date. This is a useful method of determining bull age, particularly as determination of the age of *B. indicus* bulls by estimating dentition can pose risks to operator safety because restraint facilities are often not optimal and the temperament of this type of cattle is, in general, uneasy. Unfortunately, the provision of birth date is not common, and age information is provided by branded numbers, which represent either the year of birth or of consecutive calving. Some bulls lack any form of permanent ID and are just identified by name (e.g. 13.1% of bulls submitted for a BSE in tropical Costa Rica), which complicates the certification of the sire in case of selling, purchasing or future examinations. There is a need for regulation of this matter by local or regional zebu breeders (Chacón, 2009).

Accurate recording of age for *B. indicus* bulls is important for several reasons. For example, age is positively and significantly correlated with SC and semen quality. Similarly, the breed of the bull should be recorded, as *B. indicus* genotypes differ in SC and body weight relationships, as discussed later in this chapter. There is also a higher prevalence of venereal diseases and of skeletal and testicular pathologies in aged (i.e. >7 years) bulls. Another consideration with retrained older bulls is the risk of inbreeding – a factor whose effects are probably underestimated in tropical beef cattle stocks as a result of inadequate record keeping. Age of the bulls is further important in terms of relative ranking of the bull breeding team when results of the BSEs conducted on a particular farm are consolidated. Older bulls tend to have a shorter reproductive life expectancy – owing to a number of problems, as discussed above – so they should either be culled, or if particularly valuable, be used for monitored hand mating programmes.

The bull BSE should be based on a compilation of the considerations described above, and also include a history of nutritional and reproductive management, previous illnesses, abrupt weight changes prior to examination and any available information on preceding reproductive performance of the sire. Problems such as lameness, or any systemic disease causing fever, may adversely affect mating ability and spermatogenesis. Any potential existing problems in the cattle population, such as low heat observation, low calving rate or further reproductive problems, such as abortions, should be investigated as well.

Reproductive management

The breeding system employed, i.e. single or multiple sire, bull to female ratio, continuous or seasonal breeding and physiological status of the females (e.g. primiparous, first calf or multiparous and post-partum interval) should be recorded. In multiple sire systems, if the male is subordinated by a dominant bull partner, he is likely to have few chances to access the sexually active group of females. This system has the disadvantage of favouring social dominance, especially when age is not uniform in the group, which is often the case in the tropics (Blockey, 1979; Chenoweth, 1981; McCosker *et al.*, 1989; Rodríguez *et al.*, 1993; Chacón, 2000). Dominance is not necessarily related to bull libido and fertility. In addition, dominant older sires tend to monopolize mating opportunities with females, or at least prevent less dominant animals from accessing females, even though they are more likely to have a higher prevalence of reproductive problems (McCosker *et al.*, 1989; Pérez *et al.*, 1992).

McCosker *et al.* (1989) reported low conception rates in herds located in tropical Australia in which multiple sire breeding systems used Brahman bulls of mixed ages and where up to 82% of the calving cows were impregnated by the dominant bull. Furthermore, Rodríguez *et al.* (1993) reported that the dominant zebu (Indu-Brazil) bull was responsible for most (63%) of the bull-related sexual activity in the breeding herd. If this is generally so, the general belief of most farmers in tropical areas of the effectiveness of using a bull/cow ratio of 1:25 is not well founded. In continuous mating systems, many of the cows in the herd could be either pregnant or non-oestrous, thus lowering the ratio even more. Consequently, using a multiple sire breeding system in which the bulls are of mixed ages is often not efficient and can contribute to decreased reproductive outcomes.

The physiological status of the cows is also important, especially in relation to cyclicity and fertility. Early or middle post-partum cows (<5 months) that are still feeding calves have higher probabilities of being non-oestrous, especially if they are not in good body condition. In heifers, genetic or nutritional factors may contribute to lowered cyclicity and pregnancy rates that should not be attributed to deficiencies in the bull(s).

Bulls that have been undergoing sexual rest may show semen/sperm characteristics that are typical of senility and sperm accumulation within the extragonadal system.

These include increased sperm concentration, lowered sperm motility and vigour, and evidence of sperm deterioration, such as detached sperm heads and damaged acrosomes. Conversely, bulls that have just been removed from the breeding herd in which they have been active may show semen characteristics that are typical of high rates of extragonadal sperm transport. These include, above all, relatively low sperm concentration. In either case, these findings are often temporary, emphasizing the importance of considering the reproductive history of the bull when drawing conclusions from the spermiogram.

Libido and mating behaviour

The term libido describes the willingness and eagerness of a male to mount and to attempt service of a female, while mating ability describes the physical suitability of the male to complete a service (Blockey, 1979; Chenoweth, 1981). Several methods have been developed to measure libido, mainly in *B. taurus* bulls, in which the sire's sexual response is evaluated during exposure to a restrained non-oestrous or oestrous cow for a certain interval (5–10 min). In *B. taurus* bulls, the libido score is highly correlated ($r = 0.98$) with the proportion of heifers served during a breeding season and has proved to be highly repeatable in this species ($r > 0.6$) in consecutive trials under corral conditions (Price, 1987). Libido is highly inherited in bulls (heritability, $h^2 = 0.5–0.59$) and has been shown to be significantly correlated with sexual activity in the breeding pasture (Chenoweth, 1981; Hernández *et al.*, 1991), although this does not necessarily translate into pregnancy rates. Blockey (1979) reported in Australia that at least 30% of bulls from several breeds with satisfactory semen quality had poor libido. In fact, libido is not significantly correlated with other reproductive traits, such as SC or semen quality.

When standard libido and serving capacity tests are applied to *B. indicus* bulls, they often achieve low scores in comparison with *B. taurus* bulls. This is considered to reflect differences in sexual behaviour between the species and not a genetic tendency for low libido in the former group. Given these particular differences, the application and interpretation of the results obtained after applying such tests to zebu sires are fairly difficult. For example, the sexual activity of *B. indicus* bulls seems to differ if tested under corral or grass conditions, and the repeatability of libido score results is less than that reported for *B. taurus*. Indu-Brazil bulls undergoing libido evaluations performed fewer services under corral conditions during a 30 min period than they did under grazing conditions (Hernández *et al.*, 1991). Similar results were reported in Gir bulls, where almost 50% of the bulls that failed to show interest in females under corral conditions were quite active when exposed to them on pasture (Piccinali *et al.*, 1992). Indu-Brazil bulls showed inconsistent sexual behaviour in the presence of oestrous-synchronized cows (Orihuela *et al.*, 1988), with a higher mounting frequency at night (18.00–24.00 h), as well as early in the morning (06.00–09.00 h). Similar results were reported by Rodríguez *et al.* (1993) with Gir bulls in Mexico.

As zebu bulls have a tendency to mount cows in mid oestrus only, their sexual response could be limited if tested with restrained oestrous females under corral conditions (Chenoweth, 1981). When exposed to a group of cows synchronized for oestrus, *B. indicus* bulls first established dominance over larger females and prevented passive ones from being mounted by other females (Galina *et al.*, 1987; Price, 1987; Hernández *et al.*, 1991; Piccinali *et al.*, 1992). This behaviour is seen under natural mating conditions as well, where courting activities, including butting and chin resting, can be evident for up to 48 h before the initiation of clinical signs of heat in the female (Aguilar Chavarria, 2008), which tend to persist for shorter period than in *B. taurus* females. *B. indicus* bulls have also been reported to be over attentive to individual females at the expense of servicing other eligible females. Finally, zebu bulls may be more prone to sexual inhibition when tested under corral conditions with observers at hand, even when tested with oestrous females (Galina *et al.*, 1987).These factors may combine to explain the lower frequency of services reported for these bulls during libido testing (Chenoweth and Osborne, 1975; Blockey, 1979;

Hernández *et al.*, 1991; Piccinali *et al.*, 1992), although this did not necessarily result in lowered efficiency in detecting and servicing eligible synchronized females. In general, the evaluation of libido or sex drive is not part of the routine of zebu bulls, partly because of the difficulties in implementation and interpretation, as discussed above. In addition, the extensive rearing and the fact that electroejaculation (EE) is the method of choice for semen collection in *B. indicus* impede the direct evaluation of these important variables during the BSE.

However, despite this, practitioners should not overlook the consideration of bull libido and mating ability during application of the BSE under field conditions. Much relevant information can be obtained from observations of bull conformation and locomotion, and from histories provided by bull owners and handlers. The physical and reproductive examinations conducted on the restrained bull will also reveal important clues as to his ability to physically perform his reproductive role. Unfortunately, data from the Andrology Section of UNA in Costa Rica shows that in 46.64% (1223 out of 2622) of bulls extensively reared and submitted for BSE, cattlemen could not comment on the libido status of the bulls as they had never observed them performing service (Chacón, 2009). In spite of this observation, it is considered important to include questions on the sexual behaviour of bulls when conducting the BSE. These include whether or not the bull has shown interest in oestrous females and attempted service and, if the latter was observed, whether or not the service was successful.

While potentially informative, the answers to these questions often lead to more investigations. If the bull has never been seen to attempt service despite having opportunities, further questioning would need to eliminate many of the aspects discussed above. If the situation is still unclear, the bull should be bred individually with a group of cycling cows so that libido and mating ability can be observed in a more objective way. If doubts still persist, then the actual pregnancy rate of the herd can be determined by the practitioner after allowing the bull to breed with cycling cows for a prudent period. In Costa Rica, many *B. indicus* bulls submitted for BSE and reported by farmers as low libido sires have ended up with herd pregnancy rates (diagnosed by per-rectal palpation) of 85% or higher after allowing the bull to mate individually.

General physical examination

If possible, bulls should be examined prior to restraint for general demeanour, locomotion and evident clinical signs of illness or pathology. Animals that appear to be clearly sick, depressed or in an extremely low body condition should be separated immediately and classified as unsound for breeding, and recommended for re-inspection or culling according to practitioner criteria. Although some authors suggest the use of tranquilizer drugs to allow safe access to fractious zebu bulls for conduct of the BSE, this is not necessary if undue stress (e.g. the use of electric cattle prods) is avoided and safe restraint procedures are followed. Once in the chute, the eyes should be observed, seeking for the presence of corneal opacities or any other damage that might compromise the sight capacity of the sire to identify oestrous females.

Evaluation of body condition

Scoring the body condition (or fat cover) of bulls is a subjective but useful assessment as it provides an easy, quick and cheap method correlated with changes in body weight. An additional consideration is that many beef cattle farms in tropical areas lack a scale to weigh the animals. Body condition score (BCS) provides an estimate of the fat cover on certain areas of the animal, especially the rib wall, lumbar sacral area, hip (the space between iliac crest and the ischial tuberosity), tail head and ischium tibialis muscles. *B. indicus* bulls should ideally have a BCS that provides them with a buffer when going into the breeding pasture because they can be expected to lose some weight during the mating season.

Caution should be used in relying completely on BCS in small-framed bulls that have been underfed during development, as these bulls may show an acceptable BCS as adults despite being underweight according to breed and age. A convenient method for scoring body condition in *B. indicus* bulls is the system that was used by the International Livestock Centre for Africa (ILCA, now part of the CGIAR International Livestock Research Institute, ILRI) (Nicholson and Butterworth, 1986). This method uses a five-point scale with half-point increments. A general evaluation of the bull's BCS may be conducted in the yard, but it is more accurate to do this with palpation once the animal is restrained within a chute.

Cyclical changes in the BCS of *B. indicus* bulls are often caused by seasonal climatic changes and are mainly related to the quantity and quality of pastures. In tropical Costa Rica, the lowest BCS found in different zebu breeds occurred at the end of either the dry or the rainy season (Quintero, 2000). In the latter case, excessive rainfall lowers both intake and digestibility of the pastures. In the adult bull, these seasonal variations are reflected in changes in SC and related variations in semen quality, in particular decreasing sperm production. Such changes are usually recoverable once normal nutritional conditions are re-established, but it is important to note that the recovery of body weight and BCS precedes increases in SC or semen quality (Chacón *et al.*, 2002). However, as discussed earlier, an exception may occur with young animals in which Sertoli cell numbers may be permanently reduced as a result of nutritional deprivation during the peri-pubertal period.

It should also be remembered that, even though BCS is a good indicator of the energy reserves of the bull, it does not reflect his status in relation to those micronutrients that are essential for effective reproductive function, such as vitamins (A, E) and minerals (selenium, cobalt, zinc, etc.). Unfortunately, under extensive tropical management conditions, supplementation of *B. indicus* bulls with such essential micronutrients is the exception rather than the rule.

Feet and legs

The mating ability of the bull is very dependent upon the soundness of his feet and legs, especially the rear limbs, which support about 70% of body weight during mounting. Despite this, the examination of limb conformation in *B. indicus* sires is often overlooked when bulls are evaluated for breeding soundness under tropical field conditions.

Zebu bulls should be evaluated for mobility and foot and leg soundness before being restrained, or the detection of some problems might be missed. It is advisable to observe the bull for lameness or other abnormal locomotion while he is walking and turning. This visual inspection of the locomotor system should include front, lateral and rear views of the sire.

Once restrained, closer attention can be paid to conformational aspects of the muscles and limbs, particularly in the lower limb region. The bull's inner and outer toe lengths should be similar in the normal conformation. Any difference, if found, indicates a dissimilar wearing of hooves as a result of abnormal conformation. Essentially similar pathologies for foot and leg conformation are observed in *B. indicus* and *B. taurus* bulls (Chacón *et al.*, 1999a) though some observers believe that, in general, *B. indicus* bulls are less prone to foot and leg problems. The examiner should attempt to distinguish between genetic or acquired problems. Several genetically based problems of unsound foot and leg conformation that were observed during examination of extensively reared *B. indicus* bulls in tropical Costa Rica are shown in Plates 33, 34 and 35. Clinical evidence of foot and leg problems was observed in around 1% of *B. indicus* sires reared extensively (Chacón *et al.*, 1999a), but this relatively low value should not discourage the need for a systematic and careful examination of the locomotor system in this species.

As mentioned above, abnormal foot and leg conformation can seriously impair not only the physical ability to complete service, but also the ability of the bull to locate oestrous cows in the pasture, with these factors combining to reduce herd reproductive rates. Furthermore, chronic pain associated with foot and leg problems will decrease bull libido and service activity. McCosker *et al.* (1989) reported that a Brahman bull affected by lameness may lose about 150 kg of body weight under extensive conditions.

Affected bulls generally lose body condition as well and can suffer degeneration of the seminiferous epithelia as well, so that semen quality is reduced.

Overfeeding, particularly high energy levels, is probably the most common causal factor of acquired lameness in *B. indicus* bulls. This often occurs in zebu bulls being prepared for sales and show/fairs, and can lead to a number of problems, including excessive stress on joints, laminitis and osteoarthrosis (mainly in the hip, stifle, hock and carpal/metacarpal joints). In general, the prognosis for recovery is poor once pathological signs are apparent, and such overfed bulls often achieve poor reproductive performance owing to lameness and poor mating ability, accelerated subsequent weight loss and testicular degeneration. This tendency to overfeed young bulls by pure-bred breeders is widespread and is detrimental to the reproductive performance of the bulls so raised. It should be actively discouraged by advisors and educators.

Examination of the reproductive organs

Prepuce and penis

The assessment of preputial health in zebu bulls is an important aspect of their BSE. Here, focus should be on sheath length, angle, diameter and integrity. The latter includes evidence of chronic eversion of the preputial membrane, and inspection of the orifice and cavity for signs of trauma, infection and obstruction. Some of these observations, such as preputial length, angle and diameter, can be made before the bull is restrained and while he is still relatively relaxed. Examination for possible constrictions or obstructions is best conducted once the bull has been safely restrained in the chute, preferably with immobilization of the rear leg on the side of the operator. In addition, relevant observations may be made during EE, particularly when penile protrusion occurs, although this does not occur as commonly in *B. indicus* as it does in *B. taurus* bulls.

The length and relative angle of the preputial sheath can be assessed by relating its lower part (tip of the external preputial orifice) with a horizontal line drawn at the level of the hock joint (see Plates 36 and 37). Those sheaths that hang above this line are regarded as acceptable (i.e. have a normal preputial conformation, as in Plate 36), whereas those that hang below it are considered to be abnormal. An abnormally elongated or pendulous prepuce (EP) is more likely to become irritated and traumatized, especially when preputial eversion is part of the syndrome, with posthitis being a common sequel. The defect is linked with a dysfunction or absence of the anterior preputial muscle, which elevates the front-distal part of the sheath. Failure or inadequacy of this muscle allows the distal part of the sheath to be lowered, giving it a 'V-shaped appearance' (as in Plate 37), often accompanied by a chronically prolapsed preputial mucosa. Posthitis, in turn, often leads to fibrosis and constriction of the preputial epithelium, which can then result in phimosis and associated loss of mating ability and libido. As there are genetic considerations involved, corrective surgery is usually not recommended and affected males should be regarded as candidates for culling. Some degree of recovery may follow the use of measures such as the administration of anti-inflammatory products, but relapse is frequent in the short term. It is often possible to identify susceptible bulls during the BSE before the first mating, but the problem becomes more evident as the bull ages. One consideration is that some zebu bulls may have an elongated prepuce in combination with a large cranial fold. Such bulls may or may not develop subsequent posthitis, as this can depend on their age and the type of pasture in the paddock. While it is difficult to predict whether problems will occur or not, it is advisable to note any predisposing features of the sheath at the time of evaluation.

In a survey of 2622 bulls extensively reared in tropical Costa Rica, *B. indicus* and *B. indicus* × *B. taurus* crosses had a higher prevalence of EP than *B. taurus* sires (11.6 and 14.7% versus 3.3%, respectively) (J. Chacón, unpublished). Within the *B. indicus* species (n = 1750), the overall prevalence of EP was higher in Gir strains and lowest in Nellore, with the relative percentages being Gir × Brahman

33.3%, Indu-Brazil × Brahman 23.0%, Gir 16.7%, Indu-Brazil 16.0%, Brahman 13.9%, Guzerat 11.8%, Nellore × Brahman 3.2% and Nellore 1.8% (J. Chacón, unpublished).

Similar penile problems to those observed in *B. taurus* bulls do occur in *B. indicus*, but less commonly. These include a persistent penile frenulum, which is often represented as a fibrous raphe connecting the ventral region of the distal penis to the preputial mucosa, which prevents the penis from extending normally. The length of this band-like fibrous tissue varies from a short cord-like structure to a continuous band of tissue (Plate 38). This condition seriously impairs bull mating ability by deviating the penis and impeding penetration during breeding. The defect was detected at a low level (<1%) in extensively managed *B. indicus* bulls (*n* ~3500) in Costa Rica. However, it is possible that this frequency does not reflect the actual occurrence of the defect, as many of these *B. indicus* bulls did not expose their penis during EE, making assessment difficult. Even though surgical correction is often simple to perform, culling of affected bulls should be strongly considered as there is a strong genetic basis for this condition.

Penile deviations, such as the corkscrew penis, are best diagnosed when observing attempts at natural service, when special attention should be given to penile protrusion, erection, penetration and thrust. Here it is important not to make a definitive diagnosis on the basis of observations during EE only, because this method may artificially induce a transient corkscrew deviation in some bulls, so leading to misdiagnosis. In addition, relevant information concerning penile problems and associated mating dysfunction can be obtained from the owners and handlers of the bulls.

Inspection of scrotal contents

The clinical assessment of the scrotum and its contents represents an important part of the BSE, with this being particularly relevant for *B. indicus* sires that have been raised and managed under harsh conditions. Safe evaluation of the scrotum and contents must take into account the temperament of many *B. indicus* bulls. Restraint must ensure close contact between the bull's rear legs and thighs, with the defences placed behind the bull (Plate 40). Systematic evaluation includes visual assessment of the relative scrotal length, aspect and symmetry, as well as palpation of the spermatic cord. Testicular consistency (tone or resilience) is usually determined by palpation before measuring SC (Plate 39). This evaluation should include a careful palpation of epididymal structures, including the caput, body and cauda.

From the rear of a safely restrained bull, the testicles should be pulled down within the scrotum until skin folds are no longer observed in the scrotal sac. Testicular shape and symmetry can then be assessed visually. An oval shape was the most common testicular form (79.7%) in 1750 *B. indicus* bulls inspected during BSE in tropical Costa Rica (J. Chacón, unpublished results). Although the effect of different testicular shapes (such as oval, round or elongated) on sperm production may be hypothesized, it should be emphasized that SC is the most important testicular measurement. Thus round testicles have been considered as being possibly undesirable because they might have a reduced volume of seminiferous epithelium compared with more conventionally shaped testicles with the same SC. This issue has been studied in Holstein (*B. taurus*) bulls (Bailey *et al.*, 1996), but it has yet to be studied in zebu sires.

In a survey of 1750 *B. indicus* bulls in Costa Rica, the testicles were almost always (95.2%) relatively symmetrical (J. Chacón, unpublished results), and any significant deviation from symmetry should be carefully evaluated before a prognosis is given. Non-symmetrical testicles might be the result of either testicular hypoplasia and/or atrophy. Testicular hypoplasia and degeneration are not easy to differentiate from each other, especially if chronic atrophy has occurred. Therefore, particular care should be taken in the diagnosis of testicular hypoplasia. This may be suspected if the problem is unilateral (Plate 41), if there are similar problems in related males, and if there is a lack of development of the attendant scrotum and epididymal structures. However, definitive diagnosis is reliant upon testicular histopathology. If the diagnosis is confirmed, then affected bulls should be removed from the breeding pool owing to the genetic nature of this problem.

Other pathologies that can be diagnosed during the examination of the scrotal contents include substandard SC, orchitis, epididymitis, hydrocele, and adhesions between the tunica vaginalis parietalis and the testicular capsule (Chacón, 2000). Testicular consistency (see above and Plate 39) can be ranked by palpation in *B. indicus* sires as soft, normal or hard. It should be determined by pulling and holding down the testicles while the practitioner simultaneously applies surface pressure on both gonads with his thumbs. Testicular consistency can be a good indicator of disrupted testicular function if assessed by an experienced operator. Testicles in extensively managed bulls, which are assessed as soft on palpation, are often associated with decreased semen quality, including higher levels of sperm head abnormalities (20.0 versus 13.0%, $P < 0.001$) (Chacón *et al.*, 1999a) as well as more retained proximal droplets (30.7 versus 10.3%, $P < 0.05$) (Chacón, 2001) compared with bulls whose testicles were regarded as being of normal consistency. Brahman bulls with a slight-to-moderate reduction in testicular consistency also showed more degenerative changes in the seminiferous epithelium than did control animals. These changes were characterized by the presence of cellular debris in Sertoli cells as a consequence of increased phagocytic activity (which leads to the presence of hollow areas in the seminiferous epithelium), as well as by abnormalities in sperm chromatin condensation (Chacón *et al.*, 1999b).

Those abnormalities of the testicular tunics encountered were mainly limited to acquired conditions, such as adhesions between the different layers secondary to orchitis and hydrocele.

Length of the scrotum

The length of the scrotum deserves to be discussed separately because relationships have been observed between scrotal length, the prevalence of sperm abnormalities and BSE classification (Chacón *et al.*, 1999a, 2000, 2010; Chacón, 2001). In these studies, which were carried out in tropical Costa Rica, scrotal length (SL) in breeding bulls of a number of breeds (including *B. indicus*) was classified in terms of the relative relationship between the bottom of the scrotum and the hock joint. The classification categories were: (i) short, a virtually neckless scrotum with the testicles being held close to the abdominal wall; (ii) normal, a scrotum with a distinguishable neck and with its bottom above or at the level of the hock joint; and (iii) long, the scrotum bottom below the hock joint (Plate 42). The assessment of SL should be performed before semen collection.

Regardless of breed, *B. indicus* sires ($n = 598$) with a long scrotum had more sperm nucleus abnormalities than did bulls with a normal scrotum length (19 versus 14%, respectively, $P < 0.01$) (Chacón *et al.*, 1999a); similar results were found in a second survey of bulls ($n = 302$) from different zebu breeds (32.7 versus 12.8%, respectively, $P < 0.05$) (Chacón, 2001). Consequently, zebu bulls with a long scrotum were more frequently classified as unsound for breeding compared with sires with a normal SL.

In a recent report based on a large population of breeding bulls extensively reared under tropical conditions ($n = 4231$), the overall prevalence of short, normal or long scrotum was 4.2, 84.6 and 11.2%, respectively (Chacón *et al.*, 2010). The frequency of sires with a long scrotum was higher ($P < 0.0001$) in *B. taurus* (14.6%; $n = 117/802$) than in *B. indicus* (10.8%; $n = 312/2888$) and crossbreeds (8.2%; $n = 44/541$). In particular *B. indicus* breeds, the relative prevalence of bulls with a long scrotum was: Indu-Brazil, 23.4%; Gir, 13.7%; Indu-Brazil × Brahman, 13.4%; Brahman, 7.8%; Nellore × Brahman, 2.2%; and Nellore, 2.0%. It is interesting to note that, even though the overall prevalence of bulls with a long scrotum is higher in *B. taurus* bulls, the highest frequency was in *B, indicus* sires of the Indu-Brazil genotype.

Regardless of the species, the occurrence of unsound bulls was significantly higher in sires with a long scrotum than it was in bulls of normal SL (72.4 versus 34.1%, $P < 0.0001$); the same pattern was seen for the frequency of testicular pathologies (i.e. orchitis, hydrocele, sperm granuloma and testicular atrophy) diagnosed during examination of the scrotal contents (7.2 versus 1.1%, respectively). Differences in semen quality were also found in relation to SL regardless of the species studied. Sperm motility was lower in bulls with a long scrotum than in those with a normal length scrotum (53.8 versus 63.8%, $P < 0.0001$). Moreover, bulls with a long scrotum had higher levels of abnormal acrosomes, nucleus and head defects, as well as proximal droplets compared with bulls with a normal length scrotum (29.2 versus 17.4%, $P < 0.0001$) (Chacón *et al.*, 2010).

These results suggest that spermatogenesis can be more adversely affected in bulls with a long scrotum than in those with a normal SL, although the mechanisms causing this difference are yet to be determined. A strong possibility is that this anatomical anomaly contributes to reduced thermoregulatory efficiency of the affected testicles. Riemerschmid and Quinlan (1941) and Waites (1970) have observed that a long, pendulous bull scrotum may adversely affect the requisite temperature gradient between the inguinal region and the bottom of the scrotal sac. As this gradient directly influences testicular thermoregulation mechanisms (in particular via the pampiniform plexus), it can be hypothesized that the presence of a long scrotum might, in some way, affect the normal process of heat exchange. However, it is also possible that more pendulous scrota may predispose the bull testicle to physical attrition, leading to traumatic damage to the seminiferous epithelium. Obviously, the pathogenesis of this relationship deserves to be fully studied in the future.

Scrotal circumference (SC)

As in *B. taurus* genotypes, SC is an indirect but reliable indicator of Sertoli cell number, testicular weight, volume of testicular parenchyma and semen quality in *B. indicus* bulls (Kumi-Diaka *et al.*, 1983). In addition, SC is negatively correlated with age at puberty of the offspring, and positively correlated with *in vivo* fertility of the sire (McCosker *et al.*, 1989; Morris *et al.*, 1989; Rekwot *et al.*, 1994).

Despite the evident benefits for individual and herd fertility of selecting bulls on the basis of SC, this has not resulted in its widespread acceptance as a tool for the selection of *B. indicus* sires in tropical areas (Chacón, 2000). For example, in Costa Rica, when farmers were asked to nominate their four main criteria when purchasing zebu sires, a small minority (6%) considered testicle size to be of importance (Chacón, 2009). Instead, it was evident that *B. indicus* sires were selected based mainly on phenotypic characteristics, which explains why low SC in zebu sires is, subsequent to seminal problems, the second most important cause of bulls being classified as unsound for breeding purposes (Braden, 1992; Chenoweth *et al.*, 1996; Chacón *et al.*, 1999a). This indicates the importance of strenuous extension and educational programmes to convince producers of the importance of good SC in their sires.

As SC represents such an important BSE component, it is relevant to discuss the proper methodology for its accurate assessment. The measurement of SC is usually the final step during the examination of the scrotal contents, and is one of the aspects in which most practitioners fail at the time of evaluating the bull's breeding soundness under field conditions in tropical regions (Chacón, 2009). In order to perform a reliable assessment of SC in zebu sires, the bull should be firmly and safely restrained, as stated above. Both testicles should be fully pulled down within the scrotal sac, taking care to not insert the fingers between them and thus erroneously increase the SC measurement obtained. At this point, the scrotal skin should be totally smooth with no skin folds. In the meantime, and with the opposite hand, the scrotal metal tape is placed at the widest point of the scrotum (equator) and the tape pressed until resistance is encountered. This is the moment at which SC should be read (see Plate 40).

Age and body weight are positively correlated with SC in *B. indicus* bulls (Morris *et al.*, 1989), at least until full sexual maturity is attained. Brahman adult sires raised under tropical conditions in Australia showed a correlation of SC with age and weight of 0.61 and 0.72, respectively (McCosker *et al.*, 1989), whereas Trocóniz *et al.* (1991) reported values (unadjusted) of up to 0.82 and 0.81, respectively, for Nellore bulls in Venezuela. These estimates are similar to those for beef and dairy *B. taurus* bulls in non-tropical areas (Coulter and Foote, 1979).

Breed variations lead to the high variance seen in SC between bulls of the same age and raised under similar conditions. These breed differences occur when comparing *B. indicus* with *B. taurus* bulls, but also within *B. indicus* genotypes (Chacón *et al.*, 2000), and confirm the need for using SC standards that are adjusted for both breed and age. In Costa Rica, comparison of 18–26 month old *B. indicus* and *B. taurus* bulls showed generally higher SC values in the latter group when compared with Gir, Brahman and Nellore bulls (Chacón *et al.*, 2000), though this difference was not apparent in older bulls. These findings confirm that a slower SC growth pattern tends to occur in young zebu breeds than in *B. taurus* bulls. In Florida, Brahman sires showed lower SC values than Angus and Hereford bulls at young ages (Chenoweth *et al.*, 1996). In the humid tropics of Mexico, Dios Vallejo *et al.* (1991) reported lower SCs for Nellore bulls at any age compared with Indu-Brazil, Brahman and Gir sires; similarly, Glauber *et al.* (1990) and Trocóniz *et al.* (1991) found a lower SC for Nellore than for Brahman and Guzerat bulls in Argentina and Venezuela, respectively.

The use of age-related thresholds for SC, which have been developed in semi-tropical or temperate areas with well-fed *B. taurus* or even *B. indicus* breeds, may result in a disparate proportion of zebu bulls raised in tropical environments failing to pass the BSE owing to unsatisfactory testicular size. Consequently, there is a need to establish values for SC adjusted for breed and age in grass-fed *B. indicus* bulls raised under tropical conditions, so as to enable more confident use of this trait in selection programmes for breeding bulls. As an example, Table 13.1 shows data for SC from grass-fed *B. indicus* bulls in tropical Costa Rica (Chacón *et al.*, 2000). Notice the higher SC in Indu-Brazil bulls at any age compared with Nellore, Brahman and Gir sires.

Assessment of internal organs

Transrectal palpation of the internal reproductive organs is performed before semen

Table 13.1. Mean (LSM ± SEM) scrotal circumference (SC, cm) in grass-fed *Bos indicus* bulls extensively reared in tropical Costa Rica. Adapted from Chacón *et al.* (2000); data for Brahman × Nellore bulls aged 6.1–7.0, 7.1–8.0 and >8.1 years and Brahman × Indu-Brazil bulls aged 7.1–8.0 years from author (unpublished).

Age (years)	Brahman	Nellore	Indu-Brazil	Gyr	Brahman-Nellore	Brahman-Indu-Brazil
1.5–2.0	32.2 ± 0.4 (*n* = 71)	31.0 ± 0.6 (*n* = 26)	34.2 ± 0.6 (*n* = 37)	31.6 ± 1.0 (*n* = 10)	30.5 ± 1.1 (*n* = 8)	34.5 ± 1.0 (*n* = 11)
2.1–3.0	34.4 ± 0.3 (*n* = 185)	33.2 ± 0.5 (*n* = 39)	36.1 ± 0.4 (*n* = 101)	34.3 ± 1.0 (*n* = 48)	33.2 ± 0.6 (*n* = 28)	35.5 ± 0.5 (*n* = 50)
3.1–4.0	37 ± 0.3 (*n* = 161)	35.5 ± 0.6 (*n* = 25)	37.9 ± 0.3 (*n* = 134)	35.4 ± 0.6 (*n* = 30)	33.9 ± 0.7 (*n* = 22)	37.1 ± 0.5 (*n* = 48)
4.1–5.0	37.1 ± 0.3 (*n* = 135)	36.8 ± 0.6 (*n* = 27)	39.1 ± 0.4 (*n* = 85)	38.2 ± 0.6 (*n* = 29)	35.6 ± 0.7 (*n* = 19)	37.6 ± 0.6 (*n* = 28)
5.1–6.0	38.0 ± 0.4 (*n* = 106)	37.7 ± 0.7 (*n* = 21)	39.6 ± 0.5 (*n* = 51)	39.6 ± 0.7 (*n* = 22)	35.9 ± 1.2 (*n* = 7)	39.5 ± 0.5 (*n* = 20)
6.1–7.0	38.0 ± 0.4 (*n* = 86)	39.5 ± 1.0 (*n* = 9)	40.1 ± 0.6 (*n* = 30)	39.7 ± 1.0 (*n* = 8)	39.1 ± 2.6 (*n* = 7)	39.9 ± 1.0 (*n* = 9)
7.1–8.0	38.5 ± 0.5 (*n* = 51)	NA	40.7 ± 0.7 (*n* = 22)	40.2 ± 1.0 (*n* = 7)	41.2 ± 1.2 (*n* = 5)	40.5 ± 3.2 (*n* = 4)
>8.1	37.8 ± 0.4 (*n* = 71)	NA	41.8 ± 0.7 (*n* = 21)	41.2 ± 1.1 (*n* = 12)	40.2 ± 3.6 (*n* = 5)	NA

NA, no data available.

collection. It aims to determine the soundness of the internal accessory sexual genitalia and, at the same time, helps to remove excess faecal material and to pre-stimulate the bull before EE. Again, safe restraint of the bull is essential (Plate 43). The operator introduces a lubricated gloved arm gently into the rectum as far as the wrist to locate the pelvic urethra on the floor of the pelvis. This urethral musculature surrounds the pelvic penis at this point and responds to hand stimulation with pulsatile contractions. At this stage, the operator should also evacuate as many faeces as possible from the rectum. The body of the prostate is then located at the anterior end of the urethralis muscle. The prevalence of abnormalities diagnosed during the clinical examination at this level is minimal in zebu sires (Campero *et al.*, 1988; Chacón *et al.*, 1999a).

The most common abnormality (<1% occurrence) encountered during transrectal examination of internal reproductive organs in *B. indicus* bulls is usually accessory genital disease, which is most evident as bulbourethral adenitis or seminal vesiculitis (Campero *et al.*, 1988; Chacón *et al.*, 1999a). Normal vesicular glands are usually lobulated, painless when touched, approximately symmetrical (≤20% variation is common) and anteriorly mobile. Changes in the volume of these glands can be associated with age/puberty and variations in BCS in extensively reared *B. indicus* sires (Chacón *et al.*, 2002). In general, zebu bulls are not overly discomforted by the act of transrectal palpation.

Semen collection and handling

Semen collection in *B. indicus* bulls can be performed by the use of an artificial vagina (AV), massage of the internal accessory genitalia or EE. The AV is commonly employed in *B. taurus* sires used for regular semen collection and enables evaluation of libido while semen collection is in progress (Ball *et al.*, 1983). In contrast, the use of this method in *B. indicus* bulls is mainly restricted to those sires kept at AI centres, as they need to be well trained to mount a restrained cow or a teaser. Despite intensive efforts at training, many zebu bulls never succeed in servicing an AV because most are usually reluctant to mount restrained surrogates, whether in oestrus or not (Chenoweth, 1981; Price, 1987; Piccinali *et al.*, 1992). Moreover, the nervous disposition of *B. indicus* bulls may inhibit their sexual response in the presence of the operator. Because of the nature of their sexual response, as well as their volatile temperament, which often makes close handling unpredictable, the use of an AV for semen collection is difficult, if not impossible, in *B. indicus* bulls under field conditions. Consequently, EE is the method of choice for collecting semen samples from zebu bulls under field conditions in the tropics (Chacón, 2000).

Application of electroejaculation in Bos indicus *bulls*

Adequate restraint is an important prerequisite for EE in zebu bulls, as is prior stimulation by transrectal palpation, as discussed above.

There are several aspects to the correct restraint of the male during EE. The zebu bull must first be confined such that it has some free lateral space (approximately 10–15 cm on each side) although it is incapable of horizontal movement. The latter requirement can be achieved by placing closely apposed buffers: the lower at the hock joint level and the upper at the level of the thigh. Many operators are reluctant to restrain the head or neck of zebu bulls during EE, as it often causes the bull to immediately lie down once stimulation is initiated, or even earlier. Similar responses are often obtained when squeeze head crushes are employed. If it is necessary to restrain the head, a preferred method is to secure it via the horns, if present. The floor surface plays an important role in bull stability during EE, which tends to induce involuntary extension of the rear limbs. A slippery or insecure surface, such as occurs in a number of commercially available restraint systems, will not assist the bull to achieve a secure position during EE, with the consequence that he may attempt to lie down or, at least, provide an inadequate ejaculate. Accordingly, rough floor surfaces and back stops for hind limb extension are recommended for EE in *B. indicus* bulls.

At least several minutes should be devoted to the pre-EE stimulatory transrectal palpation (Chacón, 2000). Here, a careful and gentle manipulation of the vasa deferentia, ampullae, prostate and vesicular glands can markedly improve the success of EE when using minimal electrical stimulus in zebu sires (Chacón, 2000).

The EE method is dependent upon electrical stimulation of those autonomic nerves responsible for emission and ejaculation with the aid of a longitudinal three ventral electrode probe connected to an electroejaculator powered by either a mains or portable electric source. The relevant nervous fibres derive from the mesenteric and pelvic plexus as well as the lumbar (hypogastric) and sacral (internal pudendal) nerves. These nerves control the contractions of the excurrent ducts and the vesicular glands, and the emission process into the urethra, leading to ejaculation in the bull (Ball *et al.*, 1983). Even though EE also promotes relaxation of the retractor penis muscle, it is possible for ejaculation to occur into the prepuce without penile protrusion or erection. Data on EE responses in extensively reared genotypes (*B. taurus*, n = 439; crossbred, n = 425; *B. indicus*, n = 1696) in Central America showed that intra-preputial ejaculation occurred in 28.5, 37.9 and 46.1% of cases, respectively (J. Chacón, unpublished results).

In general, *B. indicus* bulls should have little problem in performing satisfactorily when subject to EE as long as they are well restrained, relatively unstressed, submitted to competent pre-collection transrectal palpation and stimulated gradually and rhythmically by a skilled operator. The pattern of stimulation should be tailored to each bull, as the response to EE is variable among zebu bulls. Electrical stimuli should always be carefully and gradually applied while monitoring the response of the bull. In general, manually operated EE machines are, therefore, preferable to automatic machines when working with zebu bulls.

Semen can be successfully collected from most bulls by EE. According to different authors (Carroll *et al.*, 1963; Chenoweth and Osborne, 1978; Chenoweth *et al.*, 1996), failure rates vary widely (0.6 to 15%). In those cases, an important distinction must be made as to whether or not the difficulty in semen collection was due to poor technique or to some pathological condition in the bull. The former consideration can predominate with very temperamental zebu sires or when bulls are overly stressed before EE, as when electric cattle prods or dogs are used to handle the animals. Moreover, some operators tend to curtail stimulation as soon as a colour change is noted in the ejaculate, on the assumption that a representative ejaculate has been obtained. This represents poor technique. Stimulation should continue until the operator is confident that the entire sperm-rich fraction has been obtained, otherwise there will be misinterpretations of the sample obtained. If collection is unsuccessful, another attempt may be made after a short resting period (e.g. 10–15 min). If sperm are still not obtained on the second attempt, consideration should be given to more extensive investigations of the reasons for failure, with culling being the final resort.

EE tends to provide higher ejaculate volumes than does use of the AV, owing to the direct stimulation of those accessory genital organs that produce the non-sperm fractions of the ejaculate. This contributes to a lowered sperm concentration, but total sperm per ejaculate, sperm motility and fertility are not affected (Ball *et al.*, 1983; León *et al.*, 1991). It is also noteworthy that EE can induce corkscrew deviation of the penis (Carroll *et al.*, 1963), although this condition should actually be diagnosed under natural mating conditions only.

Spermiogram evaluation

Interpretation of the spermiogram poses particular challenges in the tropical areas where most *B. indicus* bulls are raised, and necessitates careful attention to aspects of bull history, environment, clinical findings, management and field work conditions. In addition, caution must be exercised when assessing semen collected by EE, as discussed above.

Sperm output

The determination of sperm output cannot be realistically achieved on the basis of a single ejaculate that has been collected in relative isolation under field conditions, when multiplying semen volume (ml) by sperm concentration (million/ml) will provide a figure that is, unfortunately, susceptible to great variability. Sperm numbers per ejaculate are related to the frequency of ejaculation of the bull in the period preceding the examination (Foster *et al.*, 1970; Everett *et al.*, 1978), a factor that cannot be controlled in the field. The first ejaculate obtained after a period of sexual inactivity can show abnormally high sperm concentration, whereas that obtained from a bull just removed from active service activity in the field may show relatively low sperm numbers. According to data obtained in tropical Costa Rica from extensively managed *B, indicus* sires of 2 years of age or older, in good condition and classified as sound for breeding, the average number of spermatozoa per ejaculate obtained from a competent EE was approximately 5 billion; this was double the estimate obtained from bulls ranked as unsound for breeding (J. Chacón, unpublished results).

Another consideration in interpreting the significance of the sperm count in *B. indicus* sires is that, as stated above, their response to EE is highly variable, and this can affect sperm numbers, especially if a partial ejaculate is obtained as a result of mishandling or incorrect stimulation of the bull during EE.

Sperm motility

While sperm motility is a prerequisite for fertilization to occur, caution should be taken when assessing and interpreting this trait in bulls examined in the field, especially when these are zebu bulls in the tropics. This is because sperm motility is highly susceptible to handling and environmental influences, and its repeatability when assessed in natural breeding bulls is low, regardless of the method of semen collection (Almquist and Cunningham, 1966; Ball *et al.*, 1983; Gipson *et al.*, 1987; León *et al.*, 1991). The use of EE may also lead to contamination of the sample with urine or other material, especially if intra-preputial ejaculation occurs during semen collection, with resulting adverse effects on sperm motility (Igboeli and Rakha, 1971a,b; Chacón, 2000). In addition, Aguilar Chavarria (2008) reported a significant correlation between motility and ejaculation frequency in zebu bulls reared in tropical areas, which means that this estimation can be affected by sexual activity in the period prior to semen collection.

These factors lead to high variability in sperm motility when assessed in *B. indicus* breeds under farm conditions, so the reliability of this parameter as a major indicator of semen quality is reduced. This leads to relatively low thresholds being adopted for acceptable sperm motility in bulls being examined for breeding soundness (Barth, 2001). Furthermore, the use of any such threshold poses problems, as sperm motility is regarded as a compensable trait (Saacke, 2008), such that sufficient numbers of motile sperm in the ejaculate can compensate for those that are immotile and thereby ensure adequate fertility. If such concerns still persist at the time of examining a semen sample from a zebu bull that was low motility, the practitioner can perform a second or even third consecutive ejaculation within a relatively short period (10–15 min) in order to verify the original finding.

It should be noted that these considerations on the variability of sperm motility, and its relatively low acceptability threshold when assessed under field conditions, should not diminish the potential importance of this trait for semen fertility. More reliance is placed on this trait when it is assessed under controlled conditions (e.g. at a bull semen AI centre) where bulls are subject to regular collection and environmental conditions are controlled.

Sperm morphology

Although relationships between sperm shape (i.e. morphology) and function (i.e. potential fertility of the sire) have been documented since the pioneering studies of Williams and Savage (1925), and later those of Lagerlöf (1934), the evaluation of sperm morphology in semen samples collected under field conditions from *B. indicus* bulls undergoing a BSE is often omitted in tropical areas (Chacón, 2009). The reasons for this omission can include a lack of competence and confidence in performing the assessment, and the absence of appropriate equipment and facilities on farm. However, such omission means that the BSE is incomplete and more prone to error than if sperm morphology assessment had been included. So it is considered of high priority to incorporate sperm morphology as part of the routine evaluation, even if the available facilities must be improvised. Awareness and training programmes for practitioners are necessary to increase the level and quality of semen assessments, of which sperm morphology is an important aspect.

In general, good microscope optics, whether bright-field or phase, are recommended for sperm morphology assessment, and this consideration includes regular cleaning, maintenance and alignment. A magnification of 1000× is recommended for this assessment, which generally requires an oil immersion objective. Most practitioners in the tropics use bright-field microscopy with a stained (e.g. eosin–nigrosin) semen smear to evaluate morphology at 400× magnification. The method of observation (i.e. phase contrast versus bright-field microscopy) and the use of stained semen smears or 'wet' mounts can influence the prevalence of observed morphological defects from the same semen sample and, thereby, the proper interpretation of the sample (Sprecher and Coe, 1996).

A recent survey was designed to compare the levels of sperm abnormalities in semen smears from *B. indicus* bulls ($n = 198$) stained with eosin–nigrosin and examined under bright-field microscopy versus samples fixed in buffered formol saline (BFS) and examined under phase contrast microscopy (both at 1000× magnification). This showed that the prevalence (%) of damaged and malformed acrosomes, abnormal sperm nuclei (diadem, simple pouches, crater) and distal cytoplasmic droplets was significantly higher ($P < 0.001$) in samples fixed in BFS and observed under phase contrast microscopy than in smears stained with eosin–nigrosin and examined under bright-field microscopy (J. Chacón, unpublished).

In another study, sperm morphology was examined in 302 semen samples from grass-fed *B. indicus* breeding bulls using either wet smears fixed with BFS or dried smears stained with carbol-fuchsin. Abnormalities of the sperm head (especially narrow at the base, and undeveloped) were more commonly detected in semen smears stained with carbol-fuchsin and examined with bright-field microscopy than in wet smears examined under phase contrast microscopy (12.0% ± 1.1 and 11.7% ± 1.3 versus 3.2% ± 1.1 and 5.3% ± 1.3, respectively; $P < 0.001$). This resulted in a further 12% of bulls being categorized as unsound for breeding (Chacón, 2001). Because the assessment of sperm morphology is almost always conducted on farm under tropical conditions, it is important that the method used to determine the different abnormalities allows reliable evaluation of these traits. The higher percentage of sperm head defects (size and shape) diagnosed when using dry smears stained with carbol-fuchsin compared with BFS-fixed wet smears emphasizes the importance of using both methods when assessing sperm morphology. This can raise logistical problems, as breeders and producers often request an immediate report on bulls being examined on farm, and good laboratory facilities are often some distance away. One practical approach to this problem would be to conduct a preliminary evaluation of sperm morphology on farm using the preferred method, and later perform a follow-up assessment at the laboratory of any samples that are considered to warrant closer study for confirmation of a diagnosis.

The assessment of the sperm morphology of *B. indicus* bulls in the field, when conducted in conjunction with the other semen and physical evaluations, should allow a prognosis to be made of the breeding potential of the bull. This prognosis can be influenced by the types and the frequencies of the different sperm abnormalities encountered. Differentiation should be attempted between those sperm abnormalities that are considered to be of genetic origin (Chenoweth, 2005), and those that are due to factors such as testicular/epididymal dysfunction, spermiostasis associated with sexual rest or post-ejaculatory environmental effects. One approach is to summarize the abnormalities considered as compensable separately from uncompensable defects (Saacke, 2008), as well as counting specific defects separately from acquired defects. This practice allows assumptions to be made about the impact of the defects on a sire's fertility and the aetiology of different abnormalities, thus facilitating temporal monitoring of particular defects. In most cases, it is difficult to set a cut-off point for the maximum percentage of sperm abnormalities allowed, because the impact on a bull's fertility is different according to the type of defects found.

The type of sperm abnormalities found in semen samples from *Bos indicus* bulls extensively managed in tropical areas resembles those reported for *Bos taurus* bulls in more temperate areas. For a detailed description of the types and prevalence of different morphological abnormalities in semen samples from extensively managed *Bos indicus* bulls and their relationships with BBSE classification and relevant clinical findings, readers are referred to Chacón, 2001.

Complementary Techniques

Ultrasonography of the scrotal contents, infrared scrotal thermography and the use of differential interference contrast (DIC) microscopy for the evaluation of sperm morphology are techniques that can add precision to bull BSE findings. Some of these techniques, such as ultrasonography, have not been widely evaluated in *B. indicus* bulls. Ultrasound has proven to be useful in detecting testicular and epididymal anomalies and also, when used transrectally, for examining anomalies of the accessory genital organs. For this, portable ultrasound machines can be used with multiple frequency probes (5.0–7.5 MHz), some of which may be used either internally or externally. With the latter approach to scrotal/testicular investigations, an adequate amount of coupling gel should be applied directly on the probe, which is first gently placed on the posterior surface of the scrotum. Good restraint, as discussed previously, is essential. Knowledge of the normal echographic picture of the testicular parenchyma and of other structures is very important in helping to interpret the findings and establish a diagnosis. Unfortunately, such information is scarce for *B. indicus* and, in most cases, it has been extrapolated from *B. taurus* bulls, although a recent publication (Chacón *et al.*, 2012a) provides relevant reference information for the ultrasound assessment of the scrotal contents of extensively reared Brahman sires.

In another study, based on a larger population of *B. indicus* ($n = 178$) and *B. taurus* ($n = 48$) bulls, the overall prevalence of bulls with hyperechoic foci within the testicular parenchyma was lower ($P < 0.01$) in left and right testicles in *B. indicus* than in *B. taurus* (44.9 and 42.8 versus 64.5 and 60.4%, respectively). Likewise, relatively older bulls (>4 and ≤7 years of age) had more lesions than younger sires in both genotypes (64.9 and 59.7 versus 32.5 and 32.5% for left and right testicles, respectively, $P < 0.0001$) (Chacón *et al.*, 2012b). The most common finding obtained during the ultrasound scrotal/testicular assessment of extensively reared *B. indicus* bulls was the presence of hyperechoic foci within the testicular parenchyma (Chacón *et al.*, 2012b). However, there was a lack of association between bull BSE classification and the general echographic appearance of the scrotal contents. Despite this, echographic assessment of the scrotal contents of extensively reared zebu bulls has the advantage of being able to reinforce other findings. For example, an oligozoospermic ejaculate from a *B. indicus* bull, which was then diagnosed as having a significant area of testicular parenchyma affected by diffuse hyperechoic foci, could be diagnosed immediately as unsound and recommended for culling, instead of requiring re-examination. It should be emphasized here that ultrasound assessment of the scrotal contents should not substitute for any of the steps of the routine BSE, but should be considered as an aid to improve the precision of ranking of the andrological status of bulls. While the numbers are low, the relatively lower occurrence of echographic lesions found in the *B. indicus* bulls than in their *B. taurus* counterparts may be related to their greater ability to withstand tropical environmental stressors, such as heat stress.

Scrotal skin temperature, and the pattern of the temperature gradient of the scrotum, as assessed by infrared thermography, has been associated with semen quality in Holstein bulls (Gábor *et al.*, 1998). The use of scrotal infrared thermography in *B. indicus* bulls to perform such assessments has been very rare to date, no doubt due to the relative cost of the equipment and the lack of appropriate infrastructure facilities on farm. However, modern devices allow the determination of temperature skin surfaces with a high sensitivity (±0.2°C), and will allow the detection of zebu bulls that have poor testicular thermoregulation. The combination of such findings with those obtained from the physical and semen assessments would undoubtedly increase the reliability of andrological diagnosis and prognosis in zebu bulls.

Another method for assessing sperm morphology, which is regarded as the current 'gold standard' for research purposes, is DIC microscopy. This technique has particular advantages for detecting certain abnormalities, such as the sperm nuclear diadem/crater defect (Saacke and Marshall, 1968; Coulter *et al.*, 1978). Furthermore, Visintin *et al.* (1984) reported that assessment of semen samples from zebu bulls at AI centres using DIC allowed the diagnosis of 5% more acrosomal abnormalities than was possible when ordinary phase contrast optics were used. Despite this, DIC usage is restricted under field conditions as the equipment is expensive and requires considerable expertise for best results. Thus, DIC tends to be limited to laboratory usage, where it can provide useful information that complements the findings of practitioners in the field.

Final Bull Ranking after the BSE

The final ranking of *B. indicus* bulls that have undergone a BSE should consider and summarize all of those aspects related to their potential breeding effectiveness under field conditions. Zebu bulls can be judged as sound or unsound for breeding, or recommended for re-evaluation at a later time, with the appropriate interval being dependent upon the particular problems encountered. For example, acquired physical and clinical problems that require treatment should be resolved before the bull is classified as sound. If resolution is not achievable, then culling is advised.

The category of 'unsound for breeding' should be used judiciously, and only when it is considered that the bull will not recover from the problem at hand, and/or that represents a significant genetic fault. The final judgement should take into account that bulls required to perform natural breeding under extensive conditions will need to be physically capable of doing the job. Acceptable thresholds for different sperm traits, such as sperm motility and morphology, may vary according to different regions, even though there is considerable conformity worldwide. In addition, differing standards for acceptable scrotal size may be adopted by different breed organizations, which might also impose breed-specific requirements that have more to do with 'type' than with reproductive capabilities.

Bulls classified as sound for breeding are those that pass the physical examination, with a scrotal circumference appropriate to breed and age standards, healthy reproductive organs and without evident semen quality problems. In Costa Rica, a maximum threshold of 15% uncompensable sperm abnormalities (head shape, nuclear defects and proximal droplets) has been instituted for a bull to be classified as sound for breeding. Levels above this value are treated with caution, and further analysis of relevant clinical findings is undertaken to decide whether the bull in question should be declared as unsound (and culled immediately) or placed in the deferred category for re-inspection at a later date. Using these criteria, the prevalence of unsound *B. indicus* bulls in Costa Rica was reported as 29.0% (Chacón *et al.*, 1999a), a level that was subsequently confirmed in another study to be 30.7% (Chacón *et al.*, 2010). In both these studies, the prevalence of unsound *B. indicus* bulls was significantly lower than in *B. taurus* and crossbreeds: 41.0% and 46.0%, respectively, $P < 0.001$ (Chacón *et al.*, 1999a); and 48.0% and 46.1%, $P < 0.0001$ (Chacón *et al.*, 2010).

In conclusion, the underuse of the BSE in *B. indicus* bulls in tropical areas is a factor that contributes to relatively poor reproductive rates of tropical beef cattle. It is considered that the relevant knowledge and techniques now available should be fully exploited to remedy this situation. To achieve this will require effective extension and education programmes for both practitioners and producers using current information, as this chapter attempts to provide.

References

Aguilar Chavarria A (2008) Tasa de concepción en monta natural de un toro Brahman con disfunción primaria de epidídimo [Conception rate under natural mating in a Brahman bull with primary epididymal dysfunction] Licentiate thesis, Universidad Nacional (UNA), Heredia, Costa Rica.

Aire TA and Akpokodje JU (1975) Development of puberty in the White Fulani (*Bos indicus*) bull calf *British Veterinary Journal* **131** 146–151.

Almquist JO and Cunningham DC (1966) Semen traits of beef bulls ejaculated frequently *Journal of Animal Science* **25** 916, abstract 153.

Aponte PM, de Rooij DG and Bastidas P (2005) Testicular development in Brahman bulls *Theriogenology* **64** 1440–1455.

Bailey TL, Monke D, Hudson RS, Wolfe DF, Carson RL and Riddell MG (1996) Testicular shape and its relationship to sperm production in mature Holstein Bulls *Theriogenology* **46** 881–887.

Ball L, Ott RS, Mortimer RG and Simons JC (1983) Manual for breeding soundness examination of bulls *Journal of the Society for Theriogenology* **12** 1–65.

Barth A (2001) Evaluation of sperm motility In *Proceedings of the Annual Meeting of the Society for Theriogenology* pp 57–60.

Blockey MA de B (1979) Observations on group mating of bulls at pasture *Applied Animal Ethology* **5** 15–34.

Braden S (1992) Andrologische Untersuchungen von Bullen in der trockenen Pazifikregion (Pacífico seco) Costa Ricas unter besonderer Berücksichtigung der Hodendegeneration [Andrological evaluation in bulls from the dry Pacific region of Costa Rica with emphasis in testicular degeneration] Doctoral thesis, Faculty of Veterinary Medicine, Hannover, Germany.

Brito LFC, Silva AEDF, Barbosa RT, Kastelic JP (2004) Testicular thermoregulation in *Bos indicus*, crossbred and *Bos taurus* bulls: relationship with scrotal, testicular vascular cone and testicular morphology, and effects on semen quality and sperm production *Theriogenology* **61** 511–528.

Campero CM, Ladds PW and Thomas AD (1988) Pathological findings in the bulbourethral glands of bulls *Australian Veterinary Journal* **65** 241–244.

Carroll EJ, Ball L and Scott JA (1963) Breeding soundness in bulls. A summary of 10,940 examinations *Journal of American Veterinary Medical Association* **142** 1105–1111.

Cartwright TC (1955) Responses of beef cattle to high ambient temperatures *Journal of Animal Science* **14** 350–362.

Cartwright, TC (1980) Prognosis of zebu cattle: research and application *Journal of Animal Science* **50** 1221–1226.

Carvalho FA, Lammoglia MAL, Simoes MJ and Randel RD (1995) Breed affects thermoregulation and epithelial morphology in imported and native cattle subjected to heat stress *Journal of Animal Science* **73** 3570–3573.

Chacón J (2000) Breeding soundness evaluation of zebu bulls. With special reference to variations in clinical parameters and sperm characteristics in sires extensively managed in the dry tropics of Costa Rica. Doctoral thesis, Swedish University of Agricultural Sciences (SLU), Uppsala, Sweden.

Chacón J (2001) Assessment of sperm morphology in zebu bulls under field conditions in the tropics *Reproduction in Domestic Animals* **36** 91–99.

Chacón J (2009) Manejo reproductivo y diagnostico andrológico del toro en ganaderías extensivas del tropico centroamericano [Breeding management and andrological diagnosis in bulls extensively reared in tropical Central America] In *Memorias de la Asociación Latinoamericana de Producción Animal (ALPA), Volume XVII, Suplemento 1, San Juan, Puerto Rico, 17–24 October 2009* pp 14–29.

Chacón J, Pérez E, Müller E, Söderquist L and Rodriguez-Martinez H (1999a) Breeding soundness evaluation of extensively managed bulls in Costa Rica *Theriogenology* **52** 221–231.

Chacón J, Müller E and Rodriguez-Martinez H (1999b) Morphological features of the seminiferous and cauda epididymides epithelia of breeding zebu bulls with normal and decreased testicular consistency *Journal of Reproduction and Development* **45** 119–128.

Chacón J, Aranda D and Pérez E (2000) Scrotal circumference in grass-fed bulls extensively reared in Costa Rica In *Proceedings of the 14th International Congress on Animal Reproduction, Stockholm Volume 1* p. 105.

Chacón J, Pérez E and Rodriguez-Martinez H (2002) Seasonal variations in testicular consistency, scrotal circumference and spermiogramme parameters of extensively reared Brahman (*Bos indicus*) bulls in the tropics *Theriogenology* **58** 41–50.

Chacón J, Jiménez A and Vargas B (2010) A proposal for categorization of scrotum length and its relationship with andrological classification in extensively managed bulls In *Abstracts of the 8th International Ruminant Reproduction Symposium Volume 1* p 45.

Chacón J, Navarro L, Vargas B and Víquez C (2012a) Ultrasonography of the scrotal contents in sound for breeding zebu bulls extensively reared in Costa Rica *Reproduction in Domestic Animals* **47** 519.

Chacón J, Navarro L, Vargas B and Víquez C (2012b) Echographic assessment of scrotal contents in extensively reared bulls *Reproduction in Domestic Animals* **47** 520.

Chenoweth PJ (1981) Libido and mating behaviour in bulls, boars and rams. A review *Theriogenology* **16** 155–177.

Chenoweth PJ (2005) Genetic sperm defects *Theriogenology* **64** 457–468.

Chenoweth PJ and Osborne HG (1975) Breed differences in the reproductive function of young beef bulls in central Queensland *Australian Veterinary Journal* **51** 405–406.

Chenoweth PJ and Osborne HG (1978) Breed differences in the response of young beef bulls to electroejaculation *Australian Veterinary Journal* **54** 333–337.

Chenoweth PJ, Chase CC Jr, Thatcher MJD, Wilcox CJ and Larsen RE (1996) Breed and other effects on reproductive traits and breeding soundness categorization in young beef bulls in Florida *Theriogenology* **46** 1159–1170.

Corrêa AB, Vale Filho VR, Corrêa GSS, Andrade VJ, Silva MA and Dias JC (2006) Características do sêmen e maturidade sexual de touros jovens da raça Tabapuã (*Bos taurus indicus*) em diferentes manejos alimentares [Semen characteristics and sexual maturity of young Tabapuã (*Bos taurus indicus*) bulls under different feeding management] *Arquivos Brasileiros do Medicina Veterinaria e Zootecnia* **58** 823–830.

Coulter GH and Foote RH (1979) Bovine testicular measurements as indicators of reproductive performance and their relationship to productive traits in cattle *Theriogenology* **11** 297–311.

Coulter GH, Oko RJ and Costerton JW (1978) Incidence and ultrastructure of "crater" defect of bovine spermatozoa *Theriogenology* **9** 165–171.

Crabo B (1988) Sperm production by farm animals under tropical conditions In *Proceedings of the 11th International Congress on Animal Reproduction and Artificial Insemination, University College Dublin, June 26–30, 1988 Volume V* pp 238–245.

Crockett JR, Koger M and Franke DE (1978) Rotational crossbreeding of beef cattle: reproduction by generation *Journal of Animal Science* **46** 1163–1169.

Da Silva RG and Casagrande JF (1976) Influence of high environmental temperatures on some characteristics of zebu bull semen In *Proceedings of the 8th International Congress on Animal Reproduction and Artificial Insemination, Krakow, July 12–16, 1976 Volume II* pp 939–942.

Dios Vallejo OO de, Cruz Méndez JM de la, Alvarez Falcón JL, Arriola HB, Ruiz Leyva P and Santos López JL (1991) Efecto de la época del año y edad en el desarrollo testicular de cuatro razas cebuinas en área tropical húmeda [Effect of season and age on testicular development of four zebu breeds in the humid tropics] *Veterinaria México* **21** 3–8.

Dunn TG and Moss GE (1992) Effects of nutrient deficiencies and excesses on reproductive efficiency of livestock *Journal of Animal Science* **70** 1580–1593.

Evans JV (1963) Adaptation to subtropical environments by zebu and British breeds of cattle in relation to erythrocyte characters *Australian Journal of Agricultural Research* **14** 559–571.

Everett RW, Bean B and Foote RH (1978) Sources of variation of semen output *Journal of Dairy Science* **61** 90–95.

Fields MJ, Burns WC and Warnick AC (1979) Age, season and breed effects on testicular volume and semen traits in young beef bulls *Journal of Animal Science* **48** 1299–1304.

Fields MJ, Hentges JF Jr and Cornelisse KW (1982) Aspects of the sexual development of Brahman versus Angus bulls in Florida *Theriogenology* **18** 17–31.

Finch VA (1986) Body temperature in beef cattle: its control and relevance to production in the tropics *Journal of Animal Science* **62** 531–542.

Fordyce G, Howit CJ, Holroyd RG, O'Rourke PK and Entwistle KW (1996) The performance of Brahman-Shorthorn and Sahiwal-Shorthorn beef cattle in the dry tropics of northern Queensland. 5: Scrotal circumference, temperament, ectoparasite resistance, and the genetics of growth and other traits in bulls *Australian Journal of Experimental Agriculture* **36** 9–17.

Foster J, Almquist J and Martig RC (1970) Reproductive capacity of beef bulls. IV Changes in sexual behavior and semen characteristics among successive ejaculations *Journal of Animal Science* **30** 244–252.

Gábor G, Sasser RG, Kastelic JP, Coulter GH, Falkay G, Mézes MM, Bozó S, Völgyi-Csík J, Bárány I and Szász F Jr (1998) Morphologic, endocrine and thermographic measurements of testicles in comparison with semen characteristics in mature Holstein–Friesian breeding bulls *Animal Reproduction Science* **51** 215–224.

Galina CS and Arthur GH (1991) Review of cattle reproduction in the tropics. Part 6 The male *Animal Breeding Abstracts* **59** 403–412.

Galina CS and Russell JM (1987) Research and publishing trends in cattle reproduction in the tropics: Part 1. A global analysis *Animal Breeding Abstracts* **55** 755–749.

Galina CS, Orihuela A and Duchateau A (1987) Reproductive physiology in zebu cattle. Unique reproductive aspects that affect their performance *Veterinary Clinics of North America Food and Animal Practice* **3** 619–632

Gipson TA, Vogt DW, Ellersieck MR and Massey JW (1987) Genetic and phenotypic parameter estimates for scrotal circumference and semen traits in young beef bulls *Theriogenology* **28** 547–555

Glauber CE, Acosta APG and Repetto IMA (1990) Circunferencia escrotal en toros *Bos indicus* y derivados [Scrotal circumference in *Bos indicus* bulls and their crosses] *Veterinaria Argentina* **7** 466–472.

Hernández Pichardo JE, Galina Hidalgo CS, Trujillo AO and Navarro-Fierro R (1991) Evaluación de la líbido de toros Cebú en pruebas en corral y en potrero [Evaluation of libido of zebu bulls tested in enclosures or on pasture] *Veterinaria México* **22** 41–45.

Igboeli G and Rakha A (1971a) Puberty and related phenomena in Angoni (short horn zebu) bulls *Journal of Animal Science* **33** 647–653.

Igboeli G and Rakha A (1971b) Seasonal changes in the ejaculate characteristics of Angoni (short horn zebu) bulls *Journal of Animal Science* **33** 651–654.

Kumi-Diaka J, Nagaratnam V and Rwuaan JS (1981) Seasonal and age-related changes in semen quality and testicular morphology of bulls in a tropical environment *Veterinary Record* **3** 13–15.

Kumi-Diaka J, Osori DIK, Njoku CO and Ogwu D (1983) Quantitative estimation of spermatogenesis in bulls (*Bos indicus*) in a tropical environment of Nigeria *Veterinary research communications* **6** 215–222.

Lagerlöf N (1934) Morphologische Untersuchungen über Veränderungen im Spermabild und in den Hoden bei Bullen mit verminderter oder aufgehobener Fertilität [Changes in the spermatozoa and in the testes of bulls with impaired or enhanced fertility] *Acta Pathologica Microbiologica Scandinavica* **Supplementum 19** 66–77.

Lamothe Zavaleta C (1990) Reproductive performance of zebu cattle in Mexico. MSc thesis, Swedish University of Agricultural Sciences, Uppsala, Sweden.

León H, Porras AA, Galina CS and Navarro-Fierro R (1991) Effect of the collection method on semen characteristics of zebu and European type cattle in the tropics *Theriogenology* **36** 349–355.

McCosker TH, Turner AF, McCool CJ, Post TB and Bell K (1989) Brahman bull fertility in a North Australian rangeland herd *Theriogenology* **32** 285–297.

Morris DL, Tyner CL, Morris PG and Forgason RL, Forgason JL, Williams JS and Young MF (1989) Correlation of scrotal circumference and age in American Brahman bulls *Theriogenology* **31** 489–453.

Müller E (1990) Andrological evaluation of bulls in the tropics In *Proceedings of the Joint IFS-SIPAR Seminar in Animal Reproduction, Paysandú, Uruguay* pp 143–150.

Murgueitio E (1990) Intensive sustainable livestock production: An alternative to tropical deforestation *AMBIO* **19** 397–400.

Navarro L, Alpízar E and Chacón J (2008) Conception rate in extensively managed beef cattle herds bred by bulls with different andrological status in the south area of tropical Costa Rica *Reproduction in Domestic Animals* **43** 166.

Nicholson MJ and Butterworth MH (1986) A Guide to Condition Scoring of Zebu Cattle. International Livestock Centre for Africa, Addis Ababa, Ethiopia.. Available at: http://www.fastonline.org/CD3WD_40/LSTOCK/002/CattlGen/condsc-zebu/condsc00.htm (accessed 12 September 2013).

Ocanto D, Patiño A, Ramos C, Escobar S and Linares T (1984) Pubertad en machos Brahman y Criollo Rio Limón bajo condiciones del llano Venezolano [Puberty in Brahman and Criollo Río Limón bulls under the conditions of the Venezuelan lowlands] In *Proceedings of the 10th International Congress on Animal Reproduction and Artificial Insemination, University of Illinois, Urbana-Champaign, June 10–14, 1984 Volume II* pp 171–173.

Ohashi OM, Sousa JS, Ribeiro HFL and Vale WG (1988) Disturbios reprodutivos em touros *B. indicus*, *B. taurus* e mesticos, criados em clima Amazonico [Reproductive disorders in *B. indicus*, *B. taurus* and cross-bred bulls reared in the Amazon region] *Pesquisa Veterinaria Brasilera* **8** 31–35

Orihuela A, Galina CS and Duchateau A (1988) Behavioural patterns of zebu bulls towards cows previously synchronised with prostaglandin $F_2\alpha$ or oestrogen under corral and field conditions *Applied Animal Behaviour Science* **21** 267–276.

Oyedipe EO, Kumi-Diaka J and Osori DIK (1981) Determination of onset of puberty in zebu bulls under tropical conditions of northern Nigeria *Theriogenology* **16** 419–431.

Pérez E, Conrad PA, Hird D, Ortuño A, Chacón J, BonDurant R and Noordhuizen J (1992) Prevalence and risk factors for *Tritrichomonas foetus* infection in cattle in northeastern Costa Rica *Journal of Preventive Veterinary Medicine* **14** 155–165.

Pexton JE, Farin PW, Gerlach RA, Sullins JL, Shoop MC and Chenoweth PJ (1989) Efficiency of single sire mating programs with beef bulls mated to estrus synchronized females *Theriogenology* **32** 705–716.

Piccinali R, Galina CS and Navarro-Fierro R (1992) Behavioural patterns of zebu bulls towards females synchronised with $PGF_2\alpha$ or oestrogens under corral and field conditions *Applied Animal Behaviour Science* **35** 125–133.

Price EO (1987) Male sexual behaviour *Veterinary Clinics of North America: Food Animal Practice* **3** 405–422.

Quintero M (2000) Evaluación andrológica en toros *Bos indicus* en diferentes periodos del año bajo condiciones del Pacífico norte costarricense [Breeding soundness evaluation in *Bos indicus* bulls in different seasons in the north Pacific of Costa Rica] Licentiate thesis, Universidad Nacional (UNA), Heredia, Costa Rica.

Rekwot PI, Oyedipe EO and Ehoche OW (1994) The effects of feed restriction and realimentation on the growth and reproductive function of Bokoloji bulls *Theriogenology* **42** 287–295.

Riek RF (1962) Studies on the reactions of animals to infestation with ticks. VI. Resistance of cattle to infestation with the tick *Boophilus microplus* (Canestrini) *Australian Journal of Agricultural Economics* **13** 532–550.

Riemerschmid G and Quinlan J (1941) Further observations on the scrotal skin temperature of the bull, with some remarks on the intra-testicular temperature *Onderstepoort Journal of Veterinary Science* **17** 123–140.

Rodríguez C, Galina CS, Gutiérrez R, Navarro R and Piccinalli [Piccinali] R (1993) Evaluación de la actividad sexual de los toros Cebú bajo condiciones de empadre múltiple con hembras sincronizadas con $PGF_2\alpha$ [Evaluation of sexual activity in zebu bulls used for multiple matings with $PGF_{2\alpha}$-synchronized females] *Ciencias Veterinarias* **15** 41–49.

Saacke RG (2008) Sperm morphology: Its relevance to compensable and uncompensable traits in semen *Theriogenology* **70** 473–478.

Saacke RG and Marshall CE (1968) Observations on the acrosomal cap of fixed and unfixed bovine spermatozoa *Journal of Reproduction and Fertility* **16** 511–514.

Sanders JO (1980) History and development of zebu cattle in the United States *Journal of Animal Science* **50** 1188–1200.

Silva-Mena C (1997) Peripubertal traits of Brahman bulls in Yucatán *Theriogenology* **48** 675–685.

Silva-Mena C, Aké-López R and Delgado-León R (2000) Sexual behavior and pregnancy rate of *Bos indicus* bulls *Theriogenology* **53** 991–1002.

Sprecher DJ and Coe PH (1996) Differences in bull spermiogramme using eosin–nigrosin stain, Feulgen stain, and phase contrast microscopy methods *Theriogenology* **45** 757–764.

Tegegne A, Entwistle KW and Mukasa-Mugerwa E (1992) Nutritional influences on growth and onset of puberty in Boran and Boran × Friesian bulls in Ethiopia *Theriogenology* **37** 1005–1015.

Trocóniz JF, Beltrán J, Bastidas H, Larreal H and Bastidas P (1991) Testicular development, body weight changes, puberty and semen traits of growing Guzerat and Nellore bulls *Theriogenology* **35** 815–827.

Turner JW (1980) Genetic and biological aspects of zebu adaptability *Journal of Animal Science* **50** 1201–1205.

Vale Filho VR do, Pinto PA, Megale F, Fonseca J and Soares LCO (1980) Fertility of the bull in Brazil. A survey of 1088 bulls and 17,945 ejaculations from *Bos taurus* and *Bos indicus* and crosses under tropical conditions In *Proceedings of the 9th International Congress on Animal Reproduction and Artificial Insemination, 16th–20th June, 1980, Madrid, Spain Volume VI* pp 545–548. Editorial Garsi, Madrid.

Vale Filho VR do, Andrade V, Bergmann J, Reis S, Mendonça R and Mourão G (1996) Breeding soundness evaluation (BSE-Z) of prospective young Nellore bulls (zebu) supplemented with pasture In *Proceedings of the 13th International Congress on Animal Reproduction and Artificial Insemination, Sydney, Australia Volume II* pp 6–9.

Visintin J, Barnabe V, Barnabe R and Viana W (1984) Seminal characteristics of zebu bulls In *Proceedings of the 10th International Congress on Animal Reproduction and Artificial Insemination, June 10–14 1984, University of Illinois at Urbana-Champaign, Illinois, USA. Volume II, Brief Communications* Paper 69.

Waites G (1970) Temperature regulation and the testis In *The Testis* Ed A Johnson, W Gomes and V Vandemark pp 251–279. Academic Press, New York.

Wildeus S and Entwistle KW (1982) Pospubertal changes in gonadal and extragonadal sperm reserves in *Bos Indicus* strain bulls *Theriogenology* **17** 655–667.

Wildeus S and Entwistle KW (1983) Spermiogram and sperm reserves in hybrid *Bos indicus* × *Bos taurus* bulls after scrotal insulation *Journal of Reproduction and Fertility* **69** 711–716.

Wildeus S and Entwistle KW (1984) Seasonal differences in reproductive characteristics of *Bos indicus* and *Bos taurus* bulls in tropical northern Australia *Tropical Animal Production* **9** 142–150.

Williams WW and Savage A (1925) Observations on the seminal micropathology of bulls *The Cornell Veterinarian* **15** 353–375.

Wolf FR, Almquist JO and Hale EB (1965) Pubertal behaviour and pubertal characteristics of beef bulls on a high nutrition allowance *Journal of Animal Science* **24** 761–765.

14 Applied Andrology in Water Buffalo

Sayed Murtaza H. Andrabi*
National Agricultural Research Centre, Park Road, Islamabad, Pakistan

Introduction

Buffalo are known for their habit of wallowing in water, especially during the hot hours of the day, and hence they are also referred to as water buffalo (hereafter mostly referred to as buffalo). The domestic buffalo (*Bubalus bubalis*) has been broadly classified on the basis of habitat as river and swamp types, each considered a subspecies. This characteristic is so marked that the buffalo can be described as a semi-aquatic mammal. The river buffalo (*B. b. bubalis*) prefers to immerse itself in running water or ponds. The swamp buffalo (*B. b. carabanesis*) likes stagnant water or mud and prepares its own wallow by digging with its horns. Keeping buffalo away from water makes them more susceptible to environmental stresses (heat) and results in poor productive and reproductive performance (Marai and Haeeb, 2010).

The world buffalo population is increasing, and in 2007 was estimated to be over 177 million head (FAO, 2007). Since 1981/2, buffalo numbers have increased by 50 million. Interest in buffalo rearing and breeding is growing in several countries where the animal has previously been neglected – and even unknown. More than 97% of the population is located in Asia, where buffalo play a prominent role in rural livestock production both providing milk and meat, and being used as working animals. In recent decades, buffalo farming has expanded widely in Mediterranean areas and in Latin America. For the manufacture of specific dairy products such as Mozzarella cheese, buffalo are reared in different regions of Australia. Buffalo are also reared for meat purposes in non-traditional countries because of the good dietary value of the meat, which contains less saturated fat than beef and pork.

Only in India and Pakistan are there well-defined buffalo breeds (Drost, 2007). Here, they are classified into five major groups: the Murrah, Gujarat, Uttar Pradesh, Central Indian and South Indian breeds. The Nili-Ravi buffalo, belonging to the Murrah group, is recognized as the highest milk-producing breed of buffalo (Cockrill, 1974). The swamp buffalo found in South-east and Far East Asia has low milk production, and is mostly used as a draft animal by small farm holders or for meat purposes (Andrabi, 2009).

The male plays an important role in any successful reproductive management programme. With the development of frozen semen technology for buffalo breeding, the

* E-mail: andrabi123@yahoo.com

demand for the best males has increased considerably. The aim of this chapter is to describe the major reproductive and andrological parameters of water buffalo bulls.

Puberty

Generally, puberty in a bull occurs when the bull calf produces sufficient sperm to successfully impregnate a female; this has been defined as the time when a bull first produces an ejaculate containing 50 million spermatozoa, of which more than 10% are progressively motile (Wolf *et al.*, 1965). The same principle has been used to define puberty in buffalo bulls (Ahmad *et al.*, 2010). The pubertal period is associated with rapid testicular growth, changes in luteinizing hormone (LH) release pattern, an increase in blood plasma testosterone and the initiation of spermatogenesis.

The onset of puberty in buffalo bulls occurs later than it does in *Bos taurus* bulls. Sexual maturity of the male buffalo is attained at about 2–3 years of age, depending on the type of breed, management practices and feeding. Appropriate feeding and management of prepubertal buffalo bulls is thought to be of value in enhancing puberty, because the first signs of sexual interest and meiotic divisions of spermatogonial cells have been found to occur as early as 9 months of age (Ali *et al.*, 1981).

The first sexual interest of a buffalo bull may coincide with the development of fertilizing capacity or may precede it by a variable period. Although bulls are fertile during the immediate post-pubertal period, the quality and quantity of semen produced increases over the subsequent months. The testis size shows a curvilinear increase in relation to age. It increases slowly between 5 and 15 months of age, rapidly between 15 and 25 months of age and slowly again between 25 and 38 months of age. The plasma testosterone concentrations are low up to 21 months of age, start to rise at 25 months and reach peak levels at 38 months of age (Ahmad *et al.*, 1984). Changes in interstitial cells during the development of the buffalo testis reveal that adult Leydig cells are visible in 3-month-old buffalo calves, but mesenchymal cells are seen from 18 months of age onwards. The percentage of adult Leydig cells reaches a maximum by 72 months of age and beyond (Rana and Bilaspuri, 2000).

Some of the salient characteristics linked to puberty in buffalo bulls are presented in Table 14.1. The ages and body weights at puberty in male buffalo are quite variable, with pubertal characteristics determined more by body weight than by age. Under good conditions, testicular spermatogenic cell divisions commence by approximately 12 months and active spermatogenesis can be seen by 15 months of age (Azmi *et al.*, 1990). However, the ejaculate contains viable spermatozoa only after 24–30 months of age, indicating that male buffaloes mature more slowly and have a longer time lag between the onset of spermatogenesis and the achievement of puberty than *B. taurus* bulls (Perera, 1999).

Seasonality

Buffalo bulls are capable of breeding throughout the year, but some seasonal fluctuation in reproductive functions is evident in most countries. Heuer *et al.* (1987) attributed 40% of the observed seasonality of buffalo fertility to the male. In river type buffalo, semen volume, sperm concentration and initial sperm motility did not differ significantly between different seasons, but there was a significant difference post freezing (Tuli and Singh, 1983). Several studies have reported better freezability and conception rate of spermatozoa harvested during the autumn/winter or peak breeding season compared with those collected and processed during the summer (dry or wet) or low breeding season (Tuli and Singh, 1983; Heuer et al., 1987; Bhavsar *et al.*, 1989; Sagdeo *et al.*, 1991; Bahga and Khokar, 1991; Younis *et al.*, 1998; Koonjaenak *et al.*, 2007b,c).

There is a possibility that a seasonal variation in the biochemical composition of seminal plasma and/or spermatozoa may occur, as it does in other farm animals (Cabrera *et al.*, 2005; Argov *et al.*, 2007; Koonjaenak *et al.*, 2007c). There are a few scattered reports available that describe differences in chemical

Table 14.1. Puberty in water buffalo.

Breed type	Age at puberty (months)	Body weight (kg)	Scrotal circumference (cm)/Testicular volume (cm^3)	Plasma/serum testosterone (pg/ml; otherwise ng/ml)	Reference
River (Egyptian)	15–17			[bc]208.4 ± 93.8	Hemeida *et al.*, 1985
River (Iraqi)	18.04	336.6	23.4		El-Wishy, 1978
River (Murrah cross)	~24			0.14 ± 0.11	Barreto *et al.*, 1996
River (Murrah)	32.4 ± 2.2	408.6 ± 53.4	24.5 ± 2.8/[a]175.9 ± 81.4		Pant *et al.*, 2003
River (Nili-Ravi)	24.9 ± 0.9				Ahmad *et al.*, 1984
River (Nili-Ravi)	22.8 ± 1.1	421 ± 191	23.8/[a]188 ± 12	3.3 ± 1.2	Ahmad *et al.*, 1989
River (Nili-Ravi)	23.6 ± 0.9				Ahmad *et al.*, 1991
River (Nili-Ravi)	~24	450 ± 40	26 ± 0.5		Ahmad *et al.*, 2010
River (Nili-Ravi)	26.5 ± 0.8	515 ± 21	26.3 ± 0.0		Asghar *et al.*, 1985
River (Nili-Ravi)	25.4	501	25.3		Heuer and Bajwa, 1986
Swamp (Australian)	30–33	>250	17–20		McCool and Entwistle, 1989
Swamp (Malaysian)	29 ± 3	380 ± 20			Nordin *et al.*, 1986
Swamp (Thai)	~20				Chantaraprateep *et al.*, 1985

[a]Values are either average or mean ± SEM; [b]value in pg/ml; [c]values are Mean ± SD.

composition of buffalo seminal plasma and spermatozoa under different climatic conditions (Singh *et al.*, 1969; Mohan *et al.*, 1979; Sidhu and Guraya, 1979), but the information given in these studies is insufficient to explain the variation in the freezability of buffalo spermatozoa during different seasons (Andrabi, 2009).

Genitalia

The male reproductive system can be divided into three components: the primary sex organ, i.e. testes; a group of accessory glands and ducts, i.e. the epididymis, vesicular glands, bulbourethral gland and prostate; and the external genitalia or copulatory organ i.e. the penis. The reproductive organs of the water buffalo bull are similar to those of the bull of domestic cattle (*B. indicus*), but the testes and scrotum are smaller and the penile sheath is less pendulous. The sheath of the penis adheres close to the body in the swamp type of buffalo. but is more pendulous in the river type. As in cattle, the testis and epididymis can be palpated through the scrotal wall, while the prostate, seminal vesicles and ampullae of the ductus deferens can be palpated per rectum (Ahmad and Noakes, 2009).

The scrotum of the river buffalo is 20–25 cm in length, has a distinct neck and is pendulous. The scrotum of the swamp buffalo is about 10 cm in length (fully extended) and the neck is not distinct (MacGregor, 1941). The testes of the swamp buffalo descend into the scrotum at 2–4 or 6 months of age, while they may be present in the scrotum at birth in the river type. The normal testes of buffalo are ovoid in shape and turgid on palpation, with a marked resonance. The testes are of unequal size, with the left usually bigger. In river buffalo, the average measurements (length × width) of the right and left testicles (including the epididymis) are 14.2 × 6.41 cm and 15.0 × 6.87 cm, respectively. Joshi *et al.* (1967) reported the length, breadth and circumference of the testes, excluding the epididymis, as 7.60 × 4.30 × 12.20 cm for the right and 7.87 × 4.33 × 12.29 cm for the left. In swamp buffalo, MacGregor (1941) reported that the average right and left testicle measurements, including the epididymis length × width) were 11.18 × 4.85 cm and 11.27 × 4.82 cm, respectively. The diameter of the seminiferous tubule in water buffalo testes ranges from 170 to 200 μm and each is 5–100 cm in length. The total length is as high as 5000 m in a pair of testes. The height of the seminiferous epithelium is about 56 μm and the interstitial nuclear diameter is close to 13 μm (Ahmad *et al.*, 2010).

During calfhood, the penis is firmly adhered within the sheath and cannot be extended. However, at about 2 to 4 months before puberty, partial protrusion occurs during mounting, followed by separation of penis from the sheath, complete erection and eventually mating and ejaculation. In river buffalo, the penis hangs clear of the abdomen by 15–30 cm, being attached thereto by a triangular fold of skin running backwards from the umbilicus. In swamp type buffalo, the sheath of the penis adheres closely to the abdomen except for the last 2–3 cm (MacGregor, 1941). The average length of the penis is 83.51 cm in the river buffalo and 56.72 cm in the swamp type (Joshi *et al.*, 1967). The prepuce of water buffalo is long and narrow, and is devoid of hairs.

Sexual Behaviour and Libido

Libido is the eagerness of the male animal to copulate with the female. Bull libido, or sex drive, is an important aspect of fertility and there is great individual variation in it. There is a large genetic component to libido (Chenoweth, 1997) although, within bulls, Landaeta-Hernandez *et al.* (2001) reported poor repeatability in test results for libido, service rate and reaction time to service, suggesting that there are other influences on mating behaviour as well, such as learning and/or environmental factors.

The sexual behaviour of the water buffalo bull is similar to, but less intense than, that of the *B. taurus* bull. Libido is suppressed during the hotter part of the day. Sniffing of the vulva or of female urine and the flehmen reaction precede mounting of the oestrous female. The occurrence of flehmen behaviour in the buffalo bull is significantly increased during oestrus compared with dioestrus (Rajanarayanan and Archunan, 2004). Mating is brief and lasts only a few seconds, and the ejaculatory thrust is less marked than in the cattle bull. After ejaculation, in contrast to the *B. taurus* bull, the buffalo bull dismounts slowly and the penis retracts gradually into the sheath (Jainudeen and Hafez, 1987). Following ejaculation and dismounting, the buffalo bull shows a sexual refractory period, but a quick return to mounting behaviour is shown by males when given an opportunity to mate a new oestrous female (Jainudeen and Hafez, 1987). Buffalo bulls usually continue to tease the same female buffalo and repeatedly mount her, perhaps within a 10–15 min period, but the interval between and number of mountings vary between males.

Good libido and proper mating ability of a buffalo bull are also desirable traits for a successful artificial insemination (AI) programme to harvest maximum semen of acceptable quality. Wide variation in libido and sexual behaviour exists between semen donor buffalo bulls (Anzar *et al.*, 1993). Bulls of the best category in terms of libido and sexual behaviour must be used to overcome the two different types of infertility related to these behavioural characteristics – delayed puberty and failure of semen production (Ahmad *et al.*, 1985).

Ejaculation time in seconds, one of the indices of libido, is higher in swamp buffalo than in river buffalo, 467.24 ± 353.9 and 112.57 ± 62.4, respectively (± SE) (Ramakrishnan *et al.*, 1989). The influence of environment (season) on libido score (0–6) in river buffalo bulls was significantly higher during the autumn/early winter peak breeding season, 4.30 ± 0.13 (± SE), than in the summer low breeding season, 3.46 ± 0.16 (± SE); similarly, reaction time (s) was significantly lower, 31.24 ± 2.27 (± SE), during the autumn peak breeding season than in the summer low breeding season, 34.28 ± 3.3 (± SE) (Younis *et al.*, 2003).

Sperm Production

Among farm animals (with the exception of the boar), the water buffalo bull has one of the shortest spermatogenic cycles. Buffalo spermatogenesis constitutes 4.57 cycles of the seminiferous epithelium (Guraya and Bilaspuri, 1976). The duration of the seminiferous epithelial cycle and of spermatogenesis are 8.6–8.7 days and 38 days, respectively (Sharma and Gupta, 1980; McCool *et al.*, 1989). In general, the frequency of cell stages in water buffalo and cattle is similar, at eight (Pawar and Wrobel, 1991).

In sexually mature river buffalo, the total epididymal sperm reserve per animal is about 36.2 billion. The efficiency of sperm production averages 14.5×10^6 sperm/g testicular parenchyma daily, with a mean of 2.02×10^9 sperm/testis. Thus, a typical buffalo bull produces about 4×10^9 sperm daily (Sharma and Gupta, 1979). This may vary according to age, season and nutrition. Pant *et al.* (2003) reported $5.2–8.4 \times 10^6$ sperm/g of testicular parenchyma daily, with a range of $2.18–3.37 \times 10^9$ total sperm daily in river buffalo bulls. In a mature swamp buffalo bull, daily sperm production is about 1.86×10^9 and epididymal sperm reserves are 9.7×10^9 (McCool and Entwistle, 1989). The overall short duration of spermatogenesis, the smaller testicular size and low rate of daily sperm production in both types of water buffalo, compared with cattle, perhaps reflects species difference in the length of the sexual season and mating behaviour (Jainudeen and Hafez, 1987).

Dhingra and Goyal (1975) explained in detail the different types of spermatogenic cells in the adult water buffalo bull. Type A spermatogonia have a spherical to ovoid nucleus with finely granulated chromatin, homogeneously dispersed in the nucleoplasm, and have one or two nucleoli adhering to the nuclear membrane. Type A_0 spermatogonia are characterized by nuclei containing finely granulated chromatin and a nucleolus attached to the nuclear envelope. The A_1 type spermatogonia, in contrast, have finely granulated chromatin with the nucleolus adhering to the nuclear membrane. The nuclei of A_2 type spermatogonia resemble those of type A_1, but contain coarse granular chromatin dispersed in the nucleoplasm. The intermediate type of spermatogonia acquire a central position of the nucleolus, but the chromatin remains coarsely granulated and non-clumped. Three classes of type B (B_1–B_3) spermatogonia are present based on the degree of clumping of the chromatin and the central position of the nucleolus. Type B_1 cells are characterized by nuclei containing a few flakes of chromatin and a centrally located nucleolus. Type B_2 cells show comparatively more clumping of chromatin than type B_1 spermatogonia, and this is dispersed at random in the nucleoplasm and along the nuclear envelope. Type B_3 spermatogonia show chromaphilic chromatin dispersed in the nucleoplasm and adhering along the nuclear membrane.

Semen Collection

The procedure for the collection of buffalo semen is by using a teaser and an artificial vagina (AV) (Fig. 14.1). Buffalo bulls are considered among the easiest of mammals to be trained to serve an AV (Presicce, 2007), provided that the temperature (39–42°C), type of inner liner (smooth or rough) and air pressure of the AV are appropriate. Buffalo bulls are less choosy than cattle bulls about the teaser or dummy animal and quickly mount a teaser or a male buffalo in their service of an AV for semen collection. Successful electroejaculation (EE) and transrectal message methods have also been used for semen collection in buffalo bulls. However, the quality of semen collected is better from using an AV than from EE or transrectal message.

In normal routine, a 30.48 cm long bovine 'Danish model' AV is used for semen collection in buffalo bulls. Adult bulls prefer a smooth inner lining to the AV, whereas senile bulls donating semen prefer an AV with rough inner lining. Two consecutive ejaculates at 10–20 min interval are collected from buffalo bulls twice a week with an AV. The semen is collected after at least two false mounts (sexual stimulation). Frequent or delayed collections result in poor semen quality (Anzar *et al.*, 1988).

Semen Characteristics

Semen colour

Buffalo bulls normally produce greyish to milky white or creamy white semen, with a slight tinge of blue (Jainudeen and Hafez, 1987; Vale, 1994). The yellowish semen produced by some *Bos* bulls because of the presence of riboflavin is rarely produced by buffalo bulls, although under some pathological conditions, buffalo bulls can produce semen of differing colour and consistency.

Ejaculate volume

The volume of the buffalo ejaculate varies from bull to bull and within each bull, depending on breed and age of bull. Significant individual variations in the semen volume of buffalo bulls have been reported by numerous workers (Table 14.2). Young bulls coming into service produce about 1–2 ml of semen, while older bulls provide up to 6 ml (Vale, 1994). In general, the volume of the semen increases

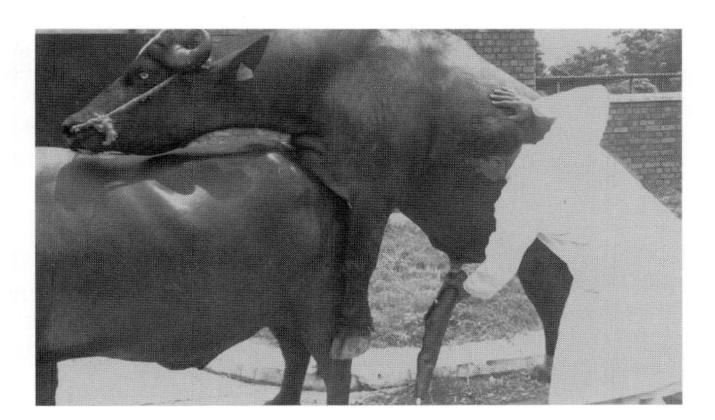

Fig. 14.1. Semen collection from a water buffalo bull, which shows bull dismounting after thrusting.

Table 14.2. Semen characteristics of water buffalo.

Breed type	Ejaculate volume (ml)	Sperm concentration ($\times 10^9$/ml)	Sperm motility (%)	Reference
River (Egyptian)	2.7–4.9	0.22–2.74	64–84	Sayed, 1958; Abdou *et al.*, 1977
River (Jafrabadi)	5.0–5.3	1.17–1.24	60–70	Kodgali, 1967; Kerur, 1971
River (Kundhi)	2.0–6.0	–	7–70	Samo *et al.*, 2005
River (Murrah)	2.5–3.6	3.82–5.90	–	Pant *et al.*, 2003
River (Murrah)	1.5–5.0	0.32–2.20	20–80	Sen Gupta *et al.*, 1963; Dabas *et al.*, 1982
River (Murrah)	2.0–8.0	0.60–1.20	70	Vale, 1994
River (Nili-Ravi)	1.5–6.0	2.50–3.70	70–90	Ishaq, 1972
River (Surti)	1.0–4.0	0.63–1.00	–	Kodgali, 1967; Kodgali *et al.*, 1972
Swamp (Malaysian)	2.9	1.06	70.7	Jainudeen *et al.*, 1982

with age, but it also depends upon the general and reproductive health of the animal and the frequency of ejaculation.

Semen pH and buffering capacity

The pH of freshly ejaculated buffalo bull semen ranges from 6.4 to 6.9, or sometimes up to 7.0. The buffering capacity of seminal plasma in buffalo bulls is much higher on the acidic side than on the alkaline side and, overall is much lower than that of cattle bulls (Singh and Sadhu, 1973).

Sperm concentration

Marked variations in the concentration of buffalo bull spermatozoa in semen have been reported (Table 14.2), with the number of spermatozoa/ml ejaculate varying from 0.2 to over 3 billion. Factors such as sexual development, age, season, nutrition and reproductive health affect the concentration of spermatozoa in ejaculates.

Semen viscosity

Viscosity is a measure of the fluidity or density of the semen. The viscosity of buffalo semen is reported to be 1.88 centipoise (0.188 Pa/s). A positive correlation between viscosity and concentration has been reported for buffalo bull semen. The density of the semen varies from 0.84 to 1.12 g/ml, and is largely due to the sperm concentration. The average value of the semen surface tension is reported to be 65.03 dyne/cm (650.03 μN/cm). Sexual development, age, season, nutrition and reproductive health all affect the concentration of spermatozoa in ejaculates. Viscosity, density and surface tension of the semen, like sperm concentration, depend upon sexual development, age, season, nutrition and reproductive health (Sidhu and Guraya, 1985, pp. 8–10).

Sperm motility

Generally, the motility of buffalo bull semen is lower than that of the *Bos* bull. The range of sperm motility found in buffalo bull semen is shown in Table 14.2.

Sperm biometry

Various sperm morphometric indices of buffalo bull spermatozoa are presented in Table 14.3.

Table 14.3. Sperm morphometric indices for water buffalo.

	Head (μm)		Acrosomal cap (μm)		Midpiece (μm)		Tail (μm)		
Breed type	Length	Width	Length	Width	Length	Width	Length	Width	Reference
River (Egyptian)	7.20	4.45	–	–	–	–	–	–	El-Azab, 1980
River (Murrah)	7.63–7.66	4.71–4.89	–	–	11.41–11.57	0.62–0.71	–	–	Pant and Mukherjee, 1972
River (Murrah)	7.59 ± 0.01	4.91 ± 0.01	3.64 ± 0.01	4.41 ± 0.01	–	–	56.14 ± 0.18	–	Roy *et al.*, 2008
River (Murrah)	7.50	4.70	–	–	11.95	0.62	42.50	0.50	Sidhu, 1974
River (Nili-Ravi)	8.3	4.5	–	–	12.2	–	54.8	–	Saeed *et al.*, 1989

Values are either average or mean ± SEM.

Biochemical Composition of Semen

Buffalo semen contains higher concentrations of fructose, acid and alkaline phosphatase activity and inorganic phosphorus, but lower concentrations of citric acid and ascorbic acid than cattle bull semen. Also in contrast to cattle bull semen, higher alkaline phosphatase in buffalo semen is concomitant with decreased motility and percentage of live cells, depressed dehydrogenase activity and a slight decrease in the rate of fructose hydrolysis. Data on the enzymic activity and other biochemical constituents of water buffalo semen are given in Tables 14.4 and 14.5.

Both acetylcholinesterase and amylase are significantly lower in buffalo than in cattle bull semen. The buffalo sperm acrosome is rich in hydrolytic enzymes, such as alkaline phosphatase and beta-glucuronidase, although acid phosphatase is localized mainly in the post-acrosomal segment. Buffalo semen

Table 14.4. Enzymatic constituents of water buffalo semen. Modified from Chauhan and Srivastava, 1973 and Andrabi, 2009.

Characteristic	Seminal plasma	Spermatozoa
Acid phosphatase	315.31 ± 22.66 BU[a]/100 ml	–
	194 ± 10 BU/100 ml	39 ± 6 BU/10^{11} cell
Aldolase	70.31 ± 27.79 SLU[b]/ml	–
Alkaline phosphatase	312.50 ± 24.04 BU/100 ml	–
	270 ± 9 BU/100 ml	63 ± 6 BU/10^{11} cell
Deoxyribonuclease	–	2007.33 ± 112.01 KU[c]/ml
Glutamic-oxaloacetic transaminase	166.72 ± 14.08 U[d]/ml	–
Glutamic-pyruvic transaminase	134.56 ± 4.57 U/ml	–
Lactic dehydrogenase	1671.5 ± 113.11 BBU[e]/ml	–

Values are either mean ± SD or mean ± SE.
Units of enzyme activity: [a]BU, Bodansky units; [b]SLU, Sibley–Lehninger units; [c]KU, Kunitz units; [d]U, (standard) units; [e]BBU, Berga-Broida units.

Table 14.5. Biochemical composition of water buffalo semen. Modified from Andrabi, 2009.

Characteristic	Seminal plasma	Spermatozoa
Alanine	0.413 mM	–
Ascorbic acid	3.9 ± 0.5 mg/100 ml	–
Aspartic acid	0.395 mM	–
Cholesterol	18.5 ± 0.9 % of total neutral lipids	117.6 ± 1.6 % of total neutral lipids
	53.67 ± 8.72 mg/100 ml	
Citric acid	444.9 ± 17.4 mg/100 ml	–
Fructose	368.12–815.71 mg/100 ml	–
Glutamic acid	4.28 mM	–
Glutathione	32.49 ± 5.10 µmol/ml (whole semen)	–
Glycine	1.34 mM	–
Glycolipids	0.581 mg/ml	0.397 mg/10^9 cells
Lactic acid	82 ± 6 mg/100 ml	167 ± 9 µg/10^{11} cells
Lipids	1.5 mg/ml or 1.75 ± 0.03 mg/ml	1.320 ± 0.030 mg/10^9 cells
Lysine	0.133 mM	–
Neutral lipids	0.439 mg/ml	0.286 mg/10^9 cells
Phospholipids	0.594 mg/ml	0.548 mg/10^9 cells
Serine	0.60 mM	–

Values are either average or mean ± SEM.

contains characteristically low levels of potassium compared with sodium. The sodium: potassium ratio in buffalo seminal plasma is 1:3.7, on average, with well over 60% of ejaculates having a ratio of 1:2.86. The zinc (Zn) concentration in seminal plasma averages 86.88 mmol/l, whereas its concentration in sperm is greater, averaging 14.3 mmol/cell. Increased motility and a decreased percentage of abnormalities are correlated with an increased Zn concentration in spermatozoa, but no relationship has been found between Zn concentration in the seminal plasma and sperm motility.

Preservation of Semen

Liquid storage

Buffalo semen can be stored at a refrigeration temperature of 5°C for up to 72 h without significant decrease in quality, provided that it is diluted in milk-based extenders or with media that have the same composition as those used for deep freezing (except for glycerol) (Dhami *et al.*, 1994; Akhter *et al.*, 2008, 2011b).

Fresh cow's milk is widely used as a diluent for the liquid storage of buffalo semen (Kumar *et al.*, 1993a). It is recommended that, before use for dilution, the milk should be heated, cooled overnight in a refrigerator, the fat layer removed and the milk then reheated in a water bath for a few minutes. After repeated cooling, the remaining fat should be removed by filtration through cotton wool. Tris(hydroxymethyl)aminomethane, egg yolk-citrate and egg yolk-lactose are also popular diluting media for the liquid storage of buffalo semen (Akhter *et al.*, 2011b). After storage of semen in milk, Tris and citrate-based diluents at 5°C for 24 h, the decrease in motility was similar in the three media, but after 48 h, only milk and Tris were able to maintain sperm motility (Kumar *et al.*, 1992). Dhami *et al.* (1994) examined the relative efficacy of Tris-, citrate- and lactose-based diluents, and found that the best sperm survival after 72 h at 5°C was in Tris buffer. Recently, Bioxcell has been found to be suitable for the liquid storage of cooled buffalo semen for up to 5 days (Akhter *et al.*, 2011b), although fertility results are not yet available on the use of Bioxcell as a medium for this purpose.

The proportion of motile spermatozoa studied at different egg yolk concentrations (1, 2.5, 5, 10 and 20%) in Tris-based diluent showed that low levels (1 and 2.5% egg yolk) are the best for 72 h storage at 5°C (Sahni and Mohan, 1990a,b). There was no improvement in viability of spermatozoa after 72 h storage at 5°C when 5, 10 and 20% egg yolk was included in milk- or Tris-based extenders. However, there was an improvement in Tris-based extender (10 or 20% egg yolk) after the addition of 0.1% L-cysteine, or of EDTA or its tetrasodium salt (Kumar *et al.*, 1993a).

The examination of different glycerol concentrations (3, 6 or 9%) in milk-, Tris- and citrate-based extenders (Kumar *et al.*, 1992) revealed that for 24 h storage at 5°C, glycerol was not required in the extender; but in the case of a longer period of storage (72 h or more), glycerol protected the motility of the spermatozoa. There was no difference between 3 and 6% glycerol for up to 24 h of storage, but beyond 24 h, the motility of spermatozoa was maintained better with 6% than with 3 or 9% glycerol in the extender.

Frozen storage

Buffer

The dilution of semen in a suitable buffer is one of the important factors affecting sperm survival during cryopreservation (Rasul *et al.*, 2000). An ideal buffer should have: (i) pH between 6 and 8, preferably 7; (ii) maximum water solubility and minimum solubility in all other solvents; (iii) minimum salt effects; (iv) minimum buffer concentration; (v) the least temperature effect; (vi) good cation interactions; (vii) greater ionic strengths; and (viii) chemical stability (Andrabi, 2009).

Development of a suitable buffering system for cryopreservation of buffalo spermatozoa has been in progress for some time, but in a hit-and-miss empirical fashion. Several studies have concentrated on the use of chemically defined buffers for buffalo semen that were originally evolved for cattle bull semen.

For example, citrate or Tris and/or citric acid or Laiciphos (IMV, France; containing Laiciphos in unknown buffer) or Biociphos (IMV, France; containing Biociphos in unknown buffer) or Bioxcell (IMV, France; containing unknown buffer and animal protein-free formulae of non-toxic cryoprotectant) have been tested as buffers for the deep freezing of water buffalo spermatozoa (Chinnaiya and Ganguli, 1980a,b; Matharoo and Singh, 1980; Ahmad *et al.*, 1986; Dhami and Kodagali, 1990; Singh *et al.*, 1990, 1991, 2000; Dhami *et al.*, 1994; Akhter *et al.*, 2010). Tris-based buffer has been reported as the most suitable. Zwitterion buffers such as Tes and Hepes have also been used for deep freezing buffalo spermatozoa, but with varying success (Oba *et al.*, 1994; Chachur *et al.*, 1997; Rasul *et al.*, 2000). A study with Bioxcell indicated that it can be an alternative to the laboratory-made Tris-based extender for the cryopreservation of buffalo semen provided that the fertility results on a large scale are found to be satisfactory (Akhter *et al.*, 2010).

From the results of the aforementioned studies, it is suggested that buffers, particularly Tris-based, may provide the most satisfactory system to improve the post-thaw freezability of buffalo spermatozoa. It is believed that Tris-based buffer has a pH nearer to the pK_a (acid dissociation constant) and is least influenced by temperature compared with other buffers. Additionally, the differences that have been found in the efficacy of different buffers suggest that buffalo spermatozoa are more prone to freezing stress than cattle bull spermatozoa, possibly owing to biochemical factors that influence membrane fluidity during cryogenic preservation (Andrabi, 2009).

Permeable cryoprotectant

Glycerol is often polyhydroxylated and capable of hydrogen bonding with water, as well as being able to permeate across the cell membrane and being non-toxic during exposure to cells in a concentration of about 1–5 mol/l – depending on the cell type and conditions of exposure (Fuller and Paynter, 2004). More specifically, the physiological actions of glycerol during the cryopreservation of spermatozoa occur by replacing the intracellular water necessary for the maintenance of cellular volume, interacting with ions and macromolecules, and depressing the freezing point of water, with the consequent lowering of electrolyte concentrations in the unfrozen fraction so that less ice forms at any given temperature (Holt, 2000; Medeiros *et al.*, 2002).

For the cryopreservation of buffalo semen, several studies have been carried out in an attempt to find the optimum concentration of glycerol (0–12% v/v) and method of glycerolization (one versus two steps) (Jainudeen and Dass, 1982; Kumar *et al.*, 1992; Nastri *et al.*, 1994; Ramakrishnan and Ariff, 1994; Abbas and Andrabi, 2002; Singh *et al.*, 2006). From these (available) studies, it is suggested that a glycerol concentration of 5–7%, added either in one step to the initial extender or in two steps to the milk-based extender, is suitable for the cryopreservation of buffalo bull spermatozoa.

Ethylene glycol could be another option for cryopreservation of buffalo spermatozoa. The permeability of ethylene glycol was found to be higher than that of glycerol in spermatozoa of different species (Gilmore *et al.*, 1995, 1998; Phelps *et al.*, 1999), resulting in lower hydraulic conductivity and subsequently a reduction in the osmotic stress to which cells are exposed during cooling and freezing (Gilmore *et al.*, 1995). Propylene glycol also has the basic properties of a cryoprotectant, i.e. it is miscible with water in all proportions, its solutions in water have profoundly depressed freezing points and (presumably) it has a low intrinsic toxicity because it is widely used in the food and pharmaceutical industries (Arnaud and Pegg, 1990). Recently, Valdez *et al.* (2003) and Rohilla *et al.* (2005) have tested ethylene glycol or propylene glycol as substitute for glycerol. Their preliminary results suggest that ethylene glycol may be used for freezing buffalo spermatozoa.

Dimethyl sulfoxide (DMSO) is a rapidly penetrating cryoprotectant with a lower molecular weight than glycerol. It may also inhibit the harmful effect of hydroxyl radicals, which appear during cell respiration and are detrimental to cells (Johnson and Nasr-Esfahani, 1994; Yu and Quinn, 1994). More recently,

Rasul *et al.* (2007) studied glycerol and/or DMSO, added either at 37 or at 4°C, as a cryoprotectant for buffalo spermatozoa. The addition of DMSO did not allow satisfactory cryopreservation of buffalo spermatozoa in Tris-citric acid based extender. The exact mechanism involved in the antagonist effect of DMSO on the cryoprotection ability of glycerol is not understood, but its lethal effect is attributed to a toxic rather than an osmotic effect (Rasul *et al.*, 2007). It is believed that, as a result of the lower molecular weight of DMSO, its penetrating ability into the cell is higher than that of glycerol.

Non-permeable cryoprotectant

Egg yolk is a common component of semen freezing extenders for most livestock species, including the buffalo. Little attention has been paid to the concentration of egg yolk necessary for freezing buffalo semen, and generally it is used at 20% in a semen extender for cryopreservation (Andrabi *et al.*, 2008). The use of egg yolk at a higher concentration may have deleterious effects when combined with the toxicity (amino acid oxidase activity) of dead spermatozoa, resulting in a lower post-thaw spermatozoal quality (Shannon, 1972). The enhanced toxicity associated with increased egg yolk is probably owing to the elevated substrate available for hydrogen peroxide formation. It should also be noted that as the yolk concentration is increased in the diluents, the pH of the medium decreases and tends towards the acidic side. This may also be a reason for the depressing effect of higher amounts of yolk on the quality of thawed spermatozoa (Sansone *et al.*, 2000).

Studies have been conducted on different concentrations (0–20% v/v) of chicken egg yolk in extender as a non-permeable cryoprotectant for buffalo semen (Sahni and Mohan, 1990a,b; Kumar *et al.*, 1994; Singh *et al.*, 1999). Yolk added at a 20% concentration yielded better results than any of the other concentrations tested. Egg yolks from duck, guinea fowl and the indigenous Indian hen (desi) in extender have been tested for improving the post-thaw quality of buffalo bull spermatozoa. Duck egg yolk has been found to be more suitable for the freezing medium than chicken egg yolk and other avian egg yolks (Andrabi *et al.*, 2008).

The contents of the egg yolk used as a component of the semen freezing extender may also influence the parameters of the sperm post thaw. It has been reported that there are different proportions of low-density lipoproteins (LDLs) in commercial hen egg yolk depending on the hybrid line selected, and the management and nutritional practices adopted (Bathgate *et al.*, 2006). The discovery of anti-cryoprotective factors and inconsistent LDL composition in whole egg yolk increased interest in the use of purified LDLs in semen freezing extender. Recently, Akhter *et al.* (2011a) determined whether LDLs extracted from the egg yolk in extender improve the freezability and fertility of buffalo bull semen. On the basis of their results, it is concluded that LDLs (at 10%) in extender can be a better alternative to fresh egg yolk for the cryopreservation of buffalo bull semen.

Polyethylene glycol (PEG) is a non-permeable cryoprotectant that may slow down the process of ice nucleation during the cryogenic process, thus protecting the cellular membrane. Another protective mechanism by which PEG operates may be its coupling with hydrophobic molecules to produce non-ionic surfactants. The effect has been studied of the addition of PEG 20 in supplementation to hen egg yolk in a freezing medium for buffalo bull semen (Cheshmedjieva *et al.*, 1996), but further studies are required to determine whether it may be a better option than other protectants for the cryopreservation of buffalo spermatozoa.

Sugars that are not capable of diffusing across a plasma membrane, such as lactose, sucrose, raffinose, trehalose or dextrans, are also added to extender as non-permeable cryoprotectants for buffalo semen cryopreservation (Ahmad and Chaudhry, 1980; Ala Ud *et al.*, 1981; Dhami and Sahni, 1993). In these instances, the sugars create osmotic pressure, inducing cell dehydration and, therefore, a lower incidence of intracellular ice formation. These sugars also interact with the phospholipids in the plasma membrane and reorganize it, which results in sperm that are better suited to surviving the cryopreservation process. However, post-thaw motilities have never achieved acceptable levels when sugar was used as the sole cryoprotectant (Andrabi, 2009).

Bacteriospermia and its Control

Even after following the standard epidemiological rule to maintain pathogen-free bull semen for AI, microbes may still exist in the semen of the bull. Bacteria in bovine semen may come from the testes or epididymis, the accessory glands, the vas deferens, the urethra, the prepuce or the penis (Thibier and Guerin, 2000). Extenders with ingredients of animal origin (egg yolk or milk) can also be a source of bacteria, and also result in the contamination of semen (de Ruigh *et al.*, 2006). Even under conditions in which great care is taken, semen may become contaminated at the time of collection or during the subsequent handling/packaging 'dip and wipe' procedure (Holt, 2000), or even during storage in the deep-frozen state (Bielanski *et al.*, 2003; Mazzilli *et al.*, 2006). Table 14.6 lists bacteria that have been isolated from fresh and thawed buffalo semen.

It is documented that semen containing substantial bacteria usually also has leucocytes and polymorphonuclear granulocytes, and this causes the production of reactive oxygen species (ROS) (Ochsendrof, 1998). Increased ROS levels impair sperm functions and, ultimately, their fertilizing capacity (Maxwell and Stojanov, 1996). The presence of microorganisms, especially bacteria, in the ejaculates, can affect fertilization directly by their adherence to spermatozoa, and impair their motility and induce the acrosome reaction (Morrell, 2006). Microbes can also have an indirect effect by producing toxins (Morrell, 2006), and those transmitted through semen can result in abortions and infections of the female genital tract (Bielanski *et al.*, 2000).

Bacteria in semen and their control via addition of antibiotics in freezing diluents may affect the viability or fertility of cryopreserved

Table 14.6. Bacteria isolated from fresh and thawed water buffalo semen. Modified from Andrabi, 2007a.

Fresh semen	Thawed semen
Acholeplasma laidlawii	*Acinetobacter calceaceticus*
Alcaligenes sp.	*Aeromonas* sp.
Bacillus sp.	*Bacillus* sp.
Chlamydia psittaci	*Campylobacter sputorum*
Corynebacterium equi	*Corynebacterium* sp.
Corynebacterium pseododiptheriticum	*Enterobacter liquefasciens*
Diphtheroids (*Corynebacterium* spp.)	*Escherichia coli*
Enterobacter sp.	*Pseudomonas aeruginosa*
Escherichia coli	*Streptococcus* sp.
Micrococcus sp.	
Mycoplasma arginini	
Mycoplasma bovigenitalium	
Mycoplasma bovis	
Proteus mirabilis	
Pseudomonas aeruginosa	
Pseudomonas fluorescens	
Pseudomonas genus	
Pseudomonas putida	
Pseudomonas testosteroni	
Shigella sp.	
Staphylococcus aureus	
Streptococcus pyogenes	
Ureaplasma spp.	

bovine spermatozoa (Thibier and Guerin, 2000; Morrell, 2006). Conventionally, benzylpenicillin (1000 IU/ml) and streptomycin sulfate (1000 μg/ml), either alone or in combination, is added to freezing diluents of buffalo bull semen (Sansone *et al.*, 2000). However, streptomycin and penicillin (SP) was not found to be an effective combination to control bacteriospermia in buffalo semen (Gangadhar *et al.*, 1986; Aleem *et al.*, 1990; Amin *et al.*, 1999; Akhter *et al.*, 2008). Ahmed and Greesh, (2001), Ahmed *et al.* (2001a, b), Hasan *et al.* (2001), Andrabi (2007a) and Akhter *et al.* (2008) found that bacteria isolated from buffalo bull semen were resistant to penicillin; SP was also deleterious to the post-thaw quality of spermatozoa. The authors concluded that gentamicin, amikacin, norfloxacin, or a combination of gentamicin, tylosin and linco-spectin, (GTLS) may be added in extender for the efficient preservation of buffalo spermatozoa. Moreover, Andrabi *et al.* (2001) reported a better conception rate from frozen–thawed semen with GTLS as diluent than with SP.

Semen Processing

During the process of cryopreservation, spermatozoa are subjected to chemical/toxic, osmotic, thermal and mechanical stresses, which are conspicuous at dilution, cooling and equilibration, or at the freezing and thawing stage. The success of semen cryopreservation depends to a notable degree on the dilution rate used. Originally, semen was diluted to protect spermatozoa during cooling, freezing and thawing, but the rate of dilution was often changed for technical reasons, such as increasing the number of females that could be inseminated with each ejaculate, or standardizing the number of spermatozoa in each dose of frozen–thawed semen. In farm animals, semen has been diluted with specific volumes of extenders or by diluting semen to a specific concentration of spermatozoa. Dilution rates of 1:1 to 1:12 have been successfully used for buffalo semen. Perhaps a better way of diluting semen for comparison purposes is based on the sperm concentration. Reports of buffalo spermatozoa with acceptable fertility were with frozen samples ranging from 120×10^6 to 30×10^6 cells/ml (Tahir *et al.*, 1981; Andrabi *et al.*, 2006).

After dilution, the semen is cooled to a temperature close to 4 or 5°C. Cooling is a period of adaptation of the spermatozoa to reduced metabolism. Extended semen is cooled slowly to avoid potential cold shock. Cold shock is believed to impair the function of membrane proteins that are necessary for structural integrity or ion metabolism. Major changes in bovine spermatozoa during this phase occur near 15 to 5°C, and do not happen below 0°C (Watson, 2000). Rapid cooling reduces the rate of fructose breakdown, oxygen uptake and ATP synthesis by the sperm, which results in the loss of energy supply and motility. Furthermore, cold shock may increase calcium uptake by sperm. It has been empirically determined that cooling cattle bull spermatozoa from body temperature to 5°C at a rate of 10°C/h has the minimum of deleterious effects (Parks, 1997). Dhami *et al.* (1992) studied the effect of cooling rates (5, 30, 60 and 120 min from 10 to 5°C versus 120 min from 28 to 5°C) on the deep freezing of buffalo semen diluted in Tris-based extender. Their results suggest that buffalo semen can be frozen successfully after 30 min of cooling at 10°C.

Equilibration is traditionally recognized as the total time during which spermatozoa remain in contact with glycerol before freezing. At this stage, glycerol penetrates into the sperm to establish a balanced intracellular and extracellular concentration. It should not be overlooked that equilibration includes the concentration balance not only of glycerol, but also of the other osmotically active extender components. Therefore, there is an interaction between the equilibration process (and possibly other cryogenic procedures as well) and the type of extender (buffer and cryoprotectant) used. Tuli *et al.* (1981) examined the equilibration of buffalo semen diluted in Tris- or citric acid-based extender for 2, 4 or 6 h. They found that post-thaw sperm survivability was better after 4 h equilibration than after 2 or 6 h. Dhami *et al.* (1996) conducted a study to determine the relative efficacy of four cooling rates (10/30°C to 5°C;

1 and 2 h each) and two equilibration periods at 5°C (0 and 2 h) for the cryopreservation of buffalo ejaculates. They concluded that the slow cooling of semen straws from 30 to 5°C for 2 h, compared with faster cooling (1 h), or with a lower initial temperature (10°C) and 2 h of equilibration at 5°C, appeared to be necessary for the successful cryopreservation of buffalo semen, as determined by sperm survivability and fertility.

Of considerable importance for cryopreservation is the cooling/freezing rate in the critical temperature range (–5°C to –50°C), which determines whether the spermatozoa will remain in equilibrium with their extracellular environment or will become progressively supercooled, with the increasing possibility of intracellular ice formation. During slow cooling, dehydration of the spermatozoa can proceed to the point of osmotic equilibrium between intracellular and extracellular space, i.e. cellular dehydration will be maximal; in contrast, raising the cooling rate too much means that dehydration is not fast enough to prevent the occurrence of intracellular ice nucleation. However, if the cooling rate is within the required values (50–100°C/min) this results in less excessive intracellular dehydration, less excessive intracellular solute concentrations and less shrinkage of the cells. Moreover, at optimum cooling/freezing rates, the spermatozoa remain vulnerable to unfavourable conditions for a shorter period of time.

Sukhato *et al.* (2001) determined the effects of freezing rate and intermediate plunge temperature (cooling at 10, 20 or 30°C/min each to –40, –80 or –120°C before being plunged into liquid nitrogen) on the post-thaw quality and fertility of buffalo spermatozoa. They found that cooling/freezing spermatozoa from 4 to –120°C, at either 20 or 30°C/min, yielded better progressive sperm motility and fertility rates. Bhosrekar *et al.* (1994) compared the conventional (over liquid nitrogen in static vapour for 10 min) and control (programmable) freezing methods for buffalo bull semen; they concluded that freezing at the rate of 17.32°C/min between +4 and –40°C with a programmable freezer, produced better quality frozen semen than the conventional method of freezing. More recently, Rasul (2000) examined the effect of freezing rates on post-thaw viability of buffalo spermatozoa extended in Tris-citric acid-based extender. The freezing rates examined from 4 to –15°C were 3°C/min or 10°C/min, whereas the freezing rates investigated from –15 to –80°C were 10°C/min, 20°C/min or 30°C/min. It was concluded that the different freezing rates tested gave similar results in terms of post-thaw spermatozoal viability, as judged by visual and computer-determined motilities, motion characteristics, plasma membrane integrity and intactness of the acrosomal ridge.

In the freeze–thaw procedure, the warming phase is just as important to the survival of spermatozoa as the freezing phase. Spermatozoa that have survived cooling to –196°C still face the challenge of warming and thawing, and thus must twice traverse the critical temperature zone, i.e. between –5 and –50°C. The thawing effect depends on whether the rate of cooling has been sufficiently high to induce intracellular freezing, or low enough to produce cell dehydration. In the former case, fast thawing is required to prevent recrystallization of any intracellular ice present in the spermatozoa. Spermatozoa thawed at a fast rate may also be exposed for a shorter time to the concentrated solute and cryoprotectant glycerol, and the restoration of the intracellular and extracellular equilibrium is more rapid than for slow thawing. Also, leaving semen straws at high temperatures for too long a time may result in pH fluctuation and, subsequently, in protein denaturation and cell death. A practical thaw for cattle bull spermatozoa packed in a 0.25 ml straw (as recommended by most AI organizations), is in a 35°C water bath for at least 30 s.

For the cryopreservation of buffalo spermatozoa in Tris-based extender, Rao *et al.* (1986) tested two thawing rates (37°C for 30 s and 75°C for 9 s). They concluded that the best value for post-thaw motility was observed in semen thawed at 37° for 30 s. Dhami *et al.* (1992) studied the effect of thawing rates (40°C for 60 s, 60°C for 15 s and 80°C for 5 s) on the post-thaw motility of buffalo spermatozoa cryopreserved in Tris-based extender. They reported that thawing at 60°C for 15 s yielded a higher sperm motility than other rates tested. In another study, Dhami *et al.* (1996) determined the thawing rates for buffalo semen; the rates investigated were 4°C for 5 min, 40°C for 1 min or 60°C for 15 s. They concluded that thawing of semen at 60°C for 15 s yielded high post-thawing spermatozoal recovery and longevity. In Pakistan, the inseminators are advised to thaw buffalo semen packed in 0.5 ml straws at 35–37 °C for 45–60 s (Anwar *et al.*, 2008).

Semen Evaluation

Motility of spermatozoa

Motility is routinely accessed by visual estimate of the percentage of progressively motile cells. A small drop of semen is placed on a dry slide maintained at 37°C, and examined at a magnification of 200× or 400×. The introduction of computer aided (also known as computer assisted) semen analysis (CASA) has enabled those working in the field to use new parameters in the assessment of buffalo sperm motility (Rasul *et al.*, 2000). In a semen sample, there can be variations in the degree of progressive movement of cells and in the lateral dislocation of sperm heads. The CASA system can evaluate parameters such as speed, direction and the beat cross frequency of spermatozoa. Fabbrocini *et al.* (1996) considered that for forward-moving buffalo bull spermatozoa, only those that traced five straight tracks and had a minimum velocity of 50 μm/s as shown by CASA were of suitable quality, whereas Rasul *et al.* (2001), who acquired a digital image at the rate of 32 frames/s, judged the minimum velocity limit for forward motile buffalo bull spermatozoa to be 30 μm/s.

Aguiar *et al.* (1994) observed 78.6 ± 5.6% motile spermatozoa in the semen of buffalo bulls. Kumar *et al.* (1993b) found that in the semen of Murrah buffalo bulls bred in India, the numbers of motile spermatozoa varied from 60.8 ± 1.5% to 69 ± 4%, and the occurrence of non-motile samples was about 30%. Samo *et al.* (2005) reported an overall motility of 67.64% in Kundhi buffalo in Sindh, Pakistan.

Sperm viability

The percentage of live spermatozoa determines the quality of the ejaculate. Semen with more than 30% dead spermatozoa initially may not be suitable for storage and freezing. Differential staining techniques have been used for the determination of live and dead spermatozoa in buffalo semen (Sansone *et al.*, 2000). Data on the ranges of sperm viability evaluated in buffalo bull semen by different workers are presented in Table 14.7.

Table 14.7. Sperm viability and sperm abnormalities evaluated in water buffalo.

Breed type	Sperm viability (%)	Sperm abnormality (%)	Reference
River (Egyptian)	79.60 ± 0.02	15.30 ± 0.10	Fayez *et al.*, 1987
River (Mehsana)	90.32 ± 0.17	8.33 ± 0.25	Bhavsar, 1987
River (Murrah)	88.61 ± 0.43	7.64 ± 0.36	Dhami, 1992
River (Murrah)	84.76 ± 0.70	4.01 ± 0.79	Gill *et al.*, 1974
River (Murrah)	86.57 ± 0.22	12.51 ± 0.11	Kumar *et al.*, 1993b
River (Murrah)	88.61 ± 1.30	8.70 ± 1.10	Sengar and Sharma, 1965
River (Nili-Ravi)	–	13.0	Ahmad *et al.*, 1987
River (Nili-Ravi)	–	14.7	Heuer *et al.*, 1982
River (Surti)	83.13 ± 0.85	10.55 ± 0.87	Dhami and Kodagali, 1989
Swamp (Malaysian)	86.5	10.3	Jainudeen *et al.*, 1982
Swamp (Thai)	–	12.1	Koonjaenak *et al.*, 2007a

Values are either average or mean ± SEM.

Abnormal spermatozoa

Abnormal spermatozoa are usually detected by staining methods and are classified as head, midpiece and tail abnormalities. In the semen of Nili-Ravi buffalo, most abnormalities were found on the sperm heads (5.78 ± 2.1%); the occurrence of midpiece abnormalities were less than 1% and that of abnormal tails varied from 3.92 ± 1.0% to 5.7 ± 0.4%. The occurrence of cytoplasmic droplets was less than 1% (Saeed *et al.*, 1990). Similar proportions of abnormalities were observed in the semen of Brazilian buffalo (Aguiar *et al.*, 1994) and Murrah buffalo (Kumar *et al.*, 1993b); the latter authors suggested that males providing semen with over 15–20% of abnormal spermatozoa should be examined for their fertility. Data on the ranges of occurrence of sperm abnormalities evaluated in buffalo bulls by different workers are presented in Table 14.7.

Acrosomal and membrane integrity

Acrosome abnormalities are examined by using the Giemsa stain technique (Ramakrishnan and Ariff, 1994) or fluoresceinated lectins (Chachur *et al.*, 1997). More than 90% of spermatozoa were observed with intact acrosomes in the semen of buffalo bulls bred in Bahia (Aguiar *et al.*, 1994) and in the semen of Murrah buffalo bulls after Giemsa staining (Kumar *et al.*, 1993b).

Recent studies of the plasma membrane integrity of buffalo sperm in terms of function and/or structure have involved the hypo-osmotic swelling test (HOST) with or without supravital staining (Rasul *et al.*, 2000; Akhter *et al.*, 2010). HOST relies on the physiological phenomenon that functionally membrane-intact sperm will swell when placed into a moderately hypo-osmotic environment (~190 mOsmol/kg), whereas membrane-damaged sperm do not. With the application of supravital stains, structurally damaged plasma membranes will allow staining of the underlying structures.

Sperm–oocyte interaction

Assessments of spermatozoa are mainly based on the examination of their motility, concentration and morphology, although these characteristics do not give any reliable indication of their fertilizing capability (Sansone *et al.*, 2000). Therefore, a method based on spermatozoon–egg interaction has been used. Di Matteo (1997) developed a simple technique to assay the capacity of buffalo spermatozoa to bind to the zona pellucida of the oocyte. As buffalo oocytes are difficult to obtain, because females of this species are slaughtered mostly because of old age or illness, bovine oocytes, either preserved in saline solution or matured *in vitro*, were used. The results showed that for a rapid evaluation of fresh or frozen–thawed buffalo spermatozoa, saline-stored bovine oocytes can be used, and gave similar results to buffalo oocytes. Zona-free hamster oocytes may be more convenient to obtain and to use for functional tests (Ramesha *et al.*, 1993). These oocytes permit entry of the spermatozoa of many mammals, including buffalo, provided that the spermatozoa have completed capacitation and the acrosomal reaction, and so can be used for assessment of their fertilizing capacity (Barnabe *et al.*, 1997).

Sperm DNA damage

Normal sperm genetic material is required for successful fertilization, as well as for further embryo and fetal development that will result in a healthy offspring. Sperm DNA contributes half of the offspring's genomic material, and abnormal DNA can lead to disarrangements in the productive process. In the post-genomic era, the rapid advance of molecular biology has resulted in numerous techniques for assessing sperm DNA fragmentation in mammalian species (Andrabi, 2007b). Of these, the assays listed in Table 14.8 have been standardized and used to evaluate the DNA integrity of buffalo bull spermatozoa. However, any wider use of these molecular techniques in buffalo andrology is still limited.

Table 14.8. Assays used to evaluate sperm DNA damage in water buffalo.

Breed type	Assay used	Reference
River (Murrah)	Sperm chromatin structure assay	Kadirvel *et al.*, 2009
River (Murrah)	Terminal deoxynucleotidyl transferase-mediated dUTP nick-end labelling assay	Kadirvel *et al.*, 2012
River (Murrah)	Comet assay	Kumar *et al.*, 2011
River (Murrah)	Sperm Chromatin Dispersion assay and acridine orange assay	Pawar and Kaul, 2011
River (Nili-Ravi)	Acridine orange assay	Andrabi *et al.*, 2011
River (Surti)	Comet assay	Selvaraju *et al.*, 2010
Swamp (Thai)	Sperm chromatin structure assay	Koonjaenak *et al.*, 2007b

References

Abbas A and Andrabi SMH (2002) Effect of different glycerol concentrations on motility before and after freezing, recovery rate, longevity and plasma membrane integrity of Nili-Ravi buffalo bull spermatozoa *Pakistan Veterinary Journal* **22** 1–4.

Abdou MSS, El-Guindi MM, El-Menoufy AA and Zaki K (1977) Some biochemical and metabolic aspects of the semen of bovines (*Bubalus bubalis* and *Bos taurus*). I. Seminal fructose in consecutive ejaculates and its relation to libido and semen quality *Zeitschrift für Tierzüchtung und Züchtungsbiologie* **94** 8–15.

Aguiar PHP, Andrade VJ, Abreu JJ and Gomez NBN (1994) Physical and morphological semen characteristics of buffaloes aged from four to eight years old. *Proceedings of the 4th World Buffalo Congress, Sao Paulo, Brazil, 27–30 June, 1994 Volume 3* pp 486–488 Ed WG Vale VH Barnabe and JCA De Mattos. International Buffalo Federation, Rome.

Ahmad K and Chaudhry RA (1980) Cryopreservation of buffalo semen *Veterinary Record* **106** 199–201.

Ahmad M, Latif M, Ahmad M, Qazi MH, Sahir N and Arslan M (1984) Age-related changes in body weight, scrotal size and plasma testosterone levels in buffalo bulls (*Bubalus bubalis*) *Theriogenology* **22** 651–656.

Ahmad M, Latif M, Ahmad M, Khan IH, Ahmad N and Anzar M (1985) Postmortem studies on infertile buffalo bulls: anatomical and microbiological findings *Veterinary Record* **117** 104–109.

Ahmad M, Ahmad KM and Khan A (1986) Cryopreservation of buffalo spermatozoa in Tris (hydroxymethylaminomethane) *Pakistan Veterinary Journal* **6** 1–3.

Ahmad M, Latif M and Ahmad M (1987) Morphological abnormalities of spermatozoa of Nili-Ravi buffalo *Buffalo Journal* **3** 153–160.

Ahmad N and Noakes D (2009) Reproduction in the buffalo In *Veterinary Reproduction and Obstetrics* 9th edition pp 824–835 Ed DE Noakes, TJ Parkinson and GCW England. Saunders Elsevier, Edinburgh.

Ahmad N, Shahab M, Khurshid S and Arslan M (1989) Pubertal development in male buffalo: longitudinal analysis of body growth, testicular size and serum profiles of testosterone and oestradiol *Animal Reproduction Science* **19** 161–170.

Ahmad N, Shahab M, Anzar M and Arslan M (1991) Changes in the behaviour and androgen levels during pubertal development of the buffalo bull *Applied Animal Behaviour Science* **32** 101–105.

Ahmad N, Umair S, Shahab M and Arslan M (2010) Testicular development and establishment of spermatogenesis in Nili-Ravi buffalo bulls *Theriogenology* **73** 20–25.

Ahmed K and Greesh M (2001) Effect of antibiotics on the bacterial load and quality of semen of Murrah buffalo bulls during preservation *Indian Journal of Animal Reproduction* **22** 79–80.

Ahmed K, Greesh M and Tripathi RP (2001a) Effect of different antibiotics on semen quality during post thaw incubation at 37°C *Indian Journal of Animal Reproduction* **22** 81.

Ahmed K, Kumar AA and Mohan G (2001b) Bacterial flora of preputial washing and semen of Murrah buffalo bulls and their antibiotic sensitivity pattern *Indian Journal of Comparative Microbiology, Immunology and Infectious Diseases* **22** 63–64.

Akhter S, Ansari MS, Andrabi SMH, Ullah N and Qayyum M (2008) Effect of antibiotics in extender on bacterial and spermatozoal quality of cooled buffalo (*Bubalus bubalis*) bull semen *Reproduction in Domestic Animals* **43** 272–278.

Akhter S, Ansari MS, Rakha BA, Andrabi SMH, Iqbal S and Ullah N (2010) Cryopreservation of buffalo (*Bubalus bubalis*) semen in Bioxcell extender *Theriogenology* **74** 951–955.

Akhter S, Ansari MS, Rakha BA, Andrabi SMH, Kahlid M and Ullah N (2011a) Effect of low density lipoproteins in extender on freezability and fertility of buffalo (*Bubalus bubalis*) bull semen *Theriogenology* **76** 759–764.

Akhter S, Ansari MS, Rakha BA, Ullah N, Andrabi SMH and Khalid M (2011b) *In vitro* evaluation of liquid-stored buffalo semen at 5°C diluted in soya lecithin based extender (Bioxcell®), Tris-citric egg yolk, skim milk and egg yolk-citrate extenders *Reproduction in Domestic Animals* **46** 45–49.

Ala Ud D, Chaudhry RA and Ahmad K (1981) Effect of different extenders on freezability of buffalo semen *Pakistan Veterinary Journal* **1** 59–61.

Aleem M, Chaudhry RA, Khan NU, Rizvi AR and Ahmed R (1990) Occurrence of pathogenic bacteria in buffalo semen *Buffalo Journal* **6** 93–98.

Ali HH, Ahmed IA and Yassen AM (1981) Sexual development and postpubertal changes in seminal characteristics of buffalo bulls *Alexandria Journal of Agricultural Research* **29** 47–57.

Amin AS, Darwish GM, Maha SZ and Hassan HM (1999) Trial to control *Chlamydia psittaci* in processed buffalo semen *Assiut Veterinary Medical Journal* **40** 319–332.

Andrabi SMH (2007a) Effects of antibiotics on motility, sperm morphology, membrane integrity, fertility and bacteriological quality of buffalo spermatozoa. PhD thesis, Quaid-i-Azam University, Islamabad. Available at: http://eprints.hec.gov.pk/2379/1/2234.htm (accessed 19 March 2011).

Andrabi SMH (2007b) Mammalian sperm chromatin structure and assessment of DNA fragmentation *Journal of Assisted Reproduction and Genetics* **24** 561–569.

Andrabi SMH (2009) Factors affecting the quality of cryopreserved buffalo (*Bubalus bubalis*) bull spermatozoa *Reproduction in Domestic Animals* **44** 552–569.

Andrabi SMH, Ahmad N, Abbas A and Anzar M (2001) Effect of two different antibiotic combinations on fertility of frozen buffalo and Sahiwal bull semen *Pakistan Veterinary Journal* **21** 166–169.

Andrabi SMH, Siddique M, Ullah N and Khan LA (2006) Effect of reducing sperm numbers per insemination dose on fertility of cryopreserved buffalo bull semen *Pakistan Veterinary Journal* **26** 17–19.

Andrabi SMH, Ansari MS, Ullah N, Anwar M, Mehmood A and Akhter S (2008) Duck egg yolk in extender improves the freezability of buffalo bull spermatozoa *Animal Reproduction Science* **104** 427–433.

Andrabi SMH, Mehmood A and Anwar M (2011) DNA integrity assay for fresh and cryopreserved buffalo bull spermatozoa In *Technical Working Group Annual Report, National Agricultural Research Centre, Islamabad*.

Anwar M, Andrabi SMH, Mehmood M and Ullah N (2008) Effect of low temperature thawing on the motility and fertility of cryopreserved water buffalo and zebu bull semen *Turkish Journal of Veterinary and Animal Sciences* **32** 413–416.

Anzar M, Ahmad M, Bakhsh M, Ahmad M and Iqbal N (1988) Effect of frequency of ejaculation on sexual behaviour and semen production ability in buffalo bulls In *Proceedings of 11th International Congress on Animal Reproduction and Artificial Insemination, University College Dublin, June 26–30, 1988 Volume 3* Paper No 225.

Anzar M, Ahmad M, Nazir M, Ahmad N and Shah IH (1993) Selection of buffalo bulls: sexual behavior and its relationship to semen production and fertility *Theriogenology* **40** 1187–1198.

Argov N, Sklan D, Zeron Y and Roth Z (2007) Association between seasonal changes in fatty-acid composition, expression of VLDL receptor and bovine sperm quality *Theriogenology* **67** 878–885.

Arnaud FG and Pegg DE (1990) Permeation of glycerol and propane-1,2-diol into human platelets *Cryobiology* **27** 107–118.

Asghar AA, Chaudhry MA and Iqbal J (1985) Productive and reproductive performance of Nili-Ravi buffaloes under optimal conditions of feeding and management In *Sixth Annual Report, Livestock Production Research Institute, Bahadurnagar, Okara, Pakistan* pp 32–33.

Azmi TI, Bongso TA, Harisah M and Basrur PK (1990) The Sertoli cell of the water buffalo – an electron microscopic study *Canadian Journal of Veterinary Research* **54** 93–98.

Bahga CS and Khokar BS (1991) Effect of different seasons on concentration of plasma luteinizing hormone and seminal quality vis-à-vis freezability of buffalo bulls (*Bubalus bubalis*) *International Journal of Biometeorology* **35** 222–224.

Barnabe RC, Barnabe VH, Ferrari S, Zogno MA, Zuge RM and Baruselli PS (1997) "*In vitro*" zona free hamster oocyte penetration test for evaluating the fertility of buffalo sperm In *Proceedings of the 5th World Buffalo Congress, Caserta, Italy, October 13–16, 1997 Volume 1* pp 869–872 Ed A Borghese, S Failla and VL Barile. International Buffalo Federation, Rome.

Barreto JB, Melo MIV, Lage AP, Marques AP and Vale VR (1996) Buffalo (*Bubalus bubalis*) sexual development: endocrinological and testicular histological aspects. II. From 10 to 24 months of age *Buffalo Bulletin* **15** 42–44.

Bathgate R, Maxwell WMC and Evans G (2006) Studies on the effect of supplementing boar semen cryopreservation media with different avian egg yolk types on *in vitro* post-thaw sperm quality *Reproduction in Domestic Animals* **41** 68–73.

Bhavsar BK (1987) Studies on seminal characteristics, biochemical constituents, freezability and fertility of Mehsana buffalo bulls. PhD thesis, Gujarat Agricultural University, Anand.

Bhavsar BK, Dhami AJ and Kodagali SB (1989) Monthly variations in freezability and fertility of Mehsana buffalo semen *Indian Journal of Dairy Science* **42** 246–250.

Bhosrekar MR, Mokashi SP, Purohit JR, Gokhale SB and Mangurkar BR (1994) Effect of glycerolization and deep freezing on the levels and release of enzymes in buffalo semen in relation to initial seminal attributes In *Proceedings of the 4th World Buffalo Congress, Sao Paulo, Brazil, 27–30 June, 1994, Volume 3*, pp 465–467 Ed WG Vale VH Barnabe and JCA De Mattos. International Buffalo Federation, Rome.

Bielanski A, Devenish J and Philips-Todd B (2000) Effect of *Mycoplasma bovis* and *Mycoplasma bovigenitalium* in semen on fertilization and association with *in vitro* produced morula and blastocyst stage embryos *Theriogenology* **53** 1213–1223.

Bielanski A, Bergeron H, Lau PC and Devenish J (2003) Microbial contamination of embryos and semen during long term banking in liquid nitrogen *Cryobiology* **46** 146–152.

Cabrera F, Gonzalez F, Batista M, Calero P, Medrano A and Gracia A (2005) The effect of removal of seminal plasma, egg yolk level and season on sperm freezability of canary buck (*Capra hircus*) *Reproduction in Domestic Animals* **40** 191–195.

Chachur MGM, Oba E and Gonzales CIM (1997) Equilibrium time influence on motility, vigour and membrane integrity of thawed buffalo semen using triladyl, glycine-egg yolk and tes extenders In *Proceedings of the 5th World Buffalo Congress, Caserta, Italy, October 13–16, 1997 Volume 1* pp 846–849 Ed A Borghese, S Failla and VL Barile. International Buffalo Federation, Rome.

Chantaraprateep P, Kamopattana M, Luengthongkum P, Usanakornkul S, Lohachit C and Ratanapany R (1985) Puberty of Thai swamp buffalo bull In *Annual Report of National Buffalo Research and Development Centre Project, Bangkok* pp 85–94.

Chauhan RAS and Srivastava RK (1973) Enzyme composition of buffalo seminal plasma *Journal of Reproduction and Fertility* **34** 165–166.

Chenoweth PJ (1997) Bull libido/serving capacity *Veterinary Clinics of North America: Food Animal Practice* **13** 331–344.

Cheshmedjieva SB, Vaisberg CN and Kolev SI (1996) On the lipid composition of buffalo spermatozoa frozen in different cryoprotective media *Bulgarian Journal of Agricultural Science* **2** 23–26.

Chinnaiya GP and Ganguli NC (1980a) Acrosomal damage of buffalo spermatozoa during freezing in extenders *Journal of Veterinary Medicine A* **27** 339–342.

Chinnaiya GP and Ganguli NC (1980b) Freezability of buffalo semen in different extenders *Journal of Veterinary Medicine A* **27** 563–568.

Cockrill WR (1974) *The Husbandry and Health of the Domestic Buffalo.* Food and Agricultural Organization of the United Nations (FAO), Rome.

Dabas YPS, Verma MC and Tripathi SS (1982) Biochemical studies on Red Dane and Murrah bull seminal plasma *Cherion* **1** 184–187.

de Ruigh L, Bosch JC, Brus MC, Landman B and Merton JS (2006) Ways to improve the biosecurity of bovine semen *Reproduction in Domestic Animals* **41** 268–274.

Dhami AJ (1992) Comparative evaluation of certain procedures in deep freezing of cattle and buffalo semen under tropical climate. PhD thesis, Indian Veterinary Research Institute, Izatnagar.

Dhami AJ and Kodagali SB (1989) Comparative analysis of physico-biochemical attributes in static and motile semen ejaculates from Surti buffalo bulls *Indian Journal of Animal Science* **59** 344–347.

Dhami AJ and Kodagali SB (1990) Freezability, enzyme leakage and fertility of buffalo spermatozoa in relation to the quality of semen ejaculates and extenders *Theriogenology* **34** 853–863.

Dhami AJ and Sahni KL (1993) Effect of extenders, additives and deep freezing on the leakage of lactic dehydrogenase from cattle and buffalo spermatozoa *Indian Journal of Animal Science* **63** 251–256.

Dhami AJ, Sahni KL and Mohan G (1992) Effect of various cooling rates (from 30° to 5°C) and thawing temperatures on the deep-freezing of *Bos taurus* and *Bos bubalis* semen *Theriogenology* **38** 565–574.

Dhami AJ, Jani VR, Mohan G and Sahni KL (1994) Effect of extenders and additives on freezability, post-thaw thermoresistance and fertility of frozen Murrah buffalo semen under tropical climate *Buffalo Journal* **10** 35–45.

Dhami AJ, Sahni KL, Mohan G and Jani VR (1996) Effects of different variables on the freezability, post-thaw longevity and fertility of buffalo spermatozoa in the tropics *Theriogenology* **46** 109–120.

Dhingra LD and Goyal HO (1975) A study on the different types of spermatogenesis in buffalo (*Bubalus bubalis*) *Acta Anatomica* **93** 219–227.

Di Matteo L (1997) Interazione eterologhe *in vitro* tra spermatozoi de *B. bubalis* ed ovociti di bovino. PhD thesis, University of Napoli Federico II, Napoli.

Drost M (2007) Bubaline versus bovine reproduction *Theriogenology* **68** 447–449.

El-Azab AI (1980) The interaction of season and nutrition on semen quality in buffalo bulls. PhD thesis, Cairo University, Cairo.

El-Wishy AB (1978) Reproduction performance of Iraqi buffaloes. Seasonal variation in sexual desire and semen characteristics *Zuchthygiens* **13** 28–32.

Fabbrocini A, Sansone G, Talev, R, Gualtieri R, Palazzo M, Formato R and Matassino D (1996) The use of MPA lectin to evaluate the functional integrity of frozen–thawed *Bubalus bubalis* spermatozoa In *Proceedings of the 13th International Congress on Animal Reproduction, Sydney, 30 June 4–July 1996 Volume 3* p 24.

FAO (2007) Global Livestock Production and Health Atlas. Food and Agricultural Organization of the United Nations, Rome. Available at: http://kids.fao.org/glipha/ (accessed 21 February 2011).

Fayez I, Marai M, Daader AH and Navsar AE (1987) Some physical and biochemical attributes of buffalo semen *Egyptian Journal of Animal Science* **25** 99–104.

Fuller B and Paynter S (2004) Fundamentals of cryobiology in reproductive medicine *Reproduction Biomed Online* **9** 680–691.

Gangadhar KS, Rao AR and Subbaiah G (1986) Effect of antibiotics on bacterial load in frozen semen of buffalo bulls *Indian Veterinary Journal* **63** 489–493.

Gill RS, Gangwar PG and Thakker OP (1974) Seminal attributes in buffalo bulls as affected by different seasons *Indian Journal of Animal Science* **44** 415–418.

Gilmore JA, McGann LE, Liu J, Gao DY, Peter AT, Kleinhans FW and Critser JK (1995) Effect of cryoprotectant solutes on water permeability of human spermatozoa *Biology of Reproduction* **53** 985–995.

Gilmore JA, Liu J, Gao DY, Peter AT and Critser JK (1998) Determination of plasma membrane characteristics of boar spermatozoa and their relevance to cryopreservation *Biology of Reproduction* **58** 28–36.

Guraya SS and Bilaspuri GS (1976) Stages of seminiferous epithelial cycle and relative duration of spermatogenic processes in the buffalo (*Bos bubalus*) *Gegenbaurs Morphologisches Jahrbuch* **122** 147–161.

Hasan S, Andrabi SMH, Muneer R, Anzar M and Ahmad N (2001) Effects of a new antibiotic combination on post-thaw motion characteristics and membrane integrity of buffalo and Sahiwal bull spermatozoa and on the bacteriological quality of their semen *Pakistan Veterinary Journal* **21** 6–12.

Hemeida NA, El-Baghdady YR and El-Fadaly (1985) Serum profiles of androstenedione, testosterone and LH from birth through puberty in buffalo male calves *Journal of Reproduction and Fertility* **74** 311–316.

Heuer C and Bajwa MA (1986) Selection of Nil-Ravi Buffalo bulls for artificial insemination. Effect of season on fertility of frozen buffalo semen *Zuchthygiene* **21** 257–262.

Heuer C, Bader H and Bajwa MA (1982) Sperm morphology of the Nili-Ravi buffaloes *Pakistan Veterinary Journal* **2** 155–160.

Heuer C, Tahir MN and Amjad H (1987) Effect of season on fertility of frozen buffalo semen *Animal Reproduction Science* **13** 15–21.

Holt WV (2000) Basic aspects of frozen storage of semen *Animal Reproduction Science* **62** 3–22.

Ishaq SM (1972) *Ninth Annual Report, Directorate of Livestock Farms, Government of the Punjab, Lahore.*

Jainudeen MR and Dass S (1982) Effect of level of glycerol, rates of freezing and thawing on the survival of buffalo spermatozoa in straws In *Animal Production and Health in the Tropics. Proceedings of First Asian–Australian Animal Science Congress, Serdang (Malaysia), 2–5 Sep 1980* pp 409–411 Ed MR Jainudeen and AR Omar. Penerbit Universiti Pertanian Malaysia, Serdang.

Jainudeen MR and Hafez ESE (1987) Cattle and water buffalo In *Reproduction in Farm Animals* 5th edition pp 297–314 Ed ESE Hafez. Lea and Febriger, Philadelphia.

Jainudeen MR, Bongso TA and Dass S (1982) Semen characteristics of the swamp buffalo (*Bubalus bubalis*) *Animal Reproduction Science* **4** 213–217.

Johnson MH and Nasr-Esfahani MH (1994) Radical solutions and cultural problems: could free oxygen radicals be responsible for the impaired development of preimplantation mammalian embryos in vitro? *Bioassays* **16** 31–38.

Joshi N, Luktuke SN and Chatterjee SN (1967) Studies on the biometry of reproductive tract and some endocrine glands of the buffalo-male *Indian Veterinary Journal* **44** 137–141.
Kadirvel G, Kumar S and Kumaresan A (2009) Lipid peroxidation, mitochondrial membrane potential and DNA integrity of spermatozoa in relation to intracellular reactive oxygen species in liquid and frozen–thawed buffalo semen *Animal Reproduction Science* **114** 125–134.
Kadirvel G, Periasamy S and Kumar S (2012) Effect of cryopreservation on apoptotic-like events and its relationship with cryocapacitation of buffalo (*Bubalus bubalis*) sperm *Reproduction in Domestic Animals* **47** 143–150.
Kerur VK (1971) Studies on semen characteristics in Gir, Kankrej, Jaffri, Surti and Murrah breeds *Indian Journal of Animal Health* **10** 119–121.
Kodagali SB (1967) Studies on semen characteristics in Gir and Jaffri breeds *Indian Veterinary Journal* **44** 773–776.
Kodagali SB, Bhavsar BK and Deshpande AD (1972) Semen characteristics of Surti buffalo bulls *Gujarat College of Veterinary Science and Animal Husbandry Magazine* **5** 15–20.
Koonjaenak S, Chanatinart V, Ekwall H and Rodriguez-Martinez H (2007a) Morphological features of spermatozoa of swamp buffalo AI bulls in Thailand *Journal of Veterinary Medicine A* **54** 169–178.
Koonjaenak S, Johannisson A, Pongpeng P, Wirojwuthikul S, Kunavongkrit A and Rodriguez-Martinez H (2007b) Seasonal variation in nuclear DNA integrity of frozen–thawed spermatozoa from Thai AI swamp buffaloes (*Bubalus bubalis*) *Journal of Veterinary Medicine A* **54** 377–383.
Koonjaenak S, Pongpeng P, Wirojwuthikul S, Johannisson A, Kunavongkrit A and Rodriguez-Martinez H (2007c) Seasonality affects post-thaw plasma membrane intactness and sperm velocities in spermatozoa from Thai AI swamp buffaloes (*Bubalus bubalis*) *Theriogenology* **67** 1424–1435.
Kumar R, Jagan Mohanarao G, Arvind and Atreja SK (2011) Freeze–thaw induced genotoxicity in buffalo (*Bubalus bubalis*) spermatozoa in relation to total antioxidant status *Molecular Biology Reports* **38** 1499–1506.
Kumar S, Sahni KL and Mohan G (1992) Effect of different levels of glycerol and yolk on freezing and storage of buffalo semen in milk, Tris and sodium citrate buffers *Buffalo Journal* **8** 151–156.
Kumar S, Sahni KL, Benjamin BN and Mohan G (1993a) Effect of various levels of yolk on deep freezing and storage of buffalo semen in different diluters without adding glycerol *Buffalo Journal* **9** 79–85.
Kumar S, Sahni KL and Bishta GS (1993b) Cytomorphological characteristics of motile and static semen of buffalo bulls *Buffalo Journal* **9** 117–127.
Kumar S, Sahni KL and Mohan G (1994) Effect of yolk, glycerol and sugars on post-thaw survival of buffalo spermatozoa in Tris dilutor *Indian Journal of Animal Science* **64** 362–364.
Landaeta-Hernández AJ, Chenoweth PJ and Berndtson WE (2001) Assessing sex-drive in young *Bos taurus* bulls *Animal Reproduction Science* **66** 151–160.
MacGregor R (1941) The domestic buffalo *Veterinary Record* **53** 443–450.
Marai IFM and Haeeb AAM (2010) Buffalo's biological functions as affected by heat stress – a review *Livestock Science* **127** 89–109.
Matharoo JS and Singh M (1980) Revivability of buffalo-spermatozoa after deep freezing the semen using various extenders *Journal of Veterinary Medicine A* **27** 385–391.
Maxwell WMC and Stojanov T (1996) Liquid storage of ram semen in the absence or presence of some antioxidants *Reproduction Fertility and Development* **8** 1013–1020.
Mazzilli F, Delfino M, Imbrogno N, Elia J and Dondero F (2006) Survival of micro-organisms in cryostorage of human sperm *Cell Tissue Banking* **7** 75–79.
McCool CJ and Entwistle KW (1989) Reproductive function in the Australian Swamp buffalo bull: age effects and seasonal effects *Theriogenology* **31** 583–594.
McCool CJ, Entwistle KW and Townsend MP (1989) The cycle of the seminiferous epithelium in the Australian swamp buffalo *Theriogenology* **31** 399–417.
Medeiros CMO, Forell F, Oliveira ATD and Rodrigues JL (2002) Current status of sperm cryopreservation: why is it better? *Theriogenology* **57** 327–344.
Mohan G, Madan ML and Razdan MN (1979) Composition of Murrah buffalo bull semen during winter and summer months in India *Tropical Agriculture (Trinidad)* **54** 21–28.
Morrell JM (2006) Update on semen technologies for animal breeding *Reproduction in Domestic Animals* **41** 63–67.
Nastri MJF, Del Sorbo C, Fabbrocini A, Fasano G and Sansone G (1994) Performances motility in cooled and freeze thawed *B. bubalis* spermatozoa at different osmotic pressures In *Proceedings of 7th*

International Symposium of Spermatology, Cairns, Queensland, Australia, 9–14 October 1994 pp 99–100 Ed M Bradley and JM Cummins. Australian Society of Reproductive Biology, East Melbourne, Victoria.

Nordin W, Bongso TA and Tan HS (1986) Relationship of semen characteristics and testosterone profiles with age in growing swamp buffalo bulls (*Bubalus bubalis*) *Buffalo Bulletin* **5** 66–67.

Oba E, Fuck EJ, Bicudo SD, Pap FO and Ohashi OM (1994) Preliminary study on different mediums for deep freezing of buffalo semen In *Proceedings of the 4th World Buffalo Congress, Sao Paulo, Brazil, 27–30 June, 1994, Volume 3* pp 579–581. International Buffalo Federation, Rome..

Ochsendrof FR (1998) Infection and reactive oxygen species *Andrologia* **31** 81–86.

Pant KP and Mukherjee DP (1972) The effect of seasons on the sperm dimensions of buffalo bulls *Journal of Reproduction and Fertility* **29** 425–429.

Pant HC, Sharma RK, Patel SH, Shukla HR, Mittal AK, Kasiraj RK, Misra AK and Prabhakar JH (2003) Testicular development and its relationship to semen production in Murrah buffalo bulls *Theriogenology* **60** 27–34.

Parks JE (1997) Hypothermia and mammalian gametes In *Reproductive Tissue Banking* pp 229–261 Ed AM Karow and JK Critser. Academic Press, San Diego.

Pawar HS and Wrobel KH (1991) The Sertoli cell of the water buffalo (*Bubalus bubalis*) during the spermatogenic cycle *Cell and Tissue Research* 265 43–50.

Pawar K and Kaul G (2011) Assessment of buffalo (*Bubalus bubalis*) sperm DNA fragmentation using a sperm chromatin dispersion test *Reproduction in Domestic Animals* **46** 964–969.

Perera BMAO (1999) Reproduction in water buffalo: comparative aspects and implications for management *Journal of Reproduction and Fertility* **54** 157–168.

Phelps MJ, Liu J, Benson JD, Willoughby CE, Gilmore JA and Critser JK (1999) Effects of Percoll separation, cryoprotective agents, and temperature on plasma membrane permeability characteristics of murine spermatozoa and their relevance to cryopreservation *Biology of Reproduction* **61** 1031–1041.

Presicce G (2007) Reproduction in the water buffalo *Reproduction in Domestic Animals* **42** 24–32.

Rajanarayanan S and Archunan G (2004) Occurrence of flehmen in male buffaloes (*Bubalus bubalis*) with special reference to estrus *Theriogenology* **61** 861–866.

Ramakrishnan P and Ariff MO (1994) Effect of glycerol level and cooling rate on post-thaw semen quality of Malaysian swamp buffalo In *Proceedings of the 4th World Buffalo Congress, Sao Paulo, Brazil, 27–30 June, 1994, Volume 3* pp 540–542 Ed WG Vale VH Barnabe and JCA De Mattos. International Buffalo Federation, Rome.

Ramakrishnan P, Adnan S, Nordin Y and Shanmugavelu S (1989) A comparison of semen characteristics of the swamp and Murrah buffaloes In *Seminar on Buffalo Genotypes for Small Farms in Asia, 15–19 May 1989, Serdang, Malaysia* pp 235–238 Ed MK Vidyadaran, TI Azmi and PK Basrur. Centre for Tropical Animal Production and Disease Studies, Universiti Pertanian Malaysia, Serdang.

Ramesha KP, Goswami SL and Das SK (1993) Zona-free hamster oocyte penetration test for assessing fertility of Murrah buffalo (*Bubalus bubalis*) bull *Buffalo Journal* **9** 259–263.

Rana BK and Bilaspuri GS (2000) Changes in interstitial cells during development of buffalo testis *The Veterinary Journal* **159** 179–185.

Rao AVN, Haranath GB, Sekharam GS and Rao JR (1986) Effect of thaw rates on motility, survival and acrosomal integrity of buffalo spermatozoa frozen in medium French straws *Animal Reproduction Science* **12** 123–129.

Rasul Z (2000) Cryopreservation of buffalo semen. PhD thesis, Quaid-i-Azam University, Islamabad.

Rasul Z, Anzar M, Jalali S and Ahmad N (2000) Effect of buffering systems on post-thaw motion characteristics, plasma membrane integrity, and acrosome morphology of buffalo spermatozoa *Animal Reproduction Science* **59** 31–41.

Rasul Z, Ahmad N and Anzar M (2001) Changes in motion characteristics, plasma membrane integrity, and acrosome morphology during cryopreservation of buffalo spermatozoa *Journal of Andrology* **22** 278–283.

Rasul Z, Ahmed N and Anzar M (2007) Antagonist effect of DMSO on the cryoprotection ability of glycerol during cryopreservation of buffalo sperm *Theriogenology* **68** 813–819.

Rohilla RK, Tuli RK and Goyal RL (2005) Comparative study of the effects of cryoprotective agents in freezing Murrah buffalo bull semen *Indian Journal of Veterinary Research* **14** 37–43.

Roy B, Nagpaul PK, Pankaj PK, Mohanty TK, Raina VS and Mishra A (2008) Studies on the biometry of sperm of riverine buffalo bulls (*Bubalus bubalis*) *Buffalo Bulletin* **27** 192–195.

Saeed A, Chaudhry RA, Khan IH and Khan NU (1989) Studies on morphology of buffalo bull semen of different age groups *Buffalo Journal* **5** 99–102.

Saeed A, Chaudhry RA, Khan IH and Khan NU (1990) Morphology of semen buffalo bulls of different age groups In *Recent Advances in Buffalo Research. Proceedings of II World Buffalo Congress, Held in India during 12–16 December 1988 Volume III. Physiology and Reproduction* pp 17–19 Ed RM Acharya, RR Lokeshwar and S Kumar. Indian Society of Buffalo Development and Indian Council of Agricultural Research, New Delhi.

Sagdeo LR, Chitnis AB and Kaikini AS (1991) Effect of seasonal variations on freezability of Surti buffalo bull semen *Indian Journal of Animal Reproduction* **12** 1–3.

Sahni KL and Mohan G (1990a) Effect of various levels of yolk on viability of buffalo semen at 37°, 5° and –196°C In *Recent Advances in Buffalo Research. Proceedings of II World Buffalo Congress, Held in India during 12–16 December 1988 Volume III. Physiology and Reproduction* pp 63–65 Ed RM Acharya, RR Lokeshwar and S Kumar. Indian Society of Buffalo Development and Indian Council of Agricultural Research, New Delhi.

Sahni KL and Mohan G (1990b) Yolk as a cryoprotectant in deep-freezing of bovine semen *Indian Journal of Animal Science* **60** 828–829.

Samo MU, Brohi NA, Kaka I, Qureshi TA and Memon MM (2005) Study on sexual behaviour and seminal quality characteristics of Kundhi buffalo bulls *Pakistan Journal of Biological Sciences* **8** 1628–1629.

Sansone G, Nastri MJF and Fabbrocini A (2000) Storage of buffalo (*Bubalus bubalis*) semen *Animal Reproduction Science* **62** 55–76.

Sayed AA (1958) Some factors affecting the semen of cattle and buffaloes. PhD thesis, Cairo University Cairo.

Selvaraju S, Nandi S, Subramani TS, Raghavendra BS, Rao SB and Ravindra JP (2010) Improvement in buffalo (*Bubalus bubalis*) spermatozoa functional parameters and fertility *in vitro*: effect of insulin-like growth factor-I *Theriogenology* **73** 1–10.

Sen Gupta BP, Misra MS and Roy A (1963) Climatic environment and reproductive behaviour of buffaloes. I. Effect of different seasons on various seminal attributes *Indian Journal of Dairy Science* **16** 150–165.

Sengar DPS and Sharma VD (1965) Studies on successive semen ejaculates of buffalo bulls I. Semen characteristics *Indian Journal of Dairy Science* **18** 54–60.

Shannon P (1972) The effect of egg yolk level and dose rate of semen diluted in caprogen In *Proceedings of the 7th International Congress on Animal Reproduction and Artificial Insemination, Munich, 6–9 June 1972* pp 1440–1442 (also in Summaries, pp. 279–280). Deutsche Gesellschaft für Züchtungskunde, Bonn.

Sharma AK and Gupta RC (1979) Estimation of daily sperm production ratio (DSPR) by quantitative testicular histology in buffalo bulls (*Bubalus bubalis*) *Archives of Andrology* **3** 147–152.

Sharma AK and Gupta RC (1980) Duration of seminiferous epithelial cycle in buffalo bulls (*Bubalus bubalis*) *Animal Reproduction Science* **3** 217–224.

Sidhu KS (1974) Morphological and histochemical studies on the spermatozoa of buffalo bull. MSc thesis, Punjab Agricultural University Ludhiana.

Sidhu KS and Guraya SS (1979) Effects of seasons on physico-chemical characteristics of buffalo (*Bubalus bubalis*) semen *Indian Journal of Animal Sciences* **49** 884–889.

Sidhu KS and Guraya SS (1985) *Buffalo Bull Semen: Morphology, Biochemistry, Physiology and Methodology*. USG Publishers and Distributors, Ludhiana.

Singh B and Sadhu DP (1973) Buffering capacity and CO, and N, content of seminal plasma of cattle and buffaloes *Indian Journal of Animal Sciences* **43** 579–583.

Singh B, Mahapatro BB and Sadhu DP (1969) Chemical composition of cattle and buffalo spermatozoa and seminal plasma under different climatic conditions *Journal of Reproduction and Fertility* **20** 175–178.

Singh J, Pangawkar GR, Biswas RK, Srivastava AK and Sharma RD (1990) Studies on lactic dehydrogenase and sorbitol dehydrogenase release in relation to deep freezing of buffalo semen in certain extenders *Theriogenology* **34** 371–378.

Singh J, Pangawkar GR, Biswas RK and Kumar N (1991) Studies on buffalo sperm morphology during various stages of freezing in certain extenders *Indian Journal of Animal Reproduction* **12** 126–129.

Singh P, Singh S and Hooda OK (1999) Effect of different level of egg yolk on freezability of buffalo semen *Haryana Veterinarians* **38** 26–28.

Singh P, Jindal JK, Singh S and Hooda OK (2000) Freezability of buffalo bull semen using different extenders *Indian Journal of Animal Reproduction* **21** 41–42.

Singh P, Singh I, Singh S and Sharma RK (2006) Initial stage glycerolization prevents the incidence of backward sperm motility during cryopreservation and increases buffalo semen freezability *Indian Journal of Animal Science* **76** 777–779.

Sukhato P, Thongsodseang S, Utha A and Songsasen N (2001) Effects of cooling and warming conditions on post-thawed motility and fertility of cryopreserved buffalo spermatozoa *Animal Reproduction Science* **67** 69–77.

Tahir MN, Bajwa MA, Latif M, Mushtaq M and Shah MH (1981) Effects of insemination dose and season on conception rates in buffaloes *Pakistan Veterinary Journal* **1** 161–163.

Thibier M and Guerin B (2000) Hygienic aspects of storage and use of semen for artificial insemination *Animal Reproduction Science* **62** 233–251.

Tuli RK and Singh M (1983) Seasonal variation in freezability of buffalo semen *Theriogenology* **20** 321–324.

Tuli RK, Mehar S and Matharoo JS (1981) Effect of different equilibration times and extenders on deep freezing of buffalo semen *Theriogenology* **16** 99–104.

Valdez CA, Bautista JAN, Rio ASD, Flor JACG and Torres EB (2003) Ethylene glycol as a cryoprotectant for freezing buffalo spermatozoa: I. Effects of concentration and equilibration time on post-thaw survival *Philippine Journal of Veterinary Medicine* **40** 37–43.

Vale WG (1994) Collection, processing and deep freezing of buffalo semen *Buffalo Journal* **10** 65–81.

Watson PF (2000) The causes of reduced fertility with cryopreserved semen *Animal Reproduction Science* **60–61** 481–482.

Wolf FR, Almquist JO and Hale EB (1965) Prepubertal behavior and pubertal characteristics of beef bulls on high nutrient allowance *Journal of Animal Science* **24** 761–765.

Younis M, Samad HA, Ahmad N and Ali CS (1998) Effect of age and breeding season on the freezability of buffalo bull semen *Pakistan Veterinary Journal* **18** 219–223.

Younis M, Samad HA, Ahmad N and Ahmad I (2003) Effects of age and season on the body weight, scrotal circumference and libido in Nili-Ravi buffalo bulls maintained at the semen production unit, Qadirabad *Pakistan Veterinary Journal* **23** 59–65.

Yu ZW and Quinn PJ (1994) Dimethyl sulphoxide: a review of its applications in cell biology *Bioscience Reports* **14** 259–281.

15 Applied Andrology in Swine

Gary C. Althouse*
University of Pennsylvania, Kennett Square, Pennsylvania, USA

Introduction

As with other livestock species, reproductive examination of the boar can be a valuable component in managing reproductive performance in a herd. Production strategies in today's swine industry can allow a boar to have a measurable impact on herd reproductive performance. Along with traditional examinations being performed on boars with suspected sub-fertility or infertility, it has become increasingly more common to examine boars that are entering or are resident in a dedicated stud providing extended semen for artificial insemination (AI) use.

Boar Selection

With the rapid incorporation of AI into the global swine industry over the past two decades, a more intensive process has evolved in the selection of boars for use in such (AI) breeding programmes. Along with visual assessment and possible performance evaluations, current genetic methodologies incorporating computerized data analysis and molecular screening provide the industry with objective information for analysing a boar's genetic merit (Safranski, 2008). Data collected on traits of interest and/or individual performance from groups of individuals of similar lineage, from paternal and maternal lines, and from their offspring, are utilized to assess a boar as a viable breeding prospect. Computational predictive outcomes on an individual boar can include estimated breeding value (EBV), expected progeny difference (EPD) and best linear unbiased predictors (BLUP). Marker assisted selection (MAS) is a molecular screening tool that can be employed to see whether an individual boar carries a particular genetic marker that is associated with a desirable or undesirable phenotypic trait. Collectively, these data provide producers with valuable information that will assist them in targeting specific improvements in their production herds.

In addition to genetic selection, a thorough inquiry into the disease status at the boar's source should be performed, preferably via a veterinarian-to-veterinarian conversation. Supportive diagnostic paperwork as part of the source herd's routine monitoring programme should be made available for review. The information obtained through this investigation, along with the strict incorporation of disease control strategies for all

* E-mail: gca@vet.upenn.edu

incoming boars, such as isolation, acclimatization and recovery, will greatly help in minimizing the introduction of infectious disease into the herd in which the boar will eventually reside.

Sexual Development and Puberty

In boars, both testes should be fully descended and present in the scrotum at birth. Absence of one or both testicles in the scrotum is indicative of cryptorchidism, a heritable condition that should preclude further consideration of the boar for future use in a breeding programme.

Sperm production is directly correlated with numbers of Sertoli cells in the testes. The proliferation of Sertoli cells in the boar occurs postnatally, in two distinct phases. The initial proliferation phase starts at birth and continues through to 1 month of age, with the second and final phase occurring at 3–4 months of age (França *et al.*, 2000).

Puberty is a temporal phase in a boar's sexual maturation in which distinct anatomical and physiological changes occur. These changes are directly under the influence of steroid (i.e. androgen) production, which originates primarily from the Leydig cells of the testis. In the prepubertal boar, there is an ill-defined penile sigmoid flexure. This results from the presence of a fibrous band known as the frenulum, which attaches the glans penis to the internal lamina of the prepuce and aids retention of the penis within the prepuce in the young boar. As androgen levels increase, penile growth and development of the sigmoid flexure ensue. Behaviourally, physical mounting activity and penile thrusting commence, and these promote the breakdown of the frenulum. Testicular synthesis of androgens increases from 70 to 170 days of age, with blood concentrations markedly increasing after 200 days of age (França *et al.*, 2000). Puberty culminates with the presence of all sperm progenitor cells in the testis (Plate 44) and of spermatozoa in the ejaculate, which may occur in domesticated large breeds as young as 4.5 to 6 months of age (Cameron, 1987).

Genotype, environment and management all contribute to boar development and puberty. Crossbred boars tend to reach sexual maturity earlier than purebred boars. Group rather than individual housing has been found to be beneficial to the boar's sexual development (Hacker *et al.*, 1994). In swine, the sexes exhibit differences in growth rates, feed utilization and nutrient needs. As such, split sex feeding is a successful management strategy that is used to optimize development of the young boar and gilt. Split sex feeding has its best benefit when applied to boars after they have reached 45–50 kg body weight.

Assessment

Physical and behavioural examination

The boar should first be visually examined while in an enclosed, secure area. Boars are normally inquisitive and will thoroughly investigate their surroundings. At this time, the boar should be assessed for body condition, symmetry, balance and skeletal conformation as it ambulates freely in the pen. Particular attention should be paid at this time to current or potential musculoskeletal conditions that may interfere with the boar's ability to approach and mount a collection dummy or female. Visual acuity in each eye is next determined by the investigator extending a hand from a perpendicular position towards the eye to elicit a blink reflex, or to elicit animal movement towards or away from the extended hand.

The reproductive genitalia (Fig. 15.1) are evaluated visually and by digital palpation. The non-pendulous scrotum is first examined for uniformity of the skin and absence of pathological indications. The testes should be freely moveable within the scrotum. Palpation of the testes should yield a uniformly firm tissue of resilient texture. In boars, the testes should be symmetrical and there should be less than 0.5 cm in diameter difference between them (Clark *et al.*, 2003). Mature boars should have a minimum testis size of 6.5 cm (width) × 10 cm (length). The cranial (caput) epididymis and cauda epididymis,

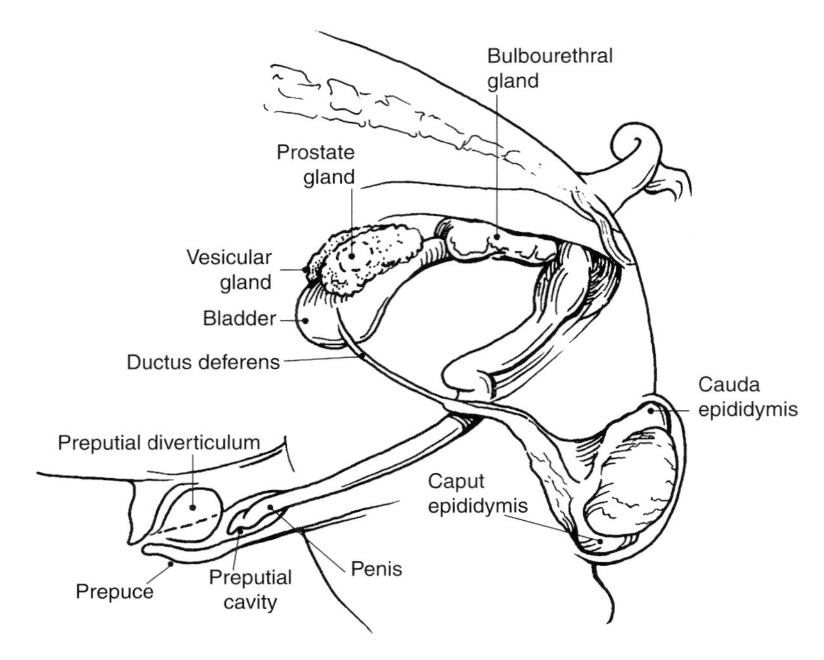

Fig. 15.1. Anatomy of the boar urogenital tract (modified from Kuster and Althouse, 2007).

located at the ventral and dorsal aspects of each testis, respectively, are easily palpable and should be assessed. Abnormalities of the testis and epididymis are only rarely found bilaterally in the boar. The prepuce, including the preputial diverticulum, is palpated along its length for normality, and examined for any abnormalities noted. In the adult boar, the external preputial orifice should readily allow the insertion of an index finger, providing for examination of the preputial cavity and diverticulum.

As AI is widely used in the global swine industry, most boars will be trained for semen collection using a dummy sow. Therefore, boars should demonstrate interest in investigating and mounting a stationary dummy sow. Field observations suggest that the majority of boars successfully trained for semen collection will have mounted the dummy sow within the first four training sessions. If a boar is to be used in a natural mating programme (e.g. pen breeding or hand mating), he should be exposed to several oestrous females in order to assess his ability to mount and perform intromission. In both cases, assessments can be made of both libido and of how well the boar 'works' the female/dummy sow.

Semen collection

Because of the widespread use of AI in swine, a large proportion of boars have been trained for semen collection using a collection dummy. The boar is introduced into a semen collection pen and allowed to freely approach the dummy. Within a few minutes, the boar will mount the collection dummy, and this is soon followed by pelvic thrusting. Good footing at the base of the collection dummy is needed at this point so that the boar maintains thrusting and exteriorizes the distal portion of the glans penis.

Semen is collected from boars using the gloved hand technique. Some glove materials have been found to be spermicidal (Ko *et al.*, 1993), so the use of gloves that are known to be non-spermicidal is required. The semen collector enters the pen with two gloves on the collecting hand. While the boar is on the collection dummy and thrusting, the collector kneels and extends the gloved hand underneath the boar and massages the sheath. Sheath massage will aid in performing a hygienic semen collection as it will facilitate evacuation of spermicidal preputial fluids, particularly from the preputial diverticulum, and aid in further exteriorizing the penis. Once the penis is exteriorized, the outer glove is discarded. A semen collection container (i.e. an insulated thermos or Styrofoam cup) is held with the free hand, and using the gloved hand, the corkscrew tip of the glans penis is grasped and digital pressure applied to 'lock' the glans within the hand. Once adequate digital pressure is applied and the penis is locked in the hand, the boar will respond positively by fully extending his penis and reducing thrusting.

Ejaculation then commences with the initial pre-sperm fraction, which is usually clear and free of sperm. The pre-sperm fraction is not collected. Sperm-rich and sperm-poor fractions follow, both of which are collected into the receptacle. A gel fraction, originating from the bulbourethral gland, is normally emitted in conjunction with the sperm-poor fraction. As this gel fraction is of minimal value, it is intentionally separated from the fluid portions of the ejaculate by placing a filter or gauze over the opening of the semen collection container. Once ejaculation is completed, the penis is visually examined for any pathology before it is released. The neat semen is taken to the laboratory for analysis, making sure that it is protected from chemical (e.g. water, detergents, alcohol, etc.), temperature (hot, cold) and ultraviolet light insults.

Although not commonly used, electroejaculation (EE) can be performed in boars when it is not possible to obtain an ejaculate using the gloved hand technique and a collection dummy. Boars selected for EE should be presented with a history of known libido and mating ability. The animals are initially placed under general anaesthesia and then positioned in lateral recumbency. Using a gloved hand, the rectum is evacuated and a boar-specific rectal probe inserted (Plate 45). The penis is exteriorized from within the prepuce using atraumatic dressing forceps (see Plate 45). Once exposed, gauze is wrapped around the penile shaft to maintain exposure during the process and to facilitate visual examination. Using an electroejaculator, a pulsating current of progressive amplitude and frequency is applied through the rectal probe until ejaculation commences. The ejaculate is collected into a sterile, warmed receptacle for protection and transport to the laboratory for further examination.

Semen evaluation

Evaluation of a boar ejaculate should include assessment of both its physical and cellular characteristics. Laboratories should be cognizant of the appropriate handling conditions in order to avoid temperature, osmotic and/or pH shock to the neat semen. Neat semen should be held and manipulated at body temperature (~38°C) during evaluation, and all equipment and materials that come into contact with the neat semen should be at a similar temperature.

Physical assessment

The ejaculate should be examined for colour, turbidity, odour and viscosity. Normal boar ejaculates will be whitish grey to white in colour. Samples will be opalescent when held up to light. With gentle agitation, mass swirling of suspended cells can be easily visualized by the naked eye. Neat semen will have a very fluid viscosity, similar to aqueous solutions. Fresh blood in the boar ejaculate (haemospermia) will colour the semen pink to bright red. A yellow or brownish colour to the ejaculate may be due to a pathological problem (i.e. infection, oxidized blood) of the reproductive tract, or to contamination with preputial fluid or urine. A characteristic odour will confirm the latter. Ejaculate volume is determined by weighing the ejaculate (1 g weight = 1 ml volume), or by using an isothermal volume measuring device. Ejaculate volume can be multiplied by sperm concentration to obtain the total sperm number in an ejaculate.

Cellular assessment

A basic evaluation of neat semen should include a detailed analysis of sperm motility, sperm morphology, sperm concentration and calculation of total sperm numbers. Table 15.1 provides suggested minimum values of neat boar semen obtained via the gloved hand technique from adult, crossbred boars that are on a 4–7 day semen collection schedule.

Motility

Estimations of sample motility should be made as soon as possible after semen collection. Accurate assessment of motility is performed by a skilled examiner using a high-quality microscope with, preferably, a heated (38°C) stage. Given the subjective nature of this assessment, accuracy is highly dependent upon sample preparation, microscope quality, technician experience and natural ability. To minimize the effects of systematic and experimenter error, sample preparation and microscopic assessment must be standardized. The most common technique employed in laboratories is to place a 7–10 μl sample of semen on a warmed (38°C) microscope slide, overlaid with a coverslip. The loaded slide is focused under a microscope at 200× magnification. If appropriately prepared, the sample should appear as a monolayer of cells, within which the motility of individual sperm can be easily ascertained. If the sample is too concentrated, a subsample can be diluted with equal (v/v) parts of semen and a thermo-equivalent isotonic diluent, before placement on the slide with a coverslip. At 200× or 400× magnification, motility should be estimated to the nearest 5% by viewing sperm activity in at least four different fields on the slide, and taking the average of these readings for the final motility estimate. From this examination, the presence of any spermatogenic progenitor cells, leucocytes, erythrocytes and bacteria should also be noted.

Table 15.1. Minimum suggested values for neat boar semen obtained via gloved hand technique from adult from crossbred boars on a 4–7 day semen collection schedule.

Semen variable	Characteristic/value
Colour	Whitish-grey to white
Ejaculate volume (ml)	216 (range 130–500)
Gross sperm motility	>75%
Sperm concentration (millions/ml)	370 (range 140–735)
Sperm morphology	<20% abnormal[a]
Turbidity	Opalescent

[a]Includes both proximal and distal cytoplasmic droplets.

Morphology

Both wet and dry mount preparations can be used for assessing boar sperm morphology (i.e. size, shape, appearance characteristics). Wet mounts are examined using microscopes that provide for their own internal contrast (e.g. phase contrast, differential interference contrast or DIC). A wet mount is prepared by immobilizing a small amount of sample on the slide using an isotonic aliphatic aldehyde solution, overlaid with a coverslip, and then viewed under oil immersion at a minimum of 1000× magnification.

A variety of contrast stains are commercially available for examining boar sperm morphology using dry mounted slides. A good contrast stain should accentuate the outline of the sperm when using a light microscope, thus allowing easier visualization and identification of normal and abnormal sperm. To make a stained slide, a small volume of sample is mixed with 7–10 μl of contrast stain using the edge of a second clean slide. In a similar manner to performing a blood smear, the edge of the second slide is used to draw the mixture across the slide to produce a thin layer that is air dried. The morphology of individual sperm is assessed under oil immersion. When using both wet and dry mount techniques, a minimum of 100 sperm should be morphologically assessed and categorized (Kuster *et al.*, 2004). Typical sperm abnormalities seen in boar semen are shown in Figs 15.2 to 15.6 inclusive. Normal boar ejaculates should exhibit less than 20% of abnormal sperm (Bach *et al.*, 1982; Althouse, 1998).

Concentration

Sperm concentration in boar semen can be determined using a variety of available techniques. If done infrequently, the most economical and practical method is with the use

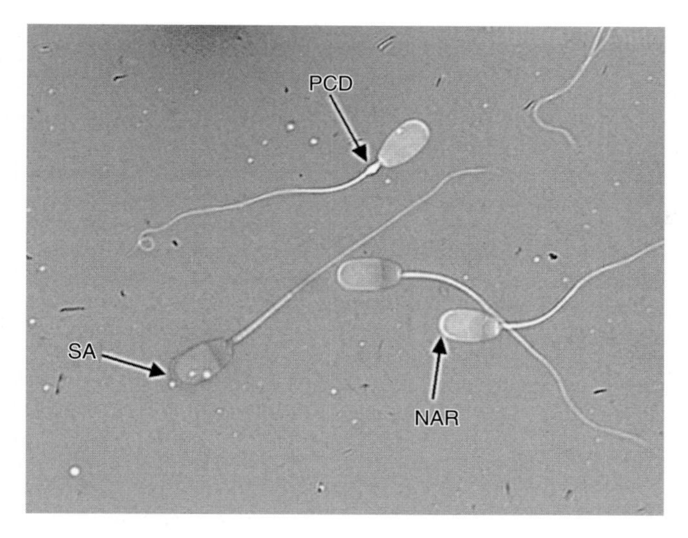

Fig. 15.2. Photomicrograph of ejaculated boar spermatozoa stained with eosin–nigrosin under 1050× magnification. NAR, normal apical ridge of acrosome; PCD, abnormal proximal cytoplasmic droplet; SA, abnormal swollen or disrupted acrosome.

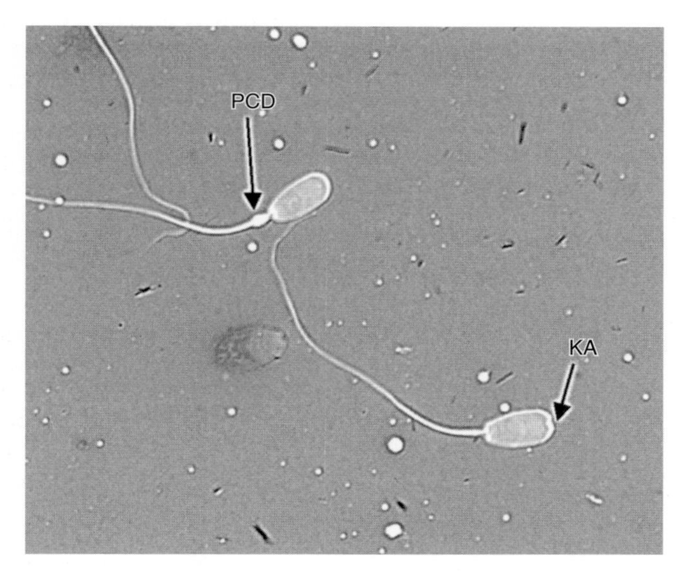

Fig. 15.3. Photomicrograph of ejaculated boar spermatozoa stained with eosin–nigrosin under 1050× magnification. KA, abnormal, knobbed acrosome; PCD, abnormal proximal cytoplasmic droplet.

of a counting chamber such as a haemocytometer (Althouse *et al.*, 1995). A portion of neat semen is diluted one part semen to 200 parts of an immobilizing solution. The sample is mixed and then a subsample charged on to each chamber of the haemocytometer. The haemocytometer is then set in a humidified chamber and the charged sample allowed to settle over a 5 min interval. The haemocytometer is placed under a microscope and the grid

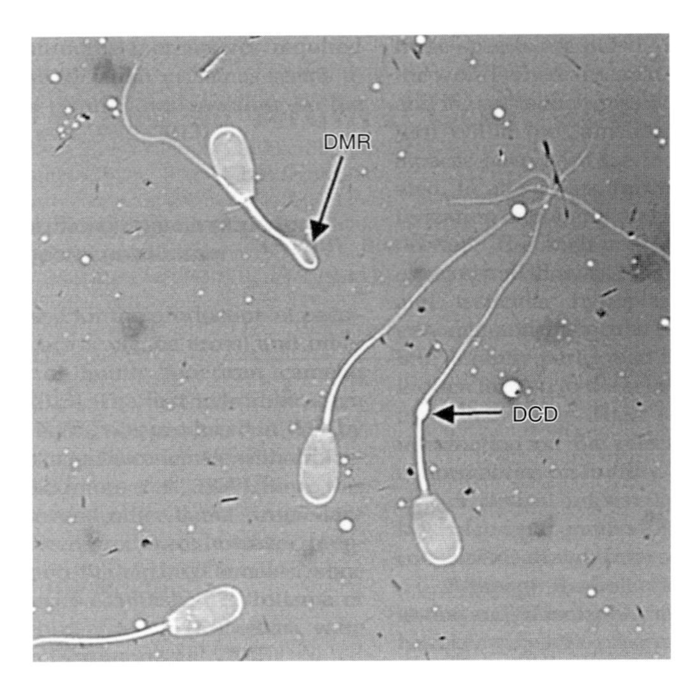

Fig. 15.4. Photomicrograph of ejaculated boar spermatozoa stained with eosin–nigrosin under 1050× magnification. DCD, abnormal distal cytoplasmic droplet; DMR, abnormal distal midpiece reflex.

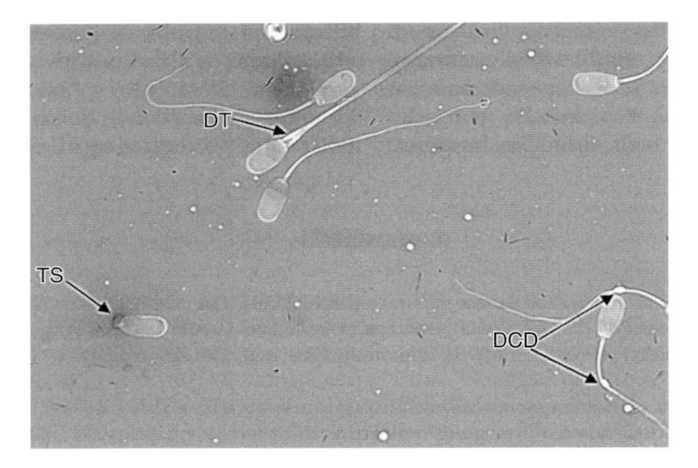

Fig. 15.5. Photomicrograph of ejaculated boar spermatozoa stained with eosin–nigrosin under 1050× magnification. DCD, abnormal distal cytoplasmic droplet; DT, abnormal double tail; TS, abnormal tail stump defect.

surface brought into focus at 200× or 400× magnification. Sperm are then counted in a minimum of five large (80 small) squares in the central grid. To aid counting, only sperm heads touching the top and left lines of each large square are included in the count. Sperm tails only or tails touching any of the lines are not counted. This counting methodology is then

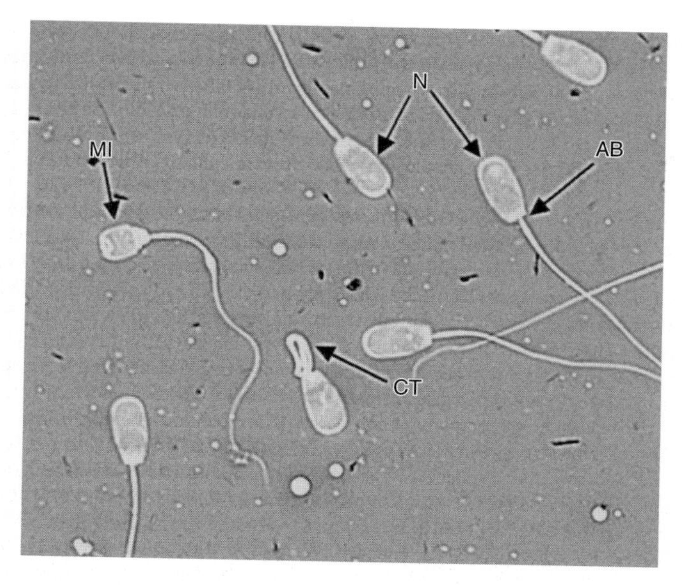

Fig. 15.6. Photomicrograph of ejaculated boar spermatozoa stained with eosin–nigrosin under 1050× magnification. AB, spermatozoa with abaxial tail attachment (considered normal in the boar); CT, abnormal coiled tail; MI, abnormal microcephalic head with distal cytoplasmic droplet; N, morphologically normal spermatozoa.

applied to the opposite chamber (or, if only a single chamber, the process is repeated), and the two counts averaged if they are within 10% of each other. The average number is then inserted into a formula supplied by the chamber distributer at the 1:200 dilution used to arrive at an estimate of the number of sperm cells/ml of gel-free semen. Appropriate training and skill are necessary in order to avoid systematic and experimenter error when using the haemacytometric technique.

In laboratories that perform frequent semen assessments, a common method of determining sperm concentration is through the measurement of sample opacity using photometry. With this technique, the number of sperm cells and other seminal plasma components all contribute to the opacity of a sample. Neat boar semen is normally too opaque for light to be transmitted through it, so a fixed dilution of a subsample of the neat semen is necessary when obtaining a measurement. Accuracy of the photometer in estimating boar sperm concentration is dependent upon several factors, including species-specific instrument calibration, the correct dilution, readings obtained within the equipment's optimum operating range, etc. It is important that manufacturer recommendations be followed when using a photometer. Using similar methodologies, more advanced photometers are available that measure sperm concentration using fluorescent probes that are excited to emit light at distinct wavelengths. The benefit of this type of instrument over traditional photometry is that only actual cells are counted. As with regular photometry, however, one needs to be aware of the need to minimize systematic and experimenter's error in order to obtain an accurate reading.

Maturation of the swine industry has brought increased implementation of computer automated (also known as computer assisted) semen analysis (CASA) systems at boar studs. Most CASA systems have the ability to assess sperm motility and determine sperm concentration with good accuracy and precision. Some of these systems offer a software option that assesses sperm morphology, the accuracy of which is currently being investigated. It should be noted, though, that inattention to sample preparation, slide choice and computer programming can all lead to erroneous results (Douglas-Hamilton *et al.*, 2005).

Supplemental semen tests

In certain circumstances, supplemental tests of the ejaculate may be warranted to aid in identifying a specific sperm dysfunction that may have an impact on fecundity or to assist in developing an appropriate medical treatment plan. Semen culture is routinely performed if infection of the urogenital tract (e.g. of the epididymides or accessory sex glands) is suspected. Given the nature of the test, the semen collector should give special consideration to hygiene and sanitation when obtaining a semen sample (Althouse, 2008). Both aerobic and anaerobic screening should be performed. In species other than the boar, biochemical assays may be used to help identify which accessory sex gland is infected. In the boar, regardless of the accessory sex gland affected, treatment choice generally remains the same, making the routine use of these assays of minimal value.

Of the readily available sperm function tests, the determination of sperm viability is the most common test performed. Sperm viability is easily and accurately determined using specific fluorophores with which sperm membrane permeability can be assessed in concert with multipurpose instrumentation, such as CASA, fluorescent photometry and flow cytometry. Flow cytometry can also be used to screen samples for other sperm variables, many of which are of an academic nature, such as membrane fluidity/capacitation, lipid peroxidation, acrosomal damage, early apoptosis, mitochondrial potential and DNA damage.

The hypo-osmotic swelling (HOS) test is a simple investigation that can be performed to assess the water permeability of intact sperm cell membranes. The HOS test is considered to be a vitality screening test, in that live spermatozoa should be able to withstand moderate hypo-osmotic stress and thus exhibit swelling when exposed to a hypotonic solution. The test can be of value when screening samples that have low motility sperm or no motile sperm upon initial evaluation, and the evaluator does not have access to instrumentation with fluorescence capabilities to determine sperm viability. Briefly, the test is performed by incubating a portion of the semen sample in a 150 mOsm solution for 30 min at 37°C (Vazquez *et al.*, 1997). After incubation, a 7–10 μl drop of a well-mixed sample is placed on a microscope slide overlaid with a coverslip. Samples are examined under 400× magnification and a minimum of 100 sperm assessed for the presence of swollen tails (Fig. 15.7), an indicator of a sperm with intact membranes.

Sperm-binding assays, such as the sperm zona pellucida binding test, can be of value in assessing an important step in the fertilization process. This detailed *in vitro* test is performed by incubating capacitated sperm with denuded oocytes at 37°C. After a period of time, samples are washed to remove non-bound and loosely bound sperm and then chemically fixed before examination under microscopy. The results of this test are limited to confirming that a boar's sperm has the ability to bind to an oocyte and do not act as a complete predictor of fertility (Collins *et al.*, 2008).

Causes of Reproductive Failure

Infectious causes

Normal spermatogenesis in the boar occurs within a defined testicular temperature range that is slightly below core body temperature. A febrile response frequently occurs in boars undergoing an acute infectious disease episode. Although the febrile response may be beneficial to activating the immune system to combat infection, an undesirable side effect is that the core testis temperature increases and spermatogenesis is disrupted. To minimize the effects of systemic illness on spermatogenesis, appropriate treatment should be implemented quickly and the animal monitored to keep core body temperature below 39.5°C. Coincidentally, vaccine-induced febrile conditions can also elicit similar results.

A number of infectious agents have been associated with reproductive problems in the boar and/or appear to play an epidemiological role in the transmission of the disease

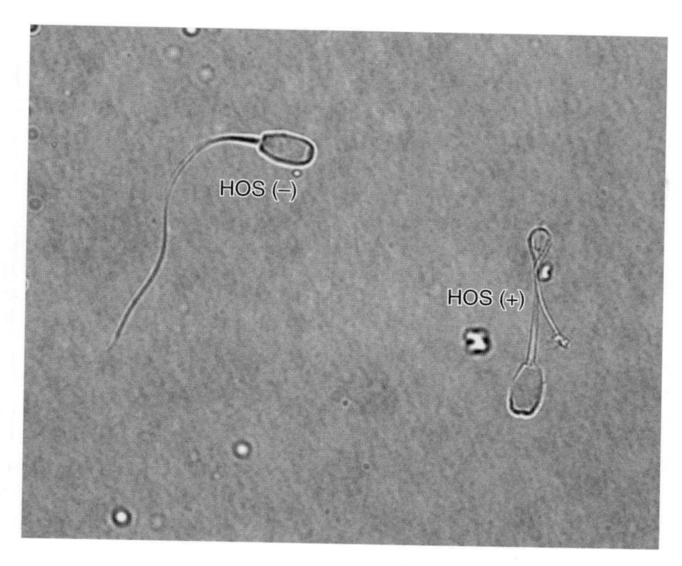

Fig. 15.7. Boar sperm undergoing a hypo-osmotic swelling (HOS) test. Positive and negative responses are shown.

(Althouse and Rossow, 2011). Pathogens that have been found to localize within the testis/epididymis and elicit an inflammatory response include the causative agents of brucellosis, Japanese encephalitis and porcine rubulavirus/blue eye disease. Several systemic pathogens have also been identified that disseminate into the urogenital tract, so leading to shedding in the semen; these pathogens include the agents of African swine fever, Aujeszky's disease/pseudorabies, classical swine fever/hog cholera, chlamydiophilosis, foot-and-mouth disease, leptospirosis, porcine circovirus type 2-associated disease, porcine reproductive and respiratory syndrome and swine vesicular disease. Porcine parvovirus, which causes infectious infertility, is also included in this category.

Non-infectious causes

Environmental

With the increased use of confinement housing to address biosecurity concerns at boar studs, the most common environmental insult associated with boar reproductive failure is exposure to high ambient or 'heat stress' temperatures. At heat stress temperatures, the thermoregulatory mechanisms of the testis are compromised, causing disruptions in sperm cell development, differentiation and proliferation. Collectively, these events lead to spermiogram alterations and reduced fecundity (Wettemann *et al.*, 1976; Stone, 1982; Malmgren and Larsson, 1984). In the boar, possible heat stress events can occur starting at temperatures of 27°C with greater than 50% relative humidity (RH), and is highly probable when ambient temperatures exceed 29°C and 50% RH (Fig. 15.8). After the onset of a heat stress event, alterations to the spermiogram (i.e. reduced viability, low sperm motility, increased abnormal sperm morphology, decrease in total sperm numbers) first appear approximately 10–14 days post exposure, and can remain present for 6-plus weeks thereafter. The duration of the altered spermiogram may extend beyond a single 44–47 day spermatogenic/epididymal maturation cycle (Swierstra, 1968) if the length and severity of the heat stress insult is prolonged. Additional factors that may contribute to eliciting a heat stress event include boar genotype (McNitt and First, 1970) and age at time of insult (Huang *et al.*, 2010).

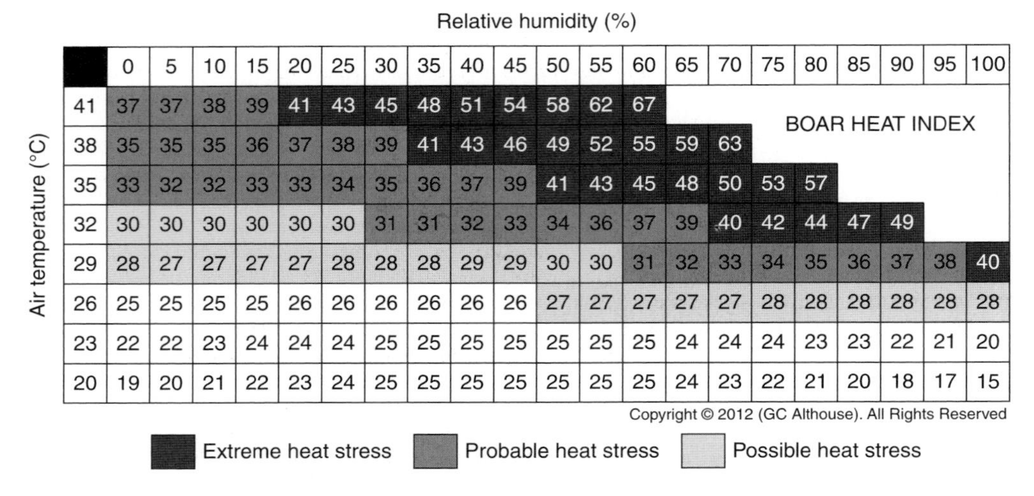

Air temperature (°C) \ Relative humidity (%)	0	5	10	15	20	25	30	35	40	45	50	55	60	65	70	75	80	85	90	95	100
41	37	37	38	39	41	43	45	48	51	54	58	62	67								
38	35	35	35	36	37	38	39	41	43	46	49	52	55	59	63						
35	33	32	32	33	33	34	35	36	37	39	41	43	45	48	50	53	57				
32	30	30	30	30	30	30	31	31	32	33	34	36	37	39	40	42	44	47	49		
29	28	27	27	27	27	28	28	28	29	29	30	30	31	32	33	34	35	36	37	38	40
26	25	25	25	25	26	26	26	26	26	26	27	27	27	27	27	28	28	28	28	28	28
23	22	22	23	24	24	24	25	25	25	25	25	25	25	24	24	24	23	23	22	21	20
20	19	20	21	22	23	24	25	25	25	25	25	25	25	24	23	22	21	20	18	17	15

Fig. 15.8. Boar heat index showing chance of a heat stress event based upon ambient temperature and relative humidity (reprinted with permission).

Producers should make a concerted effort to house boars in areas that maintain ambient temperatures below 27°C and 60% RH. Depending upon a stud's locale and the weather conditions, boar housing environmental temperatures may be best controlled using evaporative cooling cells or an HVAC (heating, ventilation and air conditioning) system. Additional measures common to enhancing conduction and convection of body heat from the working boar include programmable, thermostatically controlled mechanical ventilation (to increase air flow velocity across the boar) and whole body sprinkler systems.

Boars seem to be tolerant of low (–15 to –20°C) ambient temperatures if protected from the elements (wind, rain, snow, etc.). In confinement operations, ambient temperatures should not be allowed to go below 13°C (G.C. Althouse, unpublished). If boars have outdoor access, efforts should be made to provide for adequate protection from the elements (by provision of a windbreak, shelter, deep bedding, etc.).

Nutrition

As they are omnivores, components utilized in the formulation of boar diets can vary widely and usually favour feedstuffs that are locally available. Because of the importance of providing adequate nutrition in order to optimize boar productivity, diet cost does not tend to be an overwhelming factor when formulating a boar diet. An adult boar's nutritional needs are different from those of gestating or lactating sows, so a boar-specific diet is recommended. A typical ration for boars standing at an AI stud will contain approximately 3000 ME (metabolizable energy) kcal/kg, 16–18% crude protein (0.85% lysine), 4% fat and 5% fibre (Althouse, 2003; Table 15.2). With this type of ration, the boar will need to consume 1.5–3.0 kg feed daily with the goal of providing adequate nutrition to meet the requirements for maintenance and reproductive function. Thermoneutrality for the boar is believed to be at 20°C (Kemp *et al.*, 1989), so daily amounts of feed need to be adjusted to address changes in season/environmental conditions in order to achieve maintenance requirements. In some situations, diet formulation (i.e. energy density) may need to be modified – depending upon season – to deliver a balanced diet while reducing the total amount of feed provided. To assess whether nutritional needs are being met in individual animals, the most frequently used method is body condition scoring. Depending upon genotype and conformation, boars should be maintained at a body condition score of 3–4 (on a scale of 1–5). Overfeeding and excess body conditioning should be avoided as it can lead to foot and leg problems, which can have a negative impact on boar reproduction (Wilson, 2000).

Because cereal grains contribute substantially to the make-up of a boar's diet, vigilance in minimizing the chance of mycotoxicosis is crucial to boar health and normal spermatogenesis. Mycotoxicosis is caused by the presence of toxic compounds, known as mycotoxins, which are produced by fungi or moulds that colonize cereal crops. Mycotoxins can be produced either pre or post crop harvest. Factors that favour mycotoxin formation include crop stress, weather conditions and production practices. Mycotoxicosis in a boar is dependent upon the type of mycotoxin(s) consumed, the amount consumed and the duration of exposure. Mycotoxins of current concern to boar health include aflatoxin, ergot, fumonisin, trichothecene toxins (e.g. ochratoxin, T-2 toxin), vomitoxin (deoxynivalenol) and zearalenone. Screening of source cereal grains for these mycotoxins is recommended as a prudent measure.

Boars should have *ad libitum* access to potable water. Limited access to water, or access to only poor quality water, can affect boar health and productivity. Depending upon the region and source, periodic (e.g. at least annual) screening may be necessary to confirm the potability of the water supplied to boars.

Table 15.2. A typical ration for boars standing at an artificial insemination (AI) station (Althouse, 2003)[a].

Nutrient	Typical	Range
Calcium (%)	0.92	0.70–1.10
Fat (%)	3.9	2.9–4.67
Fibre (%)	5.0	2.8–7.5
Metabolizable energy (kcal/kg)	3,076	2,900–3,267
Protein (%)	16.2	12.0–21.5
Salt (%)	0.45	0.27–0.53
Selenium (ppm)	0.30	0.15–0.30
Total lysine (%)	0.85	0.43–1.30
Total methionine + cysteine (%)	0.49	0.46–0.71
Total phosphorus (%)	0.78	0.60–1.04
Vitamin A (IU/kg)	11,000	4,000–12,760
Vitamin D (IU/kg)	1,350	200–2,000
Vitamin E (IU/kg)	56	22–110
Zinc (ppm)	150	50–265

[a]Reprinted with permission from: Althouse GC (2003) Management of reproduction in pigs – the boar and artificial insemination. In *Animal Health and Production Compendium* 2003 edition. CAB International, Wallingford, UK.

Management

Whether in a natural mating or an AI breeding programme, efforts should be made to use boar power efficiently. The optimization of a particular boar for breeding is dependent upon the available boar power, genotype, libido, season, animal age and health. Disregarding any of these factors can lead to boar overuse or underuse, with a consequent negative impact on reproductive performance. Overuse tends to be the problem that most frequently occurs in a breeding programme. Within a 2 to 3 week period, boar overuse can lead to decreases in total sperm numbers, semen quality and subsequent fertility (Swierstra and Dyck, 1976; Hemsworth *et al.*, 1983; Cameron, 1987). The negative outcomes of boar overuse are usually observed earlier in an AI boar than a naturally mated boar because ejaculate compensability is lost with the ejaculate dilution and reduced sperm numbers that are used in an insemination dose.

Additionally, young boars (<8 months of age) exposed to overuse conditions early in their careers at stud tend to have chronic

Table 15.3. Suggested semen collection frequencies for artificial insemination (AI) boars based upon boar age.

Boar age (months)	Semen collection frequency
<8	Once every 7–10 days
8–10	Once every 7 days
10–12	Up to three times in a 14 day period
12+	Up to two times in a 7 day period

sub-fertility problems, leading to premature culling from the herd (Leman and Rodeffer, 1976). To avoid both boar overuse and underuse, appropriate boar power needs to be available (G.C. Althouse, unpublished). Under natural mating conditions, a male:female ratio of 1:15 to 1:25 should be available. For boars used exclusively in AI programmes, a male:female ratio of 1:150 to 1:250 is a desirable target. Table 15.3 provides data on suggested semen collection frequencies based upon boar age at an AI stud. At a minimum, boars should be used once every 10–14 days.

References

Althouse GC (1998) Cytoplasmic droplets on boar sperm cells *Journal of Swine Health and Production* **6** 128.

Althouse GC (2003) Management of reproduction in pigs – the boar and artificial insemination In *Animal Health and Production Compendium* 2003 edition pp 1–38 Ed G Richards. CAB International, Wallingford, UK.

Althouse GC (2008) Sanitary procedures for the production of extended semen *Reproduction in Domestic Animals* **43** 374–378.

Althouse GC and Rossow K (2011) The potential risk of infectious disease dissemination via artificial insemination in swine *Reproduction in Domestic Animals* **46** 64–67.

Althouse GC, Bruns KA, Evans LE, Hopkins SM and Hsu WH (1995) A simple technique for the purification of plasma membranes from ejaculated boar spermatozoa *Preparative Biochemistry* **25** 69–80.

Bach VS, Neundorf P, Stemmier KH, Mudra K and Ueckert H (1982) Hohe und Bewertung des anteiles anomaler Spermien beim Eber [Number of abnormal sperm in boars and its interpretation] *Monatshefte für Veterinärmedizin* **37** 463–467.

Cameron RDA (1987) Sexual development and semen production in boars *Pig News and Information* **8** 389–396.

Clark SG, Schaeffer DJ and Althouse GC (2003) B-mode ultrasonographic evaluation of paired testicular diameter of mature boars in relation to average total sperm numbers *Theriogenology* **60** 1011–1023.

Collins ED, Flowers WL, Shanks RD and Miller DJ (2008) Porcine sperm zona binding ability as an indicator of fertility *Animal Reproduction Science* **104** 69–82.

Douglas-Hamilton DH, Smith NG, Kuster CE, Vermeiden JPW and Althouse GC (2005) Capillary-loaded particle fluid dynamics: effect on estimation of sperm concentration *Journal of Andrology* **26** 115–122.

França LR, Silva Junior VA, Chiarini-Garcia H, Garcia SK and Debeljuk L (2000) Cell proliferation and hormonal changes during postnatal development of the testis in the pig *Biology of Reproduction* **63** 1629–1636.

Hacker RR, Du Z and D'Arcy CJ (1994) Influence of penning type and feeding level on sexual behavior and feet and leg soundness in boars *Journal of Animal Science* **72** 2531–2537.

Hemsworth PH, Winfield CG and Hansen C (1983) High mating frequency for boars: predicting the effect of sexual behaviour, fertility and fecundity *Animal Production* **37** 409–413.

Huang YH, Lo LL, Liu SH and Yang TS (2010) Age-related changes in semen quality characteristics and expectations of reproductive longevity in Duroc boars *Animal Science Journal* **81** 432–437.

Kemp B, den Hartog LA and Grooten HJG (1989) The effect of feeding level on semen quantity and quality of breeding boars *Animal Reproduction Science* **20** 245–254.

Ko JCH, Evans LE and Althouse GC (1993) Toxicity effects of latex gloves on boar spermatozoa *Swine Health and Production* **1** 24–26.

Kuster CE and Althouse GC (2007) Reproductive physiology and endocrinology of boars In *Current Therapy in Large Animal Theriogenology* 2nd edition pp 717–721 Ed RS Youngquist and WR Threlfall. Saunders Elsevier, St Louis.

Kuster CE, Singer RS and Althouse GC (2004) Determining sample size for the morphological assessment of sperm *Theriogenology* **61** 691–703.

Leman AD and Rodeffer HE (1976) Boar management *Veterinary Record* **98** 457–459.

Malmgren L and Larsson K (1984) Semen quality and fertility after heat stress in boars *Acta Veterinaria Scandinavica* **25** 425–435.

McNitt JI and First NL (1970) Effects of 72-hr heat stress on semen quality in boars *International Journal of Biometeorology* **14** 373–380.

Safranski TJ (2008) Genetic selection of boars *Theriogenology* **70** 1310–1316.
Stone BA (1982) Heat induced infertility of boars: the interrelationship between depressed sperm output and fertility and an estimation of the critical air temperature above which sperm output is impaired *Animal Reproduction Science* **4** 283–299.
Swierstra EE (1968) Cytology and duration of the cycle of the seminiferous epithelium of the boar; duration of spermatozoan transit through the epididymis *Anatomical Record* **161** 171–186.
Swierstra EE and Dyck GW (1976) Influence of the boar and ejaculation frequency on pregnancy rate and embryonic survival in swine *Journal of Animal Science* **42** 455–460.
Vazquez JM, Martinez EA, Martinez P, Garcia-Artiga C and Roca J (1997) Hypoosmotic swelling of boar spermatozoa compared to other methods for analyzing the sperm membrane *Theriogenology* **47** 913–922.
Wettemann RP, Wells ME, Omtvedt IT, Pope CE and Turman EJ (1976) Influence of elevated ambient temperature on reproductive performance of boars *Journal of Animal Science* **42** 664–669.
Wilson ME (2000) Nutritional effects on boar semen production In *Proceedings of the 4th International Conference on Boar Semen Preservation* pp 193–198 Ed LA Johnson and HD Guthrie. Allen Press, Inc., Lawrence, Kansas.

16 Applied Andrology in Camelids

Ahmed Tibary,[1]* Lisa K. Pearson[1] and Abelhaq Anouassi[2]
[1] *Washington State University, Pullman, Washington, USA;*
[2] *Veterinary Research Centre, Abu Dhabi, United Arab Emirates*

Introduction

The reproductive system of the male camelid presents several anatomical and physiological peculiarities compared with other domestic species. Understanding these peculiarities is a prerequisite for thorough breeding soundness evaluation, male infertility investigation and research on sperm biotechnologies of camelids.

The family Camelidae includes six species, two Old World Camelids (OWC) and four New World Camelids (NWC) (aka South American Camelids, SAC). The OWC are *Camelus dromedarius* (the dromedary or single humped camel) and *C. bactrianus* (the Bactrian or double-humped camel), which are more adapted to the deserts of central Asia, the Arabian Peninsula and Africa than to other regions. The NWC are represented by two domesticated species, the llama (*Lama glama*) and the alpaca (*Vicugna pacos*), and two wild species, the guanaco (*L. guanacoe*) and the vicuña (*V. vicugna*). Each of the domesticated species of camelid presents a variety of 'breeds' or 'types' that presents specific production characteristics (pack animal versus the production of fibre, meat or dairy products, etc). Despite great differences in their phenotypic appearance, these species share many genetic and biological characteristics. However, because of the differences in their evolution, they have developed slightly different reproductive patterns. All species of domesticated camelids are considered to be important production animals in their respective native geographic areas. Alpacas and llamas in North American and European countries, and racing camels in some countries of the Middle East, benefit from a high standard of individual veterinary medical care owing to their sentimental or economic value.

This chapter reviews the major aspects of the reproductive anatomy, physiology and common abnormalities of the male camelid. Breeding soundness examination is emphasized, with special reference to semen characteristics and factors that may affect its quality. The chapter is largely inspired by previous complete reviews by the authors (Tibary and Vaughan, 2006; Tibary *et al.*, 2007; Tibary, 2008), to which the reader is referred for more complete references.

Anatomy of the Reproductive Tract

The male reproductive organs have been well described for the llama, alpaca

* E-mail: tibary@vetmed.wsu.edu

(Collazos, 1972; Delhon and Lawzewitsch, 1987, 1994; Tibary and Anouassi, 1997a; Fowler, 1998), guanaco and vicuña (Delhon *et al.*, 1983; Urquieta and Rojas, 1990; Urquieta *et al.*, 1994), as well as the dromedary (El-Wishy, 1988; Tibary and Anouassi, 1997a) and Bactrian camel (Tibary and Anouassi, 1997a).

Scrotum and testicles

The mechanism and chronology of testicular descent remains unstudied in the Camelidae. In most males, the testes are present in the scrotum at birth, but they are usually soft and difficult to palpate. According to some authors, testicular descent is not complete until the second or third year of life in the dromedary and Bactrian camel.

The scrotum is non-pendulous and situated high in the perineal region at the level of the ischial arch. The testicles are relatively small compared with those of other domestic livestock and are directed caudo-dorsally (Tibary and Vaughan, 2006; Tibary *et al.*, 2007). Season does not seem to affect the size of the testes and scrotum in llamas and alpacas. In the vicuña, testicular size is greater in the summer than in the winter (Urquieta *et al.*, 1994). There is great seasonal variation in testicular size in the dromedary and Bactrian camel.

Testicular size

Testicular size is an important parameter in the evaluation of the breeding potential of males and can be used to predict daily sperm production.

South American camelid testicles are ovoid and exhibit age-related size variation (Table 16.1). In adult llamas, the length, width and depth are respectively 5–7, 2.5–3.5 and 3–4 cm (Fowler, 1998). In the alpaca, the length and width of the testes are 4–5 and 2.5–3, cm respectively. In the vicuña, longitudinal and transverse testes diameters are, respectively, 3.3 and 1.69 cm in summer and 2.64 and 1.50 cm in winter (Urquieta and Rojas, 1990; Urquieta *et al.*, 1994). The average individual testicular weight recorded in 3.5-year-old llamas (average body weight of 133 kg) is 24 g (Johnson, 1989).

In the dromedary, reported testicular dimensions vary greatly according to different studies. This variation is probably due to the effect of age, sexual activity and even the 'breed' or strain of camels studied. In India, the long axis and diameter of the dromedary testes range respectively from 6 to 13 cm and 3 to 6 cm. Smaller values were reported for dromedaries in Egypt. The mean testicular length and width are respectively 9.3 ± 0.7 and 4.5 ± 0.4 cm in the rainy season and 9.5 ± 0.6 and 4.5 ± 0.5 cm in the dry season in Nigeria (Djang *et al.*, 1988). The mean scrotal circumference reported for 197 dromedaries was 32.4 ± 2.4 cm

Table 16.1. Mean (range) of testicular size (length × width) and weight in alpacas, llamas and vicuñas. Adapted from Sumar, 1983; Fowler, 1998; and Bravo *et al.*, 2002.

Age (months)/ Sires (source)	Alpaca (*n* = 158)		Llama (*n* = 54)		Vicuña (*n* = 6)
	Size (cm)	Weight (g)	Size (cm)	Weight (g)	Size (cm)
6	1.0 × 0.4	0.6	2.4 × 1.4	–	0.7 × 0.3
12	2.3 × 1.5	2.9	3.4 × 2.3	5.1	1.1 × 0.7
18	2.8 × 1.9	6.6	3.5 × 2.6	14	1.5 × 0.8
24	3.3 × 2.2	9.9	3.9 × 2.3	17.4	2.1 × 1.3
30	3.6 × 2.3	13.9	4.4 × 2.5	17.8	–
36	3.6 × 2.4	13.6	4.5 × 2.7	18.2	2.5 × 1.4
Sires (Fowler *et al.*, 1998)	3.7 × 2.4	17.2	5.4 × 3.3	–	3.3 × 1.9
Sires (Sumar, 1983)	4.0 (3.2–4.8) × 2.6 (1.9–3.2)	17.7 (13.5–28.0)	–	–	–

and this was positively correlated with age. As a general rule, the average testicular length, width and depth are 9.1, 5.1 and 4.3 cm, and the average weight is 92 g for male dromedaries 3 years of age and older. In contrast to domestic NWC, camels exhibit a large seasonal variation in testicular size (Tibary and Anouassi, 1997a,c; Tibary *et al.*, 2007). Reference ranges for testicular length, depth and breadth have recently been determined for Egyptian dromedaries of various age groups (Table 16.2) (Derar *et al.*, 2012).

Reported testicular weights in adult dromedaries show large variations (32–225 g) (Tingari *et al.*, 1984; Tibary and Anouassi, 1997a). These variations may be due to type of dromedary, season and nutritional conditions (Akingbemi and Aire, 1991).

Testicular size during the breeding season (usually the coolest or rainy months of the year) is 150 to 200% of that recorded outside the breeding season (during the hottest or driest months of the year) (Tibary and Anouassi, 1997a,c). In a study in India, the effect of season on the weight of the testicle was more pronounced in dromedaries between 9 and 14 years of age than in dromedaries younger than 9 or older than 14 years. The maximum weight of the testicle was observed during the period December to March (36–225 g) compared with the periods of April to July (32–181 g) and August to November (32–191 g). Season also has a marked effect on the consistency of the testis. During the breeding season, dromedary testicles are usually turgid but they become soft during the period of sexual rest. There is a need for further characterization of testicular descent and testicular growth in various breeds of camel.

Testicular histology

The histomorphology of the camelid testis is similar to that of other domestic livestock (Flores, 1970; Delhon and Lawzewitsch, 1987). The diameter of the seminiferous tubules varies from 174 to 240 μm (Tibary and Anouassi, 1997a).

In the camel, the interstitial tissue occupies a larger area than the seminiferous tubules, particularly during the winter months, and seems to be the major factor in the seasonal variation of testicular size and weight. The increased amount of interstitial tissue observed during the breeding season is due to an increase in the number and volume of Leydig cells (Tibary and Anouassi, 1997b).

The outer diameter of the seminiferous tubules in adult dromedaries varies according to the study and has been found to be in the range of 113 to 250μm. This variability could reflect an effect of the type of dromedary studied and/or the effect of age and season. The diameter of the seminiferous tubules is greatest during the breeding season (Tibary and Anouassi, 1997a). Early studies in the dromedary indicated a complete arrest of spermatogenic activity during the summer season. However, later studies showed the presence of spermatogenesis throughout the year (Tibary *et al.*, 2007).

Blood is supplied to the testicle primarily via the testicular artery. Along its trajectory, the testicular artery gives off several branches that supply blood successively to the testicular cord envelopes, the epididymis and the testicular parenchyma. The most distinct features of the artery during its trajectory are its enlargement as it approaches the testis and the distribution of its branches. The epididymal branch arises from the upper part of the vascular cone (the pampiniform plexus) and

Table 16.2. Mean ± SEM of testicular length, depth and breadth (width, cm) in Egyptian dromedaries measured by calipers (Derar *et al.*, 2012).

	Left testicle			Right testicle		
Age (years)	Length	Width	Depth	Length	Width	Depth
1.5–3	3.7 ± 0.3	2.1 ± 0.2	1.6 ± 0.2	3.7 ± 0.3	1.9 ± 0.2	1.5 ± 0.2
4–6	6.6 ± 0.2	2.9 ± 0.1	2.6 ± 0.2	6.3 ± 0.2	2.8 ± 0.1	2.4 ± 0.2
7–12	7.8 ± 0.5	3.7 ± 0.1	3.2 ± 0.1	7.7 ± 0.5	3.5 ± 0.1	3.0 ± 0.1

runs along the body of the epididymis, which it supplies before dividing into several branches at the level of the cauda epididymis, which anastomose with the deferential artery (artery of ductus deferens). The head of the epididymis is supplied by branches of the testicular artery given off at the level of the pampiniform plexus.

Epididymis

The epididymis is composed of three distinct parts: the caput (head), corpus (body) and cauda (tail), as illustrated by Plates 46 and 47, which show a dromedary testicle and epididymis. The epididymis faces laterally along the dorsal border of the testis, with the head curving around the cranial pole of the testis. It is adherent to the cranial surface (caput) and both poles of the respective testis, and allows the corpus epididymis to form a sinus with the testis situated externally. In llamas, all three parts of the epididymis can be palpated, although in most instances only the small tail of the epididymis is easily palpable dorsal to the testis. In the dromedary, the cauda epididymis is round and well protruded – about 3–4 cm above the respective extremity of the testis. The total weight of the epididymis is approximately 10–40 g, and the weight ratio between the testis and the epididymis varies from 3:1 to 6:1. (Tibary and Anouassi, 1997a; Tibary *et al.*, 2007).

Histological and histochemical studies on llamas defined six segments of the epididymis (Delhon and Lawzewitsch, 1994). These regional differences represent different secretory functions that may play a role in the process of sperm maturation (Delhon and Lawzewitsch, 1994). On the basis of histological appearance, the dromedary epididymis is divided, into three segments: initial, middle and terminal. The middle segment is further divided into proximal, intermediate and distal parts.

Ductus deferens

In the llama, the ductus deferens is 1 mm in diameter at the junction of the epididymis. It widens to 2 mm in the abdominal cavity towards the pelvic urethra (ampulla). The length of the ductus deferens is about 40 cm (Smith *et al.*, 1994).

In the dromedary, the ductus deferens enters the abdominal cavity via the medial angle of the inguinal canal then turns caudally in the genital fold, where it becomes straight and thick, forming the ampulla; it then passes deep into the prostate and opens directly into the colliculus seminalis via an ostium ejaculatorium (ejaculatory orifice) (Tibary and Anouassi, 1997a).

Prepuce and penis

The prepuce is located in the inguinal region. It is flattened from side to side and triangular in shape when viewed laterally. The prepuce adheres to the glans penis until 2 or 3 years of age, making exteriorization of the penis impossible in young males. In the absence of sexual stimulation, the small preputial orifice (ostium praeputiale) is directed caudally. The prepuce has a well-developed muscular apparatus, consisting of the cranial, the lateral and the caudal preputial muscles. These muscles allow movement of the preputial orifice cranially during erection and mating behaviour (Tibary and Anouassi, 1997c; Fowler, 1998).

In the dromedary, the preputial skin is usually darker in colour than that in the rest of the body. It is covered with short fine hair and presents two nipples on either side of the base of the prepuce, near its caudal border. In males that are used intensively for breeding, a callus may develop at the lower part of the cranial curvature. The average length of the base, cranial and caudal borders is 28.0, 17.5 and 19.7 cm, respectively (Tibary and Anouassi, 1997a).

The camelid penis is fibroelastic. In the absence of erection, the penis is retracted into its sheath via a pre-scrotal sigmoid flexure. The length of the penis ranges from 35 to 45 cm in llamas and alpacas, and from 59 to 68 cm in camels. The penis is cylindrical, gradually decreasing in diameter from its root at the ischial arch (1.2–2 cm in the llama; 2.2 cm in the dromedary) to the neck of the glans penis (collum glandis, preputial reflection), which is 0.8–1 cm in diameter in the llama and 0.4–1 cm in the dromedary) (Tibary and Anouassi, 1997a; Fowler, 1998).

The glans penis is long (9–12 cm) and the distal tip consists of a cartilaginous process that has a slight clockwise curvature (Plate 48). The end of the urethra is located at the base of the cartilaginous process, not at the tip. The curved nature of the cartilaginous process of the camelid penis allows both penetration of the cervical rings, through combined rotational and thrusting movements, and the intrauterine deposition of semen. The retractor penis muscles (Musculus retractor penis) continue on the ventral aspect of the penis and are attached to the ventral convexity of the sigmoid flexure of the penis. There is a dorsal urethral diverticulum at the level of the pelvic symphysis that prevents passage of a catheter into the bladder (Tibary and Anouassi, 1997a; Fowler, 1998).

Accessory sex glands

The most notable feature in the anatomy of the internal genitalia of camelids is the absence of seminal vesicles (Tibary and Anouassi, 1997a). The prostate is usually described as a small H-shaped gland firmly attached to the dorsolateral aspect of the pelvic urethra near the trigone of the bladder. Its ventral aspect is slightly concave. Numerous prostatic ducts direct the secretion of this gland directly into the colliculus seminalis. The length, width and height of the prostate are respectively 3, 3 and 2 cm in the llama and 3, 2 and 2 cm in the alpaca.

The two bulbourethral glands are spherical to ovoid and located on the dorsolateral aspect of the pelvic urethra, above and just cranial to the ischial arch. The bulbourethral gland diameter is 1.5 to 2 cm in the llama and 0.8 to 1.5 cm in the alpaca (Tibary and Vaughan, 2006).

In the dromedary, the size and weight of all glands vary significantly and tend to reach a maximum at 10.5 to 15 years of age. During the breeding season, the average weights of the ampullae, prostate and bulbourethral glands are respectively 3.0, 12.0 and 3.6 g for animals between 4.5 and 6 years of age, and 4.2, 16.1 and 4.9 g for males aged 10.5 to 15 years. Size and weight of the accessory sex glands show a large seasonal variation. Maximum gland weight is recorded during the breeding season (winter and spring) and the minimum values are recorded during summer (Tibary and Anouassi, 1997a,b).

The poll glands of the dromedary and Bactrian camel

One of the most noticeable anatomical characteristics related to reproduction in camels is the presence of modified sweat glands called poll glands (Taha *et al.*, 1994). In the adult, they are oval and slightly elevated skin areas situated about 5 cm below the apex of the head, on the back of the poll, on each side of midline (Plate 49). The poll glands are identified as two dark spots, especially during the rutting season. This colour is due to oxidation of their secretions, which increase in amount during the rutting season. Activity of the poll glands is highly seasonal and closely follows testicular activity (Tibary and Anouassi, 1997b).

Reproductive Physiology

Puberty and sexual maturation

Studies on puberty in the male camelid are scarce. Reported age at puberty is variable and may reflect genetic, nutritional and climatic changes, as well as the effect of season of birth. Male llamas and alpacas can display mounting behaviour at a young age (<1 year) (Sumar, 1985). However, complete erection and intromission is only possible when the penis is completely freed from its preputial attachments. The process of preputial detachment usually begins at the age of 12–13 months and coincides with an increase in plasma testosterone concentration (Bravo and Johnson, 1994). Plasma testosterone concentrations in 11-month-old alpaca males are similar to those found in adult males (Losno and Coyotupa, 1979). In llamas, plasma testosterone is

<90 pg/ml from birth until 20 months old, and then increases to more than 1000 pg/ml. Preputial adhesions are lost in up to 10% of males by 1 year of age, in 70% of males by 2 years and in all males by 3 years of age, which corresponds to the recommended age for breeding (Sumar, 1996; Tibary and Vaughan, 2006). The variation in age at which penile preputial attachments are lost may be partially explained by plane of nutrition as there is a correlation between body size and mean testicular length, although the wide variation in testicular size at any one age or body size suggests that other factors, probably genetic, are also important (Galloway, 2000).

Appearance of a lumen in the seminiferous tubules and the presence of spermatozoa were reported respectively at 12 months and 15–18 months of age in the alpaca (Montalvo *et al.*, 1979). Other authors have reported that sperm production starts as early as 10–12 months of age in a few male alpacas and llamas, and is usually present by the age 1.5–2 years (Smith *et al.*, 1994). Testicular growth in these species is slow and maximum size is not reached until 3 years of age (Bravo and Johnson, 1994). The male llama is not mature until 2.5–3 years of age but pregnancies have resulted from younger males (Johnson, 1989). Alpacas reach sexual maturity at 5 years of age, which corresponds to a body weight of 62.5 kg. The testicles at this age should be at least 4–5 cm in length and 2.5–3 cm in width (Sumar, 1996).

In the wild, guanaco and vicuña males may reach puberty at an early age, but they live in a bachelor group until they reach maturity between 4 and 6 years of age. At this point, they start seeking a harem and become territorial. In a study of two animals, the testes of a 16-month-old vicuña male were aspermatogenic, whereas those of 2-year-old male were producing spermatozoa (Koford, 1957).

The age at puberty in the dromedary, is poorly defined. Display of sexual behaviour has been reported as early as 2 years of age, but field observations by the authors suggest that sperm production and fertilizing ability are not achieved until 3–4 years of age for this camel (Rahim, 1997). Indian dromedaries seem to be slower in reaching puberty than their Arabian or African counterparts (Khan and Kohli, 1972; Sharma, 1981; Tibary and Anouassi, 1997b; Al-Qarawi *et al.*, 2001; Tibary *et al.*, 2005, 2007).

In the traditional management of dromedary herds, males are not used for breeding until 5–6 years of age (Abdel Raouf *et al.*, 1975; Azouz *et al.*, 1992). This allows maximum fertility as both sperm production and interstitial tissue (Leydig cell) activity reach a plateau at 6 years old, which corresponds to the age of sexual maturity according to most authors. Normal sexual activity continues until 20 years old, when some males begin to show signs of senile changes in their sexual behaviour or sperm production. In the authors' experience, breeding males have been successfully used up to 25–28 years of age (Tibary and Anouassi, 1997b).

In the Bactrian camel, puberty is believed to occur around 4 years of age, though some males may enter breeding activity as early as 3 years old (Chen *et al.*, 1980). Sexual maturity is reached at an age similar to that observed in the dromedary (5–6 years) and then declines after 15 years of age, although some males continue to breed until they are 20 years old (Chen *et al.*, 1980).

Seasonality

The seasonality of reproductive activity in male camels is widely accepted based on several behavioural and endocrinological studies, as well as on the basis of reproductive patterns in the wild (Novoa, 1970; Mukasa-Mugerwa, 1981). Given the wide geographical distribution of the dromedary, the breeding season is highly variable but, in general, it coincides with the period of lower temperature, lower humidity and increased rainfall (Table 16.3) (Gombe and Oduor-Okelo, 1977; Marie, 1987; Tibary *et al.*, 2007). Seasonality in the male is evidenced by changes in sexual behaviour, morphology and function of the genital and associated organs, and endocrinological profiles (Bedrak *et al.*, 1983; Friēländer *et al.*, 1984; Marie, 1987; Azouz *et al.*, 1992; Deen *et al.*, 2005; Zia ur *et al.*, 2007; Deen, 2008; Riaz *et al.*, 2011).

Table 16.3. Reported breeding season in camels (*Camelus* spp.) by country.

Species	Country	Breeding season
C. dromedarius	Egypt	March to April
		March to May
	India	Mid-September to mid-February
		November to February
		October to March
	Israel	January to April
	Kenya	November–May (with peak in rainy seasons November–December, March–May)
	Morocco	December to May
	Nigeria	Wet season
	Pakistan	December to March
	Saudi Arabia	November to July
		November to February
	Somalia	October to March
	Sudan	October to March
	Tunisia	November to April
	Turkmenistan	Mid-January to the end of May
C. bactrianus	China	Mid-December to mid-April

Seasonality is more pronounced in the Bactrian camel, with the rutting season lasting throughout the winter months and into early spring. Dominant males demonstrate reproductive activity earlier than young or subordinate males. Individual variations exist, and each male can exhibit breeding activity lasting from 50 to 100 days (Chen *et al.*, 1980).

Spermatogenesis and sperm production

Spermatogenesis is similar to that described for other species (Tibary and Anouassi, 1997b). The cycle of the seminiferous tubule epithelium in dromedary camels and llamas consists of eight stages. Some differences exist between the two species in the percentage frequencies of cells representing each stage. In the dromedary, the most frequent stages of the seminiferous tubule epithelium cycle are stage 1 (22%) followed by stage 8 (16.4%), whereas in the llama, the most frequent cellular association is that of stage 8 (19%) followed by stage 3 (18%). The least frequent cellular association is that of stage 4 (8.2%) in the dromedary and stage 5 (5.8%) the llama. Stages 1–4 represent 54% of all the cellular associations in both species (Delhon and Lawzewitsch, 1987; Bustos-Obregon *et al.*, 1997).

The llama epididymis presents six different segments based on histological and histochemical characteristics. Three of these segments are located in the head of the epididymis (I, II, III), two are located in the body (IV and V), and the sixth includes the distal part of the body and the tail of the epididymis. The major differences among these segments are the height of the epithelial cells, the positive intraepithelial periodic acid-Schiff reaction (PAS), alkaline phosphatase activity (ALP) and lactate dehydrogenase activity (LDH).

Segment I represents a short region where efferent ductules enter the epididymal duct. Segments II and III are characterized by high epithelial cell mitotic activity and weak LDH activity. The epithelial lining increases in height from 50 μm in Segment I to 60–140 μm in Segment II, and then decreases progressively to reach an average of 67, 65, 35 and 27 μm for Segments III, IV, V and VI, respectively. Epithelial cells in Segment IV contain amylase and PAS-positive and neuraminidase-resistant secretory granules. Segment V is characterized by strong ALP and LDH activities. Segment VI is characterized by moderate ALP and high LDH activities, and is highly packed with spermatozoa. The PAS reaction is absent in all but Segment IV and is weak in Segment VI. ALP activity is present in the proximal part of Segment V and, to a lesser extent, in the distal part of Segment V and in Segment VI (Delhon and Lawzewitsch, 1994).

Epididymal transit of camelid sperm is associated with a shift in the position of the cytoplasmic droplet of the sperm, which, in the llama, becomes distal but remains present in more than 60% of the spermatozoa in the terminal segment. The cytoplasmic droplet is lost when each spermatozoon reaches the ductus deferens (Delhon and Lawzewitsch, 1994). There is a decrease in the percentage of abnormal spermatozoa (bent midpiece and tail) during epididymal transit.

Sperm production is correlated with testicular size in SAC (Table 16.4). Mean testicular length is also correlated with testicular weight and may be used as a simple means of assessing testicular size in alpacas (Galloway, 2000).

Spermatogenesis occurs throughout the year in all camelids and shows wide seasonal variation. Sperm production is affected by geographical location (including nutritional, climatic and other environmental factors), herd management, extent of domestication and social structure of the camelid group (El-Wishy, 1988; Tibary and Vaughan, 2006).

In South America, the size of testicles in the male vicuña is affected by season and is greater in summer than in winter (Urquieta and Rojas, 1990; Urquieta *et al.*, 1994). Spermatogenesis is not completely arrested during winter because all germ cell types are still observed at this time. The larger testicular size in summer has been attributed mainly to the increase in diameter of the seminiferous tubules and Leydig cells.

Table 16.4. Development of testicular function in alpacas with testicles of different sizes (Galloway, 2000).

Mean testicular length (cm)	Proportion of males (%)	% Testicular tissue producing elongated spermatids
<3	100	0
3–4	68	<10
	31	30–60
>4	36	<10
	31	10–60
	31	>60

In the male vicuña, individual testosterone values tend to be higher in summer, but some high values have also been noted in winter (Urquieta *et al.*, 1994). Seasonal changes in plasma testosterone concentrations have also been reported in male alpacas; samples collected in late summer had the highest monthly mean values, whereas those collected in winter had the lowest (Losno and Coyotupa, 1979). The higher plasma testosterone levels in vicuñas in summer are probably responsible for the behavioural changes observed in males during this period of the year. High levels of testosterone are consistently found in dominant males and young males that are trying to recruit their own harem (Urquieta *et al.*, 1994). Outside their normal habitat, llamas show a strong effect of season on sperm production, with low production in the summer (Gauly, 1997).

While many authors have not found any seasonal effects on spermatogenesis in the dromedary (Singh and Bharadwaj, 1978; Osman *et al.*, 1979), others have reported reduced spermatogenic activity during the non-breeding season (Abdel Raouf *et al.*, 1975). A steady drop of spermatogenic activity was observed in Egyptian dromedaries in the summer, manifested by an increasing number of exfoliated cells, increasing number and size of cytoplasmic vacuoles, and a decrease in sperm content. Spermatogenetic activity was at its lowest rate from June to August, and then increased in September, reaching a peak by November to January (Tingari *et al.*, 1984). Activity fell in March, and decreased steadily from April onwards. Presumably, camels living south of the equator would exhibit activity in different months, although in a similar climatic season (Tingari *et al.*, 1984).

The average daily sperm production in the dromedary was estimated at 0.751×10^9 for mature camels (8–10 years old; Ismail, 1982). The average gonadal sperm reserve and daily sperm production per gram of testicular parenchyma were estimated at $1.7–3.4 \times 10^9$ and $30.0–61 \times 10^6$, respectively

(Osman and El Azab, 1974; El Wishy and Omar, 1975; Ismail, 1982, 1988). This is a low sperm production rate compared with other species. The sperm production rate is at its highest during the breeding season (Osman and El Azab, 1974; Abdel Raouf *et al.*, 1975; Ismail, 1982; Tingari *et al.*, 1984; Osman and Plöen, 1986; Tibary *et al.*, 2007). Morphometric studies showed that testicular volume, weight of the testis, average diameter of Sertoli cells, volume of the intertubular compartment, relative and total volume and numbers of Leydig per testis, and percentage interstitial tissue in the parenchyma of the testis, were significantly higher during the winter and spring seasons (Riaz *et al.*, 2011). Sperm reserve increases with age until 10 years old and then remains constant until 15 years old (Abdel Raouf *et al.*, 1975; Ismail, 1982). A decrease in sperm production is noticed starting at 20 years old.

The epididymal sperm reserve of the dromedary averages 2.3 to 6.1 × 10^9 spermatozoa. Half to two thirds of this reserve is located in the body of the epididymis, with the head and tail contributing only 5.2–12.3 and 21–36%, respectively. The reported relative distribution of epididymal sperm differs according to the study (Akingbemi and Aire, 1991; Tibary and Anouassi, 1997b).

Epididymal transit and sperm maturation

Epididymal sperm transit is poorly studied in SAC. In the llama, as already noted, epididymal transit causes a shift in the position of the cytoplasmic droplet, which becomes distal. However, droplets were present in more than 60% of the spermatozoa in the terminal segment of the epididymis and were not completely lost until spermatozoa reached the ductus deferens (Delhon and Lawzewitsch, 1994).

In the dromedary, sperm transport through the epididymis lasts on average 4.3 days (0.22 days in the head, 2.5 days in the body and 1.5 days in the tail) (Ismail, 1982, 1988). The fertilizing capacity of the dromedary sperm is acquired during the transit in the head and body of the epididymis, which lasts about 2.9 days (Ismail, 1982). Morphological changes of the spermatozoa were observed as the spermatozoa moved through the epididymis, signifying a maturation process (Osman and Plöen, 1986). The dromedary contrasts with other domestic species in that a high number of epididymal spermatozoa are morphologically abnormal. The migration of the cytoplasmic droplet is slow and a large proportion of spermatozoa in the tail of the epididymis still have such a droplet (Osman and Plöen, 1986).

Mating Behaviour and Ejaculation

All male camelids display the flehmen response after smelling freshly excreted urine and faeces of the female. The head is lifted, the upper lip is elevated without curling and the mouth is opened slightly (Tibary and Anouassi, 1997b; Tibary, 2003). When introduced into a female herd, the male will chase a sexually receptive female and try to force her down by mounting her and putting pressure on her pelvis (llamas and alpacas) or neck (camels). Some males may protrude the penis and display rotation of the glans as they are attempting to mount. Erection and copulation occur with the female sitting in sternal recumbency. The penis is fully extended after successful vaginal penetration (Tibary and Anouassi, 1997b; Tibary, 2003).

Mating behaviour in South American Camelids (SAC)

During copulation, llamas and alpacas produce a low guttural sound known as 'orgling'. This sound is produced as air is expired through the mouth while the cheeks are inflated. Males can display all aspects of copulatory behaviour without achieving intromission (Fernandez-Baca, 1970).

The duration of copulation is variable, averaging 20–25 min and ranging from a few minutes to more than an hour (Table 16.5). There is no correlation between copulation time and conception rate (Vaughan, 2001). Factors affecting the length of copulation are species and breed, age of male and female,

Table 16.5. Mating duration in camelids.

Species	Mean (min)	Range (min)	Significant factors
Camelus bactrianus	12.2	1–20	Breed, mating frequency
Camelus dromedarius	5.5	1–22	Sire, age, season, breed, mating frequency, nutrition
Lama glama	20	5–65	Sire, age, mating frequency, female age
Vicugna pacos	25	5–50	Sire, age, mating frequency, female age

sire, season, frequency of mating and presence of other males (Bravo *et al.*, 1997c). Duration of copulation tends to be shorter for younger males and decreases in all males with increased frequency of mating (Vaughan, 2001; Vaughan *et al.*, 2003a,b; Tibary and Vaughan, 2006). Length of copulation is longer for multiparous females (21.5 min) than for maiden females (14.7 min). Also, when several males are present in the same herd, copulation is shorter than when no other males are present (8 versus 21.5 min) (Smith *et al.*, 1994).

During paddock mating, a male may attempt to breed with the same female several times until she ovulates and becomes sexually non-receptive. Male llamas with good libido can breed with up to 18 females in a day for the first 4 to 5 days after introduction into a group of sexually receptive females. Mating activity decreases sharply after the first 2 weeks, when males lose interest in all females. Some 70% of the breeding takes place in these first days, with a pregnancy rate of 50% achieved by the end of the mating period (Sumar, 1985). It has been recommended to change males every 2 weeks so that all females can be bred during a short interval of time (Sumar, 1985) or to alternate paddock mating (5 days) and male rest (2 days) to keep male libido high and allow accurate determination of the parentage of offspring. In alpacas, length of copulation and fertility are negatively affected when the number of matings per male increases to more than four times a day (Bravo and Johnson, 1994).

Alpaca males exhibit different sexual behaviour patterns if they are maintained with females all year round from when they are run in separate flocks. In the former case, they show distinct seasonal variations in activity. In the latter case, they show continuous libido and breeding capability, serving females whenever they have an opportunity (Fernandez-Baca *et al.*, 1972); higher plasma testosterone levels have been found in this situation (Losno and Coyotupa, 1979), although no related changes in fertility rate have been reported (Urquieta *et al.*, 1994). Aggressive behaviour towards other males has been observed, culminating in a direct confrontation with biting, and also chest ramming, which can lead to serious injuries, especially if the canine teeth are well developed (Tibary, 2003).

Ejaculation in camelids appears to occur throughout the entire duration of copulation (Tibary and Anouassi, 1997b; Tibary and Vaughan, 2006). Sperm is present in the ejaculate of alpacas within 5 min of the initiation of copulation (Fernandez-Baca, 1970; Kubicek, 1974; Lichtenwalner *et al.*, 1996a). In the llama, urethral pulses increase in frequency 4 min after the start of copulation and occur in clusters every minute. Each cluster lasts 20 s and is composed of four to five rapid urethral pulses followed by a tremor of the whole body. Each cluster is preceded by two repositions and 38 pelvic thrusts. The urethral pulses accompanying the whole body tremor are considered to be a single ejaculation (Lichtenwalner *et al.*, 1996a, 1997). Thus, ejaculation in the llama starts approximately 4 min after the beginning of copulation and occurs every minute (18 to 19 ejaculations/22 min). These observations (multiple ejaculation and tremors during ejaculation) have been confirmed by the pattern of ejaculation observed during semen collection using an artificial vagina (Lichtenwalner *et al.*, 1996a,b, 1997). Semen is deposited deep in the uterine horns and most likely at the papilla of the utero-tubal junction (Bravo *et al.*, 2002; Tibary and Vaughan, 2006). Intrauterine deposition of semen helps to compensate for a low sperm concentration and plays a major role in induction of ovulation.

Rutting behaviour in Old World Camelids (OWC)

The onset of the breeding season is accompanied by an increased aggressiveness of the male dromedary towards other animals in the herd (especially other males) and sometimes even towards people. Confined males show increased pacing and anxiety, and attempts to break out of the corral or pen are common. During the breeding season, male dromedaries spend most of their time guarding the herd and surveying for any strange male or female in heat. Because of this continuous stress, a net reduction of food intake and increased digestive transit (stress diarrhoea) are observed, and males tend to lose weight (up to 35%) during the breeding season, sometimes to the point of emaciation (Tibary and Anouassi, 1997b). Aggressive behaviour during the breeding season has also been described in the Bactrian camel (Chen *et al.*, 1980). In a free-roaming herd, dominant males chase each other and engage in fighting, which can lead to serious biting injuries.

Sexual behaviour in the dromedary camel is characterized by frequent exteriorization of the soft palate (also called dulaa or dulla) (Plate 50) and marking (Tibary and Anouassi, 1997b). The protrusion of the dulla occurs at an interval of 5 to 30 min and is accompanied by a loud gurgling/roaring sound. Protrusion of the soft palate becomes more frequent with increased stimulation (presence of another male or females). Some dromedaries do not exhibit complete exteriorization of the soft palate, but instead only a mere flapping; these are usually older animals or males that have had their soft palate removed surgically during their racing careers. Bactrian camels do not exteriorize the soft palate to the same extent as dromedaries during their sexual behaviour, but both Bactrian camels and dromedaries display more saliva production and frothing than usual (Plate 51).

Marking takes two major forms: (i) urine spraying; and (ii) smudging of the poll gland secretions. Plate 52 shows a male dromedary rubbing poll gland secretions on to the ground. Both types of marking behaviour have also been described in the Bactrian camel (Chen *et al.*, 1980; Tibary and Anouassi, 1997b).

Male dromedaries and Bactrian camels frequently produce a metallic sound by grinding the molars via lateral movements of the lower jaw. This sound can be produced any time, but usually replaces the gurgling and ejection of the soft palate during copulation.

As already noted for camelids in general, copulation is completed with the female sitting in the sternal position. The duration of copulation in dromedaries is variable (Table 16.5) (Abdel Rahim and El-Nazier, 1992; Agarwal *et al.*, 1995). Among factors affecting the length of copulation are breed, age, sire, season and frequency of mating (Tibary and Anouassi, 1997b; Tibary *et al.*, 2005). The duration of copulation tends to be shorter at the beginning of the breeding season and in younger males. The copulation time decreases as weather becomes warmer and the end of the breeding season approaches. Bactrian camels exhibit similar copulation behaviour. According to one study, copulation time is limited to 1–6 min in 86% of the matings, but can be as long as 10 min (Chen *et al.*, 1980).

The dromedary displays multiple distinct periods of pelvic thrusting, gluteal muscle contractions and semen discharge during copulation. Contractions of the ischiocavernous, bulbospongious and urethralis muscles surrounding the pelvic urethra may be correlated with ejaculation. Ejaculation starts within a couple of minutes of intromission and continues throughout the copulation period.

Breeding Soundness Examination

Studies on the pathology of the male camelid reproductive tract are scarce (Tibary and Anouassi, 1997d, 2002; Tibary *et al.*, 2001). A comprehensive study on abnormalities in alpacas examined 3015 breeding males and

792 male reproductive tracts at post-mortem (Sumar, 1983). The incidence of pathological conditions in breeding alpacas was 18.2% (testicular hypoplasia 10%, cryptorchidism 5.7% and ectopic testes 2.5%). In slaughterhouse material, the incidence of abnormalities was 30.2% (hypoplasia 10.8%, cryptorchidism 3%, ectopic testes 1.9%, testicular cysts 14.5%) (Sumar, 1983).

Male breeding soundness examination (BSE) is an important part of the evaluation of herd infertility and decision making in the selection or purchase of a herd sire (Tibary and Anouassi, 1997c). Most often, the evaluation of males for infertility is attempted only if a gross abnormality is seen or after a long period of unsuccessful matings occurs, which limits the ability of the clinician to reach a diagnosis in time to prevent economic loss. Examination of the male should be conducted methodically to avoid oversight of any problems that may affect reproductive performance. A standard for BSE is yet to be determined for camelids. Male evaluation should include: identification of the animal, health and reproductive history, a detailed description of the reason(s) for examination, general health examination, special examination of the genital system and an evaluation of mating ability. A complete blood count and serum biochemistry panel should be performed on all males recently introduced to the farm. New males generally represent one of the most common biosecurity breaches in a camelid operation, and it is highly recommended to test for contagious diseases upon purchase. Importantly, the BSE should also include an evaluation of semen characteristics. Accordingly, an account of both semen collection and seminal characteristics is included in this section.

History and physical examination

The history taken should include identification and age of the animal, origin and type of management, breeding record, previous health problems and reason for examination. The type of management relates mainly to the description of the herd (size, number of females and males) and housing (paddocks, individual stalls or pens). Breeding history should include information on breeding management – whether free mating or hand mating (when the male is handled with a lead and brought to the female), breeding frequency and conception rate. Diseases of systems other than the genital tract can seriously affect reproductive performance of the male. For example, lesions of the musculoskeletal system can impair the physical ability of the male to copulate, and poor conformation or weakness of the hind legs may compromise mounting ability. A prolonged febrile condition or debilitating diseases can also affect spermatogenesis. In order to identify such underlying conditions, the examiner should take a complete health history, including previous illnesses, vaccination record and recent treatments.

For alpacas and llamas, special consideration should be given to congenital and possibly inherited disorders (Table 16.6) (Tibary *et al.*, 2011a).

Evaluation and disorders of the penis and prepuce

Extreme caution should be taken when examining the rutting male camel. Sedation is often required for thorough examination of the external genitalia. A more thorough

Table 16.6. Common congenital abnormalities in alpacas and llamas that may have a genetic base.

System	Type of abnormality
Musculoskeletal	Hernias, angular limb deformities, polydactyly, syndactyly, crooked tail, choanal atresia, campylognathia (wry face), cleft palate, dwarfism, jaw misalignments
Reproductive	Segmental aplasia of reproductive organs, epididymal cysts, gonadal hypoplasia, testicular cysts, failure to induce ovulation, persistent frenulum
Senses	Cataract, blue eyes (deafness), nasolacrimal duct aplasia, gopher ears
Viscera	Atresia ani, atresia coli, ventricular septal defect

examination of the penis and prepuce in adult camels requires general anaesthesia or pudendal nerve block (Ahmed *et al.*, 2011).

Examination of the prepuce in llamas and alpacas may require restraint in lateral recumbency because the genital area is often masked by the fleece. Sedation may be required in order to exteriorize the penis or for detailed examination of an abnormally pendulous prepuce, the presence of oedema, laceration or preputial mucosa prolapse. Penile attachment to the prepuce is normal in young, prepubertal animals, but can signal the presence of adhesions or persistent frenulum in the mature male. The penis should be completely free at 3 years of age. The most common lesions observed on the penis are lacerations, pustules or abrasions. The glans penis should be examined for signs of inflammation due to traumatic balanitis or the presence of hair rings. Urolithiasis is a common problem and should be suspected in the case of dysuria.

Preputial swelling

Preputial swelling is due to local inflammation caused by contact with chemical or physical irritants, parasitic infestation, or rupture of the urethra. The condition may also be part of a large plaque of ventral oedema in some animals suffering from heat stress or other disease processes that result in hypoproteinaemia. If the preputial swelling is due to urethral rupture, the urine accumulated in the subcutaneous space should be drained and a urethrostomy performed (Tibary *et al.*, 2008).

In the dromedary, preputial swelling can be due to local inflammation caused by contact with chemical or physical irritants, parasite (tick) infestation or rupture of the urethra (Bishnoi and Gahlot, 2004). The authors have also observed a case of preputial swelling in a dromedary after breeding overuse. Preputial swelling can also be part of a large plaque of ventral oedema in some animals suffering from acute trypanosomiasis.

Preputial injuries in the dromedary are often associated with tight girth straps in riding or working animals, and can cause severe sloughing of the tissue and protrusion of the penis.

Preputial prolapse

Preputial prolapse may be a complication of masturbation behaviour or trauma, and may require replacement of the penis into the sheath and the insertion of stay sutures (Lane, 1999; Tibary, 2003; Tibary *et al.*, 2008).

Paraphimosis

Paraphimosis, in which the camelid is unable to retract the protruded penis, is seen in animals with an accumulation of dirt in the preputial opening and may lead to balanoposthitis and necrosis of the tip of the penis. In the llama, paraphimosis and balanoposthitis (swelling of the head of the penis) may be due to the presence of 'hair rings' usually acquired secondarily to entanglement in fibre of the female during mating (Tibary and Vaughan, 2006).

Cleaning and replacement of the prolapsed tip of the penis into the prepuce may require sedation of the animal and application of an ointment containing antibiotics and anti-inflammatories (such as commercial bovine mastitis medications). Systemic antibiotic and anti-inflammatory treatment may result in recovery within a week. Surgical debridement, urethrostomy and amputation of the penis may be required if severe adhesions and gangrene develop (Tibary *et al.*, 2008).

Phimosis

Phimosis (in which the camelid is unable to protrude its penis out of the preputial sheath) in post-pubertal animals may be due to a congenitally small preputial opening or to the presence of lesions (abscesses, nodules) that prevent exteriorization of the penis during copulation. In such cases, surgical correction by enlargement of the preputial orifice may be attempted. Balanoposthitis and phimosis are common in males that display overt masturbation behaviour (breeding the ground or other objects).

Phimosis may also be a complication of preputial abscessation or adhesions as a result of traumatic injuries (Plate 53). These lesions may often be complicated by dysuria, or by urethral/bladder rupture resulting in uraemia and death (Tibary *et al.*, 2008).

Urolithiasis

Urolithiasis has been reported in male camelids by several authors (Kingston and Stäempfli, 1995; Gutierrez *et al.*, 2002, 2008; Tibary *et al.*, 2008). Most of the calculi that are found in the condition have been identified in the distal urethra or at the level of the sigmoid flexure. Affected animals initially show signs of intermittent colic, which become more frequent and are accompanied by lethargy and anorexia as the condition progresses. Deterioration of the animal usually signals rupture of the bladder and peritonitis. Relief of the condition can be attempted via urethral catheterization. However, this is not always possible, especially in young alpacas and llamas, because of the penile preputial attachment that prevents exteriorization of the penis. In cases where catheterization is not possible, a urethrostomy should be performed. Recurrence of obstruction is very common, even after urethrostomy. Urolithiasis can be prevented by better nutritional management (a well-balanced ration, salt supplementation and access to fresh water), assuming that the aetiopathogenesis of the condition is similar to that seen in sheep.

In the dromedary, urethral obstruction is frequently associated with intra-urethral adhesions resulting from chronic urethritis or calculi, or from mechanical compression of the urethra with a tight girth strap in working camels. A retrospective study of 35 cases showed that 75% of the cases had extensive necrosis of the urethral mucosa associated with total blockage of urine flow by calculi or a muco-cellular mass. Symptoms of uraemia will develop if this condition is not relieved by urethrostomy within 3 days. Immediate relief of the condition is obtained by drainage of urine from the subcutaneous tissue using two small incisions (5 cm) made on each side of the prepuce after epidural anaesthesia. This drainage is followed by a urethrostomy (Tibary and Anouassi, 1997d; Tibary *et al.*, 2008).

Examination and disorders of the testes

Examination of the testicles and epididymis includes inspection, palpation, measurement and ultrasonography of the scrotum and its contents. This examination can be completed on the restrained male in the standing or sitting position. This facilitates manipulation of the organs, especially in animals suspected of having cryptorchidism, hypoplasia or hydrocele. In some cases, if the male is aggressive or the scrotum is very sensitive, sedation may be required.

Testes should be present within the scrotum at birth and visible in males by 2 years old. In older males, the scrotum may sometimes be pendulous with a longer neck. One of the testicles may be slightly more ventral than the other, but they are nearly equal in size. Absence of visible testicles in the scrotum may be due to severe testicular degeneration, ectopic testicles or cryptorchidism.

The scrotal skin is thin and smooth, but can become thick and folded in the case of severe testicular degeneration. It should be examined for bite wounds (by other males) or evidence of external parasite infestation (mange – caused by mites, tick infestation). On palpation, the normal testes are smooth, firm and resilient. The testicles become hard and fibrotic or very soft as a result of degenerative changes. The scrotal sac should be free from fluid.

Size of the testis can be evaluated by measuring its length and width (see Tables 16.1 and 16.2). However, it is important to note that accuracy of these measurements will be affected by operator, technique of measurement, body condition and age of the animal, season and pathological conditions (i.e. orchitis, haematoma, hypoplasia or degenerative atrophy).

Ultrasonography is an important diagnostic technique for the evaluation of the testicular parenchyma as well as the surrounding tissues. In llamas and alpacas, it is preferable to conduct this examination using a 7.5 or 8 MHz transducer fitted with a stand-off pad in order to avoid artefacts. In camels, a 5 MHz linear transducer is sufficient. The normal testis shows a peripheral area of homogenous tissue corresponding to the testicular parenchyma and a central echogenic area corresponding to the fibrous mediastinum testis. The epididymis is small and not easily visualized by ultrasonography (Tibary and Anouassi, 1997c, 2000). Accumulation of fluid in the vaginal cavity (hydrocele) is observed frequently and with varying degrees of intensity in heat-stressed camelids. Testicular ultrasonography allows more precise measurement of the testicles than do external calipers (Bott *et al.*, 2008). Attempts to use Doppler ultrasonography to evaluate blood flow to the testicles have been made, but the technique is not used routinely in practice.

Testicular biopsy is indicated in cases of unexplained infertility or sub-fertility, testicular asymmetry and abnormal testicular ultrasonography not consistent with haematoma or orchitis. This technique is useful for the diagnosis of spermatogenic arrest, oligospermatogenesis, hypogonadism, inflammation and neoplasia (Tibary, 2001, 2004; Tibary and Vaughan, 2006). Four techniques have been proposed and tested for testicular biopsy in camelids: (i) wedge biopsy; (ii) fine-needle aspirate (Stelletta *et al.*, 2011); (iii) Trucut® needle or split-needle spring-loaded biopsy; and (iv) needle 'core' biopsy (Johnson and Schultheiss, 1994; Heath *et al.*, 2002; Tibary, 2004). Fine-needle aspirates are sometimes difficult to interpret (Plates 56 and 57). Trucut® or split-needle spring-loaded biopsy instruments have been found to be safe and result in minimal damage to the testicular parenchyma. They also provide a satisfactory amount of tissue for examination of the spermatogenic activity of the seminiferous tubule when a 14 gauge needle is used (Tibary, 2001; Heath *et al.*, 2002; Waheed *et al.*, 2011).

Scrotal skin lesions

The most common lesions found on the scrotum are traumatic or parasitic inflammation. Scrotal trauma due to bites from other males is also common. Prognosis for reproductive soundness of the individual male depends on the extent of the injury and the time elapsed until detection. Differential diagnoses include orchitis, heat stress and hydrocele. Deep lacerations are frequently complicated by testicular haemorrhage, infection and development of scirrhous cord. These cases require urgent surgical intervention (castration). As already mentioned, parasite infestations commonly seen in this anatomical region are mange (mite) and tick infestations.

Testicular hypoplasia

Testicular hypoplasia is commonly seen in SAC. Its incidence is estimated at 10% and it may be unilateral or bilateral, as well as total or partial. The left testis seems to be more affected than the right in bilateral cases (Sumar, 1983). Histology of the testicular parenchyma shows absence of or small seminiferous tubules with absence of spermatogenesis (Tibary, 2004; Waheed *et al.*, 2011). It is suspected that testicular hypoplasia is an inherited trait in camelids; therefore, affected males should be removed from the breeding programme.

Cryptorchidism

Cryptorchidism is relatively rare in camels (El-Hariri and Deeb, 1979; Garcia Pereira *et al.*, 2004). In alpacas, an abattoir study found an incidence of 3% unilateral cryptorchidism (58.3% left testis and 41.7% right testis) in 792 animals (Sumar, 1983). Cryptorchidism was also reported in related vicuñas (three cases in a population of 60 individuals), which suggests that the condition may be hereditary (Rietschel, 1990). Bilateral cryptorchidism has been described in a *Sry*-negative XX, sex-reversal case in a llama with multiple congenital abnormalities (Drew *et al.*, 1999). Monorchism – true absence of one testicle – has also been reported in a few alpacas, and is accompanied by absence of the ipsilateral kidney (Sumar, 1989). Methods of diagnosis include history, clinical evaluation and testosterone response after injection of human chorionic gonadotrophin (hCG; 3000 IU IV for llamas and 6000 IU for camels) (Perkins *et al.*, 1996; Tibary *et al.*, 2011b). All cases of cryptorchidism seen by the authors have been intra-abdominal. The retained testicle may be visualized by transabdominal ultrasonography in the inguinal area. Cryptorchidectomy may be performed via a parainguinal laparotomy approach or a laparoscopy-assisted technique (Garcia Pereira *et al.*, 2004).

Ectopic testicles

Abnormal location of the testicles has been seen by the authors in llamas and alpacas but not in camels. An incidence of 1.9% in alpacas has been reported, with the left testicle being affected 73.3% of the time (Sumar, 1983) (Plate 54). These males are preferably removed from the breeding pool.

Testicular cysts (rete testis cysts)

Testicular cysts have been described by the authors (Tibary, 2001; Tibary *et al.*, 2001; Tibary and Vaughan, 2006) (Plate 55). Cystic dilation within the rete testis can be visualized by ultrasonography. These cysts may affect the mediastinum testis and interstitial tissue of the testicle only. The exact aetiopathogenesis of these cysts is not known but they could be due to disturbed lymphatic drainage. Other cysts may be caused by blocked efferent ductules (Bott *et al.*, 2010; Tibary *et al.*, 2011c).

Hydrocele

Hydrocele is the collection of fluid between the visceral and the parietal layers of the tunica vaginalis. It can be due to inflammatory or non-inflammatory disturbances of scrotal/testicular drainage. The scrotal sacs become pendulous and increase in size. The scrotum is not painful and the testes are usually freely palpable, as well as the fluid, which can be isolated in one area. Diagnosis is easily performed by visualization of the fluid in the scrotal sac using ultrasonography. The nature of the fluid varies from anechoic (clear serous fluid) to slightly echoic. This condition is frequently seen in hot weather and tends to resolve progressively with decreasing ambient temperature. Hydrocele can develop following obstruction of the normal blood flow in the spermatic cord. A case of hydrocele in a llama was attributed to the presence of an abscess at the level of the external inguinal ring. Long-standing hydrocele affects the thermoregulation of the testes and can decrease the quality and quantity of semen (Tibary and Anouassi, 1997d; Tibary and Vaughan, 2006).

Orchitis

Orchitis is relatively rare but may occur following septicaemia or penetrating wounds. A case of unilateral orchitis due to *Streptococcus equi* subsp. *zooepidemicus* was reported in an alpaca (Aubry *et al.*, 2000). Systemic antimicrobial treatment may be attempted but is not always successful. Castration of the affected testicle in valuable males may increase the chance of salvaging the non-affected testicle and the reproductive life of the animal. In the male dromedary, orchitis has been associated with brucellosis (Ahmed and Nada, 1993; Tibary and Anouassi, 1997d; Tibary *et al.*, 2005).

Testicular degeneration

Testicular degeneration is probably the most common cause of acquired infertility due to testicular pathology (Ahmed and Nada, 1993; Tibary *et al.*, 2005). It may result from severe heat stress, trauma or chronic inflammation of the testes and/or scrotum or secondary to severe or chronic systemic disease, fever, toxins, nutritional or hormonal imbalance, or stress. In camels, testicular degeneration has been presumed to be associated with *Trypanosoma evansi* infection (Al-Qarawi and El-Belely, 2004). The degenerated testicles are smaller than normal and either soft or hard and fibrous. Testicular degeneration leads to deterioration of semen quality (azoospermia, oligozoospermia and teratozoospermia) and with deterioration of the seminiferous epithelium (Plate 58).

Testicular neoplasia

Tumours of the testes are rare. The most commonly reported neoplasm is the seminoma (Tibary and Vaughan, 2006; Birincioğlu *et al.*, 2008). An interstitial cell tumour has been described in a cryptorchid testis in the dromedary (El-Hariri and Deeb, 1979).

Abnormalities of the epididymis

There are few reports on abnormalities of the epididymis in camelids. The prevalence of epididymitis in camels seems to be higher and associated with *Brucella* spp. seropositivity (Ahmed and Nada, 1993). Epididymal cysts have been reported in the llama (Fowler, 1998) and alpaca (Sumar, 1983; Tibary, 2001; Gray *et al.*, 2007). In alpacas, cystic structures were found in 14.5% of slaughtered animals, mainly on the anterior aspect of the head of the epididymis and near the ventral border of the testis. The majority of these cysts were 1–5 mm in diameter, although in one case the cyst was 50 mm wide (Plate 59). Bilateral epididymal aplasia has been diagnosed by the authors in two alpacas and a llama. ALP concentration of the ejaculate is not a reliable indicator of bilateral epididymal blockage in camelids (Pearson *et al.*, 2013).

Evaluation of the accessory sex glands

Evaluation of the accessory sex glands is sometimes required and is limited to ultrasonography of the prostate and bulbourethral glands (Sánchez and Correa, 1995). There are no published reports of accessory sex gland disorders. A syndrome of stranguria (difficulty in urination) and enlargement of the prostate and bulbourethral glands has been seen by the authors in juvenile males around the time of puberty. These males seem to recover within a few weeks.

Evaluation of mating ability

Mating ability of the male is best observed in the presence of a receptive female. During this evaluation the succession of the normal behavioural pattern is recorded, as well as the times needed for each step: vocalization, chasing, forcing down, mounting, intromission and duration of copulation. Behavioural problems during mating can be due to shyness, inexperience or lack of libido.

Mating behaviour includes several steps that need to be observed closely. Camelid males display the flehmen response by investigating the female perineal area or its dung pile. Following this, the male proceeds to chase the female, and attempts to mount her in the standing position. Receptive females will respond to mounting attempts by assuming a sternal sitting position for copulation.

Sexual behaviour in the male camelid is poorly studied, particularly when it comes to resolving libido problems. Lack of sexual interest can be strictly behavioural (such as in the case of a young inexperienced animal) or functional (hormonal deficiencies or systemic diseases). In young animals, inhibition of group play may reduce the ability of animals to learn normal sexual behaviour. Show animals may have a slower onset of normal sexual behaviour too. Young males may be too shy to perform, particularly in a clinic setting. The rules of thumb are to give some time to young animals and introduce them slowly to breeding in order to learn, and to act quickly on proven males if they lose interest.

In proven adult males, deterioration of libido is mainly due to systemic diseases such as infections, musculoskeletal disease, or poor general health and body condition. Males that are housed continuously with females may also lose some of their sexual interest in them. Loss of libido may be noticed in animals suffering from heat stress. Decreased libido has also been observed in males suffering from megaesophagus as orgling is inhibited. Copulation difficulties and rectal prolapse may be seen in large obese camels.

During copulation, the penis has to penetrate through the cervix and deposit semen deep into the uterine horns. Erection failure in young males may be due to insufficient development (lack of penile–preputial detachment). All males should be able to exteriorize the penis by 3 years of age. Penile exteriorization or erection failure may be due to congenital abnormalities such as persistent frenulum, short penis, or abnormal function of the retractor muscle penis. Acquired conditions that lead to failure of exteriorization of the penis are preputial stenosis or penile–preputial adhesions. These conditions are often due to severe inflammation resulting from trauma (male breeding objects on the ground, injuries inflicted by other males or preputial prolapse) (Tibary *et al.*, 2008). Erection failure has been observed in two males following a neurological syndrome due to infection with *Paraphostrongylus tenuis* (meningeal worm) (Tibary and Anouassi, 2002; Tibary and Vaughan, 2006).

Semen collection and seminal characteristics

Semen collection

The collection of semen from camelids presents many difficulties owing to the nature of their copulatory behaviour and the slow (dribbling) process of ejaculation. The main techniques used in practice are the artificial vagina, electroejaculation or post-coital aspiration from a female (Pacheco Curie, 2008).

ARTIFICIAL VAGINA. Successful collection of semen using an artificial vagina (AV) has been reported in alpacas (Vaughan *et al.*, 2003a), llamas (Ferré and Werkmeister, 1996; Lichtenwalner *et al.*, 1996b) and camels (Tibary and Anouassi, 1997c). Semen can be collected using a dummy mount fitted with the collection apparatus or by holding an AV next to a receptive female. The dummy is made of a wooden or plastic frame covered with an alpaca hide (Garnica *et al.*, 1993; Bravo *et al.*, 1997a; Vaughan *et al.*, 2003a). Males are trained to serve the AV-fitted dummy by using a receptive female as a stimulus (Garnica *et al.*, 1993; Vaughan *et al.*, 2003a). The AV used in the alpaca is made of a PVC pipe or rubber hose (25 cm long and 7 cm in diameter) fitted with an inner lining with annular constrictions (an electrical cord coiled in a 2.5 cm band, or cylindrical foam rubber) to simulate the cervical rings and encourage ejaculation (Bravo *et al.*, 1997b). The outer chamber of the AV is filled with warm water, kept at 45°C during ejaculation, and attached to the dorsal and lateral walls of the dummy (Garnica *et al.*, 1993; Ferré and Werkmeister, 1996).

An improvement in the technique of semen collection by AV has been reported in llamas (Lichtenwalner *et al.*, 1996a). A sheep AV fitted with a non-spermicidal liner and cone is held at a 30° angle by an adjustable stand inside a half-dummy mount made of fibreglass. A constant temperature (38.3–40.0°C) and pressure (60–80 mm Hg) inside the AV is maintained by a continuous flow of warm water using a peristaltic pump. Semen collection is conducted on a platform, thus allowing the system to be periodically checked from underneath. The half dummy is placed behind a receptive female in such a way that the male can see and smell her during the entire length of collection. Semen collection (i.e. the presence of spermatozoa) was successful in 87% of the attempts made using this technique. In the llamas tested, the average duration of the collection of semen with spermatozoa in the ejaculate was 31.7 ± 12 min; for semen without spermatozoa, the average duration of collection was 37.1 ± 12.2 min (Lichtenwalner *et al.*, 1996a,b, 1997). In alpacas, the average collection time was 21.6 min (Bravo *et al.*, 1997a).

In camels, a modified bovine AV is used for semen collection. The modification consists of shortening the length of the outer hard shell and the simulation of cervical rings on the protruding part of the inner liner using a coil. The artificial vagina may be held by the operator to the side of a mount female or placed in between her hocks (Tibary and Anouassi, 1997c). Recently, systems such as that described for the llama have also been utilized for camels (Al-Eknah *et al.*, 2001).

ELECTROEJACULATION. Electroejaculation (EE), using a ram probe, has been accomplished under various degrees of sedation or anaesthesia in llamas (Graham *et al.*, 1978; Merlian *et al.*, 1986), alpacas (Fernandez-Baca and Calderón, 1966) and vicuñas (Fernandez-Baca and Novoa, 1969). Response to the electrical stimulus varies among males. Erection is possible during EE in the llama but failure to obtain an ejaculate or obtaining only a few spermatozoa is common (Merlian *et al.*, 1986). Camelids take a longer time to produce an ejaculate by EE than do ruminants, and therefore stimulation should be done with extreme care, starting with a very low voltage until erection is achieved.

Recent improvements in equipment and anaesthesia protocols have made the EE technique more reliable. Anaesthesia is induced by an IM combination of xylazine, ketamine and butorphanol. The penis is exteriorized and held by an assistant. An electroejaculator probe with three linear electrodes is utilized. The electrodes should be placed at the level of the prostate, at which depth is determined by transrectal ultrasonography. It is important to encourage the male to void urine before anaesthesia and EE in order to avoid the contamination of semen with urine. Muscle tremors and posterior limb movement is not eliminated and an assistant should be available to physically control these movements (Director *et al.*, 2007; Picha *et al.*, 2011).

In llamas, EE resulted in a high volume of ejaculate and higher sperm viability and motility than in ejaculates collected using an AV (Giuliano *et al.*, 2008). EE has been reported for camels under various degrees of sedation (Al-Qarawi *et al.*, 2002). Detomidine hydrochloride (80 μg/kg IM) is the drug of choice for sedation (Jochle *et al.*, 1990). However, the authors advise against this technique as routine practice in large camels because of complications and welfare considerations (Tibary and Anouassi, 1997c).

POST-COITAL ASPIRATION. In veterinary practice, semen evaluation is generally performed on a sample aspirated from the female genital tract immediately after mating. For this purpose, the male is mated with a healthy receptive female. Semen is then aspirated using a uterine pipette fitted with a 12 ml syringe. Males should be allowed to breed with the female for at least 12 min (Tibary and Vaughan, 2006).

Semen characteristics

The physical and biological characteristics of the ejaculate vary greatly depending on the conditions of collection (e.g. method of collection, fertility and libido of male, environmental temperature) (Table 16.7). Ejaculates tend to be non-fractionated, although some parameters (e.g. concentration, percentage of live and normal sperm) may vary slightly from the beginning to the end of an ejaculate (Lichtenwalner *et al.*, 1996a; Bravo *et al.*, 2002).

VOLUME. Ejaculate volume ranges from 0.4 to 12.5 ml (Fernandez-Baca and Calderón, 1966; Quispe, 1987; Garnica *et al.*, 1993). In alpacas, ejaculates obtained by AV tend to decrease in volume with increasing frequency of use (Bravo *et al.*, 1997b). Volume of ejaculates in llamas varies between 1 and 8 ml (Baer and Hellemann, 1998). Volume of the ejaculate in camels varies between 2 and 12 ml.

COLOUR. The colour of the ejaculate depends on the sperm concentration and the proportion of accessory sex gland secretions. It is predominately cloudy grey to milky white (Quispe, 1987; Garnica *et al.*, 1993), but can sometimes be creamy white (Garnica *et al.*, 1993). Seminal plasma makes up 80–90% of the ejaculate. Individual ejaculates may be heterogeneous, with some translucent material mixed with cloudy areas (Tibary and Anouassi, 1997c).

Table 16.7. Characteristics of the ejaculate in Camelidae.[a]

Species	Collection method[b]	Volume (ml)	Sperm concentration (million/ml)	Sperm motility (%)	Sperm of normal morphology (%)
Camelus bactrianus	AV	2.5–12.5	200–1600	20–80	50–90
	EE	1–12	200–600	50–80	50–90
Camelus dromedarius	AV	2–12.5	200–1600	20–80	50–90
	EE	1–9	331–800	20–80	40–70
Lama glama	AV	0.2–8	18	20–80	40–70
	EE	0.3–12.5	20	50–95	50–80
Vicugna pacos	AV	0.4–6	82.5–250	20–80	45–75
	EE	0.2–12	10–60	20–80	–

[a]This compilation is not exhaustive and is meant to illustrate the large variation in semen parameters.
[b]AV, artificial vagina; EE, electroejaculation.

Semen collected by post-coital aspiration from the vagina may contain varying amounts of blood as a result of trauma to the female reproductive tract. Haemorrhage may be secondary to copulation and penetration of the cervix (and/or hymen in maidens) by the penis. The presence of red blood cells does not appear to kill camelid semen as in other species (Tibary, 2003).

VISCOSITY. One of the most important physical characteristics of camelid semen is its high viscosity, which makes it very difficult to handle during laboratory procedures and mixing with extenders (Vaughan *et al.*, 2003a). The viscosity of the semen is attributed to the presence of glycosaminoglycans (GAGs), among which the highest concentration is of keratin sulfate, which is secreted by bulbourethral glands (Kershaw-Young *et al.*, 2012a). The physiological role of this characteristic is not clear. The degree of viscosity depends on the individual male and on the proportion of gelatinous seminal fluid, and tends to decrease with increasing number of ejaculates on any day. The viscosity of alpaca semen collected by AV tended to decrease on the third ejaculate (Bravo *et al.*, 1997b). Complete liquefaction of alpaca semen may take several hours; in one study, the mean time for this was 23 ± 1.2 h (range 8–48 h) (Garnica et al., 1993). The liquefaction of camelid semen may be promoted by exposure to various proteolytic enzymes (trypsin, collagenase, fibrolysin, hyaluronidase) (Bravo *et al.*, 2000). All of these enzymes are effective, but they may cause damage to the spermatozoa if incubation is prolonged or the concentration is too high. A trypsin solution of 1:250 seems to be effective with minimal negative effects on spermatozoa (Bravo *et al.*, 1999, 2000). Collagenase at a concentration of 0.1% is also often used for the liquefaction of semen (Bravo *et al.*, 2000; Giuliano *et al.*, 2010; Carretero *et al.*, 2012).

SPERM CONCENTRATION. Sperm concentration is generally estimated using a haemocytometer (Tibary and Anouassi, 1997c). Electronic methods have not been investigated and may present difficulties because of the viscous nature of camelid semen. Sperm concentration is highly variable and is affected by age, method of collection and ejaculate rank (Table 16.7). Interruption of copulation results in reduced concentration of semen in the ejaculate. Sperm concentration in the ejaculate also decreases in successive ejaculations, although no difference is observed if an interval of at least 12 h is allowed between successive ejaculations (Bravo *et al.*, 1997b). A period of sexual rest should be provided before clinical evaluation, as sperm concentration tends to decrease in alpaca males maintained with females (Flores *et al.*, 2002).

It is not uncommon to obtain ejaculates without spermatozoa due to lack of adaptation to the AV, particularly in camels. This needs to be differentiated from true azoospermia or oligozoospermia as a result of testicular or epididymal abnormalities. Most testicular abnormalities are detected during routine BSE, and this prevents loss of time incurred by using a non-fertile male. The most common congenital abnormalities found on routine evaluation are testicular hypoplasia and testicular or epididymal cysts. Acquired conditions resulting in oligozoospermia or azoospermia include severe degenerative changes resulting from trauma, infection or heat stress.

PH. The pH of camelid semen is slightly alkaline, at 7.5 to 8.4 (Kubicek, 1974; Lichtenwalner *et al.*, 1996b; Baer and Hellemann, 1998; Wani *et al.*, 2011). In llamas, the pH of semen collected by AV did not differ significantly at different times during copulation. This is probably due to the fact that camelid semen is not emitted in distinct fractions (Lichtenwalner *et al.*, 1996a,b).

SPERM MOTILITY. It is important to note that sperm motility is very difficult to evaluate owing to the viscous nature of camelid semen. Individual sperm motility is very low in undiluted semen, and is best described as oscillatory. Motility is rated based on the oscillatory intensity of sperm (Garnica *et al.*, 1993; Tibary and Anouassi, 1997c; Tibary and Vaughan, 2006). Following enzymatic liquefaction, progressive sperm motility varies between 60 and 80%. It is important to note that camel semen is extremely sensitive to AV rubber liner spermatoxicity.

SPERM MORPHOLOGY. Mature camelid spermatozoa exhibit the same anatomical features as other domestic mammals. The total length of the camelid sperm cell is smaller than that of the bull, buffalo, ram, donkey and stallion, but larger than that of the boar (Table 16.8).

The head of camelid spermatozoa is described as elliptical. Lengths of the head and midpiece are shorter than those of other (large domestic) animals, while the tail is longer than that of the boar and stallion spermatozoa and shorter than that of all other animals (Merlian *et al.*, 1986). Sperm morphology is ideally evaluated using the traditional eosin–nigrosin stain; however, Diff-Quik/ Dip Quick® (Giemsa) and Spermac® have also been used in the authors' laboratory (Tibary and Anouassi, 1997c; Tibary, 2003; Tibary and Pugh, 2003; Tibary and Vaughan, 2006). Morphological evaluation is performed as for other domestic species. The morphological abnormalities should be reported according to type and location of the defect(s) observed.

All sperm abnormalities found in other livestock species can be found in camelid semen (Tibary and Anouassi, 1997c; Tibary and Vaughan, 2006). These include abnormalities of the head, midpiece, tail and proximal and distal cytoplasmic droplets. The effects of the various abnormalities on fertility have not yet been determined in camelids. In the llama, the proportion of abnormal spermatozoa in ejaculates collected by AV is highly variable and ranges from 20.9 to 96.1% (Lichtenwalner *et al.*, 1996b; Baer and Hellemann, 1998). The most common abnormalities involve the head (20 ± 19%) and the acrosome (13 ± 12%). The incidence of cytoplasmic droplets in the llama ranges from 0 to 45.4% (average 11.1 ± 12.4). The incidence of proximal droplets in four alpacas ranged from 10 to 38% (Vaughan *et al.*, 2003a). In another alpaca study, the incidence of abnormal heads, cytoplasmic droplets and abnormal tails was 9.6, 3.6 and 14.5%, respectively. The incidence of abnormalities was not affected by rank of ejaculate (Bravo *et al.*, 1997a). Computer analysis of sperm head size and shape showed significant differences both between and within ejaculates (Buendia *et al.*, 2002).

BIOCHEMICAL COMPONENTS. The biochemical composition of camelid semen is similar to that reported for other livestock species. Chloride is the main anion and calcium the main cation (Garnica *et al.*, 1993). These electrolytes could be of importance in physiological processes of the sperm cell such as motility and capacitation. Glucose is found in high concentration in the seminal plasma of alpacas and could be the main substrate for energy provision (Garnica *et al.*, 1993). Glucose concentrations are higher in younger than in adult male alpacas (Garnica *et al.*, 1993, 1995). Other components of the semen, such as lipids, phospholipids and proteins, have been quantified in the seminal plasma of the alpaca (Garnica *et al.*, 1993). The functions of these components have not yet been studied in camelids, but they may play a role in maturation and in protection of the sperm cell membrane integrity. Proteins give protection to spermatozoa against harmful effects of a high dilution rate of the ejaculate (Garnica *et al.*, 1993). In a study comparing seminal plasma characteristics between llamas and camels, pH, osmolarity, and sodium and chloride content were similar. However, camel

Table 16.8. Dimensions of the camelid spermatozoon (Tibary *et al.*, 2007).

Species/Dimension	*Camelus bactrianus*	*Camelus dromedarius*	*Lama glama*
Head length (μm)	6.0 ± 0.6	6.6 ± 0.5	5.3 ± 0.5
Head width (μm)	3.8 ± 0.1	3.9 ± 0.1	3.8 ± 0.1
Midpiece length (μm)	6.2 ± 0.7	6.8 ± 0.5	5.3 ± 1.6
Tail length (μm)	30.8 ± 1.9	37.6 ± 0.9	36.6 ± 1.8
Total length (μm)	42.9 ± 1.9	51.1 ± 0.9	49.5 ± 2.2

seminal plasma was higher in glucose, triglycerides and phosphate and lower in calcium and total protein (Wani *et al.*, 2011).

Camelid semen contains an ovulation induction factor (OIF). Seminal plasma induces ovulation in female alpacas and Bactrian camels following placement of the semen into the vagina or uterus without mating (Chen *et al.*, 1985; Xu *et al.*, 1985; Sumar, 1996) or following IM injection (Chen *et al.*, 1985; Pan *et al.*, 2001; Adams *et al.*, 2005). Although the insemination of whole semen induced ovulation in some dromedary camels, ovulation and pregnancy rates are significantly higher in females artificially inseminated following mating by a vasectomized male (Anouassi *et al.*, 1992). It has been shown recently that the OIF effect is potentiated by the presence of uterine inflammation (Adams *et al.*, 2005; Tanco *et al.*, 2011). The OIF has been isolated and identified as β nerve growth factor (β-NGF), which is conserved across multiple animal species (Ratto *et al.*, 2006, 2012; Kershaw-Young *et al.*, 2012b).

Interpretation of the spermiogram

The major problem in interpreting semen analyses in camelids is the lack of standardized methods, not only for collection but also for evaluation. Some of the physiological factors that may affect quantity and/or quality of semen, such as age, nutritional status and season, need to be critically evaluated. The study of these factors will allow clinicians to make recommendations on the frequency of use of a male based on clinical examination.

Normal males should have at least 60% morphologically normal spermatozoa (Flores *et al.*, 2002). Increased sperm abnormalities are generally associated with testicular disorders (testicular hypoplasia, testicular degeneration and heat stress). Teratozoopermia may be due to chromosomal abnormalities (chromosomal translocations) or to molecular disarrangement of spermatogenesis. These factors are not yet fully understood. Sterility has been diagnosed in one male in the authors' clinic that had abnormal sperm mitochondrial sheaths. In llamas and alpacas, heat stress affects spermatogenesis as well as sperm motility and morphology. In one study, the exposure of male llamas to an environmental temperature above 29°C decreased motility from 63.1 to 15.0%, and increased the numbers of morphologically abnormal sperm from 26.3 to 50.5% (Schwalm *et al.*, 2007).

Semen Preservation and Artificial Insemination

Semen preservation and artificial insemination have seen very little progress since the initial studies were performed in the 1960s (Tibary and Anouassi, 1997e). Except for the Bactrian camel, from which Chinese scientists have been able to collect, process and use semen to achieve high pregnancy rates, the development of artificial insemination (AI) in other camelids has been slow. Research on semen preservation and AI in camelids has regained interest in recent years, mostly as a result of interest in racing camels (Anouassi *et al.*, 1992; Tibary and Anouassi, 1997e; Skidmore *et al.*, 2013) and worldwide development of the alpaca breeding industry (Vaughan *et al.*, 2003a; Bravo *et al.*, 2013).

Semen collection and ejaculate liquefaction

Major hurdles in the development of AI are semen collection, initial ejaculate quality and the induction of liquefaction prior to the addition to extender. Only good-quality ejaculate (concentration of sperm >80 million/ml, progressive motility >70% and morphologically normal spermatozoa >70%) should be used. Despite the development of new techniques for semen collection with an AV and EE (as described earlier), ejaculate quality remains extremely variable, particularly in the dromedary camel. Semen collection with an AV in the dromedary produces a high rate of azoospermic ejaculates (Deen *et al.*, 2003). Individual variation in acceptance of the AV is also a handicap in the development of this technology. In alpacas and llamas, semen collection by AV or EE has been more reliable.

One of the most challenging factors in camelid semen processing for preservation and AI is its high viscosity (Casaretto *et al.*, 2012). Various enzymatic treatments (trypsin, hyaluronidase, amylase, collagenase) have been utilized to eliminate this viscosity and allow better mixing of the ejaculate with the extender in order to maintain its viability (Bravo *et al.*, 1999, 2000; El-Bahrawy, 2010). Collagenase (0.1% solution) is presently the most commonly used enzyme for liquefaction of llama and alpaca semen (Giuliano *et al.*, 2010; Casaretto *et al.*, 2012; Morton *et al.*, 2012; Bravo *et al.*, 2013). The ejaculate is generally incubated with the collagenase solution until liquefaction is complete. Sperm quality can then be determined from an aliquot and the rest of the liquefied ejaculate diluted with an extender. Liquefaction with 0.1% collagenase in H-TALP-BSA (HEPES-Tyrode's medium-bovine serum albumin) does not seem to increase DNA decondensation (Giuliano *et al.*, 2010; Carretero *et al.*, 2012).

Short-term semen preservation

Most of the commercially available semen extenders for ruminants (Laciphos®, Androhep®, Biociphos®, Sodium citrate-egg yolk) and equines (Lactose-egg yolk, Kenney's skim milk, INRA 96®) have been tested in one form or another on camelids (Sieme *et al.*, 1990; Tibary and Anouassi, 1997e; Vaughan *et al.*, 2003a; Deen *et al.*, 2004; Skidmore *et al.*, 2013). Egg yolk is generally added to extenders at a rate of 20% (v/v). A commercial camel semen extender (Green Buffer®-egg yolk, IMV, l'Aigle, France) was shown to preserve motility for camel and alpaca semen but is no longer available (Tibary and Anouassi, 1997e; Vaughan *et al.*, 2003a; Waheed *et al.*, 2010; Skidmore *et al.*, 2013).

Tris-buffered and egg yolk-citrate extenders seem to provide adequate preservation of viability and motility for camel sperm (Niasari-Naslaji *et al.*, 2006; Wani *et al.*, 2008). Extended semen is cooled slowly from room temperature to 4 to 8°C over 1 h. Systems used for the transport of cooled horse semen can also be used to transport cooled alpaca semen by air, thus allowing distribution over a large geographical area (Tibary and Anouassi, 1997e). Sperm viability decreased sharply after 48 h of storage in most of these extenders. A Tris-citric acid-glucose-fructose extender (SHOTOR®) was found to be comparable to Green Buffer® for the preservation of the motility of cooled Bactrian camel semen (Niasari-Naslaji *et al.*, 2006).

Lactose-egg yolk extender has been shown to maintain progressive motility and viability for up to 24 h at 4°C. Addition of catalase (500 IU/ml) may offer the advantages of reducing the effect of peroxidation and increasing the lifespan of cooled camel sperm (Medan *et al.*, 2008).

Various commercial extenders have been utilized for cooled preservation of alpaca and llama semen; these include Tris-based extender, Camel Buffer®, EDTA, Triladyl®, Biladyl®, Andromed®, lactose, Salamon's extender (Vaughan *et al.*, 2003a; Morton *et al.*, 2009; Bravo *et al.*, 2013). Biladyl® and Egg-yolk-glucose-citrate extender were shown to be superior to other commercial extenders (Morton *et al.*, 2009). Tris extender was significantly better than EDTA with or without Equex STM paste in llamas (Baer and Hellemann, 1999). Also in llamas, 11% lactose-egg yolk extender was better than Tris-citrate-fructose-egg yolk or skim-milk glucose (Giuliano *et al.*, 2012). Lactose-egg yolk extender was also shown to maintain good progressive motility of alpaca sperm for up to 24 h (Morton *et al.*, 2007).

Sperm cryopreservation

Although camelid semen cryopreservation was attempted over four decades ago, this technology remains poorly studied. The first offspring from insemination with frozen–thawed Bactrian semen was reported in 1961 (Elliot, 1961). Early attempts to freeze camel semen relied on boar or equine extenders (lactose-egg yolk-glycerol-Orvus-Equex paste) (Sieme *et al.*, 1990). Further studies on Bactrian camels used a sucrose-based extender (SYG-3: 12% sucrose, 20% egg yolk, 7% glycerol) (Chen *et al.*, 1990). SHOTOR® with 6% glycerol was shown to be superior to Green Buffer® for the cryopreservation of camel semen (Niasari-Naslaji *et al.*, 2007). The most common cryoprotectant used for camel semen cryopreservation is glycerol at concentrations varying from 4 to 7%. Most studies provide a 1 to 2 h equilibration time at 4°C before the addition of glycerol. Camel semen has been frozen in a variety of packaging, including ampoules, pellets and 0.25 and 0.5 ml straws, using the same methods as described for other species (Tibary and Anouassi, 1997e). The authors' trials with Green Buffer® with added 20% egg yolk and 5% glycerol gave excellent post-thaw motility (Tibary and Anouassi, 1997e). Similar results were obtained recently with the same extender with 6% glycerol (Morton *et al.*, 2010b). In the authors' trials, despite the excellent post-thaw quality of semen, embryo recovery rate in inseminated, superovulated females was very low (Tibary and Anouassi, 1997e).

The cryopreservation of semen from alpacas and llamas was first attempted in the late 1970s (Graham *et al.*, 1978). Llama and alpaca semen has been frozen in Tris-egg yolk-glycerol or sodium citrate-egg yolk-glycerol and other commercial semen freezing extenders (Aller *et al.*, 2003). The use of egg-yolk free commercial extender (Bociphos®) yielded poor post-thaw motility (Bürgel *et al.*, 2001). There is a lack of data on the role of factors such as extenders, individual males, seminal plasma and the degree and method of liquefaction on the freezing ability of semen. While glycerol is the main cryoprotectant used, ethylene glycol has been shown to be equally effective (Santiani *et al.*, 2005). The addition of glycerol is performed after cooling and equilibration at 4 to 5°C for 1 to 2 h. After addition of the extender, usually in one or two steps, the semen is loaded into 0.25 or 0.5 ml plastic straws. Straws are frozen by placing them on a rack at known distances (6 to 12 cm) above the surface of liquid nitrogen for 10 to 20 min. Other techniques use progressive lowering of the rack (1 cm/min) until immersion in liquid nitrogen. At the time of use, straws are thawed out by immersion in a water bath at 37°C for 30 to 40 s or 40°C for 8 s (Bravo *et al.*, 2013).

Epididymal sperm has been preserved using the same techniques described for ejaculated sperm but no fertility trials have been reported to date (Morton *et al.*, 2007, 2010c). This technique could prove valuable for the salvage of genetic material from terminal males.

Artificial insemination

AI of the camelid female is performed following the induction of ovulation. This can be induced with hCG (750–1000 IU IV in alpacas and llamas; or 1000–3000 IU IV in camels), with gonadotrophin releasing hormone (GnRH, IM; 20 μg for alpacas and llamas, 100 μg for camels) or a GnRH analogue (buserelin, IM; 8 μg for alpacas and lamas, 20 μg for camels). The ovulation induction rate is very high (90–100%) if females are selected based on maximum uterine tone and oedema and appropriate follicular size (7–12 mm for alpacas and llamas; 12–18 mm for camels) as determined by ultrasonography (Tibary and Anouassi, 1997e). Ovulation occurs between 26 and 30 h after treatment, and insemination is usually performed 24 h after treatment. The technique of insemination commonly used is similar to that for the bovine.

In camels, AI trials with cooled semen have resulted in variable pregnancy rates. The main factors affecting conception rate are sperm quality, site of deposition and inherent fertility of the male and female. Conception rates ranging from 30 to 50% have been reported with 100–300 million spermatozoa when semen is deposited in the body of the uterus (Anouassi *et al.*, 1992; Skidmore and Billah, 2006; Morton *et al.*, 2010a). Similar conception rates are achieved with as little as 8 million spermatozoa with deep horn insemination ipsilateral to the side of ovulation (Anouassi and Tibary, 2010).

AI trials with frozen–thawed semen are scarce in camelids. Excellent pregnancy rates have been achieved with a double insemination protocol using 150–300 million spermatozoa frozen in SYG-3 extender in the Bactrian camel (Chen *et al.*, 1990). In the dromedary, pregnancy rates following insemination with frozen–thawed semen are low (<10%) (Tibary and Anouassi, 1997e; Deen *et al.*, 2003). In alpacas and llamas, most of the trials published so far involve a limited number of observations and a wide variety of protocols and doses. Pregnancy rates have been extremely variable, and range from 0 to 30% for cooled semen and from 0 to 65% for frozen–thawed semen (Pacheco Curie *et al.*, 2009; Giuliano *et al.*, 2012; Bravo *et al.*, 2013).

The number of motile spermatozoa required for adequate fertilization in llamas seems to be 8 million (4–12 million depending on the study) (Bravo *et al.*, 1999, 2013).

Use of artificial insemination for hybrid production

AI has been used for the production of paco–vicuña (an alpaca × vicuña cross) and other combinations of South American camelid (Bravo *et al.*, 2013). The first hybrid between the OWC and NWC was produced in 1999 by insemination of a guanaco female with dromedary semen (Skidmore *et al.*, 1999). Since this first hybrid, several other llama–dromedary hybrids have been produced; however, pregnancy loss is quite high in bred females. None of the pregnancies established with llama or guanaco semen in a female dromedary were carried to term (Skidmore *et al.*, 2013).

Conclusion

Despite the research conducted on camelid reproduction over the last five decades, the body of knowledge of camelid andrology remains far behind that in other domestic species. This is in part due to the fact that these species are mostly found in an area of the world where research resources are scarce and in part to their peculiar reproductive pattern. Within the Camelidae family, llamas and alpacas have been the most clinically evaluated. In all species, further scientific characterization of puberty and sperm production is needed. The high rate of abnormalities in alpacas and llamas, such as testicular cysts and testicular hypoplasia, merits further genomic and molecular studies. A complete BSE is rarely performed in these species and is often limited to the selection of males based on testicular size. There is a complete lack of information on the effect of various sperm abnormalities on fertility. In camels, the paucity of clinical information is worsened by lack of proper methodology and erroneous conclusions drawn from some studies.

Although the first attempts at semen collection and AI took place in the late 1960s, there has been very little progress in the use of cryopreserved semen for AI. To date, the pregnancy rates achieved in AI trials remain too variable to make this technology acceptable, particularly in the dromedary camel. Improvements in the methods for semen collection and liquefaction will allow studies to develop better semen extenders in the future. One of the areas of research that needs to be addressed is the effect of semen dilution and processing on sperm transport and fertilization.

References

Abdel Rahim SA and El-Nazier AT (1992) Studies on the sexual behaviour of the dromedary camel In *Proceedings of the First International Camel Conference, Dubai, 2nd–6th February 1992* pp 115–118 Ed WR Allen, AJ Higgins, EG Mayhew, DH Snow and JF Wade. R & W Publications (Newmarket), Newmarket, UK.

Abdel Raouf M, el Bab MRF and Owaida M (1975) Studies on reproduction in the camel (*Camelus dromedarius*).V. Morphology of the testis in relation to age and season *Journal of Reproduction and Fertility* **43** 109–116.

Adams GP, Ratto MH, Huanca W and Jaswant S (2005) Ovulation-inducing factor in the seminal plasma of alpacas and llamas *Biology of Reproduction* **73** 452–457.

Agarwal VK, Agarwal SP, Lajja R, Rai AK and Khanna ND (1995) A quantitative assessment of sexual behaviour of male camels (*Camelus dromedarius*) *International Journal of Animal Sciences* **10** 371–374.

Ahmed AF, Al-Sobayil FA and Al-Halag MA (2011) Topographical anatomy and desensitization of the pudendal nerve in adult male dromedary camels *Theriogenology* **76** 772–777.

Ahmed WM and Nada AR (1993) Some pathological affections of testis and epididymis of slaughtered camels (*Camelus dromedarius*) *International Journal of Animal Sciences* **8** 33–36.

Akingbemi BT and Aire TA (1991) Testicular dimensions in sperm reserves in the camel (*Camelus dromedarius*) in Nigeria *Bulletin of Animal Health and Production in Africa* **39** 121–123.

Al-Eknah M, Homeida N and Al-Haider A (2001) A new method for semen collection by artificial vagina from the dromedary camel *Journal of Camel Practice and Research* **8** 127–130.

Al-Qarawi AA and El-Belely MS (2004) Intratesticular morphometric, cellular and endocrine changes in dromedary bulls exhibiting azoospermia *Veterinary Journal* **167** 194–201.

Al-Qarawi AA, Abdel-Rahman HA, El-Belely MS and El-Mougy SA (2001) Intratesticular morphometric, cellular and endocrine changes around the pubertal period in dromedary camels *Veterinary Journal* **162** 241–249.

Al-Qarawi AA, Abdel-Rahman HA, El-Mougy SA and El-Belely MS (2002) Use of a new computerized system for evaluation of spermatozoal motility and velocity characteristics in relation to fertility levels in dromedary bulls *Animal Reproduction Science* **74** 1–9.

Aller JF, Rebuffi GE, Cancino AK and Alberio RH (2003) Influence of cryopreservation on the motility, viability and fertility of llama (*Lama glama*) spermatozoa *Archivos de Zootecnia* **52** 15–23.

Anouassi A and Tibary A (2010) Effect of volume and timing of induction of ovulation on conception rate following deep horn insemination in camels (*Camelus dromedarius*) *Clinical Theriogenology* **2** 392.

Anouassi A, Adnani M and Raed EL (1992) Artificial insemination in the camel requires induction of ovulation to achieve pregnancy In *Proceedings of the First International Camel Conference, Dubai, 2nd–6th February 1992* pp 175–178 Ed WR Allen, AJ Higgins, EG Mayhew, DH Snow and JF Wade. R & W Publications (Newmarket), Newmarket, UK.

Aubry P, Swor TM, Löhr CV, Tibary A and Barrington GM (2000) Septic orchitis in an alpaca *Canadian Veterinary Journal* **41** 704–706.

Azouz A, Ateia MZ, Shawky H, Zakaria AD and Farahat AA (1992) Hormonal changes during rutting and the non-breeding season in male dromedary camels In *Proceedings of the First International Camel Conference, Dubai, 2nd–6th February 1992* pp 169–171 Ed WR Allen, AJ Higgins, EG Mayhew, DH Snow and JF Wade. R & W Publications (Newmarket), Newmarket, UK.

Baer L von and Hellemann C (1998) Semen characteristics in the llama (*Lama glama*) *Archivos de Medicina Veterinaria* **30** 171–176.

Baer L von and Hellemann C (1999) Cryopreservation of llama (*Lama glama*) semen *Reproduction in Domestic Animals* **34** 95–96.

Bedrak E, Rosenstrauch A, Kafka M and Friedlander M (1983) Testicular steroidogenisis in the camel (*Camelus dromedarius*) during the mating and the nonmating seasons *General and Comparative Endocrinology* **52** 255–264.

Birincioğlu SS, Avcı H and Aydoğan A (2008) Seminoma and cholangiocarcinoma in an 18-year-old male camel *Turkish Journal of Veterinary and Animal Sciences* **32** 141–144.

Bishnoi P and Gahlot TK (2004) Surgical disorders of reproductive tract in male camel (*Camelus dromedarius*) *Intas Polivet* **5** 275–278.

Bott [I], Rodriguez J, Sandoval S and Tibary A (2008) Relationship between testicular measurements using calipers or ultrasonography with testicular weight in alpacas (*Vicugna pacos*) *Theriogenology* **70** 576–576.

Bott I, Pearson LK, Rodriguez JS, Sandoval S, Kasimanickam RK, Sumar J and Tibary A (2010) Prevalence and pathologic features of rete testis cysts in alpacas (*Vicugna pacos*) *Clinical Theriogenology* **2** 395.

Bravo PW and Johnson LW (1994) Reproductive physiology of the male camelid *Veterinary Clinics of North America: Food Animal Practice* **10** 259–264.

Bravo PW, Flores U, Garnica J and Ordoñez C (1997a) Collection of semen and artificial insemination of alpacas *Theriogenology* **47** 619–626.

Bravo PW, Flores D and Ordoñez C (1997b) Effect of repeated collection on semen characteristics of alpacas *Biology of Reproduction* **57** 520–524.

Bravo PW, Solis P, Ordoñez C and Alarcon V (1997c) Fertility of the male alpaca: effect of daily consecutive breeding *Animal Reproduction Science* **46** 305–312.

Bravo PW, Pacheco C, Quispe G, Vilcapaza L and Ordoñez C (1999) Degelification of alpaca semen and the effect of dilution rates on artificial insemination outcome *Archives of Andrology* **43** 239–246.

Bravo PW, Ccallo M and Garnica J (2000) The effect of enzymes on semen viscosity in llamas and alpacas *Small Ruminant Research* **38** 91–95.

Bravo PW, Moscoso R, Alarcon V and Ordonez C (2002) Ejaculatory process and related semen characteristics *Archives of Andrology* **48** 65–72.

Bravo PW, Alarcon V, Baca L, Cuba Y, Ordoñez C, Salinas J and Tito F (2013) Semen preservation and artificial insemination in domesticated South American camelids. *Animal Reproduction Science* **136** 157–163.

Buendia P, Soler C, Paolicchi F, Gago G, Urquieta B, Perez-Sanchez F and Bustos-Obregon E (2002) Morphometric characterization and classification of alpaca sperm heads using the Sperm-Class AnalyzerReg. computer-assisted system *Theriogenology* **57** 1207–1218.

Bürgel H, Erhardt G and Gauly M (2001) Cryopreservation of llama (Lama glama) spermatozoa with an egg-yolk-free extender In *Progress in South American Camelids Research. Proceedings of the 3rd European Symposium and SUPREME European Seminar, Göttingen, Germany, 27–29 May 1999* pp 90–93 Ed M Gerken and C Renieri. EAAP [European Federation of Animal Science] Publication No. 105, Wageningen Pers [now Wageningen Academic Publishers], Wageningen.

Bustos-Obregon E, Rodriguez A and Urquieta B (1997) Spermatogenic cycle stages in the seminiferous epithelium of the vicuña (*Vicugna vicugna*) In *Recent Advances in Microscopy of Cells, Tissue and Organs* pp 579–584 Ed PM Motta. Delfino Editor, Rome.

Carretero MI, Giuliano SM, Casaretto CI, Gambarotta MC and Neild DM (2012) Evaluation of the effect of cooling and of the addition of collagenase on llama sperm DNA using toluidine blue *Andrologia* **44** 239–247.

Casaretto C, Martínez Sarrasague M, Giuliano S, Rubin de Celis E, Gambarotta M, Carretero I and Miragaya M (2012) Evaluation of *Lama glama* semen viscosity with a cone-plate rotational viscometer *Andrologia* **44** 335–341.

Chen BX, Yuen ZX and Pan GW (1985) Semen-induced ovulation in the Bactrian camel (*Camelus bactrianus*) *Journal of Reproduction and Fertility* **74** 335–339.

Chen BX, Zhao XX and Huang YM (1990) Freezing semen and AI in the Bactrian camel (*Camelus bactrianus*) In *Proceedings of the UCDEC Workshop 'Is it Possible to Improve the Reproductive Performance of the Camel?', 10–12 Sept., Paris* pp 285–291. Unité de Coordination pour l'Elevage Camelin, Paris.

Chen PM, Kang CL, Yuen ZX and Ge YG. (1980) Reproductive pattern of the Bactrian camel 2. Sexual behaviour *Acta Veterinaria et Zootechnica Sinica* **11** 65–76.

Collazos VGD (1972) Estudio de la irrigacion superficial del pene y prepucio de la alpaca, *Lama pacos*. BS thesis, Facultad de Medicina Veterinaria, Universidad Nacional Mayor de San Marcos, Lima.

Deen A (2008) Testosterone profiles and their correlation with sexual libido in male camels *Research in Veterinary Science* **85** 220–226.

Deen A, Sumant V and Sahani MS (2003) Semen collection, cryopreservation and artificial insemination in the dromedary camel *Animal Reproduction Science* **77** 223–233.

Deen A, Sumant V, Mamta J and Sahani MS (2004) Refrigeratory preservation of camel semen *Journal of Camel Practice and Research* **11** 137–139.

Deen A, Vyas S and Sahani MS (2005) Testosterone profiles in the camel (*C. dromedarius*) during the rutting season *Israel Journal of Veterinary Medicine* **60** 27–32.

Delhon GA and Lawzewitsch I von (1987) Reproduction in the male llama (*Lama glama*), a South American camelid. I. Spermatogenesis and organization of the intertubular space of the mature testis *Acta Anatomica* **129** 59–66.

Delhon G[A] and Lawzewitsch I von (1994) Ductus epididymidis compartments and morphology of epididymal spermatozoa in llamas *Anatomia Histologia Embryologia* **23** 217–225.

Delhon G, Zuckerberg C, Lawzewitsch I von, Larrieu E, Oporto R, Bigaztti R (1983) Estudio citologico de las gonadas de guanaco *Lama guanicoe*, Macho, en las estudios prepuperales, sexualmente maduros y seniles *Revista de la Facultad de Ciencias Veterinarias, Universidad de Buenos Aires* **1** 47–60.

Derar DR, Hussein HA and Ali A (2012) Reference values for the genitalia of male dromedary before and after puberty using caliper and ultrasonography in subtropics *Theriogenology* **77** 459–465.

Director A, Giuliano S, Trasorras V, Carretero I, Pinto M and Miragaya M (2007) Electroejaculation in llama (*Lama glama*) *Journal of Camel Practice and Research* **14** 203–206.

Djang KTF, Harun BA, Kumi-Diaka J, Yusuf H and Udomah MG (1988) Clinical and anatomical studies of the camel (*Camelus dromedarius*) genitalia *Theriogenology* **30** 1023–1031.

Drew ML, MeyersWallen VN, Acland GM, Guyer CL and Steinheimer DN (1999) Presumptive *Sry*-negative XX sex reversal in a llama with multiple congenital anomalies *Journal of the American Veterinary Medical Association* **215** 1134–1139.

El-Bahrawy KA (2010) Cryopreservation of dromedary camel semen supplemented with α-amylase enzyme *Journal of Camel Practice and Research* **17** 211–216.

El-Hariri MN and Deeb S (1979) Cryptorchidism with interstitial cell tumour in a case of camel (*Camelus dromedarius*) *Journal of the Egyptian Veterinary Medical Association* **39** 39–46.

El-Wishy AB (1988) Reproduction in the male dromedary (*Camelus dromedarius*): a review *Animal Reproduction Science* **17** 217–241.

El Wishy AB [El-Wishy AB] and Omar AA (1975) On the relation between testes size and sperm reserve in the one-humped camel (*Camelus dromedarius*) *Beiträge zur Tropischen Landwirtschaft. Veterinärmedizin* **13** 391–398.

Elliot FI (1961) Artificial insemination of Bactrian camel (*Camelus bactrianus*) *International Zoo Yearbook* **3** 94.

Fernandez-Baca S (1970) Estudios sobre la reproduccion en la alpaca, *Lama pacos Cuarto Boletin Extraordinario* **4** 33–42.

Fernandez-Baca S and Calderón W (1966) Methodos de coleccion de semen de la alpaca *Revista Facultad de Medicina Veterinaria de la Universidad Nacional Mayor de San Marcos, Lima* **18–20** 13–26.

Fernandez-Baca S and Novoa C (1969) Primeraesayo de inseminacion artificial en alpacas, *Lama pacos*, con semen de vicuña, *Vicugna vicugna Revista de la Facultad de Medicina Veterinaria de la Universidad Nacional Mayor de San Marcos, Peru* **22** 9–18.

Fernandez-Baca S, Sumar J and Novoa C (1972) Comportamiento sexual de la alpaca macho frente a la renovación de las hembras *Revista de Investigaciones Pecuarias* **1** 115–128.

Ferré LB and Werkmeister A (1996) Development of a thermo-electric artificial vagina for semen collection in camelids (preliminary results) *Revista Argentina de Producción Animal* **16** 363–365

Flores FRF (1970) *Estudio histologico del testiculo de alpacas aparentemente inaptas para la reproduccion.* BS thesis, Facultad de Medicina Veterinaria de la Universidad Nacional Mayor de San Marcos, Lima.

Flores P, García-Huidobro J, Muñoz C, Bustos-Obregón E and Urquieta B (2002) Alpaca semen characteristics previous to a mating period *Animal Reproduction Science* **72** 259–266.

Fowler ME (1998) *Medicine and Surgery of South American Camelids: Llama, Alpaca, Vicuña, Guanaco.* Blackwell Publishing, Ames.

Friеländer M, Rosenstrauch A and Bedrak E (1984) Leydig cell differentiation during the reproductive cycle of the seasonal breeder *Camelus dromedarius*: an ultrastructural analysis *General and Comparative Endocrinology* **55** 1–11.

Galloway DB (2000) The development of the testicles in alpacas in Australia In *Proceedings of the Australian Alpaca Industry Conference, Canberra, 25–27 August, 2000* pp 21–23.

Garcia Pereira FL, Allen A, Anouassi A and Tibary A (2004) Parainguinal cryptorchidectomy under general anaesthesia in a Bactrian camel (*Camelus bactrianus*) *Journal of Camel Practice and Research* **11** 103–107.

Garnica J, Achata R and Bravo PW (1993) Physical and biochemical characteristics of alpaca semen *Animal Reproduction Science* **32** 85–90.

Garnica J, Flores E and Bravo PW (1995) Citric acid and fructose concentrations in seminal plasma of the alpaca *Small Ruminant Research* **18** 95–98.

Gauly M (1997) *Seasonal Changes in Semen Characters and Serum Concentrations of Testosterone, Oestradiol-17β, Thyroxine and Triiodothyronine in Male Llamas (*Lama glama*) in Central Europe.* Fachbereich Veterinärmedizin, Justus-Liebig-Universität, Giessen.

Giuliano S, Director A, Gambarotta M, Trasorras V and Miragaya M (2008) Collection method, season and individual variation on seminal characteristics in the llama (*Lama glama*) *Animal Reproduction Science* **104** 359–369.

Giuliano S, Carretero M, Gambarotta M, Neild D, Trasorras V, Pinto M and Miragaya M (2010) Improvement of llama (*Lama glama*) seminal characteristics using collagenase *Animal Reproduction Science* **118** 98–102.

Giuliano SM, Chaves MG, Trasorras VL, Gambarotta M, Neild D, Director A, Pinto M and Miragaya MH (2012) Development of an artificial insemination protocol in llamas using cooled semen *Animal Reproduction Science* **131** 204–210.

Gombe S and Oduor-Okelo D (1977) Effect of temperature and relative humidity on plasma and gonadal testosterone concentrations in camels (*Camelus dromedarius*) *Journal of Reproduction and Fertility* **50** 107–108.

Graham EF, Schmehl MKL, Evensen BK and Nelson DS (1978) Semen preservation in non-domestic mammals In *Artificial Breeding of Non-Domestic Animals. Proceedings of a Symposium held at The Zoological Society of London, London, UK, 7–8 September, 1977* pp 153–173 Ed PF Watson. Proceedings Volume 43, Zoological Society, London.

Gray GA, Dascanio JJ, Kasimanickam R and Sponenberg DP (2007) Bilateral epididymal cysts in an alpaca male used for breeding *Canadian Veterinary Journal* **48** 741–744.

Gutierrez C, Corbera JA, Doreste F, Padrón TR and Morales M (2002) Silica urolithiasis in the dromedary camel in a subtropical climate *Veterinary Research Communications* **26** 437–442.

Gutierrez C, Corbera JA, Faye B (2008) Obstructive phosphate urolithiasis in a dromedary camel: a case report *Journal of Camel Practice and Research* **15** 77–79.

Heath AM, Pugh DG, Sartin EA, Navarre B and Purohit RC (2002) Evaluation of the safety and efficacy of testicular biopsies in llamas *Theriogenology* **58** 1125–1130.
Ismail ST (1982) Studies on the testis and epididymis of the one-humped camel (*Camelus dromedarius*). PhD thesis, Cairo University, Cairo.
Ismail STT (1988) Reproduction in the male dromedary (*Camelus dromedarius*) *Theriogenology* **29** 1407–1418.
Jochle W, Merkt H, Sieme H, Musa B and Badreldin H (1990) Sedation and analgesia with detomidine hydrocholride (Domosedan) in camelids for rectal examinations and electro-ejaculation In *Proceedings of the UCDEC Workshop 'Is it Possible to Improve the Reproductive Performance of the Camel?', 10–12 Sept., Paris* pp 263–271. Unité de Coordination pour l'Elevage Camelin, Paris.
Johnson LW (1989) Llama reproduction *Veterinary Clinics of North America: Food Animal Practice* **5** 159–182.
Johnson LW and Schultheiss PC (1994) Results of testicular biopsies in llamas In *Proceedings of the 1994 Symposium on the Health and Disease of Small Ruminants: with Supplementary Papers and an Appendix in Conjunction with the Central Veterinary Conference, Kansas City, Missouri, August 28–29, 1994* pp 54–55. American Association of Small Ruminant Practitioners, Ithaca.
Kershaw-Young CM, Evans G and Maxwell WMC (2012a) Glycosaminoglycans in the accessory sex glands, testes and seminal plasma of alpaca and ram *Reproduction, Fertility and Development* **24** 362–369.
Kershaw-Young CM, Druart X, Vaughan J and Maxwell WMC (2012b) β-Nerve growth factor is a major component of alpaca seminal plasma and induces ovulation in female alpacas *Reproduction, Fertility and Development* **24** 1093–1097.
Khan AA and Kohli IS (1972) A study on sexual behaviour of male camel (*Camelus dromedarius*). Part I *Indian Veterinary Journal* **49** 1007–1012.
Kingston JK and Stäempfli HR (1995) Silica urolithiasis in a male llama *Canadian Veterinary Journal* **36** 767–768.
Koford CB (1957) The vicuña and the puna *Ecological Monographs* **27** 153–219.
Kubicek J (1974) Samentnahme beim Alpaca durch eine Harnrohrenfistel *Zeitschrift für Tierzüchtung und Züchtungsbiologie* **90** 335–351.
Lane D (1999) Preputial prolapse in an alpaca *Canadian Veterinary Journal* **40** 260.
Lichtenwalner AB, Woods GL and Weber JA (1996a) Ejaculatory pattern of llamas during copulation *Theriogenology* **46** 285–291.
Lichtenwalner AB, Woods GL and Weber JA (1996b) Seminal collection, seminal characteristics and pattern of ejaculation in llamas *Theriogenology* **46** 293–305.
Lichtenwalner AB, Woods GL, Weber JA (1997) Pattern of emission during copulation in male llamas *Biology of Reproduction* **56** 298–298.
Losno W and Coyotupa J (1979) Testoterona sérica en alpacas macho prepúberes In *Resúmenes de Proyectos de Investigación, Período 1975–1979* p 116. Universidad Nacional Mayor de San Marcos, Lima.
Marie ME (1987) Bases endocriniennes de la fonction sexuelle chez le dromadaire (*Camelus dromedarius*) These de doctorat, l'Université Paris 6, Paris.
Medan MS, Absy G, Zeidan AE, Khalil MH, Khalifa HH, Abdel-Salaam AM and Abdel-Khalek TM (2008) Survival and fertility rate of cooled dromedary camel spermatozoa supplemented with catalase enzyme *Journal of Reproduction and Development* **54** 84–89.
Merlian CP, Sikes JD, Read BW, Boever WJ and Knox D (1986) Comparative characteristics of spermatozoa and semen from a bactrian camel, dromedary camel and llama *Journal of Zoo Animal Medicine* **10** 22–25.
Montalvo C, Cevallos E and Copaira M (1979) Estudio microscopico del parenquima testicular de la alpaca durante las estaciones del ano In *Resúmenes de de Investigación, Período 1975–1979* p 37. Universidad Nacional Mayor de San Marcos, Lima.
Morton KM, Bathgate R, Evans G and Maxwell WMC (2007) Cryopreservation of epididymal alpaca (*Vicugna pacos*) sperm: a comparison of citrate-, Tris- and lactose-based diluents and pellets and straws *Reproduction, Fertility and Development* **19** 792–796.
Morton KM, Gibb Z, Bertoldo M and Maxwell WMC (2009) Effect of diluent, dilution rate and storage temperature on longevity and functional integrity of liquid stored alpaca (*Vicugna pacos*) semen *Journal of Camelid Science* **2** 15–25.
Morton KM, Billah M and Skidmore JA (2010a) Artificial insemination of dromedary camels with fresh and chilled semen: effect of diluent and sperm dose, preliminary results In *Reproduction in Domestic Ruminants VII* p 493 [Abstract] Ed MC Lucy, JL Pate, MF Smith and TE Spencer. Published for Society of Reproduction and Fertility by Nottingham University Press, Nottingham.
Morton KM, Billah M and Skidmore JA (2010b) Pellet freezing improves post-thaw motility and viability of dromedary camel spermatozoa *Reproduction in Domestic Animimals* **45** 99.

Morton KM, Evans G and Maxwell WMC (2010c) Effect of glycerol concentration, Equex STM® supplementation and liquid storage prior to freezing on the motility and acrosome integrity of frozen–thawed epididymal alpaca (*Vicugna pacos*) sperm *Theriogenology* **74** 311–316.

Morton KM, Gibb Z, Leahy T and Maxwell WMC (2012) Effect of enzyme treatment and mechanical removal of alpaca (*Vicugna pacos*) seminal plasma on sperm functional integrity *Journal of Camelid Science* **5** 62–81.

Mukasa-Mugerwa E (1981) Reproductive performance In *The Camel* (Camelus dromedarius). *A Bibiliographical Review* pp 11–32. ILCA Monograph No 5, International Livestock Centre for Africa.

Niasari-Naslaji A, Mosaferi S, Bahmani N, Gharahdaghi AA, Abarghani A and Ghanbari A (2006) Effectiveness of a Tris-based extender (SHOTOR diluent) for the preservation of Bactrian camel (*Camelus bactrianus*) semen *Cryobiology* **53** 12–21.

Niasari-Naslaji A, Mosaferi S, Bahmani N, Gerami A, Gharahdaghi AA, Abarghani A and Ghanbari A (2007) Semen cryopreservation in Bactrian camel (*Camelus bactrianus*) using SHOTOR diluent: effects of cooling rates and glycerol concentrations *Theirogenology* **68** 618–625.

Novoa C (1970) Reproduction in Camelidae *Journal of Reproduction and Fertility* **22** 3–20.

Osman AM and El Azab EA (1974) Gonadal and epididymal sperm reserves in the camel, *Camelus dromedarius Journal of Reproduction and Fertility* **2** 425–430.

Osman DI and Plöen L (1986) Spermatogenesis in the camel (*Camelus dromedarius*) *Animal Reproduction Science* **10** 23–36.

Osman DI, Moniem KA and Tingari MD (1979) Histological observations on the testis of the camel, with special emphasis on spermatogenesis *Acta Anatomica* **104** 164–171.

Pacheco Curie JI (2008) Métodos de colección de semen en camélidos sudamericanos [Methods of semen collection in South American camelids] *REDVET (Revista Electrónica de Veterinari)* **9(5)**:050804. Available at: http://www.redalyc.org/pdf/636/63611397005.pdf (accessed 18 September 2013).

Pacheco Curie JI, Pérez Durand GM, Calle Charaja L and García Vera W (2009) Efecto del lugar y la hora de inseminación artificial sobre la fertilidad en alpacas [Effect of the place and the hour of artificial insemination on the fertility in Alpacas] *REDVET* **10(8)**:080905. Available at: http://www.veterinaria.org/revistas/redvet/n080809/080905.pdf (accessed 18 September 2013).

Pan G, Chen Z, Liu X, Li D, Xie, Q, Ling F and Fang L (2001) Isolation and purification of the ovulation-inducing factor from seminal plasma in the bactrian camel (*Camelus bactrianus*) *Theriogenology* **55** 1863–1879.

Pearson LK, Campbell AJ, Sandoval S and Tibary A (2013) Effects of vasectomy on seminal plasma alkaline phosphatase in male alpacas (*Vicugña pacos*) *Reproduction in Domestic Animals* DOI: 10.1111/rda.12199.

Perkins NR, Frazer GS and Hull BL (1996 Endocrine diagnosis of cryptorchidism in a llama *Australian Veterinary Journal* **74** 275–277.

Picha Y, Sandoval S, Pearson LK, Elzawam A, Alkar A and Tibary A (2011) Evaluation of stress response to electroejaculation in anesthetized alpacas (*Vicugna pacos*) *Clinical Theriogenology* **3** 380.

Quispe F (1987) Evaluacion de las caracteristicas fisicas del semen de la alpaca durante la epoca de empadre. BSc thesis, Facultad de Medicina Veterinaria, Universidad Nacional del Altiplano, Puno.

Rahim SEAA (1997) Studies on the age of puberty of male camels (*Camelus dromedarius)* in Saudi Arabia *The Veterinary Journal* **154** 79–83.

Ratto MH, Huanca W, Jaswant S and Adams GP (2006) Comparison of the effect of ovulation-inducing factor (OIF) in the seminal plasma of llamas, alpacas, and bulls *Theriogenology* **66** 1102–1106.

Ratto MH, Leduc Y, Valderrama X, va Staraten K, Delbaere L, Pierson R and Adams G (2012) The nerve of ovulation-inducing factor in semen *Proceedings of the National Academy of Sciences of the United States of America* **109** 15042–15047.

Riaz P, Qureshi AS, Laeeq L, Huma J, Ayesha M, Sohail H, Javed I, Zahid K and Khamas W (2011) Seasonal changes in the anatomy of testis of one-humped camel (*Camelus dromedarius*) *Journal of Camel Practice and Research* **18** 145–153.

Rietschel W (1990) Cryptorchidism in the vicugna (*Lama vicugna*) *Tierärztliche Praxis* **18** 459–461.

Sánchez AE and Correa JE (1995) Ultrasonographic description of accessory sex glands in the male llama *Archivos de Medicina Veterinaria* **27** 81–86.

Santiani A, Huanca W, Sapana R, Huanca T, Sepúlveda N and Sánchez R (2005) Effects on the quality of frozen–thawed alpaca (*Lama pacos*) semen using two different cryoprotectants and extenders *Asian Journal of Andrology* **7** 303–309.

Schwalm A, Gauly M, Erhardt G and Bergmann M (2007) Changes in testicular histology and sperm quality in llamas (*Lama glama*), following exposure to high ambient temperature *Theriogenology* **67** 1316–1323.

Sharma SS (1981) Studies of sexual physiology of stud camel (*Camelus dromedarius*) *Indian Veterinary Journal* **58** 743–744.

Sieme H, Merkt H, Musa B, Hago BEO and Willem T (1990) Liquid and deep freeze preservation of camel semen using different extenders and methods In *Proceedings of the UCDEC Workshop 'Is it Possible to Improve the Reproductive Performance of the Camel?', 10–12 Sept., Paris* pp 273–284. Unité de Coordination pour l'Elevage Camelin, Paris.

Singh UB and Bharadwaj MB (1978) Histological and histochemical studies on the testis of camel (*Camelus dromedarius*) during the various seasons and ages. Part II *Acta Anatomica* **101** 280–288.

Skidmore JA and Billah M (2006) Comparison of pregnancy rates in dromedary camels (*Camelus dromedarius*) after deep intra-uterine versus cervical insemination *Theriogenology* **66** 292–296.

Skidmore JA, Billah M, Binns, M, Short RV and Allen WR (1999) Hybridizing Old and New World camelids: *Camelus dromedarius* × *Lama guanicoe*. *Proceedings of the Royal Society, London, B* **266** 649–656.

Skidmore JA, Morton KM and Billah M (2013) Artificial insemination in dromedary camels *Animal Reproduction Science* **136** 178–186.

Smith CL, Peter AT and Pugh DG (1994) Reproduction in llamas and alpacas: a review *Theriogenology* **41** 573–592.

Stelletta C, Juyena NS, Ponce Salazar D, Ruiz J and Gutierrez G (2011) Testicular cytology of alpaca: comparison between impressed and smeared slides *Animal Reproduction Science* **125** 133–137.

Sumar J (1983) *Studies on Reproductive Pathology in Alpacas.* Faculty of Veterinary Medicine, University of Agricultural Sciences, Uppsala.

Sumar J (1985) Reproductive physiology in South American camelids In *Genetics of Reproduction in Sheep* pp 81–95 Ed RB Land and DW Robinson. Butterworths, London.

Sumar J (1989) Defectos del sistema reproductivo In *Defectos Congénitos y Hereditarios en la Alpaca – Teratologia* [Congenital and Hereditary Defects in Alpaca – Teratology] pp. 23–34 Ed J Sumar-Kalinowski. Published for Consejo Nacional de Ciencia y Tecnología, Lima by Gráfica Bellido, Lima.

Sumar JB (1996) Reproduction in llamas and alpacas *Animal Reproduction Science* **42** 1–4.

Taha AAM, Abdel-Magied EM, Abdalla MA and Abdalla AB (1994) The poll glands of the dromedary (*Camelus dromedarius*): ultrastructural characteristics *Anatomia Histologia Embryologia* **23** 269–274.

Tanco VM, Ratto MH, Lazzarotto M and Adams GP (2011) Dose-response of female llamas to ovulation-inducing factor from seminal plasma *Biology of Reproduction* **85** 452–456.

Tibary A (2001) Testicular ultrasonography and biopsy in small ruminants and lamas In *Proceedings of the Annual Conference of the Society for Theriogenology, September 12–15, 2001, Vancouver, BC, Canada* pp 369–378.

Tibary A (2003) Male reproduction In *The Complete Alpaca Book* pp 323–350 Ed E Hoffman. Bonny Doon Press, Santa Cruz.

Tibary A (2004) Testicular biopsy In *Proceedings of the North American Veterinary Conference, Orlando, Florida, January 17–21, 2004* pp 297–298. The North American Veterinary Community, Gainesville.

Tibary A (2008) Male infertility: clinical cases presentation To *International Camelid Health Conference for Veterinarians, The Ohio State University College of Veterinary Medicine, Columbus.*

Tibary A and Anouassi A (1997a) Anatomy of the male genital tract In *Theriogenology in Camelidae: Anatomy, Physiology, BSE, Pathology and Artificial Breeding* pp 17–44 Ed A Tibary. Actes Editions, Institut Agronomique et Veterinaire Hassan II, Rabat.

Tibary A and Anouassi A (1997b) Reproductive Physiology of the Male In *Theriogenology in Camelidae: Anatomy, Physiology, BSE, Pathology and Artificial Breeding* pp 49–74 Ed A Tibary. Actes Editions, Institut Agronomique et Veterinaire Hassan II, Rabat.

Tibary A and Anouassi A (1997c) Male breeding soundness examination In *Theriogenology in Camelidae: Anatomy, Physiology, BSE, Pathology and Artificial Breeding* pp 79–111 Ed A Tibary. Actes Editions, Institut Agronomique et Veterinaire Hassan II, Rabat.

Tibary A and Anouassi A (1997d) Pathology and surgery of the reproductive tract and associated organs in the male Camelidae In *Theriogenology in Camelidae: Anatomy, Physiology, BSE, Pathology and Artificial Breeding* pp 115–132 Ed A Tibary. Actes Editions, Institut Agronomique et Veterinaire Hassan II, Rabat.

Tibary A and Anouassi A (1997e) Artificial breeding and manipulation of reproduction in Camelidae In *Theriogenology in Camelidae: Anatomy, Physiology, BSE, Pathology and Artificial Breeding* pp 413–457 Ed A Tibary. Actes Editions, Institut Agronomique et Veterinarie Hassan II, Rabat.

Tibary A and Anouassi A (2000) Ultrasonography of the genital tract in camels (*Camelus dromedarius* and *Camelus bactrianus*) In *Selected Topics on Camelids* Ed TK Gahlot. The Camelids Publisher, Bikaner.

Tibary A and Anouassi A (2002) Male camelid infertility/sub-fertility: clinical cases In *Proceedings, Camelid Medicine, Surgery and Reproduction Conference* pp 427–428. The Ohio State University, Columbus.

Tibary A and Pugh D (2003) Infertility in the male lamoid In *Proceedings, Society for Theriogenology and American College of Theriogenologists Annual Conference, Columbus, Ohio, September 17–20, 2003* pp 304–312.

Tibary A and Vaughan J (2006) Reproductive physiology and infertility in male South American camelids: a review and clinical observations *Small Ruminant Research* **61** 283–298.

Tibary A, Anouassi A and Memon MA (2001) Approach to infertility diagnosis in camelids: retrospective study in alpacas, llamas and camels *Journal of Camel Practice and Research* **8** 167–179.

Tibary A, Abdelhaq A and Abdelmalek S (2005) Factors affecting reproductive performance of camels at the herd and individual level *NATO Science Series: Life and Behavioural Sciences Volume 362* pp 97–114. IOS Press, Amsterdam.

Tibary A, Anouassi A, Sghiri A and Khatir H (2007) Current knowledge and future challenges in camelid reproduction. *Society of Reproduction and Fertility Supplement* **64** 297–313.

Tibary A, Rodriguez J and Sandoval S (2008) Reproductive emergencies in camelids *Theriogenology* **70** 515–534.

Tibary A, Pearson LK and Picha Y (2011a) Congenital anomalies in crias In *Proceedings of the North American Veterinary Conference, Orlando Florida, January 15–19, 2011* pp 327–329. The North American Veterinary Community, Gainesville.

Tibary A, Picha Y, Pearson LK (2011b) Diagnosis and surgical management of cryptorchidism in camelids. *Proceedings of the North American Veterinary Conference, Orlando Florida, January 15–19, 2011* pp 330–331. The North American Veterinary Community, Gainesville.

Tibary A, Picha Y and Pearson LK (2011c) Diagnosis of testicular cysts and their significance in infertility in camelids. *Proceedings of the North American Veterinary Conference, Orlando Florida, January 15–19, 2011* p 332. The North American Veterinary Community, Gainesville.

Tingari MD, Rahma BA and Saad AHM (1984) Studies on the poll gland of the one-humped camel in relation to reproductive activity. I. Seasonal morphological and histochemical changes *Journal of Anatomy* **138** 193–205.

Urquieta B and Rojas JR (1990) Studies on the reproductive physiology of the vicuña (*Vicugna vicugna*). In *Livestock Reproduction in Latin America. Proceedings of the Final Research Coordination Meeting of the FAO/IAEA/ARCAL, Vienna* pp 407–428.

Urquieta B, Cepeda R, Caceres JE, Raggi LA and Rojas JR (1994) Seasonal variation in some reproductive parameters of male vicuña in the High Andes of northern Chile *Journal of Arid Environments* **26** 79–87.

Vaughan JL (2001) Control of ovarian follicular growth in the alpaca (*Lama pacos*). PhD thesis, Central Queensland University, North Rockhampton, Queensland.

Vaughan JL, Galloway D and Hopkins D (2003a) Artificial insemination in alpacas (*Lama pacos*) In *Publication No 03/104, RIRDC Project No AAA-1A* pp 67–73. Rural Industries Research and Development Corporation, Kingston, Australia.

Vaughan JL, Macmillan KL, Anderson GA and D'Occhio MJ (2003b) Effects of mating behaviour and the ovarian follicular state of female alpacas on conception *Australian Veterinary Journal* **81** 86–90.

Waheed MM, Ghoneim IM, Al-Eknah MM and Al-Haider AK (2010) Effect of extenders on motility and viability of chilled–stored camel spermatozoa (*Camelus dromedarius*) *Journal of Camel Practice and Research* **17** 217–220.

Waheed MM, Ghoneim IM and Hassieb MM (2011) Development and validation of testicular biopsy for diagnosis of infertility in camels (*Camelus dromedarius*) *Clinical Theriogenology* **3** 105–114.

Wani N, Billah M and Skidmore JA (2008) Studies on liquefaction and storage of ejaculated dromedary camel (*Camelus dromedarius*) semen *Animal Reproduction Science* **109** 309–318.

Wani NA, Morton KM, Billah M and Skidmor JA (2011) Biophysical and biochemical characteristics of ejaculated semen of dromedary camel (*Camelus dromedarius*) and llama (*Lama glama*) *Journal of Camel Practice and Research* **18** 97–102.

Xu YS, Wang HY, Zeng GQ, Jiang GT and Gao YH (1985) Hormone concentrations before and after semen-induced ovulation in the bactrian camel (*Camelus bactrianus*) *Journal of Reproduction and Fertility* **74** 341–346.

Zia ur R, Ahmad N, Bukhari SA, Akhtar N and Haq IU (2007) Serum hormonal, electrolytes and trace element profiles in the rutting and non-rutting one-humped male camel (*Camelus dromedarius*) *Animal Reproduction Science* **101** 172–178.

17 Applied Andrology in Endangered, Exotic and Wildlife Species

Rebecca Spindler,[1]* Tamara Keeley[2] and Nana Satake[2]
[1] *Taronga Zoo, Mosman, New South Wales, Australia;* [2] *Taronga Western Plains Zoo, Dubbo, New South Wales, Australia*

Introduction

The long-term viability of any animal population relies on the reproductive success of the constituent animals. In turn, reproductive success is intrinsically linked to an individual's health and well-being. As such, understanding reproductive physiology is essential to improving the genetic management, reproductive output and overall success of insurance populations of rare and important wildlife species. In addition, reproductive assessments should be incorporated into holistic programmes assessing the health and viability of wildlife populations. Assisted Reproductive Techniques (ARTs) can be extremely useful tools in the genetic management of these populations (Pukazhenthi *et al.*, 2006a), but should be used in a manner that will support population sustainability in the long term.

The relative ease of access to spermatozoa compared with oocytes results in greater opportunities to study andrology and, consequently, gain a better understanding of male gamete biology in a range of wildlife species. This has also led to the asymmetrical efforts and success in sperm banking compared with storing oocytes and embryos, in both domestic and wildlife species. While efforts are required to address this imbalance, the current capacity to collect, assess, cryopreserve and use spermatozoa to achieve targeted conception has great implications for the genetic management of species and conservation of biodiversity. Long-term cryopreservation of spermatozoa from wild animals, especially, can facilitate zoo-based breeding programmes by replenishing genetic material without the associated cost and anxiety of animal translocations (Wildt, 1997).

Despite this potential application, detailed knowledge of species-specific methods for studying and manipulating male reproductive activity in wildlife species is limited. There is a wealth of andrological information available on domestic species (see other chapters in this volume, and *Marshall's Physiology of Reproduction*, ed. Lamming, 1990, for summary) and humans (*Campbell's Urology*, eds Walsh *et al.*, 2002) that informs our study and control of male reproduction in wildlife species. Investigations of analogue domestic species have provided valuable launching points for investigations into wildlife reproduction and have led to the development of appropriate technology and equipment. However, it must be remembered that even closely related species often exhibit significant and biologically meaningful variations in

* E-mail: rspindler@zoo.nsw.gov.au

reproductive and endocrine parameters. A successful approach to studying or manipulating male reproduction in one species will often not work in another species without significant modification. Thus, a significant amount of effort must be exerted to develop appropriate techniques for each wildlife species, regardless of the weight of research in taxonomically close species.

Primary Challenges when Studying Wildlife Andrology

Availability of animals

Efforts to investigate wildlife andrology have focused on evaluating the semen of selected species and establishing baseline concentrations of circulating hormones (Spindler and Wildt, 2010). The numbers of animals typically included in such studies are extremely low in comparison with similar studies conducted on domestic species. This is largely due to the inability to collect longitudinal reproductive information from free-ranging animals that do not have a known life history. Investigations have relied on the availability of conspecific animals in zoos to provide baseline reproductive data, improve overall understanding of species biology and develop field techniques appropriate for the collection of samples and data from the field (Spindler and Wildt, 2010). Data collection from zoo-based animals is more laborious than it is for domestic species owing to the low numbers of animals held in each facility, the geographically dispersed locations of zoos and the diversity of environmental conditions experienced by each individual. Sourcing animals and collecting data in the field involves even greater expense and level of difficulty, and is often performed over many years, thereby introducing additional elements of variation.

Animal and researcher health and safety

Understanding the andrology of wildlife species requires close observation, physical examination, sample collection and, possibly, ultrasound or radiography of males. In most cases, this will require anaesthesia in order to avoid stress to the animal and injury to the technicians. As a result, a great deal of research is invested in developing non-invasive measures of reproduction in wildlife species. Hormone monitoring of faecal samples has greatest application in the female to determine reproductive cycling, seasonality and maturity; it also has potential to assist the reproductive management of males (Morai *et al.*, 2002; Pereira *et al.*, 2005; Hesterman and Jones, 2009). These non-invasive techniques have the added benefit of reducing the number of anaesthetic events required to gain an understanding of male reproductive physiology, and so reducing the potential impact on animal well-being.

However, most assessment of, and research into, male physiology requires the collection of semen samples and, in wildlife species, this necessitates anaesthesia. Modern anaesthetic regimes, developed to address species-specific physiology, have reduced the potential impacts for many species. The complete physiological impact of any anaesthetic regime should be considered fully before undertaking semen collection so as to ensure animal safety and well-being, as well as maximizing the chances of collecting a good-quality sample (Zambelli *et al.*, 2007).

Genetic diversity

Inbreeding depression most likely influences reproductive fitness in a variety of ways across a diverse range of species. For example, reduced genetic diversity is closely correlated with decreased testicular sperm concentrations in oldfield mice (*Peromyscus polionotus*) (Margulis and Walsh, 2002), decreased sperm quality in Cuvier's gazelles (*Gazella cuvieri*) (Roldan *et al.*, 1998), increased juvenile mortality in ungulates (Ralls *et al.*, 1979) and small mammals (Ralls and Ballou, 1982), and reduced sperm quality in clouded leopards and lions (Wildt *et al.*, 1986, 1987). Population management is essential to the reproductive health of populations and can be assisted by increased reproductive knowledge and the development of techniques to ensure gene capture and the maintenance of genetic diversity.

Environmental conditions

It is well established that reproductive success is affected by environmental stressors (Ramaley, 1981; Moberg, 1985, 2000; Lasley and Kirkpatrick, 1991; Rivier and Rivest, 1991; Pottinger, 1999; Dobson *et al.*, 2001, 2003). Wildlife species and, in particular, specialized carnivores, appear to be particularly susceptible to novel environments and stressors (Mellen, 1991; Carlstead *et al.*, 1993a,b; Jurke *et al.*, 1997; Carlstead, 2002; Moreira *et al.*, 2002; Wielebnowski, 2002; Wielebnowski *et al.*, 2002). These stressors result in the stimulation of the hypothalamic-pituitary-adrenocortical (HPA) axis. Both beneficial situations (e.g. courtship, copulation, obtaining prey, giving birth) and detrimental situations (e.g. fighting, capture, (some) transport) can elicit an adrenal response (Colborn *et al.*, 1991; Moberg, 2000). Subsequent acute (short-term) elevations in glucocorticoids can help an animal to react appropriately to the challenge through rapid energy mobilization. In contrast, a chronic (long and sustained) glucocorticoid rise can reduce fitness through a variety of mechanisms, ranging from immunosuppression to poor reproduction and offspring survival (Moberg, 1985, 2000). These potential stressors must be managed or eliminated to ensure the optimal reproduction of wildlife species.

Health and nutrition

Disease can be defined as any impairment that interferes with or modifies the performance of normal bodily functions; this includes non-infectious, infectious and parasitic diseases. Health and nutrition are intimately linked and many health issues are either directly related to poor nutrition, or are compounded by poor immune function due to poor nutrition. Owing to the tight regulations over the origin and quarantine of wildlife species, infectious diseases that directly affect reproduction are unlikely to be of concern in captivity. In free-ranging populations, diseases of concern that could be encountered include canine distemper virus, feline infectious peritonitis, genera-specific immunodeficiency viruses, rabies, herpesvirus, calicivirus, feline parvovirus (panleukopenia), feline leukaemia virus, *Leptospira interrogans*, *Chlamydia* spp., *Salmonella* spp., *Toxoplasma gondii* and *Dirofilaria immitis* (Spindler *et al.*, 2007). In many such cases, the spermatozoa of valuable males may still be used after sperm washing and the verification of disease-free status before artificial insemination (AI) (Loskutoff *et al.*, 2005).

Nutrition plays a fundamental role in overall animal demeanour, hormone production, sperm production and fecundity. For example, malnutrition alters luteinizing hormone (LH) and testosterone secretion patterns in rhesus macaques (*Macaca mulatta*) (Lado-Abeal *et al.*, 2002) and reduces overall fecundity in the marmoset (Tardif and Jaquish, 1994) and the water buffalo (Oswin-Perera, 1999). More specifically, diets lacking in specific micronutrients will have an impact on reproduction. For example, calcium influences erectile function (Mills *et al.*, 2001), spermatogenesis (Andonov and Chaldakov, 1991), oocyte maturation (Machaca and Haun, 2002), fertilization (Stricker, 1999) and embryo development (Stricker, 1999). Further, deficiency in vitamin A causes male infertility, embryonic loss, decreased neonatal survival and fetal malformations (Clagett-Dame and DeLuca, 2002), while vitamin D deficiency reduces fertility by controlling the onset of puberty and fertility in both males and females (Halloran and DeLuca, 1979). Attention to the specific dietary requirements of wildlife species must be paid to avoid risks to health and reproduction. For example, felids have a unique and specific requirement for micronutrients such as arachidonic acid, which is essential for reproduction and spermatogenesis (MacDonald *et al.*, 1984).

Ethics

All animals must be held in conditions that meet species-specific and individual needs for biological and psychological health. This ensures the best animal well-being as well as the greatest relevance of research data. As with any animal research, the benefits of the research to health, conservation or animal management must be weighed against the cost to individual animals engaged in the research. This must be undertaken by a licensed animal ethics committee that has specific knowledge of wildlife species and the associated opportunities and limitations to their study. Making these judgements about research on wildlife species must take into consideration that the impact of intervention on wildlife species may be greater than on domestic species owing to the lack of habituation to human presence and the potential risks to researchers, as described above. Conversely, these costs, if managed appropriately and humanely, are often dwarfed by the opportunity to better understand and manage the reproduction, health and even conservation status of these and other species through conducting relevant research under the highest standards and ethical conditions.

Natural and Assisted Breeding Strategies

Reproductive strategies employed by wildlife species range from asexual reproduction through promiscuity, polyandry and polygyny to monogamy. The mating strategy of the species in question will have a significant influence on male reproductive characteristics, including testis size, sperm characteristics, and even semen content. Knowledge of these strategies also allows some assumptions to be made when sampling from a species for the first time, and may also assist with developing sample collection and handling methods. For example, males in competitive breeding situations (polyandry) will tend to produce more semen, more spermatozoa and have larger testes than males with limited competition (Parker, 1998). Further, some primate, rodent, marsupial and reptile species ejaculate a coagulated 'plug' to prevent easy access to females by subsequent mates, which may cause damage to tissues if not completely expelled from the male reproductive tract during stimulated ejaculation.

Wildlife species often employ mate selection strategies that avoid inbreeding and ensure the best possible mate to avoid neonate or infant mortality. Selection tools are used to select appropriate matings in globally managed zoo programmes, and these have been demonstrated to maintain excellent levels of genetic diversity in populations over many generations (Ballou and Foose, 1996). However, the loss of genetic diversity is almost inevitable over generations. Supplementation of an insurance population with genes from wild males (using cryopreserved spermatozoa) between three and five times a year is predicted to maintain the long-term viability of even small populations of Eld's deer (*Panolia eldii*), Przewalski's horse (*Equus ferus przewalskii*) and the Sumatran tiger (*Panthera tigris sumatrae*) (Harnal *et al.*, 2002).

Species biology will often dictate that offspring gender will be skewed under specific environmental conditions. While the specific conditions triggering this imbalance may be difficult to tease apart, the impact may be resolved through the application of advanced sperm gender-sorting technologies developed using domestic species (Johnston, 1992). These technologies are still in their infancy for wildlife species. Active gender selection of offspring would resolve many issues around the management of populations with a gender imbalance (usually towards the male) that is not representative of wild populations.

Cryopreservation

The ability of human spermatozoa to withstand sub-zero temperatures was realised in the 19th century by the work of Mantegazza, although the potential applications of this technique were not recognized until the mid-20th century. The critical consideration of precise conditions and manipulations of the freezing and thawing processes to a vast number of species across a wide range of taxa has enabled long-term preservation of viable spermatozoa. Of primary concern in the development of cryopreservation methods is the avoidance of cryoinjury during the transition through solutions of altered tonicity and temperature (Mazur, 1963). These transient states include moving from isotonic solution to hypertonic solution to supercooled solution with the greatest concern for cell integrity between the temperatures of –30°C and –40°C, at which point intracellular freezing and ice crystal formation occurs.

Vitrification, whereby freezing occurs at an ultra-rapid rate in order to avoid the formation of ice crystals, is a lesser studied avenue of cryopreservation for spermatozoa, but is the preferred method of preservation for organized groups of cells, such as testicular tissues and embryos; this turns specimens into a glass-like formation (Vajta and Nagy, 2006). Frog spermatozoa are able to survive exposure to liquid air at –192°C (Luyet and Hodapp, 1938). Fowl spermatozoa were revived after partial dehydration, freezing to –76°C and storage over several months (Shaffner, 1942). Human spermatozoa have been known to be particularly resistant to the vitrification process (Isachenko *et al.*, 2003, 2012). Vitrification has been successful in recovering motile, viable spermatozoa after thawing in various species, including canids (Sanchez *et al.*, 2011), rhesus macaques (Dong *et al.*, 2009) and rainbow trout (*Oncorhynchus mykiss*) (Merino *et al.*, 2011).

The addition of glycerol was found to be beneficial to sperm cryopreservation in the mid-20th century, and since then the preferred cryopreservation methods have incorporated the use of cryoprotectant agents (CPA) to the present day. The main properties that determine an appropriate CPA include solubility, the ability to form hydrogen bonds with water and the impact on the melting point of ice with a view to reducing ice crystal formation and moderating cryoinjury. Two classes of CPA exist, penetrating and non-penetrating; penetrating substances have a molecular weight of less than 100 daltons and are able to penetrate cells. Examples of CPA routinely used in cryopreservation of spermatozoa include penetrating substances such as glycerol, dimethyl sulfoxide (DMSO) and sugars, as well as non-penetrating substances such as polyvinylpyrrolidone and polyethylene glycol.

Cryotoxicity

Sperm viability post-thaw may be considerably compromised owing to species-specific sensitivity to the CPA or concentration of the CPA used. Even within taxonomic families, species and individuals may differ markedly in their response to CPA. For example, in the domestic livestock industry, bull semen is routinely cryopreserved with commercially produced extenders and used for AI. However, using a basic semen extender developed for the domestic bull, such as a Tris-egg yolk-glycerol extender, can damage sperm DNA, plasma membranes and/or acrosomes from a related species, such as the European bison (*Bison bonasus*) (Pérez-Garnelo *et al.*, 2006) and the buffalo (*Bubalus bubalis*) (Andrabi, 2009). Other examples of animals with varying and species-specific needs include ferrets (*Mustela putorius furo*) and black-footed ferrets (*Mustela nigripes*) (Howard *et al.*, 1991), dogs and wild canids (Farstad, 1998), and cats and wild felids (Pukazhenthi *et al.*, 2006b; Herrick *et al.*, 2010).

Field applications for collecting and cryopreserving semen from free-ranging males

The ability to collect, cryopreserve and use spermatozoa from free-ranging males of any species can effectively increase the genetic diversity of a closed population, while avoiding the removal of animals from the wild, the potential stress of animal transport and the risk of disease introduction. The disadvantage of this approach is that field conditions often include an absence of electricity for cooling and centrifuging spermatozoa. Spermatozoa are clearly sensitive to high temperatures (England and Ponzio, 1996; Brinsko *et al.*, 2000; Johnston *et al.*, 2000a; Nair *et al.*, 2006; Purdy, 2006), and long-term exposure to seminal plasma has been shown to compromise sperm motility, viability and even DNA integrity (Wildt *et al.*, 1988; Centurion *et al.*, 2003; Love *et al.*, 2005). Portable instant cool packs that can be activated when required are useful in maintaining tolerable temperatures under field conditions. Further, experimentation can be useful in indicating the optimal diluent solutions that will reduce the impact of seminal plasma (Morato *et al.*, 2003). These strategies may prolong sperm viability in transit to laboratories or provide essential steps in a cryopreservation protocol. Ultimately, effective semen collection, cooling and cryopreservation techniques must be tailored to the species and the field conditions in question in order to preserve the viability of spermatozoa (Plate 60).

Taxa-specific Notes

Non-human primates

Great advances have been made in the fields of human andrology and reproductive medicine. In comparison, knowledge of non-human primates is poor and has yet to be prioritized despite a rapid decline in many species and populations (Morrell and Hodges, 1998). A diverse range of assisted reproduction protocols has been established in species that serve as models for humans. For example, the use of frozen–thawed spermatozoa has been shown to successfully produce embryos *in vitro* and result in offspring after embryo transfer in marmosets (Callitrichidae) and gorillas (*Gorilla gorilla gorilla*) (Pope *et al.*, 1997). Successful production of offspring from intracytoplasmic sperm injection has also been repeated over the past 25 years. Cryopreservation and the recovery of viable spermatozoa post thaw for AI in non-human primates has proved to be more problematic, although success has been reported in the chimpanzee (*Pan* sp.) (Gould, 1990), rhesus macaque (Wolf *et al.*, 2004) and marmoset (*Callithrix jacchus*) (Morrell *et al.*, 1998).

The varying structure of primate spermatozoa and their susceptibility to cryoinjury, particularly damage to their DNA (Li *et al.*, 2007), is of concern (Rall, 1993). As with many other wildlife species, primates show significant differences in post-thaw sperm survival between individual males, even though pre-freeze parameters are comparable. Thus, optimizing standardized protocols becomes increasingly difficult. Furthermore, a recent study in rhesus macaques found that while the morphological and motility parameters of spermatozoa post thaw were found to be of acceptable quality by visual assessment, when penetration tests through cervical mucus and hyaluronic acid gel were conducted, spermatozoa were found to have a much reduced efficiency (Tollner *et al.*, 2011).

Carnivores

Carnivores employ a wide range of reproductive strategies, often influenced by an equally wide range of ecological and social parameters. Whereas some species have received research attention, there is still a great deal of information yet to be gained when the wide diversity of species and their vulnerability to environmental perturbation are considered, as well as the lack of a domestic model that has received adequate research attention. The domestic dog and cat have provided some clues to the reproduction of their non-domestic relatives, but research on even these models has been limited compared with livestock and laboratory species.

Ursids, canids and felids are usually seasonal breeders in nature, but semen can be obtained from zoo-based males of some species throughout the year via electroejaculation (EE) (Howard *et al.*, 1986). In contrast, the spermatozoa of mustelids (Sundqvist *et al.*, 1984) or large canids (Koehler *et al.*, 1998) can only be recovered during their breeding season. Across the carnivore families, onset of puberty is variable in terms of years but is approximately proportional to the total lifespan of the species. A reproductive strategy present in males of most canid species is the presence of the bulbus glandis. This ring of erectile tissue at the base of the penis becomes engorged with blood during copulation and results in the penis being locked in the female's vagina for a prolonged period, presumably to avoid multiple matings (Kleiman, 1967). In felids, the function of the penile spines may be to enhance the stimulation of females in order to induce ovulation (Aronson and Cooper, 1967). The development of these spines is testosterone dependent and should be evaluated as part of any reproductive examination.

Semen volume and sperm concentration are highly variable across the carnivore species and may be explained largely by social structure (Swanson *et al.*, 1995; Howard, 1999). A striking attribute of spermatozoa across a wide range of carnivore families is the significant and consistent presence of morphologically abnormal spermatozoa (teratospermia). This phenomenon is related to age and nutrition of the male and to season of collection. It also has been associated with several populations known to have poor genetic diversity, including lions (*Panthera leo*) (Wildt *et al.*, 1987), cheetahs (*Acinonyx jubatus*) (Wildt *et al.*, 1983) and Florida panthers (*Felis concolor coryi*) (Barone *et al.*, 1994), although the relationship with genetics is not absolute (Pukazhenthi *et al.*, 2006c). Teratospermia alters the ability of spermatozoa to capacitate, decondense and bind to the zona pellucida of the oocyte. Both abnormal and normal-appearing spermatozoa from teratospermic ejaculates (with >60% abnormal spermatozoa) display increased sensitivity to solution tonicity and temperature change, so that the difficulties inherent in preserving the genes of the individuals and species affected are exacerbated (Pukazhenthi *et al.*, 2001b).

The collection and cryopreservation of semen from canids can be readily performed using a number of different techniques that provide good initial sperm motility post thaw (Farstad, 1996, 2000), with the exception of select species, including the red wolf (*Canis rufus*) (Goodrowe *et al.*, 1998), blue fox (*Alopex lagopus*) (Farstad, 1998) and grey wolf (*Canis lupus*) (Leibo and Songsasen, 2002). Spermatozoa are readily collected from felid species using EE (Platz *et al.*, 1978), but cryopreservation techniques yield variable success and few species have been cryobanked with the same success as canids (Howard, 1999) – possibly because felid acrosome membranes appear to be particularly cryosensitive (Pukazhenthi *et al.*, 2001a; Leibo and Songsasen, 2002). Techniques developed for felids have been successfully modified and applied to mustelids (Howard *et al.*, 1991). Ursid sperm have been collected via EE (Howard *et al.*, 2006; Chen *et al.*, 2007; Brito *et al.*, 2010) and cryopreserved successfully, including sperm of the giant panda (Spindler *et al.*, 2004; Álvarez *et al.*, 2008).

Ungulates

The six orders of ungulates span a wide range of social structures from solitary to highly gregarious. Some species in the bovid, equid and cervid families exhibit seasonal fluctuations in serum testosterone levels and have shown annual cycles in testicular regression, spermatogenesis and/or some seasonal testicular quiescence. However, animals studied in captivity may show attenuated seasonality due to absent or diminished natural cues.

ARTs are powerful tools in supporting and conserving many of the still extant species of this group, particularly in the larger, gregarious species where space constraints are often limiting. Of the many advanced ARTs available today, manipulating the male for semen collection, sample manipulation, storage, transport and insemination is often the simplest and easiest applicable set of techniques for ungulate breeding programmes. Generally, semen collection from ungulates is carried out with anaesthetized animals and EE (Watson, 1978; Howard *et al.*, 1986). Collection by artificial vagina (AV) or rectal stimulation has been successful in some species, given appropriate facilities and training opportunities (Holt, 2001).

In some wild ungulates, such as the Indian blackbuck antelope (*Antelope cervicapra*) (Sontakke *et al.*, 2009), scimitar-horned oryx (*Oryx damma*) (Roth *et al.*, 1999), Mohor gazelle (*Gazella dama mhorr*) (Holt *et al.*, 1996a) and gerenuk (*Litocranius walleri walleri*) (Penfold *et al.*, 2005), semen collection and cryopreservation, and AI using frozen semen, have been successful in producing offspring. Some of the most commonly used cryodiluents (extenders) are Berlinger, TEST, Tris, EQ (equine extender) and yolk-citrate buffers. The cryoprotectants of choice still generally contain 4–8% glycerol or 6–8% DMSO (Holt, 2001). Cryopreservation of elephant sperm has been particularly troublesome, and the low success rate has been a significant barrier to genetic management of this globally managed species. Recently, though, due to breakthroughs including the cushioned centrifuge, careful dilution techniques and the advent of directional freezing, sperm viability post thaw has been achieved (Saragusty *et al.*, 2009).

In captivity, several species exhibit a gender skew in offspring, usually towards males. Sex ratios are known to be skewed in wild populations, although evaluation of the factors that cause this is still largely unknown (Berger and Gompper, 1999). Gender skews present specific management challenges in captivity, and while research continues on the potential causes, ARTs provide an opportunity to resolve these challenges. Flow cytometric cell sorting allows for the separation, cryopreservation and use of X versus Y populations of spermatozoa (Johnson, 1995; Johnson and Welch, 1999). As with many animals, most ungulates only have 3–4% difference in the emitted fluorescence of X-bearing and Y-bearing spermatozoa after excitation, and specialized equipment is still required to orient the sperm head correctly to allow sorting. So even obtaining a small sample requires considerable time and resources. Increasingly efficient equipment that is capable of sorting cells by their morphological attributes without the use of stains may enable the potential applications of these techniques to develop in the near future.

Rodents

Knowledge of laboratory rodents such as mice, rats, hamsters, gerbils and rabbits has advanced beyond that of most other species in the field of reproductive biology; from basic physiology to endocrinology, genetic markers, proteomics and molecular signalling. Studies in non-domestic or non-laboratory rodents are still sparse, even though they are one of the most successful and widespread of the mammalian orders.

Like many wild mammalian species, most rodents respond to seasonal changes and adjust their reproduction according to photoperiod. Semen collection by EE from woodchucks (*Marmot monax*), has shown seasonally sperm-rich samples (Concannon *et al.*, 1996). Even in the blind mole rat (*Spalax ehrenbergi*), fluctuations in testosterone levels and production of spermatozoa are mediated by photoperiod (Gottreich *et al.*, 2000). Colonies of the naked mole rats (*Heterocephalus glaber*) are well known to use social suppression to control female reproduction. Social suppression of males was reported to be less rigorous where both breeding and non-breeding males produced spermatozoa (Faulkes and Abbott, 1991). The success of the order Rodentia is partially attributed to its rapid and highly adaptable breeding mechanisms (even within one genus such as the mole rats) and the wide range of breeding strategies that are governed by a multitude of environmental and physiological cues.

Many muroid rodents have evolved a falciform sperm head, where an apical hook is located on the proximal region of the head (Breed, 2004). This apical hook has been implicated in the facilitation of sperm transport in the female reproductive tract and sperm competition (Roldan *et al.*, 1992). Moore *et al.* (2002) reported interesting observations of spermatozoa from the common wood mouse (*Apodemus sylvaticus*), where over hundreds and thousands of spermatozoa formed groups or 'trains' that significantly increased sperm motility through the oviduct before dispersing for fertilization. In house mice (*Mus musculus* and *M. domesticus*), the apical hook is said to assist the formation of motile sperm bundles (Immler *et al.*, 2007).

Phylogenetic differences can also be seen in sperm physiology. Sperm chromatin undergoes reorganization during spermatogenesis into a tightly packed nucleus (Oliva, 2006). The ratios of protamine 1 to protamine 2 that enable the stabilization of the reorganized sperm nucleus are species specific. In the rodent, the ratio varies from 50:1 to 1:1.5 from the *Rattus* genus to the *Mus* genus, respectively (Corzett *et al.*, 2002). One of the greatest differences in sperm nuclear density occurs between the X and Y populations within the chinchilla (*Chinchilla laniger*) ejaculate (Johnson *et al.*, 1989). Unlike many other mammalian species, including other rodents, where the X and Y DNA difference is approximately 3–4% separation, chinchilla spermatozoa show a 7.5% separation between the

X and Y bearing spermatozoa detected by Hoechst staining (Johnson and Welch, 1999). These seemingly subtle differences, which cannot be observed by the human eye, have enabled much insight into the evolutionary differences between a highly successful order of animals and other taxa.

Cetaceans

Because of the challenges in obtaining samples from free-swimming wild cetaceans, limited research has examined seasonal changes in circulating hormone concentrations in male cetaceans. Testosterone has been detected in fatty tissues and faecal samples of free-ranging short-beaked common dolphins (*Delphinus delphis*) (Amaral, 2010) and North Atlantic right whales (*Eubalaena glacialis*) (Rolland *et al.*, 2005). Seasonal variation in serum testosterone has been quantified for captive beluga (*Delphinapterus leucas*), orca (*Orcinus orca*) and bottlenosed dolphin (*Tursiops truncatus*) (Robeck *et al.*, 2005a; Robeck and Monfort, 2006; O'Brien *et al.*, 2008).

Seasonality is observed in migratory cetaceans (e.g. most whale species) as well as in a number of dolphins and porpoises (Pomeroy, 2011). Whales predominately display a seasonal peak in mating during winter, while dolphins and porpoises are the opposite, with peak breeding observed in summer months (Pomeroy, 2011). Reproduction in whales is generally tied to annual migration patterns, with the exception of orcas, which are known to produce sperm and to breed year round (Robeck and Monfort, 2006).

Semen has been collected in several captive cetacean species, predominately using operant training and manual stimulation (Robeck *et al.*, 2004, 2009; O'Brien and Robeck, 2006; O'Brien *et al.*, 2008). EE has been used to collect semen from the bottlenosed dolphin (Fleming *et al.*, 1981). Continual spermatogenesis has been confirmed in captive orca and beluga, but beluga exhibit seasonal variations in ejaculate size and sperm concentration (O'Brien *et al.*, 2008). In captive Pacific white-sided dolphin (*Lagenorhynchus obliquidens*), spermatozoa have been confirmed only in ejaculates collected between July and October, so provide evidence of a more restricted breeding season than that observed for other cetaceans (Robeck *et al.*, 2009). Similarly, ejaculates are unobtainable or aspermic during winter months in captive bottlenosed dolphins, which corresponds with seasonal changes in serum testosterone (Schroeder and Keller, 1989).

Even though studies on sperm cryopreservation are limited by skill base and access to animals, significant progress has been made in some species. This knowledge has been directly applied to population management. For instance, offspring have been produced after AI using cryopreserved spermatozoa in Pacific white-sided dolphin, beluga and orca (Robeck *et al.*, 2004, 2009, 2010; O'Brien *et al.*, 2008). In the bottlenosed dolphin, live births have resulted from AI using liquid-stored, frozen–thawed and sex-sorted spermatozoa (Robeck *et al.*, 2005b; O'Brien and Robeck, 2006).

Reptiles

The most primitive amniotes throughout the animal kingdom are the members of the order Reptilia. Unlike fish and amphibians, this order universally uses internal fertilization and produces offspring independent of water. The differences in reptilian gonads are generally minor, even between oviparous and viviparous species (Mulaik, 1946). The reproductive tract resembles that of eutherians; however, the reproductive ducts are generally similar to those of amphibians (Risley, 1940). The primordial gonads have two defined zones – cortex and medulla. Sexual differentiation occurs with the development of one zone and the atrophy of the other under the influence of sex hormones. Relatively long periods of juvenile hermaphroditism, primarily in Chelonians, an inherent bisexual nature and adult sexual hermaphroditism occur naturally in several reptile species (Dodd, 1968).

Semen collection and histological examinations document spermiogenesis and allow semen evaluation of spermatozoa of snakes

(Esponda and Bedford, 1987; Zacariotti *et al.*, 2007), chelonians (Al-Dokhi *et al.*, 2007), caimans and common lizards (*Tropidurus itambere*) (Ferreira and Dolder, 2003). Generally, reptiles have filiform, pencil-shaped, spermatozoa (Furieri, 1970). Among the vertebrate taxa, reptilians have one of the most remarkably stable and extended periods of sperm storage naturally within the storage organ of the female reproductive tract (Birkhead and Moller, 1993). Sperm storage enables the reproductive cycles of the male and female to be asynchronous, as copulation and fertilization may be separate events. In turn, copulation frequency may be reduced, decreasing predation risk and enabling reproduction when potential mates are scarce, and increasing the opportunity for female mate choice (Birkhead and Moller, 1993). In lizards, sperm storage also allows for the selection of spermatozoa or embryos in the female reproductive tract to skew the production of offspring towards the gender with the greatest likelihood of survival (Olsson *et al.*, 2007).

With over 300 species of threatened reptilian species globally, genome resource banking is urgently required, but successful cryopreservation of reptilian spermatozoa is yet to be reported. The common cryoprotectant, glycerol, is reported to be toxic to sperm, but AI in snakes and alligators has been successful with the use of fresh semen (Watson, 1990).

Marsupials

Some marsupials have the potential to breed year round, though most display varying degrees of seasonality and semelparity. Male semelparity is characterized by complete and permanent spermatogenic failure, which in its most extreme form concludes with abrupt post-breeding season mortality (Lee *et al.*, 1977; Bradley, 1987). Marsupial male reproductive tracts are simple, comprising limited accessory sex glands, specifically one to three pairs of Cowper's glands and a carrot-shaped prostate with distinct segmentation, which varies between family groups (Rodger and Hughes, 1973; Rodger, 1976; Bedford, 2004). In marsupials, the major source of seminal fluid is the prostate. The glandular lumen of the posterior prostate produces prostatic bodies or large globular spheres, a component of the seminal fluid that appears to be similar to human prostasomes (Rodger and Hughes, 1973). Attempts to separate prostatic bodies from spermatozoa using centrifugation have failed and, although it is assumed that they may aid in the protection and viability of the spermatozoa, the role of the prostatic bodies in marsupial ejaculates is poorly understood.

Penile morphology varies among marsupial species, but is characterized by the unusual pre-penile location of the scrotum (Woolley and Webb, 1977). A number of marsupials (koala, *Phascolarctos cinereus*; southern hairy-nosed wombats, *Lasiorhinus latifrons;* kowari, *Dasyuroides byrnei*; and mulgara, *Dasycercus cristicauda*, for example) have minute epidermal penile spines (Brooks *et al.*, 1978; Woolley, 1978; Johnston *et al.*, 2000b). Penile spines are common in felids that are reflex or induced ovulators (Wildt *et al.*, 1980). To date, the koala is the only marsupial that has been identified as a reflex ovulator, but triggered by chemical rather than mechanical stimulation (Johnston *et al.*, 2000b). Across dasyurids, all species in the *Dasyurus* and *Myoictis* genera, and two of the *Antechinus* species (*A. apicalis* and *A. macdonnellensis*), have a unique penile appendage that extends from the dorsal aspect of the penis, which creates a bifid penis (Woolley and Webb, 1977). The functional significance of this appendage is unknown.

Marsupial spermatozoa undergo biochemical and functional maturation during transit through the epididymis (Harding *et al.*, 1982). Morphological changes that occur during epididymal transit include: (i) realignment of the head from perpendicular to parallel orientation in relation to the midpiece – with the exception of the koala and wombat; (ii) development of the midpiece plasma membrane and mitochondrial network; (iii) reorganization and compaction of the acrosomal matrix to the dorsal nuclear surface; and (iv) loss of the cytoplasmic droplet (Harding *et al.*, 1982; Temple-Smith, 1987). While variations exist between family groups, marsupial spermatozoa are generally longer than those of eutherians (ranging from 83 μm in the koala to 271 μm in the brown antechinus, *A. stuarii*) (Cummins and Woodall, 1985; Taggart, 1994). Marsupial spermatozoa exhibit nuclear flattening in the plane of the midpiece, and the neck of the midpiece inserts into the central region of the ventral surface of the sperm head (Hughes, 1965; Taggart, 1994). The koala and wombat have the only marsupial spermatozoa that do not exhibit nuclear flattening, as a result of the unique 'hook' shape of the head (Johnston *et al.*, 1994).

Sperm conformation may influence motility patterns as marsupial midpieces range from circular to flattened (dorsoventrally), influencing the arrangement of the mitochondria and dense outer fibres (Taggart, 1994). Possum and opossum sperm motility features sperm pairing with fusion at the head, which occurs during maturation in the epididymis (Temple-Smith, 1987; Taggart, 1994). The pairing may assist in efficient transport and aid in protecting the acrosome during sperm movement through the female reproductive tract. Although the number of spermatozoa in most marsupial ejaculates is comparable to that of some eutherian species (in the order of 10^7 or 10^8), dasyurids and American marsupials exhibit an extremely low total epididymal reserve, ranging from 10^5 to 10^6 spermatozoa (Bedford *et al.*, 1984; Taggart, 1994).

Currently, the koala is the only marsupial in which manual semen collection has been successful (Johnston *et al.*, 1997b). EE under general anaesthesia has been used successfully for the collection and examination of spermatozoa in a variety of marsupial species, including the tammar wallaby (*Macropus eugenii*) (Cummins, 1980), common (brush-tailed) possum (*Trichosurus vulpecula*) (Rodger *et al.*, 1991), koala (Johnston *et al.*, 1994), Matschie's tree kangaroo (*Dendrolagus matschiei*) (Taggart *et al.*, 1995), yellow-footed rock wallaby (*Petrogale xanthopus*) (Taggart *et al.*, 1995), eastern grey kangaroo (*Macropus giganteus*) (Johnston *et al.*, 1997a) and the common wombat (*Vombatus ursinus*) (Johnston *et al.*, 2006).

Cryopreservation studies have been conducted on several marsupial species, but freezing methodology and post-thaw motility is extremely variable between species. For example, a Tris-citrate-glucose (or fructose)-egg yolk diluent is commonly used, but the choice of cryoprotectant ranges from 4–8% glycerol for wombats and bandicoots (Taggart *et al.*, 1996), to 17.5% glycerol for possums (Molinia and Rodger, 1996) and up to 28% in southern hairy-nosed wombats (Taggart *et al.*, 1998). To date, successful sperm cryopreservation with retention of a minimum of 40% of pre-freeze motility has only been achieved in the koala, common and Southern hairy-nosed wombat (*Lasiorhinus latifrons*), brushtail possum, bandicoot (*Isoodon macrourus*) and Tasmanian devil (*Sarcophilus harrisii*) (Rodger *et al.*, 1991; Molinia and Rodger, 1996; Taggart *et al.*, 1996, 1998; Johnston *et al.*, 2006; Zee *et al.*, 2008; Keeley *et al.*, 2012). In some macropods at least, spermatozoa examined using a cryomicroscope have shown remarkable tolerance to cooling, with changes in conformation of motility profiles and the maintenance of motility down to –7°C (Holt *et al.*, 1999). Macropod spermatozoa re-warmed from –30 to 10°C showed a resumption of flagellar activity of up to 50%, but further re-warming was detrimental (Holt *et al.*, 1999).

In the koala, cryopreservation is less efficient, though chilled semen stored at 4°C has been shown to retain viability for over 8 days (Johnston *et al.*, 2000a). Under these conditions, viability and motility of approximately 50–60% has been observed over a 5 week period (N. Satake, unpublished). Even though conception has been confirmed after AI using cryopreserved spermatozoa in the brushtail possum (Johnston *et al.*, 2007b), the production of live young subsequent to AI using liquid-stored spermatozoa has been achieved in only two species, the koala (Paris *et al.*, 2005) and the tammar wallaby (Johnston *et al.*, 2003).

Monotremes

Reproductive behaviour in the short-beaked echidna (*Tachyglossus aculeatus*) features observations of extensive 'trains' of up to 11 male echidnas following females in an attempt to gain access for breeding, presumably attracted by pheromones emitted by the oestrous female (Rismiller, 1992). Males are also known to be extremely promiscuous and will mate with females still in torpor (Morrow and Nicol, 2009). In the platypus (*Ornithorhynchus anatinus*), courtship behaviour can be quite extensive and may include the male biting the female's tail and extensive chasing behaviour, all occurring in the water (Hawkins and Fanning, 1992). This behaviour is followed by several bouts of copulation that also occur in the water, but typically supported by a semi-submerged structure such as a log (Hawkins and Battaglia, 2009). Neither species breed readily in captivity.

Monotremes possess intra-abdominal testes, with the male reproductive tract including an inconspicuous disseminate prostate and a single set of Cowper's glands, which are both relatively small and therefore unlikely to provide substantial volume to the ejaculate (Carrick and Hughes, 1978; Jones *et al.*, 2004). Breeding in the short-beaked echidna is seasonal, occurring predominantly in the late Austral winter after a period of hibernation with cessation of spermatogenesis outside the breeding season (Carrick and Hughes, 1978; Jones *et al.*, 2004). Monotreme spermatozoa are filiform with a long head (50 μm in echidna, 38 μm in platypus) and overall length of approximately 100–120 μm (Carrick and Hughes, 1978; Jones *et al.*, 1992). The acrosome extends over the rostral end of approximately the first seventh of the nucleus (Carrick and Hughes, 1982; Djakiew and Jones, 1983). During epididymal maturation, short-beaked echidna spermatozoa form spheres with their rostral ends oriented towards the middle. The spermatozoa then reorient to form bundles of 20 to >100 individuals, roughly parallel to each other, a form of cooperation that may increase survival or aid in sperm competition (Djakiew and Jones, 1983; Jones *et al.*, 1992, 2004; Johnston *et al.*, 2007b).

Viable spermatozoa have been collected in the echidna by manual stimulation only (Johnston *et al.*, 2007a). All other descriptions have been obtained through the examination of reproductive tissues post-mortem. The penis of both the echidna and platypus are unique and distinctive. The platypus penis is covered with recurved spines with a bifid glans penis, which extends into two bulbous branches, each with four divided urethral ducts terminating with foliate papillae (Carrick and Hughes, 1978). In contrast, the bifid glans penis of the echidna is divided into four urethral branches, each of which terminate in 'a flower-like rosette' (Carrick and Hughes, 1978) (Plate 61). In both cases, the penis serves as a reproductive organ only, and retracts into a preputial sac adjacent to the cloaca when not in use (Carrick and Hughes, 1978; Jones *et al.*, 1992). Natural ejaculation has been observed in a captive short-beaked echidna and involves the retraction of one side (two of the four rosette openings) of the bifid penis and concurrent engorgement of the alternate side and the pooling of semen into the cups of the erect rosettes (Johnston *et al.*, 2007b). Ejaculation alternates from one side of the penis to the other, but it is unknown whether this also represents alternative contributions from the epididymides (Johnston *et al.*, 2007b).

Avids

Birds, like some reptilian and fish species, determine the sex of their offspring using the ZW sex-determination system (Smith *et al.*, 2007). So the ovum, rather than the spermatozoon determines the gender of the offspring. Almost all male non-domestic birds produce spermatozoa seasonally. Unlike mammalian species, many avian species also possess sperm storage organs, which increase the intricacy of sperm transport from the vagina to the oviduct (Bakst *et al.*, 1994). Thus, artificial insemination techniques using vaginal insemination pose more challenges. One important signalling mechanism known to control sperm motility is the regulation of intracellular calcium concentrations. Within the sperm storage organs of chickens, quails and turkeys, it is known that 17–19 mM of calcium occurs naturally, and inhibits sperm motility (Holm *et al.*, 2000). Further research must be conducted on these characteristics to provide insight into species biology and to allow for the development of ARTs.

Avian semen collection, extender composition and sperm cryopreservation techniques have been described in domestic fowl for over 50 years (Polge, 1951). In chickens, the combination of glycerol or dimethyl formamide with trehalose appears to be the most efficient cryodiluent (Terada *et al.*, 1989). However, it is unlikely that a standard cryopreservation technique for all avian species will be developed owing to the particularly wide range of membrane lipid composition and a propensity in many species for sperm membranes to become rigid during the cryopreservation process (Blesbois *et al.*, 2005). A high seminal pH may also have a negative effect on the quality of cryopreserved spermatozoa post thaw in pheasants (Saint Jalme *et al.*, 2003).

In non-domestic avian species, sperm cryopreservation methods use glycerol and DMSO as the preferred cryoprotectant. In the northern pintail (*Anas acuta*), short-term preservation in cold storage was shown to be more effective for preserving fertility in waterfowl species (Penfold *et al.*, 2001). Offspring have been produced from a number of non-domestic avian species through AI using cryopreserved spermatozoa; for example the greater sandhill crane (*Grus canadensis tabida*) (Gee *et al.*, 1985), American kestrel (*Falco sparverius*) (Gee and Sexton, 1990) and bustard (Otidae) (Samour, 2004). The importance of using species-specific protocols is clear in avian species, e.g. where modifications of cryodiluents to include a combination of dimethylacetamide and ATP benefit cryosurvival of turkey and crane spermatozoa (Blanco *et al.*, 2011). There is currently an increased need for the use of ARTs, particularly of AI and genetic resource banks (GRBs), of avian species. The lack of adequate basic knowledge of 'normal' traits and highly variable species-specific traits remain as a barrier to the efficacy of ARTs as a conservation tool (Blanco *et al.*, 2009).

Amphibians

The basic amphibian male reproductive tract consists of the testes, vasa efferentia and vasa deferentia. The primordial gonads are double structures, in which the origin of the cortex and medulla are from the peritoneum and interrenal blastema, respectively. In males, the medulla is able to act as a host to primordial germ cells for the production of testes. The male reproductive ducts of the two main amphibian orders, Urodela and Anura, are comparable. They consist of the efferent ducts of the testicular ampullae and the kidney genital segment. The ureter and vasa deferentia carry both urine and spermatozoa.

Hermaphroditism is uncommon in most amphibians; however, in *Rana* species, hermaphroditism is often observed and is a feature of individuals undergoing a transition in gender. Male reproductive cycles in amphibians are controlled by the changing ratios of basophils and acidophils within the pituitary. A change of ratio in favour of more basophils to more acidophils indicates a move into sexual quiescence, whereas an increase in basophils will indicate sexual activity (Dodd, 1968). The ease with which male reproduction can be controlled through procedures such as castration, testicular transplantation and grafting, and the injection of testicular extracts and androgens, has allowed amphibians to be useful subjects in studying reproductive control mechanisms in vertebrates.

As in most taxa, the spermatozoa of amphibians vary greatly between species. Examples of head shape may be twisted corkscrew like, cylindrical and spindle like (Asa and Phillips, 1988). Some spermatozoa may be extremely long, such as the 3 mm spermatozoa of *Discoglossus* spp. Spermatozoa of *Rana* spp., *Bufo* spp. and *Discoglossus* spp. have defining sperm morphology, including conical subacrosomes with separate sheaths and undulating tail membranes (Frazer and Glenister, 1957), which may affect cryosurvival. Anuran species spermatozoa, although having short-lived viability, have an incredible ability to withstand hypotonic environments due to the deposition of semen into fresh water during natural breeding.

Amphibians were the first taxa to be cloned in the 1950s, and protocols exist for the cloning of common laboratory species such as *Rana* and *Xenopus* (Gurdon and Byrne, 2003). Within these taxa, reproductive cloning may be a viable ART solution in the production and re-establishment of future populations of at least some of the 500 critically endangered amphibian species.

Fish

Reproductive strategies vary greatly and to the extreme in fish species, including the family Syngnathidae – seahorses and pipefish – in which the male accepts eggs from the female, fertilizes them and carries the embryos in a brood pouch. Many species are monogamous, resulting in low numbers of spermatozoa and limited possibilities for sperm competition (Watanabe *et al.*, 2000). Spermatozoa of most species of fish, whether freshwater or marine fish, are immotile within the testes and seminal plasma (Stoss, 1983). Stimulation of motility occurs when the sperm are ejaculated into the aquatic environment or the female reproductive tract, where a change in ionic/osmotic/pH balance of the surrounding fluid initiates sperm cell membrane depolarization (Morisawa *et al.*, 1983; Alavi and Cosson, 2005, 2006). Then, usually by chemoattraction, spermatozoa will be directed to the micropyle of the oocyte.

Cryopreservation of fish spermatozoa currently remains the preferred routine technique for long-term storage, though success of this technique remains low (Stoss, 1983; Zhang, 2004; Hagedorn *et al.*, 2009). The spermatozoa of freshwater species are still greatly susceptible to damage after cryopreservation (Hagedorn *et al.*, 2009), whereas the susceptibility to damage in marine species is in the range of 10–20% only (Kopeika *et al.*, 2007). Recent advances in germ cell preservation and manipulations have started to address some of the problems associated with the production of fish species. Transplantation and xenografting techniques, in combination with germ cell cryopreservation, are enabling the successful production of economically important or endangered species of fish from spermatogonial cells and surrogate recipients (Okutsu *et al.*, 2006).

Genome Resource Banking: 'The Frozen Zoo'

In the current era of the sixth great extinction, species are in decline at a rate never before witnessed (Steffen *et al.*, 2007). Traditional methods of management will not be sufficient to meet the growing demand for species management. Genetic resource banks (GRBs) hold collections of any genetic material that is viable for the potential production of embryos, whether it is made up of gametes, somatic cells or tissues. In the domestic animal production industry, this type of resource has been a well-integrated source of genetic and animal management for over 50 years, and has developed with the vast improvements in technological equipment now available. GRBs of frozen spermatozoa, and the use of such specimens in tandem with AI, are still the most practical application of ART in non-domestic animal populations (Holt *et al.*, 1996b).

The potential application of GRBs to wildlife conservation strategies was recognized in the 1970s (Watson, 1978), and has progressed significantly over ensuing decades (Holt *et al.*, 2003). The conservation of biodiversity requires the avoidance of inbreeding and minimization of the random drift effect, particularly in small populations. To maintain a GRB at an effective level, samples stored for extended periods of time must be assessed in line with breeding plans and genetic gain. Due to the difficulties in obtaining, transporting and using material from rare species, GRBs are not a commonly used tool in wildlife conservation and management. A well-managed GRB will enable efficient management of genetic variability within small insurance populations of endangered species. The storage, transfer and use of such samples will also enable close inspection and prevention of transmissible diseases, thus managing and improving the biological viability of populations.

The Frozen Ark Project, established in 2004 jointly by the Zoological Society of London, the Natural History Museum and the University of Nottingham in the UK, is now a global consortium encompassing a wide range of research institutions, each contributing knowledge and resources with respect to their focal species, tissue/cell type and techniques (Watson and Holt, 2001). Organizations that hold a wide range of captive animals, such as zoos, wildlife parks and aquariums, are incorporated into the consortium to conserve the maximum genetic materials available, and to advance research opportunities and apply research findings to propagation and management using ART.

This consortium has demonstrated great promise and there is now a need for a concerted effort to ensure that gaps in the consortium are filled. A globally linked, cooperatively managed GRB is a necessary step towards storing globally representative samples of priority wildlife species (Wildt, 1997). These samples would provide particular opportunities to facilitate genetic exchange across borders and geographic barriers, readjust gender skew in captive populations, reduce the impacts of mate incompatibility or incompetence, and reduce the risk of disease transmission (Loskutoff *et al.*, 2005). There remain logistical, political and resource barriers to the realization of such a system; however, significant progress has been made in some taxa, and our capacity and knowledge in this area continue to grow through the efforts of many scientists.

Conclusion

While females are often the rate-limiting resource for population growth, understanding male reproductive physiology is critical to understanding the impacts of environmental change and optimizing the reproductive management of wildlife species. It is already clear that the diversity of reproductive strategies found among wildlife species is extensive, and with further study, unique and fascinating mechanisms will undoubtedly be discovered. Even species within the same family often have unique strategies, sensitivities and tolerances, making the extrapolation of knowledge and techniques from one species to another fraught with difficulties and poor results. So, although generalizations across such a wide range of taxa are not advisable, this chapter has provided some information on male reproductive characteristics across wildlife species and explored some of the more interesting strategies and attempts at manipulation to manage populations and optimize reproductive health.

References

Alavi SM and Cosson J (2005) Sperm motility in fishes (I) Effects of temperature and pH: a review *Cell Biology International* **29** 101–110.

Alavi SM and Cosson J (2006) Sperm motility in fishes (II) Effects of ions and osmolality: a review *Cell Biology International* **30** 1–14.

Al-Dokhi O, Al-Wasel S and Mubarak M (2007) Ultrastructure of spermatozoa of the freshwater turtle *Mauremys caspica* (Chelonia, Reptilia) *International Journal of Zoological Research* **3** 53–63.

Álvarez M, García-Macías V, Martínez-Pastor F, Martínez F, Borragán S, Mata M, Garde J, Anel L and De Paz P (2008) Effects of cryopreservation on head morphometry and its relation with chromatin status in brown bear (*Ursus arctos*) spermatozoa *Theriogenology* **70** 1498–1506.

Amaral RS (2010) Use of alternative matrices to monitor steroid hormones in aquatic mammals: a review *Aquatic Mammals* **36** 162–171.

Andonov M and Chaldakov G (1991) Role of Ca^{2+} and cAMP in rat spermatogenesis – ultrastructural evidences *Acta Histochemica* **Supplement 41** 55–63.

Andrabi SM (2009) Factors affecting the quality of cryopreserved buffalo (*Bubalus bubalis*) bull spermatozoa *Reproduction in Domestic Animals* **44** 552–569.

Aronson L and Cooper M (1967) Penile spines in the domestic cat: their endocrine-behavior relations *Anatomical Record* **157** 71–78.

Asa CS and Phillips D (1988) Nuclear shaping in spermatids of the Thai leaf frog *Megophrys montana* *Anatomical Record* **220** 287–290.

Bakst M, Wishart G and Brillard J (1994) Oviductal sperm selection transport and storage in poultry *Poultry Science Reviews* **5** 117–143.

Ballou J and Foose T (1996) Demographic and genetic management of captive populations In *Wild Mammals in Captivity* pp 263–283 Ed D Kleiman, M Allen, K Thompson and S Lumpkin. The University of Chicago Press, Chicago.

Barone M, Roelke M, Howard J, Anderson A and Wildt D (1994) Reproductive characteristics of male Florida panthers: comparative studies from Florida, Texas, Colorado, Chile and North American zoos *Journal of Mammalogy* **75** 150–162.

Bedford JM (2004) Enigmas of mammalian gamete form and function *Biology Reviews* **79** 429–460.

Bedford JM, Rodger JC and Breed WG (1984) Why so many mammalian spermatozoa – a clue from marsupials? *Proceedings of the Royal Society of London B: Biological Sciences* **221** 221–233.

Berger J and Gompper M (1999) Sex ratios in extant ungulates: products of contemporary predation or past life histories? *Journal of Mammalogy* **80** 1084–1113.

Birkhead TR and Moller AP (1993) Sexual selection and the temporal separation of reproductive events: sperm storage data from reptiles, birds and mammals *Biological Journal of the Linnean Society* **50** 295–311.

Blanco J, Wildt D, Hofle U, Voelker W and Donoghue A (2009) Implementing artificial insemination as an effective tool for *ex situ* conservation of endangered avian species *Theriogenology* **71** 200–213.

Blanco J, Long J, Gee G, Wildt D and Donoghue A (2011) Comparative cryopreservation of avian spermatozoa: benefits of non-permeating osmoprotectants and ATP on turkey and crane sperm cryosurvival *Animal Reproduction Science* **123** 242–248.

Blesbois E, Grasseau I and Seigneurin F (2005) Membrane fluidity and the ability of domestic bird spermatozoa to survive cryopreservation *Reproduction* **129** 371–378.

Bradley AJ (1987) Stress and mortality in the red-tailed phascogale, *Phascogale calura* (Marsupialia: Dasyuridae) *General and Comparative Endocrinology* **67** 85–100.

Breed WG (2004) The spermatozoon of Eurasian murine rodents: its morphological diversity and evolution *Journal of Morphology* **26** 52–69.

Brinsko SP, Rowan KR, Varner DD and Blanchard TL (2000) Effects of transport container and ambient storage temperature on motion characteristics of equine spermatozoa *Theriogenology* **53** 1641–1655.

Brito L, Sertich P, Stull G, Rives W and Knobbe M (2010) Sperm ultrastructure, morphometry, and abnormal morphology in American black bears (*Ursus americanus*) *Theriogenology* **74** 1403–1413.

Brooks DE, Gaughwin M and Mann T (1978) Structural and biochemical characteristics of the male accessory organs of reproduction in the hairy-nosed wombat (*Lasiorhinus latifrons*) *Proceedings of the Royal Society of London B: Biological Sciences* **201** 191–207.

Carlstead K (2002) Using cross-institutional behavior surveys to study relationships between animal temperament and reproduction In *Proceedings of the Second International Symposium on Assisted Reproductive Technology for the Conservation and Genetic Management of Wildlife* pp 54–62 Ed NM Loskutoff. Henry Doorly Zoo, Omaha, Nebraska.

Carlstead K, Brown J and Strawn W (1993a) Behavioral and physiological correlates of stress in laboratory cats *Applied Animal Behaviour Science* **38** 143–158.

Carlstead K, Brown JL and Seidensticker J (1993b) Behavioral and adrenocortical responses to environmental changes in leopard cats (*Felis bengalensis*) *Zoo Biology* **12** 321–331.

Carrick FN and Hughes RL (1978) Reproduction in male monotremes *The Australian Zoologist* **20** 211–231.

Carrick FN and Hughes RL (1982) Aspects of the structure and development of monotreme spermatozoa and their relevance to the evolution of mammalian sperm morphology *Cell and Tissue Research* **222** 127–141.

Centurion F, Vazquez JM, Calvete JJ, Roca J, Sanz L, Parrilla I, Garcia EM and Martinez EA (2003) Influence of porcine spermadhesins on the susceptibility of boar spermatozoa to high dilution *Biology of Reproduction* **69** 640–646.

Chen L, Hou R, Zhang Z, Wang J, An X, Chen Y, Zheng H, Xia G and Zhang M (2007) Electroejaculation and semen characteristics of Asiatic black bears (*Ursus thibetanus*) *Animal Reproduction Science* **101** 358–364.

Clagett-Dame M and DeLuca HF (2002) The role of vitamin A in mammalian reproduction and embryonic development *Annual Review of Nutrition* **22** 347–381.

Colborn DR, Thompson DL, Roth TL, Capehart JS and White KL (1991) Responses of cortisol and prolactin to sexual excitement and stress in stallions and geldings *Journal of Animal Science* **69** 2556–2562.

Concannon P, Roberts P, Parks J, Bellezza C and Tennant B (1996) Collection of seasonally spermatozoa-rich semen by electroejaculation of laboratory woodchucks (*Marmota monax*), with and without removal of bulbourethral glands *Laboratory Animal Science* **46** 667–675.

Corzett M, Mazrimas J and Balhorn R (2002) Protamine 1:protamine 2 stoichiometry in the sperm of eutherian mammals *Molecular Reproduction and Development* **61** 519–527.

Cummins JM (1980) Decondensation of sperm nuclei of Australian marsupials: effects of air drying and of calcium and magnesium *Gamete Research* **3** 351–367.
Cummins JM and Woodall PF (1985) On mammalian sperm dimensions *Journal of Reproduction and Fertility* **75** 153–175.
Djakiew D and Jones R (1983) Sperm maturation, fluid transport, and secretion and absorption of protein in the epididymis of the echidna, Tachyglossus aculeatus *Journal of Reproduction and Fertility* **68** 445–456.
Dobson H, Tebble R, Smith RF and Ward WR (2001) Is stress really all that important? *Theriogenology* **55** 65–74.
Dobson H, Ghuman S, Prabhakar S and Smith R (2003) A conceptual model of the influence of stress on female reproduction *Reproduction* **125** 151–163.
Dodd JM (1968) Gonadal and gonadotrophic hormones in lower vertebrates In *Marshall's Physiology of Reproduction* 3rd edition pp 417–582 Ed A Parkes. Longmans Green, London.
Dong Q, Correa L and VandeVoort C (2009) Rhesus monkey sperm cryopreservation with TEST-yolk extender in the absence of permeable cryoprotectant *Cryobiology* **58** 20–27.
England GCW and Ponzio P (1996) Comparison of the quality of frozen–thawed and cooled–rewarmed dog semen *Theriogenology* **46** 165–171.
Esponda P and Bedford J (1987) Post-testicular change in the reptile sperm surface with particular reference to the snake, *Natrix fasciata Journal of Experimental Zoology* **241** 123–132.
Farstad W (1996) Semen cryopreservation in dogs and foxes *Animal Reproduction Science* **42** 251–260.
Farstad W (1998) Reproduction in foxes: current research and future challenges *Animal Reproduction Science* **51** 35–42.
Farstad W (2000) Current state in biotechnology in canine and feline reproduction *Animal Reproduction Science* **60–61** 375–387.
Faulkes CG and Abbott DH (1991) Social control of reproduction in breeding and non-breeding male naked mole-rats (*Heterocephalus glaber*) *Journal of Reproduction and Fertility* **93** 427–435.
Ferreira A and Dolder H (2003) Cytochemical study of spermiogenesis and mature spermatozoa in the lizard *Tropidurus itambere* (Reptilia, Squamata) *Acta Histochemica* **105** 339–352.
Fleming AD, Yanagimachi R and Yanagimachi H (1981) Spermatozoa of the Atlantic bottlenosed dolphin, *Tursiops truncatus Journal of Reproduction and Fertility* **63** 509–514.
Frazer JF and Glenister T (1957) Factors affecting the motility of toad spermatozoa, and their application in a rapid test for pregnancy *Journal of Physiology* **135** 49–50.
Furieri P (1970) Sperm morphology in some reptiles: Squamata and Chelonia In *Comparative Spermatology* pp 115–131 Ed B Baccetti. Academic Press, New York.
Gee G and Sexton T (1990) Cryopreservation of American kestrel semen with dimethyl sulfoxide *Zoo Biology* **3** 361–371.
Gee G, Bakst M and Sexton T (1985) Cryogenic preservation of semen from the greater sandhill crane *Journal of Wildlife Management* **49** 480–484.
Goodrowe KL, Hay MA, Platz CC, Behrns SK, Jones MH and Waddell WT (1998) Characteristics of fresh and frozen–thawed red wolf (*Canis rufus*) spermatozoa *Animal Reproduction Science* **53** 299–308.
Gottreich A, Hammel I, Yogev L and Terkel J (2000) Effect of photoperiod variation on testes and accessory sex organs in the male blind mole rat *Spalax ehrenbergi Life Sciences* **67** 521–529.
Gould K (1990) Techniques and significance of gamete collection and storage in the great apes *Journal of Medical Primatology* **19** 537–551.
Gurdon J and Byrne J (2003) The first half-century of nuclear transplantation *Proceedings of the Royal Society of London B: Biological Sciences* **100** 8048–8052.
Hagedorn M, Ricker J, McCarthy M, Meyers S, Tiersch T, Varga ZM and Kleinhans FW (2009) Biophysics of zebrafish (*Danio rerio*) sperm *Cryobiology* **58** 12–19.
Halloran BP and DeLuca HF (1979) Vitamin D deficiency and reproduction in rats *Science* **204** 73–74.
Harding HR, Woolley PA and Shorey CD (1982) Sperm ultrastructure, spermiogenesis and epididymal sperm maturation in dasyurid marsupials: phylogenetic implications In *Carnivorous Marsupials* pp 659–673 Ed M Archer. Surrey Beatty, Sydney.
Harnal V, Wildt D, Bird D, Monfort S and Ballou J (2002) Computer simulations to determine the efficacy of different genome resource banking strategies for maintaining genetic diversity *Cryobiology* **44** 122–131.
Hawkins M and Battaglia A (2009) Breeding behaviour of the platypus (*Ornithorhynchus anatinus*) in captivity *Australian Journal of Zoology* **57** 283–293.

Hawkins M and Fanning D (1992) Courtship and mating behaviour of captive platypus at Taronga Zoo In *Platypus and Echidnas* pp 106–114 Ed M Augee. The Royal Zoological Society of NSW, Sydney.

Herrick JR, Campbell M, Levens G, Moore T, Benson K, D'Agostino J, West G, Okeson DM, Coke R, Portacio SC, Leiske K, Kreider C, Polumbo PJ and Swanson WF (2010) *In vitro* fertilization and sperm cryopreservation in the black-footed cat (*Felis nigripes*) and sand cat (*Felis margarita*) *Biology of Reproduction* **82** 552–562.

Hesterman H and Jones SM (2009) Longitudinal monitoring of plasma and fecal androgens in the Tasmanian devil (*Sarcophilus harrisii*) and the spotted-tailed quoll (*Dasyurus maculatus*) *Animal Reproduction Science* **112** 334–346.

Holm L, Ekwall H, Wishart G and Ridderstrale Y (2000) Localization of calcium and zinc in the sperm storage tubules of chicken, quail and turkey using X-ray microanalysis *Journal of Reproduction and Fertility* **118** 331–336.

Holt WV (2001) Germplasm cryopreservation in elephants and wild ungulates In *Cryobanking the Genetic Resource: Wildlife Conservation for the Future?* Ed W Holt and P Watson. Taylor and Francis, London.

Holt WV, Abaigar T and Jabbour HN (1996a) Oestrous synchronization, semen preservation and artificial insemination in the Mohor gazelle (*Gazella dama mhorr*) for the establishment of a genome resource bank programme *Reproduction, Fertility and Development* **8** 1215–1222.

Holt WV, Bennett PM, Volovouev V and Watson PF (1996b) Genetic resource banks in wildlife conservation *Journal of Zoology* **238** 531–534.

Holt WV, Penfold LM, Johnston SD, Temple-Smith P, McCallum C, Shaw J, Lindemans W and Blyde D (1999) Cryopreservation of macropod spermatozoa: new insights from the cryomicroscope. *Reproduction, Fertility and Development* **11** 345–353.

Holt WV, Abaigar T, Watson PF and Wildt DE (2003) Genetic resource banks for species conservation In *Reproductive Science and Integrated Conservation* pp 267–280 Ed WV Holt, AR Pickard, JC Rodger and DE Wildt. Cambridge University Press, Cambridge.

Howard J (1999) Assisted reproductive techniques in nondomestic carnivores In *Zoo and Wild Animal Medicine: Current Therapy IV* pp 449–457 Ed M Fowler and R Miller. WB Saunders, Philadelphia.

Howard J, Bush M and Wildt D (1986) Semen collection, analysis and cryopreservation in nondomestic mammals In *Current Therapy in Theriogenology* pp 1047–1053 Ed D Morrow. WB Saunders, Philadelphia.

Howard JG, Bush M, Morton C, Morton F, Wentzel K and Wildt DE (1991) Comparative semen cryopreservation in ferrets (*Mustela putorius furo*) and pregnancies after laparoscopic intrauterine insemination with frozen–thawed spermatozoa *Journal of Reproduction and Fertility* **92** 109–118.

Howard JG, Zhang Z, Li D, Huang Y, Zhang M, Hou R, Ye Z, Li G, Zhang J, Huang S, Spindler R, Zhang H and Wildt DE (2006) Male reproductive biology in giant pandas In *Giant Pandas: Biology, Veterinary Medicine and Management* Ed DE Wildt, A Zhang, H Zhang, DL Janssen and S Ellis. Cambridge University Press, Cambridge.

Hughes RL (1965) Comparative morphology of spermatozoa from five marsupial families *Australian Journal of Zoology* **13** 533–543.

Immler S, Moore H, Breed W and Birkhead T (2007) By hook or by crook? Morphometry, competition and cooperation in rodent sperm *PLoS ONE* **2(1)**: e170.

Isachenko E, Isachenko V, Katkov I, Dessole S and Nawroth F (2003) Vitrification of mammalian spermatozoa in the absence of cryoprotectants: from past practical difficulties to present success *Reproductive Biomedicine Online* **6** 191–200.

Isachenko V, Maettner R, Petrunkina A, Sterzik K, Mallmann P, Rahimi G, Sanchez R, Risopatron J, Damjanoski I and Isachenko E (2012) Vitrification of human ICSI/IVF spermatozoa without cryoprotectants: new capillary technology *Journal of Andrology* **33** 462–468.

Johnson LA (1992) Gender preselection in domestic animals using flow cytometrically sorted sperm *Journal of Animal Science* **70** 8–18.

Johnson LA (1995) Sex preselection by flow cytometric separation of X and Y chromosome-bearing sperm based on DNA difference: a review *Reproduction, Fertility and Development* **7** 893–903.

Johnson LA and Welch GR (1999) Sex preselection: high-speed flow cytometric sorting of X and Y sperm for maximum efficiency *Theriogenology* **52** 1323–1341.

Johnson LA, Flook JP and Hawk HW (1989) Sex preselection in rabbits: live births from X and Y sperm separated by DNA and cell sorting *Biology of Reproduction* **41** 199–203.

Johnston SD, McGowan MR, Carrick FN, Cameron RDA and Tribe A (1994) Seminal characteristics and spermatozoal morphology of captive Queensland koalas (*Phascolarctos cinereus adustus*) *Theriogenology* **42** 501–511.

Johnston SD, Blyde D, Gamble J, Higgins D, Field H and Cooper J (1997a) Collection and short-term preservation of semen from free-ranging eastern grey kangaroos (*Macropus giganteus*: Macropodidae) *Australian Veterinary Journal* **75** 648–651.

Johnston SD, O'Callaghan P, McGowan MR and Phillips NJ (1997b) Characteristics of koala (*Phascolarctos cinereus adustus*) semen collected by artificial vagina *Journal of Reproduction and Fertility* **109** 319–323.

Johnston SD, McGowan MR, Phillips NJ and O'Callaghan P (2000a) Optimal physicochemical conditions for the manipulation and short-term preservation of koala (*Phascolarctos cinereus*) spermatozoa *Journal of Reproduction and Fertility* **118** 273–281.

Johnston SD, McGowan MR, O'Callaghan P, Cox R and Nicolson V (2000b) Studies of the oestrous cycle, oestrous and pregnancy in the koala (*Phascolarctos cinereus*) *Journal of Reproduction and Fertility* **120** 49–57.

Johnston SD, McGowan MR, O'Callaghan P, Cox R, Houlden B, Haig S and Taddeo G (2003) Birth of koalas (*Phascolarctos cinereus*) at Lone Pine Koala Sanctuary following artificial insemination *International Zoo Yearbook* **38** 160–172.

Johnston SM, Blyde D, McClean R, Lisle A and Holt WV (2006) An investigation into the similarities and differences governing the cryopreservation success of koala (*Phascolarctos cinereus*: Goldfuss) and common wombat (*Vombatus ursinus*: Shaw) spermatozoa *Cryobiology* **53** 218–228.

Johnston SD, Nicolson V, Madden C, Logie S, Pyne M, Roser A, Lisle AT and D'Occhio M (2007a) Assessment of reproductive status in male echidnas *Animal Reproduction Science* **97** 114–127.

Johnston SD, Smith B, Pyne M, Stenzel D and Holt WV (2007b) One-sided ejaculation of echidna sperm bundles *The American Naturalist* **170** 162–164.

Jones RC, Stone GM and Zupp J (1992) Reproduction in the male echidna In *Platypus and Echidnas* pp 115–126 Ed M Augee. The Royal Zoological Society of NSW, Sydney.

Jones RC, Djakiew D and Dacheux JL (2004) Adaptations of the short-beaked echidna *Tachyglossus aculateus* for sperm production, particularly in an arid environment *Australian Mammalogy* **26** 199–204.

Jurke M, Czekala N, Lindburg D and Millard S (1997) Fecal corticoid metabolite measurement in the cheetahs (*Acinonyx jubatus*) *Zoo Biology* **16** 133–147.

Keeley T, McGreevy PM and O'Brien JK (2012) Cryopreservation of epididymal sperm collected postmortem in the Tasmanian devil (*Sarcophilus harrisii*) *Theriogenology* **78** 315–325.

Kleiman DG (1967). Some aspects of social behavior in the Canidae *American Zoologist* **7** 365–372.

Koehler JK, Platz CCJ, Waddell W, Jones MH and Behrns S (1998) Semen parameters and electron microscope observations of spermatozoa of the red wolf, *Canis rufus Journal of Reproduction and Fertility* **114** 95–101.

Kopeika E, Kopeika J and Zhang T (2007) Cryopreservation of fish sperm *Methods in Molecular Biology* **368** 203–217.

Lado-Abeal L, Veldhuis J and Norman R (2002) Glucose relays information regarding nutritional status to the neural circuits that control the somatotropic, corticotropic and gonadotropic axes in adult male rhesus macaques *Endocrinology* **143** 403–410.

Lamming, G. (Ed) (1990) *Marshall's Physiology of Reproduction* 4th edition. Churchill Livingstone, Edinburgh.

Lasley BL and Kirkpatrick JF (1991) Monitoring ovarian function in captive and free-ranging wildlife by means of urinary and fecal steroids *Journal of Zoo and Wildlife Medicine* **22** 23–31.

Lee AK, Bradley AJ and Braithwaite RW (1977) Corticosteroid levels and male mortality in *Antechinus stuartii* In *The Biology of Marsupials* Ed B Stonehouse and D Gilmore. MacMillan, London.

Leibo SP and Songsasen N (2002) Cryopreservation of gametes and embryos of non-domestic species *Theriogenology* **57** 303–326.

Li M, Meyers S, Tollner T and Overstreet J (2007) Damage to chromosomes and DNA of rhesus monkey sperm following cryopreservation *Journal of Andrology* **28** 493–501.

Loskutoff N, Huyser C, Singh R, Walker D, Thornhill A, Morris L and Webber L (2005) Use of a novel washing method combining multiple density gradients and trypsin for removing human immunodeficiency virus-1 and hepatitis C virus from semen *Fertility and Sterility* **84** 1001–1010.

Love CC, Brinsko SP, Rigby SL, Thompson JA, Blanchard TL and Varner DD (2005) Relationship of seminal plasma level and extender type to sperm motility and DNA integrity *Theriogenology* **63** 1584–1591.

Luyet BJ and Hodapp EL (1938) Revival of Frog's spermatozoa vitrified in liquid air *Proceedings of the Society for Experimental Biology and Medicine* **39** 433–434.

MacDonald ML, Rogers QR, Morris JG and Cupps PT (1984) Effects of linoleate and arachidonate deficiencies on reproduction and spermeogenesis in the cat *Journal of Nutrition* **114** 719–726.

Machaca K and Haun S (2002) Induction of maturation-promoting factor during *Xenopus* oocyte maturation uncouples Ca^{2+} store depletion from store-operated Ca^{2+} entry *Journal of Cell Biology* **156** 75–85.
Margulis SW and Walsh A (2002) The effects of inbreeding on testicular sperm concentration in *Peromyscus polionotus Reproduction, Fertility and Development* **14** 63–67.
Mazur P (1963. Kinetics of water loss from cells at subzero temperatures and the likelihood of intracellular freezing *Journal of General Physiology* **47** 47–369.
Mellen J (1991) Factors influencing reproductive success in small captive exotic felids (*Felis* spp.): A multiple regression analysis *Zoo Biology* **10** 95–110.
Merino O, Risopatrón J, Sánchez R, Isachenko E, Figueroa E, Valdebenito I and Isachenko V (2011) Fish (*Oncorhynchus mykiss*) spermatozoa cryoprotectant-free vitrification: stability of mitochondrion as criterion of effectiveness *Animal Reproduction Science* **124** 125–131.
Mills T, Chitaley K and Lewis R (2001) Vasoconstrictors in erectile physiology *International Journal of Impotence Research* **5** 29–34.
Moberg GP (1985) Influence of stress on reproduction: measure of well-being In *Animal Stress* pp 245–267 Ed GP Moberg. Waverly Press, Baltimore.
Moberg GP (2000) Biological response to stress: implications for animal welfare In *The Biology of Animal Stress* pp 1–21 Ed GP Moberg and JA Mench. CAB International, Wallingford.
Molinia FC and Rodger JC (1996) Pellet-freezing spermatozoa of two marsupials: the tammar wallaby, *Macropus eugenii*, and the brushtail possum, *Trichosurus vulpecula Reproduction Fertility and Development* **8** 681–684.
Moore H, Dvorakova K, Jenkins N and Breed WG (2002) Exceptional sperm cooperation in the wood mouse *Nature* **41** 174–177.
Morai RN, Mucciolo RG, Gomes ML, Lacerda O, Moraes W, Moreira N, Graham LH, Swanson WF and Brown JL (2002) Seasonal analysis of semen characteristics, serum testosterone and fecal androgens in the ocelot (*Leopardus pardalis*), margay (*L. wiedii*) and tigrina (*L. tigrinus*) *Theriogenology* **57** 2027–2041.
Morato R, Wildt D and Spindler R (2003) Influence of medium-term storage on cat sperm prior to cryopreservation *Theriogenology* **59** 260.
Moreira N, Brown JL, Moraes W, Bellem A, Swanson WF and Monteiro-Filho ELA (2002) Effects of captivity conditions on reproductive cyclicity, adrenocortical activity and behavior in female tigrina, *Leopardus tigrinus* and margay, *Leopardus wiedii* In *The Second International Symposium on Assisted Reproductive Technology for the Conservation and Genetic Management of Wildlife* pp 85–89 Ed NM Loskutoff. Henry Doorly Zoo, Omaha, Nebraska.
Morisawa M, Okuno M, Suzuki K, Morisawa S and Ishida K (1983) Initiation of sperm motility in teleosts *Journal of Submicroscopical Cytology* **15** 61–65.
Morrell JM and Hodges JK (1998) Cryopreservation of non-human primate sperm: priorities for future research *Animal Reproduction Science* **53** 43–63.
Morrell JM, Nubbemeyer R, Heistermann M, Rosenbusch J, Kuderling I, Holt W and Hodges JK (1998) Artificial insemination in *Callithrix jacchus* using fresh or cryopreserved sperm *Animal Reproduction Science* **52** 165–174.
Morrow G and Nicol SC (2009) Cool sex? Hibernation and reproduction overlap in the echidna *PLoS One* **4(6):** e6070.
Mulaik D de M(1946) A comparative study of the urogenital systems of an oviparous and two ovoviviparous species of the lizard genus *Sceloporus Bulletin of the University of Utah* **37(8)** *Biology Series* **IX(7)** 3–24. Available at: http://content.lib.utah.edu/utils/getfile/collection/uspace/id/6733/filename/6739.pdf (accessed 21 September 2013).
Nair SJ, Brar AS, Ahuja CS, Sangha SP and Chaudhary KC (2006) A comparative study on lipid peroxidation, activities of antioxidant enzymes and viability of cattle and buffalo bull spermatozoa during storage at refrigeration temperature *Animal Reproduction Science* **96** 21–29.
O'Brien JK and Robeck TR (2006) Development of sperm sexing and associated assisted reproductive technology for sex preselection of captive bottlenose dolphins (*Tursiops truncatus*) *Reproduction, Fertility and Development* **18** 319–329.
O'Brien JK, Steinman KJ, Schmitt T and Robeck TR (2008) Semen collection, characterisation and artificial insemination in the beluga (*Delphinapterus leucas*) using liquid-stored spermatozoa *Reproduction, Fertility and Development* **20** 770–783.
Okutsu T, Yano A, Nagasawa K, Shikina S, Kobayashi T, Takeuchi Y and Yoshizaki G (2006) Manipulation of fish germ cell: visualization, cryopreservation and transplantation *Journal of Reproduction and Development* **52** 685–693.

Oliva R (2006) Protamines and male infertility *Human Reproduction Update* **12** 417–435.

Olsson M, Healey M, Wapstra E, Schwartz T, Lebas N and Uller T (2007) Sons are made from old stores: sperm storage effects on sex ratio in a lizard *Biology Letters* **3** 491–493.

Oswin-Perera B (1999) Reproduction in water buffalo: comparative aspects and implications for management *Journal of Reproduction and Fertility Supplement* **54** 157–168.

Paris DB, Taggart DA, Shaw G, Temple-Smith PD and Renfree MB (2005) Birth of pouch young after artificial insemination in the tammar wallaby (*Macropus eugenii*) *Biology of Reproduction* **72** 451–459.

Parker G. (1998) Sperm competition and the evolution of ejaculates: towards a theory base In *Sperm Competition and Sexual Selection* pp 3–49 Ed TR Birkhead and AP Møller. Academic Press, London.

Penfold L, Harnal V, Lynch W, Bird D, Derrickson S and Wildt D (2001) Characterization of northern pintail (*Anas acuta*) ejaculate and the effect of sperm preservation on fertility *Reproduction* **121** 267–275.

Penfold L, Monfort S, Wolfe B, Citino S and Wildt D (2005) Reproductive physiology and artificial insemination studies in wild and captive gerenuk (*Litocranius walleri walleri*) *Reproduction, Fertility and Development* **17** 707–714.

Pereira RJ, Duarte JM and Negrao JA (2005) Seasonal changes in fecal testosterone concentrations and their relationship to the reproductive behavior, antler cycle and grouping patterns in free-ranging male Pampas deer (*Ozotoceros bezoarticus bezoarticus*) *Theriogenology* **63** 2113–2125.

Pérez-Garnelo S, Oter M, Borque C, Talavera C, Delclaux M, Martínez-Nevado E, Palasz A and De la Fuente J (2006) Post-thaw viability of European bison (*Bison bonasus*) semen frozen with extenders containing egg yolk or lipids of plant origin and examined with a heterologous *in vitro* fertilization assay *Journal of Zoo and Wildlife Medicine* **37** 116–125.

Platz C, Wildt DE and Seager S (1978) Pregnancy in the domestic cat using artificial insemination with previously frozen spermatozoa *Journal of Reproduction and Fertility* **52** 279–282.

Polge C (1951) Functional survival of fowl spermatozoa after freezing at –79 degrees C *Nature* **167** 949–950.

Pomeroy P (2011) Reproductive cycles of marine mammals *Animal Reproduction Science* **124** 184–193.

Pope CE, Dresser BL, Chin NW, Liu JH, Loskutoff NM, Behnke EJ, Brown C, McRae MA, Sinoway CE, Campbell MK, Cameron KN, Owens OM, Johnson CA, Evans RR and Cedars MI (1997) Birth of a western lowland gorilla (*Gorilla gorilla gorilla*) following *in vitro* fertilization and embryo transfer *American Journal of Primatology* **41** 247–260.

Pottinger TG. (1999). The impact of stress on animal reproductive activities In *Stress Physiology in Animals* pp 130–177 Ed PHM Balm. Academic Press, Sheffield.

Pukazhenthi B, Wildt D and Howard J (2001a) Enhanced cat sperm cryopreservation *Journal of Andrology* **22(S1)** 120 [abstract].

Pukazhenthi B, Wildt D and Howard J (2001b) The phenomenon and significance of teratospermia in felids *Journal of Reproduction and Fertility Supplement* **57** 423–433.

Pukazhenthi B, Comizzoli P, Travis AR and Wildt DE (2006a) Applications of emerging technologies to the study and conservation of threatened and endangered species *Reproduction, Fertility and Development* **18** 77–90.

Pukazhenthi B, Laroe D, Crosier A, Bush LM, Spindler R, Pelican K, Bush M, Howard JG and Wildt DE (2006b) Challenges in cryopreservation of clouded leopard (*Neofelis nebulosa*) spermatozoa *Theriogenology* **66** 1790–1796.

Pukazhenthi B, Neubauer K, Jewgenow K, Howard J and Wildt D (2006c) The impact and potential etiology of teratospermia in the domestic cat and its wild relatives *Theriogenology* **66** 112–121.

Purdy PH (2006) The post-thaw quality of ram sperm held for 0 to 48 h at 5 degrees C prior to cryopreservation *Animal Reproduction Science* **93** 114–123.

Rall W (1993) Cryobiology of gametes and embryos from non-human primates In *In vitro Fertilization and Embryo Transfer in Primates* pp 223–245 Ed D Wolf, R Stouffer and R Brenner. Springer Verlag, New York.

Ralls K and Ballou J (1982) Effect of inbreeding on juvenile mortality in some small mammal species *Laboratory Animals* **16** 159–166.

Ralls K, Brugger K and Ballou J (1979) Inbreeding and juvenile mortality in small populations of ungulates *Science* **206** 1101–1103.

Ramaley JA. (1981) Stress and fertility In *Environmental Factors in Mammal Reproduction* pp 127–141 Ed D Gilmore and B Cook. Macmillan, London.

Risley PL (1940) Intersexual glands of turtle embryos following injection of male sex hormone *Journal of Morphology* **67** 439.

Rismiller PD (1992) Field observations on Kangaroo Island echidnas (*Tachyglossus aculeatus multiaculeatus*) during the breeding season In *Platypus and Echidnas* pp 101–105 Ed M Augee. The Royal Zoological Society of NSW, Sydney.

Rivier C and Rivest S (1991) Effect of stress on the activity of the hypothalamic-pituitary-gonadal axis: peripheral and central mechanisms *Biology of Reproduction* **45** 523–532.

Robeck TR and Monfort SL (2006) Characterization of male killer whale (*Orcinus orca*) sexual maturation and reproductive seasonality *Theriogenology* **66** 242–250.

Robeck TR, Steinman KJ, Gearhart S, Reidarson TR, McBain JF and Monfort SL (2004) Reproductive physiology and development of artificial insemination technology in killer whales (*Orcinus orca*) *Biology of Reproduction* **71** 650–660.

Robeck TR, Monfort SL, Calle PP, Dunn JL, Jensen E, Boehm JR, Young S and Clark ST (2005a) Reproduction, growth and development in captive beluga (*Delphinapterus leucas*) *Zoo Biology* **24** 29–49.

Robeck TR, Steinman KJ, Yoshioka M, Jensen E, O'Brien JK, Katsumata E, Gili C, McBain JF, Sweeney J and Monfort SL (2005b) Estrous cycle characterisation and artificial insemination using frozen–thawed spermatozoa in the bottlenose dolphin (*Tursiops truncatus*) *Reproduction* **129** 659–674.

Robeck TR, Steinman KJ, Greenwell M, Ramirez K, Bonn WV, Yoshioka M, Katsumata E, Dalton L, Osborn S and O'Brien JK (2009) Seasonality, estrous cycle characterization, estrus synchronization, semen cryopreservation, and artificial insemination in the Pacific white-sided dolphin (*Lagenorhynchus obliquidens*) *Reproduction* **138** 391–405.

Robeck TR, Steinman KJ, Montano GA, Katsumata E, Osborn S, Dalton L, Dunn JL, Schmitt T, Reidarson T and O'Brien JK (2010) Deep intra-uterine artificial inseminations using cryopreserved spermatozoa in beluga (*Delphinapterus leucas*) *Theriogenology* **74** 989–1001.

Rodger JC (1976) Comparative aspects of the accessory sex glands and seminal biochemistry of mammals *Comparative Biochemistry and Physiology* **55B** 1–8.

Rodger JC and Hughes RL (1973) Studies of the accessory glands of male marsupials *Australian Journal of Zoology* **21** 303–320.

Rodger JC, Cousins SJ and Mate KE (1991) A simple glycerol-based freezing protocol for the semen of a marsupial *Trichosurus vulpecula*, the common brushtail possum *Reproduction, Fertility and Development* **3** 119–125.

Roldan ERS, Gomendio M and Vitullo AD (1992) The evolution of eutherian spermatozoa and underlying selective forces: female selection and sperm competition *Biological Reviews of Cambridge Philosophical Society* **67** 551–593.

Roldan ERS, Cassinello J, Abaigar T and Gomendio M (1998) Inbreeding, fluctuating asymmetry, and ejaculate quality in an endangered ungulate *Proceedings of the Royal Society of London B: Biological Sciences* **265** 243–248.

Rolland RM, Hunt KE, Kraus SD and Wasser SK (2005) Assessing reproductive status of right whales (*Eubalaena glacialis*) using fecal hormone metabolites *General and Comparative Endocrinology* **142** 308–317.

Roth T, Bush L, Wildt D and Weiss R (1999) Scimitar-horned oryx (*Oryx dammah*) spermatozoa are functionally competent in a heterologous bovine *in vitro* fertilization system after cryopreservation on dry ice, in a dry shipper, or over liquid nitrogen vapor *Biology of Reproduction* **60** 493–498.

Saint Jalme M, Lecoq R, Seigneurin F, Blesbois E and Plouzeau E (2003) Cryopreservation of semen from endangered pheasants: the first step towards a cryobank for endangered avian species *Theriogenology* **59** 875–888.

Samour JH (2004) Semen collection, spermatozoa cryopreservation, and artificial insemination in nondomestic birds *Journal of Avian Medicine and Surgery* **18** 219–223.

Sanchez R, Risopatron J, Schulz M, Villegas J, Isachenko V, Kreinberg R and Isachenko E (2011) Canine sperm vitrification with sucrose: effect on sperm function *Andrologia* **43** 233–241.

Saragusty J, Hildebrandt T, Behr B, Knieriem A, Kruse J and Hermes R (2009) Successful cryopreservation of Asian elephant (*Elephas maximus*) spermatozoa *Animal Reproduction Science* **115** 255–266.

Schroeder JP and Keller KV (1989) Seasonality of serum testosterone levels and sperm density in *Tursiops truncatus* *Journal of Experimental Zoology* **249** 316–321.

Shaffner CS (1942) Longevity of fowl spermatozoa in frozen condition *Science* **96** 337.

Smith C, Roeszler K, Hudson Q and Sinclair A (2007) Avian sex determination: what, when and where? *Cytogenic Genome Research* **11** 165–173.

Sontakke S, Patil M, Umapathy G, Rao K and Shivaji S (2009) Ejaculate characteristics, short-term semen storage and successful artificial insemination following synchronisation of oestrus in the Indian blackbuck antelope (*Antilope cervicapra*) *Reproduction, Fertility and Development* **21** 749–756.

Spindler RE and Wildt DE (2010) Male reproduction: assessment, management, assisted breeding and fertility control In *Wild Mammals in Captivity: Principles and Techniques for Zoo Management* 2nd edition pp 429–446 Ed DG Kleiman, KV Thompson and C Kirk Baer. University of Chicago Press, Chicago.

Spindler RE, Huang Y, Howard JG, Wang P, Zhang H, Zhang G and Wildt DE (2004) Acrosomal integrity and capacitation are not influenced by sperm cryopreservation in the giant panda *Reproduction* **127** 547–556.

Spindler RE, Songsasen N and Deem S (2007) Factors affecting the reproductive success of jaguars In *Manejo e Conservação de Carnivoros Neotropicais* pp 281–304 Ed R Morato, F Rodrigues, E Eizirik, P Mangini, F Azavedo and J Marinho-Filho. Instituto Brasileiro do Meio Ambiente e dos Recursos Naturais Renováveis (IBAMA), Brasilía.

Steffen W, Crutzen PJ, McNeill JR (2007) The Anthropocene: are humans now overwhelming the great forces of nature. *Ambio* **36** 614–621.

Stoss J (1983) Fish gamete preservation and spermatozoa physiology In *Fish Physiology* pp 305–330 Ed W Hoar, D Randall and E Donaldson. Academic Press, London.

Stricker S (1999) Comparative biology of calcium signaling during fertilization and egg activation in animals *Developmental Biology* **21** 157–176.

Sundqvist C, Lukola A and Valtonen M (1984) Relationship between serum testosterone concentrations and fertility in male mink (*Mustela vison*) *Journal of Reproduction and Fertility* **70** 409–412.

Swanson W, Wildt D, Cambre R, Citino S, Quigley K, Brousset D, Morais R, Moreira N, O'Brien S and Johnson W (1995) Reproductive survey of endemic felid species in Latin American zoos: male reproductive status and implications for conservation. *Proceedings of the American Association of Zoo Veterinarians* **1** 374–380.

Taggart DA (1994) A comparison of sperm and embryo transport in the female reproductive tract of marsupial and eutherian mammals *Reproduction, Fertility and Development* **6** 451–472.

Taggart DA, Leigh CM, Schultz D and Breed WG (1995) Ultrastructure and motility of spermatozoa in macropodid and potoroidid marsupials *Reproduction, Fertility and Development* **7** 1129–1140.

Taggart DA, Leigh CM, Steele VR, Breed WG, Temple-Smith PD and Phelan J (1996) Effect of cooling and cryopreservation on sperm motility and morphology of several species of marsupial *Reproduction, Fertility and Development* **8** 673–679.

Taggart DA, Steele VR, Schultz D, Dibben R, Dibben J and Temple-Smith PD (1998) Semen collection and cryopreservation in the southern hairy-nosed wombat (*Lasiorhinus latifrons*): implications for conservation of the northern hairy-nosed wombat (*Lasiorhinus krefftii*) In *Wombats* pp 180–191 Ed RT Wells and PA Pridmore. Surrey Beatty and Sons, Chipping Norton, New South Wales.

Tardif S and Jaquish C (1994). The common marmoset as a model for nutritional impacts upon reproduction *Annals of the New York Academy of Sciences* **709** 214–215.

Temple-Smith P (1987) Sperm structure and marsupial phylogeny In *Possums and Opossums: Studies in Evolution* pp 171–193 Ed M Archer. Surrey Beatty & Sons and Royal Zoological Society of New South Wales, Sydney.

Terada T, Ashizawa K, Maeda T and Tsutsumi Y (1989) Efficacy of trehalose in cryopreservation of chicken spermatozoa *Japanese Journal of Animal Reproduction* **35** 20–25.

Tollner T, Dong Q and VandeVoort C (2011) Frozen–thawed rhesus sperm retain normal morphology and highly progressive motility but exhibit sharply reduced efficiency in penetrating cervical mucus and hyaluronic acid gel *Cryobiology* **62** 15–21.

Vajta G and Nagy ZP (2006) Are programmable freezers still needed in the embryo laboratory? Review on vitrification *Reproductive Biomedicine Online* **12** 779–796.

Walsh PC, Retik AB, Darracott Vaughan E Jr and Wein AJ (Eds) (2002) *Campbell's Urology* 8th edition. Elsevier Saunders, Philadelphia.

Watanabe S, Hara M and Watanabe Y (2000) Male internal fertilization and introsperm-like sperm of the seaweed pipefish (*Syngnathus schlegeli*) *Zoological Science* **17** 759–767.

Watson PF (1978) A review of techniques of semen collection in mammals *Symposia of the Zoological Society of London* **43** 97–126.

Watson PF (1990) Artificial insemination and the preservation of semen In *Marshall's Physiology of Reproduction* 4th edition pp 747–769 Ed G Lamming. Churchill Livingstone, Edinburgh.

Watson PF and Holt WV (2001) Organizational issues concerning the establishment of a genetic resource bank In *Cryobanking the Genetic Resource: Wildlife Conservation for the Future?* pp 113–122 Ed P Watson and W Holt. Taylor and Francis, London and New York.

Wielebnowski N (2002) Studying the impact of stress using survey information, observations & corticoid analyses In *The Second International Symposium on Assisted Reproductive Technology for the Conservation and Genetic Management of Wildlife* pp 63–72 Ed NM Loskutoff. Henry Doorly Zoo, Omaha, Nebraska.

Wielebnowski NC, Fletchall N, Carlstead K, Busso JM and Brown JL (2002) Noninvasive assessment of adrenal activity associated with husbandry and behavioral factors in the North American clouded leopard population *Zoo Biology* **21** 77–98.

Wildt DE (1997) Genome resource banking: impact on biotic conservation and society In *Tissue Banking in Reproductive Biology* pp 399–439 Ed AM Karow and J Critser. Academic Press, New York.

Wildt DE, Seager SWJ and Chakraborty PK (1980) Effect of copulatory stimuli on ovulatory and serum luteinizing hormone response in the cat *Endocrinology* **107** 1212–1217.

Wildt DE, Bush M, Howard JG, O'Brien SJ, Meltzer D, Van Dyk A, Ebedes H and Brand DJ (1983) Unique seminal quality in the South African cheetah and a comparative evaluation in the domestic cat *Biology of Reproduction* **29** 1019–1025.

Wildt DE, Howard JG, Chakraborty PK and Bush M (1986) Reproductive physiology of the clouded leopard: II. A circannual analysis of adrenal-pituitary-testicular relationships during electroejaculation or after an adrenocorticotropin hormone challenge *Biology of Reproduction* **34** 949–959.

Wildt DE, Bush M, Goodrowe KL, Packer C, Pusey AE, Brown JL, Joslin P and O'Brien SJ (1987) Reproductive and genetic consequences of founding isolated lion populations *Nature* **329** 328–331.

Wildt DE, Phillips LG, Simmons LG, Chakraborty PK, Brown JL, Howard JG, Teare A and Bush M (1988) A comparative analysis of ejaculate and hormonal characteristics of the captive male cheetah, tiger, leopard, and puma *Biology of Reproduction* **38** 245–255.

Wolf DP, Thormahlen S, Ramsey C, Yeoman RR, Fanton J and Mitalipov S (2004) Use of assisted reproductive technologies in the propagation of rhesus macaque offspring *Biology of Reproduction* **71** 486–493.

Woolley P (1978) Phallic morphology of *Dasyuroides byrnei* and *Dasycercus cristicauda* (Marsupialia; Dasyuridae) *Australian Journal of Zoology* **35** 535–540.

Woolley P and Webb SJ (1977). The penis of dasyurid marsupials In *The Biology of Marsupials* pp 307–323 Ed B Stonehouse and D Gilmore. Macmillan, London.

Zacariotti R, Grego K, Fernandes W, Sant'anna S and de Barros Vaz Guimaraes M (2007) Semen collection and evaluation in free-ranging Brazilian rattlesnakes (*Crotalus durissus terrificus*) *Zoo Biology* **26** 155–160.

Zambelli D, Cunto M, Prati F and Merlo B (2007) Effects of ketamine or medetomidine administration on quality of electroejaculated sperm and on sperm flow in the domestic cat *Theriogenology* **68** 796–803.

Zee YP, Holt WV, Gosalvez J, Allen CD, Nicolson V, Pyne M, Burridge M, Carrick FN and Johnston SD (2008) Dimethylacetamide can be used as an alternative to glycerol for the successful cryopreservation of koala (*Phascolarctos cinereus*) spermatozoa *Reproduction, Fertility and Development* **20** 724–733.

Zhang T (2004) Cryopreservation of gametes and embryos of aquatic species In *Life in the Frozen State* pp 415–435 Ed B Fuller, N Lane and EE Benson. CRC Press, London.

18 Male Animal Contraception

Scott T. Norman* and Tonya M. Collop
Charles Sturt University, Wagga Wagga, New South Wales, Australia

Introduction

Male animal contraception is considered desirable in many animal management contexts. Examples include wild and domestic animal population control (Asquith *et al.*, 2006), the control and management of individual animals, and assisting in the management of breeding programmes. As a minimum, effective male contraception should either block sperm production, or interfere with the ability of sperm to reach or to fertilize an oocyte (Hogarth *et al.*, 2011). The primary contraceptive effect may be coupled with requirements to modify secondary sex characteristics and behaviour, as the situation dictates.

Essential requirements for methods of male contraception include effectiveness and safety. More specific requirements may include options such as reversibility, and whether or not the technique allows the maintenance of male physiological processes, behaviour and social structure (Spay Neuter Task Force, 2011). When designing a suitable male contraceptive, there is a need to consider the health and safety of the animal, human health and safety, and the specific requirements of animal caretakers. This means that it is unlikely that there will be one universally acceptable method of male contraception and that a number of different options should be available (Bowen, 2008).

For many contraceptive techniques, the procedure is coupled either directly or indirectly to a reduction in the concentrations of circulating sex steroids. A reduction in circulating testosterone concentrations has negative effects on libido, male secondary sex characteristics, psychotropic effects, protein anabolism, bone structure and haematopoiesis (Neischlag *et al.*, 2004). These effects may be seen as beneficial in reducing undesirable meat qualities such as boar taint in pork, male territorial behaviour and modification of general behaviour to improve human safety while handling and training. However, in some contexts, such as for athletic animals and the production of lean beef, the reduction in sex steroids associated with some forms of male contraception may be seen as a disadvantage.

This chapter will define the need for male contraception and explore current contraceptive options, with emphasis on non-surgical techniques. Male contraception will then be reviewed in a species-specific context.

* E-mail: snorman@csu.edu.au

General Concepts

Before embarking on a review of male contraceptive techniques, it is worthwhile defining some of the terminology involved. In this review, the term 'contraception' in the male refers to any procedure that prevents spermatozoa achieving successful fertilization of the oocyte. In addition to immunological techniques, and procedures such as transection or occlusion of the ductus deferens, male contraception may involve inactivation of the gonads. Importantly, contraception commonly implies the possibility of reversibility (Munson, 2006). In contrast, castration is defined here as any process rendering a male incapable of reproduction through permanent removal, destruction or inactivation of the gonads. Castration may be achieved by surgical, chemical or possibly immunological means, but to be considered as castration in this discussion, the technique must be irreversible. Permanency is an important aspect of the definition that provides a universal understanding that castration is not a reversible procedure.

Male Contraception – Defining the Need

The application of male contraceptive techniques is justifiable only if there are benefits to one or more of the following:

1. The welfare or management of the treated animal.
2. The welfare or management of an animal population.
3. The welfare or management of another species (including humans) that may be adversely affected by the treated animals.
4. The sustainability of a habitat that might otherwise be adversely affected by the treated animals.
5. Disease control within animal and human populations.

With these factors in mind, male contraception is most commonly undertaken to control animal populations, modify undesirable behaviour, modify secondary sex characteristics to assist with animal management and assist with disease control. There are also very specific applications for male contraception, such as teaser animals (i.e. a sterile male used for the purposes of eliciting sexual behaviour in the female) for the preparation of sheep, cattle and horses to assist with the management of breeding programmes.

Population control

Population control may have different end points depending on the context. In many instances, it specifically refers to efforts to correct or to prevent excessive animal numbers. However, in animal husbandry, population control can relate to the close management of breeding in order to manipulate the timing of progeny production and/or the genetic make-up of the population.

Overpopulation can be problematic for both domestic and wild animal species. Domestic pet overpopulation is well recognized, with estimates of between 10 and 20 million unwanted dogs and cats being euthanized every year in the USA (Bowen, 2008), and similar trends noted throughout the world (RSPCA, 2012). Other undesirable consequences of pet overpopulation include the economic cost of animal control programmes to society, the potential for serious injuries being inflicted on other animals or humans, and sanitation problems in cities associated with animal faeces and urine. Depending on geographic location, there are significant overpopulation problems associated with many other animal species, including wild or feral horses, deer, elk, geese, pigs, possums, rabbits and elephants (Barfield *et al.*, 2006).

There is also an annual loss of wild or feral animals due to starvation, being killed on roads, or death associated with fighting or hunting. Stray or wild animals on roads can pose a major risk to human safety. In countries where rabies is prevalent, stray dogs are important vectors of the virus, leading to the deaths of tens of thousands of people annually (Knobel *et al.*, 2005).

Mating behaviour needs to be considered when assessing the potential for male contraception to assist with wild, or feral, animal population control (Jewgenow *et al.*, 2006). Animal populations in which monogamous breeding occurs, such as the fox, are more amenable to control by male contraceptive techniques. In contrast, contraception may be more effectively applied to the female of polygamous species such as the domestic cat (Jewgenow *et al.*, 2006).

Situations may arise in zoological collections where otherwise endangered animals become overpopulated, resulting in the need to restrict or prevent further reproduction (Bowen, 2008). Control of the genetic make-up of a zoological or production animal population, as well as the common managerial requirement to control when offspring are produced, are managed in many species by castration of excess males. This is usually complemented by the provision of suitable infrastructure to keep intact males and females separate until breeding is required. As castration is irreversible, the problem with this approach is that it permanently removes what may be genetically desirable individuals from the gene pool before they have had an opportunity to prove their value. These contexts provide the motivation to develop effective, but reversible, contraception and behaviour modification so that animals may be utilized safely, yet still retain breeding potential should they prove suitable.

Modification of behaviour

The sex steroids have profound effects on male behaviour (Senger, 2003), influencing social interactions, libido, aggression and territorial activity. Due to their anabolic, anti-inflammatory and behaviour-modifying effects, there is the possibility that sex steroids may also influence the feeling of well-being in animals, which could affect athletic performance in species such as horses and dogs. Therefore, when developing contraceptive techniques in any species, the influence of the technique on the sex steroids and behaviour of the animal needs to be considered in conjunction with the husbandry under which the animal will be maintained. With techniques such as orchidectomy, sex steroid concentrations are permanently reduced and consideration may be given as to whether it is appropriate or legal to supplement animals with exogenous steroids if the retention of male behaviour is required.

Modification of secondary sex characteristics

Secondary sex characteristics are not directly associated with the reproductive system, but distinguish gender. Male characteristics vary widely depending on the species, with manifestations as diverse as: tusks in boars, variations in vocal range, spines on the penis in cats, manes in lions, scent glands, meat taint, the presence and size of horns or antlers, pheromone production, increased muscular development and, in birds, varied plumage. When considering a method of contraception, it is necessary to determine whether or not the maintenance of secondary sex characteristics is desirable in a species. For example, boar taint in pork can adversely affect consumer acceptance, and requires contraceptive methods to reduce sex steroid production. In contrast, the loss of secondary sex characteristics in males of feral animal populations may leave them vulnerable to injury or traumatic death resulting from adverse social interactions with intact individuals, thus highlighting a need for contraception with retained production of sex steroids.

Influencing disease processes

The influence of contraception on disease processes is mainly associated with whether or not sex-steroid production is altered. There has been extensive research into the advantages and disadvantages of removing the source of sex steroids in companion animals, with orchidectomy before puberty being a specific area of investigation (Howe *et al.*, 2000, 2001). While further studies are required, there are some emerging patterns that can be used to assist with the decision of whether, and when, to castrate an animal. Despite significant benefits with regard to management and behaviour, studies are ongoing on the long-term effects of contraceptive techniques that reduce the concentrations of sex steroids. Some areas undergoing further assessment are urinary tract function, the timing of epiphyseal closure, and ligament and tendon resilience. These studies also support the need to have contraceptive options available to allow context-specific goals to be achieved.

Methods of contraception that reduce circulating concentrations of sex steroids can assist with disease control by preventing sexually transmissible disease, by reducing steroid hormone influence on specific tissues, or by modifying behaviour that may lead to traumatic injury. There are a number of disease conditions that may be positively influenced by reduced concentrations of sex steroids, such as prostatic disease in dogs (Howe, 2006), and a reduced prevalence of injuries associated with territorial behaviour in various mammals (Hart, 1974). In the domestic situation, this is particularly relevant to catfight wounds, but also has relevance in production animal and equine contexts. Examples of sexually transmitted diseases that can be reduced or prevented by contraceptive methods preventing copulation include: transmissible venereal tumour and brucellosis in dogs; campylobacteriosis and trichomoniasis in cattle; and contagious equine metritis and coital exanthema in horses.

From the aspect of human health, the control of feral animal populations can lead to a reduction in the prevalence of bite injuries and of many zoonotic diseases, such as rabies, leptospirosis, Q fever, toxoplasmosis and other parasitic infestations (Golden, 2009; Spay Neuter Task Force, 2011).

Contraception Based on Surgical Techniques

Surgical castration and contraceptive techniques in the male consist of: the destruction or removal of the gonads; occlusion or removal of a portion of the excurrent duct system (most commonly vasectomy, but also epididymectomy); methods of redirecting or restricting the penis to prevent intromission; or amputation of the penis. This section will focus on orchidectomy and vasectomy only.

Orchidectomy

Currently, the most reliable method of sterilization for male animals of most species is to physically disassociate the vascular supply from the testicles. This requires the testicles to either be removed or rendered atrophic, and can be achieved by open or closed techniques. The result is the generally desirable combination of infertility, behaviour modification and, where applicable, improvement in consumer acceptance of meat products. While these techniques are reliable and well tolerated, it should be noted that in a field situation, care should be taken with animals exhibiting conditions such as cryptorchidism or inguinal herniation. As with all surgical procedures, there may be complications associated with restraint, anaesthesia, infection, haemorrhage and cutaneous myiasis. It appears that fewer complications are encountered when the procedure is performed on younger animals, and this is reflected in the policy documents and legislature of many jurisdictions, where it is recommended that a higher level of skill and facilities is required for surgical castration as animals get older. For species in which the surgery is performed in the field, cool, dust-free environments are desirable in order to reduce wound contamination.

As a summary, there are four physical methods that are utilized to destroy or remove the testicles. These are:

- Orchidectomy via a scrotal or para-scrotal incision, which is applicable to males of most species.
- Closed constriction of the vascular supply to the testicles utilizing elastic (Elastrator®) rings, as are commonly used in young production animals.
- Closed crushing of the vascular supply to the testicles utilizing a Burdizzo® device, as used in cattle.
- Closed crushing of the vascular supply utilizing a tension banding technique

(Calicrate®), which has been used in older cattle. This technique is not recommended owing to the risk of incomplete compression of the blood supply, leading to possible life-threatening complications (Newman, 2007).

Advantages

Surgical castration is currently perhaps the most common form of contraception used for male animals, regardless of animal species or animal management context. This is true in: the cattle industry (Coetzee *et al.*, 2010); the pig industry, in which, with the exception of a few European countries and Australia, male pigs generally undergo orchidectomy at a very young age (Bonneau and Enright, 1995); dogs and cats, where orchidectomy is the most common method of male contraception (Bowen, 2008); and the horse industry (Reilly and Cimetti, 2005), where there is a requirement for both contraception and the behaviour modification of most males. The main advantage of this technique is that permanent contraception can be guaranteed when performed by an experienced operator.

Due to the widespread use of surgical castration techniques within each industry, there are generally adequate numbers of experienced operators to competently perform the procedure in a humane manner. With species variation, orchidectomy is also beneficial in reducing the prevalence of diseases such as prostatic hyperplasia, testicular tumours and sexually transmissible diseases.

Disadvantages

Surgical castration is labour intensive, can cause morbidity and mortality, is stressful to the animal (Von Waldmann *et al.*, 1994), and there are ethical and animal welfare concerns (Bonneau and Enright, 1995; Coetzee *et al.*, 2010). Importantly, as with all surgical procedures, there are inherent costs and risks associated with restraint, anaesthesia and surgery. In some production contexts, the removal of the source of sex steroids creates economic disadvantages, as intact males are usually more feed efficient and leaner than castrated animals (Bonneau and Enright, 1995; Oonk *et al.*, 1998). Because intact males have better-feed efficiency, avoiding the loss of sex steroids associated with castration may also significantly reduce the amount of biological pollutants excreted by production animals into the environment.

The adverse effects of castration on growth and efficiency can largely be reversed by the administration of anabolic steroids, although this is not an option in many countries, where their use is banned; it also does not solve the ethical and welfare concerns that are associated with surgical castration. So the challenge is to reduce management and meat quality concerns and, at the same time, maintain the advantages of intact animals (Bonneau and Enright, 1995).

In control programmes for the feral animal population, orchidectomy may not be practical and the loss of male behaviour associated with removal of the gonads may have an undesirable influence on social dynamics and territorial behaviour (Jewgenow *et al.*, 2006). This may adversely affect the well-being of individual animals, as well as interfering with the effectiveness of population control methods that require sterile but sexually active males.

Vasectomy

In veterinary terminology, the duct leading from the tail of the epididymis to the ampulla or pelvic urethra is described as the deferent duct, or ductus deferens (Senger, 2003). However, in human terminology, it is described as the vas deferens, and remnants of this terminology associated with surgical or manipulative procedures have persisted within the veterinary literature. Thus, surgical vasectomy refers to bilateral removal of a portion of the ductus deferens, rendering the animal sterile by preventing sperm from being ejaculated during copulation (Johnston *et al.*, 2001a). A technique resulting in a similar outcome, but not technically a vasectomy, is ductal occlusion.

The procedure for vasectomy in many species is well described (Boundy and Cox, 1996; Johnston *et al.*, 2001a,b). In dogs and cats, vasectomy can be performed through a

1–2 cm incision located in the inguinal region of the dog, or located cranial to the scrotum in cats (Johnston *et al.*, 2001a,b). In sheep, a 4 cm vertical incision is made over the left and right spermatic cords on the cranial surface of the neck of the scrotum (Boundy and Cox, 1996). Following skin and subcutaneous incision, the spermatic cords are identified, separated and exteriorized from the tunic using a combination of blunt and sharp dissection. Traction and manipulation of the testicle can be helpful in identifying the spermatic cord and ductus deferens. Following isolation of the ductus deferens, a segment of the ductus is removed and the proximal and distal severed ends of the ductus are ligated. Success of the procedure can be confirmed by submission of the excised tissue for histological assessment (Johnston *et al.*, 2001a). Vasectomy may also be performed laparoscopically by occlusion of a segment of ductus using bipolar forceps and electrocoagulation (Mahalingam *et al.*, 2009).

Studies have reported azoospermia to occur from 2 to 21 days in the dog following bilateral vasectomy (Pineda *et al.*, 1976; Schiff *et al.*, 2003), and within 1 week in rams (Boundy and Cox, 1996). This is relatively quick compared with cats, in which live sperm have been identified for up to 49 days following pre-scrotal vasectomy (Pineda and Dooley, 1984). Variations in the duration from vasectomy to azoospermia is associated with sperm storage capacity and the presence or absence of accessory sex glands such as the seminal vesicles (Schiff *et al.*, 2003), which may produce secretions to support sperm viability for a short duration.

Advantages

Vasectomy is a well-developed method for permanent contraception, which has been applied to pets in order to eliminate reproductive potential while retaining their testicular function. In humans, methods for successful reversal (vasovasostomy) are well documented (Bowen, 2008), and these techniques have been transferred into animal contexts. This provides the possibility of a return to fertility should genetic worthiness be identified subsequent to contraception. There is limited data from animals, but pregnancy rates following vasovasostomy in humans are estimated at 60% if the vasectomy was performed less than 5 years before reversal and 40% if performed more than 5 years from the original surgery (Barfield *et al.*, 2006).

In the veterinary context, vasectomy is a relatively quick procedure, but still requires general anaesthesia in companion animals. Vasectomy of dominant males has been suggested as a method of feral cat population control (Howe, 2006), as vasectomized dominant tomcats can prevent submissive, intact toms from inseminating non-spayed females, and can reduce receptivity in queens by inducing periods of pseudopregnancy. In sheep, vasectomy is an ideal contraceptive choice when preparing teaser rams to assist with artificial breeding programmes, as it is a quick procedure and allows the required male behaviour to remain unaltered (Boundy and Cox, 1996). There is also application for vasectomy or ductal occlusion in captive animals, and the successful vasovasostomy of a zoo animal was confirmed when a bush dog at the St Louis Zoo in Missouri sired three healthy pups (Barfield *et al.*, 2006).

Disadvantages

Despite the effectiveness of vasectomy as a contraceptive technique, most owners of domestic animals desire both contraception and behaviour modification for their pets. The persistence of male sex characteristics and behaviours after vasectomy may permit territorial fighting and androgen-dependent conditions such as prostatic disease to develop (Johnston *et al.*, 2001a). These considerations, coupled with the fact that vasectomy in dogs and cats either costs the same as or, commonly, is more expensive than castration means that it is rarely practised in domestic species (Bowen, 2008). In a wild, or feral animal context there is a significant disadvantage in vasectomy owing to the need to capture, transport, anaesthetize and release animals.

Although the testes are immunologically shielded by the blood–testis barrier, any situation such as trauma or physiological stress that may allow the production of anti-sperm antibodies (ASAs) to spermatozoa can lead to undesirable testicular pathology. These conditions can occur as a result of vasectomy. In human studies, a large proportion of vasectomized men have ASAs in their serum and this has been associated with an increased risk of epididymitis, orchitis and varicocele (Skakkebaek *et al.*, 1994). These conditions are not only problematic in their own right, but also reduce any potential for reversibility (Bowen, 2008).

Iatrogenic 'high-flanker' cryptorchidism

In cattle, there are anecdotal reports of placing rubber (Elastrator™) rings around the scrotum, distal to the testicles, to push them firmly up into the inguinal region; thereby producing an iatrogenic, bilateral, 'high-flanker' cryptorchid. While there are no published studies reporting the effectiveness of this technique, placing the testicle close to the abdominal wall will ensure a sustained increase in testicular temperature compared with the normal scrotal position. A prolonged increase in temperature should adversely affect spermatogenesis but maintain testosterone production. Risks associated with this technique include wound dehiscence, clostridial infection and variable effectiveness.

Contraception Based on Non-surgical Castration and Ductal Obstruction

Non-surgical approaches to physical destruction of the testicles have been based on injection into the testes of a variety of materials that induce tissue destruction, orchitis and fibrosis. Other targets for blocking sperm flow include the epididymides and the ductus deferens. Numerous sclerosing agents have been injected into the tail of the epididymis to induce blockage of the tubules and subsequent azoospermia. Physical devices or chemical compounds have also been used to block the ductus deferens. Basic welfare requirements for any chemical injected into the testes, epididymides, or ductus deferens should be that they are non-mutagenic, non-carcinogenic, non-teratogenic and induce minimal pain.

Ductal obstruction

Methods of obstructing the ductus deferens can be divided into two broad categories: extravasal and intravasal techniques. For blocking the epididymides, sclerosing agents have most commonly been utilized.

Extravasal methods of obstructing the ductus deferens involve the placement of a device, such as clips or a suture, around the ductus deferens in order to cause occlusion. Extravasal techniques are not commonly used in humans owing to the difficulty in removing clips, which leads to inadequate restoration of fertility, where this is needed (Barfield *et al.*, 2006). However, the requirement for reversibility may not be as relevant in animal species. Yet attempts to avoid surgery by placing clips across the scrotal skin to occlude the ductus deferens have proven ineffective and traumatic (Barfield *et al.*, 2006).

Intravasal methods of ductal obstruction include the use of injectable silicone, cylindrical plugs, spherical and polypropylene beads, threads of silicone or suture material. Unfortunately, none of these techniques can sustain effective long-term sperm obstruction without generating fibrosis or perforation. A ductus deferens valve was developed, with the goal of turning sperm flow on or off, but unfortunately reliable occlusion and problems with ductus perforation could not be overcome (Barfield *et al.*, 2006). Percutaneous injection of sclerosing chemicals into the lumen of the ductus using compounds such as ethanol, silver nitrate, acetic acid and formaldehyde has been used as a non-reversible method of ductal occlusion in rats and dogs (Freeman and Coffey, 1973). While generally effective, there is a possibility of retrograde flow of the chemical to the testis, causing testicular atrophy (Barfield *et al.*, 2006). This may or may not be a concern, depending on the context in which it is used and any need for reversibility.

A more recent male contraceptive technique that has been trialled in humans and Langur monkeys is described as reversible inhibition of sperm under guidance (RISUG). RISUG involves the injection of styrene maleic anhydride (SMA), with or without the addition of dimethyl sulfoxide (DMSO), into the ductus deferens. The product provides contraception by physically blocking the lumen of the ductus deferens, but also acts by lowering the pH in the immediate vicinity and creating charge disturbances to sperm membranes (Mishra *et al.*, 2003). This charge disturbance ensures that function is impaired in any sperm that may pass the blockage. In Langur monkeys, uniform azoospermia was achieved by the fourth month after RISUG treatment and lasted throughout the study (540 days) (Mishra *et al.*, 2003). In a human study, lasting only 6 months, partial occlusion of the vas deferens was achieved in all men by 4 months, with the occasional appearance of sperm, sperm heads or immature germ cells in the ejaculate (Barfield *et al.*, 2006). Further work is required to determine the effectiveness of RISUG in domestic species. A variation of the RISUG technique is a formulation described as 'smart' RISUG, in which iron oxide and copper particles are dispersed into the original SMA-DMSO product (Jha *et al.*, 2009). This formulation can be detected by radiology or magnetic resonance imaging (MRI) techniques, thus allowing localization if implant removal is being considered (Yan Cheng and Mruk, 2010). It also appears to provide improved spermicidal action over the original formulation of the RISUG product, as magnetic iron particles have been shown to bind proteins, and the copper particles displace molecules on the surface of sperm (Yan Cheng and Mruk, 2010). Both of these actions of the particles therefore reduce sperm viability.

Zinc arginine (ZA) is one of the primary sclerosing agents that has been trialled for ductal obstruction of the canine epididymis, with the intra-epididymal injection of 50 mg of ZA (0.5 ml/testis) resulting in azoospermia within 90 days of injection (Fahim *et al.*, 1993). ZA is used because it is considered to be non-mutagenic, non-carcinogenic, and non-teratogenic (Fahim *et al.*, 1993). Contraception by injecting ZA into the epididymis initially seemed quite promising in both dogs and cats (Pineda and Dooley, 1984; Fahim *et al.*, 1993). Unfortunately, when applied more widely in the context of animal shelters, it was associated with a high incidence of serious inflammatory responses, and is not currently considered suitable for contraception in domestic species (Bowen, 2008). Despite this, the ZA formulation has been found to be suitable for intra-testicular administration to induce testicular degeneration, and its use for this purpose will be discussed further in the next section.

Induced Testicular Degeneration

Chemical methods of inducing testicular degeneration

Testicular degeneration can be chemically induced by a number of methods. These include the direct injection of a chemical into the testicular parenchyma, the administration of a chemical by the oral or parenteral route, and, more recently, the conjugation of cytotoxic agents to gonadotrophins. This section will focus on the induction of testicular degeneration by the intra-testicular administration of chemicals.

Intra-testicular injections have been investigated as a method of inducing orchitis, seminiferous tubule degeneration and male contraception since the 1950s (Freund *et al.*, 1953). Injecting an adjuvant, such as Freund's complete adjuvant (FCA) or Bacillus Calmette–Guérin (BCG), directly into the testis incites a local inflammatory response that enables lymphoid cells to gain access to the testicular tissue, resulting in an autoimmune response. A single intra-testicular injection of FCA, or 10–25 units of BCG, resulted in severe oligospermia or azoospermia without granuloma formation or the development of circulating ASAs (Naz and Talwar, 1981). Other substances, such as glycerol formulations (Wiebe and Barr, 1984; Wiebe *et al.*, 1989) and calcium chloride (Jana and Samanta, 2007) have been trialled as contraceptive agents. However, subsequent studies utilizing glycerol have not supported an ongoing role for their use in contraception (Immegart and Threlfall, 2000). While there have been promising preliminary results from the use of calcium chloride as a contraceptive agent in dogs, there is currently no commercial product registered for such use (Jana and Samanta, 2007).

One protocol that has generated significant interest involves the intra-testicular injection of solutions containing zinc gluconate neutralized to a pH of 7 with arginine. This formulation (commonly referred to as zinc arginine, or ZA) is considered to be efficacious and safe based on contraceptive efficacy and minimal adverse reactions, especially when applied to young dogs (Tepsumethanon *et al.*, 2005). The product is quite expensive, but it was marketed for a brief time for canine contraception as Neutersol®. It has been trialled in other species as well, but with equivocal results (Brito *et al.*, 2011), possibly due to the dosing used being based on canine studies. Unfortunately, with canine use, there were also some undesirable inflammatory responses that warranted withdrawal of the product from the US market in 2011. With modifications to the formulation, the product may re-emerge commercially as Esterisol®, Zeuterin® or Testoblock®, and with a claim to have use as a method of permanent contraception in dogs after a single administration. Despite this claim, it must be recognized that severe adverse reactions may still occur in close to 4% of cases, including necrotizing reactions requiring surgical orchidectomy and debridement (Levy *et al.*, 2008). Such reactions can be minimized through careful dose placement and calculation of the dose for each testicle based on width measurements.

Physical methods of inducing testicular degeneration

Ultrasound has been utilized to interfere with sperm production or transport via the combined effect of heat and mechanical vibration (Fahim *et al.*, 1977; Fried *et al.*, 2002; Roberts *et al.*, 2002; Leoci *et al.*, 2009). Targeted structures include testicular parenchyma, the epididymides or the ductus deferentia. For the latter two structures, short-term (20 to 120 s), high-energy (3–19 W) ultrasound can be administered to induce coagulative necrosis, resulting in luminal occlusion within 2 weeks of treatment (Fried *et al.*, 2002; Roberts *et al.*, 2002). However, there is still a need for long-term studies to assess the contraceptive efficacy of this method. With current techniques and technology, there is an unacceptably high occurrence of adverse reactions such as skin burns (Fried *et al.*, 2002), though studies to accurately determine the ideal therapeutic window for both power and duration of application for different species may help overcome this problem. The testicular parenchyma of dogs and toms were treated with a high-intensity, focused ultrasound consisting of 1–2 W/cm^2 for 10–15 min administered one to three times at 2–7 day intervals (Fahim *et al.*, 1977; Leoci *et al.*, 2009). Although it is apparent that further work is necessary to clearly define treatment intensity and duration, ultrasound administration was shown to suppress spermatogenesis without affecting testosterone concentrations.

Contraception Based on Immunological Approaches

The possibilities for immunocontraception in the male include stimulation of either the humoral or cell-mediated actions of the immune system, or a combination of these. The goals are to either stimulate antibody production against molecules (most commonly hormones) important for fertility, or to activate a cell-mediated response involving cytotoxic T-lymphocytes to specifically destroy cells required for fertility (Golden, 2009).

There are two main areas in male reproductive physiology that have been investigated as potential targets for immunocontraception. First, there is the hypothalamic-pituitary-gonadal (HPG) axis, in which gonadotrophin releasing hormone (GnRH) from the hypothalamus and the gonadotroph cells of the anterior pituitary have been the primary targets. Contraceptive vaccines targeting the HPG axis have been investigated and developed for several decades, with much of the research focused on controlling female reproduction. The first commercial vaccine, which targeted GnRH, was available for use in cattle in 1990 (Hoskinson *et al.*, 1990). Hormone receptors and the gonadotrophins themselves have also been assessed as immunological targets. In the case of gonadotrophins, inactivation of follicle stimulating hormone (FSH) is of particular interest in males as it will interfere with spermatogenesis, but not with testosterone production (Yang *et al.*, 2011). This approach fulfils a niche in which contraception is achieved without interfering with either secondary sex characteristics or male behaviour.

Secondly, immunocontraceptive targets include specific gonadal and extragonadal sites. Gonadal targets include the testicular germ cells and supporting somatic Sertoli and Leydig cells (McLaughlin and Aitken, 2011). Extragonadal sites for vaccine targeting include the epididymis and its luminal content of maturing spermatozoa. An example of this approach is immunization of male monkeys with human recombinant epididymal protease inhibitor (EPPIN) to suppress sperm maturation (Yan Cheng and Mruk, 2010). The development of vaccines against sperm membrane proteins such as glycosylphosphatidylinositol (GPI)-anchored sperm-specific protein (PH-20) is also being investigated (Sabeur *et al.*, 2002), though these anti-sperm vaccines may be more suited to deployment within the female population.

Gonadal germ cells express unique antigens, some of which develop at the time of sexual maturation, long after the differentiation between self and non-self that occurs early in fetal development (Pöllänen and Cooper, 1994; Golden, 2009). Therefore, gonadal tissues contain non-self-antigens requiring protection from the body's normal defence mechanisms. The traditional view has been that complete sequestration of testicular autoantigens behind the blood–testis barrier was the only protective mechanism preventing immune responses against them (Pöllänen and Cooper, 1994). Although the inter-Sertoli-cell tight junctions certainly protect the autoantigenic germ cells in the luminal compartment of the seminiferous tubules, there is evidence that there are also autoantigens on the surface of germ cells just about to enter meiosis, but still in the basal compartment of the seminiferous tubules (Yule *et al.*, 1988; Saari *et al.*, 1996). In addition, sperm autoantigens have been identified within the epididymis at concentrations exceeding that within the testis (Pöllänen and Cooper, 1994). It is now apparent that the regulation of immune responses to these antigens involves a system of interactions requiring a balance between activation and attenuation of responses (Golden, 2009) to ensure there are few or no immune responses mounted against most of these cells or tissues. In most cases, there is more than one protective mechanism in place to shield sperm autoantigens (Verajankorva *et al.*, 1999).

Hence, the development of an immunocontraceptive targeting gonadal germ cells requires selection of a suitable target antigen within the gonad, production of an effective vaccine and development of a suitable mode of delivery to overcome whichever immunoprotective mechanism may be in place. Importantly, the development of immunity to autoantigens increases the risk of inducing autoimmune reactions. These can present complex challenges to address (Verajankorva *et al.*, 1999).

However, perhaps, the greatest obstacle to the extensive use of immunocontraceptive technology is the variability of the duration and intensity of those immune responses that are attained (Bowen, 2008). Due to the difficulty in predicting the efficacy and duration of effect, there are problems in utilizing this form of contraception in light of the definitive contraceptive requirements for the reproductive management of most species. In general, hormonal vaccines targeting the HPG axis provide only short-term contraception from 3 to 12 months duration, but, with the use of suitable adjuvants, and the strategic administration of booster vaccinations, GnRH vaccines may be able to deliver useful fertility control in some species over extended periods (Dowsett *et al.*, 1996; Miller *et al.*, 2004; Walker *et al.*, 2007; Bowen, 2008); this method would then have the potential advantage of reversibility.

Deeper understanding of the hypothalamic control of reproduction may lead to improvements in target antigen selection, and in immunization efficiency, duration and predictability of action. The recent advances in knowledge associated with kisspeptins and their receptors (Tena-Sempere, 2006; Scott *et al.*, 2010) provide an example of additional target antigens that may be investigated in the future.

GnRH and pituitary hormone vaccines

Since its initial discovery in 1971, GnRH has been considered to be the main hormone regulating and controlling gonadotrophins in the HPG axis (Schally *et al.*, 1971), and it is only in the last decade that other peptide hormones, such as kisspeptin and gonadotrophin inhibitory hormone, have been identified to also be significant contributors to HPG control. Despite the recent identification and significance of these latter peptides, GnRH still plays a critical role in reproductive control, leading to a body of research into the development of GnRH vaccines as a means of contraception (Hoskinson *et al.*, 1990; Bonneau and Enright, 1995; Dowsett *et al.*, 1996; Levy *et al.*, 2004; Golden, 2009).

In mammals, GnRH-producing neurons are generally found scattered within the hypothalamus extending from the median eminence (ME) through to the medial basal hypothalamus (MBH) and to the preoptic area (POA) (Jasoni and Porteous, 2009; Scott *et al.*, 2010). Within this distribution, the GnRH neurons are grouped into nuclei, with species variation in GnRH activity within each nuclei. For example, GnRH neurons in rodents have been identified in the ME, POA and anterior hypothalamic areas (AHA); in horses, they are primarily located in the arcuate nucleus of the MBH (Scott *et al.*, 2010); in primates and sheep, the distribution of GnRH neurons extends from the ME through to the arcuate nucleus (ARC) of the MBH (Herbison, 2006). It is only recently that the specific anatomical location and distribution of GnRH neurons in the canine hypothalamus has been described; the pattern of distribution is heavily concentrated in the MBH, including the ME and extending into the ARC (Buchholz *et al.*, 2012).

GnRH is produced in cell bodies of the hypothalamic neurons and is transported by axonal flow to the terminal synapses. These synapses lie adjacent to the vessels of the capillary plexus within the ME (Herbison, 2006). Stimulation of the GnRH neurons causes release of stored peptide from their secretory granules into the extracellular space, followed by diffusion into the capillary blood of the ME. GnRH then travels via the hypophysial portal system to the capillary plexus within the anterior pituitary. Here, a portion of the GnRH leaves the capillaries, and becomes available for binding to the pituitary gonadotrophs, stimulating the release of FSH and LH (luteinizing hormone). It is during transport within the hypophysial portal blood that GnRH can be immunologically targeted.

GnRH is a small peptide that is identical in all mammals (Senger, 2003). These two factors (small and identical) result in it producing minimal immunogenicity. However, it can be made more immunogenic by coupling it to a carrier such as keyhole limpet haemocyanin (KLH) (Miller *et al.*, 2004), or ovalbumin (Hoskinson *et al.*, 1990). The coupled GnRH peptide is called a GnRH-conjugate; this is combined with an adjuvant to create a vaccine (Hoskinson *et al.*, 1990).

In animals treated with GnRH vaccine, anti-GnRH antibody within the hypophysial portal blood binds to the newly released GnRH from the hypothalamus, thus preventing GnRH from binding to the pituitary gonadotroph cells (Miller *et al.*, 2004). Provided that sufficient specific antibodies are present in the circulating blood entering the ME, virtually all of the GnRH peptide secreted into the hypophysial portal vessels will be bound and neutralized (Donald, 2000). The binding of the antibody to GnRH neutralizes it either by preventing it from diffusing through the capillary walls owing to the size of the GnRH-antibody complex, or by masking the receptor binding site on the GnRH molecule. This neutralization of endogenous GnRH results in profound suppressive effects on the pituitary gonadotroph cells. Further, physiological responses to GnRH and its efficacy are correlated with GnRH antibody titre; animals with high titres are sterile, whereas those with lower titres are not (Levy *et al.*, 2004).

In addition to those vaccines against GnRH, others directly targeting pituitary gonadotrophins such as FSH show promise for male contraception (Delves, 2004). The main difference between the action of FSH vaccine and GnRH vaccines is the preservation of androgen-dependent functions and behaviour with the former. Immunocontraception targeting LH has been successful in dogs, with reproductive function being severely impaired for up to 12 months (Lunnen *et al.*, 1974).

Advantages

Although large-scale trials are lacking, the induction of immunity to GnRH represents a promising approach to contraception for males. It can be administered with two injections, and possibly by a single injection; no surgery is required; there are minimal adverse reactions; it has the potential to stop both spermatogenesis and androgen-related behaviour; and, in theory, it should be reversible. As the efficacy of treatment is closely correlated with antibody concentration, measurement of anti-GnRH antibody concentrations can be used as a means of assessing vaccination history and current function. Immunization with GnRH-conjugate vaccines has been reported to induce periods of sterility or infertility in dogs (Ladd *et al.*, 1994; Jung *et al.*, 2005), cats, rats (Levy *et al.*, 2004), deer (Miller *et al.*, 2004) and boars (Wagner *et al.*, 2006). The vaccines also prevent androgen-driven aggression as effectively as surgical castration for bulls (Price *et al.*, 2003), stallions, elephants (Barfield *et al.*, 2006), piglets and lambs (Levy *et al.*, 2004), though the influence on spermatogenesis has not been adequately studied in these latter species. With the development of suitable delivery methods, GnRH vaccines may be acceptable for the control of pest species (Barfield *et al.*, 2006). Specific uses of the vaccine are outlined in the section on context-specific applications of male contraception.

Disadvantages

Despite some successes, GnRH vaccine technology is still developing. For example, the same GnRH vaccine that was effective in dogs had no efficacy in cats (Ladd *et al.*, 1994). More recently, it has been suggested that these anomalies may be overcome by modifications to the injection delivery method so that the antigen is protected from rapid destruction by the animal's immune system (Miller *et al.*, 2004).

Perhaps the biggest disadvantage of GnRH vaccine contraceptive technology is the difficulty in stimulating a prolonged immune response of predictable duration. Coupled with this is the requirement for administering one or two injections for the induction of immunity, with the possible requirement for booster vaccinations. This may be impractical in some animal management contexts. However, GnRH conjugation technology has been progressing since the 1990s; a commercial product has been released in the Australian market (Hoskinson *et al.*, 1990), and further development of a single-injection product is available in the USA (Miller *et al.*, 2004).

Within a given species, there is concern that a particular vaccine may be very effective in inducing infertility in some animals, but not effective in others. This variation in efficacy has led to concern that significant application of an immunocontraceptive may result in selection of a population of non-responders with altered or restricted immunogenetic make-up and responses (Cooper and Larsen, 2006). A corollary is that the duration of action of effective contraceptive vaccines is typically quite variable, making animal management less predictable than definitive methods such as surgical castration.

Anti-GnRH vaccines combine the inhibition of fertility with the suppression of sex-steroid dependent behaviour but, if there is a need to provide contraception while maintaining sex-hormone production, further work on specific inhibition of the pituitary gonadotrophins is required. Whereas the neutralization of FSH has resulted in severe disruption of spermatogenesis (Delves, 2004), it has been shown in some species that azoospermia will not be achieved as long as LH remains active (Barfield *et al.*, 2006). More recent work targeting FSH function has focused on priming and boosting immunizations using recombinant human FSH receptor (rhFSHR) peptide. This strategy has led to decreased fertility in mice 10 weeks after vaccination (Yang *et al.*, 2011). Even though azoospermia was still not achieved, there may be uses for this technology in the suppression of animal populations.

Immunocontraception targeting sperm membrane proteins

Immunocontraceptive technology targeting sperm membrane proteins is well suited to female contraception by the induction of sperm antibody formation in the female reproductive tract. When administered to females, such an immunocontraceptive can prevent fertilization by blocking sperm–oocyte interaction. In some contexts, there is also potential for its administration to be applicable to males (Sabeur *et al.*, 2002). In general, successful immunocontraception in males targeting sperm-membrane proteins will prevent fertilization of the oocyte and, at the same time, preserve sexual behaviour and secondary sex characteristics.

Many sperm membrane proteins have been identified and assessed as targets for immunocontraception (Suri, 2004). Yet, despite encouraging results, variability in their efficacy and duration of action has made it difficult to focus on any one immunocontraceptive approach (Bowen, 2008). As an example, immunocontraception was investigated in male monkeys immunized with EPPIN (O'rand *et al.*, 2004). The study demonstrated that effective and reversible male immunocontraception in primates is an attainable goal, but the vaccine was only successful in 78% of animals. This level of contraception may be suitable in some animal population control contexts, but it would not be an acceptable result in production animal systems.

Testis-specific lactate dehydrogenase (LDH-C4) is an enzyme found on the plasma membrane of sperm flagella (Bradley *et al.*, 1997), and LDH-C4 has been targeted on the fox spermatozoon and some success reported with the vaccination of vixens (Bradley *et al.*, 1997). Immunization with LDH-C4 also prevents pregnancy in female baboons and reduces conception rates in rabbits, but not in macaques (Suri, 2004).

Fertilization antigen-1 (FA-1) is a glycoprotein that has been isolated from the postacrosomal, midpiece and tail regions of ejaculated sperm and also within the testis. Generating antibodies to FA-1 has been successful in reducing fertility in rabbits and humans (Naz *et al.*, 1984), and warrants further investigation in other species. The mechanism of action is unclear, but it does not involve the agglutination or immobilization of sperm. In rabbits, the action appears to require an intact zona pellucida, while in humans there is the suggestion that there may be interference with capacitation (Naz *et al.*, 1984).

The sperm protein PH-20 elicits a significant sperm-directed antibody response in macaques and is being developed as an antigen for rabbits, where it is a candidate for insertion into a recombinant myxoma virus (Holland *et al.*, 1997). Orally delivered, non-infectious, non-transmissible vaccines against sperm antigens are also envisaged for kangaroo contraception (Cowan and Tyndale-Biscoe, 1997). However, due to the high number of spermatozoa present in the female tract of most mammals after copulation, achieving sufficiently high titres of antibody to disable enough sperm to prevent conception remains a concern (Cowan and Tyndale-Biscoe, 1997).

Despite the potential for sperm membrane-targeted immunocontraception to be a powerful tool for managing domestic, feral and wildlife species, there is still considerable research required to translate current knowledge into reliable, commercial products. Although there has been active research in this field since the early 1970s (Gunaga *et al.*, 1970), there are no human immunocontraceptives currently available and only approximately a dozen animal applications of the technology.

Immunocontraception inducing cell-mediated immune responses

Antigens that would otherwise be too small to induce an effective immune response, such as a small decapeptide-like GnRH, can be utilized via polynucleotide vaccine technology

(Babiuk *et al.*, 1999). This technology involves the introduction of the antigen DNA via a plasmid, which then encodes the antigen within the animal tissue into which it is injected. Importantly, both humoral and cellular immune responses are induced (Babiuk *et al.*, 1999), which provides a significant advantage of plasmid vaccines over peptide vaccines. Another advantage is that adjuvants are not required to elicit an immune response, and so adverse injection-site reactions are not common (Purswell and Kolster, 2006).

T-helper cell epitopes have been studied in canines (Walker *et al.*, 2007). Totally synthetic peptide-based immunocontraceptive vaccines, in which canine distemper virus F protein epitopes were coupled to an LH releasing hormone (LHRH/GnRH) peptide, were used to inoculate dogs. In five vaccines, a strong anti-LHRH (GnRH) antibody response, with suppression of testosterone and progesterone, was observed in a majority of the animals of different breeds examined. This could provide a strategy for the development of an immunocontraceptive vaccine for widespread use in domesticated and feral canines (Walker *et al.*, 2007).

Further interest surrounding polynucleotide vaccine technology stems from the novel possibilities that it provides for vaccine delivery. Options may include bacterial or viral carriers, microparticle encapsulation or ballistic delivery to mucosal surfaces (Babiuk *et al.*, 1999).

Contraception Based on Control of Spermatogenesis

Hormonal control of hypothalamic-pituitary function

Testicular function and fertility are dependent on normal functioning of the HPG axis. Thus, an avenue for reducing fertility in males is to interfere with the secretion of gonadotrophins, most commonly by exploiting negative feedback via the administration of sex steroids. Such treatments also have the effect of reducing testicular androgen secretion, which, in most cases, is desirable. Excluding research with humans, a majority of the work on steroidal contraception of males has been conducted in dogs and rodents (England, 1997; Bowen, 2008; Wang *et al.*, 2011).

Contraception based on the use of sex steroids

The suppression of spermatogenesis by sex steroids depends mainly on inhibiting the HPG axis, which leads to reduced secretion of LH and FSH. This can be achieved at either the hypothalamic or pituitary levels. The suppression of LH inhibits testosterone production from the Leydig cells, while the suppression of FSH interferes with androgen-binding protein production and other Sertoli cell functions. There is also a suggestion of a direct adverse effect of progestogens on epididymal sperm maturation (England, 1997).

When considering this mode of contraception, it is important to recognize that sex steroid contraceptives have little influence on existing sperm; they primarily suppress sperm production and possibly sperm maturation within the tubular tract (England, 1997; Amory and Bremner, 2000). In this context, it is likely that any contraceptive regimen based on sex steroids, or on manipulation of the HPG axis, will be associated with some delay in the onset of the maximum contraceptive effect (Amory and Bremner, 2000). The duration of this delay will vary depending on the regimen used, but it will have some relationship with the duration of spermatogenesis and epididymal transport time in the species being treated.

Oestrogens have been used for male contraception in a number of species (Barfield *et al.*, 2006), although the product most commonly used, diethyl stilboestrol (DES) has been banned in a number of countries as a result of concerns with its carcinogenicity and implication in blood dyscrasias. Other orally active oestrogenic products have been used to suppress the HPG axis, including quinestrol as a contraceptive in Brandt's vole (Wang *et al.*, 2011). However, problems with palatability need to be overcome before this compound can be used on a commercial scale.

Androgens are suited to a contraceptive role owing to their significant influence on the regulation of the HPG axis, at both hypothalamic and pituitary levels (Senger, 2003). Depending on the target species, their additional androgenic and anabolic activities may, or may not, be advantageous. When administered at greater than physiological doses, testosterone has a suppressive effect on GnRH release and spermatogenesis, though its use alone has produced considerable variability in contraceptive efficacy in humans, with complete failure to suppress spermatogenesis in some men (Gao and Dalton, 2007). In addition, the administration of natural testosterone is impractical because of its rapid degradation by the liver (Amory and Bremner, 2000). Orally active androgens (those with a 17α-ethinyl group) improve ease of administration and longevity of action, but can cause liver damage and are not considered safe for long-term use (Amory and Bremner, 2000). Most current regimens in humans use testosterone esters, such as testosterone enanthate, given by IM injection on a weekly to fortnightly basis. The onset of azoospermia occurs at around 2 to 3 months, and recovery of normal sperm counts occurs 3 to 4 months after testosterone enanthate is discontinued (Amory and Bremner, 2000).

The administration of 5 mg/kg of mixed testosterone esters (comprising testosterone proprionate 0.6 mg/kg, testosterone phenylpropionate 1.2 mg/kg, testosterone isocaproate 1.2 mg/kg, testosterone decanoate 2.0 mg/kg) to male dogs resulted in a significant decline in sperm motility within 3 weeks of treatment, which persisted for 3 months (England, 1997). Daily oral administration of 50 mg methyltestosterone to male dogs for 90 days decreased daily sperm output (DSO) (Freshman *et al.*, 1990), but mean DSO was still approximately 80 million at the completion of treatment. Chronic administration of danazol, a synthetic derivative of 17α-ethinyl testosterone, to male dogs resulted in azoospermia, with the effects reversible within 60 days (Kutzler and Wood, 2006). The administration of anti-androgens, such as flutamide or cyproterone, had a slight, transient influence on spermatogenesis, but was not effective as a contraceptive (Neumann *et al.*, 1976).

The significant variability in the contraceptive efficacy of testosterone esters (Freshman *et al.*, 1990; England, 1997) and the inconsistent pharmacokinetic profile of testosterone-only regimens (Gao and Dalton, 2007), coupled with concerns about the adverse effects on long-term health associated with 17α-ethinyl group orally-active androgens (e.g. danazol), have led to investigations into combination therapy involving the use of GnRH analogues and progestins. In addition to improved efficacy, the use of combined therapy allows a reduction in the doses of hormones administered. Combination therapy utilizing progestins coupled with physiological concentrations of exogenous testosterone results in profound suppression of gonadotrophin release and has been found to be a more effective, safer strategy than the use of testosterone alone for human male contraception (Gao and Dalton, 2007). Unfortunately, the application of this method for animal use is hampered by the impractical dosing regimens required of the androgens.

An important comparative observation is that the canine appears to differ from other species in the sensitivity of the HPG to progestogen feedback (England, 1997). Gonadotrophin suppression does not occur in the canine male, even with high doses (up to 20mg/kg medroxyprogesterone acetate) of progestogens, and there are no significant effects on libido. However, combinations of progestogens and androgens appear to have a synergistic effect and may provide a clinically useful method of reversible contraception in the dog. The combination of mixed testosterone esters (5 mg/kg) and medroxyprogesterone acetate (20 mg/kg) produced a rapid and profound decrease in semen quality in the male dog (England, 1997). It was postulated that the effect of this combination of treatment on semen quality was mediated by a direct effect of the progestogen upon the epididymal phase of sperm maturation, rather than on the suppression of pituitary LH release. In contrast, the androgen caused suppression of gonadotrophins and subsequently also of spermatogenesis.

A more contemporary option for mimicking the contraceptive action of sex steroids involves the use of androgen receptor (AR) ligands. These compounds can be classified as either AR agonists or antagonists, depending on whether they (respectively) activate or inhibit the transcription of AR target genes (Gao and Dalton, 2007). The early AR ligands were based on the chemical structure of sex steroids, but more recent research has investigated the development and use of non-steroidal selective AR modulators (SARMs) (Gao and Dalton, 2007). An advantage of SARMs over natural testosterone and other products based on the sex steroid structure is that SARMs can be tailored to produce specific effects. That is, they can be designed to achieve tissue-selective modulation of AR action. This minimizes the undesirable side effects that are normally associated with steroidal androgens and can allow greater stimulation of the required action, with the knowledge that any unwanted side effects will be minimal.

The use of non-steroidal AR agonists coupled with high oral availability is still under investigation, but these compounds may provide the opportunity to develop a male contraceptive with practical applications across a number of species and contexts. Suppression of GnRH and gonadotrophins at the hypothalamic and pituitary levels requires a potent AR agonist that has good distribution in the central nervous system (CNS) and/or selectivity. AR agonist activity in peripheral tissues will allow the maintenance of normal androgen functions (Gao and Dalton, 2007), while peripheral AR antagonist activity may be suitable where loss of secondary sex characteristics and behaviour is desirable.

Contraception based on the use of GnRH agonists

Prolonged administration of high doses of GnRH agonists has been shown to induce desensitization to GnRH of the pituitary gonadotrophs in a number of species, including rodents, sheep, goats, dogs, pigs, primates, cattle and horses (Hinojosa *et al.*, 2001; D'Occhio *et al.*, 2002; Junaidi and Norman, 2003; Junaidi *et al.*, 2007). After administration of the agonist, there is an initial gonadotrophin hypersecretion, followed by a progressive loss of pituitary responsiveness to GnRH over approximately a 4 week period. Continued desensitization is due to the downregulation of pituitary cell GnRH receptors (D'Occhio *et al.*, 2002). In some species, it has also been documented that GnRH agonist administration will directly desensitize the Leydig cells to LH stimulation, resulting in a profound reduction in androgen production and spermatogenesis (Junaidi *et al.*, 2007). From a contraceptive viewpoint, the outcome is a prolonged reduction in gonadotrophin release and functional loss of testicular activity.

While this mechanism of action represents a generalization, there are clear species differences in susceptibility to the downregulatory effects of GnRH agonists (Goericke-Pesch *et al.*, 2011), with their activity ranging from being ineffective in some species, being variable in the extent and duration of action in others, to being highly effective in yet others. For example, GnRH agonists have proven effective for contraception in wild dogs and cats (Bertschinger *et al.*, 2002), with continuous GnRH agonist administration in dogs leading to a marked drop in circulating testosterone concentrations, libido and spermatogenesis (Junaidi *et al.*, 2007). In contrast, their continuous administration in bulls or red deer stags results in an increase in plasma testosterone concentrations (Junaidi *et al.*, 2007).

Advantages

GnRH agonist treatment is an effective and reversible alternative to surgical de-sexing in dogs, queens and ferrets, plus males and females of several exotic species (Goericke-Pesch *et al.*, 2011). There is ongoing research to determine the effectiveness of GnRH agonists in other species. In particular, deslorelin (Suprelorin, Peptech Animal Health) and azagly-nafarelin (Gonazon, Intervet, Inc.) implants (Bowen, 2008) and leuprolide acetate injections (Munson, 2006) have been utilized. In addition to reducing pituitary gonadotrophin release, in dogs the use of deslorelin will desensitize Leydig cells to LH. This action enhances the suppression of androgen production, thereby providing the desirable effects of preventing androgen-related behaviour and disease. For example, deslorelin administration can be very effective in the treatment of benign prostatic hyperplasia. Additionally, the application of GnRH agonists, in contrast to surgical castration, provides the opportunity to withhold further treatment in older animals if it is considered that they may benefit from the anabolic effects resulting from restored natural testosterone production.

The administration of GnRH agonists is usually a simple process of injection or implant application. Deslorelin is currently the most commonly used GnRH agonist in dogs, in which it induces reliable contraception within 6 weeks of implantation (Junaidi *et al.*, 2003), which lasts from 6 to 12 months, depending on the implant dose (4.7 or 9.4 mg implants) administered. Once the implant reaches the end of its functional lifespan and testosterone concentrations return to normal, full fertility usually occurs within the duration of one to two spermatogenic cycles. With spermatogenesis in the canine requiring approximately 61 days (Soares *et al.*, 2009), the majority of dogs will have elongated spermatids within the seminiferous tubules within 16 weeks of deslorelin removal or implant depletion (Goericke-Pesch *et al.*, 2009). No adverse effects have been noted with the long-term administration of deslorelin (Junaidi *et al.*, 2007).

Disadvantages

There is significant variation between species in their response to GnRH agonists, making it necessary to adequately research product efficacy for every species in which it is to be utilized. An historical impediment to progress with GnRH super agonists has been the cost of production and the lack of a cost-effective means for delivering adequate doses over a prolonged period (Junaidi *et al.*, 2003). However, products such as Suprelorin® are now becoming commercially viable due to improved drug delivery methods, increased use and reduced costs.

The acute increase in circulating concentrations of LH and testosterone for a short period following the initial application of GnRH agonists may induce short-term undesirable sexual behaviour, although the concentrations of LH and testosterone fall rapidly from as soon as 9 h after implantation, and then fall further to undetectable values after approximately 3 weeks (Junaidi *et al.*, 2003). Currently, GnRH agonist implants only supply 6 or 12 months of treatment, depending on the deslorelin concentration. This means that there is a need for regular re-administration if long-term contraception is required. Until longer acting products become available, in the case of domestic animals, GnRH agonist administration could easily be combined with other routine management programmes such as health checks, heartworm prevention, or vaccinations.

Contraception based on the use of GnRH antagonists

GnRH antagonists competitively occupy GnRH receptors on gonadotrophin-producing cells in the anterior pituitary, resulting in an immediate inhibition of the HPG axis without a preceding stimulatory effect as occurs with GnRH agonists (Valiente *et al.*, 2007). Current formulations require more frequent dosing than for GnRH agonists, but GnRH antagonists have the advantage of a more rapid onset of effect, which is then easily manipulated and rapidly reversible (Huirne *et al.*, 2007). GnRH antagonists are currently registered for preventing a premature LH peak during ovarian stimulation in women (Huirne *et al.*, 2007).

When considering the clinical applications of GnRH antagonists, it is important to recognize that early, or so-called first-generation, antagonists produced undesirable side effects, including allergic reactions and skin irritation. However, the more recent, third-generation, GnRH antagonists that are currently used in humans (Huirne *et al.*, 2007) have been trialled in dogs with minimal side effects (Valiente *et al.*, 2007).

Preliminary studies utilizing the third-generation GnRH antagonist, acyline, showed that a high, single subcutaneous dose of 330 μg/kg suppressed sperm production in dogs, with suppressive effects starting within 1 week of treatment and lasting until the end of the study, 6 weeks later (Valiente *et al.*, 2007). In the stallion, treatment with the GnRH antagonist, antarelix, produced significant reductions in LH, FSH, oestradiol, libido and total sperm output (Hinojosa *et al.*, 2001). Thus, while it appears that there has been some resistance to the use of GnRH antagonists resulting from the undesirable side effects noted in early studies, recent investigations utilizing third-generation drugs in dogs and horses, and their clinical use in humans, suggest that they have the potential to provide effective and reversible male contraception.

Disadvantages

No depot or long-acting formulations of GnRH antagonists have yet been developed (Herbert and Trigg, 2005). Further investigation into the pharmacokinetics of these compounds and formulation of a sustained-release delivery mechanism is required before the practical application of GnRH antagonists as male contraceptives.

Miscellaneous hormonal control of HPG axis function

Repeated IM administration of exogenous prolactin (600 mg/kg body weight weekly for 6 months) to male dogs has been found to produce severe oligospermia within 6 weeks of treatment (Shafik, 1999). In the majority of cases, azoospermia was induced within 12 weeks of treatment. Importantly, within 3 months of drug withdrawal, the sperm count normalized, mating produced pregnancy and the resultant offspring exhibited no anomalies (Shafik, 1999). It was noteworthy, in this study, that serum concentrations of testosterone, LH and FSH were not significantly affected by prolactin treatment, which led the authors to speculate that prolactin may be having a direct inhibitory effect on the testes rather than functioning via gonadotrophin downregulation.

Non-hormonal Inhibition of Hypothalamic-pituitary Function

Cytotoxin conjugates

GnRH has a high specificity and affinity for its receptors. Additionally, the internalization of GnRH receptors following ligand–receptor binding lends itself to applications that can utilize the bound receptors as carriers (Sabeur *et al.*, 2003). For example, toxins conjugated to GnRH can be used to localize cytotoxic effects to within the pituitary gonadotroph cells, thereby disrupting the HPG axis. In this application, the toxins used need to be carefully selected to ensure that the GnRH-toxin conjugate is safe in other non-GnRH receptive tissues.

As an example, pokeweed antiviral protein has been conjugated with GnRH and administered to intact male dogs (n = 4) as three daily injections (0.1 mg/kg) (Sabeur *et al.*, 2003). Serum testosterone and LH concentrations and testicular volume decreased after treatment, and the effects of male contraception persisted for 5–6 months (Sabeur *et al.*, 2003). The treatment was generally considered to have no adverse effects, with a transient (<24 h) arthralgia noted in only a few of the dogs. In particular, there was no adverse effect on serum cortisol or thyroxine concentrations, indicating the specificity of this treatment for the pituitary gonadotrophs.

Cytotoxic conjugates of FSH are also being investigated for the targeted destruction of Sertoli cells; this will be discussed in a subsequent section.

Advantages

This is a non-surgical method of contraception that includes the abolition of androgen production. It may be reversible, but further studies are needed to determine this.

Disadvantages

The long-term effects of the treatment have not been investigated; neither has the potential for individual variability in response to the treatment been studied. Nor have dose titration studies been performed, and there is also the question of how many pituitary gonadotrophs need to be damaged in order to reliably produce contraception.

Non-hormonal Control of Spermatogenesis

Non-hormonal methods for controlling spermatogenesis are attractive alternatives to hormonal methods. In production animals, consumer resistance to hormone use is becoming increasingly apparent, while for companion animals, owners may demand the perceived comfort of knowing that their pets are not being exposed to long-term hormonal administration. Currently, the long-term effects of many of the hormonal regimens are not fully understood (Yan Cheng and Mruk, 2010), and owners of the recipients may require an assured option of reversibility. There is a need for non-hormonal, reversible contraception that does not interfere with male behaviour. This requirement has stimulated significant research in humans, primates and laboratory rodents (Walden *et al.*, 2006; Mruk and Yan Cheng, 2008; Yan Cheng and Mruk, 2010), so this section provides examples of the non-hormonal control of spermatogenesis.

Chemical or physical inhibition of spermatogenesis

Inhibition of spermatogenesis can be targeted directly at germ cell division and differentiation, or indirectly via modification of Sertoli or Leydig cell function. There are many chemical agents, such as gossypol, dinitropyrroles and carbamates, that can interfere with spermatogenesis. However, many have toxic side effects, making them unacceptable as contraceptives. This section will focus on chemicals that have shown promise as contraceptive agents and either have apparently insignificant toxic effects, or at least these effects appear to be insignificant after initial studies, so that further investigation is warranted.

N-butyldeoxynojirimycin

N-butyldeoxynojirimycin (NB-DNJ) is a glucose mimetic licensed for use to treat a genetic, metabolic disorder in human medicine (Walden *et al.*, 2006). There is also a galactose analogue of the compound: *N*-butyldeoxygalactonojirimycin (NB-DGJ). Both of these imino sugars cause infertility in male mice, but studies have focused on the effects of NB-DNJ when fed at 15mg/kg daily (Walden *et al.*, 2006). Feeding mice NB-DNJ for up to 12 months resulted in infertility as a result of reduced sperm numbers (up to 56% reduction after 6 months of treatment), reduced sperm motility (<20% individual motility after 3 months of treatment), abnormal sperm nuclear morphology (90% abnormal after 5 weeks of treatment) and abnormal acrosomal phenotype. Recovery to fertility occurred within 6 to 12 weeks after cessation of feeding NB-DNJ (Walden *et al.*, 2006). Importantly, long-term NB-DNJ treatment of mice at 15mg/kg daily does not affect body or gonad weights, reproductive endocrinology, serum chemistry or animal behaviour. There was no reported reduction or increase in serum FSH, LH or testosterone concentrations, and testicular testosterone concentrations remained within normal limits. In mating trials, none of the treated mice succeeded in impregnating females despite successful copulation (Walden *et al.*, 2006). It appears that NB-DNJ is worthy of further investigation in other species in which infertility coupled with the maintenance of androgen production and male behaviour is desirable.

Neem seed oil

The effects of neem (*Azadirachta indica*) seed oil on rat sperm production have been investigated after fractionation of the oil into six different portions (Purohit *et al.*, 2008). After oral administration of the third and fourth fractions of the oil for 60 days at 9 and 10 mg/kg daily, respectively, spermatogenic arrest was confirmed histologically. The authors also found significant reductions in sperm density within the testes and cauda epididymis, plus reduced motility of epididymal sperm. It was suggested that the contraceptive activity of neem seed fractions was due to suppression of FSH and LH secretion from the anterior pituitary (Purohit *et al.*, 2008).

No adverse reactions associated with the oil fraction administration were noted.

1-formyl, 4-dichloroacteyl piperazine and gossypol

The non-steroidal, antispermatogenic molecule, 1-formyl, 4-dichloroacteyl piperazine (CDRI-84/35), has been found to cause sterility in rats by arresting spermatogenesis (Ojha *et al.*, 2006). It is postulated that the mode of action of CDRI-84/35 is through hyperstimulation and subsequent disruption of Sertoli cell enzyme function. Large-scale, long-term contraceptive studies using this compound *in vivo* have not been performed, but these initial investigations suggest that further research is warranted. By comparison, gossypol completely inhibits rat Sertoli cell enzyme activity (β-glucuronidase, γ-glutamyl transpeptidase, lactate dehydrogenase and aromatase), but produces a similar result to CDRI-84/35 of spermatogenic arrest (Ojha *et al.*, 2006). Unfortunately, gossypol can be toxic to pigs, humans and immature ruminants, so care is needed in its use (Waites *et al.*, 1998).

Bisdiamine compounds

The bisdiamine compounds are amoebicidal drugs that specifically target germinal epithelium in the testis without affecting other rapidly dividing cell types (Munson *et al.*, 2004). However, caution is necessary as serious teratogenic effects have been documented when the compounds were administered to pregnant rats (Taleporos *et al.*, 1978), so administration must be done with care to assure that only males receive the drug.

The compounds have been successfully used to induce spermatogenic arrest in dogs, grey wolves (*Canis lupus*), rats, guinea pigs, rhesus monkeys and humans (Munson *et al.*, 2004). A safety and efficacy trial in the male domestic cat demonstrated that 150 mg/kg of the bisdiamine WIN 18,446 (Fertilysin®, SAF Bulk Chemicals, Milwaukee, Wisconsin) administered daily in food causes almost complete spermatogenic arrest in all treated individuals without damage to spermatogonia (Munson *et al.*, 2004). These effects were reversible, with no adverse effects on general health, blood parameters or behaviour, although serum testosterone concentrations were significantly lower in treated than in control males. Whether this reduction was sufficient to cause loss of secondary sex characteristics is unknown. Testosterone concentrations returned to normal within 3 months of the cessation of treatment (Munson, 2006). The main limitation to this contraceptive approach is that the cost of bisdiamines is currently prohibitive for most veterinary uses (Munson, 2006).

Carica papaya

The two principal compounds, MCP I (methanol subfraction) and ECP I (ethyl acetate subfraction), isolated from the seeds of *Carica papaya*, have been assessed for contraceptive properties in male albino rats (Lohiya *et al.*, 2005). For each compound, rats were dosed at 50mg/kg daily for a period of 360 days, and assessment was continued for up to 90 days post treatment withdrawal. Both MCP I and ECP I were found to be equally effective contraceptives, resulting in no motile spermatozoa within the ejaculate after 90 days of treatment. Histology revealed that most pathology was associated with the Sertoli cells, while the Leydig cells and spermatogonial stem cells remained normal. Serum testosterone concentrations remained normal throughout the trial period. Semen parameters had commenced improvement by 90 days post treatment and no systemic toxicity was noted during the trial period. Initial research is encouraging, but further contraceptive and toxicological studies in different species are required.

Cuminum cyminum

A methanol extract of the *Cuminum cyminum* seed induced infertility in male rats proportional to the time of exposure (Gupta *et al.*, 2011). There was evidence of suppression of spermatogenesis and epididymal function. Importantly, there was also suppression of testosterone production, which may satisfy the requirement for behaviour modification. The infertility was reversible after cessation of treatment and no adverse systemic effects associated with the treatment were noted.

Indazole-3-carboxylic acid and imidazole group of compounds

There are a number of compounds in the indazole-3-carboxylic acid group that have roles in medicine and fertility control (Bhowal *et al.*, 2008; Mruk and Yan Cheng, 2008; Yan Cheng and Mruk, 2010). Included in this group are 2-(2″-chloroacetamidobenzyl)-3-(3′-indolyl) quinoline and the imidazole group of compounds. The latter compounds (such as metronidazole) are commonly used in clinical contexts to treat infections with anaerobic bacteria and flagellated protozoa (Bhowal *et al.*, 2008). The action of 2-(2″-chloroacetamidobenzyl)-3-(3′-indolyl) quinoline against flagellated cells may be a reason for it causing infertility, but it has also been reported to interfere with Sertoli cell function, reduce serum testosterone concentrations and alter seminal plasma alkaline phosphatase activity. Additionally, as functional infertility was induced by the compound within 1 to 2 weeks of treatment, there is an indication of a direct effect of the compound on spermatozoa undergoing epididymal maturation (Bhowal *et al.*, 2008). Further investigation is required, but the initial results obtained with this compound suggest that it may be suitable for further assessment as a male contraceptive agent.

Another member of the indazole-3-carboxylic acid group, adjudin (1-(2,4-dichlorobenzyl)-1*H*-indazole-3-carbohydrazide, formerly called AF-2364), has been shown to inhibit spermatogenesis by disrupting anchoring junctions at the Sertoli cell/germ cell interface (Mruk *et al.*, 2006). It has no effect on the HPG axis. In rats, rabbits and dogs, adjudin leads to germ cell loss from the seminiferous epithelium, and reversible infertility (Hu *et al.*, 2009). Once adjudin is cleared from the systemic circulation (by 24–48 h after cessation of treatment), fertility gradually returns over the duration of a spermatogenic cycle. Unfortunately, when administered systemically in current formulations at 50 mg/kg, adjudin can induce muscle atrophy and liver inflammation in male rats (Yan Cheng and Mruk, 2010). However, in recent studies, adjudin was conjugated to a recombinant FSH mutant to assist with drug delivery directly to the Sertoli cells. In this instance, the FSH mutant had no hormonal activity, but was still capable of binding to the Sertoli cell receptor and so delivered the adjudin directly to the target tissue (Mruk *et al.*, 2006). This conjugation allowed the drug to be administered to rats at the greatly reduced total dose of 0.5 μg, thereby avoiding toxicity while still inducing reversible infertility. Unfortunately, the cost associated with the production of recombinant FSH and its conjugation to adjudin is high and currently precludes commercial application of this approach to male contraception (Yan Cheng and Mruk, 2010).

Gamendazole – (*trans*-3-(1-benzyl-6-(trifluoromethyl)-1*H*-indazol-3-yl)acrylic acid – is also an analogue of the indazole-3-carboxylic acid group of compounds. The compound causes infertility in rats, which was achieved by 3 weeks in all seven rats receiving a single oral dose of gamendazole at 3 mg/kg. This was followed by a recovery of fertility, which was observed by 9 weeks in four out of these seven animals. Toxicology studies at doses of gamendazole of approximately eight times the contraceptive dose demonstrated no gross histopathological changes (e.g. inflammation, necrosis, haemorrhage or tumours) in any organ in all treated rats when compared with untreated rats. As with 2-(2″-chloroacetamidobenzyl)-3-(3′-indolyl) quinoline and adjudin, it was concluded that the Sertoli cell was the primary target of gamendazole. Importantly, gamendazole did not affect serum testosterone concentrations, though a transient increase in serum FSH was noted within 1 week of treatment (Tash *et al.*, 2008).

It is relevant to note that three out of five adult rats died when dosed with gamendazole at approximately 70 times the infertility dose (Tash *et al.*, 2008). This is in contrast to the studies on adjudin, in which no deaths were reported when rats were treated with a comparable single dose, or when consecutive doses were administered (Mruk and Yan Cheng, 2008).

Indenopyridines

Indenopyridines were initially developed as antihistamines, but subsequently were found to cause seminiferous epithelium dysfunction and testicular atrophy in mice, rats, dogs and monkeys (Hild *et al.*, 2004, 2007a; Yan Cheng and Mruk, 2010). Members of this group of chemicals appear to target the Sertoli cells, reducing their function within 3 h of administration (Hild *et al.*, 2007b), and leading to germ cell loss from the seminiferous epithelium. One member of the indenopyridine group of chemicals, CDB-4022, has been found to disrupt the microtubule network of the Sertoli cells within 24 h of administration, and interfere with adhesion proteins such as cadherin, catenin, nectin, afadin and integrin β1 (Koduri *et al.*, 2008). Additionally, CDB-4022 activates the mitogen-activated protein kinase (MAPK) pathway, interfering with mitosis, cell differentiation and cell proliferation (Kyriakis and Avruch, 2001). CDB-4022 has variably produced either a reversible effect on fertility or irreversible infertility, depending on the species (Yan Cheng and Mruk, 2010). However, irreversibility may be prevented by pretreatment or concurrent treatment with GnRH antagonists (Hild *et al.*, 2004).

Miscellaneous chemical inhibition of spermatogenesis

Intraperitoneal administration of serotonin (5-hydroxytryptamine) has been shown to produce a marked reduction in the testicular weight of rats, with concurrent reductions in testosterone and inhibin concentrations (Hedger *et al.*, 1995). While serotonin is a neurotransmitter, it is also a secretory product of mast cells, and mediates vascular (blood) flow changes during inflammation. After 4 days of serotonin administration at 10 mg/kg, the consequent reduced testicular weights and spermatogenic activity were attributed to ischaemia resulting from constriction of the testicular artery (Hedger *et al.*, 1995). The duration of the spermatogenic effects and reversibility have yet to be determined.

Contraception based on post-spermatogenic inhibition of sperm maturation and function

Methods of achieving contraception by influencing post-testicular sperm maturation and epididymal physiology are attractive options because they circumvent the problems associated with contraception that targets the inhibition of spermatogenesis. Problems with the latter technique include extended lead times to induce azoospermia (or non-functional spermatozoa), the common need for hormonal use in the treatment regimen, the possibility of non-reversibility and the possibility of behaviour modification associated with androgen suppression.

In contrast, with post-spermatogenic sperm inhibition, the onset of contraceptive benefits is dependent on epididymal transport time. In many species this duration lies between 7 to 12 days, meaning that the onset of contraception can be rapid compared with methods that depend on interference with spermatogenesis (Cooper and Yeung, 1999).

Efforts to induce infertility by reducing the epididymal transport time of sperm, or modifying epididymal secretions, have not been successful, but a number of mechanisms have been identified and studied that have direct inhibitory action on post-spermatogenic sperm function (Cooper and Yeung, 1999). There are comprehensive reviews available of investigations into post-spermatogenic sperm inhibition (Cooper and Yeung, 1999; Aitken *et al.*, 2008; Yan Cheng and Mruk, 2010), so only some examples of these mechanisms and drugs will be described in this section.

Inhibiting sperm cation channels

Calcium plays a major role in important post-ejaculatory sperm functions, being necessary for sperm motility, capacitation, the acrosome reaction and, ultimately, fertilization (Jimenez-Gonzalez *et al.*, 2006; Aitken *et al.*, 2008). Due to this role of calcium and the speed with which it can induce cellular changes, extensive mechanisms have evolved for controlling calcium activity. In addition to the many calcium-permeable channels that have been identified in spermatozoa, sperm also express unique calcium channels (cation channels of sperm, or 'CatSpers') that are specifically active at the principal piece of the sperm (Yan Cheng and Mruk, 2010). The four CatSper genes are essential for male fertility and disruption of any one results in failed hyperactivation (Navarro *et al.*, 2008; Yan Cheng and Mruk, 2010). The targeting of CatSper function has been suggested as a possible approach to male contraceptive development (Yan Cheng and Mruk, 2010). An example is the use of nifedipine, which is an anti-hypertension drug that blocks calcium influx into sperm and affects sperm membrane cholesterol (Benoff *et al.*, 1994). The calcium blockers verapamil and diltiazem are also known to affect sperm motility in rats (Yan Cheng and Mruk, 2010). The use of anti-hypertension drugs for the control of male fertility is currently not an appealing contraceptive approach in humans (Yan Cheng and Mruk, 2010), but may be suitable for animals. Further research is required if drugs that block CatSper, but not other ion channel proteins, are to be identified.

GAPDH inhibitors

Glyceraldehyde-3-phosphate dehydrogenase (GAPDH) inhibitors, are chemical agents that selectively hinder the sperm-specific GAPDH isoenzyme, leading to the suppression of sperm glycolysis (Barfield *et al.*, 2006). Unfortunately, there is also some suppression of GAPDH in liver and kidney tissue (Aitken *et al.*, 2008). These chemicals, such as α-chlorohydrin and ornidazole, induce infertility in rats when administered daily for 1 or 2 years. Whereas side effects are not observed at antifertility doses, they have been observed at higher dosages (five times the infertility dose), resulting in current GAPDH inhibitors being considered unacceptable for use in humans (Jones and Cooper, 1999; Aitken *et al.*, 2008). Synthetic GAPDH inhibitor compounds aimed at reducing side effects by improving specificity *in vitro* were ineffective *in vivo* (Barfield *et al.*, 2006). However, GAPDH inhibitors may be useful for controlling rodent populations where infertility or euthanasia may be suitable outcomes.

While gossypol is also an enzyme inhibitor, it has serious side effects, such as hypokalaemic paralysis (Aitken *et al.*, 2008). Specific contraceptive effects have, though, been isolated to the (–) enantiomer of gossypol, which may provide an avenue for investigation as to whether treatment with this enantiomer alone may avoid adverse effects (Yu, 1987; Yu and Chan, 1998).

Induced flagellar dysfunction

Flagellar (sperm tail) deformation or dysfunction is seen in sterile males of several domestic species that are otherwise healthy and sexually active (Cooper and Yeung, 1999). One notable example is the Dag defect of sperm described in the bull. Flagellar dysfunction may be induced by a number of processes or chemicals, such as carnitine inhibitors (Cooper and Yeung, 1999), possibly gossypol (Aitken *et al.*, 2008), and the c-*ros* receptor block (adult male mice in which the c-*ros* receptor gene is knocked out show abnormal epididymal development) (Cooper and Yeung, 1999). Clinical trials will be necessary to confirm the use of this approach, but the concept of inducing flagellar dysfunction (e.g. by inducing coiled sperm tails) has merit as a contraceptive technique.

Context-specific Applications of Male Contraception

Livestock

Cattle

Contraception in bulls prevents unwanted mating, allowing greater control over genetic gains through selective breeding (Newman, 2007). A combination of contraception and behaviour modification is usually necessary with bulls, as aggressive behaviour and libido can result in serious problems. Compared with entire male cattle, castrated males are less aggressive, making them safer to handle and reducing damage to fences, gates and pasture. This means that they are easier to keep in paddocks than intact males after the time that sexual maturity would be reached

(Newman, 2007). Castrated animals are also less likely to fight other animals, thereby reducing bruising and injuries to themselves and other cattle (Bonneau and Enright, 1995).

Technologies such as GnRH vaccines, vaccines targeting sperm antigens, or GnRH agonists are all the subject of ongoing research, but they are not yet reliable enough for commercial application in cattle. Currently, the most reliable method of contraception for male cattle is to physically disassociate the blood supply from the testicles. This provides the generally desirable combination of contraception and behaviour modification and can be achieved by the open or closed techniques described below. Even though these techniques are generally reliable and well tolerated, it should be noted that in a field situation, they are not suited to animals with conditions such as cryptorchidism or inguinal hernia.

SURGICAL CASTRATION. Methods of castration for male cattle have been well described and the reader is referred to a report from Meat and Livestock Australia, which outlines in detail the different techniques used and the requirements for equipment and facilities (Newman, 2007). The report can be found online at: http://www.mla.com.au/Publications-tools-and-events/Publication-details?pubid=4007. Techniques have also been summarized in the earlier section of this chapter entitled 'Orchidectomy', in the section on 'Contraception Based on Surgical Techniques'.

Particular attention should be paid to the animal welfare and veterinary legislation for the jurisdiction within which animals are surgically castrated. Variations in such legislation may include the need for some form of anaesthesia, the age categories (and restrictions to these) within which different procedures may be utilized, the type of restraint that is acceptable during the procedure (in particular, the use of electrical restraint devices), and whether or not a procedure is considered to be an act of veterinary science.

NON-SURGICAL CASTRATION. Despite a large body of research, there is a conspicuous lack of commercially viable, non-surgical contraceptive options for bulls. Most options appear to be either still in the research and development phase, or are apparently not commercially viable in comparison with traditional contraceptive techniques. Currently, GnRH vaccines offer the most promising non-surgical option in bulls, although there is also promise for the development of depot formulations of GnRH agonists, which may allow slow release of the pharmaceutical concerned over a period of 12 months or more (Herbert and Trigg, 2005).

GNRH VACCINES. Perhaps one of the earliest attempts at a commercial non-surgical contraceptive for cattle was the GnRH vaccine marketed in Australia as Vaxstrate (Hoskinson *et al.*, 1990). While this was targeted at females, it was proof of concept that an effective GnRH vaccine for cattle could be successfully produced and marketed on a commercial scale. GonaCon vaccine, a GnRH keyhole-limpet haemocyanin (KLH) conjugate, has been successfully used in beef bulls to control reproductive behaviour (Price *et al.*, 2003). After vaccination at 4 months of age and again at 12 months, males were found to have behaviours similar to those of steers. Unfortunately, studies assessing semen quality were not carried out in this trial. From this study, and from studies in other species, it can be concluded with reasonable confidence that this latter formulation of GnRH vaccine would currently be the most likely option to produce effective non-surgical contraception in bulls. However, the cattle industry demands a high degree of success to ensure good control over breeding programmes and to prevent the bruising of beef. Therefore, extensive field trials will be necessary before any recommendation can be made for commercial application.

Horses

Other than reliable fencing, surgical castration is the most widely utilized method of contraception in the stallion. Surgical castration is well described in equine surgical texts (Auer and Stick, 2012), with restraint options including standing sedation with local anaesthesia or general anaesthesia. It is notable that complications associated with surgical castration are the most common cause of malpractice claims against equine practitioners (Wilson and Quist, 1993). The equipment that is traditionally utilized includes a scalpel blade and emasculators, but a recent innovation involves the use of a drill attachment known as the Henderson attachment, which appears to reduce post-surgical complications (Reilly and Cimetti, 2005). Up to now, as with cattle, there is a lack of commercially viable, non-surgical contraceptive options for horses.

Equity™ is a GnRH vaccine available for use in mares in Australia, but it is not registered for use in stallions. GnRH vaccines (Burger *et al.*, 2006; Janett *et al.*, 2009) and GnRH antagonists (Hinojosa *et al.*, 2001), have been trialled in stallions, and have generally resulted in reduced libido and impaired semen quality, but no treatment regimens have yet succeeded in producing reliable contraception. Both GnRH vaccines (Dowsett *et al.*, 1996) and GnRH antagonists (Hinojosa *et al.*, 2001) appear to be most reliable when used in peri-pubertal stallions; more variable responses are seen in mature stallions. In the case of GnRH agonists, the consensus is that stallions are remarkably resistant to downregulation of the HPG axis (Stout and Colenbrander, 2004), with responses to treatment ranging from partial downregulation through to hyper-stimulation. Despite this variability, GnRH vaccines and GnRH antagonists, in their current forms, may have a role in disease control, for example in preventing equine viral arteritis shedding in stallions (Burger *et al.*, 2006).

Exogenous progesterone administration may help to modify stallion behaviour (Stout, 2005), but daily administration is usually required and it may not be possible for progestogens to be used in competition animals. Neither is semen production reliably reduced with progestogen treatment (Stout and Colenbrander, 2004).

Sheep and goats

In sheep and goats, open surgical castration, or scrotal banding utilizing rubber rings (Elastrator™ rings) are the most common forms of male contraception (Earl *et al.*, 2006). Vasectomy is used in specific situations where teaser rams are required for the stimulation of cyclicity in ewes (Boundy and Cox, 1996). There has been limited success with non-surgical approaches, where GnRH vaccines appear to show the most promise (Earl *et al.*, 2006). GnRH agonists have also been found to suppress sexual behaviour and testicular function (Tillbrook *et al.*, 1993), although no commercial trials appear to have been performed yet.

Pigs

The surgical castration of young male pigs is used to assist management and to reduce the production of the tryptophan breakdown product skatole. Skatole, along with androstenone, is responsible for the androgen-induced odour known as 'boar taint' in pork from some intact males (Bonneau and Enright, 1995; Oonk *et al.*, 1998). However, surgical castration is economically undesirable, as it leads to growth retardation, reduced carcass quality and higher production costs due to more inefficient feed conversion (Walstra, 1974). While there is ongoing research, currently the most feasible alternative to surgical castration seems to be the use of a GnRH vaccine (Oonk *et al.*, 1998). Importantly, immunization of young intact male pigs against GnRH is effective at inhibiting genital tract development, reducing plasma gonadotrophin and testosterone concentrations, and decreasing fat androstenone concentrations and the incidence of boar taint. The goal is to sustain testicular androgen production for as long as possible in order to maintain the production advantages of anabolics, and then to administer the immunocontraception at a time that will reduce boar taint by the time of slaughter (Bonneau and Enright, 1995). Against this strategy is the finding of a more recent study which suggested that pigs need to be vaccinated between 10 to 14 weeks of age in order to achieve effective, long-term immunocastration (Brunius *et al.*, 2011).

Camelids

The effect of GnRH vaccination has been assessed in male camels (Ghoneim *et al.*, 2012).

A Netherlands-produced GnRH-ovalbumin conjugate vaccine was utilized and, after two injections a month apart, was found to reduce serum testosterone concentrations, libido and sperm acrosome function.

Small animal population control

Dogs

Orchidectomy is the most common form of contraception used in dogs, even though the dog is probably also the species in which most non-surgical alternatives are becoming available. Options such as Esterisol™ or Zeuterin™ for chemically induced testicular degeneration, and GnRH agonists such as Suprelorin™ or Gonazon™ for reproductive suppression or other purposes (Goericke-Pesch *et al.*, 2010) are already used commercially, and there is ongoing research on GnRH vaccines (Golden, 2009).

There is a significant body of research investigating the lifetime effects of contraception on canine health and well-being, with particular focus on surgical castration. These effects may vary depending on the age at which neutering occurs and also on whether or not the contraceptive regimen preserves androgen production. The advantages and disadvantages of early-age contraceptive techniques are not addressed in this chapter; although extensive reviews on the topic have been published (e.g. Howe *et al.*, 2000, 2001). Longitudinal studies conducted after surgical castration indicate that many of the adverse health issues encountered are related to the reduced production of sex steroid (Howe *et al.*, 2001; Spain *et al.*, 2004; Spay Neuter Task Force, 2011). With this in mind, it seems reasonable to suggest that non-surgical approaches to contraception that either preserve androgen production, or provide reversible suppression of androgen production, are desirable goals.

ORCHIDECTOMY. Castration of the dog may be performed by either an open or closed technique, in which the testis is displaced cranially and exposed using a midline prescrotal skin incision. The technique is well described in various texts (e.g. Boothe, 2003; Fossum, 2007).

The complications of castration can be divided into short-term and long-term conditions. Short-term complications include scrotal swelling, haemorrhage, bruising and infection. Long-term conditions associated with castration in companion animals have been summarized as follows (Spay Neuter Task Force, 2011):

- An increased incidence of neoplasia such as haemangiosarcoma, osteosarcoma, transitional cell carcinoma and adenocarcinoma.
- An increased incidence of obesity.
- An increased incidence of autoimmune thyroiditis and hypothyroidism.
- An increased incidence of diabetes mellitus.
- An increased incidence of cranial cruciate rupture.
- An increased incidence of hip dysplasia (if castrated earlier than 5 months of age).
- A possible increased incidence of capital physeal fractures in castrated male cats that may be partially due to increased weight gain.
- Increased shyness and hiding behaviour in castrated male cats.
- A possible increased incidence of cognitive dysfunction in castrated dogs.

While the list of possible long-term complications of castration may seem lengthy, these can be alleviated, or prevented, by utilizing contraceptive techniques that preserve the production of sex steroids.

Additionally, there are multiple benefits associated with castration unrelated to population control, including:

- a reduced risk of testicular neoplasia in castrated dogs and cats (the incidence of testicular cancer in intact animals is common, though malignancy and mortality are low);
- a reduced risk of prostatitis, benign prostatic hyperplasia, prostatic cysts and squamous metaplasia of the prostate in castrated dogs; and
- a reduced risk of perineal and inguinal hernia and perineal adenoma in castrated dogs.

INDUCED TESTICULAR DEGENERATION. ZA has been extensively evaluated as a sclerosing agent suitable for the induction of testicular degeneration in dogs (Tepsumethanon *et al.*, 2005; Oliveira *et al.*, 2007; Levy *et al.*, 2008). Studies suggest that it is capable of producing irreversible atrophy of the seminiferous tubules (Oliveira *et al.*, 2007), thus making it suitable for long-term contraception. The procedure for intra-testicular injection in companion animals involves gentle cleansing of the scrotum with soap and water. Alcohol and heavy scrubbing are to be avoided. Two tuberculin syringes clearly labelled left and right, and containing the doses calculated for each testicle, have 28G needles attached. The needle is inserted into the dorsocranial aspect of each testicle and directed caudally so that the tip is estimated to be at the centre of the testis. The solution is then gently injected (Levy *et al.*, 2008). Product information sheets should be consulted, but injection volumes will range from 0.1 to 1.0 ml per testicle depending on its width. Abdominal cryptorchid animals are not suitable candidates for this procedure.

As noted previously, ZA should be used with caution as severe adverse reactions may occur in close to 4% of cases, including necrotizing reactions requiring surgical orchiectomy and debridement (Levy *et al.*, 2008). As the procedure does not require the administration of general anaesthesia, it is suggested that induced testicular degeneration is currently most suited to otherwise high-risk anaesthesia cases, or in socio-economic contexts that preclude the use of anaesthesia and surgery due to expense.

GNRH AGONISTS. Suprelorin™ (Peptech, Australia) is a deslorelin formulation for use in dogs that has been approved for use in a number of countries (Trigg *et al.*, 2006). Two different implant doses are available for either 6 (4.7 mg) or 12 (9.4 mg) months duration of contraception. The implants are placed subcutaneously, between the shoulder blades of the dog and are designed so that removal is not necessary. Implantation can be repeated safely at either 6- or 12-monthly intervals, respectively. Alternatively, implantation can be withheld if reversion to androgen production is considered beneficial to the animal's health or well-being.

Cats

The surgical castration of tomcats is a relatively quick and safe procedure when compared with the procedure in many other species and remains the most common form of sterilization in the feline. Under short-duration general anaesthesia, the spermatic cord may be ligated either using suture material, or using the spermatic cord itself – by placing an overhand knot in the cord (Fossum, 2007).

GnRH agonists have been successful in suppressing spermatogenesis and androgen-related behaviour in domestic toms (Jewgenow *et al.*, 2006; Munson, 2006; Goericke-Pesch *et al.*, 2011) and large felids (Bertschinger *et al.*, 2006). The use of these drugs appears to be safe and reliable; however, to the best of our knowledge, none are registered for use in felines. Commercial products such as Suprelorin® and Gonazon® are registered for use in other species.

Initial trials utilising a GnRH-keyhole-limpet conjugate vaccine (GonaCon®) showed a reliable induction of antibody production and contraception (Levy *et al.*, 2004). Further development of GnRH vaccine technology has the potential to produce a commercially viable product in the near future. With the utilization of novel delivery techniques, there may also be a role for polynucleotide vaccine technology to induce immunity to GnRH.

The bisdiamine compound Fertilysin® has been trialled in toms and found to be non-toxic and reliable for use as a reversible contraception method (Munson *et al.*, 2004). A qualification to its use is that it must not be used in queens owing to its potential to cause teratogenic abnormalities. Unfortunately, its current cost is likely to preclude its use as a feline contraceptive in the near future.

While methods of ductal obstruction have been trialled in toms, the unreliability of reversibility and the degree of surgical skill and time required make these methods impractical in application (Munson, 2006).

Male contraception in wildlife and feral animal populations

Many of the current population control measures for feral or native vertebrates are dominated by lethal procedures. Despite their perceived effectiveness, animal welfare considerations, risks of environmental contamination and adverse effects on non-target species dictate a need for the development and use of contraceptive population control technologies (Barfield *et al.*, 2006). Taking into account the need for public acceptance of population control methods, plus the requirement for minimal disruption to social structure (Wang *et al.*, 2011), there is also a need in some species for contraception that preserves androgen production.

One of the biggest obstacles to the implementation of male contraception in wild or feral animals is a practical means of accurately administering the treatment. Most of the methods referred to in this chapter have potential for use as a contraceptive in wild animal populations, provided that there is a practical method for safe administration. If trapping or darting is to be avoided, the requirement for practical and safe administration dictates that treatments requiring injection or implantation are not ideal options. This leaves oral or vector distribution as the most practical alternatives.

Currently there are no contraceptives available for dissemination by insect, viral or other forms of vector technology, although polynucleotide vaccines hold future promise, so the oral dissemination of contraception in extensive populations is currently the only practical alternative. Methods of contraception suitable for oral administration are limited to the bisdiamines (Fertilysin®) (Munson *et al.*, 2004), sex steroids and, possibly, SARMs. At present, significant research is required to formulate contraceptive products suitable for any given wild or feral species.

It must also be remembered that there is species variation in response to many of the proposed contraceptive techniques, requiring any product to be trialled in each species for which it is to be used. For example, treatment of male brush-tailed possums with the GnRH agonist deslorelin had no contraceptive effect (Eymann *et al.*, 2007), despite its demonstrated effectiveness in other species.

Conclusion

Contraceptive techniques in the male can be broadly categorized into surgical and non-surgical options. For most domestic and pest species of animals, surgical contraceptive techniques are currently the most commonly used. However, many non-surgical male contraceptive options have been investigated, with a number reaching the stage of commercial production. Disappointingly, a number of promising technologies appear to be stagnating for want of either adequate research interest, or the need for a multidisciplinary approach in order to develop products for specific animal and management applications. While not precluding other options, SARMs, calcium chloride, hormone–cytotoxin conjugates, GnRH antagonists, bisdiamines, indenopyridines and polynucleotide vaccines all seem worthy of further investigation, with the latter holding promise of allowing interesting means of delivery to be developed. It is exciting that GnRH agonists and GnRH vaccines are currently achieving some, albeit small, share in the market for male animal contraception.

Some of the contraceptive techniques discussed in this chapter, such as SARMs and polynucleotide vaccines, may allow the selective retention or inclusion of contraceptive elements. Therefore, as these newer technologies develop it will be necessary for stakeholders to carefully consider the attributes of the ideal contraceptive for each circumstance.

References

Aitken RJ, Hughes LM, Griffith R and Baker MA (2008) Bridging the gap between male and female fertility control; contraception-on-demand *Contraception* **78** S28–S35.

Amory JK and Bremner WJ (2000) Newer agents for hormonal contraception in the male *Trends in Endocrinology and Metabolism* **11** 61–66.

Asquith KL, Kitchener AL and Kay DJ (2006) Immunisation of the male tammar wallaby (*Macropus eugenii*) with spermatozoa elicits epididymal antigen-specific antibody secretion and compromised fertilisation rate *Journal of Reproductive Immunology* **69** 127–147.

Auer JA and Stick JA (2012) *Equine Surgery* 4th edition. Elsevier Saunders, St Louis.

Babiuk LA, Lewis J, Suradhat S, Baca-Estrada M, Foldvari M and Babiuk S (1999 Polynucleotide vaccines: potential for inducing immunity in animals *Journal of Biotechnology* **73** 131–140.

Barfield JP, Nieschlag E, and Cooper TG (2006) Fertility control in wildlife: humans as a model *Contraception* **73** 6–22.

Benoff S, Cooper GW, Hurley I, Mandel FS, Rosenfeld DL, Scholl GM, Gilbert BR and Hershlag A (1994) The effect of calcium channel blockers on sperm fertilization potential *Fertility and Sterility* **62** 606–617.

Bertschinger HJ, Trigg TE, Jöchle W and Human A (2002) Induction of contraception in some African wild carnivores by downregulation of LH and FSH secretion using the GnRH analogue deslorelin In *Fertility control in wildlife. Proceedings of the Fifth International Symposium on Fertility Control in Wildlife, held at Skukuza, The Kruger National Park, South Africa, August 2001* Ed JF Kirkpatrick, BL Lasley, WR Allen and C Doberska *Reproduction Supplement* **60** 41–52.

Bertschinger HJ, Jago M, Nöthling JO, and Human A (2006) Repeated use of the GnRH analogue deslorelin to down-regulate reproduction in male cheetahs (*Acinonyx jubatus*) *Theriogenology* **66** 1762–1767.

Bhowal S K, Lala S, Hazra A, Paira P, Banerjee S, Mondal NB and Chakraborty S (2008) Synthesis and assessment of fertility-regulating potential of 2-(2″-chloroacetamidobenzyl)-3-(3′-indolyl) quinoline in adult rats as a male contraceptive agent *Contraception* **77** 214–222.

Bonneau M and Enright WJ (1995) Immunocastration in cattle and pigs *Livestock Production Science* **42** 193–200.

Boothe HW (2003) Testes and epididymides In *Textbook of Small Animal Surgery* 3rd edition pp 1527–1529 Ed D Slatter. Elsevier Saunders, Philadelphia.

Boundy T and Cox J (1996) Vasectomy in the ram *In Practice* **18** 330–334.

Bowen RA (2008) Male contraceptive technology for nonhuman male mammals *Animal Reproduction Science* **105** 139–143.

Bradley MP, Hinds LA and Bird PH (1997) A bait-delivered immunocontraceptive vaccine for the European red fox (*Vulpes vulpes*) by the year 2002? *Reproduction, Fertility and Development* **9** 111–116.

Brito LFC, Sertich PL, Rives Knobbe WM, Piero FD and Stull GB (2011) Effects of intratesticular zinc gluconate treatment on testicular dimensions, echodensity, histology, sperm production, and testosterone secretion in American black bears (*Ursus americanus*) *Theriogenology* **75** 1444–1452.

Brunius C, Zamaratskaia G, Andersson K, Chen G, Norrby M, Madej A and Lundström K (2011) Early immunocastration of male pigs with Improvac® – effect on boar taint, hormones and reproductive organs *Vaccine* **29** 9514–9520.

Buchholz V, Norman ST and Scott C (2012) The identification and mapping of kisspeptin and gonadotrophin releasing hormone fibres and cell bodies in the canine hypothalamus. Honours thesis, Charles Sturt University, Wagga Wagga.

Burger D, Janett F, Vidament M, Stump R, Fortier G, Imboden I and Thun R (2006) Immunization against GnRH in adult stallions: effects on semen characteristics, behaviour and shedding of equine arteritis virus *Animal Reproduction Science* **94** 107–111.

Coetzee J, Nutsch A, Barbur L and Bradburn R (2010) A survey of castration methods and associated livestock management practices performed by bovine veterinarians in the United States *BMC Veterinary Research* **6**:12.

Cooper DW and Larsen E (2006) Immunocontraception of mammalian wildlife: ecological and immunogenetic issues *Reproduction* **132** 821–828.

Cooper TG and Yeung CH (1999) Recent biochemical approaches to post-testicular, epididymal contraception *Human Reproduction Update* **5** 141–152.

Cowan PE and Tyndale-Biscoe CH (1997) Australian and New Zealand mammal species considered to be pests or problems *Reproduction, Fertility and Development* **9** 27–36.

D'Occhio M, Fordyce G, Whyte TR, Jubb TF, Fitzpatrick LA, Cooper NJ, Aspden WJ, Bolam MJ and Trigg TE (2002) Use of GnRH agonist implants for long-term suppression of fertility in extensively managed heifers and cows *Animal Reproduction Science* **74** 151–162.

Delves PJ (2004) How far from a hormone-based contraceptive vaccine? *Journal of Reproductive Immunology* **62** 69–78.

Donald L (2000) Immunization against GnRH in male species (comparative aspects) *Animal Reproduction Science* **60–61** 459–469.

Dowsett KF, Knott LM, Tshewang U, Jackson AE, Bodero DAY and Trigg TE (1996) Suppression of testicular function using two dose rates of a reversible water soluble gonadotrophin releasing hormone (GnRH) vaccine in colts *Australian Veterinary Journal* **74** 228–235.

Earl ER, Waterston MM, Aughey E, Harvey MJA, Matschke C, Colston A and Ferro VA (2006) Evaluation of two GnRH-I based vaccine formulations on the testes function of entire Suffolk cross ram lambs *Vaccine* **24** 3172–3183.

England G C (1997) Effect of progestogens and androgens upon spermatogenesis and steroidogenesis in dogs *Journal of Reproduction and Fertility Supplement* **51** 123–138.

Eymann J, Herbert CA, Thomson BP, Trigg TE, Cooper DW and Eckery DC (2007) Effects of deslorelin implants on reproduction in the common brushtail possum (*Trichosurus vulpecula*) *Reproduction, Fertility and Development* **19** 899–909.

Fahim MS, Fahim Z, Harman J, Thompson I, Montie J and Hall DG (1977) Ultrasound as a new method of male contraception *Fertility and Sterility* **28** 823–831.

Fahim MS, Wang M, Sutcu MF, Fahim Z and Youngquist RS (1993) Sterilization of dogs with intra-epididymal injection of zinc arginine *Contraception* **47** 107–122.

Fossum TW (2007) *Small Animal Surgery* 3rd edition. Mosby Elsevier, St Louis Missouri.

Freeman C and Coffey DS (1973) Sterility in male animals induced by injection of chemical agents into the vas deferens *Fertility and Sterility* **24** 884–890.

Freshman JL, Olson PN, Amann RP, Carlson ED, Twedt DC and Bowen RA (1990) The effects of methyltestosterone on reproductive function in male greyhounds *Theriogenology* **33** 1057–1073.

Freund J, Lipton MM and Thompson GE (1953) Aspermatogenesis in the guinea pig induced by testicular tissue and adjuvants *The Journal of Experimental Medicine* **97** 711–726.

Fried NM, Roberts WW, Sinelnikov YD, Wright EJ and Solomon SB (2002) Focused ultrasound ablation of the epididymis with use of thermal measurements in a canine model *Fertility and Sterility* **78** 609–613.

Gao W and Dalton JT (2007) Expanding the therapeutic use of androgens via selective androgen receptor modulators (SARMs) *Drug Discovery Today* **12** 241–248.

Ghoneim IM, Waheed MM, Al-Eknah MM and El-Bahr SM (2012) Immunization against GnRH in the male camel (*Camelus dromedarius*): effects on sexual behavior, testicular volume, semen characteristics and serum testosterone concentrations *Theriogenology* **78** 1102–1109.

Goericke-Pesch S, Spang A, Schulz M, Ozalp G, Bergmann M, Ludwig C and Hoffmann B (2009) Recrudescence of spermatogenesis in the dog following downregulation using a slow release GnRH agonist implant *Reproduction in Domestic Animals* **44** 302–308.

Goericke-Pesch S, Wilhelm E, Ludwig C, Desmoulins PO, Driancourt MA and Hoffmann B (2010) Evaluation of the clinical efficacy of Gonazon implants in the treatment of reproductive pathologies, behavioral problems, and suppression of reproductive function in the male dog *Theriogenology* **73** 920–926.

Goericke-Pesch S, Georgiev P, Antonov A, Albouy M and Wehrend A (2011) Clinical efficacy of a GnRH-agonist implant containing 4.7 mg deslorelin, Suprelorin®, regarding suppression of reproductive function in tomcats *Theriogenology* **75** 803–810.

Golden T (Compiler) (2009) *Immunocontraceptive Approaches for the Sterilization of Dogs and Cats. Scientific Think Tank November 19–21, 2009, Roanoke, VA – Final Report*. Alliance for Contraception in Dogs and Cats (ACC&D), Portland. Available at: http://www.goldenbiosci.com/resources/ACCD%20Immunocontraceptive%20Think%20Tank%20Document%20Final.pdf (accessed 23 September 2013)

Gunaga KP, Sheth AR and Rao SS (1970) Immunological studies with rat testis: antigenic characterisation *Journal of Reproduction and Fertility* **23** 263–269.

Gupta RS, Saxena P, Gupta R and Kachhawa JBS (2011) Evaluation of reversible contraceptive activities of *Cuminum cyminum* in male albino rats *Contraception* **84** 98–107.

Hart BL (1974) Gonadal androgen and sociosexual behavior of male mammals: a comparative analysis *Psychological Bulletin* **81** 383–400.

Hedger MP, Khatab S, Gonzales G and de Kretser DM (1995) Acute and short-term actions of serotonin administration on the pituitary-testicular axis in the adult rat *Reproduction, Fertility and Development* **7** 1101–1109.

Herbert CA and Trigg TE (2005) Applications of GnRH in the control and management of fertility in female animals *Animal Reproduction Science* **88** 141–153.

Herbison AE (2006) Physiology of the gonadotropin-releasing hormone neuronal network In *Knobil and Neill's Physiology of Reproduction Volume 1* 3rd edition pp 1415–1482 Ed JD Neill, JRG Challis,

DM de Kretser, DW Pfaff, JAS Richards, TM Plant and PM Wassarman. Elsevier Academic Press, St Louis/San Diego/London.

Hild SA, Attardi BJ and Reel JR (2004) The ability of a gonadotropin-releasing hormone antagonist, acyline, to prevent irreversible infertility induced by the indenopyridine, CDB-4022, in adult male rats: the role of testosterone *Biology of Reproduction* **71** 348–358.

Hild SA, Marshall GR, Attardi BJ, Hess RA, Schlatt S, Simorangkir DR, Ramaswamy S, Koduri S, Reel JR and Plant TM (2007a) Development of l-CDB-4022 as a nonsteroidal male oral contraceptive: induction and recovery from severe oligospermia in the adult male cynomolgus monkey (*Macaca fascicularis*) *Endocrinology* **148** 1784–1796.

Hild SA, Reel JR, Dykstra MJ, Mann PC and Marshall GR (2007b) Acute adverse effects of the indenopyridine, CDB-4022, on the ultrastructure of Sertoli cells, spermatocytes and spermatids in rat testes: comparison to the known Sertoli cell toxicant, di-*N*-pentylphthalate (DPP) *Journal of Andrology* **28** 4 621–629.

Hinojosa AM, Bloeser JR, Thomson SRM and Watson ED (2001) The effect of a GnRH antagonist on endocrine and seminal parameters in stallions *Theriogenology* **56** 903–912.

Hogarth CA, Amor JK and Griswold MD (2011) Inhibiting vitamin A metabolism as an approach to male contraception *Trends in Endocrinology and Metabolism* **22** 136–144.

Holland MK, Andrew J, Clarke H, Walton C and Hinds LA (1997) Selection of antigens for use in a virus-vectored immunocontraceptive vaccine: PH-20 as a case study *Reproduction, Fertility and Development* **9** 117–124.

Hoskinson RM, Rigby RD, Mattner PE, Huynh VL, D'Occhio M, Neish A, Trigg TE, Moss BA, Lindsey MJ and Coleman GD (1990) Vaxstrate: an anti-reproductive vaccine for cattle *Australian Journal of Biotechnology* **4** 166–176.

Howe LM (2006) Surgical methods of contraception and sterilization *Theriogenology* **66** 500–509.

Howe LM, Slater MR, Boothe HW, Hobson HP, Fossum TW, Spann AC and Wilkie WS (2000) Long-term outcome of gonadectomy performed at an early age or traditional age in cats *Journal of the American Veterinary Medical Association* **217** 1661–1665.

Howe LM, Slater MR, Boothe HW, Hobson HP, Holcom JL and Spann AC (2001) Long-term outcome of gonadectomy performed at an early age or traditional age in dogs *Journal of the American Veterinary Medical Association* **218** 217–221.

Hu G-X, Hu L-F, Yang D-Z, Li J-W, Chen G-R, Chen B-B, Mruk DD, Bonanomi M, Silvestrini B, Yan Cheng C and Ge R-S (2009) Adjudin targeting rabbit germ cell adhesion as a male contraceptive: a pharmacokinetics study *Journal of Andrology* **30** 87–93.

Huirne JA, Homburg R and Lambalk CB (2007) Are GnRH antagonists comparable to agonists for use in IVF? *Human Reproduction* **22** 2805–2813.

Immegart HM and Threlfall WR (2000) Evaluation of intratesticular injection of glycerol for nonsurgical sterilization of dogs *American Journal of Veterinary Research* **61** 544–549.

Jana K and Samanta PK (2007) Sterilization of male stray dogs with a single intratesticular injection of calcium chloride: a dose-dependent study *Contraception* **75** 390–400.

Janett F, Stump R, Burger D and Thun R (2009) Suppression of testicular function and sexual behavior by vaccination against GnRH (Equity) in the adult stallion *Animal Reproduction Science* **115** 88–102.

Jasoni CL and Porteous RW (2009) Anatomical location of mature GnRH neurons corresponds with their birthdate in the developing mouse *Developmental Dynamics* **238** 524–531.

Jewgenow K, Dehnhard M, Hildebrandt TB and Goritz F (2006) Contraception for population control in exotic carnivores *Theriogenology* **66** 1525–1529.

Jha RK, Jha PK and Guha SK (2009) Smart RISUG: a potential new contraceptive and its magnetic field-mediated sperm interaction *International Journal of Nanomedicine* **4** 55–64.

Jimenez-Gonzalez C, Michelangeli F, Harper CV, Barratt CLR and Publicover SJ (2006) Calcium signalling in human spermatozoa: a specialized toolkit of channels, transporters and stores *Human Reproduction Update* **12** 253–267.

Johnston S, Root Kustritz MV and Olson PN (2001a) Prevention of fertility in male dogs In *Canine and Feline Theriogenology* p 307–311 Ed S Johnston, MV Root Kustritz and PN Olson. W B Saunders, Philadelphia.

Johnston S, Root Kustritz MV and Olson PN (2001b) Prevention of fertility in the tom-cat In *Canine and Feline Theriogenology* p 521–524 Ed S Johnston, MV Root Kustritz and PN Olson. W B Saunders, Philadelphia.

Jones AR and Cooper TG (1999) A re-appraisal of the post-testicular action and toxicity of chlorinated antifertility compounds *International Journal of Andrology* **22** 130–138.

Junaidi A and Norman ST (2003) Ovarian responses after the delay of ovulation using GnRH agonist implants in superstimulated goats *Jurnal Sain Veteriner* **21** 1–5.
Junaidi A, Williamson PE, Cummins JM, Martin GB, Blackberry MA and Trigg TE (2003) Use of a new drug delivery formulation of the gonadotrophin-releasing hormone analogue Deslorelin for reversible long-term contraception in male dogs *Reproduction, Fertility and Development* **15** 317–322.
Junaidi A, Williamson PE, Martin GB, Stanton PG, Blackberry MA, Cummins JM and Trigg TE (2007) Pituitary and testicular endocrine responses to exogenous gonadotrophin-releasing hormone (GnRH) and luteinising hormone in male dogs treated with GnRH agonist implants *Reproduction, Fertility and Development* **19** 891–898.
Jung M-J, Moon Y-C, Cho I-K, Yeh J-Y, Kim S-E, Chang W-S, Park S-Y, Song C-S, Kim H-Y, Park K-K, McOrist S, Choi I-S and Lee J-B (2005) Induction of castration by immunization of male dogs with recombinant gonadotropin-releasing hormone (GnRH)-canine distemper virus (CDV) T helper cell epitope p35 *Journal of Veterinary Science* **6** 21–24.
Knobel DL, Cleaveland S, Coleman PG, Fèvre EM, Meltzer MI, Miranda MEG, Shaw A, Zinsstag J and Mesli F-X (2005) Re-evaluating the burden of rabies in Africa and Asia *Bulletin of the World Health Organization* **83** 360–369.
Koduri S, Hild SA, Pessaint L, Reel JR and Attardi BJ (2008) Mechanism of action of l-CDB-4022, a potential nonhormonal male contraceptive, in the seminiferous epithelium of the rat testis *Endocrinology* **149** 1850–1860.
Kutzler M and Wood A (2006) Non-surgical methods of contraception and sterilization *Theriogenology* **66** 514–525.
Kyriakis JM and Avruch J (2001) Mammalian mitogen-activated protein kinase signal transduction pathways activated by stress and inflammation *Physiological Reviews* **81** 807–869.
Ladd A, Tson YY, Walfield AM and Thau R (1994) Development of an antifertility vaccine for pets based on active immunization against luteinizing hormone-releasing hormone *Biology of Reproduction* **51** 1076–1083.
Leoci R, Aiudi G, De Sandro SA, Silvestre Binetti FF and Lacalandra GM (2009) Ultrasound as a mechanical method for male dog contraception *Reproduction in Domestic Animals* **44(Supplement 2)** 326–328.
Levy JK, Miller LA, Cynda Crawford P, Ritchey JW, Ross MK and Fagerstone KA (2004) GnRH immunocontraception of male cats *Theriogenology* **62** 1116–1130.
Levy JK, Crawford PC, Appel LD and Clifford EL (2008) Comparison of intratesticular injection of zinc gluconate versus surgical castration to sterilize male dogs *American Journal of Veterinary Research* **69** 140–143.
Lohiya NK, Mishra PK, Pathak N, Manivannan B, Bhande SS, Panneerdoss S and Sriram S (2005) Efficacy trial on the purified compounds of the seeds of *Carica papaya* for male contraception in albino rat *Reproductive Toxicology* **20** 135–148.
Lunnen JE, Faulkner LC, Hopwood ML and Pickett BW (1974) Immunization of dogs with bovine luteinizing hormone *Biology of Reproduction* **10** 453–460.
Mahalingam A, Kumar N, Maiti SK, Sharma AK, Dimri U and Kataria M (2009) Laparoscopic sterilisation vs. open method sterilisation in dogs: a comparison of two techniques *Turkish Journal of Veterinary Animal Science* **33** 427–436.
McLaughlin EA and Aitken RJ (2011) Is there a role for immunocontraception? *Molecular and Cellular Endocrinology* **335** 78–88.
Miller LA, Rhyan J and Killian G (2004) GonaCon™, a versatile GnRH contraceptive for a large variety of pest animal problems In *Proceedings of the Twenty First Vertebrate Pest Conference, March 1–4, 2004, Visalia, California* pp 269–273 Ed RM Timm and WP Gorenze. Vertebrate Pest Conference, University of California, Davis.
Mishra PK, Manivannan B, Pathak N, Sriram S, Bhande SS, Panneerdoss S and Lohiya NK (2003) Status of spermatogenesis and sperm parameters in langur monkeys following long-term vas occlusion with styrene maleic anhydride *Journal of Andrology* **24** 501–509.
Mruk DD and Yan Cheng C (2008) Delivering non-hormonal contraceptives to men: advances and obstacles *Trends in Biotechnology* **26** 90–99.
Mruk DD, Wong CH, Silvestrini B and Yan Cheng C (2006) A male contraceptive targeting germ cell adhesion *Nature Medicine* **12** 1323–1328.
Munson L (2006) Contraception in felids *Theriogenology* **66** 126–134.
Munson L, Chassy LM and Asa C (2004) Efficacy, safety and reversibility of bisdiamine as a male contraceptive in cats *Theriogenology* **62** 81–92.

Navarro B, Kirichok Y, Chung J and Clapham DE (2008) Ion channels that control fertility in mammalian spermatozoa *International Journal of Developmental Biology* **52** 607–613.

Naz RK and Talwar GP (1981) Immunological sterilization of male dogs by BCG *International Journal of Andrology* **4** 111–128.

Naz RK, Alexander NJ, Isahakia M and Hamilton MS (1984) Monoclonal antibody to a human germ cell membrane glycoprotein that inhibits fertilization *Science* **225** 342–344.

Neischlag E, Kamischke A and Behre HM (2004) Hormonal male contraception: the essential role of testosterone In *Testosterone: Action, Deficiency, Substitution* pp 685–714 Ed E Neischlag, HM Behre and S Neischlag. Cambridge University Press Cambridge.

Neumann F, Diallo FA, Hasan SH, Schenck B and Traore I (1976) The influence of pharmaceutical compounds on male fertility *Andrologia* **8** 203–235.

Newman R (2007) *A Guide to Best Practice Husbandry in Beef Cattle: Branding, Castrating and Dehorning* Ed I Partridge. Meat and Livestock Australia (MLA), North Sydney.

O'rand MG, Widgren EE, Sivashanmugam P, Richardson RT, Hall SH, French FS, VandeVoort CA, Ramachandra SG, Ramesh V and Jagannadha RA (2004) Reversible immunocontraception in male monkeys immunized with EPPIN *Science* **306** 1189–1190.

Ojha P, Dhar JD, Dwivedi AK, Singh RL and Gupta G (2006) Effect of antispermatogenic agents on cell marker enzymes of rat Sertoli cells *in vitro Contraception* **73** 102–106.

Oliveira ECS, Moura MR, Silva J, Peixoto CA, Saraiva KLA, Sá MJC, Douglas RH and de Pinho Marques A Jr (2007) Intratesticular injection of a zinc-based solution as a contraceptive for dogs *Theriogenology* **68** 137–145.

Oonk HB, Turkstra JA, Schaaper WMM, Erkens JHF, Schuitemaker-de Weerd MH, van Nes A, Verheijden JHM and Meloen RH (1998) New GnRH-like peptide construct to optimize efficient immunocastration of male pigs by immunoneutralization of GnRH *Vaccine* **16** 1074–1082.

Pineda MH and Dooley MP (1984) Surgical and chemical vasectomy in the cat *American Journal of Veterinary Research* **45** 291–300.

Pineda MH, Reimers TJ and Faulkner LC (1976) Disappearance of spermatozoa from the ejaculates of vasectomised dogs *Journal of the American Veterinary Medical Association* **168** 502–503.

Pöllänen P and Cooper TG (1994) Immunology of the testicular excurrent ducts *Journal of Reproductive Immunology* **26** 167–216.

Price EO, Adams TE, Huxsoll CC and Borgwardt RE (2003) Aggressive behavior is reduced in bulls actively immunized against gonadotropin-releasing hormone *Journal of Animal Science* **81** 411–415.

Purohit A, Joshi VB and Vyas KB (2008) Effect of various chromatographic fractions of neem seed oil on sperm dynamics and testicular cell population dynamics of male albino rats *Pharmaceutical Biology* **46** 660–664.

Purswell BJ and Kolster KA (2006) Immunocontraception in companion animals *Theriogenology* **66** 510–513.

Reilly MT and Cimetti LJ (2005) How to use the Henderson castrating instrument and minimize castration complications In *Proceedings of the 51st Annual Convention of the American Association of Equine Practitioners, Seattle, Washington, USA, 3–7 December, 2005* pp 494–497 Ed TD Brokken. American Association of Equine Practitioners (AAEP), Lexington.

Roberts WW, Chan DY, Fried NM, Wright EJ, Nicol T, Jarrett TW, Kavoussi LR and Solomon SB (2002) High intensity focused ultrasound ablation of the vas deferens in a canine model *Journal of Urology* **167** 2613–2617.

RSPCA (2012) *RSPCA Australia National Statistics 2011–2012*. Available at: http://www.rspca.org.au/sites/default/files/website/The-facts/Statistics/RSPCA%20Australia%20National%20Statistics%202011-2012.pdf (accessed 13 September 2013).

Saari T, Jahnukainen K and Pöllänen P (1996) Autoantigenicity of the basal compartment of seminiferous tubules in the rat *Journal of Reproductive Immunology* **31** 65–79.

Sabeur K, Foristall K and Ball BA (2002) Characterization of PH-20 in canine spermatozoa and testis *Theriogenology* **57** 977–987.

Sabeur K, Ball BA, Nett TM, Ball HH and Liu IK (2003) Effect of GnRH conjugated to pokeweed antiviral protein on reproductive function in adult male dogs *Reproduction* **125** 801–806.

Schally AV, Arimura A, Kastin AJ, Matsuo H, Baba Y, Redding TW, Nair RMG, Debeljuk L and White WF (1971) Gonadotropin-releasing hormone: one polypeptide regulates secretion of luteinizing and follicle-stimulating hormones *Science* **173** 1036–1038.

Schiff JD, Li PS, Schlegel PN and Goldstein M (2003) Rapid disappearance of spermatozoa after vasal occlusion in the dog *Journal of Andrology* **24** 361–363.

Scott CJ, Setterfield CA, Caraty A and Norman ST (2010) Is there variation in the level of input of kisspeptin terminals onto GnRH neurons across the equine oestrous cycle? *Reproduction, Fertility and Development* **22(Supplement)** 129.

Senger PL (2003) *Pathways to Pregnancy and Parturition* 2nd edition. Current Concentrations Inc, Redmond, Oregon.

Shafik A (1999) Three new methods for male contraception *Asian Journal of Andrology* **1** 161–167.

Skakkebaek NE, Giwercman A and de Kretser D (1994) Pathogenesis and management of male infertility *The Lancet* **343** 1473–1479.

Soares JM, Avelar GF and Franca LR (2009) The seminiferous epithelium cycle and its duration in different breeds of dog (*Canis familiaris*) *Journal of Anatomy* **215** 462–471.

Spain CV, Scarlett JM and Houpt KA (2004) Long-term risks and benefits of early-age gonadectomy in dogs *Journal of the American Veterinary Medical Association* **224** 380–387.

Spay Neuter Task Force (2011) Basis for Position on Mandatory Spay–Neuter in the Canine and Feline. American College of Theriogenologists, Montgomery. Available at: http://www.theriogenology.org/displaycommon.cfm?an=1&subarticlenbr=59 (accessed 23 September 2013).

Stout TAE (2005) Modulating reproductive activity in stallions – a review *Animal Reproduction Science* **89** 93–103.

Stout TAE and Colenbrander B (2004) Suppressing reproductive activity in horses using GnRH vaccines, antagonists or agonists *Animal Reproduction Science* **82–83** 633–643.

Suri A (2004) Sperm-specific proteins – potential candidate molecules for fertility control *Reproductive Biology Endocrinology* **2** 10.

Taleporos P, Salgo MP and Oster G (1978) Teratogenic action of a bis(dichloroacetyl)diamine on rats: patterns of malformations produced in high incidence at time-limited periods of development *Teratology* **18** 5–15.

Tash JS, Attardi B, Hild SA, Chakrasali R, Jakkaraj SR and Georg GI (2008) A novel potent indazole carboxylic acid derivative blocks spermatogenesis and is contraceptive in rats after a single oral dose *Biology of Reproduction* **78** 1127–1138.

Tena-Sempere M (2006) GPR54 and kisspeptin in reproduction *Human Reproduction* Update **12** 631–639.

Tepsumethanon V, Wilde H and Hemachudha T (2005) Intratesticular injection of a balanced zinc solution for permanent sterilisation of dogs *Journal of the Medical Association of Thailand* **88** 686–689.

Tillbrook AJ, Galloway DB, Williams AH and Clarke IJ (1993) Treatment of young rams with an agonist of GnRH delays reproductive development *Hormones and Behaviour* **27** 5–28.

Trigg TE, Doyle AG, Walsh JD and Swangchan-uthai T (2006) A review of advances in the use of the GnRH agonist deslorelin in control of reproduction *Theriogenology* **66** 1507–1512.

Valiente C, Corrada Y, de la Sota PE, Gerez PG and Gobello C (2007) Effect of the GnRH antagonist, acyline, on canine testicular characteristics *Theriogenology* **68** 687–692.

Verajankorva E, Martikainen M, Saraste A, Sundstrom J and Pollanen P (1999) Sperm antibodies in rat models of male hormonal contraception and vasectomy *Reproduction, Fertility and Development* **11** 49–58.

Von Waldmann KH, Otto K and Bollwhan W (1994) Castration of piglets – pain sensitivity and anaesthesia *Deutsche Tierärztliche Wochenschrift* **101** 105–109.

Wagner A, Messe N, Bergmann M, Lekhkota O and Claus R (2006) Effects of estradiol infusion in GnRH immunized boars on spermatogenesis *Journal of Andrology* **27** 880–889.

Waites GM, Wang C and Griffin PD (1998) Gossypol: reasons for its failure to be accepted as a safe, reversible male antifertility drug *International Journal of Andrology* **21** 8–12.

Walden CM, Butters TD, Dwek RA, Platt FM and van der Spoel AC (2006) Long-term non-hormonal male contraception in mice using *N*-butyldeoxynojirimycin *Human Reproduction* **21** 1309–1315.

Walker J, Ghosh S, Pagnon J, Colantoni C, Newbold A, Zeng W and Jackson DC (2007) Totally synthetic peptide-based immunocontraceptive vaccines show activity in dogs of different breeds *Vaccine* **25** 7111–7119.

Walstra P (1974) Fattening of young boars: quantification of negative and positive aspects *Livestock Production Science* **1** 187–196.

Wang D, Li N, Liu M, Huang B, Liu Q and Liu X (2011) Behavioral evaluation of quinestrol as a sterilant in male Brandt's voles *Physiology and Behavior* **104** 1024–1030.

Wiebe JP and Barr KJ (1984) The control of male fertility by 1,2,3-trihydroxypropane (THP; glycerol): rapid arrest of spermatogenesis without altering libido, accessory organs, gonadal steroidogenesis, and serum testosterone, LH and FSH *Contraception* **29** 291–302.

Wiebe JP, Barr KJ and Buckingham KD (1989) Sustained azoospermia in squirrel monkey, *Saimiri sciureus*, resulting from a single intratesticular glycerol injection *Contraception* **39** 447–457.

Wilson JF and Quist CF (1993) Professional liability in equine surgery In *Equine Surgery* pp 903–914 Ed JA Auer. WB Saunders, Philadelphia.

Yan Cheng C and Mruk DD (2010) New frontiers in nonhormonal male contraception *Contraception* **82** 476–482.

Yang L-H, Li J-T, Yan P, Liu H-L, Zeng S-Y, Wu Y-Z, Liang Z-Q and He W (2011) Follicle-stimulating hormone receptor (FSHR)-derived peptide vaccine induced infertility in mice without pathological effect on reproductive organs *Reproduction, Fertility and Development* **23** 544–550.

Yu Y-W (1987) Probing into the mechanism of action, metabolism and toxicity of gossypol by studying its (+)- and (–)-stereoisomers *Journal of Ethnopharmacology* **20** 65–78.

Yu Z-H and Chan HC (1998) Gossypol as a male antifertility agent – why studies should have been continued *International Journal of Andrology* **21** 2–7.

Yule TD, Montoya GD, Russell LD, Williams TM and Tung KS (1988) Autoantigenic germ cells exist outside the blood testis barrier *Journal of Immunology* **141** 1161–1167.

19 Semen Evaluation and Handling: Emerging Techniques and Future Development

Heriberto Rodriguez-Martinez*
Linköping University, Linköping, Sweden

Introduction

The assessment of male reproductive health (clinical andrology), including the examination of the male genital organs, the ability to mate and the analysis of collected semen, is pivotal to the establishment of artificial insemination (AI) and the selection of breeding sires (Foote, 1999). Laboratory semen assessment has become, along with the selection of the most suitable males (for productivity and fertility), more and more important for the diagnostics of semen processing steps – from sperm cryopreservation to sperm sorting for chromosomal sex. For unselected animals, semen assessment was the only means to determine whether a male was potentially fertile or not. In either case, we have experienced an explosive development of *in vitro* assays to determine sperm intactness and to measure sperm function.

Despite this availability of multiple assays, conventional semen evaluation is still often restricted to determinations of sperm numbers, sperm motility and, sometimes, but rather rarely, sperm morphology. Those few sperm parameters seem sufficient, at least under commercial conditions, to disclose the degree of 'normality' of the testes (sperm numbers, morphology) and the epididymides (sperm morphology, motility). Underlying this approach is the fundamental axiom that an ejaculate must contain above a certain number of motile, morphologically 'normal' spermatozoa in order to allow minimum sperm numbers to reach the oviducts for eventual participation in the complex process of fertilization, and finally lead to the safe development of a conceptus. This axiom is further reinforced by the increasing belief that, if sperm numbers were high, and despite the presence of morphological abnormalities, the female genital tract is able to select spermatozoa by morphology and, further, that motility is pivotal both in this selection and in the interaction with the female genitalia and the counterpart gamete (Holt and Van Look, 2004; Rodriguez-Martinez *et al.*, 2005; Rodriguez-Martinez, 2007a; Holt, 2009, 2011; Holt and Lloyd, 2010).

However, more and more methods are now available for semen evaluation that not only make it possible to disclose the level of 'normality' of the male genital organs, but also the capability of spermatozoa (mostly their membranes but also their metabolomics) to interact with the surrounding fluids (the seminal plasma, or SP, female genital fluids

* E-mail: heriberto.rodriguez-martinez@liu.se

and also *in vitro* culture media), cells (epithelia, cumulus cells, oocytes) or extracellular material (the hyaluronan coating and the zona pellucida, or ZP) before fertilization. Methods are also available to disclose the status of the different organelles, and the integrity of the nuclear genome and of the available transcriptome, which are all related to the capability to initiate early embryo development. Although some of these methods, particularly those of an "omic' nature, are still restricted to the research bench, the accompanying development of relevant instruments, from computer assisted sperm analysers for motility (CASA, computer assisted sperm analysis for motility) or morphology (ASMA, computer *assisted sperm morphometry analysis*) to bench-model flow cytometers (FCs) are making assays accessible for clinical diagnostics and for semen processing for use in assisted reproduction. The reader is advised to look at some of the many reviews that have become available over the past decade, and the references therein, which cover most of the developments made in the past second half of the last century (e.g. Rodriguez-Martinez and Larsson, 1998; Graham, 2001; Katila, 2001; Rodriguez-Martinez, 2003, 2006, 2007b; Parkinson, 2004; Graham and Mocé, 2005; Guillan *et al.*, 2005; Rijsselaere *et al.*, 2005; Petrunkina *et al.*, 2007; Rodriguez-Martinez and Barth, 2007; Mocé and Graham 2008).

The relevant questions here are: (i) is current conventional semen evaluation sufficient; and (ii) have we reached a level where current spermatological laboratory methods allow us to disclose the fertility of a male and/or his ejaculate, processed or raw? Conventional analysis of semen is most relevant for the initial investigation of male infertility but, as concluded by this author 10 years ago, only of limited value if prediction of fertility is the aim (Rodriguez-Martinez, 2003). Have we now made sufficient progress towards this goal? What will the situation be in 10 or even 20 more years? The present chapter does not intend to forecast the future, but it does aim to critically review advances in semen assessment methodologies and the capacity that different assays have to prognose fertility (or, perhaps, infertility). Although far from exhaustive, the review attempts to provide a comparative view of the different species for which clinical andrology and modern seminology interact with breeding strategies and the application of assisted reproduction technologies (ARTs). Particular attention is paid to new methods for determining DNA and transcript intactness, as well as to those biomimetic *in vitro* assays that, by resembling events during sperm transport, storage and interaction with the female genital tract and the oocyte, best provide clues for sperm selection and the role of sperm sub-populations in the ejaculate. Finally, it attempts to identify those techniques that might be available for clinicians and breeders 20 years from now.

Clinical Andrology and Diagnostic Seminology: Can they Prognose Fertility – or Just the Risk Level of Never Reaching Fertile Status?

The clinical andrological examination of a sire, including the screening of his semen, aims not only to determine the normality of the function of his testes and genital tract, but also his capacity to mate. Such an examination provides an andrological diagnosis with a final estimation of his potential capacity to procreate, which, among animal breeders, is often known as a breeding soundness evaluation (BSE). The BSE (as described in accompanying chapters), combines a general clinical examination, a detailed genital evaluation, a testing of the sire's capacity to mate and laboratory evaluation of the ejaculate(s) (Kastelic and Thundathil, 2008). The BSE, mostly applied to production animal species, has helped us to identify clear-cut cases of infertility, or even of potential sub-fertility, thus facilitating the removal of sub-fertile males from both natural and assisted breeding (AI in particular).

For species other than production animals, andrological screening has also proven to be essential, especially when detailed semen evaluation has been included, to determine the causes of substandard semen/sperm quality, and to establish the risk of infertility. As such, it pursues a purely diagnostic purpose (which is the primary objective of clinical andrology), when the restricted possibilities for treatment that we have for most pathologies are considered. Diagnoses of infections (usually of the epididymides and the accessory sexual glands/urethra) are seldom followed by treatment in production animals, mainly because of the cost and relatively low rate of success of such treatments. Such infections are, however, often treated in companion and sport animals in a manner that often mimics human medical treatments. The value of these treatments is, in many respects, questionable. Hormonal treatments of, for instance, cases of primary spermatogenic failure (either diagnosable or idiopathic), are particularly discouraged, both for their relative ineffectiveness and because of their cost in domestic animals.

In species that usually have a comparatively 'less good' semen picture, for example the feline, canine or equine, semen evaluation is directed towards the identification of whether the problem is related to sperm numbers or sperm dysfunction, as 'treatment' opportunities might be present even without hormones. Low sperm numbers in the ejaculate denote a low spermatogenic output, which might be optimized by the establishment of an individual 'rhythm' of semen collections, taking care of maintaining sperm maturation thresholds and the best possible sperm output per 'abstinence' (or non-collection) period (time elapsed from last collection). Often, this procedure might improve chances of fertility by the implementation of 'controlled mating intervals'. Assessment of the type of sperm dysfunction that may be present is most relevant here so that the most appropriate type of assisted reproduction technology can be determined, e.g. low-sperm dose AI, *in vitro* fertilization (IVF) or even intracytoplasmic sperm injection (ICSI), and, if possible, applied. This is relevant for some pet breeds and particularly for horses, where sub-fertility is widespread among stallions and results with conventional IVF are, as yet, suboptimal.

The situation is far from similar in animals (mostly production animals) that have been selected for their fertility. Semen evaluation in breeding sires of production species (such as the bovine, ovine or porcine) often goes beyond the primary diagnostic purpose (early diagnosis of subfertility, so that the sire is promptly removed to avoid unnecessary costs), and extends to prognose levels of their 'expected' fertility before AI is used and their selection as top sires is decided upon. Thus, early identification and removal of the 'not-so-good' sires is a priority.

Despite this, during clinical andrological examinations, fertilizing capacity is not monitored by a simple spermiogram, and is often only assessed after breeding by natural mating or AI of a certain number of females. The recordings of fertility are, therefore, of utmost importance if a male is to be downgraded, from 'normal' fertility output to subfertility or, in serious cases, to permanent infertility (sterility). The registering of male fertility varies in stringency among species, mostly depending on the number of females involved. In some species (e.g. the equine and canine), sires breed relatively few females and, owing to the innate variation present between ejaculates and males, they yield fertility results of less analytical reliability than livestock species in which sires breed with large numbers of females. In the equine, as stallions might breed with the same females over several oestrus cycles per season, the resulting cumulative fertility outcome ought to be properly expressed (as per oestrus or mating/AI, or per season) for reliable comparisons.

The bottom line is that the larger the number of females mated or inseminated with a certain male, the higher the stringency of the fertility registry. Controlled mating or AI is used in livestock (bovines, ovines, porcines) where the ejaculate of a sire is used on a large number of females via AI and the number of bred females is high, both allowing an adequate register of male fertility via a large number of opportunities of conception. In dairy cattle, there is a generalized use of the average rates of non-return to oestrus (non-return rates, NRR) when 28, 56, 60–90 or 124 days have elapsed since the first AI. The NRR is a composite parameter, resulting from conception and gestation variables, two essential events that show the effectiveness of AI. These data must be corrected for conditioning factors using appropriate statistical methods (mixed model, repeated measures); these modify the initial means by the effects of the herd (number of animals, location of farm), the number of AIs (1st or 2nd), female characteristics (age, parity, milking stage, etc), year, season, inseminator, day of the week, etc., to avoid drawing false conclusions. In addition, sire fertility varies over time (Hallap *et al.*, 2004, 2006a), and this must be considered too. Boar studs can be subjected to a similar approach, while introducing new parameters such as prolificacy (number of piglets born) and even viability (numbers of piglets born alive).

Estimations of *in vitro* fertility depend on the type of test used to study a heterogeneous suspension of pluri-compartmental, terminally differentiated cells: the spermatozoa. Current assessment of ejaculated spermatozoa focuses almost solely on the number of spermatozoa (per unit of volume, i.e. concentration or as total per ejaculate) and their motility (sometimes including their kinematic patterns, if a CASA instrument is used), although other aspects, such as pH and ejaculate volume are also included. Few studies are published on the relation between neat semen and fertility because after collection (when it is at 35°C), semen is always extended and then either chilled or frozen (from room temperature at +20°C, or fridge temperature at +5°C, or anything in between), and used thereafter for AI; these processes introduce other variables into the assessment of semen (its extension, dilution of the SP, additive effects, cooling, time, etc) – considerations that will be discussed later. The same considerations apply to sperm numbers, which implies that each sperm AI dose contains a specific number of spermatozoa. However, as handling (including cryopreservation and other manipulations) causes cell death, shortens lifespan and modifies the fertilizing capacity of various proportions of spermatozoa, sperm numbers in an AI dose are often excessive so that breeders are safeguarded in providing fertility to their customers. Only when viable sperm numbers are decreased to possible 'threshold numbers' do we see a proven relation between sperm numbers and fertility (Tardif *et al.*, 1999; Christensen *et al.*, 2011).

Subjectively assessed sperm motility has been statistically related to fertility even for post-thawed semen in bulls (Rodriguez-Martinez, 2003) and in pigs (Cremades *et al.*, 2005). Studies in other species have yielded erratic results, with large variation between laboratories, owing to operator bias and differences in numbers of breedings/female numbers used to determine fertility (Rodriguez-Martinez, 2006). Kinematic analyses using CASA instruments have shown variable correlations between particular motility patterns, such as linearity, and fertility in the field (Bailey *et al.*, 1994; Holt *et al.*, 1997; Zhang *et al.*, 1998; Hirai *et al.*, 2001; Januskauskas *et al.*, 2001, 2003). Further, combining motility patterns with other parameters of sperm function in AI dairy sires has allowed a reasonable estimation of fertility (Januskauskas *et al.*, 2001), as exemplified in Fig. 19.1. Major constraints for conventional CASA instrumentation relate to the variability between users (Ehlers *et al.*, 2011), as well as to the low sperm numbers analysed per sample; when these factors are combined with the still high cost of the instrumentation, the wider use of CASA instruments is jeopardized (Feitsma *et al.*, 2011). There is now alternative instrumentation available (Qualisperm™, Biophos, Switzerland, Plate 62) that is based on a different principle from that used in CASA, i.e. the classical digitalization of centroid movement over time. This novel technology is based on correlation analysis of single particles (spermatozoa) in confocal volume elements. Individual spermatozoa are projected on to a pixel grid of a CMOS (complementary metal-oxide semiconductor) camera, and the algorithm analyses the number of fluctuations using a correlation function instead of trajectories (as in the CASA system). The new system benefits from a high throughput (usually four fields per minute) and analyses >2000 spermatozoa/sample; it has been thoroughly tested on several species (Tejerina *et al.*, 2008, 2009; Johannisson *et al.*, 2009).

Most often, the proportion of morphologically normal spermatozoa in the ejaculate of a bull is related to its fertility post AI (Phillips *et al.*, 2004; Al-Makhzoomi *et al.*, 2008), by reflecting, together with sperm

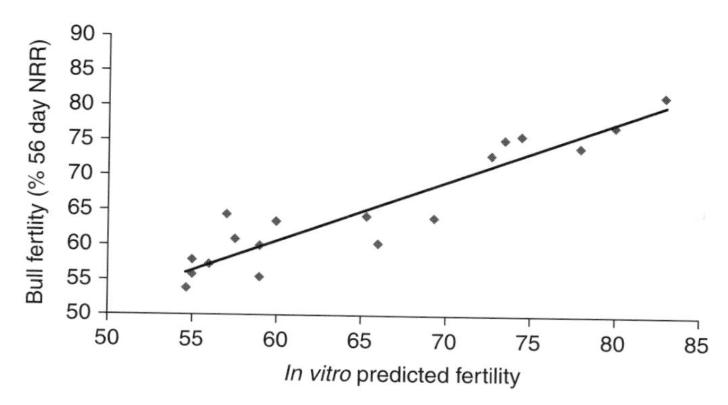

Fig. 19.1. Relationship (trend line) between 56 day non-return rates (NRRs) predicted by combining sperm parameters assessed *in vitro* with the observed field fertility (56-day NRR) in individual AI dairy bull sires (diamonds); $r = 0.90$, $P < 0.001$. Modified from Januskauskas A *et al.* (2001) *Theriogenology* **55** 947–961.

numbers and their motility, the degree of normality of spermatogenesis and sperm maturation within a cohort of sires. Continuous selection of breeding sires, and of dairy bulls in particular, for sperm quality (including morphology) has substantially decreased the incidence of sperm abnormalities in bulls of similar age within the space of 30 years (see Rodriguez-Martinez and Barth, 2007).

Morphological abnormalities are always present in any ejaculate, although various forms of abnormality differ in their impact on fertility. Some are specific defects that hamper fertilization, while others, such as the pyriform or pear-shaped sperm head deviation, impair proper embryo development (Rodriguez-Martinez and Barth, 2007), thus calling for frequent (intervals of 2 months) detailed assessments of sperm morphology in AI stud sires using wet or stained smears (Plate 63a,b). The reliability of such analyses requires large numbers of spermatozoa to be assessed per sample, i.e. 200 per wet-smear and 500 for stained sperm heads; the latter allows for determinations of defects that have clear relationships with fertility due to their uncompensable nature, such as the pyriform sperm head shape (as an expression of a defective chromatin condensation during spermiogenesis) (Al-Makhzoomi *et al.*, 2008). Software for the ASMA system has been developed since the 1980s, and has now reached an acceptable reliability for the analyses of sperm head dimensions, though other types of sperm abnormalities cannot yet be determined (Auger, 2010), and nor are there clear relationships of sperm head morphometry with fertility (Saravia *et al.*, 2007b; Gravance *et al.*, 2009).

Does the Exploration of Other Sperm Attributes, including Testing Sperm Function *In Vitro*, Prognose Fertility?

At specialized laboratories, the evaluation of semen *in vitro* goes beyond routine andrological screening. First of all, these laboratories attempt to disclose biomarkers of sperm function (e.g. intactness and reactivity of membranes, organelles, chromatin, etc.) that have proven relevant for fertilization (Silva and Gadella, 2006). Thereafter, they explore, via *in vitro* testing, how relevant the interactions are that occur between the spermatozoa and the female genital tract, and with the oocyte vestments; they also examine the process of fertilization itself, including the early development of the embryo. Finally, they seek to determine how the different outcomes relate to fertility (Rodriguez-Martinez, 2007b). A summary of current exploration methods for these sperm attributes, and of the various functional tests available, is provided in Table 19.1.

Intactness, stability and reactivity of the plasma membrane to stimuli from surrounding fluids and cells are prerequisites for the

Table 19.1. Summary of sperm attributes to be evaluated, current and possible future methods for assessment, and their individual or combined relations to fertility.

Attributes of sperm (or seminal plasma)		Current methods (with note of stain/probe/test used, where relevant; see text for further details)	Relation to fertility (*r*)		Future/possible best method/s
			Single	Combined	
Numbers		Manual counting, cytometers, Coulter counters	0.2–0.8	0.7–0.9	True volumetric counting, fluorescent staining
Motility		Operator-driven, computer assisted sperm analysis (CASA)	0.3–0.6		Computerized fluctuation analysis
Morphology		Manual (wet and stained smears), computer assisted sperm morphometry analysis (ASMA)	0.1–0.8		ASMA
Membrane	Integrity	Microscopy (Mic): phase contrast; eosin–nigrosin, HOS-t Mic/flow cytometry (Mic/FC): CFDA, SYBR-14/PI, EtdH, H33342	0.4–0.8		Cytomics (FC and imaging) -Live/Dead fixable Red Dead Cell Stain Kit
	Stability, reactivity	Mic or FC: Mero-540, Yo-PRO-1, SNARF-1, Annexin V/PI	–		Cytomics-triple stain (SNARF-1,YO-PRO-1, EtdH), HA (hyaluronan) -binding kits
Acrosome	Intactness, reactivity	Mic: wet/stained smears Mic/FC: SYBR-14/PE-PNA/PI, FITC-PNA/PSA-PI, Ca ionophores[a]	0.6–0.8[a]		Enhanced digital imaging
Mitochondria	Integrity	Mic: wet/stained smears, Rhodamine, JC-1	0.5–0.6		
	Physiological status	Mic/FC: MitoTracker Green or Deep Red	–		Cytomics-Mitotracker Deep Red
Oxidative stress	ROS (reactive oxygen species)/LPO (lipid peroxidation)	FC: H2DCFDA, MITO Sox, BODIPY probes	0.2–0.4	–	Cytomics-BODIPY probes
Chromatin/DNA intactness		Mic: Feulgen/acridine stain Mic/FC: TUNEL, COMET, SCSA,[b] SCD (Halo)	0.3–0.9[b]	–	Cytomics-SCD

Capacitation	Membrane instability, Ca^{2+} influx, tyrosine phosphorylation	Mic: Mero-540, INDO-1, FLUO, CTC, fluorescein isothiocyanate-conjugated anti-phosphotyrosine (pY) antibody	0.5–0.7	0.6–0.7	HA-binding kits, Ca ionophores
Binding	Female lining epithelium	*In vitro* tubal explant binding	0.5		Microfluidics, physiological (bicarbonate) effector challenge
	Zona pellucida (ZP)	*In vitro* ZP-binding assays	0.5		
Fertilization/ embryo development	ZP/oolemma penetration, zygote formation	IVM/IVF (*in vitro* maturation/ fertilization) penetration rate/ polyspermia	0.5–0.6		
	Cleavage/blastulation	IVF cleavage rate, morula/ blastocyst rate	0.3–0.5		
Selection		Washing and centrifugation, filtration (Sephadex, glass wool), colloid separation (Percoll, Silane), swim up/down, FC sorting	0.2–0.5	–	Silane colloid separation, HA-binding kits
‘omic analyses	Genomics	DNA decondensation rate	–	–	Gene chips
	Epigenetics	DNA methylation rate	–	–	
	Transcriptomics	Microarrays, serial analyses	–	–	Microarrays
	Proteomics	SDS, PAGE, MALDI-TOF (matrix-assisted laser desorption/ionization)	–	–	Selective protein kits
Seminal plasma	Proteomics	SDS, PAGE, MALDI-TOF	0.3–0.6	–	Fertility-associated proteins (FAPs)
	Cytokines/Chemokines	ELISA, Luminex	–	–	Selective kits

viability and fertilizing capacity of the spermatozoon. The simplest method used nowadays to study the integrity of the plasmalemma, besides microscopy using wet smears or the use of the membrane-impermeable dye eosin (such as the eosin–nigrosin test), is to expose spermatozoa to a hypo-osmotic saline solution (HOS-test or HOS-t). Cells with an intact, functional membrane change in volume by allowing water to enter the cell. Results from both the HOS-t and its subsequent development into the 'cell volumetric electronic cell sizing method' (Neild *et al.*, 2000; Petrunkina *et al.*, 2001) have provided significant correlations with fertility.

Fluorophores have been used for over two decades to assess membrane integrity, with tests ranging from the use of dyes/probes that react with cytoplasmic enzymes to those that bind with the DNA, the latter being either permeant or impermeant (Silva and Gadella, 2006). One example is the permeant acylated membrane (AM) loading dye SYBR-14, which deacylates once it enters the living cell and binds to DNA, emitting a green fluorescence when excited by blue or UV light. It is used in combination with propidium iodide (PI) or ethidium homodimer (EtdH) to stain deteriorated cells by targeting the same DNA. SYBR-14/PI or SYBR-14/EthH provides simultaneous information on the proportions of live/dead cells in the sample (Plate 64a,b). Microscopy or flow cytometry (FC) can be used for screening (Kavak *et al.*, 2003; Nagy *et al.*, 2004; Saravia *et al.*, 2007a). However, as most DNA is found in the sperm head (although there is also some in the mitochondria), the assay indicates the integrity of the plasma membrane of the head domain and, ultimately, damage in the midpiece but not in the tail. Microscopy therefore provides more information than FC, but FC compensates for this shortcoming by allowing the examination of larger numbers of sperm (Martinez-Pastor *et al.*, 2010; Hossain *et al.*, 2011a; Petrunkina and Harrison 2011).

The most advantageous use of fluorophores is when they are combined. For instance, subtle changes in the plasma membrane such as modifications in permeability can be disclosed by combining several fluorescent probes, such as SNARF-1, YO-PRO-1 and ethidium homodimer, the so-called triple stain (Peña *et al.*, 2005a, 2007), as exemplified in Plate 65. Viable, membrane-stable spermatozoa are able to pump out small probes such as YO-PRO-1, which always enters living cells. Sub-viable cells, i.e. those with incipient membrane destabilization, become permeable and unable to pump out YO-PRO-1, which then stains the DNA bright green and can easily identify the sub-viable cells before they show other signs of degeneration (Peña, 2007). Membrane permeabilization is followed by modifications of the position of phospholipids such as phosphatidylserine (PS), which can be mapped by the fluorophore combination of Annexin-V/PI (Januskauskas *et al.*, 2003; Peña *et al.* 2003, 2005a, 2007; Saravia *et al.*, 2007a). Earlier changes in the membrane, related to the destabilization of the lipid layer, can also be followed using a triple combination of fluorophores: Merocyanine-540 (Mero-540), YO-PRO-1 and Hoechst 33342. Plate 66a–b shows typical readings obtained using two different probe combination for the estimation of sperm asymmetry. This approach has proven effective in disclosing defective spermatozoa in several species, and gives good correlations with fertility (Hossain et al., 2011a).

Acrosome intactness, a prerequisite for fertilization, can be readily examined *in vitro* using phase contrast microscopy (Rodriguez-Martinez *et al.*, 1998), or it can be examined by fluorophore-linked lectins using multiparametric analysis (Nagy *et al.*, 2003, 2004). Spermatozoa can be stained with different dyes, i.e. SYBR-14/PI to identify membrane integrity, with PE-PNA(phycoerythrin conjugated peanut agglutinin) for acrosome integrity (see Plate 67), and Mitotracker Deep Red for mitochondrial intactness (Nagy *et al.*, 2003, 2004). The intactness of the approximately 100 mitochondria of sperm is indispensable for cell life and function, from motility to capacitation. Despite this, correlations between mitochondrial status and fertility have been variable, mostly owing to the changes in mitochondria functionality over time, as monitored when using the mitochondrial stain JC-1 (5,50,6,60-tetrachloro-1,10,3,30-tetraethylbenzimidazolyl-carbocyanine iodide), which can discriminate between spermatozoa with poorly and highly functional mitochondria

(Martinez-Pastor *et al.*, 2004; Hallap *et al.*, 2005b; Peña *et al.*, 2009), as exemplified in Plate 68a,b.

Besides energy, sperm mitochondria produce by-products of the metabolism of oxygen, including superoxide. This superoxide converts into the damaging hydrogen peroxide, a reactive oxygen species (ROS), which is mostly, but not completely converted to oxygen and water by the enzymes catalase or superoxide dismutase (also known as antioxidants or scavengers). A certain level of ROS is essential for sperm function, including fertilizing capacity, but only when it is kept at optimal levels by the antioxidant capacity of SP (Awda *et al.*, 2009; Mancini *et al.*, 2009; Am-in *et al.*, 2011). When excessive numbers of leucocytes are present in the ejaculate or the semen is subjected to oxidative stress (as during cooling in the absence of SP or other natural scavengers), increased ROS generation, either extrinsic (leucocytes) or intrinsic (sperm neck cytoplasm in immature or morphologically abnormal mitochondria), act to deteriorate sperm motility (Guthrie *et al.*, 2008), sperm membranes – through lipid peroxidation (LPO), and DNA integrity – via breakage and cross linking of the chromatin (Aitken and West, 1990; Koppers *et al.*, 2008). ROS levels are very variable, making their proper determination difficult, though this is possible using the probe hydroethidine. Plate 69 shows such an FC mapping of sperm ROS obtained by loading the sperm with hydroethidine and Hoechst 33258. A useful alternative, which measures oxidative stress and ROS effects indirectly, is to determine the levels of LPO in the membrane lipid bilayer by using the 5-iodoacetamidofluorescein probe family (BODIPY-C_{11}®) (Aitken *et al.*, 2007; Guthrie and Welch, 2007; Ortega Ferrusola *et al.*, 2009a). A representative scanning in stallion spermatozoa is shown in Plate 70a-d.

Mammalian spermatozoa have tightly compacted eukaryotic DNA, established by transformations during spermiogenesis in which the sperm chromatin replaces histones, first by transient proteins and then by protamines (Oliva and Castillo, 2011). The DNA strands are highly condensed by these protamines and form the basic packaging unit of sperm chromatin: the toroid. Toroids are even further compacted by intramolecular and intermolecular disulfide cross links between protamine cysteine residues (Schulte *et al.*, 2010). There are differences in compaction between species; those with less compacted chromatin (equines or canines, for instance) are theoretically more prone to DNA damage. Protamine deficiency in loosely packed chromatin can be monitored by the fluorochrome chromocanmycin-A3 (CMA3) (Tavalaee *et al.*, 2010).

Sperm chromatin can show different abnormalities related to compaction: from damage to the actual DNA physical integrity – as single- or double-stranded DNA strand breaks, nuclear protein defects interfering with histone or protamine conversion and DNA compaction – to chromatin structural abnormalities such as defective tertiary chromatin configuration. The latter scenario implies that defects will occur in the decondensation of the nucleus, thus impairing fertilization, whereas the other defects can jeopardize embryonic development as the oocyte (albeit being able to repair a limited amount of sperm DNA damage) would not be able to correct such damage (Johnson *et al.*, 2011). Sperm DNA disorders also include mutations, epigenetic modifications, base oxidation and DNA fragmentation (which is also related to sperm handling). In accord with knowledge of its increasing relevance, the evaluation of DNA integrity has increased over recent years (Barratt *et al.*, 2010).

DNA fragmentation, by being considerably present in the sperm of sub-fertile males, is considered to be the most frequent cause of paternal DNA anomalies transmitted to progeny. Damaged sperm DNA may be incorporated into the genome of the embryo, and participate in or lead to errors in DNA replication, transcription or translation during embryo development, ultimately contributing to diseases in future generations (Katari *et al.*, 2009). Moreover, DNA damage may remain in the germ line for generations, a matter of concern related to the increasing use of ICSI (which, today, is even used in animals) (Aitken *et al.*, 2009). Sperm DNA fragmentation can be studied by many techniques, including staining with the DNA fluorophore PI, which, in species where DNA compaction is not high, can present two types of staining, i.e. dimmer and brighter (Muratori *et al.*, 2008). The former relates to low sperm quality, while the latter is present among spermatozoa with good motility and morphology, without necessarily affecting their capacity to fertilize. This might not represent a problem, provided that the amount of damage is not high, because a fully mature oocyte has the capability to repair some degree of DNA damage, for instance single-stranded DNA breaks.

Several methods are used to determine DNA damage: (i) the single-cell gel electrophoresis assay (COMET); (ii) the terminal deoxynucleotidyl transferase-mediated fluorescein-dUTP nick-end labelling (TUNEL) assay; (iii) the acridine orange test (AOT); (iv) the tritium-labelled 3H-actinomycin D (3H-AMD) incorporation assay; (v) the *in situ* nick translation (ISTN) assay; (vi) the DNA breakage detection-fluorescence *in situ* hybridisations (DBD-FISH) assay; (vii) the Sperm Chromatin Dispersion test (SCD Halo); or (viii) the evaluation of the degree of induced denaturation of the DNA (the so-called 'sperm chromatin structure assay', SCSA) (Fraser, 2004; Evenson and Wixon 2006; Tamburrino *et al.*, 2012). Examples of spermatozoa examined with some of these techniques are illustrated in Plate 71a–d. Most of the above methods can use fluorescence microscopy, but SCSA and TUNEL are usually explored via FC.

Although SCSA has been extensively used, the outcome has provided conflicting relationships with fertility (Rodriguez-Martinez and Barth, 2007; Christensen *et al.*, 2011). The conflict has arisen from a single study relating the percentage of spermatozoa with denatured DNA (through an index of DNA fragmentation, DFI), to high, moderate or very low fertility values in humans, when DFI thresholds of ~0–15%, 16–29% and >30%, respectively, were recorded as in direct relation to the pregnancy outcome using ART (Evenson and Jost, 2000). Such relationships have been questioned (Collins *et al.*, 2008). A threshold for sub-fertility in bulls and boars has also been calculated (as percentage DFI), albeit using semen from young, unproven bulls with low fertility (40–60% of 90 day NRRs) and boar semen without any fertility recordings (Rybar *et al.*, 2004). These calculated threshold values (20% for bulls and 18% for boars) were much lower than those obtained with human semen. The lack of reliable fertility data when calculating these limits also leaves these values highly questionable, especially in the light of other, well-controlled studies of AI-sires with proven and varied fertility, in which neither proven stud bulls nor stud boars in AI-breeding programmes reached these high values. For instance, sperm DFI values (even as a percentage of cells with high COMP α_t values (i.e. cells outside the main population with a high proportion of single-stranded DNA over single + double-stranded DNA) in breeding sires are low (<5%) (Boe-Hansen *et al.*, 2005; De Ambrogi *et al.*, 2006; Christensen *et al.*, 2011), even in frozen–thawed semen, where the processing ought to facilitate DNA damage (Januskauskas *et al.*, 2001, 2003; Martinez-Pastor *et al.*, 2004; Hallap *et al.*, 2005a; Hernández *et al.*, 2006; Chistensen *et al.*, 2011). Conversely, equine sires unselected for sperm quality had higher DFIs and these lacked a clear relationship with equine fertility (Morrell *et al.*, 2008).

SCSA does not specifically identify the amount of DNA damage, unlike TUNEL, but rather its susceptibility to harsh treatment. The TUNEL method is more cumbersome though, and this has led to modifications of the method. For example, a TUNEL/PI procedure is now available that combines the accuracy of TUNEL and the differentiation of two sperm populations depending on PI intensity, of which one is probably participating in fertilization as the damage has no relationship with either sperm motility or morphology (Muratori *et al.*, 2008). Alternatively, the use of dithiothreitol (DTT) to decondense sperm nuclei and inclusion of a stain for dead cells provides a higher accessibility to the TdT (terminal deoxynucleotidyl transferase) enzyme of TUNEL, in combination with the detection of DNA fragmentation in live spermatozoa (Mitchell *et al.*, 2011). Taking all the above considerations into account, TUNEL appears to be a more sensitive method to predict infertility than SCSA, as determined in a recent meta-analysis (Zini *et al.*, 2008).

DTT and detergents (as sodium dodecyl sulfate, SDS) have been used to study the relative capacity of sperm nuclei to decondense *in vitro*, in an attempt to establish a method that resembles the process needed to form a male pronucleus during fertilization. The degree of decondensation can be assessed microscopically (Rodriguez *et al.*, 1985) or via FC after PI loading (Córdova-Izquierdo *et al.*, 2006), and has shown relationships with fertility in sheep and pigs, respectively. Apoptotic-like changes and the presence of apoptotic markers have been detected in species in which retained cytoplasmic droplets are common, such as in the equine (Ortega Ferrusola *et al.*, 2009a,b, 2010). However, although the presence of caspases – the primary agents of apoptosis – is clearly related to sperm storage and cooling qualities, it remains unclear as to whether their presence is biologically relevant for male fertility.

Functional Assays, Higher Discriminatory Value?

The structural integrity attributes of the sperm that have already been described are a prerequisite for sperm function, provided it is proven that they work in a concerted manner. For instance, it is logical to consider that the determination of membrane integrity in ejaculated spermatozoa is linked with sperm motility, as well as with the capacity of viable spermatozoa to undergo sperm capacitation and change their activated motility to a hyperactivated pattern, as occurs *in vivo* (Rodriguez-Martinez, 2007a). Sperm capacitation is, moreover, a stepwise process that includes the removal of epididymal and SP-adsorbed proteins, loss of membrane cholesterol, an increase in membrane fluidity due to lipid modifications, an influx of Ca^{2+} to the sperm perinuclear and neck regions and flagellum, the generation of controlled amounts of ROS, as well as the phosphorylation of protein residues (Gadella and van Gestel, 2004; Harrison and Gadella, 2005; O'Flaherty *et al.*, 2006; Tulsiani *et al.*, 2007; Fabrega *et al.*, 2011). These are steps that can be measured *in vitro* and, ultimately, associated with the fertility of the males.

The earliest stages of sperm capacitation, those related to bicarbonate and Ca^{2+} triggered phospholipid scrambling, can be mapped using Merocyanine-540 or Annexin-V, as these measure general or specific destabilization of the plasmalemma and both are related to the fertility of the semen assayed, either fresh (Harrison, 1997) or frozen–thawed (Peña *et al.*, 2004, 2007; Januskauskas *et al.*, 2005; Hallap *et al.*, 2006b), when a bicarbonate challenge has been used (Bergqvist *et al.*, 2006; Saravia *et al.*, 2007a).

Calcium influx elicits tyrosine phosphorylation, and apparently initiates sperm capacitation and hyperactivation (Suarez, 2008a,b). Therefore, mapping of low intracellular Ca^{2+} levels in spermatozoa should be a good parameter for discriminating between semen samples (or even individual boars) (Hossain *et al.*, 2011b; Waberski *et al.*, 2011). Ca^{2+} influx and displacement within the spermatozoon have been followed throughout capacitation using Indo-1 acetoxymethylester or Fluo-loaded spermatozoa (Hossain *et al.*, 2011a,b). Most changes in relation to capacitation are registered during the later stages of this process, and can be indirectly visualized by the incubation of viable spermatozoa with the fluorescent antibiotic chlortetracycline (CTC), which monitors the displacement of Ca^{2+} in the sperm head, in preparation for the occurrence of the acrosome reaction (AR) (Rodriguez-Martinez, 2007a). In AI bulls with known fertility, the proportion of non-capacitated spermatozoa (CTC stable, unreacted) in AI semen samples was significantly related to fertility (Thundathil *et al.*, 1999; Gil *et al.*, 2000). The capacitation of spermatozoa, following Ca^{2+} and bicarbonate signalling, activates adenylate cyclase and the resulting cAMP in turn activates intracellular protein kinases that enrich the equatorial subsegment in tyrosine-phosphorylated proteins (Piehler *et al.*, 2006; Jones *et al.*, 2008). There is now an anti-phosphotyrosine antibody, and FC can estimate the amount of global protein residue phosphorylation in the sperm plasma membrane, thereby providing evidence of the real-time proportion of spermatozoa that might have engaged in the process of capacitation (usually 5–10% of all cells at a given time) (Sidhu *et al.*, 2004).

In conjunction with capacitation in the oviduct, spermatozoa change their pattern of movement from a progressive activated type (acquired at ejaculation) to a non-progressive, hyperactivated type that apparently helps them to progress along the complicated tubal lumen towards the oocyte vestments, negotiate this layer and penetrate the ZP (Suarez, 2008a). As hyperactivated spermatozoa usually display a wider lateral displacement of the head than occurs during normal motility, such a variable can be determined by most CASA software. However, its relationship with fertility is low (Zhang *et al.*, 1998; Januskauskas *et al.*, 2001; Rodriguez-Martinez *et al.*, 2008).

When the AR was induced *in vitro* by exposure to homologous zona pellucidae (ZPs), or to specific proteins isolated from the ZP (Rodriguez-Martinez, 2006), it was clear that not all spermatozoa show the AR immediately upon challenge with ZP proteins, indicating that they do not respond at one time to the stimulus, as seen for capacitation (see above). Timed retrieval of spermatozoa from pig oviducts (Tienthai *et al.*, 2004) in relation to spontaneous ovulation indicated that most spermatozoa remain uncapacitated before ovulation, perhaps as a result of their sequential exposure to the bicarbonate-rich upper oviductal fluid (Rodriguez-Martinez, 2007a). Overall, the data suggest that slow responders to species-specific effectors of capacitation or the AR would probably be the ones potentially fertilizing *in vivo*, confirming the findings of an existing relation between the percentages of non-capacitated spermatozoa (stable; see above) and fertility. Easier *in vitro* AR-inducing procedures have been attempted, including exposure to progesterone (Aitken, 2006), selected glycosaminoglycans (GAGs), or the divalent cation ionophore A23187, which bypasses the physiological receptor activation/signal transduction yet promotes a massive Ca^{2+} influx similar to that recorded during capacitation and sperm–ZP binding (see Rodriguez-Martinez, 2006). In either case, rates of AR-responsiveness were significantly and positively related to fertility (as 56 day NRR post AI in cattle, or as farrowing rates in pigs). This is illustrated in Fig. 19.2 for frozen–thawed spermatozoa from individual AI dairy bull sires (diamonds) with various AI fertility in the field (Januskauskas *et al.*, 2000a,b).

Spermatozoa from human males, boars and bulls evidently contain the hyaluronan (HA) receptor CD44 in their plasma membranes (Huszar *et al.*, 2003; Tienthai *et al.*, 2003; Bergqvist *et al.*, 2006) and should thus bind to solid state HA depots (PICSI Sperm Selection device, USA) (Huszar *et al.*, 2007). The HA binding seems to trap only mature spermatozoa that are able to react with the HA and exhibit some degree of hyperactivated-like motility pattern. This technique was primarily designed to select the best spermatozoa for ICSI in humans,

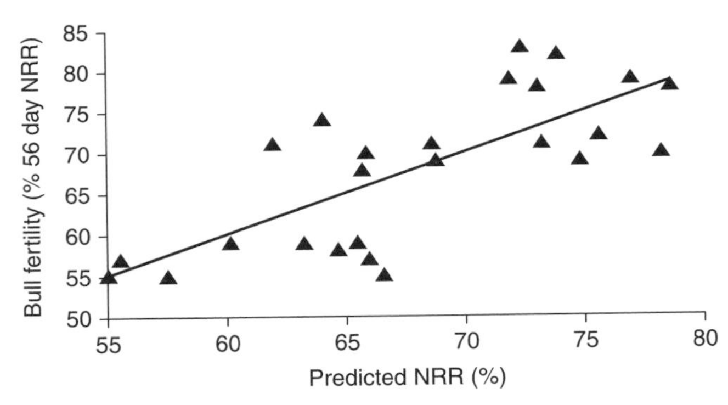

Fig. 19.2. Prediction of fertility as 56 day non-return rates (NRRs), trend line, via *in vitro* Ca-ionophore induction of the acrosome reaction (AR) in frozen–thawed spermatozoa for artificial insemination (AI) from individual dairy bull sires (triangles) with various AI fertility in the field ($r = 0.60–0.84$). Modified from Januskauskas A *et al.* (2000a) *Theriogenology* **53** 859–875.

but was later used for stallion spermatozoa (Colleoni *et al.*, 2011). The spermatozoa assessed were not only mature, but even had intact chromatin (Razavi *et al.*, 2009; Yagci *et al.*, 2010) – although they were not selected for normal morphology (Petersen *et al.*, 2010). However, the number of spermatozoa that have been explored with this method is small, and relationships between sample test results and fertility are yet to be established. An alternative to solid-state HA depots is HA-containing media, in which the results of sperm selection are quite similar, but where the fluid allows for higher sperm numbers (Parmegiani *et al.*, 2010).

Spermatozoa have also been explored for their ability to bind to the epithelium of the oviduct *in vitro*, an ability present *in vivo*, when potentially fertile spermatozoa reside in the oviductal sperm reservoir (Rodriguez-Martinez *et al.*, 2005). The binding of spermatozoa to homologous oviductal epithelial cells *in vitro* has been found to prolong their life (Lefebvre and Suarez, 1996), thus indicating that sperm co-culture with oviductal explants might mimic the capacity of a semen sample to colonize the tubal reservoir and would, therefore, indicate its potential fertilizing capacity, because uncapacitated spermatozoa are preferentially bound (Fazeli *et al.*, 1999). It should be noted here that the outcomes of such tests only provide reliable correlations when high-quality sperm samples are tested (De Pauw *et al.*, 2002), and have shown only marginal relationships with fertility (Waberski *et al.*, 2005, 2006).

The effective binding of the spermatozoon to the ZP is a critical step in the process of fertilization. The binding is species specific, only elicited by capacitated spermatozoa and precedes the AR. As ZP binding can easily be performed *in vitro*, several sperm–ZP binding tests have been designed since the 1980s, either using whole ZPs (oocytes), or hemi-ZPs (cleaved oocytes) (Rodriguez-Martinez, 2006). Even though the findings from ZP-binding tests have yielded significant correlations with AI fertility in pigs (Lynham and Harrison, 1998; Ardon *et al.*, 2005) and bulls (Zhang *et al.*, 1998), the biological significance of this assay is questioned, mainly because physiological sperm capacitation, and hence the AR, do not involve all spermatozoa at a given time. A representation of the relationship of a ZP binding (in whole oocytes) to fertility, depicting the presence of false positive and false negative cases, is shown below (Fig. 19.3).

An alternative method tested in many laboratories is the ability of presumably capacitated spermatozoa to penetrate into homologous oocytes *in vitro*, the so-called oocyte penetration test, under conditions of *in vitro* oocyte maturation (IVM) (Martínez *et al.*, 1993; Henault and Killian, 1995; Brahmkshtri *et al.*, 1999; Oh *et al.*, 2010). This test allows for

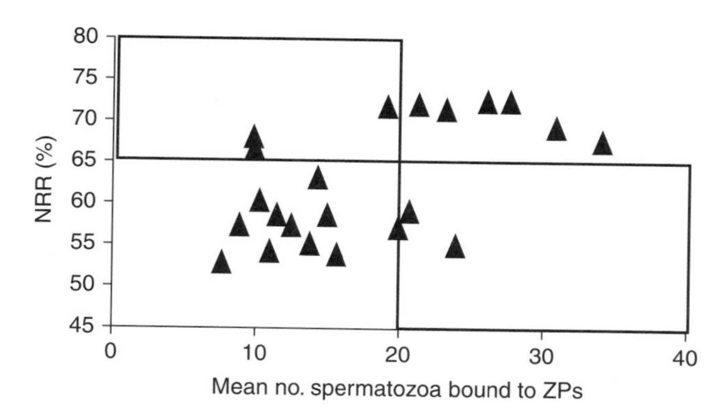

Fig. 19.3. Relationship between numbers of frozen–thawed bull spermatozoa bound to homologous zona pellucidae (ZPs) and the *in vivo* fertility (as 56 day non-return rates, or NRRs) of individual dairy bull sires (triangles); $r = 0.50$, and lines indicate threshold ZP-bound sperm numbers. Note the presence of false positives in the bottom right quadrant and false negatives in the top left quadrant. Modified from Zhang BR *et al.* (1998) *International Journal of Andrology* **21** 207–216.

screening of degrees of polyspermia, numbers of accessory spermatozoa in the ZP, decondensation ability (male pronucleus formation) and the testing of several males simultaneously (provided that each semen sample is loaded with a different fluorophore), and it also relates to fertility (Henault and Killian, 1995; Gadea, 2005). The method is simple and requires little equipment, though (i) the oocytes maturity level rules, and (ii) not all spermatozoa at a given time are capacitated and prompted to engage in penetration of the ZP. These basic timing hurdles might explain some of the variation in penetration rates found, rather than mirroring possible fertility differences among sires.

Different end points in fertilization and subsequent early embryo development can be determined using IVF, so spermatozoa of various species have been investigated for relationships between *in vitro* outcome/s and field fertility when the same semen (or males) was used for AI. In most cases, the approaches were retrospective, i.e. the fertility levels of the semen or males used were already known; only a few tests were really done prospectively, i.e. the semen was coded, used *in vitro* and the outcomes used to calculate an '*in vitro* fertility' that was subsequently compared with the 'real fertility' in the field. It soon became apparent that significant relations appeared even when the semen used had wide variations in fertility, and results could be accepted as reliable when the conditions for IVF were of a certain stringency and stability, i.e. low sperm numbers used, there were the same levels of success in a control line over time, there were no major differences between cleavage and morula/blastocyst yields, etc. Unfortunately, in most studies (see Rodriguez-Martinez, 2007b), there were only low-to-medium relations with fertility (Table 19.1), and these were lowest for morula/blastocyst rates.

Have We Found Biological Relationships with Fertility that Have Diagnostic and Prognostic Value?

As it can be seen in Table 19.1, the outcomes of most *in vitro* analyses of *individual* sperm attributes and/or functional assays (i.e. from motility to embryo development assays) show variable relationships with fertility, and these are not always statistically significant. This should not come as a surprise, because the ejaculate is not a suspension of a homogenous population of spermatozoa. Spermatozoa within an ejaculate differ morphologically, they move differently and their genome and transcriptome differ between cells. All cells, even those that were 'siblings' during spermatogenesis, are in principle different. For that reason, they neither display the same ability to interact with other cells nor have the same fertilizing capacity. The ejaculate is, therefore, considered to be a heterogeneous accumulation of various sperm cohorts – products of different spermatogenic waves that had matured along the ducti epididymides and were stored in their caudae as different cell sub-populations until mixed in the ejaculate. Moreover, the use of functional *in vitro* methods (ZP binding, IVF, etc.) bypasses some of the functional deficits of spermatozoa and aggravates the degree of uncertainty for relations with fertility *in vivo* (Barratt *et al.*, 2011).

Differences are apparent between the assays and/or attributes tested. For instance, membrane integrity evaluated via FC or fluorometry appeared to be more closely related to semen fertility than was sperm motility. This does not imply that some of the testing can be eliminated. For example, sperm motility is particularly difficult to eliminate as it embraces a number of attributes, including intactness of the plasmalemmae; it is also a simple and inexpensive procedure. Furthermore, it is important to remember that the significance of any differences encountered is a direct function of the number of spermatozoa examined (sample power), such as might be exemplified by the difference between assessing 100 spermatozoa or thousands per sample.

Prognostic strength can be also gained by carrying out different assays, even when this implies that some attributes are measured more than once. There is little inherent risk in such an approach, as the spermatozoa tested with one assay are different from all others; thus, a battery of tests is always advantageous (Rodriguez-Martinez, 2003). In line with this approach, several groups have combined the results of *in vitro* tests of the same semen samples in multiple regression analyses (see Rodriguez-Martinez and Barth, 2007), which have yielded higher correlations with fertility, even when they are retrospective (see Table 19.1). Calculation of 'predicted' fertility, in which the outcomes of various methods of semen evaluation *in vitro* were combined in multivariate analysis before the fertility of the 'donor' males was tested in the laboratory or the field, has proven valuable (Zhang *et al.*, 1999; Gil *et al.*, 2005; Ruiz-Sanchez *et al.*, 2006). Figure 19.4 depicts how *in vitro* data could be predictive of *in vivo* fertility when the results of various *in vitro* methods of evaluation of frozen–thawed semen from unproven young dairy AI bull sires (11–13 months of age) were combined to calculate their expected (potential) field fertility.

A strong relationship ($r = 0.90$) was found between predicted and real fertility. This approach enabled identification of sub-fertile bulls, whose expected and real fertility was below the limit considered for sub-fertility (a 62% non-return rate); in contrast, other young bulls that had been predicted to have satisfactory fertility had non-return rates of ≥65%. Identification of sub-fertile sires have been obtained with other bull- (Januskauskas *et al.*, 2000b; Hallap *et al.*, 2004) and boar-stud populations (Ruiz-Sanchez *et al.*, 2006). Interestingly, most sperm parameters (and to some extent even fertility) appeared to be maintained over the functional lives of the sires, provided no pathologies are acquired between measurements (Zhang *et al.*, 1997, 1998; Hallap *et al.*, 2005a, 2006a). However, intrinsic variation between ejaculates within sire was always present, which requires analyses of many ejaculates.

The spermatozoa we examine are not only different from each other, they are differently processed. They are either collected in different parts of the ejaculate (from those species with fractionated ejaculate) or they are extended, chilled, frozen, or even subjected to sperm sorting, or further prepared for IVF. All of these different scenarios result in spermatozoa with added differences in the intactness of the attributes that they need to interact with the environment (the female), including fertilization. This superimposed variation confounds further our capacity to discern the intrinsic value of *in vitro* outcomes for the estimation of potential fertility from iatrogenic implications. In other words, we need to apply a holistic view when examining a subsample of semen.

Sperm Selection: The Panacea?

Ejaculated spermatozoa are selected during their transport through the female genital tract. The proportion of abnormal spermatozoa and

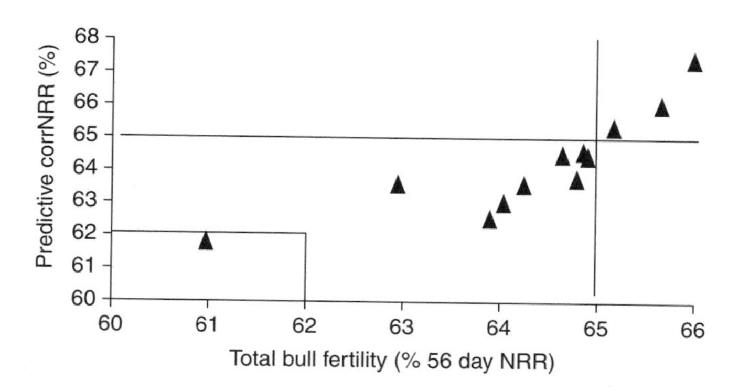

Fig. 19.4. Predictive power (as % predictive corrNRR) of *in vitro* sperm assessment (including *in vitro* fertilization, IVF) when all examinations were combined for a series of individual dairy bull sires (triangles) and related to the total fertility (as 56 day non-return rate, NRR) of the same frozen semen batches examined in the field. Lines mark boundaries for acceptable fertility. Modified from Zhang BR *et al.* (1999) *International Journal of Andrology* **22** 253–260.

of those missing some relevant sperm attribute are reduced in concert with reduced sperm numbers by the time spermatozoa approach the tubal segment in which fertilization takes place (Rodriguez-Martinez, 2007a). Which spermatozoa are selected? How are they selected? In general, selection *in vivo* should account for all attributes, known and suspected, which we consider spermatozoa to require for a 'safe' set of interactions with the female genitalia. Among these would obviously be a normal genome and morphology, indicating proper spermatogenesis, as well as a plasma membrane capable of interacting with the environment, indicating normal sperm maturation. Last, but not least, spermatozoa should display motility within a certain frame, be reactive and arrive in sufficient numbers at the oviduct in due time. We assume that a certain sub-population of spermatozoa with all these intact attributes survives the journey, thus decreasing heterogeneity for some attributes, yet leaving others, such as differential reactivity during capacitation and the presence of a haploid, unique genome and a transcriptomic load. Here, a degree of chance is most likely also involved such that a spermatozoon (or a few of them) is present at the right place and the right time to interact with the oocyte and its vestments.

Can we mimic sperm selection in the oviduct? Probably so, but only to some extent, because we still do not know if or how the female modulates all of the steps involved behind the restrictions *in vivo*. We do know, however, that ejaculated spermatozoa can be separated *in vitro* into sub-populations by differences in motility kinetics (Abaigar *et al.*, 1999; Cremades *et al.*, 2005; Peña *et al.*, 2005b, 2006; Holt *et al.*, 2007; Martinez-Pastor *et al.*, 2011), membrane integrity (Perez-Llano *et al.*, 2003), head morphometry (Núñez-Martinez *et al.*, 2005; Peña *et al.* 2005b, 2006), sustainment of handling and cryopreservation (Gil *et al.*, 2005; Peña *et al.*, 2006; Roca *et al.*, 2006), and the ability to respond to specific stimuli such as bicarbonate exposure (Holt and Harrison, 2002; Gadella and van Gestel, 2004; Holt and Van Look, 2004; Satake *et al.*, 2006), or to specific proteins isolated from the oviduct (Coy *et al.*, 2009). Several of these studies were based on individual sperm attributes, and some considered the fluids surrounding the spermatozoa, their composition and even their rheological features, with viscosity being among these. The ultimate goal would be to find a method that, by mimicking what occurs within the female genital tract, could preselect spermatozoa with primary competence for fertilization (Rodriguez-Martinez, 2006).

Methods for the *in vitro* separation of spermatozoa for robustness have been described (Rodriguez-Martinez *et al.*, 1997), with a major focus on the fact that the spermatozoa in a normal semen sample usually show a typical progressive, innate linear motility. This linearity is used to surpass natural barriers such as the cervix, where they migrate along sialic acid-rich mucus-filled deep furrows. Assays exploiting the fact that spermatozoa have an innate tendency to migrate into most media (often culture media, but also more complex preparations of varying viscosity) that are brought into contact with a semen sample (swim-over, swim-down, swim-up) have been used to mimic *in vivo* events. This simple procedure has proven useful in selecting for sperm motility and membrane integrity, essential parameters for fertilization (Rodriguez-Martinez *et al.*, 1997), and also proven valuable for fertility prognosis, because the number of viable spermatozoa post swim-up reflects the innate fertilizing capacity of the semen sample tested (Zhang *et al.*, 1998; Hallap *et al.*, 2005a,b, 2006b).

Manipulations of the viscosity, often associated with additives, of the swim-up media used have improved the results obtained, basically because they mimic the *in vivo* situation (Rodriguez-Martinez, 2007b; Hunter *et al.*, 2011). The use of oestrous cervical mucus, even heterologous mucus (e.g. bovine on ram spermatozoa), has contrasted sperm linearity and, more importantly, its relation to sperm colonization of tubal sperm reservoirs *in vivo* and to fertilizing capacity *in vitro* (Cox *et al.*, 2002, 2006; Robayo *et al.*, 2008). Hyaluronan, a component of the oviductal fluid and the cumulus cell cloud (Rodriguez-Martinez *et al.*, 2001), has proven to be an excellent additive as it increases viscosity to the right proportion *in vivo* and also selects for fertilizing capacity (Shamsuddin and Rodriguez-Martinez, 1994). As follow-ups, artificial (hyaluronate-based, not sialic acid-based) cervical mucus has also been tested, albeit with less discriminatory results (Al Naib *et al.*, 2011).

Novel methods have recently been developed using alternative multiple microfluidic flow streams for sperm self-migration that allow the sorting of motile spermatozoa (Schuster *et al.*, 2003; Chung *et al.*, 2006; Wang *et al.*, 2011). Although not suitable for the isolation of large sperm numbers, these latter methods appear to be promising when adapted for IVF, where low, quasi-physiological sperm numbers are co-incubated with oocytes (Suh *et al.*, 2003, 2006). In sum, such self-migration procedures appear to be relevant because they appear to select spermatozoa in a similar fashion to that which occurs *in vivo*. They are, however, unsuitable for procedures other than IVF. Moreover, they apparently seem to initiate capacitation-like phenomena (Shamsuddin and Rodriguez-Martinez, 1994).

Other methods have, therefore, been suggested as substitutes for use in farm animals, where a higher output of an intact population is selected (Rodriguez-Martinez *et al.*, 1997; Somfai *et al.*, 2002; Samardzija *et al.*, 2006). Examples of these methods are centrifugation through columns of adherent particles (Sephadex or glass wool) (Januskauskas *et al.*, 2005) or through discontinuous density gradients of silane-coated silica spheres (Rodriguez-Martinez *et al.*, 1997). In most species tested so far, centrifugation through a single column of species-specific formulations of colloid (based on silane-coated spheres, the so-called SLC method) has proven successful in harvesting the most robust spermatozoa from any (raw or serially processed) semen suspension, (Morrell and Rodriguez-Martinez, 2009, 2010). The selective power, which is clearly related to species differences in osmolarity and density of the colloid (Morrell *et al.*, 2011a), is equally present in different volumes of semen and sperm preparations, so that semen from species with large ejaculates can be preselected for their best sperm numbers, including a selection for morphology, motility, membrane integrity, chromatin integrity (Macías García *et al.*, 2009a,b; Morrell *et al.*, 2009a,b, 2010) and fertilizing capacity, both *in vitro* (Thys *et al.*, 2009; Colleoni *et al.*, 2011) and *in vivo* (Morrell *et al.*, 2011b).

Sperm-'omics'

The exponential advances in analytical molecular biochemistry, also named the 'omics revolution', has even involved spermatology. The 'omics revolution refers to the study of genes (genomics), and the function of their products (functional genomics) either as RNA transcripts (transcriptomics), proteins (proteomics) or the various metabolites (Aitken, 2010), thus opening possibilities for securing inventories of lipids, proteins, metabolites and RNA species and, hopefully, determining how their presence or changes in them relate to cell function. Such endeavours are made possible by the application of DNA sequencing, DNA microarrays, mass spectrometry and protein arrays, which, when proper interfaces and bioinformatic tools are available, may provide cues for sperm function. Semen is an excellent biological sample to be studied, because it is easily available and contains a majority of a pure cell type, the spermatozoon, on which all these studies can be performed (Carrell, 2008). Even SP provides a good sample, as proteomics is particularly applicable to it for diagnostics.

Sperm genomics

During fertilization, spermatozoa provide a haploid genome with intact coding regions and regulatory regions for essential genes, copies that must be intact (i.e. should not contain single- or double-stranded DNA breaks). For sperm diagnosis, examination of the DNA is of the utmost importance and accordingly has been discussed above. Here, it is noteworthy that not only DNA quality but also the packaging of the paternal genome (epigenome) is essential for normal fertility and embryonic development (Miller *et al.*, 2010; Jenkins and Carrell, 2011).

Sperm epigenomics

In addition to its direct genetic material, the spermatozoon also contributes epigenetic components (i.e. changes other than in DNA coding that can alter or regulate gene expression) that can affect early embryo development (Wu and Chu, 2008; Hales *et al.*, 2011). These include a functional centrosome for the mitosis of the newly built zygote to occur, proper packaging of the chromatin with protamines, modification of histones and the imprinting of genes. In addition, the fertilizing spermatozoon provides messenger RNAs (mRNAs) and microRNAs (miRNAs) (see below), which all contribute to the embryonic transcriptome and further regulate embryonic gene expression. The provision of a proper paternal epigenetic contribution to embryo development requires several forms of sperm epigenetic information to regulate the driving of genes towards either continued silencing or activation upon delivery to the oocyte (Jenkins and Carrell 2011). Processes such as DNA methylation, selective histone retention, sperm-specific histones with tail modifications, other chromatin-associated proteins, perinuclear theca proteins, and organization of the DNA loop domain by the sperm nuclear matrix and of sperm-born RNAs, are all included (Pacheco *et al.*, 2011; Yamauchi *et al.*, 2011).

Sperm transcriptomics

Transcription is issued by complementary DNAs (cDNAs) and their core promoters, exons and polyadenylation sites. Most protein-coding genes and a proportion of non-coding RNAs produce alternative transcripts with different exon combinations. In other words, genome-wide transcription generates a complex population of transcripts. Therefore, transcripts (not genes) are now considered the operational unit of the genome. Over the years, the protein-coding transcripts have received most attention, but more non-coding RNAs than coding transcripts have been identified. These non-coding transcripts are either housekeeping non-coding RNAs (such as ribosomal or rRNA, transfer or tRNA, small nuclear or snRNA and small nucleolar or snoRNA) or regulatory non-coding RNAs (either long non-coding or lncRNAs, short or miRNAs, small interfering or siRNAs and Piwi-associated or piRNAs) (Ponting *et al.*, 2009).

Microarray and serial analyses of gene expression assays of spermatozoa from several species have concluded that sperm mRNAs are either simple remnants of the process of spermatogenesis or are those providing the zygote with a unique set of paternal mRNAs (Krawetz *et al.*, 2011). They also seem to provide variable array signals, which correspond to the inherent variability among spermatozoa within an ejaculate, and between ejaculates and individuals. Despite this, the use of suppressive-subtraction-hybridization (Lalancette *et al.*, 2008), or of global RNA profiles of spermatozoa from fertile and infertile men (García-Herrero *et al.*, 2010), or bulls (using a cDNA collection on DNA microarrays), with different NRRs could lead to the identification of transcripts (protein kinase and ADAM5P) associated with high motility (Bissonnette *et al.*, 2009). More recently, when examined with Affymetrix bovine gene chips, high- and low-fertility bulls showed significant differences in specific transcripts associated with fertility (Feugang *et al.*, 2010). It is foreseen that microarrays will be a part of future andrological diagnostics.

Sperm proteomics

The study of protein products expressed by the genome has dramatically expanded over the past decade, owing not only to multidisciplinary methodological and instrumental developments, but also to the central role of protein interactions in cell function (Cox and Mann, 2007; Brewis and Gadella, 2010; Baker, 2011; du Plessis *et al.*, 2011). Because spermatozoa are so highly differentiated, they are advantageous cells for studying the proteomics of specific compartments, such as the membrane, which is the area of major importance for their role in interacting with the surroundings and the oocyte

(Arnold and Frohlich, 2011). However, despite this methodological development, proteomic studies of spermatozoa are still limited (Oliva *et al.*, 2009), mostly owing to difficulties in separating spermatozoa from the other cells usually contained in semen, something that has been partly solved by use of sperm separation techniques (see above) of FC sorting.

When these pre-separation issues were addressed, it led to the establishment of comprehensive sperm protein databases (Naaby-Hansen *et al.*, 1997; Duncan and Thompson, 2007; de Mateo *et al.*, 2011), with the numbers of proteins and fragments included exponentially increasing over time towards several thousands. The proteins identified so far cover the expected spectrum of function (from energy production to cell recognition), although few are accurately linked to (in)fertility, most of them being enzymes (Novak *et al.*, 2010a,b). This has called for other methods of sperm membrane isolation so that membrane domains relevant to capacitation or binding (either to the ZP or the oolemma) are more closely examined (Brewis and Gadella, 2010).

Attempts to determine how these interactions between structure and function, and further, between these and their relative impact on fertility function, require direct studies of protein function. It may even be necessary to bypass genomic studies and focus on proteomics, considering the difference between the number of protein-encoding genes (20,000–25,000), and the number of proteins present in mammals (which is 10–20 times higher). The presence of alternative splicing or posttranslational modifications in proteins (such as glycosylation, phosphorylation, proteolytic processing, lipid modification, etc.) helps to explain these basic numerical differences. Interestingly, fluids such as semen in the context of protein identification and relation to function appear to be really complex, ranging from few relevant proteins in spermatozoa towards tens or hundreds in SP (Calvete and Sanz, 2007; Duncan and Thompson, 2007). Moreover, the fact that ejaculation is fractionated in many species adds a new dimension to the action of SP proteins (and their interactions) on sperm function and on the reactivity shown by females.

Seminal Plasma (SP): The Key for Fertility?

A semen sample reflects the status of the testes, the excurrent ducts and the accessory sexual glands, and therefore we include the study of all components during the andrological examination. Semen is classically defined as a cell suspension of spermatozoa and other cells (lining cells of the excurrent ducts, epididymis or accessory glands, migrating leucocytes, and even spermatogenic cells), and of cell vesicles (epididymidosomes and prostasomes), that are suspended in an SP vehicle. The SP comprises the combined contributions of the fluids of the cauda epididymides and the accessory sexual glands. Species of mammals differ in the presence and size of the different accessory sexual glands, and this obviously leads to variations in their relative contribution to semen composition and volume, particularly for the SP. In some species, the SP represents up to 95–98% of the total semen volume (Mann and Lutwak-Mann, 1981).

Although samples and collection methods vary between species, the major difference is whether the sample is bulk or fractionated, as in many species (e.g. the stallion, canine, pig) the ejaculate is voided in spurts (also called jets), which have differing compositions, owing to the sequential emission and/or emptying of the secretions of the sexual accessory glands (Mann and Lutwak-Mann, 1981), a fact that increases semen heterogeneity. The composition and character of SP differs not only among species and among and within individuals, but even, for many species, within an ejaculate. Even though we have seen how relevant such variation is when judging 'sperm quality', we do not usually examine the SP, despite the availability of assays for specific markers (neutral α-glucuronidase for epididymal fluid, phosphatases or zinc levels for prostate fluid, or fructose for seminal vesicles (Mann and Lutwak-Mann, 1981).

Instead, most andrologists follow the classical view that the SP is no more than a vehicle for spermatozoa. In fact, SP is even regarded as deleterious for some purposes, such as sperm storage, and so it is largely removed and replaced by extenders for further handling or freezing (Rodriguez-Martinez and Barth, 2007).

However, simple components of the SP do seem to play important roles: bicarbonate modulates sperm motility and/or destabilizes the plasma membrane; zinc modulates chromatin stability. While these and other biologically active components and several hormones (e.g. oestrogens, prostaglandins) can be re-added to extenders, other SP components are more difficult to isolate and thus be added. SP proteins, which often make up to 40–60 g/l of ejaculate (in the boar, 30–60 g/l), differentially adsorb to different domains of the plasma membrane and, by affecting its capacity to interact with the environment (epithelial lining, female secretions, ZP), modulate sperm function. Moreover, components of the SP directly affect the epithelial lining, eliciting diverse signalling to the female, and its immune system in particular (reviewed by Rodriguez-Martinez *et al.*, 2008, 2009, 2010; Palacio and Martinez, 2011).

Proteomics of the Seminal Plasma (SP) in Relation to Sperm Function and Fertility

The main proteins of the SP belong to one of three groups – those carrying fibronectin type II (Fn-2) modules, spermadhesins or cysteine-rich secretory proteins (CRISPs) – and their bulk is, in most species, of vesicular gland origin (Kelly *et al.*, 2006). Table 19.2 provides an overview of the proteins present in the SP of domestic and other species with outlines of their presumed functions.

As can be seen from Table 19.2, most semen proteins in ungulates (the boar, stallion, bull, buck) are Fn-2 (originally described in the bull SP as BSPs – bovine heparin-binding proteins; Manjunath, 1984). They are now regrouped as 'binder sperm proteins' and, basically, appear in each species studied, binding to plasmalemmal phospholipids, lipoproteins, female carbohydrates and GAGs, etc., in relation to sperm capacitation (Manjunath *et al.*, 2009). While CRISP proteins are abundant in the stallion, spermadhesins dominate in pigs, where they fall within three main groups: the two heparin-binding proteins (HBPs) – alanine-glutamine-asparagine proteins (AQN-1 and -3) and alanine-tryptophan-asparagine proteins (AWNs), and the non-heparin binding porcine seminal plasma proteins (PSP-I and PSP-II) (Töpfer-Petersen *et al.*, 1998; Kelly *et al.*, 2006; Calvete and Sanz, 2007). Spermadhesins are multi-functional 12–16 kDa glycoproteins that attach to the sperm plasma membrane to various degrees at locations from the testis to the ejaculate. Collectively, they show multiple effects on spermatozoa, probably via a differential coating on the sperm membrane. AWN or AQN monomers bind directly to membrane phospholipids, followed by a secondary coating by aggregated HBPs, which stabilizes the plasma membrane over the acrosome before capacitation, mediates building of the sperm reservoir (Calvete *et al.*, 1997) and promotes sperm–ZP interaction (Rodriguez-Martinez *et al.*, 1998); these proteins are thus important to sustain sperm handling (Caballero *et al.*, 2008). PSP-I and PSP-II account for >50% of all SP proteins and not only bind to the sperm surface (Caballero *et al.*, 2006), but also display protective action on highly extended and processed spermatozoa (García *et al.*, 2006, 2008; Caballero *et al.*, 2008), including protection against cold shock (Mogielnicka-Brzozowska *et al.*, 2011). Moreover, the PSPs exhibit clear immunostimulatory activities *in vitro* and *in vivo* (see Rodriguez-Martinez *et al.*, 2009, 2010).

In the first fraction of the boar ejaculate, where spermatozoa are present, other proteins, presumably of epididymal origin, are present, such as lipocalins and an inhibitor of acrosin/trypsin (Rodriguez-Martinez *et al.*, 2009). In other species, such as the stallion, there is a similar disposition of the main SP proteins. Short Fn-2 type proteins dominate (70–80% of total protein), named HSP-1 and -2 (or SP-1 and -2). These are similar to the BSPs, and modulate capacitation (reviewed by Töpfer-Petersen *et al.*, 2005). HSP-3 is a CRISP protein associated with

Table 19.2. Overview of proteins present in the seminal plasma (SP) of domestic and other species, with their presumed functions. See text for further details.

	Protein family				Other proteins/peptides		
Species	Sperm binder proteins (Fn-2, fibronectin type II; BSPs, bovine heparin-binding proteins) (14–30 kDa)	Cysteine-rich secretory proteins (CAP-superfamily, CRISP) (20–25 kDa)	Spermadhesins (12–16 kDa)	Functions ascribed	Enzymes (phosphatases, aminopeptidases, glycosidases, hyaluronidase, PG synthase, esterases, lipases, etc.), specific peptides and growth factors, cytokines/ chemokines	Functions ascribed	Reference
Bovine	BSP-A1, A2, PDC-109		aSFP, ZB	Regulation of capacitation and sperm-binding capacity to environment and ZP (zona pellucida)	Enzymes, clusterin, osteopontin, cytokines/ chemokines	Sperm survival, fertility-associated factors, immunological modulation	Manjunath *et al.*, 2007 Calvete and Sanz, 2007 Mann and Lutwak-Mann, 1981 Jenne *et al.*, 1991 Gwathmey *et al.*, 2006 Vera *et al.*, 2003
Caprine	GSP-14, 15, 20, 22 kDa				Enzymes, cytokines/ chemokines	Sperm survival, immunological modulation	Villemure *et al.*, 2003 Druart *et al.*, 2013
Equine	HSP-1, 2, 8; EQ-12 (1–3)	CRISP1–3	HSP-7	Regulation of capacitation ZP binding (HSP-7)	Enzymes, cytokines/ chemokines	Sperm survival, regulation of coagulation, antimicrobial, immunological modulation	Töpfer-Petersen *et al.*, 2005 Alghamdi *et al.*, 2009

Continued

Table 19.2. Continued.

	Protein family				Other proteins/peptides		
Species	Sperm binder proteins (Fn-2, fibronectin type II; BSPs, bovine heparin-binding proteins) (14–30 kDa)	Cysteine-rich secretory proteins (CAP-superfamily, CRISP) (20–25 kDa)	Spermadhesins (12–16 kDa)	Functions ascribed	Enzymes (phosphatases, aminopeptidases, glycosidases, hyaluronidase, PG synthase, esterases, lipases, etc.), specific peptides and growth factors, cytokines/ chemokines	Functions ascribed	Reference
Ovine	RSVP-14, 15, 20, 22, 24		RSP	Resistance to cold shock Ion channel regulation	Enzymes, cytokines/ chemokines	Sperm survival, immunological modulation	Barrios *et al.*, 2005 Bergeron *et al.*, 2005 Palacio and Martinez, 2011
Porcine	pB1	CRISP1–3	AWN, AQN-1 and 3 PSP-I, PSP-II	Regulation of capacitation, sperm binding capacity to environment and ZP, modulation of leucocyte recruitment	Enzymes, lipocalins, inhibitor of acrosin/ trypsin, prostate specific antigen (PSA), cytokines/ chemokines	Sperm survival, fertility-associated factors, antimicrobial, immunological modulation	Rodriguez-Martinez, *et al.*, 2009 Rodriguez-Martinez *et al.*, 2011
Other (rodents, human)		CRISP1–4		Modulation of capacitation, ZP binding	Enzymes, PSA, semenogelins I–II, lactoferrin, protein C synthase, PSP-94, β-inhibin-β-microseroprotein, cytokines/ chemokines	Regulation of coagulation, antimicrobial, immunological modulation	Koppers *et al.*, 2011 Duncan and Thompson, 2007 Robertson, 2005

fertility (Hamann *et al.*, 2007), perhaps via its role as a selective protector against polymorphonuclear leucocyte (PMN)–cell binding (Troedsson *et al.*, 2010). HSP-7 is the only equine member of the spermadhesin family, and like its porcine homologue AWN-1, shows ZP-binding activity (Calvete *et al.*, 1994; Reinert *et al.*, 1996). Also in stallions, the initial ejaculate fractions contain acrosin inhibitor and prostate specific antigen (PSA), or kallikrein-like proteins (as HSP-6 and HSP-8 representing isoforms; Kareskoski *et al.*, 2011). Peptides and diverse enzymes are also components of the SP, albeit most of them are either fragment products of SP proteins, sperm-associated peptide hormones or major enzyme groups (see Mann and Lutwak-Mann, 1981; Duncan and Thompson, 2007). The epididymal lipocalin-type prostaglandin D2 synthase, present in boar and stallion SP, is an enzyme related to male fertility in several species (Barrier-Battut *et al.*, 2005, Novak *et al.*, 2010a,b). Lastly, the SP of most species contains protein compounds similar to those present in blood plasma, such as pre-albumin, albumin, α-, β- and γ-globulins, transferrin, complement factors and differential amounts of cytokines and chemokines.

In summary, SP proteins modulate several essential steps preceding fertilization – the regulation of capacitation, the establishment of the oviductal sperm reservoir, the modulation of the uterine immune response and sperm transport through the female genital tract, as well as gamete interaction and fusion. Interestingly, individual proteins from the same family appear to function in a species-specific manner. Differences in their structure, relative abundance and patterns of expression appear to determine the species-specific effects of homologous proteins (Calvete and Sanz, 2007).

Sperm transport and survival

Following mating, some spermatozoa are transported within minutes from the site of deposition in the female towards the oviduct, owing to the concerted motility of the female tract muscle (Suarez, 2008a). In animals with fractionated ejaculation, the spermatozoa are bathed in different fluids, such as the caudal epididymal fluid and the accessory gland secretion that is vented simultaneously with the corresponding spurt of ejaculation. As mentioned previously, the initial spurts of the sperm-rich fraction are acidic, and sperm proteins have been demonstrated to link themselves to acidic polysaccharides such as those in the secretions of the cervix, uterus and even the oviduct (see Rodriguez-Martinez et al., 2001). Such interactions have proven to be analogous to the binding of SP proteins; for instance, bull and stallion SP proteins have been shown to inhibit the interaction of the corresponding sperm proteins with acidic polysaccharides (Liberda *et al.*, 1998). SP proteins differentially affect sperm survival post ejaculation, and those present in the last ejaculate fractions (which are of seminal vesicle origin) have a more pronounced negative effect, perhaps in relation to the extensive presence of several proteins. In the boar, the primary secretion in the sperm peak spurts acts differently in promoting longer sperm survival (Rodriguez-Martinez *et al.*, 2008, 2009). The SP of many species contain prostasomes, known by their ability to fuse with spermatozoa and provide a mechanism for protein and lipid transfer (Burden *et al.*, 2006), thereby enhancing sperm motility and the stability of the sperm membrane.

Protein interaction with the female genital tract and the oocyte

SP proteins coat the sperm membrane during ejaculation, producing structural changes to the sperm plasma membrane in relation to capacitation, ZP recognition and fertilization. For example, AWN follows spermatozoa up to the ZP (Rodriguez-Martinez *et al.*, 1998), inhibiting sperm capacitation (Dapino *et al.*, 2009; Vadnais and Roberts, 2010), an effect that is lost when the proteins are removed from the sperm surface (Calvete *et al.*, 1997). At the same time, it has been postulated that such an initial layer of proteins might provide an anchor for aggregated spermadhesins to coat the sperm surface (Töpfer-Petersen *et al.*, 2008), thus further stabilizing the plasmalemma and preventing premature acrosomal exocytosis. The heparin-binding AQN-3, the most prominent ZP-binding protein in boar spermatozoa, remains, for example, attached to the sperm surface after capacitation (*in vitro*) and can only be recovered from the aggregating raft area of the apical ridge of the sperm head (van Gestel *et al.*, 2005, 2007).

The deposition of semen, which is to be considered foreign by the female and is therefore prone to rejection, induces a relatively rapid pro-inflammatory immunogenic response in the endometrium in order to (i) cleanse the intrauterine lumen from foreign cells, proteins and eventual pathogens, (ii) remodel the endometrium for the descending embryo/s, and (iii) secrete cytokines and growth factors that facilitate development of the embryo and placenta. Such inflammation does not occur in the oviduct, where spermatozoa find a haven prior to fertilization (Rodriguez-Martinez *et al.*, 1990, 2001, 2005). Although the presence and infiltration of granulocytes are oestrogen dependent (Robertson *et al.*, 2009a), their migration to the surface epithelium and lumen can be elicited by pro-inflammatory soluble cytokines (Sharkey *et al.*, 2007) and SP glycoproteins (spermadhesins) (Rodriguez-Martinez *et al.*, 2010). This inflammatory response is transient, dissipating within hours in normal cases, and is followed by the activation by SP of a transient state of peripheral immune tolerance by the female towards paternal antigen-bearing spermatozoa or early embryos in the oviduct (an immuno-privileged area) or in the uterus (developing embryos/fetuses and their placentae) (O'Leary *et al.*, 2004; Moldenhauer *et al.*, 2009; Robertson *et al.*, 2009a). The SP of rodents, humans and pigs contains immunoregulatory molecules, including high concentrations (100–150 ng/ml, 5–8 times higher than in blood serum) of the potent immunosuppressive transforming growth factor-β (TGF-β), a member of the multifunctional cytokine TGF family (Loras *et al.*, 1999; Robertson *et al.*, 2002); TGF-β appears to be strongly related to the induction of maternal tolerance (O'Leary *et al.*, 2004; Robertson *et al.*, 2006, 2009b; Robertson, 2007).

SP proteins and their relation to fertility

SP proteomes have been assessed in relation to reproductive outcomes (either fertility levels or (in)fertility (Drabovich *et al.*, 2011; Milardi *et al.*, 2012) in several species, and SP proteins associated with high and low fertility in bulls (Killian *et al.*, 1993) have been isolated as osteopontin (OPN) and lipocalin-type prostaglandin D synthase (Gerena *et al.*, 1998; Cancel *et al.*, 1999). The latter is always present in the sperm-rich spurts of ejaculates in species that have fractionated ejaculation, including the pig. Figure 19.5 shows the differences between portions of the boar-ejaculated SP in terms of some of the SP proteins; note the presence of epididymal lipocalins and PSP-I in the sperm-peak portion (P1) and the predominance of the other spermadhesins in the rest of the ejaculate.

OPN has been related to fertility in pigs (IVF; Hao *et al.*, 2006, 2008) and stallions (Brandon *et al.*, 1999). Some SP proteins (SP-2, SP-3, SP-4 and clusterin) have been found in higher concentrations in stallions with low fertility scores (Novak *et al.*, 2010a). SP-1 is positively (Brandon *et al.*, 1999) or negatively (Novak *et al.*, 2010a) correlated with fertility and was suggested to be homologous to a bovine fertility-associated protein, probably OPN, described by Killian *et al.* (1993). Moreover, the abundance of CRISP3 in equine SP was positively correlated with 1st-cycle conception rates (Novak *et al.*, 2010a), suggesting that this protein family might have a role in fertility, as has been proposed for rodents and humans (Koppers *et al.*, 2011). The spermadhesin PSP-I (see Fig. 19.5), seems to be negatively associated with pig fertility (Novak *et al.*, 2010b). SP cytokine levels vary among males, and variation in SP contents of TGF-β lacks a straightforward relationship with fertility (Loras *et al.*, 1999; O'Leary *et al.*, 2011). However, a female could express different levels of endogenous cytokines (relevant for embryo survival) depending on exposure to SP from different males, which might then relate to the often-observed differences in embryo survival among sires (e.g. innate fertility) (Robertson, 2007, 2010).

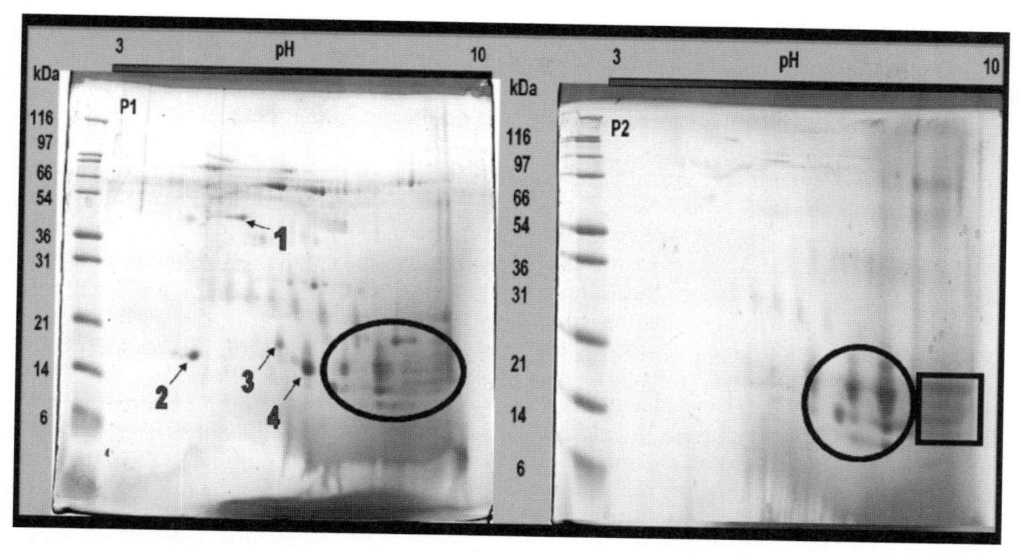

Fig. 19.5. Two-dimensional electrophoresis (2D-SDS-PAGE) of representative seminal plasma (SP) portions of the boar ejaculate depicting selected proteins. P1 is the first 10 ml of the sperm-rich fraction, also called the sperm-peak fraction; P2 is the rest of the ejaculate. (1) = actin; (2–4) = lipocalins (2, prostaglandin-D synthase (L-PGD-S)); (3–4) = epididymal secretory protein-1 (EPS-1). The circles indicate PSP-I (plasma protein I), and the square AQN-3 (alanine-glutamine-asparagine protein-3), AWN (alanine-tryptophan-asparagine proteins) and PSP-II. Lipocalins (2–4) are indicative of cauda epididymal contents. The P1 fraction shows the presence of PSP-I, but not of the rest of the spermadhesins, which are predominant in the P2 fraction.

What is Semen Analysis Going to Look Like in 20 Years?

I presume that the reader, having had the patience to reach these final pages, has come to an understanding that we have dealt with two major concepts: i.e. the *heterogeneity* of the sperm suspension and the *multitude of attributes* required for each spermatozoon to be fertile. The first recognizes the presence of large numbers of spermatozoa that are all genomically different, while the second recognizes the simultaneous presence of attributes (membrane integrity and its capacity to interact with and react to stimuli, to cite one) that have to be present alongside the functional life of a spermatozoon in order to participate in fertilization, at least theoretically. The latter also concerns the presence of a rich fluid surrounding the spermatozoa (the SP), whose signalling role in relation to fertility we are just starting to unveil. Male fertility, then, appears to be a multifactorial process in which *interactions* between the male and its semen and the female and its genital tract and oocyte/s are of central importance. Such a complex net of signalling pathways builds up an 'interactome' which must be studied in detail, with a primary focus on where it can fail and compromise fertility, but also as a diagnostic platform if defined on a steady-state base.

Therefore, any attempt to reach a diagnosis of potential (in)fertility or a prognosis for fertility levels requires a multi-parametric assessment. It is here that we need to differ on what are the intentions of the andrologist. For example, is it a diagnosis process for a male that includes its ability to breed successfully? Or is it the need to select among breeding sires with the intention of ranking these on the basis of fertility levels? One could argue that these needs are one and the same, i.e. *to prognose fertility*. We could also argue they are different. In the first case, the goal is to determine whether the male risks infertility or subfertility, a clearly *diagnostic situation* for the andrologist; this is often the stance of a BSE, especially in species where selection for breeding is not primarily related to fertility (i.e. horses). The second situation concerns breeding sires, which are often subjected to selection criteria – criteria that can include selection for fertility, a much more difficult goal to attain.

Summarizing the above considerations, the dominating scenario is that we need to achieve deeper knowledge on how spermatozoa behave and interact with their surrounding fluids, with these being the seminal plasma, extenders and fluids along the female genital tract. It is this interacting capacity that characterizes spermatozoa, with the most important variable being the intactness and fluidity of the sperm plasma membrane, which responds to stimuli and channels these stimuli into modifications of metabolism, kinematics or destabilization. Kinetics of response vary, but only those most intact spermatozoa are capable of responding appropriately. From a practitioner perspective then, the key issue is to be able to determine (i) what proportion of spermatozoa are most intact, (ii) whether these respond, and (iii) whether such screening is repeatable between ejaculates within the male.

The determination of total sperm numbers will continue to be a hallmark, and this must be quickly and accurately determined, preferably in concert with the determination of the kinematics (sperm motility) of a large number of spermatozoa (i.e. a large sample). We will probably see a further development of motility analysers based on other methods than digitalizing pictures of a microscopic field, as is currently the case with CASA. Alternatives based on a holographic on-chip imaging platform, although waiving the need for microscopes, are not suitable as relatively few spermatozoa are assessed (Su *et al.*, 2010). Other alternatives are more suitable, such as the Qualisperm™ software, which allows analyses of ten times more spermatozoa per sample, based on a technology for recording the velocity and displacement of the cells. Fluorescent probes are continuously being developed in an effort to map the intactness of the *entire* plasma membrane, which cannot be explored using conventional probes such as SYBR-14. A current example is the Live/Dead Fixable Red Dead Cell Stain Kit, an intracellular amide label (Plate 72a,b).

Even with flow cytometers becoming cheaper by the day, as well as becoming smaller and more user friendly, practitioners could benefit from the possibility of loading a supravital probe to determine live and dead spermatozoa, and observing it under a microscope that can read sperm numbers and register sperm speed and determine motility simultaneously, based on thousands of spermatozoa. Samples of these spermatozoa can even be subjected to a simplified TUNEL/PI assay (Plate 71d), which might provide a rather quick assessment of the degree of DNA damage (such as DNA fragmentation). Samples can also be stored for more detailed analyses, including transcriptome profiling (Lalancette *et al.*, 2009).

This scenario is not far away and it is relatively easy to undergo quality assurance for the instrumentation (such as a combination microscope or a bench FC), the procedures/reagents employed and the operators. The variables that still dominate here are sperm numbers, motility, and membrane and DNA intactness. A technique for monitoring DNA fragmentation in spermatozoa, the SCD Halo (Plate 71c) is now available for microscopy (Enciso *et al.*, 2006). In the SCD Halo test, semen samples can be incubated over time to monitor the dynamics of DNA fragmentation based on repeated observations; this contrasts with the current dominant tests using SCSA or TUNEL, which are performed in dedicated laboratories. Even more interesting is the presence of statistical associations between the outcome of these techniques and the use of a novel assay using birefringence (double refraction) on spermatozoa (which have natural anisotropy, and so can separate light and provide refringent images on the microscope). Birefringence is associated with the acrosome reaction, motility and the absence of DNA damage (Gianaroli *et al.*, 2008, 2010; Magli *et al.*, 2012), so this technique would improve our diagnostic capabilities. Instrumentation/software for selective digital morphological examinations will no doubt become available, provided there is consensus on their basic value. An example is the increasing use of enhanced digital imaging, which allows one to view spermatozoa at high magnifications (6000×), and thereby assess the normality of various organelles (Bartoov *et al.*, 2003). With this simple battery of variables, the clinician should have enough information to attain a more detailed BSE diagnosis.

Borderline cases will always exist and require more 'sophisticated' methods. Cytomics, in which FC is a major instrumentation/technique, will evolve, most likely following the already initiated path of multicolour analyses that enable us to test multiple attributes on individual cells using multichannelling (Hossain *et al.*, 2011a; Petrunkina and Harrison, 2011). These new methods will probably not eliminate the use of simpler methods such as microscopy for motility assessment, not even when image stream analysis (using the ImageStreamX imaging flow cytometer from Amnis, Seattle, Washington), which combines FC and imaging, is employed (Zuba-Surma *et al.*, 2007; Buckman *et al.*, 2009). An attempt has been made to demonstrate that field fertility in bull sires could be assessed simply by FC-detection of sperm concentration and viability (membrane integrity) in neat semen, or even FC determination of post-thaw viability (Christensen *et al.*, 2011). While this work reaffirmed that the number of viable spermatozoa above a certain threshold relates to the fertility of a given bull sire, the conclusion was only achievable after the application of statistical manipulation to not so clear-cut data, which indicates the need for further research.

Other methods are already under development under the banners of increased homogeneity, 'microfluidics' and reactivity markers for spermatozoa (Suh *et al.*, 2003; Lopez-Garcia *et al.*, 2008; Smith *et al.*, 2011). The rationale behind these is to mimic the conditions of the female genital tract during sperm transport and selection before fertilization, particularly as the currently available *in vitro* methods have proven to be insufficiently discriminative (Holt et *al.*, 2007). Increased homogeneity of sperm is what we intend to achieve by applying sperm selection or separation via centrifugation through silane-based colloids – similar to what is performed by the cervical sialic acid channels *in vivo*. The Inclusion of counter-fluids and sequential exposure to effectors such as bicarbonate, a relevant marker for capacitation, on to flowing spermatozoa would be easily recorded alongside events by silica-bound markers. This includes events such as the activation of adenylate cyclase and the activation of protein kinase A (PK), or changes in tail movement or in sperm protein tyrosine phosphorylation (Bailey, 2010; Baker, 2011). Alternative state-bound markers/effectors such as hyaluronan are also available, thus registering a more detailed battery of markers for sperm function in real time. Other markers, such as specific sperm receptors (CatSper, aquaporins, etc.), can be added along the fluidic line.

With developments over the past decade in either transcriptomics or proteomics, molecular analyses of the heterogeneous mixture of spermatozoa harbouring genetic diversity are warranted. As already indicated by some studies (Bissonnette *et al.*, 2009), such cellular heterogeneity provides transcript abundance of specific genes among cells. Although transcript levels vary accordingly, transcript profiles can indeed be produced for individuals yielding those ejaculates. This area is being developed at present, and we should expect major gains in transcriptomic-aided selection of breeding sires among bovine and porcine species over the next few years, based on new and more complete microarrays (containing the diversity of all transcripts present) than those currently available. The rationale here is that the genetic (transcriptomic) content is a hallmark of the quality of spermatogenesis of a given male and, therefore, depicts the potential fertility of a semen sample. In the long run, this will be developed as a diagnostic aid for the clinician (Altmäe and Salumets, 2011). The same would apply if epigenetic regulatory genes can be found to be associated with sperm motility, as has been published recently. Samples with low motility spermatozoa can then be checked by gene expression looking for specific epigenetic genes (Pacheco *et al.*, 2011). Microarray development will, no doubt, subsequently reach the clinician's desk and allow for fertility prognosis.

Alternatively, and most likely concurrently, we will probably see further development of simpler and less demanding instruments and methods for protein analysis than those available at present. Considering that most macromolecular events related to cytokine signalling by the seminal plasma, or directly by the adsorbed proteins on the sperm surface, take place at the protein level, such development is not only desirable but expected. Also likely is increased detailed determination of the proteomics of SP, not only in terms of differences among sires but, more particularly, for the differential stimulation of responses (signalling strength) by the female. An example is the differential stimulation of maternal immune tolerance to spermatozoa and the allogeneic conceptus as a measure of the differential capacity that translates into fertility differences between sires, particularly for those polytocous species in which litter size survival, or prolificacy, is important.

Conclusions

Considering that, *in vivo*, the potentially fertilizing spermatozoa are continuously subjected to several stepwise processes of sperm selection before and during fertilization, it is logical to assume that the degree of information provided by current laboratory methods used to assess sperm intactness and function would benefit from the inclusion of a sperm selection method. Use of swim-up or other cleansing methods such as differential centrifugation through colloid columns would aid in identifying a proportion of spermatozoa with those attributes of importance for colonization of the sperm reservoir from the overwhelming background population, the majority of which will not participate in fertilization, either as a natural ejaculate (from which only a small proportion of spermatozoa reaches the oviductal reservoir) or as an AI dose (in which most spermatozoa are damaged or affected by the handling and preservation procedures).

Decreasing the number of spermatozoa to be tested (*in vitro* as *in vivo*) increases our ability to separate competent spermatozoa from the bulk population. Beyond scrutiny of intactness of the relevant attributes, particularly integrity of the entire plasmalemma (using fixable stained cells), this would require a challenge with an activating effector, such as bicarbonate or HA, and a subsequent assessment of the proportion of activated spermatozoa and their longevity after activation. Both of these activators have been shown to influence fertilizing ability in direct relation to the probability of sperm–oocyte encounters and interactions.

We also need to increase our capacity to disclose the degree of DNA damage (perhaps focusing on DNA fragmentation, using simplified protocols) of those sperm subpopulations that survive handling and storage, where intrinsic ROS formation might exceed physiological levels and damage sperm DNA. Here however, the number and diversity of available assessment methods complicates the issue, and simpler, direct methods need to be designed, particularly those with a true relationship with fertility outcomes. Lastly, selective proteomic examinations of the SP are indicated in light of the readiness of certain sires to enhance maternal immunological tolerance, and these could aid in the selection of breeding males.

To conclude, it is important to remember that good clinical andrological practice should not be corrupted by the application of a never-ending plethora of new methods for assessing the details of spermatozoa – methods that lack biological significance – when the goal is diagnostic or prognostic in terms of the (in)fertility of breeding sires.

Acknowledgements

The author's studies have been made possible by grants from the Swedish Research Council FORMAS, the Swedish Farmer's Foundation for Agricultural Research (SLF) and the Swedish Research Council (VR), Stockholm, Sweden. All co-workers, past and present, are warmly acknowledged.

References

Abaigar T, Holt WV, Harrison RA and Del Barrio G (1999) Sperm subpopulations in boar (*Sus scrofa*) and gazelle (*Gazella dama mhorr*) semen as revealed by pattern analysis of computer-assisted motility assessments *Biology of Reproduction* **60** 32–41.

Aitken RJ (2006) Sperm function tests and fertility *International Journal of Andrology* **29** 69–75.

Aitken RJ (2010) Whither must spermatozoa wander? The future of laboratory seminology *Asian Journal of Andrology* **12** 99–103.

Aitken RJ and West KM (1990) Analysis of the relationship between reactive oxygen species production and leucocyte infiltration in fractions of human semen separated on Percoll gradients *International Journal of Andrology* **13** 433–451.

Aitken RJ, Wingate JK, De Iuliis GN and McLaughlin EA (2007) Analysis of lipid peroxidation in human spermatozoa using BODIPY C11 *Molecular Human Reproduction* **13** 203–11.

Aitken RJ, De Iuliis GN and McLachlan RI (2009) Biological and clinical significance of DNA damage in the male germ line *International Journal of Andrology* **32** 46–56.

Alghamdi AS, Lovaas BJ, Bird SL, Lamb GC, Rendahl AK, Taube PC and Foster DN (2009) Species-specific interaction of seminal plasma on sperm–neutrophil binding *Animal Reproduction Science* **114** 331–344.

Al-Makhzoomi A, Lundeheim N, Håård M and Rodriguez-Martinez H (2008) Sperm morphology and fertility of progeny-tested AI dairy bulls in Sweden *Theriogenology* **70** 682–691.

Al Naib A, Hanrahan JP, Lonergan P and Fair S (2011) *In vitro* assessment of sperm from bulls of high and low fertility *Theriogenology* **76** 161–167.

Altmäe S and Salumets A (2011) A novel genomic diagnostic tool for sperm quality? *Reproductive BioMedicine Online* **22** 405–407.

Am-in N, Kirkwood RN, Techakumpu M and Tantasuparuk W (2011) Lipid profiles of sperm and seminal plasma from boars having normal or low sperm motility *Theriogenology* **75** 897–903.

Ardon F, Evert M, Beyerbach M, Weitze KF and Waberski D (2005) Accessory sperm: a biomonitor of boar fertilization capacity *Theriogenology* **63** 1891–1901.

Arnold GJ and Frohlich T (2011) Dynamic proteome signatures in gametes, embryos and their maternal environment *Reproduction, Fertility and Development* **23** 81–93.

Auger J (2010) Assessing human sperm morphology: top models, underdogs or biometrics? *Asian Journal of Andrology* **12** 36–46.

Awda BJ, Mackenzie-Bell M and Buhr MM (2009) Reactive oxygen species and boar sperm function *Biology of Reproduction* **81** 553–561.

Bailey JL (2010) Factors regulating sperm capacitation *Systems in Biology and Reproductive Medicine* **56** 334–348.

Bailey JL, Robertson L and Buhr MM (1994) Relations among *in vivo* fertility, computer-analysed motility and Ca^{++} influx in bovine spermatozoa *Canadian Journal of Animal Science* **74** 53–58.

Baker MA (2011) The ‘omics revolution and our understanding of sperm cell biology *Asian Journal of Andrology* **13** 6–10.

Barratt CLR, Aitken RJ, Björndahl L, Carrell DT, de Boer P, Kvist U, Lewis SEM, Perreault SD, Perry MJ, Ramos L, Robaire B, Ward S and Zini A (2010) Sperm DNA: organization, protection and vulnerability: from basic science to clinical applications – a position report *Human Reproduction* **25** 824–838.

Barratt CLR, Mansell S, Beaton C, Tardif S and Oxenham SK (2011) Diagnostic tools in male infertility – the question of sperm dysfunction *Asian Journal of Andrology* **13** 53–58.

Barrier-Battut I, Dacheux JL, Gatti JL, Rouviere P, Stanciu C, Dacheux F and Vidament M (2005) Seminal plasma proteins and semen characteristics in relation with fertility in the stallion *Animal Reproduction Science* **89** 255–258.

Barrios B, Fernández-Juan M, Muiño-Blanco T and Cebrián-Pérez JA (2005) Immunocytochemical localization and biochemical characterization of two seminal plasma proteins that protect ram spermatozoa against cold shock *Journal of Andrology* **26** 539–549.

Bartoov B, Berkovitz A, Eltes F, Kogosovsky A, Yagoda A, Lederman H, Artzi S, Gross M and Barak Y (2003) Pregnancy rates are higher with intracytoplasmic morphologically selected sperm injection than with conventional intracytoplasmic injection *Fertility and Sterility* **80** 1413–1419.

Bergeron A, Villemure M, Lazure C and Manjunath P (2005) Isolation and characterization of the major proteins of ram seminal plasma *Molecular Reproduction and Development* **71** 461–470.

Bergqvist AS, Ballester J, Johannisson A, Hernández M, Lundeheim N and Rodriguez-Martinez H (2006) *In vitro* capacitation of bull spermatozoa by oviductal fluid and its components *Zygote* **14** 259–273.
Bissonnette N, Lévesque-Sergerie JP, Thibault C and Boissonneault G (2009) Spermatozoal transcriptome profiling for bull sperm motility: a potential tool to evaluate semen quality *Reproduction* **138** 65–80.
Boe-Hansen GB, Ersbøll AK, Greve T and Christensen P (2005) Increasing storage time of extended boar semen reduces sperm DNA integrity *Theriogenology* **63** 2006–2019.
Brahmkshtri BP, Edwin MJ, John MC, Nainar AM and Krishnan AR (1999) Relative efficacy of conventional sperm parameters and sperm penetration bioassay to assess bull fertility *in vitro Animal Reproduction Science* **54** 159–168.
Brandon CI, Heusner GL, Caudle AB and Fayrer-Hosken RA (1999) Two-dimensional polyacrylamide gel electrophoresis of equine seminal plasma proteins and their correlation with fertility *Theriogenology* **52** 863–873.
Brewis IA and Gadella BM (2010) Sperm surface proteomics: from protein lists to biological function *Molecular Human Reproduction* **16** 68–79.
Buckman C, George TC, Friend S, Sutovsky M, Miranda-Vizuete A, Ozanon C, Morrissey P and Sutovsky P (2009) High throughput, parallel imaging and biomarker quantification of human spermatozoa by ImageStream flow cytometry *Systems in Biological Reproductive Medicine* **55** 244–251.
Burden HP, Holmes CH, Persad R and Whittington K (2006) Prostasomes – their effects on human male reproduction and fertility *Human Reproduction Update* **12** 283–292.
Caballero I, Vazquez JM, García EM, Roca J, Martínez EA, Calvete JJ, Sanz L, Ekwall H and Rodriguez-Martinez H (2006) Immunolocalization and possible functional role of PSP-I/PSP-II heterodimer in highly-extended boar spermatozoa *Journal of Andrology* **27** 766–773.
Caballero I, Vazquez JM, Garcia EM, Parrilla I, Roca J, Calvete JJ, Sanz L and Martinez EA (2008) Major proteins of boar seminal plasma as a tool for biotechnological preservation of spermatozoa *Theriogenology* **70** 1352–1355.
Calvete JJ and Sanz L (2007) Insights into structure–function correlations of ungulate seminal plasma proteins In *Spermatology: Proceedings of the 10th International Symposium on Spermatology, held at El Escorial, Madrid, Spain, 17–22 September 2006 Society of Reproduction and Fertility Supplement Volume 65* pp 201–215 Ed ERS Roldan and M Gomendio. Published for Society of Reproduction and Fertility by Nottingham University Press, Nottingham.
Calvete JJ, Nessau S, Mann K, Sanz L, Sieme H, Klug E and Töpfer-Petersen E (1994) Isolation and biochemical characterization of stallion seminal-plasma proteins *Reproduction in Domestic Animals* **29** 411–426.
Calvete JJ, Ensslin M, Mburu J, Iborra A, Martínez P, Adermann K, Waberski D, Sanz L, Töpfer-Petersen E, Weitze K-F, Einarsson S and Rodriguez-Martinez H (1997) Monoclonal antibodies against boar sperm zona pellucida-binding protein AWN-1. Characterization of a continuous antigenic determinant and immunolocalization of AWN epitopes in inseminated sows *Biology of Reproduction* **57** 735–742.
Cancel AM, Chapman DA and Killian GJ (1999) Osteopontin localization in the Holstein bull reproductive tract *Biology of Reproduction* **60** 454–460.
Carrell DT (2008) Contributions of spermatozoa to embryogenesis: assays to evaluate their genetic and epigenetic fitness *Reproductive Biomedicine Online* **16** 474–484.
Christensen P, Labouriau R, Birck A, Boe-Hansen GB, Pedersen J and Borchersen S (2011) Relationships among seminal quality measures and field fertility of young dairy bulls using low-dose inseminations *Journal of Dairy Science* **94** 1744–1754.
Chung Y, Zhu X, Gu W, Smith GD and Takayama S (2006) Microscale integrated sperm sorter *Methods in Molecular Biology* **321** 227–244.
Colleoni S, Lagutina I, Lazzari G, Rodriguez-Martinez H, Galli C and Morrell JM (2011) New methods for selecting stallion spermatozoa for assisted reproduction *Journal of Equine Veterinary Science* **31** 536–541.
Collins JA, Barnhart KT and Schegel PN (2008) Do sperm DNA integrity tests predict pregnancy with *in vitro* fertilization? *Fertility and Sterility* **89** 283–231.
Córdova-Izquierdo A, Oliva JH, Lleó B, García-Artiga C, Corcuera BD and Pérez-Gutiérrez JF (2006) Effect of different thawing temperatures on the viability, *in vitro* fertilizing capacity and chromatin condensation of frozen boar semen packaged in 5 ml straws *Animal Reproduction Science* **92** 145–154.
Coy P, Gadea J, Rath D and Hunter RH (2009) Differing sperm ability to penetrate the oocyte *in vivo* and *in vitro* as revealed using colloidal preparations *Theriogenology* **72** 1171–1179.
Cox J and Mann M (2007) Is proteomics the new genomics? *Cell* **130** 395–398.

Cox JF, Zavala A, Saravia F, Rivas C, Gallardo P and Alfaro V (2002) Differences in sperm migration through cervical mucus *in vitro* relates to sperm colonization of the oviduct and fertilizing ability in goats *Theriogenology* **58** 9–18.

Cox JF, Alfaro V, Montenegro V and Rodriguez-Martinez H (2006) Computer-assisted analysis of sperm motion in goats and its relationship with sperm migration in cervical mucus *Theriogenology* **66** 860–867.

Cremades T, Roca J, Rodriguez-Martinez H, Abaigar T, Vazquez JM and Martinez EA (2005) Kinematic changes during the cryopreservation of boar spermatozoa *Journal of Andrology* **26** 610–618.

Dapino DG, Teijeiro JM, Cabada MO and Marini PE (2009) Dynamics of heparin-binding proteins on boar sperm *Animal Reproduction Science* **116** 308–317.

De Ambrogi M, Ballester J, Saravia F, Caballero I, Johannisson A, Wallgren M, Andersson M and Rodriguez-Martinez H (2006) Effect of storage in short- and long-term commercial semen extenders on the motility, plasma membrane, and chromatin integrity of boar spermatozoa *International Journal of Andrology* **29** 543–552.

de Mateo S, Castillo J, Estanyol JM, Ballesca JL and Oliva R (2011) Proteomic characterization of the human sperm nucleus *Proteomics* **11** 2714–2726.

De Pauw I, Van Soom A, Laevens H, Verberckmoes S and de Kruif A (2002) Sperm binding to epithelial oviduct explants in bulls with different non-return rates investigated with a new *in vitro* model *Biology of Reproduction* **67** 1073–1079.

Drabovich AP, Jarvi K and Diamandis EP (2011) Verification of male infertility biomarkers in seminal plasma by multiplex selected reaction monitoring assay *Molecular Cellular Proteomics* **10**: M110.004127.

Druart X, Rickard JP, Mactier S, Kohnke PL, Kershaw-Young CM, Bathgate R, Gibb Z, Crossett B, Tsikis G, Labas V, Harichaux G, Grupen CG and de Graaf SP (2013) Proteomic characterization and cross species comparison of mammalian seminal plasma *Journal of Proteomics* **91** 13–22.

du Plessis SS, Kashou AH, Benjamin DJ, Yadav SP and Agarwal A (2011) Proteomics; a subcellular look at spermatozoa *Reproductive Biology and Endocrinology* **9**:36.

Duncan MW and Thompson HS (2007) Proteomics of semen and its constituents *Proteomics Clinical Application* **1** 861–875.

Ehlers J, Behr M, Bollwein H, Beyerbach M and Waberski D (2011) Standardization of computer-assisted semen analysis using an e-learning application *Theriogenology* **76** 448–454.

Enciso M, Lopez-Fernandez C, Fernandez JL, Garcia P, Gosalvez A and Gosalvez J (2006) A new method to analyse boar sperm DNA fragmentation under bright-field or fluorescence microscopy. *Theriogenology,* **65** 308–316.

Evenson D and Jost L (2000) Sperm chromatin structure assay is useful for fertility assessment *Methods in Cell Science* **22** 169–189.

Evenson D and Wixon R (2006) Clinical aspects of sperm DNA fragmentation detection and male infertility. *Theriogenology,* **65** 979–991.

Fabrega A, Puigmule, Yeste M, Casas I, Bonet S and Pinard E (2011) Impact of epididymal maturation, ejaculation and *in vitro* capacitation on tyrosine phosphorylation patterns exhibited of boar (*Sus domesticus*) spermatozoa *Theriogenology* **76** 1356–1366.

Fazeli A, Duncan AE, Watson PF and Holt WV (1999) Sperm–oviduct interaction: induction of capacitation and preferential binding of uncapacitated spermatozoa to oviductal epithelial cells in porcine species *Biology of Reproduction* **60** 879–886.

Feitsma H, Broekhuijse MLWJ and Gadella BM (2011) Do CASA systems satisfy consumers demands? A critical analysis *Reproduction in Domestic Animals* **46(Supplement 2)** 49–51.

Feugang JM, Rodriguez-Osorio N, Kaya A, Wang H, Page G, Ostermeier GC, Topper EK and Memilli E (2010) Transcriptome analysis of bull spermatozoa: implications for male fertility *Reproductive Biomedicine Online* **21** 312–324.

Foote RH (1999) Development of reproductive biotechnologies in domestic animals from artificial insemination to cloning: a perspective *Cloning* **1** 133–142.

Fraser L (2004) Structural damage to nuclear DNA in mammalian spermatozoa: its evaluation techniques and relationship with male fertility *Polish Journal of Veterinary Science* **7** 311–321.

Gadea J (2005) Sperm factors related to *in vitro* and *in vivo* porcine fertility *Theriogenology* **63** 431–444.

Gadella BM and van Gestel RA (2004) Bicarbonate and its role in mammalian sperm function *Animal Reproduction Science* **82–83** 307–319.

García EM, Vázquez JM, Calvete JJ, Sanz L, Caballero I, Parrilla I, Gil MA, Roca J and Martinez EA (2006) Dissecting the protective effect of the seminal plasma permadhesin PSP-I/PSP-II on boar sperm functionality *Journal of Andrology* **27** 434–443.

García EM, Vázquez JM, Parrilla I, Ortega MD, Calvete JJ, Sanz L, Martínez EA, Roca J and Rodriguez-Martinez H (2008) Localization and expression of spermadhesin PSP-I/PSP-II subunits in the reproductive organs of the boar *International Journal of Andrology* **31** 408–417.

García-Herrero S, Meseguer M, Martinez-Conejero JA, Remohi J, Pellicer A and Garrido N (2010) The transcriptome of spermatozoa used in homologous intrauterine insemination varies considerably between samples that achieve pregnancy and those that do not *Fertility and Sterility* **94** 1360–1373.

Gerena RL, Irikura D, Urade Y, Eguchi N, Chapman DA and Killian GJ (1998) Identification of a fertility-associated protein in bull seminal plasma as lipocalin-type prostaglandin D synthase *Biology of Reproduction* **58** 826–833.

Gianaroli L, Magli MC, Collodel G, Moretti E, Ferraretti AP and Baccetti B (2008) Sperm head's birefringence: a new criterion for sperm selection *Fertility and Sterility* **90** 104–112.

Gianaroli L, Magli MC, Ferraretti AP, Crippa A, Lappi M, Capitani S and Baccetti B (2010) Birefringence characteristics in sperm heads allow for the selection of reacted spermatozoa for intracytoplasmic sperm injection *Fertility and Sterility* **93** 807–813.

Gil J, Januskauskas A, Håård MCh, Håård MGM, Johannisson A, Saderquist L and Rodriguez-Martinez H (2000) Functional sperm parameters and fertility of bull semen extended in Biociphos-Plus® and Triladyl® *Reproduction in Domestic Animals* **35** 69–77.

Gil MA, Roca J, Cremades T, Hernández M, Vázquez JM, Rodriguez-Martinez H and Martínez EA (2005) Does multivariate analysis of post-thaw sperm characteristics accurately estimate *in vitro* fertility of boar individual ejaculates? *Theriogenology* **64** 305–316.

Graham JK (2001) Assessment of sperm quality: a flow cytometric approach *Animal Reproduction Science* **68** 239–247.

Graham JK and Mocé E (2005) Fertility evaluation of frozen/thawed semen *Theriogenology* **64** 492–504.

Gravance CG, Casey ME and Casey PJ (2009) Pre-freeze bull sperm head morphometry related to post-thaw fertility *Animal Reproduction Science* **114** 81–88.

Guillan L, Evans G and Maxwell WMC (2005) Flow cytometric evaluation of sperm parameters in relation to fertility potential *Theriogenology* **63** 445–457.

Guthrie H and Welch G (2007) Use of fluorescence-activated flow cytometry to determine membrane lipid peroxidation during hypothermic liquid storage and freeze–thawing of viable boar sperm loaded with 4, 4-difluoro-5-(4-phenyl-1,3-butadienyl)-4-bora-3a,4a-diaza-s-indacene-3-undecanoic acid *Journal of Animal Science* **85** 1402–1411.

Guthrie HD, Welch GR and Long JA (2008) Mitochondrial function and reactive oxygen species action in relation to boar motility *Theriogenology* **70** 1209–1215.

Gwathmey TM, Ignotz GG, Mueller JL, Manjunath P and Suarez SS (2006) Bovine seminal plasma proteins PDC-109, BSP-A3, and BSP-30-kDa share functional roles in storing sperm in the oviduct *Biology of Reproduction* **75** 501–507

Hales BF, Grenier L, Lalancette C and Robaire B (2011) Epigenetic programming: from gametes to blastocyst *Birth Defects Research Part A: Clinical and Molecular Teratology* **91** 652–665.

Hallap T, Håård M, Jaakma Ü, Larsson B and Rodriguez-Martinez H (2004) Variations in quality of frozen–thawed semen from Swedish Red and White AI sires at 1 and 4 years of age *International Journal of Andrology* **27** 166–171.

Hallap T, Nagy S, Håård M, Jaakma Ü, Johannisson A and Rodriguez-Martinez H (2005a) Sperm chromatin stability in frozen–thawed semen is maintained over age in AI bulls *Theriogenology* **63** 1752–1763.

Hallap T, Nagy S, Jaakma U, Johannisson A and Rodriguez-Martinez H (2005b) Mitochondrial activity of frozen–thawed spermatozoa assessed by MitoTracker Deep Red 633 *Theriogenology* **63** 2311–2322.

Hallap T, Jaakma Ü and Rodriguez-Martinez H (2006a) Changes in semen quality in Estonian AI bulls at 3, 5 and 7 years of age *Reproduction in Domestic Animals* **41** 214–218.

Hallap T, Nagy S, Jaakma Ü, Johannisson A and Rodriguez-Martinez H (2006b) Usefulness of a triple fluorochrome combination Merocyanine 540/Yo-Pro 1/Hoechst 33342 in assessing membrane stability of viable frozen–thawed spermatozoa from Estonian Holstein AI bulls *Theriogenology* **65** 1122–1136.

Hamann H, Jude R, Sieme H, Mertens U, Töpfer-Petersen E, Distl O and Leeb T (2007) A polymorphism within the equine *CRISP3* gene is associated with stallion fertility in Hanoverian Warmblood horses *Animal Genetics* **38** 259–264.

Hao Y, Mathialagan N, Walters E, Mao J, Lai L, Becker D, Li W, Critser J and Prather RS (2006) Osteopontin reduces polyspermy during *in vitro* fertilization of porcine oocytes *Biology of Reproduction* **75** 726–733.

Hao Y, Murphy CN, Spate L, Wax D, Zhong Z, Samuel M, Mathialagan N, Schatten H and Prather RS (2008) Osteopontin improves *in vitro* development of porcine embryos and decreases apoptosis *Molecular Reproduction and Development* **75** 291–298.

Harrison RAP (1997) Sperm plasma membrane characteristics and boar semen fertility *Journal of Reproduction and Fertility Supplement* **52** 195–211.

Harrison RAP and Gadella B (2005) Bicarbonate-induced membrane processing in sperm capacitation *Theriogenology* **63** 342–51.

Henault MA and Killian GJ (1995) Effects of sperm preparation and bull fertility on *in vitro* penetration of zona-free bovine oocytes *Theriogenology* **43** 739–749.

Hernández M, Roca J, Ballester J, Vazquez JM, Martínez EA, Johannisson A, Saravia A and Rodriguez-Martinez H (2006) Differences in SCSA outcome among boars with different sperm freezability *International Journal of Andrology* **29** 583–591.

Hirai M, Boersma A, Hoeflich A, Wolf E, Foll J, Aumüller TR and Braun J (2001) Objectively measured sperm motility and sperm head morphometry in boars (*Sus scrofa*): relation to fertility and seminal plasma growth factors *Journal of Andrology* **22** 104–110.

Holt WV (2009) Is semen analysis useful to predict the odds that the sperm will meet the egg? *Reproduction in Domestic Animals* **44(Supplement 3)** 31–38.

Holt WV (2011) Mechanisms of sperm storage in the female reproductive tract: an interspecies comparison *Reproduction in Domestic Animals* **46(Supplement 2)** 68–74.

Holt WV and Harrison RAP (2002) Bicarbonate stimulation of boar sperm motility via a protein kinase A-dependent pathway: between-cell and between-ejaculate differences are not due to deficiencies in protein kinase A activation *Journal of Andrology* **23** 557–565.

Holt WV and Lloyd RE (2010) Sperm storage in the vertebrate female reproductive tract: how does it work so well? *Theriogenology* **73** 713–722.

Holt WV and Van Look KJW (2004) Concepts in sperm heterogeneity, sperm selection and sperm competition as biological foundations for laboratory tests of semen quality *Reproduction* **127** 527–535.

Holt C, Holt WV, Moore HDM, Reed HCB and Curnock RM (1997) Objectively measured boar sperm motility parameters correlate with the outcomes of on-farm inseminations: results of two fertility trials *Journal of Andrology* **18** 312–323.

Holt WV, O'Brien J and Abaigar T (2007) Applications and interpretation of computer-assisted sperm analyses and sperm sorting methods in assisted breeding and comparative research *Reproduction, Fertility and Development* **19** 709–718.

Hossain Md S, Johannisson A, Wallgren M, Nagy S, Pimenta Siqueira A and Rodriguez-Martinez H (2011a) Flow cytometry for the assessment of animal sperm integrity and functionality: state of the art *Asian Journal of Andrology* **13** 406–419.

Hossain Md S, Johannisson A, Pimenta Siqueira A, Wallgren M and Rodriguez-Martinez H (2011b) Spermatozoa in the sperm-peak-fraction of the boar ejaculate show a lower flow of Ca2+ under capacitation conditions post-thaw which might account for their higher membrane stability after cryopreservation *Animal Reproduction Science* **128** 37–44.

Hunter RH, Coy P, Gadea J and Rath D (2011) Considerations of viscosity in the preliminaries to mammalian fertilisation *Journal of Assisted Reproduction and Genetics* **28** 191–197.

Huszar G, Ozenci CC, Cayli S, Zavaczki Z, Hansch E and Vigue L (2003) Hyaluronic acid binding by human sperm indicates cellular maturity, and unreacted acrosomal status *Fertility and Sterility* **79** 1616–1624.

Huszar G, Jakab A, Sakkas D, Ozenci CC, Cayli S, Delpiano E and Ozkavukcu S (2007) Fertility testing and ICSI sperm selection by hyaluronic acid binding: clinical and genetic aspects *Reproduction Biomedicine Online* **14** 650–663.

Januskauskas A, Johannisson A, Saderquist L and Rodriguez-Martinez H (2000a) Assessment of the fertilizing potential of frozen–thawed bovine spermatozoa by calcium ionophore A23187-induced acrosome reaction *in vitro Theriogenology* **53** 859–875.

Januskauskas A, Gil J, Saderquist L and Rodriguez-Martinez H (2000b) Relationship between sperm response to glycosaminoglycans *in vitro* and non return rates of Swedish dairy AI bulls *Reproduction in Domestic Animals* **35** 207–212.

Januskauskas A, Johannisson A and Rodriguez-Martinez H (2001) Assessment of sperm quality through fluorometry and sperm chromatin structure assay in relation to field fertility of frozen–thawed semen from Swedish AI bulls *Theriogenology* **55** 947–961.

Januskauskas A, Johannisson A and Rodriguez-Martinez H (2003) Subtle membrane changes in cryopreserved bull semen in relation with sperm viability, chromatin structure, and field fertility *Theriogenology* **60** 743–758.

Januskauskas A, Lukoseviciute K, Nagy S, Johannisson A and Rodriguez-Martinez H (2005) Assessment of the efficacy of Sephadex G-15 filtration of bovine spermatozoa for cryopreservation *Theriogenology* **63** 60–78.

Jenkins TG and Carrell DT (2011) The paternal epigenome and embryogenesis: poising mechanisms for development *Asian Journal of Andrology* **13** 76–80.

Jenne DE, Lowin B, Peitsch MC, Böttcher A, Schmitz G and Tschopp J (1991) Clusterin (complement lysis inhibitor) forms a high density lipoprotein complex with apolipoprotein A-I in human plasma. *Journal of Biological Chemistry* **266** 11030–11036.

Johannisson A, Morrell JM, Thorén J, Jönsson M, Dalin AM and Rodriguez-Martinez H (2009) Colloidal centrifugation with Androcoll-E™ prolongs stallion sperm motility, viability and chromatin integrity *Animal Reproduction Science* **116** 119–128.

Johnson GD, Lalancette C, Linnemann AK, Leduc F, Boissoneault G and Krawetz SA (2011) The sperm nucleus: chromatin, RNA, and the nuclear matrix *Reproduction* **141** 21–36.

Jones R, James PS, Oxley D, Coadwell J, Suzuki-Toyota F and Howes EA (2008) The equatorial subsegment in mammalian spermatozoa is enriched in tyrosine phosphorylated proteins *Biology of Reproduction* **79** 421–431.

Kareskoski AM, Rivera del Alamo MM, Güvenc K, Reilas T, Calvete JJ, Rodriguez-Martinez H, Andersson M and Katila T (2011) Protein composition of seminal plasma in fractionated stallion ejaculates *Reproduction in Domestic Animals* **46** e79–e84.

Kastelic JP and Thundathil JC (2008) Breeding soundness evaluation and semen analysis for predicting bull fertility *Reproduction in Domestic Animals* **43(Supplement 2)** 368–373.

Katari S, Turan N, Bibikova M, Erinle O, Chalian R, Foster M, Gaughan JP, Coutifaris C and Sapienza C (2009) DNA methylation and gene expression differences in children conceived *in vitro* or *in vivo* *Human Molecular Genetics* **18** 3769–3778.

Katila T (2001) *In vitro* evaluation of frozen–thawed stallion semen: a review *Acta Veterinaria Scandinavica* **42** 199–217.

Kavak A, Johannisson A, Lundeheim N, Rodriguez-Martinez H, Aidnik M and Einarsson S (2003) Evaluation of cryopreserved stallion semen from Tori and Estonian breeds using CASA and flow cytometry *Animal Reproduction Science* **76** 205–216.

Kelly VC, Kuy S, Palmer DJ, Xu Z, Davis SR and Cooper GJ (2006) Characterization of bovine seminal plasma by proteomics *Proteomics* **6** 5826–5833.

Killian GJ, Chapman DA and Rogowski LA (1993) Fertility-associated proteins in Holstein bull seminal plasma *Biology of Reproduction* **49** 1202–1207.

Koppers A, De Iuliis G, Finnie J, McLaughlin E and Aitken R (2008) Significance of mitochondrial reactive oxygen species in the generation of oxidative stress in spermatozoa *Journal of Clinical Endocrinology and Metabolism* **93** 3199–3207.

Koppers AJ, Reddy T and O'Bryan MK (2011) The role of cysteine-rich secretory proteins in male fertility. *Asian Journal of Andrology* **13** 111–117.

Krawetz SA, Kruger A, Lalancette C, Tagett R, Anton E, Draghici S and Diamong MP (2011) A survey of small RNAs in human sperm *Human Reproduction* **26** 3401–3412.

Lalancette C, Thibault C, Bachand I, Caron N, Bissonnette N (2008) Transcriptome analysis of bull semen with extreme nonreturn rate: use of suppression-substractive hybridization to identify functional markers for fertility *Biology of Reproduction* **78** 618–635.

Lalancette C, Platts AE, Johnson GD, Emery BR, Carrell DT and Krawetz SA (2009) Identification of human sperm transcripts as candidate markers of male fertility *Journal of Molecular Medicine* **87** 735–748.

Lefebvre R and Suarez SS (1996) Effect of capacitation on bull sperm binding to homologous oviductal epithelium *Biology of Reproduction* **54** 575–582.

Liberda J, Ticha M, Zraly Z, Svecova D and Veznik Z (1998) Interaction of bull, stallion and boar seminal plasma proteins and sperms with acidic polysaccharides *Folia Biologica (Praha)* **44** 177–183.

Lopez-Garcia MD, Monson RL, Haubert K, Wheeler MB and Beebe DJ (2008) Sperm motion in a microfluidic fertilization device *Biomedical Microdevices* **10** 709–718.

Loras B, Vetele F, El Malki A, Rollet J, Soufir JC and Benahmed M (1999) Seminal transforming growth factor-β in normal and infertile men *Human Reproduction* **14** 1534–1539.

Lynham JA and Harrison RAP (1998) Use of stored pig eggs to assess boar sperm fertilizing functions *in vitro Biology of Reproduction* **58** 539–550.

Macías García B, Fernández-González L, Morrell J, Ortega-Ferrusola C, Tapia JA, Rodriguez Martínez H and Peña FJ (2009a) Single layer centrifugation through colloid positively modifies the sperm subpopulation structure of frozen–thawed stallion spermatozoa *Reproduction in Domestic Animals* **44** 523–526.

Macías García B, Morrell J, Ortega Ferrusola C, Fernández- González L, Tapia JA, Rodriguez Martínez H and Peña FJ (2009b) Centrifugation on a single layer of colloid selects the "best" quality frozen–thawed stallion spermatozoa *Animal Reproduction Science* **114** 193–202.

Magli MC, Crippa A, Muzii L, Boudjema E, Capoti A, Scaravelli G, Ferraretti AP and Gianaroli L (2012) Head birefringence properties are associated with acrosome reaction, sperm motility and morphology *Reproductive Biomedicine Online* **24** 352–359.

Mancini A, Festa R, Silvestrini A, Nicolotti N, Di Donna V, La Torre G, Pontecorvi A and Meucci E (2009) Hormonal regulation of total antioxidant capacity in seminal plasma *Journal of Andrology* **30** 534–540.

Manjunath P (1984) Gonadotropin release stimulatory and inhibitory proteins in bull seminal plasma. In *Gonadal Proteins and Peptides and their Biological Significance* pp 49–61 Ed MR Sairam and LE Atkinson. World Science Publishing, Singapore.

Manjunath P, Bergeron A, Lefebvre J and Fan J (2007) Seminal plasma proteins: functions and interaction with protective agents during semen preservation [review] In *Spermatology: Proceedings of the 10th International Symposium on Spermatology, held at El Escorial, Madrid, Spain, 17–22 September 2006 Society of Reproduction and Fertility Supplement Volume 65* pp 217–228 Ed ER Roldan and M Gomendio. Published for Society of Reproduction and Fertility by Nottingham University Press, Nottingham.

Manjunath P, Lefebvre J, Jois PS, Fan J and Wright MW (2009) New nomenclature for mammalian BSP genes *Biology of Reproduction* **80** 394–397.

Mann T and Lutwak-Mann C (1981) *Male Reproductive Function and Semen. Themes and Trends in Physiology, Biochemistry and Investigative Andrology*. Springer Verlag, Berlin.

Martínez E, Vázquez JM, Matás C, Roca J, Coy P and Gadea J (1993) Evaluation of boar spermatozoa penetrating capacity using pig oocytes at the germinal vesicle stage *Theriogenology* **40** 547–557.

Martinez-Pastor F, Johannisson A, Gil J, Kaabi M, Anel L, Paz P and Rodriguez-Martinez H (2004) Use of chromatin stability assay, mitochondrial stain JC-1, and fluorometric assessment of plasma membrane to evaluate frozen–thawed ram semen *Animal Reproduction Science* **84** 121–133.

Martinez-Pastor F, Mata-Campuzano M, Alvarez-Rodriguez M, Alvarez M, Anel L and De Paz P (2010) Probes and techniques for sperm evaluation by flow cytometry *Reproduction in Domestic Animals* **45(Supplement 2)** 67–78.

Martinez-Pastor F, Tizado EJ, Garde JJ, Anel L and De Paz P (2011) Statistical series: opportunities and challenges of sperm motility subpopulation analysis *Theriogenology* **75** 7783–795.

Milardi D, Grande G, Vincenzoni F, Messana I, Pontecorvi A, De Marinis L, Castagnola M and Marana R (2012) Proteomic approach in the identification of fertility pattern in seminal plasma of fertile men *Fertility and Sterility* **97** 67–73.

Miller D, Brinkworth M and Iles D (2010) Paternal DNA packaging in spermatozoa: more than the sum of its parts? DNA, histones, protamines and epigenetics *Reproduction* **139** 287–301.

Mitchell LA, de Iuliis GN and Aitken RJ (2011) The TUNEL assay consistently underestimates DNA damage in human spermatozoa and is influenced by DNA compaction and cell vitality: development of an improved methodology *International Journal of Andrology* **34** 2–13.

Mocé E and Graham JK (2008) *In vitro* evaluation of sperm quality *Animal Reproduction Science* **105** 104–118.

Mogielnicka-Brzozowska M, Wysocki P, Strzezek J and Kordan W (2011) Zinc-binding proteins from boar seminal plasma – isolation, biochemical characteristics and influence on spermatozoa stored at 4°C *Acta Biochimica Polonica* **58** 171–177.

Moldenhauer LM, Diener KR, Thring DM, Brown MP, Hayball JD and Robertson SA (2009) Cross-presentation of male seminal fluid antigens elicits T-cell activation to initiate the female immune response to pregnancy *Journal of Immunology* **182** 8080–8093.

Morrell JM and Rodriguez-Martinez H (2009) Biomimetic techniques for improving sperm quality in animal breeding: a review *The Open Andrology Journal* **1** 1–9.

Morrell JM and Rodriguez-Martinez H (2010) Practical applications of sperm selection techniques as a tool for improving reproductive efficiency *Veterinary Medicine International* **2011** Article ID 894767.

Morrell JM, Johannisson A, Dalin AM, Hammar L, Sandebert T and Rodriguez-Martinez H (2008) Sperm morphology and chromatin integrity in Swedish Warmblood stallions and their relationship to pregnancy rates *Acta Veterinaria Scandinavica* **50**:2.

Morrell JM, Johannisson A, Dalin AM and Rodriguez-Martinez H (2009a) Morphology and chromatin integrity of stallion spermatozoa prepared by density gradient and single layer centrifugation through silica colloids *Reproduction in Domestic Animals* **44** 512–517.

Morrell JM, Saravia F, van Wienen M, Wallgren M and Rodriguez-Martinez H (2009b) Selection of boar spermatozoa using centrifugation on a glycidoxypropyltrimethoxysilane-coated silica colloid *Journal of Reproduction and Development* **55** 547–552.

Morrell JM, Rodriguez-Martinez H and Johannisson A (2010) Single layer centrifugation of stallion spermatozoa selects the most robust spermatozoa from the rest of the ejaculate in a large sample size: data from three breeding seasons *Equine Veterinary Journal* **42** 579–585.

Morrell JM, Johannisson A and Rodriguez-Martinez H (2011a) Effect of osmolarity and density of colloid formulations on the outcome of SLC-selection of stallion spermatozoa *ISRN Veterinary Science* **2011**:128984.

Morrell JM, Mari G, Kutvölgyi G, Meurling S, Mislei B, Iacono E and Rodriguez-Martinez (2011b) Pregnancies following artificial insemination with spermatozoa from problem stallion ejaculates processed by single layer centrifugation with Androcoll-E *Reproduction in Domestic Animals* **46** 642–645.

Muratori M, Marchani S, Tamburrino L, Tocci V, Forti G and Baldi E (2008) Nuclear staining identifies two populations of human sperm with different DNA fragmentation extent and relationship with semen parameters *Human Reproduction* **23** 1035–1043.

Naaby-Hansen S, Flickinger CJ and Herr JC (1997) Two dimensional gel electrophoresic analysis of vectorially labeled surface proteins of human spermatozoa *Biology of Reproduction* **56** 771–787.

Nagy S, Jansen J, Topper E and Gadella B (2003) A triple-stain flow cytometric method to assess plasma- and acrosome-membrane integrity of cryopreserved bovine sperm immediately after thawing in presence of egg-yolk particles *Biology of Reproduction* **68** 1828–1835.

Nagy S, Hallap T, Johannisson A and Rodriguez-Martinez H (2004) Changes in plasma membrane and acrosome integrity of frozen–thawed bovine spermatozoa during a 4 h incubation as measured by multicolor flow cytometry *Animal Reproduction Science* **80** 225–35.

Neild DM, Chaves MG, Flores M, Miragaya MH, Gonzalez E and Agüero A (2000) The HOS test and its relationship to fertility in the stallion *Andrologia* **32** 351–355.

Novak S, Smith TA, Paradis F, Burwash L, Dyck MK, Foxcroft GR and Dixon WT (2010a) Biomarkers of *in vivo* fertility in sperm and seminal plasma of fertile stallions *Theriogenology* **74** 956–967.

Novak S, Ruiz-Sanchez A, Dixon WT, Foxcroft GR and Dyck MK (2010b) Seminal plasma proteins as potential markers of relative fertility in boars *Journal of Andrology* **31** 188–200.

Núñez-Martinez I, Moran JM and Peña FJ (2005) Do computer assisted derived morphometric sperm characteristics reflect DNA status in canine spermatozoa? *Reproduction in Domestic Animals* **40** 537–543.

O'Flaherty C, de Lamirande E and Gagnon C (2006) Positive role of reactive oxygen species in mammalian sperm capacitation: triggering and modulation of phosphorylation events *Free Radicals in Biology and Medicine* **41** 528–540.

Oh SA, You YA, Park YJ and Pang MG (2010) The sperm penetration assay predicts the litter size in pigs *International Journal of Andrology* **33** 604–612.

O'Leary A, Armstrong DT and Robertson SA (2011) Transforming growth factor-β (TGFβ) in porcine seminal plasma *Reproduction, Fertility and Development* **23** 748–758.

O'Leary S, Jasper MJ, Warnes GM, Armstrong DT and Robertson SA (2004) Seminal plasma regulates endometrial cytokine expression, leukocyte recruitment and embryo development in the pig *Reproduction* **128** 237–247.

Oliva R and Castillo J (2011) Proteomics and the genetics of sperm chromatin condensation *Asian Journal of Andrology* **13** 24–30.

Oliva R, de Mateo S and Estanyol JM (2009) Sperm cell proteomics *Proteomics* **9** 1004–1017.

Ortega Ferrusola C, González Fernández L, Morrel JM, Salazar Sandoval C, Macías García B, Rodriguez-Martinez H, Tapia JA and Peña FJ (2009a) Lipid peroxidation, assessed with BODIPY-C11, increases after cryopreservation of stallion spermatozoa, is stallion-dependent and is related to apoptotic-like changes *Reproduction* **138** 55–63.

Ortega-Ferrusola C, Sotillo-Galán Y, Varela-Fernández E, Gallardo-Bolaños JM, González-Fernández L, Rodriguez-Martinez H, Tapia JA and Peña FJ (2009b) Apoptotic markers can be used to forecast the freezeability of stallion spermatozoa *Animal Reproduction Science* **114** 393–403.

Ortega Ferrusola C, Gonzalez Fernandez L, Salazar Sandoval C, Macias Garcia B, Rodriguez-Martinez H, Tapia JA and Peña FJ (2010) Inhibition of the mitochondrial permeability transition pore reduces "apoptosis-like changes" during cryopreservation of equine spermatozoa *Theriogenology* **74** 458–465.

Pacheco SE, Houseman EA, Christensen BC, Marsit CJ, Kelsey KT, Sigman M and Boekelheide K (2011) Integrative DNA methylation and gene expression analyses identify DNA packaging and epigenetic regulatory genes associated with low motility sperm. *PLoS One* **6**:e20280.

Palacio JR and Martinez P (2011) Contribution of seminal plasma to the female immune regulation in embryo implantation *Advances in Neuroimmune Biology* **2** 23–30.

Parkinson TJ (2004) Evaluation of fertility and infertility in natural service bulls *The Veterinary Journal* **168** 215–229.

Parmegiani L, Cognigni GE, Ciampaglia W, Pocognoli P, March F and Filicori M (2010) Efficiency of hyaluronic acid (HA) sperm selection *Journal of Assisted Reproduction and Genetics* **27** 13–16.

Peña FJ (2007) Detecting subtle changes in sperm membranes in veterinary andrology *Asian Journal of Andrology* **9** 731–737.

Peña FJ, Johannisson A, Wallgren M and Rodriguez-Martinez H (2003) Assessment of fresh and frozen–thawed boar semen using an Annexin-V assay: a new method to evaluate sperm membrane integrity *Theriogenology* **60** 677–689.

Peña FJ, Johannisson A, Wallgren M and Rodriguez-Martinez H (2004) Effect of hyaluronan supplementation on boar sperm motility and membrane lipid architecture status after cryopreservation *Theriogenology* **61** 63–70.

Peña FJ, Johannisson A, Wallgren M and Rodriguez-Martinez H (2005a) A new and simple method to evaluate early membrane changes in frozen–thawed boar spermatozoa *International Journal of Andrology* **28** 107–114.

Peña FJ, Saravia F, García-Herreros M, Núñez I, Tapia JA, Johannisson A, Wallgren M and Rodriguez-Martinez H (2005b) Identification of sperm morphological subpopulations in two different portions of the boar ejaculate and its relation to post thaw quality *Journal of Andrology* **26** 716–723.

Peña FJ, Saravia F, Núñez-Martínez I, Johannisson A, Wallgren M and Rodriguez-Martinez H (2006) Do different portions of the boar ejaculate vary in their ability to sustain cryopreservation? *Animal Reproduction Science* **93** 101–113.

Peña FJ, Saravia F, Johannisson A, Wallgren M and Rodriguez Martinez H (2007) Detection of early changes in sperm membrane integrity pre-freezing can estimate post-thaw quality of boar spermatozoa *Animal Reproduction Science* **97** 74–83.

Peña FJ, Rodriguez-Martinez H, Tapia JA, Ortega Ferrusola C, Gonzalez Fernández L and Macías García B (2009) Mitochondria in mammalian sperm physiology and pathology: a mini-review *Reproduction in Domestic Animals* **44** 345–349.

Perez-Llano B, Yenes-Garcia P and Garcia-Casado P (2003) Four subpopulations of boar spermatozoa defined according to their response to the short hypoosmotic swelling test and acrosome status during incubation at 37°C *Theriogenology* **60** 1401–1407.

Petersen CG, Massaro FC, Mauri AL, Oliveira JB, Baruffi RL and Franco JG Jr (2010) Efficacy of hyaluronic acid binding assay in selecting motile spermatozoa with normal morphology at high magnification *Reproductive Biology and Endocrinology* **8**:149.

Petrunkina AM and Harrison RAP (2011) Cytometric solutions in veterinary andrology: developments, advantages, and limitations *Cytometry Part A* **79A** 338–348.

Petrunkina AM, Petzoldt R, Stahlberg A, Pfeilsticker J, Beyerbach M, Bader H and Töpfer-Petersen E (2001) Sperm-cell volumetric measurements as parameters in bull semen function evaluation: correlation with non-return rate *Andrologia* **33** 360–367.

Petrunkina AM, Waberski D, Günzel-Apel AR and Töpfer-Petersen E (2007) Determinants of sperm quality and fertility in domestic species *Reproduction* **134** 3–17.

Phillips NJ, McGowan MR, Johnston SD and Mayer DG (2004) Relationship between thirty post-thaw spermatozoal characteristics and the field fertility of 11 high-use Australian dairy AI-sire *Animal Reproduction Science* **81** 47–61.

Piehler E, Petrunkina AM, Ekhlasi-Hundrieser M and Töpfer-Petersen E (2006) Dynamic quantification of protein tyrosine phosphorylation of the sperm surface proteins during capacitation *in vitro Cytometry Part A* **69A** 1062–1070.

Ponting CP, Oliver PL and Reik W (2009) Evolution and functions of long noncoding RNAs *Cell* **136** 629–641.

Razavi SH, Nasr-Esfahani MH, Deemeh MR, Shayesteh M and Tavalaee M (2009) Evaluation of zeta and HA-binding methods for selection of spermatozoa with normal morphology, protamine content and DNA integrity *Andrologia* **42** 13–19.

Reinert M, Calvete JJ, Sanz L, Mann K and Töpfer-Petersen E (1996) Primary structure of stallion seminal plasma protein HSP-7, a zona-pellucida binding protein of the spermadhesin family *European Journal of Biochemistry* **242** 636–640.

Rijsselaere T, van Soom A, Thanghe S, Coryn M, Maes D and de Kruif A (2005) New techniques for the assessment of canine semen quality: a review *Theriogenology* **64** 706–719.

Robayo I, Montenegro V, Valdes C and Cox JF (2008) CASA assessment of kinematic parameters of ram spermatozoa and their relationship to migration efficiency in ruminant cervical mucus *Reproduction in Domestic Animals* **43** 393–399.

Robertson SA (2005) Seminal plasma and male factor signalling in the female reproductive tract *Cell and Tissue Research* **322** 43–52.

Robertson SA (2007) Seminal fluid signalling in the female reproductive tract: lessons from rodents and pigs *Journal of Animal Science* **85** E36–E44.

Robertson SA (2010) Immune regulation of conception and embryo implantation – all about quality control? *Journal of Reproductive Immunology* **85** 51–57.

Robertson SA, Ingman WV, O'Leary S, Sharkey DJ and Tremellen KP (2002) Transforming growth factor beta – a mediator of immune deviation in seminal plasma *Journal of Reproductive Immunology* **57** 109–128.

Robertson SA, O'Leary S and Armstrong DT (2006) Influence of semen on inflammatory modulators of embryo implantation In *Control of Pig Reproduction VII: Proceedings of the Seventh International Conference on Pig Reproduction, Kerkrade, the Netherlands, June 2005 Society of Reproduction and Fertility Supplement Volume 62* pp 231–246 Ed CJ Ashworth and RR Kraeling. Published for Society of Reproduction and Fertility by Nottingham University Press, Nottingham.

Robertson SA, Guerin LR, Bromfield JJ, Branson KM, Ahlström AC and Care AS (2009a) Seminal fluid drives expansion of the CD4+CD25+ T regulatory cell pool and induces tolerance to paternal alloantigens in mice *Biology of Reproduction* **80** 1036–1045.

Robertson SA, Guerin LR, Moldenhauer LM and Hayball JD (2009b) Activating T regulatory cells for tolerance in early pregnancy – the contribution of seminal fluid *Journal of Reproductive Immunology* **83** 109–116.

Roca J, Rodriguez-Martinez H, Vazquez JM, Bolarín A, Hernández M, Saravia F, Wallgren M and Martínez EA (2006) Strategies to improve the fertility of frozen–thawed boar semen for artificial insemination In *Control of Pig Reproduction VII: Proceedings of the Seventh International Conference on Pig Reproduction, Kerkrade, the Netherlands, June 2005 Society of Reproduction and Fertility Supplement Volume 62* pp 261–275 Ed CJ Ashworth and RR Kraeling. Nottingham University Press, Nottingham.

Rodriguez H, Ohanian C and Bustos-Obregon E (1985) Nuclear chromatin decondensation of spermatozoa *in vitro*: a method for evaluating the fertilizing ability of ovine semen *International Journal of Andrology* **8** 147–158.

Rodriguez-Martinez H (2003) Laboratory semen assessment and prediction of fertility: still utopia? *Reproduction in Domestic Animals* **38** 312–318.

Rodriguez-Martinez H (2006) Can we increase the estimative value of semen assessment? *Reproduction in Domestic Animals* **41(Supplement 2)** 2–10.

Rodriguez-Martinez H (2007a) Role of the oviduct in sperm capacitation *Theriogenology* **68** 138–146.

Rodriguez-Martinez H (2007b) State of the art in farm animal sperm evaluation *Reproduction, Fertility and Development* **19** 91–101.

Rodriguez-Martinez H and Barth AD (2007) *In vitro* evaluation of sperm quality related to *in vivo* function and fertility In *Reproduction in Domestic Ruminants VI: Proceedings of the Seventh International Symposium on Reproduction in Domestic Ruminants, Wellington, New Zealand, August 2006 Society of Reproduction and Fertility Supplement Volume 64* pp 39–54 Ed JI Juengel, JF Murray and MF Smith. Published for Society of Reproduction and Fertility by Nottingham University Press, Nottingham.

Rodriguez-Martinez H and Larsson B (1998) Assessment of sperm fertilizing ability in farm animals *Acta Agriculturae Scandinavica* **29** 12–18.

Rodriguez-Martinez H, Nicander L, Viring S, Einarsson S and Larsson K (1990) Ultrastructure of the uterotubal junction in preovulatory pigs *Anatomia, Histologia, Embryologia* **19** 16–36.

Rodriguez-Martinez H, Larsson B and Pertoft H (1997) Evaluation of sperm damage and techniques for sperm clean-up *Reproduction, Fertility and Development* **9** 297–308.

Rodriguez-Martinez H, Iborra A, Martínez P and Calvete JJ (1998) Immunoelectronmicroscopic imaging of spermadhesin AWN epitopes on boar spermatozoa bound *in vivo* to the zona pellucida *Reproduction, Fertility and Development* **10** 491–497.

Rodriguez-Martinez H, Tienthai P, Suzuki K, Funahashi H, Ekwall H and Johannisson A (2001) Oviduct involvement in sperm capacitation and oocyte development *Reproduction Supplement* **58** 129–145.

Rodriguez-Martinez H, Saravia F, Wallgren M, Tienthai P, Johannisson A, Vázquez JM, Martínez E, Roca J, Sanz L and Calvete JJ (2005) Boar spermatozoa in the oviduct *Theriogenology* **63** 514–535.

Rodriguez-Martinez H, Saravia F, Wallgren M, Roca J and Peña FJ (2008) Influence of seminal plasma on the kinematics of boar spermatozoa during freezing *Theriogenology* **70** 1242–1250.

Rodriguez-Martinez H, Kvist U, Saravia F, Wallgren M, Johannisson A, Sanz L, Peña FJ, Martinez EA, Roca J, Vazquez JM and Calvete JJ (2009) The physiological roles of the boar ejaculate In *Control of Pig Reproduction VIII: Proceedings of the Eighth International Conference on Pig Reproduction Alberta, Canada June 2009 Society of Reproduction and Fertility Supplement Volume 66* pp 1–21 Ed H Rodriguez-Martinez, JL Vallet and AJ Ziecik. Published for Society of Reproduction and Fertility by Nottingham University Press, Nottingham.

Rodriguez-Martinez H, Saravia F, Wallgren M, Martinez EA, Sanz L, Roca J, Vazquez JM and Calvete JJ (2010) Spermadhesin PSP-I/PSP-II heterodimer induces migration of polymorphonuclear neutrophils into the uterine cavity of the sow *Journal of Reproductive Immunology* **84** 57–65.

Rodriguez-Martinez H, Kvist U, Ernerudh J, Sanz L and Calvete JJ (2011) Seminal plasma proteins: what role do they play? *American Journal of Reproductive Immunology* **66(Supplement S1)** 11–22.

Ruiz-Sanchez AL, O'Donoghue R, Novak S, Dyck MK, Cosgrove JR, Dixon WT and Foxcroft GR (2006) The predictive value of routine semen evaluation and IVF technology for determining relative boar fertility *Theriogenology* **66** 736–748.

Rybar R, Faldikova L, Faldyna M, Machatkova M and Rubes J (2004) Bull and boar sperm DNA integrity evaluated by sperm chromatin structure assay in the Czech Republic *Veterinární Medicína (Veterinary Medicine – Czech)* **49** 1–8.

Samardzija M, Karadjole M, Getz I, Makek Z, Cergolj M and Dobranic T (2006) Effects of bovine spermatozoa preparation on embryonic development *in vitro Reproductive Biology and Endocrinology* **4**:58.

Saravia F, Hernández M, Wallgren MK, Johannisson A and Rodriguez-Martinez H (2007a) Cooling during semen cryopreservation does not induce capacitation of boar spermatozoa *International Journal of Andrology* **30** 485–499.

Saravia F, Núñez-Martínez I, Morán JM, Soler C, Muriel A, Rodriguez-Martinez H and Peña FJ (2007b) Differences in boar sperm head shape and dimensions recorded by computer-assisted sperm morphometry are not related to chromatin integrity *Theriogenology* **68** 196–203.

Satake N, Elliott RM, Watson PF and Holt WV (2006) Sperm selection and competition in pigs may be mediated by the differential motility activation and suppression of sperm subpopulations within the oviduct *Journal of Experimental Biology* **209** 1560–1572.

Schulte RT, Ohl DA, Sigman M and Smith GD (2010) Sperm DNA damage in male infertility: etiologies, assays, and outcomes *Journal of Assisted Reproduction and Genetics* **27** 3–12.

Schuster TG, Cho B, Keller LM, Takayama S and Smith GD (2003) Isolation of motile spermatozoa from semen samples using microfluidics *Reproductive Biomedicine Online* **7** 75–81.

Shamsuddin M and Rodriguez-Martinez H (1994) A simple, non-traumatic swim-up method for the selection of spermatozoa for *in vitro* fertilisation in the bovine *Animal Reproduction Science* **36** 61–75.

Sharkey DJ, Macpherson AM, Tremellen KP and Robertson SA (2007) Seminal plasma differentially regulates inflammatory cytokine gene expression in human cervical and vaginal epithelial cells *Molecular Human Reproduction* **13** 491–501.

Sidhu KS, Mate KE, Gunasekera T, Veal D, Hetherington L, Baker MA, Aitken RJ and Rodger JC (2004) A flow cytometric assay for global estimation of tyrosine phosphorylation associated with capacitation of spermatozoa from two marsupial species, the tammar wallaby (*Arcropus eugenii*) and brushtail possum (*Trichosurus vulpecula*) *Reproduction* **127** 95–103.

Silva P and Gadella B (2006) Detection of damage in mammalian sperm cells *Theriogenology* **65** 958–978.

Smith GD, Swain JE and Bormann CL (2011) Microfluidics for gametes, embryos, and embryonic stem cells *Seminars in Reproductive Medicine* **29** 5–14.

Somfai T, Bodo S, Nagy S, Pap AB, Ivancsics J, Baranyai B, Gocza E and Kovacs A (2002) Effect of swim up and Percoll treatment on viability and acrosome integrity of frozen–thawed bull spermatozoa *Reproduction in Domestic Animals* **37** 285–290.

Su TW, Erlinger A, Tseng A and Ozcan A (2010) Compact and light-weight automated semen analysis platform using lensfree on-chip microscopy *Analytical Chemistry* **82** 8307–8712.

Suarez SS (2008a) Regulation of sperm storage and movement in the mammalian oviduct *International Journal of Developmental Biology* **52** 455–462.

Suarez SS (2008b) Control of hyperactivation in sperm *Human Reproduction Update* **14** 647–657.

Suh RS, Phadke N, Ohl DA, Takayama S and Smith GD (2003) Rethinking gamete/embryo isolation and culture with microfluidics *Human Reproduction Update* **9** 451–461.

Suh RS, Zhu X, Phadke N, Ohl DA, Takayama S and Smith GD (2006) IVF within microfluidic channels requires total numbers and lower concentrations of sperm *Human Reproduction* **21** 477–483.

Tamburrino L, Marchiani S, Montoya M, Marino FE, Natali I, Cambi M, Forti G, Baldi E and Muratori M (2012) Mechanisms and clinical correlates of sperm DNA damage *Asian Journal of Andrology* **14** 24–31.

Tardif S, Laforest JP, Cormier N and Bailey JL (1999) The importance of porcine sperm parameters on fertility *in vivo Theriogenology* **52** 447–459.

Tavalaee M, Kiani A, Arbabian M, Deymeh MR and Hasan NEM (2010) Flow cytometry: a new approach for indirect assessment of sperm protamine deficiency *International Journal of Fertility and Sterility* **3** 177–184.

Tejerina F, Buranaamnuay K, Saravia F, Wallgren M and Rodriguez-Martinez H (2008) Assessment of motility of ejaculated, liquid-stored boar spermatozoa using computerized instruments *Theriogenology* **69** 1129–1138.

Tejerina F, Morrell J, Petterson J, Dalin A-M and Rodriguez-Martinez H (2009) Assessment of motility of ejaculated stallion spermatozoa using a novel computer-assisted motility analyzer (Qualisperm™) *Animal Reproduction* **6** 380–385.

Thundathil J, Gil J, Januskauskas A, Larsson B, Saderquist L, Mapletoft R and Rodriguez-Martinez H (1999) Relationship between the proportion of capacitated spermatozoa present in frozen–thawed semen and fertility with artificial insemination *International Journal of Andrology* **22** 366–373.

Thys M, Vanadele L, Morrell JM, Mestach J, Van Soom A, Hoogewijs M and H Rodriguez-Martinez (2009) *In vitro* fertilising capacity of frozen–thawed bull spermatozoa selected by single-layer silane (glycerolpropylsilane, GS)-coated silica colloidal centrifugation *Reproduction in Domestic Animals* **44** 390–394.

Tienthai P, Yokoo M, Kimura N, Heldin P, Sato E and Rodriguez-Martinez H (2003) Immunohistochemical localization and expression of the hyaluronan receptor CD44 in the porcine oviductal epithelium during oestrus *Reproduction* **125** 119–132.

Tienthai P, Johannisson A and Rodriguez-Martinez H (2004) Sperm capacitation in the porcine oviduct *Animal Reproduction Science* **80** 131–146.

Töpfer-Petersen E, Romero A, Varela PF, Ekhlasi-Hundrieser M, Dostàlovà Z, Sanz L and Calvete JJ (1998) Spermadhesins: a new protein family. Facts, hypotheses and perspectives *Andrologia* **30** 217–224.

Töpfer-Petersen E, Ekhlasi-Hundrieser M, Kirchhoff C, Leeb T and Sieme H (2005) The role of stallion seminal plasma proteins in fertilisation *Animal Reproduction Science* **89** 159–170.

Töpfer-Petersen E, Ekhlasi-Hundrieser M and Tsolova M (2008) Glycobiology of fertilization in the pig *International Journal of Developmental Biology* **52** 717–736.

Troedsson MHT, Doty A, Macpherson ML, Connor MC, Verstegen JP, Pozor MA and Buhi WC (2010) CRISP-3 in equine seminal plasma is involved in selective uterine sperm transport *Animal Reproduction Science* **121** 192–193.

Tulsiani DR, Zeng HT and Abou-Haila A (2007) Biology of sperm capacitation: evidence for multiple signaling pathways In *Gamete Biology: Emerging Frontiers in Fertility and Contraceptive: Proceedings of the International Congress Held at New Delhi, India February 2006 Society of Reproduction and Fertility Supplement Volume 63* pp 257–272 Ed SK Gupta, K Koyama and JF Murray. Nottingham University Press, Nottingham.

Vadnais ML and Roberts KP (2010) Seminal plasma proteins inhibit *in vitro*- and cooling-induced capacitation in boar spermatozoa *Reproduction, Fertility and Development* **22** 893–900.

van Gestel R, Brewis I, Ashton P, Helms J, Brouwers J and Gadella B (2005) Capacitation-dependent concentration of lipid rafts in the apical ridge head area of porcine sperm cells *Molecular Human Reproduction* **11** 583–590.

van Gestel RA, Brewis IA, Ashton PR, Brouwers JF and Gadella BM (2007) Multiple proteins present in purified porcine sperm apical plasma membranes interact with the zona pellucida of the oocyte *Molecular Human Reproduction* **13** 445–454.

Vera O, Vásqucz LA and Muñoz MG (2003) Semen quality and presence of cytokines in seminal fluid of bull ejaculates *Theriogenology* **60** 553–558.

Villemure M, Lazure C and Manjunath P (2003) Isolation and characterization of gelatin-binding proteins from goat seminal plasma *Reproductive and Biological Endocrinology* **1**:39.

Waberski D, Magnus F, Mendonca Ferreira F, Petrunkina AM, Weitze KF and Töpfer-Petersen E (2005) Importance of sperm-binding assays for fertility prognosis of porcine spermatozoa *Theriogenology* **63** 470–484.

Waberski D, Magnus F, Ardon F, Petrunkina AM, Weitze KF and Töpfer-Petersen E (2006) Binding of boar spermatozoa to oviductal epithelium *in vitro* in relation to sperm morphology and storage time *Reproduction* **131** 311–318.

Waberski D, Henning H and Petrunkina AM (2011) Assessment of storage effects in liquid-preserved boar semen *Reproduction in Domestic Animals* **46(Supplement 2)** 45–48.

Wang W, Liang GT, Peng YY, Liu DY and Zhou XM (2011) Effects of a microfluidic sperm sorter on sperm routine parameters and DNA integrity *Zhonghua Nan Ke Xue* **17** 301–304.

Wu TF and Chu DS (2008) Epigenetic processes implemented during spermatogenesis distinguish the paternal pronucleus in the embryo *Reproductive Biomedicine Online* **16** 13–22.

Yagci A, Murk W, Stronk J and Huszar G (2010) Spermatozoa bound to solid state hyaluronic acid show chromatin structure with high DNA chain integrity: an acridine orange fluorescence study *Journal of Andrology* **3** 566–572.

Yamauchi Y, Shaman JA and Ward WS (2011) Non-genetic contributions of the sperm nucleus to embryonic development *Asian Journal of Andrology* **13** 31–35.

Zhang BR, Larsson B, Lundeheim N and Rodriguez-Martinez H (1997) Relation between embryo development *in vitro* and 56-day non-return rates of frozen-thawed semen from dairy AI bulls *Theriogenology* **48** 221–231.

Zhang BR, Larsson B, Lundeheim N and Rodriguez-Martinez H (1998) Sperm characteristics and zona pellucida binding in relation to field fertility of frozen–thawed semen from dairy AI bulls *International Journal of Andrology* **21** 207–216.

Zhang BR, Larsson B, Lundeheim N, Håård MG and Rodriguez-Martinez H (1999) Prediction of bull fertility by combined *in vitro* assessments of frozen–thawed semen from young dairy bulls entering an AI-program *International Journal of Andrology* **22** 253–260.

Zini A, Boman JM, Belzile E and Ciampi A (2008) Sperm DNA damage is associated with an increased risk of pregnancy loss after IVF and ICSI: systematic review and meta-analysis *Human Reproduction* **23** 2663–2668.

Zuba-Surma EK, Kucia M, Abdel-Latif A, Lillard JW Jr and Ratajczak MZ (2007) The ImageStream System: a key step to a new era in imaging *Folia Histochemica Cytobiologica* **45** 279–290.

Index

Page numbers in **bold** type refer to figures and tables.